The Nature of Mathematics

Tenth Edition

Karl J. Smith
Santa Rosa Junior College

THOMSON
™
BROOKS/COLE

Australia • Canada • Mexico • Singapore • Spain
United Kingdom • United States

THOMSON

BROOKS/COLE

Mathematics Editor: John-Paul Ramin
Senior Assistant Editor: Rachael Sturgeon
Editorial Assistants: Darlene Amidon-Brent, Lisa Chow
Project Manager, Editorial Production: Janet Hill
Technology Project Manager: Rachael Sturgeon
Marketing Manager: Leah Thomson
Marketing Assistant: Jessica Perry
Advertising Project Manager: Bryan Vann
Print/Media Buyer: Karen Hunt
Production Service: Susan L. Reiland

Text Designer: Andrew Ogus
Photo Researcher: Sue C. Howard
Copy Editor: Christine M. Levesque
Technical Illustration: Lori Heckelman
Historical Note line portraits: Steven Nau/DEBORAH WOLFE LTD
Cover Designer: Roger Knox
Cover Image: © Lester Lefkowitz/CORBIS
Cover Printer: Phoenix Color Corp
Compositor: Better Graphics
Printer: R. R. Donnelley/Willard

For more information about our products
contact us at:
Thomson Learning Academic Resource Center
1-800-423-0563
For permission to use material from this text,
contact us by:
Phone 1-800-730-2214
Fax 1-800-730-2215
Web: http://www.thomsonrights.com

Library of Congress Control Number: 2003107566

Student Edition: ISBN 0-534-40023-X

Instructor's Edition: ISBN 0-534-40455-3

Brooks/Cole–Thomson Learning
10 Davis Drive
Belmont, CA 94002
USA

Asia
Thomson Learning
5 Shenton Way #01-01
UIC Building
Singapore 068808

Australia/New Zealand
Thomson Learning
102 Dodds Street
Southbank, Victoria 3006
Australia

Canada
Nelson
1120 Birchmount Road
Toronto, Ontario M1K 5G4
Canada

Europe/Middle East/Africa
Thomson Learning
High Holborn House
50/51 Bedford Row
London WC1R 4LR
United Kingdom

Latin America
Thomson Learning
Seneca, 53
Colonia Polanco
11560 Mexico D.F.
Mexico

Spain/Portugal
Paraninfo
Calle/Magallanes, 25
28015 Madrid, Spain

I dedicate this book with love
to my wife, Linda

 # Preface

Like almost every subject of human interest, mathematics is as easy or as difficult as we choose to make it. Following this Preface, I have included a *Fable*, addressed directly to you, the student. I hope you take the time to read it, and then ponder why I call it a fable.

 You will notice street sign symbols used throughout this book. I use this stop sign to mean that you should stop and pay attention to this idea, since it will be used as you travel through the rest of the book. Here are some of the other street signs you will see.

 Caution means that you will need to proceed more slowly to understand this material.

 Bump symbol means to watch out, because you are coming to some difficult material.

 WWW means you should check this out on the Web, if you have access to a computer with an Internet connection.

I also use this special font to speak to you directly out of the context of the regular textual material. I call these **author's notes**; they are comments that I might say to you if we were chatting in my office about the content in this book.

I frequently encounter people who tell me about their unpleasant experiences with mathematics. I have a true sympathy for those people, and I recall one of my elementary school teachers who assigned additional arithmetic problems as punishment. This can only create negative attitudes toward mathematics, which is indeed unfortunate. If elementary school teachers and parents have positive attitudes toward mathematics, their children cannot help but see some of the beauty of the subject. I want students to come away from this course with the feeling that mathematics can be pleasant, useful, and practical—and enjoyed for its own sake.

Since the first edition, my goal has been, and continues to be, to create a positive attitude toward mathematics. But the world, the students, and the professors are very different today than they were when I began writing this book. This is a very different book from its first printing, and this edition is very different from the previous edition. The world of knowledge is more accessible today (via the World Wide Web) than at any time in history. Supplementary help is available on the World Wide Web, and can be accessed at the following Web address:

www.mathnaturc.com

All of the Web addresses mentioned in this book are linked to the above Web address. If you have access to a computer and the World Wide Web, check out this Web address. You will find links to several search engines, history, and reference topics. Also included are a keyword search and a menu that allows you to choose a chapter and section. You will find, for each section, homework hints, a listing of essential ideas, projects, and links to related information on the Web.

This book was written for those students who need a mathematics course to satisfy the general university competency requirement in mathematics. Because of the university requirement, many students enrolling in a course that uses my book have postponed taking this course as long as possible. They dread the experience and come to class with a great deal of anxiety. Rather than simply presenting the technical details needed to proceed to the next course, I have attempted to give insight into what mathematics is, what it accomplishes, and how it is pursued as a human enterprise. However, at the same time, in this tenth edition I have included a great deal of material to help students estimate, calculate, and solve problems *outside* the classroom or textbook setting.

This book was written to meet the needs of all students and schools. How did I accomplish that goal? First, the chapters are almost independent of one another and can be covered in any order appropriate to a particular audience. Second, the problems are designed as the core of the course. There are problems that every student will find easy in order to provide the opportunity for success; there are also problems that are very challenging. Much interesting material appears in the problems, and students should get into the habit of reading (not necessarily working) all the problems whether or not they are assigned.

Level 1: mechanical or drill problems
Level 2: require understanding of the concepts
Level 3: require problem-solving skills or original thinking

What Are the Major Themes of This Book?

The major themes of this book are *problem solving* and estimation in the context of presenting the great ideas in the history of mathematics.

I believe that *learning to solve problems is the principal reason for studying mathematics*. Problem solving is the process of applying previously acquired knowledge to new and unfamiliar situations. Solving word problems in most textbooks is one form of problem solving, but students also should be faced with non-text-type problems. In the first section of this edition I introduce students to Pólya's problem-solving techniques, and these techniques are used throughout the book to solve non-text-type problems. Problem-solving examples (marked as PÓLYA'S METHOD examples) are found throughout the book. You will find problems in *each* section that require Pólya's method for problem solving, and then you can practice your problem-solving skills with problems that are marked

Level 3, Problem Solving

Students should learn the language and notation of mathematics. Most students who have trouble with mathematics do not realize that mathematics *does require hard work*. The usual pattern for most mathematics students is to open the book to the assigned page of problems and begin working. Only after getting "stuck" is an attempt made to "find it in the book." The final resort is reading the text. In this book the students are asked not only to "do math problems," but also to "experience mathematics." This means it is necessary to become involved with the **concepts** being presented, not "just get answers." In fact, the advertising slogan "Mathematics Is Not a Spectator Sport" is an invitation which suggests that the only way to succeed in mathematics is to become involved with it. Students will learn to receive mathematical ideas through listening, reading, and visualizing. They are expected to present mathematical ideas by speaking, writing, drawing pictures and graphs, and demonstrating with concrete models. There is a category of problems in each section that is designated IN YOUR OWN WORDS, which provides practice in communication skills.

A Note for Instructors

The prerequisites for this course vary considerably, as do the backgrounds of students. Some schools have no prerequisites, while other schools have an intermediate algebra prerequisite. The students, as well, have heterogeneous backgrounds. Some have little or no mathematics skills; others have had a great deal of mathematics. Even though the usual prerequisite for using this book is intermediate algebra, a careful selection of topics and chapters would allow a class with a beginning algebra prerequisite to effectively study the material.

Feel free to arrange the material in a different order from that presented in the text. I have written the chapters to be as independent of one another as possible. There is much more material than could be covered in a single course. This book can be used in classes designed for liberal arts, teacher training, finite mathematics, college algebra, or a combination of these.

Over the years many instructors from all over the country have told me that they love the material, love to teach from this book, but complain that there is just too much material in this book to cover in one, or even two, semesters. In response to these requests, I have divided some of the material into two separate volumes:

The Nature of Problem Solving in Geometry and Probability
The Nature of Problem Solving in Algebra

The first volume, *The Nature of Geometry and Probability*, includes Chapters 1, 2, 6, 7, 8, 10, 11, 12, and 15 from this text. The second volume, *The Nature of Problem Solving in Algebra*, includes Chapters 1, 3, 4, 5, 8, 9, 13, 14, 15, and 16 from this text.

One of the advantages of using a textbook that has traveled through many editions is that it is well seasoned. Errors are minimal, pedagogy is excellent, and it is easy to use; in other words, it works. For example, you will find that the sections and chapters are about the right length—each section will take about one classroom day. The problem sets are graded so that you can teach the course at different levels of difficulty, depending on the assigned problems. The problem sets are uniform in length (60 problems each), which facilitates the assigning of problems from day to day. The chapter reviews are complete and lead the students to the type of review they will need to prepare them for an examination.

I have written an extensive *Instructor's Manual* (over 460 pages) to accompany this book. It includes the complete solutions to all the problems (including the "Problem Solving" problems) as well as teaching suggestions and transparency masters.

There is also a *Student Survival and Solutions Manual* (authored by Steve Watnik) to help the students with their success. This manual has the complete solution of approximately one-half of the problems in the text.

Also available are sample tests, not only in hard-copy form, but also in electronic form for both IBM and Macintosh formats.

Changes from the Previous Edition

There are many new ideas in this edition. You will see that this edition has 17 chapters, whereas the previous edition had 12 chapters. You will find new chapters entitled *The Nature of Voting and Apportionment, The Nature of Networks and Graph Theory, The Nature of Set Theory and Counting, The Nature of Growth,* and *The Nature of Calculus.* You will also find some new sections: 1.4, Finite and Infinite; 5.8, Percents; and 11.5, The Binomial Distribution. These chapters and sections contain a great deal of new and exciting material, and speak about the changing nature of the mathematical world around us.

I've also added a prologue, "Why Math?," and an epilogue, "Why Not Math?" The prologue is designed to put the material you will find in this edition into a historical perspective, and to begin the students thinking about problem solving. The problems accompanying this prologue could serve as a pretest or diagnostic test, but I use these prologue problems to let the students know that this book will not be like other math books they may have used in the past. The epilogue is designed to tie together many parts of the book (which may or may not have been "covered" in the class) to show that there are many rooms to the mansion known as mathematics. The problems accompanying this epilogue could serve as a review to show that it would be difficult to choose a course of study in college without somehow being touched by mathematics.

The further integration of the material in the textbook and the material available online in the World Wide Web continues to be an important feature of this new edition. Without computer access, the book completely stands alone, as in the previous editions. But for those with access to the Web, there is a world of enhancement for this book. All of this material is found on our Web page **mathnature.com.**

Acknowledgments

I appreciate the suggestions of the reviewers of this edition: Rose Cavin, Chipola Junior College; Peter Chen, William Paterson University; Steven W. Davis, California State University, Fullerton; Beth Long, Pellissippi State Technical Community College; Winifred A. Mallam, Texas Women's University; Clifford D. Miller, Southwestern Assemblies of God University; Joanne V. Peeples, El Paso Community College; Susan K. Puckett, Kalamazoo Valley Community College; and Carole E. Williams, Seminole Community College.

One of the nicest things about writing a successful book is all of the letters and suggestions I've received. I would like to thank the following people who gave suggestions for previous editions of this book: Jeffery Allbritten, Brenda Allen, Richard C. Andrews, Nancy Angle, Peter R. Atwood, John August, Charles Baker, V. Sagar Bakhshi, Jerald T. Ball, Carol Bauer, George Berzsenyi, Daniel C. Biles, Jan Boal, Elaine Bouldin, Kolman Brand, Chris C. Braunschweiger, Barry Brenin, T. A. Bronikowski, Charles M. Bundrick, T. W. Buquoi, Eugene Callahan, Michael W. Carroll, Joseph M. Cavanaugh, James R. Choike, Mark Christie, Gerald Church, Robert Cicenia, Wil Clarke, Lynn Cleaveland, Penelope Ann Coe, Thomas C. Craven, Gladys C. Cummings, Ralph De Marr, Maureen Dion, Charles Downey, Mickle Duggan, Samuel L. Dunn, Beva Eastman, William J. Eccles, Gentil Estevez, Ernest Fandreyer, Loyal Farmer, Gregory N. Fiore, Robert Fliess, Richard Freitag, Gerald E. Gannon, Ralph Gellar, Gary Gislason, Mark Greenhalgh, Martin Haines, Abdul Rahim Halabieh, John J. Hanevy, Michael Helinger, Robert L. Hoburg, Caroline Hollingsworth, Scott Holm, Libby W. Holt, Peter Hovanec, M. Kay Hudspeth, Carol M. Hurwitz, James J. Jackson, Kind Jamison, Vernon H. Jantz, Charles E. Johnson, Nancy J. Johnson, Michael Jones, Martha C. Jordan, Judy D. Kennedy, Linda H. Kodama, Daniel Koral, Helen Kriegsman, Frances J. Lane, C. Deborah Laughton, William Leahey, John LeDuc, William A. Leonard, Adolf Mader, John Martin, Cherry F. May, George McNulty, Carol McVey, Valerie Melvin, Charles C. Miles, Allen D. Miller, Elaine I. Miller, John Mullen, Charles W. Nelson, Barbara Ostrick, John Palumbo, Gary Peterson, Michael Petricig, Mary Anne C. Petruska, Michael Pinter, Joan Raines, James V. Rauff, Richard Rempel, Paul M. Riggs, Jane Rood, Peter Ross, O. Sassian, Mickey G. Settle, Andrew Simoson, James R. Smart, Glen T. Smith, Donald G. Spencer, Gustavo Valadez-Ortiz, John Vangor, Arnold Villone, Clifford H. Wagner, James Walters, Barbara Williams, Stephen S. Willoughby, Jean Woody, and Bruce Yoshiwara.

Stephen DeLong, Paul McCombs, Ann Ostberg, Laurie Poe, and Steve Watnik did a superb job of checking all of the examples and checking the accuracy of the answers. I would especially like to thank Robert J. Wisner of New Mexico State for his countless suggestions and ideas over the many editions of this book; John-Paul Ramin, Craig Barth, Jeremy Hayhurst, Paula Heighton, Gary Ostedt, and Bob Pirtle of Brooks/Cole; as well as Jack Thornton, for the sterling leadership and inspiration he has been to me from the inception of this book to the present. I especially appreciate Susan Reiland for her help in countless ways, including editing, accuracy checking, and going over and above anything that could be expected from a production editor. I consider her a friend and a colleague. My appreciation also extends to technical artist Lori Heckelman, who is always there when we need her.

Finally, my thanks go to my wife, Linda, who has always been there for me. Without her this book would exist only in my dreams, and I would have never embarked as an author.

Karl J. Smith
Sebastopol, CA
smithkjs@mathnature.com

To the Student: A Fable

Once upon a time, two young ladies, Shelley and Cindy, came to a town called Mathematics. People had warned them that this was a particularly confusing town. Many people who arrived in Mathematics were very enthusiastic, but could not find their way around, became frustrated, gave up, and left town.

Shelley was strongly determined to succeed. She was going to learn her way through the town. For example, in order to learn how to go from her dorm to class, she concentrated on memorizing this clearly essential information: she had to walk 325 steps south, then 253 steps west, then 129 steps in a diagonal (southwest), and finally 86 steps north. It was not easy to remember all of that, but fortunately she had a very good instructor who helped her to walk this same path 50 times. In order to stick to the strictly necessary information, she ignored much of the beauty along the route, such as the color of the adjacent buildings or the existence of trees, bushes, and nearby flowers. She always walked blindfolded. After repeated exercising, she succeeded in learning her way to class and also to the cafeteria. But she could not learn the way to the grocery store, the bus station, or a nice restaurant; there were just too many routes to memorize. It was so overwhelming! Finally, she gave up and left town; Mathematics was too complicated for her.

Cindy, on the other hand, was of a much less serious nature. To the dismay of her instructor, she did not even intend to memorize the number of steps of her walks. Neither did she use the standard blindfold which students need for learning. She was always curious, looking at the different buildings, trees, bushes, and nearby flowers or anything else not necessarily related to her walk. Sometimes she walked down dead-end alleys in order to find out where they were leading, even if this was obviously superfluous. Curiously, Cindy succeeded in learning how to walk from one place to another. She even found it easy and enjoyed the scenery. She eventually built a building on a vacant lot in the city of Mathematics.*

* My thanks to Emilio Roxin of the University of Rhode Island for the idea for this fable.

Contents

* Optional sections.

* Optional sections.

* Optional sections.

* Optional sections.

Prologue: Why Math?
A Historical Overview

Whether you love or loathe mathematics, it is hard to deny its importance in the development of the main ideas of this world! Read the BON VOYAGE invitation on the inside front cover. The goal of this text is to help you to discover an answer to the question, "Why study math?"

The study of mathematics can be organized as a history or story of the development of mathematical ideas, or it can be organized by topic. The intended audience of this book dictates that the development should be by topic, but mathematics involves real people with real stories, so you will find this text to be very historical in its presentation. This overview rearranges the material you will encounter in the text into a historical timeline. It is not intended to be read as a history of mathematics, but rather an overview to make you want to do further investigation. Sit back, relax, and use this overview as a *starting place* to expand your knowledge about the beginnings of some of the greatest ideas in the history of the world!

We have divided this history of mathematics into seven chronological periods:

Babylonian, Egyptian, and Native American periods	3000 BC to 601 BC
Greek, Chinese, and Roman periods	600 BC to 499 AD
Hindu and Arabian period	500 to 1199
Transition period	1200 to 1599
Century of Enlightenment	1600 to 1699
Early Modern period	1700 to 1899
Modern period	1900 to present

Babylonian, Egyptian, and Native American Periods: 3000 BC to 601 BC

Mesopotamia is an ancient country located in southeast Asia between the lower Tigris and Euphrates Rivers. It is a part of modern Iraq. Mesopotamian mathematics refers to the mathematics of the ancient Babylonians, and this mathematics is sometimes referred to as Sumerian mathematics. Over 50,000 tablets

Sumerian clay tablet

from Mesopotamia have been found, and are exhibited at major museums around the world. Interesting readings about Babylon can be found by reading a history of mathematics book, such as *An Introduction to the History of Mathematics,* 6th Edition, by Howard Eves (New York: Saunders, 1990) or by looking at the many sources on the World Wide Web. You can find links to these Web sites, as

---

Babylonian, Egyptian, and Native American Periods: 3000 BC to 601 BC

	Cultural Events		Mathematical Events
3000 BC	First Dynasty of the Ancient Kingdom of Egypt	**3000 BC**	Chinese arithmetic and astronomy under Huang-ti
2800	The Great Pyramid	**2900**	Egyptian scribes use a hieratic script
2580	Cheops' Pyramid	**2850**	Egyptian simple grouping system
2500	Isis and Osiris cult in Egypt	**2200**	Mathematical tablets of Nippur: oldest example of a magic square
1900	Epic of Gilgamesh		
1700	Stonehenge	**1850**	Moscow papyrus: 25 mathematical problems
1500	First alphabets created	**1700**	Early Babylonian tablet (Plimpton 322)
1495	Obelisk of Thothmes at Karnak	**1650**	Egyptian Astrological signs
1300	Approximate beginning of Iron Age	**1350**	Rollins papyrus: elaborate mathematical problem
1250	Moses leads exodus from Egypt	**1105**	The *Chou-pei,* major Chinese text on mathematics
1200	Trojan War		
1000	Phoenicians invent alphabet		
850	Homer: *Iliad* and *Odyssey*		
753	Rome founded		

well as all Web sites in this book, by looking at the Web page for this text:

www.mathnature.com

This Web page can be your access to a world of information by using the links provided.

The mathematics of this period was very practical and it was used in construction, surveying, record keeping, and in the creation of calendars. The culture of the Babylonians reached its height about 2500 B.C., and about 1700 B.C. King Hammurabi formulated a famous code of law. In 330 B.C., Alexander the Great conquered Mesopotamia, ending the great Babylonian empire. Even though there was a great deal of political and social upheaval during this period, there was a continuity in the development of mathematics from ancient time to the time of Alexander.

The main information we have about the civilization and mathematics of the Babylonians is their numeration system, which we introduce in the text in Section 3.1 (p. 105). The Babylonian numeration system was positional with base 60. They did not have a 0 symbol, but they did represent fractions, squares, square roots, cubes, and cube roots. We have evidence that they knew the quadratic formula (Chapter 5, p. 242), and they had stated algebraic problems verbally. The base 60 system of the Babylonians led to a division of a circle into 360 equal parts that today we call degrees, and each degree was in turn divided into 60 parts that today we call seconds. The Greek astronomer Ptolemy (A.D. 85–165) used this Babylonian system, which no doubt is why we have minutes, seconds, and degree measurement today.

The Egyptian civilization existed from about 4000 B.C., and was less influenced by foreign powers than was the Babylonian civilization. Egypt was divided into two kingdoms until about 3000 B.C., when the ruler Menes unified Egypt and consequently became known as the founder of the first dynasty in 2500 B.C. This was their pyramid-building period, and the Great Pyramid of Cheops was built around 2600 B.C. (Chapter 6, pp. 329 and 333–334; see The Riddle of the Pyramids).

The Egyptians developed their own pictorial way of writing, called *hieroglyphics*, and their numeration system was consequently very pictorial (Chapter 3, p. 102).

The Egyptian numeration system is an example of a simple grouping system. Although they were able to write fractions, they used only unit fractions (p. 103). Like the Babylonians, they had not developed a symbol for zero (p. 106). Since the writing of the Egyptians was on papyrus, and not on tablets as with the Babylonians, most of the written history has been lost. Our information comes from the Rhind papyrus (p. 104), discovered in 1858 and dated to about 1700 B.C., and the Moscow papyrus (p. 104), which has been dated to about the same time period.

The mathematics of the Egyptians remained remarkably unchanged from the time of the first dynasty to the time of Alexander the Great, who conquered them in 332 B.C. The Egyptians did surveying using a unique method

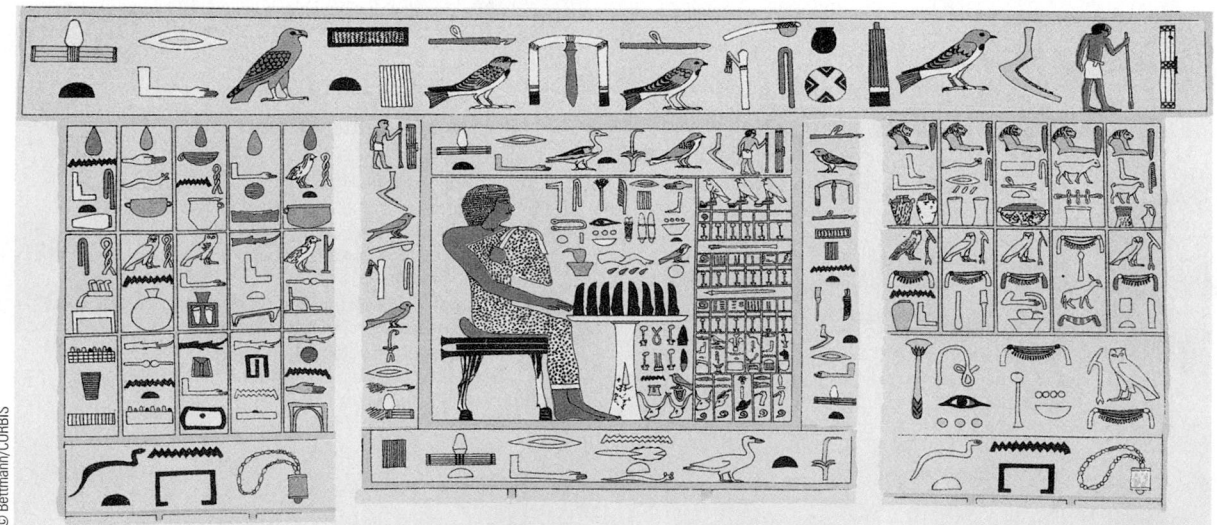

Egyptian hieroglyphics: Inscription and relief from the grave of Prince Rahdep (ca. 2800 B.C.)

© Bettmann/CORBIS

of stretching rope, so they referred to their surveyors as "rope stretchers" (p. 332). The basic unit used by the Egyptians for measuring length was the *cubit*, which was the distance from a person's elbow to the end of the middle finger. A *khet* was defined to equal 100 cubits; khets were used by the Egyptians when land was surveyed. The Egyptians did not have the concept of a variable, and all of their problems were verbal or arithmetic. Even though they solved many equations, they used the word *AHA* or *heap* in place of the variable (p. 252). For an example of an Egyptian problem, see Ahmes' dilemma (p. 15) and the statement of the problem in terms of Thoth, an ancient Egyptian god of wisdom and learning.

The Egyptians had formulas for the area of a circle (p. 364) and the volume of a cube, box, cylinder, and other figures. Particularly remarkable is their formula for the volume of a truncated pyramid of a square base, which in modern notation is

$$V = \frac{h}{3}(a^2 + ab + b^2)$$

where h is the height and a and b arc the sides of the top and bottom. Even though we are not certain the Egyptians knew of the Pythagorean theorem (p. 176), we believe they did because the rope stretchers had knots on their ropes which would form right triangles. They had a very good reckoning of the calendar, and knew that a solar year was approximately $365\frac{1}{4}$ days long. They chose as the first day of their year the day on which the Nile would flood.

The next chapter in the history of mathematics is from the Babylonian and Egyptian civilizations to the Greeks because of the conquests of Alexander the Great. Contemporaneous with the great civilizations in Mesopotamia was the great Mayan civilization in what is now Mexico. A Mayan timeline is shown in Table 1.

Table 1 Mayan Timeline

1200–1000 BC	Olmec
1800–900 BC	Early Preclassic Maya
900–300 BC	Middle Preclassic Maya
300 BC– AD 250	Late Preclassic Maya
250–600	Early Classic Maya
699–900	Late Classic Maya
900–1500	Post Classic Maya
1500–1800	Colonial period
1821–present	Mexico

Just as with the Mesopotamian civilizations, the Olmeca and Mayan civilizations lie between two great rivers, in this case the Grijalva and Papaloapan Rivers. Sometimes the Olmecas are referred to as the Tenocelome. The Olmeca culture is considered the mother culture of the Americas. What we know about the Olmecas centers around their art. We do know they were a farming community. The Maya civilization began around 2600 B.C. and gave rise to the Olmecas. A written hieroglyphic language had been developed by 700 B.C. and they had a very accurate solar calendar. The Mayan culture had developed a positional numeration system (p. 108).

You will find the influences from this period discussed throughout the book. In particular, look in the following places:

Babylonian
 Babylonian (Sumerian) numeration system, Chapter 3, p. 105
 Quadratic formula by the Babylonians, Chapter 5, pp. 242, 292 (Problem G19)

Egyptian:
 Historical Question: Thoth, Egyptian god of wisdom and learning, Chapter 1, p. 15 (Problem 58)
 Egyptian numeration system, Chapter 3, p. 102
 Rhind papyrus (1650 BC), Chapter 3, p. 103
 Ahmose, Egyptian scribe, Chapter 3, p. 104
 Moscow papyrus (1850 BC), Chapter 3, p. 104
 Historical Questions: Rhind papyrus, Chapter 3, p. 109 (Problem 36)
 Historical Questions: Yale tablet, Chapter 3, p. 109 (Problem 56)
 Egyptian fractions, Chapter 4, p. 175 (Problems 55–59)
 Egyptian knowledge of the Pythagorean theorem, Chapter 4, p. 176
 Lack of 0 in the Egyptian numeration system, Chapter 4, p. 189
 Egyptian equation-solving, Chapter 5, pp. 246–247 (Problems 59–60)
 Historical Question: Egyptian measuring rope, Chapter 6, p. 332
 The Riddle of the Pyramids, Chapter 6, p. 334
 Egyptians and the area of a circle, Chapter 17, p. 865

Others:
 Aztec nation's numeration system, Chapter 3, p. 107
 Native Americans of California (base 8 numeration system), Chapter 3, p. 115
 Yuki nation, Chapter 3, p. 115
 Aristophanes (finger counting), Chapter 3, p. 125

Greek, Chinese, and Roman Periods: 600 BC to AD 499

Greek mathematics began in 585 BC when Thales, the first of the Seven Sages of Greece (625–547 B.C.), traveled to Egypt.* The Greek civilization was most influential in our history of mathematics. So striking was their influence that

* The Seven Sages in Greek history refer to Thales of Miletus, Bias of Priene, Chilo of Sparta, Cleobulus of Rhodes, Periander of Corinth, Pittacus of Mitylene, and Solon of Athens; they were famous because of their practical knowledge about the world and how things work.

the historian Morris Kline declares, "One of the great problems of the history of civilization is how to account for the brilliance and creativity of the ancient Greeks."[†] The Greeks settled in Asia Minor, modern Greece, southern Italy, Sicily, Crete, and North Africa. They replaced the various hieroglyphic systems with the Phoenician alphabet, and with that they were able to become more literate and more capable of recording history and ideas. The Greeks had their own numeration system (p. 108). They had fractions and some irrational numbers, especially π.

The great mathematical contributions of the Greeks are Euclid's *Elements* (p. 297) and Apollonius' *Conic Sections* (p. 682). Greek knowledge developed in several centers or schools. (See Figure 1 for depiction of one of these centers of learning.) The first was founded by Thales (ca. 640–546 B.C.) and known as the Ionian in Miletus. It is reported that

[†] p. 24, *Mathematical Thought from Ancient to Modern Times* by Morris Kline (New York: Oxford University Press, 1972).

Greek, Chinese, and Roman Periods: 600 BC to AD 499

Cultural Events		Mathematical Events	
538	Persians capture Babylon	585	Thales, founder of Greek geometry
500	Pindar's *Odes*	540	The teachings of Pythagoras
480	Siddhartha, the Buddha, delivers his sermons in Deer Park	500	Sulvasutras: Pythagorean numbers
		450	Zeno: paradoxes of motion
323	Alexander the Great completes his conquest of the known world	425	Theodorus of Cyrene: irrational numbers
		384	Aristotle: logic
218	Hannibal crosses the Alps	380	Plato's Academy: logic
200	Rosetta Stone engraved	323	Euclid: geometry, perfect numbers
100	Birth of Julius Caesar	300	First use of Hindu numeration system
20	Virgil: *Aeneid*	230	Sieve of Eratosthenes
4	Birth of Christ	225	Archimedes: circle, pi, curves, series
		180	Hypsicles: number theory
▲		60	Geminus: parallel postulate
BC-AD		▲	
▼		BC-AD	
		▼	
200	Goths invade Asia Minor	50	Negative numbers used in China
324	Founding of Constantinople	75	Heron: measurements, roots, surveying
400	Augustine, *Confessions*	100	Nichomachus: number theory
476	Fall of Rome	150	Ptolemy: trigonometry
		200	Mayan calendar
		250	Diophantus: number theory, algebra
		300	Pappus: *Mathematical Collection*
		410	Hypatia of Alexandria: first woman mentioned in the history of mathematics
		480	Tsu Ch'ung-chi approximates π as 355/113

© Scala/Art Resource, NY

Figure 1
The School of Athens by Raphael, 1509. This fresco includes portraits of Raphael's contemporaries, and demonstrates the use of perspective. Note the figures in the lower right, who are, no doubt, discussing mathematics.

while he was traveling and studying in Egypt, he calculated the heights of the pyramids by using similar triangles, just as we do in Example 5 of Section 6.5 (p. 329). You can read about these two great Greek mathematicians in *Mathematical Thought from Ancient to Modern Times* by Morris Kline. You can also refer to the World Wide Web at **www.mathnature.com**.

Between 585 B.C. and 352 B.C., schools flourished, establishing the foundations for the way knowledge is organized today. Figure 2 shows each of the seven major schools, along with each school's most notable contribution. Links to textual discussion are shown within each school of thought, along with the principal person for each of these schools. Books have been written about the importance of

each of these Greek schools, and several links can be found at **www.mathnature.com**.

One of the three greatest mathematicians in the entire history of mathematics was Archimedes (287–212 B.C.). His accomplishments are truly remarkable, and you should seek out other sources about the magnitude of his accomplishments. He invented a pump (the Archimedian screw), military engines and weapons, and catapults; in addition, he used a paraboloidal mirror as a weapon by concentrating the sun's rays on the invading Roman ships. "The most famous of the stories about Archimedes is his discovery of the method of testing the debasement of a crown of gold. The king of Syracuse had ordered the crown. When it was delivered, he suspected that it was filled with baser metals

Greek Schools (585–352 B.C.)

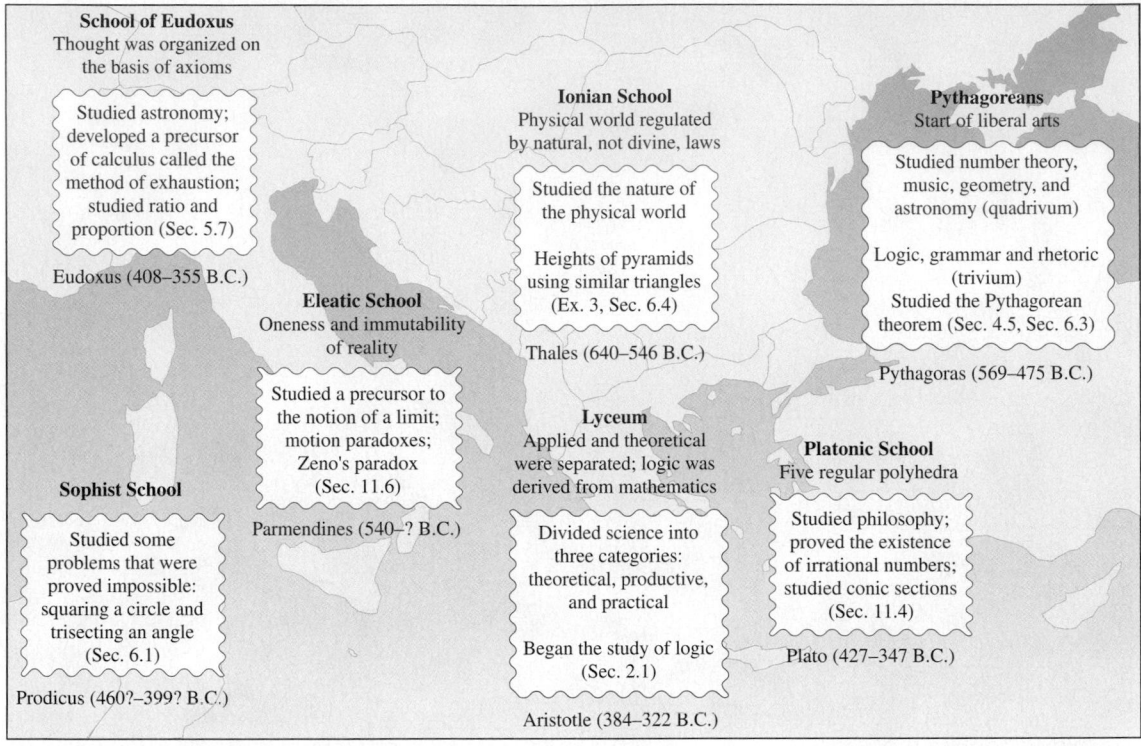

Figure 2
Greek schools from 585 B.C. to 352 B.C.

and sent it to Archimedes to devise some method of testing the contents without, of course, destroying the workmanship. Archimedes pondered the problem; one day while bathing he observed that his body was partly buoyed up by the water and suddenly grasped the principle that enabled him to solve the problem. He was so excited by this discovery that he ran out into the street naked shouting, 'Eureka!' ('I have found it!') He had discovered that a body immersed in water is buoyed up by a force equal to the weight of the water displaced, and by means of this principle was able to determine the contents of the crown."*

Although the Romans conquered the world, their mathematical contributions were minor, even though we still use Roman numerals (Section 3.1, p. 104) today in certain circumstances. Their basic unit of weight was the *as* and 1/12 of this was the *uncia,* from which we get our present-day measurements of *ounce* and *inch*, respectively. Thus, we might say their measurements were based on a duodecimal

(base 12) system. They improved on our calendar, and set up the notion of leap year every four years. The Julian calendar was adopted in 45 B.C. The Romans conquered Greece and Mesopotamia, and in 47 B.C. they set fire to the Egyptian fleet in the harbor of Alexandria. The fire spread to the city and burned the library, destroying two and a half centuries of book-collecting, including all the important knowledge of the time.

Another great world civilization existed in China, where they also developed a decimal numeration system and used a decimal system with symbols 1, 2, 3, . . . , 9, 10, 100, 1000, and 10000. Calculations were performed using small bamboo counting rods, which eventually evolved into the abacus (p. 113). Our first historical reference to the Chinese culture is the yin-yang symbol (p. 55), which has its roots in ancient cosmology. The original meaning is representative of the mountains, both the bright side and the dark side. The "yin" represents the female, or

*Ibid., pp. 105–106.

shaded, aspect, the earth, the darkness, the moon, and passivity. The "yang" represents the male, light, sun, heaven, and the active principle in nature. These words can be traced back to the Shang and Chou Dynasty (1550–1050 B.C.), but most scholars credit it to the Han Dynasty (220–206 B.C.) One of the first examples of a magic square comes from Lo River around 200 B.C., where legend tells us that the emperor Yu of the Shang dynasty received a magic square on the back of a tortoise's shell (see p. 24).

From 100 B.C. to A.D. 100 the Chinese described the motion of the planets, as well as what is the earliest known proof of the Pythagorean theorem (p. 178). The longest surviving and most influential Chinese math book is dated from the beginning of the Han Dynasty around A.D. 50. It includes measurement and area problems, proportions, volumes, and some approximations for π. Sun Zi (ca. A.D. 250) wrote his mathematical manual, which included the "Chinese remainder problem": Find n so that upon division by 3, you obtain a remainder of 2; upon division by 5 you obtain a remainder of 3; and upon division by 7 you get a remainder of 2. His solution: Add 140, 63, 30 to obtain 233, and subtract 210 to obtain 23 (see p. 203). Zhang Qiujian (ca. 450 A.D.) wrote a mathematics manual which included a formula for summing the terms of an arithmetic sequence, along with the solution to a system of two linear equations in three unknowns. The problem is the "One Hundred Fowl Problem," which is included in Problem Set 4.7 (p. 203). At the end of this historic period, the mathematician and astronomer Wang Xiaotong (ca. A.D. 626) solved cubic equations by generalization of an algorithm for finding the cube root.

Check **www.mathnature.com** for links to many excellent sites on Greek mathematics. In this text, we have included the following stories and references:

Greeks
Aristotle (384 B.C.) Logic began, Chapter 2, p. 54
Historical Question: Greek numeration system,
    Chapter 3, p. 110 (Problem 60)
Pythagoreans as a secret society, Chapter 4, p. 176
Eratosthenes created a sieve for primes, Chapter 4, p. 152
Fundamental Theorem of Arithematic, Book IX of
    Euclid's *Elements*, Chapter 4, p. 154
Hypatia (370–415), Chapter 5, p. 258
Historical Question: Diophantus' age, Chapter 5, p. 290
Euclid's *Elements* written and set the foundations for
    geometry, Chapter 6, p. 297
Apollonius (ca. 262–190 BC), Chapter 13, p. 682
Zeno's paradox, Chapter 17, pp. 862–863

Chinese
Shang dynasty emperor Yu/Lo-shu magic square,
    Chapter 1, p. 24
Yin-yang symbol (duality of nature), Chapter 2, p. 55
Historical Question: Pascal's triangle/Chinese manuscript,
    Chapter 3, pp. 109–110 (Problem 59)
Historical Question: abacus, Chapter 3, p. 113
    (Problems 44–49)
Chou Pei discovery of the Pythagorean theorem,
    Chapter 4, p. 178
One Hundred Fowl Problem/Chinese puzzle,
    Chapter 4, p. 203 (Problem 58)
Chinese Remainder problem, Chapter 4,
    p. 203 (Problem 59)

Romans
Roman numeration system, Chapter 3, p. 104

Hindu and Arabian Period: AD 500 to 1199

Much of the mathematics that we read in contemporary mathematics textbooks ignores the rich history of this period. Included on the World Wide Web are some very good sources for this period. Check our Web site **www.mathnature.com** for some links. The Hindu civilization dates back to 2000 B.C., but the first recorded mathematics was during the Śulvasūtra period from 800 B.C. to 200 A.D. In the third century, Brahmi symbols were used for 1, 2, 3, …, 9 and are significant because there was a single symbol for each number. There was no zero or positional notation at this time, but by 600 A.D. the Hindus used the Brahmi symbols with positional notation. We will discuss a numeration system that eventually evolved from these Brahmi symbols, in Section 3.2 (p. 111). For fractions, the Hindus used sexagesimal positional notation in astronomy, but in other applications they used a ratio of integers and wrote $\frac{3}{4}$ (without the fractional bar we use today). The first mathematically important period was the second period, A.D. 200–1200. The important mathematicians of this period are Āryabhata (A.D. 476–550), Brahmagupta (A.D. 598–670), Mahāvīra (9th century), and Bhāskara (1114–1185). We include some historical questions from Bhāskara and Brahmagupta in Chapter 5 (p. 262).

The Hindus developed arithmetic independently of geometry and had a fairly good knowledge of rudimentary algebra. They knew that quadratic equations had two solutions, and they had a fairly good approximation for π. Astronomy motivated their study of trigonometry. Around 1200, scientific activity in India declined, and mathematical progress ceased and did not revive until the British conquered India in the 18th century.

The Arabs invited Hindu scientists to settle in Baghdad, and when Plato's Academy closed in A.D. 529, many scholars traveled to Persia and became part of the Arab world. Omar Khayyám (1048–1122) and Nasîr-Eddin (1201–1274) worked freely with irrationals, which contrasts with the Greek idea of number. What we call Pascal's triangle dates back to this period (see Figure 1.4, p. 8). The word *algebra* comes from the Arabs in a book by the astronomer Mohammed ibn Musa al-Khwârizmî (780–850) entitled *Hisâb âl-jabr w'al mugâbalah* (see p. 216). Al-Khwârizmî solves quadratic equations and knows there are two roots, and even though they gave algebraic solutions of quadratic equations, they explained their work geometrically. The Arabs solved some cubics, but could solve only simple trigonometric problems. As stated by Morris Kline, "The Arabs made no significant advance in mathematics. What they did was absorb Greek and Hindu mathematics, preserve it, and, ultimately . . . transmit it to Europe."*

Check **www.mathnature.com** for links to many excellent sites on Hindu and Arabian mathematics. In this text, we have included the following stories and references:

Hindus:

Numeration system, Chapter 3, pp. 110–114
Historical Questions from Bhāskara, Chapter 5, p. 262 (Problems 56–58)
Historical Questions from Brahmagupta, Chapter 5, p. 262 (Problems 59–60)
Code of Manu (AD 100), Chapter 11, p. 564; this is an ethical law of classical Hinduism. It teaches that the

* *Ibid*, pp. 197.

Hindu and Arabian American periods : AD 500 to 1199

Cultural Events		Mathematical Events	
500	First plans of the Vatican Palace in Rome	630	Brahmagupta: algebra, astronomy
610	Mohammed's vision	710	Bede: calendar, finger arithmetic
697	Northern Irish submit to Roman Catholicism	750	First use of zero symbol
800	Charlemagne crowned emperor of Holy Roman Empire	810	Mohammed ibn Mûsâ al-Khwârizmî coins term 'algebra,'
832	Utrecht Psalter		Hindu numerals
	Beginning of Carolingian dynasties	850	Mahavira: arithmetic, algebra
870	First printed book	870	Iâbit ibn Qorra: algebra, magic squares, amicable numbers
871	Alfred the Great		
	Schism of the Church	900	Abû Kâmil: Algebra, Bakhshali manuscript
900	Vikings discover Greenland	976	Oldest example of written numerals in Europe
912–973	Reign of Otto I (The Great Emperor)	980	Abu'l-wefa: constructions, trig tables
950	Beginning of the Dark Ages	999	Pope Sylvester II (Gerbert): arithmetic, pi approximated as $\sqrt{8} \approx 2.83$
973–983	Emperor Otto II		
990	Development of systematic musical notation	1000	Sridhara recognizes the importance of the zero
993	First canonization of saints	1020	Al-Karkhî: algebra
1003	Leif Erickson crosses Atlantic to Vinland	1075	Game of rithmomachia
1008	World's first novel, *Tale of Genji*	1110	Omar Khayyám: cubic equations, Pascal's triangle
1028	School of Chartres	1120	Bhāskara
1050	Normans penetrate England	1125	Earliest account of a mariner's compass
1054	Macbeth defeated at Dunsinane	1150	Bhāskara: algebra
1065	Consecration of Westminster Abbey	1175	Averroës: trigonometry, astronomy
1086	Chinese use moveable type to print books		
1088	First modern university		
1096	Start of the First Crusade		
1110	Chinese invent the playing card		
1125	Commencement of troubadour music		
1154	Beginning of Plantagenet reign		
1165	Maimonides: *Mishneh Torah*		
1186	Domesday Book, tax census ordered by William the Conqueror		

caste system is divinely ordained. It also teaches the various stages through which a man is expected to pass in a successful life: student, householder, hermit, and wandering beggar. These states are only for twice-born men. Women should stay in the home under the protection and control of the chief male in the household. The code requires the cultivation of pleasantness, patience, control of mind, nonstealing, purity, control of senses, intelligence, knowledge, truthfulness, and nonirritability. The killing of cows is listed among the greatest of sins.

Arabs

Omar Khayyám, Pascal's triangle in Chapter 1, p. 12 (Example 7)

Mohammed ibn Musa al-Khwârizmî and the origin of the word *algebra*, Chapter 3, p. 216

Transition Period: 1200 to 1599

Mathematics during the Middle Ages was transitional between the great early civilizations and the Renaissance. In the 1400s the Black Death killed over 70% of the European population. The Turks conquered Constantinople, and many Eastern scholars traveled to Europe, spreading Greek knowledge as they traveled. The period from 1400 to 1600, known as the Renaissance, forever changed the intellectual outlook in Europe, and raised up mathematical thinking to new levels. Johann Gutenberg's invention of printing with movable type in 1450 changed the complexion of the world. Linen and cotton paper, which the Europeans learned about from the Chinese through the Arabians, came at precisely the right historical moment. The first printed edition of Euclid's *Elements* in a Latin translation appeared in 1482. Other early printed books were Apollonius' *Conic Sections*, Pappus' works, and Diophantus' *Arithmetica*.

The first breakthrough in mathematics was by artists who discovered mathematical perspective (see pp. 340–341 and E15). The theoretical genius in mathematical perspective was Leone Alberti (1404–1472). He was a secretary in the Papal Chancery writing biographies of the saints, but his work *Della Pictura* on the laws of perspective (1435) was a masterpiece. He said, "Nothing pleases me so much as mathematical investigations and demonstrations, especially when I can turn them into some useful practice drawing from mathematics and the principles of painting perspective and some amazing propositions on the moving of weights." He collaborated with Toscanelli, who supplied Columbus with maps for his first voyage. The best mathematician among the Renaissance artists was Albrecht Dürer (1471–1528), see p. 341. The most significant development of the Renaissance was the breakthrough in astronomical theory by Nicolaus Copernicus (1473–1543) and Johannes Kepler (1571–1630). There were no really significant new results in mathematics during this period of history. This book includes many references to the Transition Period. Some of those are listed here:

Petrus Apianus 1527 (Pascal's triangle, first in print), Chapter 1, p. 10

First recorded statement of the Fundamental Theorem of Arithmetic, Chapter 4, p. 154

Euclid's *Elements* printed in 1482, Chapter 4, p. 154; Chapter 6, p. 297

First use of the word *fraction* by Chaucer in 1321, Chapter 4, p. 170

First use of the fraction bar in 1556 by Tartaglia, Chapter 4, p. 170

Pythagorean theorem on a Moslem manuscript in 1258, Chapter 4, p. 178

First use of the root symbol, Chapter 4, p. 179

First attempts at using a decimal point, Chapter 4, p. 185

Dürer's *Designer of the Lying Woman*, Chapter 6, p. 341 (Figure 6.63)

Invention of logarithms, Chapter 8, pp. 400 and 402

John Napier (1550–1617), Chapter 8, p. 402

Tycho Brahe (1546–1601), Chapter 8, p. 402

Historical Question: Shroud of Turin, Chapter 8, p. 422 (Problem 57)

Leonardo Fibonacci (1170–1250), Chapter 9, p. 456

Girolamo Cardano (1501–1576), Chapter 11, p. 573

It is interesting to tie together some of the previous timelines to trace the history of algebra. It began around 2000 B.C. in Egypt and Babylon. This knowledge was incorporated into the mathematics of Greece between 500 B.C. and A.D. 320, as well as into the Arabian Empire and Indian mathematics around A.D. 1000. By the Transition Period, the great ideas of algebra had made their way to Europe, as shown in Figure 3.

Additional information can be found on the World Wide Web; check our Web page at **www.mathnature.com**.

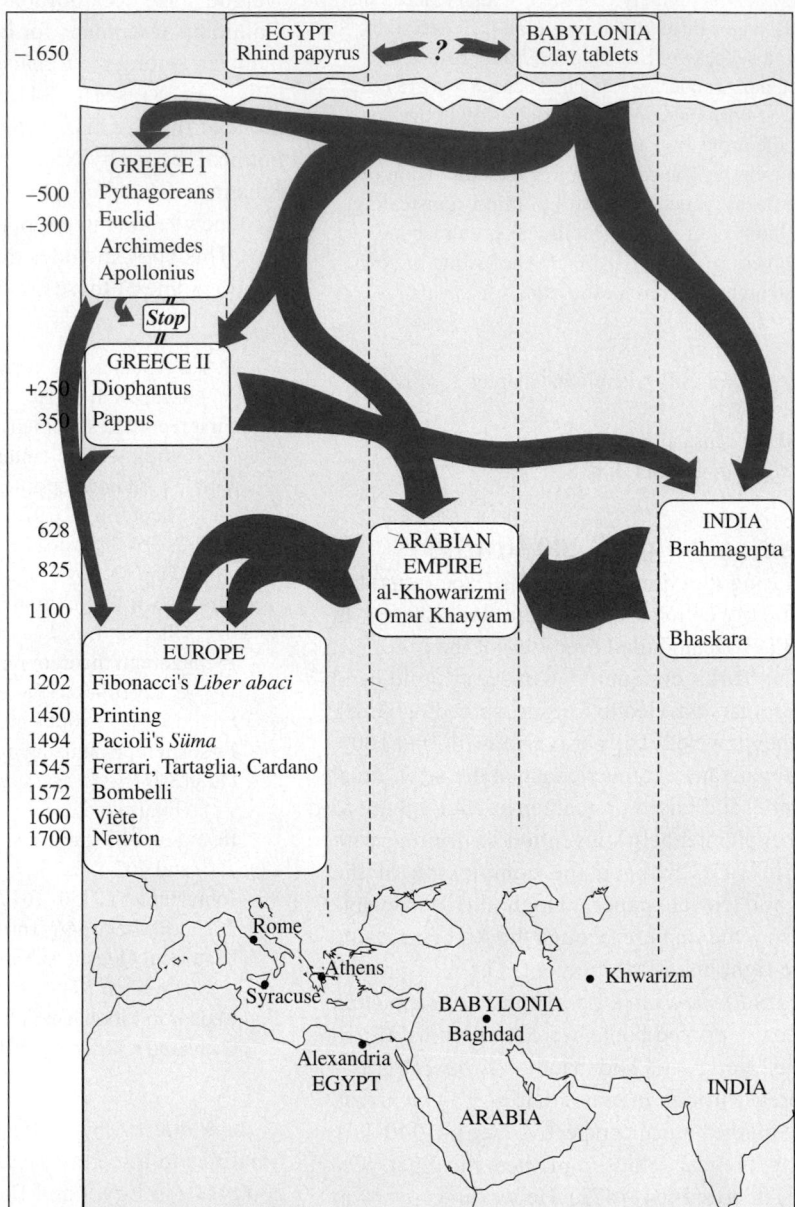

Figure 3
Mainstreams in the flow of algebra.
Additional information can be found on the World Wide Web; check our
Web page at www.mathnature.com.

Transition period: 1200 to 1599

Cultural Events

1206	Genghis Khan becomes chief prince of the Mongols
1209	Francis of Assisi initiates brotherhood
1233	Start of the Papal inquisition
1240	Amiens Cathedral rebuilt
1273	Thomas Aquinas: *Summa Theologicae*
1275	Moses de Leon: *Zohar*, major source for the cabala
1299	Florentine bankers are forbidden to use Hindu numerals
1307-21	Dante: *Divine Comedy*
1321	Chaucer
1322	The pope forbids the use of counterpoint in church music
1347-51	Approximately 75 million die of the Black Death
1364	Aztecs build Tenochtitlán
1378	Beginning of the Great Schism
1390	Chaucer: *Canterbury Tales*
1396	Metal type used for printing
1417	End of Great Schism
1420	Gutenberg and Kostner invent printer with moveable type
1429	Joan of Arc raises siege of Orleans
1435	Rogier Van der Weyden
1436	Fra Angelico begins frescoes at San Marco
1450	Florence is center of Renaissance
1454	Gutenberg prints Bible
1465	First printed music
1470	First illustrated books
1474	First book printed in English
1484	Botticelli: *Birth of Venus*
1492	Columbus discovers America
1497	Vasco da Gama rounds Cape of Good Hope
	Michelangelo: *David*
1507	Leonardo da Vinci begins *Leda and the Swan*
1513	Machiavelli: *The Prince*
1517	Luther launches Reformation
1520	Magellan discovers the straits; Luther excommunicated
1534	Henry VIII becomes head of the Church of England
1543	Publication of Copernicus' work
1558	Elizabeth crowned Queen of England
1567	Bothwell abducts Mary, Queen of Scots
1569	Tycho Brahe begins construction of a 19-foot quadrant
1573	Francis Drake sees Pacific Ocean
1583	Pope Gregory XIII creates new calendar
1588	England defeats Spanish Armada
1596	Discovery of the Marquesas

Mathematical Events

1202	Fibonacci: arithmetic, algebra, geometry, sequences, *Liber Abaci*
1250	Sacrobosco: Hindu-Arabic numerals
1260	Campanus translates Euclid
1267	Roger Bacon: *Opus*
1280	Geometry used as the basis of painting
1281	Li Yeh introduces notation for negative numbers
1303	Chu Shi-Kie: algebra, solutions of equations, Pascal's triangle
1325	Thomas Bradwardine: arithmetic, geometry, star polygons
1360	Nicole Oresme: coordinates, fractional exponents
1400	In Florence, commercial activity results in several books on mercantile arithmetic
1425	Use of perspective gives depth to Renaissance painting
1435	Ulugh Beg: trig tables
1460	Georg von Peurbach: arithmetic, table of sines
1464	Regiomontanus: establishes trigonometry
1470	First printed arithmetic book
1482	First printing of Euclid's *Elements*
1489	Johann Widmann: first use of + and − signs
1492	Pellos: use of decimal point
1505	Leonardo da Vinci: geometry, art, optics
1506	Scipione dal Ferro: cubic equations
1510	Albrecht Dürer: perspective, polyhedra, curves
1514	Dürer's *Melancholia* contains magic square
1525	Stifel: number mysticism; Rudolff; algebra, decimals
1527	Petrus Apianus; Pascal's triangle
1530	Copernicus: astronomy, trigonometry
1540	Gemma Frisius: arithmetic
1545	Tartaglia: cubic equations, arithmetic
	Ferrari: quartic equations
1550	Cardano: *Ars Magna*
	Scheubel: algebra
	Adam Riese: originator of the radical sign
1557	Robert Recorde: arithmetic, algebra, first use of = sign
1564	Galileo Galilei born
1572	Bombelli: algebra, cubic equations
1579	Viète: advocated use of decimal notation
1580	Viète: algebra, geometry, much modern notation
1583	Clavius: arithmetic, algebra, geometry
1593	Adrianus Romanus: value of π

Century of Enlightenment: 1600 to 1699

From Shakespeare and Galileo to Peter the Great and the great Bernoulli family, this period marks the growth of intellectual endeavors, and both technology and knowledge grew as never seen before in history. A great deal of the content of this book focuses on discoveries from this period of time, so instead of providing a commentary in this overview, we will simply list the references to this period in world history. Other sources and links are found on our Web page **www.mathnature.com**.

Blaise Pascal (1623–1662)/Pascal's triangle, Chapter 1, p. 9

Gottfried Leibniz (1646–1716)/logic; universal characteristic, Chapter 2, p. 56

Galileo Galilei (1564–1642)/deductive reasoning, Chapter 2, p. 72

John Napier (1550–1617)/calculating device, Chapter 3, p. 126; logarithms, Chaper 8, p. 402

Pascal's calculator (1642), Chapter 3, p. 127

Leibniz' reckoning machine 1695, Chapter 3, p. 127

Introduction of the $\times$ symbol, Chapter 4, p. 144

Leibniz proposed replacing the symbol with the dot, Chapter 4, p. 144

Marin Mersenne/Discovers first Mersenne prime, Chapter 4, p. 156

John Wallis used the $\div$ symbol, Chapter 4, p. 169

Simon Steven's use of a decimal point in 1585, Chapter 4, p. 185

Viète's use of a decimal point in 1600, Chapter 4, p. 185

Kepler's and Napier's use of a decimal point in 1616, Chapter 4, p. 185

Oughtred's use of a decimal point in 1624, Chapter 4, p. 185

Balam's use of a decimal point in 1653, Chapter 4, p. 185

Ozanam's use of a decimal point in 1691, Chapter 4, p. 185

First documented use of zero, Chapter 4, p. 188

Fermat invented a test for determining primes, Chapter 4, p. 156

René Descartes (1596–1650)/letters used for variables, Chapter 5, p. 243

Century of Enlightenment: 1600 to 1699

Cultural Events		Mathematical Events	
1600	Shakespeare: *Hamlet*	1600	Galileo: physics, astronomy, projectiles
1609	Pocahontas saves John Smith	1610	Kepler: astronomy, continuity
1611	King James Bible published	1617	Napier: logarithms, Napier's rods
1618	Beginning of the Thirty Years War	1621	Diophantus: *Arithmetica* published
1620	Pilgrims land at Plymouth	1630	Mersenne: number theory
1636	Harvard College founded, the first American college	1631	Oughtred: first table of natural logs
1643	Moliere founds *Theatre de la Comedie Française*	1635	Cavalieri: number theory
1646	Building of the Taj Mahal	1637	Descartes: *Discourse on Method*, analytic geometry
1649	Cromwell abolishes English monarchy		Fermat's Last Theorem stated in the margin of a book
1654	Coronation of Louis XIV of France	1640	Desargues: projective geometry
1659	Birth of Alessandro Scarlatti	1650	Pascal: conics, probability, computing machines, Pascal's triangle, John Wallis: algebra, imaginary numbers
1665	Great Plague in London kills 75,000		
	Newton's experiments on gravitation	1654	Pascal-Fermat correspondence begins study of probability
1668	La Fontaine: *Fables*	1658	Huygens invents the pendulum clock—theory of curves
1677	Spinoza: *Ethics*	1670	Sir Christopher Wren: architecture, imaginary numbers
1680	Stradivari makes the first cello	1678	Ceva: nature of concurrency
1683	First public museum	1680	Sir Isaac Newton: calculus, gravitation, series, hydrodynamics
1685	J. S. Bach and Handel born		
1689	Peter the Great becomes Czar of Russia	1682	Gottfried Leibniz: calculus, determinants, symbolic logic, notation, computing machines
		1690	Nicolaus Bernoulli: probability curves

Use of a colon for ratios and proportions (1633), Chapter 5, p. 263

Bernoulli family, Chapter 7, p. 372

Isaac Newton (1642–1727), Chapter 9, p. 471

Pierre de Fermat (1601–1665), Chapter 11, p. 572

Blaise Pascal (1623–1662)/beginning of probability theory, Chapter 11, p. 572

René Descartes (1596–1650)/invention of analytic geometry, Chapter 13, p. 667

René Descartes (1596–1650)/conic sections, Chapter 13, p. 682

Early Modern Period: 1700 to 1899

This period marks the dawn of modern mathematics. The Early Modern Period was characterized by experimentation and formalization of the ideas germinated in the previous century. There is so much that we could say about the period from 1700 to 1899. The mathematics that you studied in high school represents, for the most part, the ideas formulated during this period. There are a multitude of historical references to this period documented throughout the book. We list some of those references here:

George Boole (1815–1864)/Boolean algebra, Chapter 2, p. 57

Bertrand Russell (1872–1970)/formal logic, Chapter 2, p. 57

Alfred North Whitehead (1861–1947)/*Principia Mathematica*, Chapter 2, p. 57

Charles Dodgson (1832–1898)/logical proof, Chapter 2, p. 78

A. Henry Rhind/discovers Egyptian papyrus, Chapter 3, p. 104

Historical Question: duodecimal numeration system, Chapter 3, p. 119 (Problem 60)

Charles Babbage (1792–1871)/calculating machine, Chapter 3, p. 127

Work done by Fermat and Euler on factoring numbers, Chapter 4, p. 156

Colburn could do mental factorizations in the 1800s, Chapter 4, p. 156

Gauss: Any number can be written as the sum of three triangular numbers, Chapter 4, p. 184 (Problem 57)

Karl Gauss (1777–1835), Chapter 4, p. 186

π's value given as 3.2 in 1892 and as 4 in 1897, Chapter 4, p. 187

Niels Abel (1802–1829), Chapter 5, p. 217

Evariste Galois (1811–1832), Chapter 5, p. 224

Augustin Cauchy (1789–1857), Chapter 5, p. 224

Jean Fourier (1768–1830), Chapter 5, p. 224

Gregor Mendel (1822–1884), Chapter 5, p. 230

Wilhelm Weinberg (1862–1937), Chapter 5, p. 230

Henry Wadsworth Longfellow (1807–1882), Chapter 5, p. 257

Benjamin Banneker (1731–1806), Chapter 6, p. 321

Invention of the metric system, Chapter 7, p. 353

Sofia Kovalevsky (1850–1891), Chapter 7, p. 371

Sophie Germain (1776–1831), Chapter 10, p. 532

Development of the first theory of probability, Chapter 11, p. 572

Origins of statistics, Chapter 12, p. 627

Adolph Quetelet (1796–1874), Chapter 12, p. 627

Florence Nightingale (1820–1910), Chapter 12, p. 627

Leonhard Euler (1707–1783), Chapter 14, p. 711

Historical Questions (written by Euler), Chapter 14, p. 714

Gauss-Jordan elimination; Karl Gauss (1777–1855), Camille Jordan (1838–1922), Chapter 14, p. 730

James Sylvester *(American Journal of Mathematics)*, Chapter 14, p. 744

Czarina Catherine (1729–1796), Chapter 15, p. 762

Jules-Henri Poincaré (1854–1912), Chapter 15, p. 767

Felix Hausdorff (1868–1942), Chapter 15, p. 767

William Rowan Hamilton (1805–1865), Chapter 15, p. 768

Traveler's Dodecahedron problem (1771), Chapter 15, p. 771 (Problem 33)

Vilfredo Pareto (1848–1923), Chapter 16, p. 806

Marquis de Condorcet (1743–1794), Chapter 16, p. 815

Apportionment problems: Adams, Jefferson, Webster, Hamilton, and Huntington-Hill's plans, Chapter 16, pp. 836–847

Early Modern Period: 1700 to 1899

Cultural Events		Mathematical Events	
1705	Bach, *Two-part Invention no. 5*	1700	Jacob and Johann Bernoulli: applied calculus, probability
	Halley predicts return of 1682 comet	1715	Brook Taylor: series, geometry, calculus
1706	Benjamin Franklin born		of finite differences
1710	Leibniz: *Théodicée*	1720	Abraham de Moivre: probability, calculus,
1712	Last execution for witchcraft in England		complex numbers
1726	Jonathan Swift: *Gulliver's Travels*	1733	Saccheri: beginnings of analytic geometry
1738	Bach: *Mass in B Minor*	1735	Emilie de Breteuil: Newtonian studies
1740	Accession of Fredrick the Great; Israel Baal	1740	Colin Maclaurin: series, physics, higher plane curves
	Shem Toh founds Hasidism	1748	Maria Agnesi: analytic geometry
1742	Handel: *The Messiah*	1750	Leonhard Euler: number theory, applied mathematics
1756-63	The Seven Years War	1760	Compte de Buffon: connection between
1759	Voltaire: *Candide*		probability and π
1762	Rousseau: *Social Contract*	1761	Johann Peter: population statistics
1764	Paris Pantheon started	1770	Johann Lambert: irrationality of π,
1767	Isaac Watts: steam engine		non-Euclidean geometry, map projections
1776	American Declaration of Independence	1780	Lagrange: calculus, number theory
1779	Mozart: *Don Giovanni*	1790	Metric system invented
1789	French Revolution	1796	Karl Gauss: Num $= \Delta + \Delta + \Delta$
1796	Smallpox vaccine	1797	Caroline Herschel: astronomy
1799	Napoleon rules France	1799	Metric system adopted in France
1803	Robert Fulton: first steamboat	1805	Laplace: probability, differential equations, method
1804	Beethoven: *Eroica* symphony;		of least squares, integrals
	Haiti independence		Punched cards to operate Jacquard loom
1808	Goethe: *Faust, Part I*	1815	George Boole born
1810	Goya: *The Disasters of War*	1820	Sophie Germain: theory of numbers
1811	First mechanical press	1822	Feuerbach: geometry of the triangle
1812	Canned food	1824	Abel: elliptic functions, equations, series, calculus
1815	Battle of Waterloo	1825	Bolyai and Lobachevski: non-Euclidean geometry
1821	Rosetta Stone deciphered	1830	Cauchy: calculus, complex variables
1826	First photograph	1832	Babbage: calculating machines; Galois: groups,
1828	Alexander Dumas: *The Three Musketeers*		theory of equations
1830	Simon Bolivar liberates South America	1837	Trisection of an angle and duplication of the cube
1836	The first telegraph		proved impossible
1848	Karl Marx: *Communist Manifesto*	1843	Hamilton: quaternions
1851	Herman Melville: *Moby Dick*	1849	De Morgan: probability, logic
1855	Walt Whitman: *Leaves of Grass*	1850	Cayley: invariants, hyperspace, matrices
1859	Charles Darwin: *On the Origin of Species*		and determinants
1860	Gregor Mendel: genetics	1852	Byron: first programming in weaving industry
1861	American Civil War	1854	Riemann: calculus; Boole: logic, *Laws of Thought*
1862	Louis Pasteur: germ theory of infection	1855	Dirichlet: number theory
1865	Dodgson: *Alice in Wonderland*	1872	Dedekind: irrational numbers
1866	Alfred Nobel invents dynamite	1873	Brocard: geometry of the triangle
1869	Suez Canal opens	1874	Sofia Kovalevskaia wins Prix Bordin
1876	Alexander Graham Bell invents telephone	1879	Sylvester: theory of numbers, theory of invariants
1879	Thomas Edison invents lightbulb		Dodgson: Euclidean studies
1880	Rodin: *The Thinker*	1880	Georg Cantor: irrational numbers
1886	Coca-Cola bottled	1882	Lindemann: π a transcendental number
1888	Eastman develops the box camera	1886	Winifrid Merrill: first woman to receive a U.S. Ph.D.
1898	Spanish-American war		in mathematics
		1888	George Pólya born
		1890	Peano: axioms for natural numbers
		1895	Poincaré: analysis
		1896	Hadamard and Pousson: proof of prime number theorem
		1899	Hilbert: calculus

Modern Period: 1900 to Present

What we call the Modern Period includes all of the discoveries of the the last century. Students often think that all the important mathematics has been done, and there is nothing new to be discovered, but this is not true. Mathematics is alive and changing every day. There is no way a short commentary or overview can convey the richness or implications of the mathematical discoveries of this period. Historical references to this period included here are:

Modern Period: 1900 to present

Cultural Events		Mathematical Events	
1900	Freud's theories	1900	Hilbert: twenty-three famous problems
1903	First powered aircraft		Russell and Whitehead: *Principia Mathematica*, logic
1908	Henry Ford: first Model T		Cezanne orients paintings around the cone, sphere and cube
1914	World War I begins		
1917	Russian Revolution	1906	Frechet: abstract spaces
1920	U.S. women gain the right to vote	1916	Einstein: general theory of relativity
1927	Charles A. Lindbergh: solo transatlantic flight	1917	Hardy and Ramanujan; analytic number theory
1928	Penicillin	1925	Elbert Cox: first black man to receive a U.S. Ph.D. in mathematics
1929	Stock market crash		
1930	Gandhi leads march to the Salt sea	1930	Emmy Noether: algebra
1933	Hitler takes power	1931	Godel's theorem
1934	Mao heads Chinese Revolution	1934	Fields Medal established
1939	World War II begins	1946	First electronic computer; Bourbaki: *Elements*
1941	Japan bombs Pearl Harbor	1949	Marjorie Browne and Evelyn Boyd: first black women to receive U.S. Ph.D's in mathematics
1942	First controlled nuclear chain reaction		
1945	United States drops atomic bomb on Hiroshima	1950	Norbert Wiener: cybernetics
	United Nations formed	1952	John von Neumann: game theory
1947	India declares independence	1955	Homological algebra
1950	Korean War begins	1956	Turing: developed Turing Test for computer intelligence
1953	Watson and Crick discover double helix structure of DNA	1957	Datatron: first medium-priced computer ($325,000–$600,000)
1954	Birth of rock and roll	1963	Cohen: continuum hypothesis
1955	Salk polio vaccine developed	1965	John Kemeny and Thomas Kurtz develop BASIC
1957	U.S.S.R. launches first orbiting satellite: *Sputnik I*	1976	Appel and Haken solve four-color problem
1963	U.S. involvement in Vietnam War begins	1977	Apple II personal computer introduced (price: $1799)
1967	First human heart transplant	1980	Rubik's cube sweeps the world
1969	U.S. puts first man on the moon	1984	Mertens Conjecture disproved; Bieberbach Conjecture proved
1976	Physicists discover the "Charmed Quark"		
1977	Viking mission lands on Mars	1994	RSA: "unbreakable" encryption; Fermat's Last Theorem proved
1980	Smallpox declared extinct		
1981	*Voyager 2* sends back pictures from Saturn		
1987	October stock market crash		
1988	AIDS becomes worldwide epidemic		
1989	*Voyager 2* sends back pictures from Neptune		
1997	*Pathfinder* lands on Mars		

INTERNET TIMELINE; see page 130

As we enter the new millennium, we can only imagine and dream about what is to come!

One of the major themes of this text has been problem solving. The following problem set is a potpourri of problems that should give you a foretaste of the variety of ideas and concepts that we will consider in this book. Although none of these problems is to be considered routine, you might wish to attempt to work some of them before you begin, and then return to these problems at the end of your study in this book.

Prologue Problem Set

1. **IN YOUR OWN WORDS** The title of the prologue begins with "Why study math?" Take a clean sheet of paper and draw a vertical line down the center. In the left column, write down reasons why you should study math, and in the right column, write down reasons why you should not study math. Spend at least 20 minutes filling in reasons. Finally, write a summary sentence or two describing what you found.

2. **IN YOUR OWN WORDS** What are the seven chronological periods into which the prologue divided history? Which period seems the most interesting to you, and why?

3. **IN YOUR OWN WORDS** Select what you believe to be the most interesting cultural event and the most interesting mathematical event of the Egyptian, Babylonian, and Native American periods.

4. **IN YOUR OWN WORDS** Select what you believe to be the most interesting cultural event and the most interesting mathematical event of the Greek, Chinese, and Roman periods.

5. **IN YOUR OWN WORDS** Select what you believe to be the most interesting cultural event and the most interesting mathematical event of the Hindu and Arabian period.

6. **IN YOUR OWN WORDS** Select what you believe to be the most interesting cultural event and the most interesting mathematical event of the Transition period.

7. **IN YOUR OWN WORDS** Select what you believe to be the most interesting cultural event and the most interesting mathematical event of the Century of Enlightenment.

8. **IN YOUR OWN WORDS** Select what you believe to be the most interesting cultural event and the most interesting mathematical event of the Early Modern period.

9. **IN YOUR OWN WORDS** Select what you believe to be the most interesting cultural event and the most interesting mathematical event of the Modern period.

10. A long, straight fence having a pole every 8 ft is 1,440 ft long. How many fence poles are needed for the fence?

11. How many cards must you draw from a deck of 52 playing cards to be sure that at least two are from the same suit?

12. How many people must be in a room to be sure that at least four of them have the same birthday (not necessarily the same year)?

13. Find the units digit of $3^{2001} - 2^{2001}$.

14. If a year had two consecutive months with a Friday the thirteenth, which months would they have to be?

15. On Saturday evenings, a favorite pastime of the high school students in Santa Rosa, California, is to cruise certain streets. The selected routes are shown in the following illustration. Is it possible to choose a route so that all of the permitted streets are traveled exactly once?

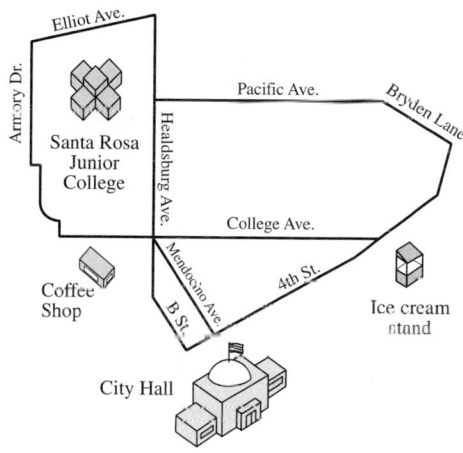

Santa Rosa street problem

16. What is the largest number that is a divisor of both 210 and 330?

17. If Ann is 3 years older than Brittany, Brittany is 2 years older than Chelsea, Deidre is 5 years older than Elysse, and Elysse is a year older than Fawn, then what can be said of Deidre's age as compared to Chelsea's age?

18. The News Clip shows a letter printed in the "Ask Marilyn" column of *Parade* magazine. How would you answer it? *Hint:* We won't give you the answer, but we will quote one line from Marilyn's answer: "So the question should be not why the smaller one yields that much, but why it yields that little."

> **Dear Marilyn,**
>
> I recently purchased a tube of caulking and it says a 1/4-inch bead will yield about 30 feet. But it says a 1/8-inch bead will yield about 96 feet —more than three times as much. I'm not a math genius, but it seems that because 1/8 inch is half of 1/4 inch, the smaller bead should yield only twice as much. Can you explain it?
>
> Norm Bean, St. Louis, Mo.

From "Ask Marilyn," by Marilyn vos Savant, *Parade Magazine*, September 27, 1992. Reprinted with permission from Parade, © 1992.

19. If the population of the world on 10/12/2002 is 6.248 billion, when do you think the world population will reach 7 billion? Calculate the date (to the nearest month) using the information that the world population reached 6 billion on October 12, 1999.

20. The Pacific 10 football conference consists of the following schools:

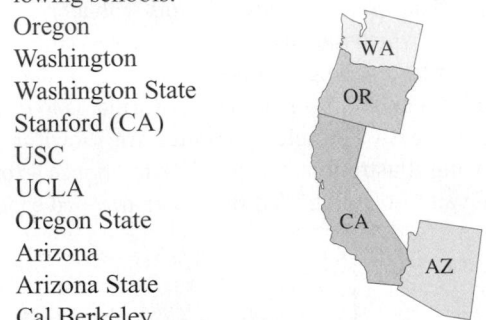

Oregon
Washington
Washington State
Stanford (CA)
USC
UCLA
Oregon State
Arizona
Arizona State
Cal Berkeley

   a. Is it possible to visit each of these schools by crossing each common state border exactly once? If so, show the path.
   b. Is it possible to start the trip in any given state and end the trip in the state in which you started?

21. If $(a, b) = a \times b + a + b$, determine the value of $((1, 2), (3, 4))$.

22. If it is known that all Angelenos are Venusians and all Venusians are Los Angeles residents, then what must necessarily be the conclusion?

23. If 1 is the first odd number, what is the 473rd odd number?

24. If $1 + 2 + 3 + \cdots + n = \dfrac{n(n + 1)}{2}$, what is the sum of the first 100,000 counting numbers beginning with 1?

25. A four-inch cube is painted red on all sides. It is then cut into one-inch cubes. What fraction of all the one-inch cubes are painted on one side only?

26. If slot machines had two arms and people had one arm, then it is probable that our number system would be based on the digits 0, 1, 2, 3, and 4 only. How would the number we know as 18 be written in such a number system?

27. If $M(a, b)$ stands for the larger number in the parentheses, and $m(a, b)$ stands for the lesser number in the parentheses, what is the value of $M(m(1, 2), m(2, 3))$?

28. If a group of 50 persons consists of 20 males, 12 children, and 25 women, how many men are in the group?

29. There are only five regular polyhedra and we have shown the patterns that give those polyhedra. Name the polyhedron obtained from each of the patterns shown in the figure.

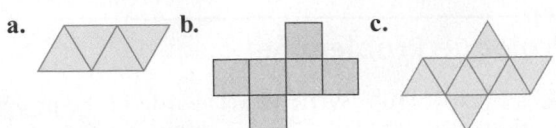
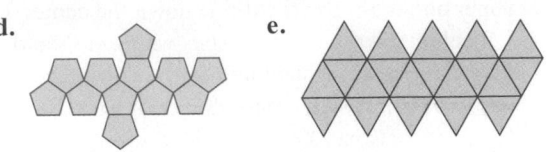

30. Jack and Jill decided to exercise together by walking around a lake. Jack walks around the lake in 16 minutes and Jill jogs around the lake in 10 minutes. If Jack and Jill start at the same time and at the same place, and continue to exercise until they return to the starting point at the same time, how long will they be exercising?

31. What is the 1,000th positive integer that is not divisible by 3?

32. A frugal man allows himself a glass of wine before dinner on every third day, an after-dinner chocolate every fifth day, and a steak dinner once a week. If it happens that he enjoys all three luxuries on March 31, what will be the date of the next steak dinner that is preceded by wine and followed by an after-dinner chocolate?

33. How many trees must be cut to make a trillion one-dollar bills? To answer this question you need to make some assumptions. Assume that a pound of paper is equal to a pound of wood, and also assume that a dollar bill weighs about one gram. This implies that a pound of wood yields about 450 dollar bills. Furthermore, estimate that an average tree has a height of 50 ft and a diameter of 12 inches. Finally, assume that wood yields about 50 lb/ft^3.

34. Estimate the volume of beer in the six-pack shown in the photograph.

35. You are given a square with sides equal to 8 inches, with two inscribed semicircles of radius 4. What is the area of the shaded region?

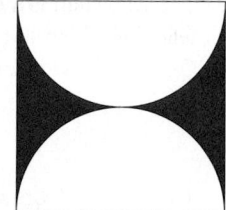

36. Critique the statement given in the News Clip.

Smoking ban
Judy Green, owner of the White Restaurant and an adamant opponent of a smoking ban, went so far as to survey numerous restaurants. She cited one restaurant that suffered a 75% decline in business after the smoking ban was activated.

37. The two small circles have radii of 2 and 3. Find the ratio of the area of the smallest circle to the area of the shaded region.

38. A large container filled with water is to be drained, and you would like to drain it as quickly as possible. You can drain the container with either one 1-in. diameter hose or two $\frac{1}{2}$-in. hoses. Which do you think would be faster (one 1-in. drain or two $\frac{1}{2}$-in. drains), and why?

39. A gambler went to the horse races two days in a row. On the first day, she doubled her money and spent $30. On the second day, she tripled her money and spent $20, after which she had what she started with the first day. How much did she start with?

40. What conclusions can you draw from the graphs that are reprinted here from the July 19, 1993, issue of *Newsweek*?

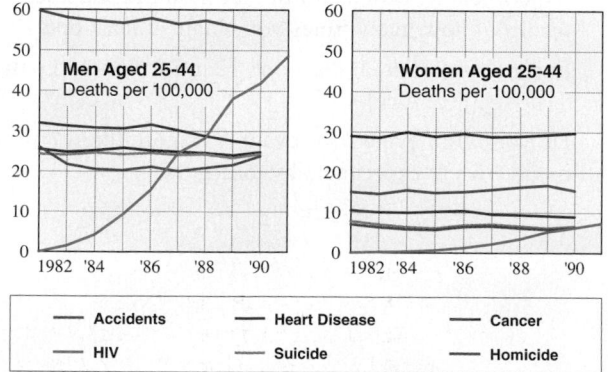

Men Aged 25-44
Deaths per 100,000

Women Aged 25-44
Deaths per 100,000

— Accidents — Heart Disease — Cancer
— HIV — Suicide — Homicide

Figures for causes other than HIV unavailable for 1991. Source: Centers for Disease Control and Prevention

41. A charter flight has signed up 100 travelers. They are told that if they can sign up an additional 25 persons, they can save $78 each. What is the cost per person if 100 persons make the trip?

42. Find $\lim\limits_{n \to \infty} \left(1 + \dfrac{1}{n}\right)^n$.

43. Suppose that it costs $450 to enroll your child in a 10-week summer recreational program. If this cost is pro-rated (that is, reduced linearly over the 10-week period), express the cost as a function of the number of weeks that have elapsed since the start of the 10-week session. Draw a graph to show the cost at any time for the duration of the session.

44. Candidates Ramirez (R), Smith (S), and Tillem (T) are running for office. According to public opinion polls, the preferences are (percentages rounded to the nearest percent):

Ranking:	38%	29%	24%	10%
1st choice	R	S	T	R
2nd choice	S	R	S	T
3rd choice	T	T	R	S

   a. Who will win the plurality vote?
   b. Who will win by the Borda count?
   c. Does a strategy exist that the voters in the 24% column could use to vote insincerely to help their first choice?

45. Suppose the percentage of alcohol in the blood t hours after consumption is given by

$$C(t) = 0.3e^{-t/2}$$

What is the rate at which the alcohol is changing with respect to time?

46. A hospital wishes to provide for its patients a diet that has a minimum of 100 g of carbohydrates, 60 g of protein, and 40 g of fats per day. These requirements can be met with two foods:

Food	Carbohydrates	Protein	Fats
A	6 g	3 g	1 g
B	2 g	2 g	2 g

It is also important to minimize costs; food A costs $0.14 per unit and food B costs $0.06 per unit. How many units of each food should be bought for each patient per day to meet the minimum daily requirements at the lowest cost?

47. Here is a map of a small village. To walk from *A* to *B*, Sarah obviously must walk at least 7 blocks (all the blocks are the same length). What is the number of shortest paths from *A* to *B*?

48. What is the smallest number of operations needed to build up to the number 100 if you start at 0 and use only two operations: doubling or increasing by 1. *Challenge*: Answer the same question for the positive integer n.

49. On June 1, 2000, the U.S. national debt was \$5.7 trillion and on that date there were 274.9 million people. How long would it take to pay off this debt if *every* person pays \$1 per day?

50. Supply the missing number in the following sequence: 10, 11, 12, 13, 14, 15, 16, 17, 20, 22, 24, ____, 100, 121, 10000.

51. If $\log_2 x + \log_4 x = \log_b x$, what is b?

52. Answer the question asked in the News Clip from the "Ask Marilyn" column of *Parade* magazine.

Dear Marilyn,

Three safari hunters are captured by a sadistic tribe of natives and forced to participate in a duel to the death. Each is given a pistol and tied to a post the same distance from the other two. They must take turns shooting at each other, one shot per turn. The worst shot of the three hunters (1 in 3 accuracy) must shoot first. The second turn goes to the hunter with 50–50 (1 in 2) accuracy. And (if he's still alive!) the third turn goes to the crack shot (100% accuracy). The rotation continues until only one hunter remains, who is then rewarded with his freedom. Which hunter has the best chance of surviving, and why?

From "Ask Marilyn," by Marilyn vos Savant, *Parade Magazine*, July 16, 1995. Reprinted with permission from Parade, © 1992.

53. How many different configurations can you see in the following figure?

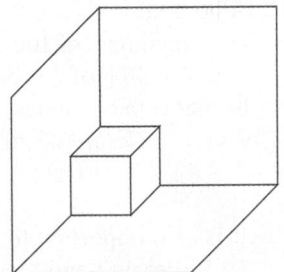

54. Five cards are drawn at random from a pack of cards that have been numbered consecutively from 1 to 104, and have been thoroughly shuffled. What is the probability that the numbers on the cards as drawn are in increasing order of magnitude?

55. The Kabbalah is a body of mystical teachings from the Qur'an (Koran). One Medieval inscription is shown:

ב	ט	ד
ז	ה	ג
ו	א	ח

This shows Hebrew characters that can be translated into numbers:

4	9	2
3	5	7
8	1	6

What can you say about this pattern of numbers?

56. What is the maximum number of points of intersection of n distinct lines?

57. The equation $P = 153{,}000e^{0.05t}$ represents the population of a city t years after 2000. What is the population in the year 2000? Show a graph of the city's population for the next 20 years.

58. The Egyptians had an interesting pictorial numeration system. Here is how you would count using Egyptian numerals:

|, ||, |||, ||||, |||||, ||||||, |||||||, ||||||||, |||||||||, ∩, ∩|, ∩||, ∩|||, . . .

Write down your age using Egyptian numerals. The symbol "|" is called a stroke, and the "∩" is called a heel bone. The Egyptians used a scroll for 100, a lotus flower for 1,000, a pointing finger for 10,000, a polliwog for 100,000, and an astonished man for the number 1,000,000. *Without* doing any research, write what you think today's date would look like using Egyptian numerals.

59. If you start with \$1 and double your money each day, how much money would you have in 30 days?

60. Consider two experiments:

Experiment *A*: Roll one die 4 times and keep a record of how many times you obtain at least one six.

Event E = {obtain at least one 6 in 4 rolls of a single die}

Experiment *B*: Roll a pair of dice 24 times and keep a record of how many times you obtain at least one 12.

Event F = {obtain at least one 12 in 24 rolls of a pair of dice}

Do you think event E or event F is more likely? You might wish to experiment by rolling dice.

1 The Nature of Problem Solving

> The idea that aptitude for mathematics is rarer than aptitude for other subjects is merely an illusion which is caused by belated or neglected beginners.
>
> J. F. HERBART

CONTENTS

OVERVIEW

There are many reasons for reading a book, but the best reason is because you want to read it. Although you are probably reading this first page because you were required to do so by your instructor, it is my hope that in a short while you will be reading this book because you *want* to read it. It was written for people who think they don't like mathematics, or people who think they can't work math problems, or people who think they are never going to use math. The common thread in this book is *problem solving*—that is, strengthening your ability to solve problems—not in the classroom, but outside the classroom. This first chapter is designed to introduce you to the nature of problem solving.

As you begin your trip through this book, I wish you a BON VOYAGE!

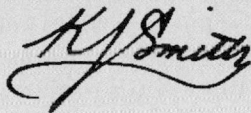

IMPORTANT IDEAS

Guidelines for Problem Solving (1.1)
Order of Operations (1.2)
Extended Order of Operations (1.3)
Laws of Exponents (1.3)
Denoting Sets (1.4)
One-to-one Correspondences and Infinite Sets (1.4)

BOOK REPORTS

Write a 500-word report on one of these books:
Mathematical Magic Show, Martin Gardner (New York: Alfred A. Knopf, 1977).
How to Solve It: A New Aspect of Mathematical Method, George Pólya (New Jersey: Princeton University Press, 1945, 1973).

LINKS

Individual Projects
Group Projects
Research Links

www.mathnature.com

1.1 PROBLEM SOLVING

A Word of Encouragement

Do you think of mathematics as a difficult, foreboding subject that was invented hundreds of years ago? Do you think that you will never be able (or even want) to use mathematics? If you answered "yes" to either of these questions, then I want you to know that I have written this book for you. I have tried to give you some insight into how mathematics is developed, and to introduce you to some of the people behind the mathematics. In this book, I will present some of the great ideas of mathematics, and then we will look at how these ideas can be used in an everyday setting to build your problem-solving abilities. *The most important prerequisite for this course is an openness to try out new ideas—a willingness to experience the suggested activities rather than to sit on the sideline as a spectator.* I have attempted to make this material interesting by putting it together differently from the way you might have had mathematics presented in the past. You will find this book difficult if you wait for the book or the teacher to give you answers—instead *be willing to guess, experiment, estimate, and manipulate,* and try out problems *without fear of being wrong!*

There is a common belief that mathematics is to be pursued only in a clear-cut logical fashion. This belief is perpetuated by the way mathematics is presented in most textbooks. Often it is reduced to a series of definitions, methods to solve various types of problems, and theorems. These theorems are justified by means of proofs and deductive reasoning. I do not mean to minimize the importance of proof in mathematics, for it is the very thing that gives mathematics its strength. But the power of the imagination is every bit as important as the power of deductive reasoning. As the mathematician Augustus De Morgan once said, "The power of mathematical invention is not reasoning but imagination."

Hints for Success

Mathematics is different from other subjects. One topic builds upon another, and you need to make sure that you understand *each* topic before progressing to the next one.

You must make a commitment to attend each class. Obviously, unforeseen circumstances can come up, but you must plan to attend class regularly. Pay attention to what your teacher says and does, and take notes. If you must miss class, write an outline of the text corresponding to the missed material, including working out each text example on your notebook paper.

You must make a commitment to daily work. Do not expect to save up and do your mathematics work once or twice a week. It will take a daily commitment on your part, and you will find mathematics difficult if you try to "get it done" in spurts. You could not expect to become proficient in tennis, soccer, or playing the piano by practicing once a week, and the same is true of mathematics. Try to schedule a regular time to study mathematics each day.

Read the text carefully. Many students expect to get through a mathematics course by beginning with the homework problems, then reading some examples, and reading the text only as a desperate attempt to find an answer. This procedure is backward; do your homework only *after* reading the text.

Writing Mathematics

The fundamental objective of education always has been to prepare students for life. A measure of your success with this book is a measure of its usefulness to you in your life. What are the basics for your knowledge "in life"? In this information age with ac-

Mathematics is one component of any plan for liberal education. Mother of all the sciences, it is a builder of the imagination, a weaver of patterns of sheer thought, an intuitive dreamer, a poet. The study of mathematics cannot be replaced by any other activity...

American Mathematical Monthly
Volume 56, 1949, p. 19

cess to a world of knowledge on the Internet, we still would respond by saying that the basics remain "reading, 'riting, and 'rithmetic." As you progress through the material in this book, we will give you opportunities to read mathematics and to consider some of the great ideas in the history of civilization, to develop your problem-solving skills ('rithmetic), and to communicate mathematical ideas to others ('riting). Perhaps you think of mathematics as "working problems" and "getting answers," but it is so much more. Mathematics is a way of thought that includes all three Rs, and to strengthen your skills you will be asked to communicate your knowledge in written form.

Journals

To begin building your skills in writing mathematics, you might keep a journal summarizing each day's work. Keep a record of your feelings and perceptions about what happened in class. Ask yourself, "How long did the homework take?" "What time of the day or night did I spend working and studying mathematics?" "What is the most important idea that I should remember from the day's lesson?" To help you with your journals, or writing of mathematics, you will find problems in this text designated **"IN YOUR OWN WORDS."** (For example, look at Problems 1–5 of the problem set at the end of this section.) There are no right answers or wrong answers to this type of problem, but you are encouraged to look at these for ideas of what you might write in your journal.

Journal Ideas

Write in your journal every day.

Include important ideas.

Include new words, ideas, formulas, or concepts.

Include questions that you want to ask later.

If possible, carry your journal with you so you can write in it anytime you get an idea.

Reasons for Keeping a Journal

It will record ideas you might otherwise forget.

It will keep a record of your progress.

If you have trouble later, it may help you diagnose areas for change or improvement.

It will build your writing skills.

Individual Research

At the end of each chapter you will find problems requiring some library research. I hope that as you progress through the course you will find one or more topics that interest you so much that you will want to do additional reading on that topic, even if it is not assigned.

Your instructor may assign one or more of these as term projects. One of the best ways for you to become aware of all the books and periodicals that are available is to log onto the Internet, or visit the library to research specific topics.

Preparing a mathematics paper or project can give you interesting and worthwhile experiences. In preparing a paper or project, you will get experience in using resources to find information, in doing independent work, in organizing your presentation, and in communicating ideas orally, in writing, and in visual demonstrations. You will broaden your background in mathematics and encounter new mathematical topics that you never before knew existed. In setting up an exhibit, you will experience the satisfaction of demonstrating what you have accomplished. It may be a way of satisfying your curiosity and your desire to be creative. It is an opportunity for developing originality, craftsmanship, and new mathematical understandings. If you are requested to do some individual research problems, here are some suggestions.

1. *Select a topic that has interest potential.* Do not do a project on a topic that does not interest you. Suggestions are given on the Web at

   www.mathnature.com

2. *Find as much information about the topic as possible.* Many of the Individual Research problems have one or two references to get you started. In addition, check the following sources:

   PERIODICALS: *The Mathematics Teacher, Teaching Children Mathematics* (formerly *Arithmetic Teacher*), and *Scientific American;* each of these has its own cumulative index; also check the *Reader's Guide.*

   SOURCE BOOKS: *The World of Mathematics* by Newman is a gold mine of ideas. *Mathematics,* a Time–Life book by David Bergamini, may provide you with many ideas. *Encyclopedias* can be consulted after you have some project ideas; however, I do not have in mind that the term project necessarily be a term paper.

   INTERNET: Use one or more search engines on the Internet for information on a particular topic. The more specific you can be in describing what you are looking for, the better the engine will be able to find material on your topic. Here are some of the search engines that you might check:

AltaVista	Ask Jeeves	Dogpile	Encuesta
Excite	Google	HotBot	Lycos
Search Internet	SavvySearch	WebCrawler	WWWomen
Yahoo			

   The two search engines that are especially useful are Dogpile and Google because they search the other search engines. Check the Web address for this book for specific computer links:

   www.mathnature.com

   If you do not have a computer or a modem, then you may need to visit your college or local library for access to this research information.

3. *Prepare and organize your material into a concise, interesting report.* Include drawings in color, pictures, applications, and examples to get the reader's attention and add meaning to your report.

4. *Build an exhibit that will tell the story of your topic.* Remember the science projects in high school? That type of presentation might be appropriate. Use models, applications, and charts that lend variety. Give your paper or exhibit a catchy, descriptive title.

5. *A **term** project cannot be done in one or two evenings.*

Group Research

Working in small groups is typical of most work environments, and being able to work with others to communicate specific ideas is an important skill to learn. At the end of each chapter is a list of suggested group projects, and you are encouraged to work with three or four others to submit a single report.

HISTORICAL NOTE

George Pólya
(1887–1985)

Born in Hungary, Pólya attended the universities of Budapest, Vienna, Göttingen, and Paris. He was a professor of Mathematics at Stanford University. Pólya's research and winning personality earned him a place of honor not only among mathematicians, but among students and teachers as well. His discoveries spanned an impressive range of mathematics, real and complex analysis, probability, combinatorics, number theory, and geometry. Pólya's *How to Solve It* has been translated into 15 languages. His books have a clarity and elegance seldom seen in mathematics, making them a joy to read. For example, here is his explanation of why he was a mathematician: "It is a little shortened but not quite wrong to say: I thought I am not good enough for physics and I am too good for philosophy. Mathematics is in between."

Guidelines for Problem Solving

We begin this study of **problem solving** by looking at the *process* of problem solving. As a mathematics teacher, I often hear the comment, "I can do mathematics, but I can't solve word problems." There *is* a great fear and avoidance of "real-life" problems because they do not fit into the same mold as the "examples in the book." Few practical problems from everyday life come in the same form as those you study in school.

To compound the problem, learning to solve problems takes time. All too often, the mathematics curriculum is so packed with content that the real process of problem solving is slighted, and because of time limitations, becomes an exercise in mimicking the instructor's steps instead of developing into an approach that can be used long after the final examination is over.

Before we build problem-solving skills, it is necessary to build certain prerequisite skills necessary for problem solving. It is my goal to develop your skills in the mechanics of mathematics, in understanding the important concepts, and finally in applying those skills to solve a new type of problem. I have segregated the problems in this book to help you build these different skills:

IN YOUR OWN WORDS	This type of problem asks you to discuss or rephrase main ideas or procedures using your own words.
Level 1 Problems	These are mechanical and drill problems, and are directly related to an example in the book.
Level 2 Problems	These problems require an understanding of the concepts and are loosely related to an example in the book.
Level 3 Problems	These problems are extensions of the examples, but generally do not have corresponding examples.
Problem Solving	These require problem-solving skills or original thinking and generally do not have direct examples in the book. These should be considered Level 3 problems.
Research Problems 	These problems require Internet research or library work. Most are intended for individual research but a few are group research projects. You will find these problems for research in the chapter summary and at the Web address for this book: **www.mathnature.com**

The model for problem solving that we will use was first published in 1945 by the great, charismatic mathematician George Pólya. His book *How to Solve It* (Princeton University Press, 1973) has become a classic. In Pólya's book you will find this problem-solving model as well as a treasure trove of strategy, know-how, rules of thumb, good advice, anecdotes, history, and problems at all levels of mathematics. His problem-solving model is as follows.

Guidelines for Problem Solving

Pay attention to boxes that look like this—they are used to tell you about important procedures that are used throughout the book.

First	You have to *understand the problem.*
Second	*Devise a plan.* Find the connection between the data and the unknown. Look for patterns, relate to a previously solved problem or a known formula, or simplify the given information to give you an easier problem.
Third	*Carry out the plan.*
Fourth	*Look back.* Examine the solution obtained.

Pólya's original statement of this procedure is reprinted in the box on the next page.

Let's apply this procedure for problem solving to the map shown in Figure 1.1; we refer to this problem as the **street problem.** Melissa lives at the YWCA (point *A*) and works at Macy's (point *B*). She usually walks to work. How many different routes can Melissa take?

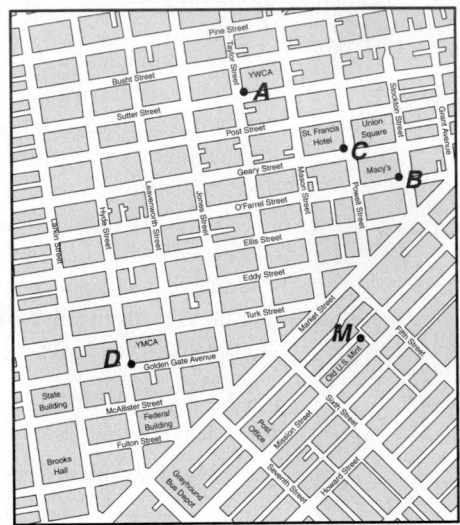

Figure 1.1 Portion of a map of San Francisco

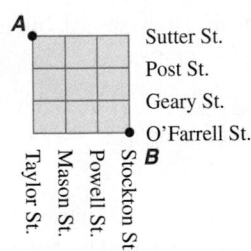

Figure 1.2 Simplified portion of Figure 1.1

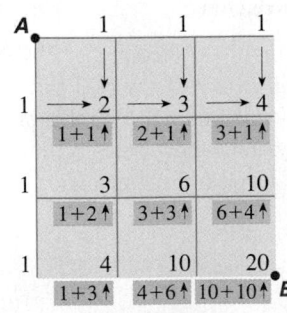

Figure 1.3 Map with solution

Where would you begin with this problem?

Step 1. *Understand the Problem.* Can you restate it in your own words? Can you trace out one or two possible paths? What assumptions are reasonable? We assume that she will not do any backtracking—that is, she always travels toward her destination. We also assume that she travels along the city streets—she cannot cut diagonally across a lot or a block.

Step 2. *Devise a Plan.* Simplify the question asked. Consider the simplified drawing shown in Figure 1.2.

Step 3. *Carry Out the Plan.* Count the number of ways it is possible to arrive at each point, or as it is sometimes called, a *vertex.*

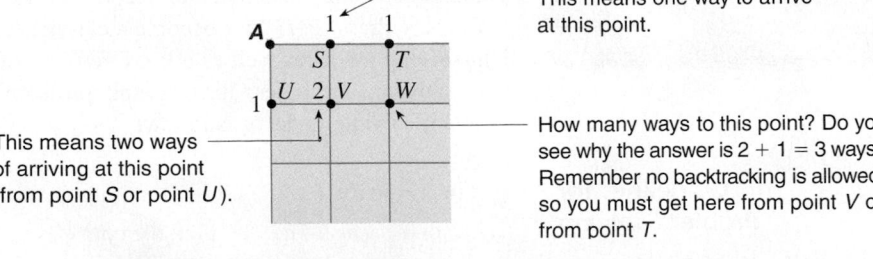

Now fill in all the possibilities on Figure 1.3, as shown by the above procedure.

Step 4. *Look Back.* Does the answer 20 different routes make sense? Do you think you could fill in all of them?

UNDERSTANDING THE PROBLEM

First

You have to understand the problem.

What is the unknown? What are the data? What is the condition? Is it possible to satisfy the condition? Is the condition sufficient to determine the unknown? Or is it insufficient? Or redundant? Or contradictory?

Draw a figure. Introduce a suitable notation.

Separate the various parts of the condition. Can you write them down?

DEVISING A PLAN

Second

Find the connection between the data and the unknown. You may be obliged to consider auxiliary problems if an immediate connection cannot be found.

Have you seen it before? Or have you seen the same problem in a slightly different form?

Do you know a related problem? Do you know a theorem that could be useful?

Look at the unknown! And try to think of a familiar problem having the same or a similar unknown.

Is the problem related to one you have solved before? Could you use it?

Could you use its result? Could you use its method? Should you introduce some auxiliary element in order to make its use possible?

Could you restate the problem? Could you restate it still differently? Go back to definitions.

If you cannot solve the proposed problem try to solve first some related problem. Could you imagine a more accessible related problem? A more general problem? A more special problem? An analogous problem? Could you solve a part of the problem? Keep only a part of the condition, drop the other part; how far is the unknown then determined, how can it vary? Could you derive something useful from the data? Could you think of other data appropriate to determine the unknown? Could you change the unknown or the data, or both if necessary, so that the new unknown and the new data are nearer to each other? Did you use all the data? Did you see the whole condition? Have you taken into account all essential notions involved in the problem?

CARRYING OUT THE PLAN

Third

Carry out your plan.

Carrying out your plan of the solution, *check each step*. Can you see clearly that the step is correct? Can you prove that it is correct?

LOOKING BACK

Fourth

Examine the solution.

Can you *check the result?* Can you check the argument?

Can you derive the result differently? Can you see it at a glance?

This is taken word for word as it was written by Pólya in 1941. It was printed in *How to Solve It* (Princeton, NJ: Princeton University Press, 1973).

EXAMPLE 1

In how many different ways could Melissa get from the YWCA (point *A*) to the St. Francis Hotel (point *C* in Figure 1.1), using the method of Figure 1.3?

Solution   Draw a simplified version of Figure 1.3, as shown below.

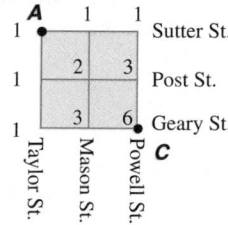

There are 6 different paths.   ◆

Problem Solving by Patterns

Let's formulate a general solution. Consider a map with a starting point *A*:

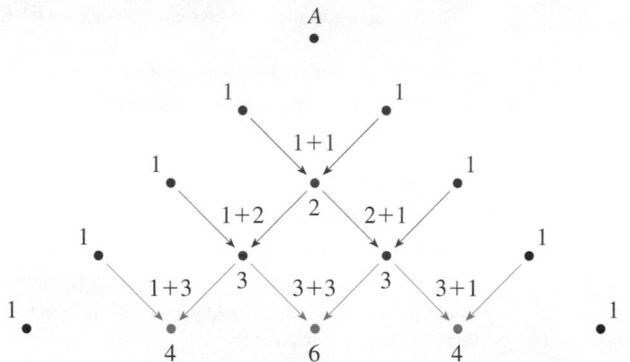

Do you see the pattern for building this figure? Each new row is found by adding the two previous numbers, as shown by the arrows. This pattern is known as **Pascal's triangle.** In Figure 1.4 the rows and diagonals are numbered for easy reference.

In the Sherlock Holmes mystery *The Final Solution*, Moriarty is a mathematician who wrote a treatise on Pascal's triangle.

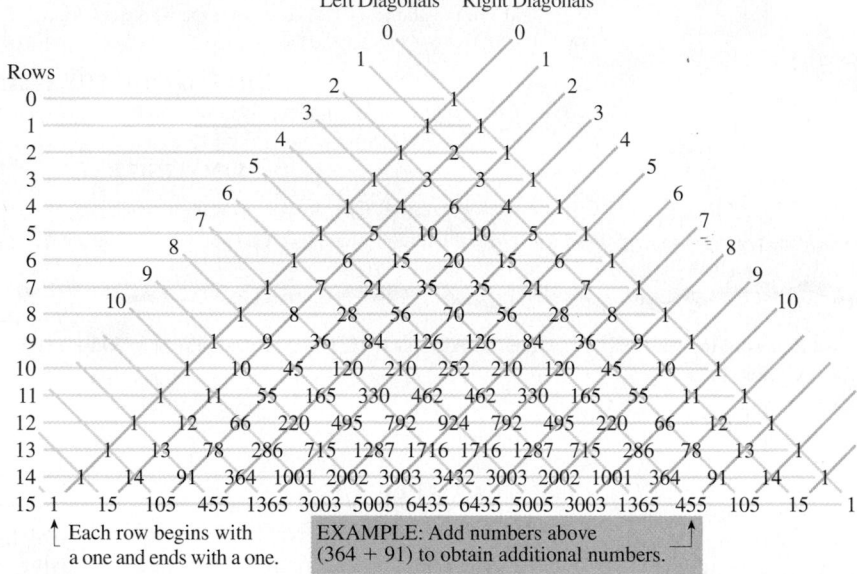

www.mathnature.com
There is an online interactive version of Pascal's triangle.

Figure 1.4 **Pascal's triangle**

How does this apply to Melissa's trip from the YWCA to Macy's? It is 3 blocks down and 3 blocks over. Look at Figure 1.4 and count out these blocks as shown in Figure 1.5.

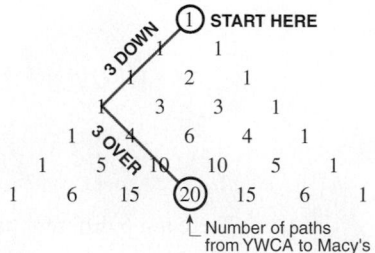

Figure 1.5 **Using Pascal's triangle to solve the street problem**

EXAMPLE 2

In how many different ways could Melissa get from the YWCA (point *A* in Figure 1.1) to the YMCA (point *D*)?

Solution   Look at Figure 1.1; from point *A* to point *D* is 7 blocks down and 3 blocks left. Use Figure 1.4 as follows:

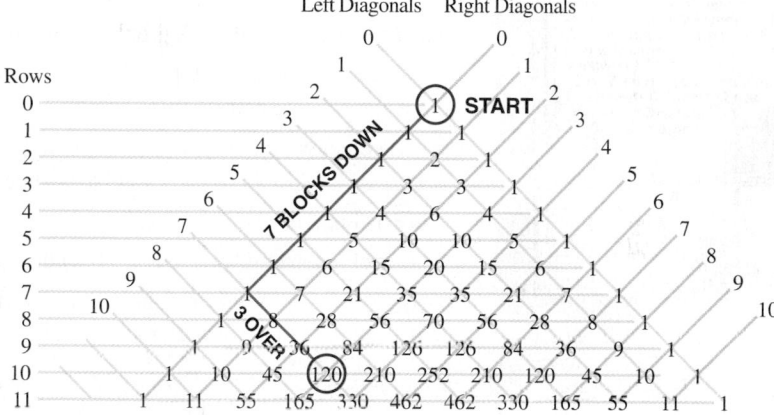

We see that there are 120 paths.   ◆

Pascal's triangle applies to the street problem only if the streets are rectangular. If the map shows irregularities (for example, diagonal streets or obstructions), then you must revert back to numbering the vertices.

EXAMPLE 3

In how many different ways could Melissa get from the YWCA (point *A*) to the Old U.S. Mint (point *M*)?

Solution   If the streets are irregular or if there are obstructions, you cannot use Pascal's triangle, but you can still count the blocks in the same fashion, as shown at the right. There are 52 paths from point *A* to point *M* (if, as usual, we do not allow backtracking).

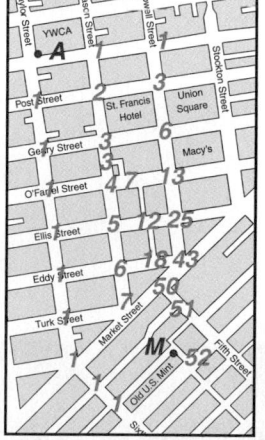

◆

Problem solving is a difficult task to master, and you are not expected to master it after one section of this book (or even after several chapters of this book). However, you must make building your problem-solving skills an ongoing process. One of the most important aspects of problem solving is to relate new problems to old problems. The problem-solving techniques outlined here should be applied when you are faced with a new problem. When you are faced with a problem similar to one you have

The title page of an arithmetic book by Petrus Apianus in 1527 is reproduced above. It was the first time Pascal's triangle appeared in print.

already worked, you can apply previously developed techniques (as we did in Examples 2 and 3). Now, because Example 4 seems to be a new type of problem, we again apply the guidelines for problem solving.

EXAMPLE 4   PÓLYA'S METHOD

A jokester tells you that he has a group of cows and chickens and that he counted 13 heads and 36 feet. How many cows and chickens does he have?

Solution   Let's use Pólya's problem-solving guidelines.

Understand the Problem. A good way to make sure you understand a problem is to attempt to phrase it in a simpler setting:

one chicken and one cow: 2 heads and 6 feet (chickens have two feet; cows have four)

two chickens and one cow: 3 heads and 8 feet

one chicken and two cows: 3 heads and 10 feet

Devise a Plan. How you organize the material is often important in problem solving. Let's organize the information into a table:

No. of chickens	No. of cows	No. of heads	No. of feet
0	13	13	52

Do you see why we started here? The problem says we must have 13 heads. There are other possible starting places (13 chickens and 0 cows, for example), but an important aspect of problem solving is to start with *some* plan.

No. of chickens	No. of cows	No. of heads	No. of feet
1	12	13	50
2	11	13	48
3	10	13	46
4	9	13	44

Carry Out the Plan. Now, look for patterns. Do you see that as the number of cows decreases by one and the number of chickens increases by one, the number of feet must decrease by two? Does this make sense to you? Remember, step 1 requires that you not just push numbers around, but that you understand what you are doing. Since we need 36 feet for the solution to this problem, we see

$$44 - 36 = 8$$

so the number of chickens must increase by an additional four. The answer is 8 chickens and 5 cows.

Look Back.

No. of chickens	No. of cows	No. of heads	No. of feet
8	5	13	36

Check: 8 chickens have 16 feet and 5 cows have 20 feet, so the total number of heads is $8 + 5 = 13$, and the number of feet is 36.   ◆

EXAMPLE 5   PÓLYA'S METHOD

If a family has 5 children, in how many different birth orders could the parents have a 3-boy, 2-girl family?

Solution   *Understand the Problem.* Part of understanding the problem might involve estimation. For example, if a family has 1 child, there are 2 possible orders (B or G). If a

family has 2 children, there are 4 orders (BB, BG, GB, GG); for 3 children, 8 orders; for 4 children, 16 orders; and for 5 children, a total of 32 orders. This means that an answer of 140 possible orders is an unreasonable answer.

Devise a Plan. You might begin by enumeration:

> BBBGG, BBGBG, BBGGB, . . .

This would seem to be too tedious. Instead, rewrite this as a simpler problem and look for a pattern.

1 child:	B ← one way	2 children:	BB ← one way
	G ← one way		BG ⎱
			⎰ two ways
			GB ⎰
			GG ← one way

3 children: BBB ← one way
BBG ⎫
BGB ⎬ three ways
GBB ⎭
BGG ⎫
GBG ⎬ three ways
GGB ⎭
GGG ← one way

Look at the possibilities:

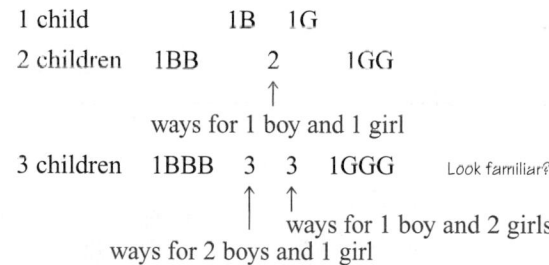

1 child   1B   1G

2 children   1BB   2   1GG
                      ↑
              ways for 1 boy and 1 girl

3 children   1BBB   3   3   1GGG   *Look familiar?*
                    ↑   ↑
                    | ways for 1 boy and 2 girls
          ways for 2 boys and 1 girl

Look at Pascal's triangle in Figure 1.4; for 5 children, look at row 5.

Carry Out the Plan.

row 5:   1   5   10   10   5   1
         ↑   ↑   ↑   ↑   ↑   ↑
      5 boys 4 boys 3 boys 2 boys 1 boy
              1 girl 2 girls 3 girls 4 girls 5 girls

They could have 3 boys and 2 girls in a total of 10 ways.

Look Back. We estimated that there are a total of 32 ways a family could have 5 children; let's sum the number of possibilities we found in carrying out the plan to see if it totals 32:

> $1 + 5 + 10 + 10 + 5 + 1 = 32$ ◆

The following example illustrates the necessity of carefully reading the question.

EXAMPLE 6

Nick and Marsha are driving from Santa Rosa, CA, to Los Angeles, a distance of 460 miles. They leave at 11:00 A.M. and average 50 mph. On the other hand, Mary and Dan leave at 1:00 P.M. in Dan's sports car. Who is closer to Los Angeles when they meet for dinner in San Luis Obispo at 5:00 P.M.?

Solution *Understand the Problem.* If they are sitting in the same restaurant, then they are all the same distance from Los Angeles. ◆

The last example of this section illustrates that problem solving may require that you change the conceptual mode.

EXAMPLE 7

If you have been reading the information in the margins, you may have noticed that Blaise Pascal was born in 1623 and died in 1662. You may also have noticed that the first time Pascal's triangle appeared in print was in 1527. How can this be?

Solution It was a reviewer of this book who brought this apparent discrepancy to my attention. The facts are all correct. How could Pascal's triangle have been in print almost 100 years before he was born? The fact is, the number pattern we call Pascal's triangle is *named after* Pascal, but was not *discovered* by Pascal. This number pattern seems to have been discovered several times, by Johann Scheubel in the 16th century, by the Chinese mathematician Nakone Genjun, and recent research has traced the triangle pattern as far back as Omar Khayyám. ◆

"Wait!" you exclaim. "How was I to answer the question in Example 7—I don't know all those facts about the triangle." You are not expected to know these facts, but you are expected to begin to think critically about the information you are given, and the assumptions you are making. It was never stated that Blaise Pascal was the first to think of or publish Pascal's triangle!

Problem Set 1.1

LEVEL 1

1. **IN YOUR OWN WORDS** In the text it was stated that "the most important prerequisite for this course is an openness to try out new ideas—a willingness to experience the suggested activities rather than to sit on the sideline as a spectator." Do you agree or disagree that this is the *most* important prerequisite? Discuss.

2. **IN YOUR OWN WORDS** What do you think the primary goal of mathematics education should be? What do you think it is in the United States? Discuss the differences between what it is and what you think it should be.

3. **IN YOUR OWN WORDS** In the chapter overview (did you read it?), it was pointed out that this book was written for people who think they don't like mathematics, or people who think they can't work math problems, or people who think they are never going to use math. Do any of those descriptions apply to you, or someone you know? Discuss.

4. **IN YOUR OWN WORDS** Discuss Pólya's problem-solving model.

5. **IN YOUR OWN WORDS** At the beginning of this section three hints for success were listed. Discuss each of these from your perspective. Are there any other hints that you might add to this list?

6. Given Pascal's triangle, as shown in Figure 1.4, write down the next two rows.

7. Describe the location of the numbers 1, 2, 3, 4, 5, . . . in Pascal's triangle.

8. Describe the location of the numbers 1, 4, 10, 20, 35, . . . in Pascal's triangle.

9. **IN YOUR OWN WORDS** In Example 2, the solution was found by going 7 blocks down and 3 blocks over. Could the solution also have been obtained by going 3 blocks over and 7 blocks down? Would this Pascal's triangle solution end up in a different location? Describe a property of Pascal's triangle that is relevant to an answer for this question.

10. If a family has 5 children, in how many ways could the parents have 2 boys and 3 girls as children?

11. If a family has 6 children, in how many ways could the parents have 3 boys and 3 girls as children?

12. If a family has 7 children, in how many ways could the parents have 4 boys and 3 girls as children?

13. If a family has 8 children, in how many ways could the parents have 3 boys and 5 girls as children?

In Problems 14–17, *what is the number of direct routes from point A to point B?*

14.

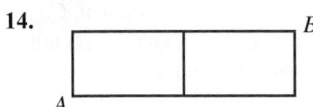

15.

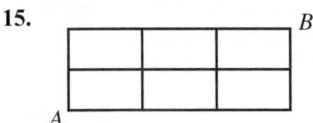

16.

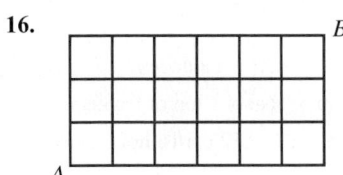

17.

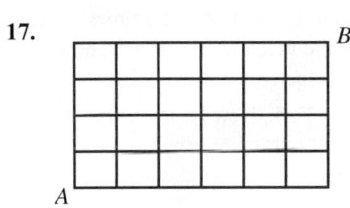

Use the map in Figure 1.6 *to determine the number of different paths from point A to the point indicated in Problems* 18–21. *Remember, no backtracking is allowed.*

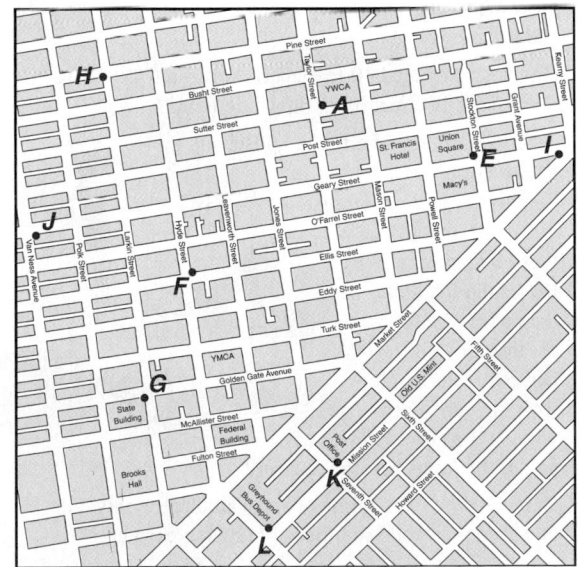

Figure 1.6 **Map of a portion of San Francisco**

18. *E*   **19.** *F*   **20.** *G*   **21.** *H*

LEVEL 2

22. If an island's only residents are penguins and bears, and if there are 16 heads and 34 feet on the island, how many penguins and how many bears are on the island?

23. Below are listed three problems. Do not solve these problems; simply tell which one you think is most like Problem 22.
   A. A penguin in a tub weighs 8 lb, and a bear in a tub weighs 800 lb. If the penguin and the bear together weigh 802 lb, how much does the tub weigh?
   B. A bottle and a cork cost $1.10, and the bottle is a dollar more than the cork. How much does the cork cost?
   C. Bob has 15 roses and 22 carnations. Carol has twice as many roses and half as many carnations. How many flowers does Carol have?

24. Ten full crates of walnuts weigh 410 pounds, whereas an empty crate weighs 10 pounds. How much do the walnuts alone weigh?

25. There are three separate, equal-size boxes, and inside each box there are two separate small boxes, and inside each of the small boxes there are three even smaller boxes. How many boxes are there all together?

26. If you expect to get 50,000 miles on each tire from a set of five tires (four and one spare), how should you rotate the tires so that each tire gets the same amount of wear, and how far can you drive before buying a new set of tires?

27. a. What is the sum of the numbers in row 1 of Pascal's triangle?
   b. What is the sum of the numbers in row 2 of Pascal's triangle?
   c. What is the sum of the numbers in row 3 of Pascal's triangle?
   d. What is the sum of the numbers in row 4 of Pascal's triangle?

28. What is the sum of the numbers in row n of Pascal's triangle?

Use the map in Figure 1.6 *to determine the number of different paths from point A to the point indicated in Problems* 29–32. *Remember, no backtracking is allowed.*

29. *I*   **30.** *J*   **31.** *K*   **32.** *L*

LEVEL 3

33. IN YOUR OWN WORDS Suppose you have a long list of numbers to add, and you have misplaced your calculator. Discuss the different approaches that could be used for adding this column of numbers.

34. IN YOUR OWN WORDS You are faced with a long division problem, and you have misplaced your calculator. You do not remember how to do long division. Discuss your alternatives to come up with the answer to your problem.

35. IN YOUR OWN WORDS You have 10 items in your grocery cart. Six people are waiting in the express lane (10 items or less); one person is waiting in the first checkout stand and two people are waiting in another checkout stand. The other checkout stands are closed. What additional information do you need in order to decide which lane to enter?

36. IN YOUR OWN WORDS You drive up to your bank and see five cars in front of you waiting for two lanes of the drive-through banking services. What additional information do you need in order to decide whether to drive through or park your car and enter the bank to do your banking?

37. A boy cyclist and a girl cyclist are 10 miles apart and pedaling toward each other. The boy's rate is 6 miles per hour, and the girl's rate is 4 miles per hour. There is also a friendly fly zooming continuously back and forth from one bike to the other. If the fly's rate is 20 miles per hour, by the time the cyclists reach each other, how far does the fly fly?

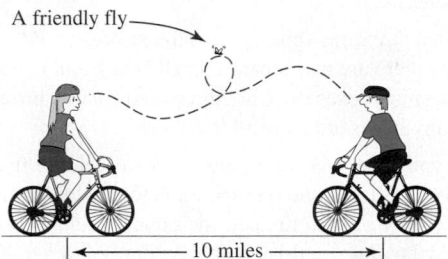

A friendly fly

|← 10 miles →|

38. Two volumes of Newman's *The World of Mathematics* stand side by side, in order, on a shelf. A bookworm starts at page i of Volume I and bores its way in a straight line to the last page of Volume II. Each cover is 2 mm thick, and the first volume is $\frac{17}{19}$ as thick as the second volume. The first volume is 38 mm thick without its cover. How far does the bookworm travel?

39. Alex, Beverly, and Cal live on the same straight road. Alex lives 10 miles from Beverly and Cal lives 2 miles from Beverly. How far does Alex live from Cal?

40. In a different language, *liro cas* means "red tomato." The meaning of *dum cas dan* is "big red barn" and *xer dan* means "big horse." What do you think the words for "red barn" are in this language?

41. Write down a three-digit number. Write the number in reverse order. Subtract the smaller of the two numbers from the larger to obtain a new number. Write down the new number. Reverse the digits again, but add the numbers this time. Complete this process for another three-digit number. Do you notice a pattern, and does your pattern work for all three-digit numbers?

42. Start with a common fraction between 0 and 1. Form a new fraction, using the following rules:

New denominator: Add the numerator and denominator of the original fraction.

New numerator: Add the new denominator to the original numerator.

Write the new fraction and use a calculator to find a decimal equivalent to four decimal places. Repeat these steps again, this time calling the new fraction the original. Continue the process until a pattern appears about the decimal equivalent. What is the decimal equivalent (correct to two decimal places)?

43. The number 6 has four divisors—namely, 1, 2, 3, and 6. List all numbers less than 20 that have exactly four divisors.

Problems 44–56 are not typical math problems but are problems that require only common sense (and sometimes creative thinking).

44. How many 3-cent stamps are there in a dozen?

45. Which weighs more—a ton of coal or a ton of feathers?

46. If you take 7 cards from a deck of 52 cards, how many cards do you have?

47. At six o'clock the grandfather clock struck 6 times. If it was 30 seconds between the first and last strokes, how long will it take the same clock to strike noon?

48. Oak Park cemetery in Oak Park, New Jersey, will not bury anyone living west of the Mississippi. Why?

49. Two U.S. coins total $0.30, yet one of these coins is not a nickel. What are the coins?

50. Two girls were born on the same day of the same month of the same year to the same parents, but they are not twins. Explain how this is possible.

51. How many outs are there in a baseball game that lasts the full 9 innings?

52. If posts are spaced 10 feet apart, how many posts are needed for 100 feet of straight-line fence?

53. A farmer has to get a fox, a goose, and a bag of corn across a river in a boat that is large enough only for him and one of these three items. If he leaves the fox alone with the goose, the fox will eat the goose. If he leaves the goose alone with the corn, the goose will eat the corn. How does he get all the items across the river?

54. Can you place ten lumps of sugar in three empty cups so that there is an odd number of lumps in each cup?

55. Six glasses are standing in a row. The first three are empty, and the last three are full of water. By handling and moving only one glass, it is possible to change this arrangement so that no empty glass is next to another empty one and no full glass is next to another full one. How can this be done?

56. Suppose you are chasing someone with a 10-mile head start, and that you are running at a rate that is one mile per hour faster than the person you are chasing. Also suppose that a fly is flying back and forth between the two of you at the rate of 20 miles per hour. How far would the fly fly (if the fly could fly nonstop all that time) by the time you reached the other person?

PROBLEM SOLVING

Each section of the book has one or more problems designated by **PROBLEM SOLVING.** *These problems may require additional insight, information, or effort to solve. True problem-solving ability comes from solving problems that "are not like the examples" but rather require independent thinking. I hope you will make it a habit to read these problems and attempt to work those that interest you, even though they may not be part of your regular class assignment.*

57. Consider the routes from *A* to *B* and notice that there is now a barricade blocking the path. Work out a general solution for the number of paths with a blockade, and then illustrate your general solution by giving the number of paths for each of the following street patterns.

a.

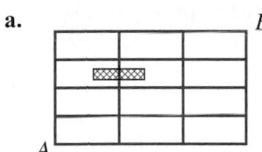

b.

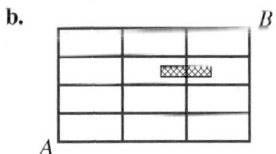

c.

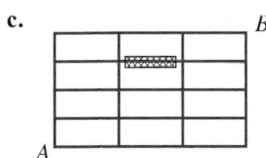

58. **HISTORICAL QUESTION** Thoth, an ancient Egyptian god of wisdom and learning, has abducted Ahmes, a famous Egyptian scribe, in order to assess his intellectual prowess. Thoth places Ahmes before a large funnel set in the ground (see Figure 1.7). It has a circular opening 1,000 ft in diameter, and its wall are quite slippery. If Ahmes attempts to enter the funnel, he will slip down the wall. At the bottom of the funnel is a sleep-inducing liquid that will instantly put Ahmes to sleep for eight hours if he touches it.* Thoth hands Ahmes two objects: a rope 1,006.28 ft in length and the skull of a chicken. Thoth says to Ahmes, "If you are able to get to the central tower and touch it, we will live in harmony for the next millennium. If not, I will detain you for further testing. Please note that

*From "The Thoth Maneuver," by Clifford A. Pickover, *Discover,* March 1996, p. 108. Clifford Pickover/© 1996. Reprinted with permission of Discover Magazine. Nenad Jakesevic and Sonja Lamut/© 1996. Reprinted with permission of Discover Magazine.

with each passing hour, I will decrease the rope's length by a foot." How can Ahmes reach the central ankh tower and touch it?

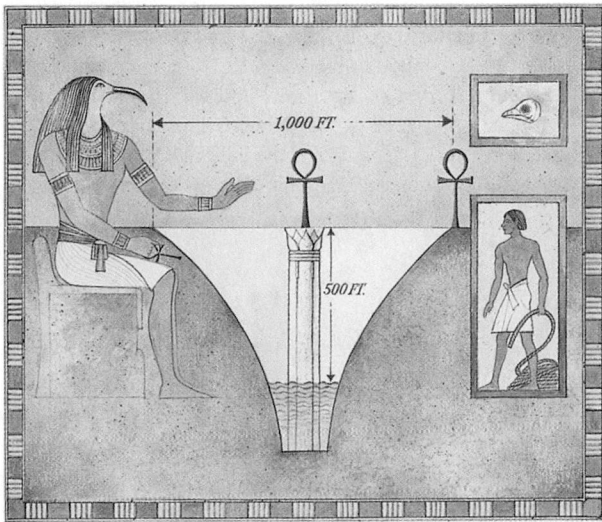

Figure 1.7 Ahmes' dilemma. Note that there are two ankh-shaped towers. One stands on a cylindrical platform in the center of the funnel. The platform's surface is at ground level. The distance from the platform surface to the liquid is 500 ft. The other ankh tower is on land, at the edge of the funnel.

59. A magician divides a deck of cards into two equal piles, counts down from the top of the first pile to the seventh card, and shows it to the audience without looking at it herself. These seven cards are replaced face down in the

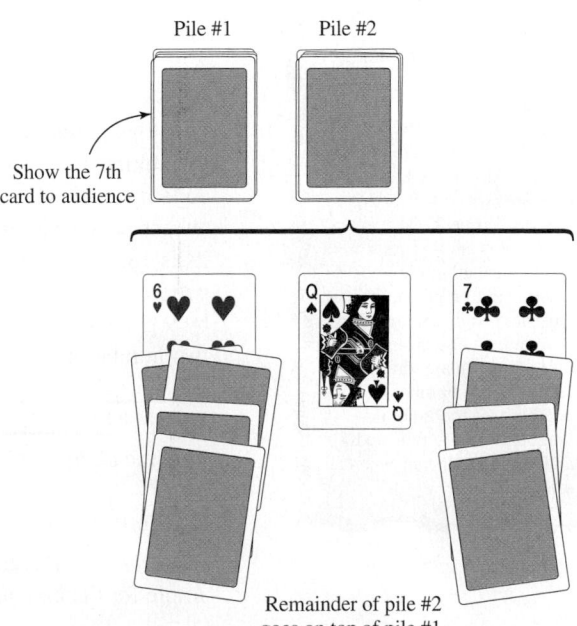

Remainder of pile #2 goes on top of pile #1.

same order on top of the first pile. She then picks up the other pile and deals the top three cards up in a row in front of her. If the first card is a six, then she starts counting with "six" and counts to ten, thus placing four more cards on this pile as shown. In turn, the magician does the same for the next two cards. If the card is a ten or a face card, then no additional cards are added. The remainder of this pile is placed on top of the first pile. Next, the magician adds the values of the three face-up cards ($6 + 10 + 7$ for this illustration), and counts down in the first deck this number of cards. That card is the card that was originally shown to the audience. Explain why this trick works.

60. A very magical teacher had a student select a two-digit number between 50 and 100 and write it on the board out of view of the instructor. Next, the student was asked to add 76 to the number, producing a three-digit sum. If the digit in the hundreds place is added to the remaining two-digit number and this result is subtracted from the original number, the answer is 23, which was predicted by the instructor. How did the instructor know the answer would be 23? *Note:* This problem is dedicated to my friend Bill Leonard of Cal State, Fullerton. His favorite number is 23.

1.2 INDUCTIVE AND DEDUCTIVE REASONING

Studying numerical patterns is one frequently used technique of problem solving. Let's begin with some simple patterns.

A Pattern of Nines

A very familiar pattern is found in the ordinary "times" tables. By pointing out patterns, teachers can make it easier for children to learn some of their multiplication tables. For example, consider the multiplication table for 9s:

$$1 \times 9 = 9$$
$$2 \times 9 = 18$$
$$3 \times 9 = 27$$
$$4 \times 9 = 36$$
$$5 \times 9 = 45$$
$$6 \times 9 = 54$$
$$7 \times 9 = 63$$
$$8 \times 9 = 72$$
$$9 \times 9 = 81$$
$$10 \times 9 = 90$$

What patterns do you notice? You should be able to see many number relationships by looking at the totals. For example, notice that the sum of the digits to the right of the equality is 9 in all the examples ($1 + 8 = 9, 2 + 7 = 9, 3 + 6 = 9$, and so on). Will this always be the case for multiplication by 9? (Consider $11 \times 9 = 99$. The sum of the digits is 18. However, notice the result if you add the digits of 18.) Do you see any other patterns? Can you explain why they "work"? This pattern of adding after multiplying by 9 generates a sequence of numbers: $9, 9, 9, \ldots$. We call this the *nine pattern*. The two number tricks described in the News Clip in the margin use this nine pattern.

EXAMPLE 1   **PÓLYA'S METHOD**

Find the *eight pattern*.

Solution We use Pólya's problem-solving guidelines for this example.

Understand the Problem. What do we mean by *eight pattern*? Do you understand the example for the *nine pattern*?

Devise a Plan. We will carry out the multiplications of successive counting numbers by 8 and if there is more than a single-digit answer, we add the digits.* What we are looking for is a pattern for these single-digit numerals.

Carry Out the Plan.

$$1 \times 8 = 8, \quad 2 \times 8 = \underbrace{16}, \quad 3 \times 8 = \underbrace{24}, \quad 4 \times 8 = \underbrace{32}, \quad 5 \times 8 = \underbrace{40}, \ldots$$

Single digits: 8 $1 + 6 = 7$ $2 + 4 = 6$ $3 + 2 = 5$ $4 + 0 = 4$

Continue with some additional terms:

$$6 \times 8 = 48 \quad \text{and} \quad 4 + 8 = 12 \quad \text{and} \quad 1 + 2 = 3$$
$$7 \times 8 = 56 \quad \text{and} \quad 5 + 6 = 11 \quad \text{and} \quad 1 + 1 = 2$$
$$8 \times 8 = 64 \quad \text{and} \quad 6 + 4 = 10 \quad \text{and} \quad 1 + 0 = 1$$
$$9 \times 8 = 72 \quad \text{and} \quad 7 + 2 = 9$$
$$10 \times 8 = 80 \quad \text{and} \quad 8 + 0 = 8$$

We now see the eight pattern: 8, 7, 6, 5, 4, 3, 2, 1, 9, 8, 7, 6, ...

Look Back. Let's do more than check the arithmetic, since this pattern seems clear. The problem seems to be asking whether we understand the concept of a *nine pattern* or an *eight pattern*. Verify that the seven pattern is 7, 5, 3, 1, 8, 6, 4, 2, 9, 7, 5, 3, 1, ...   ◆

Order of Operations

Complicated arithmetic problems can sometimes be solved by using patterns. Given a difficult problem, a problem solver will often try to *solve a simpler, but similar, problem.* The second suggestion for solving using Pólya's problem-solving procedure stated, "If you cannot solve the proposed problem, look around for an appropriate related problem (a simpler one, if possible)." For example, suppose we wish to compute the following number:

$$10 + 123,456,789 \times 9$$

Instead of doing a lot of arithmetic, let's study the following pattern:

$$2 + 1 \times 9$$
$$3 + 12 \times 9$$
$$4 + 123 \times 9$$

Do you see the next entry in this pattern? Do you see that if we continue the pattern we will eventually reach the desired expression of $10 + 123,456,789 \times 9$? Using Pólya's strategy, we begin by working these easier problems. Thus, we begin with $2 + 1 \times 9$. There is a possibility of ambiguity in calculating this number:

Left-to-right   *Multiplication first*
$2 + 1 \times 9 = 3 \times 9 = 27$ $2 + 1 \times 9 = 2 + 9 = 11$

Although either of these might be acceptable in certain situations, it is not acceptable to get two different answers to the same problem. We therefore all agree to do a problem like this by multiplying first. If we wish to change this order, we use parentheses, as in $(2 + 1) \times 9 = 27$.

> Imagination is a sort of faint perception.
>
> Aristotle

---

*Counting numbers are the numbers we use for counting—namely 1, 2, 3, 4, ... Sometimes they are also called **natural numbers.** The integers are the counting numbers, their opposites, and 0, namely . . . , −3, −2, −1, 0, 1, 2, 3, ... We assume a knowledge of these numbers.

Order of Operations

This is important! Take time looking at what this says.

> 1. First, perform any operations enclosed in parentheses.
> 2. Next, perform multiplications and divisions as they occur by working from left to right.
> 3. Finally, perform additions and subtractions as they occur by working from left to right.

Thus, the correct result for $2 + 1 \times 9$ is 11. Also,

$$3 + 12 \times 9 = 3 + 108 = 111$$
$$4 + 123 \times 9 = 4 + 1,107 = 1,111$$
$$5 + 1,234 \times 9 = 5 + 11,106 = 11,111$$

Do you see a pattern? If so, then make a prediction about the desired result. If you do not see a pattern, continue with this pattern to see more terms, or go back and try another pattern. For this example, we predict

$$10 + 123,456,789 \times 9 = 1,111,111,111$$

The most difficult part of this type of problem solving is coming up with a correct pattern. For this example, you might guess that

$$2 + 1 \times 1$$
$$3 + 12 \times 2$$
$$4 + 123 \times 3$$
$$5 + 1,234 \times 4$$

leads to $10 + (123,456,789 \times 9)$. Calculating, we find

$$2 + 1 \times 1 = 2 + 1 = 3$$
$$3 + 12 \times 2 = 3 + 24 = 27$$
$$4 + 123 \times 3 = 4 + 369 = 373$$
$$5 + 1,234 \times 4 = 5 + 4,936 = 4,941$$

If you begin a pattern and it does not lead to a pattern of answers, then you need to remember that part of Pólya's problem-solving procedure is to work both backward and forward. Be willing to give up one pattern and begin another.

We also point out that the patterns you find are not unique. One last time, we try a pattern for $10 + 123,456,789 \times 9$:

$$10 + 1 \times 9 = 10 + 9 = 19$$
$$10 + 12 \times 9 = 10 + 108 = 118$$
$$10 + 123 \times 9 = 10 + 1,107 = 1,117$$
$$10 + 1,234 \times 9 = 10 + 11,106 = 11,116$$
$$\vdots \qquad\qquad \vdots$$

We do see a pattern here (although not quite as easily as the one we found with the first pattern for this example):

$$10 + 123,456,789 \times 9 = 1,111,111,111$$

Inductive Reasoning

The type of reasoning used here and in the first sections of this book—first observing patterns and then predicting answers for more complicated problems—is called **inductive reasoning.** It is a very important method of thought and is sometimes called the *scientific method.* It involves reasoning from particular facts or individual cases to a general **conjecture**—a statement you think may be true. That is, a generalization is

B. C. reprinted by permission of Johnny Hart and Creators Syndicate.

made on the basis of some observed occurrences. The more individual occurrences we observe, the better able we are to make a correct generalization. Peter in the *B.C.* cartoon makes the mistake of generalizing on the basis of a single observation.

EXAMPLE 2   PÓLYA'S METHOD

What is the sum of the first 100 consecutive odd numbers?

Solution   We use Pólya's problem-solving guidelines for this example.

Understand the Question. Do you know what the terms mean? Odd numbers are 1, 3, 5, . . . , and *sum* means to add:

$$1 + 3 + 5 + \cdots + \ ?$$

The first thing you need to understand is what the last term will be, so you will know when you have reached 100 consecutive odd numbers.
1 + 3 is two terms.
1 + 3 + 5 is three terms.

It seems as if the last term is always one less than twice the number of terms. Thus, the sum of the first 100 consecutive odd numbers is

$$1 + 3 + 5 + \cdots + 195 + 197 + 199$$

This is one less than 2(100).

Devise a Plan   The plan we will use is to look for a pattern:

$$1 = 1 \quad \text{one term}$$
$$1 + 3 = 4 \quad \text{sum of two terms}$$
$$1 + 3 + 5 = 9 \quad \text{sum of three terms}$$

Do you see a pattern yet? If not, continue:

$$1 + 3 + 5 + 7 = 16$$
$$1 + 3 + 5 + 7 + 9 = 25$$

Carry Out the Plan. It appears that the sum of 2 terms is 2 · 2; of 3 terms, 3 · 3; of 4 terms, 4 · 4; and so on. The sum of the first 100 consecutive odd numbers is therefore 100 · 100.

Looking Back. Does 100 · 100 = 10,000 seem correct?   ◆

The numbers 2 · 2, 3 · 3, 4 · 4, and 100 · 100 from Example 2 are usually written as 2^2, 3^2, 4^2, and 100^2. The number b^2 means $b \cdot b$ and is pronounced **b squared,** and the number b^3 means $b \cdot b \cdot b$ and is pronounced **b cubed.** The process of repeated multiplication is called **exponentiation.**

Deductive Reasoning

Another method of reasoning used in mathematics is called **deductive reasoning.** This method of reasoning produces results that are *certain* within the logical system being developed. That is, deductive reasoning involves reaching a conclusion by using a formal structure based on a set of **undefined terms** and a set of accepted unproved **axioms** or **premises.** The conclusions are said to be proved and are called **theorems.**

The most useful axiom in problem solving is the principle of substitution: If two quantities are equal, one may be substituted for the other without changing the truth or falsity of the statement.

Substitution Property

Boxes that look like this are highlighting important definitions.

If $a = b$, then a may be **substituted** for b in any mathematical statement without affecting the truth or falsity of the given mathematical statement.

The simplest way to illustrate the substitution property is to use it in evaluating a formula.

EXAMPLE 3

A billiard table is 4 ft by 8 ft. Find the perimeter.

Solution   Use an appropriate formula,

$$P = 2\ell + 2w \qquad P = \text{PERIMETER}, \ell = \text{LENGTH, and } w = \text{WIDTH}$$

Substitute the known values into the formula:

$\ell = 8 \qquad w = 4$
$\downarrow \qquad\quad \downarrow$ These arrows mean substitution.

$$P = 2(8) + 2(4)$$
$$= 16 + 8$$
$$= 24$$

The perimeter is 24 ft.   ◆

"For a minute I thought we had him stymied!"

Tom Henderson, *The Saturday Evening Post* © 1960.

Problem solving depends not only on the substitution property, but also on translating statements from English to mathematical symbols. On a much more advanced level, this process is called *mathematical modeling* (we will consider some mathematical modeling later in this book). However, for now, we will simply call it **translating.** Example 4 reviews much of the terminology from your previous mathematics courses. These terms include **sum** to indicate the result obtained from addition, **difference** for the result from subtraction, **product** for the result of a multiplication, and **quotient** for the result of a division.

EXAMPLE 4

Write the word statement or phrase in symbols. Do not simplify the translated expressions.

a. The sum of seven and a number
b. The difference of a number subtracted from seven
c. The quotient of two numbers
d. The product of two consecutive numbers
e. The difference of the squares of a number and two
f. The square of the difference of two from a number
g. The sum of two times a number and six is equal to two times the sum of a number and three

Mathematics, of course, is not the *only* cornerstone of opportunity in today's world. Reading is even more fundamental as a basis for learning and for life. What is different today is the great increase in the importance of mathematics to so many areas of education, citizenship, and careers.

Everybody Counts, p. 3

Solution

a. Since *sum* indicates addition, this can be rewritten as

(SEVEN) + (A NUMBER)

Now select some variable—say, $s = $ A NUMBER.
The symbolic statement is $7 + s$.

b.   (SEVEN) − (A NUMBER)

Let $d = $ A NUMBER; then the expression is $7 - d$.

c. $$\frac{\text{A NUMBER}}{\text{ANOTHER NUMBER}}$$

If there is more than one unknown in a problem, and no relationship between those unknowns is given, more than one variable may be needed. Let $m = $ A NUMBER and $n = $ ANOTHER NUMBER. Then the expression is $\dfrac{m}{n}$.

d.   (A NUMBER)(NEXT CONSECUTIVE NUMBER)

If there is more than one unknown in a problem but a given relationship exists between those unknowns, *do not choose more variables than you need* for the problem. In this problem, a consecutive number means one more than the first number:

(A NUMBER)(NEXT CONSECUTIVE NUMBER)

NEXT CONSECUTIVE NUMBER $=$ A NUMBER $+ 1$
    This step uses the substitution property.

(A NUMBER)(A NUMBER $+ 1$)

Let $x = $ A NUMBER; then the variable expression is

$$x(x + 1)$$

e.   (A NUMBER)$^2 - 2^2$

Let $x = $ A NUMBER; then

$$x^2 - 2^2$$

This is called a *difference of squares*.

f.   (A NUMBER $- 2$)^2

Let $x = $ A NUMBER; then

$$(x - 2)^2$$

g. If "a number" is referred to more than once in a problem, it is assumed to be the same number.

$$2(\text{A NUMBER}) + 6 = 2(\text{A NUMBER} + 3)$$

Let $n = $ A NUMBER;

$$2n + 6 = 2(n + 3)$$ ◆

Problem Set 1.2

LEVEL 1

1. IN YOUR OWN WORDS Discuss the nature of *inductive* and *deductive reasoning*.

2. IN YOUR OWN WORDS Explain what is meant by the *seven pattern*.

3. IN YOUR OWN WORDS What do we mean by order of operations?

4. IN YOUR OWN WORDS What is the scientific method?

5. IN YOUR OWN WORDS What is the substitution principle?

Perform the operations in Problems 6–19.

6. a. $5 + 2 \times 6$ **b.** $7 + 3 \times 2$

7. a. $14 + 6 \times 3$ **b.** $30 \div 5 \times 2$

8. a. $3 \times 8 + 3 \times 7$ **b.** $3(8 + 7)$

9. a. $(8 + 6) \div 2$ **b.** $8 + 6 \div 2$

10. a. $12 + 6/3$ **b.** $(12 + 6)/3$

11. a. $450 + 550/10$ **b.** $\dfrac{450 + 550}{10}$

12. a. $20/2 \cdot 5$ **b.** $20/(2 \cdot 5)$

13. a. $1 + 3 \times 2 + 4 + 3 \times 6$
     b. $3 + 6 \times 2 + 8 + 4 \times 3$

14. **a.** $10 + 5 \times 2 + 6 \times 3$
    b. $4 + 3 \times 8 + 6 + 4 \times 5$

15. **a.** $8 + 2(3 + 12) - 5 \times 3$
    b. $25 - 4(12 - 2 \times 6) + 3$

16. **a.** $3 + 9 - 3 \times 2 + 2 \times 6 \div 3$
    b. $[(3 + 9) \div 3] \times 2 + [(2 \times 6) \div 3]$

17. **a.** $3 + [(9 \div 3) \times 2] + [(2 \times 6) \div 3]$
    b. $[(3 + 9) \div (3 \times 2)] + [(2 \times 6) \div 3]$

18. Does the *B.C.* cartoon illustrate inductive or deductive reasoning? Explain your answer.

B. C. reprinted by permission of Johnny Hart and Creators Syndicate.

19. Does the News Clip illustrate inductive or deductive reasoning? Explain your answer.

> The old fellow in charge of the checkroom in a large hotel was noted for his memory. He never used checks or marks of any sort to help him return coats to their rightful owners.
> Thinking to test him, a frequent hotel guest asked him as he received his coat, "Sam, how did you know this is my coat?"
> "I don't, sir," was the calm response.
> "Then why did you give it to me?" asked the guest.
> "Because," said Sam, "it's the one you gave me, sir."
>
> *Lucille J. Goodyear*

Problems 20–26 are modeled after Example 1. Find the requested pattern.

20. three pattern
21. four pattern
22. five pattern
23. six pattern
24. What is the sum of the first 25 consecutive odd numbers?
25. What is the sum of the first 50 consecutive odd numbers?

26. What is the sum of the first 1,000 consecutive odd numbers?

Rewrite the word statement or phrase in Problems 27–37, using symbols. Do not simplify; simply translate.

27. **a.** The sum of three and the product of two and four
    b. The product of three and the sum of two and four

28. **a.** The quotient of three divided by the sum of two and four
    b. Ten times the sum of four and three

29. **a.** Eight times nine plus ten
    b. Eight times the sum of nine and ten

30. **a.** The sum of three squared and five squared
    b. The square of the sum of three and five

31. **a.** The square of three increased by the cube of two
    b. The cube of three decreased by the square of two

32. **a.** The square of a number plus five
    b. Six times the cube of a number

33. **a.** The sum of the squares of 4 and 9
    b. The square of the sum of 4 and 9

34. **a.** The sum of a number and one is equal to one added to the number.
    b. Two added to a number is equal to the sum of the number and two.

35. **a.** Three times the sum of a number and four is equal to sixteen.
    b. Five times the sum of a number and one is equal to the sum of five times the number and five.

36. **a.** The sum of six times a number and twelve is equal to six times the sum of the number and two.
    b. A number times the sum of seven and the number is equal to zero.

37. **a.** The sum of a number squared and eight times the number is equal to six.
    b. The square of a number plus eight times the number is equal to the number times the sum of eight and the number.

LEVEL 2

In Problems 38–46, write a formula to express the given relation.

38. The area A of a parallelogram is the product of the base b and the height h.

39. The area A of a triangle is one-half the product of the base b and the height h.

40. The area A of a rhombus is one-half the product of the diagonals p and q.

41. The area A of a trapezoid is the product of one-half the height h and the sum of the bases a and b.

42. The volume V of a cube is the cube of the length s of an edge.

43. The volume V of a rectangular solid is the product of the length ℓ, the width w, and the height h.

44. The volume V of a cone is one-third the product of pi, the square of the radius r, and the height h.

45. The volume V of a circular cylinder is the product of pi, the square of the radius r, and the height h.

46. The volume V of a sphere is the product of four-thirds pi times the cube of the radius r.

LEVEL 3

47. Consider the following pattern:

$$9 \times 1 - 1 = 8$$
$$9 \times 21 - 1 = 188$$
$$9 \times 321 - 1 = 2,888$$
$$9 \times 4,321 - 1 = 38,888$$

Use this pattern and inductive reasoning to find the next problem and the next answer in the sequence.

48. Use Problem 47 to predict the answer to

$$9 \times 987,654,321 - 1$$

49. Use Problem 47 to predict the answer to

$$9 \times 10,987,654,321 - 1$$

50. Consider the following pattern:

$$123,456,789 \times 9 \ = 1,111,111,101$$
$$123,456,789 \times 18 = 2,222,222,202$$
$$123,456,789 \times 27 = 3,333,333,303$$

Use this pattern and inductive reasoning to find the next problem and the next answer in the sequence.

51. Use Problem 50 to predict the answer to

$$123,456,789 \times 9,000$$

52. Use Problem 50 to predict the answer to

$$123,456,789 \times 81,000$$

PROBLEM SOLVING

53. How many squares are there in Figure 1.8?

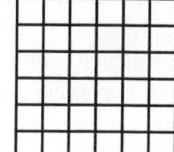

Figure 1.8 How many squares?

Hint:

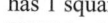

has 1 square
TOTAL: 1

has   1   2-by-2 square
      4   1-by-1 squares
TOTAL: 5 (by addition)

has   1   3-by-3 square
      4   2-by-2 squares
      9   1-by-1 squares
TOTAL: 14 (by addition)

54. How many triangles are there in Figure 1.9?

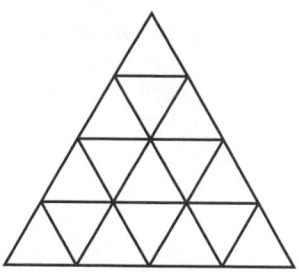

Figure 1.9 How many triangles?

55. A four-inch cube is painted red on all sides. It is then cut into one-inch cubes. What fraction of all the one-inch cubes are painted on exactly one side?

56. Problem 55 gives rise to several patterns. Suppose you consider a set of painted cubes, each of which is made up of several smaller cubes. Use patterns to fill in the blanks in the following table. The last entries (for a cube with length of edge 10 in.) have been filled in so that you can check the patterns you obtain.

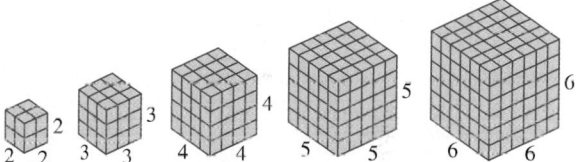

	Length of edge	Total cubes	Number of smaller cubes with the indicated number of painted faces 3	2	1	0
a.	2	8				
b.	3	27				
c.	4					
d.	5					
e.	6					
f.	7					
g.	8					
h.	9					
	10	1,000	8	96	384	512

57. You have 9 coins, but you are told that one of the coins is counterfeit and weighs just a little more than an authentic coin. How can you determine the counterfeit with 2 weighings on a two-pan balance scale? (This problem is discussed in Chapter 2.)

A **magic square** *is an arrangement of numbers in the shape of a square, with the sum of each vertical column, each horizontal row, and each diagonal all equal. The numbers of rows and columns are the same, and this number is called the* **order** *of the magic square. A* **standard magic square** *is made up of the consecutive counting numbers starting with 1. Problems 58–60 are about magic squares.*

58. The first known example of a magic square comes from China. Legend tells us that around the year 200 B.C. the emperor Yu of the Shang dynasty received the following magic square etched on the back of a tortoise's shell:

The incident supposedly took place along the Lo River, so this magic square has come to be known as the Lo-shu magic square. The even numbers are black (female numbers) and the odd numbers are white (male numbers). Translate this magic square into modern symbols.

59.

$100 REWARD

In 1987 Martin Gardner (long-time math buff and past editor of the "Mathematical Games" department of

Scientific American) offered $100 to anyone who could find a 3×3 magic square made with consecutive primes. The prize was won by Harry Nelson of Lawrence Livermore Laboratories. He produced the following simplest such square:

1,480,028,201	1,480,028,129	1,480,028,183
1,480,028,153	1,480,028,171	1,480,028,189
1,480,028,159	1,480,028,213	1,480,028,141

Prove that this is a magic square.

60.

$100 REWARD (FOR REAL)

Now, for a real $100 offer: *Find a 3×3 magic square with nine distinct square numbers.* If you find such a magic square, write to me and we will submit it to Martin Gardner to obtain our prize. Show that the following magic squares do not win the award.

a.

127^2	46^2	58^2
2^2	113^2	94^2
74^2	82^2	97^2

b.

35^2	3495^2	2958^2
3642^2	2125^2	1785^2
2775^2	2058^2	3005^2

1.3 SCIENTIFIC NOTATION AND ESTIMATION

"How Big Is the Cosmos?" There is a dynamic website demonstrating the answer to this question (go to "Powers of Ten" at **www.mathnature.com**). Figure 1.10 illustrates the size of the known cosmos. How can the human mind comprehend such numbers? Scientists often work with very large numbers. Distances such as those in Figure 1.10 are measured in terms of the distance that light, moving at 186,000 miles per second, travels in a year. In this section, we turn to patterns to see if there is an easy way to deal with very large and very small numbers. We will also discuss estimation as a problem-solving technique.

Exponential Notation

We often encounter expressions that comprise multiplication of the same numbers. For example,

$$10 \cdot 10 \cdot 10 \quad \text{or} \quad 6 \cdot 6 \cdot 6 \cdot 6 \cdot 6 \cdot 6 \cdot 6 \quad \text{or}$$

$$15 \cdot 15 \cdot 15 \cdot 15 \cdot 15 \cdot 15 \cdot 15 \cdot 15 \cdot 15 \cdot 15 \cdot 15 \cdot 15 \cdot 15 \cdot 15$$

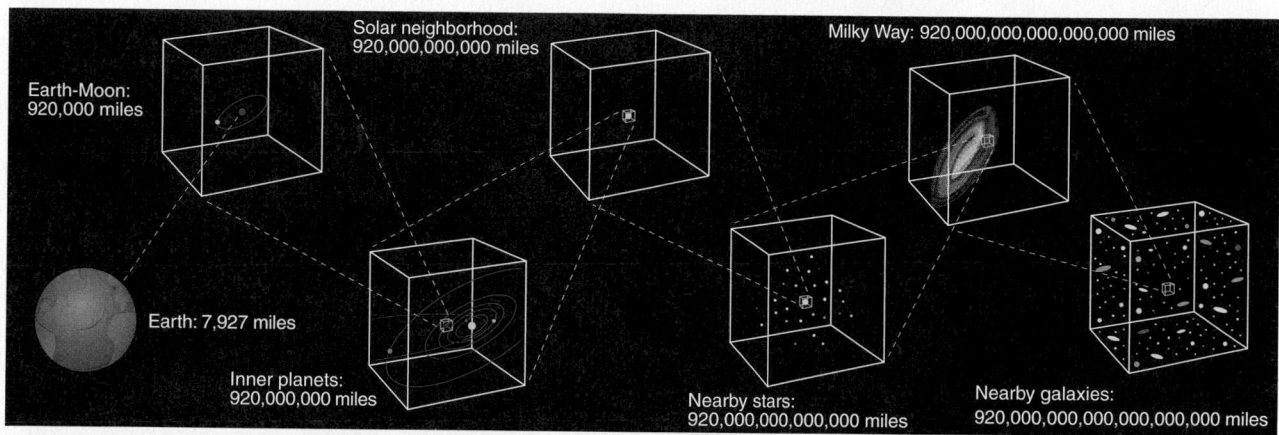

Figure 1.10 **Size of the universe.* Each successive cube is a thousand times as wide and a billion times as voluminous as the one before it. The contents and width of each one are described in the caption beneath it.**

These numbers can be written more concisely using what is called **exponential notation:**

$$10 \cdot 10 \cdot 10 = 10^3 \qquad 6 \cdot 6 \cdot 6 \cdot 6 \cdot 6 \cdot 6 \cdot 6 = 6^7$$

$$\underbrace{10 \cdot 10 \cdot 10}_{3 \text{ factors}} \qquad \underbrace{6 \cdot 6 \cdot 6 \cdot 6 \cdot 6 \cdot 6 \cdot 6}_{7 \text{ factors}}$$

$$\underbrace{15 \cdot 15 \cdot 15 \cdot 15 \cdot 15 \cdot 15 \cdot 15 \cdot 15 \cdot 15 \cdot 15 \cdot 15 \cdot 15 \cdot 15 \cdot 15}_{14 \text{ factors}} = 15^{14}$$

Exponential Notation

> For any nonzero number b and any counting number n,
>
> $$b^n = \underbrace{b \cdot b \cdot b \cdot \cdots \cdot b}_{n \text{ factors}}, \qquad b^0 = 1, \qquad b^{-n} = \frac{1}{b^n}$$
>
> The number b is called the **base,** the number n in b^n is called the **exponent,** and the number b^n is called a **power** or **exponential.**

EXAMPLE 1

Write without exponents.

a. 10^5 **b.** 6^2 **c.** 7^5 **d.** 2^{63} **e.** 3^{-2} **f.** 8.9^0

Solution

a. $10^5 = 10 \cdot 10 \cdot 10 \cdot 10 \cdot 10 = 100,000$

b. $6^2 = 6 \cdot 6 = 36$

c. $7^5 = 7 \cdot 7 \cdot 7 \cdot 7 \cdot 7$ or $16,807$

d. $2^{63} = \underbrace{2 \cdot 2 \cdot 2 \cdot \cdots \cdot 2 \cdot 2}_{63 \text{ factors}}$ or $9,223,372,036,854,775,808$

Note: You are not expected to find the form at the right; the factored form is acceptable.

CAUTION Note: 10^5 is not five multiplications, but rather five factors of 10.

*Illustration is adapted from *The Universe,* Life Nature Library.

e. $3^{-2} = \dfrac{1}{3^2} = \dfrac{1}{9}$

f. $8.9^0 = 1$ ◆

Since an exponent is an indicated multiplication, the proper procedure is first to simplify the exponent, and then to carry out the multiplication. This leads to an **extended order-of-operations agreement.**

Extended Order of Operations

> 1. First, perform any operations enclosed in parentheses.
>
> 2. Next, perform any operations that involve raising to a power.
>
> 3. Perform multiplications and divisions as they occur by working from left to right.
>
> 4. Finally, perform additions and subtractions as they occur by working from left to right.

Scientific Notation

There is a similar pattern for multiplications of any number by a power of 10. Consider the following examples, and notice what happens to the decimal point.

$$9.42 \times 10^1 = 94.2 \qquad \text{\small We find these answers by direct multiplication.}$$
$$9.42 \times 10^2 = 942.$$
$$9.42 \times 10^3 = 9{,}420.$$
$$9.42 \times 10^4 = 94{,}200.$$

Do you see the pattern? Look at the decimal point (which is included for emphasis). If we multiply 9.42×10^5, how many places to the right will the decimal point be moved?

$$9.42 \times 10^5 = 9\ 42{,}000.$$

5 places to the right

Using this pattern, can you multiply the following *without direct calculation?*

$$9.42 \times 10^{12}$$
$$= 9{,}420{,}000{,}000{,}000 \qquad \text{\small This answer is found by observing the pattern, not by direct multiplication.}$$

The pattern also extends to smaller numbers:

$$9.42 \times 10^{-1} = 0.942 \left.\vphantom{\begin{matrix}a\\a\\a\end{matrix}}\right\} \;\text{\small These numbers are found by direct}$$
$$9.42 \times 10^{-2} = 0.0942 \quad\text{\small multiplication. For example,}$$
$$9.42 \times 10^{-3} = 0.00942 \quad\text{\small } 9.42 \times 10^{-2} = 9.42 \times \tfrac{1}{100}$$
$$\text{\small } = 9.42 \times 0.01 = 0.0942$$

Do you see that the same pattern also holds for multiplying by 10 with a negative exponent? Can you multiply the following *without direct calculation?*

$$9.42 \times 10^{-6} = 0.000009\ 42$$

Moved six places to the left

These patterns lead to a useful way for writing large and small numbers, called *scientific notation.*

Scientific Notation

> The **scientific notation** for a nonzero number is that number written as a power of 10 times another number x, such that x is between 1 and 10, including 1; that is, $1 \leq x < 10$.

EXAMPLE 2

Write the given numbers in scientific notation.

a. 123,400   **b.** 0.000035   **c.** 1,000,000,000,000   **d.** 7.35

Solution

a. $123{,}400 = 1.234 \times 10^5$

b. $0.000035 = 3.5 \times 10^{-5}$

c. $1{,}000{,}000{,}000{,}000 = 10^{12}$
  Technically, this is 1×10^{12} with the 1 understood.

d. 7.35 (or 7.35×10^0) is in scientific notation.   ◆

EXAMPLE 3

Assuming that light travels at 186,000 miles per second, what is the distance (in miles) that light travels in 1 year? This is the unit of length known as a *light-year*. Give your answer in scientific notation.

Solution One year is 365.25 days $= 365.25 \times 24$ hours
$$= 365.25 \times 24 \times 60 \text{ minutes}$$
$$= 365.25 \times 24 \times 60 \times 60 \text{ seconds}$$
$$= 31{,}557{,}600 \text{ seconds}$$

Since light travels 186,000 miles each second and there are 31,557,600 seconds in 1 year, we have

$$186{,}000 \times 31{,}557{,}600 = 5{,}869{,}713{,}600{,}000 \approx 5.87 \times 10^{12}$$

Thus, light travels 5.87×10^{12} miles in 1 year.   ◆

Calculators

Throughout the book we will include calculator comments for those of you who have (or expect to have) a calculator. Calculators are classified according to their ability to perform different types of calculations, as well as by the type of logic they use to do the calculations. The problem of selecting a calculator is further complicated by the multiplicity of brands from which to choose. Therefore, choosing a calculator and learning to use it require some sort of instruction.

For most nonscientific purposes, a four-function calculator with memory is sufficient for everyday use. If you anticipate taking several mathematics and/or science courses, you will find that a scientific calculator is a worthwhile investment. These calculators use essentially three types of logic: *arithmetic, algebraic,* and *RPN.* In the previous section, we discussed the correct order of operations, according to which the correct value for

$$2 + 3 \times 4$$

is 14 (multiply first). An algebraic calculator will "know" this and will give the correct answer, whereas an arithmetic calculator will simply work from left to right and obtain the incorrect answer, 20. Therefore, if you have an arithmetic-logic calculator, you will need to be careful about the order of operations. Some arithmetic-logic calculators provide parentheses, , so that operations can be grouped, as in

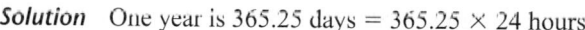

but then you must remember to insert the parentheses.

The last type of logic is RPN. A calculator using this logic is characterized by [ENTER] or [SAVE] keys and does not have an equal key [=]. With an RPN calculator, the operation symbol is entered after the numbers have been entered.

📇 **Computational Window**   ⬜ ☐ ✕

The different levels of calculators are distinguished primarily by their price.

1. **Four-function calculator with memory (under $10).** These calculators have a keyboard consisting of the numerals, the four arithmetic operations or functions (addition ⊞, subtraction ⊟, multiplication ⊠, and division ⊡, and a memory register indicated by a key marked M, STO, or M+). These calculators are often given away free as promotion and are, for the most part, obsolete.

2. **Scientific calculators ($10–$40).** These calculators include additional mathematical functions, such as square root √; trigonometric sin, cos, and tan; logarithmic log; and exponential exp. Depending on the particular brand, a scientific model may have other keys as well. Most calculators sold today are scientific calculators.

3. **Programmable/graphics calculators ($50–$250).** These calculators "remember" the sequence of steps for complex calculations. Graphics calculators can display a graph of an input equation.

Texas Instruments Incorporated

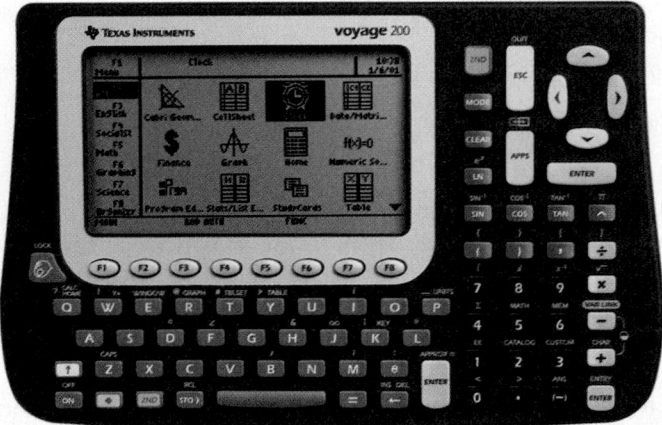

Texas Instruments Incorporated

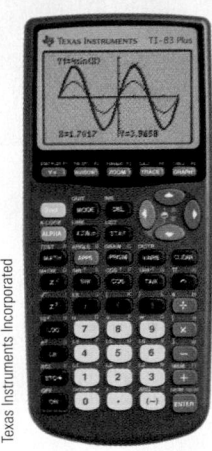

Texas Instruments Incorporated

4. **Special-purpose calculators ($40–$400).** Special-use calculators for business, statistics, surveying, medicine, and even gambling and chess are available.

These three types of logic can be illustrated by the problem $2 + 3 \times 4$:

Arithmetic logic: 3 ⊠ 4 ⊟ ⊞ 2 ⊟   Input to match order of operations

Algebraic logic: 2 ⊞ 3 ⊠ 4 ⊟   Input is the same as the problem

RPN logic: 2 ENTER 3 ENTER 4 ⊠ ⊞   Operations input last

In this book, we will illustrate the examples using algebraic logic. If you have a calculator with RPN logic, you can use your owner's manual to change the examples to RPN. We will also indicate the keys to be pushed by drawing boxes around the numerals and operational signs as shown.

Polls show that the public generally thinks that, in mathematics education, calculators are bad while computers are good. People believe that calculators will prevent children from mastering arithmetic, an important burden which their parents remember bearing with courage and pride. Computers, on the other hand, are not perceived as shortcuts to undermine school traditions, but as new tools necessary to society that children who understand mathematics must learn to use. What the public fails to recognize is that both calculators and computers are equally essential to mathematics education and have equal potential for wise use or for abuse.

Everybody Counts, p. 61

EXAMPLE 4

Show the calculator steps for $14 + 38$.

Solution Be sure to turn your calculator on, or clear the machine if it is already on. A clear button is designated by $\boxed{C}$, and the display will show 0 after the clear button is pushed. You will need to check these steps every time you use your calculator, but after a while it becomes automatic. We will not remind you of this on each example.

Press	Display		
$\boxed{1}$	1	Here we show each numeral in a single box, which means you key in one numeral at a time, as shown. But from now on, this will be shown as $\boxed{14}$.	
$\boxed{4}$	14		
$\boxed{+}$	14	Some calculators display all the keystrokes:	
$\boxed{3}\,\boxed{8}$	38	$14 + 38$	
$\boxed{=}$	52	52	◆

After completing Example 4, you can either continue with the same problem or start a new problem. If the next button pressed is an operation button, the result 52 will be carried over to the new problem. If the next button pressed is a numeral, the 52 will be lost and a new problem started. For this reason, it is not necessary to press $\boxed{C}$ to clear between problems. The button $\boxed{CE}$ is called the *clear entry* key and is used if you make a mistake keying in a number and do not want to start over with the problem. For example, if you want $2 + 3$ and accidentally push

$\boxed{2}\,\boxed{+}\,\boxed{4}$　you can then push　$\boxed{CE}\,\boxed{3}\,\boxed{=}$

to obtain the correct answer. This is especially helpful if you are in the middle of a long problem. Some models have a $\boxed{\leftarrow}$ key instead of a $\boxed{CE}$ key.

EXAMPLE 5

Show the calculator steps and display for $4 + 3 \times 5 - 7$.

Solution

Press:	$\boxed{4}$	$\boxed{+}$	$\boxed{3}$	$\boxed{\times}$	$\boxed{5}$	$\boxed{-}$	$\boxed{7}$	$\boxed{=}$ or $\boxed{ENTER}$
Display:	4	4	3	3	5	19	7	12
or	4	$4+$	$4+3$	$4+3*$	$4+3*5$	$4+3*5-$	$4+3*5-7$	12

If you have an algebraic-logic calculator, your machine will perform the correct order of operations. If it is an arithmetic-logic calculator, it will give the incorrect answer 28 unless you input the numbers using the order-of-operations agreement.　　◆

EXAMPLE 6

Repeat Example 3 using a calculator; that is, find

$$365.25 \times 24 \times 60 \times 60 \times 186{,}000$$

Solution When you press these calculator keys and then press $\boxed{=}$ the display will probably show something that looks like:

5.86971 12　or　5.86971 +12　or　5.86971E12

This display is a form of scientific notation. The 12 or +12 at the right (separated by one or two blank spaces) is the exponent on the 10 when the number in the display is written in scientific notation. That is,

5.86971 12　means　5.86971×10^{12}　　◆

Suppose you have a particularly large number that you wish to input into a calculator—say, 920,000,000,000,000,000,000 miles divided by 7,927 miles (from Figure 1.10). You can input 7,927 but if you attempt to input the larger number you will be stuck when you fill up the display (9 or 12 digits). Instead, you will need to write

$$920{,}000{,}000{,}000{,}000{,}000{,}000 = 9.2 \times 10^{20}$$

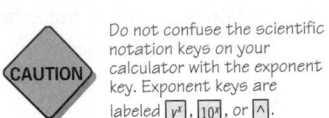

Do not confuse the scientific notation keys on your calculator with the exponent key. Exponent keys are labeled y^x, 10^x, or $\wedge$.

This may be entered by pressing an $\boxed{EE}$, $\boxed{EEx}$, or $\boxed{EXP}$ key:

$\boxed{9.2}$ $\boxed{EE}$ $\boxed{20}$ $\boxed{\div}$ $\boxed{7927}$ $\boxed{=}$ *Display:* 1.160590387E17

This means that the last cube in Figure 1.10 is about 1.2×10^{17} times larger than the earth.

Scientific notation is represented in a slightly different form on many calculators and computers, and this new form is sometimes called **floating-point form.** When representing very large or very small answers, most calculators will automatically output the answers in floating-point notation. The following example compares the different forms for large and small numbers with which you should be familiar.

EXAMPLE 7

Write each given number in scientific notation and in calculator notation.
a. 745   **b.** 1,230,000,000   **c.** 0.00573   **d.** 0.00000 06239

Solution   The form given in this example is sometimes called **fixed-point form** or **decimal notation** to distinguish it from the other forms.

	Decimal or fixed-point notation	*Scientific notation*	*Floating-point*
a.	745	7.45×10^2	7.45 02
b.	1,230,000,000	1.23×10^9	1.23 09
c.	0.00573	5.73×10^{-3}	5.73 −03
d.	0.00000 06239	6.239×10^{-7}	6.239 −07

◆

Estimation

Part of problem solving is using common sense about the answers you obtain. This is even more important when using a calculator, because there is a misconception that if a calculator or computer displays an answer, "it must be correct." Reading and understanding the problem are parts of the process of problem solving.

When problem solving, you must ask whether the answer you have found is reasonable. If I ask for the amount of rent you must pay for an apartment, and you do a calculation and arrive at an answer of $16.25, you know that you have made a mistake. Likewise, an answer of $135,000 would not be reasonable. As we progress through this course you will be using a calculator for many of your calculations, and with a calculator you can easily press the wrong button and come up with an outrageous answer. One aspect of *looking back* is using common sense to make sure the answer is reasonable. The ability to recognize the difference between reasonable answers and unreasonable ones is important not only in mathematics, but whenever you are problem solving. This ability is even more important when you use a calculator, because pressing the incorrect key can often cause outrageously unreasonable answers.

Whenever you try to find an answer, you should ask yourself whether the answer is reasonable. How do you decide whether an answer is reasonable? One way is to **estimate** an answer. Webster's *New World Dictionary* tells us that as a verb, to *estimate* means "to form an opinion or a judgment about" or to calculate "approximately." In the *1986 Yearbook* of the National Council of Teachers of Mathematics, we find:

> *The broad* mathematical context *for an estimate is usually one of the following types:*
>
> A.   *An exact value is known but for some reason an estimate is used.*
> B.   *An exact value is possible but is not known and an estimate is used.*
> C.   *An exact value is impossible.*

We will work on building your estimation skills throughout this book.

EXAMPLE 8

If your salary is $14.75 per hour, your annual salary is approximately
A. $5,000   B. $10,000   C. $15,000   D. $30,000   E. $45,000

Solution   Problem solving often requires some assumptions about the problem. For this problem, we are not told how many hours per week you work, or how many weeks per year you are paid. We assume a 40-hour week, and we also assume that you are paid for 52 weeks per year.

   Estimate: Your hourly salary is about $15 per hour.
              A 40-hour week gives us $40 \times \$15 = \$600$ per week.

              For the estimate, we calculate the wages for 50 weeks instead of 52:

              50 weeks yields $50 \times \$600 = \$30,000$.

The answer is **D**.   ◆

EXAMPLE 9

Use the map to estimate the distance from Orlando International Airport to Disney World.

Solution   Note that the scale is 20 miles to 1 in. Looking at the map, you will note that it is approximately $\frac{3}{4}$ in. from the airport to Disney World. This means that we estimate the distance to be 15 miles.   ◆

   There are two important reasons for estimation: (1) to form a reasonable opinion, or (2) to check the reasonableness of an answer. If reason (1) is our motive, we should not think it necessary to follow an estimation by direct calculation. To do so would defeat the purpose of the estimation. On the other hand, if we are using the estimate for reason (2)—to see whether an answer is reasonable—we might perform the estimate as a check on the calculated answer for Example 3 (the problem about the speed of light):

$$186,000 \text{ miles per second} \approx 2 \times 10^5 \text{ miles per second}$$

and

$$\text{one year} \approx 4 \times 10^2 \text{ days} \approx \underbrace{4 \times 10^2}_{days} \times \underbrace{2 \times 10}_{hr\ per\ day} \times \underbrace{6 \times 10}_{min\ per\ hr} \times \underbrace{6 \times 10}_{sec\ per\ hr}$$

$$= \underbrace{(4 \times 2 \times 6 \times 6) \times 10^5}_{seconds\ per\ year} \approx (10 \times 36) \times 10^5 \approx 3.6 \times 10^7 \text{ seconds}$$

   Thus, one light year is about $(2 \times 10^5)(3.6 \times 10^7) \approx 7.2 \times 10^{12}$ miles. This estimate seems to confirm the reasonableness of the answer 5.87×10^{12} we obtained in Example 3.

Laws of Exponents

In working out the previous estimation for Example 9, we used some properties of exponents that we can derive by, once again, turning to some patterns. Consider

$$10 \cdot 10 \cdot 10 \cdot 10 \cdot 10 = 10^5$$

and

$$10^2 \cdot 10^3 = (10 \cdot 10) \cdot (10 \cdot 10 \cdot 10) = 10^5$$

When we *multiply powers* of the same base, we *add* exponents. This is called the **addition law of exponents.**

$$2^3 \cdot 2^4 = (2 \cdot 2 \cdot 2) \cdot (2 \cdot 2 \cdot 2 \cdot 2)$$
$$= 2^{3+4}$$
$$= 2^7$$

Suppose we wish to raise a power to a power. We can apply the addition law of exponents. Consider

$$(2^3)^2 = 2^3 \cdot 2^3 = 2^{3+3} = 2^{2 \cdot 3} = 2^6$$
$$(10^2)^3 = 10^2 \cdot 10^2 \cdot 10^2 = 10^{2+2+2} = 10^{3 \cdot 2} = 10^6$$

When we *raise a power to a power*, we *multiply* the exponents. This is called the **multiplication law of exponents.**

A third law is needed to raise products to powers. Consider

$$(2 \cdot 3)^2 = (2 \cdot 3) \cdot (2 \cdot 3)$$
$$= (2 \cdot 2) \cdot (3 \cdot 3)$$
$$= 2^2 \cdot 3^2$$

Thus, $(2 \cdot 3)^2 = 2^2 \cdot 3^2$.

$$(3 \cdot 10^4)^2 = 3^2 \cdot (10^4)^2 = 3^2 \cdot 10^8 = 9 \cdot 10^8$$

This result, called the **distributive law of exponents,** says that to *raise a product to a power,* raise each factor to that power and then multiply.

Similar patterns can be observed for quotients. We now summarize the five laws of exponents.

Laws of Exponents

Let a and b be any nonzero real numbers, and let m and n be any integers.

Addition law:	$b^m \cdot b^n = b^{m+n}$
Multiplication law:	$(b^n)^m = b^{mn}$
Subtraction law:	$\dfrac{b^m}{b^n} = b^{m-n}$
Distributive laws:	$(ab)^m = a^m b^m \qquad \left(\dfrac{a}{b}\right)^m = \dfrac{a^m}{b^m}$

EXAMPLE 10   **PÓLYA'S METHOD**

Under $\frac{3}{4}$ impulse power the *Starship Enterprise* will travel 1 million kilometers (km) in 3 minutes.* Compare full impulse power with the speed of light, which is approximately $1.08 \cdot 10^9$ kilometers per hour (km/hr).

Solution   We use Pólya's problem-solving guidelines for this example.

Star Trek, The Next Generation (episode that first aired the week of May 15, 1993).

Understand the Problem. You might say, "I don't know anything about Star Trek," but with most problem solving in the real world, the problems you are asked to solve are often about situations with which you are unfamiliar. Finding the necessary information to understand the question is part of the process. We assume that full impulse is the the same as 1 impulse power, so that if we multiply $\frac{3}{4}$ impulse power by $\frac{4}{3}$ we will obtain $\left(\frac{3}{4} \cdot \frac{4}{3}\right) = 1$ full impulse power.

Devise a Plan. We will calculate the distance traveled (in kilometers) in one hour under $\frac{3}{4}$ power, and then will multiply that result by $\frac{4}{3}$ to obtain the distance in kilometers per hour under full impulse power.

Carry Out the Plan.

YIELD — Spend some time with this plan; it illustrates conversion of units.

$$\frac{3}{4} \text{ impulse power} = \frac{1,000,000 \text{ km}}{3 \text{ min}} \qquad \text{Given}$$

$$= \frac{10^6 \text{ km}}{3 \text{ min}} \cdot \frac{\mathbf{20}}{\mathbf{20}} \qquad \text{Multiply by } \frac{20}{20} \text{ to change}$$
$$\qquad\qquad\qquad\qquad \text{3 minutes to 60 minutes}$$
$$= \frac{10^6 \cdot 2 \cdot 10 \text{ km}}{60 \text{ min}} \qquad \text{so that we can convert to hours.}$$

$$= \frac{2 \cdot 10^7 \text{ km}}{1 \text{ hr}}$$

$$= 2 \cdot 10^7 \text{ km/hr}$$

We now multiply both sides by $\frac{4}{3}$ to find the distance under full impulse.

$$\frac{4}{3}\left(\frac{3}{4} \text{ impulse power}\right) = \frac{4}{3} \cdot 2 \cdot 10^7 \text{ km/hr}$$

$$\text{full impulse power} = \frac{8}{3} \cdot 10^7 \text{ km/hr}$$

$$\approx 2.666666667 \cdot 10^7 \text{ km/hr}$$

Comparing this to the speed of light, we see

$$\frac{\text{IMPULSE SPEED}}{\text{SPEED OF LIGHT}} = \frac{2.666666667 \cdot 10^7}{1.08 \cdot 10^9} = \frac{2.666666667}{1.08} \cdot 10^{7-9} \approx 0.025$$

Look Back. We see that full impulse power is about 2.5% of the speed of light.   ◆

Comprehending Large Numbers

We began this section by looking at the size of the cosmos. But just how large is large? Most of us are accustomed to hearing about millions, billions (budgets or costs of disasters), or even trillions (the national debt is about $6.3 trillion), but how do we really understand the magnitude of these numbers? Would you do a better job than Dennis' parents in the cartoon on the next page at explaining "How much is a million?"

A **million** is a fairly modest number, 10^6. Yet if we were to count one number per second, nonstop, it would take us about 278 hours or approximately $11\frac{1}{2}$ days to count to a million. Not a million days have elapsed since the birth of Christ (a million days is about 2,700 years). A large book of about 700 pages contains about a million letters. A million balloons (see photograph in the margin) were released at a single time. How large a room would it take to hold 1,000,000 inflated balloons?

But the age in which we live has been called the age of billions. How large is a **billion?** How long would it take you to count to a billion? Go ahead—make a guess. To get some idea about how large a billion is, let's compare it to some familiar units:

More than a million balloons were released at one time on December 5, 1985, at Disneyland by the city of Anaheim, California.

DENNIS THE MENACE® used by permission of Hank Ketcham and © 1972 by North American Syndicate.

- If you gave away $1,000 *per day,* it would take you more than 2,700 *years* to give away a billion dollars.
- A stack of a billion $1 bills would be more than 59 miles high.
- At 8% interest, a billion dollars would earn you $219,178.08 interest *per day!*
- A billion seconds ago, the movie *Godfather* came out and the videogame *Pong* was released.
- A billion minutes ago was about A.D 100.
- A billion hours ago, people had not yet appeared on earth.

But a billion is only 10^9, a mere nothing when compared with the real giants. Keep these magnitudes (sizes) in mind. Earlier in this section we noticed that a cube containing our solar neighborhood is 9.2×10^{11} miles on a side. (See Figure 1.10.) This is less than a billion times the size of the earth. (Actually, it is $9.2 \cdot 10^{11} \div 7,927 \approx 1.2 \times 10^8$.) Now we are moving beyond that comparison to some real giants.

There is an old story of a king who, being under obligation to one of his subjects, offered to reward him in any way the subject desired. Being of mathematical mind and modest tastes, the subject simply asked for a chessboard with one grain of wheat on the first square, two on the second, four on the third, and so forth. The old king was delighted with this modest request! Alas, the king was soon sorry he granted the request.

EXAMPLE 11   PÓLYA'S METHOD

Estimate the magnitude of the grains of wheat on the last square of a chessboard.

Solution   We use Pólya's problem-solving guidelines for this example.

Understand the Problem.   Each square on the chessboard has grains of wheat placed on it. To answer this question you need to know that a chessboard has 64 squares. The first square has $1 = 2^0$ grains, the next has $2 = 2^1$ grains, the next $4 = 2^2$, and so on. Thus, he needed 2^{63} grains of wheat for the last square alone. We showed this number in Example 1d.

Devise a Plan. We know (from Example 1) that $2^{63} \approx 9.22337 \times 10^{18}$. We need to find the size of a grain of wheat, and then convert 2^{63} grains into bushels. Finally, we need to state this answer in terms we can understand.

Carry Out the Plan. I went to a health food store, purchased some raw wheat, and found that there are about 250 grains per cubic inch (in.^{3}). I also went to a dictionary and found that a bushel is 2,150 in.^{3}. Thus, the number of grains of wheat in a bushel is

$$2{,}150 \times 250 = 537{,}500 = 5.375 \times 10^5$$

To find the number of bushels in 2^{63} grains we need to divide:

$$(9.22337 \times 10^{18}) \div (5.375 \times 10^5) = \frac{9.22337}{5.375} \times 10^{18-5}$$
$$\approx 1.72 \times 10^{13}$$

Look Back. This answer does not mean a thing without looking back and putting it in terms we can understand. I went to a *World Almanac* and found that in 1996 the U.S. wheat production was 2,281,763,000 bushels. To find the number of years it would take the United States to produce the necessary wheat for the last square of the chessboard, we need to divide the production into the amount needed:

$$\frac{1.72 \times 10^{13}}{2.28 \times 10^9} = \frac{1.72}{2.28} \times 10^{13-9} \approx 0.75 \times 10^4$$

This is 7,500 years! ◆

What is the name of the largest number you know? Recently we have heard about the national debt, which exceeds $6 **trillion.** Table 1.1 shows some large numbers.

Table 1.1 Some Large Numbers

1	*one:*	1
8	*byte:*	8 ← A basic unit on a computer; a string of eight binary digits
10	*ten:*	10
10^2	*hundred:*	100
10^3	*thousand:*	1,000
2^{10}	*kilobyte:*	1,024 ← Computer term for 1,024 bytes
10^6	*million:*	1,000,000 ← Dennis the Menace cartoon
2^{20}	*megabyte:*	1,048,576 ← A unit of computer storage capacity; Mb
10^9	*billion:*	1,000,000,000 ← Discussed in text
2^{30}	*gigabyte:*	1,073,741,824 ← Approximately 1,000 Mb; abbreviated Gb
10^{12}	*trillion:*	1,000,000,000,000 ← National debt $6 trillion
10^{15}	*quadrillion:*	1,000,000,000,000,000 ← Number of all words ever printed
10^{18}	*quintillion:*	1,000,000,000,000,000,000
10^{21}	*sextillion:*	1,000,000,000,000,000,000,000
		⋮
10^{63}	*vigintillion:*	1 followed by 63 zeros
		⋮
10^{100}	***googol:***	10,000,000,000,000,000,000,000,000,000,000,-000,000,000,000,000,000,000,000,000,000,000,000,000,-000,000,000,000,000,000,000,000,000,000,000

Do Things Really Change?

"Students today can't prepare bark to calculate their problems. They depend upon their slates which are more expensive. What will they do when their slate is dropped and it breaks? They will be unable to write!"

Teacher's Conference, 1703

"Students depend upon paper too much. They don't know how to write on slate without chalk dust all over themselves. They can't clean a slate properly. What will they do when they run out of paper?"

Principal's Association, 1815

"Students today depend too much upon ink. They don't know how to use a pen knife to sharpen a pencil. Pen and ink will never replace the pencil."

National Association of Teachers, 1907

"Students today depend upon store bought ink. They don't know how to make their own. When they run out of ink they will be unable to write word or ciphers until their next trip to the settlement. This is a sad commentary on modern education."

The Rural American Teacher, 1929

"Students today depend upon these expensive fountain pens. They can no longer write with a straight pen and nib (not to mention sharpening their own quills). We parents must not allow them to wallow in such luxury to the detriment of learning how to cope in the real business world, which is not so extravagant."

PTA Gazette, 1941

"Ball point pens will be the ruin of education in our country. Students use these devices and then throw them away. The American virtues of thrift and frugality are being discarded. Business and banks will never allow such expensive luxuries."

Federal Teacher, 1950

"Students today depend too much on hand-held calculators."

Anonymous, 1995

Problem Set 1.3

LEVEL 1

1. **IN YOUR OWN WORDS** What do we mean by *exponent?*

2. **IN YOUR OWN WORDS** Define *scientific notation* and discuss why it is useful.

3. **IN YOUR OWN WORDS** Do you plan to use a calculator for working the problems in this book? If so, what type of logic does it use?

4. **IN YOUR OWN WORDS** Describe the differences in evaluating exponents and using scientific notation on a calculator.

5. **IN YOUR OWN WORDS** How many classrooms would be necessary to hold 1,000,000 inflated balloons?

6. **IN YOUR OWN WORDS** What is the largest number whose name you know? Describe the size of this number.

7. **IN YOUR OWN WORDS** What is a *trillion?* Do not simply define this number, but discuss its magnitude (size) in terms that are easy to understand.

Write each of the numbers in Problems 8–11 in scientific notation and in floating-point notation (as on a calculator).

8. **a.** 3,200   **b.** 0.0004   **c.** 64,000,000,000

9. **a.** 23.79   **b.** 0.000001   **c.** 35,000,000,000

10. **a.** 5,629   **b.** 630,000   **c.** 0.00000 0034

11. **a.** googol   **b.** 1,200,300   **c.** 0.00000 123

Write each of the numbers in Problems 12–15 in fixed-point notation.

12. **a.** 7^2 **b.** 7.2×10^{10} **c.** 4.56 +3

13. **a.** 2^6 **b.** 2.1×10^{-3} **c.** 4.07 +4

14. **a.** 6^3 **b.** 4.1×10^{-7} **c.** 4.8 −7

15. **a.** 6^{-2} **b.** 3.217×10^7 **c.** 8.89 −11

Write each of the numbers in Problems 16–19 in scientific notation.

16. The velocity of light in a vacuum is about 30,000,000,000 cm/sec.

17. The wavelength of the orange-red line of krypton 86 is about 6,100 Å.

18. The distance between Earth and Mars (220,000,000 miles) when drawn to scale is 0.00000 25 in.

19. The national debt is approximately $6 trillion. It has been proposed that this number be used to define a new monetary unit, a *light buck*. That is, one light buck is the amount necessary to generate domestic goods and services at the rate of $186,000 per second.

Write each of the numbers in Problems 20–23 in fixed-point notation.

20. A kilowatt-hour is about 3.6×10^6 joules.

21. A ton is about 9.06×10^2 kilograms.

22. The volume of a typical neuron is about 3×10^{-8} cm³.

23. If the sun were a lightbulb, it would be rated at 3.8×10^{25} watts.

In Problems 24–31, first estimate your answer and then calculate the exact answer.

24. How many hours are there in 365 days?

25. How many pages are necessary to make 1,850 copies of a manuscript that is 487 pages long? (Print on one side only.)

26. If your payroll deductions are $255.83 per week and your weekly gross wages are $1,025.66, what is your net pay?

27. If your monthly salary is $1,543 per month, what is your annual salary?

28. If you are paid $6.25 per hour, what is your annual salary?

29. Carrie Dashow, the "say hello" woman, is trying to personally greet 1,000,000 people. In her first year, which ended on January 3, 2000, she had greeted 13,688 people. At this rate, how long will it take her to greet one million people?

30. If your car gets 23 miles per gallon, how far can you go on 15 gallons of gas?

31. If your car travels 280 miles and uses 10.2 gallons, how many miles per gallon did you get?

LEVEL 2

Compute the results in Problems 32–37. Leave your answers in scientific notation.

32. a. $(6 \times 10^5)(2 \times 10^3)$ **b.** $\dfrac{6 \times 10^5}{2 \times 10^3}$

33. a. $\dfrac{(5 \times 10^4)(8 \times 10^5)}{4 \times 10^6}$ **b.** $\dfrac{(6 \times 10^{-3})(7 \times 10^8)}{3 \times 10^7}$

34. a. $\dfrac{(6 \times 10^7)(4.8 \times 10^{-6})}{2.4 \times 10^5}$ **b.** $\dfrac{(2.5 \times 10^3)(6.6 \times 10^8)}{8.25 \times 10^4}$

35. a. $\dfrac{(2xy^{-2})(2^{-1}x^{-1}y^4)}{x^{-2}y^2}$ **b.** $\dfrac{x^2y(2x^3y^{-5})}{2^{-2}x^4y^{-8}}$

36. a. $\dfrac{0.00016 \times 500}{2,000,000}$ **b.** $\dfrac{15,000 \times 0.0000004}{0.005}$

37. a. $\dfrac{4,500,000,000,000 \times 0.00001}{50 \times 0.0003}$

b. $\dfrac{0.0348 \times 0.00000\,00000\,00002}{0.000058 \times 0.03}$

38. Estimate the distance from Los Angeles International Airport to Disneyland.

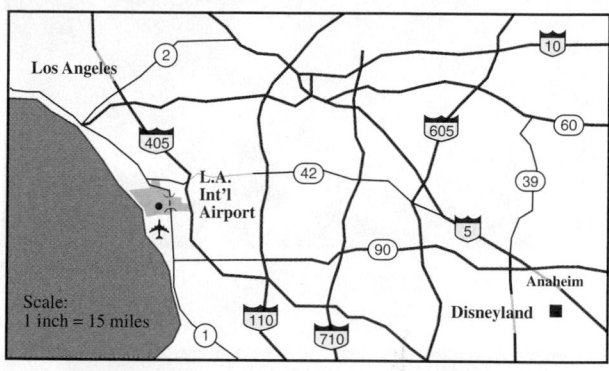

39. Estimate the distance from Fish Camp to Yosemite Village in Yosemite National Park.

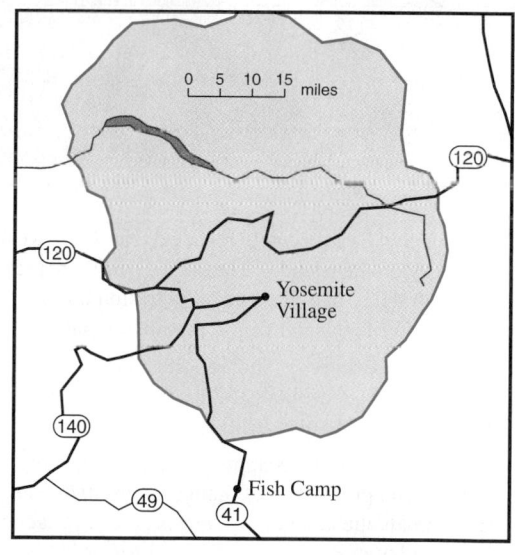

Estimate the number of items in each photograph in Problems 40–43.

40.

© Fredrik Bodin/Stock Boston

41.

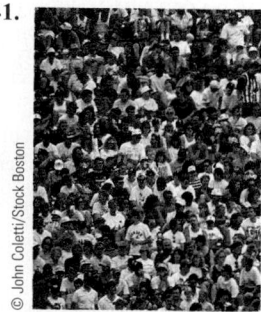

© John Coletti/Stock Boston

42.

© David Gilkey/Daily Camera, Boulder, CO

43.

44. The Associated Press recently reported that W. M. Keck Observatory on the island of Hawaii took infrared pictures of galaxy 4C41.17 at a distance of 72 trillion billion miles from Earth. Write this number in scientific notation.

45. The United Press International reported that our atmosphere weighs 5 quadrillion, 157 trillion tons. Write this number in scientific notation.

46. A box of oranges contains approximately 96 oranges. If the U.S. annual production of oranges is 186,075,000 boxes, estimate the number of oranges produced each year in the United States.

47. Approximately how high would a stack of 1 million $1 bills be? (Assume there are 233 new $1 bills per inch.)

48. Estimate how many pennies it would take to make a stack 1 in. high. Approximately how high would a stack of 1 million pennies be?

49. There are 2,260,000 grains in a pound of sugar. If the U.S. annual production of sugar is 30,000,000 tons, estimate the number of grains of sugar produced in a year in the United States. Use scientific notation. (*Note:* There are 2,000 lb per ton.)

LEVEL 3

50. a. What is the largest number you can represent on your calculator?

b. What is the largest number you can think of using only three digits?

c. Use your calculator to convert this number into scientific notation.

51. Zerah Colburn (1804–1840) toured America when he was 6 years old to display his calculating ability. He could instantaneously give the square and cube roots of large numbers. It is reported that it took him only a few seconds to find 8^{16}. Use your calculator to *help you* find this number exactly (not in scientific notation).

52. Jedidiah Buxion (1707–1772) never learned to write, but given any distance he could tell you the number of inches, and given any length of time he could tell you the number of seconds. If he listened to a speech or a sermon, he could tell the number of words or syllables in it. It reportedly took him only a few moments to mentally calculate the number of cubic inches in a right-angle block of stone 23,451,789 yards long, 5,642,732 yards wide, and 54,465 yards thick. Estimate this answer on your calculator. You will use scientific notation because of the limitations of your calculator, but remember that Jedidiah gave the *exact* answer by working the problem in his head.

53. George Bidder (1806–1878) not only possessed exceptional power at calculations but also went on to obtain a good education. He could give immediate answers to problems of compound interest and annuities. One question he was asked was: If the moon is 238,000 miles from the earth and sound travels at the rate of 4 miles per minute, how long would it be before the inhabitants of the moon could hear the Battle of Waterloo? By calculating *mentally,* he gave the answer in less than one minute! First make an estimate, and then use your calculator to give the answer in days, hours, and minutes, to the nearest minute.

54. A sheet of notebook paper is approximately 0.003 in. thick. Tear the sheet in half so that there are 2 sheets. Repeat so that there are 4 sheets. If we repeat again, there will be a pile of 8 sheets. Continue in this fashion until the paper has been halved 50 times. If it were possible to complete the process, how high would you guess the final pile would be? After you have guessed, *compute* the height.

PROBLEM SOLVING

55. If it takes one second to write down each digit, how long will it take to write down all the numbers from 1 to 1,000?

56. If it takes one second to write down each digit, how long will it take to write down all the numbers from 1 to 1,000,000?

57. Imagine that you have written down the numbers from 1 to 1,000. What is the total number of zeros you have recorded?

58. Imagine that you have written down the numbers from 1 to 1,000,000. What is the total number of zeros you have recorded?

59. a. If the entire population of the world moved to California and each person were given an equal amount of area, how much space would you *guess* that each person would have (multiple choice)?

    A. 7 in.^{2}

    B. 7 ft^2

    C. 70 ft^2

    D. 700 ft^2

    E. 1 mi^2

  b. If California is 158,600 mi^2 and the world population is 6.3 billion, calculate the answer to part **a.**

60. It is known that a person's body has about one gallon of blood in it, and that a cubic foot will hold about 7.5 gallons of liquid. It is also known that Central Park in New York has an area of 840 acres. If walls were built around the park, how tall would those walls need to be to contain the blood of all 6,300,000,000 people in the world?

© Rafael Macia/Photo Researchers

1.4   FINITE AND INFINITE

In the last section, we discussed some very large numbers. In Problem 6 of Section 1.3 we asked the question, What is the largest number whose name you know? No matter what your answer, we can find a larger number by adding one to that "largest number." Somehow we know that "there is no end," but we need to introduce some notation and terminology to tackle the "really big" numbers!

Denoting Sets

A fundamental concept in mathematics—and in life, for that matter—is the sorting of objects into similar groupings. Every language has an abundance of words that mean "a collection" or "a grouping." For example, we speak of a *herd* of cattle, a *flock* of birds, a *school* of fish, a track *team*, a stamp *collection*, and a *set* of dishes. All these grouping words serve the same purpose, and in mathematics we use the word *set* to refer to any collection of objects.

In Section 1.2, we introduced the idea of *undefined terms*. The word **set** is a perfect example of why there must be some undefined terms. Every definition requires other terms, so some *undefined terms* are necessary to get us started. To illustrate this idea, let's try to define the word *set* by using dictionary definitions:

    "*Set*: a collection of objects." What is a collection?

    "*Collection*: an *accumulation*." What is an accumulation?

    "*Accumulation*: a *collection,* a *pile,* or a *heap*." We see that the word *collection* gives us a **circular definition.** What is a *pile?*

    "*Pile*: a *heap*." What is a heap?

    "*Heap*: a *pile*."

Do you see that a dictionary leads us in circles? In mathematics, we do not allow circular definitions, and this forces us to accept some words without definition. The term *set* is undefined. Remember, the fact that we do not define *set* does not prevent us from having an intuitive grasp of how to use the word.

Sets are usually specified in one of two ways. The first is by *description,* and the other is by the *roster* method. In the **description method,** we specify the set by describing it in such a way that we know exactly which elements belong to it. An example is the set of 50 states in the United States of America. We say that this set is **well defined,** since there is no doubt that the state of California belongs to it and that the state of Germany does not—neither does the state of confusion. Lack of confusion, in fact, is necessary in using sets. The distinctive property that determines the inclusion or exclusion of a particular element is called the *defining property* of the set.

Consider the example of *the set of good students in this class.* This set is not well defined, since it is a matter of opinion whether a student is a "good" student. If we agree, however, on the meaning of the words *good students,* then the set is said to be *well defined.* A better (and more precise) formulation is usually required—for example, *the set of all students in this class who received a C or better on the first examination.* This is well defined, since it can be clearly determined exactly which students received a C or better on the first test.

In the **roster method,** the set is defined by listing the members. The objects in a set are called **members** or **elements** of the set and are said to **belong to** or **be contained in** the set. For example, instead of defining a set as *the set of all students in this class who received a C or better on the first examination,* we might simply define the set by listing its members: {Howie, Mary, Larry}.

Sets are usually denoted by capital letters, and the notation used for sets is braces. Thus, the expression

$$A = \{4, 5, 6\}$$

means that *A* is the name for the set whose members are the numbers 4, 5, and 6.

Sometimes we use braces with a defining property, as in the following examples:

{states in the United States of America}

{all students in this class who received an A on the first test}

The most common use of set terminology is to refer to certain sets of numbers. In Chapter 4, we will discuss certain sets of numbers, which we list here:

$\mathbb{N} = \{1, 2, 3, 4, \dots \}$	Set of **natural,** or **counting, numbers**
$\mathbb{W} = \{0, 1, 2, 3, 4, \dots \}$	Set of **whole numbers**
$\mathbb{Z} = \{\dots, -2, -1, 0, 1, 2, \dots \}$	Set of **integers**
$\mathbb{Q} = \{\frac{a}{b} \mid a \in \mathbb{Z}, b \in \mathbb{Z}, b \neq 0\}$	Set of **rational numbers**

The notation we used for the set of rational numbers may be new to you. If we try to list the set of rational numbers by roster, we will find that this is a difficult task (see Problem 50). A new notation called **set-builder notation** was invented to allow us to combine both the roster and the description methods. Consider:

The set of all x
$$\overbrace{\{x} \mid x \text{ is an even counting number}\}$$
↑
such that

We now use this notation for the set of rational numbers:

$$\left\{ \frac{a}{b} \;\middle|\; a \text{ is an integer and } b \text{ is a nonzero integer} \right\}$$

Read this as: "The set of all $\dfrac{a}{b}$ such that *a* is an integer and *b* is a nonzero integer."

EXAMPLE 1

Specify the given sets by roster. If the set is not well defined, say so.

a. {counting numbers between 10 and 20}

b. {distinct letters in the word *happy*}

c. {counting numbers greater than 1,000}

d. {U.S. presidents arranged in chronological order}

e. {good U.S. presidents}

Solution

a. {11, 12, 13, 14, 15, 16, 17, 18, 19}
Notice that *between* does not include the first and last numbers.

b. {*h, a, p, y*}

c. {1001, 1002, 1003, . . . }
Notice that it is sometimes impossible or impractical to write *all* the elements of a particular set using the roster method. We use *three dots* to indicate that some element have been omitted. You must be careful, however, to list enough elements so that someone looking at the set can see the intended pattern.

d. {Washington, Adams, Jefferson, . . . , Clinton, Bush}

e. Not well defined ◆

EXAMPLE 2

Specify the given sets by description.

a. {1, 2, 3, 4, 5, . . .}

b. {0, 1, 2, 3, 4, 5, . . .}

c. {. . . , −3, −2, −1, 0, 1, 2, 3, . . .}

d. {12, 14, 16, . . . , 98}

e. {4, 44, 444, 4444, . . .}

f. {*m, a, t, h, e, i, c, s*}

Solution Answers may vary.

a. Counting (or natural) numbers

b. Whole numbers

c. Integers

d. {Even numbers between 10 and 100}

e. {Counting numbers whose digits consist of fours only}

f. {Distinct letters in the word *mathematics*} ◆

If S is a set, we write $a \in S$ if a is a member of the set S, and we write $b \notin S$ if b is not a member of the set S. Thus, "$a \in \mathbb{Z}$" means that the variable a is an integer, and the statement "$b \in \mathbb{Z}, b \neq 0$" means that the variable b is a nonzero integer.

EXAMPLE 3

Let C = cities in California, a = city of Anaheim, and b = city of Berlin. Use set membership notation to describe relations among a, b, and c.

Solution $a \notin C$; $b \notin C$ ◆

If we consider a set of elements of a given set, that set is known as a **subset.** For example, consider the set of students in your class. The set of students in the class receiving a grade of C would be a subset of the set of students.

Equal and Equivalent Sets

We say that two sets are **equal** if they contain exactly the same elements. Thus, if $E = \{2, 4, 6, 8, \ldots\}$,

$$\{x \mid x \text{ is an even counting number}\} = \{x \mid x \in E\}$$

The order in which you represent elements in a set has no effect on set membership. Thus,

$$\{1, 2, 3\} = \{3, 1, 2\} = \{2, 1, 3\} = \cdots$$

Also, if an element appears in a set more than once, it is not generally listed more than a single time. For example,

$$\{1, 2, 3, 3\} = \{1, 2, 3\}$$

Another possible relationship between sets is that of *equivalence*. Two sets are **equivalent** if they have the same *number* of elements. Don't confuse this concept with equality. Equivalent sets do not need to be equal sets, but equal sets are always equivalent.

EXAMPLE 4

Which of the following sets are equivalent? Are any equal?

$$\{\circ, \triangle, \square\}, \{5, 8, 11\}, \{1, \ulcorner, \sqcap\}, \{\bullet, \odot, \star\}, \{1, 2, 3\}$$

Solution   All of the given sets are equivalent. Notice that no two of them are equal, but they all share the property of "threeness."   ◆

The number of elements in a set is often called its **cardinality.** The cardinality of the sets in Example 4 is 3; that is, the common property of the sets is the **cardinal number** of the set. The cardinality of a set S is denoted by $|S|$. Equivalent sets with four elements each have in common the property of "fourness," and thus we would say that their cardinality is 4.

EXAMPLE 5

Find the cardinality of the following.

a. $R = \{5, \triangle, Y, \pi\}$ **b.** $S = \{\quad\}$

c. $T = \{\text{states of the United States}\}$

Solution

a. The cardinality of R is 4, so we write $|R| = 4$.

b. The cardinality of S (the empty set) is 0, so we write $|S| = 0$.

c. The cardinality of T is 50 or $|T| = 50$.   ◆

Infinite Sets

Certain sets such as $\mathbb{N}$, $\mathbb{Z}$, $\mathbb{W}$, or $A = \{1000, 2000, 3000, \ldots\}$ have a common property. We call these *infinite sets*. If the cardinality of a set is 0 or a counting number, we say it is **finite.** Otherwise, we say it is **infinite.** We can also say that a set is finite if it has a cardinality less than some counting number, even though we may not know its precise cardinality. For example, we can safely assert that the set of students attending the University of Hawaii is finite even though we may not know its cardinality, because the cardinality is certainly less than a million.

You use the ideas of cardinality and of equivalent sets every time you count something, even though you don't use the term *cardinality*. For example, a set of sheep is counted by finding a set of counting numbers that has the same cardinality as the set of sheep. These equivalent sets are found by using an idea called *one-to-one correspondence*. If the set of sheep is placed into a one-to-one correspondence with a set of pebbles, the set of pebbles can then be used to represent the set of sheep.

One-to-One Correspondence

Two sets A and B are said to be in a **one-to-one correspondence** if we can find a pairing so that

1. Each element of A is paired with precisely one element of $B;$ and

2. Each element of B is paired with precisely one element of A.

EXAMPLE 6

Show which of the following sets are equivalent.
$M = \{3\}, N = \{\text{three}\}, P = \{\triangle, \square, \circ, \nabla\}, Q = \{t, h, r, e\}$

Solution Notice that

$$M = \{3\} \qquad\qquad P = \{\triangle, \square, \circ, \nabla\}$$
$$\updownarrow \qquad\qquad\qquad \updownarrow\ \updownarrow\ \updownarrow\ \updownarrow$$
$$N = \{\text{three}\} \qquad Q = \{\ t,\ \ h,\ \ r,\ \ e\}$$

so we say "M is equivalent to N" and we write $M \leftrightarrow N$. Note $P \leftrightarrow Q$. We also see that $M = N$. ◆

Since infinity is not a number, we cannot correctly say that the cardinality of the counting numbers is infinity. In the late 18th century, Georg Cantor (1845–1918) assigned a cardinal number $\aleph_0$ (pronounced "aleph-null") to the set of counting numbers. That is, $\aleph_0$ is the cardinality of the set of counting numbers

$$\mathbb{N} = \{1, 2, 3, 4, \ldots\}$$

The set

$$E = \{2, 4, 6, 8, \ldots\}$$

also has cardinality $\aleph_0$, since it can be put into a one-to-one correspondence with set $\mathbb{N}$:

$$\mathbb{N} = \{1, 2, 3, 4, \ldots, n, \ldots\}$$
$$\updownarrow \updownarrow \updownarrow \updownarrow \qquad \updownarrow$$
$$E = \{2, 4, 6, 8, \ldots, 2n, \ldots\}$$

Notice that all elements of E are elements of the set $\mathbb{N}$, and that $E \neq \mathbb{N}$. In such a case, we say that E is a **proper subset** of $\mathbb{N}$. Cantor used this property as a defining property for infinite sets.

Infinite Set

An **infinite set** is a set that can be placed in a one-to-one correspondence with a proper subset of itself.

EXAMPLE 7

Show that the set of integers $\mathbb{Z} = \{\ldots, -3, -2, -1, 0, 1, 2, 3, \ldots\}$ is infinite.

Solution   We can show that the set of integers can be placed into a one-to-one correspondence with the set of counting numbers:

$$\mathbb{N} = \{1, 2, \quad 3, 4, \quad 5, 6, \quad 7, \ldots, 2n, 2n + 1, \ldots\}$$

$$\uparrow\uparrow \quad \uparrow\uparrow \quad \uparrow\uparrow \quad \uparrow \quad\quad \uparrow \quad\quad \uparrow$$

$$\mathbb{Z} = \{0, 1, -1, 2, -2, 3, -3, \ldots, \quad n, \quad -n, \ldots\}$$

Since $\mathbb{N}$ is a proper subset of $\mathbb{Z}$, we see that the set of integers is infinite.   ◆

Example 7 not only shows that the set of integers is infinite, it also shows that the cardinality of $\mathbb{Z}$ is $\aleph_0$. If a set has cardinality $\aleph_0$, we say that the set is **countable.** Thus, both the set of counting numbers and the set of integers are countable.

Cartesian Product of Sets

There is an operation of sets called the *Cartesian product* that provides a way of generating new elements when given the elements of two sets. Suppose we have the sets

$$A = \{a, b, c\} \quad \text{and} \quad B = \{1, 2\}$$

The **Cartesian product** of set A and B, denoted by $A \times B$, is the set of all *ordered pairs* (x, y) where $x \in A$ and $y \in B$. For this example,

$$A \times B = \{(a, 1), (a, 2), (b, 1), (b, 2), (c, 1), (c, 2)\}$$

EXAMPLE 8

How many elements are in the Cartesian product of the given sets?

a. $A = \{\text{Frank, George, Hazel}\}$ and
$B = \{\text{Alfie, Bogie, Calvin, Doug, Ernie}\}$

b. $C = \{\text{U.S. Senators}\}$ and
$D = \{\text{U.S. President, U.S. Vice President, Secretary of State}\}$

Solution

a. One of the ways for finding a Cartesian product is to represent the sets as a rectangular array.

$A \times B$	Alfie, a	Bogie, b	Calvin, c	Doug, d	Ernie, e
Frank, f	(f, a)	(f, b)	(f, c)	(f, d)	(f, e)
George, g	(g, a)	(g, b)	(g, c)	(g, d)	(g, e)
Hazel, h	(h, a)	(h, b)	(h, c)	(h, d)	(h, e)

Since a Cartesian product can be arranged as an array, we see the number of elements is the product of the number of elements in the sets. That is, since $|A| = 3$ and $|B| = 5$, we see $|A \times B| = 3 \times 5 = 15$.

b. We are looking for $|C \times D|$, but we see it is not practical to form the rectangular array because $|C| = 100$ and $|D| = 3$. We still can visualize the size of the array even without writing it out, so we conclude

$$|C \times D| = |C| \times |D| = 100 \times 3 = 300 \quad\quad ◆$$

We will frequently find it convenient to use the multiplication principle illustrated in Example 8, so we restate it as a general property.

Fundamental Counting Principle

The **fundamental counting principle** gives the number of ways of performing two or more tasks. If task A can be performed in m different ways, and if, after task A is performed, a second task B can be performed in n ways, then task A followed by task B can be performed in $m \times n$ ways.

EXAMPLE 9

Classify each of the sets as finite or infinite.

a. Set of people on Earth

b. Set of license plates that can be issued using three letters followed by three numerals

c. Set of drops of water in all the oceans of the world

Solution

a. We do not know the size of this set, but we do know that there are less than 7 billion people on Earth. If the cardinality of a set is less than a finite number (and 7 billion is finite), then it must be a finite set.

b. We can use the fundamental counting principle to calculate the number of possible license plates:

$$26 \times 26 \times 26 \times 10 \times 10 \times 10 = 17,576,000$$

Note that each "26" counts the number of letters of the alphabet, and each "10" counts the number of numerals in each position. Since this is a particular number, we see that this set of license plates is finite.

c. Certainly this must be an infinite set . . . who could count the number of drops? Let's consider this a bit more carefully. What is a drop? Is that well defined? If you are told by your doctor to use two drops to dilate the pupils of your eyes, can you do that? You have a device called an "eyedropper" and you squeeze out two drops. Now, there is a finite number of drops that make up an ounce, and there are a certain number of ounces in a gallon, and a certain number of gallons in a cubic foot. The earth will fit entirely inside a cube of a certain size, and if this cube were filled with water, it would contain a finite number of gallons. The number of drops of water in all the oceans of the world is certainly less than this number, and hence the set is finite. ◆

Universal and Empty Sets

We conclude this section by considering two special sets. The first is the set that contains every element under consideration, and the second is the set that contains no elements. A **universal set,** denoted by U, contains all the elements under consideration in a given discussion; and the **empty set** contains no elements, and thus has cardinality 0. The empty set is denoted by $\{\ \ \}$ or $\varnothing$. Do not confuse the notations $\varnothing$, 0, and $\{\varnothing\}$. The symbol $\varnothing$ denotes a *set* with no elements; the symbol 0 denotes a *number;* and the symbol $\{\varnothing\}$ is a set with one element (namely, the set containing $\varnothing$).

For example, if $U = \{1, 2, 3, 4, 5, 6, 7, 8, 9\}$, then all sets we would be considering would have elements only among the elements of U. No set could contain the number 10, since 10 is not in that agreed-upon universe.

For every problem, a universal set must be specified or implied, and it must remain fixed for that problem. However, when a new problem is begun, a new universal set can be specified.

Notice that we defined *a* universal set and *the* empty set; that is, a universal set may vary from problem to problem, but there is only one empty set. After all, it doesn't matter whether the empty set contains no numbers or no people—it is still empty. The following are examples of descriptions of the empty set:

$$\{\text{living saber-toothed tigers}\} \qquad \{\text{counting numbers less than 1}\}$$

Problem Set 1.4

LEVEL 1

1. **IN YOUR OWN WORDS** Why do you think mathematics accepts the word *set* as an undefined term?
2. **IN YOUR OWN WORDS** Distinguish between equal and equivalent sets.
3. **IN YOUR OWN WORDS** What is a universal set?
4. **IN YOUR OWN WORDS** What is the empty set?
5. **IN YOUR OWN WORDS** Give three examples of the empty set.
6. **IN YOUR OWN WORDS** Give three examples of an infinite set.

Tell whether each set in Problems 7–14 is well defined. If it is not well defined, change it so that it is well defined.

7. The set of students attending the University of California
8. {Grains of sand on Earth}
9. The set of counting numbers between 3 and 4
10. The set of happy people in your country
11. The set of people with pointed ears
12. {Counting numbers less than 0}
13. {Good bets on the next race at Hialeah}
14. {Years that will be bumper years for growing corn in Iowa}

Specify the sets in Problems 15–22 by roster.

15. {Distinct letters in the word *mathematics*}
16. {Current U.S. president}
17. {Odd counting numbers less than 11}
18. {Positive multiples of 3}
19. {Distinct letters in the word *pipe*}
20. {Counting numbers greater than 150}
21. {Even counting numbers between 5 and 15}
22. {Counting numbers containing only 1s}

Specify the sets in Problems 23–28 by description.

23. {1, 2, 3, 4, 5, 6, 7, 8, 9}
24. {1, 11, 121, 1331, 14641, . . .}
25. {10, 20, 30, . . ., 100}
26. {50, 500, 5000, . . .}
27. {101, 103, 105, . . ., 169}   28. {m, i, s, p}

Write out in words the description of the sets given in Problems 29–34, and then list each set in roster form.

29. $\{x \mid x$ is an odd counting number$\}$
30. $\{x \mid x$ is a natural number between 1 and 10$\}$
31. $\{x \mid x \in \mathbb{N}, x \neq 8\}$
32. $\{x \mid x \in \mathbb{W}, x \leq 8\}$
33. $\{x \mid x \in \mathbb{W}, x < 8\}$
34. $\{x \mid x \in \mathbb{W}, x \notin E\}$ where $E = \{2, 4, 6, \ldots\}$

Find the Cartesian product of the sets given in Problems 35–38.

35. $A = \{c, d, f\}$ and $B = \{w, x\}$; find $A \times B$.
36. $A = \{c, d, f\}$ and $B = \{w, x\}$; find $B \times A$.
37. $A = \{1, 2, 3, 4, 5\}$ and $B = \{a, b, c\}$; find $A \times B$.
38. $A = \{1, 2, 3, 4, 5\}$ and $B = \{a, b, c\}$; find $B \times A$.
39. If $A = \{$letters of the alphabet$\}$ and $B = \{0, 1, 2, \ldots, 9\}$, what is $|A \times B|$?
40. If $A = \{$letters of the alphabet$\}$, what is $|A \times A|$?
41. If $C = \{500, 501, 502, \ldots, 599\}$ and $D = \{$U.S. state capitals$\}$, what is $|C \times D|$?
42. If $M = \{$distinct letters in the word *mathematics*$\}$ and $N = \{$distinct letters in the word *nevertheless*$\}$, what is $|M \times N|$?

LEVEL 2

43. Consider the sets

$$A = \{\text{distinct letters in the word } pipe\}$$
$$B = \{4\}$$
$$C = \{p, i, e\}$$
$$D = \{2 + 1\}$$
$$E = \{\text{three}\}$$
$$F = \{3\}$$

a. What is the cardinality of each set?
b. Which of the sets can be placed into a one-to-one correspondence?
c. Which of the given sets are equal?

44. Consider the sets

$$A = \{16\}$$
$$B = \{10 + 6\}$$
$$C = \{10, 6\}$$
$$D = \{2^5\}$$
$$E = \{2, 5\}$$

  a. What is the cardinality of each set?
  b. Which of the sets can be placed into a one-to-one correspondence?
  c. Which of the given sets are equal?

LEVEL 3

45. Show that there is more than one one-to-one correspondence between the sets $\{m, a, t\}$ and $\{1, 2, 3\}$.

46. Show that the following sets have the same cardinality, by placing the elements into a one-to-one correspondence.

$$\{1, 2, 3, 4, \ldots, 999, 1000\}$$
$$\{7964, 7965, 7966, 7967, \ldots, 8962, 8963\}$$

47. Show that the following sets do not have the same cardinality.

$$\{1, 2, 3, 4, \ldots, 586, 587\}$$
$$\{550, 551, 552, 553, \ldots, 902, 903\}$$

48. Do the following sets have the same cardinality?

$$\{48, 49, 50, \ldots, 783, 784\}$$
$$\{485, 487, 489, \ldots, 2053, 2055\}$$

49. Classify the following sets as finite or infinite.
  a. The set of stars in the Milky Way
  b. The set of counting numbers greater than one million
  c. The set of people who are living or have ever lived
  d. The set of grains of sand on all the beaches on Earth

50. List the set of rational numbers between 0 and 1 by roster. List the elements systematically, and do not list elements that have been previously listed.

Show that each set in Problem 51–54 has cardinality $\aleph_0$.

51. $\{-1, -2, -3, \ldots\}$

52. $\{1000, 3000, 5000, \ldots\}$

53. $\{1, 2, 4, 8, 16, \ldots\}$

54. $\{-85, -80, -75, \ldots\}$

Show that each set in Problems 55–58 is infinite by placing it in a one-to-one correspondence with a proper subset of itself.

55. $\mathbb{W}$ **56.** $\mathbb{N}$

57. $\{12, 14, 16, \ldots\}$ **58.** $\{4, 44, 444, 4444, \ldots\}$

PROBLEM SOLVING

59. Make up an example to show that

$$\aleph_0 + \aleph_0 = \aleph_0.$$

60. Show that the two line segments have the same number of points:

_____

_____

CHAPTER SUMMARY

"Numeracy is the ability to cope confidently with the mathematical demands of adult life."

Mathematics Counts

IMPORTANT TERMS

Addition law of exponents [1.3]
Axiom [1.2]
Base [1.3]
Belongs to [1.4]
Billion [1.3]
Cardinal number [1.4]
Cardinality [1.4]
Cartesian product [1.4]
Circular definition [1.4]
Conjecture [1.2]
Contained in [1.4]
Countable set [1.4]
Counting number [1.2, 1.4]
Cube [1.2]

Decimal notation [1.3]
Deductive reasoning [1.2]
Description method [1.4]
Difference [1.2]
Distributive laws of exponents [1.3]
Element [1.4]
Empty set [1.4]
Equal sets [1.4]
Equivalent sets [1.4]
Estimate [1.3]
Exponent [1.3]
Exponential [1.3]
Exponential notation [1.3]
Exponentiation [1.2]

Extended order of operations [1.3]
Finite set [1.4]
Fixed-point form [1.3]
Floating-point form [1.3]
Fundamental counting principle [1.4]
Googol [1.3]
Inductive reasoning [1.2]
Infinite set [1.4]
Laws of exponents [1.3]
Member [1.4]
Million [1.3]
Multiplication law of exponents [1.3]
Natural number [1.4]
(continued)

TYPES OF PROBLEMS

Use Pólya's method to solve a problem. [1.1]
Use Pascal's triangle as an aid to problem solving. [1.1]
Answer questions by using inductive reasoning. [1.2]
Simplify an expression using the order of operations. [1.2, 1.3]
Write out formulas for given word statements. [1.2]
Distinguish inductive from deductive reasoning. [1.2]
Translate mathematical statements. [1.2]
Write out exponential numbers without using exponents. [1.3]
Write a large or small number in scientific notation. [1.3]
Use a calculator to answer numerical questions. [1.3]
Estimate answers to numerical questions. [1.3]
Simplify numerical problems by using the laws of exponents. [1.3]
Describe the relative sizes of large and small numbers. [1.3]
Specify sets by roster and by description. [1.4]
Determine whether sets have the same cardinality by placing them in a one-to-one
    correspondence. [1.4]
Show that a given set has cardinality $\aleph_0$. [1.4]
Show that a given set is infinite. [1.4]

CHAPTER 1 REVIEW QUESTIONS

1. In your own words, describe Pólya's problem-solving model.

2. In how many ways can a person walk 5 blocks north and 4 blocks west, if the streets are arranged in a standard rectangular arrangement?

3. A chessboard consists of 64 squares, as shown in Figure 1.11. The rook can move one or more squares horizontally or vertically. Suppose a rook is in the upper left-hand corner of a chessboard. Tell how many ways the rook can reach the point marked "X". Assume that the rook always moves toward its destination.

4. Compute $111{,}111{,}111 \times 111{,}111{,}111$. Do not use direct multiplication; show all your work.

5. What is meant by "order of operations"?

6. What is scientific notation?

7. Does the story in the News Clip illustrate inductive or deductive reasoning?

Figure 1.11 Chessboard

Q: What has 18 legs and
    catches flies?
A: I don't know, what?
Q: A baseball team. What
    has 36 legs and catches
    flies?
A: I don't know that, either.
Q: Two baseball teams. If
    the United States has 100
    senators, and each state has
    2 senators, what does...
A: I know this one!
Q: Good. What does each
    state have?
A: Three baseball teams!

8. Show the calculator keys you would press as well as the calculator display for the result. Also state the brand and model of the calculator you are using.

   a. 2^{63} **b.** $\dfrac{9.22 \times 10^{18}}{6.34 \times 10^{6}}$

9. Assume that there is a $281.9 billion budget "windfall." Which of the following choices would come closest to liquidating this windfall?

   A. Buy the entire U.S. population a steak dinner.

   B. Burn one dollar per second for the next 1,000 years.

   C. Give $80,000 to every resident of San Francisco and use the remainder of the money to buy a $200 CD player for every resident of China.

10. One method for estimating the capacity of a boat is to divide the product of the length and width of the boat by 15. Write a formula to represent this idea.

11. What is wrong, if anything, with the following "Great Tapes" advertisement?

   > Instead of reading the 100 greatest books of all time, buy these beautifully transcribed books on tape. If you listen to only one 45-minute tape per day, you will complete the greatest books of all time in only one year

12. In 1995, it was reported that an iceberg separated from Antarctica. The size of this iceberg was reported equal to 7×10^{16} ice cubes. Convert this size to meaningful units.

13. The national debt in 2003 was about $6,300,000,000,000. Write this number in scientific notation. Suppose that in 2003 there were 290,000,000 people in the United States. If the debt is divided equally among these people, how much is each person's share?

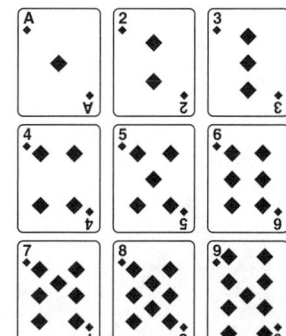

14. Rearrange the cards in the formulation to the left so that each horizontal, vertical, and diagonal line of three adds up to 15.

15. Assume that your classroom is 20 ft × 30 ft × 10 ft. If you fill this room with dollar bills (assume that a stack of 233 dollar bills is 1 in. tall), how many classrooms would it take to contain the national debt of $6,300,000,000,000?

16. The set of rational numbers is $\mathbb{Q} = \{\frac{a}{b} \mid a \in \mathbb{Z}, b \in \mathbb{N}\}$.

   a. Write out this statement in words.

   b. Give an example for the number q where $q \in \mathbb{Q}$.

17. Show that the set $F = \{5, 10, 15, 20, 25, \ldots\}$ is infinite.

18. **a.** Write the set $\{-1, -2, -3, \ldots\}$ using set-builder notation.

   b. Show that this set has cardinality $\aleph_0$.

19. Consider the following pattern:

    1 is happy.
    10 is happy because $1^2 + 0^2 = 1$, which is happy.
    13 is happy because $1^2 + 3^2 = 10$, which is happy.
    19 is happy because $1^2 + 9^2 = 82$ and $8^2 + 2^2 = 68$
            and $6^2 + 8^2 = 100$ and $1^2 + 0^2 + 0^2 = 1$,
            which is happy.

    On the other hand,
    2, 3, 4, 5, 6, 7, 8, and 9 are unhappy.
    11 is unhappy because $1^2 + 1^2 = 2$, which is unhappy.
    12 is unhappy because $1^2 + 2^2 = 5$, which is unhappy.

    Find one unhappy number as well as one happy number.

20. Suppose you could write out 7^{1000}. What is the last digit?

GROUP RESEARCH PROBLEMS

Working in small groups is typical of most work environments, and learning to work with others to communicate specific ideas is an important skill. Work with three or four other students to submit a single report based on each of the following questions.

G1. It is stated in the text that "Mathematics is alive and constantly changing. As we complete the last decade of this century, we stand on the threshold of major changes in the mathematics curriculum in the United States." Report on some of these recent changes.

  References Lynn Steen, *Everybody Counts: A Report to the Nation on the Future of Mathematics Education* (Washington, DC: National Academy Press, 1989). See also, *Curriculum and Evaluation Standards for School Mathematics* from the National Council of Teachers of Mathematics (Reston, VA: NCTM, 1989).

G2. Do some research on Pascal's triangle, and see how many properties you can discover. You might begin by answering these questions:

  a. What are the successive powers of 11?

  b. Where are the natural numbers found in Pascal's triangle?

  c. What are triangular numbers and how are they found in Pascal's triangle?

  d. What are the tetrahedral numbers and how are they found in Pascal's triangle?

  e. What relationships do the patterns in Figure 1.12 have to Pascal's triangle?

Multiples of 2   Multiples of 3

Figure 1.22 **Patterns in Pascal's triangle**

  References James N. Boyd, "Pascal's Triangle," *Mathematics Teacher,* November 1983, pp. 559–560.

  Dale Seymour, *Visual Patterns in Pascal's Triangle,* (Palo Alto, CA: Dale Seymour Publications, 1986).

  Karl J. Smith, "Pascal's Triangle," *Two-Year College Mathematics Journal,* Volume 4 (Winter 1973).

INDIVIDUAL RESEARCH PROBLEMS

Learning to use sources outside your classroom and textbook is an important skill, and here are some ideas for extending some of the ideas in this chapter. You can find references to these projects in a library or at

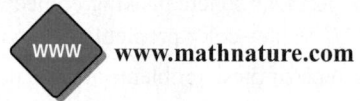

www.mathnature.com

PROJECT 1.1 What do the following people have in common?

Ralph Abernathy, civil rights leader
Harry Blackmun, Associate Justice of the U.S. Supreme Court
David Dinkins, former mayor of New York City
Art Garfunkel, folk-rock singer
Alexander Solzhenitsyn, Nobel-prize-winning novelist
J. P. Morgan, banking, steel, and railroad magnate
Michael Jordan, basketball superstar

PROJECT 1.2 Find some puzzles, tricks, or magic stunts that are based on mathematics.

PROJECT 1.3 Discover some properties about the magic square known as *Melancholia.*

PROJECT 1.4 Write a short paper about the construction of magic squares.

PROJECT 1.5 Design a piece of art based on a magic square.

PROJECT 1.6 An *alphamagic square,* invented by Lee Sallows, is a magic square so that not only do the numbers spelled out in words form a magic square, but the numbers of letters of the words also form a magic square. For example,

five	twenty-two	eighteen
twenty-eight	fifteen	two
twelve	eight	twenty-five

gives rise to two magic squares:

5	22	18
28	15	2
12	8	25

and

4	9	8
11	7	3
6	5	10

The first magic square comes from the numbers represented by the words in the alphamagic square, and the second magic square comes from the numbers of letters in the words of the alphamagic square. Find another alphamagic square.

PROJECT 1.7 Answer the question posed in Problem 59, Section 1.3 for your own state. If you live in California, then use Florida.

PROJECT 1.8 Read the article, "Mathematics at the Turn of the Millennium," by Philip A. Griffiths, *The American Mathematical Monthly,* January 2000, pp. 1–14. Briefly describe each of these famous problems:

a. Fermat's last theorem

b. Kepler's sphere packing conjecture

c. The four-color problem

Which of these problems are discussed later in this text, and where?

The objective of this article was to communicate something about mathematics to a general audience. How well did it succeed with you? Discuss.

2 The Nature of Logic

. . . Symbolic logic is Mathematics, Mathematics is Symbolic Logic, the twain are one.
J. F. Keyser

CONTENTS

OVERVIEW

In the first chapter we studied patterns that allow us to form conjectures using inductive reasoning. Inductive reasoning is the process of forming conjectures that are based on a number of observations of specific instances.

There is, however, a type of reasoning that produces *certain* results. It is based on accepting certain *premises* and then *deducing* certain inescapable conclusions. Not only much of mathematics, but also much of the reasoning we do in the world, is based on the principles of logic introduced in this chapter.

We begin with *simple statements,* which are either true or false, and then we combine those statements using certain connectives, such as *and, or, not, because, either . . . or, neither . . . nor,* and *unless.* These connected statements are called compound statements, and we discover the truth or falsity of compound statements using truth tables. The final step in our journey through this chapter is to begin with an argument, translate it into symbols, derive logical conclusions, and then translate back into English to complete a proof of a logical argument.

IMPORTANT IDEAS

Truth Table of Fundamental Operators [2.1, 2.2]
Law of Double Negation [2.2]
Law of Contraposition [2.2]
De Morgan's Laws [2.3]
Negation of a Conditional [2.3]
Direct Reasoning [2.4]
Indirect Reasoning [2.4]
Transitive Reasoning [2.4]
Relationship Between Logic and Circuits [2.6]

BOOK REPORTS

Write a 500-word report on one of these books:
Overcoming Math Anxiety, Sheila Tobias (Boston: Houghton Mifflin Co., 1978).
Mathematical Puzzles for Beginners & Enthusiasts, Geoffrey Mott-Smith (New York: Dover, 1954).
Escalante, The Best Teacher in America, Jay Mathews (New York: Holt, 1988).

LINKS

Individual Projects
Group Projects
Research Links

www.mathnature.com

2.1 | DEDUCTIVE REASONING

Terminology

There is a type of reasoning that produces results based on certain laws of logic. For example, consider the following argument.

1. If you read the *Times,* then you are well informed.
2. You read the *Times.*
3. Therefore, you are well informed.

Statements 1, 2, and 3 are called an **argument.** If you accept statements 1 and 2 of the argument as true, then you *must* accept statement 3 as true. Statements 1 and 2 are called the **hypotheses** or **premises** of the argument, and statement 3 is called the **conclusion.** Such reasoning is called **deductive reasoning** and, if the conclusion follows from the hypotheses, the reasoning is said to be **valid.**

The purpose of this chapter is to build a logical foundation to aid you, not only in your study of mathematics and other subjects, but also in your day-to-day contact with others.

Logic is a method of reasoning that accepts only those conclusions that are inescapable. This is possible because of the strict way in which every concept is defined. That is, everything must be defined in a way that leaves no doubt or vagueness in meaning. Nothing can be taken for granted, and dictionary definitions are not usually sufficient. For example, in English one often defines a sentence as "a word or group of words stating, asking, commanding, requesting, or exclaiming something; conventional unit of connected speech or writing, usually containing a subject and predicate, beginning with a capital letter, and ending with an end mark." In symbolic logic, we use the word *statement* to refer to a declarative sentence.

Statement

> A **statement** is a declarative sentence that is either true or false, but not both true and false. A **simple statement** is one without any connective.

All of the following are statements, since they are either true or false.

1. School starts on Friday the 13th.

2. $5 + 6 = 16$

3. Fish swim.

4. Mickey Mouse is president.

If the sentence is a question or a command, or if it is vague or nonsensical, then it cannot be classified as true or false; thus, we would not call it a statement. For example:

5. Go away!

6. What are you doing?

7. This sentence is false.

These are not statements by our definition, since they cannot possibly be either true or false.

 Fuzzy logic

In Eastern culture, the yin-yang symbol shows the duality of nature: male–female, light–dark, good–evil.

Difficulty in simplifying arguments may arise because of their length, the vagueness of the words used, the literary style, or the possible emotional impact of the words used. Consider the following two arguments:

I. If you use heroin, then you first used marijuana.
   Therefore, if you use marijuana, then you will use heroin.
II. If George Washington was assassinated, then he is dead.
   Therefore, if he is dead, he was assassinated.

Logically, these two arguments are exactly the same, and both are **invalid** forms of reasoning. Nearly everyone would agree that the second is invalid, but many people see the first as valid, because of the emotional appeal of the words used.

To avoid these difficulties, and to be able to simplify a complicated logical argument, we set up an *artificial symbolic language.* This procedure was first suggested by Leibniz in his search to unify all of mathematics. What we will do is invent a notational shorthand. We denote simple statements with letters such as *p, q, r, s, . . . ,* and then define certain connectives. The problem, then, is to *translate the English statements into symbolic form, simplify the symbolic form,* and then *translate the symbolic form back into English statements.*

The key to simplifying complicated logical arguments is to consider only simple statements connected by certain **operators (connectives),** each of which is defined precisely. Some of these operators are *not, and, or, neither . . . nor, if . . . then, unless,* and *because.*

A **compound statement** is formed by combining simple statements with one or more of these operators. Because of our basic definition of a statement, we see that the *truth value* of any statement is either true (T) or false (F), but we should point out that the newest chapter of logic is presently being written. This new logic, called **fuzzy logic,** defines gradations of "T" and "F" to describe real-world concepts that are, by nature, often vague. The difference between the logic we study in this chapter and fuzzy logic is something Aristotle called the **law of the excluded middle.** As we saw in Chapter 1, an object either does or does not belong to a set. There is no middle ground: The number 5 belongs fully to the set of counting numbers or it does not. This law of the excluded middle avoids the contradiction of an object that both is and is not the same object at the same time. Sets that are fuzzy break the law of the excluded middle—to some degree. The study of fuzzy logic requires concepts of probability (Chapter 11), as well as an understanding of the logic we introduce in this chapter. For purposes of this course, we accept the law of the excluded middle as an axiom, and work within a framework that forces every statement to be either true or false.

Truth Value

> The **truth value** of a *simple statement* is either true (T) or false (F).
>
> The **truth value** of a *compound statement* is true or false and depends only on the truth values of its simple component parts. It is determined by using the rules for connecting those parts with well-defined operators.

It is not sufficient to assume that we know the meanings of the operators *and, or, not,* and so on, even though they may seem obvious and simple. The strength of logic is that it does not leave any meanings to chance or to individual interpretation. In defining the truth values of these words, we will, however, try to conform to common usage.

Conjunction

If p and q represent two simple statements, then "p and q" is the compound statement using the operator called **conjunction.** The word **and** is symbolized by $\wedge$. For example,

> *I have a penny **and** a quarter in my pocket.*

When is this compound statement true? Let p and q represent two simple statements as follows:

> p: I have a penny. q: I have a quarter.

There are four logical possibilities (not necessarily all true). The four possibilities can be shown in table form:

		p	q
1. I have a penny. I have a quarter.	1.	T	T
2. I have a penny. I do not have a quarter.	2.	T	F
3. I do not have a penny. I have a quarter.	3.	F	T
4. I do not have a penny. I do not have a quarter.	4.	F	F

If p and q are both true, we would certainly say that the compound statement is true; otherwise, we say it is false. Thus, we define "$p \wedge q$" according to Table 2.1.

Table 2.1 Definition of Conjunction

p	q	$p \wedge q$
T	T	T
T	F	F
F	T	F
F	F	F

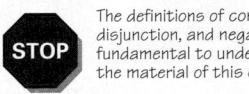

The definitions of conjunction, disjunction, and negation are fundamental to understanding the material of this chapter.

The common usage of the word *and* conforms to the technical definition of conjunction.

It is worth noting that statements p and q need not be related. For example,

> *Fish swim and there are* 100 *U.S. senators.*

is a true compound statement.

Disjunction

The operator **or,** denoted by $\vee$, is called **disjunction.** The meaning of this simple word is ambiguous, as we can see by considering the following examples.

I. *I have a penny or a quarter in my pocket.*
II. *The president is speaking in New York or in California at* 7:00 P.M. *tonight.*

Let's represent four statements as follows:

> p: I have a penny in my pocket.
> q: I have a quarter in my pocket.
> n: The president is speaking in New York at 7:00 P.M. tonight.
> c: The president is speaking in California at 7:00 P.M. tonight.

In everyday terms, what do we mean by each of these examples?

Statement I may mean:

> *I have a penny in my pocket. I have a quarter in my pocket. I have both a penny and a quarter in my pocket.*

George Boole (1815–1864)
Gottfried Leibniz advanced Aristotelian logic with his universal characteristic. However, the world took little notice of Leibniz' logic until George Boole completed his book *An Investigation of the Laws of Thought*. Boole considered various mathematical operators by separating them from the other commonly used symbols. This idea was popularized by Bertrand Russell (1872–1970) and Alfred North Whitehead (1861–1947) in their monumental *Principia Mathematica*. In this work, they began with a few assumptions and three undefined terms, and built a system of symbolic logic. From this, they then formally developed the theory of arithmetic and mathematics.

Statement II may mean:

> *The president is speaking in New York at* 7:00 P.M. *tonight.*
> *The president is speaking in California at* 7:00 P.M. *tonight.*
> It does *not* mean that he will do both.

These statements illustrate different usages of the word *or*. Now we are forced to select a single meaning for the operator, so we choose a definition that conforms to statement I. That is, "$p \lor q$" means "p or q, perhaps both." Thus, we define "$p \lor q$" according to Table 2.2.

Table 2.2 Definition of Disjunction

p	q	$p \lor q$
T	T	T
T	F	T
F	T	T
F	F	F

In logic, the statement defined in Table 2.2 is called the **inclusive or.** The second meaning of the word *or* ("p or q, but not both") is called the **exclusive or.** In this book, we will translate the exclusive or by using the operators **either . . . or.** Thus, in this book, we would translate the two examples as:

I. *I have a penny or a quarter in my pocket.*
II. *The president is speaking either in New York or in California at* 7:00 P.M. *tonight.*

Negation

The operator **not,** denoted by $\sim$, is called **negation.** Table 2.3 serves as a straightforward definition of negation. The negation of a true statement is false, and the negation of a false statement is true.

Table 2.3 Definition of Negation

p	$\sim p$
T	F
F	T

EXAMPLE 1

For the statement *t:* Otto is telling the truth, translate $\sim t$.

Solution $\sim t$:   Otto is not telling the truth.   ◆

We also can translate $\sim t$ as "It is not the case that Otto is telling the truth." You must be careful when negating statements containing the words *all, none,* or *some.* For example, write the negation of

> *All students have pencils.*

Go ahead—write it down before reading on. Did you write "No students have pencils" or "All students do not have pencils"? Remember that if a statement is false, then its negation must be true. The correct negations are:

> *Not all students have pencils.*
> *At least one student does not have a pencil.*
> *It is not the case that all students have pencils.*
> *Some students do not have pencils.*

In mathematics, the word *some* is used to mean "at least one." Table 2.4 gives some of the common negations.

Table 2.4 Negation of *All, Some,* and *No*

Statement	Negation
All	Some . . . not
Some	No
Some . . . not	All
No	Some

EXAMPLE 2

Write the negation of each statement.

a. All people have compassion.

b. Some animals are dirty.

c. Some students do not take Math 10.

d. No students are enthusiastic.

Solution

a. Some people do not have compassion.

b. No animal is dirty.

c. All students take Math 10.

d. Some students are enthusiastic.   ◆

We consider the negation of compound statements in the next section.

Order of Operations

Parentheses are used to indicate the order of operations. Thus,

$$\sim(n \wedge c) \text{ means the negation of the statement "}n \text{ and } c\text{,"}$$

and is read as "it is not the case that *n* and *c*." On the other hand,

$$\sim n \wedge c \text{ means "the negation of } n \text{ and the statement } c\text{,"}$$

which is read "not *n* and *c*."

> Part of the difficulty in translating English into logical statements is the inaccuracy of our language. For example, how can "fat chance" and "slim chance" mean the same thing?

EXAMPLE 3

Translate the given statements into words.

a. $p \wedge q$ **b.** $\sim p$ **c.** $\sim(p \wedge q)$ **d.** $\sim p \wedge q$ **e.** $\sim(\sim q)$

Let *p:* I eat spinach; *q:* I am strong.

Solution

a. A translation of $p \wedge q$ is "I eat spinach, and I am strong."

b. A translation of $\sim p$ is "I do not eat spinach."

c. A translation for $\sim(p \wedge q)$ is "It is not the case that I eat spinach and am strong."

d. A translation for $\sim p \wedge q$ is "I do not eat spinach, and I am strong."

e. A translation of $\sim(\sim q)$ is "I am not not strong" or, if we assume that "not strong" is the same as "weak," then we can translate this as "I am not weak."   ◆

We can now consider the truth value of a compound statement using parentheses.

EXAMPLE 4

Suppose n is a true statement, and c is a false statement. What is the truth value of the compound statement $(n \vee c) \wedge \sim(n \wedge c)$?

Solution   Begin with the symbolic statement, fill in the truth values given in the problem, and then simplify using the correct order of operations and the definitions in Tables 2.1–2.3.

$$(n \vee c) \wedge \sim(n \wedge c) \quad \text{Start with given statement.}$$
$$(T \vee F) \wedge \sim(T \wedge F) \quad \text{Fill in the truth values.}$$
$$T \wedge \sim F \quad \text{Simplify.}$$
$$T \wedge T$$
$$T$$

Thus, the compound statement is true. This result does not depend on the particular statements n and c. As long as n is true and c is false, the result of the compound statement is the same—namely, true.   ◆

We conclude this section by finding the truth values for the statements in Example 3.

EXAMPLE 5

Suppose that p is T and q is F. Find the truth values for the statements in Example 3.
a. $p \wedge q$ **b.** $\sim p$ **c.** $\sim(p \wedge q)$ **d.** $\sim p \wedge q$ **e.** $\sim(\sim q)$
Let p: I eat spinach; q: I am strong.

Solution
a. $p \wedge q$
    $T \wedge F$ Substitute truth values.
      F   Thus, the statement "I eat spinach, and I am strong" is false.

b. $\sim p$
    $\sim T$ Substitute.
     F   The statement "I do not eat spinach" is false.

c. $\sim(p \wedge q)$
    $\sim(T \wedge F)$ Substitute.
    $\sim F$ Definition of conjunction
     T   Definition of negation
The statement "It is not the case that I eat spinach and am strong" is true.

d. $\sim p \wedge q$
    $\sim T \wedge F$ Substitute.
    $F \wedge F$ Definition of negation
      F   Definition of conjunction
The statement "I do not eat spinach, and I am strong" is false.

If you are a sports fan, you are used to statements that may be difficult to translate. For example, "I don't think there's any chance that they won't decline the penalty." Or, "San Francisco has been able to win this season by consistently outscoring their opponents."

e. $\sim(\sim q)$
$\quad \sim(\sim F)$ Substitute.
$\quad\quad \sim(T)$ Definition of negation
$\quad\quad\quad F$ Definition of negation again

The statement "I am not weak" is false.   ◆

Problem Set 2.1

LEVEL 1

1. **IN YOUR OWN WORDS** What is an operator?

2. **IN YOUR OWN WORDS** What do we mean by *conjunction*? Include as part of your answer the definition.

3. **IN YOUR OWN WORDS** What do we mean by *disjunction*? Include as part of your answer the definition.

4. **IN YOUR OWN WORDS** What do we mean by *negation*? Include as part of your answer the definition.

5. **IN YOUR OWN WORDS** Describe the procedure for finding the truth value of a compound statement.

According to the definition, which of the examples in Problems 6–9 are statements?

6. **a.** Hickory, Dickory, Dock, the mouse ran up the clock.
   b. $3 + 5 = 9$
   c. Is John ugly?
   d. John has a wart on the end of his nose.

7. **a.** March 18, 2004, is a Monday.
   b. Division by zero is impossible.
   c. Logic is difficult.
   d. $4 - 6 = 2$

8. **a.** $6 + 9 \neq 7 + 8$
   b. Thomas Jefferson was the 23rd president.
   c. Sit down and be quiet!
   d. If wages continue to rise, then prices will also rise.

9. **a.** Dan and Mary were married on August 3, 1979.
   b. $6 + 12 \neq 10 + 8$
   c. Do not read this sentence.
   d. Do you have a cold?

Write the negation of each statement in Problems 10–19.

10. All mathematicians are ogres.

11. All dogs have fleas.

12. Some integers are negative.

13. Some people do not pay taxes.

14. No even integers are divisible by 5.

15. No triangles are squares.

16. All squares are rectangles.

17. All counting numbers are divisible by 1.

18. Some apples are rotten.

19. Some integers are not odd.

20. Let *p*: Prices will rise; *q*: Taxes will rise.
    Translate each of the following statements into symbols.
    a. Prices will rise, or taxes will not rise.
    b. Prices will rise, and taxes will not rise.
    c. Prices will rise, and taxes will rise.
    d. Prices will not rise, and taxes will rise.

21. Assume that prices rise and taxes also rise. Under these assumptions, which of the statements in Problem 20 are true?

22. Assume that prices rise and taxes do not rise. Under these assumptions, which of the statements in Problem 20 are true?

23. Let *p*: Prices will rise; *q*: Taxes will rise.
    Translate each of the following statements into words.
    a. $p \vee q$ **b.** $\sim p \wedge q$
    c. $p \vee \sim q$ **d.** $\sim p \vee \sim q$

24. Let *p*: Paul is peculiar; and *q*: Paul likes to read mathematics textbooks. Translate each of the following statements into symbols.
    a. Paul is peculiar and he likes to read mathematics textbooks.
    b. Paul likes to read mathematics textbooks or he is peculiar.
    c. Paul likes to read mathematics textbooks and he is not peculiar.
    d. Paul does not like to read mathematics textbooks and Paul is not peculiar.

25. Assume that Paul likes to read mathematics textbooks, and also that he is not peculiar. Under these assumptions, which of the statements in Problem 24 are true?

26. Assume that Paul does not like to read mathematics textbooks, and also that he is not peculiar. Under these assumptions, which of the statements in Problem 24 are true?

27. Let *p*: Paul is peculiar; and *q*: Paul likes to read mathematics textbooks. Translate each of the following statements into words.
    a. $p \wedge q$ **b.** $\sim p \wedge q$
    c. $p \vee \sim q$ **d.** $\sim p \vee \sim q$

28. Let *p:* Today is Friday; *q:* There is homework tonight. Translate each of the following statements into words.
   a. $p \wedge q$ **b.** $\sim p \wedge q$
   c. $p \vee \sim q$ **d.** $\sim p \vee \sim q$

29. Assume *p* is T and *q* is T. Under these assumptions, which of the statements in Problem 27 are true?

30. Assume *p* is F and *q* is T. Under these assumptions, which of the statements in Problem 28 are true?

Find the truth value for each of the compound statements in Problems 31–38.

31. Assume *r* is F, *s* is T, and *t* is T.
   a. $(r \vee s) \vee t$ **b.** $(r \wedge s) \wedge \sim t$

32. Assume *r* is T, *s* is T, and *t* is T.
   a. $r \wedge (s \vee t)$ **b.** $(r \wedge s) \vee (r \wedge t)$

33. Assume *p* is T, *q* is T, and *r* is F.
   a. $(p \vee q) \wedge r$ **b.** $(p \wedge q) \wedge \sim r$

34. Assume *p* is T, *q* is T, and *r* is T.
   a. $p \wedge (q \vee r)$ **b.** $(p \wedge q) \vee (p \wedge r)$

35. Assume *p* is T, *q* is T, and *r* is T.
   a. $(p \vee q) \vee (r \wedge \sim q)$ **b.** $\sim(\sim p) \vee (p \wedge q)$

36. Assume *p* is T, *q* is F, and *r* is T.
   a. $(p \wedge q) \vee (p \wedge \sim r)$ **b.** $(\sim p \vee q) \wedge \sim p$

37. Assume *p* is T and *q* is F.
   a. $\sim(p \wedge q)$ **b.** $\sim p \wedge q$

38. Assume *p* is T and *q* is T.
   a. $(\sim p \vee q) \wedge \sim p$ **b.** $(p \wedge \sim q) \vee (p \wedge q)$

LEVEL 2

Translate the statements in Problems 39–47 into symbols. For each simple statement, be sure to indicate the meanings of the symbols you use. Answers are not unique.

39. W. C. Fields is eating, drinking, and having a good time.

40. Sam will not seek and will not accept the nomination.

41. Jack will not go tonight, and Rosamond will not go tomorrow.

42. Fat Albert lives to eat and does not eat to live.

43. The decision will depend on judgment or intuition, and not on who paid the most.

44. The successful applicant for the job will have a B.A. degree in liberal arts or psychology.

45. The winner must have an A.A. degree in drafting or three years of professional experience.

46. **a.** Dinner includes soup and salad, or the vegetable of the day.
   b. Dinner includes soup, and salad or the vegetable of the day.

47. **a.** Marsha finished the sign and table, or a pair of chairs.
   b. Marsha finished the sign, and the table or a pair of chairs.

In Problems 48–57, find the truth value when p is T, q is F, and r is F.

48. $(p \wedge q) \wedge r$ **49.** $p \vee (q \wedge r)$

50. $(p \vee q) \wedge \sim(p \vee \sim q)$ **51.** $(p \wedge \sim q) \vee (r \wedge \sim q)$

52. $\sim(\sim p) \wedge (q \vee p)$ **53.** $(r \wedge p) \vee (q \wedge r)$

54. $\sim(r \wedge q) \wedge (q \vee \sim q)$ **55.** $(p \wedge q) \vee (p \wedge r)$

56. $(p \wedge q) \vee [(p \vee \sim q) \vee (\sim r \wedge p)]$

57. $(q \vee \sim q) \wedge [(p \wedge \sim q) \vee (\sim r \vee r)]$

PROBLEM SOLVING

58. Prove the law of double negation. That is, prove that for any proposition *p*, $\sim(\sim p)$ has the same truth value as *p*.

59. Smith received the following note from Melissa: "Dr. Smith, I wish to explain that I was really joking when I told you that I didn't mean what I said about reconsidering my decision not to change my mind." Did Melissa change her mind or didn't she?

60. Their are three errers in this item. See if you can find all three.

2.2

TRUTH TABLES AND THE CONDITIONAL

Constructing Truth Tables

The connectives *and, or,* and *not* introduced in the previous section are called the **fundamental operators.** To go beyond these and consider additional operators, we need a device known in logic as a **truth table.** A truth table shows how the truth values of compound statements depend on the fundamental operators. Tables 2.1, 2.2, and 2.3 from the previous section should be memorized. They are summarized in Table 2.5 on page 62.

NOT AN ENTRANCE

NOT AN EXIT

DEPT. OF LOGIC

Robert Mankoff, *Saturday Review*, June 11, 1977.
© 1977 by Robert Mankoff.

Table 2.5 Truth Table of the Fundamental Operators

p	q	Conjunction $p \wedge q$	Disjunction $p \vee q$	Negation $\sim p$	Negation $\sim q$
T	T	T	T	F	F
T	F	F	T	F	T
F	T	F	T	T	F
F	F	F	F	T	T

EXAMPLE 1

Construct a truth table for the compound statement:

Alfie did not come last night and did not pick up his money.

Solution First identify the operators: *not . . . and . . . not.* Next, choose variables to represent the simple statements. Let

p: Alfie came last night.   q: Alfie picked up his money.

Translate the English statement into symbols: $\sim p \wedge \sim q$.

Finally, construct a truth table. List all possible combinations of truth values for the simple statements (see columns A and B).

A	B
p	q
T	**T**
T	**F**
F	**T**
F	**F**

Insert the truth values for $\sim p$ and $\sim q$ (see columns C and D).

A	B	C	D
p	q	$\sim p$	$\sim q$
T	T	**F**	**F**
T	F	**F**	**T**
F	T	**T**	**F**
F	F	**T**	**T**

Finally, insert the truth values for $\sim p \wedge \sim q$ (see column E).

A	B	C	D	E
p	q	$\sim p$	$\sim q$	$\sim p \wedge \sim q$
T	T	F	F	**F**
T	F	F	T	**F**
F	T	T	F	**F**
F	F	T	T	**T**

The only time the compound statement is true is when *both p and q are false.* ◆

There is a story about a logic professor who was telling her class that a double negative is known to mean a negative in some languages and a positive in others (as in English). She continued by saying that there is no spoken language in which a double positive means a negative. Just then she heard from the back of the room a sarcastic "Yeah, yeah."

EXAMPLE 2

Construct a truth table for $\sim(\sim p)$.

Solution

p	$\sim p$	$\sim(\sim p)$
T	F	T
F	T	F

◆

Notice from Example 2 that $\sim(\sim p)$ and p have the same truth values. (See also Problem 58 of Section 2.1.) If two statements have the same truth values, one can replace the other in any logical expression. This means that the double negation of a statement is the same as the original statement.

Law of Double Negation

$\sim(\sim p)$ may be replaced by p in any logical expression.

EXAMPLE 3

Use the law of double negation to rewrite the following statement made by an Iraqi official and reported on a national news report (December 3, 2002): "All that the U.S. has said about Iraq is a false lie."

Solution   If we assume that a lie is not the truth, then a "false lie" is the negation of "not the truth." This means a correct equivalent translation is, "All that the U.S. has said about Iraq is the truth."

◆

The following example provides a truth-table solution for Problem 58 of the previous problem set.

EXAMPLE 4

Construct a truth table to determine when the following statement is true.

$$\sim(p \wedge q) \wedge [(p \vee q) \wedge q]$$

Solution   We begin as before and move from left to right with parentheses taking precedence, focusing our attention on no more than two columns at a time (refer to Table 2.5 to find the correct entries).

A	B	C	D	E	F	G
p	q	$p \wedge q$	$\sim(p \wedge q)$	$p \vee q$	$(p \vee q) \wedge q$	$\sim(p \wedge q) \wedge [(p \vee q) \wedge q]$
T	T	T	F	T	T	F
T	F	F	T	T	F	F
F	T	F	T	T	T	T
F	F	F	T	F	F	F

The compound statement is true only when p is false and q is true.   ◆

Conditional

We can now use truth tables to prove certain useful results and to introduce some additional operators. The first one we'll consider is called the **conditional.** The statement "if p, then q" is called a *conditional statement.* It is symbolized by $p \rightarrow q$, where p is called the **antecedent,** and q is called the **consequent.** There are several ways of using a conditional, as illustrated by the following examples.

1. We use "if–then" to indicate a *logical* relationship—one in which the consequent follows logically from the antecedent:

   If $\sim(\sim p)$ has the same truth value as p, then p can replace $\sim(\sim p)$.

THERE'S TOO MUCH UNCERTAINTY ABOUT OUR RELATIONSHIP!

if

2. We can use "if–then" to indicate a causal relationship.

If John drops that rock, then it will land on my foot.

3. We can use "if–then" to report a decision on the part of the speaker.

If John drops that rock, then I will hit him.

4. We can use "if–then" when the consequent follows from the antecedent by the very definition of the words used.

If John drives a Geo, then John drives a car.

5. Finally, we can use "if–then" to make a *material implication.* There is no logical, causal, or definitional relationship between the antecedent and consequent; we use the expression simply to convey humor or emphasis.

If John gets an A on that test, then I'm a monkey's uncle.

The consequent is obviously false, and the speaker wishes to emphasize that the antecedent is also false.

Our task is to state a definition of the conditional that applies to all of these *if–then* statements. We will approach the problem by asking under what circumstances a given conditional would be false. Let's consider another example.

Suppose I make you a promise:

"If I receive my check tomorrow, then I will pay you the $10 that I owe you."

If I keep my promise, let's agree to say the statement is true; if I don't, then it is false. Let

$p:$ I receive my check tomorrow.
$q:$ I will pay you the $10 that I owe you.

We symbolize the promise by $p \rightarrow q$. There are four possibilities:

	p	q	
Case 1:	T	T	*I receive my check tomorrow, and I pay you the* $10. In this case, the promise, or conditional, is true.
Case 2:	T	F	*I receive my check tomorrow, and I do not pay you the* $10. In this case, the conditional is false, since I did not fulfill my promise.
Case 3:	F	T	*I do not receive my check tomorrow, but I pay you the* $10. In this case, I certainly didn't break my promise, so the conditional is true.
Case 4:	F	F	*I do not receive my check tomorrow, and I do not pay you the* $10. Here, again, I did not break my promise, so the conditional is true.

Actually, the promise was not tested for cases 3 and 4, since I didn't receive my check. Assume the principle of "innocent until proven guilty." The only time that I will have broken my promise is in case 2. The test for a conditional is to determine when it is false. In symbols,

$p \rightarrow q$ is false whenever $p \wedge {\sim}q$ is true (case 2)

or

$p \rightarrow q$ is true whenever ${\sim}(p \wedge {\sim}q)$ is true.

Construct a truth table for ${\sim}(p \wedge {\sim}q)$.

A	B	C	D	E
p	q	$\sim q$	$p \wedge \sim q$	$\sim(p \wedge \sim q)$
T	T	F	F	T
T	F	T	T	F
F	T	F	F	T
F	F	T	F	T

We use this truth table to define the conditional $p \to q$ as shown in Table 2.6.

STOP

Add this to your list of important definitions.

Table 2.6 Definition of Conditional

p	q	$p \to q$
T	T	T
T	F	F
F	T	T
F	F	T

The following examples illustrate the definition of the conditional.

EXAMPLE 5

Make up an example illustrating each of the four cases for the conditional.

Solution   Examples can vary.

a. Case 1:   T → T   *If 7 < 14, then 7 + 2 < 14 + 2.*
This is true, since both component parts are true.

b. Case 2:   T → F   *If 7 + 5 = 12, then 7 + 10 = 15.*
This is false, since the antecedent is true, but the consequent is false.

c. Case 3:   F → T   *If 7 + 5 = 15, then 8 + 2 = 10.*
This is true, since the antecedent is false.

d. Case 4:   F → F   *If 7 + 5 = 25, then 7 = 20.*
This is true, since the antecedent is false.   ◆

Example 5 shows that the conditional, in mathematics, does not mean that there is any cause-and-effect relationship. *Any* two statements can be joined with the if-then connective, and the result must be true or false.

EXAMPLE 6   PÓLYA'S METHOD

The following sentence is found on a tax form:
> If you do not itemize deductions on Schedule A and you have charitable contributions, then complete the worksheet on page 14 and enter the allowable part on line 36b.

Use symbolic form to analyze this sentence.

Solution   We use Pólya's problem-solving guidelines for this example.

Understand the Problem.   We need to rephrase, or analyze, this statement so that we can determine an action to take under all possible circumstances.

Devise a Plan.   We will translate the statement into symbolic form. It could then be analyzed using a truth table.

Carry Out the Plan. First identify the operators:

If you do **not** itemize deductions on Schedule A **and** you have charitable contributions, **then** complete the worksheet on page 14 **and** enter the allowable part on line 36b.

Next, assign variables to simple statements:

d: You itemize deductions on Schedule A.

c: You have charitable contributions.

w: You complete the worksheet on page 14.

b: You enter the allowable part on line 36b.

Rewrite the sentence, making substitutions for the variables:

If not d **and** c, **then** (w **and** b).

Complete the translation into symbols:

$$(\sim d \land c) \to (w \land b)$$

Look Back.

We check, for example, to see whether the statement is true if d, c, w, and b are all true. To check all possible circumstances, you would construct a truth table with 16 rows.

$$(\sim d \land c) \to (w \land b)$$
$$(\sim T \land T) \to (T \land T)$$
$$(F \land T) \to T$$
$$F \to T$$
$$T$$

The given statement is true in this case.   ◆

Translations for the Conditional

The *if* part of a conditional need not be stated first. All of the following statements have the same meaning:

Conditional translation	*Example*
If p, then q.	If you are 18, then you can vote.
q, if p.	You can vote, if you are 18.
p, only if q.	You are 18 only if you can vote.
All p are q.	All 18-year-olds can vote.

In addition to these translations for the conditional, there are related statements.

Converse
Inverse
Contrapositive

Given the conditional $p \to q$, we define:

the **converse** is $q \to p$;
the **inverse** is $\sim p \to \sim q$;
the **contrapositive** is $\sim q \to \sim p$.

EXAMPLE 7

Write the converse, inverse, and contrapositive of the statement: *If it is a 300Z, then it is a car.*

Solution Let *p:* It is a 300Z; *q:* It is a car. The given statement is symbolized as $p \rightarrow q$.

Converse: $q \rightarrow p$	If it is a car, then it is a 300Z.
Inverse: $\sim p \rightarrow \sim q$	If it is not a 300Z, then it is not a car.
Contrapositive: $\sim q \rightarrow \sim p$	If it is not a car, then it is not a 300Z. ◆

As you can see from Example 7, not all these statements are equivalent in meaning. The contrapositive and the original statement always have the same truth values, as do the converse and the inverse. We see this in Table 2.7, and summarize it in the following box.

Law of Contraposition

A conditional may always be replaced by its contrapositive without having its truth value affected.

Table 2.7 Truth table for variations of the conditional

				Statement	Converse	Inverse	Contrapositive
p	*q*	*~p*	*~q*	$p \rightarrow q$	$q \rightarrow p$	$\sim p \rightarrow \sim q$	$\sim q \rightarrow \sim p$
T	T	F	F	T	T	T	T
T	F	F	T	F	T	T	F
F	T	T	F	T	F	F	T
F	F	T	T	T	T	T	T

EXAMPLE 8

Assume that the following statement is true:

$p \rightarrow \sim q$ If you obey the law, then you will not go to jail.

Write the converse, inverse, and contrapositive.

Solution We note that *p:* You obey the law; *q:* You go to jail.

Converse: $\sim q \rightarrow p$	If you do not go to jail, then you obey the law.
Inverse: $\sim p \rightarrow q$	Note: $\sim(\sim q)$ is replaced by q (double negation).
	If you do not obey the law, then you will go to jail.
Contrapositive: $q \rightarrow \sim p$	If you go to jail, then you did not obey the law. *This has the same truth value as the given statement.* ◆

Problem Set 2.2

LEVEL 1

1. **IN YOUR OWN WORDS** What is a truth table?
2. **IN YOUR OWN WORDS** What is a conditional? Discuss.
3. **IN YOUR OWN WORDS** What is the law of double negation?

4. **IN YOUR OWN WORDS** What is the law of contraposition?

Construct a truth table for the statements given in Problems 5–22.

5. $\sim p \vee q$
6. $\sim p \wedge \sim q$
7. $\sim(p \wedge q)$
8. $\sim r \vee \sim s$

9. $\sim(\sim r)$

10. $(r \wedge s) \vee \sim s$

11. $p \wedge \sim q$

12. $\sim p \vee \sim q$

13. $(\sim p \wedge q) \vee \sim q$

14. $(p \wedge \sim q) \wedge p$

15. $p \vee (p \rightarrow q)$

16. $(p \wedge q) \rightarrow p$

17. $[p \wedge (p \vee q)] \rightarrow p$

18. $[p \vee (p \wedge q)] \rightarrow p$

19. $(p \vee q) \vee r$

20. $(p \wedge q) \wedge \sim r$

21. $[(p \vee q) \wedge \sim r] \wedge r$

22. $[p \wedge (q \vee \sim p)] \vee r$

Write the converse, inverse, and contrapositive of the statements in Problems 23–28.

23. $\sim p \rightarrow \sim q$

24. $\sim r \rightarrow t$

25. $\sim t \rightarrow \sim s$

26. If you break the law, then you will go to jail.

27. I will go Saturday if I get paid.

28. If you brush your teeth with Smiles toothpaste, then you will have fewer cavities.

LEVEL 2

Translate the sentences in Problems 29–36 into if–then form.

29. All triangles are polygons.

30. All prime numbers greater than 2 are odd numbers.

31. All good people go to heaven.

32. Everything happens to everybody sooner or later if there is time enough. (G. B. Shaw)

33. We are not weak if we make a proper use of these means which the God of Nature has placed in our power. (Patrick Henry)

34. All useless life is an early death. (Goethe)

35. All work is noble. (Thomas Carlyle)

36. Everything's got a moral if only you can find it. (Lewis Carroll)

First decide whether each simple statement in Problems 37–40 is true or false. Then state whether the given compound statement is true or false.

37. If $5 + 10 = 16$, then $15 - 10 = 3$.

38. The moon is made of green cheese only if Mickey Mouse is president.

39. If $1 + 1 = 10$, then the moon is made of green cheese.

40. $3 \cdot 2 = 6$ only if water runs uphill.

Tell which of the statements in Problems 41–44 are true.

41. Let p: $2 + 3 = 5$; q: $12 - 7 = 5$.
   a. $\sim p \vee q$
   b. $\sim p \wedge \sim q$
   c. $\sim(p \wedge q)$

42. Let p: 2 is prime; q: 1 is prime.
   a. $\sim p \vee \sim q$
   b. $\sim(\sim p)$
   c. $(p \wedge q) \vee \sim q$

43. Let p: $5 + 8 = 10$; q: $4 + 4 = 8$.
   a. $(p \wedge \sim q) \wedge p$
   b. $p \vee (p \rightarrow q)$
   c. $(p \wedge q) \rightarrow p$

44. Let p: $1 + 1 = 2$; q: $9 - 3 = 5$.
   a. $(p \rightarrow \sim q) \rightarrow (q \rightarrow \sim p)$
   b. $(p \rightarrow q) \rightarrow (\sim q \rightarrow \sim p)$

Translate the statements in Problems 45–50 into symbolic form.

45. If the qualifying person is a child and not your dependent, enter this child's name.

46. If the amount on line 31 is less than \$26,673 and a child lives with you, turn to page 27.

47. If line 32 is \$86,025 or less, multiply \$2,500 by the total number of exemptions claimed on line 6e.

48. If married and filing a joint return, enter your spouse's earned income.

49. If you are a student or disabled, see line 6 of instructions.

50. If the income on line 1 was reported to you on Form W-2 and the "Statutory employee" box on that form was checked, see instructions for Schedule C, line 1, and check here.

51. Repeat Example 6, except this time assume that you do not complete the worksheet on page 14, but all other statements are true.

52. Under what conditions will the statement in Example 6 be true?

LEVEL 3

In Problems 53–58, fill the blanks with a symbolic statement that follows from the given statement.

53. The applicant for the position must have a two-year college degree in drafting or five years of experience in the field. Let
   a: You are an applicant for the position.
   q: You are qualified for the position.
   e: You have a two-year college degree in drafting.
   f: You have five years of experience in the field.
   a. $a \rightarrow$ _____
   b. $(a \wedge e) \rightarrow$ _____
   c. $(a \wedge f) \rightarrow$ _____

54. To qualify for a loan, the applicant must have a gross income of at least \$35,000 if single or combined income of \$50,000 if married. Let
   q: You qualify for a loan.
   m: You are married.
   i: You have an income of at least \$35,000.
   b: Your spouse has an income of at least \$35,000.
   a. $(\sim m \wedge \sim i) \rightarrow$ _____
   b. $(\sim m \wedge i) \rightarrow$ _____
   c. $[m \wedge (i \wedge b)] \rightarrow$ _____

55. To qualify for the special fare, you must fly on Monday, Tuesday, Wednesday, or Thursday, and you must stay over a Saturday evening. Let

 q: You qualify for the special fare.
 m: You fly on Monday.
 t: You fly on Tuesday.
 w: You fly on Wednesday.
 h: You fly on Thursday.
 s: You stay over a Saturday evening.

 _____ → *q*

56. This contract is noncancellable by tenant for 60 days. Let

 t: You are a tenant.
 d: It is within 60 days.
 c: This contract can be canceled.
 $(t \wedge d) \to$ _____

57. The tenant agrees to lease the premises for 12 months beginning on September 1 and at a monthly rental charge of $800. Let

 t: You are a tenant.
 m: You lease the premises for 12 months.
 s: You will begin on September 1.
 p: You will pay $800 per month.
 $t \to$ _____

58. Utilities, except for water and garbage, are paid by the tenant. Let

 t: You are a tenant.
 w: You pay water.
 g: You pay garbage.
 u: You pay the other utilities.

 $t \to$ _____

> **PROBLEM SOLVING**

59. If Apollo can do anything, could he make an object that he could not lift?

60. Decide about the truth or falsity of the following statement:

 If wishes were horses, then beggars could ride.

2.3 | OPERATORS AND LAWS OF LOGIC

Biconditional, Implication, and Logical Equivalence

In the previous section we took great care to point out that a statement $p \to q$ and its converse $q \to p$ do not have the same truth values. However, it may be the case that $p \to q$ *and also* $q \to p$. In this case, we write

$$p \leftrightarrow q$$

and call this operator the **biconditional.** To determine the truth values of the biconditional, we construct a truth table for $(p \to q) \wedge (q \to p)$.

p	q	$p \to q$	$q \to p$	$(p \to q) \wedge (q \to p)$
T	T	T	T	T
T	F	F	T	F
F	T	T	F	F
F	F	T	T	T

This leads us to define the biconditional so that it is true only when both p and q are true or when both p and q are false (that is, whenever they have the same truth values). This definition is shown in Table 2.8 on page 70.

Signpost in New York City

Table 2.8 Definition of Biconditional

p	q	$p \leftrightarrow q$
T	T	T
T	F	F
F	T	F
F	F	T

In mathematics and logic, $p \leftrightarrow q$ is translated in several ways, all of which have the same meaning:

Biconditional translation

1. p if and only if q.
2. q if and only if p.
3. If p then q, and conversely.
4. If q then p, and conversely.

EXAMPLE 1

Rewrite the following as one statement:

1. *If a polygon has three sides, then it is a triangle.*
2. *If a polygon is a triangle, then it has three sides.*

Solution A polygon is a triangle if and only if it has three sides. ◆

The set of logical possibilities for which a given statement is true is called its **truth set.** A logical statement in which the conclusion is equivalent to the premise is called a **tautology.** This means that a compound statement is a tautology if you obtain only Ts on a truth table.

EXAMPLE 2

Is $(p \vee q) \to (\sim q \to p)$ a tautology?

Solution

p	q	$p \vee q$	$\sim q$	$\sim q \to p$	$(p \vee q) \to (\sim q \to p)$
T	T	T	F	T	T
T	F	T	T	T	T
F	T	T	F	T	T
F	F	F	T	F	T

Thus, the given statement is a tautology. ◆

If a conditional is a tautology, as in Example 2, then it is called an **implication** and is symbolized by $\Rightarrow$. That is, Example 2 can be written

$$(p \vee q) \Rightarrow (\sim q \to p)$$

The implication symbol $p \Rightarrow q$ is pronounced "p implies q."

A biconditional statement $p \leftrightarrow q$ that is also a tautology (that is, always true) is a **logical equivalence,** written $p \Leftrightarrow q$ and read "p is logically equivalent to q."

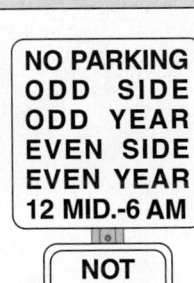

NO PARKING
ODD SIDE
ODD YEAR
EVEN SIDE
EVEN YEAR
12 MID.-6 AM

NOT
A
THRU
WAY

Many drivers must throw up their hands in utter confusion when they come upon this wacky sign in Medford, Massachusetts. Local police say the sign is really easy to decipher: You can't park on the odd side of the street—the side in which the house numbers are odd—from midnight to 6 A.M. in an odd-numbered year such as 2001. Likewise, parking on the even side is taboo in an even year. The reason for this zany arrangement is to keep one side clear of snow for emergency vehicle use.

EXAMPLE 3

Show that $(p \rightarrow q) \Leftrightarrow \sim(p \wedge \sim q)$.

Solution   We begin by constructing a truth table for

$$(p \rightarrow q) \leftrightarrow \sim(p \wedge \sim q).$$

p	q	$p \rightarrow q$	$\sim q$	$p \wedge \sim q$	$\sim(p \wedge \sim q)$	$(p \rightarrow q) \leftrightarrow \sim(p \wedge \sim q)$
T	T	T	F	F	T	T
T	F	F	T	T	F	T
F	T	T	F	F	T	T
F	F	T	T	F	T	T

Since all possibilities are true (it is a tautology), we see it is a logical equivalence and we write

$$(p \rightarrow q) \Leftrightarrow \sim(p \wedge \sim q)$$

$\blacklozenge$

Notice the difference between the symbols $\rightarrow$ and $\Rightarrow$ as well as $\leftrightarrow$ and $\Leftrightarrow$. The conditional ($\rightarrow$) and biconditional ($\leftrightarrow$) are used as logical connectives and the truth values for these may be true or false. If, because of the statements being connected, the truth table gives *only true values,* then we use the symbols $\Rightarrow$ and $\Leftrightarrow$ in place of $\rightarrow$ and $\leftrightarrow$ and call them implication ($\Rightarrow$) and logical equivalence ($\Leftrightarrow$).

Laws of Logic

We can use the idea of logical equivalence to write two previously stated laws:

Law of double negation: $\sim(\sim p) \Leftrightarrow p$

Law of contraposition: $(\sim q \rightarrow \sim p) \Leftrightarrow (p \rightarrow q)$

You will be asked to prove these laws in the problem set. Other important laws are developed in the next section. Two additional laws, called **De Morgan's laws,** will be considered in this section.

De Morgan's Laws

$$\sim(p \vee q) \Leftrightarrow \sim p \wedge \sim q$$
$$\sim(p \wedge q) \Leftrightarrow \sim p \vee \sim q$$

There is a strong tie between set theory (Chapter 10) and symbolic logic, so it is no coincidence that there are De Morgan's laws both in set theory and in symbolic logic. We prove the second one here and leave the first for the problem set. The proof, by truth table, is shown:

p	q	$p \wedge q$	$\sim(p \wedge q)$	$\sim p$	$\sim q$	$\sim p \vee \sim q$	$\sim(p \wedge q) \leftrightarrow (\sim p \vee \sim q)$
T	T	T	F	F	F	F	T
T	F	F	T	F	T	T	T
F	T	F	T	T	F	T	T
F	F	F	T	T	T	T	T

Since the last column is all Ts, we can write $\sim(p \wedge q) \Leftrightarrow \sim p \vee \sim q$.

Negation of a Compound Statement

De Morgan's laws can be used to write the negation of a compound statement using
conjunction and disjunction.

EXAMPLE 4

Write the negation of the compound statements.

a. John went to work or he went to bed.

b. Alfie didn't come last night and didn't pick up his money.

Solution Begin by writing the statement in symbolic form, then find the negation, and
finally use the rules of logic to simplify before translating back into English.

a. Let w: John went to work; b: John went to bed.

The symbolic form is:	$w \vee b$
Negation:	$\sim(w \vee b)$
Simplify (De Morgan's law):	$\sim w \wedge \sim b$

Translate back into English:

   John did not go to work and he did not go to bed.

b. Let p: Alfie came last night; q: Alfie picked up his money.

Symbolic form:	$\sim p \wedge \sim q$
Negation:	$\sim(\sim p \wedge \sim q)$
Simplify (De Morgan's law):	$\sim(\sim p) \vee \sim(\sim q)$
Simplify (law of double negation):	$p \vee q$
Translate:	Alfie came last night or he picked up his money.	◆

Sometimes it is necessary to find the **negation of a conditional,** $p \rightarrow q$. It can be
shown (see Problem 39) that

$$(p \rightarrow q) \Leftrightarrow (\sim p \vee q)$$

Therefore, the negation of $p \rightarrow q$ is equivalent to the negation of $\sim p \vee q$:

$$\sim(p \rightarrow q) \Leftrightarrow \sim(\sim p \vee q)$$

$$\Leftrightarrow \sim(\sim p) \wedge \sim q \quad \text{De Morgan's law}$$

$$\Leftrightarrow p \wedge \sim q \qquad \text{Law of double negation}$$

Negation of a Conditional

$$\sim(p \rightarrow q) \Leftrightarrow (p \wedge \sim q)$$

EXAMPLE 5 PÓLYA'S METHOD

The officer said to the detective, "If John was at the scene of the crime, then he knows
that Jean could not have done it." "No, Colombo," answered the detective, "that is not
correct. John was at the scene of the crime and Jean did it." Was the detective's statement a correct negation of Colombo's statement?

Solution We use Pólya's problem-solving guidelines for this example.

Understand the Problem. We are asking for the negation of a given statement, which we
then wish to compare with the proposed negation.

Devise a Plan. Translate the given statement into symbolic form, find the negation, simplify, and then translate back into English.

Carry Out the Plan.

Let *p:* John was at the scene of the crime; *q:* Jean committed the crime. Then $p \rightarrow \sim q$ is a symbolic statement of the given statement, "If John was at the scene of the crime, then he knows that Jean could not have done it." We now form the negation and then simplify:

$$\sim(p \rightarrow \sim q) \Leftrightarrow p \wedge \sim(\sim q) \quad \text{Negation of a conditional}$$

$$\Leftrightarrow p \wedge q \qquad \text{Law of double negation}$$

Translate back into English: John was at the scene of the crime and Jean committed the crime.

Look Back. The detective's negation was correct. ◆

Miscellaneous Operators

Occasionally, we encounter other operators, and it is necessary to formulate precise definitions of these additional operators. As Table 2.9 shows, they are all defined in terms of our previous operators.

Use this table for reference.

Table 2.9 Additional Operators

p	*q*	Either *p* or *q* $(p \vee q) \wedge \sim(p \wedge q)$	Neither *p* nor *q* $\sim(p \vee q)$	*p* unless *q* $\sim q \rightarrow p$	*p* because *q* $(p \wedge q) \wedge (q \rightarrow p)$	No *p* is *q* $p \rightarrow \sim q$
T	T	F	F	T	T	F
T	F	T	F	T	F	T
F	T	T	F	T	F	T
F	F	F	T	F	F	T

EXAMPLE 6

Show that the statement "*p* because *q*" as defined in Table 2.9 is equivalent to conjunction.

Solution According to Table 2.9, "*p* because *q*" means $(p \wedge q) \wedge (q \rightarrow p)$ and conjunction means $p \wedge q$.

p	*q*	$(p \wedge q) \wedge (q \rightarrow p)$	$p \wedge q$	$(p \wedge q) \wedge (q \rightarrow p) \leftrightarrow (p \wedge q)$
T	T	T	T	T
T	F	F	F	T
F	T	F	F	T
F	F	F	F	T
		↑ From Table 2.9	↑ From Table 2.5	

The last column is all Ts, so we can write $(p \wedge q) \wedge (q \rightarrow p) \Leftrightarrow p \wedge q$. ◆

Problem Set 2.3

LEVEL 1

1. **IN YOUR OWN WORDS** Discuss the difference between the conditional and the biconditional.

2. **IN YOUR OWN WORDS** Discuss the procedure for finding the negation of compound statements.

3. **IN YOUR OWN WORDS** Discuss when you use the symbols ↔ and ⇔.

4. **IN YOUR OWN WORDS** Discuss when you use the symbols → and ⇒.

5. **IN YOUR OWN WORDS** Make up five good English statements, each using one of the additional operators shown in Table 2.9.

Use truth tables in Problems 6–13 to determine whether the given compound statement is a tautology.

6. $(p \wedge q) \vee (p \rightarrow \sim q)$

7. $(p \vee q) \wedge (q \rightarrow \sim p)$

8. $(\sim p \rightarrow q) \rightarrow p$

9. $(\sim q \rightarrow p) \rightarrow q$

10. $(p \wedge q) \leftrightarrow (p \vee q)$

11. $(p \rightarrow q) \leftrightarrow (\sim p \vee q)$

12. $(p \vee q) \leftrightarrow (q \vee p)$

13. $(p \vee \sim q) \leftrightarrow (\sim p \vee q)$

Verify the indicated definition in Problems 14–17 from Table 2.9 using a truth table.

14. either p or q

15. neither p nor q

16. p unless q

17. no p is q

Translate the statements in Problems 18–27 into symbols. For each simple statement, indicate the meanings of the symbols you use.

18. Neither smoking nor drinking is good for your health.

19. I will not buy a new house unless all provisions of the sale are clearly understood.

20. To obtain the loan, I must have an income of $85,000 per year.

21. I am obligated to pay the rent because I signed the contract.

22. I cannot go with you because I have a previous engagement.

23. No man is an island.

24. Either I will invest my money in stocks or I will put it in a savings account.

25. Be nice to people on your way up 'cause you'll meet 'em on your way down. (Jimmy Durante)

26. No person who has once heartily and wholly laughed can be altogether irreclaimably bad. (Thomas Carlyle)

27. If by the mere force of numbers a majority should deprive a minority of any clearly written constitutional right, it might, in a moral point of view, justify revolution. (Abraham Lincoln)

Use the tautology $(p \rightarrow q) \Leftrightarrow (\sim p \vee q)$ to write each statement in Problems 28–33 in an equivalent form.

28. If I go, then I paid $100.

29. If the cherries have turned red, then they are ready to be picked.

30. We will not visit New York or visit the Statue of Liberty.

31. Melissa will not watch Jay Leno or the NBC late-night orchestra.

32. The sun is shining or I will not go to the park.

33. The money is available or I will not take my vacation.

LEVEL 2

34. Show that the definition for *neither p nor q* could also be $\sim p \wedge \sim q$.

35. Prove the law of double negation by using a truth table.

36. Prove the law of contraposition by using a truth table.

37. Prove De Morgan's law:
$$\sim(p \vee q) \Leftrightarrow \sim p \wedge \sim q$$

38. In the text we used laws of logic to prove that
$$\sim(p \rightarrow q) \Leftrightarrow (p \wedge \sim q)$$
Use a truth table to prove this result.

39. Prove $(p \rightarrow q) \Leftrightarrow (\sim p \vee q)$.

Write the negation of the compound statements in Problems 40–53.

40. $p \rightarrow q$

41. $p \rightarrow \sim q$

42. $\sim p \rightarrow q$

43. $\sim p \rightarrow \sim q$

44. John went to Macy's or Sears.

45. Jane went to the basketball game or to the soccer game.

46. Tim is not here and he is not at home.

47. Sally is not on time and she missed the boat.

48. If I can't go with you, then I'll go with Bill.

49. If you're out of Schlitz, you're out of beer.

50. If $x + 2 = 5$, then $x = 3$.

51. If $x - 5 = 4$, then $x = 1$.

52. If $x = -5$, then $x^2 = 25$.

53. $2x + 3y = 8$ if $x = 1$ and $y = 2$.

PROBLEM SOLVING

54. To qualify for a loan of $200,000, an applicant must have a gross income of $72,000 if single, or $100,000 combined income if married. It is also necessary to have assets of at least $50,000. Write these statements symbolically, and decide whether Liz, who is single and has assets of $125,000, satisfies the conditions for obtaining a loan. She has an income of $58,000.

55. An airline advertisement states, "OBTAIN 40% OFF REGULAR FARE." Read the fine print:

> You must purchase your tickets between January 5 and February 15 and fly round trip between February 20 and May 3. You must also depart on a Monday, Tuesday, or Wednesday, and return on a Tuesday, Wednesday, or Thursday. You must also stay over a Saturday night.

Write out these conditions symbolically.

56. The contract states, "No alterations, redecorating, tacks, or nails may be made in the building, unless written permission is obtained." Write this statement symbolically.

57. The contract states, "The tenant shall not let or sublet the whole or any portion of the premises to anyone for any purpose whatsoever, unless written permission from the landlord is obtained." Write this statement symbolically.

58. One day in a foreign country I met three politicians. Now, all of the politicians of this country belonged to one of two political parties. The first was the Veracious party, consisting of persons who could tell only the truth. The other party, called the Deceit Party, consisted of persons who were chronic liars. I asked these politicians to which party they belonged. The first said something I did not hear. The second remarked, "He said he belonged to the Veracious Party." The third said "You're a liar!" To which party did the third politician belong?

59. Translate into symbolic form: *Either Alfie is not afraid to go, or Bogie and Clyde will have lied.*

60. A man is about to be electrocuted but is given a chance to save his life. In the execution chamber are two chairs, labeled 1 and 2, and a jailer. One chair is electrified; the other is not. The prisoner must sit on one of the chairs, but before doing so, he may ask the jailer one question, to which the jailer must answer yes or no. The jailer is a consistent liar or else a consistent truth teller, but the prisoner does not know which. Knowing that the jailer either deliberately lies or faithfully tells the truth, what question should the prisoner ask?

2.4 THE NATURE OF PROOF

Sidney Harris, *American Scientist* magazine © 1977. Reprinted by permission.

"I think you should be more explicit here in step two."

One of the greatest strengths of mathematics is its concern with the logical proof of its propositions. Any logical system must start with some undefined terms, definitions, and postulates or axioms. We have seen several examples of each of these. From here, other assertions can be made. These assertions are called theorems, and they must be proved using the rules of logic. In this section, we are concerned not so much with "proving mathematics" as with investigating the nature of proof in mathematics since the idea of proof does occupy a great portion of a mathematician's time.

In Chapter 1 we discussed inductive reasoning. Experimentation, guessing, and looking for patterns are all part of inductive reasoning. After a conjecture or generalization has been made, it needs to be proved using the rules of logic. After the conjecture is proved, it is called a **theorem.** Often several years will pass from the conjectural stage to the final proved form. Certain definitions must be made, and certain axioms or postulates must be accepted. Perhaps the proof of the conjecture will require the results of some previous theorems.

A **syllogism** is a form of reasoning in which two statements or premises are made and a *logical conclusion* is drawn from them. In this book, we will consider three types of syllogisms: direct reasoning, indirect reasoning, and transitive reasoning.

Direct Reasoning

The simplest type of syllogism is **direct reasoning.** This type of argument consists of two *premises,* or *hypotheses,* and a *conclusion.* For example,

$p \rightarrow q$	If you receive an A on the final, then you will pass the course.
p	You receive an A on the final.
$\therefore q$	Therefore, you pass the course.

The three dot symbol $\therefore$ is used to symbolize the word *therefore,* which is used to separate the conclusion from the premises. We can use a truth table to prove direct reasoning. We begin by noting that the argument form can be rewritten as

$$[(p \rightarrow q) \wedge p] \rightarrow q$$

p	q	$p \rightarrow q$	$(p \rightarrow q) \wedge p$	$[(p \rightarrow q) \wedge p] \rightarrow q$
T	T	T	T	T
T	F	F	F	T
F	T	T	F	T
F	F	T	F	T

Since the argument is always true, we can write $[(p \rightarrow q) \wedge p] \Rightarrow q$, which proves the reasoning form called direct reasoning. It is also called **modus ponens, law of detachment,** or **assuming the antecedent.**

Direct Reasoning

STOP

> Major premise: $p \rightarrow q$
> Minor premise: p
> Conclusion: $\therefore q$

EXAMPLE 1

Use direct reasoning to formulate a conclusion for each of the given arguments.

a. If you play chess, then you are intelligent.
You play chess.

b. If $x + 2 = 3$, then $x = 1$.
$x + 2 = 3$.

c. If you are a logical person, then you will understand this example.
You are a logical person.

Solution

a. You are intelligent.

b. $x = 1$.

c. You understand this example. ◆

Indirect Reasoning

The following syllogism illustrates what we call **indirect reasoning.**

$p \rightarrow q$ If you receive an A on the final, then you will pass the course.
$\underline{\sim q}$ You did not pass the course.
$\therefore \sim p$ Therefore, you did not receive an A on the final.

We can prove this is valid by using a truth table (see Problem 6) or by using direct reasoning as follows:

$$[(\sim q \rightarrow \sim p) \wedge \sim q] \Rightarrow \sim p \quad \text{Direct reasoning}$$

$$[(p \rightarrow q) \wedge \sim q] \Rightarrow \sim p \quad \text{Law of contraposition}$$

This type of reasoning is also called **modus tollens** or **denying the consequent.**

Indirect Reasoning

STOP

> Major premise: $p \rightarrow q$
> Minor premise: $\sim q$
> Conclusion: $\therefore \sim p$

EXAMPLE 2

Formulate a conclusion for each statement by using indirect reasoning.

a. If the cat takes the rat, then the rat will take the cheese.
The rat does not take the cheese.

b. If x is an even number, then $3x$ is an even number.
$3x$ is not an even number.

c. If you received an A on the test, then I am Napoleon.
I am not Napoleon.

Solution

a. The cat does not take the rat.

b. x is not an even number.

c. You did not receive an A on the test. ◆

Transitive Reasoning

Sometimes we must consider some extended arguments. Transitivity allows us to reason through several premises to some conclusion. The argument form is given in the box.

Transitive Reasoning

Premise:	$p \rightarrow q$
Premise:	$q \rightarrow r$
Conclusion:	$\therefore p \rightarrow r$

We can prove transitivity by using a truth table. Notice that for three statements we need a truth table with eight possibilities.

p	q	r	$p \rightarrow q$	$q \rightarrow r$	$p \rightarrow r$	$(p \rightarrow q) \wedge (q \rightarrow r)$	$[(p \rightarrow q) \wedge (q \rightarrow r)] \rightarrow (p \rightarrow r)$
T	T	T	T	T	T	T	T
T	T	F	T	F	F	F	T
T	F	T	F	T	T	F	T
T	F	F	F	T	F	F	T
F	T	T	T	T	T	T	T
F	T	F	T	F	T	F	T
F	F	T	T	T	T	T	T
F	F	F	T	T	T	T	T

All Ts

Since transitivity is always true, we may write

$$[(p \rightarrow q) \wedge (q \rightarrow r)] \Rightarrow (p \rightarrow r)$$

EXAMPLE 3

Formulate a conclusion for each argument.

a. If you attend class, then you will pass the course.
If you pass the course, then you will graduate.

b. If you graduate, then you will get a good job.
If you get a good job, then you will meet the right people.
If you meet the right people, then you will become well known.

c. If $x + 2x + 3 = 9$, then $3x + 3 = 9$.
If $3x + 3 = 9$, then $3x = 6$.
If $3x = 6$, then $x = 2$.

Solution

a. If you attend class, then you will graduate.

b. The transitive law can be extended to several premises:
If you graduate, then you will become well known.

Notice that we can apply the transitive law to both parts **a** and **b:** If you attend class, then you will become well known.

c. If $x + 2x + 3 = 9$, then $x = 2$.   ◆

Logical Proof

The point of studying these various argument forms is to acquire the ability to apply them to longer and more involved arguments.

EXAMPLE 4

Form a valid conclusion using all these statements. We number the premises for easy reference.

> 1. If I receive a check for $500, then we will go on vacation.
> 2. If the car breaks down, then we will not go on vacation.
> 3. The car breaks down.

Solution   Begin by changing the argument into symbolic form.

1. $c \rightarrow v$ where   *c:* I receive a $500 check.
                             v: We will go on vacation.

2. $b \rightarrow \sim v$ *b:* The car breaks down.

3. b

Next, simplify the argument. For this example, we rearrange the premises. Notice the numbers to help you keep track of the premises.

2.	$b \rightarrow \sim v$	Given
1′.	$\sim v \rightarrow \sim c$	Contrapositive of the first premise
∴	$b \rightarrow \sim c$	Transitive
3.	b	Given
∴	$\sim c$	Direct reasoning

Finally, we translate the conclusion back into words: *I did not receive a check for* $500.   ◆

EXAMPLE 5   PÓLYA'S METHOD

Form a valid conclusion using all the statements.*

> 1. All unripe fruit is unwholesome.
> 2. All these apples are wholesome.
> 3. No fruit grown in the shade is ripe.

*From Lewis Carroll, *Symbolic Logic and The Game of Logic* (New York: Dover Publications, 1958).

Solution   We use Pólya's problem-solving guidelines for this example.

Understand the Problem. The forms in this argument are not exactly like those we are used to seeing. We recall that the statement *all p is q* can be translated as *if p then q.*

Devise a Plan. The procedure we will use is to (step 1) translate into symbols, (step 2) simplify using logical arguments, and finally (step 3), translate the symbolic form back into English.

Carry Out the Plan.

Step 1:

1. $\sim r \to (\sim w)$ where   *r:*   This fruit is ripe.

2. $a \to w$ *w:*   This fruit is wholesome.

3. $s \to (\sim r)$ *a:*   This fruit is an apple.

                                      s: This fruit is grown in the shade.

Note: The sentence "No *p* is *q*" is translated as $p \to (\sim q)$.

Step 2:
Let (1'), $w \to r$, replace (1) by the law of contraposition.
Let (3'), $r \to \sim s$, replace (3) by the law of contraposition.
Rearranging the premises, we have:

2. $a \to w$
1'. $\dfrac{w \to r}{\therefore a \to r}$ Transitive

3'. $\dfrac{r \to (\sim s)}{a \to (\sim s)}$ Transitive

Step 3:

Conclusion: *All these apples were not grown in the shade.* If we assume that $\sim s$ represents grown in the sun, then the conclusion can be stated more simply: *All these apples were grown in the sun.*

Look Back. Does this conclusion seem reasonable?   ◆

Fallacies

Sometimes *invalid arguments* are given. The remainder of this section is devoted to some of the more common **logical fallacies.**

Fallacy of the Converse

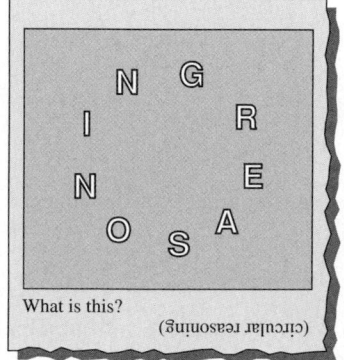

What is this?

(circular reasoning)

EXAMPLE 6
_____

Show that the following argument is not valid.

> If a person reads the *Times,* then she is well informed.
> This person is well informed.
> Therefore, this person reads the *Times.*

Solution   This argument has the following form:

$p \to q$
$\dfrac{q}{\therefore p}$

By considering the associated truth table, we can test the validity of our argument.

p	q	$p \to q$	$(p \to q) \wedge q$	$[(p \to q) \wedge q] \to p$
T	T	T	T	T
T	F	F	F	T
F	T	T	T	F
F	F	T	F	T

We see that the result is not always true; thus the argument is invalid.   ◆

If $p \to q$ were replaced by $q \to p,$ the argument in the preceding example would become valid. That is, the argument would be valid if the direct statement and the converse had the same truth values, which in general they do not. For this reason the argument is sometimes called the **fallacy of the converse** or the fallacy of **assuming the consequent.**

We can often show that a given argument is invalid by finding a **counterexample.** In the preceding example we found a counterexample by looking at the truth table. The entry in the third row is false, so the argument can be shown to be false in the case in which p is false and q is true. In terms of this example, a person could never see the *Times* (p false) and still be well informed (q true).

Fallacy of the Inverse
Consider the following argument:

> If a person reads the *Times,* then he is well informed.
> This person does not read the *Times.*
> Therefore, this person is not well informed.

As we have seen in Example 6, a person who reads the *Tribune* might also be well informed. This line of reasoning is called the **fallacy of the inverse** (sometimes also called the fallacy of **denying the antecedent**). A truth table for

$$[(p \to q) \wedge (\sim p)] \to (\sim q)$$

shows that the fallacy of the inverse is not valid. (The truth table is left as a problem.)

EXAMPLE 7

Test the validity of the following argument.

> If a person goes to college, he will make a lot of money.
> You do not go to college.
> Therefore, you will not make a lot of money.

Solution   This argument has the following form:

$$p \to q$$
$$\underline{\sim p\quad\quad}$$
$$\therefore \sim q$$

We recognize this reasoning as the fallacy of the inverse (the fallacy of denying the antecedent).   ◆

False Chain Pattern
In certain areas of this country, it is thought that thunderstorms cause milk to sour. This belief is an example of the fallacy we will call the **false chain pattern.** It can be shown as follows:

> Hot, humid weather favors thunderstorms.
> Hot, humid weather favors bacterial growth, which causes milk to sour.
> Therefore, thunderstorms cause milk to sour.

The false chain pattern is illustrated by

$$p \to q$$
$$\underline{p \to r}$$
$$\therefore q \to r$$

and can be proved invalid by construction of an appropriate truth table.

The three types of common fallacies that we have discussed can be summarized as follows:

Fallacy of the Converse (assuming the consequent)	*Fallacy of the Inverse* (denying the antecedent)	*False Chain Pattern*
$p \to q$	$p \to q$	$p \to q$
$\underline{q}$	$\underline{\sim p}$	$\underline{p \to r}$
$\therefore p$	$\therefore \sim q$	$\therefore q \to r$

Study these fallacies, and notice how they differ from direct reasoning, indirect reasoning, and transitivity.

Problem Set 2.4

LEVEL 1

1. **IN YOUR OWN WORDS** Explain what we mean by direct reasoning.

2. **IN YOUR OWN WORDS** Explain what we mean by indirect reasoning.

3. **IN YOUR OWN WORDS** Explain what we mean by transitive reasoning.

4. **IN YOUR OWN WORDS** What do we mean by logical fallacies?

5. **IN YOUR OWN WORDS** What is a syllogism?

6. Prove $[(p \to q) \wedge \sim q] \to \sim p$ by constructing a truth table. What is the name we give to this type of reasoning?

7. By constructing a truth table, show that

$$[(p \to q) \wedge (p \to r)] \to (q \to r)$$

   is an invalid argument. What is the name of this fallacy?

8. Prove that $[(p \to q) \wedge \sim p] \to \sim q$ is an invalid argument. What is the name of this fallacy?

9. **IN YOUR OWN WORDS** There are similarities between Pólya's problem-solving method and the steps a detective will go through in solving a case. Rewrite each of these detective problem-solving steps using the language of Pólya's problem-solving method. (See Problems 55–56 below for examples of detective-type problems.)

   Understand the case.
   What are you looking for?

   Investigate the case.
   Have you solved a similar case?
   What are the facts?

   Analyze the facts/data.
   What information is important?
   What information is not important?
   What pieces of information do not seem to fit together logically?
   Which data are inconsistent with the given information?

   Reexamine the facts.
   Do the facts support the solution?
   Can we obtain a conviction?

Determine whether each argument in Problems 10–13 is valid or invalid. Give reasons for your answer.

10. **a.** $p \to q$
    $\underline{\sim q}$
    $\therefore \sim p$

    b. $p \to q$
    $\underline{q}$
    $\therefore p$

11. **a.** $p \vee q$
    $\underline{\sim p}$
    $\therefore q$

    b. $p \vee q$
    $\underline{\sim q}$
    $\therefore p$

12. **a.** $p \to q$
    $\underline{\sim p}$
    $\therefore \sim q$

    b. $p \to q$
    $\underline{p}$
    $\therefore q$

13. **a.** $p \to \sim q$
    $\underline{q}$
    $\therefore \sim p$

    b. $\sim p \to q$
    $\underline{\sim p}$
    $\therefore q$

Determine whether each argument in Problems 14–36 is valid or invalid. If valid, name the type of reasoning and if invalid, determine the error in reasoning.

14. If I inherit $1,000, I will buy you a cookie.
    I inherit $1,000.
    Therefore, I will buy you a cookie.

15. If a^2 is even, then a must be even.
    a is odd.
    Therefore, a^2 is odd.
    Note: Assume that if a number is odd, then it is not even.

16. All snarks are fribbles.
    All fribbles are ugly.
    Therefore, all snarks are ugly.

17. All cats are animals.
    This is not an animal.
    Therefore, this is not a cat.

18. If I don't get a raise in pay, I will quit.
    I don't get a raise in pay.
    Therefore, I quit.

19. If Fermat's Last Theorem is ever proved, then my life is
    complete.
    Fermat's Last Theorem was proved in 1994.
    Therefore, my life is complete.

20. If Congress appropriates the money, the project can be
    completed.
    Congress appropriates the money.
    Therefore, the project can be completed.

21. If Alice drinks from the bottle marked
    "poison," she will become sick.
    Alice does not drink from a bottle that is marked "poison."
    Therefore, she does not become sick.

22. Blue-chip stocks are safe investments.
    Stocks that pay a high rate of interest are safe investments.
    Therefore, blue-chip stocks pay a high rate of interest.

23. If Al understands logic, then he enjoys this sort of
    problem.
    Al does not understand logic.
    Therefore, Al does not enjoy this sort of problem.

24. If Al understands a problem, it is easy.
    This problem is not easy.
    Therefore, Al does not understand this problem.

25. If Mary does not have a little lamb, then she has a
    big bear.
    Mary does not have a big bear.
    Therefore, Mary has a little lamb.

26. If $2x - 4 = 0$, then $x = 2$.
    $x \neq 2$
    Therefore, $2x - 4 \neq 0$.

27. If Todd eats Krinkles cereal, then he has "extra energy."
    Todd has "extra energy."
    Therefore, Todd eats Krinkles cereal.

28. If Ron uses Slippery oil, then his car is in good running
    condition.
    Ron's car is in good running condition.
    Therefore, Ron uses Slippery oil.

29. If you get a fill-up of gas, you will get a free car wash.
    You get a fill-up of gas.
    Therefore, you will get a free car wash.

30. If the San Francisco 49ers lose, then the Dallas
    Cowboys win.
    If the Dallas Cowboys win, then they will go to the
    Super Bowl.
    Therefore, if the San Francisco 49ers lose, then the
    Dallas Cowboys will go to the Super Bowl.

31. If Missy uses Smiles toothpaste, then she has
    fewer cavities.
    Therefore, if Missy has fewer cavities, then she uses
    Smiles toothpaste.

32. All mathematicians are eccentrics.
    All eccentrics are rich.
    Therefore, all mathematicians are rich.

33. (Let x be an integer and y a nonzero integer for
    this argument.)
    If Q is a rational number, then $Q = \dfrac{x}{y}$, where $\dfrac{x}{y}$ is a
    reduced fraction.
    $Q \neq \dfrac{x}{y}$, where $\dfrac{x}{y}$ is a reduced fraction.
    Therefore, Q is not a rational number.

34. If you like beer, you'll like Bud.
    You don't like Bud.
    Therefore, you don't like beer.

35. No students are enthusiastic.
    You are enthusiastic.
    Therefore, you are not a student.

36. If the crime occurred after 4:00 A.M., then Smith could
    not have done it.
    If the crime occurred at or before 4:00 A.M., then Jones
    could not have done it.
    The crime involved two persons, if Jones did not commit
    the crime.
    Therefore, if Smith committed the crime, it involved
    two persons.

LEVEL 2

*In Problems 37–54, form a valid conclusion, using all the
premises given for each argument. Give reasons.*

37. If you learn mathematics, then you are intelligent.
    If you are intelligent, then you understand human nature.

38. If I am idle, then I become lazy.
    I am idle.

39. If we go to the concert, then we are enlightened.
    We are not enlightened.

40. If you climb the highest mountain, then you feel great.
    If you feel great, then you are happy.

41. $a = 0$ or $b = 0$
    $a \neq 0$

42. If $a \cdot b = 0$, then $a = 0$ or $b = 0$.
    $a \cdot b = 0$

43. If we interfere with the publication of false information,
    we are guilty of suppressing the freedom of others.
    We are not guilty of suppressing the freedom of others.

44. If a nail is lost, then a shoe is lost.
If a shoe is lost, then a horse is lost.
If a horse is lost, then a rider is lost.
If a rider is lost, then a battle is lost.
If a battle is lost, then a kingdom is lost.

Beetle Bailey © 1974 by King Features Syndicate, Inc.

45. If I eat that piece of pie, I will get fat.
I will not get fat.

46. If 2 divides a positive integer N and if N is greater than 2, then N is not a prime number.
N is a prime number.

47. If we win first prize, we will go to Europe.
If we are ingenious, we will win first prize.
We are ingenious.

48. If I am tired, then I cannot finish my homework.
If I understand the homework, then I can finish my homework.

LEVEL 3

49. Babies are illogical.
Nobody is despised who can manage a crocodile.
Illogical persons are despised.

50. All hummingbirds are richly colored.
No large birds live on honey.
Birds that do not live on honey are dull in color.

51. No ducks waltz.
No officers ever decline to waltz.
All my poultry are ducks.

52. Everyone who is sane can do logic.
No lunatics are fit to serve on a jury.
None of your sons can do logic.

53. If the government awards the contract to Airfirst Aircraft Company, then Senator Firstair stands to earn a great deal of money.
If Airsecond Aircraft Company does not suffer financial setbacks, then Senator Firstair does not stand to earn a great deal of money.
The government awards the contract to Airfirst Aircraft Company.

54. If you go to college, then you get a good job.
If you get a good job, then you make a lot of money.
If you do not obey the law, then you do not make a lot of money.
You go to college.

PROBLEM SOLVING

55. "The Case of the Dead Professor"*
The detective, Columbo, had just arrived at the scene of the crime and found that the professor had been at the lab working for hours. He seemed to have electrocuted himself and ended up blowing the fuses for the whole building. Later in the night, the janitor came to clean up the lab and found the professor's body. Columbo suspected the janitor, but when he was questioned, he vehemently denied killing the professor. He said, "I came to work late. When I got off the elevator, I found the professor dead with his head on the lab table. I wish I could tell you more, but that is all I know."
Then Columbo said, "I think you can tell us more at headquarters." Downtown under grueling interrogation, the janitor confessed. What made Columbo suspect the janitor?

56. "The Case of the Tumbled Tower"*
Dwayne got up at 6:00 A.M. and was watching the sun rise from his bedroom window. After the sun came up, he started working on his toothpick tower in his room. The tower was very fragile. While he was working on his tower, his little brother came into the room bugging him. Dwayne's brother wanted to be more like Dwayne and he wanted to build something out of toothpicks, too. He was very jealous of Dwayne.
In the afternoon, Dwayne went out to buy candy at the candy store and he left his little brother home (even though he was supposed to be babysitting). On his way out of the house, he heard on the radio that there was going to be a slight westerly wind coming. He then left the house, forgetting that he had left his bedroom window open. When he returned and saw that he had left the window open, he thought that the wind had blown the tower over. But then he remembered something and said that his little brother must have knocked over the tower. What did he remember?

*Problems 55 and 56 were adapted from "Solving the Mystery," by Frances R. Curicio and J. Lewis McNeece, *The Mathematics Teacher,* November 1993, pp. 682–685.

Form a valid conclusion using all of the statements in Problems 57–60.

57. Nobody who really appreciates Beethoven fails to keep silent while the *Moonlight Sonata* is being played. Guinea pigs are hopelessly ignorant of music. No one who is hopelessly ignorant of music ever keeps silent while the *Moonlight Sonata* is being played.

58. No kitten that loves fish is unteachable. No kitten with a tail will play with a gorilla. Kittens with whiskers always love fish. No teachable kitten has green eyes. No kittens without whiskers have tails.

59. When I work a logic problem without grumbling, you may be sure it is one that I can understand. These problems are not arranged in regular order, like the problems I am used to. No easy problem ever makes my head ache. I can't understand problems that are not arranged in regular order, like those I am used to. I never grumble at a problem unless it gives me a headache.

60. Every idea of mine that cannot be expressed as a syllogism is really ridiculous. None of my ideas about rock stars is worth writing down. No idea of mine that fails to come true can be expressed as a syllogism. I never have any really ridiculous idea that I do not at once refer to my lawyer. All my dreams are about rock stars. I never refer any idea of mine to my lawyer unless it is worth writing down.

2.5 | PROBLEM SOLVING USING LOGIC

In Chapter 1 we laid the foundation for problem solving. The key to building problem-solving skills is to encounter problem solving in a variety of contexts. We will now apply another method of proof to solving some logic puzzles that have been around for a long time but that nevertheless continue to challenge the lay reader. Example 1 gives an analysis of Problem 57 of Problem Set 1.2.

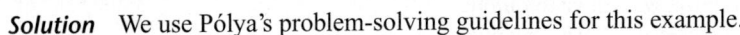

EXAMPLE 1   **PÓLYA'S METHOD**

ODD BALL PROBLEM   You are given nine steel balls of the same size and color. One of the nine balls is slightly heavier in weight; the others all weigh the same. Using a two-pan balance, what is the minimum number of weighings necessary to find the ball of different weight?

Solution   We use Pólya's problem-solving guidelines for this example.

Understand the Problem. Do you understand the terminology? Are you familiar with a two-pan balance?

Devise a Plan. One of the techniques for Pólya's method is to guess and test. That is the method we will use here. We will find a solution by trial and error (that is, guess), and then test it. Next we will see whether we can find a solution with fewer weighings. We will accomplish this by first solving some simpler problems.

Carry Out the Plan. Suppose we had two steel balls. Then one weighing would suffice. What about three steel balls? One weighing would still suffice. How? Put one ball on each pan. If it doesn't balance, then you have found the heavier one. If it balances, then the one not weighed is the heavier one. Does this give you an idea for the nine balls?

Let's try two weighings.

1. Divide the 9 steel balls into 3 groups of 3. First weighing: Balance 3 balls against 3 balls.

    a. The weighing either balances or doesn't balance (law of excluded middle).

    b. If it balances, then the heavier one is in the group not weighed. If it doesn't balance, then take the group with the heavier ball.

2. Divide the 3 steel balls from the heavier group. Second weighing: Balance 1 ball against 1 ball.
   a. The weighing either balances or doesn't balance. (Why?)
   b. If it doesn't balance, then the heavier ball is the one that tips the scale. If it does balance, then the heavier ball is the one not weighed.

Look Back. Two weighings is the solution.   ◆

EXAMPLE 2   PÓLYA'S METHOD

LIAR PROBLEM   On the South Side, a member of the mob had just knocked off a store. Since the boss had told them all to lay low, he was a bit mad. He decided to have a talk with the boys.

From the boys' comments, can you help the Boss figure out who committed the crime? Assume that the Boss is telling the truth.

Solution   We use Pólya's problem-solving guidelines for this example.

Understand the Problem.

Let a: Alfie did it.

   b: Bogie did it.

   c: Clyde did it.

   d: Dirty Dave did it.

   f: Fingers did it.

Translate the sentences into symbolic form:

Alfie said:	$b \vee c$
Bogie said:	$\sim(f \vee b)$
Clyde said:	$[\sim(b \vee c)] \wedge \sim[\sim(f \vee b)]$
Dirty Dave said:	$[(b \vee c) \wedge (f \vee b)] \vee [\sim(b \vee c) \wedge \sim(f \vee b)]$
Fingers said:	$\sim\{[(b \vee c) \wedge (f \vee b)] \vee [\sim(b \vee c) \wedge \sim(f \vee b)]\}$

Assume that a is true, and check all these statements. Next, assume that b is true, and check all the statements. Do the same for $c, d,$ and f. Since we assume the boss is telling the truth, we look for those cases (if any) in which three are truthful and two are lying.

Devise a Plan.   We can summarize our analysis by using a matrix (a rectangular array) with vacant cells for all possible pairings of the elements in each set.

	a	b	c	d	f
a					
b					
c					
d					
f					

Let us assume that the *a* at the left means that we are making the assumption that Alfie did it. Then we place a 1 or a 0 in each position to indicate the truth or falsity, respectively, of each witness' statement with the assumption that person *x* (as listed at the left) is guilty.

Carry Out the Plan.

Witness statements:

Conclusion:		a	b	c	d	f
Alfie did it:	a	0	1	0	1	0
Bogie did it:	b	1	0	0	1	0
Clyde did it:	c	1	1	0	0	1
Dave did it:	d	0	1	0	1	0
Fingers did it:	f	0	0	1	0	1

Next we assume that Bogie did it, and we fill in the second row of the matrix. We continue until the matrix is complete, as shown. Now if we take the boss's statement as the major premise, we see that in only one case are there three truthful mobsters and two liars. Thus we can say that Clyde knocked off the store.

Look Back. Assume that Clyde knocked off the store and check each of the mob's comments to see that in this case there are three truthful mobsters and two liars.   ◆

EXAMPLE 3   **PÓLYA'S METHOD**

PRISONER PROBLEM   During an ancient war three prisoners were brought into a room. In the room was a large box containing three white hats and two black hats. Each man was blindfolded, and one of the hats was placed on his head. The men were lined up, one behind the other, facing the wall. The blindfold of the man farthest from the wall was removed, and he was permitted to look at the hats of the two men in front of him. If he knew (not guessed) the color of the hat on his head, he would be freed. However, he was unable to tell. The blindfold was then taken from the head of the next man, who could see only the hat of the one man in front of him. This man had the same chance for freedom, but he, too, was unable to tell the color of his hat. The remaining man then told the guards the color of the hat he was wearing and was released. What color hat was he wearing, and how did he know?

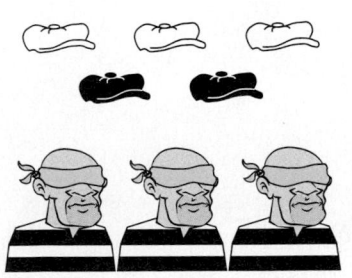

Solution   We use Pólya's problem-solving guidelines for this example.

Understand the Problem.
The situation looks like this:

Devise a Plan. We consider what the last man in line could see, and then decide which of these possibilities fit the conditions of what the middle man in the line saw.

Carry Out the Plan.

1. What did man C see? There are four possibilities:

	I	II	III	IV
A	white	white	black	black
B	white	black	white	black

If he had seen possibility IV, he would have known that he had on a white hat. (Why?) He did not know, so it must be possibility I, II, or III.

2. What did man B see? If he had seen a black hat, then he would have known that his hat was white. (Why?) This rules out possibility III.

3. Since both of the remaining possibilities, I and II, have a white hat on the front man, man A *knew* he had a white hat.

Look Back. Reread the question to see whether all the conditions of the problem are satisfied.   ◆

EXAMPLE 4   PÓLYA'S METHOD

REPORTER'S ERROR PROBLEM   Emor D. Nilap, a news reporter who was a little backward, was sent to cover a billiards tournament. Since he wanted to do a good job, he rounded up some human-interest facts to make his story a little more interesting. He gathered the following information:

1. The men competing were Pat, Milt, Dick, Joe, and Karl.
2. Milt had once beaten the winner at tennis.
3. Pat and Karl frequently played cards together.
4. Joe's finish ahead of Karl was totally unexpected.
5. The man who finished fourth left the tournament after his match and did not see the last two games.
6. The winner had not met the man who came in fifth prior to the day of the tournament.
7. The winner and the runner-up had never met until Joe introduced them just before the final game.

When Emor returned to his office, he found that he had forgotten the order in which the men had finished but he was able to figure it out from the information provided. Can you figure it out?

Solution   We use Pólya's problem-solving guidelines for this example.

Understand the Problem. We need to find the order in which the men finished the tournament.

Devise a Plan. The procedure we will use is to form a matrix of all possibilities and then use the rules of logic to eliminate the parts that are impossible; what remains is the solution.

Carry Out the Plan.
From (1) we are able to make the following matrix of the possible solutions:

Pat	Milt	Dick	Joe	Karl
1	1	1	1	1
2	2	2	2	2
3	3	3	3	3
4	4	4	4	4
5	5	5	5	5

The steps are listed with the "work" shown after the last step.

a. We are given seven premises. We begin by the process of elimination; that is, we cross out those positions that it would be impossible for the men to occupy. The order in which we work the problem is not unique.

b. From (7), Joe is not the winner or the runner-up. Thus we cross out 1 and 2 under Joe's name and label those places *b,* as shown below.

c. By (5), Joe did not finish fourth; cross out and label *c.*

d. Joe finished ahead of Karl, by (4); therefore Karl was not first, second, or third. By the same premise, Joe could not have been last. (Why?)

e. Therefore Joe finished third, which means that Pat, Milt, and Dick are not in third place.

f. By (2), Milt was not the winner.

g. By (7), Milt was not the runner-up.

h. By (6), Milt was not fifth.

i. Therefore Milt finished fourth, which means that Pat, Dick, and Karl are not in fourth place.

j. Therefore Karl finished fifth, which means that the others did not finish fifth.

k. By (6) and the fact that Karl finished last, along with (3), we see that Pat could not have finished first.

ℓ. Therefore Pat is second, which means that Dick was not second.

m. Therefore Dick finished first, and the problem is finished.

Pat	Milt	Dick	Joe	Karl
$\cancel{1}^k$	$\cancel{1}^f$	$\boxed{1}^m$	$\cancel{1}^b$	$\cancel{1}^d$
$\boxed{2}^\ell$	$\cancel{2}^g$	$\cancel{2}^\ell$	$\cancel{2}^b$	$\cancel{2}^d$
$\cancel{3}^e$	$\cancel{3}^e$	$\cancel{3}^e$	$\boxed{3}^e$	$\cancel{3}^d$
$\cancel{4}^i$	$\boxed{4}^i$	$\cancel{4}^i$	$\cancel{4}^c$	$\cancel{4}^i$
$\cancel{5}^j$	$\cancel{5}^h$	$\cancel{5}^j$	$\cancel{5}^d$	$\boxed{5}^j$

Look Back. We have found that Dick finished first, Pat second, Joe third, Milt fourth, and Karl last. Reread each of the statements to make sure this solution does not cause any inconsistencies. ◆

EXAMPLE 5   PÓLYA'S METHOD

TURKEY PROBLEM   A man I know once owned a number of turkeys. One day, one of his gobblers flew over the man's fence and laid an egg on a neighbor's property. To whom did the egg belong—to the man who owned the gobbler, to the gobbler, or to the neighbor?

Solution   We use Pólya's problem-solving guidelines for this example.

Understand the Problem. To understand the question, you must understand the terminology. Since a turkey gobbler is a male turkey, it could not lay an egg. You must be careful to use common sense along with the rules of logic when answering puzzle problems. ◆

Problem Set 2.5

PROBLEM SOLVING

1. **The Hat Game**   Harry, Larry, and Moe are sitting in a circle so that they see each other. The game is played by a judge placing either a black or a white hat on each person's head so that each person can see the others' hats, but not the hat on his or her own head. Now when the judge says "GO," all players who see a black hat must raise their hands. The first player to deduce (not guess) the color of his own hat wins the game. Now, the judge puts a black hat on each player's head, and then said, "GO." All players raised their hands and after a few moments Moe said he knew his hat was black. How did he deduce this?

2. **Broken Window Problem**   Three children were playing baseball; their names were Alice, Bob, and Cole. One of them hit a home run and broke your expensive plate glass window, and you went out to question the children. Each child made two statements.

   Alice: "Cole didn't do it. Bob did it."
   Bob: "I didn't do it. Cole did it."
   Cole: "I didn't do it. Alice did it."

   Now if you know that one of the three always tells the truth, one always lies, and the other tells the truth half the time, can you point to the guilty party?

3. **Bear Problem**   A fox, hunting for a morsel of food, spotted a huge bear about 100 yards due east of him. Before the hunter could become the hunted, the crafty fox ran due north for 100 yards but then realized the bear had not noticed him. Thus he stopped and remained hidden. At this point the bear was due south of the fox. What was the color of the bear?

4. **The Marble Players**   Four boys were playing marbles; their names were Gary, Harry, Iggy, and Jack. One of the boys had 9 marbles, another had 15, and each of the other two had 12. Their ages were 3, 10, 17, and 18, but not respectively. Gary shot before Harry and Jack. Jack was older than the boy with 15 marbles. Harry had fewer than 15 marbles. Jack shot before Harry. Iggy shot after Harry. If Iggy was 10 years old, he did not have 15 marbles. Gary and Jack together had an even number of marbles. The youngest boy was not the one with 15 marbles. If Harry had 12 marbles, he wasn't the youngest. The 10-year-old shot after the 17-year-old. In what order did they shoot, and how old was each boy?

5. **Flower Garden**   I visited a beautiful flower garden yesterday and counted exactly 50 flowers. Each flower was either red or yellow, and the flowers were not all the same color.

My friend made the following observation: No matter which two flowers you might have picked, at least one was bound to be red. From this can you determine how many were red and how many were yellow?

6. **Sock Problem**   I have a habit of getting up before the sun rises. My socks are all mixed up in the drawer, which contains 10 black and 20 blue socks. I reach into the drawer and grab some socks in the dark. How many socks do I need to take from the drawer to be sure that I have a matched pair?

7. **Whodunit?**   Daniel Kilraine was killed on a lonely road, two miles from Pontiac, at 3:30 A.M. on March 3, 1997. Otto, Curly, Slim, Mickey, and The Kid were arrested a week later in Detroit and questioned. Each of the five made four statements, three of which were true and one of which was false. One of these men killed Kilraine. Whodunit? Their statements were:*

   Otto: "I was in Chicago when Kilraine was murdered. I never killed anyone. The Kid is the guilty man. Mickey and I are pals."

   Curly: "I did not kill Kilraine. I never owned a revolver in my life. The Kid knows me. I was in Detroit the night of March 3rd."

   Slim: "Curly lied when he said he never owned a revolver. The murder was committed on March 3rd. One of us is guilty. Otto was in Chicago at the time."

   Mickey: "I did not kill Kilraine. The Kid has never been in Pontiac. I never saw Otto before. Curly was in Detroit with me on the night of March 3rd."

   The Kid: "I did not kill Kilraine. I have never been in Pontiac. I never saw Curly before. Otto lied when he said I am guilty."

8. **Coin Problem.**   Suppose you are given 12 coins, one of which is counterfeit and weighs a little more or less than the real coins. Using a two-pan balance scale, what is the minimum number of weighings necessary to find the counterfeit coin? (If you are interested in a general solution to this type of problem, see T. H. O'Beirne's book, *Puzzles and Paradoxes,* Oxford University Press, New York, 1965, Chapters 2 and 3.)

---

*Reprinted with permission of The Macmillan Company from *Introduction to Logic* (2nd ed.), by Irving Copi. Copyright © 1961 by The Macmillan Company.

9. **Mixed Bag Problem**   Suppose I have three bags, one with two peaches, another with two plums, and a third mixed bag with one peach and one plum. Now I give the bags (in mixed-up order) to Alice, Betty, and Connie. I tell the three to look into their bags and that I want each to make a statement about the contents of her bag, but I want them to lie. Here is what they say:

> *Alice:*   I have two peaches.
> *Betty:*   I have two plums.
> *Connie:*   I have one peach and one plum.

Now here is the game. I want you to develop a strategy of asking one of the three to reach into their bag and pull out one fruit and show it to you. The fruit is then returned to the bag and you ask another to do the same thing. Continue until you can deduce which bag is the mixed bag. What is the minimum possible number of necessary moves? Explain.

10. **Rectangle Tangle**   Fill in each blank with a digit so that every statement is true.*

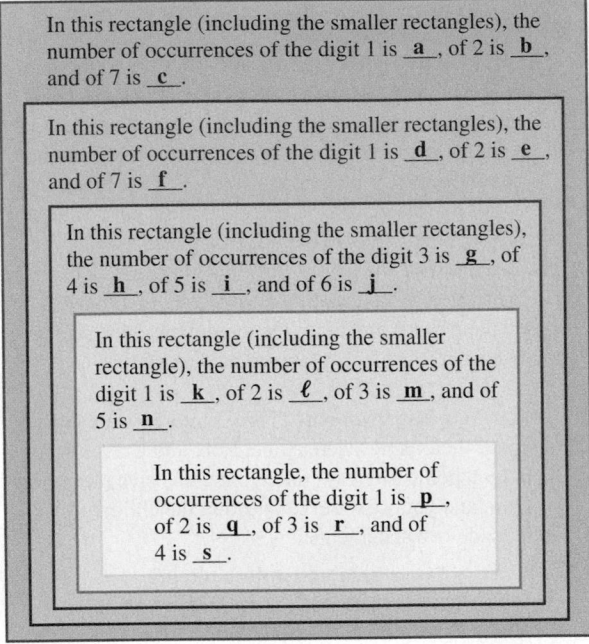

In this rectangle (including the smaller rectangles), the number of occurrences of the digit 1 is __a__ , of 2 is __b__ , and of 7 is __c__ .

In this rectangle (including the smaller rectangles), the number of occurrences of the digit 1 is __d__ , of 2 is __e__ , and of 7 is __f__ .

In this rectangle (including the smaller rectangles), the number of occurrences of the digit 3 is __g__ , of 4 is __h__ , of 5 is __i__ , and of 6 is __j__ .

In this rectangle (including the smaller rectangle), the number of occurrences of the digit 1 is __k__ , of 2 is __ℓ__ , of 3 is __m__ , and of 5 is __n__ .

In this rectangle, the number of occurrences of the digit 1 is __p__ , of 2 is __q__ , of 3 is __r__ , and of 4 is __s__ .

*By Guney Mentes in *Games,* August 1996, p. 43.

2.6 | LOGIC CIRCUITS

One of the applications of symbolic logic that is easiest to understand is that of an electrical **circuit** (see Figure 2.1). Consider the following symbols that we use for the parts of a circuit.

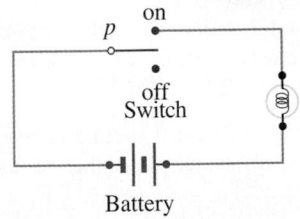

Figure 2.1 **Schematic diagram showing how a light might be connected to a switch**

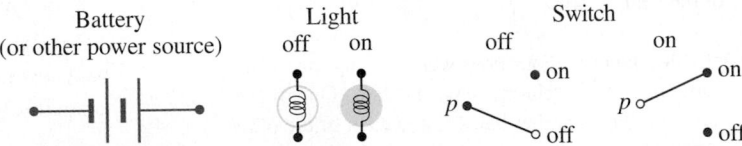

There are two types of circuits for connecting two switches together; the first is called a **series circuit,** and the other is called a **parallel circuit.**

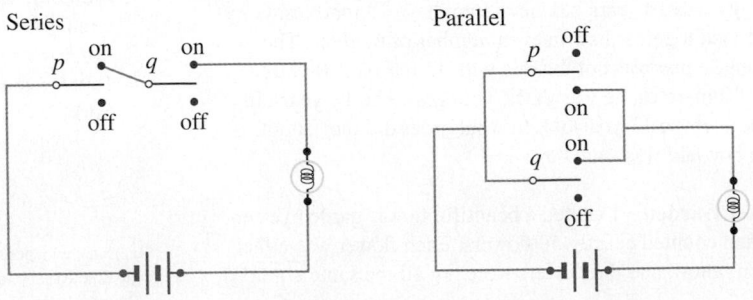

EXAMPLE 1   PÓLYA'S METHOD

How can symbolic logic be used to describe circuits?

Solution   We use Pólya's problem-solving guidelines for this example.

Understand the Problem. When two switches are connected in *series,* current will flow only when both switches are closed. When they are connected in *parallel,* current will flow if either switch is closed.

Devise a Plan. Let switch p be considered as a logical proposition that is either true or false. If p is true, then the switch will be considered closed (current flows), and if p is false, then the switch will be considered open (current does not flow).

Carry Out the Plan. A series circuit of two switches p and q can be specified as a logical statement $p \wedge q$. A parallel circuit can be specified as a logical statement $p \vee q$.

Look Back. Each circuit has four possible states, as shown.

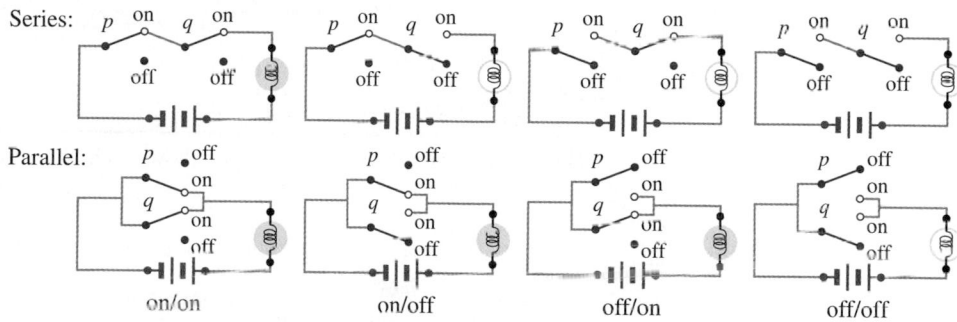

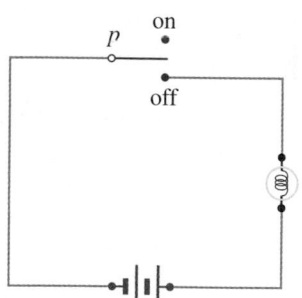

Figure 2.2 Negation circuit

The three fundamental operators are conjunction, disjunction, and negation. In Example 1 we saw that a series circuit can be represented as a conjunction, and a parallel circuit as a disjunction. The circuit for negation is relatively easy, as is shown in Figure 2.2.

It is often necessary to combine circuits. We symbolically represent these as **gates.** The series circuit is called an **AND-gate,** the parallel circuit is called an **OR-gate,** and the negation circuit is called a **NOT-gate.** The notation for each of these is shown in Figure 2.3.

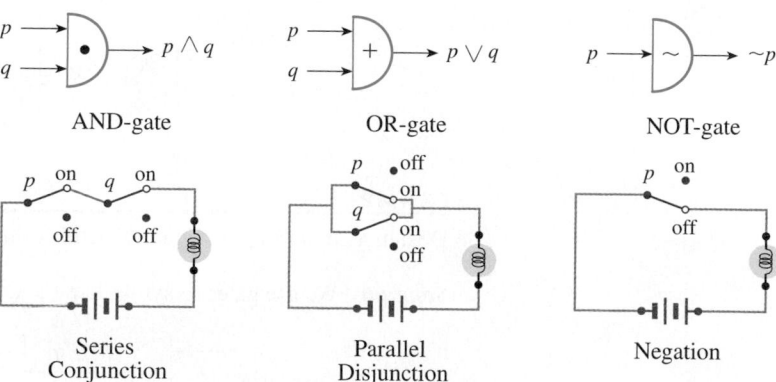

Figure 2.3 Circuits and gates

p	q	$p \land q$	$p \lor q$	$\sim p$
T; on	T; on	T; light on	T; light on	F; light off
T; on	F; off	F; light off	T; light on	F; light off
F; off	T; on	F; light off	T; light on	T; light on
F; off	F; off	F; light off	F; light off	T; light on

Notice that the NOT-gate has a single input (proposition p) and a single output (proposition $\sim p$). That is, current will flow out of the NOT-gate in those cases in which the light is on for the negation circuit. Similarly, the AND-gate and OR-gate have two switches, p and q. These are symbolized as two inputs, p and q. The single output stands for the light.

We can construct logic circuits to simulate logical truth tables. For example, suppose we wish to find the truth values for $\sim(p \lor q)$. We should design the following circuit (remember that parentheses indicate the operation to be performed first):

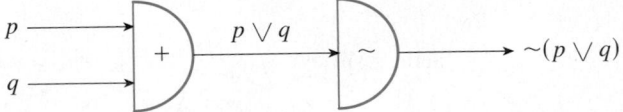

EXAMPLE 2

Design a circuit for $\sim p \land (\sim q)$. Show both the circuit, and the simplified gate diagram.

Solution   For the circuit we begin with a truth table:

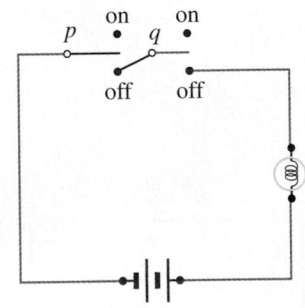

p	q	$\sim p$	$\sim q$	$\sim p \land \sim q$
T	T	F	F	**F**
T	F	F	T	**F**
F	T	T	F	**F**
F	F	T	T	**T**

The light must be on only when both switches are off:

We can simulate this circuit by drawing the gate diagram:

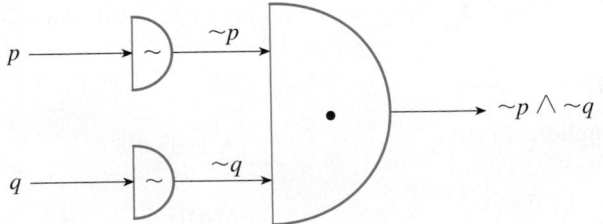

EXAMPLE 3

Design a circuit that will find the truth values for $(p \lor q) \land q$.

Solution   We use gates to symbolize $(p \lor q) \land q$:

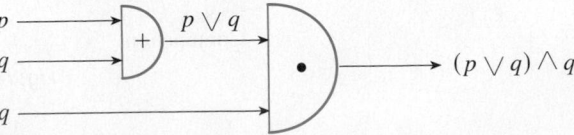

Next we must have some way of determining when the output values of $(p \vee q) \wedge q$ are true and when they are false. We can do this by again connecting a light to the circuit. When $(p \vee q) \wedge q$ is true, the light should be on; when it is false, the light should be off. We represent this connection as follows:

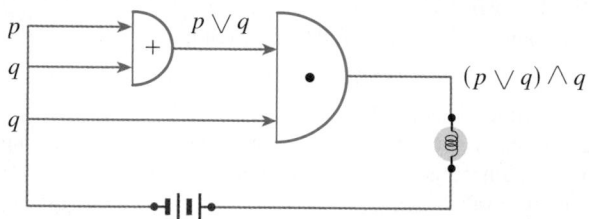

Problem Set 2.6

LEVEL 1

1. IN YOUR OWN WORDS Explain when the light on a circuit is on and when it is off.

2. IN YOUR OWN WORDS When will the light be on for the following circuits?
  a. series circuit   **b.** parallel circuit

3. IN YOUR OWN WORDS What do the following circuit symbols mean?

  a.

  b.

  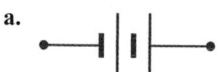

  c.

  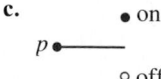

4. IN YOUR OWN WORDS What do the following circuit symbols mean?

  a.

  b.

  c.

  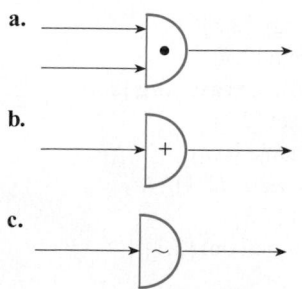

Using switches, design a circuit that would find the truth values for the statements in Problems 5–16 (answers are not unique).

5. $p \wedge q$	**6.** $p \vee q$	**7.** $\sim p \wedge q$
8. $p \wedge \sim q$	**9.** $\sim(p \vee q)$	**10.** $\sim(p \wedge q)$
11. $p \to q$	**12.** $q \to p$	**13.** $p \to \sim q$
14. $q \to \sim p$	**15.** $\sim(p \to \sim q)$	**16.** $\sim(q \to p)$

LEVEL 2

Using AND*-gates,* OR*-gates, and* NOT*-gates, design circuits that would find the truth values for the sentences in Problems 17–22.*

17. $p \wedge \sim q$	**18.** $(p \wedge q) \vee r$
19. $(p \wedge q) \vee (p \wedge r)$	**20.** $\sim(\sim p \vee q)$
21. $\sim[(p \wedge q) \vee r]$	**22.** $(\sim p \wedge q) \vee (p \vee \sim q)$

LEVEL 3

23. The conditional $p \to q$ was defined to be equivalent to $\sim(p \wedge \sim q)$. Use gates to represent this conditional.

24. The cost of the circuit is often a factor in work with switching circuits. The circuit you constructed in Problem 23, by using the definition of the conditional, required three gates. Construct a truth table for $p \to q$ and $\sim p \vee q$. What do you notice about the truth values of the result? Design circuits for $p \to q$ by using only one OR-gate and one NOT-gate.

Using the results of Problems 23 and 24, design circuits for the conditional statements given in Problems 25–27.

25. $\sim p \rightarrow q$ **26.** $p \rightarrow \sim q$ **27.** $\sim q \rightarrow \sim p$

PROBLEM SOLVING

28. a. Suppose an engineer designs the following circuit:

$$\sim\{\sim[(p \vee q) \wedge (\sim p)] \wedge (\sim p)\}$$

What will the circuit look like?

b. If each gate costs 2¢ and a company is going to manufacture one million items using this circuit, how much will the circuits cost?

c. Use truth tables to find the following.

  i. $\sim\{\sim[(p \vee q) \wedge \sim p] \wedge (\sim p)\}$
  ii. $p \vee q$
  iii. What do you notice about the final column of each of these truth tables?

d. Draw a simpler circuit that will output the same values as those in part **a.**

e. If the company is going to manufacture one million items using this simpler circuit, how much will the circuits cost? How much will the company save by using this simpler circuit rather than the more complicated one?

29. Alfie, Bogie, and Clyde are the members of a Senate committee. Design a circuit that will output *yea* or *nay* depending on the way the majority of the committee members voted.

30. Suppose that the Senate committee of Problem 29 has five members. Design a circuit that will output the result of the majority of the committee's vote.

CHAPTER SUMMARY

"Students should be able to construct cogent arguments in support of their claims."

NCTM Standards

IMPORTANT TERMS

And [2.1]
AND-gate [2.6]
Antecedent [2.2]
Argument [2.1]
Assuming the antecedent [2.4]
Assuming the consequent (fallacy) [2.4]
Because [2.3]
Biconditional [2.3]
Circuit [2.6]
Compound statement [2.1]
Conclusion [2.1]
Conditional [2.2]
Conjunction [2.1]
Connective [2.1]
Consequent [2.2]
Contrapositive [2.2]
Converse [2.2]
Counterexample [2.4]
Deductive reasoning [2.1]
De Morgan's laws [2.3]
Denying the antecedent (fallacy) [2.4]
Denying the consequent [2.4]
Direct reasoning [2.4]

Disjunction [2.1]
Either . . . or [2.1]
Exclusive *or* [2.1]
Fallacy [2.4]
Fallacy of the converse [2.4]
Fallacy of the inverse [2.4]
False chain pattern [2.4]
Fundamental operators [2.2]
Fuzzy logic [2.1]
Gates [2.6]
Hypothesis [2.1]
Implication [2.3]
Inclusive *or* [2.1]
Indirect reasoning [2.4]
Invalid argument [2.1]
Inverse [2.2]
Law of contraposition [2.2]
Law of detachment [2.4]
Law of double negation [2.2]
Law of the excluded middle [2.1]
Logic [2.1]
Logical equivalence [2.3]
Logical fallacy [2.4]
Modus ponens [2.4]

Modus tollens [2.4]
Negation [2.1]
Negation of a conditional [2.3]
Neither . . . nor [2.3]
No *p* is *q* [2.3]
Not [2.1]
NOT-gate [2.6]
Operator [2.1]
Or [2.1]
OR-gate [2.6]
Parallel circuit [2.6]
Premise [2.1]
Series circuit [2.6]
Simple statement [2.1]
Statement [2.1]
Syllogism [2.4]
Tautology [2.3]
Theorem [2.4]
Transitive reasoning [2.4]
Truth set [2.3]
Truth table [2.2]
Truth value [2.1]
Unless [2.3]
Valid argument [2.1]

TYPES OF PROBLEMS

Determine whether a sentence is a statement. [2.1]
Write the negation of *all, some,* and *not.* [2.1]
Find truth value of simple and compound statements. [2.1]
Translate statements into symbolic form. [2.1–2.3]
Translate symbolic form into verbal statements. [2.2–2.3]
Construct a truth table for a given symbolic form. [2.2]
Apply the definition of the conditional. [2.2]
Write the converse, inverse, and contrapositive for a given statement. [2.2]
Determine whether a given symbolic statement is true or false. [2.2]
Decide whether a given statement is a tautology. [2.3]
Write an implication as a disjunction. [2.3]
Write the negation of a compound statement. [2.3]
Given certain real-life premises, reach conclusions. [2.3, 2.4]
Be able to recognize, state, and prove valid forms of reasoning, namely, direct reasoning, indirect reasoning, and transitive reasoning. [2.4]
Prove logical statements. [2.3, 2.4]
Determine whether a given argument is valid or invalid. [2.4]
Classify valid forms of reasoning and recognize common fallacies. [2.4]
Find a valid conclusion for a given argument. [2.4]
Solve logical puzzles. [2.5]
Design a circuit to simulate truth values of a given logical statement. [2.6]
Use gates to design a circuit. [2.6]

CHAPTER 2 REVIEW QUESTIONS

1. **a.** What is a logical statement?

   b. What is a tautology?

   c. What is the law of contraposition?

2. Complete the following truth table.

p	q	$\sim p$	$p \wedge q$	$p \vee q$	$p \rightarrow q$	$p \leftrightarrow q$

Construct truth tables for the statements in Problems 3–6.

3. $\sim(p \wedge q)$ 4. $[(p \vee \sim q) \wedge \sim p] \rightarrow \sim q$

5. $[(p \wedge q) \wedge r] \rightarrow p$ 6. $\sim(p \wedge q) \leftrightarrow (\sim p \vee \sim q)$

7. State and prove the principle of direct reasoning.

8. Give an example of indirect reasoning.

9. Give an example of a common fallacy of logic, and show why it is a fallacy.

10. Is the following valid?
$$p \rightarrow q$$
$$\frac{\sim q}{\therefore \sim p}$$

Can you support your answer?

11. Write the negation of each of the following statements.

   a. All birds have feathers.

   b. Some apples are rotten.

   c. No car has two wheels.

   d. Not all smart people attend college.

   e. If you go on Tuesday, then you cannot win the lottery.

12. Which of the following statements are true?

   a. If $111 + 1 = 1,000$, then I'm a monkey's uncle.

   b. If $6 + 2 = 10$ and $7 + 2 = 9$, then $5 + 2 = 7$.

   c. If $6 + 2 = 10$ or $5 + 2 = 7$, then $7 + 2 = 9$.

   d. If $5 + 2 = 7$ and $7 + 2 \neq 9$, then $6 + 2 = 10$.

   e. If the moon is made of green cheese, then $5 + 7 = 57$.

13. Let p: P is a prime number; q: $P + 2$ is a prime number. Translate the following statements into verbal form.

   a. $p \rightarrow q$ b. $(p \vee q) \wedge [\sim(p \wedge q)]$

14. Consider this statement: "All computers are incapable of self-direction."

   a. Translate this statement into symbolic form.

   b. Write the contrapositive of the statement.

15. Translate into symbols and identify the argument.

   If there are a finite number of primes, then there is some natural number, greater than 1, that is not divisible by any prime.
   Every natural number greater than 1 is divisible by a prime number.
   Therefore, there are infinitely many primes.

16. a. Using switches, design a circuit that would find the truth values for the statement $\sim p \vee \sim q$.

   b. Using AND-gates, OR-gates, or NOT-gates, design a circuit that would find the truth values for $[(p \wedge q) \wedge p]$.

In Problems 17–19, form a valid conclusion using all the premises given in each problem.

17. If I attend to my duties, I am rewarded.
   If I am lazy, I am not rewarded.
   I am lazy.

18. All organic food is healthy.
   All artificial sweeteners are unhealthy.
   No prune is nonorganic.

19. All squares are rectangles.
   All rectangles are quadrilaterals.
   All quadrilaterals are polygons.

20. **Table Puzzle**

   The mathematics department of a very famous two-year college consists of four people: Josie, who is department chairman, Maureen, Terry, and Warren. For department meetings they always sit in the same seats around a square table. Their hobbies are (in alphabetical order) baking, gardening, hiking, and surfing. Josie, whose hobby is gardening, sits on Terry's left. Maureen sits at the hiker's right. Warren, who faces Terry, is not the baker.

   Make a drawing showing who sits where, and state the hobby of each person.

GROUP RESEARCH PROBLEMS

Working in small groups is typical of most work environments, and learning to work with others to communicate specific ideas is an important skill. Work with three or four other students to submit a single report based on each of the following questions.

G3. Write a symbolic statement for each of the following verbal statements. (1) Either Donna does not like Elmer because Elmer is bald, or Donna likes Frank and George because they are handsome twins. (2) Either neither you nor I am honest, or Hank is a liar because Iggy did not have the money.

G4. Three schools have a track meet and enter one person in each of the events. The number of events is unknown, and so is the scoring system—except that the winner of each event scores a certain number of points, second place scores fewer points, and third place scores fewer still. Georgia won with 22, and Alabama and Florida tied with 9 each. Florida won the high jump. Who won the mile run? *

G5. Consider the following question: "Of all possible collections of states that yield 270 or more electoral votes—enough to win a presidential election—which collection has the smallest geographical area?" *Hint:* Let O be the set consisting of the optimal collection of states. Your group should choose a state and *prove* mathematically that the state is *not* in O or prove that the state *is* in O.

> **References** This problem is found in the article "Proof by Contradiction and the Electoral College," by Charles Redmond, Michael P. Federici, and Donald M. Platte in *The Mathematics Teacher,* November 1998.
> The U.S. Electoral College Calculator:
> **http://www.archives.gov/federal_register/electoral_college/calculator.html**
> National Archives and Records Administration Federal Register
> **http://www.archives.gov/federal_register/electoral_college.html**

G6. Five cabbies have been called to pick up five fares at the Hilton Towers. On arrival, they find that their passengers are slightly intoxicated. Each man has a different first and last name, a different profession, and a different destination; in addition, each man's wife has a different first name. Unable to determine who's who and who's going where, the cabbies want to know: **Who is the accountant? What is Winston's last name? Who is going to Elm Street?** Use only the following facts to answer these questions:
   1. Sam is married to Donna.
   2. Barbara's husband gets into the third cab.
   3. Ulysses is a banker.
   4. The last cab goes to Camp St.
   5. Alice lives on Denver Street.
   6. The teacher gets into the fourth cab.
   7. Tom gets into the second cab.
   8. Eve is married to the stock broker.
   9. Mr. Brown lives on Denver St.
   10. Mr. Camp gets into the cab in front of Connie's husband.
   11. Mr. Adams gets into the first cab.
   12. Mr. Duncan lives on Bourbon St.

(*continued*)

13. The lawyer lives on Anchor St.
14. Mr. Duncan gets into the cab in front of Mr. Evans.
15. The lawyer is three cabs in front of Victor.
16. Mr. Camp is in the cab in front of the teacher.

G7. Consider the apparatus shown in Figure 2.4. Notice that there are 12 chutes (numbered 1 to 12), and if you drop a ball into the chute it will slide down the tube until it reaches an AND-GATE or an OR-GATE. If two balls reach an AND-GATE, then one ball will pass through, but if only one reaches an AND-gate, it will not pass through. If one or two balls reach an OR-GATE, then one ball will pass through. The object is to obtain a reward by having a ball reach the location called REWARD. What is the fewest number of balls that can be released in order to gain the reward?

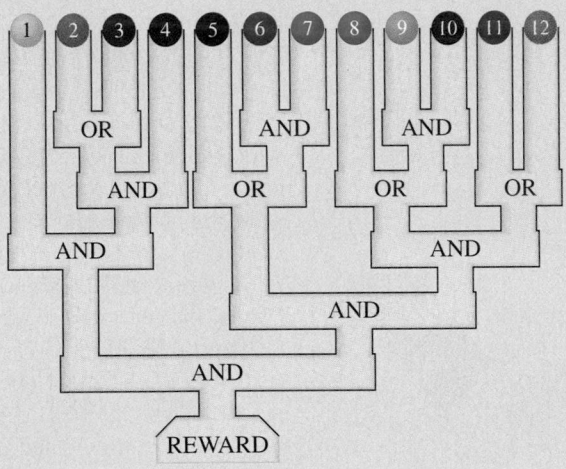

Figure 2.4 **Reward game**

INDIVIDUAL RESEARCH PROBLEMS

Learning to use sources outside your classroom and textbook is an important skill, and here are some ideas for extending some of the ideas in this chapter. You can find references to these projects in a library or at

   **www.mathnature.com**

PROJECT 2.1 What do the following people have in common?

Ira Glasser, executive director of the ACLU
Bram Stoker, author of *Dracula*
David Robinson, basketball star
Ed Thorpe, inventor of programmed-trading on Wall Street
Clifford Brown, 1950s jazz trumpeter

PROJECT 2.2 Deck of Cards Puzzle from *Games* magazine. See
www.mathnature.com, Section 2.1 for a statement of this puzzle.

PROJECT 2.3 Do some research to explain the differences between the words *necessary* and *sufficient.*

PROJECT 2.4 Sometimes statements p and q are described as *contradictory, contrary,* or *inconsistent.* Consult a logic text, and then define these terms using truth tables.

PROJECT 2.5 Between now and the end of the course, look for logical arguments in newspapers, periodicals, and books. Translate these arguments into symbolic form. Turn in as many of them as you find. Be sure to indicate where you found each argument.

PROJECT 2.6 Suppose a prisoner must make a choice between two doors. One door leads to freedom and the other door is booby-trapped so that it leads to death. The doors are labeled as follows:

Door 1:   This door leads to freedom and the other door leads to death.

Door 2:   One of these doors leads to freedom and the other of these doors leads to death.

If exactly one of the signs is true, which door should the prisoner choose? Give reasons for your answer.

PROJECT 2.7 Convention Problem   A mathematician attended a convention of fifty men and women scientists. The mathematician observed that if any two of them were picked at random, at least one of the two would be male. From this information, is it possible to deduce what percentage of the attendees were women?

PROJECT 2.8 Flower Problem   Three students visited a very patriotic garden of red, white, and blue flowers, but in addition there were some yellow flowers. One student observed that if any four flowers were picked, one of them would be red. Another observed that if any four were picked, at least one of them would be blue. The third shouted that if four were picked, one would be yellow. Does this necessarily mean that if any four were picked, one would be white?

PROJECT 2.9 Baseball Problem   Nine men play the positions on a baseball team. See the Web site for a description of this problem.

PROJECT 2.10 Build a simple device which will add single digit numbers.

3 The Nature of Numeration Systems

To err is human, but to really foul things up requires a computer.
ANONYMOUS

CONTENTS

OVERVIEW

It is nearly impossible to graduate from college today and not have some computer skills. We are all affected by computers, and even though we do not need to understand all about what makes them operate any more than we need to know about an internal combustion engine to drive a car, it is worthwhile to learn just a little about how they do what they do.

We study numeration systems in this chapter so that we can better understand our own numeration system, as well as gain some insight into the internal workings of a computer. In addition to looking at computers today, we take a peek back into the history of computers.

Some knowledge about computers and how they work may help you to understand how better to deal with computers in your day-to-day life.

IMPORTANT IDEAS

Properties of numeration systems [3.1]
Decimal numeration system [3.2]
Converting between numeration systems [3.3]
Use a binary numeration system to store "data" in a computer [3.4]
The history of calculating devices [3.5]
Distinguish between computer hardware and software [3.5]

BOOK REPORTS

Write a 500-word report on one of these books:
How Computers Work, Ron White (Emeryville, CA: Ziff-Davis Press, 1993).
Nerds 2.0.1, Stephen Segaller (New York: TV Books, 1998).

LINKS

Individual Projects

Group Projects

Research Links

www.mathnature.com

3.1 | EARLY NUMERATION SYSTEMS

"Do you know your numbers?" is a question you might hear one child asking another. At an early age, we learn our numbers as we learn to count, but if we are asked to define what we mean by *number,* we are generally at a loss. There are many different kinds of numbers, and one type of number is usually defined in terms of more primitive types of numbers. The word *number* is taken as one of our primitive (or undefined) words, but the following definition will help us distinguish between a number and the symbol that we use to represent a number.

Number/Numeral

A **number** is used to answer the question "How many?" and usually refers to numbers used to count objects:

$$\{1, 2, 3, 4, 5, 6, 7, 8, 9, 10, \dots\}$$

This set of numbers is called the set of **counting numbers.**

A **numeral** is a symbol used to represent a number, and a **numeration system** consists of a set of basic symbols and some rules for making other symbols from them, the purpose being the identification of all numbers.

The invention of a precise and "workable" numeration system is one of the greatest inventions of humanity. It is certainly equal to the invention of the alphabet, which enabled humans to carry the knowledge of one generation to the next. It is simple for us to use the symbol 17 to represent this many objects:

● ● ● ● ● ● ● ● ● ● ● ● ● ● ● ● ●

However, this is not the first and probably will not be the last numeration system to be developed; it took centuries to arrive at this stage of symbolic representation. Here are some of the ways that 17 has been written:

Tally: |||| |||| |||| || Egyptian: ∩ | | | | | | | Roman: XVII

Mayan: .. ___   Linguistic: *seventeen, seibzehn, dix-sept*

> There is a story of a woman who decided she wanted to write to a prisoner in a nearby state penitentiary. However, she was puzzled over how to address him, since she knew him only by a string of numerals. She solved her dilemma by beginning her letter: "Dear 5944930, may I call you 594?"

The concept represented by each of these symbols is the same, but the symbols differ. The concept or idea of "seventeenness" is called a *number;* the symbol used to represent the concept is called a *numeral.*

Three of the earliest civilizations known to use numerals were the Egyptian, Babylonian, and Roman. We shall examine these systems for two reasons. First, it will help us to more fully understand our own numeration (decimal) system; second, it will help us to see how the ideas of these other systems have been incorporated into our system.

Egyptian Numeration System

Perhaps the earliest type of written numeration system developed was a **simple grouping system.** The Egyptians used such a system by the time of the first dynasty, around 2850 B.C. The symbols of the Egyptian system were part of their hieroglyphics and are shown in Table 3.1.

Any number is expressed by using these symbols *additively*; each symbol is repeated the required number of times, but with no more than nine repetitions. This additively is called the **addition principle.**

Table 3.1 Egyptian Hieroglyphic Numerals

Our Numeral (decimal)	Egyptian Numeral	Descriptive Name
1	|	Stroke
10	∩	Heel bone
100	ϩ	Scroll
1,000	⚘	Lotus flower
10,000	⌐	Pointing finger
100,000	⌒	Polliwog
1,000,000	⚲	Astonished man

"No, no! 𓃀 as in 𓃀𓆓𓏏𓅱𓀀."
Alan Dunn, *The New Yorker Magazine,* 1952.

EXAMPLE 1

Write $\ell\,\&\,\&\,{}^{99}_{9}\,\cap\cap\,|||||$ using decimal numerals.

Solution

$$\ell\,\&\,\&\,{}^{99}_{9}\,\cap\cap\,||||| = 10{,}000 + 2 \times 1{,}000 + 3 \times 100 + 4 \times 10 + 5 \times 1$$
$$= 12{,}345$$

The position of the individual symbols is not important. For example,

$$\ell\,\&\,\&\,{}^{99}_{9}\,\cap\cap\,||||| = \&\,\&\,{}^{99}_{9}\,\ell\,\cap\cap\,|||||$$
$$= |||||\,\ell\,\&\,\cap\cap\cap\cap\,{}^{99}_{9}\,\& $$

The Egyptians had a simple **repetitive-type** numeration system. Addition and subtraction were performed by repeating the symbols and regrouping.

EXAMPLE 2

Use Egyptian arithmetic to find

a. $245 + 457$ **b.** $142 - 67$

Solution

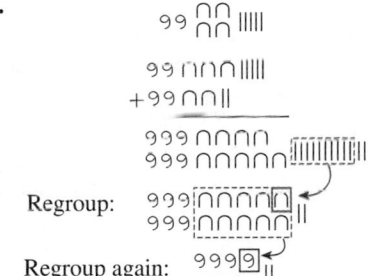

In Egyptian arithmetic, multiplication and division were performed by successions of additions that did not require the memorization of a multiplication table and could easily be done on an abacus-type device (see links on this topic at **www.mathnature.com**).

The Egyptians also used unit fractions (that is, fractions of the form $1/n$, for some counting number n). These were indicated by placing an oval over the numeral for the denominator. Thus,

$$\underset{|||}{\frown} = \frac{1}{3} \qquad \underset{||||}{\frown} = \frac{1}{4} \qquad \underset{\cap}{\frown} = \frac{1}{10} \qquad \underset{\cap\cap\cap\,|||}{\frown} = \frac{1}{33}$$

The fractions $\frac{1}{2}$ and $\frac{2}{3}$ were exceptions:

$$\underset{||}{\frown} \text{ or } \sqsubset\ = \frac{1}{2} \qquad \phi \text{ or } \varphi = \frac{2}{3}$$

Fractions that were not unit fractions (with the exception of $\frac{2}{3}$ just mentioned) were represented as sums of unit fractions. For example,

$$\frac{2}{7} \quad \text{could be expressed as} \quad \frac{1}{4} + \frac{1}{28}$$

Repetitions were not allowed, so $\frac{2}{7}$ would not be written as $\frac{1}{7} + \frac{1}{7}$. Why so difficult? We don't know, but the Rhind papyrus (see Figure 3.1) includes a table for all such decompositions for odd denominators from 5 to 101.

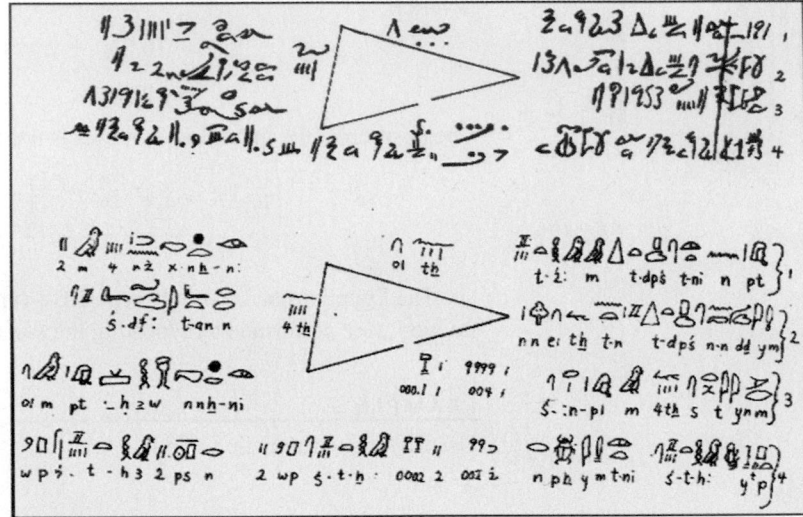

Figure 3.1 **This is a portion of the Rhind papyrus showing the area of a triangle.**

Roman Numeration System

The Roman numeration system was used in Europe until the 18th century, and is still used today in outlining, in numbering certain pages in books, and for certain copyright dates. The Roman numerals were selected letters from the Roman alphabet, as shown in Table 3.2. The Roman numeration system is *repetitive,* like the Egyptian system, but has two additional properties, called the *subtraction principle* and the *multiplication principle.*

Table 3.2 Roman Numerals

Decimal Numeral	1	5	10	50	100	500	1,000
Roman Numeral	I	V	X	L	C	D	M

Subtraction Principle

One property of the Roman numeration system is based on the **subtraction principle.** Reading from the left, we add the value of each numeral unless its value is smaller than the value of the numeral to its right, in which case we subtract it from that number. Only the numbers 1, 10, 100, . . . , can be subtracted, and only from the next two higher numbers. For example,

    I, II, III, IIII or IV, V, VI, VII, VIII, VIIII or IX, X, XI, . . .

are the successive counting numbers. Instead of IIII we write IV, which (using the subtraction principle) means $5 - 1 = 4$. The I can be subtracted from V or X, and the X can be subtracted from L or C; thus

    XL   means $50 - 10 = 40$

The C can be subtracted only from D or M. For example,

    DC means $500 + 100 = 600$ but   CD means $500 - 100 = 400$

Multiplication Principle

A bar is placed over a Roman numeral or a group of numerals to mean 1,000 times the value of the Roman numerals themselves. When one symbol (such as a bar) is used to multiply the value of an associated symbol, it is called the **multiplication principle.** For example,

$$\overline{V} \text{ means } 5 \times 1,000 = 5,000 \quad \text{and} \quad \overline{XL} \text{ means } 40 \times 1,000 = 40,000$$

EXAMPLE 3

Write each Roman numeral as a decimal numeral.

a. CCLXIII **b.** MXXIV **c.** MCMXCIX **d.** $\overline{\text{CDCLIX}}$

Solution

a. CCLXIII $= 100 + 100 + 50 + 10 + 1 + 1 + 1 = 263$

b. MXXIV $= 1,000 + 10 + 10 + (5 - 1) = 1,024$

c. MCMXCIX $= 1,000 + (1,000 - 100) + (100 - 10) + (10 - 1) = 1,999$

d. $\overline{\text{CDCLIX}} = (500 - 100) \times 1,000 + 100 + 50 + (10 - 1) = 400,159$ ◆

EXAMPLE 4

Write each decimal numeral as a Roman numeral.

a. 49 **b.** 379 **c.** 345,123 **d.** 2,003

Solution

a. $49 = (50 - 10) + (10 - 1) = \text{XLIX}$

b. $379 = 100 + 100 + 100 + 50 + 20 + (10 - 1) = \text{CCCLXXIX}$

c. $345,123 = 345 \times 1,000 + 123$

$$= [100 + 100 + 100 + (50 - 10) + 5] \times 1,000 + (100 + 20 + 3)$$

$$= \overline{\text{CCCXLV}}\text{CXXIII}$$

d. $2,003 = 2,000 + 3 = \text{MMIII}$ ◆

Babylonian Numeration System

The Babylonian numeration system differed from the Egyptian in several respects. Whereas the Egyptian system was a simple grouping system, the Babylonians employed a much more useful **positional system.** Since they lacked papyrus, they used mostly clay as a writing medium, and thus the Babylonian cuneiform was much less pictorial than the Egyptian system. They employed only two wedge-shaped characters, which date from 2000 B.C. and are shown in Table 3.3.

Notice from the table that, for numbers 1 through 59, the system is *repetitive.* However, unlike in the Egyptian system, the position of the symbols is important. The ◁ symbol *must* appear to the left of any ▼s to represent numbers smaller than 60. For numbers larger than 60, the symbols ◁ and ▼ are written to the left of ◁ (as in a positional system), and now take on a new value that is 60 times larger than the original value. That is,

$$\blacktriangledown \,\,{}^{◁◁}_{◁}\,\, \overset{\blacktriangledown\blacktriangledown\blacktriangledown}{\blacktriangledown\blacktriangledown} \quad \text{means} \quad (1 \times 60) + 35$$

Study the following example.

Table 3.3 Babylonian Cuneiform Numerals

Our Numeral (decimal)	Babylonian Numeral
1	▼
2	▼▼
9	▼▼▼▼▼▼▼▼▼
10	◁
59	◁◁◁ ▼▼▼▼▼ ◁◁ ▼▼▼▼

HISTORICAL NOTE

There are many clay tablets available to us (more than 50,000), and the work of interpreting them is continuing. Most of our knowledge of the tablets is less than 130 years old and has been provided by Professors Otto Neugebauer and F. Thureau-Dangin.

EXAMPLE 5

Write each of the Babylonian numerals in decimal form.

a. ▼▼▼ ◁◁ ◁ ▼▼▼▼▼▼   **b.** ▼▼▼▼▼ ◁ ▼   **c.** ◁◁ ▼▼▼ ◁ ▼▼▼▼▼▼

Solution **a.** ▼▼▼ ◁◁ ◁ ▼▼▼▼▼▼ $= (3 \times 60) + 59 = 239$

b. ▼▼▼▼▼ ◁ ▼ $= (5 \times 60) + 11 = 311$

c. ◁◁ ▼▼▼ ◁ ▼▼▼▼▼▼ $= (23 \times 60) + 16 = 1{,}396$ ◆

This system is called a *sexagesimal system* and uses the principle of position. However, the system is not fully positional, because only numbers larger than 60 use the position principle; numbers within each basic 60-group are written by a simple grouping system. A true positional sexagesimal system would require 60 different symbols.

The Babylonians carried their positional system a step further. If any numerals were to the left of the second 60-group, they had the value of 60×60 or 60^2. Thus,

$$
▼▼ ◁◁ ▼▼▼ ◁◁ ▼▼ = (2 \times 60^2) + (45 \times 60) + 24
$$
$$
= 7{,}200 + 2{,}700 + 24
$$
$$
= 9{,}924
$$

The Babylonians also made use of a *subtractive symbol*, ⌐ . That is, 38 could be written

◁◁◁ ▼▼▼▼▼▼▼   or   ◁◁◁◁ |▼▼

EXAMPLE 6

Change each Babylonian numeral to decimal form.

a. ▼▼▼ ◁◁ |▼   **b.** ▼▼ ◁◁ ▼▼▼▼ ◁ ▼▼▼   **c.** ◁◁ ◁◁◁ ▼▼▼

Solution

a. $(3 \times 60) + 20 - 1 = 199$

b. $(2 \times 60^2) + (23 \times 60) + 16 = 7{,}200 + 1{,}380 + 16 = 8{,}596$

c. Note the ambiguity; this might be 54 or it might be

$(20 \times 60) + 34 = 1{,}234$ ◆

EXAMPLE 7

Write 4,571 as a Babylonian numeral.

Solution
$$
1 \times 60^2 = 3{,}600
$$
$$
16 \times 60 = 960
$$
$$
11 \times 1 = \underline{11}
$$
$$
4{,}571
$$

Thus, ▼ ◁ ▼▼▼ ◁ ▼ is the Babylonian numeral.   ◆

Although this positional numeration system was in many ways superior to the Egyptian system, it suffered from the lack of a zero or place-holder symbol. For example, how is the number 60 represented? Does ▼▼ mean 2 or 61? In Example 6c the value of the number had to be found from the context. (Scholars tell us that such ambiguity can be resolved only by a careful study of the context.) However, in later Babylon, around 300 B.C., records show that there is a zero symbol, an idea that was later used by the Hindus.

Arithmetic with the Babylonian numerals is quite simple, since there are only two symbols. A study of the Babylonian arithmetic will be left for the reader (see Problem 6).

Properties of Numeration Systems

A *numeration system* consists of a set of basic symbols. Because the set of counting numbers is infinite and the set of symbols in any numeration system is finite, some rules for using the set of symbols are necessary. Different historical systems use one or more of these properties:

A *simple grouping* system invents a symbol to represent the number of objects in a predetermined group of objects. For example, the symbol

3	is used to represent	● ● ●
V	is used to represent	● ● ● ● ●
or ∩	is used to represent	● ● ● ● ● ● ● ● ● ●

A *positional system* reuses a symbol to represent different numbers of objects by changing the position in its symbolic (numerical) representation. For example,

in 35 the "3" is used to represent

● ● ● ● ● ● ● ● ● ●
● ● ● ● ● ● ● ● ● ●
● ● ● ● ● ● ● ● ● ●

whereas in 53 the "3" is used to represent

● ● ●

The *addition principle* means that the various values of individual symbols (numerals) are added, as in ∩∩∩ |||| meaning

$$10 + 10 + 10 + 1 + 1 + 1 + 1$$

The *subtraction principle* means that certain values are subtracted, as in IX meaning $10 - 1 = 9$. The *multiplication principle* means that certain symbolic values are multiplied, as in $\overline{V}$ meaning $1{,}000 \times 5$.

A *repetitive system* reuses the same number over and over in an additive manner, as in XXX meaning $10 + 10 + 10$.

Other Historical Systems

One of the richest areas in the history of mathematics is the study of historical numeration systems. In this book, we discuss the Egyptian numeration system because it illustrates a simple grouping system, the Babylonian numeration system because it illustrates a positional numeration system, and the Roman system because it is still in limited use today.

Other numeration systems of interest from a historical point of view are the Mayan, Greek, and Chinese numeration systems. These are summarized in Table 3.4.

Table 3.4 Historical Numeration Systems

Numeration System	Classification	Selected Numerals							
		1	2	5	10	50	100	500	1,000
Decimal system	positional (10s)	1	2	5	10	50	100	500	1,000
Egyptian system	grouping								
Roman system	grouping	I	II	V	X	L	C	D	M
Babylonian system (also called Sumerian)	positional (60s)								
Greek	grouping	α	β	ϵ	ι	ν	ρ	ϕ	α'
Mayan	positional (vertical)								
Chinese	positional (vertical)								

Problem Set 3.1

LEVEL 1

1. **IN YOUR OWN WORDS** Explain the difference between *number* and *numeral*. Give examples of each.

2. **IN YOUR OWN WORDS** Discuss the similarities and differences between a simple grouping system and a positional system. Give examples of each.

3. **IN YOUR OWN WORDS** What do you regard as the shortcomings and contributions of the Egyptian numeration system?

4. **IN YOUR OWN WORDS** What do you regard as the shortcomings and contributions of the Roman numeration system?

5. **IN YOUR OWN WORDS** What do you regard as the shortcomings and contributions of the Babylonian numeration system?

6. **IN YOUR OWN WORDS** Discuss addition and subtraction for Babylonian numerals. Show examples.

7. Tell which of the named properties apply to the Egyptian, Roman, and Babylonian numeration systems.
   a. grouping system
   b. positional system
   c. repetitive system
   d. additive system
   e. subtractive system
   f. multiplicative system

Write each numeral in Problems 8–35 as a decimal numeral. By decimal numeral, we mean the system we use in everyday arithmetic.

8.

9.

10.

11.

12.

13.

14.

15.

16.

17.

18.

19.

20. XLVIII

21. DCCIX

22. MCMXCVII

23. MMI

24. $\overline{\text{DL}}$

25. $\overline{\text{CD}}$

26. $\overline{\text{VMMDC}}$

27. $\overline{\text{IXDCCXII}}$

28.

29.

30.

31.

32.

33.

34.

35.

The Rhind papyrus contains many problems. Two are symbolized on the following simulated papyrus. The one in Problem 37 is an 18th century Mother Goose rhyme. Answer the question posed in each problem.

36.

> In each of 7 houses are 7 cats.
>
> Each cat kills 7 mice.
>
> Each mouse would have eaten 7 spelt (wheat).
>
> Each ear of spelt would have produced 7 kehat of grain.
>
> Query: How much grain is saved by the 7 houses' cats?

37.

> As I was going to St. Ives
>
> I met a man with seven wives.
>
> Every wife had seven sacks,
>
> Every sack had seven kits.
>
> Kits, cats, sacks, and wives,
>
> How many were there going to St. Ives?

LEVEL 2

Write each of the numerals in Problems 38–43 in the following systems:

a. Egyptian   **b.** Roman   **c.** Babylonian

38. 47   **39.** 75   **40.** 258

41. 521   **42.** 852   **43.** 2,001

44. Kathy had a dream in which she was selling roses in an international market. She started out with 143 roses. Then her first customer, an Egyptian, bought ∩||||| of them. Soon afterward, a Babylonian asked to buy ◁◁◁ ▼▼▼▼▼▼ roses; later XIV roses were bought by (you guessed it) a Roman. How many roses did she have left to sell at the end of her dream?

45. In a strange mix-up, several people (none of whom spoke the same language) needed to pool their resources to survive the scorching desert heat. Eric had 45 bottles of water; Dimetrius had ∩∩∩ |||| bottles; Sparticus had ◁◁◁◁◁ ▼▼▼▼ bottles. How many bottles of water did they have all together?

Perform the indicated operations in Problems 46–51.

46.

$$\begin{array}{r} \text{𓏤}\;∩∩\;|| \\ ∩∩\;|| \\ +\;99\;∩∩∩∩\;||||| \\ \hline \end{array}$$

47.

$$\begin{array}{r} 99\;∩|\;| \\ -\;\;∩∩\;|||| \\ \hline \end{array}$$

48.

$$\begin{array}{r} 99\;∩∩||\; \\ -\;9\;∩∩∩∩\;||||| \\ \hline \end{array}$$

49.

$$\begin{array}{r} ◁◁◁\;▼▼▼▼ \\ +\;◁\;\begin{smallmatrix}▼▼▼\\▼▼▼\end{smallmatrix} \\ \hline \end{array}$$

50.

$$\begin{array}{r} ◁◁\;▼▼▼▼ \\ -\;◁\;\begin{smallmatrix}▼▼▼\\▼▼▼\end{smallmatrix} \\ \hline \end{array}$$

51.

$$\begin{array}{r} ▼◁\;▼▼ \\ ◁◁◁\;▼▼▼▼ \\ -\;◁\;\;▼▼▼ \\ \hline \end{array}$$

PROBLEM SOLVING

52. In Example 3c, we wrote 1999 as MCMXCIX, which, according to the National Institute of Standards and Technology, is the preferred representation. Which of the following does not also represent 1999?
A. MCMXCXI   B. MCMXCVIIII
C. MDCCCCLXXXXVIIII   D. MIM

53. a. What is the largest number that uses each of the seven Roman numerals exactly once?

    b. What is the smallest number that uses each of the seven Roman numerals exactly once?

54. Which year in the millennium (1001 to 2000) is written with the most Roman numerals?

55. What is the largest number that begins with the given indicated symbol?
a. I   **b.** V   **c.** X   **d.** L   **e.** C

56. The Yale tablet from the Babylonian collection, written between 1900 B.C. and 1600 B.C., contains the following algebra problem:

    The length of a rectangle exceeds the width by 7, and the area is 60. What are the dimensions of the rectangle?

Answer the question in the Babylonian numeration system.

57. Answer the following question from the Yale tablet (see Problem 56):

    The length of a rectangle exceeds the width by 10 and the area is 600. What are the dimensions of the rectangle?

Give your answer using Babylonian notation.

58. IN YOUR OWN WORDS The people from the Long Lost Land had only the following four symbols: □, △, ▽, and ○. Write out the first 20 numbers. Find □ + ▽ and △ × △. (Use your imagination to invent a system to answer these questions.)

59. IN YOUR OWN WORDS In Chapter 1 we introduced Pascal's triangle. The reproduction in Figure 3.2 (page 110) is from a 14th century Chinese manuscript. Even though we have not discussed these ancient Chinese numerals, see if you can reconstruct the basics of their numeration system. (Use your imagination.)

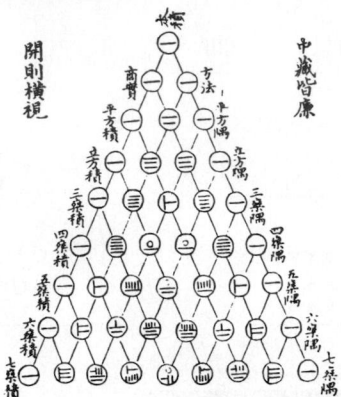

Figure 3.2 **Chinese triangle**

60. IN YOUR OWN WORDS The Ionic Greek numeration system (approximately 3000 B.C.) counts as follows:

$$\alpha, \beta, \gamma, \delta, \epsilon, \Im, \zeta, \eta, \theta, \iota, \iota\alpha, \iota\beta, \iota\gamma, \iota\epsilon, \iota\Im, \iota\zeta, \theta\eta,$$
$$\iota\theta, \kappa, \kappa\alpha, \dots$$

Other numbers are: λ (for 30), μ (for 40), ν (for 50), ξ (for 60), o (for 70), π (for 80), $\mathbb{Q}$ (for 90), ρ (for 100), σ (for 200), τ (for 300), υ (for 400), ϕ (for 500), χ (for 600), ψ (for 700), ω (for 800), and Π (for 900). Try to reconstruct the basics of their numeration system. (Use your imagination.)

3.2 HINDU–ARABIC NUMERATION SYSTEM

The numeration system in common use today (the one we have been calling the decimal system) has ten symbols—namely, 0, 1, 2, 3, 4, 5, 6, 7, 8, and 9. The selection of ten digits was no doubt a result of our having ten fingers (digits).

The symbols originated in India in about 300 B.C. However, because the early specimens do not contain a zero or use a positional system, this numeration system offered no advantage over other systems then in use in India.

The date of the invention of the zero symbol is not known. The symbol did not originate in India but probably came from the late Babylonian period via the Greek world.

By the year 750 A.D. the zero symbol and the idea of a positional system had been brought to Baghdad and translated into Arabic. We are not certain how these numerals were introduced into Europe, but it is likely that they came via Spain in the 8th century. Gerbert, who later became Pope Sylvester II in 999, studied in Spain and was the first European scholar known to have taught these numerals. Because of their origins, these numerals are called the **Hindu–Arabic numerals.** Since ten basic symbols are used, the Hindu–Arabic numeration system is also called the *decimal numeration system,* from the Latin word *decem,* meaning "ten."

Although we now know that the decimal system is very efficient, its introduction met with considerable controversy. Two opposing factions, the "algorists" and the "abacists," arose. Those favoring the Hindu–Arabic system were called algorists, since the symbols were introduced into Europe in a book called (in Latin) *Liber Algorismi de Numero Indorum,* by the Arab mathematician al-Khowârizmî. The word *algorismi* is the origin of our word *algorithm.* The abacists favored the status quo—using Roman numerals and doing arithmetic on an abacus. The battle between the abacists and the algorists lasted for 400 years. The Roman Catholic Church exerted great influence in commerce, science, and theology. The church criticized those using the "heathen" Hindu–Arabic numerals and consequently kept the world using Roman numerals until 1500. Roman numerals were easy to write and learn, and addition and subtraction with them were easier than with the "new" Hindu–Arabic numerals. It seems incredible that our decimal system has been in general use only since about the year 1500.

Decimal System—Grouping by Tens

Let's examine the Hindu–Arabic or **decimal numeration system** a little more closely:

1. It uses ten symbols, called digits.

2. Larger numbers are expressed in terms of powers of 10.

3. It is positional.

Consider how we count objects:

■ ■ ■ ■ ■ ■ ■ ■ ■ ■
1 2 3 4 5 6 7 8 9 ?

At this point we could invent another symbol as the Egyptians did (you might suggest 0, but remember that 0 represents no objects), or we could reuse the digit symbols by repeating them or by altering their positions. That is, we agree to use **ten,** written 10, to mean 1 group of

■■■■■■■■■■

The symbol 0 was invented as a place-holder to show that the 1 here is in a different position from the 1 representing ■. We continue to count:

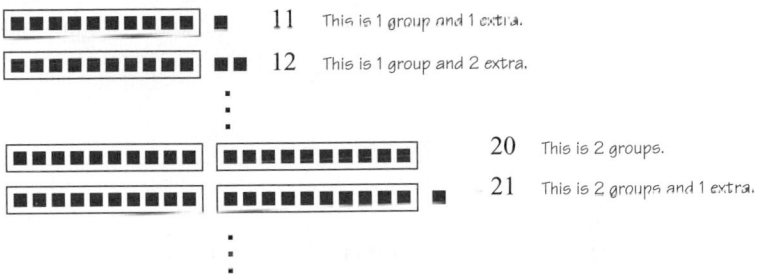

We continue in the same fashion until we have 9 groups and 9 extra. What's next? It is 10 groups or a group of groups:

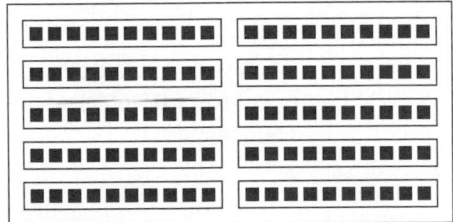

We call this group of groups a $10 \cdot 10$ or 10^2 or a **hundred.** We again use position and repeat the symbol 1 with still different meaning: 100.

EXAMPLE 1

What does 134 mean?

Solution

134 means that we have 1 group of 100, 3 groups of 10, and 4 extra. In symbols,

We denote this more simply by writing:

These represent the number in each group.

$$(1 \times 10^2) + (3 \times 10) + 4$$

These are the names of the groups.

This leads us to the meaning *one hundred, three tens, four ones.*   ◆

Expanded Notation

The representation, or the meaning, of the number 134 in Example 1 is called **expanded notation,** or expanded form.

EXAMPLE 2

Write 52,613 in expanded form.

Solution $52,613 = 50,000 + 2,000 + 600 + 10 + 3$

$$= 5 \times 10^4 + 2 \times 10^3 + 6 \times 10^2 + 1 \times 10 + 3$$ ◆

EXAMPLE 3

Write $4 \times 10^8 + 9 \times 10^7 + 6 \times 10^4 + 3 \times 10 + 7$ in decimal form.

Solution You can use the order of operations and multiply out the digits, but you should be able to go directly to decimal form if you remember what place-value means:

$$49\ 00600\ 37 = 490,060,037$$

Notice that there were no powers of 10^6, 10^5, 10^3, or 10^2.   ◆

A period, called a **decimal point** in the decimal system, is used to separate the fractional parts from the whole parts. The positions to the right of the decimal point are fractions:

$$\frac{1}{10} = 10^{-1}, \quad \frac{1}{100} = 10^{-2}, \quad \frac{1}{1,000} = 10^{-3}$$

To complete the pattern, we also sometimes write $10 = 10^1$ and $1 = 10^0$.

EXAMPLE 4

Write 479.352 using expanded notation.

Solution

$$479.352 = 400 + 70 + 9 + 0.3 + 0.05 + 0.002$$

$$= 400 + 70 + 9 + \frac{3}{10} + \frac{5}{100} + \frac{2}{1,000}$$

$$= 4 \times 10^2 + 7 \times 10^1 + 9 \times 10^0 + 3 \times 10^{-1} + 5 \times 10^{-2} + 2 \times 10^{-3} \quad ◆$$

Problem Set 3.2

LEVEL 1

1. **IN YOUR OWN WORDS** Illustrate the meaning of 123 by showing the appropriate groupings.
2. **IN YOUR OWN WORDS** Illustrate the meaning of 145 by showing the appropriate groupings.
3. **IN YOUR OWN WORDS** Illustrate the meaning of 1,134 by showing the appropriate groupings.
4. **IN YOUR OWN WORDS** What is expanded notation?
5. **IN YOUR OWN WORDS** Define b^n for b any nonzero number and n any counting number.
6. **IN YOUR OWN WORDS** Define b^{-n} for b any nonzero number and n any counting number.

Give the meaning of the numeral 5 in each of the numbers in Problems 7–10.

7. 805   8. 508   9. 0.00567   10. 58,000,000

Write the numbers in Problems 11–27 in decimal notation.

11. **a.** 10^5 **b.** 10^3 12. **a.** 10^6 **b.** 10^4
13. **a.** 10^{-4} **b.** 10^{-3} 14. **a.** 10^{-2} **b.** 10^{-6}
15. **a.** 5×10^3 **b.** 5×10^2
16. **a.** 8×10^{-4} **b.** 7×10^{-3}
17. **a.** 6×10^{-2} **b.** 9×10^{-5}
18. **a.** 5×10^{-6} **b.** 2×10^{-9}
19. $1 \times 10^4 + 0 \times 10^3 + 2 \times 10^2 + 3 \times 10^1 + 4 \times 10^0$
20. $6 \times 10^1 + 5 \times 10^0 + 0 \times 10^{-1} + 8 \times 10^{-2} + 9 \times 10^{-3}$
21. $5 \times 10^5 + 2 \times 10^4 + 1 \times 10^3 + 6 \times 10^2 + 5 \times 10^1 + 8 \times 10^0$
22. $6 \times 10^7 + 4 \times 10^3 + 1 \times 10^0$
23. $7 \times 10^6 + 3 \times 10^{-2}$
24. $6 \times 10^9 + 2 \times 10^{-3}$
25. $5 \times 10^5 + 4 \times 10^2 + 5 \times 10^1 + 7 \times 10^0 + 3 \times 10^{-1} + 4 \times 10^{-2}$
26. $3 \times 10^3 + 2 \times 10^1 + 8 \times 10^0 + 5 \times 10^{-1} + 4 \times 10^{-2} + 2 \times 10^{-4}$
27. $2 \times 10^4 + 6 \times 10^2 + 4 \times 10^{-1} + 7 \times 10^{-3} + 6 \times 10^{-4} + 9 \times 10^{-5}$

Write each of the numbers in Problems 28–43 in expanded notation.

28. 741   29. 728,407
30. 0.096421   31. 27.572
32. 47.00215   33. 521
34. 6,245   35. 2,305,681
36. 428.31   37. 5,245.5
38. 0.00000 527   39. 100,000.001
40. 893.0001   41. 8.00005
42. 678,000.01   43. 57,285.9361

LEVEL 2

One of the oldest devices used for calculation is the abacus, as shown in Figure 3.3. Each rod names one of the positions we use in counting. Each bead on the bottom represents one unit in that column and each bead on the top represents 5 units in that column. The number illustrated in Figure 3.3 is 1,734. What number is illustrated by the drawings in Problems 44–49?

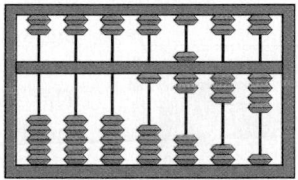

$10^6\ 10^5\ 10^4\ 10^3\ 10^2\ 10^1\ 10^0$

Figure 3.3 **Abacus**

44.   45.

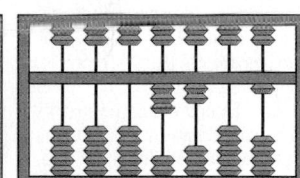

$10^6\ 10^5\ 10^4\ 10^3\ 10^2\ 10^1\ 10^0$ $10^6\ 10^5\ 10^4\ 10^3\ 10^2\ 10^1\ 10^0$

46.   47.

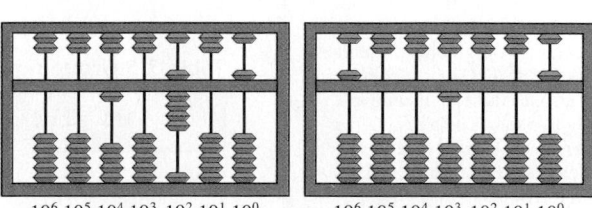

$10^6\ 10^5\ 10^4\ 10^3\ 10^2\ 10^1\ 10^0$ $10^6\ 10^5\ 10^4\ 10^3\ 10^2\ 10^1\ 10^0$

48.   49.

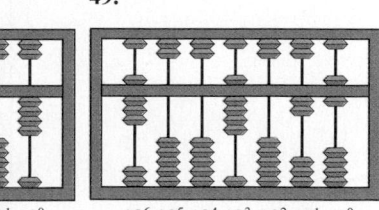

$10^6\ 10^5\ 10^4\ 10^3\ 10^2\ 10^1\ 10^0$ $10^6\ 10^5\ 10^4\ 10^3\ 10^2\ 10^1\ 10^0$

Sketch an abacus to show the numbers given in Problems 50–57.

50. 132   **51.** 849   **52.** 3,214   **53.** 9,387

54. 1,998   **55.** 2,001   **56.** 3,000,400   **57.** 8,007,009

PROBLEM SOLVING

58. Can you find a pattern?

0, 1, 2, 10, 11, 12, 20, 21, 22, 100, . . .

59. Can you find a pattern?

0, 1, 2, 3, 4, 10, 11, 12, 13, 14, 20, 21, . . .

60. Notice the following pattern for multiplication by 11:

$$14 \times 11 = 154 \qquad 51 \times 11 = 561$$

↑   ↑

Insert the sum of the original two digits between those digits.

Use expanded notation to show why this pattern "works."

3.3   DIFFERENT NUMERATION SYSTEMS

In the previous section we discussed the Hindu–Arabic numeration system and grouping by tens. However, we could group by twos, fives, twelves, or any other counting number. In this section, we summarize numeration systems with bases other than ten. This not only will help you understand our own numeration system, but will give you insight into the numeration systems used with computers, namely, base 2 **(binary),** base 8 **(octal),** and base 16 **(hexadecimal).**

Number of Symbols

The number of symbols used in a particular base depends on the method of grouping for that base. For example, in base ten the grouping is by tens, and in base five the grouping is by fives. Suppose we wish to count ■■■■■■■■■■■ in various bases. Let's look for patterns in Table 3.5. Note the use of the subscript following the numeral to keep track of the base in which we are working.

Do you see any patterns? Suppose we wish to continue this pattern. Can we group by elevens or twelves? We can, provided new symbols are "invented." For base eleven (or higher bases), we use the symbol T to represent ■■■■■■■■■■. For base twelve (or higher bases), we use E to stand for ■■■■■■■■■■■. For bases larger than twelve, other symbols can be invented.

Table 3.5 Grouping in Various Bases

Base	Symbols	Method of Grouping	Notation
two	0, 1	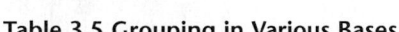	1011_{two}
three	0, 1, 2		102_{three}
four	0, 1, 2, 3		23_{four}
five	0, 1, 2, 3, 4		21_{five}
six	0, 1, 2, 3, 4, 5		15_{six}
seven	0, 1, 2, 3, 4, 5, 6		14_{seven}
eight	0, 1, 2, 3, 4, 5, 6, 7		13_{eight}
nine	0, 1, 2, 3, 4, 5, 6, 7, 8		12_{nine}
ten	0, 1, 2, 3, 4, 5, 6, 7, 8, 9		11_{ten}

When photographs are sent back from space, they are sent using a binary numeration system. The images are not usually recorded on photographic film, but instead the image is broken up into tiny dots, called *pixels*. For example, a photograph might be divided into 1,000 pixels horizontally and 500 pixels vertically. Each pixel is then assigned a number representing its brightness — 0 for pure white and 63 for pure black. These numbers are sent back as six-digit binary numbers 000000 to 111111. A computer translates these digits into a photograph. The photo shown here is the scarred face of Triton, Neptune's largest moon. The image was sent by *Voyager 2* on August 24, 1989.

For example, $2T_{twelve}$ means that there are two groupings of twelve and T (ten) extra:

■■■■■■■■■■■■ ■■■■■■■■■■■■ ■■■■■■■■■■

We continue with the pattern from Table 3.5 by continuing beyond base ten in Table 3.6.

Table 3.6 Grouping in Various Bases

Base	Symbols	Method of Grouping	Notation
eleven	$0, 1, 2, \ldots, 8, 9, T$	■■■■■■■■■■■	10_{eleven}
twelve	$0, 1, 2, \ldots, 8, 9, T, E$	■■■■■■■■■■■	E_{twelve}
thirteen	$0, 1, 2, \ldots, 8, 9, T, E, U$	■■■■■■■■■■■	$E_{thirteen}$
fourteen	$0, 1, 2, \ldots, 9, T, E, U, V$	■■■■■■■■■■	$E_{fourteen}$

Do you see more patterns? Can you determine the number of symbols in the **base b system?** Remember that b stands for some counting number greater than 1. Look for a pattern:

$$\textbf{base } \boldsymbol{b}\textbf{, } \boldsymbol{b}\textbf{ symbols} \begin{cases} \text{base two, two symbols} \\ \text{base three, three symbols} \\ \text{base four, four symbols} \\ \text{base five, five symbols} \\ \quad \vdots \\ \text{base ten, ten symbols} \\ \text{base eleven, eleven symbols} \\ \quad \vdots \\ \text{base twenty, twenty symbols} \\ \quad \vdots \end{cases}$$

Change from Base *b* to Base 10

To change from base b to base ten, we write the numerals in expanded notation. The resulting number is in base ten.

EXAMPLE 1

Change each number to base ten.

a. 1011.01_{two} **b.** 1011.01_{four} **c.** 1011.01_{five}

Solution

a. $1011.01_{two} = 1 \times 2^3 + 0 \times 2^2 + 1 \times 2^1 + 1 \times 2^0 + 0 \times 2^{-1} + 1 \times 2^{-2}$
$= 8 + 0 + 2 + 1 + 0 + 0.25 = 11.25$

b. $1011.01_{four} = 1 \times 4^3 + 0 \times 4^2 + 1 \times 4^1 + 1 \times 4^0 + 0 \times 4^{-1} + 1 \times 4^{-2}$
$= 64 + 0 + 4 + 1 + 0 + 0.0625 = 69.0625$

c. $1011.01_{five} = 1 \times 5^3 + 0 \times 5^2 + 1 \times 5^1 + 1 \times 5^0 + 0 \times 5^{-1} + 1 \times 5^{-2}$
$= 125 + 0 + 5 + 1 + 0 + 0.04 = 131.04$ ◆

HISTORICAL NOTE

A study of the Native Americans of California uncovered the use of a wide variety of number bases. Several of these bases are discussed by Barnabus Hughes in an article entitled "California Indian Arithmetic" in *The Bulletin of the California Mathematics Council* (Winter 1971/1972). According to Professor Hughes, the Yukis, who lived north of Willits, used both the quaternary (base four) and the octonary (base eight) systems. Instead of counting on their fingers, these Native Americans enumerated the spaces between the fingers.

The sketch above shows a Karok woman because the Yukis were nearly eliminated at the beginning of the 20th century and consequently there are no known photographs of the Yukis. The Native Americans in the western corner of California use the decimal system, while those in the rest of the state use the quinary (base five) system for small numbers, but switch to a vigesimal (base twenty) system for large numbers.

Change from Base 10 to Base *b*

To see how to change from base ten to any other valid base, let's again look for a pattern:

To change from base ten to base two, group by twos.
To change from base ten to base three, group by threes.
To change from base ten to base four, group by fours.
To change from base ten to base five, group by fives.

The groupings from this pattern are summarized in Table 3.7.

Table 3.7 Place-value Chart

Base	Place Value					
2	$2^5 = 32$	$2^4 = 16$	$2^3 = 8$	$2^2 = 4$	$2^1 = 2$	$2^0 = 1$
3	$3^5 = 243$	$3^4 = 81$	$3^3 = 27$	$3^2 = 9$	$3^1 = 3$	$3^0 = 1$
4	$4^5 = 1,024$	$4^4 = 256$	$4^3 = 64$	$4^2 = 16$	$4^1 = 4$	$4^0 = 1$
5	$5^5 = 3,125$	$5^4 = 625$	$5^3 = 125$	$5^2 = 25$	$5^1 = 5$	$5^0 = 1$
7	$7^5 = 16,807$	$7^4 = 2,401$	$7^3 = 343$	$7^2 = 49$	$7^1 = 7$	$7^0 = 1$
8	$8^5 = 32,768$	$8^4 = 4,096$	$8^3 = 512$	$8^2 = 64$	$8^1 = 8$	$8^0 = 1$
10	$10^5 = 100,000$	$10^4 = 10,000$	$10^3 = 1,000$	$10^2 = 100$	$10^1 = 10$	$10^0 = 1$
12	$12^5 = 248,832$	$12^4 = 20,736$	$12^3 = 1,728$	$12^2 = 144$	$12^1 = 12$	$12^0 = 1$

The next example shows how we can interpret this grouping process in terms of a simple division.

EXAMPLE 2   PÓLYA'S METHOD

Convert 42 to base two.

Solution   We use Pólya's problem-solving guidelines for this example.

Understand the Problem. Using Table 3.7, we see that the largest power of two smaller than 42 is 2^5, so we begin with $2^5 = 32$:

$$42 = 1 \times 2^5 + 10$$
$$10 = 0 \times 2^4 + 10$$
$$10 = 1 \times 2^3 + 2$$
$$2 = 0 \times 2^2 + 2$$
$$2 = 1 \times 2^1 + 0$$

We could now write out 42 in expanded notation.

Devise a Plan. Instead of carrying out the steps by using Table 3.7, we will begin with 42 and carry out repeated division, saving each remainder as we go.

Carry Out the Plan. We are changing to base 2, so we do repeated division by 2:

$$\begin{array}{r} 21 \\ 2\overline{)42} \end{array}$$ **r. 0 ← save remainder**

Next we need to divide 21 by 2, but instead of rewriting our work we work our way up:

$$\begin{array}{r} 10 \\ 2\overline{)21} \\ 2\overline{)42} \end{array}$$ **r. 1 ← save all remainders**
r. 0 ← save remainder

Continue by doing repeated division.

Stop when you get a zero here

$$
\begin{array}{r}
0 \quad \text{r. 1} \\
2\overline{)1} \quad \text{r. 0} \\
2\overline{)2} \quad \text{r. 1} \\
2\overline{)5} \quad \text{r. 0} \\
2\overline{)10} \quad \text{r. 1} \\
2\overline{)21} \quad \text{r. 0} \\
2\overline{)42}
\end{array}
$$

Answer is found by reading down.

Thus, $42 = 101010_{two}$.

Look Back. $101010_{two} = 1 \times 2^5 + 1 \times 2^3 + 1 \times 2 = 32 + 8 + 2 = 42$ ◆

EXAMPLE 3

Write 42 in **a.** base three **b.** base four

Solution

a. Begin with $3^3 = 27$ (from Table 3.7):

$$
\begin{array}{lll}
42 = 1 \times 3^3 + 15 & \text{or} & \quad 0 \quad \text{r. 1} \\
15 = 1 \times 3^2 + 6 & & 3\overline{)1} \quad \text{r. 1} \\
6 = 2 \times 3^1 + 0 & & 3\overline{)4} \quad \text{r. 2} \\
0 = 0 \times 3^0 & & 3\overline{)14} \quad \text{r. 0} \\
& & 3\overline{)42}
\end{array}
$$

Thus, $42 = 1120_{three}$.

b. Begin with $4^2 = 16$ (from Table 3.7):

$$
\begin{array}{lll}
42 = 2 \times 4^2 + 10 & \text{or} & \quad 0 \quad \text{r. 2} \\
10 = 2 \times 4^1 + 2 & & 4\overline{)2} \quad \text{r. 2} \\
2 = 2 \times 4^0 & & 4\overline{)10} \quad \text{r. 2} \\
& & 4\overline{)42}
\end{array}
$$

Thus, $42 = 222_{four}$. ◆

How would you change from base ten to base seven? To base eight? To base *b*? We see from the above pattern that we group by *b*'s or perform repeated division by *b*.

EXAMPLE 4   PÓLYA'S METHOD

Suppose you need to purchase 1,000 name tags and can buy them by the gross (144), the dozen (12), or individually. The name tags cost $0.50 each, $4.80 per dozen, and $50.40 per gross. How much should you order to minimize the cost?

Solution   We use Pólya's problem-solving guidelines for this example.

Understand the Problem. If you purchase 1,000 tags individually, the cost is $0.50 \times 1,000 = \$500$. This is not the least cost possible, because of the bulk discounts. We need to find the maximum number of gross, the number of dozens, and then purchase the remainder individually.

Devise a Plan. We will proceed by repeated division by 12, which we recognize as equivalent to changing the number to base twelve.

Carry Out the Plan. Change 1,000 to base 12:

$$
\begin{array}{r}
0 \quad \text{r. } 6 \\
12\overline{)6} \quad \text{r. } 11, \text{ or } E \text{ in base twelve} \\
12\overline{)83} \quad \text{r. } 4 \\
12\overline{)1{,}000}
\end{array}
$$

Thus, $1{,}000 = 6E4_{twelve}$ so you must purchase 6 gross, 11 dozen, and 4 individual tags.

Look Back. The cost is $6 \times \$50.40 + 11 \times \$4.80 + 4 \times \$0.50 = \$357.20$. As you can see, this is considerably less expensive than purchasing the individual name tags. ◆

Converting Between Numeration Systems

To convert from a base b numeration system to the Hindu–Arabic (decimal) numeration system, write the base b numeral in expanded form and then simplify.

To convert from the Hindu–Arabic numeration system to a base b numeration system, perform repeated division by b and then read the remainders from the top down.

Problem Set 3.3

LEVEL 1

1. IN YOUR OWN WORDS Explain how to change from base eight to base ten.

2. IN YOUR OWN WORDS Explain how to change from base sixteen to base ten.

3. IN YOUR OWN WORDS Explain how to change from base ten to base eight.

4. IN YOUR OWN WORDS Explain how to change from base ten to base sixteen.

5. IN YOUR OWN WORDS Explain how to change from base two to base eight.

6. IN YOUR OWN WORDS Discuss the binary (base two) numeration system.

7. Count the number of people in the indicated base.

a. base ten   **b.** base five   **c.** base thirteen
d. base eight   **e.** base two   **f.** base twelve

8. Count the number of people in the indicated base.

a. base ten   **b.** base five   **c.** base three
d. base eight   **e.** base two   **f.** base nine

In Problems 9–14, write the numbers in expanded notation.

9. a. 643_{eight} **b.** 5387.9_{twelve}

10. a. $111\,011\,000_{two}$ **b.** 750_{eight}

11. a. 110111.1001_{two} **b.** 5411.1023_{six}

12. a. 64200051_{eight} **b.** 1021.221_{three}

13. a. 323000.2_{four} **b.** 234000_{five}

14. a. 3.40231_{five} **b.** 2033.1_{four}

Change the numbers in Problems 15–30 to base ten.

15. 527_{eight} **16.** 527_{twelve}

17. 1101.11_{two} **18.** $25TE_{twelve}$

19. 431_{five} **20.** 65_{eight}

21. 1011.101_{two} **22.** 11101000110_{two}

23. 573_{twelve} **24.** 4312_{eight}

25. 2110_{three} **26.** 4312_{five}

27. 537.1_{eight}

28. 3721_{eight}

29. 5742_{eight}

30. 101111001_{two}

31. Change 628 to base four.

32. Change 724 to base five.

33. Change 427 to base twelve.

34. Change 256 to base two.

35. Change 615 to base eight.

36. Change 412 to base five.

37. Change 615 to base two.

38. Change 5,133 to base twelve.

39. Change 795 to base seven.

40. Change 512 to base two.

41. Change 4,731 to base twelve.

42. Change 52 to base three.

43. Change 76 to base four.

44. Change 602 to base eight.

45. Change 2,000 to base eight.

46. Change 2,001 to base two.

LEVEL 2

Use number bases to answer the questions given in Problems 47–59.

47. Change 52 days to weeks and days.

48. Change 158 hours to days and hours.

49. Change 55 inches to feet and inches.

50. Change 39 ounces to pounds and ounces.

51. Change 500 into gross, dozens, and units.

52. Change $4.59 into quarters, nickels, and pennies.

53. Using only quarters, nickels, and pennies, what is the minimum number of coins needed to make $0.84?

54. Suppose you have two quarters, four nickels, and two pennies. Use base five to write a numeral to indicate your financial status.

55. A bookstore ordered 9 gross, 5 dozen, and 4 pencils. Write this number in base twelve and in base ten.

56. Change $8.34 into the smallest number of coins consisting of quarters, nickels, and pennies.

57. Change 44 days to weeks and days.

58. Change 54 months to years and months.

59. Change 29 hours to days and hours.

PROBLEM SOLVING

60. The *duodecimal numeration system* refers to the base twelve system, which uses the symbols 0, 1, 2, 3, 4, 5, 6, 7, 8, 9, *T, E*. Historically, a numeration system based on 12 is not new. There were 12 tribes in Israel and 12 Apostles of Christ. In Babylon, 12 was used as a base for the numeration system before it was replaced by 60. In the 18th and 19th centuries, Charles XII of Sweden and Georg Buffon (1707–1788) advocated the adoption of the base twelve system. Even today there is a Dozenal (Duodecimal) Society of America that advocates the adoption of this system. According to the Society's literature, no one "who thought long enough—three to 17 minutes—to grasp the central idea of the duodecimal system ever failed to concede its superiority." Study the duodecimal system from 3 to 17 minutes, and comment on whether you agree with the Society's statement.

3.4 | BINARY NUMERATION SYSTEM

Base Two

For a computer to be of use in calculating, the human mind had to invent a way of representing numbers and other symbols in terms of electronic devices. In fact, it is the simplest of electronic devices, the switch, that is at the heart of communication between machines and human beings. A switch can be only "on" or "off," as illustrated in Figure 3.4.

Figure 3.4 **Two-state device. Light bulbs serve as a good example of two-state devices. The 1 is symbolized by "on," and the 0 is symbolized by "off."**

Decimal		*Binary*
0	↔	0
1	↔	1
2	↔	10
3	↔	11
4	↔	100
5	↔	101
6	↔	110
7	↔	111
8	↔	1000
9	↔	1001
10	↔	1010
11	↔	1011
12	↔	1100
13	↔	1101
14	↔	1110
15	↔	1111
16	↔	10000
17	↔	10001
.		.
.		.
.		.
31	↔	11111
32	↔	100000
.		.
.		.
.		.

Figure 3.5 Binary and decimal numeration. Can you see the pattern?

We can see that it would be easy to represent the numbers "one" and "zero" by "on" and "off," but that won't get us very far. The key idea that will allow us to represent every symbol we need is that of the **binary numeration system.** Binary numerals represent the numbers 1, 2, 3, . . . using only the symbols 1 and 0, as shown in Figure 3.5. Recall that expanded notation provides the link between the binary numeration and decimal numeration systems, and that for a binary numeration system the grouping is by twos.

EXAMPLE 1

What does 1110_{two} mean?

Solution

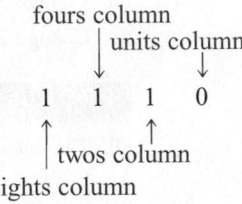

We see this means: $\quad 1 \times 8 + 1 \times 4 + 1 \times 2 + 0 = 8 + 4 + 2 = 14 \quad$ ◆

EXAMPLE 2

Change 10111_{two} to base 10.

Solution

$$10111_{two} = 1 \times 16 + 0 \times 8 + 1 \times 4 + 1 \times 2 + 1 \times 1$$
$$= 16 + 4 + 2 + 1 = 23 \qquad ◆$$

To change a number from decimal to binary, it is necessary to find how many units (1 if odd, 0 if even), how many twos, how many fours, and so on are contained in that given number. This can be accomplished by repeated division by 2.

EXAMPLE 3

Change 47 to a binary numeral.

Solution

$$
\begin{array}{r}
0 \quad \text{r. 1} \\
2\overline{)\,1} \quad \text{r. 0} \\
2\overline{)\,2} \quad \text{r. 1} \\
2\overline{)\,5} \quad \text{r. 1} \\
2\overline{)11} \quad \text{r. 1} \\
2\overline{)23} \quad \text{r. 1} \\
2\overline{)47}
\end{array}
$$

There is one thirty-two.
There are zero sixteens.
There is one eight.
There is one four.
There is one two.
There is one unit.

If you read down the remainders, you obtain the binary numeral 101111. *Check:* 101111 means

$$1 \times 32 + 0 \times 16 + 1 \times 8 + 1 \times 4 + 1 \times 2 + 1 \times 1 = 47 \qquad ◆$$

ASCII Code

In the computer, each switch, called a **bit** (from <u>bi</u>nary dig<u>it</u>), represents a 1 or 0, depending on whether it is on or off. A group of eight of these switches is lined up to form a **byte,** which can represent any number from 0 to $11111111_{two} = 255_{ten}.$

TODAY ONLY

MARKED DOWN TO

SCHOCHET

Of course, when we use computers we need to represent more than numerals. There must be a representation for every letter and symbol that we wish to use. For example, we will need to represent statements such as

$$x = 3.45 - y$$

The code that has been most commonly used is the American Standard Code for Information Interchange, developed in 1964 and is called **ASCII** (pronounced "ask-key") **code.** A partial list of this code is shown in Table 3.8.

From this table you can see that the word *cat* would be represented by the ASCII code 99 97 116. In the machine, this would be represented by binary numerals, each one in a different byte. Thus, cat is represented as shown:

Symbol	ASCII Code	Binary	Byte
c	99	01100011	off on on off off off on on
a	97	01100001	off on on off off off off on
t	116	01110100	off on on on off on off off

The computer is able to store large amounts of information by having a large number of bytes, each one with an *address:*

ADDRESS **1**	INFORMATION
ADDRESS **2**	INFORMATION
ADDRESS **3**	IS
ADDRESS **4**	STORED
ADDRESS **5**	IN
ADDRESS **6**	EACH
ADDRESS **7**	NUMBERED
ADDRESS **8**	LOCATION
ADDRESS **9**	INFORMATION
ADDRESS **10**	INFORMATION

Table 3.8 Partial Listing of ASCII Code. The entire ASCII code has 127 symbols.

ASCII Code	Symbol	ASCII Code	Symbol	ASCII Code	Symbol
32	Space	56	8	91	[
33	!	57	9	92	\
34	"	58	:	93	]
35	#	59	;	94	^
⋮	⋮	⋮	⋮	95	–
48	0	65	A	96	`
49	1	66	B	97	a
50	2	67	C	98	b
51	3	⋮	⋮	⋮	⋮
52	4	87	W	120	x
53	5	88	X	121	y
54	6	89	Y	122	z
55	7	90	Z	⋮	⋮

A computer has billions of such addresses. For example, suppose we want to store the information, *3 cats.* We need six locations or addresses:

Address	Information	
1000	00110011	← **Code for 3**
1001	00100000	← **Code for space**
1010	01100011	← **Code for *c***
1011	01100001	← **Code for *a***
1100	01110100	← **Code for *t***
1101	01110011	← **Code for *s***

During the process of solving a problem, the storage unit will contain not only the set of instructions that specify the sequence of operations required to complete the problem, but also the input data. Notice that an address number has nothing to do with the contents of that address. The early computers stored these individual bits with switches, vacuum tubes, or cathode ray tubes, but computers today use much more sophisticated means for storing bits of information. These advances in storing information have enabled computers to use codes that use many more than eight bits. For example, a location might look like the following:

Address 00101   0001111100011011011000000001101

Binary Arithmetic

The binary numeration system has only two symbols, 0 and 1, which makes counting in the system simple. Arithmetic is particularly easy in the binary system, since the only arithmetic "facts" one needs are the following:

Addition

+	0	1
0	0	1
1	1	10

Multiplication

×	0	1
0	0	0
1	0	1

You simply must remember when adding $1 + 1$ in binary that you put down 0 and "carry" the 1, as shown in the following example.

EXAMPLE 4

Carry out the following operations:

a. $11_{two} + 101_{two}$

b. $11_{two} \times 101_{two}$

Solution

a.
$$\begin{array}{r} 11_{two} \\ + \, 101_{two} \\ \hline 1000_{two} \end{array}$$
Note $1 + 1 = 10$, carry 1;
$1 + 1 = 10$, carry 1 again;
$1 + 1 = 10$ for the last time

b.
$$\begin{array}{r} 11_{two} \\ \times \, 101_{two} \\ \hline 11 \\ 00 \\ 11 \\ \hline 1111_{two} \end{array}$$

◆

Problem Set 3.4

LEVEL 1

1. **IN YOUR OWN WORDS** Describe Figure 3.4 on page 119.

2. **IN YOUR OWN WORDS** Describe a procedure for converting a decimal number into binary representation.

What decimal number is represented by the lightbulbs shown in Problems 3–6?

3.

4.

5.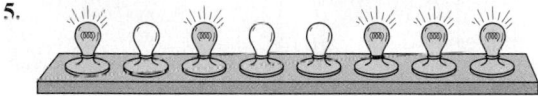

6.

Write each number given in Problems 7–18 as a decimal numeral.

7. 1101_{two} 8. 1001_{two} 9. 1011_{two}

10. 1111_{two} 11. 11101_{two} 12. 10111_{two}

13. 11011_{two} 14. 11111_{two} 15. 1100011_{two}

16. 1110111_{two} 17. 10111000_{two} 18. 11111111_{two}

Write each number given in Problems 19–30 as a binary numeral.

19. 13   20. 15   21. 35   22. 46

23. 51   24. 63   25. 64   26. 256

27. 128   28. 615   29. 795   30. 803

LEVEL 2

Write the ASCII codes for the capital letters in Problems 31–34.

31. DO   32. PRINT   33. END   34. SAVE

Write the words for the ASCII codes given in Problems 35–38.

35. 72 65 86 69   36. 70 85 78

37. 83 84 85 68 89   38. 72 65 82 68

Perform the indicated operations in Problems 39–46.

39. 11_{two}
    $+ 10_{two}$

40. 1011_{two}
    $+ 111_{two}$

41. 110_{two}
    $+ 111_{two}$

42. 1101_{two}
    $+ 1100_{two}$

43. 101_{two}
    $- 11_{two}$

44. 11011_{two}
    $- 10110_{two}$

45. 111_{two}
    $\times 101_{two}$

46. 10110_{two}
    $\times 1001_{two}$

LEVEL 3

47. **IN YOUR OWN WORDS** There is a slogan among computer operators: *GIGO (garbage in, garbage out)*. This phrase reflects the fact that a computer will do exactly what it is told to do. Consider the following humorous story.

    A very large computer system has been created for the military. It was built and staffed by the best computer people in the country. "The system is now ready to answer questions," said the spokesperson for the project. A four-star general bit off the end of a cigar, looked whimsically at his comrades, and said, "Ask the machine if there will be war or peace." The machine replied: YES. "Yes *what*?" bellowed the general. The operator typed in this question, and the machine answered: YES, SIR!

    In terms of what you know about logic, why did the computer answer the first question YES correctly?

48. **IN YOUR OWN WORDS** The **octal numeration system** refers to the base eight system, which uses the symbols 0, 1, 2, 3, 4, 5, 6, 7. Describe a process for converting octal numbers to decimal and decimal to octal.

Computer programmers often use octal numerals (see Problem 48) instead of binary numerals because it is easy to convert from binary to octal directly. Table 3.9 shows this conversion.

Table 3.9 Octal Binary Equivalence

Octal	Binary
0	000
1	001
2	010
3	011
4	100
5	101
6	110
7	111

A binary number 1101110111 is separated into three-digit groupings by starting with the right end of the number and supplying leading zeros at the left if necessary: 001 101 110 111. *The binary groups are then replaced by their octal equivalents:*

$$001_{two} = 1_{eight} \qquad 101_{two} = 5_{eight}$$
$$110_{two} = 6_{eight} \qquad 111_{two} = 7_{eight}$$

and the binary number is converted to its octal equivalent: 1567. *Conversely, an octal number can be expanded to a binary number using the same table of equivalents:*

$$5307_{eight} = 101\ 011\ 000\ 111_{two}$$

Use this information for Problems 49–59.

Convert the numbers in Problems 49–53 to the binary system.

49. a. 5_{eight} **b.** 6_{eight}

50. a. 14_{eight} **b.** 56_{eight}

51. a. 167_{eight} **b.** 624_{eight}

52. a. 7045_{eight} **b.** 3062_{eight}

53. a. 5700_{eight} **b.** 04320_{eight}

Convert the numbers in Problems 54–59 to the octal system.

54. a. 101_{two} **b.** 100_{two}

55. a. 011_{two} **b.** 001_{two}

56. $000\ 000\ 111\ 111\ 101\ 000_{two}$

57. $100\ 000\ 000\ 101\ 110\ 111_{two}$

58. $111\ 111\ 011\ 101\ 010\ 001_{two}$

59. $100\ 101\ 011\ 001\ 010\ 111_{two}$

PROBLEM SOLVING

60. Magic number cards You will need to use a pair of scissors for this problem. Construct the five cards shown at the right.

Cut out those parts of the cards that are white. Ask someone to select a number between 1 and 31. Hold the cards face up on top of one another, and ask the person if his or her number is on the card. If the answer is yes, place the card on the table face up with the word *yes* at the upper left corner. If the answer is no, place the card face up with the word *no* at the upper left corner. Repeat this procedure with each card, placing the cards on top of one another on the table. After you have gone through the five cards, turn the entire stack over. The chosen number will be seen. Explain why this trick works.

Card 1

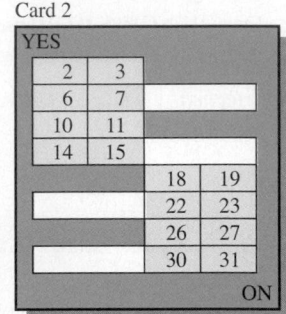

Card 2

Card 3

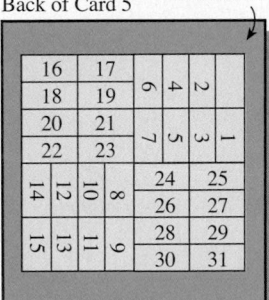

Card 4

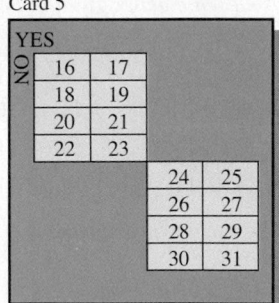

"YES" is on back of this corner

Back of Card 5

Card 5

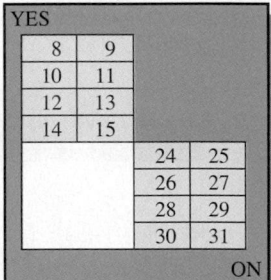

3.5 | HISTORY OF CALCULATING DEVICES

One of the biggest surprises of our time has been the impact of the personal computer on everyone's life. Few people foresaw that computers would jump the boundaries of scientific and engineering communities and create a revolution in our way of life in the last half of the 20th century. No one expected to see a computer sitting on a desk, much less on desks in every type of business, large and small, and even in our homes.

"It says, 'don't fold, spindle,
or mutilate.'"

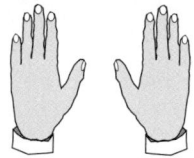

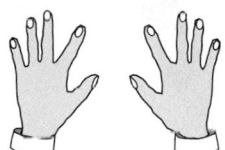

Today no one is ready to face the world without some knowledge of computers. What has created this change in our lives is not only the advances in technology that have made computers small and affordable, but the tremendous imagination shown in developing ways to use them.

In this section we will consider the historical achievements leading to the easy availability of calculators and computers.

First Calculating Tool

The first "device" for arithmetic computations is finger counting. It has the advantages of low cost and instant availability. You are familiar with addition and subtraction on your fingers, but here is a method for multiplication by 9. Place both hands as shown in the margin. To multiply 4×9, simply bend the fourth finger from the left as shown:

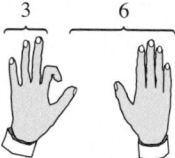

The answer is read as 36, the bent finger serving to distinguish between the first and second digits in the answer. What about 36×9? Separate the third and fourth fingers from the left (as shown in the margin), since 36 has 3 tens. Next, bend the sixth finger from the left, since 36 has 6 units. Now the answer can be read directly from the fingers:

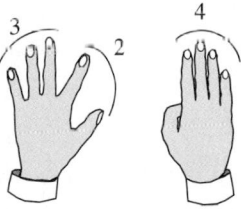

The answer is 324.

Aristophanes devised a complicated finger-calculating system in about 500 B.C., but it was very difficult to learn. Figure 3.6 shows an illustration from a manual published about two thousand years later, in 1520.

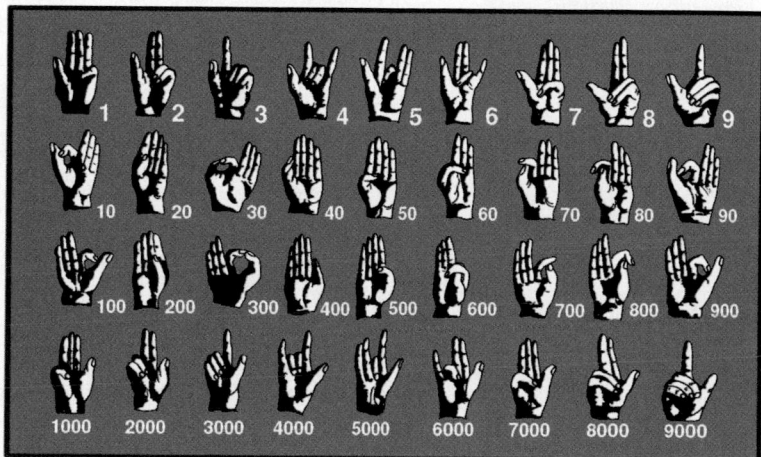

Figure 3.6 **Finger calculation was important in the Middle Ages.**

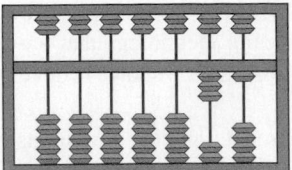

Figure 3.7 **Abacus: The number shown is 31.**

Early Calculating Devices

Numbers can be represented by stones, slip knots, or beads on a string. These devices evolved into the abacus, as shown in Figure 3.7. Abaci (plural of abacus) were used thousands of years ago, and are still used today. In the hands of an expert, they even rival calculators for speed in performing certain calculations. An abacus consists of rods that contain sliding beads, four or five in the lower section, and two in the upper that equal one in the lower section of the next higher denomination. The abacus is useful today for teaching mathematics to youngsters. One can actually see the "carry" in addition and the "borrowing" in subtraction.

In the early 1600s, John Napier invented a device similar to a multiplication table with movable parts, as shown in Figure 3.8. These rods are known as *Napier's rods* or *Napier's bones*.

Figure 3.8 **Napier's rods (1617)**

A device used for many years was a *slide rule* (see Figure 3.9), which was also invented by Napier. The answers given by a slide rule are only visual approximations and do not have the precision that is often required.

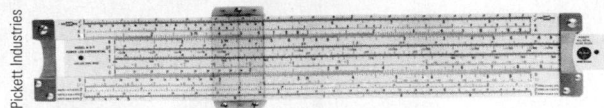

Figure 3.9 **Slide rule**

Mechanical Calculators

The 17th century saw the beginnings of calculating machines. When Pascal was 19, he began to develop a machine to add long columns of figures (see Figure 3.10). He built several versions, and since all proved to be unreliable, he considered this project a failure; but the machine introduced basic principles that are used in modern calculators.

Figure 3.10 **Pascal's calculator (1642)**

The next advance in mechanical calculators came from Germany in about 1672, when the great mathematician Gottfried Leibniz studied Pascal's calculators, improved them, and drew up plans for a mechanical calculator. In 1695 a machine was finally built (see Figure 3.11), but this calculator also proved to be unreliable.

Figure 3.11 **Leibniz' reckoning machine (1695)**

In the 19th century, an eccentric Englishman, Charles Babbage, developed plans for a grandiose calculating machine, called a "difference engine," with thousands of gears, ratchets, and counters (see Figure 3.12).

Figure 3.12 **Babbage's calculating machines**

Four years later in 1826, even though Babbage had still not built his difference engine, he began an even more elaborate project—the building of what he called an "analytic engine." This machine was capable of an accuracy to 20 decimal places, but it, too, could not be built because the technical knowledge to build it was not far enough advanced. Much later, IBM built both the difference and analytic engines based on Babbage's design, but using modern technology, and they work perfectly.

Hand-Held Calculators

In the past few years pocket calculators have been one of the fastest growing items in the United States. There are probably two reasons for this increase in popularity. Most people (including mathematicians!) don't like to do arithmetic, and a good calculator is very inexpensive. It is assumed that you have access to a calculator to use with this book, and the basics of their use were introduced in Chapter 1.

First Computers

The devices discussed thus far in this section would all be classified as calculators. With some of the new programmable calculators, the distinction between a calculator and a computer is less well defined than it was in the past. Stimulated by the need to

HISTORICAL NOTE

Charles Babbage **(1792–1871)** Babbage continued for 40 years trying to build his "engines." He was constantly seeking financial help to subsidize his projects. However, his eccentricities often got him into trouble. For example, he hated organ-grinders and other street musicians, and for this reason people ridiculed him. He was a no-nonsense man and would not tolerate inaccuracy. Had he been willing to compromise his ideas of perfection with the technical skill available, he might have developed a workable machine.

make ballistics calculations and to break secret codes during World War II, researchers made great advances in calculating machines. During the late 1930s and early 1940s John Vincent Atanasoff, a physics professor at Iowa State University, and his graduate student Clifford E. Berry built an electronic digital computer, but could not get it to work properly. The first working electronic computers were invented by J. Presper Eckert and John W. Mauchly at the University of Pennsylvania. All computers now in use derive from the original work they did between 1942 and 1946. They built the first fully electronic digital computer called ENIAC (Electronic Numerical Integrator and Calculator). The ENIAC filled a space 30 × 50 feet, weighed 30 tons, had 18,000 vacuum tubes, cost $487,000 to build, and used enough electricity to operate three 150-kilowatt radio stations. Vacuum tubes generated a lot of heat and the room where computers were installed had to be kept in a carefully controlled atmosphere; still, the tubes had a substantial failure rate, and in order to "program" the computer, many switches and wires had to be adjusted.

There is some controversy over who actually invented the first computer. In the early 1970s there was a lengthy court case over a patent dispute between Sperry Rand (who had acquired the ENIAC patent) and Honeywell, who represented those who originally worked on the ENIAC project. The judge in that case ruled that "between 1937 and 1942, Atanasoff . . . developed and built an automatic electronic digital computer for solving large systems of linear equations . . . . Eckert and Mauchly did not themselves invent the automatic electronic digital computer, but instead derived that subject matter from one Dr. John Vincent Atanasoff." On the other hand, others believe that a court of law is not the place for deciding questions about the history of science and give Eckert and Mauchly the honor of inventing the first computer. In 1980, the Association for Computing Machinery honored them as founders of the computer industry.

The UNIVAC I, built in 1951 by the builders of the ENIAC, became the first commercially available computer. Unlike the ENIAC, it could handle alphabetic data as well as numeric data. The invention of the transistor in 1947 and solid-state devices in the 1950s provided the technology for smaller, faster, more reliable machines.

Present-Day Computers

In 1958, Seymour Cray developed the first **supercomputer,** sometimes called the *Cray* computer, which could handle at least 10 million instructions per second. It is used in major scientific research and military defense installations. The newest version of this computer, the X-1, operates a million times faster than the first supercomputer. In 2002, the world's fastest computer, "The Earth Simulator," was put into service in Tokyo, Japan (see Figure 3.13). It is reported that it can perform at a speed in excess of 40TFLOPS (40 trillion operations per second).

Throughout the 1960s and early 1970s, computers continued to become faster and more powerful; they became a part of the business world. The "user" often never saw the machines. A "job" would be submitted to be "batch processed" by someone trained to run the computer. The notion of "time sharing" developed, so that the computer could handle more than one job, apparently simultaneously, by switching quickly from one job to another. Using a computer at this time was often frustrating; an incomplete job would be returned because of an error and the user would have to resubmit the job. It often took days to complete the project.

The large computers, known as *mainframes,* were followed by the **minicomputers,** which took up less than 3 cubic feet of space and no longer required a controlled atmosphere. These were still used by many people at the same time, though often directly through the use of terminals.

Keep in mind some of the deception that goes on in the name of technology. People sometimes think that "if a calculator or a computer did it, then it must be correct." This is false, and the fact that a "machine" is involved has no effect on the validity of the results. Have you ever been told that "The computer requires...," or "There is nothing we can do; it is done by computer," or "The computer won't permit it." What they really mean is they don't want to instruct the *programmer* to do it.

© AP/Wide World Photos

Figure 3.13 **The "Earth Simulator" supercomputer**

The term **personal computer** was first used by Steward Brand in *The Whole Earth Catalog* (1968), which, oddly enough, was before any such computers existed. In 1976, Steven Jobs and Stephen Wozniak built the Apple, the first personal computer to be commercially successful. This computer proved extremely popular. Small businesses could afford to purchase these machines that, with a printer attached, took up only about twice the space of a typewriter. Technology "buffs" could purchase their own computers. This computer was designed so that innovations that would improve or enlarge the scope of performance of the machine could be added with relatively little difficulty. Owners could open up their machines and install new devices. One such device, the **mouse,** was introduced to the world by Douglas Engelbart in 1968. This increased contact between the user and the machine produced an atmosphere of tremendous creativity and innovation. In many cases, individual owners brought their own machines to work to introduce their superiors to the potential usefulness of the personal computer.

In the fast-moving technology of the computer world, the invention of language to describe it changes as fast as teenage slang. Today there are battery-powered **laptops,** weighing about 4 pounds, which are more powerful than the huge ENIAC. In 1989 a pocket computer weighing only 1 pound was introduced. The most recent trend has been to link together several personal computers so that they can share software and different users can easily access the same documents. These are called LANs, or **local area networks.** The most widely known networks are the **Internet,** which was first described by Paul Baran of the RAND Corp. in 1962, and the **World Wide Web** (www), developed by Tim Berners-Lee based on Baran's ideas and released in 1992. The growth of the Internet and the World Wide Web could not have been predicted. Figure 3.14 (page 130) shows some of the significant events in the growth of the Internet from 1962 to the present. In 1962 there were 4 hosts, and in 1992 there were 727,000. By January 1, 1996, there were almost 10 million. In addition to educational access through most schools, you can access the Internet through services such as MSN, America Online, and Earthlink.

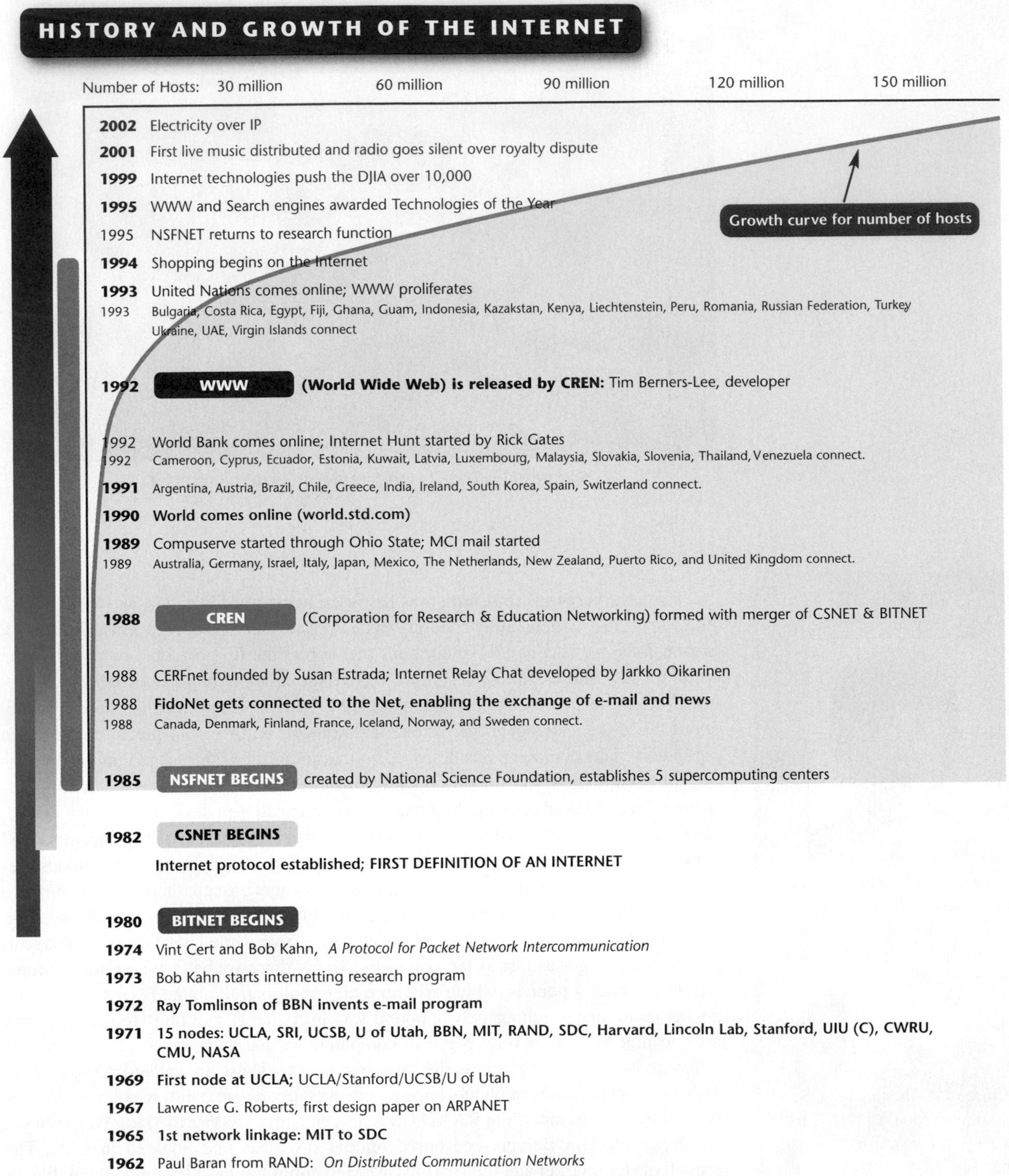

HISTORY AND GROWTH OF THE INTERNET

Number of Hosts: 30 million   60 million   90 million   120 million   150 million

2002 Electricity over IP

2001 First live music distributed and radio goes silent over royalty dispute

1999 Internet technologies push the DJIA over 10,000

1995 WWW and Search engines awarded Technologies of the Year

1995 NSFNET returns to research function

1994 Shopping begins on the Internet

1993 United Nations comes online; WWW proliferates

1993 Bulgaria, Costa Rica, Egypt, Fiji, Ghana, Guam, Indonesia, Kazakstan, Kenya, Liechtenstein, Peru, Romania, Russian Federation, Turkey, Ukraine, UAE, Virgin Islands connect

1992   **WWW**   **(World Wide Web) is released by CREN:** Tim Berners-Lee, developer

1992 World Bank comes online; Internet Hunt started by Rick Gates

1992 Cameroon, Cyprus, Ecuador, Estonia, Kuwait, Latvia, Luxembourg, Malaysia, Slovakia, Slovenia, Thailand, Venezuela connect.

1991 Argentina, Austria, Brazil, Chile, Greece, India, Ireland, South Korea, Spain, Switzerland connect.

1990 **World comes online (world.std.com)**

1989 Compuserve started through Ohio State; MCI mail started

1989 Australia, Germany, Israel, Italy, Japan, Mexico, The Netherlands, New Zealand, Puerto Rico, and United Kingdom connect.

1988   **CREN**   (Corporation for Research & Education Networking) formed with merger of CSNET & BITNET

1988 CERFnet founded by Susan Estrada; Internet Relay Chat developed by Jarkko Oikarinen

1988 **FidoNet gets connected to the Net, enabling the exchange of e-mail and news**

1988 Canada, Denmark, Finland, France, Iceland, Norway, and Sweden connect.

1985   **NSFNET BEGINS**   created by National Science Foundation, establishes 5 supercomputing centers

1982   **CSNET BEGINS**

Internet protocol established; FIRST DEFINITION OF AN INTERNET

1980   **BITNET BEGINS**

1974 Vint Cert and Bob Kahn, *A Protocol for Packet Network Intercommunication*

1973 Bob Kahn starts internetting research program

1972 Ray Tomlinson of BBN invents e-mail program

1971 15 nodes: UCLA, SRI, UCSB, U of Utah, BBN, MIT, RAND, SDC, Harvard, Lincoln Lab, Stanford, UIU (C), CWRU, CMU, NASA

1969 First node at UCLA; UCLA/Stanford/UCSB/U of Utah

1967 Lawrence G. Roberts, first design paper on ARPANET

1965 1st network linkage: MIT to SDC

1962 Paul Baran from RAND: *On Distributed Communication Networks*

Growth curve for number of hosts

Figure 3.14 **The growth and history of the Internet**

Computer Hardware

The machine, or computer itself, is referred to as **hardware.** A personal computer might sell for under $1,000, but many extras, called **peripherals,** can increase the price by several thousand dollars. In this section we will familiarize you with the various components of a typical personal computer system that you might find in a home, classroom, or office. We will also discuss the link between the electronic signals of the computer and the human mind that enables us to harness the power and speed of the computer.

The basic parts of a computer are shown in Figure 3.15.

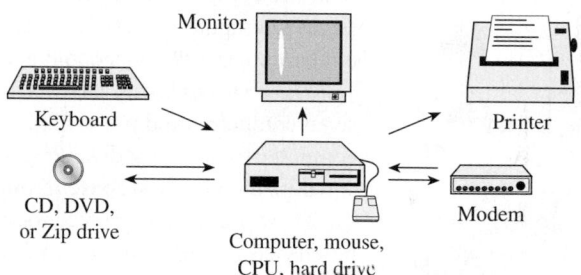

Figure 3.15 **Input and output with a computer**

We will focus our attention in this section on the terminology you will need to know to purchase a computer. The central unit of a computer contains a small microchip, called the *central processing unit*, where all processing takes place, and the computer memory, where the information is stored while the computer is completing a given task. The microchip and the unit that contains this chip are collectively referred to as the CPU. In 1972, Intel developed the first 8-bit microprocessor called the 8008, replaced successively by the 8080, 80386, 80486, Pentium, 80686, and Pentium II, III, and IV chips. In 1979, Motorola introduced the 68000 chip, followed by the 68020 and 68030 chips, both used by the Macintosh computer.

Although you do not need to understand electronic circuitry, a simple conceptual model of the inside of a computer can be very useful. The memory of a computer can be thought of as rows and rows of little storage boxes, each one with an address—yes, an address, just like your address. These boxes are of two types: **ROM** (read-only memory) and **RAM** (random-access memory). ROM is very special: You can never change what is in it and it is not erased when the computer is turned off. It is programmed by the manufacturer to contain routines for certain processes that the computer must always use. The CPU can fetch information from it, but cannot send new results back. It is well protected because it is at the heart of the machine's capabilities. On the other hand, programs and data are stored in RAM while the computer is completing a task. If the power to the computer is interrupted for any time, however brief, everything in RAM is lost forever.

Your computer will also have memory. When referring to memory, we used to talk in terms of a thousand bytes, called a kilobyte (KB or K), or in terms of a thousand kilobytes, called a megabyte (MB or MEG). However, today, we refer to gigabytes (GB); one gigabyte is about one billion bytes. Your computer will come with 128, 256, or more MEG of RAM. You will also need a **hard drive,** which is used as additional memory to store the programs you purchase. Hard disks in sizes of 20, 40, 60 or more gigabytes are commonly installed in microcomputers.

To use a computer, you will need **input** and **output devices.** A **keyboard** and a *mouse* are the most commonly used input devices. The most common output devices are a **monitor** and a **printer.** One of the monitor's main functions is to enable you to keep track of what is going on in the computer. The letters or pictures seen on a monitor are displayed by little dots called **pixels,** just as they are on a television. As the number of dots is increased, we say that the **resolution** of the monitor increases. The better the resolution, the easier on your eyes.

Computer Software

Communication with computers has become more sophisticated, and when a computer is purchased it often comes with several *programs* that allow the user to communicate with the computer. A **computer program** is a set of step-by-step directions that instruct a computer how to carry out a certain task. Programs that allow the user to carry out particular tasks on a computer are referred to as **software.**

The business world has welcomed the advances made possible by the use of personal computers and powerful programs, known as **software packages,** that enable the computer to do many diverse tasks. The most widely used computer applications are **word processing, database management,** and **spreadsheets.**

There are several sources for obtaining software. *Selfware* refers to software that is purchased at a store, *shareware* is try-before-you-buy software. Some software is available on the Internet free of charge (*freeware* or *public domain software*). The most widely used software is from Microsoft, which owns *Windows, Word, Internet Explorer,* and *Excel.* Bill Gates (1956-   ) is the cofounder of Microsoft and is known as the world's wealthiest person. In 1975, Ed Roberts built the first personal computer (the Altair 8800) in Albuquerque, New Mexico; it was featured on the cover of *Popular Electronics,* and was named because the word Altair was used in an episode of *Star Trek.* Bill Gates and his partner, Paul Allen, were students at Harvard at the time, and traveled to New Mexico to look at this new machine. The story about the beginning of the huge Microsoft empire is told by Stephen Segaller in *Nerds 2.0.1:* "He [Bill Gates] kept postponing and postponing actually writing the code [for the Altair]. He said, 'I know how to write it, I have a design in my head, I'll get it done, don't worry about it, Paul.' Four days before he was due to go back to Harvard he checked into a hotel and he was incommunicado for three days. Bill came back three or four days later with this huge sheet of paper. He'd written 4K of code in three days, and typed the whole thing in, got it working, and went back to school, just barely. It was really one of the most amazing displays of programming I've ever seen."

Uses of Computers

Today, computers are used for a variety of purposes (see Figure 3.16). **Data processing** involves large collections of data on which relatively few calculations need to be made. For example, a credit card company could use the system to keep track of the balances of its customers' accounts. **Information retrieval** involves locating and displaying material from a description of its content. For example, a police department may want information about a suspect, or a realtor may want listings that meet the criteria of a client. **Pattern recognition** is the identification and classification of shapes, forms, or relationships. For example, a chess-playing program will examine the chessboard carefully to determine a good next move. The computer carries out these tasks by accepting information, performing mathematical and logical operations with the information, and then supplying the results of the operations as new information. Finally, a computer can be used for **simulation** of a real situation. For example, a prototype of a new airplane can be tested under various circumstances to determine its limitations.

By using a **modem,** installed internally or attached as a peripheral, and a program called a **communications package,** a computer can transfer information to and from

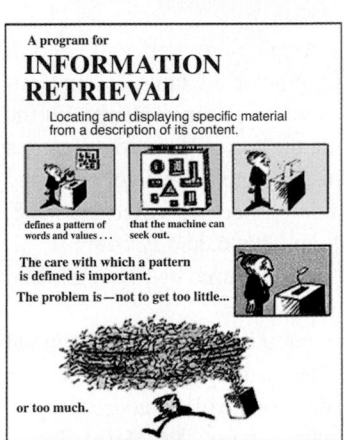

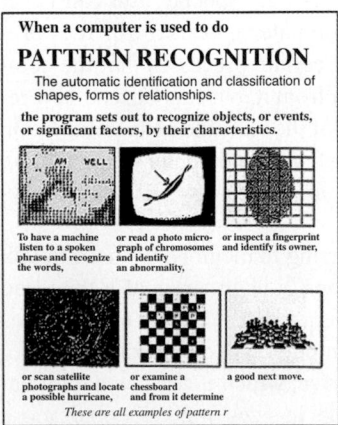

From *A Computer Glossary* © 1968 International Business Machines Corporation.

Figure 3.16 **Common computer uses**

other computers directly, via satellite, or over a phone line. If you send software or data from your computer to another, this is called **uploading;** if you get software or data from another computer, this is called **downloading.** A person or company may set up a computer specifically to be accessed by modem. The phone number is then made available to a private list of people or to the public. Some companies specialize in making databases available commercially. There is usually a charge to use these services based on the time you are connected, or the time you are **online.** Networks have been designed specifically to enable people to send messages to each other via computer. Such networks, known as *electronic mail* or **e-mail,** have become commonplace. Individuals or user groups set up computers accessible by modem to share information about a common interest. These are called **bulletin boards** or **chat rooms.** Often, though not always, these are free and the only charge to the user is the cost of the telephone call.

Misuses of Computers

Most companies and bulletin boards require a **password** or special procedures to be able to go online with their computers. A major problem for certain computer installations has been to make sure that it is impossible for the wrong person to get into their computer. There have been many instances where people have "broken into" computers just as a safecracker breaks into a safe. They have been clever enough to discover how to access the system. A few of these people have been caught and prosecuted. A security break of this nature can be very serious. By transferring funds in a bank computer, someone can steal money. By breaking into a military installation, a spy could steal secrets or commit sabotage. By breaking into a university's computers, one could change grades or records. **Computer abuse** includes *illegal use* (as described above, but also includes illegal copying of programs, called *pirating*) and abuse caused by *ignorance.* Ignorance of computers leads to the assumption that the output data are always correct or that the use of a computer is beyond one's comprehension.

With the tremendous advances in computer technology, financial investors and analysts can simulate various possible outcomes and select strategies that best suit their goals. Designers and engineers use computer-aided design, known as CAD. Using computer graphics, sculptors can model their work on a computer that will rotate the image so that the proposed work can be viewed from all sides. Computers have been used to help the handicapped with learning certain skills and by taking over certain functions. Computers can be used to "talk" for those without speech. For those without sight, there are artificial vision systems that translate a visual image to a tactile "image" that can be "seen" on their back. A person who has never seen a candle burn can learn the shape and dynamic quality of a flame.

Along with the advances in computer technology came new courses, and even new departments, in the colleges and universities. New specialties developed involving the design, use, and impact of computers on our society. The feats of the computer captured the imagination of some who thought that a computer could be designed to be as intelligent as a human mind and the field known as **artificial intelligence** was born. Many researchers worked on natural language translation with very high hopes initially, but the progress has been slow. Progress continues in "expert systems," programs that involve the sophisticated strategy of pushing the computer ever closer to human-like capabilities. The quest has produced powerful uses for computers.

What does the future hold? Earlier editions of this book attempted to make predictions of the future, and in every case those predictions became outdated during the life of the edition. Instead of making a prediction, we invite you to visit a computer store.

***Stephen Hawking* (1942–)**
Stephen Hawking, a physicist known the world over for his study of the origins of the universe, has had a degenerative disease for many years. It has left him unable to speak, although his mind remains as brilliant as ever. He communicates with his colleagues and family and even gives speeches using a computer with a voice synthesizer. There has recently been a movie featuring his life and his accomplishments, and in 1992 he even appeared in an episode of *Star Trek: The Next Generation.* He is one of the great thinkers of our time.

Problem Set 3.5

LEVEL 1

1. **IN YOUR OWN WORDS** Describe some of the computing devices that were used before the invention of the electronic computer.

2. **IN YOUR OWN WORDS** Distinguish between hardware and software.

3. **IN YOUR OWN WORDS** What does the CPU in a computer do?

4. **IN YOUR OWN WORDS** What are input and output devices?

5. **IN YOUR OWN WORDS** What is the difference between ROM and RAM?

6. **IN YOUR OWN WORDS** Give at least one example of each of the indicated computer functions: data processing, information retrieval, pattern recognition, and simulation.

7. **IN YOUR OWN WORDS** What are five common uses for a computer at home? Describe the purpose of each.

8. **IN YOUR OWN WORDS** What are five common uses for a computer in the office? Describe the purpose of each.

9. **IN YOUR OWN WORDS** It has been said that "computers influence our lives increasingly every year, and the trend will continue." Do you see this as a benefit or a detriment to humanity? Explain your reasons.

10. **IN YOUR OWN WORDS** A heated controversy rages about the possibility of a computer actually thinking. Do you believe that is possible? Do you think a computer can eventually be taught to be truly creative?

Use finger multiplication to do the calculations shown in Problems 11–14.

11. **a.** 3×9 **b.** 7×9 **c.** 6×9

12. **a.** 5×9 **b.** 8×9 **c.** 9×9

13. **a.** 27×9 **b.** 48×9 **c.** 56×9

14. **a.** 35×9 **b.** 47×9 **c.** 68×9

15. Arrange the development of the following computers in chronological order of their first appearance:

Altair, Cray, ENIAC, Apple, and UNIVAC

There were many names mentioned in this section. Give a brief description for each person named in Problems 16–33, telling how they were related to the development of computers.

16. Paul Allen
17. Aristophanes
18. John Atanasoff
19. Charles Babbage
20. Paul Baran
21. Clifford Berry
22. Steward Brand
23. Seymour Cray
24. J. Presper Eckert
25. Douglas Engelbart
26. Bill Gates
27. Steven Jobs
28. Tim Berners-Lee
29. Gottfried Leibniz
30. John Mauchly
31. John Napier
32. Blaise Pascal
33. Stephen Wozniak

LEVEL 2

Problems 34–45 list a specific task. Decide whether a computer should be used, and if so, is it because of the computer's speed, complicated computations, repetition, or some other reason?

34. Guiding a missile to its target

35. Controlling the docking of a spacecraft

36. Tabulating averages for a teacher's classes

37. Identifying the marital status indicated on tax returns reporting income more than $100,000

38. Sorting the mail according to zip code

39. Assembling automobiles on an assembly line

40. Turning on and off the lights in a public office building

41. Typing and printing a term paper

42. Teaching a person to solve quadratic equations

43. Teaching a person to play the piano

44. Teaching a person how to repair a flat tire

45. Compiling a weather report and forecast

Problems 46–50 are multiple-choice questions designed to get you to begin thinking about computer use and abuse.

46. As part of a patient monitoring system in a local hospital, the computer (a) constantly monitors the patient's pulse and respiration rate; (b) sounds an alarm if the rates fall outside a preset range; and (c) notifies the nearest nursing station if either or both of its systems fail to function properly. The computer's capabilities are suited to these tasks because the tasks call for:
   A. Speed
   B. Repetition
   C. Program modification
   D. Simulation
   E. Data processing

47. Computers are capable of organizing large amounts of data. Which of the following situations would call for a computer having that capability?
   A. Printing the names and addresses of all employees
   B. Turning lights on and off at specified times
   C. Running a statistical analysis of family incomes as taken from the 2000 census
   D. Monitoring the temperature and humidity in a large office building
   E. All of the above

48. Karlin Publishing purchased a program for its computer that would enable it to streamline the printing process and decrease the time it takes to print each page. The program was too fast for Karlin's printer, so the company hired Shannon Foley to eliminate the problem. The human function required by Shannon for this job was:
   A. Performing calculations
   B. Modifying the hardware
   C. Interpreting output
   D. Modifying the software
   E. None of the above

49. Melissa knew that if she flunked mathematics she would be ineligible to play in her team's last five basketball games. When she found out that one of her classmates had been able to change someone else's grade on the school computer, she had this same person change her math grade from F to C. This example of computer abuse was:
   A. Invasion of privacy
   B. Stealing funds
   C. Pirating software
   D. Falsifying information
   E. Stealing information

50. One example of computer abuse is categorized as assuming that computers are largely incomprehensible. Which of the following situations would indicate this abuse?
   A. A secretary refuses promotion because he will have to interact with the company's computer.
   B. A clerk inputs the wrong information into the computer to defraud the company.
   C. A student changes the grade of a classmate.
   D. A hacker (computer hobbyist) copies a copyrighted program for three of her friends.
   E. A credit card customer wants to receive his bill on the 25th of the month instead of on the 10th, but when he calls to have it changed he is told that there is nothing that can be done since that is the way the computer was programmed.

LEVEL 3

51. **IN YOUR OWN WORDS** There is a great deal of concern today about **invasion of privacy** by computer. With more and more information about all of us being kept in computerized databases, there is the increasing possibility that a computer may be used to invade our privacy. Discuss this issue.

52. **IN YOUR OWN WORDS** Have you ever been told that something cannot be changed because "that is the way the computer does it"? Discuss this issue.

53. **IN YOUR OWN WORDS** Have you ever been told that something is right because the computer did it? For example, suppose I have my computer print, THE VALUE OF π IS 3.141592. Is this statement necessarily correct? What confidence do you place on computer results? What confidence do you think any of us should place on computer results? Discuss.

54. **IN YOUR OWN WORDS** Definitions for "thinking" are given. *According to each of the given definitions,* answer the question, "Can computers think?"
   a. To remember
   b. To subject to the process of logical thought
   c. To form a mental picture of
   d. To perceive or recognize
   e. To have feeling or consideration for
   f. To create or devise
   g. To have the ability to learn

55. **IN YOUR OWN WORDS** What do you mean by "thinking" and "reasoning"? Try to formulate these ideas as clearly as possible, and then discuss the following question: "Can computers think or reason?"

56. **IN YOUR OWN WORDS** In his book *Future Shock,* Alvin Toffler divided humanity's time on earth into 800 lifetimes. The 800th lifetime, in which we now live, has produced more knowledge than the previous 799 combined, and this has been made possible by computers. How have computers made this possible? What impact has this had on our lives? Do people actually know more today because they have ready access to vast amounts of knowledge?

57. **IN YOUR OWN WORDS** In his article "Toward an Intelligence Beyond Man's" (*Time,* February 20, 1978), Robert Jastrow claimed that by the 1990s computer intelligence would match that of the human brain. He quoted Dartmouth President Emeritus John Kemeny as saying we would "see the ultimate relation between man and computer as a symbiotic union of two living species, each completely dependent on the other for survival." He called the computer a new form of life. Now, with the advantage of a historical perspective, do you think his predictions came to pass? Now that the 1990s have passed, do you agree or disagree with the hypothesis that a computer could be a "new form of life"?

PROBLEM SOLVING

58. **IN YOUR OWN WORDS** One of the things that can go wrong when we use a computer is due to *machine error.* Such errors can be caused by an electrical surge or a power failure. Talk to some users of computers and gather some anecdotal information about computer problems due to machine errors.

59. **IN YOUR OWN WORDS** It is possible that a computer can make a mistake due to a *programming error.* There are three types of programming errors: *syntax errors, run-time errors,* and *logic errors.* Do some research and write a paragraph about each of these types of programming errors.

60. **IN YOUR OWN WORDS** One category of possible computer errors includes *data errors* or *input errors.* What is meant by this type of error?

CHAPTER SUMMARY

"The availability of low-cost calculators, computers, and related new technology has already dramatically changed the nature of business, industry, government, sciences, and social sciences."

NCTM Standards

IMPORTANT TERMS

Addition principle [3.1]
Artificial intelligence [3.5]
ASCII code [3.4]
Base *b* numeration system [3.3]
Binary numeration system [3.3; 3.4]
Bit [3.4]
Bulletin board [3.5]
Byte [3.4]
Chat rooms [3.5]
Communications package [3.5]
Computer abuse [3.5]
Computer program [3.5]
Counting number [3.1]
Data processing [3.5]
Database manager [3.5]
Decimal numeration system [3.2]
Decimal point [3.2]
Download [3.5]
e-mail [3.5]
Expanded notation [3.2]
Hard drive [3.5]
Hardware [3.5]

Hindu–Arabic numerals [3.2]
Hundred [3.2]
Information retrieval [3.5]
Input [3.5]
Input device [3.5]
Internet [3.5]
Keyboard [3.5]
Laptop [3.5]
Minicomputers [3.5]
Modem [3.5]
Monitor [3.5]
Mouse [3.5]
Multiplication principle [3.1]
Network [3.5]
Number [3.1]
Numeral [3.1]
Numeration system [3.1]
Octal numeration system [3.3; 3.4]
Online [3.5]
Output [3.5]
Output device [3.5]
Password [3.5]

Pattern recognition [3.5]
Peripheral [3.5]
Personal computer [3.5]
Pixel [3.5]
Positional system [3.1]
Printer [3.5]
Program [3.5]
RAM [3.5]
Repetitive system [3.1]
Resolution [3.5]
ROM [3.5]
Simple grouping system [3.1]
Simulation [3.5]
Software [3.5]
Software package [3.5]
Spreadsheet [3.5]
Subtraction principle [3.1]
Supercomputer [3.5]
Ten [3.2]
Upload [3.5]
Word processing [3.5]
World Wide Web [3.5]

TYPES OF PROBLEMS

Know the principal properties, advantages, and disadvantages of the Egyptian, Babylonian, and Roman numeration systems. [3.1]

Write decimal numerals for numbers written in the Egyptian, Babylonian, and Roman numeration systems. [3.1]

Write decimal numerals in the Egyptian, Babylonian, and Roman numeration systems. [3.1]

Perform addition and subtraction in the Egyptian and Babylonian numeration systems. [3.1]

Give the meaning of a particular numeral in the Hindu–Arabic numeration system. [3.2]

Write the decimal representation for a number written in expanded notation. [3.2]

Write a decimal numeral in expanded notation. [3.2]

Use an abacus to illustrate the meaning of a decimal number. [3.2]

Count objects in various number bases. [3.3]

Write numbers in various bases in expanded notation. [3.3]

Change numbers from base b to base 10. [3.3]
Change numbers from base 10 to base b. [3.3]
Solve applied problems by using number bases. [3.3]
Change a binary numeral to a decimal numeral. [3.4]
Use the binary numeration system to represent a number. [3.4]
Use the binary numeration system and the ASCII code to represent a word. [3.4]
Add, subtract, and multiply using binary numeration systems. [3.4]
Convert from binary to octal and from octal to binary numeration systems. [3.4]
Know some principal events in the history of computers. [3.5]
Know some principal events in the history of the Internet. [3.5]
Know the principal uses for a computer. [3.5]
Know the principal computer abuses. [3.5]

CHAPTER 3 REVIEW QUESTIONS

1. **IN YOUR OWN WORDS** What do we mean when we say that a numeration system is positional? Give examples.
2. **IN YOUR OWN WORDS** Is addition easier in a positional system or in a grouping system? Discuss and show examples.
3. **IN YOUR OWN WORDS** What are some of the characteristics of the Hindu–Arabic numeration system?
4. **IN YOUR OWN WORDS** Briefly discuss some of the events leading up to the invention of the computer.
5. **IN YOUR OWN WORDS** Discuss some computer abuses.
6. **IN YOUR OWN WORDS** Briefly describe each of the given computer terms.
   a. hardware   **b.** software   **c.** word processing   **d.** modem
   e. e-mail   **f.** RAM   **g.** computer bulletin board

Write the numbers given in Problems 7–10 in expanded notation.

7. one billion   8. 436.20001   9. 523_{eight} 10. 1001110_{two}
11. Write $4 \times 10^6 + 2 \times 10^4 + 5 \times 10^0 + 6 \times 10^{-1} + 2 \times 10^{-2}$ in decimal notation.

Write the numbers in Problems 12–15 in base ten.

12. 11101_{two} 13. 1111011_{two} 14. 122_{three} 15. 821_{twelve}

Write the numbers in Problems 16–18 in base two.

16. 12   17. 52   18. 2003   19. one million
20. **a.** Write 1,331 in base twelve.
    b. Write 100 in base five.

GROUP RESEARCH

Working in small groups is typical of most work environments, and learning to work with others to communicate specific ideas is an important skill. Work with three or four other students to submit a single report based on each of the following questions.

G8. Invent an original numeration system.

G9. Organize a debate. One side represents the algorists and the other side the abacists. The year is 1400. Debate the merits of the Roman numeration system and the Hindu–Arabic numeration system.

> **Reference:** Barbara E. Reynolds, "The Algorists vs. The Abacists: An Ancient Controversy on the Use of Calculators," *The College Mathematics Journal,* Vol. 24, No. 3, May 1993, pp. 218–223. Includes additional references.

G10. Organize a debate. The issue: "Resolved: Computers can think."

G11. In a now famous paper, Alan Turing asked, "What would we ask a computer to do before we would say that it could think?" In the 1950s Turing devised a test for "thinking" that is now known as the **turing test.** Dr. Hugh Loebner, a New York philanthropist, has offered $100,000 for the first machine that fools a judge into thinking it is a person. In 1991, the Computer Museum in Boston held a contest in which 10 judges at the museum held conversations on terminals with eight respondents around the world, including six computers and two humans. The conversations of about 15 minutes each were limited to particular subjects, such as wine, fishing, clothing, and Shakespeare, but in a true turing test, the questions could involve any topic. Work as a group to decide the questions you would ask. Do you think a computer will ever be able to pass the test?

> **References:** Betsy Carpenter, "Will Machines Ever Think?" *U.S. News & World Report,* October 17, 1988, pp. 64–65.
>
> Stanley Wellborn, "Machines That Think," *U.S. News & World Report,* December 5, 1983, pp. 59–62.

G12. Construct an exhibit on ancient computing methods. Some suggestions for your exhibit are charts of sample computations by ancient methods, pebbles, tally sticks, tally marks in sand, Roman number computations, abaci, Napier's bones, and old computing devices. You should consider answering the following questions as part of your exhibit: How do you multiply with Roman numerals? What is the scratch system? What is the lattice method of computation? What changes in our methods of long multiplication and long division have taken place over the years? How did the old computing machines work? Who invented the slide rule?

INDIVIDUAL RESEARCH PROBLEMS

Learning to use sources outside your classroom and textbook is an important skill, and here are some ideas for extending some of the ideas in this chapter. You can find references to these projects in a library or at **www.mathnature.com.**

PROJECT 3.1 What do the following people have in common?
    Eamon de Valera, Prime minister and president of the Republic of Ireland
    Tom Lehrer, songwriter-parodist
    Edmund Husserl, the "Father of Phenomenology"
    Frank Ryan, quarterback for the Cleveland Browns

PROJECT 3.2 Write a paper discussing the Egyptian method of multiplication.

PROJECT 3.3 What are some of the significant events in the development of mathematics? Who are some of the famous people who have contributed to mathematical knowledge?

PROJECT 3.4 Is it possible to have a numeration system with a base that is negative?

PROJECT 3.5 "I became operational at the HAL Plant in Urbana, Ill., on January 12, 1997," the computer HAL declares in Arthur C. Clarke's 1968 novel, *2001: A Space Odyssey.* Now that time has passed and many advances have been made in computer technology between 1968 and today, write a paper showing the similarities and differences between HAL and the computers of today.

PROJECT 3.6 Software bugs (for example, the Y2K Millennium Bug) can have devastating effects. Write a paper on some famous software bugs, and some of the problems that they have caused.

PROJECT 3.7 Build a working model of Napier's rods.

PROJECT 3.8 Write a paper and prepare a classroom demonstration on the use of an abacus. Build your own device as a project.

PROJECT 3.9 Write a paper regarding the invention of the first electronic computer.

PROJECT 3.10 Visit a computer store, talk to a salesperson about the available computers, and then write a paper on your experiences.

PROJECT 3.11 Write a history of the hand-held calculator.

PROJECT 3.12 Find out what local, state, and federal governments have stored in their computers about you and your family. Find out what you can see and what others can see. This will provide you with an interesting intellectual journey, if you wish to take it.

4 The Nature of Numbers

> In most sciences one generation tears down what another has built, and what one has established another undoes. In mathematics alone each generation builds a new story to the old structure.
>
> HERMAN HANKEL

CONTENTS

OVERVIEW

Before we can do mathematical work, we need to have some building blocks for our journey. Those building blocks are sets of numbers. We assume that you know about some of these sets of numbers, but to form a common basis for the rest of the textbook, we will discuss some sets of numbers and their properties in this chapter.

The first set of numbers we encounter as children is the set of counting numbers. As we build more and more complex sets of numbers in this chapter, we move from the counting numbers to the integers (which include the counting numbers, zero, and their opposites), to the fractions, which we characterize as numbers whose decimal representations terminate or repeat. Even though this is a very useful set of numbers, there are certain applications (finding the length of a diagonal of a square, for example) that require numbers that cannot be written as terminating or repeating decimals. This set is called the set of irrational numbers.

IMPORTANT IDEAS

Properties of numbers: closure property [4.1]; commutative property [4.1]; associative property [4.1]; distributive property [4.1]; identity [4.6]; inverse [4.6]; groups [4.6]; fields [4.6]

Relationships among the following sets of numbers: natural numbers [4.1]; whole numbers [4.3]; integers [4.3]; rational numbers [4.4]; irrational numbers [4.5]; real numbers [4.6]

Meanings of fundamental operations: addition, multiplication [4.1]; subtraction [4.1]; division [4.2]; know the difference between a square root and an irrational number [4.5]

Rules of divisibility [4.2]

Operations among the following sets of numbers: integers [4.3]; rationals [4.4]; irrationals [4.5]; reals [4.6]

Least number of divisors [4.2]

Fundamental theorem of arithmetic [4.2]

Fundamental property of fractions [4.4]

Pythagorean theorem [4.5]

Elementary operations [4.7]

BOOK REPORTS

Write a 500-word report on one of these books:
Estimation and Mental Computation, Harold L. Schoen and Marilyn J. Zweng (Reston, VA: Yearbook of the National Council of Teachers of Mathematics, 1986).
The Man Who Knew Infinity, S. Kanigel (New York: Charles Scribners, 1991).

LINKS

WWW

Individual Projects

Group Projects

Research Links

www.mathnature.com

4.1 | NATURAL NUMBERS

The most basic set of numbers used by any society is the set of numbers used for counting:

$$\mathbb{N} = \{1, 2, 3, 4, 5, 6, 7, 8, 9, 10, 11, \ldots\}$$

This set of numbers is called the set of **counting numbers** or **natural numbers.** Let's assume that you understand what the numbers in this set represent, and you understand the operation of **addition,** $+$. That is, we assume, without definition, knowledge of the operation of addition of natural numbers.

There are a few self-evident properties of addition for this set of natural numbers. They are called "self-evident" because they almost seem too obvious to be stated explicitly. For example, if you jump into the air, you expect to come back down. That assumption is well founded in experience and is also based on an assumption that jumping has certain undeniable properties. But astronauts have found that some very basic assumptions are valid on earth and false in space. Recognizing these assumptions (properties, axioms, laws, or postulates) is important.

HISTORICAL NOTE

The word *add* comes from the Latin word *adhere,* which means "to put to." Widman first used "$+$" and "$-$" signs in 1489 when he stated, "What is $-$, that is minus, what is $+$, that is more." The symbol "$+$" is believed to be a derivation of the Latin *et* ("and").

Closure Property

When we add or multiply any two natural numbers, we know that we obtain a natural number. This "knowing" is an assumption based on experiences (inductive reasoning), but we have actually experienced only a small number of cases for all the possible sums and products of numbers. The scientist—and the mathematician in particular—is very skeptical about making assumptions too quickly. The assumption that the sum or product of two natural numbers is a natural number is given the name *closure* and is referred to as the **closure property.** The property is phrased in terms of sets and operations. Think of a set as a "box"; there is a label on the box—say, addition. If *all* additions of numbers in the box have answers that are already *in* the box, then we say the set is **closed** for addition. If there is at least one answer that is not contained in the box, then the set is said to be **not closed** for that operation.

A closed box

Closure for $+$ in $\mathbb{N}$

> Let $\mathbb{N}$ be the set of natural (or counting) numbers. Let a and b be any natural numbers. Then
>
> $$a + b \text{ is a natural number.}$$
>
> We say $\mathbb{N}$ is **closed for addition.**

EXAMPLE 1

Is the set $A = \{1, 2, 3, 4, 5, 6, 7, 8, 9, 10\}$ closed for addition?

Solution
The set A is *not closed* for addition because

$$5 + 7 = 12 \quad \text{and} \quad 12 \notin A$$

$5 + 7 = 12$; 12 is not in the box; the box is not closed.

Note: The fact that $5 + 3 = 8$ is in the set does not change the fact that A is not closed for addition.

A box that is not closed
$A = \{1, 2, 3, 4, 5, 6, 7, 8, 9, 10\}$

You need find only one *counterexample* to show that a property does not hold.

Commutative and Associative Properties

The word commute *can mean to travel back and forth from home to work; this back-and-forth idea can help you remember that the commutative property applies if you read from left to right or from right to left.*

Another self-evident property of the natural numbers concerns the order in which they are added. It is called the **commutative property for addition** and states that the *order* in which two numbers are added makes no difference; that is (if we read from left to right),

$$a + b = b + a$$

for any two natural numbers a and b. The commutative property allows us to rearrange numbers; it is called a *property of order.* Together with another property, called the *associative property,* it is used in calculation and simplification.

The word associate *can mean connect, join, or unite; with this property you associate two of the added numbers.*

The **associative property for addition** allows us to group numbers for addition. Suppose you wish to add three numbers—say, 2, 3, and 8:

$$2 + 3 + 8$$

To add these numbers you must first add two of them and then add this sum to the third. The associative property tells us that, no matter which two numbers are added first, the final result is the same. If parentheses are used to indicate the numbers to be added first, then this property can be symbolized by

$$(2 + 3) + 8 = 2 + (3 + 8)$$

The parentheses indicate the numbers to be added first. This associative property for addition holds for *any* three or more natural numbers.

Add the column of numbers in the margin. How long does it take? Five seconds is long enough if you use the associative and commutative properties for addition:

$$(9 + 1) + (8 + 2) + (7 + 3) + (6 + 4) + 5 = 10 + 10 + 10 + 10 + 5 = 45$$

However, it takes much longer if you don't rearrange (commute) and regroup (associate) the numbers:

$$
\begin{aligned}
(9 + 8) + (7 + 6 + 5 + 4 + 3 + 2 + 1) &= (17 + 7) + (6 + 5 + 4 + 3 + 2 + 1) \\
&= (24 + 6) + (5 + 4 + 3 + 2 + 1) \\
&= (30 + 5) + (4 + 3 + 2 + 1) \\
&= (35 + 4) + (3 + 2 + 1) \\
&= (39 + 3) + (2 + 1) \\
&= (42 + 2) + 1 \\
&= 44 + 1 \\
&= 45
\end{aligned}
$$

The numbers in the margin:

$$
\begin{array}{r}
9 \\
8 \\
7 \\
6 \\
5 \\
4 \\
3 \\
2 \\
+\ 1 \\
\hline
\end{array}
$$

The properties of associativity and commutativity are not restricted to the operation of addition. For example, these properties also hold for $\mathbb{N}$ and multiplication, as we will now discuss.

Multiplication is defined as repeated addition.

Multiplication

> For $a \neq 0$, **multiplication** is defined as follows:
> $$a \times b \text{ means } \underbrace{b + b + b + \cdots + b}_{a \text{ addends}}$$
> If $a = 0$, then $0 \times b = 0$.

We now consider the property of **closure for multiplication.**

Closure for · in ℕ

Let ℕ be the set of natural (or counting) numbers. Let a and b be any natural numbers. Then

ab is a natural number.

We say ℕ is **closed for multiplication.**

EXAMPLE 2

Is the set $B = \{0, 1\}$ closed for multiplication?

Solution   The set B is *closed* for the operation of multiplication, because all possible products are in B:

$$0 \times 0 = 0 \qquad 0 \times 1 = 0 \qquad 1 \times 0 = 0 \qquad 1 \times 1 = 1 \qquad \blacklozenge$$

We can now consider commutativity and associativity for multiplication in the set ℕ of natural numbers:

Commutativity: $2 \times 3 \overset{?}{=} 3 \times 2$

Associativity: $(2 \times 3) \times 4 \overset{?}{=} 2 \times (3 \times 4)$

The question mark above the equal sign signifies that we should not assume the conclusion (namely, that the expressions on both sides are equal) until we check the arithmetic. Even though we can check these properties for particular natural numbers, it is impossible to check them for *all* natural numbers, so we accept the following axioms.

Commutative and Associative Properties

For any natural numbers a, b, and c:

Commutative properties	***Associative properties***
Addition: $a + b = b + a$	Addition: $(a + b) + c = a + (b + c)$
Multiplication: $ab = ba$	Multiplication: $(ab)c = a(bc)$

To distinguish between the commutative and associative properties, remember the following:

1. When the *commutative property* is used, the *order* in which the elements appear from left to right is changed, but the grouping is not changed.

2. When the *associative property* is used, the elements are *grouped* differently, but the order in which they appear is not changed.

3. If *both* the order and the grouping have been changed, then both the commutative and associative properties have been used.

We are not confined to ℕ when discussing the associative and commutative properties (or any of the properties, for that matter). Nor are we restricted to addition and multiplication for our operations. Indeed, it is often fun to form your own "group" of numbers and see whether the properties hold for your group under your designated operation.

> There still remain three studies suitable for free man. Arithmetic is one of them.
>
> Plato

Distributive Property

Are there properties in ℕ that involve both operations? Consider an example.

©MarkAntman/The Image Works

EXAMPLE 3

Suppose you are selling tickets for a raffle, and the tickets cost $2 each. You sell 3 tickets on Monday and 4 tickets on Tuesday. How much money did you collect?

Solution I You sold a total of $3 + 4 = 7$ tickets, which cost $2 each, so you collected $2 \times 7 = 14$ dollars. That is,

$$2 \times (3 + 4) = 14$$

Solution II You collected $2 \times 3 = 6$ dollars on Monday and $2 \times 4 = 8$ dollars on Tuesday for a total of $6 + 8 = 14$ dollars. That is,

$$(2 \times 3) + (2 \times 4) = 14$$

Since these solutions are equal, we see

$$2 \times (3 + 4) = (2 \times 3) + (2 \times 4) \qquad \blacklozenge$$

Do you suppose this would be true if the tickets cost a dollars and you sold b tickets on Monday and c tickets on Tuesday? Then the equation would be

$$a \times (b + c) = (a \times b) + (a \times c)$$

or simply

$$a(b + c) = ab + ac$$

This example illustrates the **distributive property.**

Distributive Property for Multiplication Over Addition

$$a(b + c) = ab + ac$$

STOP

It there is one property to remember, it is this one!

In the set $\mathbb{N}$ of natural numbers, is addition distributive over multiplication? We wish to check

$$3 + (4 \times 5) \stackrel{?}{=} (3 + 4) \times (3 + 5)$$

Checking:

$$3 + (4 \times 5) = 3 + 20 = 23 \quad \text{and} \quad (3 + 4) \times (3 + 5) = 7 \times 8 = 56$$

Thus, addition is not distributive over multiplication in the set of natural numbers.

The distributive property can also help to simplify arithmetic. Suppose you wish to multiply 9 by 71. You can use the distributive property to do the following mental multiplication:

$$9 \times 71 = 9 \times (70 + 1) = (9 \times 70) + (9 \times 1) = 630 + 9 = 639$$

This allows you to do the problem quickly and simply in your head.

Definition of Subtraction

Since these properties hold for the operations of addition and multiplication, we might reasonably ask whether they hold for other operations. *Subtraction* is defined as the opposite of addition.

Subtraction

The operation of **subtraction** is defined in terms of addition:

$$a - b = x \quad \text{means} \quad a = b + x$$

From the definition of subtraction, 2 − 3 = □ means 2 = □ + 3 so we need to find a number that, when added to 3, gives the result 2.

To test the commutative property for subtraction, we check a particular example:

$$3 - 2 \overset{?}{=} 2 - 3$$

Now, $3 - 2 = 1$, but $2 - 3$ doesn't even exist in the set of natural numbers. Therefore, the commutative property does not hold for subtraction in $\mathbb{N}$.

Furthermore, to provide the result of the operation of subtraction for $2 - 3$, we must find a number that when added to 3 gives the result 2. But there is *no such natural number.* Thus, the set of natural numbers is *closed* for addition and multiplication, but is *not closed* for subtraction. In Section 4.3, we will add elements to the set of natural numbers to create the set of *integers*, which will be closed for subtraction as well as for addition and multiplication.

To make sure you understand the properties discussed in this section, this section's problem set focuses on the properties of closure, commutativity, associativity, and distributivity rather than on the set of natural numbers and the operations of addition, multiplication, and subtraction. Since you are so familiar with the set of natural numbers and with these operations, you could probably answer questions about them without much reflection on the concepts involved. Therefore, in the problem set we work with some operations other than addition, multiplication, and subtraction. When we refer to a table, rows are horizontal and columns are vertical.

Problem Set 4.1

LEVEL 1

IN YOUR OWN WORDS *Explain each of the words or concepts in Problems 1–9.*

1. Natural number
2. Multiplication
3. Subtraction
4. Closure for addition
5. Commutativity
6. Associativity
7. Distributivity
8. Closure for multiplication
9. Contrast commutativity and associativity.

Use the definition of multiplication to show what each expression in Problems 10–15 means.

10. **a.** $2 \cdot 3$ **b.** $3 \cdot 2$ 11. **a.** $3 \cdot 4$ **b.** $4 \cdot 3$
12. **a.** $5 \cdot 2$ **b.** $2 \cdot 5$ 13. **a.** $2 \cdot 184$ **b.** $184 \cdot 2$
14. **a.** $3 \cdot 145$ **b.** $145 \cdot 3$ 15. **a.** xy **b.** yx

In Problems 16–26, classify each as an example of the commutative property, the associative property, or both.

16. $3 + 5 = 5 + 3$
17. $2 + 3 + 5 = 2 + 5 + 3$
18. $2 + (3 + 5) = (2 + 3) + 5$
19. $6 + (2 + 3) = (6 + 2) + 3$
20. $6 + (2 + 3) = (6 + 3) + 2$
21. $6 + (2 + 3) = 6 + (3 + 2)$
22. $6 + (2 + 3) = (2 + 3) + 6$
23. $(4 + 5)(6 + 9) = (4 + 5)(9 + 6)$
24. $(4 + 5)(6 + 9) = (6 + 9)(4 + 5)$
25. $(3 + 5) + (2 + 4) = (3 + 5) + (4 + 2)$
26. $(3 + 5) + (2 + 4) = (3 + 4) + (5 + 2)$

27. "Isn't this one just too sweet, dear?" asked the wife as she tried on a beautiful diamond ring. "No," the husband replied. "It's just too dear, sweet." Does this story remind you of the associative or the commutative property?

28. Is the operation of putting on your shoes and socks commutative?

29. In the English language, the meanings of certain phrases can be very different depending on the association of the words. For example,

> (MAN EATING) TIGER

is not the same as

> MAN (EATING TIGER)

Decide whether each of the following groups of words is associative.
a. HIGH SCHOOL STUDENT
b. SLOW CURVE SIGN
c. BARE FACTS PERSON
d. RED FIRE ENGINE
e. TRAVELING SALESMAN JOKE
f. BROWN SMOKING JACKET

30. Think of three nonassociative word triples as shown in Problem 29.

LEVEL 2

31. IN YOUR OWN WORDS Why do you think *addition* of natural numbers was left undefined? Try to write a definition. Look in one or more dictionaries. What problems do you find with these definitions?

32. Consider the set $A = \{1, 4, 7, 9\}$ with an operation $*$ defined by the table.

$*$	1	4	7	9
1	9	7	1	4
4	7	9	4	1
7	1	4	7	9
9	4	1	9	7

$a * b$ means find the entry in row a and column b; for example, $7 * 9 = 9$ (the entry in row 7 and column 9). Find each of the following.

a. $7 * 4$ b. $9 * 1$ c. $1 * 7$ d. $9 * 9$
e. $4 * 7$ f. $1 * 9$ g. $7 * 1$ h. $7 * 7$

33. Consider the set $F = \{1, -1, i, -i\}$ with an operation $\times$ defined by the table.

$\times$	1	-1	i	$-i$
1	1	-1	i	$-i$
-1	-1	1	$-i$	i
i	i	$-i$	-1	1
$-i$	$-i$	i	1	-1

$a \times b$ means find the entry in row a and column b; for example, $-1 \times (-i) = i$ (the entry in row -1 and column $-i$). Find each of the following.

a. $-1 \times i$ b. $i \times i$ c. $-i \times i$
d. $-i \times 1$ e. $1 \times i$ f. $-i \times -i$

34. Consider the set A and the operation $*$ from Problem 32. Is the set A closed for the operation of $*$? Give reasons for your answer.

35. Consider the set F and the operation of $\times$ from Problem 33. Is the set F closed for the operation of $\times$? Give reasons for your answer.

36. Consider the set A and the operation $*$ from Problem 32. Does the set A satisfy the given property for the operation of $*$? Give reasons for your answer.
a. Associative b. Commutative

37. Consider the set F and the operation of $\times$ from Problem 33. Does the set F satisfy the given property for the operation of $\times$? Give reasons.
a. Associative b. Commutative

38. Let a be the process of putting on a shirt; let b be the process of putting on a pair of socks; and let c be the process of putting on a pair of shoes. Let $\star$ be the operation of "followed by."
a. Is $\star$ commutative for $\{a, b, c\}$?
b. Is $\star$ associative for $\{a, b, c\}$?

39. Consider the set $\mathbb{N}$ of natural numbers and an operation $\leftarrow$ which means "select the first of the two." That is,
$$4 \leftarrow 3 = 4; 3 \leftarrow 4 = 3; 5 \leftarrow 7 = 5; 6 \leftarrow 6 = 6$$
Is the set $\mathbb{N}$ closed for the operation of $\leftarrow$? Give reasons.

40. Consider the set $\mathbb{N}$ of natural numbers and an operation $\leftarrow$ defined in Problem 39.
a. Is $\leftarrow$ associative for $\mathbb{N}$? Give reasons.
b. Is $\leftarrow$ commutative for $\mathbb{N}$? Give reasons.

41. Consider the operation $\bullet$ defined by the following table.

$\bullet$	$\square$	$\triangle$	$\circ$
$\square$	$\circ$	$\square$	$\triangle$
$\triangle$	$\square$	$\triangle$	$\circ$
$\circ$	$\triangle$	$\square$	$\circ$

a. Find $\square \bullet \triangle$.
b. Find $\triangle \bullet \circ$.
c. Does $\circ \bullet \square = \square \bullet \circ$? Is the set commutative for $\bullet$?
d. Does $(\circ \bullet \triangle) \bullet \triangle = \circ \bullet (\triangle \bullet \triangle)$?

42. Let $\downarrow$ mean "select the smaller number" and $\rightarrow$ mean "select the second of the two." Is $\rightarrow$ distributive over $\downarrow$ in the set of natural numbers, $\mathbb{N}$?

43. Let $\downarrow$ mean "select the smaller number" and $\rightarrow$ mean "select the second of the two." Is $\downarrow$ distributive over $\rightarrow$ in the set of natural numbers, $\mathbb{N}$?

44. Do the following problems mentally using the distributive property.
a. 6×82 b. 8×41 c. 7×49

45. Do the following problems mentally using the distributive property.
a. 5×99 b. 4×88 c. 8×52

46. Is the set $\{0, 1\}$ commutative for the operation of multiplication? Give reasons.

47. Is the set $\{0, 1\}$ associative for the operation of multiplication? Give reasons.

48. Is the set $\{-1, 0, 1\}$ commutative for the operation of multiplication? Give reasons.

49. Is the set of even natural numbers closed for the operation of addition? Give reasons.

50. Is the set of even natural numbers closed for the operation of multiplication? Give reasons.

51. Is the set of odd natural numbers closed for the operation of multiplication? Give reasons.

52. Is the set of odd natural numbers closed for the operation of addition? Give reasons.

LEVEL 3

53. Let $S = \{1, 2, 3, \ldots, 9, 10\}$. Define an operation $\odot$ as $a \odot b = 0 \cdot a + 1 \cdot b$. Is S closed for the operation of $\odot$?

54. Let $S = \{1, 2, 3, \ldots, 99, 100\}$. Define an operation $\oplus$ as $a \oplus b = 2a + b$. Check the commutative and associative properties.

55. Let $S = \{1, 2, 3, \ldots, 999, 1000\}$. Define an operation $\otimes$ as $a \otimes b = 2(a + b)$. Check the commutative and associative properties.

56. Consider a soldier facing in a given direction (say, north). Let us denote "left face" by ℓ, "right face" by r, "about face" by a, and "stand fast" by f. (The element f means "don't move from your present position." It does not mean "return to your original position.") Then we define $H = \{\ell, r, a, f\}$ and an operation $\star$ meaning "followed by." Thus, $\ell \star \ell = a$ means "left face" followed by "left face" and is the same as the single command "about face." Complete the following table.

$\star$	ℓ	r	a	f
ℓ	a			
r				
a				
f				

57. Is the set H from Problem 56 closed for the operation of $\star$?

58. Does the set H and operation $\star$ from Problem 56 satisfy the following properties? Give reasons.
   a. Associative   **b.** Commutative

59. Mensa is an association for people with high IQs. An advertisement for the organization offers the following challenge: "Take this instant test to see if you're a genius."

> *Put the appropriate plus or minus signs between the numbers, in the correct places, so that the sum total will equal 1.*
> 0 1 2 3 4 5 6 7 8 9 = 1

The advertisement also states, "This problem stumps 45% of the Mensa members who try it. And they *all* have IQs in the top 2% nationwide. See if you can pass this test. If you can do it, you might have what it takes to join us. To find out, . . . write to American Mensa, 2626 East 14th St., Brooklyn, NY 11235."

60. The Vanishing Leprechaun Puzzle* The puzzle shown in Figure 4.1 consists of three pieces and was originally published by W. A. Elliott Company, 212 Adelaide St. W., Toronto, Canada M5H 1W7.

   If we place the pieces together as shown at the top, we see 15 leprechauns. However, if we *commute* pieces A and B as shown at the bottom, we count 14 leprechauns!

   Clearly, from Figure 4.1 we see

   $AB \neq BA$

Can you explain where the vanishing leprechaun went?

A   B

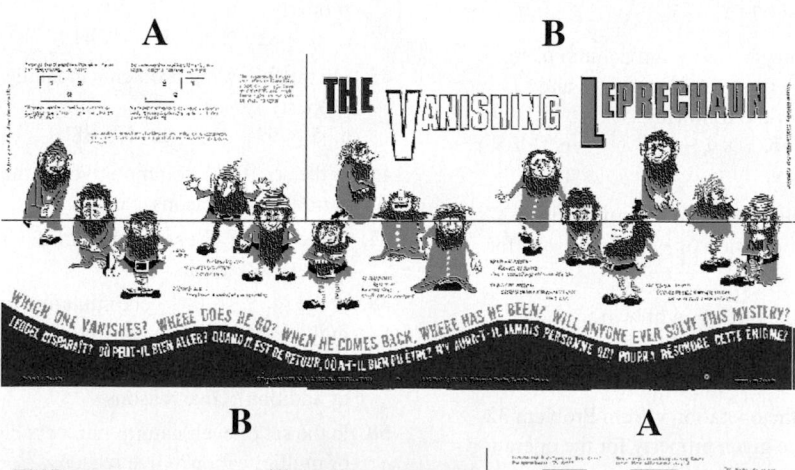

B   A

Figure 4.1 Leprechaun Puzzle: How many leprechauns?

4.2 | PRIME NUMBERS

Divisibility

A set of numbers that is important, not only in algebra but in all of mathematics, is the set of prime numbers. To understand prime numbers, you must first understand the idea of divisibility, along with some new terminology and notation.

The natural number 10 is divisible by 2, since there is a natural number 5 so that $10 = 2 \cdot 5$; it is not divisible by 3, since there is no natural number k such that $10 = 3 \cdot k$. This leads us to the following definition.

Divisibility

> If m and d are natural numbers, and if there is a natural number k so that $m = d \cdot k$, we say that **d is a divisor of m, d is a factor of m, d divides m,** and **m is a multiple of d.** We denote this relationship by $d|m$.

 **CAUTION** Do not confuse this notation with the notation sometimes used for fractions: "5/30" means 5 divided by 30, which is a fraction; "5|30" means "5 divides 30," which is a statement.

That is, $5|30$ is read "5 divides 30" and means that there exists some natural number k—namely, 6—such that $30 = 5 \cdot k$.

EXAMPLE 1

Tell whether each of the following is true or false, and give the meaning of each.

a. $7|63$ **b.** $8|104$ **c.** $14|2$ **d.** $6|15$

Solution

a. $7|63$ is read "7 divides 63" and is true since we can find a natural number k—namely, 9—such that $63 = 7 \cdot k$.

b. $8|104$ is true, since $104 = 8 \cdot 13$.

c. $14|2$ is false because we can find no natural number k so that $2 = 14 \cdot k$. We write $14 \nmid 2$ to say that 14 does not divide 2.

d. $6|15$ is false, because we can find no natural number k so that $15 = 6 \cdot k$.   ◆

It is easy to see that 1 divides every natural number m, since $m = 1 \cdot m$. Also, by the commutative property of multiplication, $m = m \cdot 1$; thus every natural number m divides itself. We have proved the following theorem.

Least Number of Divisors

CAUTION This property is basic to understanding what follows.

> Every natural (counting) number greater than 1 has at least two distinct divisors, itself and 1.

Consider the number 341,592. Is this number divisible by 2? By 3? By 4? By 5? You may know some ways of answering these questions without the necessity of actually doing the division.

EXAMPLE 2   PÓLYA'S METHOD

Find a rule for the divisibility of any number M by 2.

Solution   We use Pólya's problem-solving guidelines for this example.

Understand the Problem. You might already know the rule for divisibility by 2. It says that *if the last digit of the number is even, then the number is divisible by 2.* That is, if the number M ends in 0, 2, 4, 6, or 8, it is divisible by 2. This example asks us to show why this rule "works."

Devise a Plan. We will begin with a simpler example, say, 341,592. We write the number in expanded notation.

$$341{,}592 = 3 \times 10^5 + 4 \times 10^4 + 1 \times 10^3 + 5 \times 10^2 + 9 \times 10^1 + 2$$

The question to answer is "When will this number be divisible by 2?" Associate all the digits except the last one:

$$341{,}592 = \underbrace{3 \times 10^5 + 4 \times 10^4 + 1 \times 10^3 + 5 \times 10^2 + 9 \times 10^1}_{\text{This is divisible by 2.}} + 2$$

The associated part is *always* divisible by 2, since $2|10$, $2|10^2$, $2|10^3$, $2|10^4$, and $2|10^5$.

Carry Out the Plan. For the number M we see that since $2|10$, $2|10^2$, . . . , $2|10^n$, the divisibility of M by 2 depends solely on whether 2 divides the last digit.

Look Back. We see that $2|341{,}592$ since $2|2$. Also $2|838$ since $2|8$, and $2\nmid839$ since $2\nmid9$. ◆

Similar rules apply for divisibility by 4 or 8. You might even expect to try the same type of rule for divisibility by 3, but this situation is not quite so easy, as shown by the following example.

EXAMPLE 3   **PÓLYA'S METHOD**

Find a rule for divisibility by 3.

SOLUTION   We use Pólya's problem-solving guidelines for this example.

Understand the Problem. Try a simple example. We see that $3|84$ since we can find a natural number—namely, 28—so that $84 = 3 \cdot k$. We also note that $3\nmid4$, so the same type of rule that worked for divisibility by 2 will not work for 3.

Devise a Plan. We will once again look at the expanded notation.

Consider a simpler problem—say, the divisibility of 341,592 by 3:

$$341{,}592 = 3 \times 10^5 + 4 \times 10^4 + 1 \times 10^3 + 5 \times 10^2 + 9 \times 10^1 + 2$$

The plan is to make each product divisible by 3. We do this by adding and subtracting 1 from each term containing 10^b, where b is a natural number.

Carry Out the Plan.

$$341{,}592 = 3 \times (10^5 - 1 + 1) + 4 \times (10^4 - 1 + 1) + 1 \times (10^3 - 1 + 1)$$
$$+ 5 \times (10^2 - 1 + 1) + 9 \times (10^1 - 1 + 1) + 2$$

We now use the distributive and associative properties to rewrite this expression:

$$341{,}592 = [3 \times (10^5 - 1) + 3] + [4 \times (10^4 - 1) + 4] + [1 \times (10^3 - 1) + 1]$$
$$+ [5 \times (10^2 - 1) + 5] + [9 \times (10^1 - 1) + 9] + 2$$
$$= \underbrace{[3(10^5 - 1) + 4(10^4 - 1) + 1(10^3 - 1) + 5(10^2 - 1) + 9(10^1 - 1)]}_{\text{This is divisible by 3.}}$$

$$\underbrace{+ [3 + 4 + 1 + 5 + 9 + 2]}_{\text{This is the sum of the digits.}}$$

Notice what we have done:

$$10^1 - 1 = 9$$
$$10^2 - 1 = 99$$
$$10^3 - 1 = 999$$
$$10^4 - 1 = 9{,}999$$
$$10^5 - 1 = 99{,}999$$

These are all divisible by 3, and hence we see that if

$$3|(3 + 4 + 1 + 5 + 9 + 2), \text{ then } 3|341{,}592$$

Checking, we see that $3|24$ (sum of digits), so $3|341{,}592$. Furthermore, since $3|(10^n - 1)$ for any natural number n, we see that the divisibility of a number N by 3 depends on whether 3 divides the sum of the digits.

Look Back. If the sum of the digits of a number N is divisible by 3, then N is divisible by 3. ◆

Some of the more common rules of divisibility are shown in Table 4.1.

Table 4.1 Rules of Divisibility for a Natural Number *N*

N is divisible by	*Test*
1	all *N*
2	if the last digit is divisible by 2.
3	if the sum of the digits is divisible by 3.
4	if the number formed by the last two digits is divisible by 4.
5	if the last digit is 0 or 5.
6	if the number is divisible by 2 and by 3.
8	if the number formed by the last three digits is divisible by 8.
9	if the sum of the digits is divisible by 9.
10	if the last digit is 0.
12	if the number is divisible by 3 and by 4.

Since every natural number greater than 1 has at least two divisors, can any number have more than two?

Checking: 2 has exactly two divisors: 1, 2
        3 has exactly two divisors: 1, 3
        4 has more than two divisors: 1, 2, and 4

Thus, some numbers (such as 2 and 3) have exactly two divisors, and some (such as 4 and 6) have more than two divisors. Do any natural numbers have fewer than two divisors?

We now state a definition that classifies each natural number according to the number of divisors it has.

Prime Number

> A **prime number** is a natural number that has exactly two divisors.
> A natural number that has more than two divisors is called a **composite number.**

We see that 2 is prime, 3 is prime, 4 is composite (since it is divisible by three natural numbers), 5 is prime, 6 is composite (since it is divisible by 1, 2, 3, and 6). Note that every natural number greater than 1 is either prime or composite. The number 1 is neither prime nor composite.

One method for finding primes smaller than some given number was first used by a Greek mathematician, Eratosthenes, more than 2,000 years ago. The technique is known as the **sieve of Eratosthenes.** Suppose we wish to find the primes less than 100. We prepare a table of natural numbers 1–100, as shown in Table 4.2.

Table 4.2 Finding Primes Using the Sieve of Eratosthenes

X́	②	③	X́	⑤	X́	⑦	X́	X́9	X́0
⑪	X́2	⑬	X́4	X́5	X́6	⑰	X́8	⑲	X́0
X́1	X́2	㉓	X́4	25	X́6	X́7	X́8	㉙	X́0
㉛	X́2	X́3	X́4	35	X́6	㊲	X́8	X́9	X́0
㊶	X́2	㊸	X́4	X́5	X́6	㊼	X́8	X́9	X́0
X́1	X́2	㊳	X́4	55	X́6	X́7	X́8	㊾	X́0
㊱	X́2	X́3	X́4	65	X́6	㊲	X́8	X́9	X́0
㊸	X́2	⑦3	X́4	X́5	X́6	X́7	X́8	㉇	X́0
X́1	X́2	㊳	X́4	85	X́6	X́7	X́8	㊾	X́0
X́1	X́2	X́3	X́4	95	X́6	㊲	X́8	X́9	1X́0

Cross out 1, since it is not classified as a prime number.
Draw a circle around 2, the smallest prime number. Then cross out every following multiple of 2, since each is divisible by 2 and thus is not prime.
Draw a circle around 3, the next prime number. Then cross out each succeeding multiple of 3. Some of these numbers, such as 6 and 12, will already have been crossed out because they are also multiples of 2.
Circle the next open prime, 5, and cross out all subsequent multiples of 5.
The next prime number is 7; circle 7 and cross out multiples of 7. Since 7 is the largest prime less than $\sqrt{100} = 10$, we now know that all the remaining numbers are prime.

The process is a simple one, since you do not have to cross out the multiples of 3 (for example) by checking for divisibility by 3 but can simply cross out every third number. Thus, anyone who can count can find primes by this method. Also, notice that in finding the primes under 100, we had crossed out all the composite numbers by the time we crossed out the multiples of 7. That is, to find all primes less than 100: (1) Find the largest prime smaller than or equal to $\sqrt{100} = 10$ (7 in this case); (2) cross out multiples of primes up to and including 7; and (3) all the remaining numbers in the chart are primes.

This result generalizes. If you wish to find all primes smaller than n:

1. Find the *largest* prime less than or equal to $\sqrt{n}$.

2. Cross out the multiples of primes less than or equal to $\sqrt{n}$.

3. All the remaining numbers in the chart are primes.

Phrasing this another way, if n is composite, then one of its factors must be less than or equal to $\sqrt{n}$. That is, if $n = ab$, then it can't be true that *both a and b* are greater than $\sqrt{n}$ (otherwise $ab > \sqrt{n}\sqrt{n} = n = ab$, so $ab > ab$ is a contradiction). Thus, one of the factors must be less than or equal to $\sqrt{n}$.

Prime Factorization

Prime numbers are fundamental to many mathematical processes. In particular, we use prime numbers in working with rational numbers later in this chapter. You will need to understand *prime factorization, greatest common factor,* and *least common multiple.*

The operation of **factoring** is the reverse of the operation of multiplying. For example, multiplying 3 by 6 yields $3 \cdot 6 = 18$ and this answer is unique (only one answer is possible). In the reverse process, called factoring, you are given the number 18 and asked for numbers that can be multiplied together to give 18. This process is *not* unique; we list several different factorizations of 18:

$$18 = 1 \cdot 18 = 18 \cdot 1 = 2 \cdot 9 = 9 \cdot 2 = 1 \cdot 1 \cdot 2 \cdot 9 = 3 \cdot 6 = 2 \cdot 3 \cdot 3 = \cdots$$

There are, in fact, infinitely many possibilities. We make some agreements, so the process gives a unique answer:

1. We will not consider the order in which the factors are listed as important. That is, $2 \cdot 9$ and $9 \cdot 2$ are considered the same factorization.

2. We will not consider 1 as a factor when writing out any factorizations. That is, prime numbers do not have factorizations.

3. Recall that we are working in the set of natural numbers; thus, $18 = 36 \cdot \frac{1}{2}$ and $18 = (-2)(-9)$ are *not* considered factorizations of 18.

With these agreements, we have greatly reduced the possibilities:

$$18 = 2 \cdot 9 = 3 \cdot 6 = 2 \cdot 3^2$$

These are the only three possible factorizations. Notice that the last factorization contains only prime factors; thus it is called the **prime factorization** of 18.

It should be clear that, if a number is composite, it can be factored into two natural numbers greater than 1. Each of these two numbers will be prime or composite. If both are prime, then we have a prime factorization. If one or more is composite, we repeat the process, and continue until we have written the original number as a product of primes. It is also true that this representation is unique. This is one of the most important results in arithmetic, and it carries the impressive title **fundamental theorem of arithmetic.**

Fundamental Theorem of Arithmetic

STOP As you might guess from the name, this result is important for the following material.

Every natural number greater than 1 is either a prime or a product of primes, and its prime factorization is unique (except for the order in which the factors appear).

EXAMPLE 4

Find the prime factorizations: **a.** 385 **b.** 1,400

Solution

a. One of the easiest ways to find the prime factors of a number is to try division by each of the prime numbers in order: 2, 3, 5, 7, ... The rules of divisibility in Table 4.1 may help. For this example, we need to check primes up to $\sqrt{385} \approx 19$. If none of the primes up to 19 is a factor of 385, then 385 is prime. We see by inspection that 385 is not divisible by 2 or 3. It is divisible by 5, so

$$385 = 5 \cdot 77$$

Since 77 is composite $(77 = 7 \cdot 11)$, we write

$$385 = 5 \cdot 7 \cdot 11$$

Finding prime factorizations is a process that is used in a multitude of mathematical applications.

Many people prefer to write these using a **factor tree**.

$$385$$
$$5 \cdot 77$$
$$5 \cdot 7 \cdot 11$$

b. Using a factor tree, we may begin with *any* factors of 1,400:

$$1,400$$
$$10 \cdot 140$$
$$2 \cdot 5 \cdot 10 \cdot 14$$
$$2 \cdot 5 \cdot 2 \cdot 5 \cdot 2 \cdot 7$$

We write the prime factorization using exponents:

$$1,400 = 2^3 \cdot 5^2 \cdot 7$$ ◆

The answer to Example 4 is written in what is called **canonical form.** The *canonical representation* of a number is the representation of that number as a product of primes using exponential notation with the factors arranged in order of increasing magnitude.

EXAMPLE 5

Find the canonical representation of 3,465.

Solution   We use a factor tree.

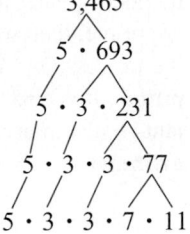

$$3,465$$
$$5 \cdot 693$$
$$5 \cdot 3 \cdot 231$$
$$5 \cdot 3 \cdot 3 \cdot 77$$
$$5 \cdot 3 \cdot 3 \cdot 7 \cdot 11$$

The prime factorization in canonical form is $3^2 \cdot 5 \cdot 7 \cdot 11$. ◆

HISTORICAL NOTE

A statement equivalent to the fundamental theorem of arithmetic is found in Book IX of Euclid's *Elements*. This work is not only the earliest known major Greek mathematical book, but it is also the most influential textbook of all time. It was composed around 300 B.C. and first printed in 1482. Except for the Bible, no other book has been through so many printings. Euclid, the first professor of mathematics at the Museum of Alexandria, was the author of at least ten other books.

Greatest Common Factor

Suppose we look at the set of factors common to a given set of numbers:

Factors of 18:   {1, 2, 3,   6, 9, 18}
                                              Common factors: {1, 2, 3, 6}
Factors of 12:   {1, 2, 3, 4, 6, 12}

The *greatest common factor* is the largest number in the set of common factors.

Greatest Common Factor

> The **greatest common factor (g.c.f.)** of a set of numbers is the largest number that divides (evenly) into each of the numbers in the given set.

The procedure for finding the greatest common factor involves the canonical form of the given numbers. For example, suppose we want to find the greatest common factor

Factors of:
24: {1, 2, 3, 4, 6, 8, 12, 24}
30: {1, 2, 3, 5, 6, 10, 15, 30}
Common factors:
{1, 2, 3, 6}
g.c.f. = 6

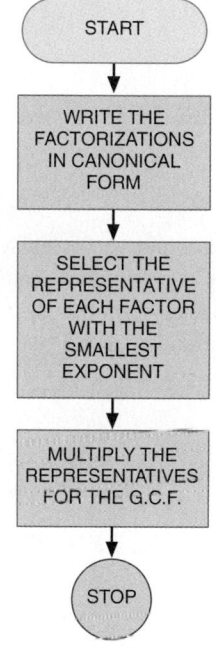

Figure 4.2 Procedure for finding the g.c.f.

Multiples of:
24: {24, 48, 72, 96, 120, . . .}
30: {30, 60, 90, 120, . . .}
Common multiples: {120, 240, . . .}
l.c.m. = 120

Least Common Multiple

of 24 and 30. We could, of course, find all factors, then the intersection (common factors), and finally the greatest one in that set. Instead, find the canonical representation of each:

$$24 = 2^3 \cdot 3 \quad = \quad 2^3 \cdot 3^1 \cdot 5^0$$
$$30 = 2 \cdot 3 \cdot 5 \quad = \quad 2^1 \cdot 3^1 \cdot 5^1$$

Select one representative from each of the columns in the factorization. The representative we select when finding the g.c.f. is the one with the smallest exponent. The g.c.f. is the product of these representatives.

$$\text{g.c.f.} = 2^1 \cdot 3^1 \cdot 5^0$$

The procedure for finding the g.c.f. is summarized in Figure 4.2.

EXAMPLE 6

Find the greatest common factor of the given sets of numbers.

a. 300, 144   **b.** 15, 28   **c.** 3150, 588, 280

Solution

a. $300 = 2^2 \cdot 3^1 \cdot 5^2$
$144 = 2^4 \cdot 3^2 \cdot 5^0$
$\text{g.c.f.} = 2^2 \cdot 3^1 \cdot 5^0$
$= 4 \cdot 3 \cdot 1 = 12$

b. $15 = 2^0 \cdot 3^1 \cdot 5^1 \cdot 7^0$
$28 = 2^2 \cdot 3^0 \cdot 5^0 \cdot 7^1$
$\text{g.c.f.} = 2^0 \cdot 3^0 \cdot 5^0 \cdot 7^0$
$= 1 \cdot 1 \cdot 1 \cdot 1 = 1$

c. $3150 = 2^1 \cdot 3^2 \cdot 5^2 \cdot 7^1$
$588 = 2^2 \cdot 3^1 \cdot 5^0 \cdot 7^2$
$280 = 2^3 \cdot 3^0 \cdot 5^1 \cdot 7^1$
$\text{g.c.f} = 2^1 \cdot 3^0 \cdot 5^0 \cdot 7^1$
$= 14$ ◆

If the greatest common factor of two numbers is 1, we say that the numbers are **relatively prime.** Notice that 15 and 28 are relatively prime, but they themselves are not prime. It is possible for relatively prime numbers to be composite numbers.

Least Common Multiple

The greatest common factor is the largest number in the intersection of the factors of a set of given numbers. On the other hand, the *least common multiple* is the smallest number in the intersection of the multiples of a set of given numbers.

> The **least common multiple (l.c.m.)** of a set of numbers is the smallest number that each of the numbers in the set divides into evenly.

An algorithm for finding the least common multiple is very much like the one for the g.c.f. The process, as before, begins by finding the canonical representations of the numbers involved. For example, to obtain the l.c.m. of the numbers 24 and 30, write each in canonical form:

$$24 = 2^3 \cdot 3^1 \cdot 5^0 \qquad 30 = 2^1 \cdot 3^1 \cdot 5^1$$

For the l.c.m., we choose the representative of each factor with the largest exponent. The l.c.m. is the product of these representatives.

$$\text{l.c.m.} = 2^3 \cdot 3^1 \cdot 5^1 = 120$$

The procedure for finding the least common multiple is shown in Figure 4.3 (page 156).

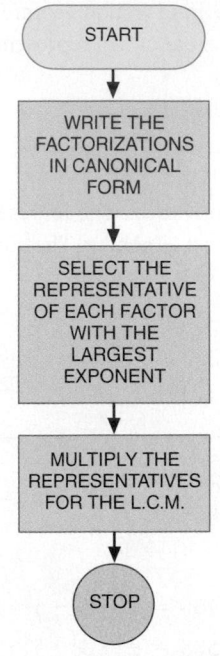

Figure 4.3 Procedure for finding the l.c.m.

EXAMPLE 7

Find the least common multiple of 300, 144, and 108.

Solution

$$300 = 2^2 \cdot 3^1 \cdot 5^2$$
$$144 = 2^4 \cdot 3^2 \cdot 5^0$$
$$108 = 2^2 \cdot 3^3 \cdot 5^0$$
$$\text{l.c.m.} = 2^4 \cdot 3^3 \cdot 5^2 = 16 \cdot 27 \cdot 25 = 10{,}800$$

This means that the *smallest* number that 300, 144, and 108 *all* divide into evenly is 10,800. ◆

In Pursuit of Primes

The method of Eratosthenes gives a finite list of primes, but it is not very satisfactory to use if we wish to determine whether a given number n is a prime. For centuries, mathematicians have tried to find a formula that would yield *every* prime. Let's try to find a formula that results in giving only primes. A possible candidate is

$$n^2 - n + 41$$

If we try this formula for $n = 1$, we obtain $1^2 - 1 + 41 = 41$.

For $n = 2$: $2^2 - 2 + 41 = 43$, a prime
For $n = 3$: $3^2 - 3 + 41 = 47$, a prime

So far, so good. That is, we are obtaining only primes. Continuing, we keep finding only primes for n up to 40:

For $n = 40$: $40^2 - 40 + 41 = 1{,}601$, a prime

Inductively, we might conclude that the formula yields only primes, but the next value provides a counterexample:

For $n = 41$: $41^2 - 41 + 41 = 41^2 = 1{,}681$, not a prime!

A more serious attempt to find a prime number formula was made by Pierre de Fermat, who tried the formula

$$2^{2^n} + 1$$

For $n = 1$: $2^{2^1} + 1 = 5$, a prime
For $n = 2$: $2^{2^2} + 1 = 2^4 + 1 = 17$, a prime
For $n = 3$: $2^{2^3} + 1 = 2^8 + 1 = 257$, a prime
For $n = 4$: $2^{2^4} + 1 = 2^{16} + 1 = 65{,}537$, a prime
For $n = 5$: $2^{2^5} + 1 = 2^{32} + 1 = 4{,}294{,}967{,}297$

Is 4,294,967,297 a prime? The answer is not easy. See the Historical Note in the margin. It turns out this number is not prime! It is divisible by 641. Whether this formula generates any other primes is unknown.

In 1644, the French priest and number theorist Marin Mersenne (1588–1648) stated without proof that the number

$$2^{251} - 1$$

is composite. In the 19th century, mathematicians finally proved Mersenne correct when they discovered that this number was divisible by both 503 and 54,217. Mersenne did discover, however, that

$$2^{257} - 1$$

is a prime number.

HISTORICAL NOTE

David Hilbert (1862–1943)
In 1900, David Hilbert delivered an address before the International Congress of Mathematicians in Paris. Instead of solving a problem, Hilbert presented a list of 23 problems. The mathematical world went to work on these problems and, even today, is still working on some of them. In the process of solving these problems, entire new frontiers of mathematics were opened. Hilbert's 10th problem was solved in 1970 by Yuri Matyasievich, who built upon the work of Martin Davis, Hilary Putnam, and Julia Robinson. A consequence of Hilbert's 10th problem is the fact that there must exist a polynomial with integer coefficients that will produce only prime numbers.

In 1970, a young Russian named Matyasievich discovered several explicit polynomials (such as $n^2 - n + 41$) of this sort that generate only prime numbers, but all of those he discovered are too complicated to reproduce here. The largest known prime number at that time was

$$2^{11,213} - 1$$

which was discovered at the University of Illinois through the use of number theory and computers. The mathematicians were so proud of this discovery that the following postmark was used on the university's postage meter:

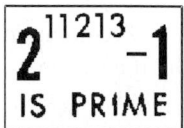 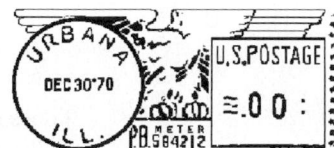

Courtesy of Donald B. Gillies, University of Illinois

Some other large prime numbers and the dates of their discovery are shown in Table 4.3.

Table 4.3 Some Large Primes

Prime Number	Date of Discovery	Source
$2^{257} - 1$	1644	Marin Mersenne
$2^{11,213} - 1$	December 1970	University of Illinois
$2^{19,937} - 1$	March 1971	Bryant Tuckerman
$2^{21,701} - 1$	October 1978	Laura Nickel and Curt Knoll
$2^{23,209} - 1$	January 1979	Curt Knoll
$2^{86,243} - 1$	May 1983	David Slowinski
$2^{216,091} - 1$	November 1985	Amdahl Benchmark Center
$2^{756,839} - 1$	February 1992	David Slowinski and Paul Gage
$2^{858,433} - 1$	April 1994	David Slowinski and Paul Gage
$2^{1,398,269} - 1$	November 1996	Joel Armengaud
$2^{2,976,221} - 1$	December 1997	Gordon Spence
$2^{3,021,377} - 1$	1998	Clarkson and Kurowski
$2^{6,972,593} - 1$	1999	Hajratwala and Kurowski
$2^{13,466,917} - 1$	2001	Cameron and Kurowski

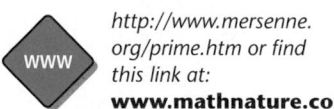

http://www.mersenne. org/prime.htm or find this link at:
www.mathnature.com

If you are interested in finding out more about the search for large primes, you might wish to check the Great Internet Mersenne Prime Search, or as it is usually known, GIMPS. GIMPS is a worldwide project coordinated by George Woltman, who wrote a program for the PC to find primes. The hunt for record prime numbers used to be the exclusive domain of supercomputers, but today, by using thousands of individual machines, it is possible to collectively surpass even the most powerful computers. The last five primes on this list were found using Woltman's program.

Infinitude of Primes

Table 4.3 shows some very large primes. Is there a largest prime? If there is no largest prime, then there must be infinitely many primes.

EXAMPLE 8   PÓLYA'S METHOD

Show that there is no largest prime.

Solution   We use Pólya's problem-solving guidelines for this example.

Understand the Problem. Is there a prime larger than the largest known prime shown in Table 4.3? If we find one, then we are finished. If we can't find one, is there a way we can still show it is not the largest prime?

Devise a Plan. We will consider a simpler problem. Suppose we believe that 19 is the largest prime. The task is to show that there exists a prime larger than 19. There are two ways to proceed. We could simply find a larger prime—say, 23. But what if we can't find a larger one? Without actually finding the largest prime, we will proceed by showing that 19 can't be the largest prime. Consider the number $M = (2 \cdot 3 \cdot 5 \cdot 7 \cdot 11 \cdot 13 \cdot 17 \cdot 19) + 1$. This number is larger than 19. Is it a prime?

Carry Out the Plan. According to our assumption, M must be composite, since it is larger than 19. But if it is composite, it has a prime divisor. Check all the primes:

2 does not divide M since $2|(2 \cdot 3 \cdot 5 \cdot 7 \cdot 11 \cdot 13 \cdot 17 \cdot 19)$, and thus 2 does not divide 1 more than this number.

3 does not divide M for the same reason.

Repeat for every known prime.

Thus, if M is not divisible by any known prime, then either it must be prime or there is a prime divisor larger than 19. In either case, we have found a prime larger than 19.

Look Back. If *anyone* claims to be in possession of the largest prime, we need only carry out an argument like the above to find a larger prime. Thus, we are saying that there are infinitely many primes, since it is impossible to have a largest prime. ◆

Problem Set 4.2

LEVEL 1

1. **IN YOUR OWN WORDS** What is a prime number?

2. **IN YOUR OWN WORDS** Describe a process for finding a prime factorization.

3. **IN YOUR OWN WORDS** What is the canonical representation of a number?

4. **IN YOUR OWN WORDS** What does g.c.f. mean, and what is the procedure for finding the g.c.f. of a set of numbers?

5. **IN YOUR OWN WORDS** What does l.c.m. mean, and what is the procedure for finding the l.c.m. of a set of numbers?

6. **IN YOUR OWN WORDS** Compare and contrast finding the g.c.f. and l.c.m. of a set of numbers.

Which of the numbers in Problems 7–10 are prime?

7. a. 59	**b.** 57	**c.** 1	**d.** 1,997
8. a. 63	**b.** 73	**c.** 79	**d.** 1,999
9. a. 43	**b.** 97	**c.** 171	**d.** 2,001
10. a. 91	**b.** 87	**c.** 111	**d.** 2,003

Are the statements in Problems 11–14 true or false?

11. a. $6|48$	**b.** $7|48$	**c.** $8|48$	**d.** $9|48$
12. a. $6\!\!\not|39$	**b.** $5\!\!\not|30$	**c.** $16\!\!\not|576$	**d.** $3|7,823$

13. **a.** $15|5$ **b.** $5|83,410$
    c. $2|628,174$ **d.** $10|148,729,320$

14. **a.** $15|4,814$ **b.** $17|255$
    c. $9|7,823,572$ **d.** $10\!\!\not|148,729,320$

15. Find all prime numbers less than or equal to 300.

16. Determine the largest prime you need to consider to be sure that you have excluded, in the sieve of Eratosthenes, all primes less than or equal to:
    a. 200   **b.** 500
    c. 1,000   **d.** 1,000,000

Write the prime factorization for each of the numbers in Problems 17–20. If the number is prime, so state.

17. a. 24	**b.** 30	**c.** 300	**d.** 144
18. a. 108	**b.** 740	**c.** 699	**d.** 123

19. a. 120   **b.** 90   **c.** 75   **d.** 975

20. a. 490   **b.** 4,752   **c.** 143   **d.** 51

Find the canonical representation for each of the numbers in Problems 21–36.

21. 83   **22.** 97

23. 127   **24.** 113

25. 377   **26.** 151

27. 105   **28.** 187

29. 67   **30.** 229

31. 315   **32.** 111

33. 567   **34.** 568

35. 2,869   **36.** 793

Find the g.c.f. and l.c.m. of the sets of numbers in Problems 37–44.

37. {60, 72}   **38.** {95, 1425}

39. {12, 54, 171}   **40.** {11, 13, 23}

41. {9, 12, 14}   **42.** {3, 6, 15, 54}

43. {75, 90, 120}   **44.** {85, 100, 240}

LEVEL 2

45. Bill and Sue both work at night. Bill has every sixth night off and Sue has every eighth night off. If they are both off tonight, how many nights will it be before they are both off again at the same time?

46. Two movie theaters, UAI and UAII, start their movies at 7:00 P.M. The movie at UAI takes 75 minutes and the movie at UAII takes 90 minutes. If the shows run continuously, when will they again start at the same time?

47. We used a sieve of Eratosthenes in Table 4.2 by arranging the first 100 numbers into 10 rows and 10 columns. Repeat the sieve process for the first 100 numbers by arranging the numbers in the following patterns.

a. By 6:
```
1   2   3   4   5   6
7   8   9  10  11  12
13  14  15  ...
```

b. By 7:
```
1   2   3   4   5   6   7
8   9  10  11  12  13  14
15  16  17  ...
```

c. By 21:   1   2   3   4   5   6   ...

As you are using these sieves, look for patterns. Describe some of the patterns you notice. Do you think that any of these sieves are better than the one shown in Table 4.2? Why or why not?

48. Use the sieve in Problem 47a to make a conjecture about primes and multiples of 6.

49. Lucky Numbers   Set up a sieve similar to the one illustrated in Table 4.4. What are the lucky numbers less than 100?

Table 4.4 Lucky Numbers

1	2̸	3	4̸	5̸	6̸
7	8̸	9	1̸0̸	1̸1̸	1̸2̸
13	1̸4̸	15	1̸6̸	1̸7̸	1̸8̸
19	2̸0̸	21	2̸2̸	2̸3̸	24
25	2̸6̸	27	2̸8̸	2̸9̸	30
31	3̸2̸	33	3̸4̸	3̸5̸	36
37	3̸8̸	39	40	4̸1̸	42
43	4̸4̸	45	4̸6̸	4̸7̸	4̸8̸

Start counting with 1 *each time.*
Cross out every second number (shown as /).
The next uncrossed number is 3, so cross out every 3rd number that remains (shown as ×).
The next uncrossed number is 7, so cross out every 7th number that remains (shown as −).
Continue in the same fashion. The numbers that are not crossed out are called *lucky numbers.*

LEVEL 3

50. Pairs of consecutive odd numbers that are primes are called *prime twins.* For example, 3 and 5, 11 and 13, and 41 and 43 are prime twins. Can you find any others?

51. Three consecutive odd numbers that are primes are called *prime triplets.* It is easy to show that 3, 5, and 7 are the only prime triplets. Can you explain why this is true?

52. HISTORICAL QUESTION   In 1742 the mathematician Christian Goldbach observed that every even number (except 2) seemed representable as the sum of two primes. Goldbach could not prove this result, known today as *Goldbach's conjecture,* so he wrote to his friend, the world-famous mathematician Leonhard Euler. Euler was unable to prove or disprove this conjecture, and it remains unsolved to this day. Write the following numbers as the sum of two primes (the first three are worked for you):

4 = 2 + 2	6 = 3 + 3	8 = 5 + 3
10 =	12 =	14 =
16 =	18 =	20 =
40 =	80 =	100 =

To date, this conjecture has not been proved, but the Russian mathematician L. Schnirelmann (1905–1938) proved that every positive integer can be represented as the sum of not more than 300,000 primes. That may seem like a long way off from Goldbach's conjecture, but at least

300,000 is a finite number! Later, another mathematician, I. M. Vinogradoff, proved that there exists a number N such that all numbers larger than N can be written as the sum of, at most, four primes.

53. Let $S = \{1, 2, 3, 5, 6, 10, 15, 30\}$, and define an operation $\mathbf{M}$ meaning *least common multiple*. For example,

$$5 \; \mathbf{M} \; 10 = 10, \qquad 5 \; \mathbf{M} \; 6 = 30, \qquad 10 \; \mathbf{M} \; 30 = 30$$

a. Is S closed for the operation of $\mathbf{M}$?
b. Is S associative for $\mathbf{M}$?
c. Is S commutative for $\mathbf{M}$?

54. Use an argument similar to the one in the text to show that 23 is not the largest prime.

PROBLEM SOLVING

55. In the text, we showed that 19 is not the largest prime by considering

$$M = 2 \cdot 3 \cdot 5 \cdot 7 \cdot 11 \cdot 13 \cdot 17 \cdot 19 + 1$$

Now, M is either prime or composite. If it is prime, then since it is larger than 19, we have a prime larger than 19. If it is composite, it has a prime divisor larger than 19. In either case, we find a prime larger than 19. Show that this number M does not always generate primes. That is,

$2 + 1 = 3$, a prime
$2 \cdot 3 + 1 = 7$, a prime
$2 \cdot 3 \cdot 5 + 1 = 31$, a prime

Find an example in which the product of consecutive primes plus 1 does not yield a prime.

56. What is the smallest natural number that is divisible by the first 20 counting numbers?

57. Some primes are 1 more than a square. For example, $5 = 2^2 + 1$. Can you find any other primes p so that $p = n^2 + 1$?

58. Some primes are 1 less than a square. For example, $3 = 2^2 - 1$. Can you find any other primes p so that $p = n^2 - 1$?

59. **HISTORICAL QUESTION** The Pythagoreans studied numbers to find certain mystical properties in them. Certain numbers they studied were called *perfect numbers*. A *perfect number* is a natural number that is equal to the sum of all its divisors that are less than the number itself. A divisor that is less than the number itself is called a *proper divisor*. The proper divisors of 6 are $\{1, 2, 3\}$ and $1 + 2 + 3 = 6$, so 6 is a perfect number. It is not hard to show that 6 is the smallest perfect number. On the other hand, 24 is not perfect, since its proper divisors are $\{1, 2, 3, 4, 6, 8, 12\}$, which have the sum $1 + 2 + 3 + 4 + 6 + 8 + 12 = 36$. The Pythagoreans discovered the first four perfect numbers. Fourteen centuries later the fifth perfect number was discovered. There are only 33 known perfect numbers. The 32nd perfect number is

$$2^{N-1}(2^N - 1)$$

where N is the largest known prime (see Table 4.3). It is known that all even perfect numbers are of the form shown for the 33rd perfect number. Show that if $N = 5$, the resulting number is perfect.

60. **HISTORICAL QUESTION** The Pythagoreans studied numbers that they called *amicable* or *friendly*. A pair of numbers is *friendly* if each number is the sum of the proper divisors of the other (a proper divisor includes the number 1, but not the number itself). The Pythagoreans discovered that 220 and 284 are friendly. The proper divisors of 220 are

$$\{1, 2, 4, 5, 10, 11, 20, 22, 44, 55, 110\}$$

and

$$1 + 2 + 4 + 5 + 10 + 11 + 20 + 22 + 44 + 55 + 110 = 284$$

Also, the proper divisors of 284 are

$$\{1, 2, 4, 71, 142\}$$

and

$$1 + 2 + 4 + 71 + 142 = 220$$

The next pair of friendly numbers was found by Pierre de Fermat: 17,296 and 18,416. In 1638 the French mathematician René Descartes found a third pair, and the Swiss mathematician Leonhard Euler found more than 60 pairs. In 1866, a 16-year-old Italian schoolboy, Nicolo Pagonini, found another relatively small pair of friendly numbers that had been overlooked by the great mathematicians. He found the pair of numbers 1,184 and 1,210. Show that 1,184 and 1,210 are friendly.

4.3 | INTEGERS

Historically, an agricultural-type society would need only natural numbers, but what about a subtraction such as

$$5 - 5 = ?$$

Certainly society would have a need for a number representing $5 - 5$, so a new number, called **zero**, was invented, so that $5 = 5 + 0$ (remember the definition of subtrac-

B.C. cartoon reprinted by permission of Johnny Hart and Creators Syndicate.

tion). If this new number is annexed to the set of natural numbers, a set called the set of **whole numbers** is formed:

$$\mathbb{W} = \{0, 1, 2, 3, 4, \ldots\}$$

This one annexation to the existing numbers satisfied society's needs for several thousand years.

However, as society evolved, the need for bookkeeping advanced, and eventually the need to answer this:

Can we annex new numbers to the set $\mathbb{W}$ so that it is possible to carry out all subtractions?

The numbers that need to be annexed are the opposites of the natural numbers. The opposite of 3, which is denoted by -3, is the number that when added to 3 gives 0. If we add these opposites to the set $\mathbb{W}$ we have the following set:

$$\mathbb{Z} = \{\ldots, -3, -2, -1, 0, 1, 2, 3, \ldots\}$$

This set is known as the set of **integers.** It is customary to refer to certain subsets of $\mathbb{Z}$ as follows:

1. Positive integers: $\{1, 2, 3, 4, \ldots\}$

2. Zero: $\{0\}$

3. Negative integers: $\{-1, -2, -3, \ldots\}$

Now with this new enlarged set of numbers, are we able to carry out all possible additions, subtractions, and multiplications? Before we answer this question, let's review the process by which we operate within the set of integers. It is assumed that you have had an algebra course, so the following summary is intended only as a review.

You might recall that the process for describing the operations with integers requires the notion of *absolute value,* which represents the distance of a number from the origin when plotted on a number line. We give an algebraic definition.

Absolute Value

This definition may be difficult for you to understand; stop for a few moments to make sure you understand what it says.

The **absolute value** of x, denoted by $|x|$, is defined as

$$|x| = \begin{cases} x, & \text{if } x \geq 0 \\ -x, & \text{if } x < 0 \end{cases}$$

EXAMPLE 1

Find the absolute value of the numbers: **a.** $|5|$ **b.** $|-5|$ **c.** $|-(-3)|$

Solution

a. $|5| = 5$, since $5 \geq 0$ **b.** $|-5| = 5$, since $-5 < 0$ and $-(-5) = 5$

c. $|-(-3)| = |3| = 3$, since $3 \geq 0$ ◆

Addition of Integers

If one (or both) of the integers is 0, then we use the identity property to write $x + 0 = 0 + x = x$, for all x. We could introduce the addition of nonzero integers in terms of number lines, and we note that if the numbers we're adding have the same sign, the result is the same as the sum of the absolute values, except for a plus or a minus sign (since their directions on a number line are the same). If we're adding numbers with different signs, their directions are opposite, so the net result is the difference of the absolute values with a sign to indicate final position. This is summarized by the following procedure for adding integers.

Addition of Integers

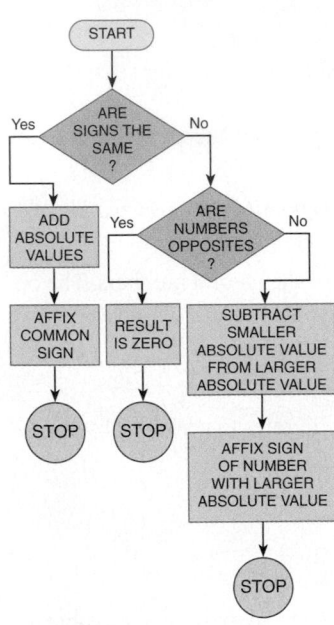

CAUTION

You may feel you already know how to add integers, but this is one of the fundamental ideas of mathematics.

To add nonzero integers x and y, look at the signs of x and y:

1. If the signs are the same:

$$\left.\begin{array}{l}\text{POSITIVE} + \text{POSITIVE} = \text{POSITIVE} \\ \text{NEGATIVE} + \text{NEGATIVE} = \underbrace{\text{NEGATIVE}}_{\text{Sign part}}\end{array}\right\}\underbrace{|x| + |y|}_{\text{Whole-number part}}$$

2. If the signs are different:

$$\left.\begin{array}{l}\text{POSITIVE} + \text{NEGATIVE} \\ \text{NEGATIVE} + \text{POSITIVE}\end{array}\right\} = \underbrace{\left\{\begin{array}{l}\text{Sign of the larger} \\ \text{absolute value}\end{array}\right.}_{\text{Sign part}}\underbrace{\left\{\begin{array}{l}\text{Subtract smaller} \\ \text{absolute value from} \\ \text{larger absolute value}\end{array}\right.}_{\text{Whole-number part}}$$

Notice that all integers consist of two parts: a sign part and a whole-number part.

EXAMPLE 2

Add the integers:

a. $41 + 13$ **b.** $-41 + (-13)$ **c.** $41 + (-13)$ **d.** $-41 + 13$

Solution

a. POSITIVE + POSITIVE: $41 + 13 = 54$ Add absolute values.

b. NEGATIVE + NEGATIVE: $-41 + (-13) = -54$ Add absolute values.

c. POSITIVE + NEGATIVE: $41 + (-13) = 28$ Subtract absolute values.

d. NEGATIVE + POSITIVE: $-41 + 13 = -28$ Subtract absolute values.   ◆

⊞ Calculator Window   `_ □ ✕`

When using a calculator, you must distinguish between a negative sign, as in -5, and a subtraction sign, as in $8 - 5$. For entering negative numbers into a calculator, you'll find a key marked `+/−` or `CHS` or `(−)`. These keys change the sign of a number. For example, $5 + (-6)$ is entered as

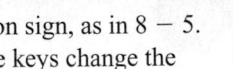

EXAMPLE 3

Indicate the sequence of keys to enter $(-8) + (-5)$ into a calculator.

Solution

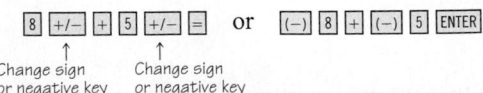

   ↑   ↑
Change sign   Change sign
or negative key   or negative key

   ◆

Notice that a calculator has separate keys for subtraction `−` and opposite `+/−` or `(−)`. The `+/−` or `(−)` key changes the sign of a number to the opposite of its present sign. For example, $5 - (-2)$ would be entered as

$\boxed{5}\ \boxed{-}\ \boxed{2}\ \boxed{+/-}\ \boxed{=}$ or $\boxed{5}\ \boxed{-}\ \boxed{(-)}\ \boxed{2}\ \boxed{\text{ENTER}}$

Because the calculator assumes that all numbers entered are positive, a negative number is obtained by taking the opposite of a positive.

Multiplication of Integers

For whole numbers, multiplication is defined as repeated addition, since we say that $5 \cdot 4$ means

$$\underbrace{4 + 4 + 4 + 4 + 4}_{\text{5 addends}}$$

However, we cannot do this for the integers, since $(-5) \cdot 4$ or

$$\underbrace{4 + 4 + 4 + \cdots + 4}_{}$$

-5 addends does not make sense

Even though you may remember how to multiply integers, we consider four patterns.

POSITIVE · POSITIVE We know how to multiply positive numbers since these are natural numbers. *The product of two positive numbers is a positive number.*

POSITIVE · NEGATIVE Consider, for example, $3 \cdot (-4)$. We look at the pattern:

$$3 \cdot 4 = 12$$
$$3 \cdot 3 = 9$$
$$3 \cdot 2 = 6$$
$$3 \cdot 1 = 3$$
$$3 \cdot 0 = 0$$

What comes next? **Answer this question before reading further.**

$$3 \cdot (-1) = -3$$
$$3 \cdot (-2) = -6$$
$$3 \cdot (-3) = -9$$
$$3 \cdot (-4) = -12$$

Do you know how to continue? Try building a few more such patterns using different numbers. What did you discover about the product of a positive number and a negative number? *The product of a positive number and a negative number is a negative number.*

NEGATIVE · POSITIVE Since we now know how to multiply a positive by a negative, and if we assume the commutative property holds, the result here must be the same for a negative times a positive. *The product of a negative number and a positive number is a negative number.*

NEGATIVE · NEGATIVE Consider the example $-3 \cdot (-4)$. Let's build another pattern.

$$-3 \cdot 4 = -12$$
$$-3 \cdot 3 = -9$$
$$-3 \cdot 2 = -6$$
$$-3 \cdot 1 = -3$$
$$-3 \cdot 0 = 0$$

What comes next? **Answer this question before reading further.**

$$-3 \cdot (-1) = 3$$
$$-3 \cdot (-2) = 6$$
$$-3 \cdot (-3) = 9$$
$$-3 \cdot (-4) = 12$$

Thus, as the pattern indicates: *The product of two negative numbers is a positive number.* We summarize our discussion in the following box.

Multiplication of Integers

CAUTION   Take a few moments with this idea and make sure you know how to multiply integers. Can you explain this idea to someone else?

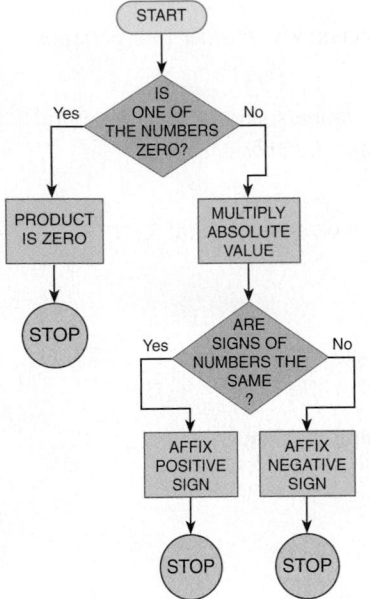

To multiply x and y, look at the signs of x and y:

$$x \cdot 0 = x \quad \text{for all numbers } x$$

$$
\left.
\begin{array}{l}
\text{POSITIVE} \cdot \text{POSITIVE} \ = \text{POSITIVE} \\
\text{POSITIVE} \cdot \text{NEGATIVE} = \text{NEGATIVE} \\
\text{NEGATIVE} \cdot \text{POSITIVE} \ = \text{NEGATIVE} \\
\text{NEGATIVE} \cdot \text{NEGATIVE} = \text{POSITIVE}
\end{array}
\right\}
\underbrace{|x|}_{\text{Sign part}} \cdot \underbrace{|y|}_{\text{Whole-number part}}
$$

EXAMPLE 4

Multiply the given integers.

   a. $(41)(13)$ **b.** $(-41)(-13)$ **c.** $(41)(-13)$ **d.** $(-41)(13)$

Solution

   a. POSITIVE · POSITIVE: $(41)(13) = 533$ positive

   b. NEGATIVE · NEGATIVE: $(-41)(-13) = 533$ positive

   c. POSITIVE · NEGATIVE: $(41)(-13) = -533$ negative

   d. NEGATIVE · POSITIVE: $(-41)(13) = -533$ negative   ◆

Subtraction of Integers

What about subtracting negative numbers? Negative already indicates "going back." Does subtraction of a negative indicate "going ahead?" Consider the following pattern:

$$4 - 4 = 0$$
$$4 - 3 = 1$$
$$4 - 2 = 2$$
$$4 - 1 = 3$$
$$4 - 0 = 4$$

Stop and look for patterns:

$$4 - (-1) = 5$$
$$4 - (-2) = 6$$
$$4 - (-3) = 7$$

Guided by these results, we make the following procedure for subtraction of integers.

Subtraction

STOP   This is saying that we do not subtract integers, but rather we change subtraction problems to addition by adding the opposite.

$$x - y = x + (-y)$$

To subtract, add the opposite (of the number being subtracted).

EXAMPLE 5

Subtract the given integers.

   a. $41 - 13$ **b.** $41 - (-13)$ **c.** $-41 - 13$ **d.** $-41 - (-13)$

Solution

   a. POSITIVE − POSITIVE: $41 \underset{\text{subtraction}}{-} 13 = 41 \underset{\text{addition}}{+} (\underset{\text{opposite}}{-13}) = 28$

                                     complete the addition

b. POSITIVE − NEGATIVE: $41 - (-13) = 41 + 13 = 54$

c. NEGATIVE − POSITIVE $-41 - 13 = -41 + (-13) = -54$

d. NEGATIVE − NEGATIVE: $-41 - (-13) = -41 + 13 = -28$ ◆

Division of Integers

Let's take an overview of what has been done in this chapter. We began with the *natural numbers,* which are closed for addition and multiplication. Next, we defined subtraction and created a situation where it was impossible to subtract some numbers from others. After looking at the prime numbers and factorization, we then "created" another set (called the *integers*) that includes not only the natural numbers, but also zero and the opposite of each of its members.

Since the subtraction of integers is defined in terms of addition, we can easily show that the integers are closed for subtraction. You are asked to do this in Problem 57. We now will define division and then ask the question, "Is the set of integers closed for division?"

Division is defined as the opposite operation of multiplication.

Division

If a, b, and z are integers, where $b \neq 0$, then **division** $a \div b$ is written as $\dfrac{a}{b}$ and is defined in terms of multiplication:

$$\frac{a}{b} = z \quad \text{means} \quad a = bz$$

Since division is defined in terms of multiplication, the rules for dividing integers are identical to those for multiplication. We summarize the procedure for $x \div y$, but first we must make sure $y \neq 0$, because division by zero is not defined.

$$\left.\begin{array}{l} \text{POSITIVE} \div \text{POSITIVE} = \text{POSITIVE} \\ \text{POSITIVE} \div \text{NEGATIVE} = \text{NEGATIVE} \\ \text{NEGATIVE} \div \text{POSITIVE} = \text{NEGATIVE} \\ \text{NEGATIVE} \div \text{NEGATIVE} = \text{POSITIVE} \end{array}\right\} \underbrace{|x|}$$

Sign part   Whole-number part

EXAMPLE 6

Divide the given integers.

a. $12 \div 6$ **b.** $-18 \div 2$ **c.** $10 \div (-2)$ **d.** $-65 \div (-13)$

Solution

a. POSITIVE ÷ POSITIVE: $\dfrac{12}{6} = 2$ **b.** NEGATIVE ÷ POSITIVE: $\dfrac{-18}{2} = -9$

c. POSITIVE ÷ NEGATIVE: $\dfrac{10}{-2} = -5$ **d.** NEGATIVE ÷ NEGATIVE: $\dfrac{-65}{-13} = 5$ ◆

Notice that for $\dfrac{a}{b}$, we require $b \neq 0$. Why do we not allow **division by zero?** We consider two possibilities.

1. Division of a nonzero number by zero:

$$a \div 0 \quad \text{or} \quad \frac{a}{0} = x$$

HISTORICAL NOTE

Srinivasa Ramanujan
(1887–1920)
Ramanujan, who lived only 33 years, was a mathematical prodigy of great originality. He was largely self-taught, but was "discovered" in 1913 by the eminent British mathematician G. H. Hardy. Hardy brought Ramanujan to Cambridge, and in 1918 Ramanujan became the first Indian to become a Fellow of the Royal Society. An often-told story about Hardy and Ramanujan is that when Hardy visited Ramanujan in the hospital, he came in a taxi bearing the number 1729. He asked Ramanujan if there was anything interesting about this number. Without hesitation, Ramanujan said there was: It is the smallest positive integer that can be represented in two different ways as a sum of two cubes:

$$1{,}729 = 1^3 + 12^3 = 9^3 + 10^3$$

What does this mean? Is there such a number x so that this makes sense? We see that any number x would have to be such that $a = 0 \cdot x$. But $0 \cdot x = 0$ for all x, and since $a \neq 0$, we see that such a situation is impossible. That is, $a \div 0$ does not exist.

2. Division of zero by zero:

$$0 \div 0 \quad \text{or} \quad \frac{0}{0} = x$$

What does this mean? Is there such a number x? We see that *any* x makes this true, since $0 \cdot x = 0$ for all x. But this leads to certain absurdities, for example:

If $\dfrac{0}{0} = 2$, then this checks since $0 \cdot 2 = 0$; also

if $\dfrac{0}{0} = 5$, then this checks since $0 \cdot 5 = 0$.

NEVER DIVIDE BY 0.

CAUTION

Abraham Lincoln used a biblical reference (Mark 3:25) to initiate his campaign in 1858. He said, "A house divided against itself cannot stand." I offer a corollary to Lincoln's statement: "A house divided by itself is one (provided, of course, that the house is not zero)."

But since both 2 and 5 are equal to the *same* number, we would conclude that $2 = 5$. This is absurd, so we say that division of zero by zero is excluded (or that it is *indeterminate*).

Is the set of integers closed for division? Certainly we can find many examples in which an integer divided by an integer is an integer. Does this mean that the set of integers is closed? What about $1 \div 2$ or $4 \div 5$? These numbers do not exist in the set of integers; thus, the set is *not* closed for division. Now, as long as society has no need for such division problems, the question of inventing new numbers will not arise. However, as the need to divide 1 into 2 or more parts arises, some new numbers will have to be invented so that the set will be closed for division. We'll do this in the next section. The problem with inventing such new numbers is that it must be done in such a way that the properties of the existing numbers are left unchanged. That is, closure for addition, subtraction, and multiplication must be retained.

Problem Set 4.3

LEVEL 1

1. **IN YOUR OWN WORDS** Explain how to add integers.

2. **IN YOUR OWN WORDS** Explain how to subtract integers.

3. **IN YOUR OWN WORDS** Explain how to multiply integers.

4. **IN YOUR OWN WORDS** Explain how to divide integers.

5. **IN YOUR OWN WORDS** Explain the difference between $0 \div 5$ and $5 \div 0$.

6. **IN YOUR OWN WORDS** Why is division by 0 not defined?

Evaluate each absolute value expression in Problems 7–8.

7. **a.** $|30|$ **b.** $|-30|$ **c.** $|0|$ **d.** $-|30|$
    e. $|8| - |8|$

8. **a.** $|18|$ **b.** $|-18|$ **c.** $-|-18|$ **d.** $|-(-18)|$
    e. $|-18| + |18|$

Simplify the expressions in Problems 9–37.

9. **a.** $5 + 3$ **b.** $-5 + 3$

10. **a.** $4 + (-7)$ **b.** $-2 + (-4)$

11. **a.** $7 + 3$ **b.** $-9 + 5$

12. **a.** $-10 + 4$ **b.** $-8 + (-10)$

13. **a.** $-15 + 8$ **b.** $|-14 + 2|$

14. **a.** $|8 + (-10)|$ **b.** $-38 + (-14)$

15. **a.** $10 - 7$ **b.** $7 - 10$

16. **a.** $6 - (-4)$ **b.** $0 - (-15)$

17. **a.** $3(-6)$ **b.** $-5(4)$

18. **a.** $\dfrac{-4}{-2}$ **b.** $\dfrac{-6}{-3}$

19. **a.** $14(-5)$ **b.** $-14(-5)$

20. **a.** $-5(8-12)$ **b.** $[-5(8)]-12$

21. **a.** $\dfrac{-12}{4}$ **b.** $\dfrac{-63}{-9}$

22. **a.** $\dfrac{12}{-4}$ **b.** $(-6)\dfrac{14}{-2}$

23. **a.** $\dfrac{-528}{-4}$ **b.** $(-1)^3$

24. **a.** $(-1)^4$ **b.** $10\left(\dfrac{-8}{-2}\right)$

25. **a.** $7(-8)$ **b.** $-5(15)$

26. **a.** $-5(-6)$ **b.** $\dfrac{-42}{3}$

27. **a.** -2^2 **b.** $(-2)^2$

28. **a.** $(-3)^2$ **b.** -3^2

29. **a.** $162+(-12)$ **b.** $-12+[(-4)+(-3)]$

30. **a.** $-46-(-46)$ **b.** $|7-(-3)|$

31. **a.** $|-5-(-10)|$ **b.** $|5-(-5)|$

32. **a.** $-7-(-18)$ **b.** $62-(-112)$

33. **a.** $-4-8$ **b.** $31+(-16)$

34. **a.** $-14-21$ **b.** $-9+16+(-11)$

35. **a.** $|-8|+(-8)$ **b.** $-9-(4-5)$

36. **a.** $-(23+14)$ **b.** $-7-(6-4)$

37. **a.** $-6-(-6)$ **b.** $-18-5$

LEVEL 2

Simplify the expressions in Problems 38–50.

38. **a.** $-8+7+16$
    b. $14+(-10)-8-11$

39. **a.** $6+(-8)-5$
    b. $-2(-3)+(-1)(6)$

40. **a.** $-5(7)-(-9)$
    b. $-2(3)-(-8)$

41. **a.** $6-(-2)$
    b. $-5-(-3)$

42. **a.** $-15-(-6)$
    b. $-4+|6-8|$

43. **a.** $\dfrac{-32}{-8}-5-(-7)$
    b. $\dfrac{-15}{-5}-4-(-8)$

44. **a.** $-3-[(-6)-4]$
    b. $5+(-19)+|15|$

45. **a.** $|-3|-[-(-2)]$
    b. $15-(-7)$

46. **a.** $[-54\div(-9)]\div 3$
    b. $-54\div[(-9)\div 3]$

47. **a.** $[48\div(-6)]\div(-2)$
    b. $48\div[(-6)\div(-2)]$

48. **a.** $15-(-3)-|4-11|$
    b. $-12+(-7)-10-14$

49. **a.** $-5(2)+(-3)(-4)-6(-7)$
    b. $-8(3)-6(-4)-2(-8)$

50. **a.** $1(-2)(3)(-4)(5)(-6)(7)(-8)(9)(-10)$
    b. $1+(-2)+3+(-4)+5+(-6)+7+(-8)$
    $+9+(-10)$

LEVEL 3

51. Perform the indicated operations. Let k be a natural number.
    a. 1^6 **b.** 1^{67} **c.** 1^{2001}
    d. 1^{2k} **e.** 1^{2k+1}

52. Perform the indicated operations. Let k be a natural number.
    a. 2^2 **b.** 2^3 **c.** 2^4
    d. 2^5
    e. Is 2^{2k+1} positive or negative?

53. Perform the indicated operations. Let k be a natural number.
    a. $(-1)^6$ **b.** $(-1)^{67}$ **c.** $(-1)^{2001}$
    d. $(-1)^{2k}$ **e.** $(-1)^{2k+1}$

54. Perform the indicated operations. Let k be a natural number.
    a. $(-2)^2$ **b.** $(-2)^3$ **c.** $(-2)^4$
    d. $(-2)^5$
    e. Is $(-2)^{2k+1}$ positive or negative?

55. **a.** State the commutative property.
    b. Is $\mathbb{Z}$ commutative for addition? Give reasons.
    c. Is $\mathbb{Z}$ commutative for subtraction? Give reasons.
    d. Is $\mathbb{Z}$ commutative for multiplication? Give reasons.
    e. Is $\mathbb{Z}$ commutative for division? Give reasons.

56. **a.** State the associative property.
    b. Is $\mathbb{Z}$ associative for addition? Give reasons.
    c. Is $\mathbb{Z}$ associative for subtraction? Give reasons.
    d. Is $\mathbb{Z}$ associative for multiplication? Give reasons.
    e. Is $\mathbb{Z}$ associative for division? Give reasons.

57. Show that the set $\mathbb{Z}$ is closed for subtraction.

PROBLEM SOLVING

58. Find a finite subset of $\mathbb{Z}$ that is closed for multiplication.

59. Multiply $1{,}234{,}567 \times 9{,}999{,}999$
    a. using a calculator   **b.** using patterns

60. Four Fours   B.C. apparently has a mental block against fours, as we can see from the cartoon.

See if you can handle fours by writing the numbers from 1 to 10 using four 4s, operation symbols, or possibly grouping symbols for each. Here are the first three completed for you:

$$\frac{4}{4} + 4 - 4 = 1$$

$$\frac{4}{4} + \frac{4}{4} = 2$$

$$\frac{4 + 4 + 4}{4} = 3$$

B.C. cartoon reprinted by permission of Johnny Hart and Creators Syndicate.

4.4 RATIONAL NUMBERS

Historically, the need for a closed set for division came before the need for closure for subtraction. We need to find some number k so that

$$1 \div 2 = k$$

As we saw in Section 3.1, the ancient Egyptians limited their fractions by requiring the numerators to be 1. The Romans avoided fractions by the use of subunits; feet were divided into inches and pounds into ounces, and a twelfth part of the Roman unit was called an *uncia*.

However, people soon felt the practical need to obtain greater accuracy in measurement and the theoretical need to close the number system with respect to the operation of division. In the set $\mathbb{Z}$ of integers, some divisions are possible:

$$\frac{10}{-2}, \quad \frac{-4}{2}, \quad \frac{-16}{-8}, \quad \cdots$$

However, certain others are not:

$$\frac{1}{2}, \quad \frac{-16}{5}, \quad \frac{5}{12}, \quad \cdots$$

Just as we extended the set of natural numbers by creating the concept of opposites, we can extend the set of integers. That is, the number $\frac{5}{12}$ is defined to be that number obtained when 5 is divided by 12. This new set, consisting of the integers as well as the quotients of integers, is called the set of *rational numbers*.

Rational Number

> The set of **rational numbers,** denoted by $\mathbb{Q}$, is the set of all numbers of the form
>
> $$\frac{a}{b}$$
>
> where a and b are integers, and $b \neq 0$.

HISTORICAL NOTE

The symbol ÷ was adopted by John Wallis (1616–1703) and was used in Great Britain and in the United States [but not on the European continent, where the colon (:) was used]. In 1923, the National Committee on Mathematical Requirements stated: "Since neither ÷ nor : as signs of division play any part in business life, it seems proper to consider only the needs of algebra, and to make more use of the fractional form and (where the meaning is clear) of the symbol "/" and to drop the symbol ÷ in writing algebraic expressions."
—From *Report of the National Committee on Mathematical Requirements* under the auspices of the Mathematical Association of America, Inc. (1923), p. 81.

Notice that a rational number has fractional form. In arithmetic you learned that if a number is written in the form $\frac{a}{b}$ it means $a \div b$ and that a is called the **numerator** and b the **denominator.** Also, if a and b are both positive, $\frac{a}{b}$ is called

a **proper fraction** if $a < b$;

an **improper fraction** if $a > b$; and

a **whole number** if b divides evenly into a.

It is assumed that you know how to perform the basic operations with fractions, but we will spend the next few pages reviewing those operations.

Fundamental Property

If the greatest common factor of the numerator and denominator of a given fraction is 1, then we say the fraction is in lowest terms or **reduced.** If the greatest common factor is not 1, then divide both the numerator and denominator by this greatest common factor using the **fundamental property of fractions.**

Fundamental Property of Fractions

> If $\frac{a}{b}$ is any rational number and x is any nonzero integer, then
>
> $$\frac{a \cdot x}{b \cdot x} = \frac{x \cdot a}{x \cdot b} = \frac{a}{b}$$

STOP The fundamental property works only for products (factors) and not for sums.

That is, given some fraction that you wish to simplify:

1. Find the g.c.f. of the numerator and denominator (this is x in the fundamental property).

2. Use the fundamental property to simplify the fraction.

EXAMPLE 1

Reduce the given fractions.

a. $\dfrac{24}{30}$ **b.** $\dfrac{300}{144}$

In this book, we agree to leave all fractional answers in reduced form.

Solution

a. First, find the greatest common factor: $24 = 2^3 \cdot 3^1 \cdot 5^0$
$$30 = 2^1 \cdot 3^1 \cdot 5^1$$
$$\text{g.c.f.} = 2^1 \cdot 3^1 \cdot 5^0 = 6$$

Next, use the fundamental property to simplify the fraction:

Thus, $\dfrac{24}{30} = \dfrac{6 \cdot 2^2}{6 \cdot 5} = \dfrac{2^2}{5} = \dfrac{4}{5}$

↑
g.c.f.

b. $300 = 2^2 \cdot 3^1 \cdot 5^2$

$144 = 2^4 \cdot 3^2 \cdot 5^0$

$\text{g.c.f.} = 2^2 \cdot 3^1 \cdot 5^0 = 12$

$\dfrac{300}{144} = \dfrac{12 \cdot 5^2}{12 \cdot 2^2 \cdot 3} = \dfrac{5^2}{2^2 \cdot 3} = \dfrac{25}{12}$

Note that $\frac{25}{12}$ is reduced because the g.c.f. of the numerator and denominator is 1. Notice that a reduced fraction may be an improper fraction.   ◆

Operations with Rationals

Operations with Rational Numbers

If $\dfrac{a}{b}$ and $\dfrac{c}{d}$ are rational numbers, then

Addition: $\dfrac{a}{b} + \dfrac{c}{d} = \dfrac{ad}{bd} + \dfrac{bc}{bd} = \dfrac{ad + bc}{bd}$

Subtraction: $\dfrac{a}{b} - \dfrac{c}{d} = \dfrac{ad}{bd} - \dfrac{bc}{bd} = \dfrac{ad - bc}{bd}$

Multiplication: $\dfrac{a}{b} \times \dfrac{c}{d} = \dfrac{ac}{bd}$

Division: $\dfrac{a}{b} \div \dfrac{c}{d} = \dfrac{ad}{bc}$ $(c \neq 0)$

You will note that both addition and subtraction require that we first obtain *common denominators*. That process requires a multiplication of fractions, so we begin with an example reviewing multiplication of rational numbers.

EXAMPLE 2

Multiply the given rational numbers.

a. $\dfrac{1}{3} \times \dfrac{2}{5}$ **b.** $\dfrac{2}{3} \times \dfrac{-4}{7}$ **c.** $-5 \times \dfrac{-2}{3}$ **d.** $3\frac{1}{2} \times 2\frac{3}{5}$ **e.** $\dfrac{3}{4} \times \dfrac{2}{3}$

Solution

a. $\dfrac{1}{3} \times \dfrac{2}{5} = \boxed{\dfrac{1 \times 2}{3 \times 5}} = \dfrac{2}{15}$

↑
This step is often done in your head

b. $\dfrac{2}{3} \times \dfrac{-4}{7} = \dfrac{-8}{21}$

c. When multiplying a whole number and a fraction, write the whole number as a fraction, and then multiply:

$$-5 \times \frac{-2}{3} = \frac{-5}{1} \times \frac{-2}{3} = \frac{10}{3}$$

d. When multiplying mixed numbers, write the mixed numbers as improper fractions, and *then* multiply:

$$3\tfrac{1}{2} \times 2\tfrac{3}{5} = \frac{7}{2} \times \frac{13}{5} = \frac{91}{10}$$

e. For most complicated fractions, the best procedure is to reduce *before* multiplying rather than after, as illustrated here.

$$\frac{\overset{1}{\cancel{3}}}{\underset{2}{\cancel{4}}} \times \frac{\overset{1}{\cancel{2}}}{\underset{1}{\cancel{3}}} = \frac{1 \times 1}{2 \times 1} = \frac{1}{2}$$

◆

EXAMPLE 3   **PÓLYA'S METHOD**

Justify the rule for division of rational numbers.

Solution   We use Pólya's problem-solving guidelines for this example.

Understand the Problem. Given $\dfrac{a}{b} \div \dfrac{c}{d}$ where $c \neq 0$. Note that we don't say $b \neq 0$ and $d \neq 0$, because the definition of rational number excludes these possibilities, but does not exclude $c = 0$, so this is the condition that must be stated. The example asks us to show where the rule for division as stated in the previous box comes from. Let

$$\frac{a}{b} \div \frac{c}{d} = \square$$

We are looking for the value of $\square$. To understand what we are doing here, look at a more familiar problem:

$$\frac{2}{3} \div \frac{4}{5} = \square \quad \text{means} \quad \frac{4}{5} \times \square = \frac{2}{3}$$

What do we put into the box to get the answer? Do it in two steps. First multiply $\frac{4}{5}$ by $\frac{5}{4}$ to obtain 1, and *then* multiply by $\frac{2}{3}$ to obtain the result that makes the equation true:

$$\frac{4}{5} \times \boxed{\frac{5}{4} \times \frac{2}{3}} = \frac{2}{3}$$

Thus, $\dfrac{2}{3} \div \dfrac{4}{5} = \boxed{\dfrac{5}{4} \times \dfrac{2}{3}}$. This seems to suggest that we "invert" the fraction we are dividing by, and then multiply.

Devise a Plan. We will use the definition of division, and then the fundamental property of fractions to multiply the numerator and denominator by the same number.

Carry Out the Plan.

$$\frac{a}{b} \div \frac{c}{d} = \frac{\dfrac{a}{b}}{\dfrac{c}{d}} \qquad \text{Write the division using fractional notation.}$$

$$= \frac{\dfrac{a}{b} \times \dfrac{d}{c}}{\dfrac{c}{d} \times \dfrac{d}{c}} \qquad \text{Multiply numerator and denominator by } \frac{d}{c}; \text{ fundamental property of fractions}$$

$$= \frac{\dfrac{ad}{bc}}{\dfrac{cd}{dc}} \qquad \text{Multiplication of the "big" fractions}$$

$$= \frac{\dfrac{ad}{bc}}{1} \qquad \text{Reduce the fraction } \frac{cd}{dc}.$$

$$= \frac{ad}{bc}$$

Look Back. The result here, $\dfrac{a}{b} \div \dfrac{c}{d} = \dfrac{ad}{bc}$, checks with the entry in the box. ◆

EXAMPLE 4

Divide the given rational numbers.

a. $\dfrac{-3}{4} \div \dfrac{-4}{7}$ **b.** $\dfrac{4}{3} \div \dfrac{8}{9}$

Solution

a. $\dfrac{-3}{4} \div \dfrac{-4}{7} = \dfrac{-3}{4} \times \dfrac{7}{-4} = \dfrac{-21}{-16} = \dfrac{21}{16}$ **b.** $\dfrac{4}{3} \div \dfrac{8}{9} = \dfrac{\overset{1}{\cancel{4}}}{\underset{1}{\cancel{3}}} \times \dfrac{\overset{3}{\cancel{9}}}{\underset{2}{\cancel{8}}} = \dfrac{1 \times 3}{1 \times 2} = \dfrac{3}{2}$ ◆

To carry out addition or subtraction of fractions, you must find the **least common denominator.** The least common denominator is the same as the least common multiple.

EXAMPLE 5

Simplify the given expressions.

a. $\dfrac{5}{24} + \dfrac{7}{30}$ **b.** $\dfrac{19}{300} + \dfrac{55}{144} + \dfrac{25}{108}$ **c.** $\dfrac{7}{18} - \dfrac{-5}{24}$ **d.** $\dfrac{-1}{15} - \dfrac{27}{50}$

Solution

This example is intentionally long. Stick with it because it was designed to illustrate many of the procedures of this section.

a. The denominators are 24 and 30, so we find the l.c.m. of these numbers.

$$24 = 2^3 \cdot 3^1 \cdot 5^0$$
$$30 = 2^1 \cdot 3^1 \cdot 5^1$$
$$\text{l.c.m.} = 2^3 \cdot 3^1 \cdot 5^1 = 120$$

$$\frac{5}{24} = \frac{5}{24} \cdot \frac{5}{5} = \frac{25}{120}$$

We are simply multiplying each fraction by the identity 1, since $\frac{5}{5}$ and $\frac{4}{4}$ are both equal to 1.

$$+ \frac{7}{30} = \frac{7}{30} \cdot \frac{4}{4} = \frac{28}{120}$$

$$\frac{53}{120}$$

The answer is in reduced form, since 53 and 120 are relatively prime.

b. If the numbers are complicated, it is easier to work in factored form. In Section 4.2 we found the l.c.m. of 300, 144, and 108 to be $2^4 \cdot 3^3 \cdot 5^2$. Thus,

$$\frac{19}{300} = \frac{19}{2^2 \cdot 3 \cdot 5^2} \cdot \frac{2^2 \cdot 3^2}{2^2 \cdot 3^2} = \frac{19 \cdot 2^2 \cdot 3^2}{2^4 \cdot 3^3 \cdot 5^2}$$ This is the step during which we multiply each fraction by 1.

$$\frac{55}{144} = \frac{5 \cdot 11}{2^4 \cdot 3^2} \cdot \frac{3 \cdot 5^2}{3 \cdot 5^2} = \frac{3 \cdot 5^3 \cdot 11}{2^4 \cdot 3^3 \cdot 5^2}$$

$$+ \frac{25}{108} = \frac{5^2}{2^2 \cdot 3^3} \cdot \frac{2^2 \cdot 5^2}{2^2 \cdot 5^2} = \frac{2^2 \cdot 5^4}{2^4 \cdot 3^3 \cdot 5^2}$$

Adding: $$\frac{19 \cdot 2^2 \cdot 3^2}{2^4 \cdot 3^3 \cdot 5^2} + \frac{3 \cdot 5^3 \cdot 11}{2^4 \cdot 3^3 \cdot 5^2} + \frac{2^2 \cdot 5^4}{2^4 \cdot 3^3 \cdot 5^2}$$

$$= \frac{684}{2^4 \cdot 3^3 \cdot 5^2} + \frac{4{,}125}{2^4 \cdot 3^3 \cdot 5^2} + \frac{2{,}500}{2^4 \cdot 3^3 \cdot 5^2} = \frac{7{,}309}{2^4 \cdot 3^3 \cdot 5^2}$$

Now 7,309 is not divisible by 2, 3, or 5; thus, 7,309 and $2^4 \cdot 3^3 \cdot 5^2$ are relatively prime, and the solution is complete:

$$\frac{7{,}309}{2^4 \cdot 3^3 \cdot 5^2} \quad \text{or} \quad \frac{7{,}309}{10{,}800}$$

c.
$$18 = 2^1 \cdot 3^2$$

$$24 = 2^3 \cdot 3^1$$

$$\text{l.c.m.} = 2^3 \cdot 3^2 = 72$$

$$\frac{7}{18} = \frac{7}{18} \cdot \frac{4}{4} = \frac{28}{72}$$

$$- \frac{-5}{24} = \frac{5}{24} \cdot \frac{3}{3} = \frac{15}{72}$$

$$\frac{43}{72}$$

d.
$$15 = \qquad 3^1 \cdot 5^1$$

$$50 = 2^1 \cdot \qquad 5^2$$

$$\text{l.c.m.} = 2^1 \cdot 3^1 \cdot 5^2 = 150$$

$$\frac{-1}{15} = \frac{-1}{15} \cdot \frac{10}{10} = \frac{-10}{150}$$

$$+ \; -\frac{27}{50} = \frac{-27}{50} \cdot \frac{3}{3} = \frac{-81}{150}$$ Notice with subtraction it is usually easier to add the opposite.

$$\frac{-91}{150}$$

◆

The set $\mathbb{Q}$ is closed for the operations of addition, subtraction, multiplication, and nonzero division. As an example, we will show that the rationals are closed for addition. We need to show that, given any two elements of $\mathbb{Q}$, their sum is also an element of $\mathbb{Q}$. Suppose

$$\frac{x}{y} \text{ and } \frac{w}{z} \text{ are any two rational numbers.}$$

By definition of addition,

$$\frac{x}{y} + \frac{w}{z} = \frac{xz + wy}{yz}$$

We now need to show that $\dfrac{xz + wy}{yz}$ is a rational number. Since x and w are integers and y and z are nonzero integers, we know from closure of the integers for multiplication that xz and wy are also integers. Since the set of integers is closed for addition, we know that $xz + wy$ is also an integer. This means that, since yz is a nonzero integer,

$$\frac{xz + wy}{yz} \text{ is a rational number.}$$

Thus, the rational numbers are closed for addition.

Problem Set 4.4

LEVEL 1

1. **IN YOUR OWN WORDS** What does it mean for a fraction to be reduced? Describe a process for reducing a fraction.

2. **IN YOUR OWN WORDS** Describe the process for multiplying fractions.

3. **IN YOUR OWN WORDS** Describe the process for dividing fractions.

4. **IN YOUR OWN WORDS** Describe the process for adding fractions.

5. **IN YOUR OWN WORDS** Use algebra to show where the formula for subtracting fractions comes from.

Completely reduce the fractions in Problems 6–17.

6. **a.** $\dfrac{3}{9}$ **b.** $\dfrac{6}{9}$ 7. **a.** $\dfrac{2}{10}$ **b.** $\dfrac{3}{12}$

8. **a.** $\dfrac{4}{12}$ **b.** $\dfrac{6}{12}$ 9. **a.** $\dfrac{14}{7}$ **b.** $\dfrac{38}{19}$

10. **a.** $\dfrac{92}{20}$ **b.** $\dfrac{72}{15}$ 11. **a.** $\dfrac{42}{14}$ **b.** $\dfrac{16}{24}$

12. **a.** $\dfrac{18}{30}$ **b.** $\dfrac{70}{105}$ 13. **a.** $\dfrac{50}{400}$ **b.** $\dfrac{140}{420}$

14. **a.** $\dfrac{150}{1,000}$ **b.** $\dfrac{2,500}{10,000}$ 15. **a.** $\dfrac{78}{455}$ **b.** $\dfrac{75}{500}$

16. **a.** $\dfrac{240}{672}$ **b.** $\dfrac{5,670}{12,150}$ 17. **a.** $\dfrac{2,431}{3,003}$ **b.** $\dfrac{47,957}{54,808}$

Perform the indicated operations in Problems 18–37. (Recall that negative exponents are sometimes used to denote fractions. For example, $\frac{1}{7} = 7^{-1}$.)

18. **a.** $\dfrac{2}{3} + \dfrac{7}{9}$ **b.** $\dfrac{-5}{7} + \dfrac{4}{3}$

19. **a.** $\dfrac{-12}{35} - \dfrac{8}{15}$ **b.** $2^{-1} + 3^{-1}$

20. **a.** $3 + 3^{-1}$ **b.** $2 + 2^{-1}$

21. **a.** $\dfrac{7}{9} - \dfrac{2}{3}$ **b.** $\dfrac{4}{7} - \dfrac{-5}{9}$

22. **a.** $\dfrac{-3}{5} - \dfrac{-6}{9}$ **b.** $\dfrac{6}{7} \div \dfrac{-3}{7}$

23. **a.** $\dfrac{5}{3} \div \dfrac{7}{12}$ **b.** $\dfrac{-2}{9} \div \dfrac{6}{7}$

24. **a.** $\dfrac{2}{3} \times 9$ **b.** $6 \times \dfrac{5}{9}$

25. **a.** $3^{-1} + 5^{-1}$ **b.** $2^{-1} + 5^{-1}$

26. **a.** $\dfrac{1}{2} + \dfrac{1}{3} + \dfrac{1}{5}$ **b.** $2^{-1} + 3^{-1} + 5^{-1}$

27. **a.** $\dfrac{2}{3} \times \dfrac{5}{7}$ **b.** $\dfrac{-1}{8} \times \dfrac{2}{-5}$

28. **a.** $\dfrac{4}{9} \div \dfrac{2}{3}$ **b.** $\dfrac{105}{-11} \div \dfrac{-15}{33}$

29. **a.** 7×7^{-1} **b.** -14×14^{-1}

30. **a.** 8×8^{-1} **b.** -12×12^{-1}

31. a. $6 \div 6^{-1}$ **b.** $-5 \div 5^{-1}$

32. a. $\dfrac{1}{10} \cdot \dfrac{-2}{5}$ **b.** $\dfrac{-5}{18} \cdot \dfrac{9}{25}$

33. a. $\left(\dfrac{3}{7} \cdot \dfrac{3}{5}\right) \div \dfrac{1}{2}$ **b.** $\dfrac{2}{-3} \cdot \dfrac{-2}{15}$

34. a. $\dfrac{4}{5}\left(\dfrac{17}{95}\right) + \dfrac{4}{5}\left(\dfrac{78}{95}\right)$ **b.** $\dfrac{4}{5}\left(\dfrac{17}{95} + \dfrac{78}{95}\right)$

35. a. $\dfrac{2}{3} - \dfrac{7}{12}$ **b.** $\dfrac{-7}{24} + \dfrac{-13}{16}$

36. a. $\dfrac{28}{9} - \dfrac{4}{27}$ **b.** $\dfrac{1}{-8} + \dfrac{1}{-7}$

37. a. $-5 + (-5)^2$ **b.** $-3 + (-3)^2$

LEVEL 2

Perform the indicated operations in Problems 38–48.

38. a. $\dfrac{-2}{15} + \dfrac{3}{5} + \dfrac{7}{12}$ **b.** $-2\frac{3}{5} + 4\frac{1}{8} - 7\frac{1}{10}$

39. a. $\dfrac{\frac{1}{2} + \frac{-2}{3}}{\frac{5}{6} - \frac{-3}{5}}$ **b.** $\dfrac{\frac{1}{3} - \frac{-1}{4}}{\frac{7}{8} - \frac{3}{16}}$

40. a. $\dfrac{2^{-1} + 3^{-2}}{2^{-1} + 3^{-1}}$ **b.** $\dfrac{6 + 2^{-1}}{\frac{1}{2} + \frac{1}{3}}$

41. a. $\dfrac{-3}{4} \cdot \dfrac{119}{200} + \dfrac{-3}{4} \cdot \dfrac{81}{200}$ **b.** $\dfrac{-3}{4}\left(\dfrac{119}{200} + \dfrac{81}{200}\right)$

42. a. $3 + 3^{-1} + 3^{-2}$ **b.** $5^{-1} + 5^{-2} + 5$

43. $\dfrac{11}{144} + \dfrac{17}{300} + \dfrac{7}{108}$ **44.** $\dfrac{11}{108} - \dfrac{7}{144} + \dfrac{23}{300}$

45. $\dfrac{7}{60} - \dfrac{19}{90} + \dfrac{21}{51}$ **46.** $\dfrac{14}{90} + \dfrac{7}{60} - \dfrac{11}{50}$

47. $\dfrac{143}{210} + \dfrac{15}{124} + \dfrac{11}{1,085}$ **48.** $\dfrac{15}{484} - \dfrac{5}{234} + \dfrac{27}{200}$

LEVEL 3

49. Show that the set $\mathbb{Q}$ of rationals is closed for subtraction.

50. Show that the set $\mathbb{Q}$ of rationals is closed for nonzero division.

51. Is the set $\mathbb{Z}$ of integers closed for addition, subtraction, multiplication, and nonzero division? Explain.

52. Is $\mathbb{Q}$, the set of rationals, associative for addition? Explain.

53. Is $\mathbb{Q}$, the set of rationals, commutative for addition? Explain.

54. Is $\mathbb{Q}$, the set of rationals, associative and/or commutative for multiplication? Explain.

PROBLEM SOLVING

A unit fraction (a fraction with a numerator of 1) is sometimes called an Egyptian fraction. On page 103, we said that the Egyptians expressed their fractions as a sum of distinct (different) unit fractions. How might the Egyptians have written the fractions in Problems 55–58?

55. $\frac{3}{4}$ **56.** $\frac{47}{60}$

57. $\frac{67}{120}$ **58.** $\frac{1}{17}$

59.

> *A quantity and its two-thirds and its half and its one-seventh together make 33. Find the quantity.*
>
> *apyrus*
> *answer:* $14 + \frac{1}{4} + \frac{1}{56} + \frac{1}{97} + \frac{1}{194} + \frac{1}{388} + \frac{1}{679} + \frac{1}{776}$

Recall that the Egyptians used only unit fractions to represent numbers. Is the answer given on the papyrus correct?

60. The sum and difference of the same two squares may be primes, as in this example:

$$9 - 4 = 5 \quad \text{and} \quad 9 + 4 = 13$$

Can the sum and difference of the same two primes be squares? Can you find more than one example?

4.5 IRRATIONAL NUMBERS

Pythagorean Theorem

We have been considering numbers as they relate to practical problems. However, numbers can be appreciated for their beauty and interrelationships. The Pythagoreans, to our knowledge, were among the first to investigate numbers for their own sake.

   Much of the Pythagoreans' lifestyle was embodied in their beliefs about numbers. They considered the number 1 the essence of reason; the number 2 was identified with opinion; and 4 was associated with justice because it is the first number that is the product of equals (the first perfect squared number, other than 1). Of the numbers greater than 1, odd numbers were masculine and even numbers were feminine; thus, 5 represented marriage, since it was the union of the first masculine and feminine numbers ($2 + 3 = 5$).

The Pythagoreans were also interested in special types of numbers that had mystical meanings: perfect numbers, friendly numbers, deficient numbers, abundant numbers, prime numbers, triangular numbers, square numbers, and pentagonal numbers. Other than the prime numbers we have already considered, the perfect square numbers are probably the most interesting. They are called **perfect squares** because they can be arranged into squares (see Figure 4.4). They are found by squaring the natural numbers.

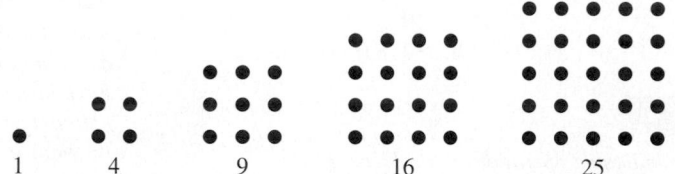

Figure 4.4 Square numbers 1, 4, 9, 16, and 25. Other square numbers (not pictured) are 36, 49, 64, 81, 100, 121, 144, 169, ...

The Pythagoreans discovered the famous property of **square numbers** that today bears Pythagoras' name. They found that if they constructed any right triangle and then constructed squares on each of the legs of the triangle, the area of the larger square was equal to the sum of the areas of the smaller squares (see Figure 4.5).

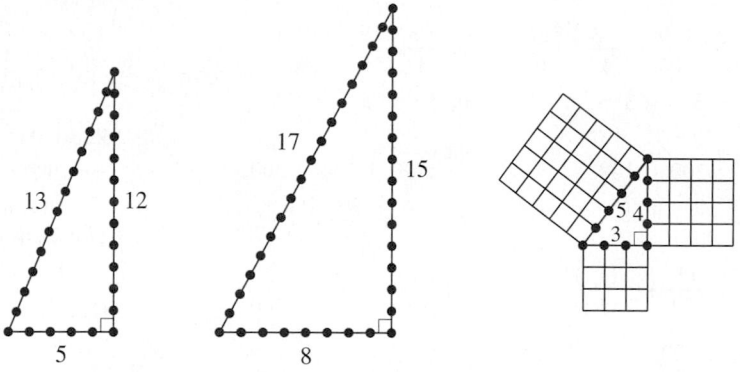

Figure 4.5 Relationships of sides of right triangles

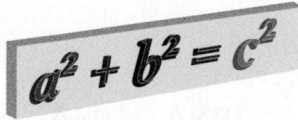

$$a^2 + b^2 = c^2$$

Today we state the **Pythagorean theorem** algebraically by saying that, if a and b are the lengths of the **legs** (or sides) of a right triangle, and c is the length of the **hypotenuse** (the longest side), then the square of the length of the hypotenuse is equal to the sum of the squares of the lengths of the other two sides.

The following result, called the Pythagorean theorem, is one of the most famous (and important) results in all of mathematics.

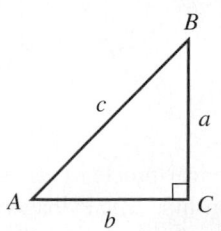

Pythagorean Theorem

For a right triangle ABC, with sides of length a, b, and hypotenuse c,

$$a^2 + b^2 = c^2$$

Also, if $a^2 + b^2 = c^2$ for a triangle with sides a, b, and c, then $\triangle ABC$ is a right triangle.

B.C. cartoon reprinted by permission of Johnny Hart and Creators Syndicate.

Square Roots

The Pythagoreans were overjoyed with this discovery, but it led to a revolutionary idea in mathematics—one that caused the Pythagoreans many problems.

Legend tells us that one day while the Pythagoreans were at sea, one of their group came up with the following argument. Suppose each leg of a right triangle is 1, then

$$a^2 + b^2 = c^2$$
$$1^2 + 1^2 = c^2$$
$$2 = c^2$$

If we denote the number whose square is 2 by $\sqrt{2}$, we have $\sqrt{2} = c$. The symbol $\sqrt{2}$ is read "square root of two." This means that $\sqrt{2}$ is that number such that, if multiplied by itself, is exactly equal to 2; i.e., $\sqrt{2} \times \sqrt{2} = 2$.

Square Root

> The **square root** of a nonnegative number n is a number so that its square is equal to n. In symbols, the **positive square root of n**, denoted by $\sqrt{n}$, is defined as that number for which
>
> $$\sqrt{n}\,\sqrt{n} = n$$

EXAMPLE 1

What is $\sqrt{3}$?
The definition says that it is the number with the property that $\sqrt{3}\,\sqrt{3} = 3$.

Use the definition of square root to find the indicated products.

a. $\sqrt{3} \times \sqrt{3}$ **b.** $\sqrt{4} \times \sqrt{4}$ **c.** $\sqrt{5} \times \sqrt{5}$ **d.** $\sqrt{15} \times \sqrt{15}$

e. $\sqrt{16} \times \sqrt{16}$ **f.** $\sqrt{144} \times \sqrt{144}$ **g.** $\sqrt{200} \times \sqrt{200}$

Solution

a. 3   **b.** 4   **c.** 5   **d.** 15   **e.** 16   **f.** 144   **g.** 200   ◆

Some square roots are rational. For example, $\sqrt{4} \times \sqrt{4} = 4$ and we also know $2 \times 2 = 4$, so it seems that $\sqrt{4} = 2$. But wait! We also know $(-2) \times (-2) = 4$, so isn't it just as reasonable to say $\sqrt{4} = -2$? Mathematicians have agreed that the square root symbol may be used only to denote positive numbers, so that $\sqrt{4} = 2$ and NOT -2.

What about square roots of numbers that are not perfect squares? Is $\sqrt{2}$, for example, a rational number? Remember, if $\sqrt{2} = a/b$, where a/b is some fraction so that

$$\frac{a}{b} \cdot \frac{a}{b} = 2$$

then it is *rational*. The Pythagoreans were among the first to investigate this question. Now remember that, for the Pythagoreans, mathematics and religion were one; they

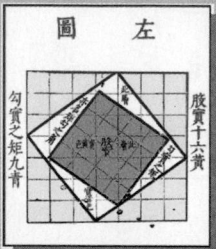
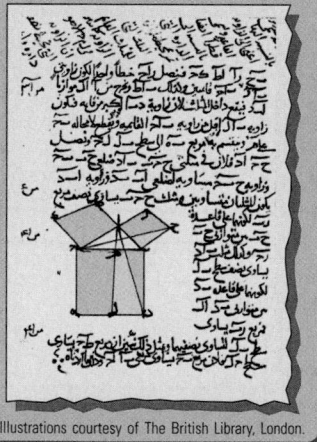
asserted that all natural phenomena could be expressed by whole numbers or ratios of whole numbers. Thus, they believed that $\sqrt{2}$ must be some whole number or fraction (ratio of two whole numbers). Suppose we try to find such a rational number:

$$\frac{7}{5} \times \frac{7}{5} = \frac{49}{25} = 1.96 \quad \text{or, try again:} \quad \frac{707}{500} \times \frac{707}{500} = \frac{499,849}{250,000} = 1.999396$$

We are "getting closer" to 2, but we are still not quite there, so we really get down to business and use a calculator:

$$\boxed{2} \ \boxed{\sqrt{}} \quad \textit{Display:} \quad 1.414213562$$

If you square this number, do you obtain 2? Notice that the last digit of this multiplication will be 4; what should it be if it were the square root of 2? Even if we use a computer to find the following possibility for $\sqrt{2}$, we see that its square is still not 2:

1.41421356237309504880168872420969807856967187537694807317766797379907324784

Can you give a brief argument showing why that can't be $\sqrt{2}$? Project 4.12 asks for a *proof* that $\sqrt{2}$ is not rational. Such a number is called an *irrational number*. The set of **irrational numbers** is the set of numbers whose decimal representations do not terminate nor do they repeat.

It can be shown that not only $\sqrt{2}$ is irrational, but also $\sqrt{3}$, $\sqrt{5}$, $\sqrt{6}$, $\sqrt{7}$, $\sqrt{8}$, $\sqrt{10}$; in fact, the square root of any whole number that is not a perfect square is irrational. Also the cube root of any whole number that is not a perfect cube, and so on, is an irrational number. The **number π**, which is the ratio of the circumference of any circle to its diameter, is also not rational. The number π cannot be written in exact decimal form (which is why we use the symbol "π"), but if you press the $\boxed{\pi}$ key on your calculator, you will see the approximation 3.141592654. In everyday work, you will use irrational numbers when finding the circumference and area of a circle.

Another commonly used irrational number is the **number e,** which is defined in Chapter 8, and is related to natural growth or decay. Like π, the number e cannot be written in exact decimal form, but if you press the $\boxed{e^{\wedge}}$ $\boxed{1}$ keys on your calculator, you will see the approximation 2.718281828.

EXAMPLE 2

Approximate e^2 and e^{-3} using your calculator.

Solution

This example is a check to make sure you know how to use your calculator. Check the outputs of your calculator with those shown here. Look for the $\boxed{e^x}$ key on your calculator. Some calculators require input of the value of x first, whereas others require input of the exponent after the e^x key is pressed:

$$e^2 \approx 7.389056099 \qquad e^{-3} \approx 0.0497870684 \qquad \blacklozenge$$

We also use irrational numbers when applying the Pythagorean theorem. Since the Pythagorean theorem asserts $a^2 + b^2 = c^2$, then

$$c = \sqrt{a^2 + b^2}$$

Also, if you wish to find the length of one of the legs of a right triangle, say, a, when you know both b and c, you can use the formula

$$a = \sqrt{c^2 - b^2}$$

EXAMPLE 3

If a 13-ft ladder is placed against a building so that the base of the ladder is 5 ft away from the building, how high up does the ladder reach?

Solution   Consider Figure 4.6. Since h is one of the legs, use the formula

$$a = \sqrt{c^2 - b^2}$$

$$h = \sqrt{13^2 - 5^2}$$ Substitute the lengths of the hypotenuse and the known leg.

unknown

$$= \sqrt{169 - 25} = \sqrt{144} = 12$$

Thus, the ladder reaches 12 ft up the side of the building.

13 ft. ladder

5 ft.

Figure 4.6 **Ladder problem** ◆

The Pythagorean theorem is of value only when dealing with a right triangle. Carpenters often make use of this property when they want to construct a right angle. That is, if $a^2 + b^2 = c^2$, then an angle of the triangle must be a right angle.

EXAMPLE 4

A carpenter wants to make sure that the corner of a room is square (is a right angle). If she measures out sides (legs) of 3 ft and 4 ft, how long should she make the diagonal (hypotenuse) in order to make sure the corner is square?

Solution   The triangle (corner of the room) is shown in Figure 4.7. The hypotenuse is the unknown, so use the formula

$$c = \sqrt{a^2 + b^2}$$

Thus,

$$c = \sqrt{3^2 + 4^2} = \sqrt{9 + 16} = \sqrt{25} = 5$$

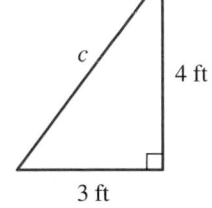

c

4 ft

3 ft

Figure 4.7 **Building a right angle**

If she makes the diagonal 5 ft long, then by the Pythagorean theorem, the angle is a right angle, which forces the corner of the room to be square.   ◆

Your answers to both Examples 3 and 4 are rational. Suppose the result is irrational. You can either leave the result in **radical form** or approximate the result, as shown in Example 5.

EXAMPLE 5

Suppose you need to attach several guy wires to your TV antenna, as shown in Figure 4.8 (page 180). If one guy wire is attached 20 ft away from the base of an antenna, and the height of the wire on the antenna is 30 ft, what is the exact length of one guy wire, and what is the length to the nearest foot?

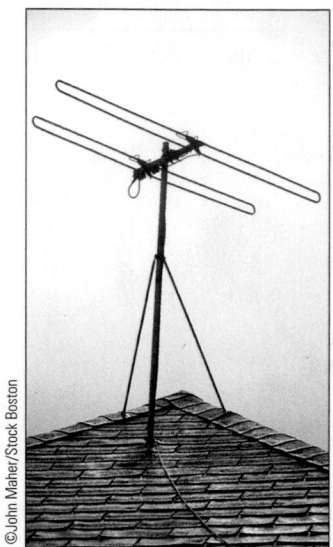

Figure 4.8 TV antenna

Solution The length of the guy wire is the length of the hypotenuse of a right triangle. Thus,

$$c = \sqrt{a^2 + b^2} = \sqrt{20^2 + 30^2} = \sqrt{400 + 900} = \sqrt{1{,}300}$$

The exact length of the guy wire is $\sqrt{1{,}300}$ and is irrational since 1,300 is not a perfect square. What is this length as an approximate rational number? Because

$$30^2 = 900 \quad \text{and} \quad 40^2 = 1{,}600$$

and 1,300 is between 900 and 1,600 we know $\sqrt{1{,}300}$ is between 30 and 40. For a better approximation, we can use a calculator:

$$\boxed{1300} \ \boxed{\sqrt{}} \qquad \textit{Display:} \ 36.05551275$$

The guy wire is 36 ft (to the nearest ft). If the application will not allow any length less than $\sqrt{1{,}300}$, then, instead of rounding, take the *next larger foot*. ◆

Operations with Square Roots

There are times when a square root is irrational and yet we do not want a rational approximation. In such cases, we will need to know certain **laws of square roots** and when a square root is simplified.

Laws of Square Roots

You will not be able to deal properly with radicals if you do not understand the material in this box.

Let a and b be positive numbers. Then:

1. $\sqrt{0} = 0$ 2. $\sqrt{a^2} = a$ 3. $\sqrt{ab} = \sqrt{a}\sqrt{b}$ 4. $\sqrt{\dfrac{a}{b}} = \dfrac{\sqrt{a}}{\sqrt{b}}$

5. A square root is *simplified* if:

   The **radicand** (the number under the radical sign) has no factor with an exponent larger than 1 when it is written in factored form.

   The radicand is not written as a fraction or by using negative exponents.

   There are no square root symbols used in the denominators of fractions.

EXAMPLE 6

Simplify $\sqrt{8}$.

Solution

Step 1. Factor the radicand: $\sqrt{8} = \sqrt{2^3}$

Step 2. Write the radicand as a product of as many factors with exponents of 2 as possible; if there is a remaining factor, it will have an exponent of 1:

$$\sqrt{2^3} = \sqrt{2^2 \cdot 2^1}$$

Step 3: Use Law 3 for square roots: $\sqrt{2^2 \cdot 2^1} = \sqrt{2^2} \cdot \sqrt{2^1}$

Step 4: Use Law 2 for square roots: $\sqrt{2^2} \cdot \sqrt{2^1} = 2\sqrt{2}$ ◆

Notice that the simplified form in Example 6 still contains a radical, so $\sqrt{8}$ is an irrational number. We call $2\sqrt{2}$ the *exact* simplified representation for $\sqrt{8}$ and the calculator representation 2.828427125 is an *approximation*. The whole process of simplifying radicals depends on factoring the radicand and separating out the square factors, and is usually condensed as shown by the following example.

EXAMPLE 7

Simplify the radical expression and assume that the variables are positive. If the expression is simplified, so state.

a. $\sqrt{441}$ **b.** $3\sqrt{2{,}100}$ **c.** $\sqrt{(x+y)^2}$ **d.** $\sqrt{x^2+y^2}$

e. $\sqrt{\dfrac{2}{5}}$ **f.** $\dfrac{7\sqrt{2}}{\sqrt{6}}$ **g.** $\sqrt{0.1}$ **h.** $\dfrac{4}{\sqrt{20}}$

i. $\sqrt{\dfrac{5x^2}{27y}}$ **j.** $\sqrt{2^2-4(1)(-5)}$ **k.** $3+\sqrt{5}$

l. $\dfrac{3+\sqrt{5}}{3}$

m. $\dfrac{6+3\sqrt{5}}{3}$ **n.** $\dfrac{-(-2)+\sqrt{(-2)^2-4(2)(-1)}}{2(2)}$

Solution There are many parts to this example, and you should review them carefully because they lay the groundwork for future sections in this book.

a. If you do not see any factors that are square numbers, you can use a factor tree:
$$\sqrt{441}=\sqrt{3^2\cdot 7^2}=3\cdot 7=21$$

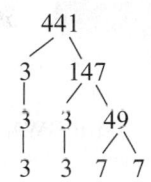

b. $3\sqrt{2{,}100}=3\sqrt{10^2\cdot 3\cdot 7}=3(10\sqrt{3\cdot 7})=30\sqrt{21}$

c. $\sqrt{(x+y)^2}=x+y$

d. $\sqrt{x^2+y^2}$ is simplified; remember that we are looking for square factors, not square terms.

e. $\sqrt{\dfrac{2}{5}}=\dfrac{\sqrt{2}}{\sqrt{5}}\cdot\dfrac{\sqrt{5}}{\sqrt{5}}=\dfrac{\sqrt{10}}{5}$

f. This example came from a *Peanuts* cartoon, and the solution shown in the cartoon is correct.

Peanuts cartoon © 1979 United Feature Syndicate, Inc. Reprinted by permission.

g. $\sqrt{0.1}=\sqrt{\dfrac{1}{10}}=\dfrac{\sqrt{1}}{\sqrt{10}}\cdot\dfrac{\sqrt{10}}{\sqrt{10}}=\dfrac{\sqrt{10}}{10}$ This is also written as $\frac{1}{10}\sqrt{10}$ or $0.1\sqrt{10}$.

h. $\dfrac{4}{\sqrt{20}}=\dfrac{4}{2\sqrt{5}}\cdot\dfrac{\sqrt{5}}{\sqrt{5}}=\dfrac{2\sqrt{5}}{5}$

i. $\sqrt{\dfrac{5x^2}{27y}}=\sqrt{\dfrac{5x^2}{3^2\cdot 3y}}=\dfrac{x\sqrt{5}}{3\sqrt{3y}}\cdot\dfrac{\sqrt{3y}}{\sqrt{3y}}=\dfrac{x\sqrt{5\cdot 3y}}{3(3y)}=\dfrac{x\sqrt{15y}}{9y}$

j. $\sqrt{2^2-4(1)(-5)}=\sqrt{24}=2\sqrt{6}$

k. $3+\sqrt{5}$ is simplified.

l. $\dfrac{3+\sqrt{5}}{3}$ is simplified.

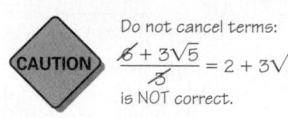

Do not cancel terms:

$\dfrac{\cancel{6} + 3\sqrt{5}}{\cancel{6}} = 2 + 3\sqrt{5}$

is NOT correct.

Confirm this numerically with

$\dfrac{6 + 3\sqrt{5}}{3} \approx 4.236$

$2 + 3\sqrt{5} \approx 8.708$

But

$2 + \sqrt{5} \approx 4.236$ from Example 7m.

m. $\dfrac{6 + 3\sqrt{5}}{3} = \dfrac{3(2 + \sqrt{5})}{3} = 2 + \sqrt{5}$

n. $\dfrac{-(-2) + \sqrt{(-2)^2 - 4(2)(-1)}}{2(2)} = \dfrac{2 + \sqrt{4 + 4(2)}}{4} = \dfrac{2 + \sqrt{12}}{4}$

$= \dfrac{2 + 2\sqrt{3}}{4} = \dfrac{2(1 + \sqrt{3})}{4} = \dfrac{1 + \sqrt{3}}{2}$ ◆

Problem Set 4.5

LEVEL 1

1. **IN YOUR OWN WORDS** What is the Pythagorean theorem?

2. **IN YOUR OWN WORDS** Explain the two meanings of the square root symbol—as an operation and as a number.

3. **IN YOUR OWN WORDS** Discuss the exact and decimal approximations for an irrational number.

4. **IN YOUR OWN WORDS** What does it mean for a square root to be simplified?

5. **IN YOUR OWN WORDS** A computer approximation for $\sqrt{2}$ is

1.41421356237309504880168872420969807856967187537694

Give a brief argument showing why that can't be $\sqrt{2}$.

6. **IN YOUR OWN WORDS** What do you think is meant by a triangular number?

7. Write the following limerick in symbols, and then tell whether it is true or false:*

> A dozen, a gross, and a score
> Plus three times the square root of four
> Divided by seven
> Plus five times eleven
> Is nine squared and not a bit more.

Use the definition of square root to find the indicated products in Problems 8–11. Assume that the variables are positive.

8. **a.** $\sqrt{6} \times \sqrt{6}$ **b.** $\sqrt{7} \times \sqrt{7}$
   c. $\sqrt{9} \times \sqrt{9}$ **d.** $\sqrt{14} \times \sqrt{14}$

9. **a.** $\sqrt{30} \times \sqrt{30}$ **b.** $\sqrt{36} \times \sqrt{36}$
   c. $\sqrt{807} \times \sqrt{807}$ **d.** $\sqrt{169} \times \sqrt{169}$

10. **a.** $\sqrt{400} \times \sqrt{400}$ **b.** $\sqrt{2.5} \times \sqrt{2.5}$
    c. $\sqrt{2.4} \times \sqrt{2.4}$ **d.** $\sqrt{0.25} \times \sqrt{0.25}$

11. **a.** $\sqrt{a} \times \sqrt{a}$ **b.** $\sqrt{xy} \times \sqrt{xy}$
    c. $2\sqrt{b} \times 2\sqrt{b}$ **d.** $5\sqrt{w} \times 8\sqrt{w}$

*From *Omni*, March 1995, "Games" department, p. 104.

Classify each number in Problems 12–17 as rational or irrational. If it is rational, write it without a square root symbol. If it is irrational, approximate it with a rational number correct to the nearest thousandth.

12. **a.** $\sqrt{9}$ **b.** $\sqrt{25}$ **c.** e **d.** π

13. **a.** $\sqrt{10}$ **b.** $\sqrt{30}$ **c.** π^2 **d.** e^2

14. **a.** $\sqrt{36}$ **b.** $\sqrt{50}$ **c.** $\sqrt{e}$ **d.** $\dfrac{\pi}{2}$

15. **a.** $\sqrt{169}$ **b.** $\sqrt{400}$ **c.** e^π **d.** π^e

16. **a.** $\sqrt{500}$ **b.** $\sqrt{1,000}$ **c.** $\dfrac{\pi}{6}$ **d.** $e\pi$

17. **a.** $\sqrt{1,024}$ **b.** $\sqrt{1,936}$ **c.** $\sqrt{\pi}$ **d.** $\sqrt{\dfrac{\pi}{2}}$

Simplify the expressions in Problems 18–37. Assume that the variables are positive.

18. **a.** $-\sqrt{16}$ **b.** $-\sqrt{144}$ **c.** $\sqrt{125}$ **d.** $\sqrt{96}$

19. **a.** $\sqrt{1,000}$ **b.** $\sqrt{2,800}$ **c.** $\sqrt{2,240}$ **d.** $\sqrt{4,410}$

20. **a.** $3\sqrt{75}$ **b.** $2\sqrt{90}$ **c.** $5\sqrt{48}$ **d.** $3\sqrt{96}$

21. **a.** $\sqrt{\dfrac{1}{2}}$ **b.** $\sqrt{\dfrac{1}{3}}$ **c.** $\sqrt{\dfrac{3}{5}}$ **d.** $\sqrt{\dfrac{3}{7}}$

22. **a.** $-\sqrt{0.1}$ **b.** $-\sqrt{0.4}$ **c.** $\sqrt{0.75}$ **d.** $\sqrt{0.05}$

23. **a.** $\dfrac{1}{\sqrt{2}}$ **b.** $\dfrac{-1}{\sqrt{3}}$ **c.** $\dfrac{2}{\sqrt{5}}$ **d.** $\dfrac{5}{\sqrt{10}}$

24. **a.** $\sqrt{(a + b)^2}$ **b.** $\sqrt{a^2 + b^2}$

25. **a.** $\sqrt{x^2 + 4}$ **b.** $\sqrt{(x + 2)^2}$

26. **a.** $\sqrt{5^2 - 4(3)(2)}$ **b.** $\sqrt{7^2 - 4(5)(2)}$

27. **a.** $\sqrt{10^2 - 4(5)(-5)}$ **b.** $\sqrt{12^2 - 4(3)(12)}$

28. **a.** $\sqrt{6^2 - 4(3)(-2)}$ **b.** $\sqrt{2^2 - 4(1)(-1)}$

29. **a.** $\dfrac{6 + 2\sqrt{5}}{2}$ **b.** $\dfrac{8 - 4\sqrt{3}}{4}$

30. **a.** $\dfrac{12 - 3\sqrt{2}}{6}$ **b.** $\dfrac{6 - 2\sqrt{5}}{4}$

31. **a.** $\dfrac{3 - 9\sqrt{x}}{3}$ **b.** $\dfrac{9 + 3\sqrt{x}}{-3}$

32. **a.** $\dfrac{3}{\sqrt{x}}$ **b.** $\dfrac{-7}{\sqrt{y}}$

33. a. $\sqrt{\dfrac{4x^2}{25y}}$ **b.** $\sqrt{\dfrac{5y}{16x}}$

34. $\dfrac{-7 + \sqrt{7^2 - 4(2)(3)}}{2(2)}$

35. $\dfrac{-(-2) - \sqrt{(-2)^2 - 4(6)(-3)}}{2(6)}$

36. $\dfrac{-10 - \sqrt{10^2 - 4(3)(6)}}{2(3)}$

37. $\dfrac{-(-12) + \sqrt{(-12)^2 - 4(1)(-1)}}{2(1)}$

LEVEL 2

38. How far from the base of a building must a 26-ft ladder be placed so that it reaches 10 ft up the wall?

39. How high up on a wall does a 26-ft ladder reach if the bottom of the ladder is placed 10 ft from the base of the building?

40. If a carpenter wants to make sure that the corner of a room is square and measures out 5 ft and 12 ft along the walls, how long should he make the diagonal?

41. If a carpenter wants to be sure that the corner of a building is square and measures out 6 ft and 8 ft along the sides, how long should she make the diagonal?

42. A television antenna is to be erected and held by guy wires. If the guy wires are 15 ft from the base of the antenna and the antenna is 10 ft high, what is the exact length of each guy wire? What is the length of each guy wire to the nearest foot? If three guy wires are to be attached, how many feet of wire should be purchased if it can't be bought by a fraction of a foot?

43. What is the exact length of the hypotenuse if the legs of a right triangle are 2 in. each?

44. What is the exact length of the hypotenuse if the legs of a right triangle are 3 ft each?

45. An empty lot is 400 ft by 300 ft. How many feet would you save by walking diagonally across the lot instead of walking the length and width?

46. A diagonal brace is to be placed in the wall of a room. The height of the wall is 8 ft and the wall is 20 ft long. What is the exact length of the brace? What is the length of the brace to the nearest foot?

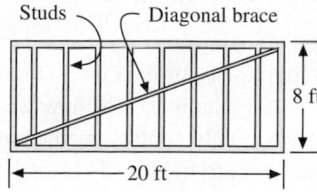

47. A balloon rises at a rate of 4 ft per second when the wind is blowing horizontally at a rate of 3 ft per second. After three seconds, how far away from the starting point, in a direct line, is the balloon?

LEVEL 3

48. Consider a square inch as shown in Figure 4.9.

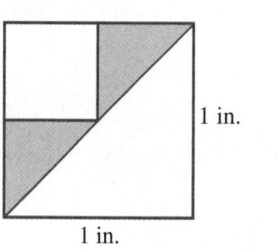 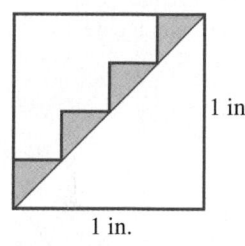

1 in.

a. Two stairs   **b.** Four stairs

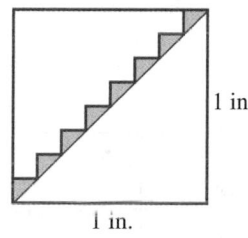

1 in.

c. Eight stairs

Figure 4.9 **Find the length of steps**

  a. Find the total length of the segments making up the stairs in each part of Figure 4.9.
  b. Find the length of the diagonal of each square.
  c. It seems that if we continue the progression started with the three parts of Figure 4.9, at some point the stairs will become indistinguishable from the diagonal. This seems to say that your answers for parts **a** and **b** should be the same (if there are enough stairs). Discuss.

49. Find an irrational number between 1 and 3.

50. Find an irrational number between 0.53 and 0.54.

51. Find an irrational number between $\frac{1}{11}$ and $\frac{1}{10}$.

52. Without using a radical symbol, write an irrational number using only 2s and 3s.

53. Suppose three squares of uniform thickness are made of gold plate and you are offered either the large square or the two smaller ones. Which choices would you make for each of the squares having sides whose lengths are given below?
  a. 1 in. and 1 in. or 2 in.
  b. 3 in. and 4 in. or 5 in.
  c. 4 in. and 5 in. or 7 in.
  d. 5 in. and 7 in. or 9 in.
  e. 10 in. and 11 in. or 15 in.

PROBLEM SOLVING

54. The Pythagorean theorem tells us that the sum of the squares of the lengths of the legs of a right triangle is equal to the square of the length of the hypotenuse. Verify that the theorem is true by tracing the squares and fitting them onto the pattern of the upper two squares shown in Figure 4.10. Next, cut the squares along the lines and re-arrange the pieces so that the pieces all fit into the large square in Figure 4.10.

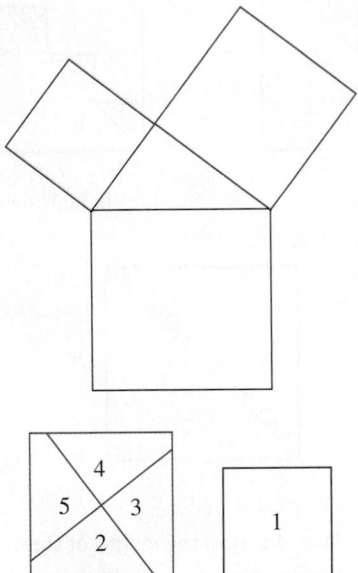

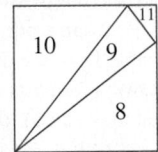 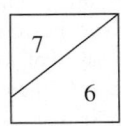

Figure 4.10 **Pythagorean theorem squares**

55. Repeat Problem 54 for the following squares.

56. A man wanted to board a plane with a 5-ft long steel rod, but airline regulations say that the maximum length of any object or parcel checked on board is 4 ft. Without bending or cutting the rod, or altering it in any way, how did the man check it through without violating the rule?

57. HISTORICAL QUESTION The Historical Note on page 186 introduces the great mathematician Karl Gauss. Gauss kept a scientific diary containing 146 entries, some of which were independently discovered and published by others. On July 10, 1796, he wrote

What do you think this meant? Illustrate with some numerical examples.

58. What mathematical property is illustrated in Figure 4.11?

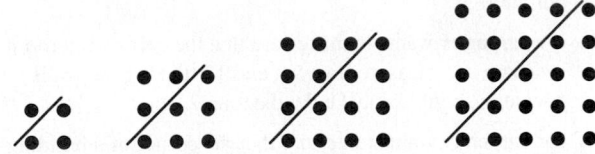

Figure 4.11 **Can you discover this mathematical property?**

59. How are the square numbers embedded in Pascal's triangle?

60. How are the triangular numbers embedded in Pascal's triangle?

4.6 **GROUPS, FIELDS, and REAL NUMBERS**

Definition of Real Numbers

You are familiar with the rational numbers (fractions, for example) and the irrational numbers (π or square roots of certain numbers, for example), and now we wish to consider the most general set of numbers to be used in elementary mathematics. This set consists of the annexation of the irrational numbers to the set of rational numbers and is called the set of *real numbers*.

Real Numbers

The set of real numbers is the most common set used in elementary math.

> The set of **real numbers,** denoted by $\mathbb{R}$, is defined as the union of the set of rationals and the set of irrationals.

Decimal Representation

Let's consider the decimal representation of a real number. If a number is *rational,* then its decimal representation is either **terminating** or **repeating.**

EXAMPLE 1

Find the decimal representation of each of the given rational numbers.

a. $\dfrac{1}{4}$ **b.** $\dfrac{5}{8}$ **c.** $\dfrac{58}{10}$ **d.** $\dfrac{2}{3}$ **e.** $\dfrac{1}{6}$ **f.** $\dfrac{5}{11}$ **g.** $\dfrac{1}{7}$

Solution

a. $\dfrac{1}{4} = 0.25$ [1] [÷] [4]   This is a terminating decimal.

b. $\dfrac{5}{8} = 0.625$ [5] [÷] [8]   This is a terminating decimal.

c. $\dfrac{58}{10} = 5.8$ [58] [÷] [10]   This is a terminating decimal.

d. $\dfrac{2}{3} = 0.666\ldots$ [2] [÷] [3]   *Display:* .6666666667
This is a repeating decimal.

Notice that calculators always represent decimals as terminating decimals, so you need to *interpret* the calculator display as a repeating decimal. If you look at the long division, you can see that the division never terminates:

$$
\begin{array}{r}
.666\ldots \\
3\overline{)2.000\ldots} \\
\underline{1.8} \\
20 \\
\underline{18} \\
2\ldots
\end{array}
$$

e. $\dfrac{1}{6} = 0.166\ldots$ [1] [÷] [6]   This is a repeating decimal.

f. $\dfrac{5}{11} = 0.4545\ldots$ [5] [÷] [11]   This is a repeating decimal.

g. $\dfrac{1}{7} \approx 0.143$ [1] [÷] [7]   *Display:* .1428571429

It may happen that you do not recognize a pattern by looking at the calculator display. For most of our work, an approximation of the result will suffice. Can you use long division to show that the decimal representation *must* terminate (have a 0 remainder) or *must* repeat? For $\frac{1}{7}$ it repeats after six digits.   ◆

When a decimal repeats, we sometimes use an overbar to indicate the numerals that repeat. For Example 1,

One digit repeats: $\frac{2}{3} = 0.\overline{6}, \quad \frac{1}{6} = 0.1\overline{6},$

Two digits repeat: $\frac{5}{11} = 0.\overline{45},$

Six digits repeat: $\frac{1}{7} = 0.\overline{142857}$

Courtesy of Patrick J. Boyle

Real numbers that are *irrational* have decimal representations that are *nonterminating* and *nonrepeating:*

$$\sqrt{2} = 1.414213\ldots \qquad \pi = 3.141592\ldots \qquad e = 2.71828\ldots$$

In each of these examples, the numbers exhibit no repeating pattern and are irrational. Other decimals that do not terminate or repeat are also irrational:

$$0.12345678910111213\ldots \qquad 0.10110111011110111110\ldots$$

We now have some different ways to classify real numbers:

1. Positive, negative, or zero
2. A rational number or an irrational number
   a. If the decimal representation terminates, it is rational.
   b. If the decimal representation repeats, it is rational.
   c. If it has a nonterminating and nonrepeating decimal, it is irrational.

We have illustrated the procedure for changing from a fraction to a decimal: Divide the numerator by the denominator. To reverse the procedure and to change from a terminating decimal representation of a rational number to a fractional representation, use expanded notation. Recall that $10^{-1} = \frac{1}{10}$, $10^{-2} = \frac{1}{100}$, $10^{-3} = \frac{1}{1,000}$, . . . , $10^{-n} = \frac{1}{10^n}$. Thus 0.5 means $5 \times 10^{-1} = 5 \cdot \frac{1}{10} = \frac{5}{10} = \frac{1}{2}$.

EXAMPLE 2

Change the terminating decimals to fractional form.

a. 0.123   **b.** 56.28   **c.** 0.3479

Solution   Write each in expanded notation.

a. $0.123 = 1 \times 10^{-1} + 2 \times 10^{-2} + 3 \times 10^{-3}$

$$= \frac{1}{10} + \frac{2}{100} + \frac{3}{1,000} = \frac{123}{1,000}$$

b. $56.28 = \dfrac{5,628}{100} = \dfrac{1,407}{25}$ The steps shown in part **a** can often be done mentally.

c. $0.3479 = \dfrac{3,479}{10,000}$ Make sure the fraction is reduced.   ◆

It is assumed that you can carry out the basic operations with real numbers written in decimal form. You are asked to carry out the basic operations of addition, subtraction, multiplication, and division of decimal fractions.

Real Number Line

If we consider a line and associate the numbers 0 and 1 with two points situated so that the 1 is to the right of 0, we call the distance between these points a **unit distance.** Next, if we mark off equal distances to the right and associate the successive points with the natural numbers, and mark equal distances to the left and associate those points successively with the opposites of the natural numbers, we have drawn a **number line,** as shown in Figure 4.12.

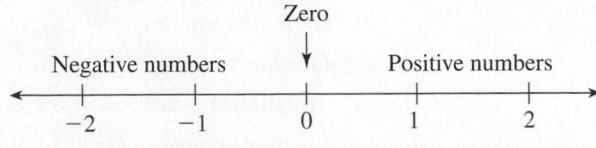

Figure 4.12 A number line

A mnemonic for remembering π:

"Yes, I have a number."

Notice that the number of letters in the words gives an approximation for π (the comma is the decimal point).

Spend some time studying these relationships.

If we now associate points on the number line with rational numbers, it appears that the number line is just about "filled up" by the rationals. The reason for this feeling of "fullness" is that the rationals form what is termed a **dense set.** That is, between every two rationals we can find another rational (see Figure 4.13).

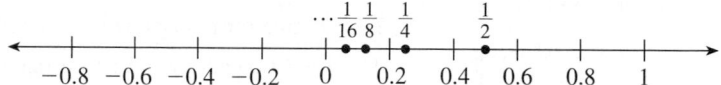

Figure 4.13 A number line with rationals

If we plot *all* the points of this dense set called the rationals, are there still any "holes"? In other words, is there any room left for any of the irrationals? We have shown that $\sqrt{2}$ is irrational. We can show that there is a place on the number line representing this length by using the Pythagorean theorem, as shown in Figure 4.14.

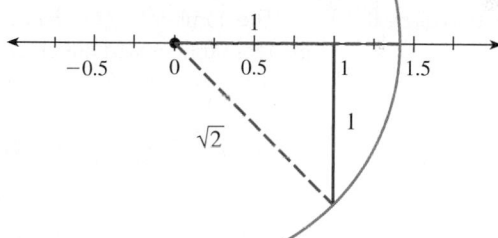

Figure 4.14 Finding an irrational "hole" on a dense number line with rationals plotted

We could show that other irrationals have their places on the number line (see Question G16 of the Group Research at the end of the chapter as well as Figure 4.15). These points (corresponding to both the rational and irrational numbers) when plotted on a line form what is is known as the **real number line.**

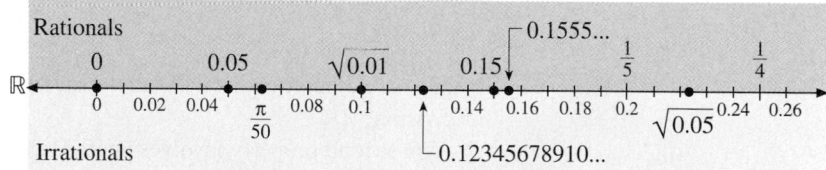

Figure 4.15 Real number line showing some rationals and some irrationals

The relationships among the various sets of numbers we have been discussing are shown in Figure 4.16.

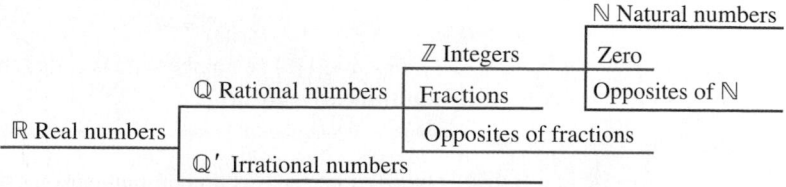

Figure 4.16 Classifications within the set of real numbers, $\mathbb{R}$

Properties of Real Numbers

At the beginning of this chapter, we stated some properties of natural numbers that apply to sets of real numbers as well. We repeat them here for easy reference. Let a, b, and c be real numbers; we denote this by a, b, $c \in \mathbb{R}$. Then,

	Addition	**Multiplication**
Closure:	$(a + b) \in \mathbb{R}$	$ab \in \mathbb{R}$
Associative:	$(a + b) + c = a + (b + c)$	$(ab)c = a(bc)$
Commutative:	$a + b = b + a$	$ab = ba$
Distributive for multiplication over addition:		$a(b + c) = ab + ac$

There are two additional properties that are important in the set of real numbers: the identity and inverse properties.

In Section 3.1, we mentioned that the development of the concept of zero and a symbol for it did not take place at the same time as the development of the natural numbers. The Greeks were using the letter "oh" for zero as early as A.D. 150, but the predominant system in Europe was the Roman numeration system, which did not include zero. It was not until the 15th century, when the Hindu–Arabic numeration system finally replaced the Roman system, that the zero symbol came into common usage.

The number 0 (zero) has a special property for addition that allows it to be added to any real number without changing the value of that number. This property is called the **identity property for addition** of real numbers.

Identity for Addition

> There exists in $\mathbb{R}$ a number 0, called **zero,** so that
> $$0 + a = a + 0 = a$$
> for any $a \in \mathbb{R}$. The number zero is called the **identity for addition** or the **additive identity.**

Remember that when you are studying algebra, you are studying ideas and not just rules about specific numbers. A mathematician would attempt to isolate the *concept* of an identity. First, does an identity property apply for other operations?

Multiplication	**Subtraction**	**Division**
$\square \times a = a \times \square = a$	$\triangle - a = a - \triangle = a$	$\nabla \div a = a \div \nabla = a$
$\uparrow$ same number	$\uparrow$ same number	$\uparrow$ same number

Is there a real number that will satisfy any of the blanks for multiplication, subtraction, or division?

The second property involves another special, important number in $\mathbb{R}$, namely, the number 1 (one). This number has the property that it can multiply any real number without changing the value of that number. This property is called the **identity property for multiplication** of real numbers.

Identity for Multiplication

> There exists in $\mathbb{R}$ a number 1, called **one,** so that
> $$1 \times a = a \times 1 = a$$
> for any $a \in \mathbb{R}$. The number one is called the **identity for multiplication** or the **multiplicative identity.**

Notice that there is no real number that satisfies the identity property for subtraction or division. There may be identities for other operations or for sets other than the set of real numbers.

In Section 4.3, we spoke of opposites when we were adding and subtracting integers. Recall the property of opposites:

$$5 + (-5) = 0 \qquad -128 + 128 = 0 \qquad a + (-a) = 0$$

When opposites are added, the result is zero, the identity for addition. This idea, which can be generalized, is called the **inverse property for addition.**

Inverse Property for Addition

For each $a \in \mathbb{R}$, there is a unique number $(-a) \in \mathbb{R}$, called the **opposite** (or **additive inverse**) of a, so that

$$a + (-a) = -a + a = 0$$

Recall that the product of a number and its reciprocal is 1, the identity for multiplication. The reciprocal of a number, then, is the multiplicative inverse of the number, as we will now show.

Inverse for multiplication

$$5 \times \square = \square \times 5 = 1$$

$$5 \times \frac{1}{5} = \frac{1}{5} \times 5 = 1$$

Since $\frac{1}{5} \in \mathbb{R}$, $\frac{1}{5}$ is an inverse of 5 for multiplication.

Inverse for multiplication

$$-128 \times \triangle = \triangle \times (-128) = 1$$

$$-128 \times \frac{1}{-128} = \frac{1}{-128} \times (-128) = 1$$

Since $\frac{1}{-128} \in \mathbb{R}$, $\frac{1}{-128}$ is an inverse of -128 for multiplication.

To show this inverse property for multiplication in a general way, we seek to find a replacement for the box for each and every real number a: $a \times \square = \square \times a = 1$

Does the inverse property for multiplication hold for every real number a? No, because if $a = 0$, then $0 \times \square = \square \times 0 = 1$ does not have a replacement for the box in $\mathbb{R}$. However, the inverse property for multiplication holds for all *nonzero* replacements of a, and we adopt this condition as part of the inverse property for multiplication of real numbers.

Inverse Property for Multiplication

For *each* number $a \in \mathbb{R}$, $a \neq 0$, there exists a number $a^{-1} \in \mathbb{R}$, called the **reciprocal** (or **multiplicative inverse**) of a, so that

$$a \times a^{-1} = a^{-1} \times a = 1$$

EXAMPLE 3

Given the set $A = \{1, 0, 1\}$. Does this set A have an element that satisfies the inverse property for multiplication?

Solution Before we can talk about the inverse property, we need to find the identity for the operation, in this case, multiplication. Without an identity for multiplication, we cannot have an inverse for multiplication. The operation of multiplication has the identity 1. Check to see that *each* element of A has an inverse:

Member of set	Operation	Inverse	Identity for answer
-1	$\times$	-1	1
0	$\times$	none, but need to check only nonzero elements	
1	$\times$	1	1

Since every nonzero element in the set has an inverse, we say the inverse property for multiplication is satisfied. ◆

We have now introduced several properties that we can apply to a given set with a given operation. If every element of a set satisfies the closure, associative, identity, and inverse properties for a particular operation, then we call that set a *group*. We use the symbol ∘ to stand for any operation. This operation might be $+$, $\times$, or any other *given* operation.

The idea of using a variable to represent an operation will take some extra effort to understand.

Group

The idea of a group is an important unifying idea in more advanced mathematics.

Let $\mathbb{S}$ be any set, let ∘ be any operation, and let *a, b,* and *c* be any elements of $\mathbb{S}$. We say that $\mathbb{S}$ is a **group** for the operation of ∘ if the following properties are satisfied:

1. *Closure:* $(a \circ b) \in \mathbb{S}$

2. *Associative:* $(a \circ b) \circ c = a \circ (b \circ c)$

3. *Identity:* There exists a number $\text{I} \in \mathbb{S}$ so that
$$x \circ \text{I} = \text{I} \circ x = x \text{ for } every \ x \in \mathbb{S}$$

4. *Inverse:* For *each* $x \in \mathbb{S}$, there exists a corresponding $x^{-1} \in \mathbb{S}$ so that
$$x \circ x^{-1} = x^{-1} \circ x = \text{I, where I is the identity element in } \mathbb{S}$$

Furthermore, $\mathbb{S}$ is called a **commutative group** (or **Abelian group**) if the following property is satisfied:

5. *Commutative:* $a \circ b = b \circ a$

A commutative group is called Abelian *to honor the mathematician Niels Abel (see Historical Note on page 217).*

EXAMPLE 4

Is the set $\mathbb{N}$ of natural numbers a group for multiplication?

Solution We have already studied the properties of $\mathbb{N}$, so we can form hasty conclusions:

1. Closure: The product of any two natural numbers is a natural number.

2. Associative: Yes

3. Identity: Yes, namely, 1

4. Inverse: No, since there is no number □ in $\mathbb{N}$ so that
$$3 \times \square = \square \times 3 = 1$$

All we need to do to show that a property doesn't hold is to come up with one counterexample. We might also note that an inverse exists, namely $\frac{1}{3}$, but $\frac{1}{3} \notin \mathbb{N}$.

Therefore, $\mathbb{N}$ does not form a group for multiplication. ◆

This is a lengthy example, but if you stick with it, you will have a better understanding of the ideas of this section.

EXAMPLE 5   PÓLYA'S METHOD

Let us partition the set of natural numbers into two sets, E (even) and O (odd). Consider the operations of addition $(+)$ and multiplication $(\times)$ in the set $\{E, O\}$. Does the set form a group for either or both of these operations?

Solution We use Pólya's problem-solving guidelines for this example.

Understand the Problem. In this example, we are considering a set consisting of two elements E and O (never mind that each of these elements is also a set). How could we possibly define addition? $E + O$ means that we take *any* even number and add to it *any* odd number, and then ask whether the result is even, odd, or something else; in this case we conclude that the answer must be an odd number, so we write $E + O = O$. Consider the operations of addition and multiplication:

Addition	*Multiplication*
even + even = even	even × even = even
even + odd = odd	even × odd = even
odd + even = odd	odd × even = even
odd + odd = even	odd × odd = odd

These operations can be summarized in table format:

+	E	O		×	E	O
E	E	O		E	E	E
O	O	E		O	E	O

Devise a Plan. We will check the properties for each of these operations one at a time. If and when we find a counterexample for one of the group properties, we will have the conclusion that it is not a group for that operation.

Carry Out the Plan.	*Addition*	*Multiplication*
1. Closure:	Yes	Yes

Every entry in the tables is either an E or O, both of which are in the set.

2. Associative:	Yes	Yes

We prove this by checking all possibilities (there are several). We show a few here:

$(E + E) + E = E + (E + E)$	$(E \times E) \times E = E \times (E \times E)$
$(E + E) + O = E + (E + O)$	$(E \times E) \times O = E \times (E \times O)$
$(E + O) + E = E + (O + E)$	$(E \times O) \times E = E \times (O \times E)$
$\vdots$	$\vdots$

3. Identity:	Yes, it is E.	Yes, it is O.
4. Inverse:	Yes	No

Inverse of E is E	E does not have an inverse
since $E + E = E$	because $E \times ? = O$
Inverse of O is O	
since $O + O = E$	

Conclusion: The set $\{E, O\}$ is a *group* for +, but not for ×.

5. Commutative:	Yes	Yes

The tables are symmetric with respect to the principal diagonal. That is, $E + O = O + E$ and $E \times O = O \times E$.

Look Back. The set $\{E, O\}$ is a commutative group for addition.   ◆

Example 5 showed that a given set was a commutative group for one operation but not for another. If a set is a commutative group for two operations, and *also* satisfies the distributive property, it is called a *field*. We define a field in terms of the set $\mathbb{R}$ and the operations of + and ×.

Field

CAUTION

The field properties summarize the main properties used with the set of real numbers.

A **field** is a set $\mathbb{R}$, with two operations $+$ and $\times$ satisfying the following properties for any elements $a, b, c \in \mathbb{R}$:

	Addition, +	*Multiplication, ×*
Closure:	1. $(a + b) \in \mathbb{R}$	2. $ab \in \mathbb{R}$
Associative:	3. $(a + b) + c = a + (b + c)$	4. $(a \times b) \times c = a \times (b \times c)$
Identity:	5. There exists $0 \in \mathbb{R}$ so that $0 + a = a + 0 = a$ for every element a in $\mathbb{R}$.	6. There exists $1 \in \mathbb{R}$ so that $1 \times a = a \times 1 = a$ for every element a in $\mathbb{R}$.
Inverse:	7. For each $a \in \mathbb{R}$, there is a unique number $(-a) \in \mathbb{R}$ so that $a + (-a) = (-a) + a = 0$	8. For each $a \in \mathbb{R}$, $a \neq 0$, there is a unique number $\dfrac{1}{a} \in \mathbb{R}$ so that $a \times \dfrac{1}{a} = \dfrac{1}{a} \times a = 1$
Commutative:	9. $a + b = b + a$	10. $ab = ba$

Distributive for multiplication over addition:

$$11.\ a \times (b + c) = a \times b + a \times c$$

The set of real numbers is a field, but there are other fields. In our definition of a field we used $\mathbb{R}$ and the operations of addition and multiplication for the sake of understanding, but for the mathematician, a field is defined as a set with *any two* operations satisfying the 11 stated properties.

Problem Set 4.6

LEVEL 1

1. **IN YOUR OWN WORDS** What is the distinguishing characteristic between the rational and irrational numbers?

2. **IN YOUR OWN WORDS** A segment of length $\sqrt{2}$ is shown in Figure 4.4. Describe a process you might use to draw a segment of length $\sqrt{3}$.

3. **IN YOUR OWN WORDS** Explain the identity property.

4. **IN YOUR OWN WORDS** Explain the inverse property.

5. What is a group?

6. What is a field?

7. Tell whether each number is an element of $\mathbb{N}$ (a natural number), $\mathbb{Z}$ (an integer), $\mathbb{Q}$ (a rational number), $\mathbb{Q}'$ (an irrational number), or $\mathbb{R}$ (a real number). You may need to list more than one set for each answer.
   a. 7   **b.** 4.93   **c.** 0.656656665 . . .
   d. $\sqrt{2}$ **e.** 3.14159   **f.** $\frac{17}{43}$
   g. $0.00\overline{27}$ **h.** $\sqrt{9}$ **i.** 0

8. Tell whether each number is an element of $\mathbb{N}$ (a natural number), $\mathbb{Z}$ (an integer), $\mathbb{Q}$ (a rational number), $\mathbb{Q}'$ (an irrational number), or $\mathbb{R}$ (a real number). Since these sets are not all disjoint, you may need to list more than one set for each answer.
   a. 19   **b.** 6.48   **c.** 1.868686868 . . .
   d. $\sqrt{8}$ **e.** π **f.** $0.00\overline{12}$
   g. $\sqrt{16}$ **h.** $\sqrt{1,000}$ **i.** $\frac{1}{7}$

Express each of the numbers in Problems 9–12 as a decimal.

9. **a.** $\dfrac{3}{2}$ **b.** $\dfrac{7}{10}$ **c.** $\dfrac{3}{5}$ **d.** $\dfrac{27}{15}$

10. **a.** $\dfrac{5}{6}$ **b.** $\dfrac{2}{7}$ **c.** $\dfrac{3}{25}$ **d.** $2\dfrac{1}{6}$

11. **a.** $\dfrac{2}{3}$ **b.** $2\dfrac{2}{13}$ **c.** $\dfrac{15}{3}$ **d.** $\dfrac{12}{11}$

12. **a.** $\dfrac{-4}{5}$ **b.** $-\dfrac{2}{3}$ **c.** $-\dfrac{17}{6}$ **d.** $\dfrac{-14}{5}$

Change the terminating decimals in Problems 13–20 to fractional form.

13. a. 0.5   **b.** 0.8   **14. a.** 0.25   **b.** 0.75

15. a. 0.45   **b.** 0.234   **16. a.** 0.111   **b.** 0.52

17. a. 98.7   **b.** 0.63   **18. a.** 0.24   **b.** 16.45

19. a. 15.3   **b.** 6.95   **20. a.** 0.64   **b.** 6.98

Carry out the operations with decimal forms in Problems 21–25.

21. a. $6.28 - 3.101$ **b.** $-6.824 + 1.32$
    c. $1.36 + 0.541$ **d.** $6.31 - 12.62$

22. a. $-4.2 - 0.921$ **b.** $8.23 + (-0.005)$
    c. $-6.03 \times (-4.6)$ **d.** 5.002×9.009

23. a. -0.44×0.298 **b.** $-10.5(6.23)$
    c. $3.72 \div 0.3$ **d.** $-5.95 \div -7.00$

24. a. $13.06 \div 0.02$ **b.** $8 \div 4.002$
    c. $0.5(6.2 + 3.4)$ **d.** $0.25(5.03 - 4.005)$

25. a. $5.2 \times 2.3 - 4.5$ **b.** $5.2 - 2.3 \times 4.5$
    c. $8.2 + 2.8 \times 23$ **d.** $8.2 \times 2.8 + 23$

Identify each of the properties illustrated in Problems 26–35.

26. $5 + 7 = 7 + 5$

27. $5 \cdot 1 = 1 \cdot 5$

28. $5 \cdot \frac{1}{5} = 1$

29. $3(4 + 8) = 3(4) + 3(8)$

30. $a + (10 + b) = (a + 10) + b$

31. $a + (10 + b) = (10 + b) + a$

32. mustard + catsup = catsup + mustard

33. (red + blue) + yellow = red + (blue + yellow)

34. $15 + [a + (-a)] = 15 + 0$

35. $\dfrac{x^2 + x - 1}{x^2 - 1} = \dfrac{x^2 + x - 1}{x^2 - 1} \cdot \dfrac{\frac{1}{x^2}}{\frac{1}{x^2}}$

LEVEL 2

Check whether each of the sets and operations in Problems 36–44 form a group.

36. $\mathbb{N}$ for $+$ **37.** $\mathbb{N}$ for $-$ **38.** $\mathbb{W}$ for $+$

39. $\mathbb{W}$ for $\times$ **40.** $\mathbb{Z}$ for $+$ **41.** $\mathbb{Z}$ for $\times$

42. $\mathbb{Q}$ for $+$ **43.** $\mathbb{Q}$ for $\times$ **44.** $\mathbb{Q}$ for $\div$

45. a. Given the set $\{1, 2, 3, 4\}$ and the operation $\times$, construct a multiplication table showing all possible answers for the numbers in the set.
    b. Given the set $\{1, 2, 3, 4\}$ and the operation $*$ defined by $a * b = 2a$, construct a table for $*$ showing all possible answers for numbers in the set.
    c. Verify as many of the field properties as possible for the operations of $\times$ and $*$.

LEVEL 3

Find a rational number and an irrational number between each of the given pairs of numbers in Problems 46–50.

46. 1 and 10   **47.** 2 and 3

48. 3 and 4   **49.** 4.5 and 4.6

50. 8.00 and 8.01

51. List the rationals between 0 and 1 by roster.

Let $\circ$ be an arbitrary operation in Problems 52–59. Describe the operation $\circ$ for each problem.

52. $5 \circ 3 = 8;\ 7 \circ 2 = 9;\ 9 \circ 1 = 10;\ 8 \circ 2 = 10;\ \ldots$

53. $5 \circ 3 = 15;\ 7 \circ 2 = 14;\ 9 \circ 1 = 9;\ 8 \circ 2 = 16;\ \ldots$

54. $5 \circ 3 = 2;\ 7 \circ 2 = 5;\ 9 \circ 1 = 8;\ 8 \circ 2 = 6;\ \ldots$

55. $1 \circ 9 = 11;\ 2 \circ 7 = 10;\ 9 \circ 0 = 10;\ 9 \circ 8 = 18;\ \ldots$

56. $8 \circ 0 = 1;\ 5 \circ 4 = 21;\ 1 \circ 0 = 1;\ 5 \circ 6 = 31;\ \ldots$

57. $4 \circ 6 = 20;\ 8 \circ 2 = 20;\ 7 \circ 9 = 32;\ 6 \circ 8 = 28;\ \ldots$

58. $4 \circ 7 = 1;\ 4 \circ 5 = 3;\ 7 \circ 3 = 11;\ 12 \circ 9 = 15;\ \ldots$

59. $4 \circ 7 = 17;\ 5 \circ 6 = 26;\ 6 \circ 4 = 37;\ 2 \circ 8 = 5;\ \ldots$

PROBLEM SOLVING

60. Symmetrics of a Square Cut out a small square and label it as shown in Figure 4.17.

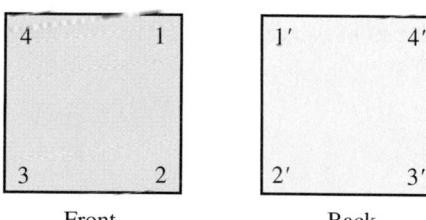

Front   Back

Figure 4.17 **Square construction**

Be sure that 1 is in front of $1'$, 2 is in front of $2'$, 3 is in front of $3'$, and 4 is in front of $4'$. We will study certain *symmetries* of this square—that is, the results that are obtained when a square is moved around according to certain rules that we will establish.

    Hold the square with the front facing you and the 1 in the top right-hand corner as shown. This is called the *basic position.*

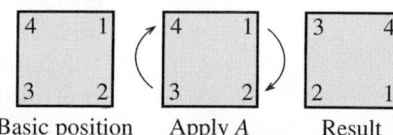

Basic position   Apply *A*   Result

    Now rotate the square 90° clockwise so that 1 moves into the position formerly held by 2 and so that 4 and 3 end up on top. We use the letter *A* to denote this rotation of the square. That is, *A* indicates a clockwise rotation of the square through 90°. Other symmetries can be obtained similarly according to Table 4.5. You should be able to tell how each of the results in the table was found. Do this before continuing with the problem.

We now have a set of elements: {*A, B, C, D, E, F, G, H*}. We must define an operation that combines a pair of these symmetries. Define an operation ★ which means "followed by." Consider, for example, *A* ★ *B*: Start with the basic position, apply *A, followed by B* (without returning to basic position) to obtain the result shown:

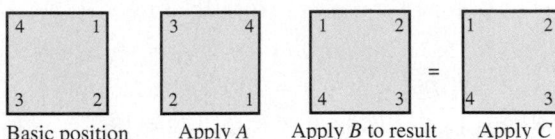

Thus, we say *A* ★ *B* = *C*, meaning "*A* followed by *B* is the same as applying the single element *C*."

a. Complete a table for the operation ★ and the set {*A, B, C, D, E, F, G, H*}.
b. Is the set closed for ★?
c. Is the set associative for ★?
d. Is the set commutative for ★?
e. Does the set have an identity for ★?
f. Does the inverse property hold for the set and the operation ★?
g. Is the set of symmetries of a square with the operation ★ a group?

Table 4.5   Symmetries of a square

Element	Description	Result
A	90° clockwise rotation	3 4 / 2 1
B	180° clockwise rotation	2 3 / 1 4
C	270° clockwise rotation	1 2 / 4 3
D	360° clockwise rotation	4 1 / 3 2
E	Flip about a **horizontal** line through the middle of the square	3' 2' / 4' 1'
F	Flip about a **vertical** line through the middle of the square	1' 4' / 2' 3'
G	Flip along a line drawn from **upper** left to lower right	4' 3' / 1' 2'
H	Flip along a line drawn from **lower** left to upper right	2' 1' / 3' 4'

4.7 DISCRETE MATHEMATICS

Discrete mathematics is that part of mathematics that deals with sets of objects that can be counted or processes that consist of a sequence of individual steps. That is, a *discrete* set is one that is *finite* (the cardinality of the set is a natural number). Computers have enhanced the importance of discrete mathematics because many natural phenomena are idealized in terms of discrete (or countable) steps or numbers of objects. For example, television sets today are termed *digital* TVs, which means that the screen is divided into a large number of cells, and each cell is associated with a number designating its color, brightness, or other characteristics of that picture in that cell. CDs and digital music utilize the same ideas from discrete mathematics. Figure 4.18 shows a famous image that has been *digitized* by computer and used on a cover of a *Finite Mathematics* textbook.

When you hear the word *algebra*, you probably think of the algebra you encountered in high school. The dictionary says that *algebra* is a generalization of arithmetic. Most people speak of algebra in the singular, but to a mathematician **algebra** refers to a structure, a set of symbols that operate according to certain agreed-upon properties. A mathematical dictionary reveals that we can speak of an *algebra with a unit element, a simple algebra,* an *algebra of* subsets, *a commutative algebra,* or an *algebra over a field.* In the next chapter we consider the algebra studied in high school, but in this section we will look at a different, discrete, algebra that focuses on the definition of operations and on the properties we have studied in this chapter.

Clock Arithmetic

We now consider a mathematical system based on a 12-hour clock (see Figure 4.19). We'll need to define some operations for this set of numbers, and we'll use the way we tell time as a guide to our definitions. For example, if you have an appointment at 4:00 P.M. and you are two hours late, you arrive at 6:00 P.M. On the other hand, if your appointment is at 11:00 A.M. and you're two hours late, you arrive at 1:00 P.M. That is, on a 12-hour clock,

$$4 + 2 = 6 \quad \text{and} \quad 11 + 2 = 1$$

Using the clock as a guide, we define an operation called "clock addition," which is different from ordinary addition because it consists of a *finite* set, $\{1, 2, 3, 4, 5, 6, 7, 8, 9, 10, 11, 12\}$, and is *closed* for addition, as you can see by looking at Table 4.6.

Cover image courtesy of Professor Leon D. Harmon, Case Western Reserve University.

Figure 4.18 **A digitized image of the *Mona Lisa***

Figure 4.19 **A 12-hour clock**

Table 4.6 Addition on a 12-Hour Clock

+	1	2	3	4	5	6	7	8	9	10	11	12
1	2	3	4	5	6	7	8	9	10	11	12	1
2	3	4	5	6	7	8	9	10	11	12	1	2
3	4	5	6	7	8	9	10	11	12	1	2	3
4	5	6	7	8	9	10	11	12	1	2	3	4
5	6	7	8	9	10	11	12	1	2	3	4	5
6	7	8	9	10	11	12	1	2	3	4	5	6
7	8	9	10	11	12	1	2	3	4	5	6	7
8	9	10	11	12	1	2	3	4	5	6	7	8
9	10	11	12	1	2	3	4	5	6	7	8	9
10	11	12	1	2	3	4	5	6	7	8	9	10
11	12	1	2	3	4	5	6	7	8	9	10	11
12	1	2	3	4	5	6	7	8	9	10	11	12

EXAMPLE 1

Use a 12-hour clock to find the following sums.

a. $7 + 5$ **b.** $9 + 5$ **c.** $4 + 11$ **d.** $12 + 3$

Solution

a. $7 + 5 = 12$ **b.** $9 + 5 = 2$ **c.** $4 + 11 = 3$ **d.** $12 + 3 = 3$ ◆

Subtraction, multiplication, and division may be defined as they are in ordinary arithmetic.

Elementary Operations

STOP Note that these definitions of the basic operations are consistent with the usual definitions.

Subtraction: $a - b = x$ means $a = b + x$

Multiplication: $a \times b = ab$ means $\underbrace{b + b + \cdots + b}_{a \text{ addends}}$

Zero multiplication: If $a = 0$, then $a \times b = 0 \times b = 0$.

Division: $a \div b = \dfrac{a}{b} = x$ means $a = bx$ has an inverse for multiplication.

EXAMPLE 2

Carry out the given operations on a 12-hour clock by using the definitions for the elementary operations.

a. $4 - 9$ **b.** 4×9 **c.** $4 \div 7$ **d.** $\dfrac{4}{9}$

Solution

a. $4 - 9 = x$ means $4 = 9 + x$.

That is, what number when added to 9 produces 4? From Table 4.6, we see $x = 7$.

b. 4×9 means $9 + 9 + 9 + 9 = 12$ (from Table 4.6).

c. $4 \div 7 = t$ means $4 = 7t$.

That is, what number can be multiplied by 7 to obtain the result 4? We can proceed by trial and error:

$$7 \times 1 = 7; \quad 7 \times 2 = 2; \quad 7 \times 3 = 9; \quad 7 \times 4 = 4; \quad \ldots$$

We see $t = 4$.

d. $\dfrac{4}{9}$ means $4 \div 9 = s$ or $4 = 9s$.

We need to find the number that, when multiplied by 9, produces 4. Once again, proceed by checking each possibility:

$$9 \times 1 = 9; \quad 9 \times 2 = 6; \quad 9 \times 3 = 3; \quad 9 \times 4 = 12;$$
$$9 \times 5 = 9; \quad 9 \times 6 = 6; \quad 9 \times 7 = 3; \quad 9 \times 8 = 12;$$
$$9 \times 9 = 9; \quad 9 \times 10 = 6; \quad 9 \times 11 = 3; \quad 9 \times 12 = 12$$

Even though this process is tedious, it is complete because we are dealing with a finite set, and since we have checked all possibilities we see there is no such number. We say $\dfrac{4}{9}$ does not exist. ◆

We see from Example 2 that it might be worthwhile to construct a multiplication table. However, since the addition and multiplication tables for a 12-hour clock are rather large, we shorten our arithmetic system by considering a clock with fewer than 12 hours.

Modulo Five Arithmetic

Consider a mathematical system based on a 5-hour clock, numbered 0, 1, 2, 3, and 4, as shown in Figure 4.20. Clock addition on this 5-hour clock is the same as on an ordinary clock, except that the only numbers in this set are $\{0, 1, 2, 3, 4\}$.

We define the operations of addition, subtraction, multiplication, and division just as we did for a 12-hour clock. Since subtraction and division are defined in terms of addition and multiplication, respectively, we need only the two operation tables that are shown in Table 4.7.

Figure 4.20 **A 5-hour clock**

Table 4.7 Addition and Multiplication Tables

+	0	1	2	3	4
0	0	1	2	3	4
1	1	2	3	4	0
2	2	3	4	0	1
3	3	4	0	1	2
4	4	0	1	2	3

×	0	1	2	3	4
0	0	0	0	0	0
1	0	1	2	3	4
2	0	2	4	1	3
3	0	3	1	4	2
4	0	4	3	2	1

Instead of speaking about "arithmetic on the 5-hour clock," mathematicians usually speak of "modulo 5 arithmetic." The set $\{0, 1, 2, 3, 4\}$, together with the operations defined in Table 4.7, is called a **modulo 5** or **mod 5** system. Suppose it is 4 o'clock on a 5-hour clock. What time will the clock show 9 hours later? We could write

$$4 + 9 = 3 \quad \text{and} \quad 2 + 1 = 3,$$

thus $4 + 9 = 2 + 1$. Since we do not wish to confuse this with ordinary arithmetic in which

$$4 + 9 \neq 2 + 1$$

we use the following notation:

$$4 + 9 \equiv 2 + 1, (\text{mod } 5)$$

which is read "$4 + 9$ is congruent to $2 + 1$, mod 5." We define *congruence* as follows.

Congruence Mod m

> The real numbers a and b are **congruent modulo m,** written $a \equiv b, (\text{mod } m)$, if a and b differ by a multiple of m.

EXAMPLE 3

Decide whether each statement is true or false.

a. $3 \equiv 8, (\text{mod } 5)$ **b.** $3 \equiv 53, (\text{mod } 5)$ **c.** $3 \equiv 19, (\text{mod } 5)$

Solution

a. $3 \equiv 8, (\text{mod } 5)$ because $8 - 3 = 5$, and 5 is a multiple of 5.

b. $3 \equiv 53, (\text{mod } 5)$ because $53 - 3 = 50$, and 50 is a multiple of 5.

c. $3 \not\equiv 19, (\text{mod } 5)$ because $19 - 3 = 16$, and 16 is not a multiple of 5. ◆

Another way to determine whether two numbers are congruent mod m is to divide each by m and check the remainders. If the remainders are the same, then the numbers are congruent mod m. For example, $3 \div 5$ gives a remainder 3, and $53 \div 5$ gives a remainder of 3, so $3 \equiv 53, (\text{mod } 5)$.

EXAMPLE 4

Solve each equation for x. In other words, find a replacement for x that makes each equation true.

a. $4 + 9 \equiv x, (\text{mod } 5)$ **b.** $15 + 92 \equiv x, (\text{mod } 5)$

c. $2 + 4 \equiv x, (\text{mod } 5)$ **d.** $2 - 4 \equiv x, (\text{mod } 5)$

e. $7 \times 5 \equiv x, (\text{mod } 7)$ **f.** $3 - 5 \equiv x, (\text{mod } 12)$

Solution

a. $4 + 9 = 13 \equiv 3, (\text{mod } 5)$ **b.** $15 + 92 = 107 \equiv 2, (\text{mod } 5)$

c. $2 + 4 = 6 \equiv 1, (\text{mod } 5)$

d. $2 - 4 \equiv 7 - 4, (\text{mod } 5)$ Since $2 \equiv 7, (\text{mod } 5)$, we can replace 2 by 7.

             $\equiv 3, (\text{mod } 5)$

e. $7 \times 5 = 35 \equiv 0, (\text{mod } 7)$

f. $3 - 5 \equiv 15 - 5, (\text{mod } 12)$ Since $3 \equiv 15, (\text{mod } 12)$

             $\equiv 10, (\text{mod } 12)$ ◆

Notice that every whole number is congruent modulo 5 to exactly one element in the set $I = \{0, 1, 2, 3, 4\}$, which contains all possible remainders when dividing by 5. Since the number of elements in I is rather small, we can easily solve equations by trying the numbers 0, 1, 2, 3, and 4. In modulo 7 try numbers in the set $\{0, 1, 2, \ldots, 6\}$ and in modulo 10 try numbers in the set $\{0, 1, 2, \ldots, 9\}$. Consider the following example.

EXAMPLE 5

The manager of a TV station hired a college student, U. R. Stuck, as an election-eve runner. The request for reimbursement turned in is shown in the margin. The station manager refused to pay the bill, since the mileage was not honestly recorded. How did he know? *Note:* The answer has nothing to do with the illegible digits in the mileage report, and you should also assume that the mileage is rounded to the nearest mile.

Solution Let $x =$ the number of miles in one round trip. For the reimbursement request, we see there are 8 trips, so the total mileage must be $8x$. Since the last digits are legible, we see the difference is 5, (mod 10). Thus,

$$8x \equiv 5, (\text{mod } 10)$$

We solve this by checking each possible value:

    $x = 0$: $8(0) \equiv 0, (\text{mod } 10)$
    $x = 1$: $8(1) \equiv 8, (\text{mod } 10)$
    $x = 2$: $8(2) = 16 \equiv 6, (\text{mod } 10)$
    $x = 3$: $8(3) = 24 \equiv 4, (\text{mod } 10)$
    $x = 4$: $8(4) = 32 \equiv 2, (\text{mod } 10)$
    $x = 5$: $8(5) = 40 \equiv 0, (\text{mod } 10)$
    $x = 6$: $8(6) = 48 \equiv 8, (\text{mod } 10)$
    $x = 7$: $8(7) = 56 \equiv 6, (\text{mod } 10)$
    $x = 8$: $8(8) = 64 \equiv 4, (\text{mod } 10)$
    $x = 9$: $8(9) = 72 \equiv 2, (\text{mod } 10)$

These are the only possible values for x in modulo 10, and none gives a mileage reading for 5, (mod 10) in the units digit. Therefore, Stuck did not report the mileage honestly.   ◆

REQUEST FOR REIMBURSEMENT FOR MILEAGE	
NAME	U.R. Stuck
ADDRESS	1234 Fifth St.
S.S. NO.	576-38-4459
ENDING MILEAGE	14▨2.8
BEGINNING MILEAGE	14,8▨3
NO. OF TRIPS	8

EXAMPLE 6   PÓLYA'S METHOD

Suppose you are planning to buy paper to cover some shelves. You need to cover 1 shelf 100 inches long, and in order to minimize the amount of waste you need to cover at least 30 11-inch shelves, but not more than 50 11-inch shelves. The paper can be purchased in multiples of 36 inches. How much paper should you buy to minimize the waste?

Solution   We use Pólya's problem-solving guidelines for this example.

Understand the Problem. Suppose I need to cover 30 shelves; then I will need $30 \times 11 + 100$ inches of paper. Since $30 \times 11 + 100 = 430$ inches and the paper comes in 36-inch lengths, I can find $430 \div 36 = 11.94$. This means the *minimum* purchase is $12 \times 36 = 432$ inches of paper. On the other hand, if I need to cover 50 shelves, then I will need $50 \times 11 + 100 = 650$ inches; $650 \div 36 = 18.05$. This requires 19 units of length 36 inches: $36 \times 19 = 684$ inches.

Devise a Plan. We know the minimum purchase and we also know the maximum purchase. We need to find a solution that specifies a general solution (for any number of shelves between 30 and 50). Since the paper comes in multiples of 36 inches, we must buy $36x$ inches of paper, where x represents the number of multiples we buy. Suppose we need to cover k shelves at 11 inches and an extra shelf at 100 inches for a total of $11k + 100$ inches. We will write this as a congruence modulo 11 and then solve for x.

Carry Out the Plan. The amount of shelf paper required is $11k + 100$, which means

$$36x = 11k + 100$$

$$36x \equiv 100, \ (\mathrm{mod}\ 11)$$

$$3x \equiv 1, \ (\mathrm{mod}\ 11) \quad \text{Note } 36 \equiv 3, \text{ (mod 11) and } 100 \equiv 1, \text{ (mod 11)}.$$

Since we are working in modulo 11, and if we assume no waste, we can check all possibilities (since the set of possibilities is *finite*, namely, 0, 1, 2, 3, 4, 5, 6, 7, 8, 9, and 10; all these congruences are modulo 11):

$x = 0$: $3(0) = 0 \not\equiv 1$	$x = 1$: $3(1) = 3 \not\equiv 1$	$x = 2$: $3(2) = 6 \not\equiv 1$
$x = 3$: $3(3) = 9 \not\equiv 1$	$x = 4$: $3(4) = 12 \equiv 1$	$x = 5$: $3(5) = 15 \not\equiv 1$
$x = 6$: $3(6) = 18 \not\equiv 1$	$x = 7$: $3(7) = 21 \not\equiv 1$	$x = 8$: $3(8) = 24 \not\equiv 1$
$x = 9$: $3(9) = 27 \not\equiv 1$	$x = 10$: $3(10) = 30 \not\equiv 1$	

The solution is $x \equiv 4, \ (\mathrm{mod}\ 11)$.

Look Back. We know that the minimum is 12 multiples of the 36-inch paper, and the maximum is 19 multiples. To *minimize the waste* we see that $x \equiv 4, \ (\mathrm{mod}\ 11)$ must be $x = 4, 15, 26, \ldots$. This answer means that to minimize the waste we should buy 15 multiples of the 36-inch paper:

$$36(15) = 540$$

inches of paper. This amount allows us to cover the 100-inch board and 40 11-inch boards.   ◆

Group Properties for a Modulo System

It appears that some modulo systems "behave" like ordinary algebra and others do not. The distinction is found by determining which systems form a field. We assume that modulo arithmetic follows the usual order of operations. Let's first explore the group properties for the set $I = \{0, 1, 2, 3, 4\}$ and the operation of addition, modulo 5.

Closure for $+$:   The set I of elements modulo 5 is closed with respect to addition. That is, for any pair of elements, there is a unique element that represents their sum, and that is also a member of the original set.

Associative for $+$:   Addition of elements modulo 5 satisfies the associative property. That is,

$$(a + b) + c \equiv a + (b + c)$$

There was a young fellow
named Ben
Who could only count
modulo 10
He said when I go
Past my last little toe
I shall have to start
over again.

for all elements *a, b,* and *c* ∈ *I.* As a specific example, we evaluate 2 + 3 + 4 in two ways:

$$(2 + 3) + 4 = 9 \equiv 4, \ (\text{mod } 5)$$
$$2 + (3 + 4) = 9 \equiv 4, \ (\text{mod } 5)$$

Identity for +:   The set *I* of elements modulo 5 includes an identity element for addition. That is, the set contains an element 0 such that the sum of any given element and zero is the given element. In modulo 5,

$$0 + 0 \equiv 0, 1 + 0 \equiv 1, 2 + 0 \equiv 2, 3 + 0 \equiv 3, 4 + 0 \equiv 4$$

Inverse for +:   Each element in arithmetic modulo 5 has an inverse with respect to addition. That is, for each element *a* ∈ *I,* there exists a unique element *a'* ∈ *I* such that $a + a' \equiv a' + a \equiv 0$. The element *a'* is said to be the *inverse* of *a.* Specifically (in mod 5):

	elements in the set	identity
	↓	↓
The inverse of 0 is 0:	0 + 0	≡ 0
The inverse of 1 is 4:	1 + 4	≡ 0
The inverse of 2 is 3:	2 + 3	≡ 0
The inverse of 3 is 2:	3 + 2	≡ 0
The inverse of 4 is 1:	4 + 1	≡ 0
	↑	
	inverses	

Since the closure, associative, identity, and inverse properties are satisfied for addition modulo 5, we conclude that it is a *group*.

EXAMPLE 7

Is the set {1, 2, 3, 4, 5} a group for multiplication modulo 6?

Solution   This set has an identity element 1 for multiplication. However, the inverse property is not satisfied (in mod 6):

	elements in the set	identity
	↓	↓
The inverse of 1 is 1:	1 × 1	≡ 1
The inverse of 2 does not exist:	2 × ?	≡ 1

$$2 \times 1 \equiv 2, 2 \times 2 \equiv 4, 2 \times 3 \equiv 0,$$
$$2 \times 4 \equiv 2, 2 \times 5 \equiv 4$$

None of the possible products gives the answer 1.

However, the best way to proceed is to check the closure property first. Note that $2 \times 3 = 6 \equiv 0,(\text{mod } 6)$, which is not in the set, so the set is not closed.

The set is *not* a group for multiplication modulo 6.   ◆

EXAMPLE 8

Is the set *I* = {0, 1, 2, 3, 4} a field for + and × modulo 5?

Solution   We have shown (pages 199–200) that four of the 11 field properties are satisfied.

Commutative for +: Addition in arithmetic modulo 5 satisfies the commutative property. That is,

$$a + b \equiv b + a$$

where a and b are any elements in I. Specifically, we see that the entries in Table 4.7 are symmetric with respect to the principal diagonal.

We can now say that I is a commutative group for addition. We continue by verifying other properties for multiplication.

Closure for ×: The set I is closed for multiplication, as we can see from Table 4.7.

Associative for ×: This property is satisfied, and the details are left for you to verify.

Identity for ×: The identity element for multiplication is 1, since (in mod 5):

$$0 \times 1 \equiv 0, 1 \times 1 \equiv 1, 2 \times 1 \equiv 2, 3 \times 1 \equiv 3, 4 \times 1 \equiv 4$$

Inverse for ×: The inverse for multiplication can be checked by finding the inverse of each element (mod 5).

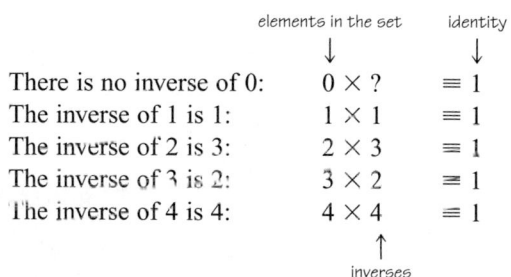

There is no inverse of 0:
The inverse of 1 is 1:
The inverse of 2 is 3:
The inverse of 3 is 2:
The inverse of 4 is 4:

Since every *nonzero* element has an inverse for multiplication, the inverse property for multiplication in a field is satisfied. We note, however, that the set I is *not* a group because the element 0 does not have an inverse for multiplication.

Commutative for ×: The set is commutative for multiplication, as we can see by looking at Table 4.7.

Distributive for × *over* +: We need to check $a(b + c) \equiv ab + ac$.

Check some particular examples:

$$2(3 + 4) = 2(7) = 14 \equiv 4, \text{(mod 5)}$$
$$2(3) + 2(4) = 6 + 8 = 14 \equiv 4, \text{(mod 5)}$$

Thus, $2(3 + 4) \equiv 2(3) + 2(4)$. Also,

$$4(1 + 2) \equiv 4(1) + 4(2)$$

and

$$3(2 + 4) \equiv 3(2) + 3(4)$$

These examples seem to imply that the distributive property holds.

The set I is a field for the operations of addition and multiplication, modulo 5. ◆

Problem Set 4.7

1. **IN YOUR OWN WORDS** What do we mean by clock arithmetic?

2. **IN YOUR OWN WORDS** How can the operations of addition and multiplication be defined for a 24-hour clock?

3. **IN YOUR OWN WORDS** Discuss the meaning of the definition of congruence modulo m.

4. Define precisely the concept of congruence modulo m.

Perform the indicated operations in Problems 5–14 using arithmetic for a 12-hour clock.

5. **a.** $9 + 6$ **b.** $5 - 7$

6. **a.** 5×3 **b.** 2×7

7. **a.** $7 + 10$ **b.** $7 - 9$

8. **a.** 6×7 **b.** $1 \div 5$

9. **a.** $5 + 7$ **b.** $4 - 8$

10. **a.** $2 - 6$ **b.** 9×3

11. **a.** 4×8 **b.** 2×3

12. **a.** $1 \div 12$ **b.** $10 + 6$

13. **a.** $3 \times 5 - 7$ **b.** $7 + 3 \times 2$

14. **a.** $5 \times 2 - 11$ **b.** $5 \times 8 + 5 \times 4$

Which of the statements in Problems 15–20 are true?

15. **a.** $5 + 8 \equiv 1$, (mod 6)
    b. $4 + 5 \equiv 1$, (mod 7)

16. **a.** $5 \equiv 53$, (mod 8)
    b. $102 \equiv 1$, (mod 2)

17. **a.** $47 \equiv 2$, (mod 5)
    b. $108 \equiv 12$, (mod 8)

18. **a.** $5,670 \equiv 270$, (mod 365)
    b. $2,001 \equiv 39$, (mod 73)

19. **a.** $1,996 \equiv 0$, (mod 1,996)
    b. $246 \equiv 150$, (mod 6)

20. **a.** $126 \equiv 1$, (mod 7)
    b. $144 \equiv 12$, (mod 144)

Perform the indicated operations in Problems 21–28.

21. **a.** $9 + 6$, (mod 5)   **b.** $7 - 11$, (mod 12)

22. **a.** 4×3, (mod 5)   **b.** $1 \div 2$, (mod 5)

23. **a.** $5 + 2$, (mod 4)   **b.** $2 - 4$, (mod 5)

24. **a.** 6×6, (mod 8)   **b.** $5 \div 7$, (mod 9)

25. **a.** $4 + 3$, (mod 5)   **b.** $6 - 12$, (mod 8)

26. **a.** $2 \div 3$, (mod 7)   **b.** 121×47, (mod 121)

27. **a.** $7 + 41$, (mod 5)   **b.** $4 - 5$, (mod 11)

28. **a.** 62×4, (mod 2)   **b.** $7 \div 12$, (mod 13)

Solve each equation for x in Problems 29–37. Assume that k is any natural number.

29. **a.** $x + 3 \equiv 0$, (mod 7)
    b. $4x \equiv 1$, (mod 5)

30. **a.** $x + 5 \equiv 2$, (mod 9)
    b. $4x \equiv 1$, (mod 6)

31. **a.** $x - 2 \equiv 3$, (mod 6)
    b. $3x \equiv 2$, (mod 7)

32. **a.** $5x \equiv 2$, (mod 7)
    b. $7x + 1 \equiv 3$, (mod 11)

33. **a.** $x^2 \equiv 1$, (mod 4)
    b. $x \div 4 \equiv 5$, (mod 9)

34. **a.** $x^2 \equiv 1$, (mod 5)
    b. $4 \div 6 \equiv x$, (mod 13)

35. **a.** $4k \equiv x$, (mod 4)
    b. $4k + 2 \equiv x$, (mod 4)

36. $2x^2 - 1 \equiv 3$, (mod 7)

37. $5x^3 - 3x^2 + 70 \equiv 0$, (mod 2)

38. Assume that today is Monday (day 2). Determine the day of the week it will be at the end of each of the following periods. (Assume no leap years.)
    a. 24 days   **b.** 155 days
    c. 365 days   **d.** 2 years

39. Assume that today is Friday (day 6). Determine the day of the week it will be at the end of each of the following periods. (Assume no leap years.)
    a. 30 days   **b.** 195 days
    c. 390 days   **d.** 3 years

40. Your doctor tells you to take a certain medication every 8 hours. If you begin at 8:00 A.M., show that you will not have to take the medication between midnight and 7:00 A.M.

41. Suppose you make six round trips to visit a sick aunt and wish to record your mileage to the nearest mile. You forget the original odometer reading, but you do remember that the units digit has increased by 8 miles. What are the possible distances between your house and your aunt's house?

42. Suppose you are planning to purchase some rope. You need between 15 and 20 pieces that are 7 inches long and one piece that is 80 inches long. The rope can be purchased in multiples of 12 inches. How much rope should you buy to minimize waste?

43. If you know that your aunt in Problem 41 lives somewhere between 10 and 15 miles from your house, how far exactly is her house, given the information in Problem 41?

44. Is the set $\{0, 1, 2, 3, 4, 5\}$ a group for addition modulo 6?

45. Is the set {0, 1, 2, 3, 4, 5, 6} a group for addition modulo 7?

46. Is the set {0, 1, 2, 3} a field for addition and multiplication modulo 4?

Problems 47–52 involve the set {0, 1, 2, 3, 4, 5, 6, 7, 8, 9, 10} and addition and multiplication modulo 11.

47. Make a table for addition and multiplication.

48. Is the set a group for addition?

49. Is the set a group for multiplication?

50. Is the set a commutative group for addition?

51. Is the set a commutative group for multiplication?

52. Is the set a field for the operations of addition and multiplication?

PROBLEM SOLVING

An International Standard Book Number (ISBN) is used to identify books. The ISBN number for the 9th edition of The Nature of Mathematics *was 0-534-36890-5. The first digit, 0, indicates the book is published in an English-speaking country. The next three digits, 534, identify the publisher (Brooks/Cole), and the next five digits identify the particular book). The last digit is a check digit, which is used as follows. Multiply the first digit by 10, the next digit by 9, and so on:*

$$10(0) + 9(5) + 8(3) + 7(4) + 6(3) + 5(6) + 4(8)$$
$$+ 3(9) + 2(0) + 1(5) = 209$$

This number is congruent to 0, (mod 11). This is not by chance. The check digit, 5, is chosen so that this sum is 0, (mod 11). In other words, the check digit x for the book with ISBN number 0-534-34015-x is found by considering

$$10(0) + 9(5) + 8(3) + 7(4) + 6(3) + 5(4) + 4(0)$$
$$+ 3(1) + 2(5) = 148 \equiv 5$$

This means that for check digit x, $5 + x \equiv 0$, (mod 11). The one-digit solution to this equation is $x = 6$, so the check digit is 6. What is the check digit for a book with the given ISBN in Problems 53–54?

53. a. 0-534-13728-*x*   **b.** 0-691-02356-*x*

54. a. 0-8028-1430-*x*   **b.** 9-68-7270-82-*x*

55. What are the possible check digits for ISBNs? What is the check digit for the book with ISBN 0-13-330325-X? What do you think the X stands for in this ISBN?

56. If it is now 2 P.M., what time will it be 99,999,999,999 hours from now?

57. Write a schedule for 12 teams so that each team will play every other team once and no team will be idle.

58. One Hundred Fowl Problem (an old Chinese puzzle)   A man buys 100 birds for $100. A rooster is worth $10, a hen is worth $3, and chicks are worth $1 a pair. How many roosters, hens, and chicks did he buy if he bought at least one of each type?

59. Chinese Remainder Problem   A band of 17 pirates decided to divide their doubloons into equal portions. When they found that they had 3 coins remaining, they agreed to give them to their Chinese cook, Wun Tu. But 6 of the pirates were killed in a fight. Now when the treasure was divided equally among them, there were 4 coins left that they considered giving to Wun Tu. Before they could divide the coins, there was a shipwreck and only 6 pirates, the coins, and the cook were saved. This time equal division left a remainder of 5 coins for the cook. Now Wun Tu took advantage of his culinary position to concoct a poison mushroom stew so that the entire fortune in doubloons became his own. What is the smallest number of coins that the cook would have finally received?*

60. What is the next smallest number of coins that would satisfy the conditions of Problem 59?

*This problem is from Sun Zi (ca. A.D. 250) who wrote a mathematical manual during the Three Kingdoms period in China.

4.8 | CRYPTOGRAPHY

Cryptography is the art of writing or deciphering messages in code. Work in cryptography combines theoretical foundations with practical applications. It involves, at some level, ways to generate random-looking sequences and to detect and evaluate nonrandom effects. You might think of governments, covert operations, and spies when you think of cryptography, but there are many applications outside government. Cable television companies, whose signals are easily intercepted from satellite relays, use encryption schemes to prevent unauthorized use of their transmissions. Banks encrypt financial transactions and records in their computers for purposes of authenticity and integrity— that is, to make sure that the sender and contents are really as they appear—as well as for privacy. Cryptography figured prominently in a recent movie on the war against the drug cartel in Colombia. A cryptologist needs skills in communications,

Games *magazine, December 1985*

engineering, speech research, signals processing, and the design of specialized computers.

Simple Codes

Simple codes can be formed by replacing one letter by another. You may have seen this type of code in a children's magazine, or as a puzzle problem in a newspaper. The television show *Wheel of Fortune* uses a variation of this simple code-breaking skill. The cartoon at the left is from a popular game magazine. Codes such as this can easily be broken by using some logic and a knowledge of our language. Here is an analysis of how this code could be broken.

> RAE WA XLHH XDL WAYLTSAT XDNX VM DL KALZS'X
>
> PAIIEXL IR LGLPEXVAS, V FVHH SLYLT YAXL MAT DVI
>
> NWNVS!

We have color-coded the solution by showing the new material at each step in color.

1. The hint for this puzzle was "A four-letter word that starts and ends with the same letter is often THAT." From this, we conclude that the coded word XDNX is *that*, so that X is T, D is H, N is A. Fill in these letters above the letters of the puzzle:

```
       T     TH            THAT       H         T
   RAE WA XLHH XDL WAYLTSAT XDNX VM DL KALZS'X
       T      T                    T      H
   PAIIEXL IR LGLPEXVAS, V FVHH SLYLT YAXL MAT DVI
    A  A
   NWNVS!
```

2. The three letter word XDL must be the word *the*, so fill in E for L:

```
                 TH
       TE    THE      E    THAT    HE    E    T
   RAE WA XLHH XDL WAYLTSAT XDNX VM DL KALZS'X
        TE   E E  T              E E    TE      H
   PAIIEXL IR LGLPEXVAS, V FVHH SLYLT YAXL MAT DVI
    A  A
   NWNVS!
```

3. The single letter V is not A since N is A, so we guess that it is I:

```
       TE    THE      E    THAT I HE    E    T
   RAE WA XLHH XDL WAYLTSAT XDNX VM DL KALZS'X
        TE   E E TI  I I    E E    TE      HI
   PAIIEXL IR LGLPEXVAS, V FVHH SLYLT YAXL MAT DVI
    A  AI
   NWNVS!
```

4. The coded word XLHH begins with TE_ _; since the last two letters are the same, we assume that H is the letter L:

```
       TELL THE      E    THAT I HE    E    T
   RAE WA XLHH XDL WAYLTSAT XDNX VM DL KALZS'X
        TE   E E TI   I ILL E E    TE      HI
   PAIIEXL IR LGLPEXVAS, V FVHH SLYLT YAXL MAT DVI
    A  AI
   NWNVS!
```

5. The coded word FVHH ends with _ILL; assume F is W. Also we see the coded word VM which is I_, so we also assume that the coded M is decoded as F:

TELL THE   E   THAT I F HE   E   T
RAE WA XLHH XDL WAYLTSAT XDNX VM DL KALZS'X

     TE   E E T I   I W ILL E E   TE F   HI
PAIIEXL IR LGLPEXVAS, V F VHH SLYLT YAXL MAT DVI

A   AI
NWNVS!

6. Now, observing the context of the cartoon and making some assumptions about what a prisoner might be saying, we make some trial-and-error guesses: He wants a pardon from the governor, so we count letters and fill in the blanks for the word *governor:*

O GO TELL THE GOVERNOR THAT IF HE OE NT
RAE WA XLHH XDL WAYLTSAT XDNX VM DL KALZS'X

O TE   E E TION I WILL NEVER VOTE FOR III
PAIIEXL IR LGLPEXVAS, V FVHH SLYLT YAXL MAT DVI

AG A IN
NWNVS!

7. We can now fill in the completed deciphered message. Note that spaces translate as spaces:

YOU GO TELL THE GOVERNOR THAT IF HE DOESN'T COMMUTE MY EXECUTION, I WILL NEVER VOTE FOR HIM AGAIN!

Modular Codes*

One common code is based on modular arithmetic, and it is known as a **modular code.** Rather than simply substituting one letter for another (which is an easily broken code), more sophisticated codes can be developed. To **encrypt** a message means to scramble it by something called an **encoding key.** The result is a secret or coded message, called **ciphertext.** The coded message is then unscrambled using a secret **decoding key.** This coding procedure requires that both the sender and the receiver know, and conceal, the encoding and decoding keys. Suppose, for example, that we wish to send the following secret message:

THE FBI HAS BED BUGS.

Suppose we select some encoding key—say, multiply by 2. Then we code the message according to some given modular number, say 29, as shown in Figure 4.21 (page 206). Notice that we use 29 instead of 0; this is still considered a modulo 29 system since $29 \equiv 0, (\text{mod } 29)$.

*This code requires Section 4.7.

On October 11, 1988, Mark S. Manasse of Digital Corporation's Systems and Arjen K. Lenstra of the University of Chicago linked over a dozen users of some 400 computers on three continents to find the factors of a 100-digit number.

Several of the most secure cipher systems invented in the past decade are based on the fact that large numbers are extremely difficult to factor, even using the most powerful computers for long period of time. The accomplishment of factoring a 100-digit number "is likely to prompt cryptographers to reconsider their assumptions about cipher security," Lenstra said in a telephone interview. ... Using larger numbers makes the work of cryptographers more cumbersome.

The Mathematics Teacher, Jan. 1990, p. 70.

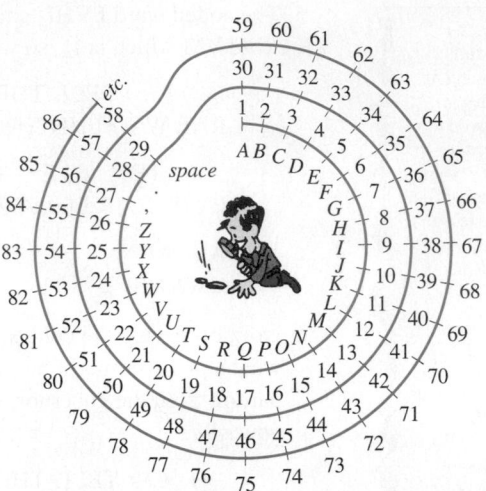

Figure 4.21 Modular 29 code

We look up the numerical value for each letter, space, and punctuation mark according to Figure 4.21.

Message:

20	8		5	29	6		2	9	29	8		1		19	29	2		5		4	29	2		21	7	19	28
T	H	E		F	B	I		H	A	S		B	E	D		B	U	G	S	.

Note that the "code" 20-8-5-29-6-2-9-29-8-1-19-29-2-5-4-29-2-21-7-19-28 would be "easy" to break. Why?

Now we encode by multiplying each of these numbers by the encoding key; then we *modulate,* or write each of these answers modulo 29:

Message:

20	8	5	29	6	2	9	29	8	1	19	29	2	5	4	29	2	21	7	19	28
T	H	E		F	B	I		H	A	S		B	E	D		B	U	G	S	.

Numerical value: 20- 8- 5-29- 6- 2- 9-29- 8 -1 -19-29-2- 5- 4-29- 2- 21- 7- 19- 28

Encode (encoding key 2):   40-16- 10-58-12- 4- 18-58-16- 2- 38-58- 4- 10- 8-58- 4- 42- 14- 38- 56

Modulate: 11-16- 10-29-12- 4- 18-29-16- 2- 9-29- 4- 10- 8-29- 4- 13- 14- 9- 27

Coded message   K P J   L D R   P B I   D J H   D M N I   ,

The coded message is KPJ LDR PBI DJH DMNI,

The decoding key is the inverse of the encoding key. Since the encoding key is to multiply by 2, the decoding key is to divide by 2.

EXAMPLE 1

Decode the following message for which the encoding key is to multiply by 2: NBXXMC RI B CAXX FBK,

Solution   First use Figure 4.21 to find the numerical value:

14-2-24-24-13-3-29-18-9-29-2-29-3-1-24-24-29-6-2-11-27

The decoding key is to divide by 2, so we look at the code and write the odd numbers so that they are evenly divisible by 2, (mod 29):

14- 2- 24- 24- 42- 32- 58- 18- 38- 58- 2- 58- 32- 30- 24- 24- 58- 6- 2- 40- 56

Decoding key (divide by 2):   7- 1- 12- 12- 21- 16- 29- 9- 19- 29- 1- 29- 16- 15- 12- 12- 29- 3- 1- 20- 28

Decoded message (Figure 4.21):   G- A- L- L- U- P- - I- S- - A- P- O- L- L- - C- A- T- .

The decoded message is: GALLUP IS A POLL CAT.   ◆

It is easy to multiply two large prime numbers to obtain a large number as the answer. But the reverse process—factoring a large number to determine its components—presents a formidable challenge. The problem appears so hard that the difficulty of factoring underlies the so-called RSA method of encrypting digital information. An international team of computer scientists recently spent 8 months finding the factors of a 129-digit number that was suggested 17 years ago as a test of the security of the RSA cryptographic scheme. This effort demonstrates the strength of the RSA cryptosystem. However, this effort also demonstrates that significantly larger numbers may be necessary in the future to ensure security.

Science News, May 7, 1994

Unbreakable Codes

In 1970, mathematicians Whitfield Diffie and Martin Hellman showed a way to make the keys public. Suppose a code has two keys, an encoding key and a decoding key, and also suppose it is impossible to compute one key from the other in the sense that no person or computer would be able to do it; then this would constitute an unbreakable code. Here is the way such a code works. Everyone owns a unique pair of keys, one of which remains private, but the other is public in the sense that it is listed in a readily available directory. To send a message, you look up the public key for the person to whom the message is to be sent. You use the public key to scramble the message. The receiver then uses their private key to decode the message with total secrecy and privacy.

Three mathematicians, Rivest, Shamir, and Adleman, created a public key algorithm known as RSA, from their initials. This method depends on ideas of prime numbers and factoring studied in this chapter. Consider two prime numbers, say, 11 and 7. Now if I hand you their product, 77, this product can be made public, while the factors 11 and 7 remain secret. To *encode* you need the number 77, but to *decode* you need the factors. Now, *if* the factors are large enough (say, 200 digits each), then the product is so large that it can never be factored, and the result is an unbreakable code. *

Why would we need an unbreakable public-key code? Some applications include protecting money transfers from tampering, shielding sensitive business data from competitors, and protecting computer software from viruses. The government has published a *digital signature algorithm* (DSA), which depends on a single very large prime number.

*RSA Laboratories, Redwood City, CA, tests its ability to create difficult ciphers by establishing a series of cryptographic contests. Its first challenge required 140 days to solve. The coded message was "Strong cryptography makes the world a better place." If you are interested in this type of challenge, check out www.rsa.com for additional information.

Problem Set 4.8

LEVEL 1

Number the letters of the alphabet from 1 to 26; code a comma as 27, period as 28, and space as 29. Encode the messages in Problems 1–4.

1. NEVER SAY NEVER.
2. THE EAGLE HAS LANDED.
3. YOU BET YOUR LIFE.
4. MY BANK BALANCE IS NEGATIVE.

Number the letters of the alphabet from 1 to 26 and code a blank as 29. Decode the messages in Problems 5–8.

5. 1-18-5-29-23-5-29-8-1-22-9-14-7-29-6-21-14-29-25-5-20
6. 9-29-12-15-22-5-29-13-1-20-8-5-13-1-20-9-3-19
7. 6-1-9-12-21-18-5-29-20-5-1-3-8-5-19-29-19-21-3-3-5-19-19
8. 19-17-21-1-18-5-29-13-5-1-12-19-29-13-1-11-5-29-18-15-21-14-4-29-16-5-15-16-12-5

Give the decoding key for the encoding keys in Problems 9–14.

9. Multiply by 8.
10. Divide by 6.
11. Times 20, add 2.
12. Divide by 4, minus 3.
13. Multiply by 4 and add 2, then double the result.
14. Multiply by 3 and subtract 3, then divide the result by 2.

LEVEL 2

Use Figure 4.21 to encode or decode the messages in Problems 15–23.

15. Encoding key: Multiply by 3.
    NEVER SAY NEVER.
16. Encoding key: Multiply by 5.
    THE EAGLE HAS LANDED.
17. Encoding key: Multiply by 2 and add 5.
    YOU BET YOUR LIFE.
18. Encoding key: Multiply by 4 and subtract 10.
    MY BANK BALANCE IS NEGATIVE.
19. Encoding key: Multiply by 3.
    XEJSBQ LEJSBQ ,. C RCGG UEQZ
20. Encoding key: Multiply by 2 and add 10.
    LJMQKVTSSKQJ.SJKITJ,ZKJULECSJ.IJSKGTKITJTE
    STSJSETTM
21. Encoding key: multiply by 2 and subtract 7.
    SKAXBVIXAVXVGKQDV.WSYQCLT

22. Encoding key: multiply by 2 and subtract 11.
ALL PERSONS BY NATURE DESIRE KNOWLEDGE.

23. Encoding key: Multiply by 3 and add 5.
SHOW ME A DROPOUT FROM A DATA PROCESSING
SCHOOL AND I WILL SHOW YOU A
NINCOMPUTER.

PROBLEM SOLVING

*The ciphers in Problems 24–27 are taken from Games
Magazine, December 1985.*

24. L FPHWDJ QJPQFJ TDP MJJQ BPKR. WDJU HVJ
IPTHVBR TDP DHXJG'W KPW WDJ KSWR WP ALWJ
QJPQFJ WDJNRJFXJR. (*Hint:* Ciphertext pattern
QJPQFJ often represents PEOPLE.)

25. GHJWHY NDW LOGKL TGRLBK, UBLRGXV, GHV
XYOZLD WH DZL DWR VWS ZL RXBOJ G UGH
MWX GOO LYGLWHZHSL. (*Hint:* A three-letter word
after a series of words set off by commas is often AND.)

26. GRAPE MA QLMVFGCB, BLLCRQU XRUX
GPMRQUB, TFQBMPQMEK BTXLYNEL DLBM
BXFVB PUPRQBM LPTX FMXLG. (*Hint:* Ciphertext
B represents S. Note its high frequency as a first and last
letter. *Bonus hint:* The fifth word is *not* THAT.)

27. MZDGBWV-MVCWZ WTZ HR ZHMAD
XHKEBWTEG NZHLTUCG TUCWV CELTZHEXCEB
RHZ PCWBSCZ GBWBTHE. (*Hint:* The five vowels, A
to U, are represented by C, H, K, T, and W, but not necessarily in that order.)

*Cryptic arithmetic is a type of mathematics puzzle in which
letters have been replaced by the digits of numbers. Replace
each letter by a digit (the same digit for the same letter
throughout; different digits for different letters), and the arithmetic will be performed correctly. Break the codes in Problems 28–30.*

28.	SEND	**29.**	THIS	**30.**	DAD	
	+ MORE		IS		SEND	
	MONEY		+ VERY		+ MORE	
			EASY		MONEY	

CHAPTER SUMMARY

"More new jobs will require more postsecondary mathematics education."

A Challenge of Numbers

IMPORTANT TERMS

Abelian group [4.6]
Absolute value [4.3]
Addition [4.1]
Additive identity [4.6]
Additive inverse [4.6]
Algebra [4.7]
Associative property [4.1]
Canonical form [4.2]
Ciphertext [4.8]
Closed for addition [4.1]
Closed for multiplication [4.1]
Closed set [4.1]
Closure property [4.1]
Commutative group [4.6]
Commutative property [4.1]
Composite number [4.2]
Congruent modulo *m* [4.7]
Counting numbers [4.1]
Cryptography [4.8]
Decoding key [4.8]

Denominator [4.4]
Dense set [4.6]
Discrete mathematics [4.7]
Distributive property [4.1]
Divides [4.2]
Divisibility [4.2]
Division [4.3, 4.7]
Division by zero [4.3]
Divisor [4.2]
e [4.5]
Encoding key [4.8]
Encrypt [4.8]
Factor [4.2]
Factor tree [4.2]
Factoring [4.2]
Field [4.6]
Fundamental property
    of fractions [4.4]
Fundamental theorem
    of arithmetic [4.2]

g.c.f. [4.2]
Greatest common factor [4.2]
Group [4.6]
Hypotenuse [4.5]
Identity for addition [4.6]
Identity for multiplication [4.6]
Identity property [4.6]
Improper fraction [4.4]
Integers [4.3]
Inverse [4.6]
Inverse property [4.6]
Irrational number [4.5]
Laws of square roots [4.5]
l.c.m. [4.2]
Least common denominator [4.4]
Least common multiple [4.2]
Leg [4.5]
Modular codes [4.8]
Modulo 5 [4.7]
Multiple [4.2]

Multiplication [4.1, 4.7]
Multiplicative identity [4.6]
Multiplicative inverse [4.6]
Natural numbers [4.1]
Number line [4.6]
Numerator [4.4]
One [4.6]
Opposites [4.6]
Perfect square [4.5]
π [4.5]
Positive square root [4.5]
Prime factorization [4.2]

Prime number [4.2]
Proper fraction [4.4]
Pythagorean theorem [4.5]
Radical form [4.5]
Radicand [4.5]
Rational number [4.4]
Real number line [4.6]
Real numbers [4.6]
Reciprocal [4.6]
Reduced fraction [4.4]
Relatively prime [4.2]
Repeating decimal [4.6]

Rules of divisibility [4.2]
Sieve of Eratosthenes [4.2]
Square number [4.5]
Square root [4.5]
Subtraction [4.1; 4.7]
Terminating decimal [4.6]
Unit distance [4.6]
Whole numbers [4.3; 4.4]
Zero [4.3, 4.6]
Zero multiplication [4.7]

TYPES OF PROBLEMS

Demonstrate the definition of multiplication. [4.1]
Determine whether a given set with a given operation is closed. [4.1]
Recognize and distinguish the commutative and associative properties. [4.1]
Apply the distributive property with a variety of operations. [4.1]
Determine whether a given number is prime or not. [4.2]
Tell whether one number divides another number. [4.2]
Find the prime factorization and write the answer in canonical form. [4.2]
Find the least common multiple of a set of numbers. [4.2]
Find the greatest common factor of a set of numbers. [4.2]
Show that there is no largest prime number. [4.2]
Use problem-solving techniques to solve applied problems. [4.2–4.8]
Find the absolute value of a number. [4.3]
Carry out operations with integers. [4.3]
Reduce fractions using the fundamental property of fractions. [4.4]
Carry out operations with fractions. [4.4]
Use the definition of square root to simplify radical expressions. [4.5]
Classify numbers as rational or irrational. [4.5]
Determine into which of the following sets that a given number belongs: $\mathbb{N}$ (natural numbers), $\mathbb{Z}$ (integers), $\mathbb{Q}$ (rational numbers), $\mathbb{Q}'$ (irrational numbers), or $\mathbb{R}$ (real numbers). [4.6]
Express a rational number as a decimal. [4.6]
Express a terminating decimal as a fraction. [4.6]
Use the order of operations to simplify real numbers. [4.6]
Find a rational number or an irrational number between each of a given pair of numbers. [4.6]
Recognize and distinguish examples of the closure, associative, commutative, identity, and inverse properties. [4.6]
Know and be able to describe each of the field properties. [4.6]
Carry out operations in modular arithmetic. [4.7]
Solve modular equation. [4.7]
Decide if a given set and operation forms a group. [4.7]
Decide if a given set and two operations forms a field. [4.7]
Encode and decode simple phrases. [4.8]
Break a simple code. [4.8]

CHAPTER 4 REVIEW QUESTIONS

Perform the indicated operations in Problems 1–8.

1. $-4 + 5(-3)$

2. $\dfrac{4}{7} + \dfrac{5}{9}$

3. $\dfrac{7}{30} + \dfrac{5}{42} + \dfrac{5}{99}$

4. $-\sqrt{10} \cdot \sqrt{10}$

5. $\left(\dfrac{11}{12} + 2\right) + \dfrac{-11}{12}$

6. $\dfrac{3^{-1} + 4^{-1}}{6}$

7. $\dfrac{-7}{9} \cdot \dfrac{99}{174} + \dfrac{-7}{9} \cdot \dfrac{75}{174}$

8. $\dfrac{-3 + \sqrt{3^2 + 4(2)(3)}}{2(2)}$

Reduce each fraction in Problems 9–13. If it is reduced, so state.

9. $\dfrac{8}{3}$

10. $\dfrac{16}{18}$

11. $\dfrac{100}{825}$

12. $\dfrac{184}{207}$

13. $\dfrac{1{,}209}{2{,}821}$

Find the decimal representation for each number in Problems 14–18, and classify it as rational or irrational.

14. $\dfrac{3}{8}$

15. $\dfrac{7}{3}$

16. $\dfrac{3}{7}$

17. $\dfrac{125}{20}$

18. $\dfrac{3}{13}$

Find the prime factorization of each number given in Problems 19–23. If it is prime, so state.

19. 89

20. 101

21. 349

22. 1,001

23. 6,825

Solve the equations in Problems 24–26 for x.

24. $\dfrac{x}{5} \equiv 2,\ (\text{mod } 8)$

25. $2x \equiv 3,\ (\text{mod } 7)$

26. $2x^2 + 7x + 1 \equiv 0,\ (\text{mod } 2)$

27. Find the greatest common factor for the set of numbers {49, 1001, 2401}.

28. Find the least common multiple for the set of numbers {49, 1001, 2401}.

29. If $a \odot b = a \times b + a + b$, what is the value of $(1 \odot 2) \odot (3 \odot 4)$?

30. If $a \uparrow b$ stands for the larger number of the pair and $a \downarrow b$ stands for the lesser number of the pair, what is the value of $(1 \downarrow 2) \uparrow (2 \downarrow 3)$?

Define each of the operations given in Problems 31–33.

31. multiplication

32. subtraction

33. division

34. IN YOUR OWN WORDS Explain why we cannot divide by zero.

35. Find an irrational number between 34 and 35.

36. What is a field? Describe each property.

37. Consider the set $\mathbb{N}$ and let $a, b \in \mathbb{N}$. Let $⫯$ be an operation defined by $a ⫯ b = $ g.c.f. of a and b. Is $\mathbb{N}$ a commutative group for the operation of $⫯$?

38. **Do you weigh too much?**   In the United States, a fat person is defined as anyone with a "body mass index" (or BMI) of 27.6 or higher. The World Health Organization defines a fat person as one with a BMI of 25 or higher. To determine BMI, multiply your weight in pounds by 703, and then divide that result by your height (in inches) squared. Write this in formula form using w for weight in pounds and h for height in inches. Calculate the BMI (rounded to the nearest tenth) for each of these people:

    a. 5 ft 5 in., 165 lb   **b.** 6 ft, 185 lb

    c. What is the weight (rounded to the nearest pound) for a 5-ft, 6-in. tall person who wants a BMI of 25?

39. The News Clip is from the column "Ask Marilyn," *Parade Magazine,* April 15, 1990. Answer the question asked in the article.

    > There are 1,000 tenants and 1,000 apartments. The first tenant opens every door. The second closes every other door. The third tenant goes to every third door, opening it if it is closed and closing it if it is open. The fourth tenant goes to every fourth door, closing it if it is open and opening it if it is closed. This continues with each tenant until the 1,000th tenant closes the 1,000th door. Which doors are open?

    Anita Mueller, Arnold, MD

40. Suppose you are building a stairway (such as shown in Figure 4.22). Assume that the vertical rise is 8 ft, the horizontal run is 12 ft, and the maximum rise for each step is 8 inches. How many steps are necessary? What is the total length of the segments making up the stairs, and what is the length of the diagonal?

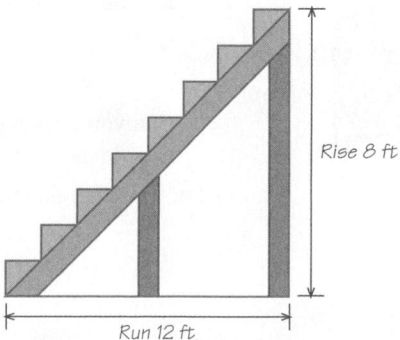

Figure 4.22 **Building a stairway**

GROUP RESEARCH

Working in small groups is typical of most work environments, and learning to work with others to communicate specific ideas is an important skill. Work with three or four other students to submit a single report based on each of the following questions.

G13. With only a straightedge and compass, use a number line and the Pythagorean theorem to construct a segment whose length is $\sqrt{2}$. Measure the segment as accurately as possible, and write your answer in decimal form. *Do not use a calculator or any tables.* Now, continue your work to construct segments whose lengths are $\sqrt{3}$, $\sqrt{4}$, $\sqrt{5}$, ...

G14. Four Fours   Write the numbers from 1 to 100 (inclusive) using exactly four fours. See Problem 60, Problem Set 4.3, to help you get started.

G15. Pythagorean Theorem   Write out three different proofs of the Pythagorean theorem.

G16. Modular Art   Many interesting designs such as those shown here can be created using patterns based on modular arithmetic. Prepare a report for class presentation based on the article "Using Mathematical Structures to Generate Artistic Designs" by Sonia Forseth and Andrea Price Troutman, *The Mathematics Teacher,* May 1974, pp. 393–398. Another source is "Mod Art: The Art of Mathematics" by Susan Morris, *Technology Review,* March/April 1979.

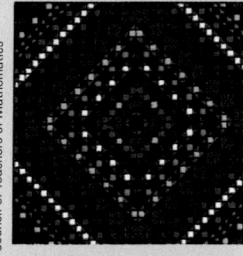

G17. The Babylonians estimated square roots using the following formula:

$$\text{If } n = a^2 + b, \text{ then } \sqrt{n} \approx a + \frac{b}{2a}$$

For example, if $n = 11$, then $n = 9 + 2$, so that $n = 11$, $a = 3$, and $b = 2$.
↑
perfect square

This Babylonian approximation for $\sqrt{11}$ is found by

$$\sqrt{11} \approx 3 + \frac{2}{2(3)} \approx 3.3333\ldots$$

With a calculator, we obtain $\sqrt{11} \approx 3.31662479$. Write a paper about this approximation method. Here are some questions you might consider:

a. Can b be negative?

b. Consider the following possibilities:

$$|b| < a^2 \qquad |b| = a^2 \qquad |b| > a^2$$

Can you formulate any conclusions about the appropriate hypotheses for the Babylonian approximation?

c. Put this formula into a historical context.

Individual Research Problems

Learning to use sources outside your classroom and textbook is an important skill, and here are some ideas for extending some of the ideas in this chapter. You can find references to these projects in a library or at

www.mathnature.com

PROJECT 4.1 What do the following people have in common?

Corazon Aquino, former president of the Philippines
Leon Trotsky, revolutionary
Carole King, singer-songwriter
Heloise (Poncé Cruse Evans), columnist, *Hints from Heloise*
Florence Nightingale, pioneer in professional nursing

PROJECT 4.2 In the text we tried some formulas that might have generated only primes, but, alas, they failed. Below are some other formulas. Show that these, too, do not generate only primes.

    a. $n^2 + n + 41$ **b.** $n^2 - 79n + 1,601$ **c.** $2n^2 + 29$
    d. $9n^2 - 489n + 6,683$ **e.** $n^2 + 1$, n an even integer

PROJECT 4.3 For what values of n is $11 \cdot 14^n + 1$ a prime?

PROJECT 4.4 A formula that generates all prime numbers is given by David Dunlop and Thomas Sigmund in their book *Problem Solving with the Programmable Calculator* (Englewood Cliffs, NJ: Prentice-Hall, 1983). The authors claim that the formula $\sqrt{1 + 24n}$ produces every prime number except 2 and 3, but give no proof or reference to a proof. Create a table, and give an argument to support or find a counterexample to disprove their claim.

PROJECT 4.5 The largest known prime, $2^{13,466,917} - 1$, is a number that has 4,053,946 digits. A number this large is hard to comprehend. Write a paper making the size of this number meaningful to a nonmathematical reader.

PROJECT 4.6 Investigate some of the properties of primes not discussed in the text. Why are primes important to mathematicians? Why are primes important in mathematics? What are some of the important theorems concerning primes?

PROJECT 4.7 Find the one composite number in the following set:

$$31$$
$$331$$
$$3331$$
$$33331$$
$$333331$$
$$3333331$$
$$33333331$$
$$333333331$$

PROJECT 4.8 Celebrate the millennium

    a. Consider the product $1 \cdot 2 \cdot 3 \cdot \cdots \cdot 1{,}998 \cdot 1{,}999 \cdot 2{,}000$. What is the last digit?

    b. Consider the product $1 \cdot 2 \cdot 3 \cdot \cdots \cdot 1{,}998 \cdot 1{,}999 \cdot 2{,}000$. From the product, cross out all even factors as well as all multiples of 5. What is the last digit?

PROJECT 4.9 We mentioned that the Egyptians wrote their fractions as sums of unit fractions. Show that every positive fraction less than 1 can be written as a sum of unit fractions.

PROJECT 4.10 The Egyptians had a very elaborate and well-developed system for working with fractions. Write a paper on Egyptian fractions.

PROJECT 4.11 Form a group using a geoboard. Go to **www.mathnature.com** for some ideas about writing this paper.

PROJECT 4.12 Prove that $\sqrt{2}$ is irrational.

PROJECT 4.13 Write a paper or prepare an exhibit illustrating the Pythagorean theorem.

PROJECT 4.14 Write a paper on the symmetries of a cube. Go to **www.mathnature.com** for some ideas about writing this paper.

PROJECT 4.15 What is a Diophantine equation?

PROJECT 4.16 Prepare an exhibit on cryptography. Include devices or charts for writing and deciphering codes, coded messages, and illustrations of famous codes from history. For example, codes are found in literature in *Before the Curtain Falls* by J. Rives Childs, *The Gold Bug* by Edgar Allan Poe, and *Voyage to the Center of Earth* by Jules Verne.

PROJECT 4.17 Write a paper on the importance of cryptography for the Internet.

5 The Nature of Algebra

The fact that algebra has its origin in arithmetic, . . . led Sir Isaac Newton to designate it "Universal Arithmetic," a designation which, vague as it is, indicates its character better than any other.

GEORGE E. CHRYSTAL

CONTENTS

OVERVIEW

It has been said that algebra is the greatest labor-saving device ever invented. Even though you have probably had algebra sometime in your past, the essential algebraic ideas you will need for the rest of this text are covered or reviewed in this chapter. We begin with the basic building blocks of algebra—namely, polynomials and operations with polynomials. There are four main processes in algebra: **simplify** (Section 5.1), **factor** (Section 5.2), **evaluate** (Section 5.3), and **solve** (Sections 5.4, 5.5, and 5.7). Some applications of these ideas are introduced in the last four sections of this chapter. We look at an application from genetics, take a peek at computer spreadsheets, and solve application problems with numbers, ratios, proportions, percents, distance, and Pythagorean relationships.

IMPORTANT IDEAS

Binomial product (FOIL) [5.1]
Binomial theorem [5.1]
Procedure for factoring trinomials [5.2]
Difference of squares [5.2]
Evaluate an expression [5.3]
Use a spreadsheet [5.3]
Equation properties [5.4]
Zero-product rule [5.4]
Quadratic formula [5.4]
Linear inequalities [5.5]
Addition property of inequality [5.5]
Multiplication property of inequality [5.5]
Procedure for problem solving [5.6]
Property of proportions [5.7]
Procedure for solving proportions [5.7]
Change forms: fraction/decimal/percent [5.8]
Be able to estimate answers to percent problems [5.8]
Solve applied percent problems [5.8]
Guidelines for problem solving [5.9]
Problem-solving examples: genetics [5.3], linear and quadratic equations [5.4], linear inequalities [5.5], number relationships [5.6], distance relationships [5.6], Pythagorean theorem applications [5.6], proportion problems [5.7], percent problems [5.8]

BOOK REPORT

Write a 500-word report on this book:
Hypatia's Heritage, Margaret Alic (Boston: Beacon Press, 1986). This book is a history of women in science from antiquity through the 19th century. Since most of recorded history has been male-dominated, history books reflect this male bias and have ignored the history of women. This book is a rediscovery of women in science.

LINKS

www Individual Projects

Group Projects

Research Links

www.mathnature.com

5.1 | POLYNOMIALS

Many people think of algebra as simply a high school mathematics course in which variables are manipulated. This chapter reviews many of these procedures; however, the word **algebra** refers to a structure, or a set of axioms that forms the basis for what is accepted and what is not. For example, in the previous chapter we defined a field, which involves a set, two operations, and 11 specified properties. If you studied Section 4.7, you investigated an algebra with a finite number of elements. As you study the ordinary algebra presented in this chapter, you should remember this is only one of many possible algebras. Additional algebras are often studied in more advanced mathematics courses.

In this chapter we will review much of the algebra you have previously studied. There are four main processes in algebra which we will review in this chapter: *simplify, factor, evaluate,* and *solve.* Recall that a **term** is a number, a variable, or the product of numbers and variables. Thus, $10x$ is one term, but $10 + x$ is not (because the terms 10 and x are connected by addition and not by multiplication). A fundamental notion in algebra is that of a **polynomial,** which is a term or the sum of terms. We classify polynomials by the number of terms and by degree:

A polynomial with one term is called a **monomial.**

A polynomial with two terms is called a **binomial.**

A polynomial with three terms is called a **trinomial.**

There are other words that could be used for polynomials with more than three terms, but this classification is sufficient. To classify by degree, we recall that the **degree of a term** is the number of *variable* factors in that term. Thus, $3x$ is first-degree, $5xy$ is second-degree, 10 is zero-degree, $2x^2$ is second-degree, and $9x^2y^3$ is fifth-degree. The **degree of a polynomial** is the largest degree of any of its terms. A first-degree term is sometimes called **linear** and a second-degree term is sometimes referred to as **quadratic.** The numerical part of a term, usually written before the variable part, is called the **numerical coefficient.** In $3x,$ it is the number 3, in $5xy$ it is the number 5, and in $9x^2y^3$ it is the number 9.

EXAMPLE 1

Classify each polynomial by number of terms and by degree. If the expression is not a polynomial, so state.

a. x **b.** $x^2 + 5x - 7$ **c.** $x^3 - \dfrac{2}{x}$ **d.** $x^2y^3 - xy^2$ **e.** 5

Solution

a. Monomial, degree 1

b. Trinomial, degree 2

c. Not a polynomial because it involves division by a variable. Notice that $x^3 - \dfrac{x}{2}$ *is a* polynomial because it can be written as $x^3 - \frac{1}{2}x$ and fractional coefficients are permitted.

d. Binomial, degree 5

e. Monomial, degree 0 ◆

When writing polynomials, it is customary to arrange the terms from the highest-degree to the lowest-degree term. If terms have the same degree they are usually listed in alphabetical order.

Simplification

When working with polynomials, it is necessary to simplify algebraic expressions. The key ideas of simplification are *similar terms* and the *distributive property.* Terms that differ only in the numerical coefficients are called **like terms** or **similar terms.**

EXAMPLE 2

Simplify the given algebraic expressions.

a. $-12x - (-5)x$ **b.** $-3x - 6x + 2x$

c. $2x + 3y + 5x - 2y$ **d.** $5xy^2 - xy^2 - 4x^2y$

e. $(4x - 5) + (5x^2 + 2x - 3)$ **f.** $(4x - 5) - (5x^2 + 2x - 3)$

Solution

a. $-12x - (-5x) = -12x + 5x = -7x$ Add the opposite of $(-5x)$.

b. $-3x - 6x + 2x = -7x$

c. $2x + 3y + 5x - 2y = 7x + y$ Note the similar terms.

d. $5xy^2 - xy^2 - 4x^2y = 4xy^2 - 4x^2y$

e. $(4x - 5) + (5x^2 + 2x - 3) = 5x^2 + (4x + 2x) + (-5 - 3)$
$$= 5x^2 + 6x - 8$$

f. Recall that to subtract a polynomial, you subtract *each* term of that polynomial:
$$(4x - 5) - (5x^2 + 2x - 3) = 4x - 5 - 5x^2 - 2x - (-3)$$
$$= -5x^2 + 2x - 2 \qquad \blacklozenge$$

Simplify Polynomials

> To **simplify** a polynomial means to carry out all operations (according to the order of operations agreement) and to write the answer in a form with the highest-degree term first, with the rest of the terms arranged by decreasing degree. If there are two terms of the same degree, arrange those terms alphabetically.

Remember from beginning algebra that $-x = (-1)x,$ so to subtract a polynomial you can do it as shown in Example 2f—by subtracting *each* term—or you can think of it as an application of the distributive property:

$$(4x - 5) - (5x^2 + 2x - 3) = 4x - 5 + (-1)(5x^2 + 2x - 3)$$
$$= 4x - 5 + (-1)(5x^2) + (-1)(2x) + (-1)(-3)$$
$$= -5x^2 + 2x - 2$$

The distributive property is also important in multiplying polynomials.

EXAMPLE 3

Simplify the given algebraic expressions.

a. $3x(x^2 - 1)$ **b.** $(2x - 3)(x + 1)$ **c.** $(x + 2)(x^2 + 5x - 2)$

d. $(2x + 1)^3$

Solution

In each case, we will distribute the expression on the left: $3x$, $(2x-3)$, $(x+2)$, and $(2x+1)$, respectively. You will find your work easier to read if you work down with the equal signs aligned rather than across your paper. And finally, we use the laws of exponents (Section 1.3, p. 32) to simplify expressions such as $x(x^2) = x^1x^2 = x^{1+2} = x^3$.

a. $3x(x^2 - 1) = 3x(x^2) + 3x(-1)$ This step is usually done mentally, and is not written down.

$\qquad\qquad = 3x^3 - 3x$

b. $(2x - 3)(x + 1) = (2x - 3)(x) + (2x - 3)(1)$ Mental step

$\qquad\qquad\qquad = 2x^2 - 3x + 2x - 3$ Distributive property

$\qquad\qquad\qquad = 2x^2 - x - 3$ Add similar terms.

c. $(x + 2)(x^2 + 5x - 2) = (x + 2)(x^2) + (x + 2)(5x) + (x + 2)(-2)$

$\qquad\qquad\qquad\qquad = x^3 + 2x^2 + 5x^2 + 10x - 2x - 4$

$\qquad\qquad\qquad\qquad = x^3 + 7x^2 + 8x - 4$

d. $(2x + 1)^3 = (2x + 1)(2x + 1)(2x + 1)$ Definition of cube

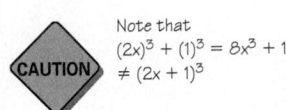
Note that
$(2x)^3 + (1)^3 = 8x^3 + 1$
$\neq (2x + 1)^3$

$\qquad\qquad = (2x + 1)[(2x + 1)(2x) + (2x + 1)(1)]$

$\qquad\qquad = (2x + 1)[4x^2 + 2x + 2x + 1]$

$\qquad\qquad = (2x + 1)(4x^2 + 4x + 1)$

$\qquad\qquad = (2x + 1)(4x^2) + (2x + 1)(4x) + (2x + 1)(1)$

$\qquad\qquad = 8x^3 + 4x^2 + 8x^2 + 4x + 2x + 1$

$\qquad\qquad = 8x^3 + 12x^2 + 6x + 1$ ◆

Shortcuts with Products

It is frequently necessary to multiply binomials, and even though we use the distributive property, we want to be able to carry out the process quickly and efficiently in our heads. Consider the following example, which leads us from multiplication using the distributive property to an efficient process that is usually called **FOIL.**

EXAMPLE 4

Simplify:

a. $(2x + 3)(4x - 5)$ **b.** $(5x - 3)(2x + 3)$ **c.** $(4x - 3)(3x - 2)$

Solution

a. $\mathbf{(2x + 3)(4x - 5)} = \mathbf{(2x + 3)(4x)} + \mathbf{(2x + 3)(-5)}$ Distributive property

$\qquad\qquad = \underbrace{8x^2}_{\substack{\uparrow \\ \text{Product of} \\ \text{first terms}}} + \underbrace{12x + (-10x)}_{\substack{\uparrow \\ \text{Sum of products of} \\ \text{inner terms and outer terms}}} + \underbrace{(-15)}_{\substack{\uparrow \\ \text{Product of} \\ \text{last terms}}}$

$\qquad\qquad = 8x^2 + 2x - 15$

b. $(5x - 3)(2x + 3) = \underbrace{10x^2}_{\substack{\uparrow \\ \text{FIRST terms}}} + \underbrace{(15x - 6x)}_{\substack{\uparrow \\ \text{OUTER terms and} \\ \text{INNER terms}}} + \underbrace{(-9)}_{\substack{\uparrow \\ \text{LAST terms}}}$

Mentally:

$\qquad\qquad = 10x^2 + 9x - 9$

c. $(4x - 3)(3x - 2) = 12x^2 - 17x + 6$ Mentally   ◆

You are encouraged to carry out the binomial multiplication mentally, as shown in Example 4c. To help you remember the process, we sometimes call this binomial multiplication FOIL to remind you

First terms + **O**uter terms + **I**nner terms + **L**ast terms

FOIL
Binomial Product

(Do this binomial multiplication mentally.)

$$(ax + b)(cx + d) = \underbrace{acx^2}_{\substack{\uparrow \\ \textbf{F}\text{irst terms}}} + \underbrace{(ad + bc)x}_{\substack{\textbf{O}\text{uter} \\ + \\ \textbf{I}\text{nner}}} + \underbrace{bd}_{\substack{\uparrow \\ \textbf{L}\text{ast terms}}}$$

EXAMPLE 5

Simplify (mentally):

a. $(2x - 3)(x + 3)$ **b.** $(x + 3)(3x - 4)$ **c.** $(5x - 2)(3x + 4)$

Solution

a. $(2x - 3)(x + 3) = \underset{\underset{\textbf{F}}{\uparrow}}{\underline{2x^2}} + \underset{\textbf{O}+\textbf{I}}{\underline{3x}} \underset{\underset{\textbf{L}}{\uparrow}}{\underline{-9}}$

b. $(x + 3)(3x - 4) = 3x^2 + 5x - 12$

c. $(5x - 2)(3x + 4) = 15x^2 + 14x - 8$ ◆

A second shortcut involves raising a binomial to an integral power—for example, $(2x + 1)^3$ or $(a + b)^8$. We look for a pattern with the following example.

EXAMPLE 6   **PÓLYA'S METHOD**

Expand (multiply out) the expression $(a + b)^8$.

Solution   We use Pólya's problem-solving guidelines for this example.

Understand the Problem. We could begin by using the definition of 8th power and the distributive property:

$$(a + b)^8 = \underbrace{(a + b)(a + b) \cdots (a + b)}_{8 \text{ factors of } a + b}$$

However, we soon see that this is too lengthy to complete directly.

Devise a Plan. We will consider a pattern of successive powers; that is, consider $(a + b)^n$ for $n = 0, 1, 2, \ldots$.

Carry Out the Plan. We begin by actually doing the multiplications:

$n = 0$: $(a + b)^0 = \qquad\qquad\qquad 1$

$n = 1$: $(a + b)^1 = \qquad\qquad 1 \cdot a + 1 \cdot b$

$n = 2$: $(a + b)^2 = \qquad\quad 1 \cdot a^2 + 2 \cdot ab + 1 \cdot b^2$

$n = 3$: $(a + b)^3 = \qquad 1 \cdot a^3 + 3 \cdot a^2b + 3 \cdot ab^2 + 1 \cdot b^3$

$n = 4$: $(a + b)^4 = 1 \cdot a^4 + 4 \cdot a^3b + 6 \cdot a^2b^2 + 4 \cdot ab^3 + 1 \cdot b^4$

$\vdots \qquad\qquad\qquad\qquad \vdots$

First, ignore the coefficients (shown in color) and focus on the variables:

$(a + b)^1$: $\quad a \quad\quad b$

$(a + b)^2$: $\quad a^2 \quad\quad ab \quad\quad b^2$

$(a + b)^3$: $\quad a^3 \quad\quad a^2b \quad\quad ab^2 \quad\quad b^3$

$(a + b)^4$: $\quad a^4 \quad\quad a^3b \quad\quad a^2b^2 \quad\quad ab^3 \quad\quad b^4$

$\vdots \quad\quad\quad\quad\quad \vdots$

Do you see a pattern? As you read from left to right, the powers of a decrease and the powers of b increase. Note that the sum of the exponents for each term is the same as the original exponent:

$(a + b)^n$: $\ a^n b^0 \quad a^{n-1}b^1 \quad a^{n-2}b^2 \cdots a^{n-r}b^r \cdots a^2 b^{n-2} \quad a^1 b^{n-1} \quad a^0 b^n$

Next, consider the numerical coefficients (shown in color):

$(a + b)^0$: $\quad\quad\quad\quad\quad\quad 1$

$(a + b)^1$: $\quad\quad\quad\quad\quad 1 \quad\quad 1$

$(a + b)^2$: $\quad\quad\quad\quad 1 \quad\quad 2 \quad\quad 1$

$(a + b)^3$: $\quad\quad\quad 1 \quad\quad 3 \quad\quad 3 \quad\quad 1$

$(a + b)^4$: $\quad\quad 1 \quad\quad 4 \quad\quad 6 \quad\quad 4 \quad\quad 1$

$\vdots \quad\quad\quad\quad\quad\quad \vdots$

Do you see the pattern? Recall Pascal's triangle from Section 1.1 (see Figure 5.1).

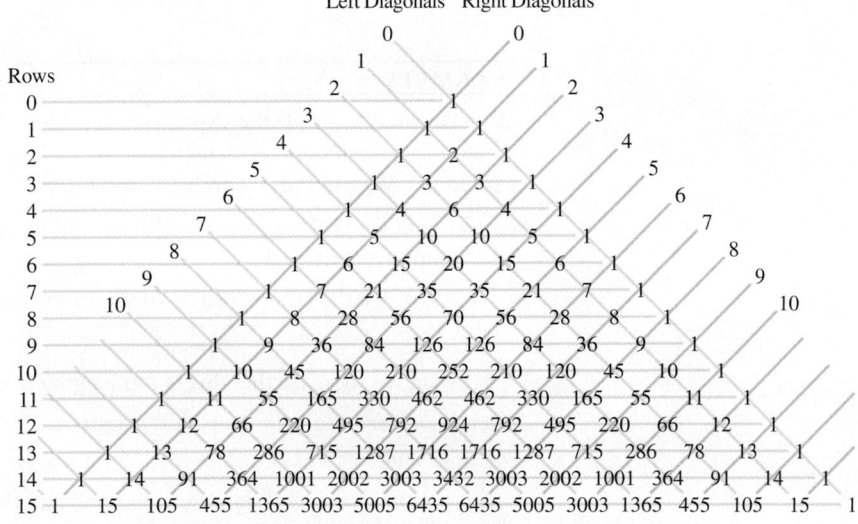

Figure 5.1 **Pascal's triangle**

To find $(a + b)^8$, we look at the 8th row of Pascal's triangle for the coefficients to complete the product:

$(a + b)^8 = a^8 + 8a^7b + 28a^6b^2 + 56a^5b^3 + 70a^4b^4 + 56a^3b^5 + 28a^2b^6 + 8ab^7 + b^8$

Look Back. Does this pattern seem correct? We have verified the pattern directly for $n = 0, 1, 2, 3$, and 4. With a great deal of algebraic work, you can verify by direct multiplication that the pattern checks for $(a + b)^5$. ◆

Binomial Theorem

The pattern we discovered for $(a + b)^8$ is a very important theorem in mathematics. It is called the **binomial theorem.** The difficulty in stating this theorem is in relating the coefficients of the expansion to Pascal's triangle. We write $\binom{4}{2}$ to represent the number in row 4, diagonal 2 of Pascal's triangle (see Figure 5.1). We see

$$\binom{4}{2} = 6 \qquad \binom{6}{4} = 15 \qquad \binom{5}{3} = 10 \qquad \binom{15}{7} = 6{,}435$$

Binomial Theorem

🛑 **STOP** Spend some time studying this theorem. There is a lot of notation here, and you should make sure you understand what it means.

For any positive integer n,

$$(a + b)^n = \binom{n}{0}a^n + \binom{n}{1}a^{n-1}b + \binom{n}{2}a^{n-2}b^2 + \cdots + \binom{n}{n-1}ab^{n-1} + \binom{n}{n}b^n$$

where $\binom{n}{r}$ is the number in the nth row, rth diagonal of Pascal's triangle.

Pascal's triangle is efficient for finding the numerical coefficients for exponents that are relatively small, as shown in Figure 5.1. However, for larger exponents we will need some additional work, which will be presented in Chapter 10. To **expand** a polynomial is to carry out the operations to simplify the expression.

EXAMPLE 7

Expand $(x - 2y)^4$.

Solution In this example, let $a = x$ and $b = -2y$, and look at row 4 of Pascal's triangle for the coefficients.

$$(a + b)^4 = a^4 + 4a^3b + 6a^2b^2 + 4ab^3 + b^4 \quad \text{Binomial theorem, } n = 4$$
$$(x - 2y)^4 = x^4 + 4x^3(-2y) + 6x^2(-2y)^2 + 4x(-2y)^3 + (-2y)^4$$
$$= x^4 - 8x^3y + 24x^2y^2 - 32xy^3 + 16y^4 \qquad \blacklozenge$$

Polynomials and Areas

We assume that you know the area formulas for squares and rectangles:

Area of a square: $A = s^2$ Area of rectangle: $A = \ell w$

⚠️ **YIELD** This is EASY if you cut out these figures and use them as manipulatives. Rearrangement of strips is hard to show in a book, but if you try it with cutout pieces, you will like it!

Areas of squares and rectangles are often represented as trinomials. Consider the area represented by the trinomial $x^2 + 3x + 2$. This expression is made up of three terms: x^2, $3x$, and 2. Translate each of these terms into an area:

Now you must rearrange these pieces into a rectangle. Think of them as cutouts and move them around. There is only one way (not counting order) to fit them into a rectangle.

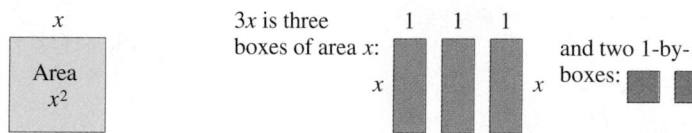

or put together it looks like:

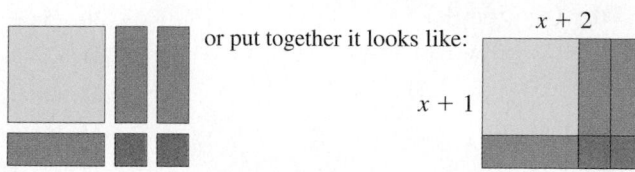

Note: The rectangle with dimensions $x + 2$ by $x + 1$ has the same area as one with dimensions $x + 1$ by $x + 2$.

Thus, $(x + 2)(x + 1) = x^2 + 3x + 2$. These observations provide another way (besides the distributive property) for verifying the shortcut method of FOIL.

EXAMPLE 8

Find the product $(x + 3)(x + 1)$ both algebraically and geometrically.

Solution

Algebraic: $(x + 3)(x + 1) = x^2 + 4x + 3$

Geometric: Draw a rectangle with sides that measure $x + 3$ and $x + 1$:

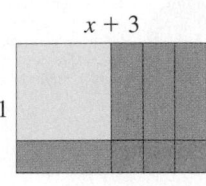

There is one square: x^2
four rectangles: $4x$
three units: 3
Thus, $(x + 3)(x + 1) = x^2 + 4x + 3$ ◆

Problem Set 5.1

LEVEL 1

1. **IN YOUR OWN WORDS** What is a polynomial?

2. **IN YOUR OWN WORDS** What is the degree of a polynomial?

3. **IN YOUR OWN WORDS** Discuss the process for adding and subtracting polynomials.

4. **IN YOUR OWN WORDS** Discuss the process of multiplying polynomials using the distributive property.

5. **IN YOUR OWN WORDS** Discuss the process of multiplying binomials using FOIL.

6. **IN YOUR OWN WORDS** Discuss the process of multiplying binomials using areas.

7. **IN YOUR OWN WORDS** What is the binomial theorem?

Simplify each expression in Problems 8–22. Classify each answer by number of terms and degree.

8. $(x + 3) + (5x - 7)$

9. $(2x - 4) - (3x + 4)$

10. $(x + y + 2z) + (2x + 5y - 4z)$

11. $(x - y - z) + (2x - 5y - 3z)$

12. $(5x^2 + 2x - 5) + (3x^2 - 5x + 7)$

13. $(2x^2 - 5x + 4) + (3x^2 - 2x - 11)$

14. $(x + 2y - 3z) - (x - 5y + 4z)$

15. $(x^2 + 4x - 3) - (2x^2 + 9x - 6)$

16. $(3x - x^2) - 5(2 - x) - (2x + 3)$

17. $3(x - 5) - 2(x + 8)$

18. $6(x + 1) - 2(x + 1)$

19. $3(2x^2 + 5x - 5) + 2(5x^2 - 3x + 6)$

20. $2(4x^2 - 3x + 2) - 3(x^2 - 5x - 8)$

21. $2(x + 3) - 3(x^2 - 3x + 1) + 4(x - 5)$

22. $3(x - 1) - 2(x^2 + 4x + 5) - 5(x + 8)$

In Problems 23–29, multiply mentally.

23. **a.** $(x + 3)(x + 2)$ **b.** $(y + 1)(y + 5)$
    c. $(z - 2)(z + 6)$ **d.** $(s + 5)(s - 4)$

24. **a.** $(x + 1)(x - 2)$ **b.** $(y - 3)(y + 2)$
    c. $(a - 5)(a - 3)$ **d.** $(b + 3)(b - 4)$

25. **a.** $(c + 1)(c - 7)$ **b.** $(z - 3)(z + 5)$
    c. $(2x + 1)(x - 1)$ **d.** $(2x - 3)(x - 1)$

26. **a.** $(x + 1)(3x + 1)$ **b.** $(x + 1)(3x + 2)$
    c. $(2a + 3)(3a - 2)$ **d.** $(2a + 3)(3a + 2)$

27. **a.** $(x + y)(x + y)$ **b.** $(x - y)(x - y)$
    c. $(x + y)(x - y)$ **d.** $(a + b)(a - b)$

28. **a.** $(5x - 4)(5x + 4)$ **b.** $(3y - 2)(3y + 2)$
    c. $(a + 2)^2$ **d.** $(b - 2)^2$

29. **a.** $(x + 4)^2$ **b.** $(y - 3)^2$
    c. $(s + t)^2$ **d.** $(u - v)^2$

LEVEL 2

Simplify the expressions in Problems 30–37.

30. $(5x + 1)(3x^2 - 5x + 2)$

31. $(2x - 1)(3x^2 + 2x - 5)$

32. $(3x - 1)(x^2 + 3x - 2)$

33. $(5x + 1)(x^3 - 2x^2 + 3x)$

34. $(5x + 1)(3x^2 - 5x + 2) - (x^3 - 4x^2 + x - 4)$

35. $3(3x^2 - 5x + 2) - 4(x^3 - 4x^2 + x - 4)$

36. $(x - 2)(2x - 3) - (x + 1)(x - 5)$

37. $(2x - 3)(3x + 2) + (x + 2)(x + 3)$

Find each product in Problems 38–43 both algebraically and geometrically.

38. $(x + 2)(x + 4)$ **39.** $(x + 1)(x + 4)$

40. $(x + 2)(x + 5)$ **41.** $(x + 3)(x + 4)$

42. $(2x + 3)(3x + 2)$ **43.** $(2x + 1)(2x + 3)$

Use the binomial theorem to expand each binomial given in Problems 44–51.

44. $(x + 1)^3$ **45.** $(x - 1)^3$

46. $(x + y)^5$ **47.** $(x + y)^6$

48. $(x - y)^7$ **49.** $(x - y)^8$

50. $(5x - 2y)^3$ **51.** $(2x - 3y)^4$

LEVEL 3

52. Write out the first three terms in the expansion $(x + y)^{12}$.

53. Write out the last three terms in the expansion $(x + y)^{14}$.

54. The number of desks in one row is $5d + 2$. How many desks are there in a room of $2d - 1$ rows if they are arranged in a rectangular arrangement?

55. An auditorium has $6x + 2$ seats in each row and $51x - 7$ rows. If the chairs are in a rectangular arrangement, how many seats are there in the auditorium?

56. Each apartment in a building rents for $800 - d$ dollars per month. What is the monthly income from $6d + 12$ units?

57. If a boat is traveling at a rate of $6b + 15$ miles per hour for a time of $10 - 2b$ hours, what is the distance traveled by the boat?

PROBLEM SOLVING

58. Binomial products can be used to do mental calculations. For example, to multiply a pair of two-digit numbers whose tens digits are the same and whose units digits add up to 10, mentally multiply the units digits and then multiply the tens digit of the first number by one more than the tens digit of the second. For example,

$$\begin{array}{c} \longrightarrow 8 \times 9 = 72 \\ \uparrow \qquad \downarrow \\ 84 \times 86 = \overline{7\,2} \quad 2\,4 \\ \downarrow \qquad \downarrow \qquad \uparrow \\ \longrightarrow 4 \times 6 = 24 \end{array}$$

Mentally multiply the given numbers.

a. 62×68 **b.** 57×53 **c.** 63×67

d. 95^2 **e.** 75^2

59. Suppose the given numbers for a mental calculation (see Problem 58) are $10x + y$ and $10x + z$. Notice that these two numbers have the same tens digit. Also assume that $y + z = 10$, which says that the units digits of the two numbers sum to 10. Algebraically show why the mental calculation described in Problem 58 "works."

60. Devise a procedure for mentally multiplying three-digit numbers with the same first two digits and units digits that add up to 10.

5.2 | FACTORING

We have called numbers that are multiplied *factors*. In Section 4.2 we defined a *prime number* and the *prime factorization* of numbers. The same process can be applied to algebraic expressions. To **factor** an expression means to write it in factored form. That is, the word "factor" is sometimes a noun and sometimes a verb. In this section, we look at the process of factoring and complete our discussion of this topic in Section 5.4 when we use factoring to solve quadratic equations. Factoring is also used extensively in algebra, so to understand some of the algebraic processes, it is necessary to understand factoring.

The approach we take in this section is different from that you will normally see in an algebra course. So often algebra is learned by brute force and memorization or symbol manipulation, but our development uses a geometric visualization that may help in your understanding not only of algebraic processes, but of geometric ones as well.

Using Areas to Factor

In the previous section, we showed how areas can be used to understand multiplication of binomials. We now use areas to factor polynomials.

EXAMPLE 1

Factor $2x^2 + 7x + 6$ using areas.

Solution　First draw the areas for the terms:

Rearrange to form a rectangle:

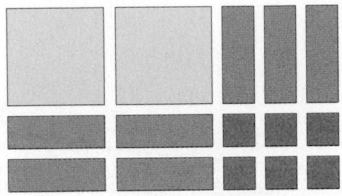

Push these pieces together to form a single rectangle:

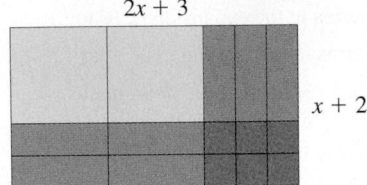

Thus, $2x^2 + 7x + 6 = (2x + 3)(x + 2)$.　◆

The following example shows how to use areas when factoring a polynomial with a subtraction, as well as with a polynomial that is not factorable.

EXAMPLE 2

Factor the following polynomials, if possible, by using areas.

a. $x^2 + 2x - 3$　　**b.** $x^2 + 5x + 8$

Solution

a. First draw the rectangles for the terms:

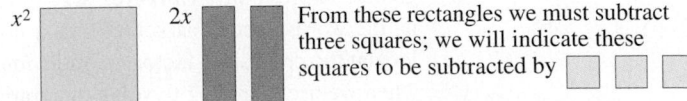

From these rectangles we must subtract three squares; we will indicate these squares to be subtracted by

First arrange the positive pieces and then place the negative pieces (the gray ones) on top of the positive pieces:

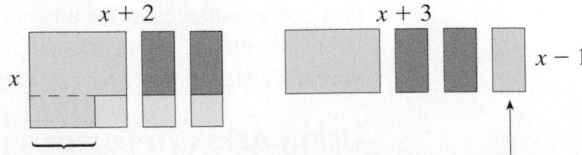

This is $x - 1$; move this piece over here

Thus, $x^2 + 2x - 3 = (x + 3)(x - 1)$.

b.

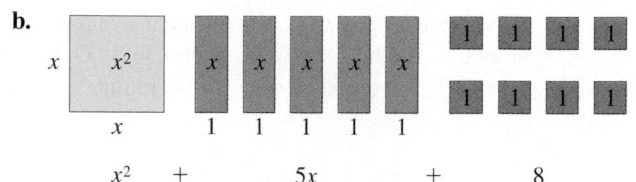

$$x^2 \quad + \quad\quad 5x \quad\quad + \quad\quad 8$$

There is no way of arranging all the pieces to form a rectangle.

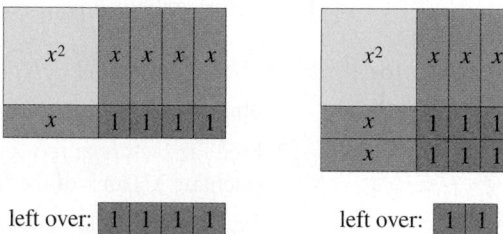

left over: 1 1 1 1   left over: 1 1

We see that this trinomial is not factorable. ◆

Using Algebra to Factor

By thinking through the steps for factoring trinomials by using areas, we can develop a process for algebraic factoring. Consider $x^2 + 3x + 2$ and compare the process below with the discussion at the beginning of this section. We consider the first term and the last term of the given trinomial.

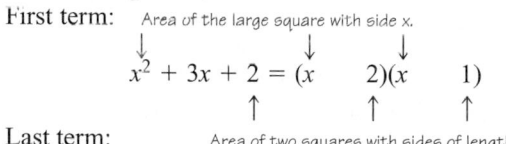

First term: Area of the large square with side x.
$$x^2 + 3x + 2 = (x \quad 2)(x \quad 1)$$
Last term: Area of two squares with sides of length 1.

There may be several ways of rearranging the areas for these first two steps (first term and last term). What you want to do is to rearrange them so that they form a rectangle. By looking at the binomial product and recalling the process we called FOIL, you see that the sum of the outer product and the inner product, which gives the middle term of the trinomial, will be the factorization that gives a rectangular area:

outer + inner   outer product + inner product

$$x^2 + 3x + 2 = (x + 2)(x + 1)$$

This method of factorization is called FOIL.

EXAMPLE 3

Factor $2x^2 + 7x + 6$ using FOIL.

Solution   This is the same as Example 1, but now we use FOIL.

First term: $2x^2 + 7x + 6 = (2x \quad)(x \quad)$

Last term: There are several possibilities:
$$(2x + 6)(x + 1)$$
$$(2x + 1)(x + 6)$$
$$(2x + 2)(x + 3)$$
$$(2x + 3)(x + 2)$$

These are equivalent to the different arrangements of "pieces" to form the rectangle in the area method shown in Example 1. Only one will form a rectangle and that is precisely the one that will give the middle term of the trinomial, namely, $7x$:

$$
\begin{array}{c}
\text{outer product} \\
+ \\
\text{inner product}
\end{array}
$$

Middle term: $2x^2 + 7x + 6 = (2x + 3)(x + 2)$

This factorization is unique (except for the order of the factors).   ◆

Procedure for Factoring Trinomials

This is a very common process; make sure you thoroughly understand how to factor trinomials.

To factor a trinomial:

1. Find the factors of the second-degree term, and set up the binomials.

2. Find the factors of the constant term, and consider all possible binomials (mentally). Think of the factors that will form a rectangle.

3. Determine the factors that yield the correct middle term. If no pair of factors produces the correct full product, then the trinomial is not factorable using integers.

This factoring approach is called FOIL.

EXAMPLE 4

Factor, if possible.

a. $2x^2 + 11x + 12$ **b.** $x^2 + 3x + 5$

Solution

a. $2x^2 + 11x + 12 = (2x \quad)(x \quad)$

Try (mentally): 1, 12; 12, 1; 2, 6; 6, 2; 3, 4; and 4, 3. Use the pair that gives the middle term, $11x$.

$$2x^2 + 11x + 12 = (2x + 3)(x + 4)$$

b. $x^2 + 3x + 5 = (x \quad)(x \quad)$

Try (mentally): 1, 5; and 5, 1. Since neither of these gives the middle term, $3x$, we say that the trinomial is not factorable.   ◆

Common Factoring

If several terms share a factor, then that factor is called a **common factor.** For example, the binomial $5x^2 + 10x$ has three common factors: 5, x, and $5x$. To factor this sum of two terms, we must change it to a product, and this can be done several ways:

$$5x^2 + 10x = 5(x^2 + 2x)$$
$$5x^2 + 10x = x(5x + 10)$$
$$5x^2 + 10x = 5x(x + 2)$$

The last of these possibilities has the greatest common factor as a factor, and is said to be **completely factored.**

When combining common factoring with trinomial factoring, the procedure will be easiest if you look for common factors first.

EXAMPLE 5

Completely factor $6x^3 - 21x^2 - 12x$.

Solution

$$6x^3 - 21x^2 - 12x = 3x(2x^2 - 7x - 4) \quad \text{Common factor first}$$
$$= 3x(2x \quad)(x \quad) \quad \text{Now use FOIL, first terms}$$
$$= 3x(2x + 1)(x - 4) \quad \text{Last terms}$$

◆

Difference of Squares

The last type of factorization we will consider is called a **difference of squares.** Suppose we start with one square, a^2:

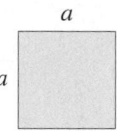

a

a

From this square, we wish to subtract another square, b^2:

This gray square should be smaller than the first ($a^2 > b^2$). Place this square (since it is gray) on top of the larger square:

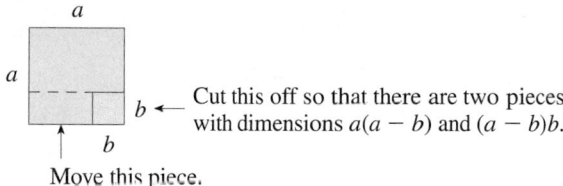

a

a

b ← Cut this off so that there are two pieces
with dimensions $a(a - b)$ and $(a - b)b$.

b

Move this piece.

Rearrange these two pieces by moving the smaller one from the bottom and positioning it vertically at the right:

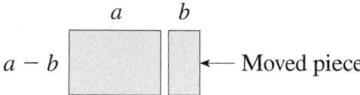

a   *b*

$a - b$ ← Moved piece

The dimensions of this new arrangement are $(a - b)$ by $(a + b)$. Thus,

$$a^2 - b^2 = (a - b)(a + b)$$

Difference of Squares

🛑 STOP

You need to remember this formula.

$$a^2 - b^2 = (a - b)(a + b)$$

EXAMPLE 6

Completely factor each expression, if possible.

a. $x^2 - 36$ **b.** $8x^2 - 50$ **c.** $6x^4 - 18x^2 - 24$

Solution

a. $x^2 - 36 = (x - 6)(x + 6)$ Difference of squares

b. $8x^2 - 50 = 2(4x^2 - 25)$ Common factor first
$$= 2(2x - 5)(2x + 5) \quad \text{Difference of squares}$$

c. $6x^4 - 18x^2 - 24 = 6(x^4 - 3x^2 - 4)$ Common factor
$$= 6(x^2 - 4)(x^2 + 1) \quad \text{FOIL}$$
$$= 6(x - 2)(x + 2)(x^2 + 1) \quad \text{Difference of squares}$$

◆

Problem Set 5.2

LEVEL 1

1. **IN YOUR OWN WORDS** Outline a procedure for factoring.

2. **IN YOUR OWN WORDS** What is a difference of squares?

If possible, completely factor the expressions in Problems 3–36.

3. $10xy - 6x$
4. $5x + 5$
5. $8xy - 6x$
6. $6x - 2$
7. $x^2 - 4x + 3$
8. $x^2 - 2x - 3$
9. $x^2 - 5x + 6$
10. $x^2 - x - 6$
11. $x^2 - 7x + 12$
12. $x^2 - 7x - 8$
13. $x^2 - x - 30$
14. $x^2 + 9x + 14$
15. $x^2 - 2x - 35$
16. $x^2 - 6x - 16$
17. $3x^2 + 7x - 10$
18. $2x^2 + 7x - 15$
19. $2x^2 - 7x + 3$
20. $3x^2 - 10x + 3$
21. $3x^2 - 5x - 2$
22. $6y^2 - 7y + 1$
23. $2x^2 + 9x + 4$
24. $7x^2 + 4x - 3$
25. $3x^2 + x - 2$
26. $3x^2 + 7x + 2$
27. $5x^3 + 7x^2 - 6x$
28. $8x^3 + 12x^2 + 4x$
29. $7x^4 - 11x^3 - 6x^2$
30. $3x^4 + 3x^3 - 36x^2$
31. $x^2 - 64$
32. $x^2 - 169$
33. $25x^2 + 50$
34. $16x^2 - 25$
35. $x^4 - 1$
36. $x^8 - 1$

LEVEL 2

Factor each expression in Problems 37–48 by using areas.

37. $x^2 + 5x + 6$
38. $x^2 + 7x + 12$
39. $x^2 + 4x + 3$
40. $x^2 + 5x + 4$
41. $x^2 + 6x + 8$
42. $x^2 + 2x + 2$
43. $x^2 - 1$
44. $x^2 - 4$
45. $x^2 + x - 2$
46. $x^2 + x - 6$
47. $x^2 - x - 2$
48. $x^2 - x - 6$

LEVEL 3

49. The area of a rectangle is $x^2 - 2x - 143$ square feet. What are the dimensions of the figure?

50. What are the dimensions of a rectangle whose area is $x^2 - 4x - 165$ square feet?

51. What is the time needed to travel a distance of $6x^2 + 5x - 4$ miles if the rate is $3x + 4$ mph?

52. If an auditorium has $x^2 - 50x - 600$ seats arranged in a rectangular fashion, what is the number of rows and how many seats are there in each row?

Factor each expression in Problems 53–58, if possible.

53. $(x + 2)(x + 4) + (5x + 6)(x - 1)$
54. $(2x + 1)(3x + 2) + (3x + 5)(x - 2)$
55. $x^6 - 13x^4 + 36x^2$
56. $x^6 - 26x^4 + 25x^2$
57. $20x^2y^2 + 17x^2yz - 10x^2z^2$
58. $12x^2y^2 + 10x^2yz - 12x^2z^2$

PROBLEM SOLVING

59. Pick any three consecutive integers—for example, 4, 5, and 6. The square of the middle term is 1 more than the product of the first and third; for example, 25 is 1 more than $4(6) = 24$. Prove this is true for any three consecutive integers.

60. Illustrate the property in Problem 59 geometrically.

5.3 EVALUATION, APPLICATIONS, AND SPREADSHEETS

If $x = a$, then x and a name the same number; x may then be replaced by a in any expression, and the value of the expression will remain unchanged. When you replace variables by given numerical values and then simplify the resulting numerical expression, the process is called *evaluating an expression*. That is, to **evaluate** an expression means to replace the variable (or variables) with given values, and then to simplify the resulting numerical expression.

EXAMPLE 1

Evaluate $a + cb$, where $a = 2$, $b = 11$, and $c = 3$.

Solution $a + cb$ *Remember: cb means c **times** b.*

Step 1. Replace each variable with the corresponding numerical value. You may need additional parentheses to make sure you don't change the order of operations.

$$a + c \cdot b$$
$$\downarrow \ \downarrow \ \downarrow\downarrow \ \downarrow$$
$$2 + 3(11) \leftarrow \text{Parentheses are necessary so that the product cb is not changed to 311.}$$

Step 2. Simplify:

$$2 + 3(11) = 2 + 33 \quad \text{Multiplication before addition}$$
$$= 35$$

◆

EXAMPLE 2

Evaluate the following where $a = 3$ and $b = 4$.

a. $a^2 + b^2$ **b.** $(a + b)^2$

Solution

a. $a^2 + b^2 = 3^2 + 4^2$

$$= 9 + 16$$

$$= \mathbf{25} \quad \text{Remember the order of operations: Multiplication comes first, and } 3^2 = 3 \cdot 3, \ 4^2 = 4 \cdot 4, \text{ which is multiplication.}$$

b. $(a + b)^2 = (3 + 4)^2$

$$= 7^2$$

$$= \mathbf{49} \quad \text{Order of operations; parentheses first.}$$

Notice that $a^2 + b^2 \neq (a + b)^2$.

◆

Remember that a particular variable is replaced by a single value when an expression is evaluated. You should also be careful to write capital letters differently from lowercase letters, because they often represent different values. This means that you should not assume that $A = 3$ just because $a = 3$. On the other hand, it is possible that other variables *might* have the value 3. For example, just because $a = 3$, do not assume that another variable—say, t—cannot also have the value $t = 3$.

EXAMPLE 3

Let $a = 1$, $b = 3$, $c = 2$, and $d = 4$. Find the value of the given capital letters.

a. $G = bc - a$ **b.** $H = 3c + 2d$ **c.** $I = 3a + 2b$ **d.** $R = a^2 + b^2d$

e. $S = \dfrac{2(b + d)}{2c}$ **f.** $T = \dfrac{3a + bc + b}{c}$

After you have found the value of a capital letter, write it in the box that corresponds to its numerical value. This exercise will help you check your work.

	37	9	5	14	6	

Solution

a. $G = bc - a$
$\quad = 3(2) - 1$
$\quad = 6 - 1$
$\quad = 5$

b. $H = 3c + 2d$
$\quad = 3(2) + 2(4)$
$\quad = 6 + 8$
$\quad = 14$

c. $I = 3a + 2b$
$\quad = 3(1) + 2(3)$
$\quad = 3 + 6$
$\quad = 9$

d. $R = a^2 + b^2d$
$\quad = 1^2 + 3^2(4)$
$\quad = 1 + 9(4)$
$\quad = 37$

e. $S = \dfrac{2(b + d)}{2c}$

$\quad = \dfrac{2(3 + 4)}{2(2)}$

$\quad = \dfrac{2(7)}{4}$

$\quad = \dfrac{7}{2}$

f. $T = \dfrac{3a + bc + b}{c}$

$\quad = \dfrac{3(1) + 3(2) + 3}{2}$

$\quad = \dfrac{3 + 6 + 3}{2}$

$\quad = \dfrac{12}{2}$

$\quad = 6$

After you have filled in the appropriate boxes, the result is

37	9	5	14	6
R	I	G	H	T

♦

In algebra, variables are usually represented by either lowercase or capital letters. However, in other disciplines, variables are often represented by other symbols or combinations of letters. For example, I recently took a flight on Delta Airlines and a formula $VM = \sqrt{A} \times 3.56$ was given as an approximation for the distance you can see from a Delta jet (or presumably any other plane). The article defined VM as the distance you can view in miles when flying at an altitude of A feet. For this example VM is interpreted as a single variable, and not as V *times* M as it normally would be in algebra.

An Application from Genetics

This application is based on the work of Gregor Mendel (1822–1884), an Austrian monk, who formulated the laws of heredity and genetics. Mendel's work was later amplified and explained by a mathematician, G. H. Hardy (1877–1947), and a physician, Wilhelm Weinberg (1862–1937). For years Mendel taught science without any teaching credentials because he had failed the biology portion of the licensing examination! His work, however, laid the foundation for the very important branch of biology known today as genetic science.

Assume that traits are determined by *genes,* which are passed from parents to their offspring. Each parent has a pair of genes, and the basic assumption is that each offspring inherits one gene from each parent to form the offspring's own pair. The genes are selected in a random, independent way. In our examples, we will assume that the researcher is studying a trait that is both easily identifiable (such as color of a rat's fur) and determined by a pair of genes consisting of a *dominant gene,* denoted by *A,* and a *recessive gene,* denoted by *a.* The possible pairings are called *genotypes:*

AA is called *dominant,* or homozygous.

Aa is called *hybrid,* or heterozygous; genetically,
the genotype *aA* is the same as *Aa.*

aa is called *recessive.*

The physical appearance is called the *phenotype:*

Genotype *AA* has phenotype *A.*

Genotype *Aa* has phenotype *A* (since *A* is dominant).

Genotype *aA* has phenotype *A.*

Genotype *aa* has phenotype *a.*

In genetics, a square called a *Punnett square* is used to display genotype. For example, suppose two individuals with genotypes *Aa* are mated, as represented by the following Punnett square:

Parent 2

		A	a
Parent 1	A	*AA*	*Aa*
	u	*aA*	*aa*

We see the result is *AA* + *Aa* + *aA* + *aa* = *AA* + 2*Aa* + *aa*. This reminds us of the binomial product

$$(p + q)^2 = p^2 + 2pq + q^2$$

Let's use binomial multiplication to find the genotypes and phenotypes of a particular example. In population genetics, we are interested in the percent, or frequency, of genes of a certain type in the entire population under study. In other words, imagine taking the two genes from each person in the population and putting them into an imaginary pot. This pot is called the *gene pool* for the population. Geneticists study the gene pool to draw conclusions about the population.

EXAMPLE 4

Suppose a certain population has two eye color genes: *B* (brown eyes, dominant) and *b* (blue eyes, recessive). Suppose we have an isolated population in which 70% of the genes in the gene pool are dominant *B,* and the other 30% are recessive *b.* What fraction of the population has each genotype? What percent of the population has each phenotype?

Solution   Let $p = 0.7$ and $q = 0.3$. Since p and q give us 100% of all the genes in the gene pool, we see that $p + q = 1$. Since

$$(p + q)^2 = p^2 + 2pq + q^2$$

we can find the percents:

genotype *BB:* $p^2 = (0.7)^2 = 0.49,$ so 49% have *BB* genotype

genotype *bB* or *Bb:* $2pq = 2(0.7)(0.3) = 0.42,$ so 42% have this genotype

genotype *bb:* $q^2 = (0.3)^2 = 0.09,$ so 9% have *bb* genotype

Check genotypes: $0.49 + 0.42 + 0.09 = 1.00$

As for the phenotypes, we look only at outward appearances, and since brown is dominant, *BB, bB,* and *Bb* all have brown eyes; this accounts for 91%, leaving 9% with blue eyes.   ◆

the devices themselves, the electronics and mechanics are referred to as hardware, but...

the directions that make the hardwar are known as

SOFTWARE

A computer's programs, plus the procedure for their use.

(PROGRAM
A set of instructions for performing computer operations.)

Source: *A Computer Glossary*, © 1968 International Business Machines Corporation.

Spreadsheets

In Section 3.5, we discussed the notion of computer software and mentioned that one of the most important computer applications is in using something called a *spreadsheet*. A **spreadsheet** is a computer program used to manipulate data and carry out calculations, or chains of calculations. If you have access to a computer and software such as Excel, Lotus 1-2-3, or Quattro-Pro, you might use that software in conjunction with this section. However, it is not necessary to have this software (or even access to a computer) to be able to study variables and the evaluation of formulas using the ideas of a spreadsheet. In fact, your first inclination when reading this might be to skip over this and say to yourself, "I don't know anything about a spreadsheet, so I will not read this. Besides, my instructor is not requiring this anyway." However, regardless of whether this is assigned, chances are that sooner or later you will be using a spreadsheet.

One of the most interesting, and important, new ways of representing variables is as a **cell,** or a "box."

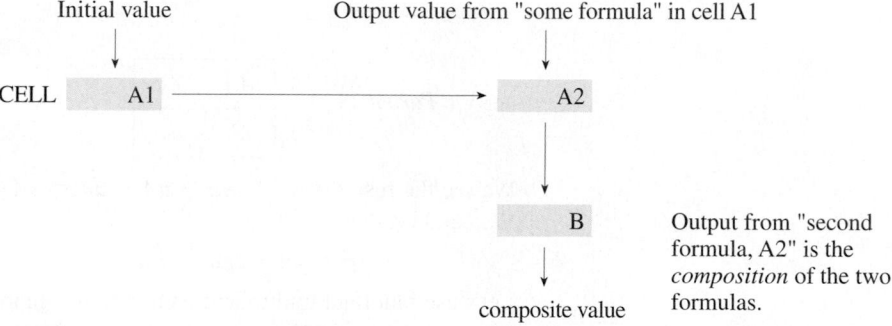

The information in a spreadsheet is stored in a rectangular array of *cells*. The content of each cell can be a number, text, or a formula. The power of a spreadsheet is that a cell's numeric value can be linked to the content of another cell. For example, it is possible to define the content of one cell as the sum of the contents of two other cells. Furthermore, if the value of a cell is changed anywhere in the spreadsheet, all values dependent on it are recalculated and the new values are displayed immediately.

Instead of designating variables as letters (such as $x, y, z, \ldots$) as we do in algebra, a spreadsheet designates variables as cells (such as B2, A5, Z146, . . .). If you type something into a cell, the spreadsheet program will recognize it as text if it begins with a letter, and as a number if it begins with a numeral. It also recognizes the usual mathematical symbols of $+$, $-$, $*$ (for $\times$), $/$ (for $\div$), and $\wedge$ (for raising to a power). Parentheses are used in the usual fashion as grouping symbols. To enter a formula, you must begin with $+$, $@$, or $=$, depending on the position in the spreadsheet. Compare some algebraic and spreadsheet evaluations:

Algebra	*Comment*	*Spreadsheet*	*Comment*
$3(x + y)$	Variables x and y	$+3*(A1 + A2)$	Variables are in contents of cells A1 and A2.
$x^2 + 2x - 5$	Variable is x	$+B3\wedge2+2*B3-5$	Variable is the contents of cell B3. Begins with $+$ to indicate that it is a formula.
$\dfrac{A + B}{2}$	Formula for the average of variables A and B	$+(A3 + A4)/2$	Variables are the contents of cells A3 and A4.

EXAMPLE 5

Translate each formula into spreadsheet notation.

a. $x + \dfrac{y}{2}$ **b.** $\dfrac{6x^2 + y}{2x}$ **c.** $-3^2 + (-5)^2 + z^3 + (z + 3)^2$

Solution

a. $+A1 + A2/2$ where the value of x is the content of cell A1 and the value of y is the content of cell A2.

b. $+(6*A1\wedge2 + A2)/(2*A1)$ where the value of x is the content of cell A1 and the value of y is in cell A2.

c. $+(-3\wedge2) + (-5)\wedge2 + A3\wedge3 + (A3 + 3)\wedge2$ where the value of z is the content of cell A3.   ◆

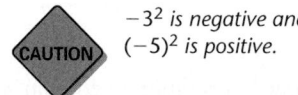

-3^2 is negative and $(-5)^2$ is positive.

For our purposes, we will assume that a spreadsheet program has an almost unlimited number of rows and columns. We will represent a typical spreadsheet as follows:

	A	B	C	D	E	F	G	H	I	J	K
1											
2											
3											
4											
5											

As an example, we will consider the way in which a spreadsheet program could be used to set up an electronic checkbook. We might fill in the spreadsheet as follows:

	A	B	C	D	E	F
1	DESCRIPTION	DEBIT	DEPOSIT	BALANCE		
2	Beginning balance					
3				+D2-B3+C3		
4				+D3-B4+C4		
5				+D4-B5+C5		

After some entries are filled in, the spreadsheet might look like the following:

	A	B	C	D	E	F
1	DESCRIPTION	DEBIT	DEPOSIT	BALANCE		
2	Beginning balance			1000.00		
3	School bookstore	250.00		750.00		
4	Paper route		100.00	850.00		
5	Ski trip	300.00		550.00		

The power of a spreadsheet derives from the way variables are referenced by cells. For example, if you go back to the spreadsheet and enter a beginning balance of $2,500 in cell D2, *all* the other entries *automatically* change:

	A	B	C	D	E	F
	Spreadsheet Application				_ □ ✕	
1	DESCRIPTION	DEBIT	DEPOSIT	BALANCE		
2	Beginning balance			2500.00		
3	School bookstore	250.00		2250.00		
4	Paper route		100.00	2350.00		
5	Ski trip	300.00		2050.00		

Once the spreadsheet has been set up, the user will enter information in column A and, depending on whether a check has been written or a deposit made, make an entry in either column B or column C. The entries in column D (beginning with cell D3) are automatically calculated by the spreadsheet program. Empty cells are assumed to have the value 0.

It should be clear that each cell from D3 downward needs to contain a different formula. In this example, the column letters and operations of the formula remain unchanged, but each row number is increased by 1 from the cell above. If each of these formulas had to be entered by hand, one by one, it is obvious that setting up a spreadsheet would be very time-consuming. This is not the case, however, and a typical spreadsheet program allows the user to copy the formula from one cell into another cell and at the same time *automatically* change its formula references. Thus, with a single command, each cell in column D is given the correct formula. We will call this command **replicating** a formula or cell. Formulas replicated down a column have their row numbers incremented, and formulas replicated across a row have the column letters incremented.

EXAMPLE 6

Consider the following spreadsheet with the indicated formulas.

	A	B	C	D	E	F	G	H	I	J	K
	Spreadsheet Application									_ □ ✕	
1	x	y = 10x	x + y	x^2							
2		+10*A2	+A2+B2	+A2^2							
3	+A2+1	+10*A3	+A3+B3	+A3^2							
4											
5											

a. Describe what the spreadsheet would show if cells A3 . . . D3 were replicated in rows 4 and 5.

b. Describe what the spreadsheet would show if column D were replicated in column E.

Solution

a.

	A	B	C	D	E	F	G	H	I	J	K
Spreadsheet Application											
1	x	y = 10x	x + y	x^2							
2		+10*A2	+A2+B2	+A2^2							
3	+A2+1	+10*A3	+A3+B3	+A3^2							
4	+A3+1	+10*A4	+A4+B4	+A4^2							
5	+A4+1	+10*A5	+A5+B5	+A5^2							

b.

	A	B	C	D	E	F	G	H	I	J	K
Spreadsheet Application											
1	x	y = 10x	x + y	x^2	x^2						
2		+10*A2	+A2+B2	+A2^2	+B2^2						
3	+A2+1	+10*A3	+A3+B3	+A3^2	+B3^2						
4											
5											

EXAMPLE 7

Given that cell A2 has the value 12, show what the spreadsheet shown in the solution for Example 6a would look like.

Solution

	A	B	C	D	E	F	G	H	I	J	K
Spreadsheet Application											
1	x	y = 10x	x + y	x^2							
2	12	120	132	144							
3	13	130	143	169							
4	14	140	154	196							
5	15	150	165	225							

We might once again remind you of the power of a computer spreadsheet. Note that the *entire* answer shown in Example 7 would *immediately* be filled in as soon as you fill in the number 12 in cell A2. If you now go back and reenter another number into cell A2, the entire spreadsheet would *immediately* change because every cell is ultimately defined in terms of the content of cell A2 in this example spreadsheet.

EXAMPLE 8

Suppose we look at a spreadsheet and see the entries 1, 1, 2, 3, 5, 8, 13, 21, 34, 55 in successive entries in column A. Suppose that 1 is entered into cell A1 and 1 into cell A2. What is the formula to be placed into cell A3?

Solution   Cell A3 should contain the formula $+A1 + A2$. Note that if this formula were replicated down column A through cell A10, the given numbers would be shown.

	A	B	C
Spreadsheet Application			
1	1		
2	1		
3	2		
4	3		
5	5		
6	8		
7	13		
8	21		
9	34		
10	55		
11			

The real power of a spreadsheet program lies not in its ability to perform calculations, but rather in its ability to answer "what-if" types of questions. For instance, for the set of numbers in Example 8, what if the number in cell A1 is divided by A2, and then the number in cell A2 is divided by A3, and so on for 50 such numbers? In the problem set you are asked to use a spreadsheet to detect a pattern for these quotients.

EXAMPLE 9

If $100 is deposited in an account that pays 5% interest compounded yearly, then at the end of the first year, the account will contain $100 + (0.05)*100 = \$105$. At the end of two years, the account will contain $105 + (0.05)*105 = \$110.25$. Suppose cell A1 contains the value 100. Then what formula must be placed in cell A2 if it is to contain the amount in the account at the end of the first year?

Solution   Cell A2 should contain the formula $A1 + (0.05)*A1$. Note that if this formula were replicated down column A to cell A11, the column would show the amount in the account at the end of each year through 10 years.   ◆

To take advantage of the "what-if" power of a spreadsheet, the previous example could be set up to allow for any interest rate. This can be done in the following way:

	A	B	C	D	E	F	G	H	I	J	K
Spreadsheet Application										_ □ ×
1	Interest rate =										
2	YEAR NUMBER	BALANCE									
3	0										
4	+A3+1	+B3+B1*B3									
5											

Note that when a number is inserted into cell B1, that number will act as the interest rate, and the number inserted into cell B3 will act as the amount of deposit. Therefore, we see that B1 is the variable representing the interest rate and B3 is the variable representing the beginning balance. If row 4 is *replicated* into rows 5 to 10, we will then obtain the data for the next 7 years. The problem with this replication, however, is that the reference to cell B1 will change as the replication takes place down column B. This difficulty is overcome by using a special character that holds a column or a row constant. The symbol we will use for this purpose is #. (Many spreadsheets use the symbol $ for this purpose.) Thus, we would change the formula in cell B4 above to

$$+B3+\#B\#1*B3$$

to mean that we want not only column B to remain constant, but also we want row 1 to remain unchanged when this entry is replicated. Note that # applies only to the character directly following its placement. We show the first 7 rows (4 years) of such a spreadsheet for which the rate is 4% and the initial deposit is $1,000.

	A	B	C	D	E
Spreadsheet Application				_ □ ×	
1	Interest rate =	0.04			
2	YEAR NUMBER	BALANCE			
3	0	1000.00			
4	1	1040.00			
5	2	1081.60			
6	3	1124.86			
7	4	1169.86			

Problem Set 5.3

LEVEL 1

1. **IN YOUR OWN WORDS** What is a variable?
2. **IN YOUR OWN WORDS** What is a spreadsheet?
3. **IN YOUR OWN WORDS** A colleague of mine speculated over lunch a few years ago:

   "Someday there will be just one programming language. IBM and Macintosh formats will merge. The trend is just the same as it is with the languages we speak in the world. Someday the entire world will speak English."

   Comment on this statement.

4. **IN YOUR OWN WORDS** Although the terms *program* and *software* can often be interchanged, there is a subtle difference in their usage. Discuss. What is meant when the term *user-friendly* is used to describe software?

In Problems 5–12, write each expression in spreadsheet notation. Let x be in cell A1, y in A2, and z in A3.

5. **a.** $\frac{2}{3}x^2$ **b.** $5x^2 - 6y^2$

6. **a.** $3x^2 - 17$ **b.** $14y^2 + 12x^2$

7. **a.** $12(x^2 + 4)$ **b.** $\frac{15x + 7}{2}$

8. **a.** $\frac{3x + 1}{12}$ **b.** $3x + \frac{1}{12}$

9. **a.** $(5 - x)(x + 3)^2$ **b.** $6(x + 3)(2x - 7)^2$

10. **a.** $(x + 1)(2x - 3)(x^2 + 4)$
    b. $(2x - 3)(3x^2 + 1)$

11. **a.** $\frac{1}{4}x^2 - \frac{1}{2}x + 12$ **b.** $\frac{2}{3}x^2 + \frac{1}{3}x - 17$

12. **a.** $1 - \frac{x}{yz}$ **b.** $\frac{1 - x}{yz}$

In Problems 13–18, write each spreadsheet expression in ordinary algebraic notation. Let cell A1 represent the variable x, A2 the variable y, A3 the variable z; B1 is a, B2 is b, and B3 is c.

13. **a.** +4*A1+3
    b. +5*A1^2 − 3*A1+4

14. **a.** +36*A3^2 − 13*A3+2
    b. +13*A2^2+(15/2)

15. **a.** +(5/4)*A1+14^2
    b. +(5/4*A1+14)^2

16. **a.** +5*A1^2 − 3*A2+4
    b. +4*(A1^2+5)*(3*A1^2 − 3)^2

17. **a.** +A1/A2*A3   **b.** +A1/(A2*A3)

18. **a.** +B2+B1*#B#3   **b.** +A1 + #B2^2

In Problems 19–42, let w = 2, x = 1, y = 2, and z = 4. Find the values of the given capital letters.

19. $A = x + z + 8$ 20. $B = 5x + y - z$

21. $C = 10 - w$ 22. $D = 3z$

23. $E = 25 - y^2$ 24. $F = w(y - x + wz)$

25. $G = 5x + 3z + 2$ 26. $H = 3x + 2w$

27. $I = 5y - 2z$ 28. $J = 2w - z$

29. $K = wxy$ 30. $L = x + y^2$

31. $M = (x + y)^2$ 32. $N = x^2 + 2xy + y^2 + 1$

33. $P = y^2 + z^2$ 34. $Q = w(x + y)$

35. $R = z^2 - y^2 - x^2$ 36. $S = (x + y + z)^2$

37. $T = x^2 + y^2z$ 38. $U = \frac{w + y}{z}$

39. $V = \frac{3wyz}{x}$ 40. $W = \frac{3w + 6z}{xy}$

41. $X = (x^2z + x)^2z$ 42. $Y = (wy)^2 + w^2y + 3x$

LEVEL 2

43. This problem will help you check your work in Problems 19–42. Fill in the capital letters from Problems 19–42 to correspond with their numerical values (the letter O has been filled in for you). Some letters may not appear in the boxes. When you are finished, darken all the blank spaces to separate the words in the secret message. Notice that some of the blank spaces have also been filled in to help you.

13	5	19	21	3	11	13	14	2	49	22	50
17	7	21	23	19	11	21	13	17	21	49	17
5	13	3	O	11		49	13	48	2	10	19
12	21	48	2	8	21	26	21	48	21	11	22
2	10	48	21	10	17	21	12	.			

Suppose that a spreadsheet contains the following values:

📊 **Spreadsheet Application**			_ □ ✕	
	A	B	C	D	E
1	6	-4	2	3	
2					
3					
4					
5					

Determine the value of cell E1 *if it contains the formula given in Problems 44–45.*

44. a. +A1+B1   **b.** +2*A1+3*B1
   c. +C1*(A1+3*B1)
   d. +A1+ ··· +D1 or @sum(A1 ... D1)

45. a. +A1 − B1/C1   **b.** +(A1 − B1)/C1
   c. +A1/C1*D1
   d. +(A1+ ··· +D1)/2 or @sum(A1 ... D1)/2

Draw a spreadsheet like the one shown and fill in the values of the missing cells assuming that the formula given in Problems 46–47 has been entered in cell C1 and replicated across row 1.

🔲 Spreadsheet Application			_ □ ✕	
	A	B	C	D	E
1	1	3			
2					
3					
4					
5					

46. a. +A1 + B1   **b.** +A1 − B1
47. a. +#A#1 + B1   **b.** +#A#1*B1

In Problems 48–49, consider the following spreadsheet, in which the formula for each cell except B1 is displayed. Determine the value of each cell given the value in cell B1.

🔲 Spreadsheet Application			_ □ ✕	
	A	B	C	D	E
1	+C2+1		+B2+1		
2	+C3+1	+A3+1	+C1+1		
3	+A2+1	+A1+1	+B1+1		
4					
5					

48. a. 1   **b.** −10   **49. a.** 0   **b.** 100

50. Show how you might use a spreadsheet to calculate the balance at the end of each year if $1,000 is deposited into an account paying 8% compounded yearly.

51. The owner of an auto dealership would like to create a spreadsheet in which sales performance information regarding each salesperson can be tabulated. She would like to use the following format:

🔲 Spreadsheet Application			_ □ ✕	
	A	B	C	D	E
1	NAME	SALES	COST	PROFIT	COMMISSION
2	John Adams	100,000	80,000	20,000	1,600
3					
4					
5					

Column A is to contain the name of each salesperson; column B, the gross sales; column C, the cost; column D, the profit; and column E, the commission, calculated at 8% of the profit. Construct a spreadsheet for 20 employees.

52. A certain population has two eye color genes; *B* (brown eyes, dominant) and *b* (blue eyes, recessive). Suppose we have an isolated population in which 75% of the genes in the gene pool are dominant *B,* and the other 25% are recessive *b.* What percent of the population has each genotype? What percent of the population has each phenotype?

53. A certain population has two fur color genes; *B* (black, dominant) and *b* (brown, recessive). Suppose we have an isolated population in which 65% of the genes in the gene pool are dominant *B,* and the other 35% are recessive *b.* What percent of the population has each genotype? What percent of the population has each phenotype?

54. A population of self-pollinating pea plants has two genes; *T* (tall, dominant) and *t* (short, recessive). Suppose we have an isolated population in which 50% of the genes in the gene pool are dominant *T,* and the other 50% are recessive *t.* What percent of the population has each genotype? What percent of the population has each phenotype?

55. When flowers known as "four-o'clocks" are crossed, there is an incomplete dominance. A four-o'clock is red if the genotype is *rr* and white if it is *ww.* If *rr* is crossed with *ww,* a hybrid *rw* results. Suppose a population of 20% red four-o'clocks is mixed with a population of 80% white four-o'clocks. What percent of the population has each genotype? What percent of the population has each phenotype?

PROBLEM SOLVING

56. What do you notice about the nine cells of the spreadsheet given in Problems 48–49?

57. A certain population has two fur color genes: *B* (black, dominant) and *b* (brown, recessive). Suppose you look at a population that is 25% brown and 75% black. Estimate the percentages in the gene pool that are *B* and *b.*

58. A population of self-pollinating pea plants has two genes: *T* (tall, dominant) and *t* (short, recessive). Suppose you look at a population that is 36% short. Estimate the percentages in the gene pool that are *T* and *t.*

59. Earlobes can be characterized as attached or free hanging. Free hanging (*F*) are dominant, and attached (*f*) are recessive. Survey your class to determine the percentage that are attached. Let q^2 be this number between 0 and 1; this represents the *ff* genotype. Estimate the gene pool for your class.

60. There are three genes in the gene pool for blood, *A*, *B*, and *O*. Two of these three are present in a person's blood: *A* and *B* dominate *O*, whereas *A* and *B* are codominant. This gives the following possibilities:

Genotype	Phenotype
AA	type *A* blood
AO	type *A* blood
AB	type *AB* blood
BO	type *B* blood
BB	type *B* blood
OO	type *O* blood

Let *p*, *q*, and *r* represent the frequencies of the genes *A*, *B*, and *O*, respectively, in the blood gene pool. Suppose a certain population has 20% type *A*, 30% type *B*, and 50% type *O*. Construct a Punnett square and find $(p + q + r)^2$ to answer the following questions. What percent of the population has each genotype? What percent of the population has each phenotype?

Pay attention to the difference between an expression and an equation. Also, note that "to solve" is not the same as "to simplify."

5.4 | EQUATIONS

Even though there are many aspects of algebra that are important to the scientist and mathematician, the ability to solve simple equations is important to the lay person and can be used in a variety of everyday applications.

An **equation** is a statement of equality. There are three types of equations: *true, false,* and *open.* An *open equation* is one with a variable.

A *true equation* is an equation without a variable such as

$$2 + 3 = 5$$

A *false equation* is an equation without a variable such as

$$2 + 3 = 15$$

Our focus is on *open equations,* those equations with a variable. The values that make an open equation true are said to **satisfy** the equation and are called the **solutions** or **roots** of the equation. To **solve** an open equation is to find all replacements for the variable(s) that make the equation true. There are three types of open equations. Those that are always true, as in

$$x + 3 = 3 + x$$

are called *identities.* Those that are always false, as in

$$x + 3 = 4 + x$$

are called *contradictions.* Most open equations are true for some replacements of the variable and false for other replacements, as in

$$2 + x = 15$$

These are called *conditional equations.* Generally, when we speak of equations we mean conditional equations. Our concern when solving equations is to find the numbers that satisfy a given equation, so we look for things to do to equations to make the solutions or roots more obvious. Two equations with the same solutions are called **equivalent equations.** An equivalent equation may be easier to solve than the original equation, so we try to get successively simpler equivalent equations until the solution is obvious. There are certain procedures you can use to create equivalent equations. In this section, we will discuss solving the two most common types of equations you will encounter: *linear* and *quadratic.*

Linear equations:	$ax + b = 0$	$(a \neq 0)$
Quadratic equations:	$ax^2 + bx + c = 0$	$(a \neq 0)$

Linear Equations

To solve a linear equation, you can use one or more of the following equation properties.

Equation Properties

CAUTION

These properties lead to a procedure for solving equations.

Addition property	Adding the same number to both sides of an equation results in an equivalent equation.
Subtraction property	Subtracting the same number from both sides of an equation results in an equivalent equation.
Multiplication property	Multiplying both sides of a given equation by the same nonzero number results in an equivalent equation.
Division property	Dividing both sides of a given equation by the same nonzero number results in an equivalent equation.

When these properties are used to obtain equivalent equations, ***the goal is to isolate the variable on one side of the equation,*** as illustrated in Example 1. You can always check the solution to see whether it is correct; substituting the solution into the original equation will verify that it satisfies the equation. Notice how the field properties are used when solving these equations.

EXAMPLE 1

Solve the given equations.

a. $x + 15 = 25$ **b.** $x - 36 = 42$ **c.** $3x = 75$ **d.** $\dfrac{x}{5} = -12$ **e.** $15 - x = 0$

Solution

CAUTION

The goal is to isolate the variable on one side of the equal sign.

a.
$$x + 15 = 25 \qquad \text{Given equation}$$
$$x + 15 - 15 = 25 - 15 \qquad \text{Subtract 15 from both sides.}$$
$$x = 10 \qquad \text{Carry out the simplification.}$$

The root (solution) of this simpler equivalent equation is now obvious (it is 10). We often display the answer in the form of this simpler equation, $x = 10$, with the variable isolated on one side.

CAUTION

Perform the opposite operation to find a simpler equivalent equation.

b.
$$x - 36 = 42 \qquad \text{Given equation}$$
$$x - 36 + 36 = 42 + 36 \qquad \text{Add 36 to both sides.}$$
$$x = 78 \qquad \text{Simplify.}$$

c. $3x = 75 \qquad \text{Given}$
$$\frac{3x}{3} = \frac{75}{3} \qquad \text{Divide both sides by 3.}$$
$$x = 25 \qquad \text{Simplify.}$$

d. $\dfrac{x}{5} = -12 \qquad \text{Given}$

$$5\left(\frac{x}{5}\right) = 5(-12) \qquad \text{Multiply both sides by 5.}$$
$$x = -60 \qquad \text{Simplify.}$$

e.
$$15 - x = 0 \qquad \text{Given}$$
$$15 - x + x = 0 + x \qquad \text{Add x to both sides.}$$
$$15 = x \qquad \text{Simplify.}$$ ◆

The equation $15 = x$ is the same as $x = 15$. This is a general property of equality called the **symmetric property of equality:** If $a = b$, then $b = a$.

While Example 1 illustrates the basic properties of equations, you will need to solve more complicated linear equations. In the following examples, some of the steps are left for mental calculations.

EXAMPLE 2

Find the root of the given equations.

a. $5x + 3 - 4x = 6 + 9$

b. $6x + 3 - 5x - 7 = 11(-2)$

c. $5x + 2 = 4x - 7$

d. $4x + x = 20$

e. $3(m + 4) + 5 = 5(m - 1) - 2$

An equation is like a mystery thriller,
It grips you once you've begun it.
You are the sleuth who stalks the killer,
X represents "whodunit."

I'M SICK OF BEING AN UNKNOWN!

The scene of the crime must first be cleared,
The suspects called into session;
You look for clues to prove your case,
Till you wring from X a confession.

Tom Sampson
Blakelack High School

Solution

a.
$$5x + 3 - 4x = 6 + 9 \qquad \text{Given}$$
$$x + 3 = 15 \qquad \text{Simplify.}$$
$$x = 12 \qquad \text{Subtract 3 from both sides.}$$

b.
$$6x + 3 - 5x - 7 = 11(-2) \qquad \text{Given}$$
$$x - 4 = -22 \qquad \text{Simplify.}$$
$$x = -18 \qquad \text{Add 4 to both sides.}$$

c.
$$5x + 2 = 4x - 7 \qquad \text{Given}$$
$$x + 2 = -7 \qquad \text{Subtract 4x from both sides.}$$
$$x = -9 \qquad \text{Subtract 2 from both sides.}$$

d.
$$4x + x = 20 \qquad \text{Given}$$
$$5x = 20 \qquad \text{Simplify.}$$
$$x = 4 \qquad \text{Divide both sides by 5.}$$

e.
$$3(m + 4) + 5 = 5(m - 1) - 2 \qquad \text{Given}$$
$$3m + 12 + 5 = 5m - 5 - 2 \qquad \text{Simplify (distributive property).}$$
$$3m + 17 = 5m - 7 \qquad \text{Simplify.}$$
$$17 = 2m - 7 \qquad \text{Subtract 3m from both sides.}$$
$$24 = 2m \qquad \text{Add 7 to both sides.}$$
$$12 = m \qquad \text{Divide both sides by 2.}$$ ◆

Quadratic Equations

To solve quadratic equations, you must first use the equation properties to write the equation in the form

$$ax^2 + bx + c = 0$$

There are two commonly used methods for solving quadratic equations. The first uses factoring, and the second uses the quadratic formula. Both of these methods require that you apply the linear equation properties to obtain a 0 on one side. Next, look to see whether the polynomial is factorable. If so, use the **zero-product rule** to set each factor equal to 0, and then solve each of those equations. If the polynomial is not factorable, then use the quadratic formula.

Zero-Product Rule

Note that you must first have a zero on one side.

If $A \cdot B = 0$, then $A = 0$ or $B = 0$, or $A = B = 0$.

If the product of two numbers is 0, then at least one of the factors must be 0.

With a quadratic equation, first obtain a 0 on one side.

EXAMPLE 3

Solve each equation.

a. $x^2 = x$ **b.** $x(x - 8) = 4(x - 9)$

Solution

a.
$$x^2 = x \quad \text{Given}$$
$$x^2 - x = 0 \quad \text{Subtract x from both sides.}$$
$$x(x - 1) = 0 \quad \text{Factor, if possible.}$$
$$x = 0, \quad x - 1 = 0 \quad \text{Zero-product rule; set each factor equal to 0.}$$
$$x = 1 \quad \text{Solve each of the resulting equations.}$$

The equation has two roots, $x = 0$ and $x = 1$. Usually you will set each factor equal to 0 and solve them mentally.

b.
$$x(x - 8) = 4(x - 9) \quad \text{Given}$$
$$x^2 - 8x = 4x - 36 \quad \text{Simplify.}$$
$$x^2 - 12x + 36 = 0 \quad \text{Subtract 4x from both sides; add 36 to both sides.}$$
$$(x - 6)(x - 6) = 0 \quad \text{Factor.}$$
$$x = 6 \quad \text{Set each factor equal to 0 and mentally solve. Since the factors are the same, there is one root, 6. In such a case we say the root of 6 has } \textit{multiplicity } \text{two.} \quad \blacklozenge$$

HISTORICAL NOTE

As early as 2000 B.C., the Babylonians had a well-developed algebra. It did not have the symbolism we associate with modern-day algebra but was written out in words. Even without the symbolism, however, the Babylonians did solve quadratic equations by using a general formula.

If the quadratic expression is not easily factorable (after you obtain a 0 on one side), then you can use the **quadratic formula,** which is derived in most high school algebra books.

Quadratic Formula

This is one of the all-time "BIGGIES" in algebra. You should remember it.

If $ax^2 + bx + c = 0$, $a \neq 0$, then
$$x = \frac{-b \pm \sqrt{b^2 - 4ac}}{2a}$$

EXAMPLE 4

Solve the given equations.

a. $2x^2 + 4x + 1 = 0$ **b.** $x^2 = 6x - 13$

Solution

a. Note that $2x^2 + 4x + 1 = 0$ has a 0 on one side and also that the left-hand side does not easily factor, so we will use the quadratic formula. We begin by (mentally) identifying $a = 2$, $b = 4$, and $c = 1$.

$$x = \frac{-(4) \pm \sqrt{(4)^2 - 4(2)(1)}}{2(2)} \quad \text{Substitute for } a, b, \text{ and } c \text{ in the quadratic formula.}$$

$$= \frac{-4 \pm \sqrt{8}}{4} \quad \text{Simplify under the square root.}$$

$$= \frac{-4 \pm 2\sqrt{2}}{4} \quad \text{Simplify radical.}$$

$$= \frac{2(-2 \pm \sqrt{2})}{4} \quad \text{Factor a 2 out of the numerator so that we can reduce the fraction. This step is usually done mentally.}$$

$$= \frac{-2 \pm \sqrt{2}}{2}$$

b.
$$x^2 = 6x - 13 \quad \text{Given}$$
$$x^2 - 6x + 13 = 0 \quad \text{Obtain a 0 on one side.}$$

$$x = \frac{-(-6) \pm \sqrt{(-6)^2 - 4(1)(13)}}{2(1)}$$

$$= \frac{6 \pm \sqrt{-16}}{2}$$

The square root of a negative number is not defined in the set of real numbers. Thus, since we are working in the set of real numbers, we say there is no real value. ◆

Since most of us have access to a calculator, we often estimate the roots of quadratic equations with radicals as rational (decimal) approximations. We illustrate this with the next calculator example.

EXAMPLE 5

Solve $5x^2 + 2x - 2 = 0$ and approximate the roots to the nearest hundredth.

Solution   From the quadratic formula, where $a = 5$, $b = 2$, and $c = -2$:

$$x = \frac{-b \pm \sqrt{b^2 - 4ac}}{2a}$$

$$= \frac{-2 \pm \sqrt{2^2 - 4(5)(-2)}}{2(5)}$$

$$= \frac{-2 \pm \sqrt{44}}{2(5)}$$

$$= \frac{-1 \pm \sqrt{11}}{5} \approx 0.46, -0.86$$ ◆

Calculator Window

We can use calculators to help us solve quadratic equations. Since there are many types of calculators, we can only offer some suggestions, and you will need to check your owner's manual.

Solve $5x^2 + 2x - 2 = 0$ using an *algebraic calculator*. To approximate these roots:

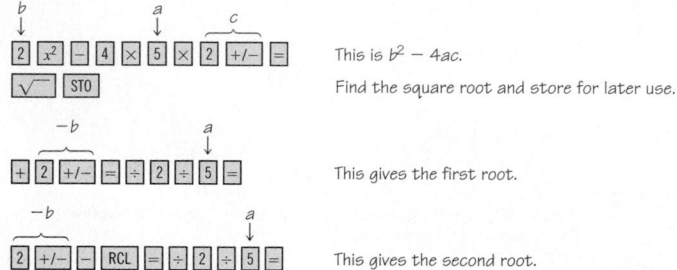

This is $b^2 - 4ac$.

Find the square root and store for later use.

This gives the first root.

This gives the second root.

Some of the steps shown here could be combined because these are simple numbers. These steps give the numerical approximation for a quadratic equation with real roots. For this quadratic equation the roots are (to four decimal places) 0.4633 and −0.8633.

Since you will have occasion to use the quadratic formula over and over again, and since many calculators have programming capabilities, this is a good time to consider writing a simple program to give the real roots for a quadratic equation. First write the equation in the form $ax^2 + bx + c = 0$, input the *a, b,* and *c* values into the calculator as *A, B,* and *C*. The program will then output the two real values (if they exist). Each brand of *graphing calculator* is somewhat different, but it is instructive to illustrate the general process. Press the ⌈PRGM⌉ key. You will then be asked to name the program; we call our program QUAD. Next, input the formula for the two roots (from the quadratic formula). Finally, display the answer:

$:(-B+\sqrt{(B^2 - 4AC)})/(2A)$

:Disp Ans

$:(-B - \sqrt{(B^2 - 4AC)})/(2A)$

:Disp Ans

For Example 5, input the *A, B,* and *C* values as follows:

⌈5⌉ ⌈STO→⌉ ⌈A⌉ ⌈2⌉ ⌈STO→⌉ ⌈B⌉ ⌈−2⌉ ⌈STO→⌉ ⌈C⌉ ⌈PRGM⌉ QUAD

Then run the program for the DISPLAY: .4633249581
                                        −.8633249581

Finally, today many calculators have a ⌈SOLVE⌉ key and the only requirement for solving the equation is to check your owner's manual for the correct format. For Example 5, input

solve $(5x^2 + 2x - 2 = 0, x)$

which gives the solution as

$$x = \frac{-(\sqrt{11} + 1)}{5} \quad \text{or} \quad x = \frac{\sqrt{11} - 1}{5}$$

Note that the form here is equivalent to (but not the same as) $x = \dfrac{-1 \pm \sqrt{11}}{5}$ obtained in Example 5.

FERMAT'S LAST THEOREM

Pierre de Fermat
(1601–1665)

Pierre de Fermat was a lawyer by profession, but he was an amateur mathematician in his spare time. He became Europe's finest mathematician, and he wrote well over 3,000 mathematical papers and notes. However, he published only one, because he did them just for fun. Every theorem that Fermat said he proved has subsequently been verified—with one defying solution until 1993! This problem, known as Fermat's Last Theorem (see Group Research Project G20 and Individual Research Project 5.5), was written by Fermat in the margin of a book:

To divide a cube into two cubes, a fourth power, or in general any power whatever above the second, into powers of the same denomination, is impossible, and I have assuredly found an admirable proof of this, but the margin is too narrow to contain it.

Many of the most prominent mathematicians since his time have tried to prove or disprove this conjecture, and on June 23, 1993, during the third of a series of lectures at a conference held at the Newton Institute in Cambridge, it was reported that British mathematician Andrew Wiles of Princeton had proved a theorem for which Fermat's Last Theorem is a corollary.

Fermat's Last Theorem has been mentioned in literary works ranging from *Sherlock Holmes* to *Star Trek: The Next Generation*. In 1983, a 23-year-old German mathematician, Gerd Faltings, proved that the number of possible exceptions is finite, and in 1988 there were reports in the *Los Angeles Times* and *The Chronicle of Higher Education* that a Japanese mathematician, Yoichi Miyaoka, proved the famous theorem, but alas, there were errors. In 1993 when Andrew Wiles made front-page headlines when he announced that he had a proof of this theorem, many mathematicians remained skeptical. After a short-lived glitch when an error in his calculations was discovered, he was vindicated and his proof was checked and verified. There are many references to this great discovery, and perhaps the best is on videotape as part of the *NOVA* series on the Public Broadcasting System. The title is *The Proof* and it was first aired in 1997.

Problem Set 5.4

LEVEL 1

1. **IN YOUR OWN WORDS** Describe a procedure for solving first-degree equations.

2. **IN YOUR OWN WORDS** Describe a procedure for solving second-degree equations.

Solve the equations in Problems 3–23.

3. **a.** $x - 5 = 10$ **b.** $6 = x - 2$

4. **a.** $8 + x = 4$ **b.** $18 + x = 10$

5. **a.** $\dfrac{x}{4} = 8$ **b.** $\dfrac{x}{-4} = 11$

6. **a.** $4x = 12$ **b.** $-x = 5$

7. **a.** $13x = 0$ **b.** $-\frac{1}{2}x = 0$

8. **a.** $A + 13 = 18$ **b.** $5 = 3 + B$

9. **a.** $-5X = -1$ **b.** $6 = C - 4$

10. **a.** $2D + 2 = 10$ **b.** $15E - 5 = 0$

11. **a.** $16F - 5 = 11$ **b.** $6 = 5G - 24$

12. **a.** $4(H + 1) = 4$ **b.** $5(I - 7) = 0$

13. **a.** $\dfrac{J}{5} = 3$ **b.** $\dfrac{2K}{3} = 6$

14. **a.** $\dfrac{3L}{4} = 5$ **b.** $1 = \dfrac{2M}{3} + 7$

15. **a.** $7 = \dfrac{2N}{3} + 11$ **b.** $-5 = \dfrac{2P + 1}{3}$

16. **a.** $\dfrac{2 - 5Q}{3} = 4$ **b.** $\dfrac{5R - 1}{2} = 5$

17. $5(6S - 81) = -3(15 + 5S)$

18. $3T + 3(T + 2) + (T + 4) + 11 = 0$

19. $3(U - 3) - 2(U - 12) = 18$

20. $6(V - 2) - 4(V + 3) = 10 - 42$

21. $5(W + 3) - 6(W + 5) = 0$

22. $6(Y + 2) = 4 + 5(Y - 4)$

23. $5(Z - 2) - 3(Z + 3) = 9$

LEVEL 2

24. This problem should help you check your work in Problems 8–23. Fill in the capital letters from Problems 8–23 to correspond with their numerical values (the letter O has been filled in for you). Some letters may not appear in the boxes. When you are finished, darken all the blank spaces

to separate the words in the secret message. Notice that one of the blank spaces has also been filled in to help you.

12	−3	0	7	8	−7	−8	$\frac{11}{5}$	O		2	$\frac{20}{3}$	$\frac{1}{3}$	−9	$\frac{4}{5}$		11
−15	7	$\frac{20}{3}$	$\frac{20}{3}$	−1	3	−6	4	O		3	2	−3	$\frac{1}{3}$	4	$\frac{20}{3}$	−28
−1	5	8	8	7	8	−3	−5	−28		3	$\frac{1}{6}$	$\frac{1}{6}$	7	−6	12	
1	7	−6	4	7	−6	6	$\frac{1}{4}$	$\frac{1}{3}$	$\frac{11}{5}$	$\frac{11}{5}$	O	$\frac{11}{5}$	8	!	!	

Solve the equations in Problems 25–46.

25. $x^2 = 10x$

26. $x^2 = 14x$

27. $5x + 66 = x^2$

28. $15x^2 + 4x = 4$

29. $x^3 = 4x$

30. $x^3 = x$

31. $4x(x - 9) = 9(1 - 4x)$

32. $4(9x - 1) = 9x(4 - x)$

33. $x^2 + 7x + 2 = 0$

34. $x^2 - 3x + 1 = 0$

35. $x^2 - 5x - 3 = 0$

36. $x^2 - 6x + 9 = 0$

37. $x^2 - 6x + 7 = 0$

38. $x^2 - 6x + 6 = 0$

39. $3x^2 + 5x - 4 = 0$

40. $2x^2 - x + 3 = 0$

41. $4x^2 + 2x = -5$

42. $6x^2 = 13x - 6$

43. $3x^2 = 11x + 4$

44. $6x^2 = 17x + 3$

45. $6x^2 = 5x$

46. $9x^2 = 2x$

Use a calculator to obtain solutions correct to the nearest hundredth in Problems 47–54.

47. $x^2 + 4 = 3\sqrt{2}\,x$

48. $x^2 - 4\sqrt{3}\,x + 9 = 0$

49. $\sqrt{2}\,x^2 + 2x - 3 = 0$

50. $4x^2 - 2\sqrt{5}\,x - 2 = 0$

51. $0.02x^2 + 0.831x + 0.0069 = 0$

52. $68.38x^2 - 4.12x - 198.41 = 0$

53. $x^2 - 11.001x + 24.098 = 0$

54. $x^2 + 4.09x = 0.078$

LEVEL 3

55. Young's Rule for calculating a child's dosage for medication is

$$\text{CHILD'S DOSE} = \frac{\text{AGE OF CHILD}}{\text{AGE OF CHILD} + 12} \times \text{ADULT DOSE}$$

a. If an adult's dose of a particular medication is 100 mg, what is the dose for a 10-year-old child?

b. If a 12-year-old child's dose of a particular medication is 10 mg, what is the adult's dose?

56. Fried's Rule for calculating an infant's dosage for medication is

INFANT'S DOSE

$$= \frac{\text{AGE OF INFANT IN MONTHS}}{150} \times \text{ADULT DOSE}$$

a. If an adult's dose of a particular medication is 50 mg, what is the dosage for a 10-month-old infant?

b. If a 15-month-old infant is to receive 7.5 mg of a medication, what is the equivalent adult dose?

PROBLEM SOLVING

57. Let $a = b = 1$. Consider

$a + b = c$	Given.
$(a + b)^2 = c(a + b)$	Multiply both sides by $(a + b)$.
$a^2 + 2ab + b^2 = ac + bc$	Expand by multiplication.
$a^2 + 2ab - ac = bc - b^2$	Subtract ac and b^2 from both sides.
$a^2 + ab - ac = -ab - b^2 + bc$	Subtract ab from both sides.
$a(a + b - c) = -b(a + b - c)$	Common factor.
$a = -b$	Divide both sides by $a + b - c$.

But since $a = b = 1$, we see $a = -1$. What is wrong here?

58. An approximation for π can be obtained from

$$\frac{\pi^2}{6} = 1 + \frac{1}{2^2} + \frac{1}{3^2} + \frac{1}{4^2} + \cdots$$

a. Solve for π.

b. Find an approximation for π using the first 20 terms.

c. Compare your answer to part **b** with an approximation of π you probably used in grade school, namely, $\frac{22}{7}$.

59. **HISTORICAL QUESTION** Al-Khwârizmî (see Historical Note on p. 216) solved the equation

$$x^2 + 10x = 39$$

a. Solve this equation.

b. Here is a translation of his solution:

. . . a square and 10 roots are equal to 39 units. The question therefore in this type of equation is about as follows: What is the square which combined with ten of its roots will give a sum total of 39? The manner of solving this type of equation is to take one-half of the roots just mentioned. Now the roots in the problem before us are 10. Therefore take 5, which multiplied by itself gives 25, an amount which you add to 39 giving 64. Having taken then the square root of this which is 8, subtract from it half the roots, 5, leaving 3. The number three therefore represents one root of this square, which itself, of course is 9. Nine therefore gives the square.

Follow these directions to find a solution. Show each step using modern algebra notation.

60. HISTORICAL QUESTION

a. Consult an algebra textbook and look up the procedure of *completing the square*. Solve Al-Khwârizmî's equation in Problem 59 by completing the square.

b. Do you think the translation of the solution in Problem 59b best describes what you did in solving the equation of Problem 59a, or what you did in solving the equation in Problem 60a?

5.5   INEQUALITIES

Comparison Property

The techniques of the previous section can also be applied to quantities that are not equal. If we are given any two numbers x and y, then obviously either

$$x = y \quad \text{or} \quad x \neq y$$

If $x \neq y$, then either

$$x < y \quad \text{or} \quad x > y$$

This property is called the **comparison property.** *

Comparison Property

For any two numbers x and y, exactly one of the following is true:

1. $x = y$ x is equal to y (the same as)
2. $x > y$ x is greater than y (larger than)
3. $x < y$ x is less than y (smaller than)

Solving Linear Inequalities

This means that if two quantities are not exactly equal, we can relate them with a greater-than or a less-than symbol (called an **inequality symbol**). The solution of

$$x < 3$$

has more than one value, and it becomes very impractical to write "The answers are $2, 1, -110, 0, 2\frac{1}{2}, 2.99, \ldots$." Instead, we relate the answer to a number line, as shown in Figure 5.2. The fact that 3 is not included (3 is not less than 3) in the solution set is indicated by an open circle at the point 3.

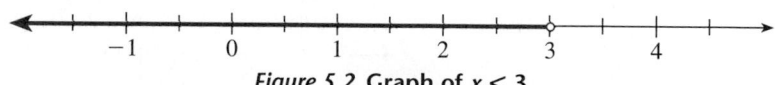

Figure 5.2 Graph of $x < 3$

If we want to include the endpoint $x = 3$ with the inequality $x < 3$, we write $x \leq 3$ and say "x is less than or equal to 3." We define two additional inequality symbols:

$$x \geq y \quad \text{means } x > y \text{ or } x = y$$
$$x \leq y \quad \text{means } x < y \text{ or } x = y$$

A *statement of order,* called an **inequality,** refers to statements that include one or more of the following relationships:

less than ($<$)   less than or equal to ($\leq$)

greater than ($>$)   greater than or equal to ($\geq$)

---

*Sometimes this is called the trichotomy property.

EXAMPLE 1

Graph the solution sets for the given inequalities.

a. $x \leq 3$ **b.** $x > 5$ **c.** $-2 \leq x$

Solution

a. Notice that in this example the endpoint is included because with $x \leq 3$, it is possible that $x = 3$. This is shown as a solid dot on the number line:

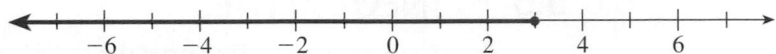

b. $x > 5$

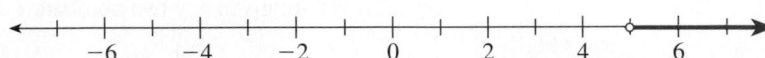

c. $-2 \leq x$

Although you can graph this directly, you will have less chance of making a mistake when working with inequalities if you rewrite them so that the variable is on the left; that is, reverse the inequality to read $x \geq -2$. Notice that the direction of the arrow has been changed, because the symbol requires that the arrow always point to the smaller number. When you change the direction of the arrow, we say that you have *changed the order of the inequality.* For example, if $-2 \leq x$, then $x \geq -2$, and we say the order has been changed. The graph of $x \geq -2$ is shown on the following number line:

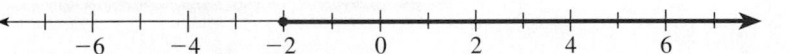

On a number line, if $x < y$, then x is to the left of y. Suppose the coordinates x and y are plotted as shown in Figure 5.3.

Figure 5.3 **Number line showing two coordinates, x and y**

If you add 2 to both x and y, you obtain $x + 2$ and $y + 2$. From Figure 5.4, you see that $x + 2 < y + 2$.

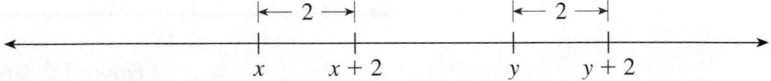

Figure 5.4 **Number line with 2 added to both x and y**

If you add some number c, there are two possibilities, as shown in Figure 5.5.

$c > 0$ ($c > 0$ is read "c is positive")

$c < 0$ ($c < 0$ is read "c is negative")

c positive:

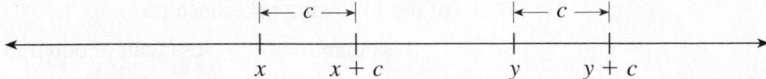

c negative:

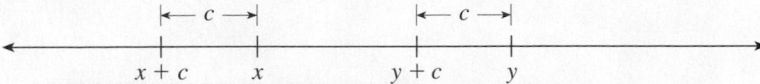

Figure 5.5 **Adding positive and negative values to x and y**

If $c > 0$, then $x + c$ is still to the left of $y + c$. If $c < 0$, then $x + c$ is still to the left of $y + c$, as shown in Figure 5.5. In both cases, $x < y$, which justifies the following property.

Addition Property of Inequality

If $x < y$, then

$$x + c < y + c$$

Also, if $x \leq y$, then $x + c \leq y + c$
if $x > y$, then $x + c > y + c$
if $x \geq y$, then $x + c \geq y + c$

Because this **addition property of inequality** is essentially the same as the addition property of equality, you might expect that there is also a multiplication property of inequality. We would hope that we could multiply both sides of an inequality by some number c without upsetting the inequality. Consider some examples. Let $x = 5$ and $y = 10$, so that $5 < 10$.

Let $c = 2$: $5 \cdot 2 < 10 \cdot 2$
$10 < 20$ True

Let $c = 0$: $5 \cdot 0 < 10 \cdot 0$
$0 < 0$ False

Let $c = -2$: $5(-2) < 10(-2)$
$-10 < -20$ False

You can see that you cannot multiply both sides of an inequality by a constant and be sure that the result is still true. However, if you restrict c to a positive value, then you can multiply both sides of an inequality by c. On the other hand, if c is a negative number, then the order of the inequality should be reversed. This is summarized by the **multiplication property of inequality.**

Multiplication Property of Inequality

Positive multiplication ($c > 0$)

If $x < y$, then

$$cx < cy$$
↑
Order unchanged

Also for $c > 0$,

if $x \leq y$, then $cx \leq cy$
if $x > y$, then $cx > cy$
if $x \geq y$, then $cx \geq cy$

Negative multiplication ($c < 0$)

If $x < y$, then

$$cx > cy$$
↑
Order reversed

Also for $c < 0$,

if $x \leq y$, then $cx \geq cy$
if $x > y$, then $cx < cy$
if $x \geq y$, then $cx \leq cy$

The same properties hold for positive and negative division. We can summarize with the following statement, which tells us how to **solve an inequality.**

Solution of Linear Inequalities

The procedure for solving linear inequalities is the same as the procedure for solving linear equations except that, if you multiply or divide by a negative number, you reverse the order of the inequality symbol.

In summary, given $x < y$, $x \leq y$, $x > y$, or $x \geq y$:

STOP   *Spend some time with this summary.*

The **inequality symbols are the** *same* if you

1. Add the same number to both sides.
2. Subtract the same number from both sides.
3. Multiply both sides by a positive number.
4. Divide both sides by a positive number.

This works the same as with equations.

STOP   *Here is where inequalities differ from equations.*

The **inequality symbols are** *reversed* if you

1. Multiply both sides by a negative number.
2. Divide both sides by a negative number.
3. Interchange the x and the y.

This is where inequalities differ from equations.

EXAMPLE 2

Solve: **a.** $-x \geq 2$ **b.** $\dfrac{x}{-3} < 1$ **c.** $5x - 3 \geq 7$

Solution

a. $-x \geq 2$

$x \leq -2$ *Multiply both sides by -1 and remember to reverse the order of the inequality.*

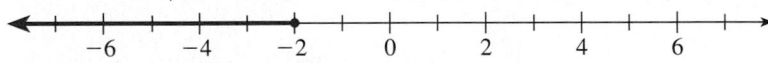

b. $\dfrac{x}{-3} < 1$

$x > -3$ *Multiply both sides by -3 and reverse the order of the inequality.*

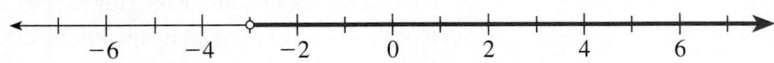

c. $5x - 3 \geq 7$

$5x - 3 + 3 \geq 7 + 3$

$5x \geq 10$

$\dfrac{5x}{5} \geq \dfrac{10}{5}$

$x \geq 2$

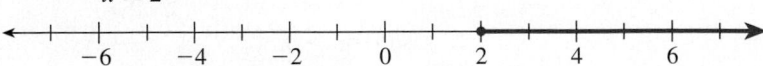

Problem Set 5.5

LEVEL 1

1. **IN YOUR OWN WORDS** What is the comparison property?

2. **IN YOUR OWN WORDS** Describe a procedure for solving a first-degree inequality.

Graph the solution sets in Problems 3–17.

3. $x < 5$ **4.** $x \geq 6$ **5.** $x \geq -3$

6. $x \leq -2$ **7.** $4 \geq x$ **8.** $-1 < x$

9. $\dfrac{x}{2} > 3$ **10.** $4 < \dfrac{x}{2}$ **11.** $-2 > \dfrac{x}{-4}$

12. $x < 50$ **13.** $x \geq 100$ **14.** $x \geq -125$

15. $x \leq -75$ **16.** $45 \geq x$ **17.** $-40 < x$

Solve the inequalities in Problems 18–41.

18. $x + 7 \geq 3$ **19.** $x - 2 \leq 5$ **20.** $x - 2 \geq -4$

21. $10 < 5 + y$ **22.** $-4 < 2 + y$ **23.** $-3 < 5 + y$

24. $2 > -s$ **25.** $-t \leq -3$ **26.** $-m > -5$

27. $5 \leq 4 - y$ **28.** $3 > 2 - x$ **29.** $5 \geq 1 - w$

30. $2x + 6 \leq 8$ **31.** $3y - 6 \geq 9$ **32.** $3 > s + 9$

33. $2 < 2s + 8$ **34.** $4 \leq a + 2$ **35.** $3 > 2b - 13$

36. $3s + 2 > 8$ **37.** $5t - 7 \geq 8$ **38.** $7u - 5 \leq 9$

39. $9 - 2v < 5$ **40.** $5 - 3w > 8$ **41.** $2 - x \geq 3x + 10$

LEVEL 2

Solve the inequalities in Problems 42–51.

42. $7 - 5A < 2A + 7$ **43.** $B > 3(1 + B)$

44. $3C > C + 19$ **45.** $2(D + 7) > 2 - D$

46. $5E - 4 < 3E - 6$ **47.** $3(3F - 2) > 4F - 3$

48. $4G - 1 > 3(G + 2)$ **49.** $5(4 + H) < 3(H + 1)$

50. $2 - 3I > 7(1 - I)$ **51.** $7(J - 2) + 5 \leq 3(2 + J)$

52. Suppose that seven times a number is added to 35 and the result is positive. What are the possible numbers?

53. If the opposite of a number must be greater than twice the number, what are the possible numbers?

54. Suppose that three times a number is added to 12 and the result is negative. What are the possible numbers?

55. If the opposite of a number must be less than 5, what are the possible numbers satisfying this condition?

PROBLEM SOLVING

56. If a number is four more than its opposite, what are the possible numbers?

57. If a number is six less than twice its opposite, what are the possible numbers?

58. If a number is less than four more than its opposite, what are the possible numbers?

59. If a number is less than six minus twice its opposite, what are the possible numbers?

60. Current postal regulations state that no package may be sent if its combined length, width, and height exceed 72 in. What are the possible dimensions of a box to be mailed with equal height and width if the length is four times the height?

5.6 | ALGEBRA IN PROBLEM SOLVING

One of the goals of problem solving is to be able to apply techniques that you learn in the classroom to situations outside the classroom. However, a first step is to learn to solve contrived textbook-type word problems to develop the problem-solving skills you will need outside the classroom.

We will now rephrase Pólya's problem-solving guidelines in a setting that is appropriate to solving word problems. This procedure is summarized in the following box.

Procedure for Problem Solving in Algebra

First: You have to *understand the problem.* This means you must read the problem and note what it is all about.
Focus on processes rather than numbers. You cannot work a problem you do not understand. A sketch may help in understanding the problem.

Second: *Devise a plan.* **Write down a verbal description of the problem using operation signs and an equal or inequality sign.** Note the following common translations.

(continued)

**Procedure for Problem
Solving in Algebra
(continued)**

Symbol	Verbal Description
=	is equal to; equals; are equal to; is the same as; is; was; becomes; will be; results in
+	plus; the sum of; added to; more than; greater than; increased by; combined; total
−	minus; the difference of; the difference between; is subtracted from; less than; smaller than; decreased by; is diminished by
×	times; product; is multiplied by; twice (2 ×); triple (3 ×); of (as in 40% of 300)
÷	divided by; quotient of; ratio of; proportional to

Third: *Carry out the plan.*   In the context of word problems, we need to proceed deductively by carrying out the following steps.

Choose a variable.   If there is a single unknown, choose a variable. If there are several unknowns, you can use the substitution property to reduce the number of unknowns to a single variable. Later we will consider word problems with more than one unknown.

Substitute.   Replace the verbal phrase for the unknown with the variable.

Solve the equation.   This is generally the easiest step. Translate the symbolic statement (such as $x = 3$) into a verbal statement. Probably no variables were given as part of the word problem, so $x = 3$ is not an answer. Generally, word problems require an answer stated in words. Pay attention to units of measure and other details of the problem.

Fourth: **Look back.**   Be sure your answer makes sense by checking it with the original question in the problem. **Remember to answer the question that was asked.**

In this section, we will focus on common types of word problems that are found in most textbooks. You might say, "I want to learn how to become a problem solver, and textbook problems are not what I have in mind; I want to do *real* problem solving." But to become a problem solver, you must first learn the basics, and there is good reason why word problems are part of a textbook. We start with these problems *to build a problem-solving* **procedure** *that can be expanded to apply to problem solving in general.*

Number Relationships

The first type of word problem we consider involves number relationships. These are designed to allow you to begin thinking about the *procedure* to use when solving word problems.

EXAMPLE 1

If you add 10 to twice a number, the result is 22. What is the number?

Solution   Read the problem carefully. Make sure you know what is given and what is wanted. Next, write a verbal description (without using variables), using operation signs and an equal sign, but still using the key words. This is called *translating* the problem.

$$10 + 2(\text{A NUMBER}) = 22$$

When there is a single unknown, choose a variable.

IMPORTANT: Do not BEGIN by choosing a variable; choose a variable only AFTER you have translated the problem.

With more complicated problems you will not know at the start what the variable should be.

Let $n = $ A NUMBER.

Use the substitution property:

$$10 + 2(\underset{\downarrow}{\text{A NUMBER}}) = 22$$
$$10 + 2n \;\;\;\;= 22$$

Solve the equation and check the solution in the original problem to see if it makes sense:

$$10 + 2n = 22$$
$$2n = 12$$
$$n = 6$$

Check: Add 10 to twice 6 and the result is 22. State the solution to a word problem in words: The number is 6.   ◆

The second type of number problem involves **consecutive integers.** If $n = $ AN INTEGER, then

$$\text{THE SECOND CONSECUTIVE INTEGER} = n + 1, \quad \text{and}$$
$$\text{THE THIRD CONSECUTIVE INTEGER} = n + 2$$

Also, if $E = $ AN EVEN INTEGER, then

$$E + 2 = \text{THE NEXT CONSECUTIVE EVEN INTEGER}$$

and if $F = $ AN ODD INTEGER, then

$$F + 2 = \text{THE NEXT CONSECUTIVE ODD INTEGER}$$

Notice that you add 2 when you are writing down consecutive evens or consecutive odds.

EXAMPLE 2

Find three consecutive integers whose sum is 42.

Solution   Read the problem. Write down a verbal description of the problem, using operation signs and an equal sign:

$$\text{INTEGER} + \text{NEXT INTEGER} + \text{THIRD INTEGER} = 42$$

This problem has three variables, but they are related. If

$$x = \text{INTEGER, then}$$
$$x + 1 = \text{NEXT INTEGER}$$
$$x + 2 = \text{THIRD INTEGER}$$

Substitute the variables into the verbal equation:

$$\underset{\downarrow}{\text{INTEGER}} + \underset{\downarrow}{\text{NEXT INTEGER}} + \underset{\downarrow}{\text{THIRD INTEGER}} = 42$$
$$x \;\;\;\; + \;\;\;\; x + 1 \;\;\;\; + \;\;\;\; x + 2 \;\;\;\; = 42$$

Solve the equation:

$$x + x + 1 + x + 2 = 42$$
$$3x + 3 = 42$$
$$3x = 39$$
$$x = 13$$

Check: $13 + 14 + 15 = 42$

The integers are 13, 14, and 15. ◆

Distance Relationships

The first example of a problem with several variables that we will consider involves distances. The relationships may seem complicated, but if you draw a figure and remember that the total distance is the sum of the separate parts, you will easily be able to analyze this type of problem.

EXAMPLE 3   PÓLYA'S METHOD

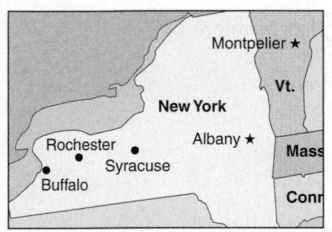

The drive from Buffalo to Albany is 210 miles across northern New York State. On this route, you pass Rochester and then Syracuse before reaching the capital city of Albany. It is 20 miles less from Buffalo to Rochester than from Syracuse to Albany, and 10 miles farther from Rochester to Syracuse than from Buffalo to Rochester. How far is it from Rochester to Syracuse?

Solution   First, you should begin by examining the problem. Remember that you cannot solve a problem you do not understand. It must make sense before mathematics can be applied to it. All too often poor problem solvers try to begin solving the problem too soon. Take your time when trying to understand the problem. Start at the beginning of the problem, and make a sketch of the situation, as shown in Figure 5.6. The cities are located in the order sketched on the line.

Buffalo	Rochester	Syracuse	Albany
B	R	S	A

Figure 5.6 **Distance problem**

Second, devise a plan. The cities are located in the order sketched in Figure 5.6, but how can that fact give us an equation? To have an equation, you must find an equality. Which quantities are equal? The distances from Buffalo to Rochester, from Rochester to Syracuse, and from Syracuse to Albany must add up to the distance from Buffalo to Albany. That is, the sum of the parts must equal the whole distance—so start there. This is what we mean when we say *translate*.

TRANSLATE.

(DIST. B TO R) + (DIST. R TO S) + (DIST. S TO A) = (DIST. B TO A)

Notice that there appear to be four variables. We now use the substitution property to *evolve*.

EVOLVE.

With this step we ask whether we know the value of any quantity in the equation. The first sentence of the problem tells us that the total distance is 210 miles, so

(DIST. B TO R) + (DIST. R TO S) + (DIST. S TO A) = **(DIST. B TO A)**

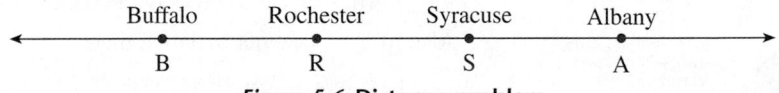

↓ Buffalo to Albany = 210

(DIST. B TO R) + (DIST. R TO S) + (DIST. S TO A) = **210**

Also, part of evolving the equation is to use substitution for the relationships that are given as part of the problem. We now translate the other pieces of given information by adding to the smaller distance in each case.

$$(\text{DIST. B TO R}) + 20 = (\text{DIST. S TO A})$$

$$(\text{DIST. R TO S}) = (\text{DIST. B TO R}) + 10$$

We now use substitution to let the equation evolve into one with a single unknown.

Pólya's method now tells us to carry out the plan. We use substitution on the above equations:

$$(\text{DIST. B TO R}) + (\text{DIST. R TO S}) + (\text{DIST. S TO A}) = 210$$

$$\uparrow \qquad\qquad \uparrow$$

$$(\text{DIST. B TO R} + 10) \qquad (\text{DIST. B TO R}) + 20$$

$$(\text{DIST. B TO R}) + [(\text{DIST. B TO R}) + 10] + [(\text{DIST. B TO R}) + 20] = 210$$

This equation now has a single variable, so we let

$$d = \text{DIST. B TO R}$$

and substitute into the equation:

$$d + [d + 10] + [d + 20] = 210$$

SOLVE:

$$3d + 30 = 210$$

$$3d - 180$$

$$d = 60$$

The equation is solved. Does that mean that the answer to the problem is "$d = 60$"? No, the question asks for the distance from Rochester to Syracuse, which is $d + 10$. So now interpret the solution and **answer** the question: The distance from Rochester to Syracuse is 70 miles.

Notice that the steps we used above can be summarized as **translate, evolve, solve,** and **answer.** Pólya's procedure requires that we look back. To be certain that the answer makes sense in the original problem, you should always check the solution:

$$60 + 70 + 80 \stackrel{?}{=} 210$$

$$210 = 210 \; \checkmark$$

◆

Remember this summary.

STOP

EXAMPLE 4   PÓLYA'S METHOD

Once upon a time (about 450 B.C.), a Greek named Zeno made up several word problems that became known as Zeno's paradoxes. This problem is not a paradox, but was inspired by one of Zeno's problems. Consider a race between Achilles and a tortoise. The tortoise has a 100-meter head start. Achilles runs at a rate of 10 meters per second, whereas the tortoise runs 1 meter per second (it is an extraordinarily swift tortoise). How long does it take Achilles to catch up with the tortoise?

Solution   We use Pólya's problem-solving guidelines for this example.

Understand the Problem.   Before you begin, make sure you understand the problem. It is often helpful to draw a figure or diagram to help you understand the problem. The situation for this problem is shown in Figure 5.7 on page 256.

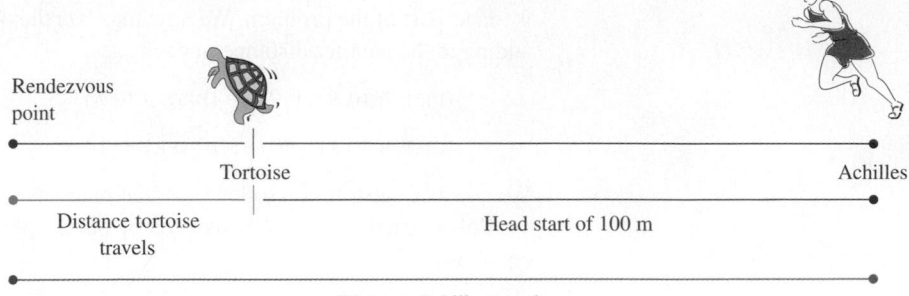

Rendezvous point

Tortoise

Achilles

Distance tortoise travels

Head start of 100 m

Distance Achilles travels

Figure 5.7 **Achilles and tortoise problem**

Devise a Plan. The plan we will use is *translate, evolve, solve,* and *answer.*

Carry Out the Plan.

TRANSLATE.

(ACHILLES' DISTANCE TO RENDEZVOUS) = (TORTOISE'S DISTANCE TO RENDEZVOUS) + (HEAD START)

EVOLVE.   The equation we wrote has three unknowns. Our goal is to evolve this equation into one with a single unknown so that we can choose that as our variable. The evolution of the equation requires that we use the substitution property to replace the unknowns with known numbers or with other expressions that, in turn, will lead to an equation with one unknown. We will begin this problem by substituting the known number.

(ACHILLES' DISTANCE TO RENDEZVOUS) = (TORTOISE'S DISTANCE TO RENDEZVOUS) + (HEAD START)
                                                                                ↓
(ACHILLES' DISTANCE TO RENDEZVOUS) = (TORTOISE'S DISTANCE TO RENDEZVOUS) + 100

There are still two unknowns, which we can change by using the following distance–rate–time formulas:

ACHILLES' DISTANCE = (ACHILLES' RATE)(TIME TO RENDEZVOUS)

TORTOISE'S DISTANCE = (TORTOISE'S RATE)(TIME TO RENDEZVOUS)

These values are now substituted into the equation:

(ACHILLES' DISTANCE TO RENDEZVOUS) = (TORTOISE'S DISTANCE TO RENDEZVOUS) + 100

(A's RATE)(TIME TO RENDEZVOUS) = (T's RATE)(TIME TO RENDEZVOUS) + 100

There are now three unknowns, but the values for two of *these* unknowns are given in the problem.

(A's RATE)(TIME TO RENDEZVOUS) = (T's RATE)(TIME TO RENDEZVOUS) + 100
        ↓   ↓
   10 (TIME TO RENDEZVOUS) =   1 (TIME TO RENDEZVOUS) + 100

The equation now has a single unknown, so let

t = TIME TO RENDEZVOUS

The last step in the evolution of this equation is to substitute the variable:

$10t = t + 100$

SOLVE.

$$9t = 100$$

$$t = \frac{100}{9}$$

ANSWER.   It takes Achilles $11\frac{1}{9}$ seconds to catch up with the tortoise.

Look Back.   Does this answer make sense? how long does it take a person to run 100 meters? The problem says Achilles runs at 10 m/sec, so at that rate it would take 10 seconds. The answer seems about right.   ◆

Pythagorean Relationships

Many word problems are concerned with relationships involving a right triangle. If two sides of a right triangle are known, the third can be found by using the Pythagorean theorem. Remember, if a right triangle has sides a and b and hypotenuse c, then

$$a^2 + b^2 = c^2$$

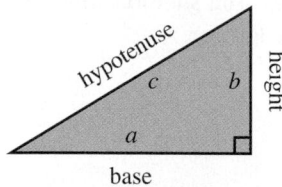

base

EXAMPLE 5

If the area of a right triangle is one square unit, and the height is two units longer than the base, find the lengths of the sides of the triangle to the nearest thousandth.

Solution   Draw a picture to help you understand the relationships.

TRANSLATE.

$$\text{AREA} = \frac{1}{2}\,(\text{BASE})(\text{HEIGHT})$$

EVOLVE.   The area is known, so begin by substituting 1 for AREA,

$$\text{AREA} = \frac{1}{2}\,(\text{BASE})(\text{HEIGHT})$$
$$\downarrow$$
$$1 \;\;= \frac{1}{2}\,(\text{BASE})(\text{HEIGHT})$$

Next, the problem tells us that HEIGHT = BASE + 2. Therefore,

$$1 = \frac{1}{2}\,(\text{BASE})(\text{HEIGHT})$$
$$\downarrow$$
$$1 = \frac{1}{2}\,(\text{BASE})(\text{BASE} + 2)$$

SOLVE.   There is now a single unknown, so we choose a variable for the unknown. Let $b = \text{BASE}$. Then

$$1 = \tfrac{1}{2}b(b + 2)$$
$$2 = b(b + 2)$$
$$2 = b^2 + 2b$$
$$0 = b^2 + 2b - 2$$

Apply the quadratic formula:

$$b = \frac{-2 \pm \sqrt{2^2 - 4(1)(-2)}}{2(1)}$$
$$= \frac{-2 \pm \sqrt{12}}{2}$$
$$= \frac{-2 \pm 2\sqrt{3}}{2}$$
$$= -1 \pm \sqrt{3}$$

Since the base cannot be negative, we have

$$\text{BASE} = -1 + \sqrt{3} \approx 0.7320508$$
$$\text{HEIGHT} = \text{BASE} + 2 = -1 + \sqrt{3} + 2 = 1 + \sqrt{3} \approx 2.7320508$$

The Pythagorean theorem is necessary for finding the third side. When finding the hypotenuse, be sure not to work with the approximate values. **You should round only once in a problem, and that is when you are finding your answer.**

$$(\text{HYPOTENUSE})^2 = (-1 + \sqrt{3})^2 + (1 + \sqrt{3})^2$$
$$= 1 - 2\sqrt{3} + 3 + 1 + 2\sqrt{3} + 3$$
$$= 8$$

It follows that

$$\text{HYPOTENUSE} = \pm\sqrt{8} = 2\sqrt{2} \approx 2.8284271$$

Positive value only, since hypotenuse represents a distance

ANSWER. State the answer to the specified number of decimal places. The lengths of the sides of the triangle are 0.732, 2.732, and 2.828. ◆

Have you ever wondered why sidewalks, pipes, or tracks have expansion joints every few feet? The next example may help you to understand why; it considers the unlikely situation in which 1-mile sections of pipe are connected together.

EXAMPLE 6   PÓLYA'S METHOD

A 1-mile-long pipeline connects two pumping stations. Special joints must be used along the line to provide for expansion and contraction due to changes in temperature. However, if the pipeline were actually one continuous length of pipe fixed at each end by the stations, then expansion would cause the pipe to bow. Approximately how high would the middle of the pipe rise if the expansion were just 1 inch over the mile?

Solution We use Pólya's problem-solving guidelines for this example.

Understand the Problem. First, understand the problem. Draw a picture as shown in Figure 5.8.

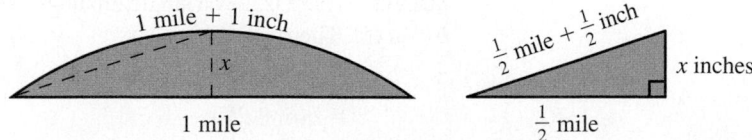

Figure 5.8 Pipeline problem

Before beginning the solution to this example, try to guess the answer. Consider the following choices for the rise in pipe:

- A. 1 inch
- B. 1 foot
- C. 1 yard
- D. 5 yards
- E. 1 mile

Go ahead, choose one of these. This is one problem for which the answer was not intuitive for the author. I guessed incorrectly when I first considered this problem.

Devise a Plan. For purposes of solution, notice from Figure 5.8 that we are assuming that the pipe bows in a circular arc. A triangle would produce a reasonable approximation since the distance x should be quite small compared to the total length. Since a right triangle is used to model this situation, the Pythagorean theorem may be used. The method we will use is to *translate, evolve, solve,* and *answer.*

Carry Out the Plan.

TRANSLATE.

$$(\text{SIDE})^2 + (\text{HEIGHT})^2 = (\text{HYPOTENUSE})^2$$

EVOLVE. The side is one-half the length of pipe, so it is 0.5 mile. Also, since the expansion is 1 inch, one-half the arc would have length 0.5 mile + 0.5 inch.

$$(\text{SIDE})^2 + (\text{HEIGHT})^2 = (\text{HYPOTENUSE})^2$$
$$\downarrow \qquad\qquad\qquad \downarrow$$
$$(0.5 \text{ mile})^2 + (\text{HEIGHT})^2 = (0.5 \text{ mile} + 0.5 \text{ in.})^2$$

There is a single unknown, so let $h = \text{HEIGHT}$. Also, notice that there is a mixture of units. Let us convert all measurements to inches. We know that

$$1 \text{ mile} = 5{,}280 \text{ ft} = 5{,}280(12 \text{ in.}) = 63{,}360 \text{ in.}$$
$$\tfrac{1}{2} \text{ mile} = 31{,}680 \text{ in.}$$

$$(0.5 \text{ mile})^2 + (\text{HEIGHT})^2 = (0.5 \text{ mile} + 0.5 \text{ in.})^2$$
$$\downarrow \qquad\qquad\qquad \downarrow$$
$$(31{,}680)^2 + h^2 = (31{,}680.5)^2$$

SOLVE.

$$h^2 = (31{,}680.5)^2 - (31{,}680)^2$$
$$h = \sqrt{(31{,}680.5)^2 - (31{,}680)^2}$$
$$\approx 177.99$$

ANSWER. The pipe rises about 180 in. in the middle.

Look Back. The solution, 177.99 in., is approximately 14.8 ft. This is an extraordinary result if you consider that the pipe expanded only *1 inch*. The pipe would bow approximately 14.8 ft at the middle. This answer does not seem correct, but in re-examining the steps, we see that it is! This answer is paradoxical or, at least, counterintuitive. ◆

Problem Set 5.6

LEVEL 1

1. **IN YOUR OWN WORDS** Outline a procedure for solving word problems in algebra.

2. **IN YOUR OWN WORDS** Explain what is meant by *translate, evolve, and solve.*

*The first step in the problem-solving procedure is NOT choosing a variable, but rather translating the main idea into symbols. We called this **devising a plan**. Carry out the first step ONLY for Problems 3–25.*

3. If you add seven to twice a number, the result is seventeen. What is the number?

4. If you subtract twelve from twice a number, the result is six. What is the number?

5. If you multiply a number by five and then subtract negative ten, the difference is negative thirty. What is the number?

6. If 6 is subtracted from three times a number, the difference is twice the number. What is the number?

7. Find two consecutive integers whose sum is 117.

8. Find two consecutive even integers whose sum is 94.

9. The sum of three consecutive integers is 105. What are the integers?

10. The sum of four consecutive integers is 74. What are the integers?

11. A house and a lot are appraised at $212,400. If the value of the house is five times the value of the lot, how much is the house worth?

12. A cabinet shop produces two types of custom-made cabinets. If the cost of one type of cabinet is four times the cost of the other, and the total price for one of each type of cabinet is $4,150, how much does each cabinet cost?

13. To stimulate his daughter in the pursuit of problem solving, a math professor offered to pay her $8 for every equation correctly solved and to fine her $5 for every incorrect solution. At the end of the first 26 problems of this problem set, neither owed any money to the other. How many problems did the daughter solve correctly?

14. A professional gambler reported that at the end of the first race at the track he had doubled his money. He bet $30 on the second race and tripled the money he came with. He bet $54 on the third race and quadrupled his original bankroll. He bet $72 on the fourth race and lost it, but still had $48 left. With how much money did he start?

15. A 10-ft pole is to be erected and held in the ground by four guy wires attached at the top. The guy wires are attached to the ground at a distance of 15 ft from the base of the pole. What is the exact length of each guy wire? How much wire should be purchased if it cannot be purchased in fractions of a foot?

16. A diagonal brace is to be placed in the wall of a room. The height of the wall is 8 ft, and the wall is 14 ft long. What is the exact length of the brace, and what is the length of the brace to the nearest foot?

17. In traveling from Jacksonville to Miami, you pass through Orlando and then through Palm Beach. It is 10 miles farther from Orlando to Palm Beach than it is from Jacksonville to Orlando. The distance between Jacksonville and Orlando is 90 miles more than the distance from Palm Beach to Miami. If it is 370 miles from Jacksonville to Miami, how far is it from Jacksonville to Orlando?

18. The drive from New Orleans to Memphis is 90 miles shorter than the drive from Memphis to Cincinnati, but 150 miles farther than the drive from Cincinnati to Detroit. If the total highway distance of a New Orleans–Memphis–Cincinnati–Detroit trip is 1,140 miles, find the length of the Cincinnati–Detroit leg of the trip.

19. Traveling from San Antonio to Dallas, you first pass through Austin and then Waco before reaching Dallas, a total distance of 280 miles. From Austin to Waco is 30 miles farther than from San Antonio to Austin, and also 20 miles farther than from Waco to Dallas. How far is it from Waco to Dallas?

20. Two persons are to run a race, but one can run 10 meters per second, whereas the other can run 6 meters per second. If the slower runner has a 50-meter head start, how long will it be before the faster runner catches the slower runner, if they begin at the same time?

21. If the rangefinder on the *Enterprise* shows a shuttlecraft 4,500 km away, how long will it take to catch the shuttle if the shuttle travels at 12,000 kph and the *Enterprise* is traveling at 15,000 kph?

22. A speeding car is traveling at 80 mph when a police car starts pursuit at 100 mph. How long will it take the police car to catch up to the speeding car? Assume that the speeding car has a 2-mile head start and that the cars travel at constant rates.

23. Two people walk daily for exercise. One is able to maintain 4.0 mph and the other only 3.5 mph. The slower walker has a mile head start when the other begins, yet they finish together. How far did each walk?

24. Two joggers set out at the same time from their homes 21 miles apart. They agree to meet at a point somewhere in between in an hour and a half. If the rate of one is 2 mph faster than the rate of other, find the rate of each.

25. Two joggers set out at the same time in opposite directions. If they were to maintain their normal rates for four hours, they would be 68 miles apart. If the rate of one is 1.5 mph faster than the rate of the other, find the rate of each.

LEVEL 2

Solve Problems 26–33. Because you are practicing a **procedure,** *you must show all of your work. You wrote down a verbal description for these problems in Problems 3–10.*

26. If you add seven to twice a number, the result is seventeen. What is the number?

27. If you subtract twelve from twice a number, the result is six. What is the number?

28. If you multiply a number by five and then subtract negative ten, the difference is negative thirty. What is the number?

29. If 6 is subtracted from three times a number, the difference is twice the number. What is the number?

30. Find two consecutive integers whose sum is 117.

31. Find two consecutive even integers whose sum is 94.

32. The sum of three consecutive integers is 105. What are the integers?

33. The sum of four consecutive integers is 74. What are the integers?

*Solve Problems 34–48. Because you are practicing a **procedure**, you must show all of your work. You wrote down a verbal description for these problems in Problems 11–25.*

34. A house and a lot are appraised at $212,400. If the value of the house is five times the value of the lot, how much is the house worth?

35. A cabinet shop produces two types of custom-made cabinets. If the cost of one type of cabinet is four times the cost of the other, and the total price for one of each type of cabinet is $4,150, how much does each cabinet cost?

36. To stimulate his daughter in the pursuit of problem solving, a math professor offered to pay her $8 for every equation correctly solved and to fine her $5 for every incorrect solution. At the end of the first 26 problems of this problem set, neither owed any money to the other. How many problems did the daughter solve correctly?

37. A professional gambler reported that at the end of the first race at the track he had doubled his money. He bet $30 on the second race and tripled the money he came with. He bet $54 on the third race and quadrupled his original bankroll. He bet $72 on the fourth race and lost it, but still had $48 left. With how much money did he start?

38. A 10-ft pole is to be erected and held in the ground by four guy wires attached at the top. The guy wires are attached to the ground at a distance of 15 ft from the base of the pole. What is the exact length of each guy wire? How much wire should be purchased if it cannot be purchased in fractions of a foot?

39. A diagonal brace is to be placed in the wall of a room. The height of the wall is 8 ft, and the wall is 14 ft long. What is the exact length of the brace, and what is the length of the brace to the nearest foot?

40. In traveling from Jacksonville to Miami, you pass through Orlando and then through Palm Beach. It is 10 miles farther from Orlando to Palm Beach than it is from Jacksonville to Orlando. The distance between Jacksonville and Orlando is 90 miles more than the distance from Palm Beach to Miami. If it is 370 miles from Jacksonville to Miami, how far is it from Jacksonville to Orlando?

41. The drive from New Orleans to Memphis is 90 miles shorter than the drive from Memphis to Cincinnati, but 150 miles farther than the drive from Cincinnati to Detroit.

If the total highway distance of a New Orleans–Memphis–Cincinnati–Detroit trip is 1,140 miles, find the length of the Cincinnati–Detroit leg of the trip.

42. Traveling from San Antonio to Dallas, you first pass through Austin and then Waco before reaching Dallas, a total distance of 280 miles. From Austin to Waco is 30 miles farther than from San Antonio to Austin, and also 20 miles farther than from Waco to Dallas. How far is it from Waco to Dallas?

43. Two persons are to run a race, but one can run 10 meters per second, whereas the other can run 6 meters per second. If the slower runner has a 50-meter head start, how long will it be before the faster runner catches the slower runner, if they begin at the same time?

44. If the rangefinder on the *Enterprise* shows a shuttlecraft 4,500 km away, how long will it take to catch the shuttle if the shuttle travels at 12,000 kph and the *Enterprise* is traveling at 15,000 kph?

45. A speeding car is traveling at 80 mph when a police car starts pursuit at 100 mph. How long will it take the police car to catch up to the speeding car? Assume that the speeding car has a 2-mile head start and that the cars travel at constant rates.

46. Two people walk daily for exercise. One is able to maintain 4.0 mph and the other only 3.5 mph. The slower walker has a mile head start when the other begins, yet they finish together. How far did each walk?

47. Two joggers set out at the same time from their homes 21 miles apart. They agree to meet at a point somewhere in between in an hour and a half. If the rate of one is 2 mph faster than the rate of other, find the rate of each.

48. Two joggers set out at the same time in opposite directions. If they were to maintain their normal rates for four hours, they would be 68 miles apart. If the rate of one is 1.5 mph faster than the rate of the other, find the rate of each.

LEVEL 3

49. The area of a right triangle is 17.5 cm^2. One leg is 2 cm longer than the other. What is the length of the shortest side?

50. The hypotenuse of a right triangle is 13.0, and one leg is 6.0 units shorter than the other. Find the dimensions of the figure (rounded to the nearest tenth).

51. Find the base and height of a triangle with area 3.0 square feet if its base is 2.0 feet (rounded to the nearest tenth) longer than its height.

52. Find the base and height of a triangle (rounded to the nearest tenth) with area 75.0 in.^{2} if its base is 10.0 in. longer than its height.

53. The annual rate, r, compounded annually, it takes for 1 dollar to grow to A dollars in 2 years is given by the formula $A = (1 + r)^2$. Find the rate necessary for a dollar to double in 2 years.

54. If P dollars is invested at an annual rate r compounded annually, at the end of 2 years it will have grown to an amount A according to the formula $A = P(1 + r)^2$. At what rate will \$1,000 grow to \$1,500 in 2 years?

PROBLEM SOLVING

55. A 1-mile-long pipeline connects two pumping stations. Special joints must be used along the line to provide for expansion and contraction due to changes in temperature. However, if the pipeline were actually one continuous length of pipe fixed at each end by the stations, then expansion would cause the pipe to bow. Approximately how high would the middle of the pipe rise if the expansion were just one-half inch over the mile?

56. **HISTORICAL QUESTION** (from Bhaskara, ca. A.D. 1120) "In a lake the bud of a water lily was observed, one cubit above the water, and when moved by the gentle breeze, it sunk in the water at two cubits' distance." Find the depth of the water.

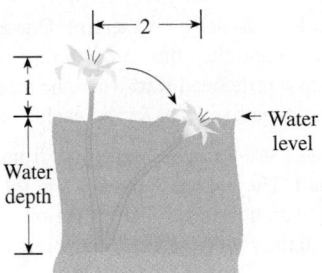

57. **HISTORICAL QUESTION** (from Bhaskara, ca. A.D. 1120) "One third of a collection of beautiful water lilies is offered to Mahadev, one-fifth to Huri, one-sixth to the Sun, one-fourth to Devi, and the six which remain are presented to the spiritual teacher." Find the total number of lilies.

58. **HISTORICAL QUESTION** (from Bhaskara, ca. A.D. 1120) "One-fifth of a hive of bees flew to the Kadamba flower; one-third flew to the Silandhara; three times the difference of these two numbers flew to an arbor, and one bee continued flying about, attracted on each side by the fragrant Keteki and the Malati." Find the number of bees.

59. **HISTORICAL QUESTION** (from Brahmagupta, ca. A.D. 630) "A tree one hundred cubits high is distant from a well two hundred cubits; from this tree one monkey climbs down the tree and goes to the well, but the other leaps in the air and descends by the hypotenuse from the high point of the leap, and both pass over an equal space." Find the height of the leap.

60. **HISTORICAL QUESTION** "Ten times the square root of a flock of geese, seeing the clouds collect, flew to the Manus lake; one-eighth of the whole flew from the edge of the water amongst a multitude of water lilies; and three couples were observed playing in the water." Find the number of geese.

5.7 | RATIOS, PROPORTIONS, AND PROBLEM SOLVING

Ratios

Ratios and proportions are powerful problem-solving tools in algebra. Ratios are a way of comparing two numbers or quantities—for example, the compression ratio of a car, the gear ratio of a transmission, the pitch of a roof, the steepness of a road, or a player's batting average. A *ratio* expresses a size relationship between two sets and is defined as the quotient of two numbers. It is written using the word *to,* a colon, or a fraction; that is, if the ratio of men to women is **5 to 4,** this could also be written as $5 : 4$ or $\frac{5}{4}$.

Ratio

$\dfrac{a}{b}$ is called the **ratio** of a to b. The two parts a and b are called its terms.

We will emphasize the idea that a ratio can be written as a fraction (or as a quotient of two numbers). Since a fraction can be reduced, a ratio can also be reduced.

EXAMPLE 1

Reduce the given ratios to lowest terms.

a. 4 to 52 **b.** 15 to 3 **c.** $1\frac{1}{2}$ to 2 **d.** $1\frac{2}{3}$ to $3\frac{3}{4}$

Solution

a. A ratio of 4 to 52

$$\frac{4}{52} = \frac{1}{13}$$

A ratio of 1 to 13

b. A ratio of 15 to 3

$$\frac{15}{3} = 5$$

Write this as $\frac{5}{1}$ because a ratio compares two numbers.
A ratio of 5 to 1

c. A ratio of $1\frac{1}{2}$ to 2

$$\frac{1\frac{1}{2}}{2} = 1\frac{1}{2} \div 2$$

$$= \frac{3}{2} \times \frac{1}{2}$$

$$= \frac{3}{4}$$

A ratio of 3 to 4

d. A ratio of $1\frac{2}{3}$ to $3\frac{3}{4}$

$$\frac{1\frac{2}{3}}{3\frac{3}{4}} = 1\frac{2}{3} \div 3\frac{3}{4}$$

$$= \frac{5}{3} \div \frac{15}{4}$$

$$= \frac{5}{3} \times \frac{4}{15}$$

$$= \frac{4}{9} \quad \textbf{A ratio of 4 to 9} \quad \blacklozenge$$

Proportions

A **proportion** is a statement of equality between ratios. In symbols,

$$\frac{a}{b} \qquad = \qquad \frac{c}{d}$$
$$\uparrow \qquad \uparrow \qquad \uparrow$$
"a is to b" **"as"** **"c is to d"**

The notation used in some books is $a : b :: c : d$. Even though we won't use this notation, we will use words associated with this notation to name the terms:

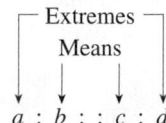

$$a : b :: c : d$$

In the more common fractional notation, we have

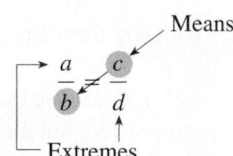

EXAMPLE 2

Read each proportion, and name the means and the extremes.

a. $\dfrac{2}{3} = \dfrac{10}{15}$ **b.** $\dfrac{m}{5} = \dfrac{3}{8}$

Solution

a. $\dfrac{2}{3} = \dfrac{10}{15}$ *Read:*   Two is to three as ten is to fifteen.

   Means: 3 and 10   *Extremes:*   2 and 15

b. $\dfrac{m}{5} = \dfrac{3}{8}$ *Read:* m is to five as three is to eight.

   Means: 5 and 3   *Extremes:* m and 8   ◆

The following property is fundamental to our study of proportions and percents.

Property of Proportions

If the product of the means equals the product of the extremes, then the ratios form a proportion.

Also, if the ratios form a proportion, then the product of the means equals the product of the extremes.

In symbols,

$$\frac{a}{b} = \frac{c}{d}$$

$$\underbrace{b \times c}_{\uparrow} = \underbrace{a \times d}_{\uparrow}$$

product of means = product of extremes

EXAMPLE 3

Tell whether each pair of ratios forms a proportion.

a. $\dfrac{3}{4}, \dfrac{36}{48}$ **b.** $\dfrac{5}{16}, \dfrac{7}{22}$

Solution

a. *Means*   *Extremes*   **b.** *Means*   *Extremes*

   4×36 3×48 16×7 5×22

          $144 = 144$ $112 \neq 110$

   Thus,   Thus,

   $$\frac{3}{4} = \frac{36}{48}$$ $$\frac{5}{16} \neq \frac{7}{22}$$

   They form a proportion.   They do not form a proportion.   ◆

You can use the cross-product method not only to see whether two fractions form a proportion, but also to compare the sizes of two fractions. For example, if a and c are whole numbers, and b and d are counting numbers, then the following property can be used to compare the sizes of two fractions:

If $ad = bc$, then $\dfrac{a}{b} = \dfrac{c}{d}$.

If $ad < bc$, then $\dfrac{a}{b} < \dfrac{c}{d}$.

If $ad > bc$, then $\dfrac{a}{b} > \dfrac{c}{d}$.

EXAMPLE 4

Insert $=$, $<$, or $>$ as appropriate.

a. $\dfrac{3}{4}$ ___ $\dfrac{18}{24}$ **b.** $\dfrac{1}{2}$ ___ $\dfrac{1}{3}$ **c.** $\dfrac{2}{3}$ ___ $\dfrac{7}{8}$ **d.** $\dfrac{3}{5}$ ___ $\dfrac{4}{7}$

Solution

a. $\dfrac{3}{4}$ ___ $\dfrac{18}{24}$ **b.** $\dfrac{1}{2}$ ___ $\dfrac{1}{3}$ **c.** $\dfrac{2}{3}$ ___ $\dfrac{7}{8}$ **d.** $\dfrac{3}{5}$ ___ $\dfrac{4}{7}$

    $3(24)$? $4(18)$ $1(3)$? $2(1)$ $2(8)$? $3(7)$ $3(7)$? $5(4)$

       72 ? 72 3 ? 2 16 ? 21 21 ? 20

           $=$ $>$ $<$ $>$ ◆

Comparing decimal numbers can sometimes be confusing; for example, which is larger,

$$0.6 \quad \text{or} \quad 0.58921?$$

The larger number is 0.6. To see this, simply write each decimal with the same number of places by affixing trailing zeros, and then compare. That is, write

    0.60000
    0.58921

Now, it is easy to see that $0.60000 > 0.58921$ because 60,000 hundred thousandths is larger than 58,921 hundred thousandths.

EXAMPLE 5

Insert $<$ or $>$ as appropriate.

a. 0.28 ___ 0.3 **b.** 0.001 ___ 0.01

c. 0.005 ___ 0.00482 **d.** 3 ___ 0.98712

Solution

a. 0.28 ___ 0.3 **b.** 0.001 ___ 0.01

    $0.28 < 0.30$ since $28 < 30$ $0.001 < 0.010$ since $1 < 10$

c. 0.005 ___ 0.00482 **d.** 3 ___ 0.98712

    $0.00500 > 0.00482$ $3.00000 > 0.98712$ ◆

Solving Proportions

The usual setting for a proportion problem is that three of the terms of the proportion are known and one of the terms is unknown. It is always possible to find the missing term by solving an equation. However, when given a proportion, we first use the property of proportions (which is equivalent to multiplying both sides of the equation by the same number).

EXAMPLE 6

Find the missing term of each proportion.

a. $\dfrac{3}{4} = \dfrac{w}{20}$ **b.** $\dfrac{3}{4} = \dfrac{27}{y}$ **c.** $\dfrac{2}{x} = \dfrac{8}{9}$ **d.** $\dfrac{t}{15} = \dfrac{3}{5}$

Solution **a.** $\dfrac{3}{4} = \dfrac{w}{20}$

PRODUCT OF MEANS $=$ PRODUCT OF EXTREMES

$$4w = 3(20)$$

$$w = \dfrac{3(\overset{5}{\cancel{20}})}{\underset{1}{\cancel{4}}}$$

Solve the equation by dividing both sides by 4. Notice that 4 is the number opposite the unknown:
$$\dfrac{3}{4} = \dfrac{w}{20}$$

$$w = 15$$

b. $\dfrac{3}{4} = \dfrac{27}{y}$

PRODUCT OF MEANS $=$ PRODUCT OF EXTREMES

$$4(27) = 3y$$

$$\dfrac{4(\overset{9}{\cancel{27}})}{\underset{1}{\cancel{3}}} = y$$

Divide both sides by 3; notice that 3 is the number opposite the unknown:
$$\dfrac{3}{4} = \dfrac{27}{y}$$

$$36 = y$$

c. $\dfrac{2}{x} = \dfrac{8}{9}$

PRODUCT OF MEANS $=$ PRODUCT OF EXTREMES

$$8x = 2(9)$$

$$x = \dfrac{\overset{1}{\cancel{2}}(9)}{\underset{4}{\cancel{8}}}$$

Divide both sides by 8; notice that 8 is the number opposite the unknown:
$$\dfrac{2}{x} = \dfrac{8}{9}$$

$$x = \dfrac{9}{4}$$

d. $\dfrac{t}{15} = \dfrac{3}{5}$

PRODUCT OF MEANS $=$ PRODUCT OF EXTREMES

$$3(15) = 5t$$

$$\dfrac{3(\overset{3}{\cancel{15}})}{\underset{1}{\cancel{5}}} = t$$

Divide both sides by 5; notice that 5 is the number opposite the unknown:
$$\dfrac{t}{15} = \dfrac{3}{5}$$

$$9 = t \qquad \blacklozenge$$

Notice that the unknown term can be in any one of four positions, as illustrated by the four parts of Example 6. But even though you can find the missing term of a proportion (called **solving the proportion**) by the technique used in Example 6, it is easier to think in terms of **the cross-product divided by the number opposite the unknown.** This method is easier than actually solving the equation because it can be done quickly using a calculator, as shown in the following examples.

Procedure for Solving Proportions

> 1. Find the product of the means or the product of the extremes, whichever does not contain the unknown term.
> 2. Divide this product by the number that is opposite the unknown term.

EXAMPLE 7

Solve the proportion for the unknown term.

a. $\dfrac{5}{6} = \dfrac{55}{y}$ **b.** $\dfrac{5}{b} = \dfrac{3}{4}$ **c.** $\dfrac{2\frac{1}{2}}{5} = \dfrac{a}{8}$

Solution

a. $\dfrac{5}{6} = \dfrac{55}{y}$

$y = \dfrac{6 \times 55}{5}$ ← Product of the means
← Number opposite the unknown

$= \dfrac{6 \times \overset{11}{\cancel{55}}}{\underset{1}{\cancel{5}}}$ You can cancel to simplify many of these problems.

$= 66$ $\boxed{6}\boxed{\times}\boxed{55}\boxed{\div}\boxed{5}\boxed{=}$ 66

Proportions are quite easy to solve if you have a calculator. You can multiply the cross-terms and divide by the number opposite the variable all in one calculator sequence.

b. $\dfrac{5}{b} = \dfrac{3}{4}$

$b = \dfrac{5 \times 4}{3}$ ← Product of the extremes
← Number opposite the unknown

$= \dfrac{20}{3}$ or $6\frac{2}{3}$ $\boxed{5}\boxed{\times}\boxed{4}\boxed{\div}\boxed{3}\boxed{=}$
Display: 6.66666667   Interpret this as $6\frac{2}{3}$.

Notice that your answers don't have to be whole numbers. This means that the correct proportion is

$$\dfrac{5}{6\frac{2}{3}} = \dfrac{3}{4}$$

c. $\dfrac{2\frac{1}{2}}{5} = \dfrac{a}{8}$

$a = \dfrac{2\frac{1}{2} \times 8}{5}$ ← Product of the extremes
← Number opposite the unknown

$= \dfrac{\frac{5}{2} \times \frac{8}{1}}{5}$

$= \dfrac{20}{5}$

$= 4$

◆

Many applied problems can be solved using a proportion. Whenever you are working an applied problem, you should estimate an answer so that you will know whether the result you obtain is reasonable.

When setting up a proportion with units, be sure that like units occupy corresponding positions, as illustrated in Examples 8–11. The proportion is obtained by applying the sentence "a is to b as c is to d" to the quantities in the problem.

EXAMPLE 8   PÓLYA'S METHOD

If 4 cans of cola sell for $1.89, how much will 6 cans cost?

Solution We use Pólya's problem-solving guidelines for this example.

Understand the Problem. We see that 4 cans sell for $1.89 and 8 cans sell for $2($1.89) = $3.78. We need to find the cost for 6 cans, which must be somewhere between $1.89 and $3.78.

Devise a Plan. There are many possible plans for solving this problem. We will form a proportion.

Carry Out the Plan. Solve "4 cans is to $1.89 as 6 cans is to what?"

$$\frac{4 \text{ cans}}{1.89 \text{ dollars}} = \frac{6 \text{ cans}}{x \text{ dollars}}$$

$$x = \frac{1.89 \times 6}{4}$$

$$= 2.835$$

Look Back. We see that 6 cans will cost $2.84.   ◆

EXAMPLE 9

If a 120-mile trip took $8\frac{1}{2}$ gallons of gas, how much gas is needed for a 240-mile trip?

Solution "120 miles is to $8\frac{1}{2}$ gallons as 240 miles is to how many gallons?"

$$\frac{120}{8\frac{1}{2}} = \frac{240}{x}$$

miles ↓   miles ↓

↑ gallons   ↑ gallons

$$x = \frac{8\frac{1}{2} \times \overset{2}{\cancel{240}}}{\underset{1}{\cancel{120}}} = 17$$

The trip will require 17 gallons.   ◆

EXAMPLE 10

If the property tax on a $375,000 home is $1,910, what is the tax on a $450,000 home?

Solution   "$375,000 is to $1,910 as $450,000 is to what?"

$$\overset{\overset{\text{value}}{\downarrow}}{\underset{\underset{\text{tax}}{\uparrow}}{\frac{375,000}{1,910}}} = \overset{\overset{\text{value}}{\downarrow}}{\underset{\underset{\text{tax}}{\uparrow}}{\frac{450,000}{x}}}$$

$$x = \frac{1,910 \times 450,000}{375,000}$$

You can do the arithmetic by canceling or on a calculator:

$$x = \frac{1,910 \times \overset{\overset{18}{\cancel{450}}}{\cancel{450,000}}}{\underset{\underset{15}{\cancel{375}}}{\cancel{375,000}}}$$

$$= 2,292$$

The tax is $2,292.   ◆

EXAMPLE 11

In a can of mixed nuts, the ratio of cashews to peanuts is 1 to 6. If a given machine releases 46 cashews into a can, how many peanuts should be released?

Solution   First, understand the problem. This problem is comparing two quantities, cashews and peanuts, so consider writing a proportion.

Translate.
$$\frac{\text{NUMBER OF CASHEWS}}{\text{NUMBER OF PEANUTS}} = \overset{\overset{\text{This is the given ratio.}}{\downarrow}}{\frac{1}{6}}$$

Evolve.   The NUMBER OF CASHEWS = 46, so by substitution

$$\frac{46}{\text{NUMBER OF PEANUTS}} = \frac{1}{6}$$

Let p = NUMBER OF PEANUTS; then the equation is

$$\frac{46}{p} = \frac{1}{6}$$

Solve.

$$6(46) = p$$
$$276 = p$$

Answer.   Thus, 276 peanuts should be released.   ◆

Problem Set 5.7

LEVEL 1

1. **IN YOUR OWN WORDS** What do we mean by ratios and proportions?

2. **IN YOUR OWN WORDS** How does the property of proportions relate to solving equations?

3. **IN YOUR OWN WORDS** How does the procedure for solving proportions relate to solving equations?

4. **IN YOUR OWN WORDS** Describe a process for setting up a proportion, given an applied problem.

Write the ratios given in Problems 5–10 as simplified ratios.

5. A cement mixture calls for 60 pounds of cement for 3 gallons of water. What is the ratio of cement to water?

6. What is the ratio of water to cement in Problem 5?

7. About 106 baby boys are born for every 100 baby girls. Write this as a simplified ratio of males to females.

8. What is the ratio of girls to boys in Problem 7?

9. If you drive 279 miles on $15\frac{1}{2}$ gallons of gas, what is the simplified ratio of miles to gallons?

10. If you drive 151.7 miles on 8.2 gallons of gas, what is the simplified ratio of miles to gallons?

Tell whether each pair of ratios in Problems 11–14 forms a proportion.

11. **a.** $\dfrac{7}{1}, \dfrac{21}{3}$ **b.** $\dfrac{6}{8}, \dfrac{9}{12}$ **c.** $\dfrac{3}{6}, \dfrac{5}{10}$

12. **a.** $\dfrac{85}{18}, \dfrac{42}{9}$ **b.** $\dfrac{403}{341}, \dfrac{13}{11}$ **c.** $\dfrac{20}{70}, \dfrac{4}{14}$

13. **a.** $\dfrac{3}{4}, \dfrac{75}{100}$ **b.** $\dfrac{2}{3}, \dfrac{67}{100}$ **c.** $\dfrac{5}{3}, \dfrac{7\frac{1}{3}}{4}$

14. **a.** $\dfrac{3}{2}, \dfrac{5}{3\frac{1}{2}}$ **b.** $\dfrac{5\frac{1}{5}}{7}, \dfrac{4}{5}$ **c.** $\dfrac{1}{3}, \dfrac{33\frac{1}{3}}{100}$

Insert =, <, or > as appropriate in Problems 15–18.

15. **a.** $\dfrac{1}{6}$ —— $\dfrac{1}{8}$ **b.** $\dfrac{1}{4}$ —— $\dfrac{1}{3}$ **c.** $\dfrac{1}{5}$ —— $\dfrac{1}{8}$

16. **a.** $\dfrac{25}{5}$ —— $\dfrac{10}{2}$ **b.** $\dfrac{14}{15}$ —— $\dfrac{42}{45}$ **c.** $\dfrac{11}{16}$ —— $\dfrac{7}{12}$

17. **a.** 0.8 —— 0.8001 **b.** 0.8 —— 0.7999
    c. 8 —— 2.81 **d.** 2.8 —— 2.88

18. **a.** π —— 3.1416 **b.** $\sqrt{2}$ —— 1.4142
    c. $\sqrt{3}$ —— 1.7320508 **d.** $\sqrt{4}$ —— 2.0000

Solve each proportion in Problems 19–42 for the item represented by a letter.

19. $\dfrac{5}{1} = \dfrac{A}{6}$ 20. $\dfrac{1}{9} = \dfrac{4}{B}$ 21. $\dfrac{C}{2} = \dfrac{5}{1}$

22. $\dfrac{7}{D} = \dfrac{1}{8}$ 23. $\dfrac{12}{18} = \dfrac{E}{12}$ 24. $\dfrac{12}{15} = \dfrac{20}{F}$

25. $\dfrac{G}{24} = \dfrac{14}{16}$ 26. $\dfrac{4}{H} = \dfrac{3}{15}$ 27. $\dfrac{2}{3} = \dfrac{I}{24}$

28. $\dfrac{4}{5} = \dfrac{3}{J}$ 29. $\dfrac{3}{K} = \dfrac{2}{5}$ 30. $\dfrac{L}{18} = \dfrac{5}{6}$

31. $\dfrac{7\frac{1}{5}}{9} = \dfrac{M}{5}$ 32. $\dfrac{4}{2\frac{2}{3}} = \dfrac{3}{N}$ 33. $\dfrac{P}{4} = \dfrac{4\frac{1}{2}}{6}$

34. $\dfrac{5}{2} = \dfrac{Q}{12\frac{3}{5}}$ 35. $\dfrac{5}{R} = \dfrac{7}{12\frac{3}{5}}$ 36. $\dfrac{1\frac{1}{3}}{\frac{1}{9}} = \dfrac{S}{2\frac{2}{3}}$

37. $\dfrac{33}{2\frac{1}{5}} = \dfrac{3\frac{3}{4}}{T}$ 38. $\dfrac{U}{1\frac{1}{2}} = \dfrac{\frac{1}{2}}{\frac{3}{4}}$ 39. $\dfrac{\frac{1}{5}}{\frac{2}{3}} = \dfrac{\frac{3}{4}}{V}$

40. $\dfrac{\frac{3}{5}}{\frac{1}{2}} = \dfrac{X}{\frac{2}{3}}$ 41. $\dfrac{9}{Y} = \dfrac{1\frac{1}{2}}{3\frac{2}{3}}$ 42. $\dfrac{Z}{2\frac{1}{3}} = \dfrac{1\frac{1}{2}}{4\frac{1}{5}}$

LEVEL 2

43. If 4 melons sell for $0.52, how much would 7 melons cost?

44. If a 121-mile trip took $5\frac{1}{2}$ gallons of gas, how many miles can be driven with a full tank of 13 gallons?

45. If a family uses $3\frac{1}{2}$ gallons of milk per week, how much milk will this family need for four days?

46. If 2 quarts of paint are needed for 75 ft of fence, how many quarts are needed for 900 ft of fence?

47. If Jack jogs 3 miles in 40 minutes, how long will it take him (to the nearest minute) to jog 2 miles at the same rate?

48. If Jill jogs 2 miles in 15 minutes, how long will it take her (to the nearest minute) to jog 5 miles at the same rate?

49. A moderately active 140-pound person will use 2,100 calories per day to maintain that body weight. How many calories per day are necessary to maintain a moderately active 165-pound person?

50. This problem will help you check your work in Problems 19–42. Fill in the capital letters from Problems 19–42 to correspond with their numerical values in the boxes. For example, if

$$\frac{W}{7} = \frac{10}{14} \quad \text{then} \quad W = \frac{7 \times 10}{14} = 5$$

Now find the box or boxes with number 5 in the corner and fill in the letter *W*. This has already been done for you. (The letter O has also been filled in for you.) Some letters may not appear in the boxes. When you are finished filling in the letters, darken all the blank spaces to separate the words in the secret message. Notice that one of the blank spaces has also been filled in to help you.

24	5	$\frac{1}{4}$	20	16	32		3	9		36	15	8	4	$\frac{2}{3}$	7
									O						
5	16	15	15	12	1	2	56		1	36	$\frac{1}{4}$	8	56	15	22
W								O							
13	30	32	32	16	32	$\frac{1}{4}$	7	22		1	12	$\frac{1}{2}$	16	2	24
									O						
25	16	2	56	16	2	21	6	8	9	9		9	32	!	!
											O				

51. If $\dfrac{V}{T} = \dfrac{V'}{T'}$, find V' when $V = 175$, $T = 300$, and $T' = 273$.

52. If $\dfrac{PV}{T} = \dfrac{P'V'}{T'}$, find V' when $V = 12$, $P = 2$, $T = 300$, $P' = 8$, and $T' = 400$.

53. The *pitch* of a roof is the ratio of the rise to the half-span. If a roof has a rise of 8 feet and a span of 24 feet, what is the pitch?

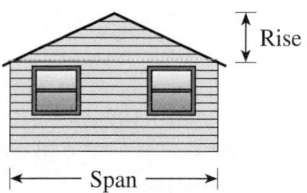

54. What is the pitch of a roof (see Problem 53) with a 3-foot rise and a span of 12 feet?

55. You've probably seen advertisements for posters that can be made from any photograph. If the finished poster will be 2 ft by 3 ft, it's likely that part of your original snapshot will be cut off. Suppose that you send in a photo that measures 3 in. by 5 in. If the shorter side of the enlargement will be 2 ft, what size should the longer side of the enlargement be so that the entire snapshot is shown in the poster?

56. Suppose you wish to make a scale drawing of your living room, which measures 18 ft by 25 ft. If the shorter side of the drawing is 6 in., how long is the longer side of the scale drawing?

57. If the property tax on a $180,000 home is $1,080, what is the tax on a $130,000 home?

58. To dilute a medication you can use the following formula:

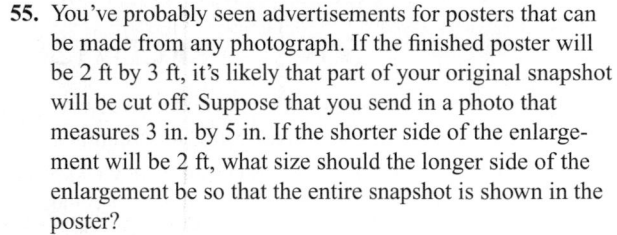

$$\frac{\text{PERCENT OF DILUTE SOLUTION}}{\text{PERCENT OF ORIGINAL SOLUTION}} = \frac{\text{AMOUNT OF STRONG SOLUTION NEEDED}}{\text{AMOUNT OF DILUTED SOLUTION WANTED}}$$

How many units of a 10% solution are needed to prepare 500 units of a 2% solution?

59. At a certain hamburger stand, the owner sold soft drinks out of two 16-gallon barrels. At the end of the first day, she wished to increase her profit, so she filled the soft-drink barrels with water, thus diluting the drink served. She repeated the procedure at the end of the second and third days. At the end of the fourth day, she had 10 gallons remaining in the barrels, but they contained only 1 pint of pure soft drink. How much pure soft drink was served in the four days?

60. Answer the question in the following *Peanuts* cartoon strip.

Peanuts © 1979. Reprinted by permission of United Feature Syndicate, Inc.

5.8 | PERCENTS

We frequently use percents, and you have, no doubt, worked with percents in your previous mathematics classes. However, we include this section to review percents and the equivalence of the decimal, fraction, and percent forms.

Percent

> **Percent** is the ratio of a given number to 100. This means that a percent is the numerator of a fraction whose denominator is 100.

Since a percent is a ratio, percents can easily be written in fractional form.

EXAMPLE 1

Write the following percents as simplified fractions.

a. "Sale 75% OFF"   **b.** "SALARIES UP 6.8%"

Solution

a. 75% means a "ratio of 75 to 100": $\dfrac{75}{100} = \dfrac{3}{4}$

b. 6.8% means a "ratio of 6.8 to 100":

$$6.8 \div 100 = 6\tfrac{8}{10} \div 100$$
$$= 6\tfrac{4}{5} \div 100$$
$$= \dfrac{\overset{17}{\cancel{34}}}{5} \times \dfrac{1}{\underset{50}{\cancel{100}}}$$
$$= \dfrac{17}{250}$$

◆

Fractions/Decimals/Percents

Percents can also be written as decimals. Since a percent is a ratio of a number to 100, we can divide by 100 by moving the decimal point two places.

Procedure for Changing a Percent to a Decimal

> To express a percent as a decimal, shift the decimal point two places to the *left* and delete the % symbol. If the percent involves a fraction, write the fraction as a decimal; *then* shift the decimal point.

EXAMPLE 2

Write each percent in decimal form.

a. 8 percent   **b.** 6.8%   **c.** $33\tfrac{1}{3}\%$ **d.** $\tfrac{1}{2}\%$

Solution

a. 8%

↑_____ If a decimal point is not shown, it is always
understood to be at the right of the whole number.

0.08 %

↑_↓

Shift the decimal point two places to the left;
add zeros as placeholders, if necessary. Delete percent symbol.

Answer: 8% = 0.08

b. 6.8%

Think: 6.8%

Shift two places, add placeholders as necessary, and delete percent symbol.

Answer: 6.8% = 0.068

c. $33\frac{1}{3}\%$

Think: $33\frac{1}{3}\%$

← Decimal point is understood.

Answer: $33\frac{1}{3}\% = 0.33\frac{1}{3}$ or 0.333 . . .

d. $\frac{1}{2}\%$

$\frac{1}{2} = 0.5$, so $\frac{1}{2}\% = 0.5\%$ *Think:* 00.5%

Answer: 0.5% = 0.005

♦

As you can see from the examples, every number can be written in three forms: fraction, decimal, and percent. Even though we discussed changing from fraction to decimal form earlier in the text, we'll review the three forms in this section. The procedure for changing from one form to another is given in Table 5.1.

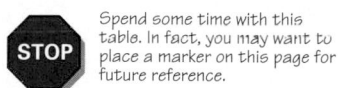

Spend some time with this table. In fact, you may want to place a marker on this page for future reference.

Table 5.1 Fraction/Decimal/Percent Conversion Chart

From \ To	Fraction	Decimal	Percent
Fraction		Divide the numerator (top) by the denominator (bottom). Write as a terminating or as a repeating decimal (bar notation).	First change the fraction to a decimal by carrying out the division to two decimal places and writing the remainder as a fraction. *Then* move the decimal point two places to the right, and affix a percent symbol.
Terminating decimal	Write the decimal without the decimal point, and multiply by the decimal name of the last digit (rightmost digit).		Shift the decimal point two places to the *right,* and affix a percent symbol.
Percent	Write as a ratio to 100 and reduce the fraction. If the percent involves a decimal, first write the decimal in fractional form, and then multiply by $\frac{1}{100}$. If the percent involves a fraction, delete the percent symbol and multiply by $\frac{1}{100}$.	Shift the decimal point two places to the *left,* and delete the percent symbol. If the percent involves a fraction, first write the fraction as a decimal, and then shift the decimal point.	

EXAMPLE 3

Write each fraction in percent form.

a. $\frac{5}{8}$ **b.** $\frac{5}{6}$

Solution

a. Look under the fraction heading in Table 5.1, and follow the directions for changing a fraction to a percent.

Step 1. $\frac{5}{8} = 0.625$

Step 2. $0.625 = 62.5\%$

Two places; add zeros as placeholders as necessary; add percent symbol.

Answer: 62.5%

$$\begin{array}{r} 0.625 \\ 8\overline{)5.000} \\ \underline{48} \\ 20 \\ \underline{16} \\ 40 \\ \underline{40} \\ 0 \end{array}$$

b. *Step 1.* $\frac{5}{6} = 0.83\frac{1}{3}$

This can also be written as $0.833\ldots$ or $0.8\bar{3}$.

Step 2. $0.83\frac{1}{3} = 83\frac{1}{3}\%$

The decimal point is understood.

Answer: $83\frac{1}{3}\%$

$$\begin{array}{r} 0.83 \\ 6\overline{)5.000} \\ \underline{48} \\ 20 \\ \underline{18} \\ 2 \end{array}$$

Carry out the division two places and save the remainder as a fraction: $\frac{2}{6} = \frac{1}{3}$

← Remainder

EXAMPLE 4

Write decimal forms as percents and fractions.

a. 0.85   **b.** 2.485

Solution

a. *Step 1.* $0.85 = 85\%$

Two places   Decimal understood

Step 2. 85% means $\frac{85}{100} = \frac{17}{20}$

b. *Step 1.* $2.485 = 248.5\%$

Step 2. 248.5% means $248.5 \times \dfrac{1}{100} = 248\frac{1}{2} \times \dfrac{1}{100}$

$$= \frac{497}{2} \times \frac{1}{100}$$

$$= \frac{497}{200} \quad \text{or} \quad 2\frac{97}{200}$$

Estimation

Percent problems are very common. We conclude this section with a discussion of estimation and some percent calculations.

The first estimation method is the *unit fraction-conversion method,* which can be used to estimate the common percents of 10%, 25%, $33\frac{1}{3}\%$, and 50%. To estimate the

Unit fraction comparison

Percent	Fraction
10%	$\frac{1}{10}$
25%	$\frac{1}{4}$
$33\frac{1}{3}$%	$\frac{1}{3}$
50%	$\frac{1}{2}$

size of a part of a whole quantity, which is sometimes called a **percentage,** rewrite the percent as a fraction and mentally multiply, as shown by the following example.

$$50\% \text{ of } 800: \quad \tfrac{1}{2} \times 800 = 400; \quad \textit{THINK: } 800 \div 2 = 400$$

$$25\% \text{ of } 1{,}200: \quad \tfrac{1}{4} \times 1{,}200 = 300; \quad \textit{THINK: } 1{,}200 \div 4 = 300$$

$$33\tfrac{1}{3}\% \text{ of } 600: \quad \tfrac{1}{3} \times 600 = 200; \quad \textit{THINK: } 600 \div 3 = 200$$

$$10\% \text{ of } 824: \quad \tfrac{1}{10} \times 824 = 82.4; \quad \textit{THINK: } 824 \div 10 = 82.4$$

(Move the decimal point one place to the left.)

If the numbers for which you are finding a percentage are not as "nice" as those given here, you can estimate by rounding the number, as shown in Example 5.

EXAMPLE 5

Estimate the following percentages:

a. 25% of 312 **b.** 50% of 843 **c.** $33\frac{1}{3}$% of 1,856 **d.** 25% of 43,350

Solution

a. Estimate 25% of 312 by rounding 312 so that it is easily divisible by 4:

$$\tfrac{1}{4} \times 320 = 80 \quad \text{Find } 320 \div 4 = 80.$$

b. Estimate 50% of 843 by rounding 843 so that it is easily divisible by 2:

$$\tfrac{1}{2} \times 840 = 420 \quad \text{Find } 840 \div 2 = 420.$$

c. Estimate $33\frac{1}{3}$% of 1,856 by rounding 1,856 so that it is *easily* divisible by 3:

$$\tfrac{1}{3} \times 1{,}800 = 600 \quad \text{Find } 1{,}800 \div 3 = 600$$

d. Estimate 25% of 43,350 by rounding 43,350 so that is is *easily* divisible by 4:

$$\tfrac{1}{4} \times 44{,}000 = 11{,}000 \qquad\qquad\qquad\qquad \blacklozenge$$

A second estimation procedure uses a multiple of a unit fraction. For example,

Think of 75% as $\frac{3}{4}$, which is $3 \times \frac{1}{4}$.

Think of $66\frac{2}{3}$% as $\frac{2}{3}$, which is $2 \times \frac{1}{3}$.

Think of 60% as $\frac{6}{10}$, which is $6 \times \frac{1}{10}$.

EXAMPLE 6

Estimate the following percentages.

a. 75% of 943 **b.** $66\frac{2}{3}$% of 8,932 **c.** 60% of 954 **d.** 80% of 0.983

Solution

a. 75% of 943 $\approx \frac{3}{4} \times 1{,}000 = 3(\frac{1}{4} \times 1{,}000) = 3(250) = 750$

b. $66\frac{2}{3}$% of 8,932 $\approx \frac{2}{3} \times 9{,}000 = 2(\frac{1}{3} \times 9{,}000) = 2(3{,}000) = 6{,}000$

c. 60% of 954 $\approx \frac{6}{10} \times 1{,}000 = 6(\frac{1}{10} \times 1{,}000) = 6(100) = 600$

d. 80% of 0.983 $\approx \frac{8}{10} \times 1 = 0.8 \qquad\qquad\qquad\qquad \blacklozenge$

The Percent Problem

Many percentage problems are more difficult than these. The following quotation was found in a recent publication: "An elected official is one who gets 51 percent of the vote cast by 40 percent of the 60 percent of voters who registered." Certainly, most of us will have trouble understanding the percent given in this quotation; but you can't pick up a newspaper without seeing dozens of examples of ideas that require some understanding of percents. A difficult job for most of us is knowing whether to multiply or divide by the given numbers. In this section, I will provide you with a sure-fire method for knowing what to do. The first step is to understand what is meant by **the percent problem.**

The Percent Problem

Study this percent problem. If you learn this, you get a written guarantee for correctly working percent problems.

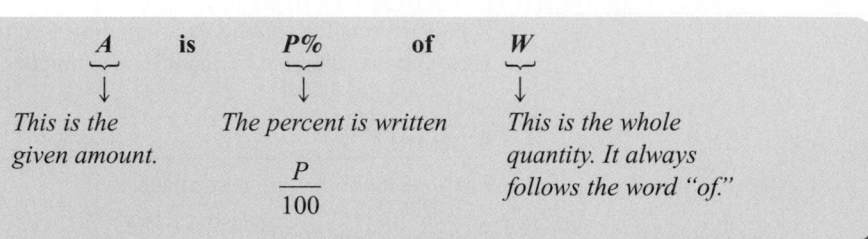

A	is	$P\%$	of	W
↓		↓		↓
This is the given amount.		*The percent is written* $\dfrac{P}{100}$		*This is the whole quantity. It always follows the word "of."*

The percent problem won't always be stated in this form, but notice that three quantities are associated with it:

1. The *amount*—sometimes called the **percentage**
2. The *percent*—sometimes called the **rate**
3. The *whole quantity*—sometimes called the **base**

Read these three steps— SLOWLY!

Now, regardless of the form in which you are given the percent problem, follow these steps to write a proportion:

1. Identify the *percent* first; it will be followed by the symbol % or the word *percent*. Write it as a fraction:

$$\frac{P}{100}$$

2. Identify the *whole quantity* next; it is preceded by the word *of*. It is the denominator of the second fraction in the proportion:

$$\frac{P}{100} = \frac{}{W} \quad \leftarrow \text{This is the quantity following the word "of."}$$

3. The remaining number is the partial amount; it is the numerator of the second fraction in the proportion:

$$\frac{P}{100} = \frac{A}{W} \quad \leftarrow \text{This is the last quantity to be inserted into the proportion.}$$

EXAMPLE 7

For each of the following cases, identify the percent, the whole quantity, and the amount (the percentage or part), and then write a proportion.

a. What number is 18% of 200?   **b.** 18% of 200 is what number?

c. 150 is 12% of what number?   **d.** 63 is what percent of 420?

e. 18% of what number is 72?   **f.** 120 is what percent of 60?

Solution

	Percent P (%)	*Whole* W ("of")	*Amount* A (part)	*Proportion* $\dfrac{P}{100} = \dfrac{A}{W}$
a. What number is 18% of 200?	18	200	unknown	$\dfrac{18}{100} = \dfrac{A}{200}$
b. 18% of 200 is what number?	18	200	unknown	$\dfrac{18}{100} = \dfrac{A}{200}$
c. 150 is 12% of what number?	12	unknown	150	$\dfrac{12}{100} = \dfrac{150}{W}$
d. 63 is what percent of 420?	unknown	420	63	$\dfrac{P}{100} = \dfrac{63}{420}$
e. 18% of what number is 72?	18	unknown	72	$\dfrac{18}{100} = \dfrac{72}{W}$
f. 120 is what percent of 60?	unknown	60	120	$\dfrac{P}{100} = \dfrac{120}{60}$

Regardless of the arrangement of the question, identify P first.

Second, identify the number following the word "of."

This number is identified last.

◆

Since there are only three letters in the proportion

$$\frac{P}{100} = \frac{A}{W}$$

there are three types of percent problems. These possible types were illustrated in Example 7. To answer a question involving a percent, write a proportion and then solve the proportion. Try solving each proportion in Example 7. The answers are:
a. $A = 36$; **b.** $A = 36$; **c.** $W = 1{,}250$; **d.** $P = 15$; **e.** $W = 400$; **f.** $P = 200$.

EXAMPLE 8

In a certain class there are 500 points possible. The lowest C grade is 65% of the possible points. How many points are equal to the lowest C grade?

Solution

What is 65% of 500 points?

$$\frac{65}{100} = \frac{A}{500} \qquad \leftarrow \text{The whole amount follows the word "of."}$$

$$A = \frac{65 \times \overset{5}{\cancel{500}}}{\underset{1}{\cancel{100}}} = 325$$

Check by estimation: 65% of 500 $\approx 6(\frac{1}{10} \times 500) = 300$.

The lowest C grade is 325 points.

◆

EXAMPLE 9

If your monthly salary is $4,500 and 21% is withheld for taxes and Social Security, how much money will be withheld from your check on payday?

Solution   How much is 21% of $4,500?

$$\frac{21}{100} = \frac{A}{4,500}$$

$$A = \frac{21 \times 4,500}{100} = 945$$

Check by estimation: 21% of $4,500 \approx 2(\frac{1}{10} \times 4,500) = 900$.

The withholding is $945.   ◆

EXAMPLE 10

You make a $25 purchase, and the clerk adds $2.25 for sales tax. This doesn't seem right to you, so you want to know what percent tax has been charged.

Solution   What percent of $25 is $2.25?

$$\frac{P}{100} = \frac{2.25}{25}$$

$$\frac{100 \times 2.25}{25} = P$$

$$9 = P \qquad \textit{Check by estimation:} \ 9\% \ of \ 25 \approx \frac{1}{10} \times 25 = 2.50.$$

The tax charged was 9%.   ◆

EXAMPLE 11

Your neighbors tell you that they paid $4,437 in taxes last year, and this amounted to 29% of their total income. What was their total income?

Solution   29% of total income is $4,437.

$$\frac{29}{100} = \frac{4,437}{W}$$

$$\frac{100 \times 4,437}{29} = W$$

$$15,300 = W$$

Check by estimation:

$$29\% \ of \ 15,300 \approx 3(\tfrac{1}{10} \times 15,000) = 3(1,500) = 4,500.$$

Since $4,500 is an estimate for $4,437, we conclude the result is correct. Their total income was $15,300.   ◆

EXAMPLE 12

If 45 is increased to 105, what is the percent increase?

Solution   The amount of increase is $105 - 45 = 60$, so as a percent problem we have "60 is what percent of 45?"

$$\frac{P}{100} = \frac{60}{45}$$

$$P = \frac{100 \times 60}{45} \approx 133.\overline{33}$$

The percent increase is $133\frac{1}{3}\%$.   ◆

WARNING! You must be careful not to add percents. For example, suppose you have $100 and spend 50%. How much have you spent, and how much do you have left?

    Amount Spent   Remainder
        $50 $50

Now, suppose you spend 50% of the remainder. How much have you spent, and how much is left?

    New Spending   Old Spending   Remainder
        $25 $50 $25

This means you have spent $75 or 75% of your original bankroll. A common ERROR is to say "50% spending + 50% spending = 100% spending." **Remember, if you add percents, you often obtain incorrect results.**

EXAMPLE 13

A newspaper headline proclaimed:

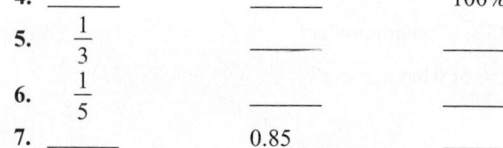

Teen drug use soars 105%

WASHINGTON – Teen drug use rose 105% between 1995 and 1997.

A national survey showed that between 1995 and 1996 youth drug use rose 30%, but between 1996 and 1997 usage soared 75%.

Over the two-year period, the rise of 105% was attributed to...

What is wrong with this headline?

Solution

We are not given all the relevant numbers, but consider the following possibility:
    Suppose there are 100 drug users, so a rise of

    100 to 130 is a 30% increase
    130 to 227 is a 75% increase
    100 to 227 is a 127% increase, NOT 105%

Remember, adding percents can give faulty results.   ◆

Problem Set 5.8

LEVEL 1

In Problems 1–21, change the given form into the two missing forms.

Fraction	Decimal	Percent
1. _____	0.75	_____
2. _____	0.2	_____
3. _____	_____	40%

	Fraction	Decimal	Percent
4.	_____	_____	100%
5.	$\frac{1}{3}$	_____	_____
6.	$\frac{1}{5}$	_____	_____
7.	_____	0.85	_____
8.	_____	_____	60%

	Fraction	Decimal	Percent
9.	$\frac{3}{8}$	_____	_____
10.	_____	_____	45%
11.	_____	_____	120%
12.	$\frac{2}{3}$	_____	_____
13.		0.05	_____
14.	_____	_____	$6\frac{1}{2}\%$
15.	$\frac{1}{6}$	_____	_____
16.	$\frac{5}{6}$	_____	_____
17.	_____	_____	$22\frac{2}{9}\%$
18.	_____	0.35	_____
19.	_____	0.175	_____
20.	$\frac{1}{12}$	_____	_____
21.	_____	0.0025	_____

Estimate the percentages in Problems 22–28.

22. a. 50% of 2,000   **b.** 25% of 400

23. a. 10% of 95,000   **b.** 10% of 85.6

24. a. 50% of 9,985   **b.** $33\frac{1}{3}\%$ of 3,600

25. a. 25% of 819   **b.** 25% of 790

26. a. 75% of 1,058   **b.** 75% of 94

27. a. 40% of 93   **b.** 90% of 8,741

28. a. $66\frac{2}{3}\%$ of 8,600   **b.** $66\frac{2}{3}\%$ of 35

Write each sentence in Problems 29–46 as a proportion, and then solve to answer the question.

29. What number is 15% of 64?

30. What number is 120% of 16?

31. 14% of what number is 21?

32. 40% of what number is 60?

33. 10 is what percent of 5?

34. What percent of $20 is $1.20?

35. 4 is what percent of 5?

36. 2 is what percent of 5?

37. What percent of 12 is 9?

38. What percent of 5 is 25?

39. 49 is 35% of what number?

40. 3 is 12% of what number?

41. 120% of what number is 16?

42. 21 is $66\frac{2}{3}\%$ of what number?

43. 12 is $33\frac{1}{3}\%$ of what number?

44. What is 8% of $2,425?

45. What is 6% of $8,150?

46. 400% of what number is 150?

LEVEL 2

47. If 11% of the 180 million adult Americans live in poverty, how many adult Americans live in poverty?

48. If 6.2% of the 180 million adult Americans are unemployed, how many adult Americans are unemployed?

49. If the sales tax is 6% and the purchase price is $181, what is the amount of tax?

50. If the sales tax is 5.5% and purchase price is $680, what is the amount of tax?

51. If you were charged $151 in taxes on a $3,020 purchase, what percent tax were you charged?

52. If 20 is increased to 25, what is the percent increase?

53. If 80 is decreased to 48, what is the percent decrease?

54. Government regulations require that, for certain companies to receive federal grant money, 15% of the total number of employees must meet minority requirements. If a company employs 390 people, how many minority people should be employed to meet the minimum requirements?

55. If Brad's monthly salary is $8,200, and 32% is withheld for taxes and Social Security, how much money is withheld each month?

56. A certain test is worth 125 points. How many points (rounded to the nearest point) are needed to obtain a score of 75%?

57. If Carlos answered 18 out of 20 questions on a test correctly, what was his percentage right?

58. If Wendy answered 15 questions correctly and obtained 75%, how many questions were on the test?

59. Shannon Sovndal received an 8% raise, which amounted to $100 per month. What was his old wage, and what will his new wage be?

60. An advertisement for a steel-belted radial tire states that this tire delivers 15% better gas mileage. If the present gas mileage is 25.5 mpg, what mileage would you expect if you purchased these tires? Round your answer to the nearest tenth of a mile per gallon.

5.9 | MODELING UNCATEGORIZED PROBLEMS

One major criticism of studying the common types of word problems presented in most textbooks is that students can fall into the habit of solving problems by using a template or pattern and be successful in class without ever developing independent problem-solving abilities. Even though problem solving may require algebraic skills, it also requires many other skills; a goal of this book is to help you develop a general problem-solving ability so that when presented with a problem that does not fit a template, you can apply techniques that will lead to a solution.

Most problems that do not come from textbooks are presented with several variables, along with one or more relationships among those variables. In some instances, insufficient information is given; in others there may be superfluous information or inconsistent information. A common mistake when working with these problems is to assume that you know which variable to choose as the unknown in the equation *at the beginning of the problem*. This leads to trying to take too much into your memory at the start, and as a result it is easy to get confused.

Now you are in a better position to practice Pólya's problem-solving techniques because you have had some experience in carrying out the second step of the process, which we repeat here for convenience.

Guidelines for Problem Solving

First: You have to *understand the problem.*

Second: *Devise a plan.*   Find the connection between the data and the unknown. Look for patterns, relate to a previously solved problem or to a known formula, or simplify the given information to give you an easier problem.

Third: *Carry out the plan.*

Fourth: *Look back.*   Examine the solution obtained.

Remember that the key to becoming a problem solver is to develop the skill to solve problems *that are not like the examples shown in the text.* For this reason, the problems in this section will require that you go beyond copying the techniques developed in the first part of this chapter.

EXAMPLE 1   **PÓLYA'S METHOD**

In California, state funding of education is based on the average daily attendance (ADA). Develop a formula for determining the ADA at your school.

Solution   We use Pólya's problem-solving guidelines for this example.

Most attempts at mathematical modeling come from a real problem that needs to be solved. There is no "answer book" to tell you when you are correct. There is often the need to do additional research to find needed information, and the need *not* to use certain information that you have to arrive at a solution.

Understand the Problem. What is meant by ADA? The first step might be a call to your school registrar to find out how ADA is calculated. For example, in California it is

This means that 525 WSCH is 525/525 = 1 ADA or 1,050 WSCH is

$$\frac{1,050}{525} = 2 \; ADA$$

In other words, we divide the WSCH by 525 to find the ADA.

based on weekly student contact hours (WSCH), and 1 ADA = 525 WSCH. Thus, you might begin by writing the formula

$$\frac{WSCH}{525} = TOTAL \; ADA$$

Devise a Plan. Mathematical modeling requires that we check this formula to see whether it properly models the TOTAL ADA at your school. How is the WSCH determined? In California, the roll sheets are examined at two census dates, and WSCH is calculated as the average of the census numbers. In mathematical modeling, you often need formulas that are not specified as part of the problem. In this case, we need a formula for calculating an average. We use the **mean:**

$$WSCH = \frac{Census \; 1 + Census \; 2}{2}$$

We now have a second attempt at a formula for TOTAL ADA:

$$\frac{Census \; 1 + Census \; 2}{2} \div 525 = TOTAL \; ADA$$

Carry Out the Plan. Does this properly model the TOTAL ADA? Perhaps, but suppose a taxpayer or congressperson brings up the argument that funding should take into account absent people, because not all classes have 100% attendance each day. You might include an *absence factor* as part of the formula. In California, the agreed absence factor for funding purposes is 0.911. How would we incorporate this into the TOTAL ADA calculation?

$$\frac{Census \; 1 + Census \; 2}{2} \cdot \frac{Absence \; factor}{525} = TOTAL \; ADA$$

This formula could be simplified to

$$TOTAL \; ADA = 0.0008676190476(c_1 + c_2)$$

where c_1 and c_2 are the numbers of students on the roll sheets on the first and second census dates, respectively.

Look Back. In fact, a Senate bill in California states this formula as

$$\frac{Census \; 1 \; WSCH + Census \; 2 \; WSCH}{2} \cdot \frac{0.911 \; Absence \; factor}{525} = TOTAL \; ADA$$

This is equivalent to the formula we derived.   ◆

EXAMPLE 2   PÓLYA'S METHOD

Suppose you wish to record a 3-hour movie on your VCR and want to obtain the best quality recording on a single videotape. Explain how you could do this. There are three modes for recording on your VCR: SP (standard play; tape will record 2 hours), LP (long play; tape will record 4 hours), and SLP (super long play; tape will record 6 hours). The best quality recording is on SP, medium quality on LP, and least quality on SLP. *

Solution   We use Pólya's problem-solving guidelines for this example.

Understand the Problem. If we record the movie at the SP speed, we will need more than one videotape. If we record the movie at the LP speed, we will be able to record the

*The idea for this example is from an article by Gregory N. Fiore in *The Mathematics Teacher,* October 1988.

movie, but an empty 1-hour space would be left at the end, and we would be sacrificing quality by recording the entire movie at the LP speed. It seems that the solution is to begin in the LP mode and at some "crucial moment" switch to the SP mode with the goal that the 3-hour movie should just fill the tape.

Devise a Plan. We begin with some assumptions. In reality, the actual lengths of VCR tapes are not exact. We will assume, however, that the times specified in the problem are exact. That is, a tape will play for exactly 2, 4, or 6 hours depending on the chosen speed. We also assume a linear relationship between recording time and the amount of tape used. Some experimentation will show us that the number of counts-per-hour on a built-in counter varies and is not linear, so we cannot use this meter on our VCR to determine when to make the switch.

Let us begin by drawing a picture of our problem:

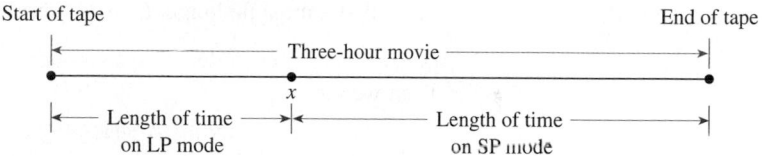

Let x be the length of time that we record on the LP mode. Then $3 - x$ is the length of time that we record on the SP mode.

How do we represent the fraction of the tape that is recorded in the LP mode?

If $x = 3$, then $\frac{3}{4}$ of the tape is used; remember that 4 hours of movies can be recorded in the LP mode.

If $x = 2$, then $\frac{2}{4}$ of the tape is used.

If $x = 1$, then $\frac{1}{4}$ of the tape is used.

Thus, the model seems to be

$$\frac{\text{TIME RECORDING IN THE LP MODE}}{4} = \frac{x}{4}$$

How do we represent the fraction of the tape that is recorded in the SP mode? Since this mode fills the tape in 2 hours, we see that the fraction of the tape recorded in the SP mode is

$$\frac{\text{TIME RECORDING IN THE SP MODE}}{2} = \frac{3 - x}{2}$$

Since we want to fill the entire tape with the movie, both fractions must add up to 1:

$$\frac{x}{4} + \frac{3 - x}{2} = 1$$

Carry Out the Plan. Simplify this equation by multiplying both sides by 4.

$$x + 2(3 - x) = 4$$
$$x + 6 - 2x = 4$$
$$x = 2$$

The mode should be changed after 2 hours.

Look Back. If we record for 2 hours on the LP mode, we have used up $\frac{1}{2}$ of the tape. We then switch to the SP mode and record for 1 hour, which uses up the other half of the tape. ◆

EXAMPLE 3   PÓLYA'S METHOD

An inlet pipe on a swimming pool can be used to fill the pool in 24 hours. The drain pipe can be used to empty the pool in 30 hours. If the pool is half-filled and then the drain pipe is accidentally opened, how long will it take to fill the pool?

Swimming pool
(cross-section)

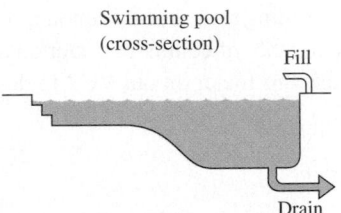

Fill

Drain

Solution   We use Pólya's problem-solving guidelines for this example.

Understand the Problem. Draw a picture, if necessary. Water comes into the pool until it is half full. Then water begins draining out of the pool at the same time it is coming in. Is there a solution? What if the drain pipe alone emptied the pool in 20 hours? In that case, the pool would never be more than half full, and instead we could ask when it would be empty. Estimate an answer. Since it takes less time to fill than to drain, the inlet pipe is working faster than the drain, so it will fill. How long? Well, you might guess that it must be longer than 24 hours.

Devise a Plan. Let us denote the amount of water drained as the amount of work done. Then we see

WORK DONE BY INLET PIPE $-$ WORK DONE BY DRAIN $=$ ONE JOB COMPLETED

But wait, is this correct? The drain works only after the pool is half full. One of Pólya's suggestions is to work a simpler problem. How long does it take to fill the first half of the pool? That's easy. Since the pool fills in 24 hours, it must be half full in 12 hours. Now, we can focus on filling the second half of the pool, so we come back to the guide we wrote down above.

Another step we can take when devising a plan is to ask whether there is an appropriate formula. In this case,

WORK DONE BY INLET PIPE $=$ (RATE OF INLET PIPE)(TIME TO FILL)

$$= \frac{1}{24}(\text{TIME}) \quad \text{\small TIME TO FILL} = \text{\small TIME TO DRAIN}$$

WORK DONE BY DRAIN $=$ (RATE OF DRAIN)(TIME TO DRAIN)

$$= \frac{1}{30}(\text{TIME})$$

Carry Out the Plan. Substitute these formulas into the guide (for the second half of the pool):

WORK DONE BY INLET PIPE $-$ WORK DONE BY DRAIN $=$ ONE-HALF JOB COMPLETED

$$\downarrow \qquad\qquad\qquad \downarrow \qquad\qquad\qquad \downarrow$$

$$\frac{1}{24}(\text{TIME}) \quad - \quad \frac{1}{30}(\text{TIME}) \quad = \quad \frac{1}{2}$$

Let $t = $ TIME   *Remember that this is the time required to fill the second half of the pool.*

$$\frac{1}{24}t - \frac{1}{30}t = \frac{1}{2}$$

$$(120)\frac{1}{24}t - (120)\frac{1}{30}t = (120)\frac{1}{2} \quad \text{\small Multiply both sides by 120.}$$

$$5t - 4t = 60$$

$$t = 60$$

It will take 12 hours to fill the first half of the pool, but it will take 60 more hours to fill the second half. Thus, it will take 72 hours to fill the pool.

Look Back. Does this answer make sense? In 72 hours, $\frac{72}{24} = 3$ swimming pools could be filled. The drain is working for 60 hours, so this is equivalent of $\frac{60}{30} = 2$ swimming pools drained. Let's see: $3 - 2 = 1$ swimming pool, so it checks.   ◆

We have refrained from working so-called age problems because they seem to be a rather useless type of word problem—one that occurs only in algebra books. They do, however, give us a chance to practice our problem-solving skill, so they are worth considering. By the way, the precedent for age problems in algebra books goes back over 2,000 years (see Problem 60 about Diophantus). The age problem given in the next example was found in the first three frames of a *Peanuts* cartoon.

Peanuts © 1972. Reprinted by permission of United Feature Syndicate, Inc.

EXAMPLE 4   **PÓLYA'S METHOD**

A man has a daughter and a son. The son is three years older than the daughter. In one year the man will be six times as old as the daughter is now, and in ten years he will be fourteen years older than the combined ages of his children. What is the man's present age?

Solution   We use Pólya's problem-solving guidelines for this example.

Understand the Problem. There are three people: a man, his daughter, and his son. We are concerned with their ages now, in one year, and in ten years.

Devise a Plan. Does this look like any of the problems we have solved? We need to distinguish among the persons' ages now and some time in the future. Let us begin by translating the known information:

Ages now:

> MAN'S AGE NOW
>
> DAUGHTER'S AGE NOW
>
> SON'S AGE NOW

We are given that

First equation: SON'S AGE NOW = DAUGHTER'S AGE NOW + 3

Ages in one year:

> MAN'S AGE NOW + 1
>
> DAUGHTER'S AGE NOW + 1
>
> SON'S AGE NOW + 1

We are given that

> MAN'S AGE NOW + 1 = 6(DAUGHTER'S AGE NOW)

This is the same as

Second equation: MAN'S AGE NOW = 6(DAUGHTER'S AGE NOW) − 1

Ages in ten years:

> MAN'S AGE NOW + 10
>
> DAUGHTER'S AGE NOW + 10
>
> SON'S AGE NOW + 10

We are given that

MAN'S AGE NOW + 10 = (SON'S AGE NOW + 10) + (DAUGHTER'S AGE NOW + 10) + $\overset{\uparrow}{14}$

14 years older

Subtract 10 from both sides:

Third equation: MAN'S AGE NOW = SON'S AGE NOW + DAUGHTER'S AGE NOW + 24

The plan is to begin with the third equation and evolve this into an equation we can solve.

Carry Out the Plan.

MAN'S AGE NOW = SON'S AGE NOW + DAUGHTER'S AGE NOW + 24

First equation: SON'S AGE NOW = (DAUGHTER'S AGE NOW + 3)

Second equation: MAN'S AGE NOW = 6(DAUGHTER'S AGE NOW) − 1

Substitute (as shown by the arrows):

6(DAUGHTER'S AGE NOW) − 1 = (DAUGHTER'S AGE NOW + 3) + DAUGHTER'S AGE NOW + 24

There is now one unknown, so let d = DAUGHTER'S AGE NOW. Then

$$6d - 1 = d + 3 + d + 24$$
$$6d - 1 = 2d + 27$$
$$4d = 28$$
$$d = 7$$

From the second equation, MAN'S AGE NOW = 6(7) − 1 = 41. The man's present age is 41.

Look Back. From the first equation, SON'S AGE NOW = 7 + 3 = 10. In one year the man will be 42, and this is 6 times the daughter's present age. In ten years the man will be 51 and the children will be 20 and 17; the man will be 14 years older than 20 + 17 = 37.   ◆

EXAMPLE 5   PÓLYA'S METHOD

To number the pages of a bulky volume, the printer used 4,001 digits. It was estimated that the volume has 1,000 pages. Is this a good estimate? How many actual pages has the volume?

Solution   We use Pólya's problem-solving guidelines for this example.

Understand the Problem. Page 1 uses one digit; page 49 uses two digits; page 125 uses three digits. Assume that the first page is page 1 and that the pages are numbered consecutively.

Devise a Plan. This does not seem to fit any of the problem types we have considered in this book. Can you reduce this to a simpler problem? If the book has exactly 9 numbered pages, how many digits does the printer use? If the book has exactly 99 numbered pages, how many digits does the printer use? The plan is to count the number of digits on single-digit pages, then those on double-digit pages, then on triple-digit pages.

Carry Out the Plan.

Single-digit pages:	9 digits total
Double-digit pages:	2(90) = 180 digits total

Do you see why this is not 2(99)? Remember, you are counting 90 pages (namely, from page 10 to page 99).

Triple-digit pages:	3(900) = 2,700 digits total

We have now accounted for 2,889 digits. Since we are looking for 4,001 digits and since we can estimate four-digit pages to have $4(9,000) = 36,000$ digits, we need to know how many pages past 999 are necessary. Let x be the total number of pages. Then the number of digits on four-digit pages is

Four-digit pages: $4(x - 999)$

Now we can solve for x by using the following equation:

$$9 + 180 + 2,700 + 4(x - 999) = 4,001$$
$$4x + 2,889 - 3,996 = 4,001$$
$$4x - 1,107 = 4,001$$
$$4x = 5,108$$
$$x = 1,277$$

Look Back. The book has 1,277 pages.   ◆

Problem Set 5.9

LEVEL 1

1. **IN YOUR OWN WORDS** Discuss the difference between solving word problems in textbooks and problem solving outside the classroom.

2. **IN YOUR OWN WORDS** In the first problem set of this book, we asked you to discuss Pólya's problem-solving model. Now we ask the same question again to see whether your perspective has changed at all. Discuss Pólya's problem-solving model.

3. The product of two consecutive positive even numbers is 440. What are the numbers?

4. The product of two consecutive positive odd numbers is 255. What are the numbers?

5. The sum of an integer and its reciprocal is $\frac{5}{2}$. What is the integer?

6. The sum of a number and twice its reciprocal is 3. What are the numbers?

LEVEL 2

Read each of the given problems and then select the best of the given answers in Problems 7–34.

7. The Hidden Valley Summer Camp is taking a field trip to the Exploratorium. There are ten children who are 12 years old or younger, six teens whose ages range from 13 years to 18 years, and three chaperones. The field trip includes transportation and lunch. The costs of the field trip are given:

$150	Bus and driver
$2.50/person	Lunch
$5.00/child	Admission ⎫
$6.00/teens	Admission ⎬ $20 discount for 25
$7.00/adult	Admission ⎭ persons or more

The bus driver will eat lunch, but will not visit the Exploratorium. What is the total cost for the field trip?
   A. $307   B. $287   C. $314   D. $294

8. An animal shelter has two purebred Labradors among the 48 dogs available for adoption. It is also known that 7/8 of the dogs are mongrels. How many of the dogs are purebreds?
   A. 42   B. 8   C. 4   D. 6

9. Shannon needs to have a term paper printed on a laser printer at Kinko's. The paper is 30 pages long (including a cover page and a bibliography page). Kinko's charges $10.00 for each 30 minutes (with a minimum charge of one hour), plus 10¢/sheet, and a binding fee of $1.25. He will also copy the original at a cost of 7¢/page. If the laser printing takes 45 minutes, what are the total charges?
   A. $26.35   B. $21.35   C. $30.00   D. $16.35

10. Hannah is selling snow cones to earn money for a train trip. She sold 23 cones for $1.25 each and spent $4.95 + $0.35 tax. In addition, she received $3.50 in tips. What is the total profit for Hannah's endeavor?
   A. $27.00   B. $26.95   C. $32.25   D. $23.45

11. When Melissa went to purchase a purse from a street vendor, she negotiated a purchase price that was 15% less than the $125.00 marked price, plus 7% sales tax on the purchase price. If she has $325.00 in her checking account, how much did she pay for the purse?
   A. $113.69   B. $211.31   C. $20.06   D. $304.94

12. Linda finds that her $560 car payment is 28% of her gross monthly income. She knows that $510 is deducted from her paycheck each month, for income tax and Social Security. What is Linda's gross monthly income (rounded to the nearest dollar)?
   A. $2,000   B. $157   C. $1,490   D. $930

13. Jane figures that her monthly car insurance payment of $180 is equal to 35% of the amount of her monthly auto loan payment. What is her total combined monthly expense for auto loan pament and insurance (rounded to the nearest dollar)?
    A. $577   B. $694   C. $514   D. $63

14. Ben wants to finance the purchase of a guitar that is priced at $1,500. He contacts ABC Finance Company; they believe the value of the guitar to be worth 10% less than that, and their policy is to lend only 80% of the value of the guitar. What is the down payment?
    A. $420 $1,320   C. $1,080   D. $1,350

15. Søren's major bills are a variable-rate mortgage, car payments ($325/mo), food ($525/mo), and utilities ($215/mo). Last year his monthly mortgage payment was $2,320. This year, his mortgage payment increased by 3%. How much (rounded to the nearest dollar) will his annual mortgage payment increase?

    A. $3,385   B. $2,390   C. $835   D. $70

16. Theron bought an HDTV for $1,280 retail price and paid 7% state tax. He also received a rebate of 5% of the retail price. What was the total amount that he paid, including tax, for the HDTV?
    A. $1,369.60   B. $1,305.60
    C. $1,433.60   D. $1,340.20

17. A swimming pool holds 25,500 gallons of water. It will be filled by a pair of 0.5-in.-diameter hoses, each of which supplies 2.50 gallons of water per minute. How many hours will it take for the hoses to fill the pool?
    A. 170   B. 83   C. 85   D. 125

18. Gary, a gardener, gives the following estimate for an installation. Each manifold will cost $85 to install, and each one will have three sprinkler heads at a cost of $9.50 each. Overhead adds 10% to the total bill. What is the total cost for this installation, consisting of four manifolds?
    A. $113.50   B. $499.40   C. $454.00   D. $103.95

19. A 40-member committee voted on whether to send a certain bill to the governing board. One-fourth voted in favor, 3/5 voted against, and the rest abstained. How many abstained?
    A. 6   B. 24   C. 34   D. 7

20. In Karl's math class with 28 students, 1/7 are receiving an A grade, 1/4 are B's, 11 are C's, and the rest are D's or F's. How many are D's or F's?
    A. 4   B. 7   C. 6   D. 22

21. Alice paid $350 for her flight from San Francisco to Denver, but Bobbie paid $278 for the same flight. Cal, on the other hand, paid 8% less than Alice. How much was paid for the three seats (rounded to the nearest dollar)?
    A. $950   B. $656   C. $1,006   D. $978

22. Doug's monthly salary was $1,250 when he received a 5% raise. Six months later, he received another 2% raise. What is his annual salary, after receiving both raises?
    A. $1,337.50   B. $1,338.75
    C. $16,050.00   D. $16,065.00

23. The top three women in the 2002 WNBA playoffs scored a total of 360 points. The leader, Tamika Whitmore (New York Liberty), scored 13 and 14 more points than the next two scorers. The points scored by these three players are:
    A. 100, 114, 146   B. 115, 116, 129
    C. 100, 114, 129   D. 316, 330, 345

24. Brett Favre (Green Bay Packers) threw 551 passes during the 2002 season. He ran with the ball 25 times. Of the 551 passes, 194 were incomplete, 16 were intercepted, and the rest were completed. What percent of the passes were complete?
    A. 61.9%   B. 57.4%   C. 35.2%   D. 38.1%

25. In 2002, Barry Bonds (San Francisco Giants) had 403 plate appearances, with 149 hits and 46 home runs. He was paid $104,895/game. If he played 143 games, how much was his annual salary (rounded to the nearest hundred dollars)?
    A. $14,999,985   B. $918,971
    C. $105,493   D. $15,000,000

26. An investor has $2,500 at 2.5%, $21,300 at 3.2%, and $8,540 at 3.8%. If this investor is in a 38% tax bracket, how much interest income, after taxes is realized?
    A. $1,068.62   B. $32,340
    C. $662.54   D. $406.08

27. An hourly salary of $25 is equivalent to what annual salary? Assume a 40-hour work week.
    A. $50,000   B. $52,000
    C. $41,600   D. $52,850

28. An annual salary of $83,000 is equivalent to what hourly salary? Assume a 40-hour work week.
    A. $38.00   B. $38.90   C. $39.90   D. $41.50

29. How many prime numbers are divisible by 2?
    A. 0   B. 1   C. 2   D. 3

30. How many prime numbers are less than 100?
    A. 23   B. 24   C. 25   D. 26

31. How far up a wall does a 10-ft ladder reach if the bottom is placed 3 ft from the base of the wall? (Answer to the nearest foot.)
    A. 8   B. 9   C. 10   D. 11

32. In order to secure a 10-ft TV antenna, three guy wires are needed. If the guy wires are 12 ft from the base of the antenna, how many feet of wire should be purchased if the packages available are:
    A. 20 ft   B. 50 ft   C. 75 ft   D. 100 ft

33. The number of distinct (different) factors of 100 is:
A. 2   B. 4   C. 8   D. 9

34. The number of distinct (different) prime factors of 100 is:
A. 2   B. 4   C. 8   D. 9

35. **IN YOUR OWN WORDS** Look back over your method of solution for any of Problems 7–34. Did you answer any of those by inspection? That is, discuss the role of estimation and common sense in problem solving.

36. **IN YOUR OWN WORDS** In this chapter we have been discussing a problem-solving technique. Look back over your method of solution for any of Problems 7–34 that you have answered. Did you use the method discussed in the text or did you use other procedures? Discuss your own "method of attack" that you *actually* used.

37. Suppose your monthly salary is $7,415 with the following deductions. (Answers should be to the nearest cent.)
a. What is the federal tax (38%)?
b. What is the state tax (6.2)%?
c. What is the retirement deduction (10%)?
d. What are the miscellaneous deductions (4.5%)?
e. What is your take-home salary?

38. Suppose your stock market portfolio was $251,000 at the start of 2000. By the start of 2002, the value was $112,950.
a. What is the percent loss in the two-year period?
b. What is the percent gain necessary for your portfolio to regain its value in the next two years?

39. The Standard Oil and the Sears buildings have a combined height of 2,590 ft. The Sears Building is 318 ft taller. What is the height of each of these Chicago towers?

40. The combined area of New York and California is 204,192 square miles. The area of California is 108,530 square miles more than that of New York. Find the land area of each state.

41. What is the sum of the first 100 consecutive even numbers?

42. A survey of 100 persons at Better Widgets finds that 40 jog, 25 swim, 16 cycle, 15 both swim and jog, 10 swim and cycle, 8 jog and cycle, and 3 persons do all three. How many people at Better Widgets do not take part in any of these three exercise programs?

43. A survey of executives of Fortune 500 companies finds that 520 have MBA degrees, 650 have business degrees, and 450 have both degrees. How many executives with MBAs have nonbusiness degrees?

44. An inlet pipe on a swimming pool can be used to fill the pool in 24 hours. The drain pipe can be used to empty the pool in 30 hours. If the pool is two-thirds filled and then the drain pipe is accidentally opened, how long will it take to fill the pool?

45. An inlet pipe on a swimming pool can be used to fill the pool in 36 hours. The drain pipe can be used to empty the pool in 40 hours. If the pool is half filled and then the drain pipe is accidentally opened, how long will it take to fill the pool?

46. Solve the videotape example for taping a movie of length ℓ where $2 < \ell \le 4$. (If $\ell \le 2$, then the solution is obvious—just use the SP mode.)

47. Reconsider Problem 46 if you are recording the movie and are eliminating the commercials from the taping process. Suppose that the movie is ℓ hours long and that there are c minutes of commercials each hour.

48. A swimming pool is 15 ft by 30 ft and an average of 5 ft in depth. It takes 25 minutes longer to fill than to drain the pool. If it can be drained at a rate of 15 ft³/min faster than it can be filled, what is the drainage rate?

PROBLEM SOLVING

The following problems do not match the examples in this book, but all the problem-solving techniques necessary to answer these questions have been covered in this book. These problems are designed to test your problem-solving abilities.

49. After paying a real estate agent a commission of 6% on the selling price, $2,525 in other costs, and $65,250 on the mortgage, you receive a cash settlement of $142,785. What was the selling price of your house?

50. An investor has $100,000 to invest and can choose between insured savings at 8.5% interest or annuities paying 12.25%. If the investor wants to earn an annual income of $9,850, how much should be invested in each?

51. The longest rod that will just fit inside a rectangular box, if placed diagonally top to bottom, is 17 inches. The box is 1 inch shorter and 3 inches longer than it is wide. How much must you cut off the rod so that it will lie flat in the bottom of the container? What are the dimensions of the box?

52. A canoeist rows downstream in 1.5 hours and back upstream in 3 hours. What is the rate of the current if the distance rowed is 9 miles in each direction?

53. A plane makes a 630-mile flight with the wind in 2.5 hours; the return flight against the wind takes 3 hours. Find the wind speed.

54. A plane makes an 870-mile flight in $3\frac{1}{3}$ hours against a strong head wind, but returns in 50 minutes less with the wind. What is the plane's speed without the wind?

55. A tour agency is booking a tour and has 100 people signed up. The price of a ticket is $2,500 per person. The agency has booked a plane seating 150 people at a cost of $240,500. Additional costs to the agency are incidental fees of $500 per person. For each $5 that the price is lowered, a new person will sign up. How much should be charged for a ticket so that the income and expenses will be the same? The lower price will be charged to all participants, even those already signed up.

56. Imagine that you have written down the numbers from 1 to 1,000,000. What is the total number of zeros you have written down?

57. Suppose you have a large bottle with a canary inside. The bottle is sealed, and it's on a scale. The canary is standing on the bottom of the bottle. Then the canary starts to fly around inside of it. Does the reading on the scale change?

58. Suppose it were possible to count all of the individual strands of hair on your head. Also suppose that it is possible to do that for any number of people. For example, if you have 4,890 hairs on your head and I have 1,596 hairs, then the product of the number of hairs on both our heads is 7,804,440. How many hairs are there in the product for all persons in New York City at midnight on New Year's Eve, December 31, 2000?

59. An ancient society called a number *sacred* if it could be written as the sum of the squares of two counting numbers.
   a. List five sacred numbers.
   b. Is the product of every two sacred numbers sacred?

60. **HISTORICAL QUESTION** Diophantus wrote the following puzzle about himself in *Anthologia Palatina:*

> *Here lies Diophantus. The wonder behold—*
> *Through art algebraic, the stone tells how old:*
> *"God gave him his boyhood one-sixth of this life,*
> *One-twelfth more as youth while whiskers grew rife;*
> *And then yet one-seventh their marriage begun;*
> *In five years there came a bouncing new son.*
> *Alas, the dear child of master and sage*
> *Met fate at just half his dad's final age.*
> *Four years yet his studies gave solace from grief;*
> *Then leaving scenes earthly he, too, found relief."*

What is Diophantus' final age?

CHAPTER SUMMARY

"Algebra is generous, she often gives more than is asked of her."

D'Alembert

IMPORTANT TERMS

Addition property [5.4]
Addition property of inequality [5.5]
Algebra [5.1]
Binomial [5.1]
Binomial theorem [5.1]
Cell [5.3]
Common factor [5.2]
Comparison property [5.5]
Completely factored [5.2]
Consecutive integers [5.6]
Degree [5.1]
Difference of squares [5.2]
Division property [5.4]
Equation [5.4]
Equation properties [5.4]
Equivalent equations [5.4]
Evaluate [5.3]
Expand [5.1]

Factor [5.1]
FOIL [5.1]
Inequality [5.5]
Inequality symbol [5.5]
Like terms [5.1]
Linear [5.1]
Linear equation [5.4]
Monomial [5.1]
Multiplication property [5.4]
Multiplication property of inequality [5.5]
Multiplicity [5.4]
Numerical coefficient [5.1]
Percent [5.8]
Percent problem [5.8]
Polynomial [5.1]
Property of proportions [5.7]
Proportion [5.7]
Quadratic [5.1]

Quadratic equation [5.4]
Quadratic formula [5.4]
Ratio [5.7]
Replication [5.3]
Root [5.4]
Satisfy [5.4]
Similar terms [5.1]
Simplify [5.1]
Solution [5.4]
Solve a proportion [5.7]
Solve an equation [5.4]
Solve an inequality [5.5]
Spreadsheet [5.3]
Subtraction property [5.4]
Symmetric property of equality [5.4]
Term [5.1]
Trinomial [5.1]
Zero-product rule [5.4]

TYPES OF PROBLEMS

Simplify an algebraic expression. [5.1]
Multiply binomials mentally. [5.1]
Expand a binomial using the binomial theorem. [5.1]
Factor a polynomial. [5.2]
Evaluate an algebraic expression. [5.3]
Write spreadsheet expressions in ordinary algebraic notation. [5.3]
Use a spreadsheet to evaluate (including the replicate command). [5.3]
Use binomials to answer genetic applications. [5.3]
Solve equations (both linear and quadratic). [5.4]
Use a calculator to solve an equation. [5.4]
Solve linear inequalities. [5.5]
Solve applied problems—in particular, those involving number, distance, and
    Pythagorean relationships. [5.6]
Set up and simplify ratios. [5.7]
Compare the size of fractions, decimals, and radicals. [5.7]
Solve proportion problems. [5.7]
Be able to change forms: fractions/decimals/percents. [5.8]
Estimate percents. [5.8]
Solve applied percent problems. [5.8]
Use problem-solving methods to answer uncategorized questions. [5.9]

CHAPTER 5 REVIEW QUESTIONS

1. **a.** What is algebra?
   b. Briefly name and describe each of the four major processes of algebra.
2. **Simplify** the given expressions.
   a. $(x - 1)(x^2 + 2x + 8)$ **b.** $x^2(x^2 - y) - xy(x^2 - 1)$
3. **Evaluate** the given expressions for $x = 2$ and $y = 3$.
   a. $x^3 - y^3$ **b.** $(x - y)(x^2 + xy + y^2)$
4. **Factor** the given expressions.
   a. $3x^2 - 27$ **b.** $x^2 - 5x - 6$
5. **Solve** the given equations.
   a. $8x - 12 = 0$ **b.** $8x - 12 = -2x^2$
6. Factor $x^2 + 5x + 6$ using areas.

Solve the equations in Problems 7–15.

7. **a.** $2x + 5 = 13$ **b.** $3x + 1 = 7x$

8. **a.** $\dfrac{2x}{3} = 6$ **b.** $2x - 7 = 5x$

9. **a.** $\dfrac{P}{100} = \dfrac{3}{20}$ **b.** $\dfrac{25}{W} = \dfrac{80}{12}$

10. $x^2 = 4x + 5$ 11. $4x^2 + 1 = 6x$

12. $3 < -x$ 13. $2 - x \geq 4$

14. $3x + 2 \leq x + 6$ 15. $14 > 5x - 1$

16. Fill in $=$, $>$, or $<$:

   a. $\dfrac{2}{3}$ ___ $\dfrac{67}{100}$ **b.** $\dfrac{3}{4}$ ___ $\dfrac{75}{100}$ **c.** $\dfrac{23}{27}$ ___ $\dfrac{92}{107}$

   d. 0.05 ___ 0.1 **e.** 0.99 ___ 0.909

17. In a recipe for bread, flour and water are to be mixed in a ratio of 1 part water to 5 parts flour. If $2\frac{1}{2}$ cups of water is called for, how much flour (to the nearest tenth cup) is needed?

18. The sum of four consecutive even integers is 100. Find the integers.

19. A population of self-pollinating pea plants has two genes: T (tall, dominant) and t (short, recessive). Suppose we have an isolated population in which 52% of the genes in the gene pool are dominant T, and the other 48% are recessive t. What fraction of the population has each genotype? What fraction of the population has each phenotype?

20. The airline distance from Chicago to San Francisco is 1,140 miles farther than the flight from New York to Chicago, and 540 miles shorter than the distance from San Francisco to Honolulu. What is the length of each leg of the 4,980-mile New York–Chicago–San Francisco–Honolulu flight?

GROUP RESEARCH PROJECTS

Working in small groups is typical of most work environments, and learning to work with others to communicate specific ideas is an important skill. Work with three or four other students to submit a single report based on each of the following questions.

G18. Is $\sqrt{2}$ rational or irrational? Give a convincing argument to support your answer.

G19. HISTORICAL QUESTION The Babylonians solved the quadratic equation $x^2 + px = q$ ($q > 0$) without the benefit of algebraic notation. Tablet 6967 at Yale University finds a positive solution to this equation to be

$$x = \sqrt{\left(\frac{p}{2}\right)^2 + q} - \frac{p}{2}$$

Using modern algebraic notation, show that this result is correct.

G20. HISTORICAL QUESTION In 1907, the University of Göttingen offered the Wolfskehl Prize of 100,000 marks to anyone who could prove Fermat's Last Theorem, which seeks any replacements for $x, y,$ and z such that $x^n + y^n = z^n$ (where n is greater than 2 and $x, y,$ and z are counting numbers). In 1937, the mathematician Samuel Drieger announced that 1324, 731, and 1961 solved the equation. He would not reveal n—the power—but said that it was less than 20. That is,

$$1,324^n + 731^n = 1,961^n$$

However, it is easy to show that this cannot be a solution *for any n.* See if you can explain why by investigating some patterns for powers of numbers ending in 4 and 1.

G21. Prove that there are infinitely many integers such that the sum of the digits in their square equals the sum of the digits in their cube.

G22. JOURNAL PROBLEM (From *Journal of Recreational Mathematics,* Vol. II, #2) Translate the following message: Wx utgtuz f pbkz tswx wlx xwozm pbkzr, f exbmwo cxlzm xm ts jzszmfi fsv cxlzm lofwzgzm tswx wlx cxlzmr xe woz rfnz uzsxntsfwtxs fkxgz woz rzpxsu tr tncxrrtkiz, fsu T ofgz frrbmzuiv exbsu fs funtmfkiz cmxxe xe wotr kbw woz nfmjts tr wxx sfmmxl wx pxswfts tw. *Ctzmmz Uz Ezmnfw*

INDIVIDUAL RESEARCH PROBLEMS

Learning to use sources outside your classroom and textbook is an important skill, and here are some ideas for extending some of the ideas in this chapter. You can find references to these projects in a library or at

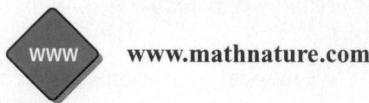

www.mathnature.com

PROJECT 5.1 What do the following people have in common?

Carl T. Rowan, columnist for the *Washington Post*
Lewis Carroll (Charles Dodgson), author of *Alice in Wonderland*
Christopher Wren, architect of St. Paul's Cathedral in London
Virginia Wade, tennis player, Wimbledon champion
Lawrence Leighton Smith, conductor and pianist

PROJECT 5.2 Write a paper on the relationship between geometric areas and algebraic expressions.

PROJECT 5.3 Write out a derivation of the quadratic formula.

PROJECT 5.4 Find any replacements for *x, y,* and *z* such that $x^n + y^n = z^n$, where *n* is greater than 2 and where *x, y,* and *z* are counting numbers. Write a history of this problem, known as Fermat's Last Theorem.

6 The Nature of Geometry

Geometry is the science created to give understanding and mastery of the external relations of things; to make easy the explanation and description of such relations and the transmission of this mastery.

G. B. HALSTED

CONTENTS

OVERVIEW

Geometry, or "earth measure," was one of the first branches of mathematics. Both the Egyptians and the Babylonians needed geometry for construction, land measurement, and commerce. They both discovered the Pythagorean theorem, although it was not proved until the Greeks developed geometry formally. This formal development utilizes deductive logic (which was briefly discussed in Chapter 2), beginning with certain assumptions, called **axioms** or **postulates.** Historically, the first axioms that were accepted seemed to conform to the physical world. In this chapter, we look at that body of mathematics known as *geometry.*

IMPORTANT IDEAS

Euclidean postulates [6.1]
Terminology associated with polygons, angles, and parallel/perpendicular lines [6.2]
Sum of the measures of angles in a triangle [6.3]
Exterior angles of a triangle [6.3]
Isosceles triangle property [6.3]
Similar triangle theorem [6.4]
Pythagorean theorem [6.5]
Golden rectangle and golden ratio [6.6]
Lobachevskian postulate [6.9]
Comparison of major two-dimensional geometries [6.9]

BOOK REPORTS

Write a 500-word report on one of the following books:
Sphereland, a Fantasy About Curved Spaces and an Expanding Universe, Dionys Burger (New York: Thomas Crowell Company, 1965).
Mathematical Recreations and Essays, W. W. R. Ball and H. S. M. Coxeter (New York: Macmillan, 1962).

LINKS

Individual Projects

Group Projects

Research Links

www.mathnature.com

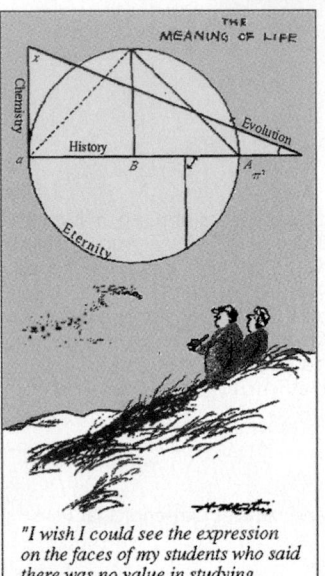

"I wish I could see the expression on the faces of my students who said there was no value in studying geometry."

© 1971. Reprinted by permission of *Saturday Review* and Henry Martin.

Early civilizations observed from nature certain simple shapes such as triangles, rectangles, and circles. The study of geometry began with the need to measure and understand the properties of these simple shapes.

6.1 GEOMETRY

Greek (Euclidean) Geometry

When we refer to Euclidean geometry, we are talking about the geometry known to the Greeks as summarized in a work known as Euclid's *Elements*. This 13-volume set of books collected all the material known about geometry and organized it into a logical deductive system. It is the most widely used and studied book in history, with the exception of the Bible. An overview of the history of mathematics, and of geometry in particular, is presented in the preface of this book, and it includes the geometry of both the Egyptians and the Greeks. We do not know very much about the life of Euclid except that he was the first professor of mathematics at the University of Alexandria (which opened in 300 B.C.).

Geometry involves **points** and sets of points called **lines, planes,** and **surfaces.** Certain concepts in geometry are called **undefined terms.** For example, what is a line? You might say, "I know what a line is!" But try to define a line. Is it a set of points? Any set of points? What is a point?

1. A point is something that has no length, width, or thickness.

2. A point is a location in space.

Certainly these are not satisfactory definitions because they involve other terms that are not defined. We will therefore take the terms *point, line,* and *plane* as undefined.

We often draw physical models or pictures to represent these concepts; however, we must be careful not to try to prove assertions by looking at pictures, since a picture may contain hidden assumptions or ambiguities. For example, consider Figure 6.1. If you look at Figure 6.1a, you might call it a square. If you look at Figure 6.1b, you might say it is something else. But what if we have in mind a cube, as shown in Figure 6.1c?

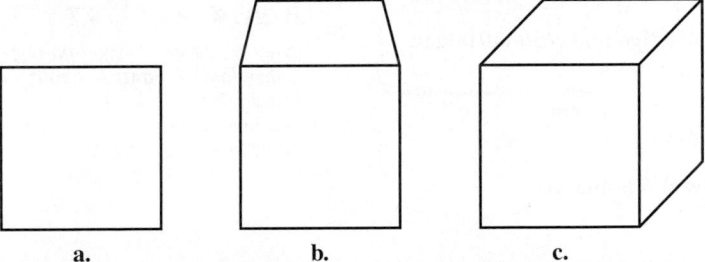

a.   b.   c.

Figure 6.1 Three views of the same object. What do you see? A square? A cube? Figures can be ambiguous (not clear, or with hidden meaning).

Even if you view this object as a cube, do you see the same cube as everyone else? Is the fly in Figure 6.2 inside or outside the cube? Thus, although we may use a figure to help us understand a problem, we cannot prove results by this technique.

Geometry can be separated into two categories:

1. Traditional (which is the geometry of Euclid)

2. Transformational (which is more algebraic than the traditional approach)

When Euclid was formalizing traditional geometry, he based it on five postulates which have come to be known as Euclid's postulates. A **postulate** or **axiom** is a statement accepted without proof. In mathematics, a result that is proved on the basis of some agreed-upon postulates is called a **theorem.**

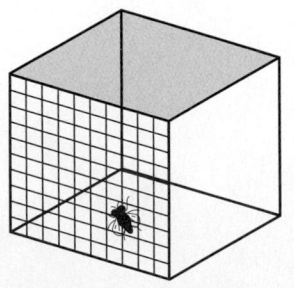

Figure 6.2 Is the fly on the cube or in the cube? On which face?

Euclid's Postulates

1. A straight line can be drawn from any point to any other point.
2. A straight line extends infinitely far in either direction.
3. A circle can be described with any point as center and with a radius equal to any finite straight line drawn from the center.
4. All right angles are equal to each other.
5. Given a straight line and any point not on this line, there is one and only one line through that point that is parallel to the given line.*

The first four of these postulates were obvious and noncontroversial, but the fifth one was different. This fifth postulate looked more like a theorem than a postulate. It was much more difficult to understand than the other four postulates, and for more than 20 centuries mathematicians tried to derive it from the other postulates or to replace it by a more acceptable equivalent. Two straight lines in the same plane are said to be **parallel** if they do not intersect.

Today we can either accept it as a postulate (without proof) or deny it. If it is denied, it turns out that no contradiction results; in fact, if it is not accepted, other geometries called **non-Euclidean geometries** result. If it is accepted, then the geometry that results is consistent with our everyday experiences and is called **Euclidean geometry.**

Let's look at each of Euclid's postulates. The first one says that a straight line can be drawn from any point to any other point.

To connect two points, you need a device called a **straightedge** (a device that we assume has no markings on it; you will use a ruler, but not to measure, when you are treating it as a straightedge). The portion of the line that connects points A and B in Figure 6.3 is called a **line segment.** We write this line segment $\overline{AB}$ (or $\overline{BA}$). We contrast this notation by $\overleftrightarrow{AB}$, which is used to name the line passing through the points A and B. We use the symbol $|\overline{AB}|$ for the length of segment $\overline{AB}$.

The second postulate says that we can draw a straight line. This seems straightforward and obvious, but we should point out that we will indicate a line by putting arrows on each end. If the arrow points in only one direction, the figure is called a **ray.** We write $\overrightarrow{AB}$ or $\overleftarrow{BA}$ for the ray with endpoint A passing through B, as shown in Figure 6.4. If we consider a point on a line, that point separates the line into parts: two **half-lines** and the point itself, as shown in Figure 6.5.

Figure 6.3 A line segment

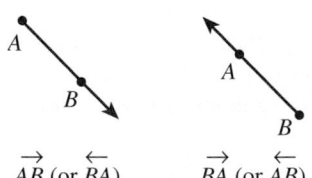

$\overrightarrow{AB}$ (or $\overleftarrow{BA}$) $\overrightarrow{BA}$ (or $\overleftarrow{AB}$)
Figure 6.4 Two rays

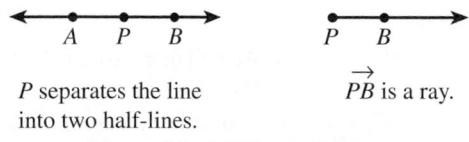

P separates the line into two half-lines. $\overrightarrow{PB}$ is a ray.

Figure 6.5 Half-lines and rays

To construct a line segment of length equal to the length of a given line segment, we need a device called a **compass.** Figure 6.6 shows a compass, which is used to mark off and duplicate lengths, but not to measure them. If objects have exactly the same size and shape, they are called **congruent.**

We can use a straightedge and compass to **construct** a figure so that it meets certain requirements. To *construct a line segment congruent to a given line segment,* copy

*The fifth postulate stated here is the one usually found in high school geometry books. It is sometimes called Playfair's axiom and is equivalent to Euclid's original statement as translated from the original Greek by T. L. Heath: "If a straight line falling on two straight lines makes the interior angle on the same side less than two right angles, the two straight lines, if produced infinitely, meet on that side on which the angles are less than the two right angles."

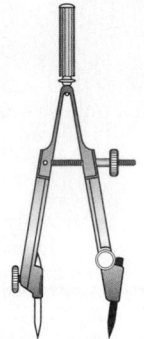

Pointer   Pencil
Figure 6.6 **A compass**

• *P*

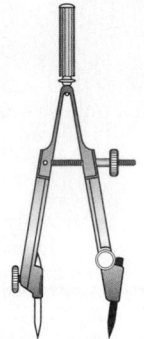

Figure 6.9 **Given a point and a line**

a segment AB on any line ℓ. First fix the compass so that the pointer is on point A and the pencil is on B, as shown in Figure 6.7a. Then, on line ℓ, choose a point C. Next, without changing the compass setting, place the pointer on C and strike an arc at D, as shown in Figure 6.7b.

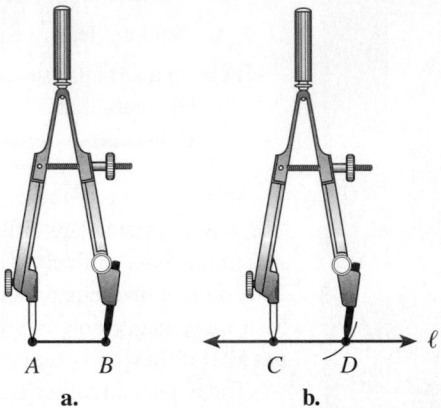

a.   b.

Figure 6.7 **Constructing a line segment**

Euclid's third postulate leads us to a second construction. The task is to construct a circle, given its center and radius. These steps are summarized in Figure 6.8.

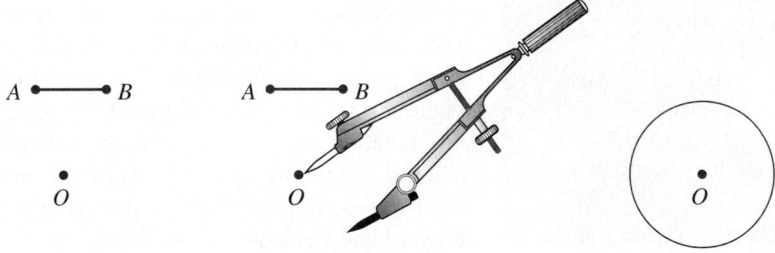

a. Given, a point and a radius of length $\overline{AB}$.

b. Set the legs of the compass on the ends of radius $\overline{AB}$; move the pointer to point O without changing the setting.

c. Hold the pointer at point O and move the pencil end to draw the circle.

Figure 6.8 **Construction of a circle**

We will demonstrate the fourth postulate in the next section when we consider angles.

The final construction of this section will demonstrate the fifth postulate. The task is to construct a line through a point P parallel to a given line ℓ, as shown in Figure 6.9. First, draw any line through P that intersects ℓ at a point A, as shown in Figure 6.10a.

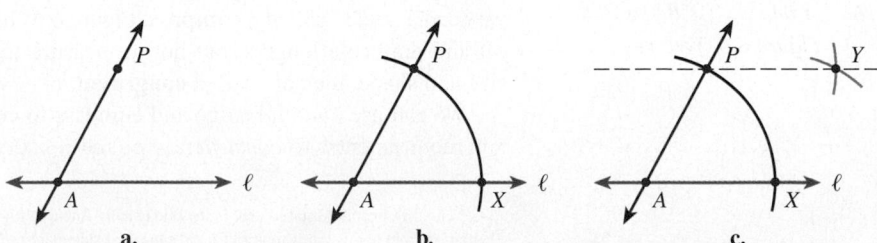

a.   b.   c.

Figure 6.10 **Construction of a line parallel to a given line through a given point**

Now draw an arc with the pointer at A and radius $\overline{AP}$, and label the point of intersection of the arc and the line X, as shown in Figure 6.10b. With the same opening of the compass, draw an arc first with the pointer at P and then with the pointer at X. Their point of intersection will determine a point Y. Draw the line through both P and Y. This line is parallel to ℓ.

Transformational Geometry

We now turn our attention to the second category of geometry. **Transformational geometry** is quite different from traditional geometry in that it deals with the study of *transformations.* A **transformation** is the passage from one geometric figure to another by means of reflections, translations, rotations, contractions, or dilations. For example, given a line L and a point P, as shown in Figure 6.11, we call the point P' the **reflection** of P about the line L if $\overline{PP'}$ is perpendicular to L and is also bisected by L. Each point in the plane has exactly one reflection point corresponding to a given line L. A reflection is called a *reflection transformation,* and the line of reflection is called the **line of symmetry.** The easiest way to describe a line symmetry is to say that if you fold a paper along its line of symmetry, then the figure will fold onto itself to form a perfect match, as shown in Figure 6.12.

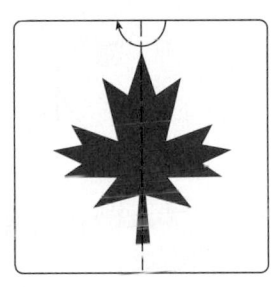
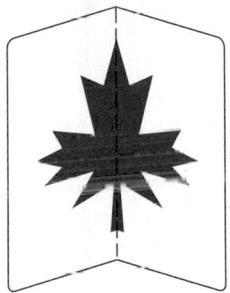

Figure 6.12 **Line symmetry on the Maple Leaf of Canada**

Many everyday objects exhibit a line of symmetry. From snowflakes in nature, to the Taj Mahal in architectural design, to many flags and logos, we see examples of lines of symmetry (see Figure 6.13).

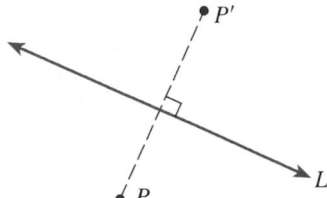

Escher's *Drawing Hands*

See www.mathnature.com
for some Escher links.

WWW

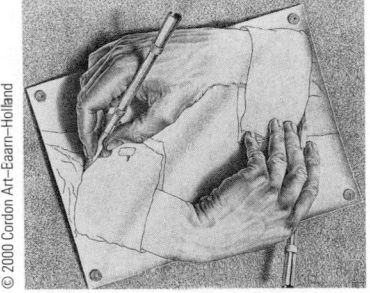

Figure 6.11 **A reflection**

Snowflakes

Taj Mahal

Chrysler logo

Figure 6.13 **Examples of line symmetry**

Other transformations include *translations, rotations, dilations,* and *contractions,* which are illustrated in Figure 6.14.

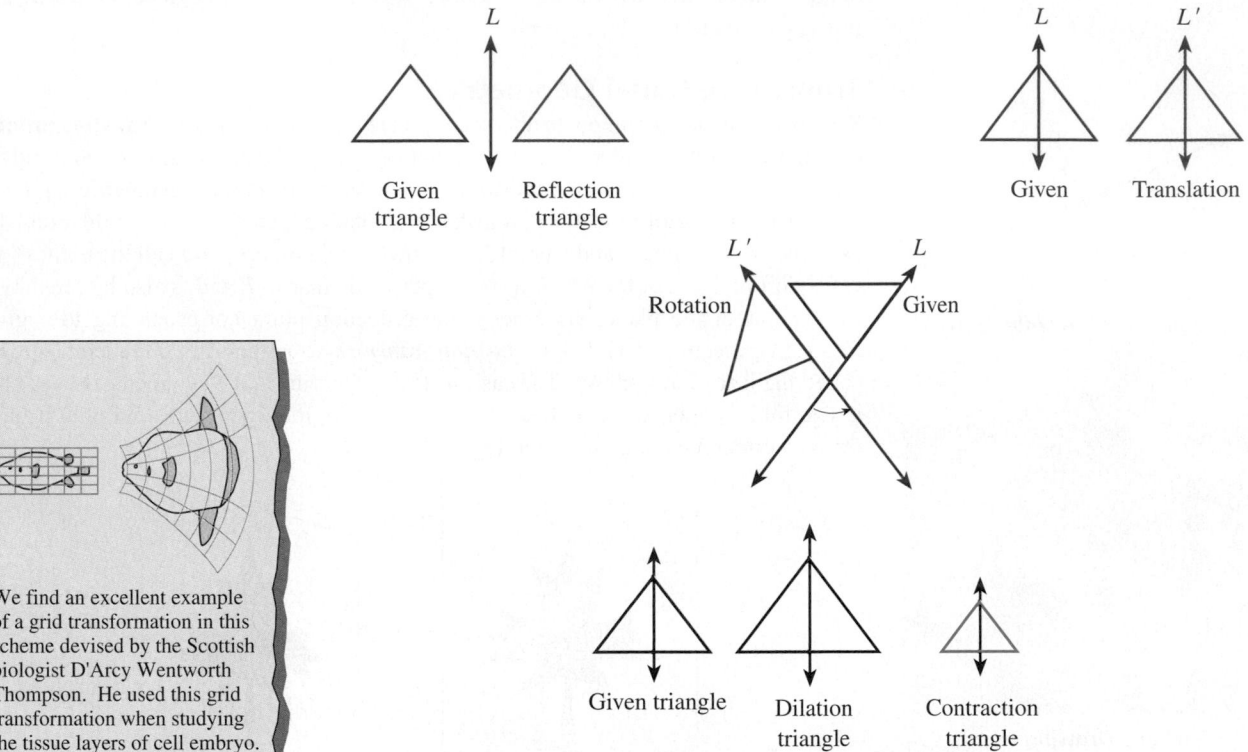

We find an excellent example of a grid transformation in this scheme devised by the Scottish biologist D'Arcy Wentworth Thompson. He used this grid transformation when studying the tissue layers of cell embryo.

Figure 6.14 **Transformations of a fixed geometric figure**

Geometry is also concerned with the study of the relationships between geometric figures. A primary relationship is that of *congruence.* A second relationship is called **similarity.** Two figures are said to be **similar** if they have the same shape, although not necessarily the same size. These ideas are considered in Sections 6.3 and 6.4.

If we were to develop this text formally, we would need to be very careful about the statement of our postulates. The notion of mathematical proof requires a very precise formulation of all postulates and theorems. For example, Euclid based many proofs about congruence on a postulate that said, "Things that coincide with one another are equal to one another" (Common Notions, Book I, *Elements*). This statement is not formulated precisely enough to be used as justification for some of the things Euclid did with congruence. Even the explanation we gave for congruence is not precise enough to be used in a formal course. However, since we are not formally developing geometry, we simply accept the general drift of the statement. Remember, though, that you cannot base a mathematical proof on general drift. On the other hand, there are some facts that we *know to be true* that tell us that certain properties are impossible. For example, if someone claims to be able to trisect an angle with a straightedge and compass, we know *without looking at the construction* that the construction is wrong. This can be frustrating to someone who believes that he or she has accomplished the impossible. In fact, it was so frustrating to Daniel Wade Arthur that he was motivated

Calligraphers sometimes experiment with some inversion transformations.

John Langdon has been creating such word designs for about 15 years, and the two shown here are from his book Wordplay. *The following inversion was found in the April 1992 issue of* OMNI *magazine.*

"THE WHOLE OF SCIENCE IS NOTHING MORE THAN A REFINEMENT OF EVERY-DAY THINKING."
ALBERT EINSTEIN

science

"ALL THAT SCIENCE CAN ACHIEVE IS A PERFECT KNOWLEDGE AND A PERFECT UNDERSTANDING OF THE NATURAL AND MORAL FORCES."
HERMAN LUDWIG FERDINAND VON HELMHOLTZ

Here is a favorite of mine from Langdon's book:

Three calligraphy illustrations from an article by Scott Morris, *OMNI Magazine* April 1992, p. 4.

www See www.mathnature.com for one student's attempt at trisecting an angle.

to take out a paid advertisement in the *Los Angeles Times,* which is reproduced in the News Clip.

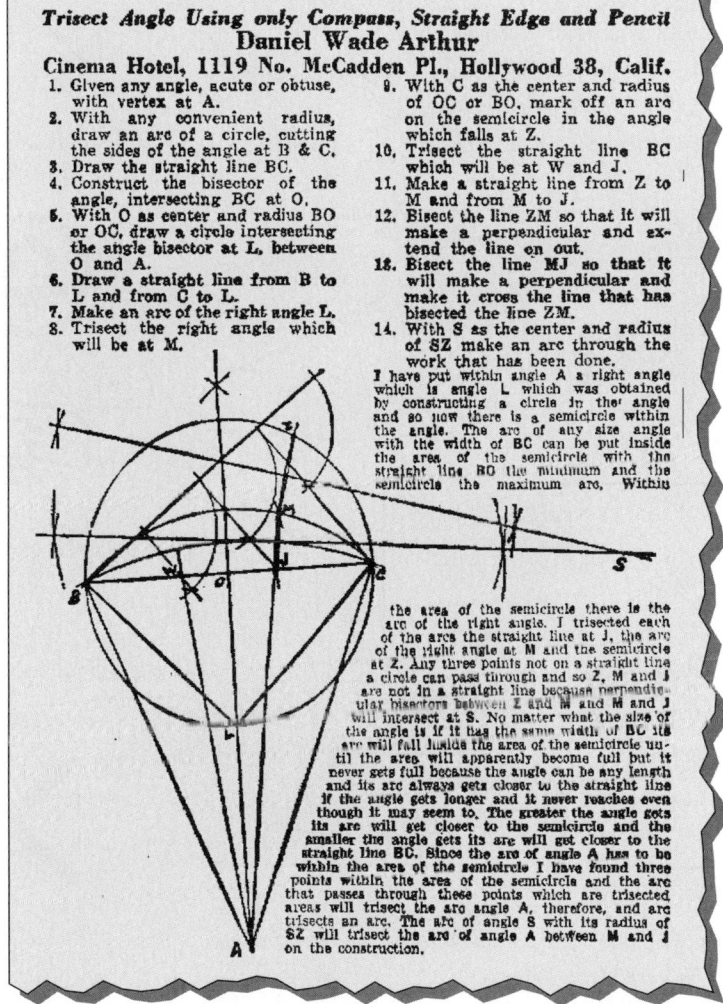

Advertisement by Daniel Wade Arthur, from the *Los Angeles Times.*

Problem Set 6.1

LEVEL 1

1. Is the woman in the figure a young woman or an old woman?

2. IN YOUR OWN WORDS Why do you think Problem 1 is included in this problem set? How does this question relate to working problems in geometry?

3. IN YOUR OWN WORDS Describe a procedure for constructing a line segment congruent to a given segment.

4. IN YOUR OWN WORDS Describe a procedure for constructing a circle with a radius congruent to a given segment.

5. **IN YOUR OWN WORDS** Discuss what it means to be an undefined term.

6. **IN YOUR OWN WORDS** Describe line symmetry.

7. **IN YOUR OWN WORDS** What is the difference between an axiom and a theorem?

8. **IN YOUR OWN WORDS** What are the two categories into which geometry is usually separated?

Construct a segment congruent to each segment given in Problems 9–12.

9. A line segment

10. A line segment

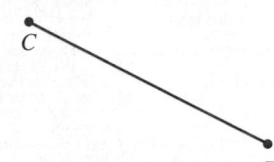

11. A line segment

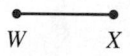

12. A line segment

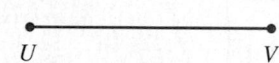

In Problems 13–16, construct a circle with given radius.

13. Circle with given radius

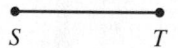

14. Circle with given radius

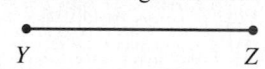

15. Circle with given radius

16. Circle with given radius

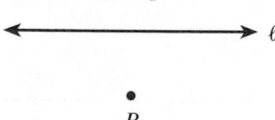

In Problems 17–20, construct a line through the given point parallel to the given line.

17. Line through *P* parallel to ℓ

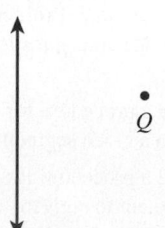

18. Line through *Q* parallel to *m*

19. Line through *R* parallel to *n*

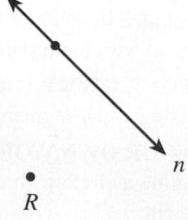

20. Line through *S* parallel to *k*

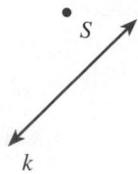

Find at least one line of symmetry for each of the illustrations in Problems 21–24.

21.

22.

23.

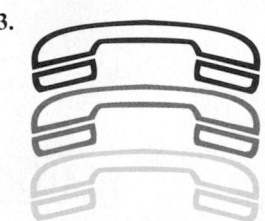

24.

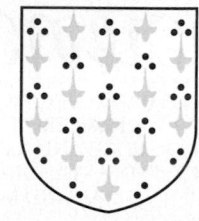

Find at least one line of symmetry for each of the state flags given in Problems 25–28. If a line symmetry is not possible, so state.

25.

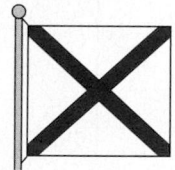

26.

27.

28.

LEVEL 2

Use the given illustration to draw the figures requested in Problems 29–36.

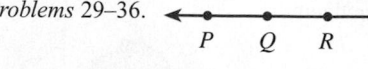

29. $\overline{PQ}$

30. $\overline{RS}$

31. $\overrightarrow{PQ}$

32. $\overleftrightarrow{RS}$

33. $\overrightarrow{PQ}$

34. $\overline{SR}$

35. $\overleftrightarrow{PQ}$

36. $\overrightarrow{RS}$

Which of the pictures in Problems 37–45 illustrate a line symmetry?

37. Chambered nautilus

38. Butterfly

39. *Winged Lion,* China, 7th–8th century A.D.

40. *Dodecahedron,* drawn by Leonardo da Vinci

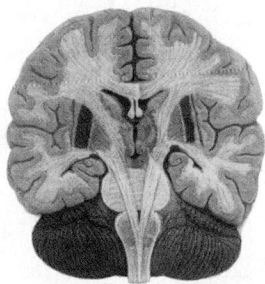

41. Human brain

42. Restaurant at Los Angeles International Airport

43. Human circulatory system

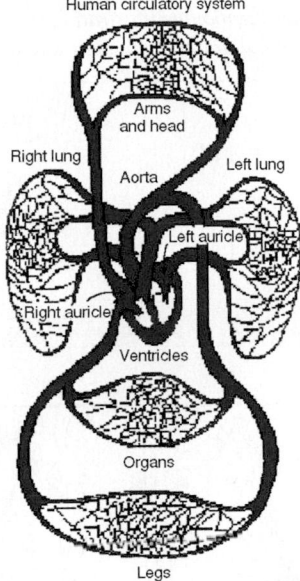

44. Empire State Building New York City

45. *Kaiser* porcelain vase

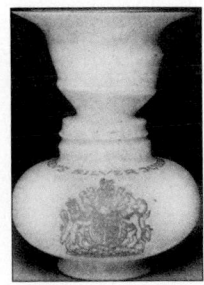

LEVEL 3

In Problems 46–47 label each cartoon as illustrating a translation, reflection, rotation, dilation, or a contraction.

46.

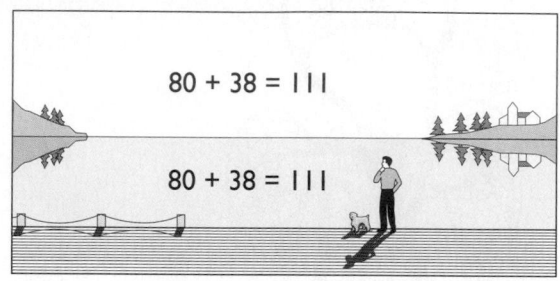

47.

Study the patterns shown here. When folded they will form cubes spelling CUBE. Letter each pattern in Problems 48–53. Answers are not unique.

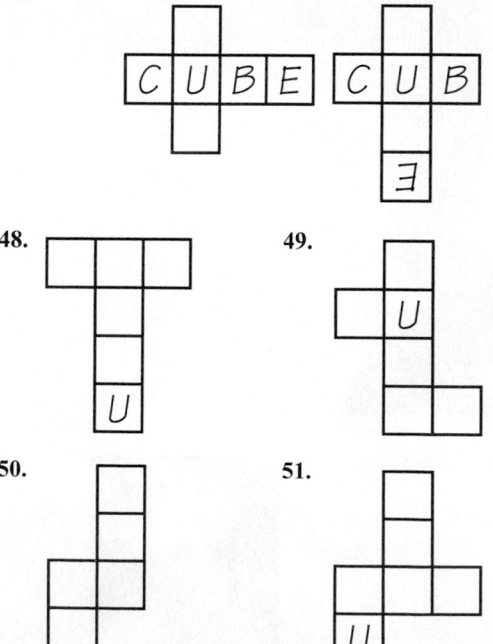

48.

49.

50.

51.

52.

53.

Start with the given cube in Problems 54–58. Then rotate the cube 90° in the directions indicated by the arrows, and select the cube marked A, B, C, or D that most closely matches the result.

54.
Rotation

A   B   C   D

55.
Rotation

A   B   C   D

56.
Rotation

A   B   C   D

57.
Rotation

A   B   C   D

58.
Rotation

A   B   C   D

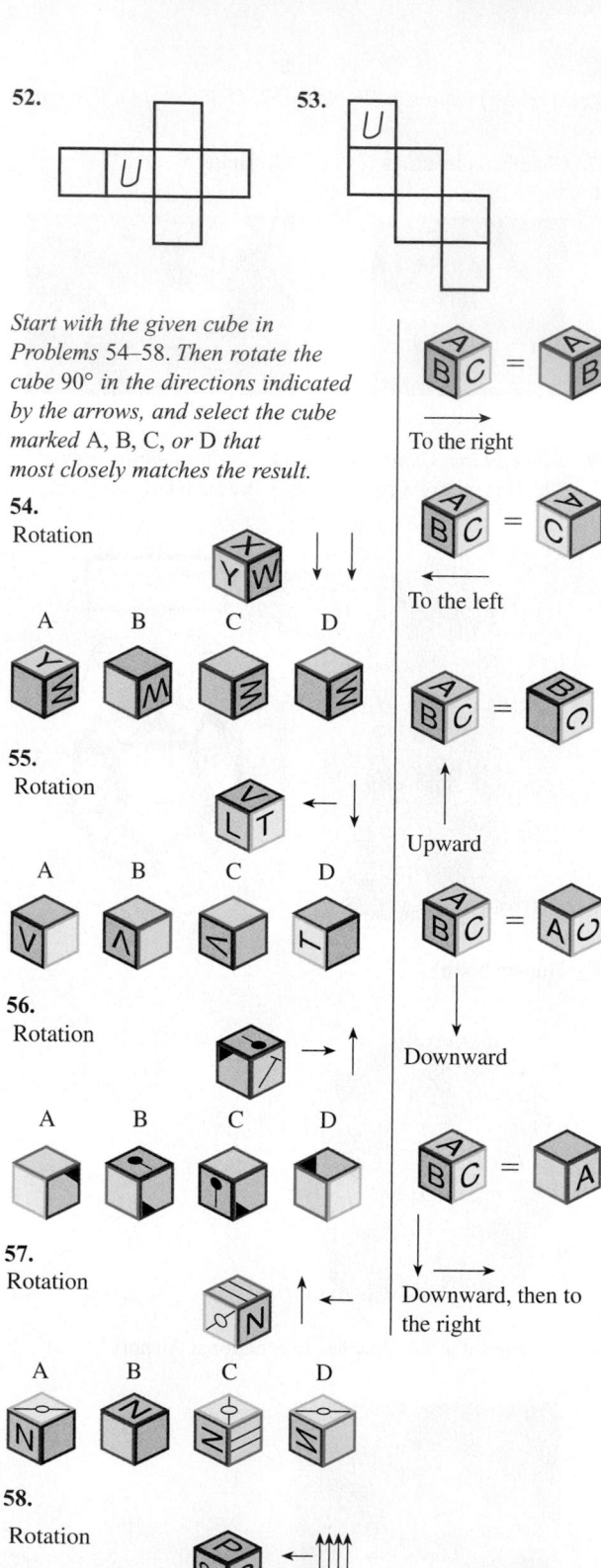

59. The letters of the alphabet can be sorted into the following categories: (1) FGJLNPQRSZ; (2) BCDEK; (3) AMTUVWY; and (4) HIOX. What defines these categories?

60. The "Mirror Image" illustration shows a problem condensed from *Games* magazine. Answer the questions asked in that problem.

Mirror Image
by Diane Dawson

In the Fabulous Kingdom, anything can happen—and usually does. When we held a looking glass up to this page, we noticed a few incongruities between it and its "reflection" on the facing page. Can you spot thirty *substantial* differences here, with or without the help of a mirror? (Since both illustrations were hand-drawn there are bound to be hairline differences—these should be ignored.)

"Mirror Image" by Diane Dawson, reprinted from *Games Magazine*, May/June 1979.

Mirror Image
by Diane Dawson

In the Fabulous Kingdom, anything can happen—and usually does. When we held a looking glass up to this page, we noticed a few incongruities between it and its "reflection" on the facing page. Can you spot thirty *substantial* differences here, with or without the help of a mirror? (Since both illustrations were hand-drawn there are bound to be hairline differences—these should be ignored.)

6.2

POLYGONS AND ANGLES

When the Euclidean geometry introduced in the preceding section is studied in high school as an entire course, it is usually presented in a *formal* manner using definitions, axioms, and theorems. The development of this chapter is *informal,* which means that we base our results on observations and intuition. We begin by assuming that you are familiar with the ideas of *point, line,* and *plane.*

Polygons

A **polygon** is a geometric figure that has three or more straight sides, all of which lie on a flat surface or plane so that the starting point and the ending point are the same. Polygons can be classified according to their number of sides, as shown in Figure 6.15. We say a polygon is **regular** if its sides are the same length.

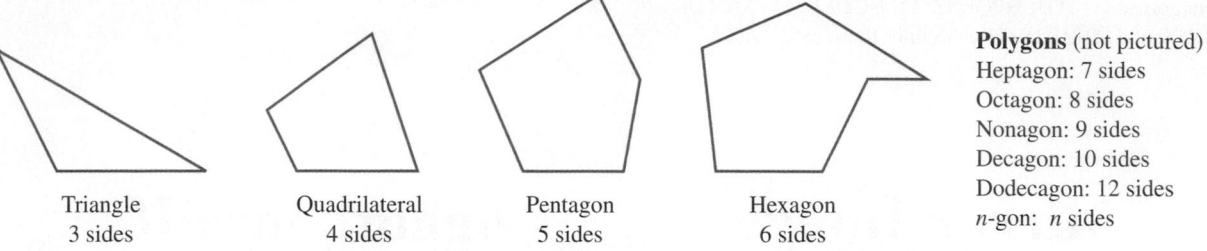

Polygons (not pictured)
Heptagon: 7 sides
Octagon: 8 sides
Nonagon: 9 sides
Decagon: 10 sides
Dodecagon: 12 sides
n-gon: *n* sides

Triangle
3 sides

Quadrilateral
4 sides

Pentagon
5 sides

Hexagon
6 sides

Figure 6.15 **Polygons classified according to number of sides**

Angles

A connecting point of two sides is called a **vertex** (plural **vertices**) and is usually designated by a capital letter. An **angle** is composed of two rays or segments with a common endpoint. The angles between the sides of a polygon are sometimes also denoted by a capital letter, but other ways of denoting angles are shown in Figure 6.16.

> **CAUTION** This notation for angles is used throughout all of mathematics.

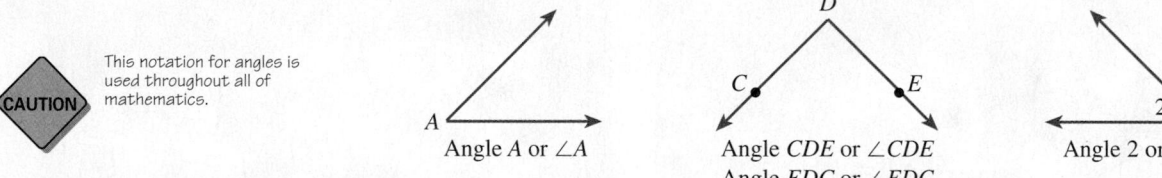

Angle *A* or ∠*A*

Angle *CDE* or ∠*CDE*
Angle *EDC* or ∠*EDC*

Angle 2 or ∠2

Figure 6.16 **Ways of denoting angles**

EXAMPLE 1

Locate each of the following angles in Figure 6.17.

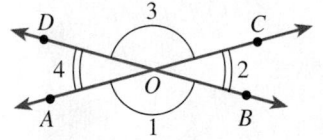

Figure 6.17

Solution

a. ∠*AOB*   **b.** ∠*COB*   **c.** ∠*BOA*

d. ∠*DOC*   **e.** ∠3   **f.** ∠4   ◆

To construct an angle with the same size as a given angle *B*, first draw a ray from *B′*, as shown in Figure 6.18a. Next, mark off an arc with the pointer at the vertex of the given angle and label the points *A* and *C*. Without changing the compass, mark off a similar arc with the pointer at *B′*, as shown in Figure 6.18b. Label the point *C′* where this arc crosses the ray from *B′*. Place the pointer at *C* and set the compass to the distance from *C* to *A*. Without changing the compass, put the pointer at *C′* and strike an arc to make a point of intersection *A′* with the arc from *C′*, as shown in Figure 6.18c. Finally, draw a ray from *B′* through *A′*.

Two angles are said to be **equal** if they describe the same angle. If we write *m* in front of an angle symbol, we mean the measure of the angle rather than the angle itself. Notice in Example 1 that parts **a** and **c** name the *same angle,* so ∠*AOB* = ∠*BOA*. Also notice the single and double arcs used to mark the angles in Example 1; these are used to denote angles with equal measure, so *m*∠*COB* = *m*∠*AOD* and *m*∠*COD* = *m*∠*AOB*, but ∠*COD* ≠ ∠*AOB* (since they are not the same angle).

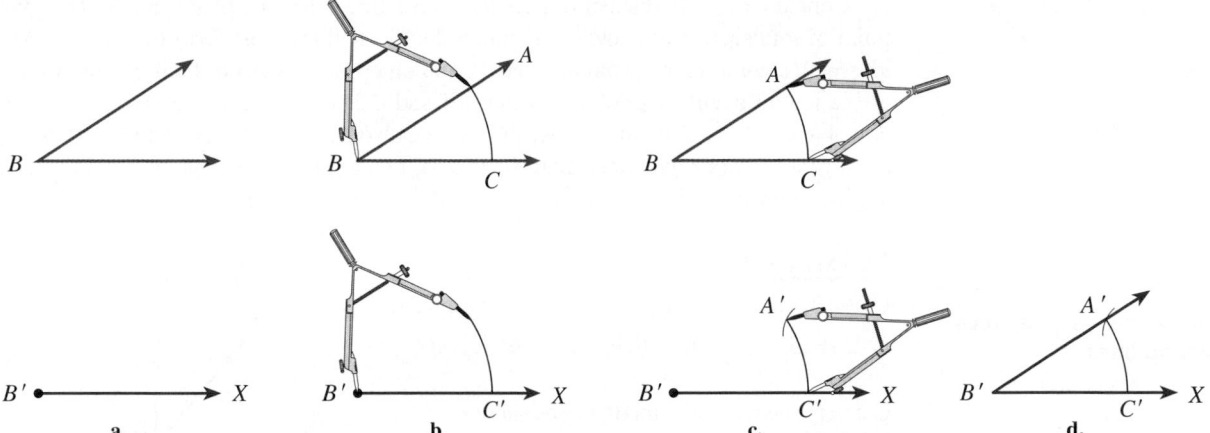

Figure 6.18 **Construction of an angle congruent to a given angle**

Denoting an angle by a single letter is preferred except in the case (as shown by Example 1) where several angles share the same vertex.

Angles are usually measured using a unit called a **degree,** which is defined to be $\frac{1}{360}$ of a full revolution. The symbol ° is used to designate degrees. To measure an angle, you can use a **protractor,** but in this book the angle whose measures we need will be labeled as in Figure 6.19.

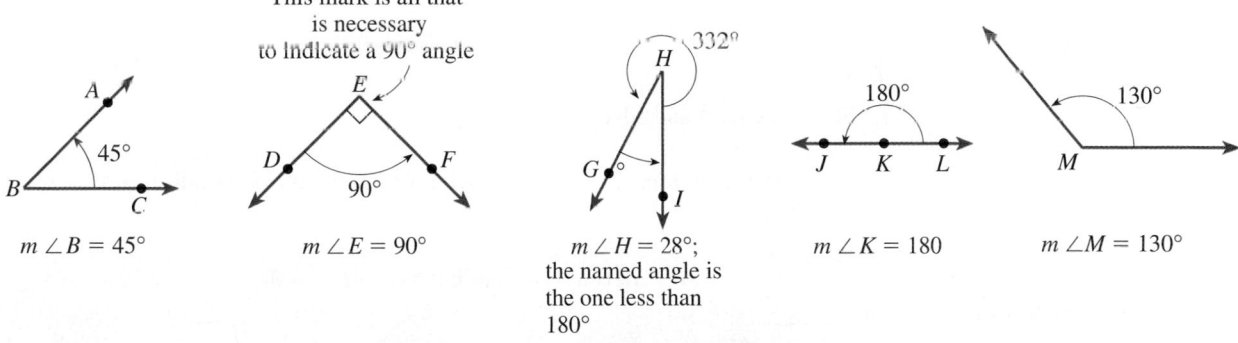

Figure 6.19 **Labeling angles**

Table 6.1 Types of Angles

Angle Measure	Classification
Less than 90°	**Acute**
Equal to 90°	**Right**
Between 90° and 180°	**Obtuse**
Equal to 180°	**Straight**

Angles are sometimes classified according to their measures, as shown in Table 6.1. Experience leads us to see the plausibility of Euclid's fourth postulate that all right angles are congruent to one another.

EXAMPLE 2

Label the angles *B, E, H, K,* and *M* in Figure 6.19 by classification.

Solution

∠*B* is acute; ∠*E* is right; ∠*H* is acute; ∠*K* is straight; ∠*M* is obtuse.   ◆

STOP

Remember these names for types of angles.

Two angles with the same measure are said to be **congruent.** If the sum of the measures of two angles is 90°, they are called **complementary angles,** and if the sum is 180°, they are called **supplementary angles.**

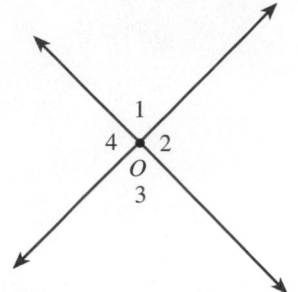

Figure 6.20 **Angles formed by intersecting lines**

Consider any two distinct (different) intersecting lines in a plane, and let O be the point of intersection as shown in Figure 6.20. These lines must form four angles. Angles with a common ray, common vertex, and on opposite sides of their common sides are called **adjacent angles.** We say that $\angle 1$ and $\angle 2$, $\angle 2$ and $\angle 3$, $\angle 3$ and $\angle 4$, as well as $\angle 4$ and $\angle 1$ are pairs of adjacent angles. We also say that $\angle 1$ and $\angle 3$ as well as $\angle 2$ and $\angle 4$ are pairs of **vertical angles**—that is, two angles for which each side of one angle is a prolongation through the vertex of a side of the other.

EXAMPLE 3

Classify the named angles shown in Figure 6.21.

a. $\angle AOB$ **b.** $\angle BOC$ **c.** $\angle DAO$

Classify the named pairs of angles shown in Figure 6.21.

d. $\angle AOB$ and $\angle BOC$ **e.** $\angle BOC$ and $\angle OPE$

f. $\angle AOB$ and $\angle COP$ **g.** $\angle FPE$ and $\angle EPO$

Solution

a. Acute

b. Obtuse

c. Right

d. Supplementary and adjacent

e. Congruent

f. Vertical

g. Supplementary and adjacent

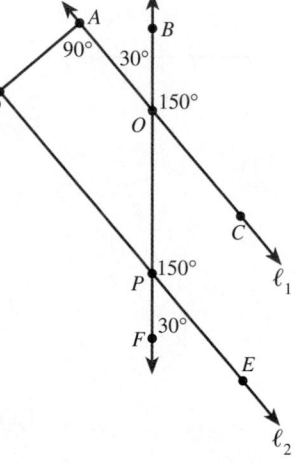

Figure 6.21 **Classifying angles**

A polygon with four sides is a **quadrilateral.** Some other classifications are given in the following box.

Classification of Quadrilaterals

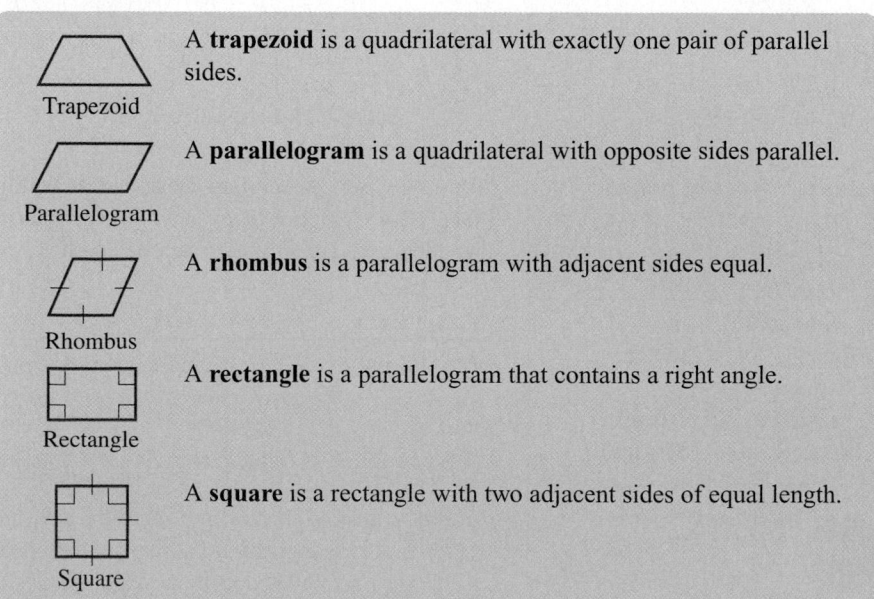

A **trapezoid** is a quadrilateral with exactly one pair of parallel sides.

Trapezoid

A **parallelogram** is a quadrilateral with opposite sides parallel.

Parallelogram

A **rhombus** is a parallelogram with adjacent sides equal.

Rhombus

A **rectangle** is a parallelogram that contains a right angle.

Rectangle

A **square** is a rectangle with two adjacent sides of equal length.

Square

Angles with Parallel Lines

Consider three lines arranged similarly to those shown in Example 3. Suppose that two of the lines, say, ℓ_1 and ℓ_2, are **parallel** (that is, they lie in the same plane and never intersect), and also that a third line ℓ_3 intersects the parallel lines at points P and Q, as shown in Figure 6.22. The line ℓ_3 is called a **transversal.**

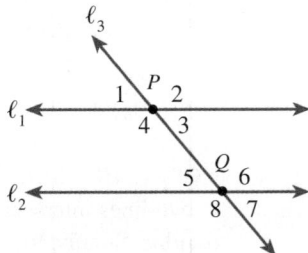

Figure 6.22 **Parallel lines cut by a transversal**

We make some observations about angles:

Vertical angles are congruent.

Alternate interior angles are pairs of angles whose interiors lie between the parallel lines, but on opposite sides of the transversal, each having one of the lines for one of its sides. Alternate interior angles are congruent.

Alternate exterior angles are pairs of angles that lie outside the parallel lines, but on opposite sides of the transversal, each with one side adjacent to each parallel. Alternate exterior angles are congruent.

Corresponding angles are two nonadjacent angles whose interiors lie on the same side of the transversal such that one angle lies between the parallel lines and the other does not. Corresponding angles are congruent.

EXAMPLE 4

Consider Figure 6.23.

a. Name the vertical angles.   **b.** Name the alternate interior angles.

c. Name the corresponding angles.   **d.** Name the alternate exterior angles.

Solution

a. The vertical angles are: $\angle 1$ and $\angle 3$; $\angle 2$ and $\angle 4$; $\angle 5$ and $\angle 7$; $\angle 6$ and $\angle 8$

b. The alternate interior angles are: $\angle 4$ and $\angle 6$; $\angle 3$ and $\angle 5$

c. The corresponding angles are: $\angle 1$ and $\angle 5$; $\angle 2$ and $\angle 6$; $\angle 3$ and $\angle 7$; $\angle 4$ and $\angle 8$

d. The alternate exterior angles are: $\angle 1$ and $\angle 7$; $\angle 2$ and $\angle 8$ ◆

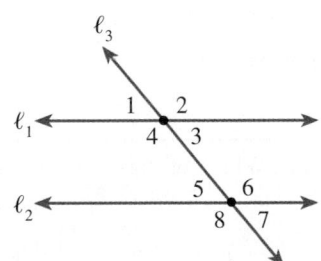

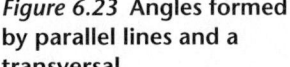

Figure 6.23 **Angles formed by parallel lines and a transversal.**

To summarize the results from Example 4, notice that the following angles are congruent, written $\simeq$:

$$\angle 1 \simeq \angle 3 \simeq \angle 5 \simeq \angle 7 \quad \text{and} \quad \angle 2 \simeq \angle 4 \simeq \angle 6 \simeq \angle 8$$

Also, all pairs of adjacent angles are supplementary.

EXAMPLE 5

Find the measures of the eight numbered angles in Figure 6.23, where ℓ_1 and ℓ_2 are parallel. Assume that $m\angle 5 = 50°$.

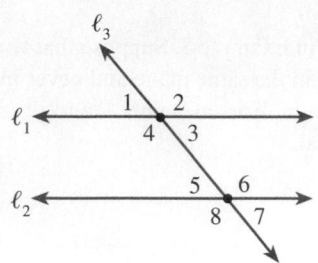

Solution

$m\angle 1 = 50°$ Corresponding angles

$m\angle 2 = 180° - 50° = 130°$ $\angle 1$ and $\angle 2$ are supplementary.

$m\angle 3 = 50°$ $\angle 3$ and $\angle 1$ are vertical angles.

$m\angle 4 = 130°$ Vertical angles

$m\angle 5 = 50°$ Given

$m\angle 6 = 130°$ Supplementary angles

$m\angle 7 = 50°$ Vertical angles

$m\angle 8 = 130°$ Supplementary angles ◆

Perpendicular Lines

If two lines intersect so that the adjacent angles are equal, then the lines are **perpendicular.** Simply, lines that intersect to form angles of 90° (right angles) are called perpendicular lines. In Figure 6.24, lines ℓ_3 and ℓ_4 intersect to form a right angle, and therefore they are perpendicular lines.

In a diagram on a printed page, any line that is parallel to the top and bottom edge of the page is considered **horizontal.** Lines that are perpendicular to a horizontal line are considered to be **vertical.** In Figure 6.25, line ℓ_5 is a horizontal line and line ℓ_6 is a vertical line.

Figure 6.24 **Perpendicular lines**

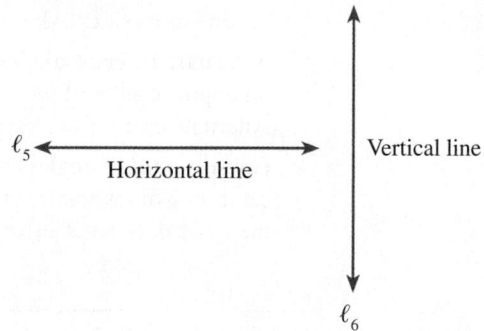

Figure 6.25 **Horizontal and vertical lines**

EXAMPLE 6

The diagram in Figure 6.26 shows lines in the same plane. Which of the given statements are true?

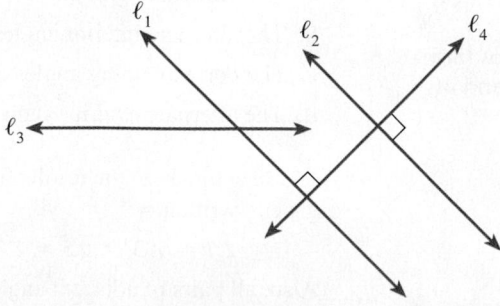

Figure 6.26

a. Lines ℓ_1 and ℓ_2 are parallel and horizontal.

b. Lines ℓ_4 and ℓ_2 are intersecting and not perpendicular.

c. Lines ℓ_2 and ℓ_3 are intersecting lines.

Solution

a. Lines ℓ_1 and ℓ_2 are parallel, but they are not horizontal. Thus, the statement (which is a conjunction) is false.

b. Lines ℓ_4 and ℓ_2 are both intersecting and perpendicular. The statement is false.

c. Lines ℓ_2 and ℓ_3 are intersecting. Recall that the arrows on ℓ_3 indicate that it goes on without end in both directions, so it will intersect ℓ_2. This statement is true.   ◆

Problem Set 6.2

LEVEL 1

1. **IN YOUR OWN WORDS** What is an angle?

2. **IN YOUR OWN WORDS** Distinguish between equal angles and congruent angles.

3. **IN YOUR OWN WORDS** Distinguish between a half-line and a ray.

4. **IN YOUR OWN WORDS** What is a quadrilateral? Describe five different classifications of quadrilaterals.

5. **IN YOUR OWN WORDS** Distinguish between horizontal and vertical lines.

6. **IN YOUR OWN WORDS** Describe right angles, acute angles, and obtuse angles.

7. **IN YOUR OWN WORDS** Describe adjacent, vertical, and corresponding angles.

8. **IN YOUR OWN WORDS** Describe parallel lines.

Name the polygons in Problems 9–14 according to the number of sides.

9. **a.** ▫   **b.**

10. **a.** ⬡   **b.**

11. **a.** △   **b.**

12. **a.**   **b.**

13. **a.** ▱   **b.**

14. **a.**   **b.**

Determine whether each sentence in Problems 15–19 is true or false.

15. **a.** Every square is a rectangle.
    b. Every square is a parallelogram.

16. **a.** Every square is a rhombus.
    b. Every rhombus is a square.

17. **a.** Every square is a quadrilateral.
    b. A parallelogram is a rectangle.

18. **a.** A rectangle is a parallelogram.
    b. A trapezoid is a quadrilateral.

19. **a.** A quadrilateral is a trapezoid.
    b. A parallelogram is a trapezoid.

For the quadrilaterals named in Problems 20–24, answer "yes" or "no" to indicate whether each of the following properties are satisfied:

a. opposite sides parallel
b. opposite sides have equal length
c. opposite angles have equal measure
d. interior angles are right angles
e. diagonals have equal length

20. rectangle   21. square
22. parallelogram   23. trapezoid
24. rhombus

LEVEL 2

Using only a straightedge and a compass, reproduce the angles given in Problems 25–30.

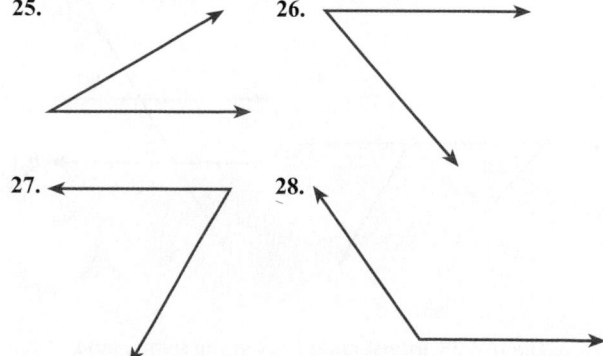

25.   26.

27.   28.

29. 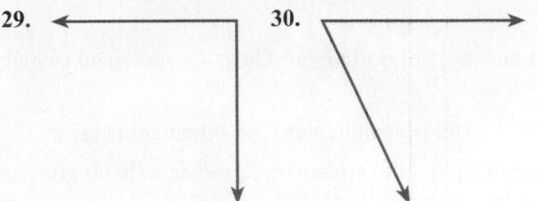 **30.**

In Problems 31–35, classify the requested angles shown in Figure 6.27, where ℓ_1 and ℓ_2 are parallel.

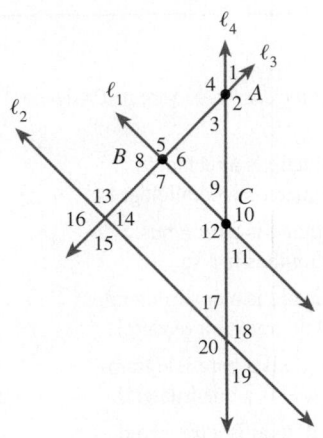

Figure 6.27

31. a. $\angle BAC$ if $m\angle 1$ is $30°$ **b.** $\angle ABC$ if $m\angle 5$ is $90°$
32. a. $\angle 18$ if $m\angle 17$ is $105°$ **b.** $\angle 19$ if $m\angle 11$ is $70°$
33. a. $\angle 10$ if $m\angle 11$ is $90°$ **b.** $\angle 16$ if $m\angle 15$ is $30°$
34. a. $\angle CBA$ if $m\angle 16$ is $120°$ **b.** $\angle BCA$ if $m\angle 19$ is $110°$
35. a. $\angle 1$ if $m\angle 2$ is $130°$ **b.** $\angle 5$ if $m\angle 15$ is $88°$

In Problems 36–43, classify the pairs of angles shown in Figure 6.27.

36. $\angle 2$ and $\angle 4$ **37.** $\angle 13$ and $\angle 14$
38. $\angle 9$ and $\angle 12$ **39.** $\angle 9$ and $\angle 10$
40. $\angle 9$ and $\angle 17$ **41.** $\angle 9$ and $\angle 11$
42. $\angle 12$ and $\angle 18$ **43.** $\angle 7$ and $\angle 13$

Each of the four diagrams in Figure 6.28 shows lines in the same plane. Classify each of the statements in Problems 44–49 as true or false.

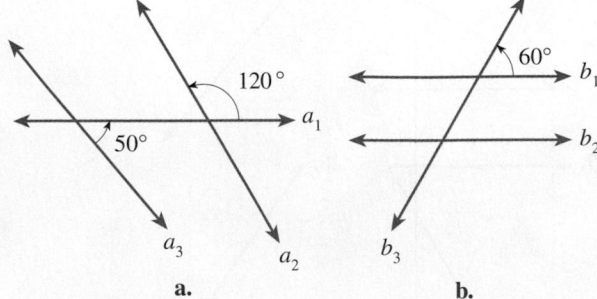

a.   **b.**
Figure 6.28 Intersecting lines in the same plane

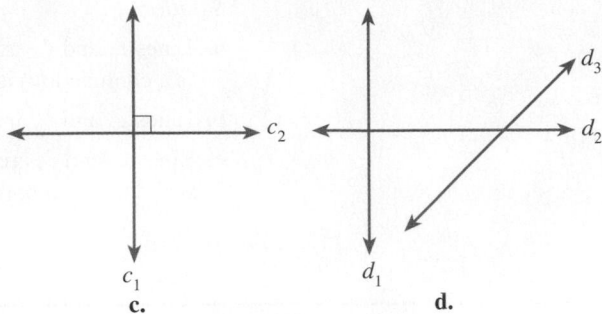

c.   **d.**

Figure 6.28 Continued

44. The lines a_1 and a_3 are both intersecting and parallel.
45. The lines b_1 and b_2 are vertical.
46. The lines c_1 and c_2 are not intersecting.
47. The lines d_1 and d_3 are intersecting.
48. The lines c_2 and d_2 are horizontal.
49. The lines a_2 and a_3 are both vertical and parallel.

In Problems 50–55, find the measures of all the angles in Figure 6.29.

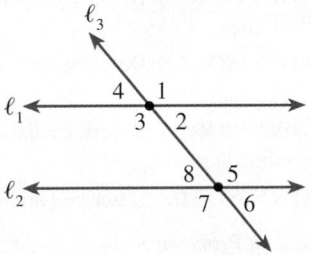

Figure 6.29 ℓ_1 is parallel to ℓ_2

50. Given $m\angle 7 = 110°$ **51.** Given $m\angle 2 = 65°$
52. Given $m\angle 6 = 19°$ **53.** Given $m\angle 1 = 153°$
54. Given $m\angle 5 = 120°$ **55.** Given $m\angle 3 = 163°$

LEVEL 3

56. The legs of a picnic table form a triangle where $\overline{AC}$ and $\overline{BC}$ have the same length, as shown in Figure 6.30. If $m\angle ACB = 90°$, find the measures of angles x and y so that the top of the table will be parallel to the ground.

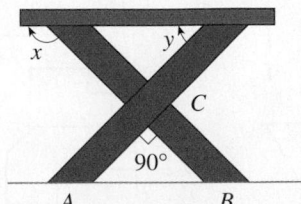

Figure 6.30 Picnic table

57. Repeat Problem 56 where $m\angle ACB = 85°$.
58. Repeat Problem 56 where $m\angle ACB = 92°$.

59. The first illustration in the accompanying figure shows a cube with the top cut off. Use solid lines and shading to depict seven other different views of a cube with one side cut off.

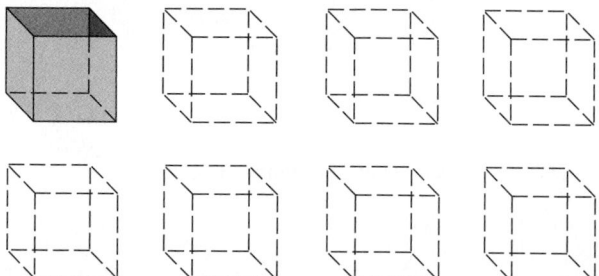

60. Allow me to start you on a journey in Golygon City. You can take a similar trip in New York, Tokyo, or almost any large city whose streets form a grid of squares. Here are your directions. Stroll down a city block, and at the end turn left or right. Walk two more blocks, turn left or right, then walk another three blocks, turn once more, and so on. Each time you turn, you must walk straight one block farther than before. If after a number of turns you arrive at

your starting point, you have traced a golygon, as shown in Figure 6.31. A *golygon* consists of straight-line segments that have lengths (measured in miles, meters, or whatever unit you prefer) of one, two, three, and so on, units. Draw some golygons. Can you make a conjecture about golygons?*

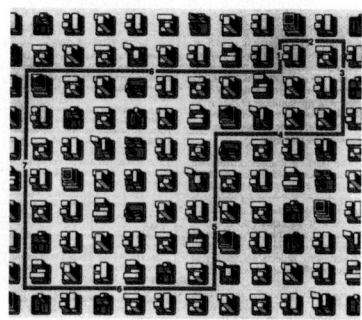

Figure 6.31 **A sample golygon**

*From "Mathematical Recreations," by A. K. Dewdney, *Scientific American,* July 1990, p. 118. Copyright © 1990 by Scientific American, Inc. All rights reserved. Illustration by Slim Films.

6.3 | TRIANGLES

Terminology

One of the most frequently encountered polygons is the **triangle.** Every triangle has six parts: three sides and three angles. We name the sides by naming the endpoints of the line segments, and we name the angles by identifying the vertex (see Figure 6.32).

A triangle and the parts of a triangle are essential ideas.

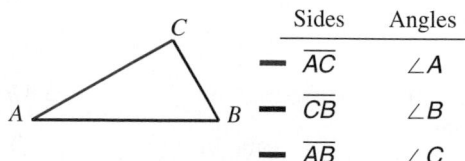

Sides	Angles
$\overline{AC}$	$\angle A$
$\overline{CB}$	$\angle B$
$\overline{AB}$	$\angle C$

Figure 6.32 **A standard triangle showing the six parts**

Triangles are classified both by sides and by angles:

By Sides

Scalene: no equal sides

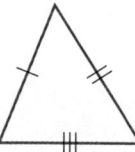

Scalene triangle

By Angles

Acute: three acute angles

Acute triangle

(continued)

By Sides
Isosceles: two equal sides

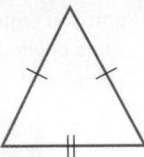

Isosceles triangle

By Angles
Right: one right angle

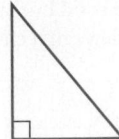

Right triangle

Equilateral: three equal sides

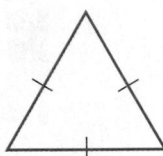

Equilateral triangle

Obtuse: one obtuse angle

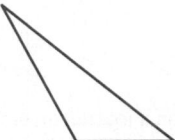

Obtuse triangle

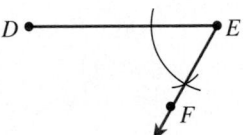

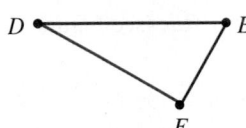

Figure 6.33 **Constructing congruent triangles**

We say that two triangles are **congruent** if they have the same size and shape. Suppose that we wish to construct a triangle with vertices D, E, and F, congruent to $\triangle ABC$ as shown in Figure 6.32. We would proceed as follows (as shown in Figure 6.33):

1. Mark off segment $\overline{DE}$ so that it is congruent to $\overline{AB}$. We write this as $\overline{DE} \simeq \overline{AB}$.

2. Construct angle E so that it is congruent to angle B. We write this as $\angle E \simeq \angle B$.

3. Mark off segment $\overline{EF} \simeq \overline{BC}$.

You can now see that, if you connect points D and F with a straightedge, the resulting $\triangle DEF$ has the same size and shape as $\triangle ABC$. The procedure we used here is called SAS, meaning we constructed two sides and an *included angle* (an angle between two sides) congruent to two sides and an included angle of another triangle. We call these **corresponding parts.** There are other procedures for constructing congruent triangles; some of these are discussed in Problem Set 6.3. For this example, we say $\triangle ABC \simeq \triangle DEF$. From this we conclude that all six corresponding parts are congruent.

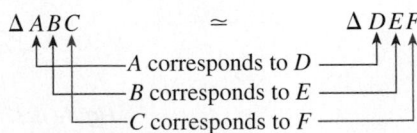

EXAMPLE 1

Name the corresponding parts of the given triangles.

a. $\triangle ABC \simeq \triangle A'B'C'$

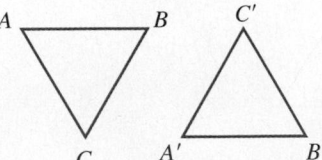

b. $\triangle RST \simeq \triangle UST$

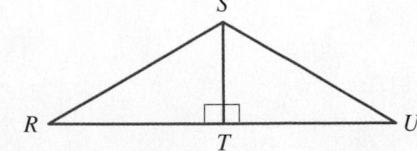

Solution

a. $\overline{AB}$ corresponds to $\overline{A'B'}$

$\overline{AC}$ corresponds to $\overline{A'C'}$

$\overline{BC}$ corresponds to $\overline{B'C'}$

$\angle A$ corresponds to $\angle A'$

$\angle B$ corresponds to $\angle B'$

$\angle C$ corresponds to $\angle C'$

b. $\overline{RS}$ corresponds to $\overline{US}$

$\overline{RT}$ corresponds to $\overline{UT}$

$\overline{ST}$ corresponds to $\overline{ST}$

$\angle R$ corresponds to $\angle U$

$\angle RTS$ corresponds to $\angle UTS$

$\angle RST$ corresponds to $\angle UST$ ◆

Angles of a Triangle

One of the most basic properties of a triangle involves the sum of the measures of its angles. To discover this property for yourself, place a pencil with an eraser along one side of any triangle as shown in Figure 6.34a.

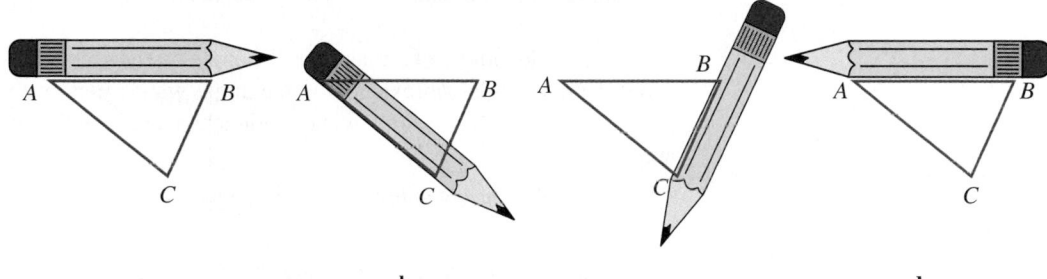

a.   **b.**   **c.**   **d.**

Figure 6.34 **Demonstration that the sum of the measures of the angles in a triangle is 180°**

Now rotate the pencil to correspond to the size of $\angle A$ as shown in Figure 6.34b. You see your pencil is along $\overline{AC}$. Next, rotate the pencil through $\angle C$, as shown in Figure 6.34c. Finally, rotate the pencil through $\angle B$. Notice that the pencil has been rotated the same amount as the sum of the angles of the triangle. Also notice that the orientation of the pencil is exactly reversed from the starting position. This leads us to the following important theorem.

Sum of the Measures of Angles in a Triangle

STOP

You will frequently need to use this property.

The sum of the measures of the angles in any triangle is 180°.

EXAMPLE 2

Find the missing angle measure in the triangle in Figure 6.35.

Solution

Let x represent the missing angle measure.

$$65 + 82 + x = 180$$
$$147 + x = 180$$
$$x = 33$$

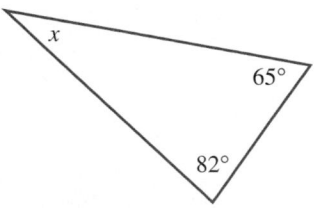

Figure 6.35 **What is *x*?**

The missing angle's measure is 33°.   ◆

EXAMPLE 3

Find the measures of the angles of a triangle if it is known that the measures are x, $2x - 15$, and $3(x + 17)$ degrees.

Solution Using the theorem for the sum of the measures of angles in a triangle, we have

$$x + (2x - 15) + 3(x + 17) = 180$$
$$x + 2x - 15 + 3x + 51 = 180 \quad \text{Eliminate parentheses.}$$
$$6x + 36 = 180 \quad \text{Combine similar terms.}$$
$$6x = 144 \quad \text{Subtract 36 from both sides.}$$
$$x = 24 \quad \text{Divide both sides by 6.}$$

Now find the angle measures:

$$x = 24$$
$$2x - 15 = 2(24) - 15 = 33$$
$$3(x + 17) = 3(24 + 17) = 123$$

The angles have measures of 24°, 33°, and 123°. ◆

An **exterior angle** of a triangle is the angle on the other side of an extension of one side of the triangle. An example is the angle whose measure is marked as x in Figure 6.36. Notice that the following relationships are true for any $\triangle ABC$ with exterior angle x:

$$m\angle A + m\angle B + m\angle C = 180° \quad \text{and} \quad m\angle C + x = 180°$$

Thus,

$$m\angle A + m\angle B + m\angle C = m\angle C + x$$
$$m\angle A + m\angle B = x \quad \text{Subtract } m\angle C \text{ from both sides.}$$

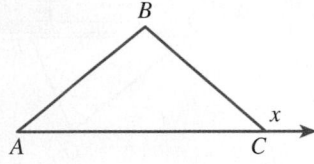
Figure 6.36 **Exterior angle *x***

Exterior Angles of a Triangle

The measure of the exterior angle of a triangle equals the sum of the measures of the two opposite interior angles.

EXAMPLE 4

Find the value of x in Figure 6.37.

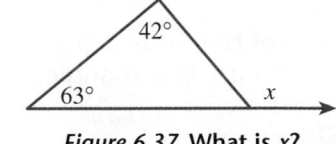
Figure 6.37 **What is *x*?**

Solution

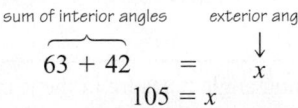

$$\underbrace{63 + 42}_{\text{sum of interior angles}} = \underset{x}{\underset{\downarrow}{\text{exterior angle}}}$$
$$105 = x$$

The measure of the exterior angle is 105°. ◆

Isosceles Triangle Property

In an isosceles triangle, there are two sides of equal length and the third side called its **base.** The angle included by its legs is called the **vertex angle** and the angles that include the base are called **base angles.**

There is an important theorem in geometry that is known as the **isosceles triangle property.**

Isosceles Triangle Property

If two sides of a triangle have the same length, then angles opposite them are congruent.

In other words, if a triangle is isosceles, then the base angles have equal measure. The converse is also true; namely, if two angles of a triangle are congruent, the sides opposite them have equal length.

EXAMPLE 5

Give a reasonable argument to prove that if a triangle is equiangular, it is also equilateral.

Solution   If $\triangle ABC$ is equiangular, then $m\angle A = m\angle B = m\angle C$. Since $m\angle A = m\angle B$, from the converse of the isosceles triangle property we have $|\overline{BC}| = |\overline{AC}|$. Again, since $m\angle B = m\angle C$, we have $|\overline{AC}| = |\overline{AB}|$. Thus, we see that all three sides have the same length, and consequently $\triangle ABC$ is equilateral.   ◆

Problem Set 6.3

LEVEL 1

1. **IN YOUR OWN WORDS** What is a triangle?

2. **IN YOUR OWN WORDS** What is the sum of the measures of the angles of a triangle?

3. **IN YOUR OWN WORDS** Explain the notation $\triangle ABC \cong \triangle DEF$.

4. **IN YOUR OWN WORDS**
   Explain why the musical instrument called a "triangle" is not a good example of a geometric triangle.

Name the corresponding parts of the triangles in Problems 5–10. Single and double marks are used to indicate segments of equal length.

5.

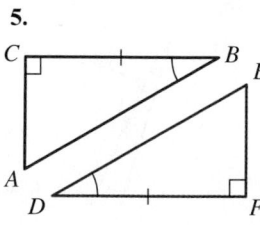

6.

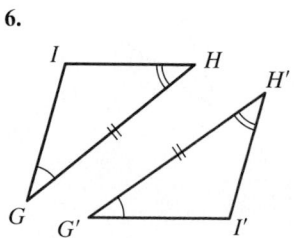

7.

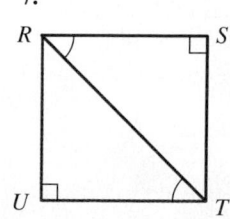

8.

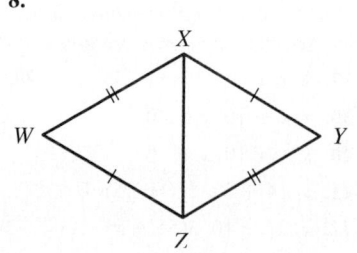

9.

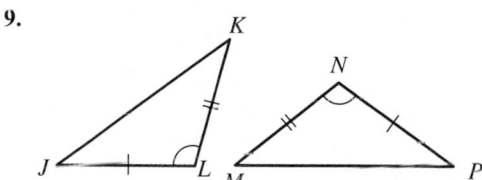

10.

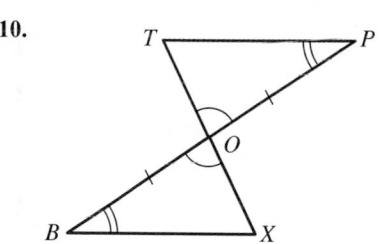

In Problems 11–16, find the measure of the third angle in each triangle.

11.   12.

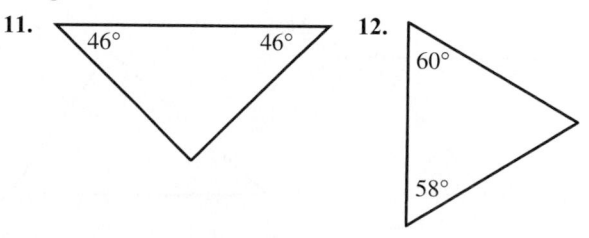

13.   14.

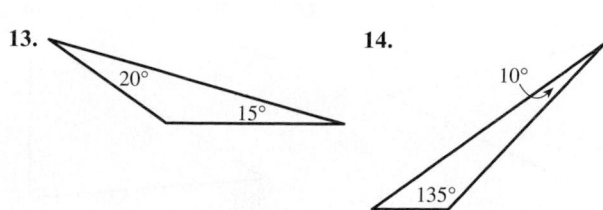

15.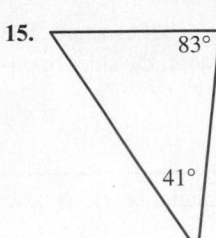

16.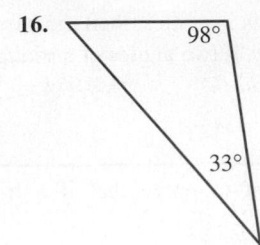

Find the measure of the indicated exterior angle in each of the triangles in Problems 17–22.

17.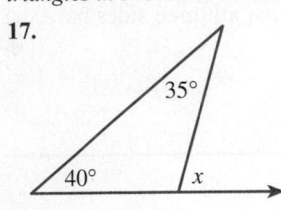

18.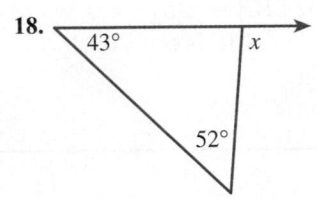

19.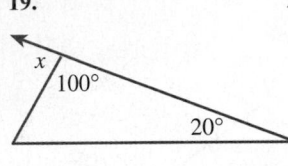

20.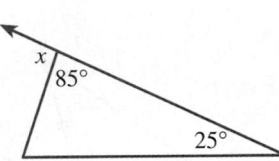

21.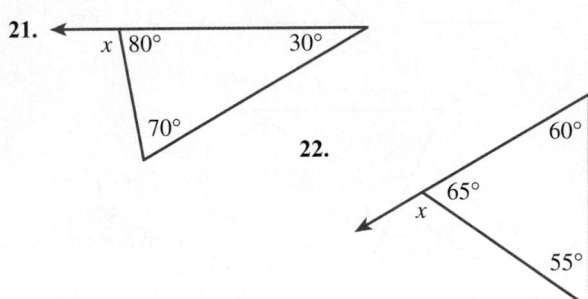

22.

Using only a straightedge and a compass, reproduce the triangles given in Problems 23–28.

23.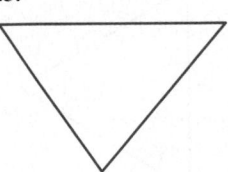

24.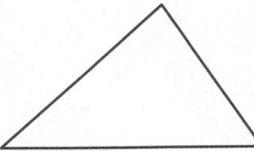

25.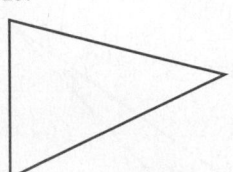

26.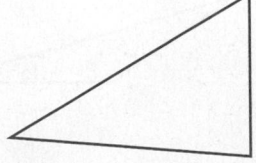

27.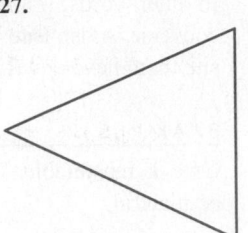

28.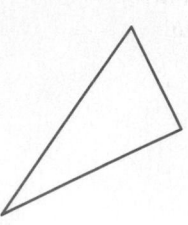

LEVEL 2

29. IN YOUR OWN WORDS Define an *isosceles right triangle.*

30. IN YOUR OWN WORDS Explain why the sum of the lengths of any two sides of a triangle must be greater than the length of the third side.

Use algebra to find the value of x in each of the triangles in Problems 31–36. Notice that the measurement of the angle is not necessarily the same as the value of x.

31.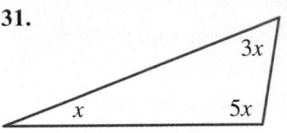

32.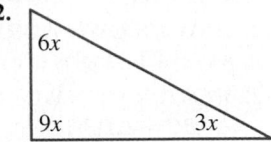

33.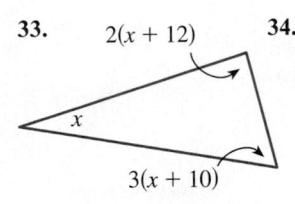

34.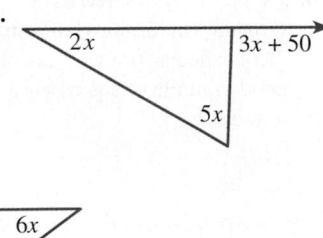

35.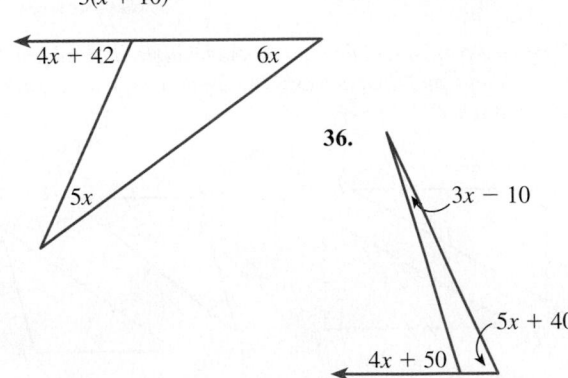

36.

Find the numerical measures of the angles of the triangle whose angle measures are given in Problems 37–42.

37. x, x, x

38. $x, x, 2x$

39. $x, x + 10, x + 20$

40. $x, x - 10, x + 10$

41. $x, 14 + 3x, 3(x + 25)$

42. $x, 3x - 10, 3(55 - x)$

Draw an example of each of the triangles described in Problems 43–50. If you think the figure cannot exist, write "impossible."

43. acute scalene

44. acute isosceles

45. acute equilateral

46. right scalene

47. right isosceles

48. right equilateral

49. obtuse scalene

50. obtuse isosceles

51. In the text we constructed congruent triangles by using SAS. Reproduce the triangle shown in Problem 25 by using SSS. This means to construct the triangle by using the lengths of the three sides.

52. In the text we constructed congruent triangles by using SAS. Reproduce the triangle shown in Problem 26 by using SSS. This means to construct the triangle by using the lengths of the three sides.

53. In the text we constructed congruent triangles by using SAS. Reproduce the triangle shown in Problem 27 by using ASA. This means to construct the triangle by using a side included between two angles.

54. In the text we constructed congruent triangles by using SAS. Reproduce the triangle shown in Problem 28 by using ASA. This means to construct the triangle by using a side included between two angles.

55. IN YOUR OWN WORDS Give a reasonable argument proving that complements of the same angle are equal.

56. IN YOUR OWN WORDS Give a reasonable argument proving that if a triangle is equilateral, it is also equiangular.

57. Show that *the sum of the measures of the interior angles of any triangle is 180° by carrying out the following steps.*
   a. Draw three triangles: one with all acute angles, one with a right angle, and a third with an obtuse angle. For example, one of these triangles might look like the following triangle:
   b. Tear apart the angles of each triangle you've drawn. (See the tear marks in the figure.)

   c. Place the pieces together to form a straight angle.

58. Show that *the sum of the measures of the interior angles of any quadrilateral is 360° by carrying out the following steps:*
   a. Draw any quadrilateral, as illustrated (but draw your quadrilateral so it has a different shape from the one shown here).

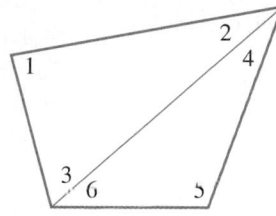

   b. Divide the quadrilateral into two triangles by drawing a diagonal (a line segment connecting two nonadjacent vertices). Label the angles of your triangles as shown in the figure.
   c. What is the sum $m\angle 1 + m\angle 2 + m\angle 3$?
   d. What is the sum $m\angle 4 + m\angle 5 + m\angle 6$?
   e. The sum of the measures of the angles of the quadrilateral is

   $$(m\angle 1 + m\angle 2 + m\angle 3) + (m\angle 4 + m\angle 5 + m\angle 6)$$

   What is this sum?
   f. Do you think this argument will apply for *any* quadrilateral?

59. Look at Problem 58. What is the sum of the measures of the interior angles of any pentagon?

60. Look at Problem 58. What is the sum of the measures of the interior angles of any octagon?

6.4   SIMILAR TRIANGLES

Congruent figures have exactly the same size and shape. However, it is possible for figures to have exactly the same shape without necessarily having the same size. Such figures are called *similar figures*. In this section we will focus on **similar triangles.** If $\triangle ABC$ is similar to $\triangle DEF$, we write

$$\triangle ABC \sim \triangle DEF$$

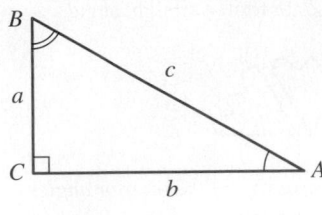

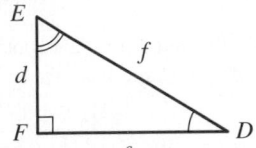

Figure 6.38 Similar triangles

Similar triangles are shown in Figure 6.38. Since these figures have the same shape, we talk about **corresponding angles** and **corresponding sides.** The corresponding angles of similar triangles are those angles that have equal measure. The corresponding sides are those sides that are opposite equal angles.

$m\angle A = m\angle D$, so these are corresponding angles.

$m\angle B = m\angle E$, so these are corresponding angles.

$m\angle C = m\angle F$, so these are corresponding angles.

Side $\overline{BC}$ is opposite $\angle A$ and side $\overline{EF}$ is opposite $\angle D$, so we say that $\overline{BC}$ corresponds to $\overline{EF}$.

$\overline{AC}$ corresponds to $\overline{DF}$.

$\overline{AB}$ corresponds to $\overline{DE}$.

Even though corresponding angles are equal, corresponding sides do not need to have the same length. If they do have the same length, the triangles are congruent. However, when they are not the same length, we can say they are proportional. From Figure 6.38 we see that the lengths of the sides are labeled a, b, c and d, e, f. When we say the sides are proportional, we mean

Primary Ratios: **Reciprocals:**

$$\frac{a}{b} = \frac{d}{e} \quad \frac{a}{c} = \frac{d}{f} \quad \frac{b}{c} = \frac{e}{f} \qquad \frac{b}{a} = \frac{e}{d} \quad \frac{c}{a} = \frac{f}{d} \quad \frac{c}{b} = \frac{f}{e}$$

We summarize with an important property of similar triangles called the **similar triangle theorem.**

Similar Triangle Theorem

This result is used in many applications.

> Two triangles are similar if two angles of one triangle are congruent to two angles of the other triangle. If the triangles are similar, then their corresponding sides are proportional.

EXAMPLE 1

Identify pairs of triangles that are similar in Figure 6.39.

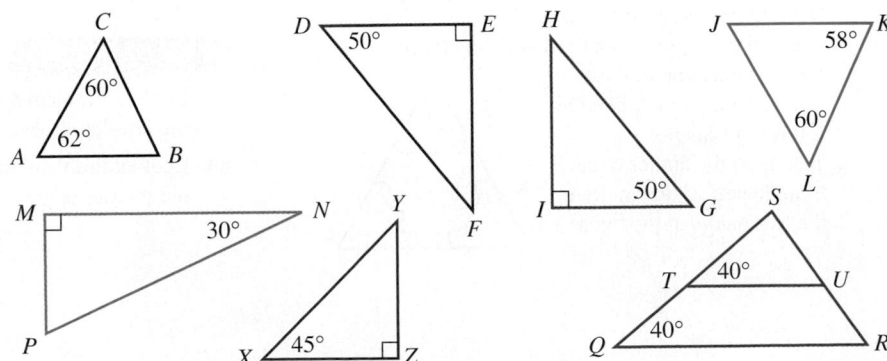

Figure 6.39 Which of these triangles are similar?

Solution $\triangle ABC \sim \triangle JKL$; $\triangle DEF \sim \triangle GIH$; $\triangle SQR \sim \triangle STU$ ◆

EXAMPLE 2

Given the similar triangles in Figure 6.40, find the unknown lengths marked b' and c'.

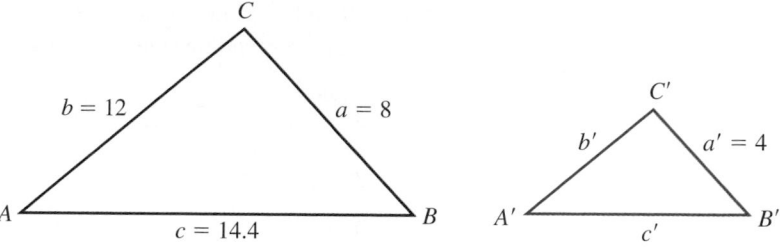

Figure 6.40 Given △*ABC* ~ △*A'B'C'*

Solution Since corresponding sides are proportional (other proportions are possible), we have

$$\frac{a}{b} = \frac{a'}{b'} \qquad\qquad \frac{a}{c} = \frac{a'}{c'}$$

$$\frac{8}{12} = \frac{4}{b'} \qquad\qquad \frac{8}{14.4} = \frac{4}{c'}$$

$$b' = \frac{4 \times 12}{8} \qquad\qquad c' = \frac{14.4 \times 4}{8}$$

$$b' = 6 \qquad\qquad c' = 7.2 \qquad\qquad\blacklozenge$$

Identifying similar triangles is simplified even further if we know that the triangles are right triangles, because then the triangles are similar if one of the acute angles of one triangle has the same measure as an acute angle of the other.

EXAMPLE 3   PÓLYA'S METHOD

Find the height of a tree that is difficult to measure directly.

Solution We use Pólya's problem-solving guidelines for this example.

Understand the Problem. We need to find the height of some tree without measuring it directly.

Devise a Plan. We assume that it is a sunny day, and will measure the height of the tree by measuring its shadow on the ground. For reference, we also measure the length of a shadow of an object of known height (say, our own height, a meterstick, or a yardstick). We will then use similar triangles and proportions to find the height of the tree.

Carry Out the Plan. Suppose that a tree and a yardstick are casting shadows as shown in Figure 6.41. If the shadow of the yardstick is 3 yards long and the shadow of the tree is 12 yards long, we use similar triangles to estimate *h,* the height of the tree, if we know that $m\angle S = m\angle S'$.

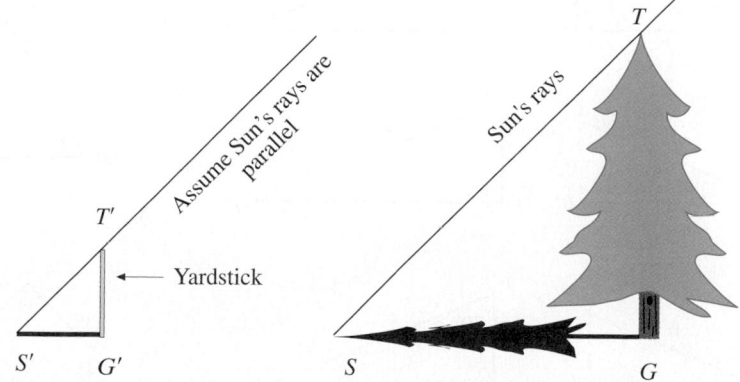

Figure 6.41 Finding the height of a tall object by using similar triangles

---

Since $\angle G$ and $\angle G'$ are right angles, and since $m\angle S = m\angle S'$, $\triangle SGT \sim \triangle S'G'T'$. Therefore, corresponding sides are proportional.

$$\frac{1}{3} = \frac{h}{12}$$ *You solved proportions like this in Chapter 5.*

$$h = \frac{1 \times 12}{3}$$

$$h = 4$$

Look Back. The tree is 4 yards, or 12 ft, tall.   ◆

There is a relationship between the sizes of the angles of a right triangle and the ratios of the lengths of the sides. In a right triangle, the side opposite the right angle is called the **hypotenuse.** Each of the acute angles of a right triangle has one side that is the hypotenuse; the other side of that angle is called the **adjacent side.** In $\triangle ABC$ with right angle at C, as shown in Figure 6.42,

the hypotenuse is c,

the side adjacent to $\angle A$ is b,

the side adjacent to $\angle B$ is a.

We also talk about an **opposite side.** The side opposite $\angle A$ is a, and the side opposite $\angle B$ is b.

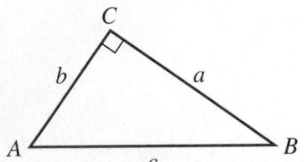

Figure 6.42 A right triangle

Problem Set 6.4

LEVEL 1

1. **IN YOUR OWN WORDS** Contrast congruent and similar triangles.

2. **IN YOUR OWN WORDS** What does it mean when we say the corresponding sides of two congruent triangles are proportional?

In Problems 3–8, tell whether it is possible to conclude that the triangles are similar.

3.
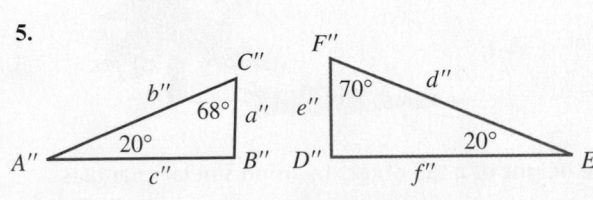

4.

5.

6.
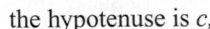

7.
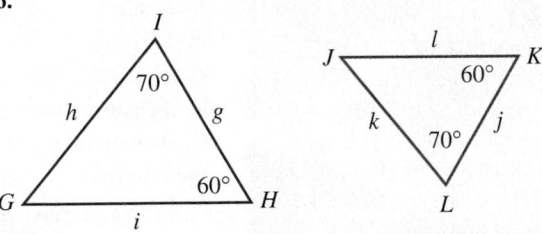

8.
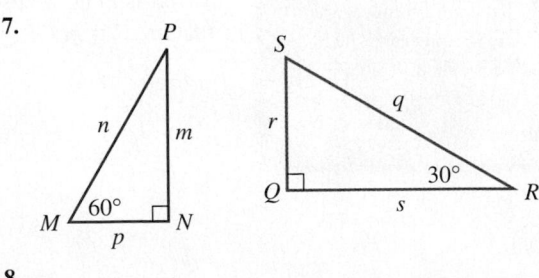

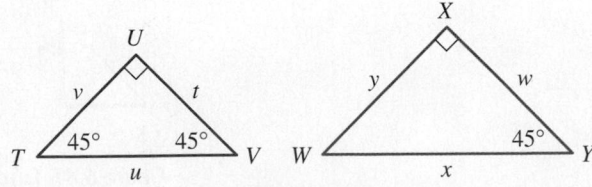

List the measures for all six angles for the figures given in Problems 9–14.

9.
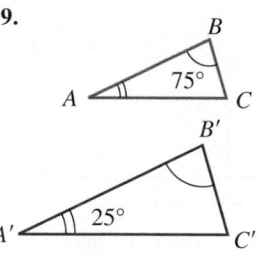

10.
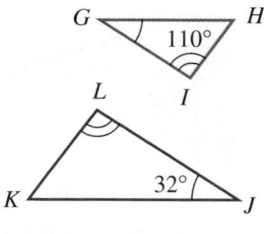

11.
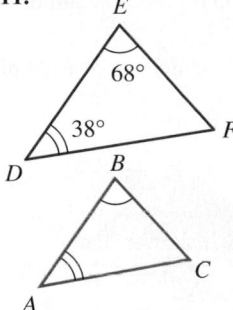

12.
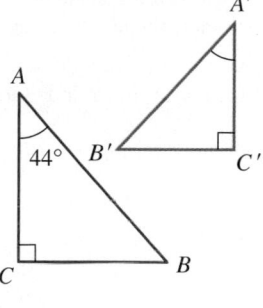

13.
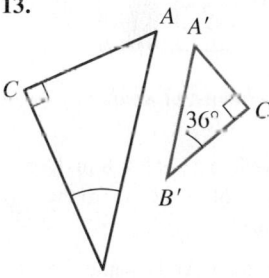

14.
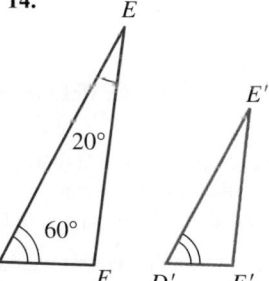

List the lengths of all six sides for the figures given in Problems 15–20.

15.
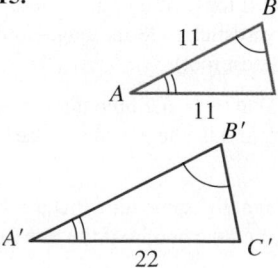

16.
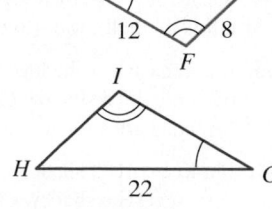

17.
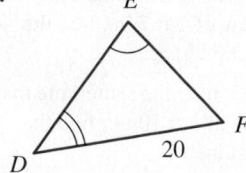

18.
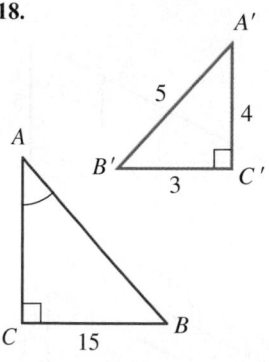

19.
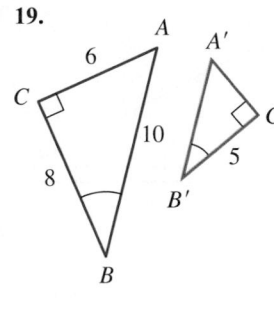

20.
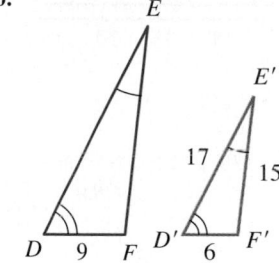

Given two similar triangles, as shown in Figure 6.43, find the unknown lengths in Problems 21–28.

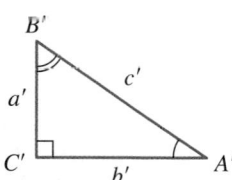

Figure 6.43 Similar triangles

21. $a = 4$, $b = 8$; find c.

22. $a' = 7$, $b' = 3$; find c'.

23. $a = 4$, $b = 8$, $a' = 2$; find b'.

24. $b = 5$, $c = 15$, $b' = 3$; find c'.

25. $c = 6$, $a = 4$, $c' = 8$; find a'.

26. $a' = 7$, $b' = 3$, $a = 5$; find b.

27. $b' = 8$, $c' = 12$, $c = 4$; find b.

28. $c' = 9$, $a' = 2$, $c = 5$; find a.

Each figure in Problems 29–34 contains two similar triangles. Find the unknown measure indicated by a variable. Answer to the nearest tenth.

29.
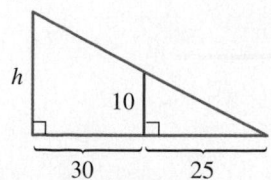

30.
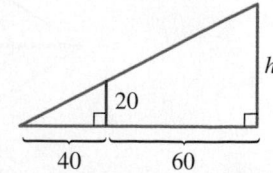

31. **32.**

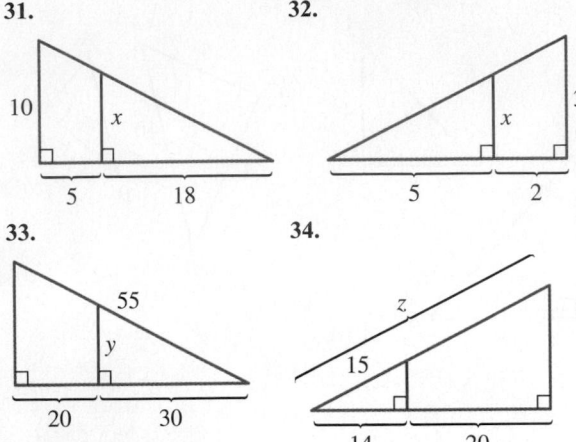

33. **34.**

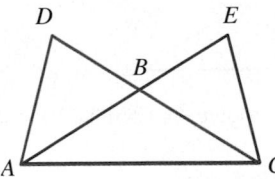

LEVEL 2

35. Given $\overline{AC}$ is perpendicular to $\overline{MB}$, and $\triangle ABC$ is an equilateral triangle. Show that $\triangle ABM \sim \triangle CBM$.

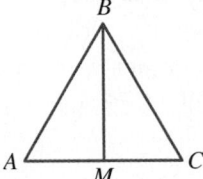

36. Given $m\angle D = m\angle E$. Show that $\triangle ABD \sim \triangle CBE$.

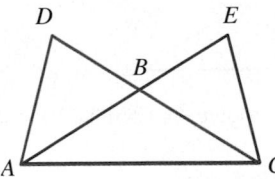

37. Given that $\overline{CT}$ bisects both $\angle ACO$ and $\angle ATO$. Show that $\triangle CAT \sim \triangle COT$.

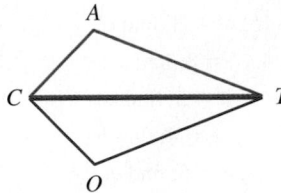

38. Given that $\overline{AW}$ bisects $\angle R_1AR_2$ and $\angle R_1OR_2$. Show that $\triangle AOR_1 \sim \triangle AOR_2$.

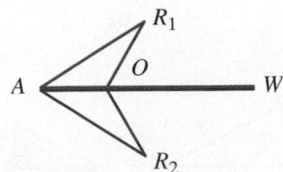

39. Use similar triangles and a proportion to find the length of the lake shown in Figure 6.44.

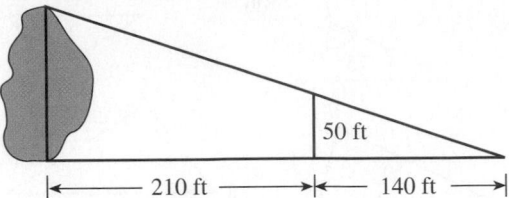

Figure 6.44 **Determining the length of a lake**

40. Suppose the distances in Problem 39 are changed as follows: 150 ft instead of 210 ft and 90 ft instead of 140 ft. What is the length of this lake?

41. Use similar triangles and a proportion to find the height of the house shown in Figure 6.45.

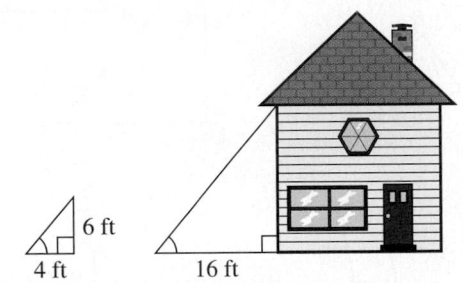

Figure 6.45 **Determining the height of a building**

42. Suppose the 6 ft distance in Problem 41 is 5 ft 8 in. Use this information to find the height of the house (to the nearest inch) shown in Figure 6.45.

43. A building casts a shadow 75 ft long. At the same time, the shadow cast by a vertical yardstick is 5 ft long. How tall is the building?

44. Suppose the shadow in Problem 43 is 75 ft 3 in. How tall is the building (to the nearest inch)?

45. A bell tower casts a shadow 45 ft long. At the same time, the shadow cast by a vertical yardstick is 23 in. long. How tall is the bell tower (to the nearest foot)

46. If a tree casts a shadow of 12 ft at the same time that a 6-ft person casts a shadow of $2\frac{1}{2}$ ft, find the height of the tree (to the nearest foot).

47. If a tree casts a shadow of 10 ft at the same time that a 5-ft person casts a shadow of 3 ft, find the height of the tree (to the nearest foot).

48. If a tree casts a shadow of 8 ft 3 in. at the same time that a 5-ft 10-in. person casts a shadow of 2 ft 7 in., find the height of the tree (to the nearest inch).

49. If a tree casts a shadow of 4 ft 5 in. at the same time that a 5-ft 9-in. person casts a shadow of 3 ft 10 in., find the height of the tree (to the nearest inch).

50. If lines are drawn on a map, a triangle can be formed by the cities of New York City, Washington, D.C., and Buffalo, New York. On the map, the distance between New York City and Washington, D.C., is 2.1 cm, New York to Buffalo is 2.5 cm, and from Buffalo to Washington, D.C., it is approximately 2.85 cm. If the actual straight-line distance from Buffalo to Washington is 285 miles, how far is it between the other pairs of cities?

51. If lines are drawn on a map, a triangle can be formed by the cities of New Orleans, Louisiana; Denver, Colorado; and Chicago, Illinois. On the map, the distance between New Orleans and Denver is 10.8 cm, from Chicago to New Orleans is 9.5 cm, and from Chicago to Denver it is 10.2 cm. If the actual straight-line distance from Chicago to New Orleans is 950 miles, how far is it between the other pairs of cities?

52. Suppose a 6-ft person wishes to determine the height of a bridge above the bottom of a canyon, as shown in Figure 6.46.

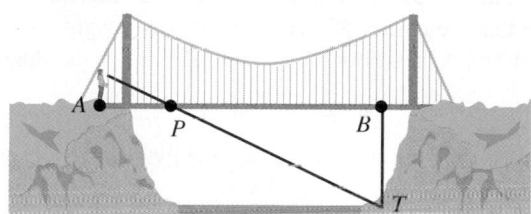

Figure 6.46 **Height of a bridge**

To do this, this person stands at one end of the bridge (point A) and looks down to a point directly below the other end (point B). With the help of a companion, point P is determined to form two triangles.
a. Identify the two triangles.
b. Are the triangles similar? Why or why not?
c. Find the distance to the bottom of the canyon, if
$$|\overline{AP}| = 10 \text{ ft} \quad \text{and} \quad |\overline{PB}| = 40 \text{ ft}$$

53. Suppose a 6-ft person wishes to determine the height of a footbridge connecting two buildings, as shown in Figure 6.47.

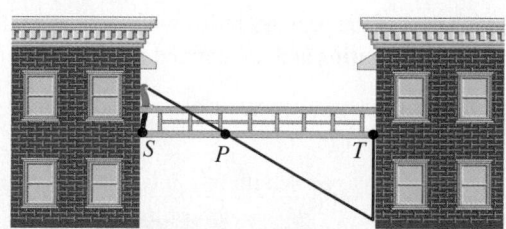

Figure 6.47 **Height of a footbridge**

To do this, this person stands at one end of the bridge (point S) and looks down to a point directly below the other end (point T). With the help of a companion, point P is determined to form two triangles.
a. Identify the two triangles.

b. Are the triangles similar? Why or why not?
c. Find the height of the footbridge, if
$$|\overline{SP}| = 10 \text{ ft} \quad \text{and} \quad |\overline{PT}| = 35 \text{ ft}$$

LEVEL 3

54. Given $|\overline{AB}| = |\overline{BC}|$ and M is the midpoint of $\overline{AC}$. Show that $\triangle ABM \sim \triangle CBM$.

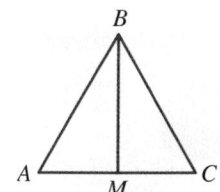

55. Given $m\angle D = m\angle E$ and $|\overline{AB}| = |\overline{BC}|$. Show that $\triangle ADC \sim \triangle CEA$.

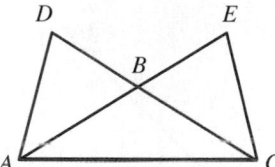

56. A useful theorem that uses proportionality, but not triangles, is the following:

If two lines connect the endpoints of parallel segments of different lengths, then a line through the intersection of the two lines connecting those endpoints divides the parallel segments proportionally. (See Figure 6.48.)

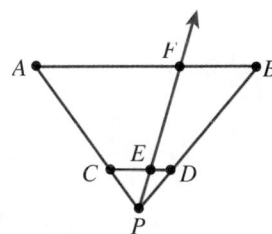

Figure 6.48 **Suppose $\overline{AB} \parallel \overline{CD}$; then**
$$\frac{|\overline{AF}|}{|\overline{BF}|} = \frac{|\overline{CE}|}{|\overline{DE}|}$$

Using this result, divide the following line segment into a 2-to-3 ratio:

_____

57. Divide the segment given in Problem 56 into two parts in a 3-to-7 ratio.

58. Present an argument showing that if two triangles are equilateral, then they are similar triangles.

59. For any right triangle ABC (right angle at C), drop a perpendicular from point C to base $\overline{AB}$ at the point D. Show that the two triangles thus formed are similar.

60. Present an argument showing that two triangles similar to a third triangle are similar to each other.

6.5 | RIGHT-TRIANGLE TRIGONOMETRY

An important theorem from geometry, the **Pythagorean theorem,** has an important algebraic representation and is important in our study of triangles.

Pythagorean Theorem

 STOP   You should be familiar with this theorem.

For any right triangle with sides of length a and b and hypotenuse of length c,

$$a^2 + b^2 = c^2$$

Also, if a, b, and c are the lengths of the sides of a triangle so that

$$a^2 + b^2 = c^2$$

then the triangle is a right triangle.

A correctly labeled right triangle is shown in Figure 6.49. In a right triangle, the sides that are not the hypotenuse are sometimes called **legs.**

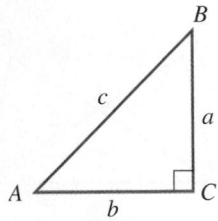

Figure 6.49 Right triangle

EXAMPLE 1

A carpenter wants to make sure that the corner of a closet is square (a right angle). If she measures out sides of 3 feet and 4 feet, how long should she make the diagonal (hypotenuse)?

Solution   The length of the hypotenuse is the unknown, so use the Pythagorean theorem:

$$c = \sqrt{a^2 + b^2}$$
$$= \sqrt{3^2 + 4^2} \quad \text{The sides are 3 and 4.}$$
$$= \sqrt{9 + 16}$$
$$= \sqrt{25}$$
$$= 5$$

She should make the diagonal 5 feet long.   ◆

Of interest to us in this section is the value found by forming the ratios of the lengths of the sides. There are six possible ratios for the triangle shown in Figure 6.49.

Primary ratios	*Reciprocal ratios:*
$\dfrac{a}{b}, \ \dfrac{a}{c}, \ \dfrac{b}{c}$	$\dfrac{b}{a}, \ \dfrac{c}{a}, \ \dfrac{c}{b}$

In this section, we study the relationship of the primary ratios with the angles A and B. These ratios are called the **trigonometric ratios** and are defined in the following box.

Trigonometric Ratios

STOP   This is the essential definition of this section. It also forms the basis for a trigonometry course.

In a right triangle ABC with right angle at C,

sin A (pronounced "sine of A") is the ratio $\dfrac{\text{length of opposite side of } A}{\text{length of hypotenuse}}$

cos A (pronounced "cosine of A") is the ratio $\dfrac{\text{length of adjacent side of } A}{\text{length of hypotenuse}}$

tan A (pronounced "tangent of A") is the ratio $\dfrac{\text{length of opposite side of } A}{\text{length of adjacent side of } A}$

EXAMPLE 2

Given a right triangle with sides of length 5 and 12, find the trigonometric ratios for the angles A and B. Show your answers in both common fraction and decimal fraction form, with decimals rounded to four places.

Solution

First use the Pythagorean theorem to find the length of the hypotenuse.

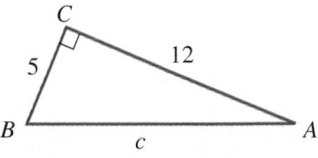

$$c = \sqrt{5^2 + 12^2}$$

$$= \sqrt{25 + 144}$$

$$= \sqrt{169}$$

$$= 13$$

$$\sin A = \frac{5}{13} \approx 0.3846; \quad \cos A = \frac{12}{13} \approx 0.9231; \quad \tan A = \frac{5}{12} \approx 0.4167$$

$$\sin B = \frac{12}{13} \approx 0.9231; \quad \cos B = \frac{5}{13} \approx 0.3846; \quad \tan B = \frac{12}{5} = 2.4 \quad \blacklozenge$$

EXAMPLE 3

Find the cosine, sine, and tangent of $45°$.

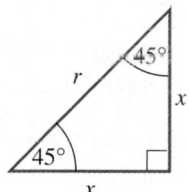

Figure 6.50

Solution
If one of the angles of a right triangle is $45°$, then the other acute angle must also be $45°$ (because the sum of the angles of a triangle is $180°$). Furthermore, since the base angles have the same measure, the triangle is isosceles (see Figure 6.50). By the Pythagorean theorem,

$$x^2 + x^2 = r^2$$

$$2x^2 = r^2$$

$$\sqrt{2}\,x = r \qquad \text{Note that } x \text{ is positive.}$$

$$\cos \theta = \frac{x}{r} \qquad \sin \theta = \frac{x}{r} \qquad \tan \theta = \frac{x}{x}$$

$$= \frac{x}{\sqrt{2}\,x} \qquad\qquad\qquad = 1$$

$$= \frac{1}{\sqrt{2}} \cdot \frac{\sqrt{2}}{}$$

Thus, $\cos 45° = \sin 45° = \dfrac{\sqrt{2}}{2}$, and $\tan 45° = 1$. $\quad \blacklozenge$

Tables of ratios for different angles are available (see Table 6.2 on page 328). Most calculators have keys for the sine, cosine, and tangent ratios. Check with your calculator owner's manual to see how to use your calculator for these problems.

Table 6.2 Trigonometric Ratios

Degrees	sin x	cos x	tan x	Degrees	sin x	cos x	tan x
1	0.0175	0.9998	0.0175	46	0.7193	0.6947	1.0355
2	0.0349	0.9994	0.0349	47	0.7314	0.6820	1.0724
3	0.0523	0.9986	0.0524	48	0.7431	0.6691	1.1106
4	0.0698	0.9976	0.0699	49	0.7547	0.6561	1.1504
5	0.0872	0.9962	0.0875	50	0.7660	0.6428	1.1918
6	0.1045	0.9945	0.1051	51	0.7771	0.6293	1.2349
7	0.1219	0.9925	0.1228	52	0.7880	0.6157	1.2799
8	0.1392	0.9903	0.1405	53	0.7986	0.6018	1.3270
9	0.1564	0.9877	0.1584	54	0.8090	0.5878	1.3764
10	0.1736	0.9848	0.1763	55	0.8192	0.5736	1.4281
11	0.1908	0.9816	0.1944	56	0.8290	0.5592	1.4826
12	0.2079	0.9781	0.2126	57	0.8387	0.5446	1.5399
13	0.2250	0.9744	0.2309	58	0.8480	0.5299	1.6003
14	0.2419	0.9703	0.2493	59	0.8572	0.5150	1.6643
15	0.2588	0.9659	0.2679	60	0.8660	0.5000	1.7321
16	0.2756	0.9613	0.2867	61	0.8746	0.4848	1.8040
17	0.2924	0.9563	0.3057	62	0.8829	0.4695	1.8807
18	0.3090	0.9511	0.3249	63	0.8910	0.4540	1.9626
19	0.3256	0.9455	0.3443	64	0.8988	0.4384	2.0503
20	0.3420	0.9397	0.3640	65	0.9063	0.4226	2.1445
21	0.3584	0.9336	0.3839	66	0.9135	0.4067	2.2460
22	0.3746	0.9272	0.4040	67	0.9205	0.3907	2.3559
23	0.3907	0.9205	0.4245	68	0.9272	0.3746	2.4751
24	0.4067	0.9135	0.4452	69	0.9336	0.3584	2.6051
25	0.4226	0.9063	0.4663	70	0.9397	0.3420	2.7475
26	0.4384	0.8988	0.4877	71	0.9455	0.3256	2.9042
27	0.4540	0.8910	0.5095	72	0.9511	0.3090	3.0777
28	0.4695	0.8829	0.5317	73	0.9563	0.2924	3.2709
29	0.4848	0.8746	0.5543	74	0.9613	0.2756	3.4874
30	0.5000	0.8660	0.5774	75	0.9659	0.2588	3.7321
31	0.5150	0.8572	0.6009	76	0.9703	0.2419	4.0108
32	0.5299	0.8480	0.6249	77	0.9744	0.2250	4.3315
33	0.5446	0.8387	0.6494	78	0.9781	0.2079	4.7046
34	0.5592	0.8290	0.6745	79	0.9816	0.1908	5.1446
35	0.5736	0.8192	0.7002	80	0.9848	0.1736	5.6713
36	0.5878	0.8090	0.7265	81	0.9877	0.1564	6.3138
37	0.6018	0.7986	0.7536	82	0.9903	0.1392	7.1154
38	0.6157	0.7880	0.7813	83	0.9925	0.1219	8.1444
39	0.6293	0.7771	0.8098	84	0.9945	0.1045	9.5144
40	0.6428	0.7660	0.8391	85	0.9962	0.0872	11.4300
41	0.6561	0.7547	0.8693	86	0.9976	0.0698	14.3007
42	0.6691	0.7431	0.9004	87	0.9986	0.0523	19.0812
43	0.6820	0.7314	0.9325	88	0.9994	0.0349	28.6362
44	0.6947	0.7193	0.9657	89	0.9998	0.0175	57.2898
45	0.7071	0.7071	1.0000	90	1.0000	0.0000	undefined

EXAMPLE 4

Find the trigonometric ratios by using either Table 6.2 or a calculator.

a. sin 45° **b.** cos 32° **c.** tan 19°

Solution

a. sin 45° ≈ 0.7071 from Table 6.2; by calculator, press $\boxed{\text{sin}}\ \boxed{45}$.*

Compare with Example 3: $\sin 45° = \dfrac{\sqrt{2}}{2} \approx 0.7171$.

b. cos 32° ≈ 0.8480 from Table 6.2; press $\boxed{\text{cos}}\ \boxed{32}$.

c. tan 19° ≈ 0.3443 from Table 6.2; press $\boxed{\text{tan}}\ \boxed{19}$.   ◆

Trigonometric ratios are useful in a variety of situations, as illustrated in the next example.

EXAMPLE 5

The angle from the ground to the top of the Great Pyramid of Cheops is 52° if a point on the ground directly below the top is 351 ft away (see Figure 6.51). What is the height of the pyramid?

Solution

From Figure 6.51 we see that for height h

$$\tan 52° = \frac{h}{351}$$

Solving for h by multiplying both sides by 351, we obtain

$$h = 351 \tan 52°$$

Now we need to know the ratio for tan 52°. By Table 6.2, tan 52° ≈ 1.2799, so $h \approx 351(1.2799) = 449.2449$ or about 449 ft.

By calculator, press $\boxed{351}\ \boxed{\times}\ \boxed{\text{tan}}\ \boxed{52}\ \boxed{=}$.

Display: 449.2595129 or about 449 ft.   ◆

Notice that the calculator and table answers in Example 5 are not identical, because both the table and the calculator give approximations of the exact value of tan 52°.

We can also use right-triangle trigonometry to find one of the acute angles if we know the trigonometric ratio. For example, suppose we know (as we do from Example 3) that

$$\tan \theta = 1$$

Also suppose that we do not know (as we do from Example 3) the angle θ. In other words, we ask, "What is the angle θ?" We answer by saying, "θ is the angle whose tangent is 1." In mathematics, we call this the **inverse tangent** and we write

$$\theta = \tan^{-1}1$$

To find the angle θ, we turn to a calculator. Find the button labeled $\boxed{\text{tan}^{-1}}$, and press †

$$\boxed{\text{tan}^{-1}}\ \boxed{1}\ \boxed{=}$$

The display is 45, which means $\theta = 45°$.

We now define the **inverse trigonometric ratios** for a right triangle.

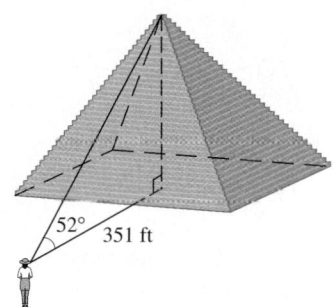

Figure 6.51 **Calculating the height of the Great Pyramid of Cheops**

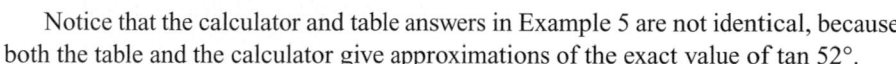

---

*Or $\boxed{45}\ \boxed{\text{sin}}$, depending on brand and model of calculator.

†On some calculators, you press $\boxed{1}\ \boxed{\text{tan}^{-1}}\ \boxed{=}$ or $\boxed{\text{tan}^{-1}}\ \boxed{1}\ \boxed{\text{ENTER}}$. Also make sure your calculator is set on DEGREE MODE.

Inverse Trigonometric Ratios

This definition is used to find angles in a right triangle when the sides are known.

For θ an acute angle in a right triangle,

$$\cos^{-1}\left(\frac{\text{ADJ}}{\text{HYP}}\right) = \theta \qquad \sin^{-1}\left(\frac{\text{OPP}}{\text{HYP}}\right) = \theta \qquad \tan^{-1}\left(\frac{\text{OPP}}{\text{ADJ}}\right) = \theta$$

We return to Example 2.

EXAMPLE 6

Given a right triangle with sides of length 5 and 12, find the measures of the angles of this triangle.

Solution
From Example 2, we see the hypotenuse is 13.
Also, from Example 2,

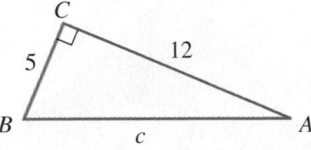

$$\sin A = \frac{5}{13} \approx 0.3846$$

$$\sin B = \frac{12}{13} \approx 0.9231$$

Our task here is to find the measures of angles A and B. What is A? We might say, "A is the measure of the angle whose sine is $\frac{5}{13}$." This is the inverse sine, and we write

$$A = \sin^{-1}\frac{5}{13} \approx 22.6°$$

Similarly,

$$B = \sin^{-1}\frac{12}{13} \approx 67.4°$$

Note: There is usually more than one ratio that can be used to find a particular angle. For this example, we could also have used

$$A = \cos^{-1}\frac{12}{13} \approx 22.6° \quad \text{or} \quad A = \tan^{-1}\frac{5}{12} \approx 22.6°$$

and

$$B = \cos^{-1}\frac{5}{13} \approx 67.4°, \quad \text{or} \quad B = \tan^{-1}\frac{12}{5} \approx 67.4° \qquad \blacklozenge$$

Right-triangle trigonometry can be used in a variety of applications. One of the most common has to do with an observer looking at an object.

Nancy and Sluggo cartoon reprinted by permission of United Feature Syndicate, Inc.

Remember to measure from the horizontal.

The **angle of depression** is the acute angle measured down from a horizontal line to the line of sight:

On the other hand, if we take the mountain climber's viewpoint, and measure from a horizontal up to the line of sight, we call this angle the **angle of elevation.**

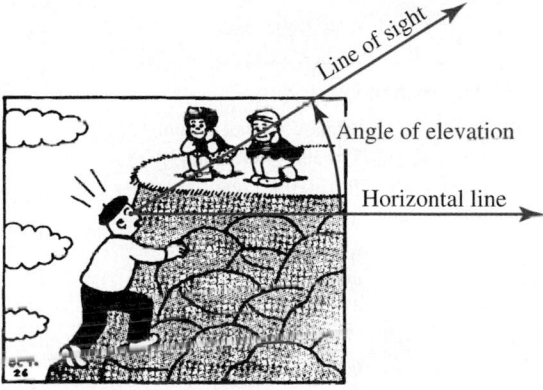

EXAMPLE 7

The angle of elevation to the top of a tree from a point on the ground 42 ft from its base is 33°. Find the height of the tree (to the nearest foot).

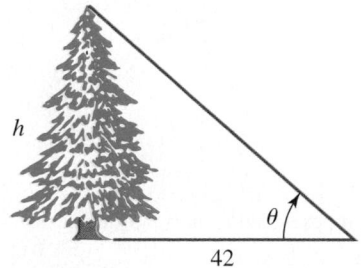

Solution Let θ = angle of elevation;

$\quad\quad h$ = height of tree.

Then,

$$\tan \theta = \frac{h}{42}$$

$$h = 42 \tan 33°$$

$$\approx 27.28$$

The tree is 27 ft tall. ◆

Problem Set 6.5

LEVEL 1

1. **IN YOUR OWN WORDS** What is the Pythagorean theorem?

2. **IN YOUR OWN WORDS** What is a sine?

3. **IN YOUR OWN WORDS** What is a cosine?

4. **IN YOUR OWN WORDS** What is a tangent?

5. **HISTORICAL QUESTION** The mathematics of the early Egyptians was practical and centered around surveying, construction, and record keeping. They used a simple device to aid surveying—a rope with 12 equal divisions marked by knots, as shown in Figure 6.52.

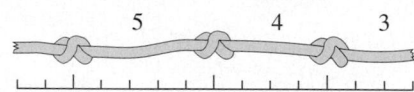

Figure 6.52 Egyptian measuring rope

When the rope was stretched and staked so that a triangle was formed with sides 3, 4, and 5, the angle formed by the shorter sides was a right angle. This method was extremely useful in Egypt, where the Nile flooded the rich lands close to the river each year. The lands needed to be resurveyed when the waters subsided.

Which of the following ropes would form right triangles?

Rope *A:* 30 knots (sides 5, 12, and 13)

Rope *B:* 9 knots (sides 2, 3, and 4)

Use the right triangle in Figure 6.53 to answer the questions in Problems 6–16.

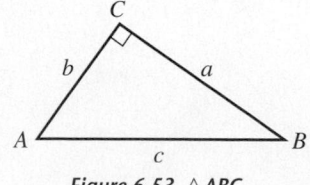

Figure 6.53 △ABC

6. What is the side opposite ∠*A*?

7. What is the side opposite ∠*B*?

8. What is the side adjacent ∠*A*?

9. What is the side adjacent ∠*B*?

10. What is the hypotenuse?

11. What is sin *A*? 12. What is sin *B*?

13. What is cos *A*? 14. What is cos *B*?

15. What is tan *A*? 16. What is tan *B*?

Find the trigonometric ratios in Problems 17–28 by using either Table 6.2 or a calculator. Round your answers to four decimal places.

17. sin 56° 18. sin 15° 19. sin 61°

20. sin 18° 21. cos 54° 22. cos 8°

23. cos 90° 24. cos 34° 25. tan 24°

26. tan 52° 27. tan 75° 28. tan 89°

Find the angles (to the nearest degree) for the information given in Problems 29–37.

29. $\sin^{-1} \dfrac{1}{2}$ 30. $\cos^{-1} 1$ 31. $\tan^{-1} 1$

32. $\cos^{-1} \dfrac{\sqrt{2}}{2}$ 33. $\tan^{-1} \sqrt{3}$ 34. $\sin^{-1} 0$

35. $\tan^{-1} 1.5$ 36. $\sin^{-1} 0.35$ 37. $\cos^{-1} 0.8$

LEVEL 2

In Problems 38–47, find the sine, cosine, and tangent for the angle A.

38.

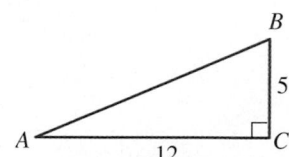

39.

40.

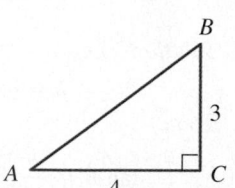

41.

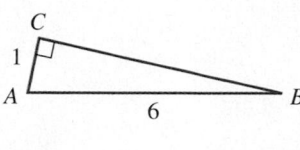

42.

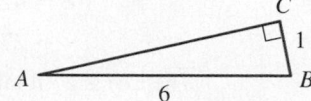

43.

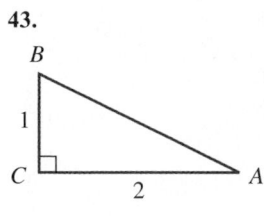

44.

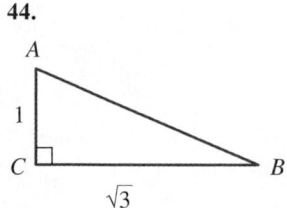

45.

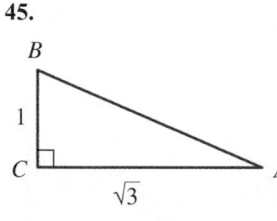

46.

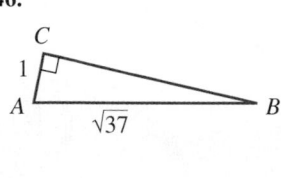

47.

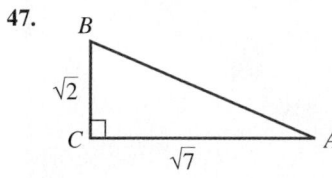

48. If the angle from the horizontal to the top of a building is 38° and the horizontal distance from its base is 90 ft, what is the height of the building (to the nearest foot)?

49. If the angle from the horizontal to the top of a tower is 52° and the horizontal distance from its base is 85 ft, what is the height of the tower (to the nearest foot)?

50. A building's angle of elevation from a point on the ground 30 m from its base is 35°. Find the height of the building (to the nearest meter).

51. From a cliff 150 m above the shoreline, a ship's angle of depression is 37°. Find the distance of the ship (to the nearest meter) from a point directly below the observer.

52. From a police helicopter flying at 1,000 ft, a stolen car is sighted at an angle of depression of 71°. Find the distance of the car (to the nearest ft) from a point directly below the helicopter.

53. A 16-ft ladder on level ground is leaning against a house. If the ladder's angle of elevation is 52°, how far above the ground (to the nearest inch) is the top of the ladder?

54. Find the height of the Barrington Space Needle (to the nearest foot) if its angle of elevation at 1,000 ft from a point on the ground directly below the top is 58.15°.

55. The world's tallest chimney is the stack at the International Nickel Company. Find its height if its angle of elevation at 1,000 ft from a point (to the nearest foot) on the ground directly below the top stack is 51.36°.

56. The angle of elevation of the top of the Great Pyramid of Khufu (or Cheops) from a point on the ground 351 ft from a point directly below the top is 52.0°. Find the height of the pyramid (to the nearest foot).

PROBLEM SOLVING

57. In a 30°–60°–90° triangle, the length of the side adjacent to the 30° angle is $\sqrt{3}$ times the length of the side opposite the 30° angle. Use this information to find the exact values for cos 30°, sin 30°, and tan 30°.

58. What is the radius of the largest circle you can cut from a rectangular poster board with measurements 11 in. by 17 in.? (See Figure 6.54.)

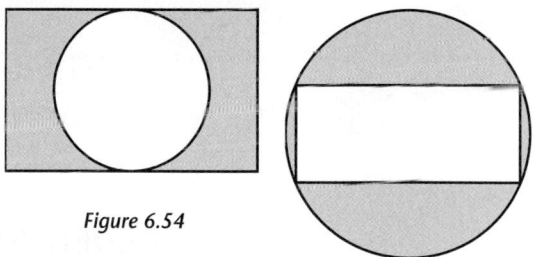

Figure 6.54

Figure 6.55

59. What is the width of the largest rectangle with length 16 in. you can cut from a circular piece of cardboard having a radius of 10 in.? (See Figure 6.55.)

60. a. If the distance from the earth to the sun is 92.9 million miles, and the angle formed between Venus, the earth, and the sun is 47° (as shown in the illustration), find the distance from the sun to Venus (to the nearest hundred thousand miles).

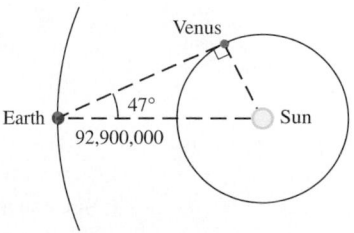

b. Find the distance from the earth to Venus (to the nearest hundred thousand miles).

HISTORICAL NOTE

The angle of elevation for a pyramid is the angle between the edge of the base and the slant height, the line from the apex of the pyramid to the midpoint of any side of the base. It is the maximum possible ascent for anyone trying to climb the pyramid to the top. In an article, "Angles of Elevation to the Pyramids of Egypt," in *The Mathematics Teacher* (February 1982, pp. 124–127), author Arthur F. Smith notes that the angle of elevation of these pyramids is either about 44° or 52°. Why did the Egyptians build pyramids using these angles of elevation? (Project 6.7, on page 350.)

Smith states that (according to Kurt Mendelssohn, *The Riddle of the Pyramids*; New York: Praeger Publications, 1974) Egyptians might have measured long horizontal distances by means of a circular drum with some convenient diameter such as one cubit. The circumference would then have been π cubits. In order to design a pyramid of convenient and attractive proportions, the Egyptians used a 4:1 ratio for the rise relative to revolutions of the drum. Smith then shows that the angle of elevation of the slant height is

$$\tan^{-1}\frac{4}{\pi} \approx 51.9°$$

If a small angle of elevation was desired (as in the case of the Red Pyramid), a 1:3 ratio might have been used. In that case, the angle of elevation of the slant height is

$$\tan^{-1}\frac{3}{\pi} \approx 43.7°$$

Arctan $\frac{3}{\pi} \approx 43.7°$

$4n$

$2\pi n$ πn πn πn

Adapted from "Angles of Elevation of the Pyramids of Egypt," in *The Mathematics Teacher*, February 1982, pp. 124–127.

6.6   GOLDEN RECTANGLES

Certain rectangles hold some special interest because of the relationship between their height and width. Consider a rectangle with height h and width w, and consider the proportion

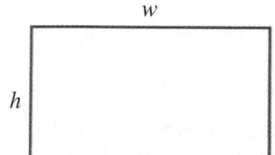

$$\frac{h}{w} = \frac{w}{h + w}$$

This relationship is called the **divine proportion.** If $h = 1$, we can solve the resulting equation for w:

$$\frac{1}{w} = \frac{w}{1 + w}$$

$$1 + w = w^2 \qquad \text{Product of means = product of extremes}$$

$$w^2 - w - 1 = 0 \qquad \text{Simplify.}$$

$$w = \frac{1 \pm \sqrt{(-1)^2 - 4(1)(-1)}}{2} = \frac{1 \pm \sqrt{5}}{2}$$

Quadratic formula

Since w is a length, we disregard the negative value to find

$$w = \frac{1 + \sqrt{5}}{2} \approx 1.618033989$$

This number is called the **golden ratio,** and is denoted by τ (pronounced *tau*).

HISTORICAL NOTE

The divine proportion has many interesting properties. It was derived by the 15th century mathematician Luca Pacioli. He defined it by dividing a line segment into two parts so that the length of the smaller part is proportional to the length of the larger part as the length of the larger part is to the length of the entire segment. *Would you mind repeating that?* Consider a segment of unit length and divide it into two parts with lengths x and $1 - x$. Then the divine proportion is

$$\frac{1 - x}{x} = \frac{x}{1}$$

If you can solve this equation for x you will find $x = 1/\tau$.

George Markowsky points out in his article, "Misconceptions About the Golden Ratio," that the first time the term *golden section* was seen in print (in German) was in a book by Martin Ohm, and it first appeared in English in the 1875 edition of *Encyclopedia Britannica.*

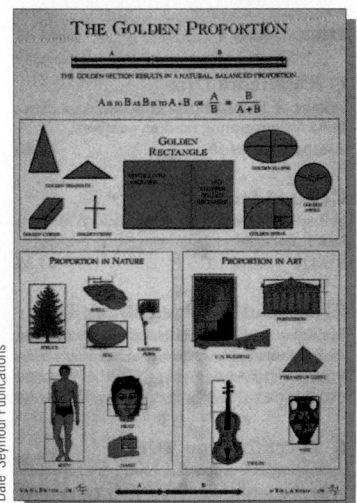

A rectangle that satisfies this proportion for finding the golden ratio is called a **golden rectangle** and can easily be constructed using a straightedge and a compass. Consider the proportion

$$\frac{h}{w} = \frac{w}{h + w}$$

which means the ratio of the height (h) to the width (w) is the same as the ratio of the width (w) to the sum of its height and width. To draw such a rectangle, we can begin with *any* square *CDHG,* as shown in Figure 6.56. This square is shown with its interior shaded. Now divide the square into two equal parts, as shown by the dashed segment labeled $\overline{AB}$. Set your compass so that it measures the length of $\overline{AC}$. Draw an arc, with center at A and radius equal to the length of $\overline{AC}$, so that it intersects the extension of side $\overline{HD}$; label this point E. Now draw side $\overline{EF}$. The resulting rectangle *EFGH* is a golden rectangle; *CDEF* is also a golden rectangle.

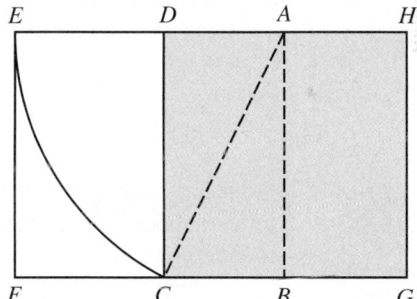

Figure 6.56 **Constructing a golden rectangle *EFGH***

There are many interesting properties associated with the golden ratio τ. Consider the pattern of numbers

$$1, 1, 2, 3, 5, 8, 13, 21, 34, 55, 89, 144, \ldots$$

which is formed by adding the first two terms for the third term, and then continues by adding successive pairs of numbers. Suppose we consider the ratios of the successive terms of the sequence:

$$\frac{1}{1} = 1.0000; \quad \frac{2}{1} = 2.0000; \quad \frac{3}{2} = 1.5000; \quad \frac{5}{3} \approx 1.667; \quad \frac{8}{5} = 1.6000;$$

$$\frac{13}{8} = 1.625; \quad \frac{21}{13} \approx 1.615; \quad \frac{34}{21} \approx 1.619; \quad \frac{55}{34} \approx 1.618; \quad \frac{89}{55} \approx 1.618$$

If you continue to find these ratios, you will notice that the sequence oscillates about a number approximately equal to 1.618, which is τ.

Suppose we repeat this same procedure, except we start with *any* two nonzero numbers, say 4 and 7:

$$4, 7, 11, 18, 29, 47, 76, 123, 199, 322, \ldots$$

Next, form the ratios of the successive terms:

$$\frac{7}{4} = 1.750; \quad \frac{11}{7} \approx 1.571; \quad \frac{18}{11} \approx 1.636; \quad \frac{29}{18} \approx 1.611; \quad \frac{47}{29} \approx 1.621; \ldots$$

These ratios are oscillating about the same number, τ!

EXAMPLE 1   **PÓLYA'S METHOD**

Find some numbers with the property that the number and its reciprocal differ by 1.

Solution   We use Pólya's problem-solving guidelines for this example.

Understand the Problem. We know, for example, that if x is such a number, then $\frac{1}{x} + 1$ must be equal to the original number.

Devise a Plan. We let x be any nonzero number. Then we need to set $\frac{1}{x} + 1$ equal to x, and solve the resulting equation.

Carry Out the Plan.

$$x = \frac{1}{x} + 1$$

$$x^2 = 1 + x \qquad \text{Multiply both sides by } x.$$

$$x^2 - x - 1 = 0$$

$$x = \frac{1 \pm \sqrt{(-1)^2 - 4(1)(-1)}}{2} = \frac{1 \pm \sqrt{5}}{2}$$

Look Back. We recognize the positive value as τ:

$$\tau = \frac{1 + \sqrt{5}}{2} \approx 1.618033989$$

Use your calculator to find the reciprocal $\frac{1}{\tau} = \frac{\sqrt{5} - 1}{2} \approx 0.618033989$. ◆

It has been said that many everyday rectangular objects have a length-to-width ratio of about 1.6:1, as illustrated in Figure 6.57. Psychologists have tested individuals to determine the rectangles they find most pleasing; the results are those rectangles whose length-to-width ratios are near the golden ratio. George Markowsky, on the other hand, in a recent article "Misconceptions About the Golden Ratio," suggests that with rectangles with greatly different length-to-width ratios, the golden rectangle is the most pleasing; but when confronted with rectangles with ratios "close" to the golden ratio, subjects are unable to select the "best" rectangle.

If we observe many works of art, we can see evidence of golden rectangles. Whether the artist had such rectangles in mind is open to speculation, but we can see golden rectangles in the work of Albrecht Dürer, Leonardo da Vinci, George Bellows, Pieter Mondriaan, and Georges Seurat (see Figure 6.58).

Figure 6.57 A 15-oz box of Kellogg's Sugar Corn Pops is 30 cm by 19 cm, for a ratio of 1.6; a l-lb box of C&H sugar is 17 cm by 10 cm, for a ratio of 1.7.

©1989 The Metropolitan Museum of Art

Figure 6.58 La Parade by the French impressionist Georges Seurat

Parthenon

The Parthenon in Athens has been used as an example of a building with a height-to-width ratio that is almost equal to the golden ratio.

EXAMPLE 2   **PÓLYA'S METHOD**

If the Parthenon is 101 feet wide, what is its height (to the nearest foot) if we assume the dimensions are in a golden ratio?

Solution   We use Pólya's problem-solving guidelines for this example.

Understand the Problem. First, understand the problem. Since the Parthenon is built to satisfy the golden ratio, the height h and the width w satisfy the following proportion:

$$\frac{h}{w} = \frac{w}{h + w}$$

Devise a Plan. The width is 101 feet, so

$$\frac{h}{101} = \frac{101}{h + 101}$$

We will solve this equation for h.

Carry Out the Plan. There is only one unknown, which is written in variable form, so we now solve the equation for h:

$$h(h + 101) = 101^2$$
$$h^2 + 101h - 101^2 = 0$$
$$h = \frac{-101 \pm \sqrt{101^2 - 4(1)(-101^2)}}{2}$$
$$= \frac{-101 \pm 101\sqrt{1 + 4}}{2}$$
$$\approx 62.4 \qquad \text{Disregard the negative solution, since distances are nonnegative.}$$

Look Back. The Parthenon is about 62 feet high.   ◆

Many studies of the human body itself involve the golden ratio (remember $\tau \approx$ 1.62). Figure 6.59 shows a drawing of an idealized athlete.

David (1501–1504) by Michelangelo illustrates many golden ratios.

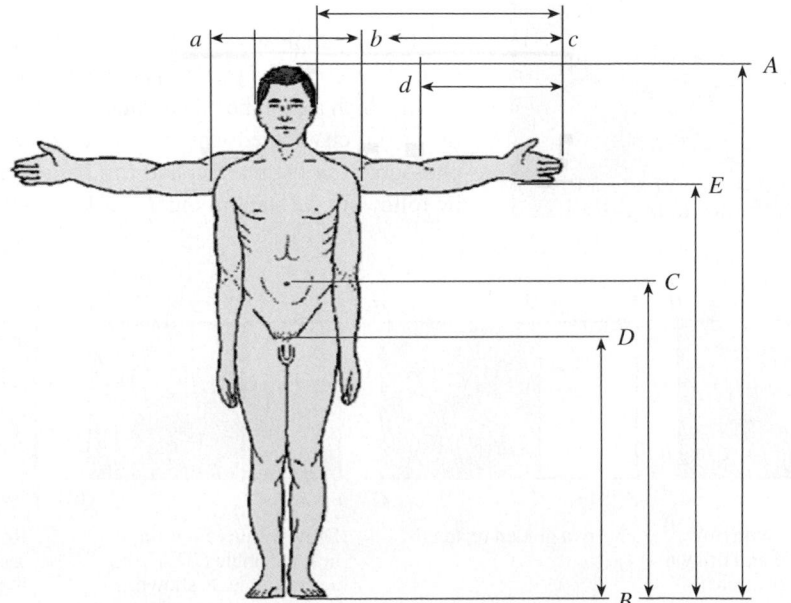

Figure 6.59 **Proportions of the human body**

Let AC = distance from the top of the head to the navel;
CB = distance from the navel to the floor; and
AB = height.

Then, $\dfrac{CB}{AC} \approx \tau$ and $\dfrac{AB}{CB} \approx \tau$.

Also, let ab = shoulder width and bc = arm length. Then

$$\frac{bc}{ab} \approx \tau$$

See if you can find other ratios on the human body that approximate τ. Figure 6.60 shows a study of the human face by da Vinci, in which the rectangles approximate golden rectangles.

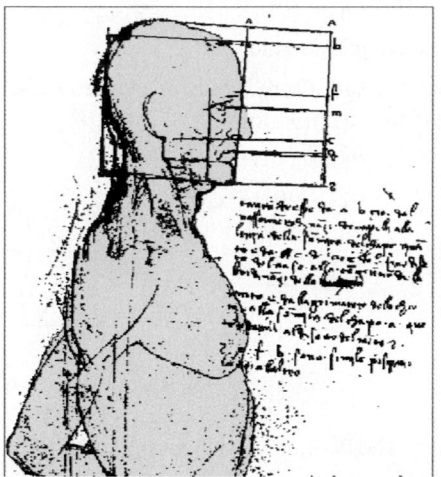

Dynamic symmetry of a human face (Leonardo da Vinci)

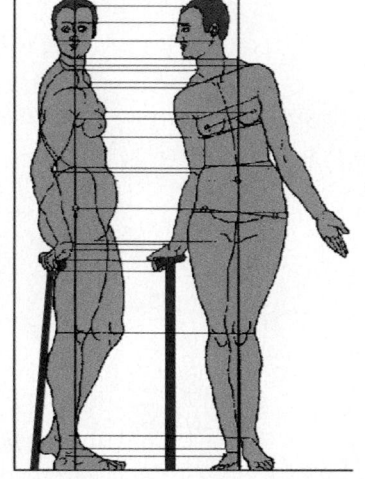

Proportions of the human body by Albrecht Dürer

© VEB Bibliographisches Institute Leipzig. Reprinted by permission.

Figure 6.60 **Golden rectangles used in art**

© Susan Van Etten/Stock Boston

The last application of the golden rectangle we will consider in this section is related to the manner in which a chambered nautilus grows. The spiral of the chambered nautilus can be seen in the photograph in the margin, and can be constructed by following the steps in Figure 6.61.

In *The Chambered Nautilus,* Oliver Wendell Holmes wrote:

"Year after year beheld the silent toll
That spread his lustrous coil;
Still, as the spiral grew,
He left the past year's dwelling for the new."

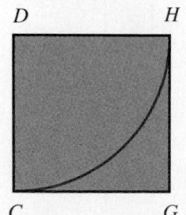

Begin with any square and draw a quarter circle as shown.

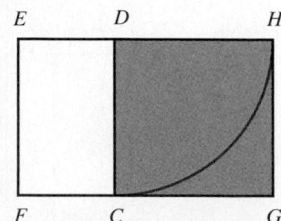

Form a golden rectangle.

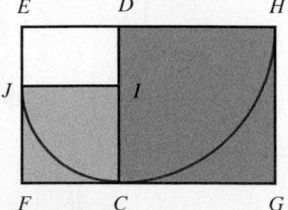

Draw a square within the new rectangle *CDEF*; draw a semicircle as shown.

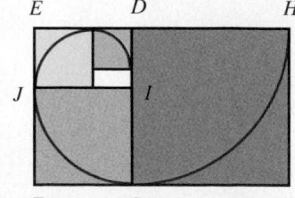

Repeat the process. The resulting curve is called a logarithmic spiral.

Figure 6.61 **A spiral is constructed using a golden rectangle.**

Problem Set 6.6

LEVEL 1

1. **IN YOUR OWN WORDS** What is a golden ratio?
2. **IN YOUR OWN WORDS** What is a golden rectangle?
3. **IN YOUR OWN WORDS** What is the divine proportion?
4. **IN YOUR OWN WORDS** What is τ?
5. **a.** Pick any two nonzero numbers. Construct a a set of numbers, in order, by adding these two numbers to find the next number. Continue by adding the previous two numbers to form your list of terms.
   b. Form the ratios of successive terms, and show that after a while they oscillate around the golden ratio.
6. Repeat Problem 5 for two other nonzero numbers.
7. Start with 1 and 3, and add these numbers to get the next number. Continue by adding two successive numbers to get the next number until you have ten terms, in order, on your list. Next, find the ratios of the successive terms. How do these compare with τ?
8. If the Parthenon in Greece is 60 ft tall (at the apex) and 97 ft wide, find the ratio of width to height and compare it with τ.
9. The Great Pyramid of Giza has dimensions as follows: height, $h = 481$ ft; base, $b = 756$ ft; and slant height, $s = 612$ ft. Is the ratio of any of these dimensions related to τ?

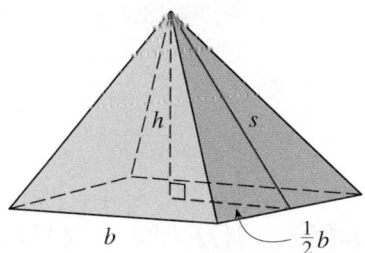

10. The American Pyramid of the Sun at Teotihuacán, Mexico, has a height of 216 ft, a base of 700 ft, and a slant height of 411 ft. Does the ratio of any of these measurements approximate τ?
11. How closely does the ratio of the dimensions of a polo field (160 yd by 300 yd) approximate τ?
12. How closely does the ratio of the dimensions of a basketball court (36 ft by 78 ft) approximate τ?

LEVEL 2

13. **a.** What are the length and width of a standard index card?
    b. Write the ratio of length to width as a decimal and compare it with τ.
14. **a.** What are the length and width of a standard brick?
    b. Write the ratio of length to width as a decimal and compare it with τ.

15. **a.** What are the length and width of this textbook?
    b. Write the ratio of length to width as a decimal and compare it with τ.

Use Figure 6.59 to find the requested measurements in Problems 16–18 using your own body as the model.

16. **a.** $CB \div AC$ **b.** $bc \div ab$
17. **a.** $AB \div CB$ **b.** $ac \div bc$
18. **a.** $AD \div AE$ **b.** $dc \div bd$
19. A plant grows for two months and then adds a new branch. Each new branch grows for two months, and then adds another branch. After the second month, each branch adds a new branch every month. Assume that the growth begins in January.

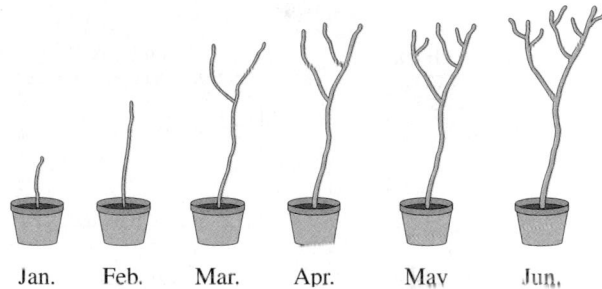

Jan.   Feb.   Mar.   Apr.   May   Jun.

   a. How many branches will there be in March?
   b. How many branches will there be in June?
   c. How many branches will there be after 12 months?
   d. Form a sequence of ratios of successive terms by considering the number of branches for the first 12 months. How do these numbers compare with τ?

Leo Moser studied the effect that two face-to-face panes of glass have on light reflected through the panes. If a ray is unreflected, it has just one path through the glass. If it has one reflection, it can be reflected two ways. For two reflections, it can be reflected three ways. Use this information in Problems 20–21.

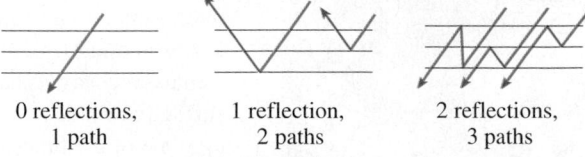

0 reflections,   1 reflection,   2 reflections,
   1 path   2 paths   3 paths

20. **a.** Show the possible paths for three reflections.
    b. Show the possible paths for four reflections.
    c. Make a conjecture about the number of paths for n reflections.
21. Form a sequence of ratios of the number of successive paths. How do these numbers compare with τ?
22. If a rectangle is to have sides with lengths in the golden ratio, what is the width if the height (the shorter side) is two units?

23. If a window is to be 5 feet wide, how high should it be, to the nearest tenth of a foot, to be a golden rectangle?

24. If a canvas for a painting is 18 inches wide, how high should it be, to the nearest inch, to be the divine proportion?

25. A photograph is to be printed on a rectangle in the divine proportion. If it is 9 cm high, how wide is it, to the nearest centimeter?

26. If the Parthenon is 60 ft high, what is its width, to the nearest foot, if we assume the building conforms to the golden ratio?

PROBLEM SOLVING

27. "When is an open book a closed book?" * If the length-to-width ratio of a book remains unchanged when that book is opened, then that book is said to be in the librarian's ratio.

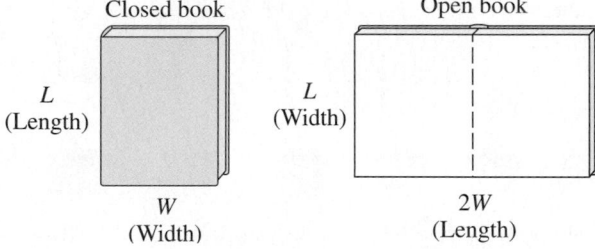

Closed book

L
(Length)

W
(Width)

Open book

L
(Width)

$2W$
(Length)

*My thanks to Monte J. Zerger of Friends University in Wichita, Kansas, for this problem. I found it in Vol. 18 of *The Journal of Recreational Mathematics*.

(Note that the length, L, of the closed book becomes the width of the opened book, and twice the width of the closed book becomes the length of the opened book.) Find an approximate and an exact representation for the constant L/W.

28. Does this book satisfy the librarian's ratio (see Problem 27)?

29. a. Show that the solutions to the equation
$$x^2 - x - 1 = 0$$
are $\tau = \dfrac{1 + \sqrt{5}}{2}$ and $\bar{\tau} = \dfrac{1 - \sqrt{5}}{2}$.

  b. Is there a relationship between the decimal representations of these numbers? If so, explain.

30. If you form a list of numbers where the first two numbers are ones, and then find successive terms (in order) by adding these two numbers to obtain the next, you will find

$$1, 1, 2, 3, 5, 8, 13, \ldots$$

Now, we see that the first number on the list is 1, and the seventh number is 13. It is known that the nth number on this list is

$$\frac{\tau^n - (-\bar{\tau})^n}{}$$

for τ defined in Problem 29. Use this formula to find the requested terms.

a. $n = 5$ **b.** $n = 10$ **c.** $n = 20$

6.7 | PROJECTIVE AND NON-EUCLIDEAN GEOMETRIES

Projective Geometry

As Europe passed out of the Middle Ages and into the Renaissance, artists were at the forefront of the intellectual revolution. No longer satisfied with flat-looking scenes, they wanted to portray people and objects as they looked in real life. The artists' problem was one of dimension, and dimension is related to mathematics, so many of the Renaissance artists had to solve some original mathematics problems. How could a flat surface be made to look three-dimensional?

One of the first (but rather unsuccessful) attempts at portraying depth in a painting is shown in Figure 6.62, Duccio's *Last Supper*. Notice that the figures are in a boxed-in room. This technique is characteristic of the period, and was an attempt to make perspective easier to define.

Artists finally solved the problem of perspective by considering the surface of the picture to be a window through which the object was painted. This technique was pioneered by Paolo Uccello (1397–1475), Piero della Francesca (1416–1492), Leonardo

Opera del Duomo, Siena

Figure 6.62 **Duccio's *Last Supper* illustrates perspective that is incorrect.**

B.C. cartoon reprinted by permission of Johnny Hart and Creators Syndicate.

da Vinci (1452–1519), and Albrecht Dürer (1471–1528). As the lines of vision from the object converge at the eye, the picture captures a cross section of them, as shown in Figure 6.63. The mathematical study of vanishing points and perspective is part of a branch of mathematics called **projective geometry,** the study of those properties of geometric configurations that do not change (are invariant) under projections.

Figure 6.63 **Albrecht Dürer's** *Designer of the Lying Woman* **shows how the problem of perspective can be overcome. The point in front of the artist's eye fixes the point of viewing the painting. The grid on the window corresponds to the grid on the artist's canvas.**

Non-Euclidean Geometry

Euclid's so-called fifth postulate (see Section 6.1) caused problems from the time it was stated. Several different formulations of this postulate are given in the margin. It somehow doesn't seem like the other postulates but rather, like a theorem that should be proved. In fact, this postulate even bothered Euclid himself, since he didn't use it until he had proved his 29th theorem. Many mathematicians tried to find a proof for this postulate.

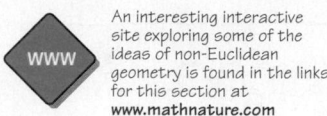

An interesting interactive site exploring some of the ideas of non-Euclidean geometry is found in the links for this section at www.mathnature.com

One of the first serious attempts to prove Euclid's fifth postulate was made by Girolamo Saccheri (1667–1733), an Italian Jesuit. Saccheri's plan was simple. He constructed a quadrilateral, later known as a **Saccheri quadrilateral,** with base angles A and B right angles, and with sides $\overline{AC}$ and $\overline{BD}$ the same length, as shown in Figure 6.64.

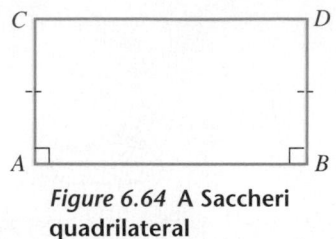

***Figure 6.64* A Saccheri quadrilateral**

As you may know from high school geometry, the summit angles C and D are also right angles. However, this result uses Euclid's fifth postulate. Now it is also true that *if* the summit angles are right angles, *then* Euclid's fifth postulate holds. The problem, then, was to establish the fact that angles C and D are right angles. Here is the plan:

1. Assume that the angles are obtuse and deduce a contradiction.

2. Assume that the angles are acute and deduce a contradiction.

3. Therefore, by the first two steps, the angles must be right angles.

4. From step 3, Euclid's fifth postulate can be deduced.

It turned out not to be as easy as he thought, because he was not able to establish a contradiction. He gave up the search because, he said, it "led to results that were repugnant to the nature of a straight line."

Saccheri never realized the significance of what he had started, and his work was forgotten until 1889. However, in the meantime, Johann Lambert (1728–1777) and Adrien-Marie Legendre (1752–1833) similarly investigated the possibility of eliminating Euclid's fifth postulate by proving it from the other postulates.

By the early years of the 19th century, three accomplished mathematicians began to suspect that the parallel postulate was independent and could not be eliminated by deducing it from the others. The great mathematician Karl Gauss, whom we've mentioned before, was the first to reach this conclusion, but since he didn't publish this finding of his, the credit goes to two others. In 1811, an 18-year-old Russian named Nikolai Lobachevski pondered the possibility of a "non-Euclidean" geometry—that is, a geometry that did not assume Euclid's fifth postulate. In 1829, he published his ideas in an article entitled "Geometrical Researches on the Theory of Parallels." The postulate he used was subsequently named after him.

The Lobachevskian Postulate

The summit angles of a Saccheri quadrilateral are acute.

This axiom, in place of Euclid's fifth postulate, leads to a geometry that we call **hyperbolic geometry.** If we use the plane as a model for Euclidean geometry, what model could serve for hyperbolic geometry? A rough model for this geometry can be

seen by placing two trumpet bells together as shown in Figure 6.65a. It is called a **pseudosphere** and is generated by a curve called a *tractrix* (as shown in Figure 6.65b). The tractrix is rotated about the line *AB*. The pseudosphere has the property that, through a point not on a line, there are many lines parallel to a given line.

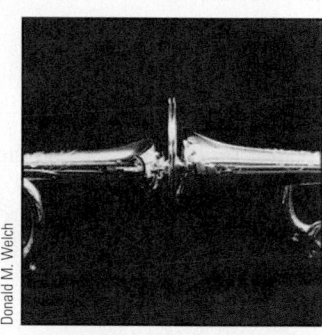

Two trumpets placed end to end serve as a physical model of a pseudosphere.

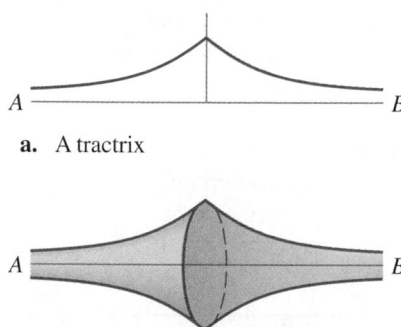

a. A tractrix

b. A tractrix rotated about the line *AB*

Figure 6.65 **A tractrix can be used to generate a pseudosphere.**

Georg Riemann (1826–1866), who also worked in this area, pointed out that, although a straight line may be extended indefinitely, it need not have infinite length. It could instead be similar to the arc of a circle, which eventually begins to retrace itself. Such a line is called *re-entrant*. An example of a re-entrant line is found by considering a great circle on a sphere. A **great circle** is a circle on a sphere with a diameter equal to the diameter of the sphere. With this model, a Saccheri quadrilateral is constructed on a sphere with the summit angles obtuse. The resulting geometry is called **elliptic geometry.** The shortest path between any two points on a sphere is an arc of the great circle through those points; these arcs correspond to line segments in Euclidean geometry. In 1854, Riemann showed that, with some other slight adjustments in the remaining postulates, another consistent non-Euclidean geometry can be developed. Notice that the fifth, or parallel, postulate fails to hold because any two great circles on a sphere must intersect at two points (see Figure 6.66).

We have not, by any means, discussed all possible geometries. We have merely shown that the Euclidean geometry that is taught in high school is not the only possible model. A comparison of some of the properties of these geometries is shown in Table 6.3 on page 344.

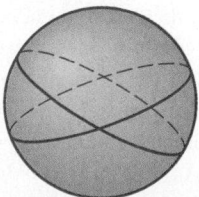

Figure 6.66 **A sphere showing the intersection of two great circles**

Table 6.3 Comparison of Major Two-Dimensional Geometries

Euclidean Geometry	*Hyperbolic Geometry*	*Elliptic Geometry*
Euclid (about 300 B.C.)	Gauss, Bolyai, Lobachevski (ca. 1830)	Riemann (ca. 1850)
Given a point not on a line, there is one and only one line through the point parallel to the given line.	Given a point not on a line, there are an infinite number of lines through the point that do not intersect the given line.	There are no parallels.

A representative line in each geometry is shown in color for each model, and the shaded portion showing a Saccheri quadrilateral is shown directly below the respective models.

Geometry is on a plane:	Geometry is on a pseudosphere:	Geometry is on a sphere:
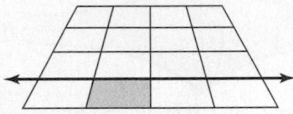	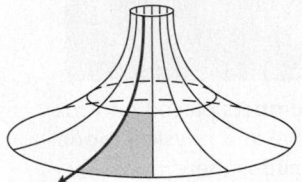	
Lines are infinitely long.	Lines are infinitely long.	Lines are finite in length.
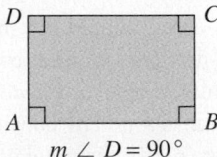	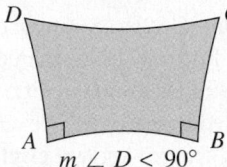	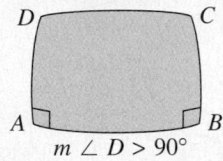
$m \angle D = 90°$	$m \angle D < 90°$	$m \angle D > 90°$
The sum of the angles of a triangle is 180°.	The sum of the angles of a triangle is less than 180°.	The sum of the angles of a triangle is more than 180°.

Problem Set 6.7

LEVEL 1

1. **IN YOUR OWN WORDS** What is a non-Euclidean geometry?

2. **IN YOUR OWN WORDS** Compare the major two-dimensional geometries.

3. **IN YOUR OWN WORDS** Why do you think Euclidean geometry remains so prevalent today even though we know there are other valid geometries?

4. Associate each of the given names with one of the following geometries: Euclidean, hyperbolic, elliptic.
   a. Karl Gauss
   b. Georg Riemann
   c. Euclid
   d. Janos Bolyai
   e. Nikolai Lobachevski

Name the geometry in which each of the statements in Problems 5–12 is possible.

5. The sum of the measures of the angles of a triangle is 180°.

6. The sum of the measures of the angles of a triangle is greater than 180°.

7. The summit angles of a Saccheri quadrilateral are right angles.

8. No line can be drawn through a given point parallel to a given line.

9. Lines have finite length.

10. Was used to measure distances in the construction of the pyramids in ancient Egypt.

11. The summit angles of a Saccheri quadrilateral are acute.

12. The measures of the base angles of a Saccheri quadrilateral are greater than 90°.

Which of the figures in Problems 13–16 are Saccheri quadrilaterals?

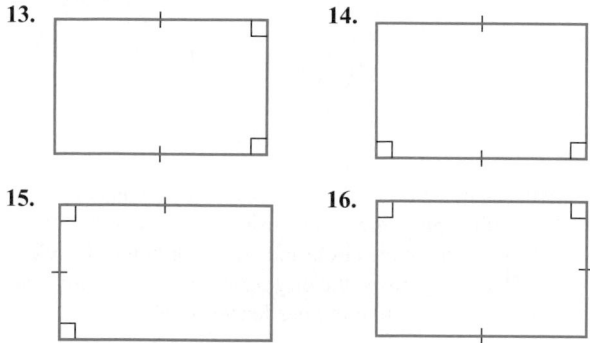

13.   14.

15.   16.

LEVEL 2

In Problems 17–20, choose one of the following statements to complete the sentence, and discuss the reasoning for your selection.

17. **IN YOUR OWN WORDS** Artists of the Renaissance discovered how to draw a three-dimensional-looking world on a flat canvas by . . .
    A. . . . reading about projective geometry from a classic book on the subject by Leonardo da Vinci.
    B. . . . discovering the use of a vanishing point for parallel lines.
    C. . . . using non-Euclidean geometry to accurately represent spherical (or three-dimensional) points on a flat surface (the canvas).

18. **IN YOUR OWN WORDS** Non-Euclidean geometry . . .
    A. . . . has no practical applications.
    B. . . . has helped us understand the difference between provable facts and assumptions.
    C. . . . was a false step in helping us understand the nature of the world in which we live.

19. **IN YOUR OWN WORDS** The knowledge that the sum of the measures of the angles of a triangle is 180° is . . .
    A. . . . known because of Figure 6.34, p. 315.
    B. . . . certain, so the foundations of geometry would be destroyed if it were proved false.
    C. . . . sometimes used to decide whether the geometry is on a plane, a pseudosphere, or a sphere.

20. **IN YOUR OWN WORDS** Euclid's treatment of the fifth axiom . . .
    A. . . . shows that he was really smart.
    B. . . . is now known to be wrong.
    C. . . . shows that it is important to question the nature of the assumptions that are made.

21. On a globe, locate San Francisco, Miami, and Detroit. Connect these cities with the shortest paths to form a triangle. Next, use a protractor to measure the angles. What is their sum?

22. On a globe, locate Tokyo, Seattle, and Honolulu. Connect these cities with the shortest paths to form a triangle. Next, use a protractor to measure the angles. What is their sum?

LEVEL 3

23. Consider the spheres shown in Figure 6.67.

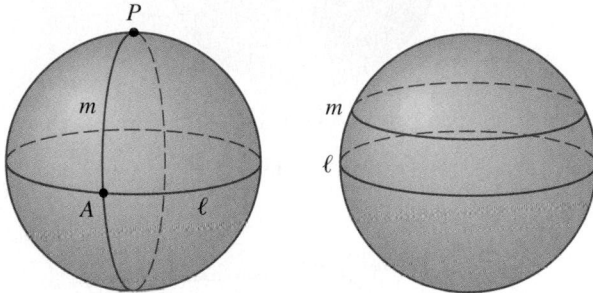

Figure 6.67 A "line" on a sphere is a great circle.

   a. Explain what is meant by a *great circle* on a sphere.
   b. Do one or both of the lettered curves in the figure at the right seem to be lines?
   c. In Euclidean geometry, we say that a line has no endpoints. Does this property apply to lines on a sphere?

24. **IN YOUR OWN WORDS** In Figure 6.67, the sphere at the left shows two lines. In Euclidean geometry, we know that two lines are either parallel or cross at exactly one point. Discuss this property in relation to what you observe in Figure 6.67.

25. **IN YOUR OWN WORDS** In Figure 6.67, the lines on the sphere at the right appear to be parallel. In Table 6.3, we said that there are no parallels on a sphere. What's wrong with our reasoning?

26. Walking north or south on the earth is defined as walking along the meridians, and walking east or west is defined as walking along parallels. There are points on the surface of the earth from which it is possible to walk 300 ft south, then a mile east, then 300 ft north, and be right back where you started. Find one such point.

27. **IN YOUR OWN WORDS** In Lobachevskian geometry, the sum of the measures of the angles of a triangle is less than 180°. Discuss the following statement: In Lobachevskian geometry, the sum of the measures of the angles of a quadrilateral is less than 360°.

28. **IN YOUR OWN WORDS** In Euclidean geometry, if two triangles are similar, then they may or may not also be congruent. In Lobachevskian geometry, if two triangles are similar, do you think that they must also be congruent?

29. Find two additional points satisfying the conditions stated in Problem 26.

© 2000 Cordon Art–Baarn–Holland

Escher: *Circle Limit III*

30. Escher's work *Circle Limit III* is a tessellation (see Section 15.3) based on Lobachevskian geometry. Consider the following model. Draw a circle.

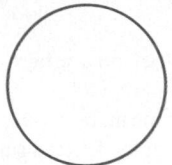

Now draw a "perpendicular circle," which is a circle whose tangents to the original circle at the points of intersection

are perpendicular. The two circles are called **orthogonal circles.**

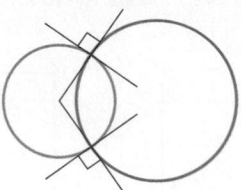

Next, set up another model of Lobachevskian geometry. Points of the plane are points inside the circle. Lines are both diameters of the circle and arcs of orthogonal circles as well as being inside the original circle. On the circle you have drawn, draw several lines for this model.

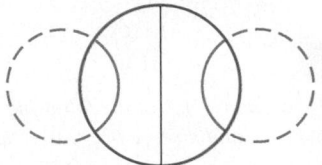

Circle Limit III, the Escher work shown here, is an example that uses this model. The white arcs through the backbones of the fish are lines in this model.

CHAPTER 6 SUMMARY

"Today's society expects schools to ensure that all students have an opportunity to become mathematically literate . . ."

NCTM Standards

IMPORTANT TERMS

Acute angle [6.2]
Acute triangle [6.3]
Adjacent angles [6.2]
Adjacent side [6.4]
Alternate exterior angles [6.2]
Alternate interior angles [6.2]
Angle [6.2]
Angle of depression [6.5]
Angle of elevation [6.5]
Axiom [6.1]
Base angles [6.3]
Base of a triangle [6.3]
Bolyai-Lobachevski geometry [6.7]
Compass [6.1]

Complementary angles [6.2]
Congruent [6.1]
Congruent angles [6.2]
Congruent triangles [6.3]
Construct [6.1]
Corresponding angles [6.2, 6.4]
Corresponding parts [6.3]
Corresponding sides [6.4]
Cosine [6.5]
Degree [6.2]
Divine proportion [6.6]
Elliptic geometry [6.7]
Equal angles [6.2]
Equilateral triangle [6.3]

Euclidean geometry [6.1]
Euclid's postulates [6.1]
Exterior angle [6.3]
Golden ratio [6.6]
Golden rectangle [6.6]
Great circle [6.7]
Half-line [6.1]
Horizontal line [6.2]
Hyperbolic geometry [6.9]
Hypotenuse [6.4]
Inverse tangent [6.5]
Inverse trigonometric ratios [6.5]
Isosceles triangle [6.3]
Isosceles triangle property [6.3]

Legs of a triangle [6.5]
Line [6.1]
Line segment [6.1]
Line of symmetry [6.1]
Lobachevskian postulate [6.7]
Non-Euclidean geometries [6.1, 6.7]
Obtuse angle [6.2]
Obtuse triangle [6.3]
Opposite side [6.4]
Parallel lines [6.1, 6.2]
Parallelogram [6.2]
Perpendicular lines [6.2]
Plane [6.1]
Point [6.1]
Polygon [6.2]
Postulate [6.1]
Projective geometry [6.7]
Protractor [6.2]
Pseudosphere [6.7]

Pythagorean theorem [6.5]
Quadrilateral [6.2]
Ray [6.1]
Rectangle [6.2]
Reflection [6.1]
Region [6.7]
Regular polygon [6.2]
Rhombus [6.2]
Right angle [6.2]
Right triangle [6.3]
Saccheri quadrilateral [6.7]
Scalene triangle [6.3]
Similar figures [6.1]
Similar triangle theorem [6.3]
Similar triangles [6.4]
Similarity [6.1]
Sine [6.5]
Square [6.2]
Straight angle [6.2]

Straightedge [6.1]
Sum of the measures of angles in a triangle [6.3]
Supplementary angles [6.2]
Surface [6.1]
Symmetry (line) [6.1]
Tangent [6.5]
Theorem [6.1]
Transformation [6.1]
Transformational geometry [6.1]
Transversal [6.2]
Trapezoid [6.2]
Triangle [6.3]
Trigonometric ratios [6.5]
Undefined terms [6.1]
Vertex (pl. vertices) [6.2]
Vertex angle [6.3]
Vertical angles [6.2]
Vertical line [6.2]

TYPES OF PROBLEMS

Construct line segments. [6.1]
Construct circles, given the radius. [6.1]
Construct parallel lines. [6.1]
Find a line of symmetry for a given piece of art. [6.1]
Decide whether a given picture is symmetric. [6.1]
Visualize objects in three dimensions. [6.1]
Classify polygons with three to twelve sides. [6.2]
Construct an angle congruent to a given angle. [6.2]
Classify angles. [6.2]
Classify quadrilaterals. [6.2]
Identify vertical, horizontal, intersecting, and parallel lines. [6.2]
Name the corresponding parts of congruent triangles. [6.3]
Find the measure of the third angle of a triangle. [6.3]
Find the measure of the exterior angles of a triangle. [6.3]
Construct a triangle congruent to a given triangle. [6.3]
Classify triangles and use the terminology associated with triangle classifications. [6.3]
Use algebra to find the measures of angles in a triangle. [6.3]
Decide whether a pair of given triangles is similar. [6.4]
List all six angles for a given pair of triangles. [6.4]
List all six sides of a given pair of triangles. [6.4]
Given a right triangle, find the length of a missing side. [6.4]
Given similar triangles, find the length of one of the sides. [6.4]
Show that a given pair of triangles is similar. [6.4]
Solve applied problems using similar triangles. [6.4]
Evaluate a trigonometric ratio. [6.5]
Find the sine, cosine, and tangent for a given angle. [6.5]
Find $\sin^{-1}x$, $\cos^{-1}x$, and $\tan^{-1}x$. [6.5]
Solve applied problems using triangles. [6.5]

Know the terminology associated with right triangles. [6.5]
Work applied problems involving the golden ratio. [6.6]
Decide whether a figure is a Saccheri quadrilateral. [6.7]

CHAPTER 6 REVIEW QUESTIONS

1. If a 4-cm cube is painted green and then cut into 64 1-cm cubes, how many of those cubes will be painted on
   a. 4 sides   **b.** 3 sides   **c.** 2 sides   **d.** 1 side   **e.** 0 sides

2. Which of the following illustrate at least one line of symmetry? If a figure illustrates symmetry, show a line of symmetry. If it does not, tell why.

   a.    **b.**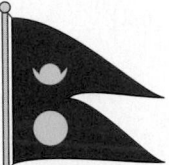

   c.    **d.**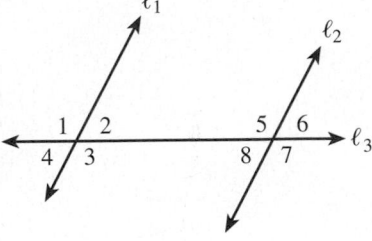

3. Suppose ℓ_1 and ℓ_2 are parallel lines. Classify the pairs of angles shown in the figure.
   a. $\angle 1$ and $\angle 5$ **b.** $\angle 5$ and $\angle 6$
   c. $\angle 2$ and $\angle 4$ **d.** $\angle 2$ and $\angle 8$
   e. $\angle 1$ and $\angle 7$ **f.** $\angle 1$ and $\angle 3$
   g. Can you classify any of the lines ℓ_1, ℓ_2, or ℓ_3 as perpendicular, horizontal, or vertical?

4. If an angle is 49°, then . . .
   a. . . . what is its complement?   **b.** . . . what is its supplement?
   c. . . . if it is an angle in a right triangle, what is the size of the other acute angle?

5. If $(3x + 20)°$, $(2x - 40)°$, and $(x - 16)°$ are angles of a triangle, then what is x?

6. If a box is 10 in. wide, what is its height (to the nearest inch) if we assume the dimensions are in a golden ratio?

7. In a right triangle with one leg of length 12 in. and a hypotenuse of 13 in., what is the length of the other leg?

8. Find the value of the given trigonometric ratios. Round your answers to four decimal places.
   a. sin 59°   **b.** tan 0°   **c.** cos 18°   **d.** tan 82°

9. The world's most powerful lighthouse is on the coast of Brittany, France, and is about 160 ft tall. Suppose you are in a boat just off the coast, as shown in Figure 6.68. Determine your distance (to the nearest foot) from the base of the lighthouse if $\angle B = 12°$.

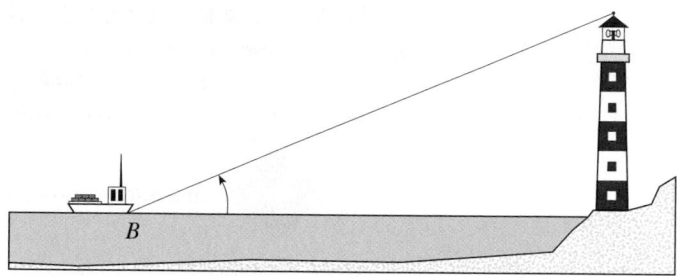

Figure 6.68 **Distance from ship to shore**

10. What is a Saccheri quadrilateral? Discuss why this quadrilateral leads to different kinds of geometries.

GROUP RESEARCH

Working in small groups is typical of most work environments, and learning to work with other students to communicate specific ideas is an important skill. Work with three or four other students to submit a single report based on each of the following questions.

G23. In the figure at the right, there are eight square rooms making up a maze. Each square room has two walls that are mirrors and two walls that are open spaces. Identify the mirrored walls, and then solve the maze by showing how you can pass through all eight rooms consecutively without going through the same room twice. If that is not possible, tell why.

Adapted from "The Amazing Mirror Maze" by Walter Wick in *Games Magazine*, September/October 1981, p. 25

G24. In Figures 6.1 and 6.2 we considered different views of a cube. Figure 6.69 shows a cube with a dot in the middle of each face.

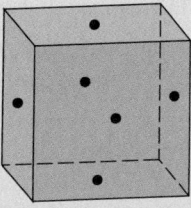

Figure 6.69 **Draw a cube so that each dot is in the center of a face of the cube.**

Figure 6.70 **Dots in a cube**

Draw a cube around the dots in Figure 6.70 so that each dot is in the middle of a face.

G25. Place a dollar bill across the top of two glasses that are at least 3.5 in. apart. Now, describe how you can place a $.50 piece in the middle of the dollar bill without having it fall.

INDIVIDUAL RESEARCH PROBLEMS

Learning to use sources outside your classroom and textbook is an important skill, and here are some ideas for extending some of the ideas in this chapter. You can find references to these projects in a library or at

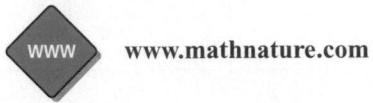

www.mathnature.com

PROJECT 6.1 What do the following people have in common?

    Eamon de Valera, Prime Minister and President of the Republic of Ireland
    Tom Lehrer, songwriter–parodist
    Edmund Husserl, the "Father of Phenomenology"
    Frank Ryan, quarterback for the Cleveland Browns

PROJECT 6.2 Write a report on optical illusions.

PROJECT 6.3 Create an aestheometry design beginning with an angle.

PROJECT 6.4 Create an aestheometry design beginning with a circle.

PROJECT 6.5 Write a report on Ramsey theory.

PROJECT 6.6 Write a paper discussing the nature of an unsolved problem as compared with an impossible problem. As part of your paper, discuss the problems of trisecting an angle and squaring a circle.

PROJECT 6.7 The Historical Note on page 334 asks the question, "Why did the Egyptians build the pyramids using a slant height angle of about 44° or 52°?" Write a paper answering this question.

PROJECT 6.8 Do some research on the length-to-width ratios of the packaging of common household items. Form some conclusions. Find some examples of the golden ratio in art. Do some research on dynamic symmetry.

PROJECT 6.9 In Example 2 of Section 6.6, we assumed the width of the Parthenon to be 101 ft and found the height to be 62.4 ft (assuming the golden ratio). If you worked Problem 8 of Section 6.6, you assumed the height to be 60 ft and found the width using the golden ratio to be 97 ft. Are the numbers from Example 2 and from Problem 8 consistent? Can you draw any conclusions?

PROJECT 6.10 The German artist Albrecht Dürer (1471–1528) is not only a Renaissance artist, but he is also somewhat of a mathematician. Do some research on the mathematics of Dürer.

PROJECT 6.11 Write a paper on perspective. How are three-dimensional objects represented in two dimensions?

PROJECT 6.12 The discovery and acceptance of non-Euclidean geometries had an impact on all of our thinking about the nature of scientific truth. Can we ever know truth in general? Write a paper on the nature of scientific laws, the nature of an axiomatic system, and the implications of non-Euclidean geometries.

7 The Nature of Measurement

Neglect of mathematics works injury to all knowledge, since he who is ignorant of it cannot know the other sciences or the things of this world.

ROGER BACON

CONTENTS

OVERVIEW

Numbers are used to count and to measure. In counting, the numbers are considered exact unless the result has been rounded. In this chapter, we study measurement in more detail.

Dimension refers to those properties called length, area, and volume. A figure having length only is said to be *one-dimensional*. A figure having area is said to be *two-dimensional,* and an object having volume is said to be *three-dimensional*.

In this chapter, we use both the United States and the metric measurement systems, not as they relate to each other, but as independent systems used to measure the size of objects in our world. The goal of this chapter is to give you the ability to both measure and estimate the size, weight, capacity, and temperature of objects.

IMPORTANT IDEAS

Understand the structure of the metric system; length (meter), capacity (liter), and mass (gram). [7.1–7.4]

Know the agreement about accuracy of measurements used in this book. [7.1]

Estimate the distance around a closed figure (perimeter and circumference), and be able to find it by measurement. [7.1]

Know the area formulas for rectangles, squares, parallelograms, triangles, trapezoids, and circles. [7.2]

Know (and estimate) the size of an acre. [7.2]

Know the volume formulas for boxes (parallelepipeds), right rectangular prisms, right circular cylinders, pyramids, right circular cones, and spheres. [7.3]

Know the units for measuring capacity. [7.3]

Know the relationship between volume and capacity. [7.3]

Know the Fahrenheit and Celsius temperatures for both water freezing and water boiling. [7.4]

BOOK REPORTS

Write a 500-word report on one of these books:
Flatland, a Romance of Many Dimensions, by A Square, Edwin A. Abbott (New York: Dover Publications, original edition printed in 1880). It can also be found on the Web at: **http://www.alcyone.com/max/lit/flatland/**
Flatterland: Like Flatland, Only More So, Ian Stewart (Cambridge, MA: Perseus Publishing, 2002).

LINKS

Individual Projects

Group Projects

Research Links

www.mathnature.com

7.1 | PERIMETER

Measuring Length

What is your height? To answer that question, you must take a measurement.

Measure of Length

MARVIN'S LESSONS IN GEOMETRY

TRAPEZOIDS & TRIANGLES

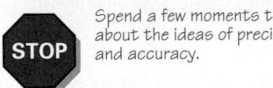

Marvin's Windows and Doors

Spend a few moments thinking about the ideas of precision and accuracy.

> To measure an object is to assign a number to its size. The number representing its linear dimension, as measured from end to end, is called its **measure** or **length.**

Measurement is never exact, and you therefore need to decide how **precise** the measure should be. For example, the measurement might be to the nearest inch, nearest foot, or even nearest mile. The precision of a measurement depends not only on the instrument used but also on the purpose of your measurement. For example, if you are measuring the size of a room to lay carpet, the precision of your measurement might be different than if you are measuring the size of an airport hangar.

The **accuracy** refers to your answer. Suppose that you measure with an instrument that measures to the nearest tenth of a unit. You find one measurement to be 4.6 and another measurement to be 2.1. If, in the process of your work, you need to multiply these numbers, the result you obtain is

$$4.6 \times 2.1 = 9.66$$

This product is calculated to two decimal places, but it does not seem quite right that you obtain an answer that is more accurate (two decimal places) than the instrument you are using to do your measurements (one decimal place). In this book, we will require that the accuracy of your answers not exceed the precision of the measurement. This means that after the calculations are completed, the final answer should be rounded. The principle we will use is stated in the following box.

Accuracy of Measurements Used in This Book

The answers we give will conform to this statement.

> All measurements are as precise as given in the text. If you are asked to make a measurement, the precision will be specified. Carry out all calculations without rounding. After you obtain a final answer, round this answer to be as accurate as *the least precise* measurement.

This means that, to avoid round-off error, you should round only once (at the end). This is particularly important if you are using a calculator, which will display 8, 10, 12, or even more decimal places.

You will also be asked to *estimate* the size of many objects in this chapter. As we introduce different units of measurement, you should remember some reference points so that you can make intelligent estimates. Many comparisons will be mentioned in the text, but you need to remember only those that are meaningful for you to estimate other sizes or distances. You will also need to choose appropriate units of measurement. For example, you would not measure your height in yards or miles, or the distance to New York City in inches.

The most common system of measurement used in the world is what we call the **metric system** (see Figure 7.1). There have been numerous attempts to make the metric system mandatory in the United States. Figure 7.1 shows that the United States is the only major country not using the metric system. Today, big business is supporting the drive toward metric conversion, and it appears inevitable that the metric system will eventually come into use in the United States. In the meantime, it is important that we understand how to use both the United States and metric systems.

Cleveland

400 MILES

640 KILOMETERS

Standard Units of Length

Figure 7.1 The metric world

The most difficult problem in changing from the **United States system** (the customary system of measurement in the United States) to the metric system is not mathematical, but psychological. Many people fear that changing to the metric system will require complex multiplying and dividing and the use of confusing decimal points. For example, in a recent popular article, James Collier states:

> *For instance, if someone tells me it's 250 miles up to Lake George, or 400 out to Cleveland, I can pretty well figure out how long it's going to take and plan accordingly. Translating all of this into kilometers is going to be an awful headache. A kilometer is about 0.62 miles, so to convert miles into kilometers you divide by six and multiply by ten, and even that isn't accurate. Who can do that kind of thing when somebody is asking me are we almost there, the dog is beginning to drool and somebody else is telling you you're driving too fast?*
>
> *Of course, that won't matter, because you won't know how fast you're going anyway. I remember once driving in a rented car on a superhighway in France, and every time I looked down at the speedometer we were going 120. That kind of thing can give you the creeps. What's it going to be like when your wife keeps shouting, "Slow down, you're going almost 130"?*
>
> *But if you think kilometers will be hard to calculate . . . .*

The author of this article has missed the whole point. Why are kilometers hard to calculate? How does he know that it's 400 miles to Cleveland? He knows because the odometer on his car or a road sign told him. Won't it be just as easy to read an odometer calibrated to kilometers or a metric road sign telling him how far it is to Cleveland?

The real advantage of using the metric system is the ease of conversion from one unit of measurement to another. How many of you remember the difficulty you had in learning to change tablespoons to cups? Or pints to gallons?

In this book we will work with both the U.S. and the metric measurement systems. You should be familiar with both and be able to make estimates in both systems. The following box gives the standard units of length.

U.S. System	Metric System
inch (in.)	**meter (m)**
foot (ft)	centimeter (cm; $\frac{1}{100}$ m)
yard (yd)	kilometer (km; 1,000 m)
mile (mi)	

To understand the size of any measurement, you need to see it, have experience with it, and take measurements using it as a standard unit. The basic unit of measurement for the U.S. system is the inch; it is shown in Figure 7.2. You can remember that an inch is about the distance from the joint of your thumb to the tip of your thumb. The basic unit of measurement for the metric system is the meter; it is also shown in Figure 7.2. You can remember that a meter is about the distance from your left ear to the tip of the fingers on the end of your outstretched right arm.

STOP You should commit the approximate size of both the inch and the centimeter to memory.

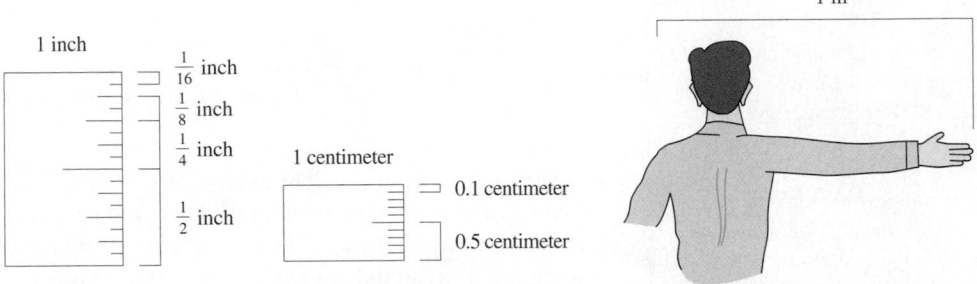

a. Basic unit of length in the U.S. system is the inch.

b. In the metric system, a distance comparable to the inch is the centimeter.

c. Basic unit of length in the metric system is the meter. In the U.S. system, a comparable unit is the yard.

Figure 7.2 **Standard units of measurement for length**

For the larger distances of a mile and a kilometer, you will need to look at maps, or the odometer of your car. However, you should have some idea of these distances. *

It might help to have a visual image of certain prefixes as you progress through this chapter. Greek prefixes **kilo-, hecto-,** and **deka-** are used for measurements larger than the basic metric unit, and Latin prefixes **deci-, centi-,** and **milli-** are used for smaller quantities (see Figure 7.3). As you can see from Figure 7.3, a centimeter is $\frac{1}{100}$ of a meter; this means that 1 meter is equal to 100 centimeters.

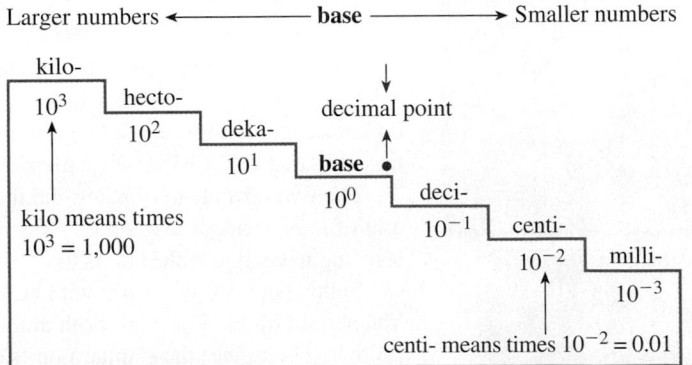

Greek prefixes for larger quantities   **Latin prefixes for smaller quantities**

Figure 7.3 **Metric prefixes**

---

*We could tell you that a mile is 5,280 ft or that a kilometer is 1,000 m, but to do so does not give you any feeling for what these distances really are. You need to get into a car and watch the odometer to see how far you travel in going 1 mile. Most cars in the United States do not have odometers set to kilometers, and until they do it is difficult to measure in kilometers. You might, however, be familiar with a 10-kilometer race. It takes a good runner about 30 minutes to run 10 kilometers and an average runner about 45 minutes. You can walk a kilometer in about 6 minutes.

Now we will measure given line segments with different levels of precision. We will consider two different rulers, one marked to the nearest centimeter and another marked to the nearest $\frac{1}{10}$ centimeter.

EXAMPLE 1

Measure the given segment

a. to the nearest centimeter.

b. to the nearest $\frac{1}{10}$ centimeter.

B ─────────────────────

Solution

a.

B ──────────────────── End of B is nearer to 5 than to 6.

| 1 | 2 | 3 | 4 | 5 | 6 | 7 | 8 |

b. B ──────────────────── B is 5.3 cm long.

| 1 | 2 | 3 | 4 | 5 | 6 | 7 | 8 |

◆

Perimeter

One application of both measurement and geometry involves finding the distance around a polygon. This distance is called the *perimeter* of the polygon.

Perimeter

Remember this term.

STOP

> The **perimeter** of a polygon is the sum of the lengths of the sides of that polygon.

Following are some formulas for finding the perimeters of the most common polygons.

An **equilateral triangle** is a triangle with all sides equal in length.

PERIMETER = 3(SIDE)

$$P = 3s$$

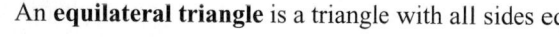

A **rectangle** is a quadrilateral with angles that are all right angles.

PERIMETER = 2(LENGTH) + 2(WIDTH)

$$P = 2\ell + 2w$$

A **square** is a rectangle with all sides equal length.

PERIMETER = 4 (SIDE)

$$P = 4s$$

Equilateral triangle

Rectangle

Square

EXAMPLE 2

Find the perimeter of each polygon.

a.

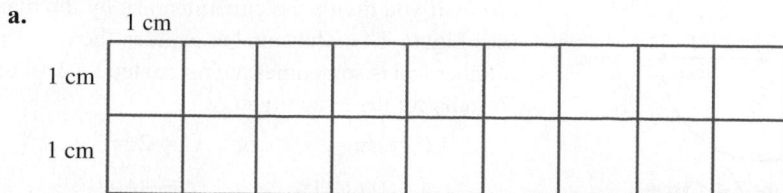

1 cm

1 cm

1 cm

1 cm

b.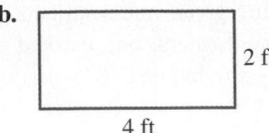

c.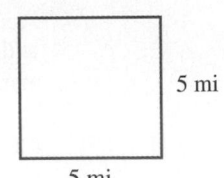

d.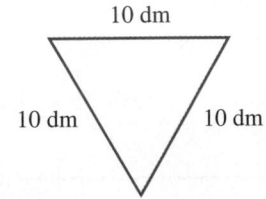

e.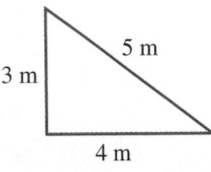

Solution

a. Rectangle is 2 cm by 9 cm, so

$$P = 2\ell + 2w$$
$$= 2(9) + 2(2)$$
$$= 18 + 4 = 22 \text{ cm}$$

b. Rectangle is 2 ft by 4 ft, so

$$P = 2\ell + 2w$$
$$= 2(4) + 2(2) = 8 + 4$$
$$= 12 \text{ ft}$$

c. Square, so

$$P = 4s = 4(5) = 20 \text{ mi}$$

d. Equilateral triangle, so

$$P = 3s = 3(10) = 30 \text{ dm}$$

e. Triangle (add lengths of sides), so

$$P = 3 + 4 + 5 = 12 \text{ m}$$

◆

EXAMPLE 3

Suppose you have enough material for 70 ft of fence and want to build a rectangular pen 14 ft wide. What is the length of this pen?

Solution

$$\text{PERIMETER} = 2(\text{LENGTH}) + 2(\text{WIDTH}) \qquad \text{This is the formula for perimeter.}$$
$$\downarrow \qquad\qquad\qquad \downarrow$$
$$70 \quad = 2(\text{LENGTH}) + 2(14) \qquad \text{Fill in the given information.}$$

Let $\ell = \text{LENGTH OF PEN}$

$$70 = 2\ell + 28$$
$$42 = 2\ell$$
$$21 = \ell$$

The pen will be 21 ft long. ◆

Circumference

A **circle** is the set of all points in a plane a given distance, called the **radius**, from a given point, called the **center.** Although a circle is not a polygon, sometimes we need to find the distance around a circle. This distance is called the **circumference.** For *any circle*, if you divide the circumference by the diameter, you will get the *same number* (see Figure 7.4). This number is given the name π **(pi)**. The number π is an irrational number and is sometimes approximated by 3.14 or $\frac{22}{7}$. We need this number π to state a formula for the circumference C:

$$C = d\pi \qquad \text{or} \qquad C = 2\pi r$$
$$d = \text{DIAMETER} \qquad\qquad r = \text{RADIUS}$$

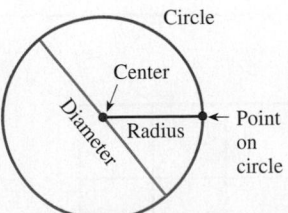

Figure 7.4 Circle

For a circle, the **diameter** is twice the radius. This means that, if you know the radius and want to find the diameter, you simply multiply by 2. If you know the diameter and want to find the radius, divide by 2.

EXAMPLE 4

Find the distance around each figure.

a.

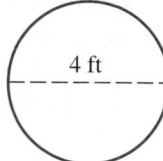

b.

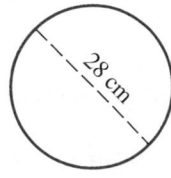

c.

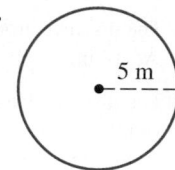

d. $\frac{1}{2}$ of a circle

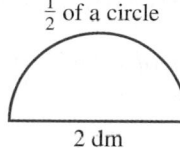

e.

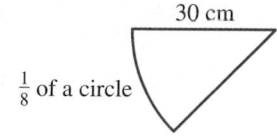

Solution

a. $C = 4\pi$

A decimal approximation (symbolized by $\approx$) is

$C \approx 4(3.14) = 12.56$

The circumference is 4π ft, which is about 13 ft.

b. $C = 28\pi$; this is about 88 cm.

c. $C = 2\pi(5) = 10\pi$; this is about 31 m.

d. This is half of a circle (called a **semicircle**); thus, the curved part is half of the circumference ($C = 2\pi \approx 6.28$), or about 3.14, which is added to the diameter:

$3.14 + 2 = 5.14$

The distance around the figure is about 5 dm.

e. This is one-eighth of a circle. The curved part is one-eighth of the circumference $[C = 2\pi(30) \approx 188.5]$, or about 23.6, which is added here to the radius (on both sides): $23.6 + 30 + 30 \approx 84$ cm.   ◆

Many calculators have a single key marked π. If you press it, the display shows an approximation correct to several decimal places (the number of places depends on your calculator):

Press: [π]   *Display:*   3.141592654

If your calculator doesn't have a key marked π, you may want to use this approximation to obtain the accuracy you want. In this book, the answers are found by using the π key on a calculator and then rounding the answer.

Problem Set 7.1

LEVEL 1

1. **IN YOUR OWN WORDS** Contrast precision and accuracy.

2. **IN YOUR OWN WORDS** What is the agreement about the accuracy of answers in this book?

3. **IN YOUR OWN WORDS** State the perimeter formulas for a square, rectangle, equilateral triangle, and regular pentagon.

4. **IN YOUR OWN WORDS** What is the formula for the circumference of a circle?

5. **IN YOUR OWN WORDS** Define π.

From memory, and without using any measuring devices, draw a line segment with approximate length as indicated in Problems 6–8.

6. **a.** 1 in.   **b.** $\frac{1}{2}$ in.   **c.** 5 cm

7. **a.** 2 in.   **b.** $\frac{1}{4}$ in.   **c.** 1 cm

8. **a.** 3 in.   **b.** 10 cm   **c.** 3 cm

Pick the best choices in Problems 9–33 by estimating. Do not measure. For metric measurements, do not attempt to convert to the U.S. system. The hardest part of the transition to the metric system is the transition to thinking in metrics.

9. The length of this math textbook is about
   A. 9 in.   B. 9 cm   C. 2 ft

10. The length of a car is about
    A. 1 m   B. 4 m   C. 10 m

11. The length of a dollar bill is about
    A. 3 in.   B. 6 in.   C. 9 in.

12. The width of a dollar bill is about
    A. 1.9 cm   B. 6.5 cm   C. 0.65 m

13. The perimeter of a dollar bill is
    A. 18 in.   B. 6 in.   C. 46 in.

14. The perimeter of a five-dollar bill is
    A. 18 cm   B. 6 cm   C. 46 cm

15. The length of a new pencil is
    A. 4 in.   B. 18 in.   C. 7 in.

16. The diameter of a new pencil is
    A. 0.25 cm   B. 0.25 in.   C. 0.25 ft

17. The circumference of an automobile tire is
    A. 60 cm   B. 60 in.   C. 1 m

18. The perimeter of this textbook is
    A. 34 in.   B. 34 cm   C. 11 in.

19. The perimeter of a VISA credit card is
    A. 30 in.   B. 30 cm   C. 1 m

20. The perimeter of a sheet of notebook paper is
    A. 1 cm   B. 10 cm   C. 1 m

21. The perimeter of the screen of a console TV set is
    A. 30 in.   B. 100 cm   C. 100 in.

22. The perimeter of a classroom is
    A. 100 ft   B. 100 m   C. 100 yd

23. The distance from your home to the nearest grocery store is most likely to be
    A. 1 cm   B. 1 m   C. 1 km

24. Your height is closest to
    A. 5 ft   B. 10 ft   C. 25 in.

25. An adult's height is most likely to be about
    A. 6 m   B. 50 cm   C. 170 cm

26. The distance around your waist is closest to
    A. 10 in.   B. 36 in.   C. 30 cm

27. The distance from floor to ceiling in a typical home is about
    A. 2.5 m   B. 0.5 m   C. 4.5 m

28. The length of a diagonal on a typical computer monitor is about
    A. 14 cm   B. 14 in.   C. 31 in.

29. The length of a 100-yard football field is
    A. 100 m
    B. more than 100 m
    C. less than 100 m

30. The distance from San Francisco to New York is about 3,000 miles. This distance is
    A. less than 3,000 km
    B. more than 3,000 km
    C. about 3,000 km

31. Suppose someone could run the 100-meter dash in 10 seconds flat. At the same rate, this person should be able to run the 100-yard dash in
    A. less than 10 sec   B. more than 10 sec   C. 10 sec

32. The prefix *centi-* means
    A. one thousand
    B. one-thousandth
    C. one-hundredth

33. The prefix *kilo-* means
    A. one thousand
    B. one-thousandth
    C. one-hundredth

Measure the segments given in Problems 34–36 with the indicated precision.

34. _____
    a. to the nearest centimeter
    b. to the nearest $\frac{1}{10}$ centimeter
    c. to the nearest inch
    d. to the nearest eighth of an inch

35.
  a. to the nearest centimeter
  b. to the nearest $\frac{1}{10}$ centimeter
  c. to the nearest inch
  d. to the nearest eighth of an inch

36. ──────────────
  a. to the nearest centimeter
  b. to the nearest $\frac{1}{10}$ centimeter
  c. to the nearest inch
  d. to the nearest eighth of an inch

LEVEL 2

Find the perimeter, circumference, or distance around the figures given in Problems 37–51, by using the appropriate formula. Round approximate answers to two decimal places.

37.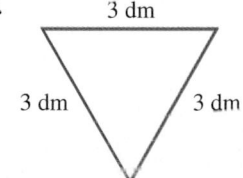
5 in. / 4 in.

38.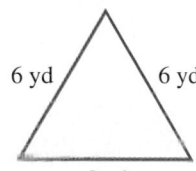
12 ft / 3 ft

39.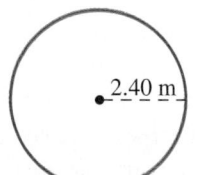
3 dm / 3 dm / 3 dm

40.
6 yd / 6 yd / 6 yd

41.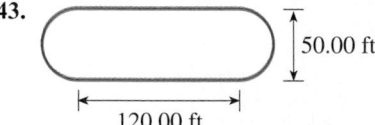
2.40 m

42.
50.00 ft

43.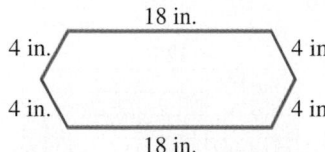
50.00 ft
120.00 ft

44.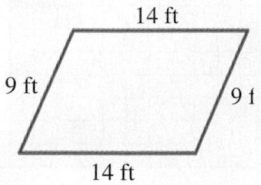
18 in.
4 in. 4 in.
4 in. 4 in.
18 in.

45.
14 ft
9 ft 9 ft
14 ft

46.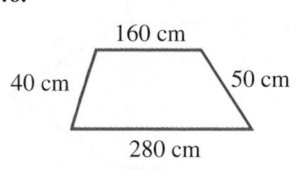
160 cm
40 cm 50 cm
280 cm

47.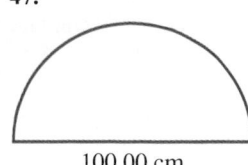
100.00 cm

48.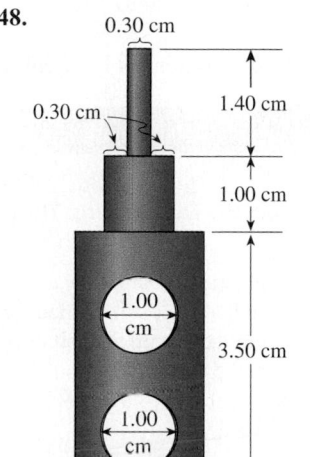
0.30 cm
0.30 cm
1.40 cm
1.00 cm
1.00 cm
1.00 cm
3.50 cm
1.00 cm
1.70 cm

49.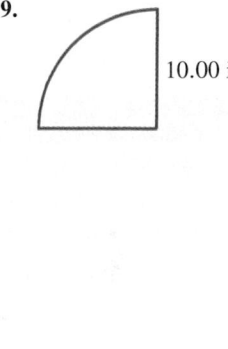
10.00 in.

50.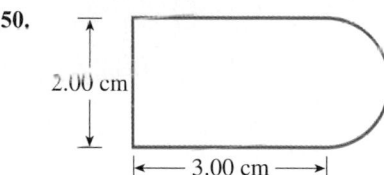
2.00 cm
3.00 cm

51.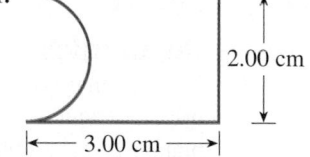
2.00 cm
3.00 cm

52. What is the width of a rectangular lot that has a perimeter of 410 ft and a length of 140 ft?

53. What is the length of a rectangular lot that has a perimeter of 750 m and a width of 75 m?

54. Find the dimensions of a rectangle with a perimeter of 54 cm if the length is 5 cm less than three times the width.

LEVEL 3

55. The perimeter of $\triangle ABC$ is 117 in. Find the lengths of the sides.

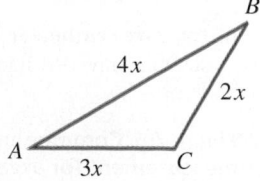
B
4x
2x
A 3x C

56. The perimeter of this pentagon is 280 cm. Find the lengths of the sides.

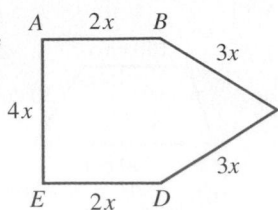

57. One of the first recorded units of measurement is the **cubit.**

Musée du Louvre, by M. Chuzeville, Paris.

Figure 7.5 **An ancient cubit, from the Louvre, Paris. The original measures 52.5 cm.**

If we define a *cubit* as the distance from your elbow to your fingertips, then we may find that the cubit defined for your body is a little different from the Egyptian cubit on display at the Louvre (see Figure 7.5). How does your cubit compare with this Egyptian cubit?

58. Find your own metric measurements.

Women	*Men*
a. height	height
b. bust	chest
c. waist	waist
d. hips	seat
e. waist to hemline	neck

59. In the sixth chapter of Genesis, the dimensions of Noah's ark are given as 300 cubits long, 50 cubits wide, and 30 cubits high. Use the Egyptian cubit in Figure 7.5 to convert these measurements to the nearest meter.

60. Suppose that we fit a band tightly around the earth at the equator. We wish to raise the band so that it is uniformly supported 6 ft above the earth at the equator.

  a. Guess how much extra length would have to be added to the band (not the supports) to do this.

  b. Calculate the amount of extra material that would be needed.

Equator

7.2   AREA

Rectangles

Suppose that you want to carpet your living room. The price of carpet is quoted as a price per square yard. A square yard is a measure of **area.** To measure the area of a plane figure, you fill it with **square units.** (See Figure 7.6.)

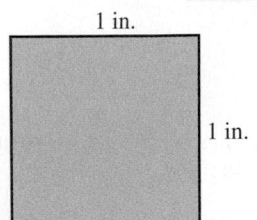

a. 1 **square inch** (actual size); abbreviated 1 sq in. or 1 in.^{2}

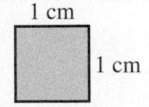

b. 1 **square centimeter** (actual size); abbreviated 1 sq cm or 1 cm^2

Figure 7.6 **Common units of measurement for area**

EXAMPLE 1

What is the area of the shaded region?

a.

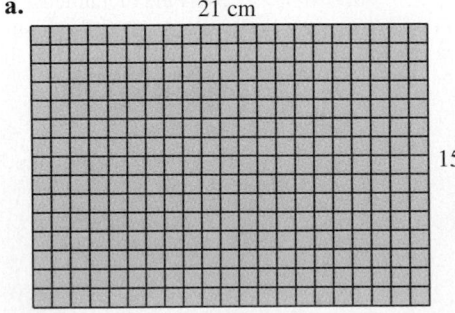

21 cm

15 cm

b.

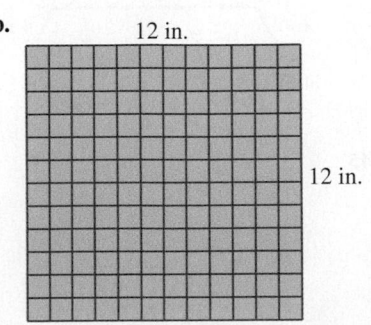

12 in.

12 in.

Solution   To find the area, count the number of square units in each region.

a. You can cout the number of square centimeters in the shaded region; there are 315 squares. Also notice:

Across   Down

21 cm × 15 cm = 21 × 15 cm × cm = 315 cm²

b. The shaded region is a **square foot.** You can count 144 square inches inside the region. Also notice:

Across   Down

12 in. × 12 in. = 144 in.²   ◆

As you can see from Example 1, the area of a rectangular or square region is the product of the distance across (length) and the distance down (width).

Area of Rectangles and Squares

Many believe the push for a U.S. conversion to the metric system has passed. However, a publication of The American Association for the Advancement of Science carried the following report: "A renewed U.S. Government commitment to the metric system of measurement promises to help American companies in international markets and to provide the opportunity to restart our commitment to the use of the metric system in science and mathematics classrooms."

Rectangles
AREA = LENGTH × WIDTH

$A = \ell w$

Length, ℓ

Width, w

Squares
AREA = SIDE × SIDE
= (SIDE)²

$A = s^2$

Side, s

Side, s

EXAMPLE 2

How many square feet are there in a square yard?

Solution
Since 1 yd = 3 ft, we see from Figure 7.7 that

$1 \text{ yd}^2 = (3 \text{ ft})^2$
$= 9 \text{ ft}^2$

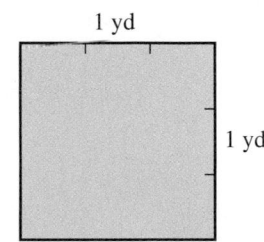

1 yd

1 yd

Figure 7.7 **1 yd² = 9 ft²**   ◆

Parallelograms

A **parallelogram** is a quadrilateral with two pairs of parallel sides, as shown in Figure 7.8. To find the area of a parallelogram, we can estimate the area by counting the number of square units inside the parallelogram (which may require estimation of partial square units), or we can show that the formula for the area of a parallelogram is the same as the formula for the area of a rectangle.

Figure 7.8 **Parallelograms**

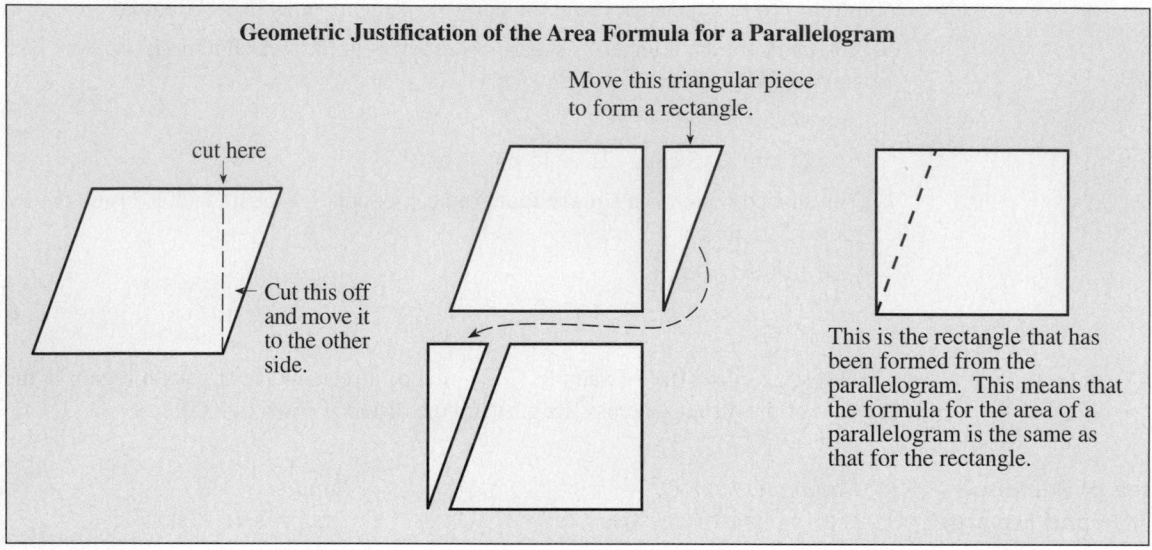

Geometric Justification of the Area Formula for a Parallelogram

Move this triangular piece to form a rectangle.

cut here

Cut this off and move it to the other side.

This is the rectangle that has been formed from the parallelogram. This means that the formula for the area of a parallelogram is the same as that for the rectangle.

Area of Parallelogram

Parallelograms

AREA = BASE × HEIGHT

$A = bh$

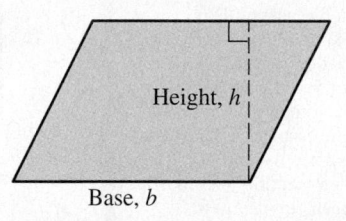

Height, h

Base, b

EXAMPLE 3

Find the area of each shaded region.

a.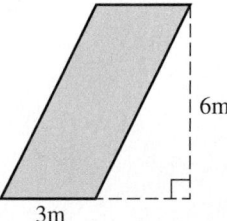

6m

3m

b.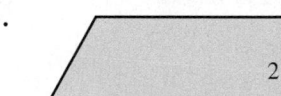

2 in.

5 in.

Solution

a. $A = 3 \text{ m} \times 6 \text{ m}$

$= 18 \text{ m}^2$

b. $A = 2 \text{ in.} \times 5 \text{ in.}$

$= 10 \text{ in.}^2$ ◆

Triangles

You can find the area of a triangle by filling in and approximating the number of square units, by rearranging the parts, or by noticing that *every* triangle has an area that is exactly half that of a corresponding parallelogram.

Geometric Justification of the Area Formula for a Triangle

These triangles have the same area.

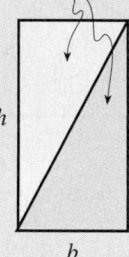

These triangles have the same area.

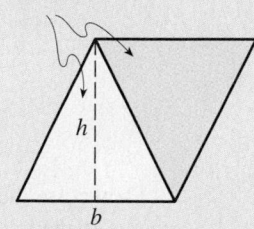

These triangles have the same area.

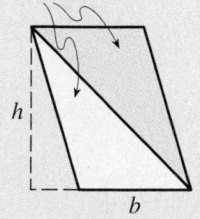

We can therefore state the following result.

Area of Triangles

Triangles

AREA $= \frac{1}{2} \times$ BASE $\times$ HEIGHT

$A = \frac{1}{2}bh$

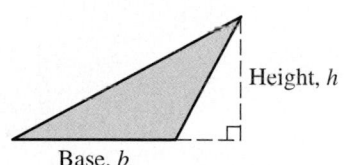

EXAMPLE 4

Find the area of each shaded region

a.

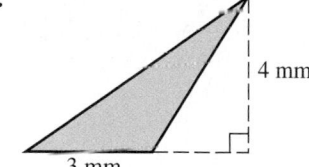

4 mm

3 mm

b.

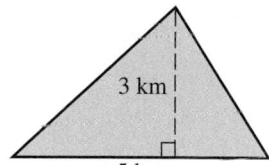

3 km

5 km

Solution

a. $A = \frac{1}{2} \times 3 \text{ mm} \times 4 \text{ mm}$

$= 6 \text{ mm}^2$

b. $A = \frac{1}{2} \times 5 \text{ km} \times 3 \text{ km}$

$= \frac{15}{2} \text{ km}^2$ or $7\frac{1}{2} \text{ km}^2$ ◆

Trapezoids

We can also find the area of a trapezoid by finding the area of triangles. A **trapezoid** is a quadrilateral with two sides parallel. These sides are called the *bases,* and the perpendicular distance between the bases is the height, as shown in Figure 7.9.

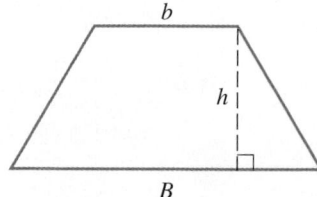

b

h

B

Figure 7.9 **Trapezoid**

The area formula can be found as the sum of the areas of triangle I and triangle II:

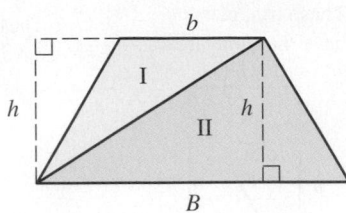

Area of triangle I: $\frac{1}{2}bh$

Area of triangle II: $\frac{1}{2}Bh$

Total area: $\frac{1}{2}bh + \frac{1}{2}Bh$

If we use the distributive property, we obtain the area formula for a trapezoid, as shown in the following box.

Area of Trapezoids

> ### *Trapezoids*
>
> $A = \frac{1}{2}h(b + B)$
>
>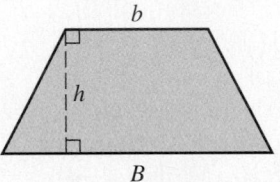

EXAMPLE 5

Find the area of each shaded region.

a.

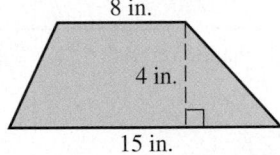

b.

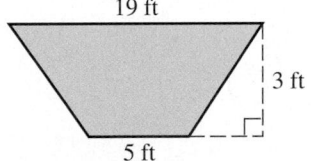

Solution

a. $h = 4;\quad b = 8;\quad B = 15$

$A = \frac{1}{2}(4)(8 + 15)$

$\quad = 2(23)$

$\quad = 46$

The area is 46 in.^{2}.

b. $h = 3;\quad b = 19;\quad B = 5$

$A = \frac{1}{2}(3)(19 + 5)$

$\quad = \frac{3}{2}(24)$

$\quad = 36$

The area is 36 ft^2.   ◆

Circles

The last of our area formulas is for the area of a circle. Historically, we know from the Rhind papyrus that the Egyptians knew of the formula for the area of a circle. We state this formula using modern notation.

Area of Circles

> ### *Circles*
>
> $A = \pi r^2$
>
>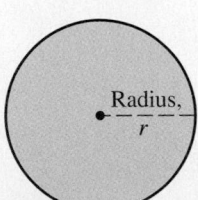
>
> Radius, *r*

Even though it is beyond the scope of this course to derive a formula for the area of a circle, we can give a geometric justification that may appeal to your intuition.

Geometric Justification of the Area Formula for a Circle

Consider a circle with radius r.

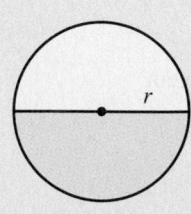

$C = 2\mathrm{p}r$

Cut the circle in half:

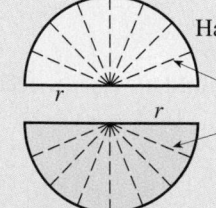

Half of the circumference is $\mathrm{p}r$.

Cut along the dashed lines so that each half lays flat when it is opened up, as shown below.

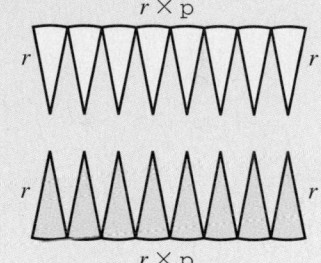

Fit these two pieces together:

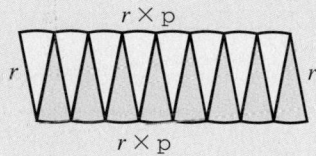

Therefore, it looks as if the area of a circle of radius r is about the same as the area of a rectangle of length $\mathrm{p}r$ and width r— that is, $\mathrm{p}r^2$.

EXAMPLE 6

Find the area of each shaded region to the nearest tenth unit.

a.

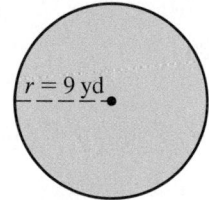

$r = 9$ yd

b.

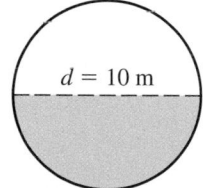

$d = 10$ m

Solution

a. On a calculator:

$$9^2\pi \approx 254.4690049$$

To the nearest tenth, the area is 254.5 yd^2.

b. Notice that the shaded portion is only half the area of the circle. On a calculator:

$$5^2\pi/2 \approx 39.26990817$$

To the nearest tenth, the area is 39.3 m^2.   ◆

Sometimes we measure area using one unit of measurement and then we want to convert the result to another.

EXAMPLE 7

Suppose your living room is 12 ft by 15 ft and you want to know how many square yards of carpet you need to cover this area.

Solution
Method I. $A = 12 \text{ ft} \times 15 \text{ ft}$
$= 180 \text{ ft}^2$
$= 180 \times 1 \text{ ft}^2$
$= 180 \times \left(\frac{1}{9} \text{ yd}^2\right)$ Since 1 yd = 3 ft,
$= 20 \text{ yd}^2$ 1 yd² = (1 yd) × (1 yd)
                                      1 yd² = (3 ft) × (3 ft)
                                      1 yd² = 9 ft²
                                      $\frac{1}{9}$ yd² = 1 ft²

Method II. Change feet to yards to begin the problem:
$12 \text{ ft} = 4 \text{ yd}$ and $15 \text{ ft} = 5 \text{ yd}$
$A = 4 \text{ yd} \times 5 \text{ yd}$
$= 20 \text{ yd}^2$ ◆

If the area is large, as with property, a larger unit is needed. This unit is called an *acre*.

Acre

> An **acre** is 43,560 ft².
>
> To convert from ft² to acres, divide by 43,560.

ESTIMATION HINT:
Real estate brokers estimate an acre as 200 ft by 200 ft or 40,000 ft².

When working with acres, you usually need a calculator to convert square feet into acres, as shown in Example 8.

EXAMPLE 8

How many acres are there in a rectangular property measuring 363 ft by 180 ft?

Solution

AREA $= 363 \text{ ft} \times 180 \text{ ft}$
$= 65,340 \text{ ft}^2$ ESTIMATE $65,340 \approx 65,000$

To change ft² to acres, divide by 43,560. $65,000 \div 40,000 = 65 \div 40$
AREA $= 65,340 \text{ ft}^2$ $= 13 \div 8$
$= (65,340 \div 43,560) \text{ acres}$ $\approx 12 \div 8 = 1.5$
$= 1.5 \text{ acres}$ *Press*:   [65340] [÷] [43560] [=]   ◆

Problem Set 7.2

LEVEL 1

1. **IN YOUR OWN WORDS** What do we mean by length?

2. **IN YOUR OWN WORDS** What do we mean by area?

3. **IN YOUR OWN WORDS** How do you find the area of a circle? Contrast with the procedure for finding the circumference of a circle.

Estimate the area of each figure in Problems 4–7 to the nearest square centimeter.

4. a. 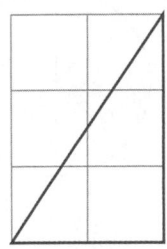   **b.**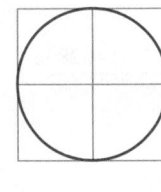

c.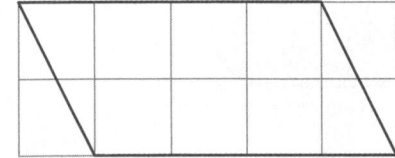

5. a. 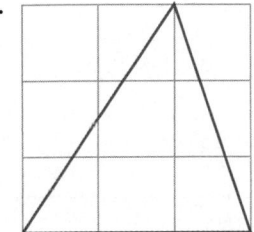   **b.**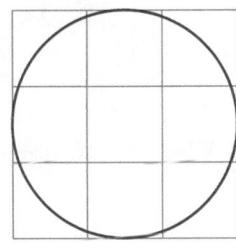

c.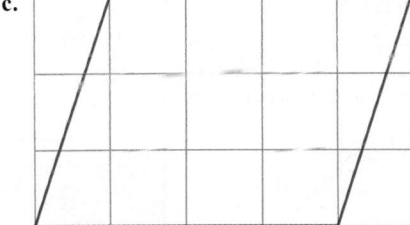

6.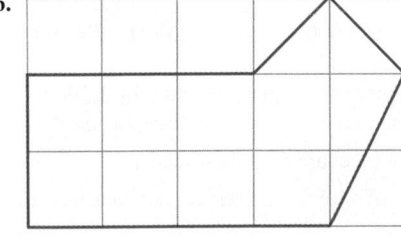

7.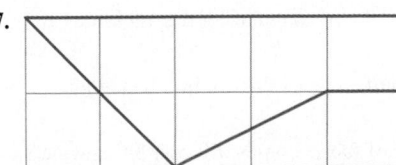

Pick the best choices in Problems 8–18 by estimating. Do not measure. For metric measurements, do not attempt to convert to the U.S. system.

8. The area of a dollar bill is
  A. 18 in.   B. 6 in.²   C. 18 in.²

9. The area of a five-dollar bill is
  A. 100 cm   B. 10 cm²   C. 100 cm²

10. The area of the front cover of this textbook is
  A. 70 in.²   B. 70 cm²   C. 70 in.

11. The area of a VISA credit card is
  A. 8 in.   B. 8 in.²   C. 8 cm²

12. The area of a sheet of notebook paper is
  A. 90 cm²   B. 10 in.²   C. 600 cm²

13. The area of a sheet of notebook paper is
  A. 90 in.²   B. 10 cm²   C. 600 in.²

14. The area of the screen of a console TV set is
  A. 4 ft²   B. 19 in.²   C. 100 in.²

15. The area of a classroom is
  A. 100 ft²   B. 1,000 ft²   C. 0.5 acre

16. The area of the floor space of the Superdome in New Orleans is
  A. 1 mi²   B. 1,000 m²   C. 10 acres

17. The area of the bottom of your feet is
  A. 1 m²   B. 400 in.²   C. 400 cm²

18. It is known that the area of the skin covering your entire body is about 100 times the area that you will find if you trace your hand on a sheet of paper. Using this estimate, the area of your skin is
  A. 300 in.²   B. 3,000 in.²   C. 3,000 cm²

LEVEL 2

Find the area of each shaded region in Problems 19–40. (Assume that given measurements are exact, and round approximate answers to the nearest tenth of a square unit.)

19.
3 in.
5 in.

20.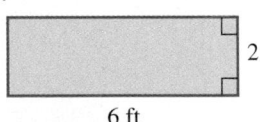
2 ft
6 ft

21.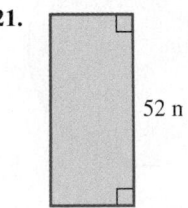
52 n
23 m

22.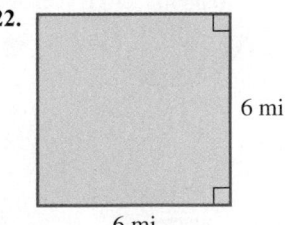
6 mi
6 mi

23.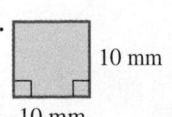
10 mm
10 mm

24.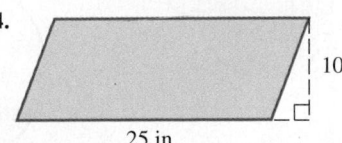
10 in.
25 in.

25.

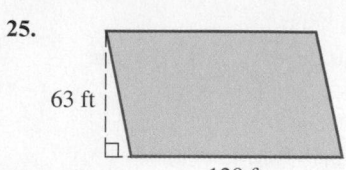

63 ft
120 ft

26.

3 m
9 m

27.

13 dm
21 dm

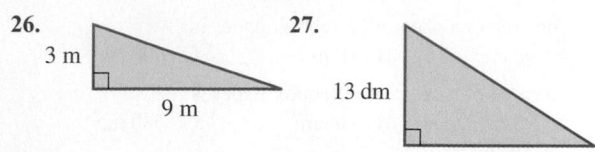

28.

8 ft   10 ft

29.

160 cm
32 cm   30 cm   50 cm
280 cm

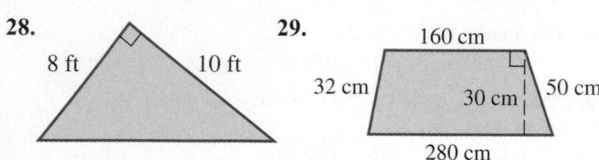

30.

8 ft
5 ft   7 ft
14 ft

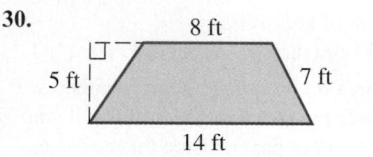

31.

4 in.
5 in.
9 in.
6 in.

32.

13 mm
15 mm
10 mm
14 mm
11 mm

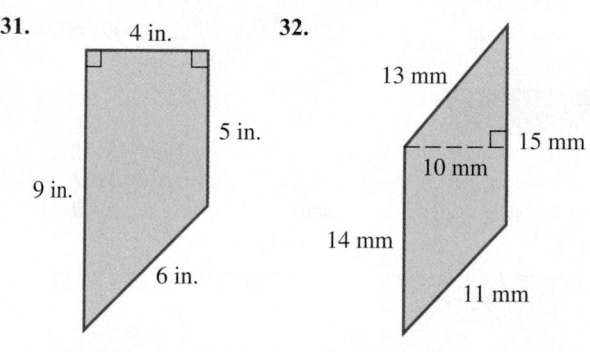

33.

20 in.

34.

30 in.

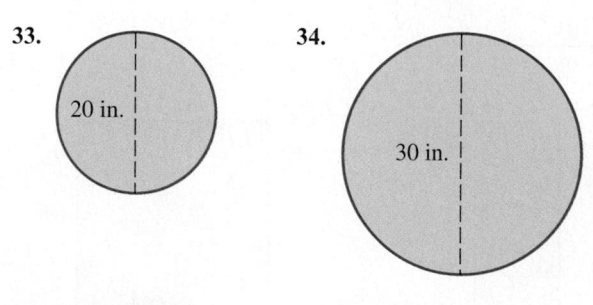

35.

10 in.

36.

r = 5 dm

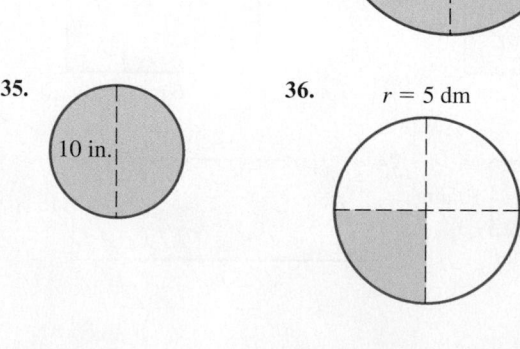

37.

d = 28 in.

38.   r = 30 cm

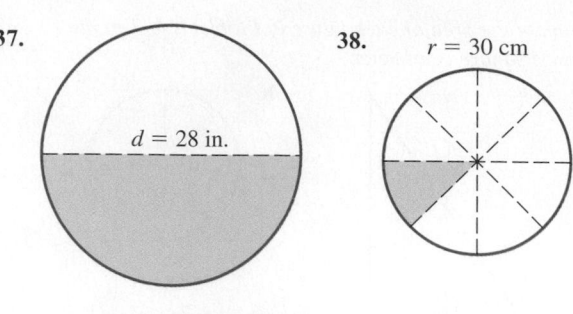

39.

2 cm
3 cm

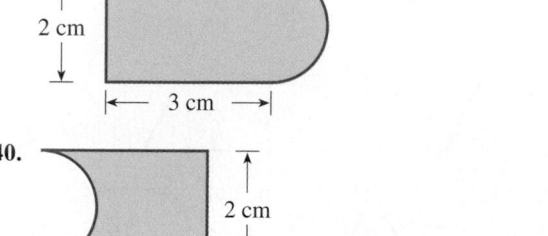

40.

2 cm
3 cm

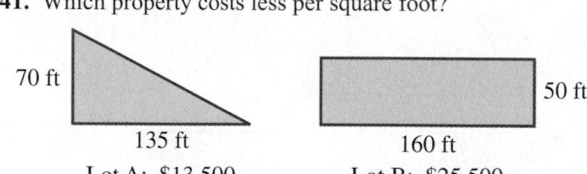

41. Which property costs less per square foot?

70 ft
135 ft
Lot A: $13,500

50 ft
160 ft
Lot B: $25,500

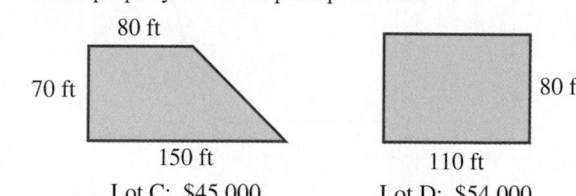

42. Which property costs less per square foot?

80 ft
70 ft
150 ft
Lot C: $45,000

80 ft
110 ft
Lot D: $54,000

43. If a rectangular piece of property is 750 ft by 1,290 ft, what is the acreage (to the nearest tenth of an acre)?

44. How many square feet are there in $4\frac{1}{2}$ acres?

45. What is the area of a television screen that measures 12 in. by 18 in.?

46. What is the area of a rectangular lot that measures 185 ft by 75 ft?

47. What is the area of a piece of $8\frac{1}{2}$-in. by 11-in. typing paper?

48. If a certain type of fabric comes in a bolt 3 feet wide, how long a piece must be purchased to have 24 square feet?

49. What is the cost of seeding a rectangular lawn 100 ft by 30 ft if 1 pound of seed costs $5.85 and covers 150 square feet? Use estimation to decide whether your answer is reasonable.

50. a. If a mini pizza has a 6-in. diameter, what is the number of square inches (to the nearest square inch)?
   b. If a small pizza has a 10-in. diameter, what is the number of square inches (to the nearest square inch)?
   c. If a medium pizza has a 12-in. diameter, what is the number of square inches (to the nearest square inch)?
   d. If a large pizza has a 14-in. diameter, what is the number of square inches (to the nearest square inch)?

 LEVEL 3

*If the lengths of three sides of a triangle are known, then the following formula, known as **Hero's (or Heron's)** formula is sometimes used:*

$$A = \sqrt{s(s - a)(s - b)(s - c)}$$

where a, b, and c are the given lengths of the sides, and s = 0.5(a + b + c). Use this formula to find the areas (correct to the nearest square foot) in Problems 51–53.

51. $a = 5$ ft, $b = 8$ ft, $c = 10$ ft

52. $a = 180$ ft, $b = 200$ ft, $c = 350$ ft

53.

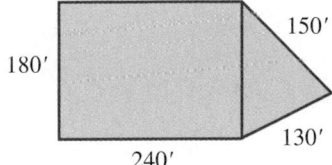

54. What is the area of a square whose diagonal is equal to 7 in.?

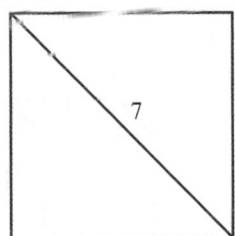

55. What is the area of the regular pentagon with a side equal to 10 in.?

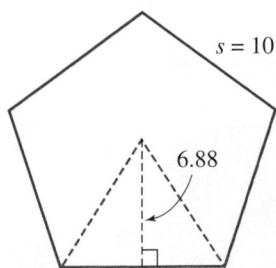

56. What is the area (to the nearest square inch) of the regular hexagon with a side equal to 10 in.?

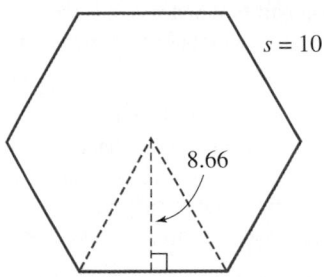

57. What is the area (to the nearest square inch) of a regular octagon with a side equal to 10 in.?

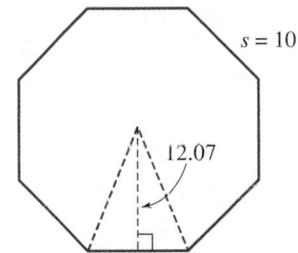

PROBLEM SOLVING

Find the area (to the nearest square inch) of the shaded region contained in the 10-in. squares in Problems 58–59. Assume that the arcs intersect the midpoints of the sides.

58.

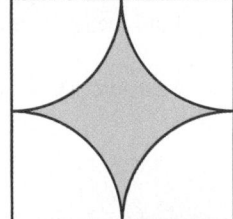

59.

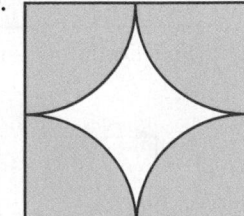

60. Figure 7.10 illustrates a strange and interesting relationship. The square in part **a** has an area of 64 cm² (8 cm by 8 cm). When *this same figure* is cut and rearranged as shown in part **b,** it appears to have an area of 65 cm². Where did this "extra" square centimeter come from?

[*Hint:* Construct your own square 8 cm on a side, and then cut it into the four pieces as shown. Place the four pieces together as illustrated. Be sure to do your measuring and cutting very carefully. Satisfy yourself that this "extra" square centimeter has appeared. Can you explain this relationship?]

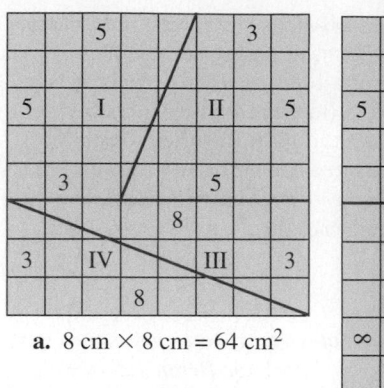

a. 8 cm × 8 cm = 64 cm²

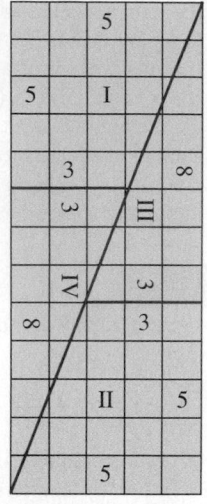

b. 13 cm × 5 cm = 65 cm²

Figure 7.10 Extra square cm?

7.3 SURFACE AREA, VOLUME, AND CAPACITY

Surface Area

Suppose you wish to paint a box whose edges are each 3 ft, and you need to know how much paint to buy. To determine this you need to find the sum of the areas of all the faces—this is called the **surface area.**

EXAMPLE 1

Find the amount of paint needed for a box with edges 3 ft.

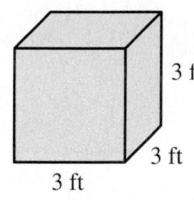

Solution A box (cube) has 6 faces of equal area.

Number of faces of cube
↓
6 × 9 ft² = 54 ft²
Area of each face

You need enough paint to cover 54 ft². ◆

EXAMPLE 2

Find the outside surface area.

a.

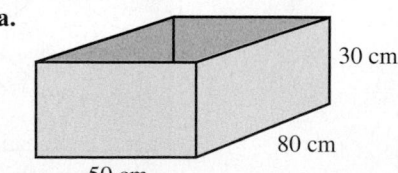

30 cm

80 cm

50 cm

b.

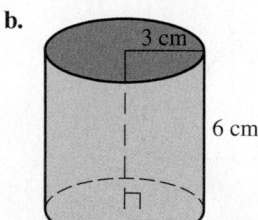

3 cm

6 cm

Solution

a. Notice that there is no top on the box. We find the sum of the areas of all the faces:

Front: $30 \times 50 = 1{,}500$
Back: $1{,}500$ *Same as front*
Side: $80 \times 30 = 2{,}400$
Side: $2{,}400$ *Sides are the same size.*
Bottom: $80 \times 50 = 4{,}000$
Total: $11{,}800 \text{ cm}^2$

b. Notice that the can has a bottom, but no lid. To find the surface area, find the area of a circle (the bottom) and think of the sides of the can as being "rolled out." The length of the resulting rectangle is the circumference of the can and the width of the rectangle is the height of the can.

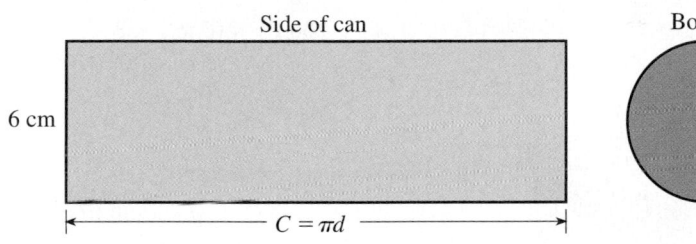

Side: $A = \ell w = (\pi d)w = \pi(6)(6) = 36\pi$

Bottom: $A = \pi r^2 = \pi(3)^2 = 9\pi$

Surface area: $36\pi + 9\pi = 45\pi \approx 141.37167$

The surface area is about 141 cm^2.   ◆

EXAMPLE 3   PÓLYA'S METHOD

You want to paint 200 ft of a three-rail fence, and need to know how much paint to purchase.

Solution   We use Pólya's problem-solving guidelines for this example.

Understand the Problem. There is some additional information you need to gather before you can answer this question.

You want to paint 200 ft of a three-rail fence that is made up of three boards, each 6 inches wide.

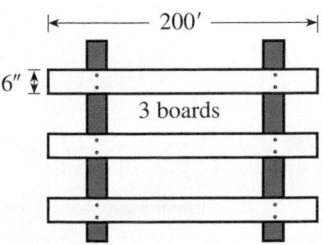

You want to know:

a. the number of square feet on one side of the fence.

b. the number of square feet to be painted if the posts and edges of the boards comprise 100 ft².

c. the number of gallons of paint to purchase if each gallon covers 325 ft².

Devise a Plan. The paint coverage is found on the can (325 ft² for this problem). The next thing we need to know is the number of square feet to be painted. We estimate the number of square feet to be painted on the posts and the edges to be 100 ft². Now, calculate the number of square feet to be painted and divide by 325 (the number of square feet per gallon).

Carry Out the Plan.

a. We first calculate A, the number of square feet to be painted on one side of the fence.

$$A = (200 \text{ ft}) \times (6 \text{ in.}) \times 3$$
$$= (200 \text{ ft}) \times (\tfrac{1}{2} \text{ ft}) \times 3$$
$$= 300 \text{ ft}^2$$

b. AMOUNT TO BE PAINTED $= 2($AMOUNT ON ONE SIDE$) +$ EDGES AND POSTS
$$= 2(300 \text{ ft}^2) + 100 \text{ ft}^2$$
$$= 700 \text{ ft}^2$$

c. $\left(\begin{array}{c}\text{NUMBER OF SQUARE} \\ \text{FEET PAINTED}\end{array}\right) = \left(\begin{array}{c}\text{NUMBER OF SQUARE} \\ \text{FEET PER GALLON}\end{array}\right)\left(\begin{array}{c}\text{NUMBER OF} \\ \text{GALLONS}\end{array}\right)$

$$700 = 325\left(\begin{array}{c}\text{NUMBER OF} \\ \text{GALLONS}\end{array}\right)$$

$$2.15 \approx (\text{NUMBER OF GALLONS}) \quad \text{\textit{Divide both sides by 325.}}$$

Look Back. If paint must be purchased by the gallon (as implied by the question), the amount to purchase is 3 gallons. ◆

Volume

To measure area, we covered a region with square units and then found the area by using a mathematical formula. A similar procedure is used to find the amount of space inside a solid object, which is called its **volume**. We can imagine filling the space with **cubes**. A **cubic inch** and a **cubic centimeter** are shown in Figure 7.11.

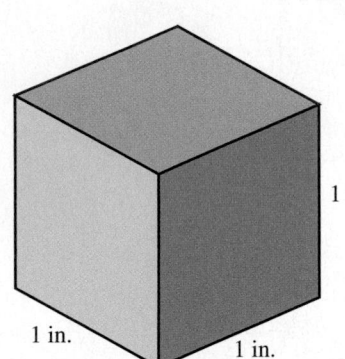

a. 1 cubic inch (1 cu in. or 1 in.³)

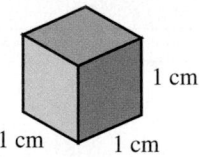

b. 1 cubic centimeter (cu cm, cc, or 1 cm³)

***Figure 7.11* Common units used for measuring volume**

Volume of a Cube

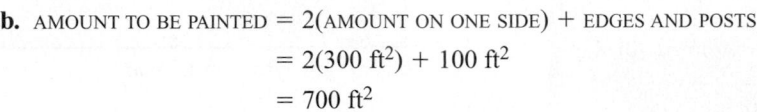

The volume V of a cube with edge s is found by

$$\text{VOLUME} = \text{EDGE} \times \text{EDGE} \times \text{EDGE}$$
$$= (\text{EDGE})^3$$

or

$$V = s^3$$

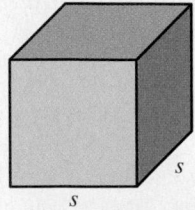

If the solid is not a cube but is a box (called a **rectangular parallelepiped**) with edges of different lengths, the volume can be found similarly.

EXAMPLE 4

Find the volume of a box that measures 4 ft by 6 ft by 4 ft.

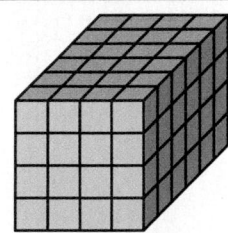

Figure 7.12 **What is the volume?**

Solution
There are 24 cubic feet on the bottom layer of cubes. Do you see how many layers of cubes will fill the solid? (See Figure 7.12.) Since there are four layers with 24 cubes in each, the total is

$$4 \times 24 = 96$$

The volume is 96 ft^3.   ◆

The volume V of a box (parallelepiped) with edges ℓ, w, and h is

VOLUME = LENGTH × WIDTH × HEIGHT

or

$$V = \ell \times w \times h$$
$$= \ell w h$$

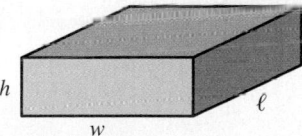

EXAMPLE 5

Find the volume of each solid.

a.

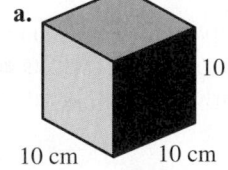

10 cm
10 cm   10 cm

b.

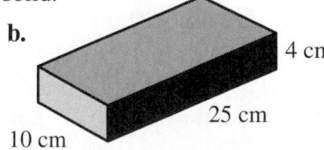

4 cm
25 cm
10 cm

c.

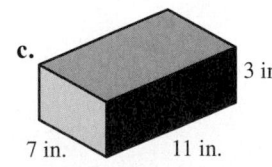

3 in.
7 in.   11 in.

Solution

a. $V = s^3$

$= (10 \text{ cm})^3$

$= (10 \times 10 \times 10) \text{ cm}^3$

$= 1{,}000 \text{ cm}^3$

b. $V = \ell w h$

$= (25 \text{ cm})(10 \text{ cm})(4 \text{ cm})$

$= (25 \times 10 \times 4) \text{ cm}^3$

$= 1{,}000 \text{ cm}^3$

c. $V = \ell w h$

$= (11 \text{ in.})(7 \text{ in.})(3 \text{ in.})$

$= 231 \text{ in.}^3$ ◆

Sometimes the dimensions for the volume we are finding are not given in the same units. In such cases, you must convert all units to a common unit. The common conversions are as follows:

1 ft = 12 in.	To convert feet to inches, multiply by 12. To convert inches to feet, divide by 12.
1 yd = 3 ft	To convert yards to feet, multiply by 3. To convert feet to yards, divide by 3.
1 yd = 36 in.	To convert yards to inches, multiply by 36. To convert inches to yards, divide by 36.

EXAMPLE 6

Suppose you are pouring a rectangular driveway with dimensions 24 ft by 65 ft. The depth of the driveway is 3 in. and concrete is ordered by the yard. By a "yard" of concrete, we mean a cubic yard. You cannot order part of a yard of concrete. How much concrete should you order?

Solution   There are three different measurements in this problem: inches, feet, and yards. Since we want the answer in cubic yards, we will convert all of these measurements to yards:

$$65 \text{ ft} = (65 \div 3) \text{ yd} = \tfrac{65}{3} \text{ yd} \qquad \textit{This is the length, } \ell.$$

$$24 \text{ ft} = (24 \div 3) \text{ yd} = 8 \text{ yd} \qquad \textit{This is the width, w.}$$

$$3 \text{ in.} = (3 \div 36) \text{ yd} = \tfrac{3}{36} \text{ yd} = \tfrac{1}{12} \text{ yd} \qquad \textit{This is the height (depth), h.}$$

$$V = \ell w h$$

$$= \tfrac{65}{3}(8)(\tfrac{1}{12}) \text{ yd}^3 \qquad \textit{Think of 8 as } \tfrac{8}{1}.$$

$$= \frac{65 \times 8 \times 1}{3 \times 12}$$

$$= \frac{130}{9}$$

$$= 14\tfrac{4}{9}$$

You must order 15 cubic yards of concrete.   ◆

Capacity

One of the most common applications of volume involves measuring the amount of liquid a container holds, which we refer to as its **capacity.** For example, if a container is 2 ft by 2 ft by 12 ft, it is fairly easy to calculate the volume:

$$2 \times 2 \times 12 = 48 \text{ ft}^3$$

But this still doesn't tell us how much water the container holds. The capacities of a can of cola, a bottle of milk, an aquarium tank, the gas tank in your car, and a swimming pool can all be measured by the amount of fluid they can hold.

Standard Units of Capacity

U.S. System	Metric System
gallon (gal)	**liter (L)**
quart (qt; $\tfrac{1}{4}$ gal)	kiloliter (kl; 1,000 L)
ounce (oz; $\tfrac{1}{128}$ gal)	milliliter (ml; $\tfrac{1}{1,000}$ L)
cup (c; 8 oz)	

HISTORICAL NOTE

Do you know where the expression "the whole 9 yards" comes from? It is related to concrete. A standard-size "cement mixer" has a capacity of 9 cubic yards of concrete. Thus, a job requiring the mixer's full capacity demands "the whole 9 yards."

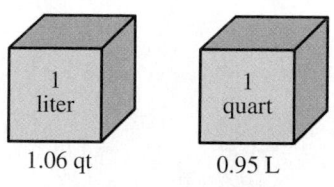

Figure 7.13 Standard capacities

You should remember some of these references for purposes of estimation. For example, remember that a can of Coke is 355 ml, and the size of a liter of milk is about the same as a quart of milk. A cup of coffee is about 300 ml and a spoonful of medicine is about 5 ml.

Most containers of liquid that you buy have capacities stated in both milliliters and ounces, or quarts and liters (see Figure 7.13). Some of these size statements are listed in Table 7.1. The U.S. Bureau of Alcohol, Tobacco, and Firearms has made metric bottle sizes for liquor mandatory, so the half-pint, fifth, and quart have been replaced by 200-ml, 750-ml, and 1-L sizes. A typical dose of cough medicine is 5 ml, and 1 kl is 1,000 L, or about the amount of water one person would use for all purposes in two or three days.

Table 7.1 Capacities of Common Grocery Items, as Shown on Labels

Item	U.S. Capacity	Metric Capacity
Milk	$\frac{1}{2}$ gal	1.89 L
Milk	1.06 qt	1 L
Budweiser	12 oz	355 ml
Coke	67.6 oz	2 L
Hawaiian Punch	1 qt	0.95 L
Del Monte pickles	1 pt 6 oz	651 ml

Since it is common practice to label capacities in both U.S. and metric measuring units, it will generally not be necessary for you to make conversions from one system to another. But if you do, it is easy to remember that a liter is just a little larger than a quart, just as a meter is a little larger than a yard.

To measure capacity, you use a measuring cup.

EXAMPLE 7

Measure the amount of liquid in the measuring cup in Figure 7.14, both in the U.S. system and in the metric system.

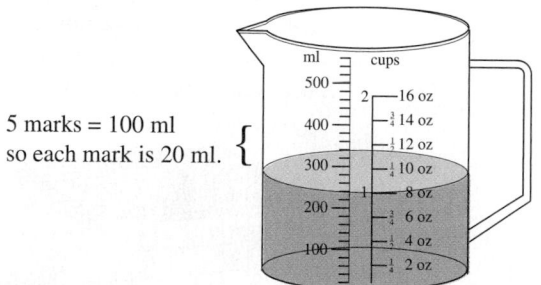

5 marks = 100 ml
so each mark is 20 ml.

Figure 7.14 Standard measuring cup with both metric and U.S. measurements

Solution

Metric: 240 ml   U.S.:   About 1 c or 8 oz   ◆

Some common relationships among volume and capacity measurements in the U.S. system are shown in Figure 7.15 on page 376.

In the U.S. system of measurement, the relationship between volume and capacity is not particularly convenient. One gallon of capacity occupies 231 in.³. This means that, since the box in part **c** of Example 5 has a volume of 231 in.³, we know that it will hold exactly 1 gallon of water.

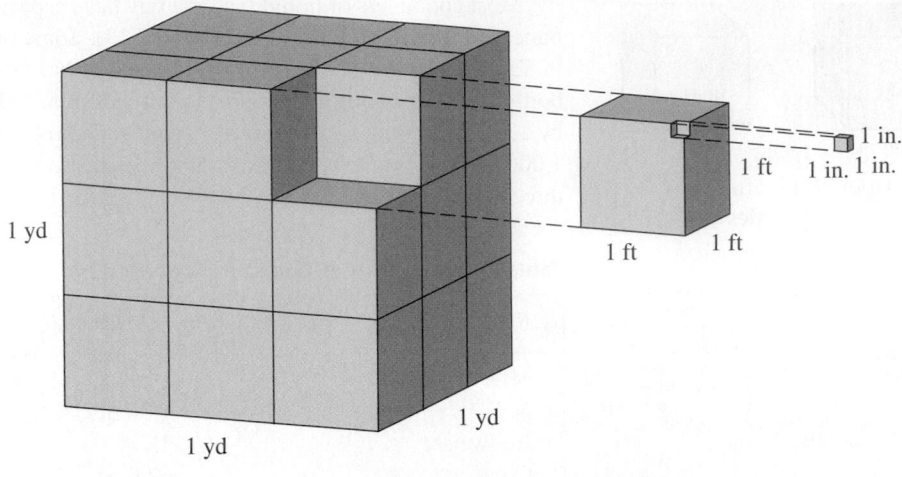

Volume:	1 yd³ = 27 ft³	1 ft³ = 1,728 in.³	1 in.³
Capacity:	about 200 gallons	7.48 gallons	1 gal = 231 in.³

Figure 7.15 **U.S. measurement relationship between volume and capacity**

To find the capacity of the 2-ft by 2-ft by 12-ft box mentioned earlier, we must change 48 ft³ to cubic inches:

ESTIMATE   Since 1 ft³ ≈ 7.5 gal,
48 ft³ ≈ 50 ft³
   ≈ (50 × 7.5) gal
   = 375 gal

$$48 \text{ ft}^3 = 48 \times (1 \text{ ft}) \times (1 \text{ ft}) \times (1 \text{ ft})$$
$$= 48 \times 12 \text{ in.} \times 12 \text{ in.} \times 12 \text{ in.}$$
$$= 82{,}944 \text{ in.}^3 \qquad \text{A calculator would help here.}$$

Since 1 gallon is 231 in.³, the final step is to divide 82,944 by 231 to obtain approximately 359 gallons.

The relationship between volume and capacity in the metric system is easier to remember. One cubic centimeter is one-thousandth of a liter. Notice that this is the same as a milliliter. For this reason, you will sometimes see cc used to mean cm³ or ml. These relationships are shown in Figure 7.16.

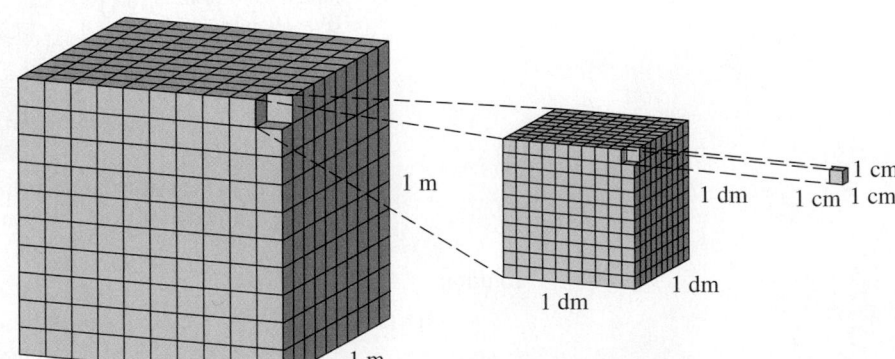

Volume:	1 m³ = 1,000 dm³	1 dm³ = 1,000 cm³	1 cm³ = 1 cc
Capacity:	1,000 L = 1 kl	1 L = 1,000 ml	1 ml = 1 cc

Figure 7.16 **Metric measurement relationship between volume and capacity**

Relationship Between Volume and Capacity

1 liter = 1,000 cm^3
1 gallon = 231 in.^{3};　1 ft^3 ≈ 7.48 gal

EXAMPLE 8

How much water would each of the following containers hold?

a.

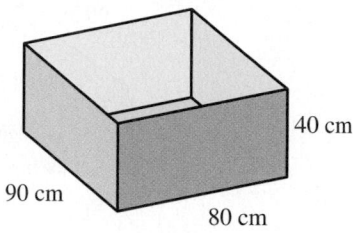

90 cm
80 cm
40 cm

b.

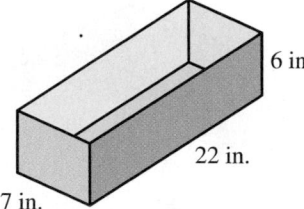

6 in.
22 in.
7 in.

ESTIMATE FOR part **b***:*
Container is approximately
$\frac{1}{2}$ ft × 2 ft × $\frac{1}{2}$ ft = $\frac{1}{2}$ ft^3 ≈
($\frac{1}{2}$ × 7.5) gal = 3.75 gal

Solution

a. $V = 90 \text{ cm} \times 80 \text{ cm} \times 40 \text{ cm}$
   $= 288,000 \text{ cm}^3$

Since each 1,000 cm^3 is 1 liter,

$$\frac{288,000}{1,000} = 288$$

This container would hold 288 liters.

b. $V = 7 \text{ in.} \times 22 \text{ in.} \times 6 \text{ in.}$
   $= 924 \text{ in.}^3$

Since each 231 in.^{3} is 1 gallon,

$$\frac{924}{231} = 4$$

This container would hold
4 gallons.　◆

EXAMPLE 9

An ecology swimming pool is advertised as being 20 ft × 25 ft × 5 ft. How many gallons will it hold?

Solution

$$V = 20 \text{ ft} \times 25 \text{ ft} \times 5 \text{ ft} = 2,500 \text{ ft}^3$$

Since 1 ft^3 ≈ 7.48 gal, the swimming pool contains

$$2,500 \times 7.48 = 18,700 \text{ gallons}　◆$$

Problem Set 7.3

LEVEL 1

1. **IN YOUR OWN WORDS** Contrast length, area, and volume.

2. **IN YOUR OWN WORDS** What do we mean by surface area?

3. **IN YOUR OWN WORDS** Contrast volume and capacity.

4. **IN YOUR OWN WORDS** Compare the sizes of a cubic inch and a cubic centimeter.

5. **IN YOUR OWN WORDS** Compare the sizes of a quart and a liter.

6. **IN YOUR OWN WORDS** Compare a meter and a yard.

Find the outside surface area in Problems 7–18. Note that some boxes have tops and others do not.

7.

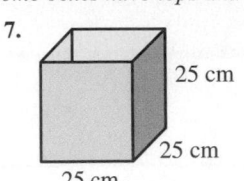

25 cm
25 cm
25 cm
25 cm

8.

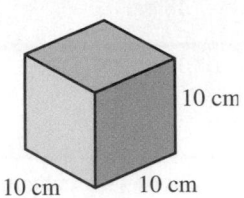

10 cm
10 cm
10 cm

9.

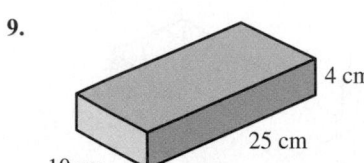

4 cm
25 cm
10 cm

10.

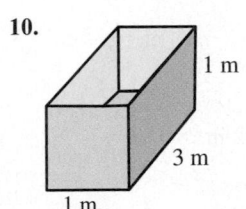

1 m
3 m
1 m

11.

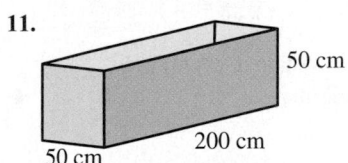

50 cm
200 cm
50 cm

12.

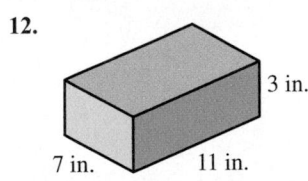

3 in.
7 in.   11 in.

13.

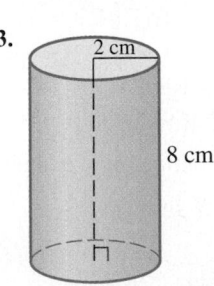

2 cm
8 cm

14.

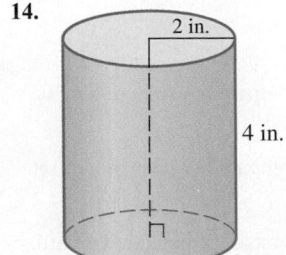

2 in.
4 in.

15.

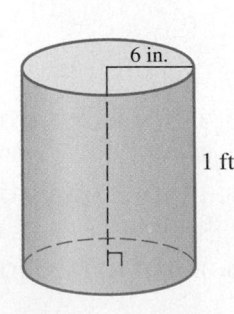

6 in.
1 ft

16.

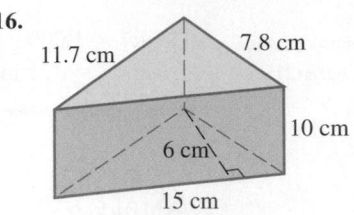

11.7 cm   7.8 cm
10 cm
6 cm
15 cm

17.

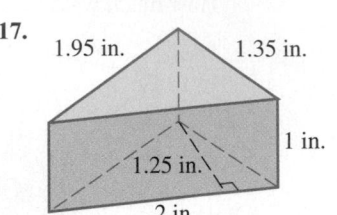

1.95 in.   1.35 in.
1 in.
1.25 in.
2 in.

18.

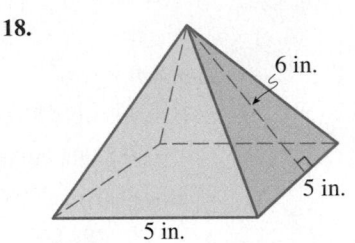

6 in.
5
5 in.
5 in.

In Problems 19–20, find the volume of each solid by counting the number of cubic centimeters in each box.

19.

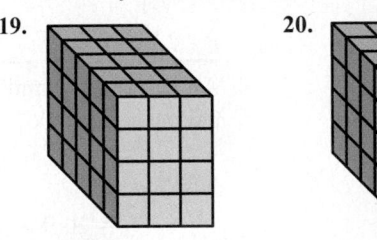

20.

21.

Find the volume of each solid in Problems 21–28.

21.

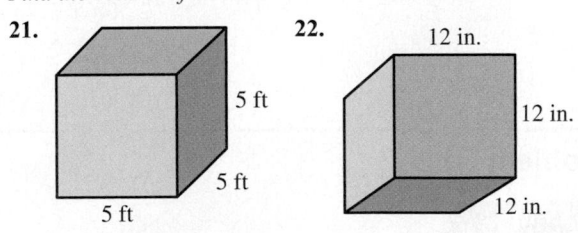

5 ft
5 ft
5 ft

22.

12 in.
12 in.
12 in.

23.

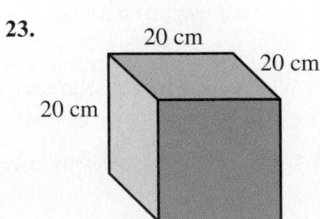

20 cm
20 cm
20 cm

24.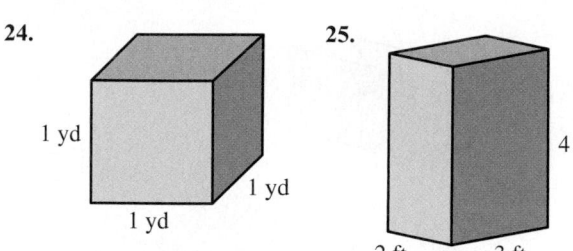

1 yd
1 yd
1 yd

25.

4 ft
2 ft   3 ft

26.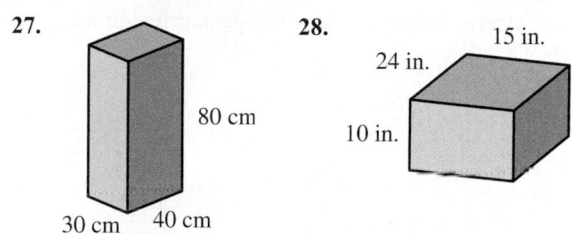

20 mm
2 mm
10 mm

27.

80 cm
30 cm   40 cm

28.

15 in.
24 in.
10 in.

Measure each amount given in Problems 29–32.

29.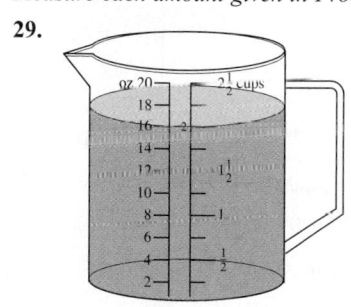

oz 20   2½ cups
18
16   2
14
12   1½
10
8   1
6
4   ½
2

a. in cups
b. in ounces

30.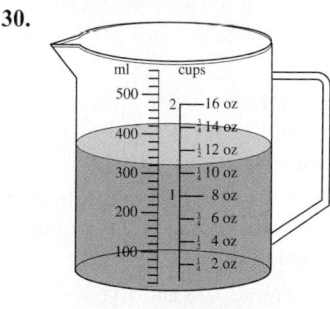

ml   cups
500 — 2 — 16 oz
            ¾ 14 oz
400 —   ½ 12 oz
300 —   ¼ 10 oz
200 — 1 — 8 oz
            ¾ 6 oz
100 —   ½ 4 oz
            ¼ 2 oz

a. in ounces
b. in milliliters

31.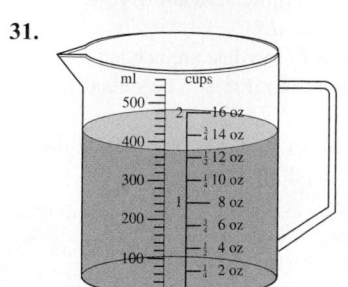

ml   cups
500 — 2 — 16 oz
            ¾ 14 oz
400 —   ½ 12 oz
300 —   ¼ 10 oz
200 — 1 — 8 oz
            ¾ 6 oz
100 —   ½ 4 oz
            ¼ 2 oz

a. in ounces
b. in milliliters

32.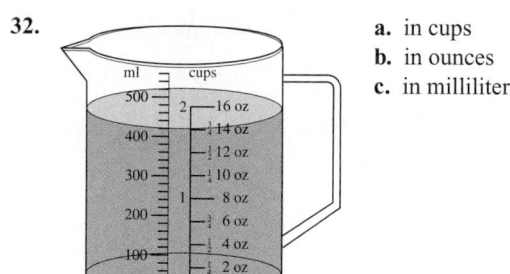

ml   cups
500 — 2 — 16 oz
            ¾ 14 oz
400 —   ½ 12 oz
300 —   ¼ 10 oz
200 — 1 — 8 oz
            ¾ 6 oz
100 —   ½ 4 oz
            ¼ 2 oz

a. in cups
b. in ounces
c. in milliliters

33. Measure the amount, in milliliters, contained in each of the given containers.

a.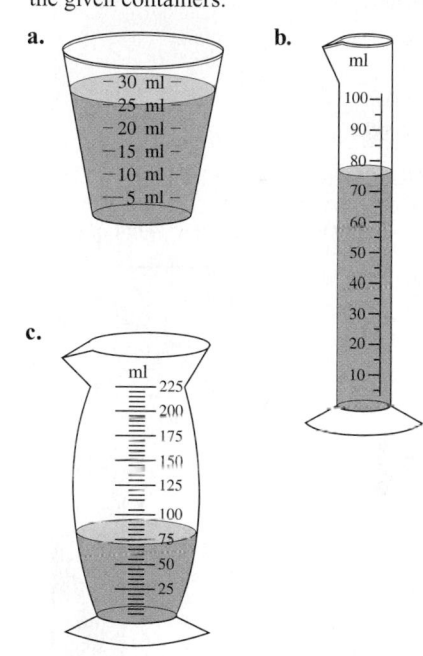

— 30 ml —
— 25 ml —
— 20 ml —
—15 ml —
—10 ml —
— 5 ml —

b.

ml
100 —
90 —
80 —
70 —
60 —
50 —
40 —
30 —
20 —
10 —

c.

ml   225
200
175
150
125
100
75
50
25

The ability to estimate capacities is an important skill to develop. Without measuring, pick the best answer in Problems 34–44.

34. An average cup of coffee is about
A. 250 ml   B. 750 ml   C. 1 L

35. If you want to paint a small bookshelf, how much paint would you probably need?
A. 5 ml   B. 500 ml   C. 5 L

36. A six-pack of beer would contain about
A. 2 ml   B. 200 ml   C. 2 L

37. The dose of a strong cough medicine might be
A. 2 ml   B. 200 ml   C. 2 L

38. A glass of water served at a restaurant is about
A. 200 ml   B. 2 ml   C. 2 L

39. Enough gas to fill your car's empty tank would be about
A. 15 L   B. 200 ml   C. 70 L

40. You order some champagne for yourself and one companion. You would most likely order
A. 2 ml   B. 700 ml   C. 20 L

41. 50 kl of water would be about enough for
  A. taking a bath
  B. taking a swim
  C. supplying the drinking water for a large city

42. The prefix *centi-* means
  A. one thousand
  B. one-thousandth
  C. one-hundredth

43. The prefix *milli-* means
  A. one thousand
  B. one-thousandth
  C. one-hundredth

44. The prefix *kilo-* means
  A. one thousand
  B. one-thousandth
  C. one-hundredth

What is the capacity for each of the containers in Problems 45–52? (Give answers in the U.S. system to the nearest tenth of a gallon or in metric to the nearest tenth of a liter.)

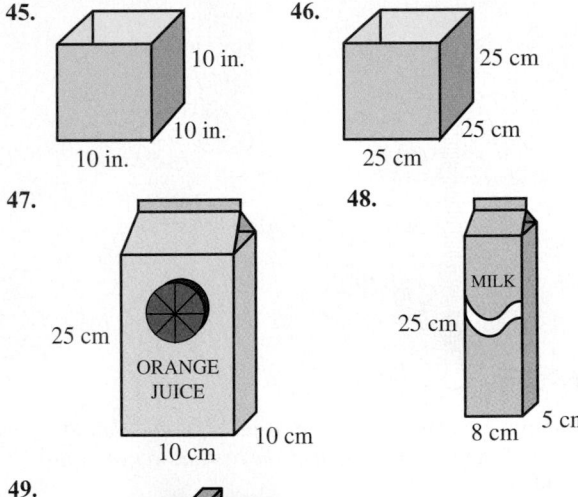

45. 10 in. ⋅ 10 in. ⋅ 10 in.

46. 25 cm ⋅ 25 cm ⋅ 25 cm

47. ORANGE JUICE — 25 cm, 10 cm, 10 cm

48. MILK — 25 cm, 8 cm, 5 cm

49. BENZENE — 30 cm, 20 cm, 15 cm

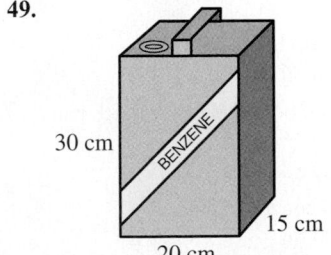

50.

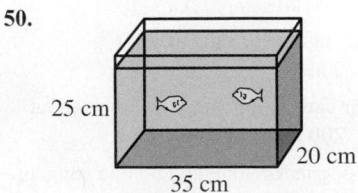
25 cm, 35 cm, 20 cm

51.

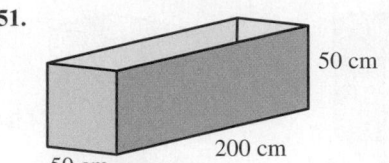
50 cm, 50 cm, 200 cm

52.

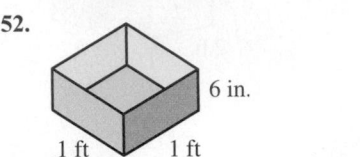
6 in., 1 ft, 1 ft

LEVEL 2

53. The exterior dimensions of a refrigerator/freezer are shown in Figure 7.17.

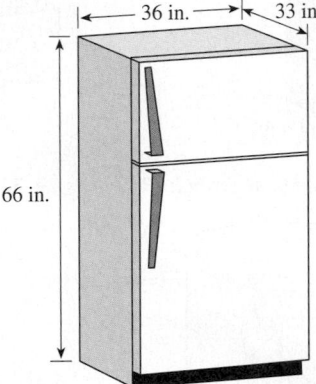
36 in., 33 in., 66 in.

Figure 7.17 **Refrigerator/freezer**

  a. How many cubic feet are contained within the refrigerator/freezer?
  b. If it is advertised as a 19-cu-ft refrigerator/freezer, how much space is taken up by the motor, insulation, and so on?

54. The exterior dimensions of a freezer are 48 inches by 36 inches by 24 inches, and it is advertised as being 27.0 cu ft. Is the advertised volume correctly stated?

55. Suppose that you must order concrete for a sidewalk 50 ft by 4 ft to a depth of 4 in. How much concrete is required? (Answer to the nearest $\frac{1}{2}$ cubic yard.)

56. Use the plot plan shown in Figure 7.18 and give your answers to the nearest *cubic yard.**
  a. How many cubic yards of sawdust are needed for preparation of the lawn area if it is to be spread to a depth of 6 inches?
  b. How many cubic yards of chips are necessary if they are to be placed to a depth of 3 inches?

*In practice, you would not round to the nearest yard, but rather would round up to ensure that you had enough material. However, for consistency in this book, we will round according to the rules developed in the first chapter.

c. How much gravel is necessary if it is to be laid to a depth of 3 inches?

d. Suppose that you wish to pave the area labeled "Concrete." How much concrete is needed if it is to be poured to a depth of 4 inches?

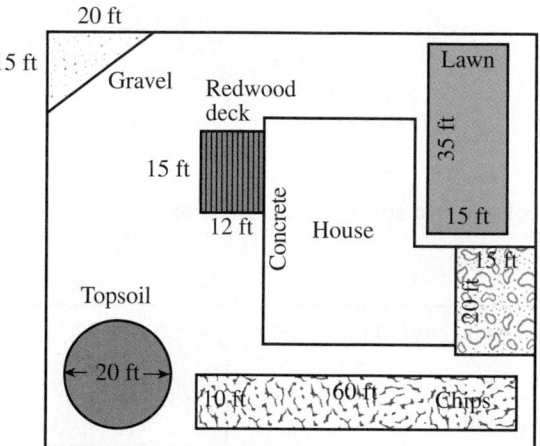

Figure 7.18 **Plot plan**

57. How much water will a 7-m by 8-m by 2-m swimming pool contain (in kiloliters)?

58. How much water will a 21-ft by 24-ft by 4-ft swimming pool contain (rounded to the nearest gallon)?

PROBLEM SOLVING

59. The total human population of the earth is about 6.2×10^9.

a. If each person has the room of a prison cell (50 sq ft), and if there are about 2.8×10^7 sq ft in a square mile, how many people could fit into a square mile?

b. How many square miles would be required to accommodate the entire human population of the earth?

c. If the total land area of the earth is about 5.2×10^7 sq mi, and if all the land area were divided equally, how many acres of land would each person be allocated (1 sq mi = 640 acres)?

60. a. Guess what percentage of the world's population could be packed into a cubical box measuring $\frac{1}{2}$ mi on each side. [*Hint:* The volume of a typical person is about 2 cu ft.]

b. Now calculate the answer to part **a,** using the earth's population as given in Problem 59.

7.4 MISCELLANEOUS MEASUREMENTS

Comparisons

In this chapter we have been discussing measurement.

Length
If you are measuring length, you will use linear measures:

in.; ft; yd; mi   cm; m; km

Area
For area, you will use square measures:

in.²; ft²; yd²; mi²   cm²; m²; km²

Volume
For volumes, you will use cubic measures:

in.³; ft³   cm³ (or cc); m³; km³

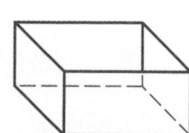

In the previous section we found the volume and the capacity of boxes; now we can extend this to other solids. In Figure 7.19 (page 382) we give some of the more common solids, along with the volume formulas. We will use B in each case to signify the area of the base and h for the height.

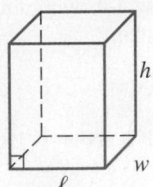
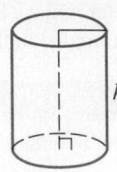
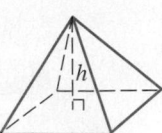
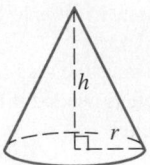
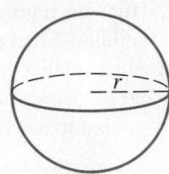

Right rectangular prism	Right circular cylinder	Pyramid	Right circular cone	Sphere
$V = Bh$	$V = Bh$	$V = \frac{1}{3}Bh$	$V = \frac{1}{3}Bh$	$V = \frac{4}{3}\pi r^3$
$S = 2\ell w + 2wh + 2\ell h$	$S = 2\pi r^2 + 2\pi rh$	$S = s^2 + s\sqrt{s^2 + 4h^2}$	$S = \pi r\sqrt{r^2 + h^2} + \pi r^2$	$S = 4\pi r^2$

Figure 7.19 **Common solids, with accompanying volume and surface area formulas**

EXAMPLE 1

Find the volume of the accompanying solid to the nearest cubic unit.

Solution   This is a right circular cylinder. Hence,

$$V = Bh$$
$$= \pi(6.2)^2(15)$$
$$\approx 1811.4423$$

Notice that $B = \pi r^2$, where $r = 6.2$; the height of the cylinder is 15 ($h = 15$).

The volume is 1,811 cm^3.   ◆

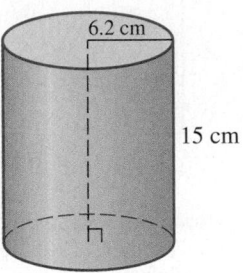

EXAMPLE 2

Find the volume of the accompanying solid to the nearest cubic unit.

Solution   This is a right triangular prism. Hence,

$$V = Bh$$
$$= \frac{1}{2}(15)(6)(10)$$
$$= 450$$

Notice that $B = \frac{1}{2}ba$, where $a = 6$ (height of triangle), $b = 15$ (base of triangle), and $h = 10$ (height of prism).

The volume is 450 m^3.   ◆

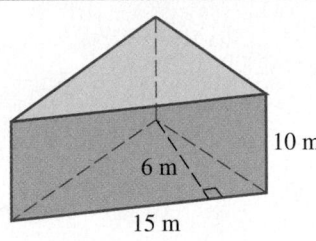

EXAMPLE 3

Find the volume of the accompanying solid to the nearest cubic unit.

Solution   This is a pyramid. Hence,

$$V = \frac{1}{3}Bh$$
$$= \frac{1}{3}(5)^2(3)$$
$$= \frac{25}{3}(3)$$
$$= 25$$

Notice that the base is a square so $B = s^2 = 5^2$.

The volume is 25 m^3.   ◆

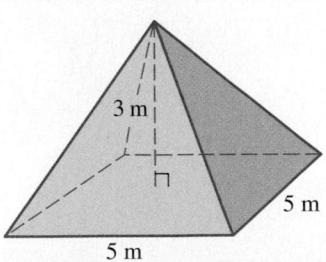

EXAMPLE 4

Find the surface area (to the nearest square unit), and also find the volume of the accompanying solid (to the nearest cubic unit).

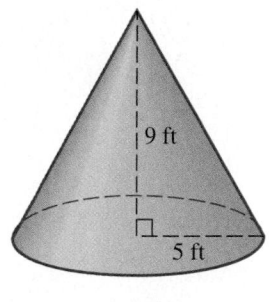

9 ft

5 ft

Solution   This is a right circular cone. Hence,

$$S = \pi r \sqrt{r^2 + h^2} + \pi r^2$$

$$= \pi(5)\sqrt{5^2 + 9^2} + 5^2\pi \quad \text{Notice that } r = 5 \text{ and } h = 9.$$

$$= 5\pi\sqrt{106} + 25\pi$$

$$\approx 240.26 \qquad \text{You can use the } \pi \text{ key on your calculator.}$$

Also,

$$V = \frac{1}{3}Bh$$

$$= \frac{1}{3}\pi(5)^2 9 \quad \text{Notice that } B = \pi r^2 \text{ and } r = 5.$$

$$= 75\pi$$

$$\approx 235.6$$

The surface area is 240 ft^2 and the volume is 236 ft^3. ◆

EXAMPLE 5

Find the surface area and volume of the sphere. Give your answer to the nearest whole unit, paying attention to the unit.

8.2 cm

Solution   For a sphere,

$$S = 4\pi r^2 \quad \text{Notice that } r = 8.2 \text{ cm.}$$

$$\approx 844.96$$

Also,

$$V = \frac{4}{3}\pi r^3$$

$$= \frac{4}{3}\pi(8.2)^3$$

$$\approx 2{,}309.564878$$

The surface area is 845 cm^2 and the volume is 2,310 cm^3. ◆

The following example compares the areas and volumes of similarly shaped figures.

EXAMPLE 6

a. Compare the area of a square whose side is tripled with the area of the original square.

b. Compare the area of a circle after its radius is doubled.

c. Compare the volume of a sphere after its radius is doubled.

d. Compare the volume of a cube when the length of its side is multiplied by 5.

Solution

a. Let s be the length of the side of the square; then the area is $A = s^2$. If the side is tripled to $3s$, then the area is $A = (3s)^2 = 9s^2$. We see that the new area is *nine times as large* as the original area.

b. Let r be the radius of the original circle; then the area is $A = \pi r^2$. If the radius is doubled to $2r$, then the area is $A = \pi(2r)^2 = 4\pi r^2$. We see that the new area is *increased four-fold*.

c. Let r be the radius of the original sphere; then the area is $A = \frac{4}{3}\pi r^3$. If the radius is doubled to $2r$, then the area is $A = \frac{4}{3}\pi(2r)^3 = \frac{4}{3}\pi(8r^3) = \frac{32}{3}\pi r^3$. We see that the volume is *eight times as large.*

d. Use patterns based on parts **a–c** to *guess* the result without calculation. Since the formula cubes the variable that is multiplied by 5, we guess the volume is $5^3 = 125$ times as large as the original volume. ◆

The measurements of length, area, volume, and capacity were discussed in the previous section. Two additional measurements we need to consider are mass and temperature.

Mass

The **mass** of an item is the amount of matter it comprises. The **weight** of an item is the heaviness of the matter.* The U.S. and metric units of measurement for mass or weight are given in the following box. Notice that, in the U.S. measurement system, an ounce is used as a weight measurement; this is not the same use of an ounce as a capacity measurement that we used earlier. The basic unit of measurement for mass in the metric system is the **gram,** which is defined as the mass of 1 cm³ of water, as shown in Figure 7.20.

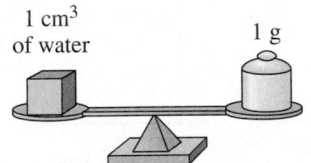
Figure 7.20 **Mass of 1 g**

Standard Units of Weight

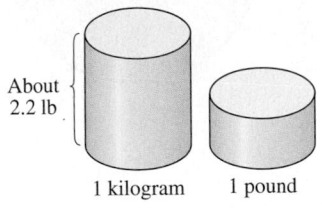

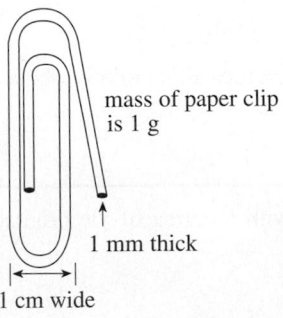

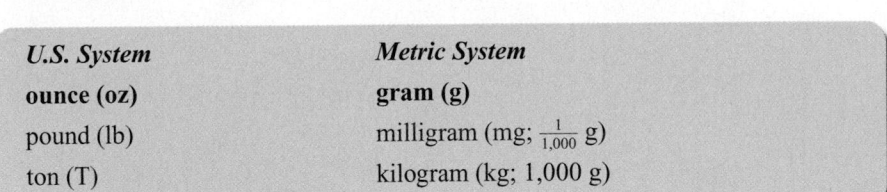

U.S. System	Metric System
ounce (oz)	**gram (g)**
pound (lb)	milligram (mg; $\frac{1}{1,000}$ g)
ton (T)	kilogram (kg; 1,000 g)

A paper clip weighs about 1 g, a cube of sugar about 3 g, and a nickel about 5 g. This book weighs about 1 kg, and an average-size person weighs from 50 to 100 kg. The weights of some common grocery items are shown in Table 7.2.

Table 7.2 Weights of Common Grocery Items, as Shown on Labels

Item	U.S. Weight	Metric Weight
Kraft cheese spread	5 oz	142 g
Del Monte tomato sauce	8 oz	227 g
Campbell cream of chicken soup	$10\frac{3}{4}$ oz	305 g
Kraft marshmallow cream	11 oz	312 g
Bag of sugar	5 lb	2.3 kg
Bag of sugar	22 lb	10 kg

*Technically, weight is the force of gravity acting on a body. The mass of an object is the same on the moon as on Earth, whereas the weight of that same object would be different on the moon and on Earth, since they have different forces of gravity. For our purposes, you can use either word, *mass* or *weight,* because we are weighing things only on Earth.

We use a scale to measure weight. To weigh items, or ourselves, in metric units, we need only replace our U.S. weight scales with metric weight scales. As with other measures, we need to begin to think in terms of metric units, and to estimate the weight of various items. The multiple-choice questions in Problem Set 7.4 are designed to help you do this.

Temperature

The final quantity of measure that we'll consider in this chapter is **temperature,** which is the degree of hotness or coldness.

Standard Units of Temperature

The U.S. system has one common temperature unit:	*The metric system has one common temperature unit:*
Fahrenheit (°F)	**Celsius (°C)**

To work with temperatures, it is necessary to have some reference points.

U.S. Temperature		*Metric Temperature*	
Water freezes:	32°F	Water freezes:	0°C
Water boils:	212°F	Water boils:	100°C

We are usually interested in measuring temperature in three areas: atmospheric temperature (usually given in weather reports), body temperature (used to determine illness), and oven temperature (used in cooking). The same scales are used, of course, for measuring all of these temperatures. But notice the difference in the ranges of temperatures we're considering. The comparisons for Fahrenheit and Celsius are shown in Figure 7.22 on page 386.

> **Go Metric—**
> **Be a liter bug!**
>
> *Mike Keedy, Purdue University*

Converting Units

As we have mentioned, one of the advantages (if not the chief advantage) of the metric system is the ease with which you can remember and convert units of measurement. The three basic metric units are related are shown in Figure 7.21.

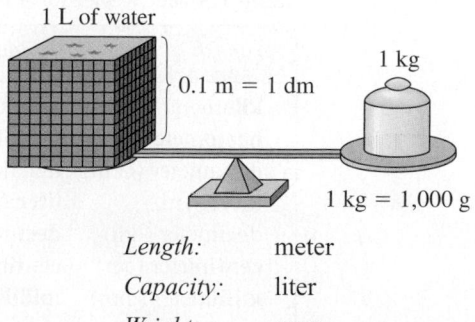

Length:	meter
Capacity:	liter
Weight:	gram

Figure 7.21 **Relationship among meter, liter, and gram**

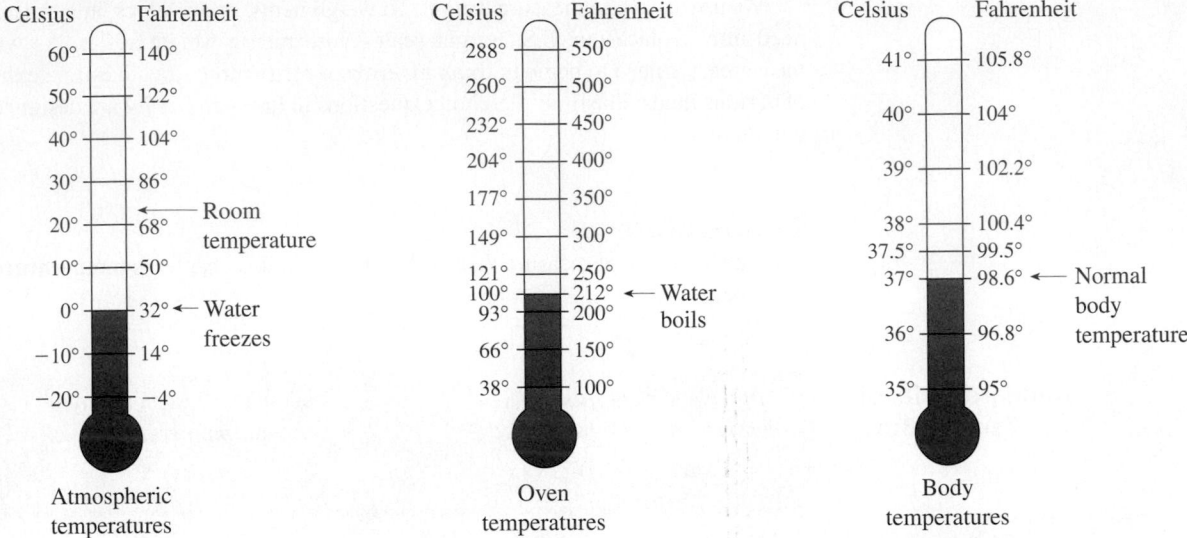

Figure 7.22 **Temperature comparisons between Celsius and Fahrenheit**

These units are combined with certain prefixes:

$$milli\text{- means } \frac{1}{1,000} \qquad centi\text{- means } \frac{1}{100} \qquad kilo\text{- means } 1,000$$

Other metric units are used less frequently, but they should be mentioned:

$$deci\text{- means } \frac{1}{10} \qquad deka\text{- means } 10 \qquad hecto\text{- means } 100$$

A listing of metric units (in order of size) is given in Table 7.3.

Table 7.3 Metric Measurements

Length	Capacity	Weight	Meaning	Memory Aid
kilometer (km)	**kilo**liter (kl)	**kilo**gram (kg)	1,000 units	**K**arl
hectometer (hm)	**hecto**liter (hl)	**hecto**gram (hg)	100 units	**H**as
dekameter (dkm)	**deka**liter (dkl)	**deka**gram (dkg)	10 units	**D**eveloped
meter (m)	**liter** (L)	**gram** (g)	1 unit	**M**y
decimeter (dm)	**deci**liter (dl)	**deci**gram (dg)	0.1 unit	**D**ecimal
centimeter (cm)	**centi**liter (cl)	**centi**gram (cg)	0.01 unit	**C**raving for
millimeter (mm)	**milli**liter (ml)	**milli**gram (mg)	0.001 unit	**M**etrics

Basic Unit: (row "meter (m) | liter (L) | gram (g) | 1 unit | My")

To convert from one metric unit to another, you **simply move the decimal point.** For example, if the height of this book is 0.235 m, then it is also

Larger prefixes
$\begin{cases} 0.000235 \text{ km} \\ 0.00235 \text{ hm} \\ 0.0235 \text{ dkm} \end{cases}$
←Three decimal places
←Two decimal places
←One decimal place

Smaller prefixes
$\begin{cases} 2.35 \text{ dm} \\ 23.5 \text{ cm} \\ 235 \text{ mm} \end{cases}$
←One decimal place
←Two decimal places
←Three decimal places

EXAMPLE 7

Write each metric measurement using each of the other prefixes.

a. 43 km   **b.** 60 L   **c.** 14.1 cg

Solution

a. All prefixes are smaller, so the numbers of units become larger:

43 km	**Given**
430 hm	One place
4,300 dkm	Two places
43,000 m	Three places
430,000 dm	Four places
4,300,000 cm	Five places
43,000,000 mm	Six places

b. Larger prefixes, smaller numbers:

60 L	**Given**
6 dkl	One place
0.6 hl	Two places
0.06 kl	Three places

Smaller prefixes, larger numbers:

60 L	**Given**
600 dl	One place
6,000 cl	Two places
60,000 ml	Three places

c. Larger prefixes:

14.1 cg	**Given**
1.41 dg	One place
0.141 g	Two places
0.0141 dkg	Three places
0.00141 hg	Four places
0.000141 kg	Five places

Smaller prefix:

14.1 cg	**Given**
141 mg	One place

◆

To make a single conversion, use the pattern illustrated by Example 7 and simply count the number of decimal places.

EXAMPLE 8

Make the indicated conversions.

a. 287 cm to km   **b.** 1.5 kl to L   **c.** 4.8 kg to g

Solution

a. From cm to km is five places; a larger prefix implies a smaller number, so move the decimal point to the left:

$$287 \text{ cm} = 0.00287 \text{ km}$$

b. From kl to L is three places; a smaller prefix implies a larger number, so move the decimal point to the right:

$$1.5 \text{ kl} = 1,500 \text{ L}$$

c. From kg to g is three places; a smaller prefix implies a larger number, so move the decimal point to the right:

$$4.8 \text{ kg} = 4,800 \text{ g}$$ ◆

Problem Set 7.4

LEVEL 1

1. **IN YOUR OWN WORDS** Discuss the merits of the metric system as opposed to the U.S. measurement system. Why do you think the metric system has not yet been adopted in the United States?

2. **IN YOUR OWN WORDS** Discuss the relationships among measuring length, capacity, and weight.

3. **IN YOUR OWN WORDS** Explain how we change units within the metric system.

Name the metric unit you would use to measure each of the quantities in Problems 4–9.

4. **a.** The distance from New York to Chicago
   b. The distance around your waist

5. **a.** Your height
   b. The height of a building

6. **a.** The capacity of a wine bottle
   b. The amount of gin in a martini

7. **a.** The capacity of a car's gas tank
   b. The amount of water in a swimming pool

8. **a.** The weight of a pencil
   b. The weight of an automobile

9. **a.** The outside temperature
   b. The temperature needed to bake a cake

Without measuring, pick the best choice in Problems 10–29 by estimating.

10. A hamburger patty would weigh about
    A. 170 g   B. 240 mg   C. 2 kg

11. A can of carrots at the grocery store most likely weighs about
    A. 40 kg   B. 4 kg   C. 0.4 kg

12. A newborn baby would weigh about
    A. 490 mg   B. 4 kg   C. 140 kg

13. You have invited 15 people for Thanksgiving dinner. You should buy a turkey that weighs about
    A. 795 mg   B. 4 kg   C. 12 kg

14. Water boils at
    A. 0°C   B. 100°C   C. 212°C

15. If it is 32°C outside, you would most likely find people
    A. ice skating   B. water skiing

16. If the doctor says that your child's temperature is 37°C, your child's temperature is
    A. low   B. normal   C. high

17. You would most likely broil steaks at
    A. 120°C   B. 500°C   C. 290°C

18. John tells you he weighs 150 kg. If John is an adult, he is
    A. underweight   B. about average
    C. overweight

19. A kilogram is ____?____ a pound.
    A. more than   B. about the same as
    C. less than

20. The prefix used to mean 1,000 is
    A. centi-   B. milli-   C. kilo-

21. The prefix used to mean $\frac{1}{1,000}$ is
    A. centi-   B. milli-   C. kilo-

22. The prefix used to mean $\frac{1}{100}$ is
    A. centi-   B. milli-   C. kilo-

23. 15 kg is a measure of
    A. length   B. capacity
    C. weight   D. temperature

24. 28.5 m is a measure of
    A. length   B. capacity
    C. weight   D. temperature

25. 6 L is a measure of
    A. length   B. capacity
    C. weight   D. temperature

26. 38°C is a measure of
A. length   B. capacity
C. weight   D. temperature

27. 7 ml is a measure of
A. length   B. capacity
C. weight   D. temperature

28. 68 km is a measure of
A. length   B. capacity
C. weight   D. temperature

29. 14.3 cm is a measure of
A. length   B. capacity
C. weight   D. temperature

Write each measurement given in Problems 30–37 using all of the metric prefixes.

	kilometer (km)	hectometer (hm)	dekameter (dkm)	meter (m)	decimeter (dm)	centimeter (cm)	millimeter (mm)
30.				9			
32.					6		
34.	4						
36.				1	5	0	

	kiloliter (kℓ)	hectoliter (hℓ)	dekaliter (dkℓ)	liter (ℓ)	deciliter (dℓ)	centiliter (cℓ)	milliliter (mℓ)
31.						6	3
33.			3	5			
35.				8			
37.		3	1				

LEVEL 2

In Problems 38–42 make the indicated conversions.

38. a. 1 cm = _____ mm   **b.** 1 cm = _____ m
    c. 1 cm = _____ km   **d.** 1 mm = _____ cm

39. a. 1 mm = _____ m   **b.** 1 mm = _____ km
    c. 1 m = _____ mm   **d.** 1 m = _____ cm

40. a. 1 m = _____ km   **b.** 1 ml = _____ dl
    c. 1 ml = _____ L   **d.** 1 ml = _____ kl

41. a. 1 L = _____ ml   **b.** 1 L = _____ dl
    c. 1 L = _____ kl   **d.** 1 kl = _____ ml

42. a. 1 kl = _____ dl   **b.** 1 kl = _____ L
    c. 1 g = _____ mg   **d.** 1 g = _____ cg

Find the surface areas of the solids in Problems 43–46 correct to the nearest square unit. Note that some do not have tops.

43.
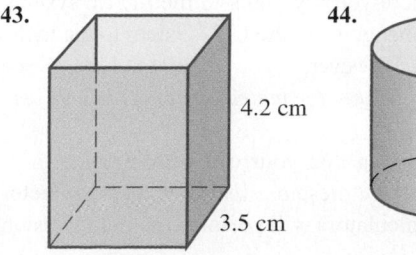
3.5 cm   4.2 cm   3.5 cm

44.
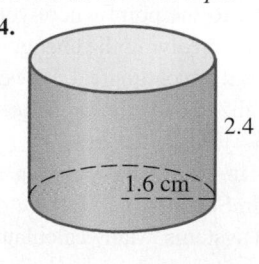
2.4   1.6 cm

45.
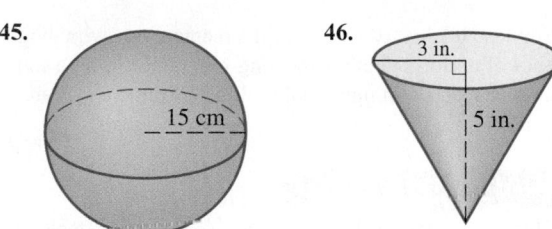
15 cm

46.
3 in.
5 in.

Find the volumes of the solids in Problems 47–50 correct to the nearest unit.

47.
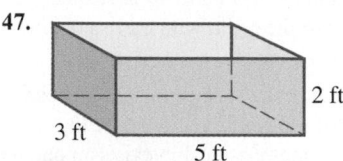
2 ft   3 ft   5 ft

48.
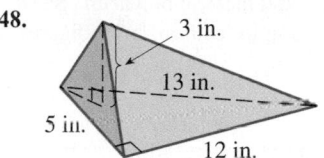
3 in.   13 in.   5 in.   12 in.

49.
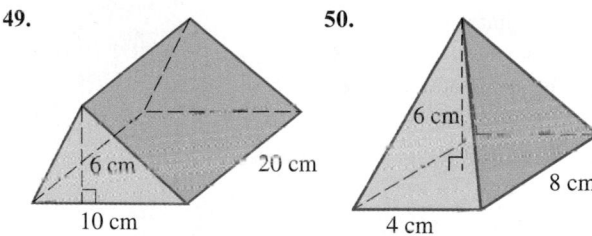
6 cm   20 cm   10 cm

50.
6 cm   8 cm   4 cm

Find the surface areas and volumes of the solids in Problems 51–52 correct to the nearest unit.

51.
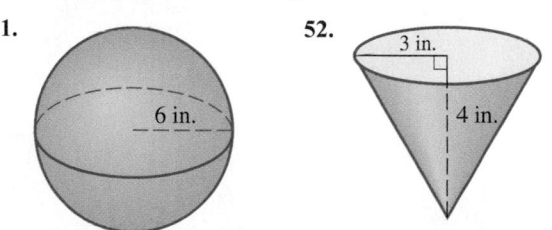
6 in.

52.
3 in.
4 in.

53. If the length of a rectangle is doubled, and the width is tripled, what effect does that have on the area?

54. If the length of a box is doubled, the width is tripled, and the height is doubled, what effect does that have on the volume?

55. If the radius of a sphere is multiplied by three, what effect does that have on the surface area?

56. If the height of a square pyramid is doubled, what effect does that have on the volume?

LEVEL 3

57. If the diameter of a circle is doubled, what effect does that have on the area?

58. Suppose it takes 10 hours to fill a rectangular swimming pool of uniform depth. How long will it take to fill a similarly shaped swimming pool if all of the dimensions are doubled?

PROBLEM SOLVING

59. Show how a spreadsheet could be used to print a table of temperature conversions between the Fahrenheit and Celsius scales. Your table should consist of a column showing Celsius temperatures from 0 to 40 in steps of 5 degrees, and a column to the right, with corresponding Fahrenheit temperatures.

60. A polyhedron is a simple closed surface in space whose boundary is composed of polygonal regions (see Figure 7.23). A rather surprising relationship exists among the number of vertices, edges, and faces of polyhedra. See if you can discover it by looking for patterns in the figures and filling in the blanks.

	Triangular pyramid	Quadrilateral pyramid	Pentagonal pyramid	Regular tetrahedron

	Regular hexahedron (cube)	Regular octahedron	Regular dodecahedron	Regular icosahedron

Figure 7.23 Some common polyhedra

Figure	Number of: Faces	Vertices	Edges
a. Triangular pyramid	4	4	6
b. Quadrilateral pyramid	5	____	8
c. Pentagonal pyramid	____	6	10
d. Regular tetrahedron	4	4	____

Figure	Number of: Faces	Vertices	Edges
e. Cube	____	____	12
f. Regular octahedron	____	6	____
g. Regular dodecahedron	____	____	30
h. Regular icosahedron	____	____	30

7.5 | U.S.–METRIC CONVERSIONS

This section may be used for reference. The emphasis in this chapter is on everyday use of the metric system. However, certain specialized applications require more precise conversions than we've considered. On the other hand, you don't want to become bogged down with arithmetic to the point where you say "nuts to the metric system."

The most difficult obstacle involving the change from the U.S. system to the metric system is not mathematical but psychological. However, if you understood the presentation in this chapter, you realized that *working within the metric system is much easier than working within the U.S. system.*

With these ideas firmly in mind, and realizing that your everyday work with the metric system is discussed in Sections 7.1–7.4, we present a list of conversion factors between these measurement systems. Many calculators will perform these conversions for you; check your owner's manual.

From *The Metrics Are Coming! The Metrics Are Coming!* by R. Cardwell. © 1975 by Dorrance & Company. Reprinted by permission.

Length Conversions

U.S. to Metric				Metric to U.S.		
When You Know	**Multiply by**	**To Find**		**When You Know**	**Multiply by**	**To Find**
in.	2.54	cm		cm	0.39370	in.
ft	30.48	cm		m	39.37	in.
ft	0.3048	m		m	3.28084	ft
yd	0.9144	m		m	1.09361	yd
mi	1.60934	km		km	0.62137	mi

Capacity Conversions

U.S. to Metric				Metric to U.S.		
When You Know	**Multiply by**	**To Find**		**When You Know**	**Multiply by**	**To Find**
tsp	4.9289	ml		ml	0.20288	tsp
tbsp	14.7868	ml		ml	0.06763	tbsp
oz	29.5735	ml		ml	0.03381	oz
c	236.5882	ml		ml	0.00423	c
pt	473.1765	ml		ml	0.00211	pt
qt	946.353	ml		ml	0.00106	qt
qt	0.9464	L		L	1.05672	qt
gal	3.7854	L		L	0.26418	gal

Weight Conversions

U.S. to Metric				Metric to U.S.		
When You Know	**Multiply by**	**To Find**		**When You Know**	**Multiply by**	**To Find**
oz	28.3495	g		g	0.0352739	oz
lb	453.59237	g		g	0.0022046	lb
lb	0.453592	kg		kg	2.2046226	lb
T	907.18474	kg				

Temperature Conversions

U.S. to Metric	Metric to U.S.
1. Subtract 32 from degrees Fahrenheit. 2. Multiply this result by $\frac{5}{9}$ to get Celsius.	1. Multiply Celsius degrees by $\frac{9}{5}$. 2. Add 32 to this result to get Fahrenheit.

CHAPTER SUMMARY

"Many arts there are which beautify the mind of people; of all other none do more garnish and beautify it than those arts which are called mathematical."

H. Billingsley

IMPORTANT TERMS

Accuracy [7.1]
Acre [7.2]
Area [7.2]
Capacity [7.3]
Celsius [7.4]
Center of a circle [7.1]
Centi- [7.1]
Circle [7.1]
Circumference [7.1]
Cone [7.4]
Cubic unit [7.3]
Cup [7.3]
Cylinder [7.4]
Deci- [7.1]
Deka- [7.1]
Diameter [7.1]
Equilateral triangle [7.1]
Fahrenheit [7.4]
Foot [7.1]
Gallon [7.3]

Gram [7.4]
Hecto- [7.1]
Inch [7.1]
Kilo- [7.1]
Length [7.1]
Liter [7.3]
Mass [7.4]
Measure [7.1]
Meter [7.1]
Metric system [7.1]
Mile [7.1]
Milli- [7.1]
Ounce [7.3, 7.4]
Parallelepiped [7.3]
Parallelogram [7.2]
Perimeter [7.1]
Pi (π) [7.1]
Pound [7.4]
Precision [7.1]

Prism [7.4]
Pyramid [7.4]
Quart [7.3]
Radius [7.1]
Rectangle [7.1]
Rectangular parallelepiped [7.3]
Semicircle [7.1]
SI system [7.1]
Sphere [7.4]
Square [7.1]
Square unit [7.2]
Surface area [7.3]
Temperature [7.4]
Ton [7.4]
Trapezoid [7.2]
United States system [7.1]
Volume [7.3]
Weight [7.4]
Yard [7.1]

TYPES OF PROBLEMS

Distinguish between the concepts of precision and accuracy. [7.1]
Be able to measure length in both the United States and metric measurement systems. [7.1]
Estimate lengths; choose an appropriate unit for measuring a given length. [7.1]
Find the perimeter, circumference, or distance around a given figure. [7.1]
Solve applied problems involving length. [7.1]
Be able to measure area in both the United States and metric measurement systems. [7.2]
Estimate areas; choose an appropriate unit for measuring a given area. [7.2]
Find the area of a given figure. [7.2]
Solve applied problems involving area. [7.2]
Find the surface area of an object. [7.3]
Estimate volumes; choose an appropriate unit for measuring a given volume. [7.3, 7.4]
Find the volume of a given solid. [7.3, 7.4]
Solve applied problems involving volume. [7.3, 7.4]
Estimate capacities. [7.3]
Find the capacity of a given container. [7.3]
Solve applied problems involving capacity. [7.3]

Measure the amount of a liquid. [7.3]

Be able to measure mass (weight) in both the United States and metric measurement systems. [7.4]

Estimate weights; choose an appropriate unit for measuring a given mass. [7.4]

Be able to measure temperature in both the United States and metric measurement systems. [7.4]

Be able to measure volume in both the United States and metric measurement systems. [7.3, 7.4]

Estimate temperatures. [7.4]

Change units within the metric system. [7.4]

Change units within the United States system. [7.4]

Change units between the metric and United States systems. [7.5]

CHAPTER 7 REVIEW QUESTIONS

1. **a.** Without any measuring device, draw a segment approximately 10 cm long.
   b. Measure this segment to the nearest tenth cm.
   c. What is the width of your index finger in both the metric and U.S. systems?

2. In Figure 7.24,
   a. What is the distance around (to the nearest inch)?
   b. What is the area (to the nearest square inch)?

Figure 7.24

3. In Figure 7.25,
   a. What is the volume?
   b. What is the capacity of the box?
   c. What is the outside surface area of the box?

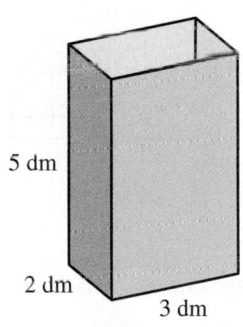

Figure 7.25

4. For a sphere with radius 1 ft,
   a. What is the exact surface area?
   b. What is the surface area (to the nearest tenth)?
   c. What is the exact volume?
   d. What is the volume (to the nearest tenth)?

5. **a.** What are the units of measurement in both the U.S. and metric measurement systems that you would use to measure the height of a building?
   b. Estimate the temperature on a hot summer day in both the U.S. and metric measurement systems.
   c. What does the prefix *milli-* mean?
   d. What does the measurement 8.6 ml measure?
   e. How many centimeters are equivalent to 10 km?

6. A rectangular pool is 80 ft by 30 ft with a depth of 3 ft.
   a. What is the surface area of the top of the pool?
   b. What is the volume of the pool?
   c. What is the capacity of the pool (to the nearest gallon)?

7. How many square yards of carpet are necessary to carpet an 11-ft by 16-ft room? (Assume that you cannot purchase part of a square yard.)

8. Consider the following advertisement from Round Table Pizza.

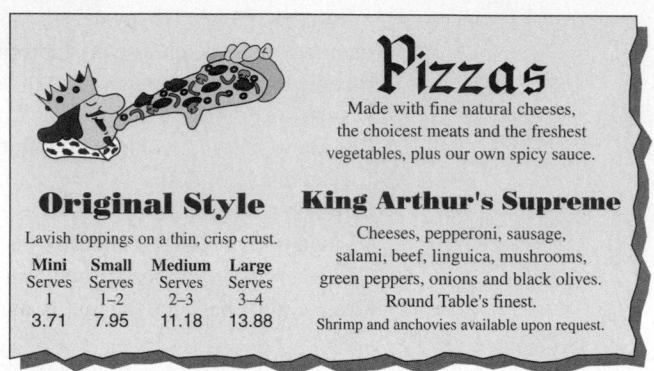

Courtesy of Round Table Pizza

a. If you order a pizza, what size should you order if you want the best price per square inch? (That is, compare the price per square inch for the various pizza sizes; assume the diameters for the four sizes are 6 in., 10 in., 12 in., and 14 in.)

b. Suppose you were the owner of a pizza restaurant and were going to offer a 16-in. diameter pizza. What would you charge for the pizza if you wanted the price to be comparable with the prices of the other sizes?

9. The Pythagorean theorem (Chapter 6) states that, for a right triangle with sides a and b and hypotenuse c,

$$a^2 + b^2 = c^2$$

In this chapter, we saw that a^2 is the area of a square of length a. We can, therefore, illustrate the Pythagorean theorem for a triangle with sides 3, 4, and 5, as shown in Figure 7.26.

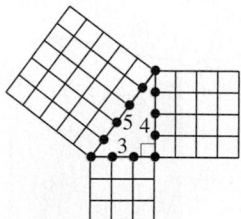

a. Illustrate the Pythagorean theorem for a right triangle with sides 5, 12, and 13.

b. On the other hand, if you draw a triangle with sides measuring 2, 3, and 4 units, you would find that

Figure 7.26 **Geometric interpretation of the Pythagorean theorem**

$$2^2 + 3^2 = 4 + 9 = 13 \text{ and } 4^2 = 16 \text{ so}$$
$$2^2 + 3^2 \neq 4^2.$$

Also notice that such triangles are not right triangles. Is it possible to form a right triangle with sides of 25, 312, and 313?

10. The distributive law states that

$$ab + ac = a(b + c)$$

In this chapter, we say ab and ac represent areas of two rectangles, one with sides a and b, and the other with sides a and c. Give a geometric justification for the distributive property.

GROUP RESEARCH

Working in small groups is typical of most work environments, and being able to work with others to communicate specific ideas is an important skill to learn. Work with three or four other students to submit a single report based on each of the following questions.

G26. Suppose a house has an 8-ft ceiling in all rooms except the living room, which has a 10-ft cathedral ceiling. Approximately how many marbles would fit into this house?

G27. Suppose you wish to build a spa on a wood deck. The deck is to be built 4 ft above level ground. It is to be 50 ft by 30 ft and is to contain a spa that is circular with a 14-ft diameter. The spa is 4 ft deep.

    a. How much water will the spa contain, and how much will it weigh? Assume that the spa itself weighs 550 lb.

    b. Draw plans for the wood deck.

    c. Draw up a materials list.

    d. Estimate the cost for this installation.

G28. Here is a simple formula for finding Pythagorean triples (numbers a, b, and c that satisfy the Pythagorean theorem). It was given to me in an elevator by my friend Bert Liberi (who is also a great mathematician). If m is any natural number greater than 1, then

$$\frac{1}{m} + \frac{1}{m + 2} = \frac{a}{b}$$

The reduced fraction $\frac{a}{b}$ will have the property that the set $\{a, b, c\}$ is a Pythagorean triple. For example, if $m = 2$, then

$$\frac{1}{2} + \frac{1}{2 + 2} = \frac{1}{2} + \frac{1}{4} = \frac{3}{4}$$

Thus, the first two numbers of the triple are 3 and 4. For the third number in the triple, we find

$$c = \sqrt{3^2 + 4^2} = \sqrt{9 + 16} = \sqrt{25} = 5$$

so the set is $\{3, 4, 5\}$. Find ten sets of Pythagorean triples.

INDIVIDUAL RESEARCH PROBLEMS

Learning to use sources outside your classroom and textbook is an important skill, and here are some ideas for extending some of the ideas in this chapter. You can find references to these projects in a library or at

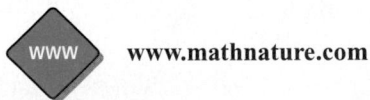

www.mathnature.com

PROJECT 7.1 HISTORICAL QUESTION In the Boston Museum of Fine Arts is a display of carefully made stone cubes found in the ruins of Mohenjo-Daro of the Indus. The stones are a set of weights that exhibit the binary pattern, 1, 2,

4, 8, 16, ... The fundamental unit displayed is just a bit lighter than the ounce in the U.S. measurement system. The old European standard of 16 oz for 1 pound may be a relic of the same idea. Write a paper showing how a set of such stones can successfully be used to measure any reasonable given weight of more than one unit.

PROJECT 7.2 Write a paper on Pythagorean triplets (numbers a, b, and c that satisfy the Pythagorean theorem).

PROJECT 7.3 Construct models for the regular polyhedra.

PROJECT 7.4 What solids occur in nature? Find examples of each of the five regular solids.

8 The Nature of Growth

Little can be understood of even the simplest phenomena of nature without some knowledge of mathematics, and the attempt to penetrate deeper into the mysteries of nature compels simultaneous development of the mathematical processes.
J. W. A. YOUNG

CONTENTS

OVERVIEW

One of the most important ideas in understanding the nature of the world around us is to understand applications of growth and decay. Ross Honsberber, a contemporary mathematics professor, said in his book *Mathematical Morsels* (Washington, DC: Mathematical Association of America, p. vii), "Mathematics abounds in bright ideas. No matter how long and hard one pursues her, mathematics never seems to run out of exciting surprises." Nothing illustrates that concept more than the study of growth and decay, which involves the understanding of both exponential and logarithmic equations. In this chapter we investigate both of these important mathematical ideas.

IMPORTANT IDEAS

A logarithm is an exponent. That is, $\log_b A$ is the exponent on a base b that gives the result A. [8.1]
State and use the change of base theorem. [8.1]

Fundamental properties of logarithms. [8.2]

Log of both sides theorem. [8.2]

Laws of logarithms—that is, the additive, subtractive, and multiplicative laws. [8.2]

BOOK REPORTS

Write a 500-word report on one of these books:
The Mathematical Experience, Philip J. Davis and Reuben Hersh (Boston: Houghton Mifflin, 1981).
Ethnomathematics: A Multicultural View of Mathematical Ideas, Marcia Ascher (Pacific Grove, CA: Brooks/Cole), 1991.

LINKS

Individual Projects

Group Projects

Research Links

www.mathnature.com

8.1 | EXPONENTIAL EQUATIONS

The measurement of growth and decay often involves the study of relatively large or relatively small quantities. Difficulties with scaling measurements is often one of our primary concerns with describing and measuring figures and data. Recall from Section 1.3 that large and small numbers are often represented in exponential form and scientific notation. In this section, we investigate solving equations known as *exponential equations*.

Exponential Equation

> An equation of the form $b^x = N$ in which an unknown value is included as part of the exponent is called an **exponential equation**.

An **exponential** is an expression of the form b^x; we begin by using a calculator to evaluate some exponentials. The number b is called the *base*. In Chapter 1, we defined b^x for integer values of x. In more advanced courses, b^x is defined for all real numbers. We will use a calculator to approximate these values. We will also frequently approximate two irrational numbers

$$\pi \approx 3.1416 \quad \text{and} \quad e \approx 2.7183$$

EXAMPLE 1

Evaluate the given exponentials (correct to two decimal places). Show the calculator steps.

a. 2^8 **b.** $3^{2.5}$ **c.** $4^{\sqrt{2}}$ **d.** π^3 **e.** e^2 **f.** e^π **g.** π^e

Press each of these on your calculator because these evaluations set the groundwork for the rest of this chapter.

Solution

Given	Keys Pressed	Evaluation	Approximation
a. 2^8	2 ∧ 8 =	256 (exact)	256.00
b. $3^{2.5}$	3 ∧ 2.5 =	15.58845727	15.59
c. $4^{\sqrt{2}}$	4 ∧ √ 2 =	7.102993301	7.10
d. π^3	π ∧ 3 =	31.00627668	31.01
e. e^2	e ∧ 2 =	7.389056099	7.39

Note: If you want to evaluate e, find e^1:

	e ∧ 1 =	2.718281828	2.72
f. e^π	e ∧ π =	23.14069263	23.14
g. π^e	π ∧ e ∧ 1 =	22.45915772	22.46 ◆

Let's solve the exponential equation $2^x = 14$. To solve an equation means to find the replacement(s) for the variable that make the equation true. You might try certain values:

$x = 1$: $2^x = 2^1 = 2$ *Too small*

$x = 2$: $2^x = 2^2 = 4$ *Too small*

$x = 3$: $2^x = 2^3 = 8$ *Still too small*

$x = 4$: $2^x = 2^4 = 16$ *Too big*

It seems as if the number you are looking for is between 3 and 4. Our task in this section is to find both an approximate as well as an exact value for x. To answer this problem, we need some preliminary information.

Definition of Logarithm

Understanding this development is essential to understanding the meaning of logarithm.

The solution of the equation $2^x = 14$ seeks an x-value. What is this x-value? We express the idea in words:

x is the exponent on a base 2 that gives the answer 14

This can be abbreviated as

x = exp of 14 to the base 2

We further shorten this notation to

$x = \exp_2 14$

This statement is read, "x is the exponent on a base 2 that gives the answer 14." It appears that the equation is now solved for x, but this is simply a notational change. The expression "exponent of 14 to the base 2" is called, for historical reasons, "the log of 14 to the base 2." That is,

$x = \exp_2 14$ and $x = \log_2 14$

Both equations mean exactly the same thing. This leads us to the following definition of logarithm.

Definition of Logarithm

Spend some time with this definition.

> For positive b and A, $b \neq 1$
> $$x = \log_b A \quad \text{means} \quad b^x = A$$
> x is called the **logarithm** and A is called the **argument.**

The statement $x = \log_b A$ should be read as "x is the log (exponent) on a base b that gives the value A." *Do not forget that a logarithm is an exponent.*

EXAMPLE 2

Write in logarithmic form: **a.** $5^2 = 25$ **b.** $\frac{1}{8} = 2^{-3}$ **c.** $\sqrt{64} = 8$

Solution

a. In $5^2 = 25$, 5 is the base and 2 is the exponent, so we write

$$2 = \log_5 25$$

Remember, the logarithmic expression "solves" for the exponent.

b. With $\frac{1}{8} = 2^{-3}$, the base is 2 and the exponent is -3:

$$-3 = \log_2 \tfrac{1}{8}$$

c. With $\sqrt{64} = 8$, the base is 64 and the exponent is $\frac{1}{2}$ (since $\sqrt{64} = 64^{1/2}$):

$$\tfrac{1}{2} = \log_{64} 8$$ ◆

EXAMPLE 3

Tell what each expression means, and then rewrite in exponential form.

a. $\log_{10} 100$ **b.** $\log_{10} \frac{1}{1,000}$ **c.** $\log_3 1$

Solution

a. $\log_{10} 100$ is the exponent on a base 10 that gives 100. We see that the exponent is 2, so we write $\log_{10} 100 = 2$; the base is 10 and the exponent is 2, so $10^2 = 100$.

b. $\log_{10} \frac{1}{1,000}$ is the exponent on a base 10 that gives $\frac{1}{1,000}$; this exponent is -3. The base is 10 and the exponent is -3, so we write

$$10^{-3} = \frac{1}{1,000}$$

c. $\log_3 1$ is the exponent on a base 3 that gives 1; this exponent is 0. The base is 3 and the exponent is 0, so we write

$$3^0 = 1 \qquad \blacklozenge$$

EXAMPLE 4

Solve for x: a. $3^x = 5$ b. $10^x = 2$ c. $e^x = 0.56$

Solution

a. $x = \log_3 5$ b. $x = \log_{10} 2$ c. $x = \log_e 0.56$ $\qquad \blacklozenge$

In elementary work, the most commonly used base is 10, so we call a logarithm to the base 10 a **common logarithm,** and agree to write it without using a subscript 10. That is, $\log x$ is a *common logarithm.* A logarithm to the base e is called a **natural logarithm** and is denoted by $\ln x$. The expression $\ln x$ is often pronounced "ell en x" or "lon x."

Logarithmic Notations

> **Common logarithm:** $\log x$ means $\log_{10} x$
> **Natural logarithm:** $\ln x$ means $\log_e x$

The solution for the equation $10^x = 2$ is $x = \log 2$, and the solution for the equation $e^x = 0.56$ is $x = \ln 0.56$.

Evaluating Logarithms

STOP

To **evaluate** a logarithm means to find a numerical value for the given logarithm. Calculators have, to a large extent, eliminated the need for logarithm tables. You should find two logarithm keys on your calculator. One is labeled ⎡LOG⎤ for common logarithms, and the other is labeled ⎡LN⎤ for natural logarithms.

EXAMPLE 5

Use a calculator to evaluate: a. $\log 5.03$ b. $\ln 3.49$ c. $\log 0.00728$

Solution

Use your own calculator to verify these answers because the number of digits shown may vary. Calculator answers are more accurate than were the old table answers, but it is important to realize that any answer (whether from a table or a calculator) is only as accurate as the input numbers. However, in this book we will not be concerned with significant digits, but instead will use all the accuracy our calculator gives us, rounding only once (if requested) at the end of the problem.

a. $\log 5.03 \approx 0.7015679851$ $\qquad\qquad$ b. $\ln 3.49 \approx 1.249901736$

c. $\log 0.00728 \approx -2.137868621$ $\qquad\qquad\qquad\qquad\qquad\qquad \blacklozenge$

Example 5 shows fairly straightforward evaluations, since the problems involve common or natural logarithms and because your calculator has both $\boxed{\text{LOG}}$ and $\boxed{\text{LN}}$ keys. However, suppose we wish to evaluate a logarithm to some base *other than* base 10 or base *e*. The first method uses the definition of logarithm (as in Example 3), and the second method uses what is called the **change of base theorem.** Before we state this theorem, we consider its plausibility with the following example.

EXAMPLE 6

Evaluate the given expressions.

a. $\log_2 8, \dfrac{\log 8}{\log 2},$ and $\dfrac{\ln 8}{\ln 2}$ **b.** $\log_3 9, \dfrac{\log 9}{\log 3},$ and $\dfrac{\ln 9}{\ln 3}$

Solution

a. From the definition of logarithm, $\log_2 8 = x$ means $2^x = 8$ or $x = 3$. Thus, $\log_2 8 = 3$. By calculator,

$$\frac{\log 8}{\log 2} \approx \frac{0.903089987}{0.3010299957} \approx 3; \quad \text{also,} \frac{\ln 8}{\ln 2} \approx \frac{2.079441542}{0.6931471806} \approx 3$$

b. $\log_3 9 = x$ means $3^x = 3^2$, so that $x = \log_3 9 = 2$. By calculator,

$$\frac{\log 9}{\log 3} \approx \frac{0.9542425094}{0.4771212547} \approx 2 \quad \text{and} \quad \frac{\ln 9}{\ln 3} \approx \frac{2.197224577}{1.098612289} \approx 2 \qquad \blacklozenge$$

You no doubt noticed that the answers to each part of Example 6 are the same. This result is summarized with the following theorem, which is proved on page 411 (in Section 8.2).

Change of Base Theorem

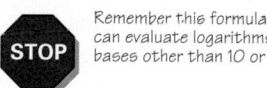

Remember this formula so you can evaluate logarithms to bases other than 10 or *e*.

$$\log_a x = \frac{\log_b x}{\log_b a}$$

EXAMPLE 7

Evaluate (round to the nearest hundredth): **a.** $\log_7 3$ **b.** $\log_3 3.84$

Solution

a. $\log_7 3 = \dfrac{\log 3}{\log 7} \approx \underbrace{\dfrac{0.4771212547}{0.84509804} \approx 0.5645750341}_{\text{This is all done by calculator and not on your paper.}} \approx 0.56$

b. $\log_3 3.84 = \dfrac{\log 3.84}{\log 3} \approx \underbrace{\dfrac{0.5843312244}{0.4771212547} \approx 1.224701726}_{\text{Calculator work}} \approx 1.22 \qquad \blacklozenge$

We now return to the problem of solving

$$2^x = 14 \quad \text{\footnotesize Given equation}$$

$$x = \log_2 14 \quad \text{\footnotesize Solution}$$

We call $\log_2 14$ the **exact solution** for the equation, and Example 8 finds an approximate solution.

EXAMPLE 8

Solve $2^x = 14$ (correct to the nearest hundredth).

Solution

We use the definition of logarithm and the change of base theorem to write

$$x = \log_2 14 = \underbrace{\frac{\log 14}{\log 2}}_{\text{calculator work}} \approx 3.807354922 \approx 3.81$$

◆

Exponential Equations

We now turn to solving *exponential equations.* Exponential equations will fall into one of three types:

	Base:	10 (common log)	e (natural log)	b (arbitrary base)
	Example:	$10^x = 5$	$e^{-0.06x} = 3.456$	$8^x = 156.8$

The following example illustrates the procedure for solving each type of exponential equation.

EXAMPLE 9

Solve the following exponential equations:

a. Base 10 (common log): $10^x = 5$

b. Base e (natural log): $e^{-0.06x} = 3.456$

c. Arbitrary base: $8^x = 156.8$

Solution

Regardless of the base, we use the definition of logarithm to solve an exponential equation.

a. $10^x = 5$

   $x = \log 5$ Definition of logarithm; this is the exact answer.

   ≈ 0.6989700043 Approximate answer

b. $e^{-0.06x} = 3.456$

   $-0.06x = \ln 3.456$ Definition of logarithm

   $x = \dfrac{\ln 3.456}{-0.06}$ Exact answer; this can be simplified to $x = -\frac{50}{3} \ln 3.456$.

   ≈ -20.66853085 Approximate answer

c. $8^x = 156.8$

   $x = \log_8 156.8$ Definition of logarithm; exact answer

   ≈ 2.43092725 Do not forget the change of base theorem: $\log_8 156.8 = \frac{\log 156.8}{\log 8}$.

Note: Many people will solve this by "taking the log of both sides":

$$8^x = 156.8$$

$$\log 8^x = \log 156.8$$

$$x \log 8 = \log 156.8 \quad \text{This property of logarithms will be developed in the next section.}$$

$$x = \frac{\log 156.8}{\log 8}$$

$$\approx 2.43092725$$

Did you notice that this result is the same? It simply involves several extra steps and some additional properties of logarithms. It is rather like solving quadratic equations by completing the square each time instead of using the quadratic formula. You can see that, before calculators, there were good reasons to avoid representations such as $\log_8 156.8$. Whenever you *see* an expression such as $\log_8 156.8$, you *know* how to calculate it: $\log 156.8/\log 8$. ◆

We now consider some more general exponential equations. We will follow the general procedure illustrated in Example 9. Some algebraic steps will be required to put the equations into the correct form. That is, before we use the definition of logarithm, we first solve for the exponential form.

EXAMPLE 10

Solve: **a.** $\dfrac{10^{5x+3}}{5} = 39$ **b.** $1 = 2e^{-0.000425x}$ **c.** $8 \cdot 6^{3x+2} = 1{,}600$

Solution
Note that, in each case, we use the definition of logarithm.

a. $\dfrac{10^{5x+3}}{5} = 39$ *Given*

$10^{5x+3} = 195$ Solve for the exponential (the base, along with its exponent) by multiplying both sides by 5.

$5x + 3 = \log 195$ Definition of logarithm: $5x + 3$ is the exponent on a base 10 that gives 195.

$5x = \log 195 - 3$ We now seek to solve the linear equation for x; subtract 3 from both sides.

$x = \dfrac{\log 195 - 3}{5}$ Divide both sides by 5; this is the exact solution.

≈ -0.1419930777 Use a calculator to find an approximate answer.

b. $1 = 2e^{-0.000425x}$ *Given*

$\dfrac{1}{2} = e^{-0.000425x}$ Solve for the exponential by dividing both sides by 2.

$-0.000425x = \ln \dfrac{1}{2}$ Definition of logarithm

$x = \dfrac{\ln 0.5}{-0.000425}$ Solve for x by dividing both sides by -0.000425. This is the exact solution.

$\approx 1{,}630.934542$ Use a calculator to approximate root.

c. $8 \cdot 6^{3x+2} = 1{,}600$ *Given*

$6^{3x+2} = 200$ Solve for the exponential by dividing both sides by 8.

$3x + 2 = \log_6 200$ Definition of logarithm

$3x = \log_6 200 - 2$ Solve for x; first subtract 2 from both sides.

$x = \dfrac{\log_6 200 - 2}{3}$ Divide both sides by 3; this is the exact solution.

≈ 0.3190157417 Use a calculator to approximate root (you will need the change of base theorem to evaluate). ◆

Example 10b illustrates exponential decay. Growth and decay problems are common examples of exponential equations. We will consider growth and decay applications in Section 8.3.

In Chapter 1 we considered large numbers. The following example involves large numbers, volumes, and exponential equations.

EXAMPLE 11   PÓLYA'S METHOD

The earth will fit into a cube that is 12,740 km on a side. How many generations of an organism that doubles each generation and takes up one μm^3 of space would it take to fill up this earth-sized cube? Note that the symbol μm stands for the length of a **micrometer,** which is defined to be one-millionth of a meter.

Solution   We use Pólya's problem-solving guidelines for this example.

Understand the Question.   Do you know what a μm is? It is one-millionth of a meter, so $(1\ \mu m)^3$ equals the size of a cube whose dimensions are:

$$1\ \mu m \times 1\ \mu m \times 1\ \mu m$$

Devise a Plan.   The plan we will use is first to find the number of organisms of size $1\ \mu m^3$ that are contained in this earth-sized cube, and then to find the number of generations by solving an exponential equation.

Carry Out the Plan.

$$V = s^3 \qquad \text{Where } s \text{ is the length of a side of a cube}$$
$$= (12{,}740\ \text{km})^3 \qquad \text{This is the earth-sized cube.}$$
$$\approx (2.07 \times 10^{12}\ \text{km}^3)$$
$$= (2.07 \times 10^{12})(1{,}000\ \text{m})^3$$
$$= 2.07 \times 10^{21}\ \text{m}^3$$
$$= (2.07 \times 10^{21})(1{,}000{,}000\ \mu m)^3 \qquad \text{1 m = 1,000,000 } \mu m$$
$$= 2.07 \times 10^{39}\ \mu m^3$$

Since the organism doubles each generation, we want to find n so that

$$2^n = 2.07 \times 10^{39}$$
$$n = \log_2(2.07 \times 10^{39}) \approx 130.6$$

It would take about 131 generations.

Looking Back.   Since n is the number of doublings, we find

$$2^{131} \approx 2.72 \times 10^{39}$$

which is a number of the appropriate magnitude.   ◆

Problem Set 8.1

LEVEL 1

1. **IN YOUR OWN WORDS** What is the definition of logarithm?

2. **IN YOUR OWN WORDS** What is a common logarithm? What is the notation used for a common logarithm?

3. **IN YOUR OWN WORDS** What is a natural logarithm? What is the notation used for natural logarithm?

4. **IN YOUR OWN WORDS**
   a. What does $\log N$ mean?
   b. What does $\ln N$ mean?
   c. What does $\log_b N$ mean?

5. **IN YOUR OWN WORDS** How do you use your calculator to evaluate the following?
   a. a common logarithm
   b. a natural logarithm
   c. a logarithm to a base b

6. **IN YOUR OWN WORDS** What is an exponential equation?

7. **IN YOUR OWN WORDS** What are the three types of exponential equations?

8. **IN YOUR OWN WORDS** Outline a procedure for solving exponential equations.

Write the equations in Problems 9–12 in logarithmic form.

9. **a.** $64 = 2^6$ **b.** $100 = 10^2$ **c.** $m = n^p$

10. **a.** $1{,}000 = 10^3$ **b.** $81 = 9^2$ **c.** $\dfrac{1}{e} = e^{-1}$

11. **a.** $\dfrac{1}{10} = 10^{-1}$ **b.** $36 = 6^2$ **c.** $s = t^n$

12. **a.** $125 = 5^3$ **b.** $9 = \left(\dfrac{1}{3}\right)^{-2}$ **c.** $a = b^c$

Use the definition of logarithm in Problems 13–18 to simplify each expression.

13. **a.** $\log_{10}10$ **b.** $\log_{10}1{,}000$ **c.** $\log_{10}10^{-5}$
14. **a.** $\log 100$ **b.** $\log 0.1$ **c.** $\log 10^{-3}$
15. **a.** $\log_5 5$ **b.** $\log_5 25$ **c.** $\log_5 5^{-3}$
16. **a.** $\log_e e$ **b.** $\log_e e^2$ **c.** $\log_e e^{-4}$
17. **a.** $\log_b b$ **b.** $\log_b b^3$ **c.** $\log_b b^{-6}$
18. **a.** $\log 10^n$ **b.** $\ln e^n$ **c.** $\log_b b^n$

Solve each exponential equation in Problems 19–22. Give the exact value for x.

19. **a.** $4^x = 5$ **b.** $5^x = 8$ **c.** $6^x = 4.5$
20. **a.** $\pi^x = 10$ **b.** $(\sqrt{2})^x = 5$ **c.** $e^x = 9$
21. **a.** $10^x = 15$ **b.** $10^x = 2.5$ **c.** $10^x = 45$
22. **a.** $e^x = 6$ **b.** $e^x = 1.8$ **c.** $e^x = 34.2$

Evaluate the given expressions in Problems 23–30 (to two decimal places).

23. **a.** $\log 4.27$ **b.** $\log_b b^2$ **c.** $\log_t t^3$
24. **a.** $\log 1.08$ **b.** $\log_e e^4$ **c.** $\log_\pi \sqrt{\pi}$
25. **a.** $\log 71{,}600$ **b.** $\log_3 9$ **c.** $\log_{19} 1$
26. **a.** $\log 18.9$ **b.** $\log_2 32$ **c.** $\log_7 1$
27. **a.** $\log 0.042$ **b.** $\log 0.321$ **c.** $\log 0.0532$
28. **a.** $\ln 2.27$ **b.** $\ln 16.77$ **c.** $\ln 7.3$
29. **a.** $\ln 10$ **b.** $\ln 100$ **c.** $\ln 1{,}000$
30. **a.** $\log e$ **b.** $\log e^2$ **c.** $\log e^3$

Write each expression in Problems 31–36 in terms of common logarithms, and then give a calculator approximation (correct to four decimal places).

31. $\log_3 45$ 32. $\log_5 91$ 33. $\log_6 10$
34. $\log_5 304$ 35. $\log_4 3.05$ 36. $\log_2 1{,}513$

Write each expression in Problems 37–42 in terms of natural logarithms, and then give a calculator approximation (correct to four decimal places).

37. $\log_2 0.0056$ 38. $\log_{8.3} 105$
39. $\log_8 10$ 40. $\log_\pi e^2$
41. $\log_8 e$ 42. $\log_{1.08} 5{,}450$

LEVEL 2

Solve the exponential equations in Problems 43–56. Show the approximation you obtain with your calculator without rounding.

43. **a.** $2^x = 128$ **b.** $3^x = 243$
44. **a.** $125^x = 25$ **b.** $216^x = 36$
45. **a.** $4^x = \frac{1}{16}$ **b.** $27^x = \frac{1}{81}$
46. **a.** $8^x = 3$ **b.** $64^x = 5$
47. **a.** $e^x = 4$ **b.** $e^x = 25$
48. **a.** $10^x = 42$ **b.** $10^x = 0.0234$
49. **a.** $10^{5x} = 5$ **b.** $10^{3x} = 0.45$
50. **a.** $10^{5-3x} = 0.041$ **b.** $10^{2x-1} = 515$
51. **a.** $e^{1-2x} = 3$ **b.** $e^{1-5x} = 25$
52. **a.** $e^{-x} = 8$ **b.** $10^{-x} = 125$

LEVEL 3

53. $2 \cdot 3^x + 7 = 61$
54. $3 \cdot 5^x + 30 = 105$
55. $\left(1 + \dfrac{0.08}{360}\right)^{360x} = 2$
56. $\left(1 + \dfrac{0.055}{12}\right)^{12x} = 2$
57. Solve $P = P_0 e^{rt}$ for t
58. Solve $I = I_0 e^{-rt}$ for r

59. If the earth's radius is approximately 3,963 mi, and if a hypothetical perpetual water-making machine pours out 1 gallon in the first minute and then doubles its capacity each minute, in which minute would this hypothetical machine pour out a single quantity of water that would be enough to fill the earth with water?

60. If the weight of the earth is 5.9×10^{21} metric tons, and if its weight were cut in half each minute, how long would it take for the earth to weigh less than one g? *Note:* A metric ton is one million grams.

8.2

LOGARITHMIC EQUATIONS

Fundamental Properties

In 1989, an earthquake measuring 7.1 on the Richter scale struck San Francisco during the World Series. Over 3,000 people were injured and 67 persons were killed. What is the amount of energy released (in ergs) by this earthquake? How would you answer this question?

Where would you begin? With a little research, you could find information on the Richter scale, which was developed by Gutenberg and Richter. The formula relating the energy E (in ergs) to the magnitude of the earthquake, M, is given by

$$M = \frac{\log E - 11.8}{1.5}$$

This equation is called a *logarithmic equation*, and the topic of this section is to solve such equations. To answer the question, we first solve for $\log E$, and then we use the definition of logarithm to write this as an exponential equation.

$$M = \frac{\log E - 11.8}{1.5} \qquad \text{Given}$$

$$1.5M = \log E - 11.8 \qquad \text{Multiply both sides by 1.5.}$$

$$1.5M + 11.8 = \log E \qquad \text{Add 11.8 to both sides.}$$

$$10^{1.5M+11.8} = E \qquad \text{1.5M + 11.8 is the exponent on a base 10 that gives } E \\ \text{(definition of logarithm).}$$

We can now answer the question. Since $M = 7.1$,

$$E = 10^{1.5M+11.8} = 10^{1.5(7.1)+11.8} \approx 2.818 \times 10^{22} \text{ ergs}$$

However, to solve certain logarithmic equations, we must first develop some properties of logarithms.

We begin with two *fundamental properties of logarithms*. If you understand the definition of logarithm, you can see these two properties are self-evident, so we call these the **Grant's tomb properties** of logarithms.

Fundamental Properties of Logarithms

1. $\log_b b^x = x$
2. $b^{\log_b x} = x \qquad x > 0$

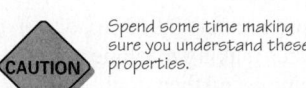 Spend some time making sure you understand these properties.

To justify property 1, remember the definition of logarithm:

$$b^M = N \quad \text{means} \quad \log_b N = M$$

and let $M = x$ and $N = b^x$. Then since $b^x = b^x$, we have $\log_b b^x = x$.

To justify property 2, remember that $\log_b x$ is the exponent on a base b that gives x, so property 2 shows exactly that fact.

EXAMPLE 1

Evaluate the given expressions.

a. $\log 10^3$ **b.** $\ln e^2$ **c.** $10^{\log 8.3}$ **d.** $e^{\ln 4.5}$ **e.** $\log_8 8^{4.2}$

Solution

a. $\log 10^3 = 3$ log 10³ is the exponent on 10 that gives 10³; obviously it is 3.

b. $\ln e^2 = 2$ ln e² is the exponent on e that gives e²; obviously it is 2.

c. $10^{\log 8.3} = 8.3$ Consider log 8.3; what is this? It is the exponent on a base 10 that gives the answer 8.3.

d. $e^{\ln 4.5} = 4.5$ Consider ln 4.5; what is this? It is the exponent on a base e that gives the answer 4.5.

e. $\log_8 8^{4.2} = 4.2$ What is the exponent on 8 that gives 8⁴·²? Obviously, it is 4.2.   ◆

Logarithmic Equations

A **logarithmic equation** is an equation for which there is a logarithm on one or both sides. The key to solving logarithmic equations is the following theorem, which we will call the **log of both sides theorem.**

Log of Both Sides Theorem

If A, B, and b are positive real numbers with $b \neq 1$, then

$$\log_b A = \log_b B \quad \text{is equivalent to} \quad A = B$$

The proof of this theorem is not difficult, and it depends on the two fundamental properties of logarithms given in the previous subsection.

Basically, all logarithmic equations in this book fall into one of four types:

Type I: The unknown is the logarithm; $\log_2 \sqrt{3} = x$.

Type II: The unknown is the base; $\log_x 6 = 2$.

Type III: The logarithm of an unknown is equal to a number; $\ln x = 5$.

Type IV: The logarithm of an unknown is equal to the logarithm of a number; $\log_5 x = \log_5 72$.

The following example illustrates the procedure for solving each type of logarithmic equation.

EXAMPLE 2

Solve the following logarithmic equations:

a. Type I: $\log_2 \sqrt{3} = x$ **b.** Type II: $\log_x 6 = 2$

c. Type III: $\ln x = 5$ **d.** Type IV: $\log_5 x = \log_5 72$

Solution

a. Type I: $\log_2 \sqrt{3} = x$. If the logarithmic expression does not contain a variable, you can use your calculator to evaluate. Remember, this type was evaluated in Section 8.1. If it is a common logarithm (base 10), use the $\boxed{\text{LOG}}$ key; if it is a natural logarithm (base e), use the $\boxed{\text{LN}}$ key; if it has another base, use the change of base theorem:

$$\log_a N = \frac{\log N}{\log a} \quad \text{or} \quad \log_a N = \frac{\ln N}{\ln a}$$

For this example, we see

$$x = \log_2 \sqrt{3} = \frac{\log \sqrt{3}}{} \approx 0.7924812504$$

b. Type II: $\log_x 6 = 2$. If the unknown is the base, then use the definition of logarithm to write an equation that is not a logarithmic equation.

$$\log_x 6 = 2 \qquad \text{Given}$$
$$x^2 = 6 \qquad \text{Definition of logarithm}$$
$$x = \pm\sqrt{6} \qquad \text{Solve quadratic equation.}$$

When solving logarithmic equations, make sure your answers are in the domain of the variable. Remember that, by definition, the base must be positive. For this example, $x = -\sqrt{6}$ is not in the domain, so the solution is $x = \sqrt{6}$.

c. Type III: The third and fourth types of logarithmic equations are the most common, and both involve the logarithm of an unknown quantity on one side of an equation. For the third type, use the definition of logarithm (and a calculator for an approximate solution, if necessary).

$$\ln x = 5$$
$$e^5 = x \qquad \text{5 is the exponent on } e \text{ that gives } x.$$

This is the exact solution. An approximate solution is $x \approx 148.4131591$.

d. Type IV: When a logarithm occurs on both sides, use the log of both sides theorem. *Make sure the log on both sides has the same base:*

$$\log_5 x = \log_5 72$$
$$x = 72 \qquad\qquad\qquad\qquad\qquad\qquad\qquad \blacklozenge$$

Example 2 illustrates the procedures for solving logarithmic equations, but most logarithmic equations are not as easy as those in Example 2. Usually, you must do some algebraic simplification to put the problem into the form of one of the four types of logarithmic equations. You might also have realized that Type IV is a special case of Type III. For example, to solve

$$\log_5 x = \log_3 72$$

which looks like Example 2d, we see that we cannot use the log of both sides theorem because the bases are not the same. We can, however, treat this as a Type III equation by using the definition of logarithm to write

$$x = 5^{\log_3 72}$$

This can be evaluated using a calculator. However, you may find it easier to visualize if we write

$$\log_5 x = \log_3 72 \approx 3.892789261$$

so that 3.892789261 is the exponent on a base 5 that gives x. In other words,

$$x \approx 5^{3.892789261} \approx 525.9481435$$

Laws of Logarithms

To simplify logarithmic expressions, we remember that a logarithm is an exponent and the laws of exponents correspond to the **laws of logarithms,** shown in the next box.

The proofs of these laws of logarithms are easy. The additive law of logarithms comes from the additive law of exponents:

$$b^x b^y = b^{x+y}$$

Let $A = b^x$ and $B = b^y$, so that $AB = b^{x+y}$. Then from the definition of logarithm, these three equations are equivalent to

$$x = \log_b A, \quad y = \log_b B, \quad \text{and} \quad x + y = \log_b(AB)$$

Laws of Logarithms

First Law (Additive)

Second Law (Subtractive)

Third Law (Multiplicative)

If A, B, and b are positive numbers, p is any real number, and $b \neq 1$:

$$\log_b(AB) = \log_b A + \log_b B$$ The log of the product of two numbers is the sum of the logs of those numbers.

$$\log_b\left(\frac{A}{B}\right) = \log_b A - \log_b B$$ The log of the quotient of two numbers is the log of the numerator minus the log of the denominator.

$$\log_b A^p = p \log_b A$$ The log of the pth power of a number is p times the log of that number.

Therefore, by putting these pieces together, we have

$$\log_b(AB) = x + y = \log_b A + \log_b B$$

Similarly, for the subtractive law of logarithms,

$$\frac{A}{B} = \frac{b^x}{b^y}$$

$$= b^{x-y} \qquad \text{Subtractive law of exponents}$$

$$\text{Thus,} \quad x - y = \log_b\left(\frac{A}{B}\right) \qquad \text{Definition of logarithm}$$

which means

$$\log_b\left(\frac{A}{B}\right) = \log_b A - \log_b B \qquad \text{Since } x = \log_b A \text{ and } y = \log_b B$$

The proof of the multiplicative law of logarithms follows from the multiplicative law of exponents and you are asked to do this in the problem set. We can also prove this multiplicative law by using the additive law of logarithms for p a positive integer:

$$\log_b A^p = \log_b \underbrace{(A \cdot A \cdot A \cdots \cdot A)}_{p \text{ factors}} \qquad \text{Definition of } A^p$$

$$= \underbrace{\log_b A + \log_b A + \log_b A + \cdots + \log_b A}_{p \text{ terms}} \qquad \text{Additive law of logarithms}$$

$$= p \log_b A$$

When logarithms were used for calculations, the laws of logarithms were used to expand an expression such as $\log\left(\dfrac{6 \cdot 45.62^2}{84.2}\right)$. Calculators have made such problems obsolete. Today, logarithms are important in solving equations, and the procedure for solving logarithmic equations requires that we take an algebraic expression involving logarithms and write it as a single logarithm. We might call this *contracting* a logarithmic expression.

EXAMPLE 3

Write each statement as a single logarithm.

a. $\log x + 5 \log y - \log z$ **b.** $\log_2 3x - 2 \log_2 x + \log_2(x + 3)$

Solution

Be sure the bases are the same before you use the laws of logarithms.

a. $\log x + 5 \log y - \log z = \log x + \log y^5 - \log z$ Third law

$$= \log xy^5 - \log z \qquad \text{First law}$$

$$= \log \frac{xy^5}{z} \qquad \text{Second law}$$

b. $\log_2 3x - 2\log_2 x + \log_2(x + 3) = \log_2 3x - \log_2 x^2 + \log_2(x + 3)$

$$= \log_2 \frac{3x(x + 3)}{x^2}$$

$$= \log_2 \frac{3(x + 3)}{x} \qquad \blacklozenge$$

EXAMPLE 4

Solve: $\log_8 3 + \frac{1}{2}\log_8 25 = \log_8 x$.

Solution

The goal here is to make this look like a Type IV logarithmic equation so that there is a single log expression on both sides.

$$\log_8 3 + \tfrac{1}{2}\log_8 25 = \log_8 x$$

$$\log_8 3 + \log_8 25^{1/2} = \log_8 x \qquad \text{\small Third law of logarithms}$$

$$\log_8 3 + \log_8 5 = \log_8 x \qquad \text{\small $25^{1/2} = (5^2)^{1/2} = 5$}$$

$$\log_8(3 \cdot 5) = \log_8 x \qquad \text{\small First law of logarithms}$$

$$15 = x \qquad \text{\small Log of both sides theorem}$$

The solution is 15. (Check to be sure 15 is in the domain of the variable.) $\blacklozenge$

When solving logarithmic equations, you must be mindful of extraneous solutions. The reason for this is that the logarithm function requires that the arguments be positive, but when solving an equation, we may not know the signs of the arguments. For example, if you solve an equation involving $\log x$ and obtain two answers, for example $x = 3$ and $x = -4$, then the value $x = -4$ must be extraneous because $\log(-4)$ is not defined.

EXAMPLE 5

Solve: **a.** $\log 15 + 2 = \log(x + 250)$ **b.** $\log 5 + \log(2x^2) = \log x + \log 15$

Solution

Use the laws of logarithms to combine the log statements. We have chosen two examples that are quite similar, but whose solutions require slightly different procedures. For part **a,** rewrite the expression so that all parts involving logarithms are on one side. In part **b,** rewrite the expression so that all logarithms involving the variable are on one side.

a. $\log 15 + 2 = \log(x + 250)$

$$2 = \log(x + 250) - \log 15 \qquad \text{\small Subtract log 15 from both sides.}$$

$$2 = \log \frac{x + 250}{15} \qquad \text{\small Second law of logarithms}$$

Type III: Use definition of a logarithm.

$$10^2 = \frac{x + 250}{15} \qquad \text{\small Definition of logarithm}$$

$$1{,}500 = x + 250 \qquad \text{\small Multiply both sides by 15.}$$

$$1{,}250 = x \qquad \text{\small Subtract 250 from both sides.}$$

The solution is $x = 1{,}250$.

b. $\log 5 + \log(2x^2) = \log x + \log 15$

$\qquad \log(2x^2) - \log x = \log 15 - \log 5$ Subtract log x and log 5 from both sides.

$$\log \frac{2x^2}{x} = \log \frac{15}{5}$$ Second law of logarithms

$$\log(2x) = \log 3$$

$$2x = 3$$ Log of both sides theorem

$$x = \frac{3}{2}$$

The solution is $x = \frac{3}{2}$.

 Type IV: Use log of both sides theorem.

EXAMPLE 6

Prove: $\log_a x = \dfrac{\log_b x}{\log_b a}$

Proof
Let $y = \log_a x$.

$$a^y = x$$ Definition of logarithm

$$\log_b a^y = \log_b x$$ Log of both sides theorem

$$y \log_b a = \log_b x$$ Third law of logarithms

$$y = \frac{\log_b x}{\log_b a}$$ Divide both sides by $\log_b a$ ($\log_b a \neq 0$).

Thus, by substitution, $\log_a x = \dfrac{\log_b x}{\log_b a}$.

EXAMPLE 7   PÓLYA'S METHOD

The largest number that you can write with three digits is 9^{9^9}. Write this number as a power of 10.

Solution   We use Pólya's problem-solving guidelines for this example.

Understand the Question. Do you know what 9^{9^9} means? Try using your calculator to find this number.

First try: $9\char`^9\char`^9 \approx 1.966270505 \times 10^{77}$
           What the calculator is giving is $(9^9)^9$. Is this the largest possible number with three digits?

Second try: $9\char`^(9\char`^9)$ produces an *overflow*, so we know this number is larger than $(9\char`^9)\char`^9$.

Devise a Plan. The plan we will use is first to find this number as a power of 10.

Carry Out the Plan.

$$9\char`^(9\char`^9) = 10^n$$ For some number n

$$9^{387,420,489} = 10^n$$ Use a calculator for 9^9.

$$n = \log 9^{387,420,489}$$ Definition of logarithm

$$= 387,420,489 \log 9$$ Property of logarithms

$$\approx 369,693,099.6$$ Calculator approximation

This means $9^{9^9} \approx 10^{369,693,099.6}$.

Problem Set 8.2

LEVEL 1

1. **IN YOUR OWN WORDS** What is a logarithmic equation?

2. **IN YOUR OWN WORDS** What are the four types of logarithmic equations?

3. **IN YOUR OWN WORDS** Outline a procedure for solving logarithmic equations.

Classify each of the statements in Problems 4–20 as true or false. If it is false, explain why you think it is false.

4. log 500 is the exponent on 10 that gives 500.

5. A common logarithm is a logarithm in which the base is 2.

6. A natural logarithm is a logarithm in which the base is 10.

7. In $\log_b N$, the exponent is N.

8. To evaluate $\log_5 N$, divide log 5 by log N.

9. If $2 \log_3 81 = 8$, then $\log_3 6,561 = 8$.

10. If $2 \log_3 81 = 8$, then $\log_3 81 = 4$.

11. If $\log_{1.5} 8 = x$, then $x^{1.5} = 8$.

12. $\ln \dfrac{x}{2} = \dfrac{\ln x}{2}$

13. $\log_b(A + B) = \log_b A + \log_b B$

14. $\log_b AB = (\log_b A)(\log_b B)$

15. $\dfrac{\log_b A}{\log_b B} = \log_b \dfrac{A}{B}$

16. $\dfrac{\log A}{\log B} = \dfrac{\ln A}{\ln B}$

17. $\dfrac{\log_b A}{\log_b B} = \log_b(A - B)$

18. $\dfrac{\log_b A}{\log_b B} = \log_b A - \log_b B$

19. $\log_b N$ is negative when N is negative.

20. log N is negative when $N > 1$.

Find a simplified value for x in Problems 21–31 by inspection. Do not use a calculator.

21. **a.** $e^{\ln 23}$ **b.** $10^{\log 3.4}$ **c.** $4^{\log_4 x}$

22. **a.** $\log 10^{4.2}$ **b.** $\ln e^3$ **c.** $\log_6 6^x$

23. **a.** $\log_b b^x$ **b.** $b^{\log_b x}$

24. **a.** $\log_5 25 = x$ **b.** $\log_2 128 = x$

25. **a.** $\log_3 81 = x$ **b.** $\log_4 64 = x$

26. **a.** $\log \frac{1}{10} = x$ **b.** $\log 10,000 = x$

27. **a.** $\log 1,000 = x$ **b.** $\log \frac{1}{1,000} = x$

28. **a.** $\log x = 5$ **b.** $\log_x e = 1$

29. **a.** $\ln x = 2$ **b.** $\ln x = 3$

30. **a.** $\ln x = 4$ **b.** $\ln x = \ln 14$

31. **a.** $\ln 9.3 = \ln x$ **b.** $\ln 109 = \ln x$

Contract the expressions given in Problems 32–35. That is, use the properties of logarithms to write each expression as a single logarithm with a coefficient of 1.

32. **a.** $\log 2 + \log 3 + \log 4$
    b. $\log 40 - \log 10 - \log 2$
    c. $2 \ln x + 3 \ln y - 4 \ln z$

33. **a.** $3 \ln 4 - 5 \ln 2 + \ln 3$
    b. $3 \ln 4 - 2 \ln(2 + 2)$
    c. $3 \ln 4 - 5(\ln 2 + \ln 3)$

34. **a.** $\ln 3 - 2 \ln 4 + \ln 8$
    b. $\ln 3 - 2 \ln(4 + 8)$
    c. $\ln 3 - 2(\ln 4 + \ln 8)$

35. **a.** $\log(x^2 - 9) - \log(x + 3)$
    b. $\log(x^2 - x - 6) - \log(x + 2)$
    c. $\ln(x^2 - 4) - \ln(x + 2)$

LEVEL 2

Solve the equations in Problems 36–54 by finding the exact solution.

36. **a.** $\frac{1}{2}x - 2 = 2$ **b.** $\frac{1}{2} \log x - \log 100 = 2$

37. **a.** $3 + 2x = 11$ **b.** $\ln e^3 + 2 \log x = 11$

38. **a.** $\frac{1}{2}x = 3 - x$ **b.** $\frac{1}{2}\log_b x = 3 \log_b 5 - \log_b x$

39. **a.** $x - 2 = 2$ **b.** $\log 10^x - 2 = \log 100$

40. **a.** $1 = x - 1$ **b.** $\ln e = \ln \dfrac{\sqrt{2}}{x} - \ln e$

41. **a.** $1 = \frac{3}{2} - x$ **b.** $\log 10 = \log \sqrt{1,000} - \log x$

42. **a.** $3 - x = 1$ **b.** $\ln e^3 - \ln x = 1$

43. **a.** $0 + x = 2$ **b.** $\ln 1 + \ln e^x = 2$

44. $\log(\log x) = 1$ 45. $\ln[\log(\ln x)] = 0$

46. $x^2 5^x = 5^x$ 47. $x^2 3^x = 9(3^x)$

48. $\log x = 1.8 + \log 4.8$ 49. $\ln x = 1.8 - \ln 4.8$

50. $\ln x - \ln 8 = 12$ 51. $\log x + \log 8 = 12$

52. $\log 2 = \frac{1}{4} \log 16 - x$

53. $\log_8 5 + \frac{1}{2} \log_8 9 = \log_8 x$

54. $\log x + \log(x - 3) = 2$

LEVEL 3

55. An advertising agency conducted a survey and found that the number of units sold, N, is related to the amount a spent on advertising (in dollars) by the following formula:

$$N = 1,500 + 300 \ln a \quad (a \geq 1)$$

a. How many units are sold after spending $1,000?
b. How many units are sold after spending $50,000?
c. How much needs to be spent (to the nearest hundred dollars) to sell 4,000 units?

56. The pH of a substance measures its acidity or alkalinity. It is found by the formula

$$pH = -\log[H^+]$$

where $[H^+]$ is the concentration of hydrogen ions in an aqueous solution given in moles per liter.

a. What is the pH (to the nearest tenth) of a lemon for which $[H^+] = 2.86 \times 10^{-4}$?

b. What is the pH (to the nearest tenth) of rainwater for which $[H^+] = 6.31 \times 10^{-7}$?

c. If the pH of a tested substance is 8.1, what is the hydrogen content? Give your answer in scientific notation.

57. The Richter scale for measuring earthquakes was developed by Gutenberg and Richter. It relates the energy E (in ergs) to the magnitude of the earthquake M by the formula

$$M = \frac{\log E - 11.8}{1.5}$$

a. A small earthquake is one that releases 15^{15} ergs of energy. What is the magnitude (to the nearest hundredth) of such an earthquake on the Richter scale?

b. A large earthquake is one that releases 10^{25} ergs of energy. What is the magnitude (to the nearest hundredth) of such an earthquake on the Richter scale?

c. How much energy is released in an 8.0 earthquake?

d. Solve for E.

58. The "forgetting curve" for memorizing nonsense syllables is given by

$$R = 80 - 27 \ln t \quad (t \ge 1)$$

where R is the percentage who remember the syllables after t seconds.

a. In how many seconds would only 10% ($R = 10$) of the students remember?

b. Solve for t.

PROBLEM SOLVING

59. The "learning curve" describes the rate at which a person learns certain tasks. If a person sets a goal of typing N words per minute (wpm), the length of time t (in days) to achieve this goal is given by

$$t = -62.5 \ln\left(1 - \frac{N}{80}\right)$$

a. According to this formula, what is the maximum number of words per minute?

b. Solve for N.

60. Prove the multiplicative law of logarithms using the multiplicative law of exponents. That is, prove $\log_b A^p = p \log_b A$.

8.3 APPLICATIONS OF GROWTH AND DECAY

In populations, bank accounts, and radioactive decay, where the *rate of change* is held constant, the average growth rate is a constant percent of the current value. In calculus, a **growth formula** is derived; it is summarized in the following box.

Growth/Decay Formula

This is one of the most useful formulas from mathematics.

STOP

Exponential **growth** or **decay** can be described by the equation

$$A = A_0 e^{rt}$$

where r is the annual growth/decay rate, t is the time (in years), A_0 is the amount present initially (present value), and A is the future value. If r is positive, this formula models growth, and if r is negative, the formula models decay.

Note: You can use this formula as long as the units of time are the same. That is, if the time is measured in days, then the growth/decay rate is a daily growth/decay rate, as illustrated by Example 4 on page 415.

Population Growth

EXAMPLE 1

On October 12, 1999, the world population reached 6 billion. If we assume a growth rate of 1.5%, when will the population reach 7 billion?

Solution

$$7 = 6e^{0.015t}$$ Growth formula

$$\frac{7}{6} = e^{0.015t}$$ Solve for the exponential.

$$0.015t = \log_e \frac{7}{6} \quad \text{or} \quad 0.015t = \ln \frac{7}{6}$$ Definition of logarithm

$$t = \frac{\ln 7/6}{0.015}$$ Divide both sides by 0.015.

$$t \approx 10.27671199$$ Approximate answer by calculator

This means that we should pass the 7 billion mark in 2009. ◆

Suppose we do not know the growth rate but have some population data. Consider the following example, assuming an exponential growth model.

EXAMPLE 2

Phoenix, Arizona, had a population of 789,794 in 1980 and 983,403 in 1990. Predict the population in Phoenix in 1994. Compare this calculated number with the actual 1994 population of 1,048,949.

Solution

We use the growth/decay formula:

$$A = A_0 e^{rt}$$

$$983,403 = 789,794 e^{r(10)}$$ Given: $A_0 = 789,794$ in 1980, $A = 983,403$ in 1990, so $t = 10$.

$$\frac{983,403}{789,794} = e^{10r}$$ Divide both sides by 789,794.

$$10r = \ln\left(\frac{983,403}{789,794}\right)$$ 10r is the exponent on the base e that gives the number 983,403/789,794.

$$r = \frac{1}{10} \ln\left(\frac{983,403}{789,794}\right)$$ Divide both sides by 10.

$$\approx 0.0219246854$$ Calculator evaluation

In 1994, we use the population in 1990 and calculate A using the above value for r:

$$A = A_0 e^{rt}$$

$$= 983,403 \, e^{r(4)}$$

$$\approx 1,073,540.933$$ Use $r \approx 0.0219246854$.

Thus, the predicted 1994 population is 1,073,541. Since the actual population, 1,048,949, is a bit less than the predicted number, we conclude that the growth rate between 1990 and 1994 has decreased a bit from the growth rate in 1980–1990. ◆

EXAMPLE 3

According to the Centers for Disease Control in Atlanta, Georgia, at the end of January 1992, the total number of AIDS-related U.S. deaths for all ages was 209,693. At that time, it was predicted that by January 1996 there would be from 400,000 to 450,000 cumulative deaths from the disease. Assuming these numbers are correct, estimate the cumulative number of AIDS-related deaths at the end of January 2002.

Solution We assume that the growth rate will remain constant over the years of our study and also that the growth takes place continuously. From the end of January 1992 to January 1996 is 4 years, so $t = 4$. Furthermore, we will work with the more conservative estimate of cumulative deaths. Consider

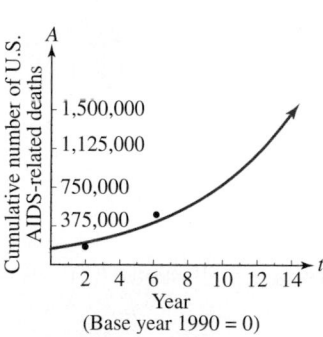

$$A = A_0 e^{rt} \qquad \text{Growth formula}$$

$$400{,}000 = 209{,}693 e^{4r} \qquad \text{\textit{t} = 4; substitute known values.}$$

$$\frac{400{,}000}{209{,}693} = e^{4r} \qquad \text{Solve for the exponential.}$$

$$4r = \ln\!\left(\frac{400{,}000}{209{,}693}\right) \qquad \text{Definition of logarithm}$$

$$r = \frac{1}{4}\ln\!\left(\frac{400{,}000}{209{,}693}\right) \qquad \text{Divide both sides by 4.}$$

$$\approx 0.1614549977 \qquad \text{Approximate answer by calculator}$$

A

Cumulative number of U.S.
AIDS-related deaths

1,500,000
1,125,000
750,000
375,000

2 4 6 8 10 12 14 t
Year
(Base year 1990 = 0)

Figure 8.1 Cumulative deaths from AIDS, 1992–2002

Thus, $A = 209{,}693 e^{0.1614549977t}$. At the end of January 2002, we see that $t = 10$ years:

$$A = 209{,}693 e^{0.1614549977(10)} \approx 1{,}053{,}839$$

A graph is shown in Figure 8.1. ◆

Radioactive Decay

We now consider another application involving decay. Radioactive materials decay over time. Each substance has a different decay rate. In the following example, when we say the decay rate of neptunium-239 is 31%, we imply that the rate is negative—that is, growth implies positive rate and decay implies negative rate; or, saying it another way, positive rate is growth and negative rate is decay.

EXAMPLE 4

If 100.0 mg of neptunium-239 (^{239}Np) decays to 73.36 mg after 24 hours, find the value of r in the growth/decay formula for t expressed in days.

Solution
Since $A = 73.36$, $A_0 = 100.0$, and $t = 1$ (day), we have

$$A = A_0 e^{rt} \qquad \text{Growth/decay formula}$$

$$73.36 = 100 e^{r(1)} \qquad \text{Substitute known values.}$$

$$0.7336 = e^r \qquad \text{Divide both sides by 100.}$$

$$r = \log_e 0.7336 \qquad \text{Definition of logarithm}$$

$$r = \ln 0.7336 \approx -0.309791358$$

Thus, the daily decay rate is approximately 31%. ◆

We often specify the radioactive decay in terms of what is called the **half-life.** This is the time that it takes for a particular radioactive substance to decay to half of its original amount.

EXAMPLE 5

Carbon-14, used for archaeological dating, has a half-life of 5,730 years. Find the decay rate for carbon-14.

Solution

If 100 mg of carbon-14 is present, we are given that in 5,730 years there will be 50 mg present. We use the growth/decay formula:

$$A = A_0 e^{rt} \qquad \text{Growth/decay formula}$$

$$50 = 100 e^{5,730r} \qquad \text{Substitute the given values.}$$

$$\frac{1}{2} = e^{5,730r} \qquad \text{Note that you can start a half-life problem with this formula; the amount present at the start is not relevant.}$$

$$5,730r = \ln 0.5 \qquad \text{Definition of logarithm}$$

$$r \approx -1.209680943E - 4 \qquad \text{Calculator evaluation}$$

Use this calculator value of r for carbon-14 calculations. ◆

We can use the decay rate for carbon-14 to date an artifact, as illustrated by the following example.

EXAMPLE 6

An archaeologist has found a fossil in which the ratio of ^{14}C to ^{12}C is 20% of the ratio found in the atmosphere. Approximately how old is the fossil?

Solution As the radioactive isotope carbon-14 (denoted by ^{14}C) decays, it changes to a stable form of carbon, called carbon-12. Once again, we use the growth/decay formula with the ratio of ^{14}C to ^{12}C to be 20% as follows:

$$A = A_0 e^{rt} \qquad \text{Growth/decay formula}$$

$$\frac{A}{A_0} = e^{rt} \qquad \text{Divide both sides by } A_0.$$

$$0.20 = e^{rt} \qquad \text{The given ratio is 20\% and the half-life is 5,730 years.}$$

$$rt = \ln 0.20 \qquad \text{Definition of logarithm}$$

$$t = \frac{\ln 0.20}{r} \qquad \text{Divide both sides by } r \text{ (approximate using the result from Example 5).}$$

$$\approx 13,304.64798$$

The fossil is approximately 13,000 yr old. ◆

Logarithmic Scales

Growth and decay examples are exponential models, but as we have seen, a logarithm is an exponent and is therefore directly related to growth and decay. A **logarithmic scale** is a scale in which logarithms are used to make data more manageable by expanding small variations and compressing large ones.

For example, prior to 1935, *seismographs* were used to record the amount of earth movement generated by an earthquake's seismic wave; this movement was recorded on a *seismogram*. The *amplitude* of a seismogram is the vertical distance between the peak or valley of the recording of the seismic wave and a horizontal line formed if there is no earth movement. This amplitude is measured using a very small unit, **micrometer** (denoted by μm), which is one-millionth of a meter. This small movement on a seismograph is used to measure a very large amount of energy released by an earthquake, and it must be done in such a way that the location of the seismograph relative to the earthquake's location (called the *epicenter*) was not relevant. You have, no doubt, heard of the well-known *Richter scale* used today as a means of measuring the magnitude, *M*, of an earthquake.

In 1935, Charles F. Richter, a seismologist at the California Institute of Technology, declared that the magnitude M of an earthquake with amplitude of A on a seismograph was

$$M = \log \frac{A}{A_0}$$

where A_0 is the amplitude of a "standard earthquake." This number M is called the **Richter number** or **Richter scale** to denote the size of an earthquake. Richter measured a large number of extremely small southern California earthquakes, and *defined* $\log A_0$ to be -1.7 for a seismograph located 20 km from the epicenter. Using the properties of logarithms, we know

$$\begin{aligned} M &= \log A - \log A_0 \\ &= \log A - (-1.7) \\ &= \log A + 1.7 \end{aligned}$$

If the seismograph is located 300 km from the epicenter, then $\log A_0$ is *defined* to be -4.0, so

$$\begin{aligned} M &= \log A - \log A_0 \\ &= \log A + 4.0 \end{aligned}$$

Since the calculation of the magnitude depends on the distance of the seismograph from the epicenter, and since for a particular earthquake the distance of the seismograph from the epicenter is not known, an actual earthquake is usually measured at three different locations in order to determine the epicenter. The Richter scale ratings for some well-known quakes are shown in Table 8.1, but it should be noted that if you search the Web you will find that earthquakes are common. In 1998 there were over 10 quakes recorded with a magnitude of over 7.0.

Table 8.1 Magnitudes of Major World Earthquakes

Date	Location	Magnitude
1906	San Francisco	8.3
1906	Valparaiso	8.6
1915	Avezzano	7.5
1920	Gansu, China	8.6
1933	Japan	8.9
1946	Honshu	8.4
1950	Assam, India	8.7
1960	Chile	9.5
1971	San Fernando	6.6
1985	Mexico	8.1
1989	Loma Prieta	7.1
1994	Northridge	6.6
1998	Balleny Islands	8.3
2003	Colima, Mexico	7.8

EXAMPLE 7

A seismograph 300 km from the epicenter of an earthquake recorded a maximum amplitude of 4.9×10^3 μm. Find the earthquake's magnitude. What would be the Richter scale reading for a seismograph 20 km from the epicenter of the same earthquake?

Solution

$$\begin{aligned} M &= \log A + 4.0 \\ &= \log(4.9 \times 10^3) + 4.0 \\ &\approx 7.7 \quad \text{\small Note: It is customary to give Richter scale readings rounded to the nearest tenth.} \end{aligned}$$

For a seismograph 20 km from the epicenter,

$$M = \log(4.9 \times 10^3) + 1.7 \approx 5.4 \qquad \blacklozenge$$

Why is the Richter scale called logarithmic? The reason is that if you increase the magnitude by 1, then the quake is 10 times stronger; if you increase the magnitude by 2, then the quake is $10^2 = 100$ times stronger; if you increase the magnitude by 3, then the quake is 10^3 times stronger. We illustrate an increase of magnitude by 4 with the following example.

EXAMPLE 8

Compare the strengths of earthquakes with magnitudes 4 and 8.

Solution Let M_1 and M_2 be the two given magnitudes. Then

$$M_1 - M_2 = (\log A_1 - \log A_0) - (\log A_2 - \log A_0)$$
$$= \log A_1 - \log A_2$$
$$= \log \frac{A_1}{A_2}$$

For this problem, in particular, we have

$$8 - 4 = \log \frac{A_1}{A_2}$$
$$4 = \log \frac{A_1}{A_2}$$
$$10^4 = \frac{A_1}{A_2}$$
$$A_1 = 10{,}000 A_2$$

A doubling of the magnitude from 4 to 8 means that the stronger earthquake's amplitude is $10^4 = 10{,}000$ times stronger. ◆

The amount of earth movement from an earthquake is measured by the amount of energy released by the earthquake. Consider the following example and compare with Example 8.

EXAMPLE 9

Compare the amount of earth movement (the energy released) by earthquakes of magnitudes of 4 and 8.

Solution Use the formula

$$M = \frac{\log E - 11.8}{1.5}$$

from Section 8.2. We solve for E:

$$1.5M = \log E - 11.8 \qquad \text{Multiply both sides by 1.5.}$$
$$1.5M + 11.8 = \log E \qquad \text{Add 11.8 to both sides.}$$
$$E = 10^{1.5M + 11.8} \qquad \text{Definition of logarithm}$$

Let $M_1 = 4$ and $M_2 = 8$ with corresponding energies E_1 and E_2, respectively. We now compare these energies:

$$\frac{E_1}{E_2} = \frac{10^{1.5(4)+11.8}}{10^{1.5(8)+11.8}} = 10^{-6} \qquad \text{Same bases, subtract exponents.}$$
$$E_1 = 10^{-6} E_2 \qquad \text{Multiply both sides by } E_2.$$
$$10^6 E_1 = E_2 \qquad \text{Multiply both sides by } 10^6.$$

The earthquake with magnitude 8 releases a million times more energy than an earthquake with magnitude 4. ◆

A second example of a logarithmic scale is the decibel rating used for measuring the intensity of sounds. To measure the intensity of sound, we need to understand that a sound is a vibration received by the ear and processed by the brain. We can place a listening device in the path of the sound and measure the amount of energy on that device per unit of area per second. This listening device acts like an eardrum, but the problem is that experiments have shown that humans perceive loudness on the basis of the ratio

Table 8.2 Decibel Ratings

Sound	dB rating
threshold recording	0
studio	20
whisper	25
quiet room	30
conversation	60
traffic	70
train	100
orchestra	110
rock music	115
threshold of pain	120
rocket	125

of intensities of two different sounds. For this reason, the unit of measurement for measuring sounds, called the **decibel,** in honor of Alexander Graham Bell, the inventor of the telephone, is defined as a ratio of the intensity of one sound, I, and another sound, $I_0 \approx 10^{-16}$ watt/cm^2, the intensity of a barely audible sound for a person with normal hearing.

The issue is further complicated by the fact that a human ear can hear an incredible range of sounds. A painful sound is 10^{14} (100 trillion) times more intense than a barely audible sound. This leads us to define the number of decibels, D, by

$$D = 10 \log \frac{I}{I_0}$$

for a sound of intensity I. The decibel rating, abbreviated dB, for various sounds is shown in Table 8.2.

EXAMPLE 10

The background sound of a study room was measured to be 10^{-10} watts/cm^2. Find the decibel rating for the room.

Solution   We are given $I = 10^{-10}$ and $I_0 = 10^{-16}$, so

$$D = 10 \log \frac{10^{-10}}{10^{-16}}$$

$$= 10 \log 10^6 \quad \text{For division, subtract exponents.}$$

$$= 60 \quad \text{Grant's tomb property: } \log 10^6 = 6. \qquad \blacklozenge$$

A scale for loudness of sounds begins at 0 dB (threshold of hearing) and extends to the threshold of pain, 120 dB. Each increase of 10 decibels is perceived as a doubling of loudness. A sound of 70 dB is twice as loud as a 60-dB sound.

A link for a sound test of decibel changes can be found at the following site:
www.mathnature.com

EXAMPLE 11

The noise in a classroom varies from 50 dB to 62 dB. Find the corresponding variation in intensities.

Solution   We are given $D_1 = 50$ and $D_2 = 62$, and we wish to compare I_1 and I_2:

$$D_2 - D_1 = 10 \log \frac{I_2}{I_0} - 10 \log \frac{I_1}{I_0}$$

$$= 10 \log I_2 - 10 \log I_0 - 10 \log I_1 + 10 \log I_0$$

$$= 10(\log I_2 - \log I_1)$$

$$= 10 \log \frac{I_2}{I_1}$$

$$62 - 50 = 10 \log \frac{I_2}{I_1} \quad \text{Substitute given values.}$$

$$1.2 = \log \frac{I_2}{I_1}$$

$$10^{1.2} = \frac{I_2}{I_1} \quad \text{Definition of logarithm}$$

$$15.85 \approx \frac{I_2}{I_1} \quad \text{Calculator approximation of } 10^{1.2}$$

$$I_2 \approx 16 I_1$$

The louder room is about 16 times noisier than the quiet room. $\qquad \blacklozenge$

Problem Set 8.3

LEVEL 1

1. **IN YOUR OWN WORDS** What is the growth/decay formula?

2. **IN YOUR OWN WORDS** What do we mean by half-life?

3. **IN YOUR OWN WORDS** What is a logarithmic scale?

4. **IN YOUR OWN WORDS** What is a micrometer?

5. **IN YOUR OWN WORDS** According to the World POPClock, the world population on 7/1/98 was 5,918,624,368 and on 7/1/99 it was 5,996,215,340. Use these numbers to find the annual growth rate for this period. Use this growth rate to project the world population on the day you work this problem. Check the World POP-Clock for the day's population and compare with your answer. Comment on the difference. Here is the Web address:

   http://www.census.gov/cgi-bin/ipc/popclockw
   As usual, you can access this Web address through the address for this text:
   www.mathnature.com

6. **IN YOUR OWN WORDS** In July 1990, the world population was about 5.3 billion. If we assume a growth rate of 0.98%, when will the population reach 6 billion? Compare this with the actual time 6 billion was reached (October 1999). Answer to the nearest month.

Find the following information for each of the cities described in Problems 7–14.
a. *Growth rate from 1980 to 1990*
b. *Predict the population for 1994.*
c. *Compare the prediction with the 1994 actual, and comment on possible reasons for discrepancies, if any.*
d. *Predict the population for the year 2005.*

City	1980	1990	1994
7. Jacksonville	540,920	636,070	665,070
8. Sacramento	275,919	369,365	373,964
9. Nashville	455,651	488,374	504,505
10. El Paso	425,259	515,342	579,307
11. Honolulu	365,048	365,272	385,881
12. Boston	562,994	574,283	547,725
13. St. Paul	270,230	272,235	262,071
14. Shreveport	206,989	198,518	196,982

15. Example 3 calculates the growth rate of AIDS-related deaths and uses this information to predict the cumulative number of AIDS-related deaths at the end of January 2002. Repeat this example for the number of deaths at the turn of the century, January 2001.

16. Example 3 calculates the growth rate of AIDS-related deaths and uses this information to predict the cumulative number of AIDS-related deaths at the end of January 2002

by using the conservative estimate of 400,000 in 1996. Repeat this example using the less conservative estimate of 450,000.

17. If the half-life of cesium-137 is 30 years, find the decay constant, r.

18. If the half-life of plutonium-238 is 86 years, find the decay constant, r.

19. Find the half-life (to the nearest year) of strontium-90 if $r = -0.0246$.

20. Find the half-life (to the nearest year) of krypton if $r = -0.0641$.

21. A seismograph 300km from the epicenter of an earthquake recorded a maximum amplitude of 5.1×10^2 μm. Find this earthquake's magnitude.

22. A seismograph 20 km from the epicenter of an earthquake recorded a maximum amplitude of 8.2×10^5 μm. Find this earthquake's magnitude.

23. Compare the strengths of earthquakes with magnitudes 4 and 6.

24. Compare the strengths of earthquakes with magnitudes 7 and 8.

Find the number of decibels for the power of the sounds given in Problems 25–30. Round to the nearest decibel.

25. A whisper, 10^{-13} watts/cm^2

26. A traffic jam, 10^{-8} watts/cm^2

27. A rock concert, 5.23×10^{-6} watts/cm^2

28. A conversation, 3.16×10^{-10} watts/cm^2

29. A rocket engine, 2.53×10^{-5} watts/cm^2

30. According to the *Guinness Book of World Records,* the world's loudest shout, by Skipper Kenny Leader, was 10^{-5}W/cm^2.

31. The energy released by an earthquake is approximated by

$$\log E = 11.8 + 1.5M$$

   What is the energy released by the 1906 San Francisco quake that measured 8.3 on the Richter scale? This energy, it is estimated, would be sufficient to provide the entire world's food requirements for a day.

LEVEL 2

32. A bacteria culture had a population of 10 million at 10:00 A.M., and by 2:00 P.M. had grown to 18 million.
   a. Predict the population at 6:00 P.M. that same day.
   b. When will the population double in size?

33. Between 1980 and 1990, the population of Los Angeles grew from 2,968,528 to 3,485,557, but by 1994 had dropped to 3,448,613. Predict the population (rounded to the nearest thousand) of Los Angeles in 2004 using the given assumptions.

a. Using the 1980 to 1990 growth rate
b. Using the 1990 to 1994 growth rate
c. Using the 1980 to 1994 growth rate

34. The Dead Sea Scrolls were written on parchment at about 100 B.C. What percentage of ^{14}C originally contained in the parchment remained when the scrolls were discovered in 1947?

35. Tests of an artifact discovered at the Debert site in Nova Scotia show that 28% of the original ^{14}C is still present. What is the probable age of the artifact?

36. The half-life of ^{234}U, uranium-234, is 2.52×10^5 yr. If 97.3% of the uranium in the original sample is present, what length of time (to the nearest thousand years) has elapsed?

37. The half-life of ^{22}Na, sodium-22, is 2.6 yr. If 15.5 g of an original 100-g specimen remains, how much time has elapsed (to the nearest year)?

38. How much more energy was released by the Loma Prieta quake (Richter scale 7.1) than the Northridge quake (Richter scale 6.6)?

39. How much more energy was released by the 1960 Chile quake (Richter scale 9.5) than the 1906 San Francisco quake (Richter scale 8.3)?

40. An artifact was found and tested for its carbon-14 content. If 12% of the original carbon-14 was still present, what is its probable age (to the nearest 100 years)?

41. An artifact was found and tested for its carbon-14 content. If 85% of the original carbon-14 was still present, what is its probable age (to the nearest 100 years)?

42. The radioactive substance neptunium-139 decays to 73.36% of its original amount after 24 hours. How long (to the nearest hour) would it take for 43% of the original neptunium to be present? What is the half-life of neptunium-139?

43. The 1989 World Series San Francisco quake was initially reported to have magnitude 7.0, but later this was revised to 7.1. How much energy released corresponds to this increase in magnitude?

44. Compare the amount of earth movement (energy released) by earthquakes of magnitudes of 3 and 6.

45. Compare the amount of earth movement (energy released) by earthquakes of magnitudes 7 and 8.

LEVEL 3

46. A certain artifact is tested by carbon dating and found to contain 73% of its original carbon-14. As a cross-check, it is also dated using radium, and was found to contain 32% of the original amount. Assuming the dating procedures were accurate, what is the half-life of radium?

47. The radioactive isotope gallium-67 (symbol ^{67}Ga) used in

the diagnosis of malignant tumors has a half-life of 46.5 hours. If we start with 100 mg of ^{67}Ga, what percentage is lost between the 30th and 35th hours? Is this the same as the percentage lost in any other 5-hour period?

If an object at temperature B is surrounded by air at temperature A, it will gradually cool so that the temperature T, t minutes later, is given by Newton's Law of Cooling:

$$T = A + (B - A)10^{-kt}$$

The constant k depends on the particular object, and can be found by

$$k = \frac{1}{t} \log \frac{B - A}{T - A}$$

Use this information for Problems 48–51.

48. You draw a tub of hot water ($k = 0.01$) for a bath. The water is 100°F when drawn and the room is 72°F. If you are called away to the phone, what is the temperature of the water 20 minutes later when you get in?

49. You take a batch of chocolate-chip cookies from the oven (250°F) when the room temperature is 74°F. If the cookies cool for 10 minutes and $k = 0.075$, what is the temperature of the cookies?

50. It is known that the temperature of a given object falls from 120°C to 70°C in an hour when placed in 20°C air. What is the temperature of the object after 30 minutes?

51. The temperature of an object is initially 100°C. In air of 22°C it cools to 45° in 30 minutes. What is the temperature in 40 minutes?

52. The atmospheric pressure P in pounds per square inch (psi) is given by

$$P = 14.7e^{-0.21a}$$

where a is the altitude above sea level (in miles). If a city has an atmospheric pressure of 13.23 psi, what is its altitude? (Recall that 1 mi = 5,280 ft.)

53. The atmospheric pressure P in pounds per square inch (psi) is approximated by $P = 14.7e^{-0.21a}$, where a is the altitude above sea level in miles. If the atmospheric pressure of Denver is 11.9 psi, estimate Denver's altitude. (Recall that 1 mi = 5,280 ft.)

54. The atmospheric pressure P in pounds per square inch (psi) is approximated by $P = 14.7e^{-0.21a}$, where a is the altitude above sea level in miles. If the pressure gauge in a small plane shows 10.2 psi, estimate the plane's altitude in feet. (Recall that 1 mi = 5,280 ft.)

55. A satellite has an initial radioisotope power supply of 50 watts (W). The power output in watts is given by

$$P = 50e^{-t/250}$$

where t is the time in days. Solve for t to find the time when the power supply is 30 W.

56. A satellite has an initial radioisotope power supply of 50 watts. The power output in watts is given by the equation $P = 50e^{-t/250}$, where t is the time in days. If the satellite will operate if there is at least 10 watts of power, how long would we expect the satellite to operate?

PROBLEM SOLVING

57. **HISTORICAL QUESTION** The Shroud of Turin (see Figure 8.2) is a rectangular linen cloth kept in the Chapel of the Holy Shroud in the cathedral of St. John the Baptist in Turin, Italy. It shows the image of a man whose wounds correspond with the biblical accounts of the crucifixion.

In 1389, Pierre d'Arcis, the Bishop of Troyes, wrote a memo to the Pope, accusing a colleague of passing off "a certain cloth, cunningly painted" as the burial shroud of Jesus Christ. Despite this early testimony of forgery, this so-called Shroud of Turin has survived as a famous relic. In 1988, a small sample of the Shroud of Turin was taken and scientists from Oxford University, the University of Arizona, and the Swiss Federal Institute of Technology were permitted to test it. It was determined that 92.3% of the Shroud's original ^{14}C still remained. According to this information, how old was the Shroud in 1988?

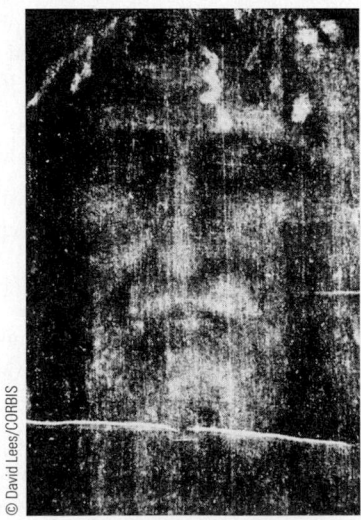

Figure 8.2 **Shroud of Turin**

58. Construct a logarithmic scale. On a sheet of paper, draw a line 10 in. long and mark it in tenths and hundredths. Place the edge of a second sheet along this line and use the first as a ruler to mark off logs on the second. At 0, mark 1, since log 1 = 0. Mark 2 at 0.301 because log 2 ≈ 0.301. Mark 3 at 0.477, and so on until you have something similar to Figure 8.3.

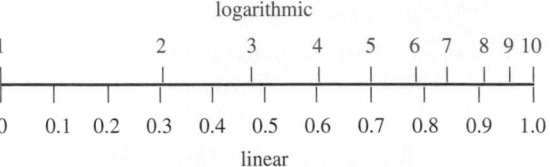

Figure 8.3 **Logarithmic scale**

Fill in the scale between 1 and 2 for every tenth, and the others at least at the halves. Using two of these log scales, you can make a simple slide rule that you can use to multiply and divide. Explain how to do this and illustrate it to at least one other person.

59. In 1986, it was determined that the *Challenger* disaster was caused by failure of the primary O-rings. Linda Tappin gives a formula in "Analyzing Data Relating to the *Challenger* Disaster," *The Mathematics Teacher*, Vol. 87, No. 6 (Sept. 1994, pp. 423–426) that relates the temperature x (in degrees Fahrenheit) around the O-rings and the expected number y of eroded or leaky primary O-rings:

$$y = \frac{6e^{5.085-0.1156x}}{1 + e^{5.085-0.1156x}}$$

a. What is the predicted number of eroded or leaky O-rings at a temperature of 75°F?

b. What is the predicted number of eroded or leaky O-rings at a temperature of 32°F?

60. The Arrhenius function is used to relate the viscosity η of a fluid (the fluid's internal friction, which is what makes it resist a tendency to flow) to its absolute temperature T:

$$\frac{1}{\eta} = Ae^{-E/(RT)}$$

where A is a constant specific to that fluid and R is the ideal gas constant. Solve this equation for T. The resulting formula is one you could use to investigate the viscosity of different grades of motor oil at different temperatures.

CHAPTER SUMMARY

Mathematics abounds in bright ideas. No matter how long and hard one pursues her, mathematics never seems to run out of exciting surprises."

Ross Honsberger

IMPORTANT TERMS

Addition law of logarithms [8.2]
Argument [8.1]
Change of base theorem [8.1]
Common logarithm [8.1]
Decay formula [8.3]
Decibel [8.3]
Evaluate [8.1]
Exact solution [8.1]

Exponential [8.1]
Exponential equation [8.1]
Grant's tomb properties [8.2]
Growth formula [8.3]
Half-life [8.3]
Laws of logarithms [8.2]
Log of both sides theorem [8.2]
Logarithm [8.1]

Logarithmic equation [8.2]
Logarithmic scale [8.3]
Micrometer [8.1, 8.3]
Multiplication law of logarithms [8.2]
Natural logarithm [8.1]
Richter number [8.3]
Richter scale [8.3]
Subtraction law of logarithms [8.2]

TYPES OF PROBLEMS

Know the definition of a logarithm. [8.1]
Evaluate logarithms. [8.1]
Use the Grant's tomb properties to simplify logarithmic expressions. [8.1, 8.2]
Solve exponential equations. [8.1]
Solve logarithmic equations. [8.2]
Solve applied problems of growth and decay. [8.3]

CHAPTER 8 REVIEW QUESTIONS

Simplify each expression in Problems 1–5 without using a calculator or computer.

1. $\log 100 + \log \sqrt{10}$
2. $\ln e + \ln 1 + \ln e^{542}$
3. $\log_8 4 + \log_8 16 + \log_8 8^{2.3}$
4. $10^{\log 0.5}$
5. $\ln e^{\log 1,000}$

Evaluate each expression in Problems 6–8 rounded to two decimal places.

6. **a.** $\log 8.43$ **b.** $\log 9,760$
7. **a.** $\ln 2$ **b.** $\ln 0.125$
8. **a.** $\log_2 10$ **b.** $\log_\pi \dfrac{1}{\pi}$

Solve each equation in Problems 9–12 rounded to calculator accuracy.

9. $10^x = 85$ 10. $e^x = 500$
11. $435^x = 890$ 12. $e^{3x+1} = 45$

Give the exact solution for each equation in Problems 13–17.

13. $\log_6 x = 4$

14. $2^{3x-1} = 6$

15. $10^{2x} = 5$

16. $\log(x + 1) = 2 + \log(x - 1)$

17. $3 \ln \dfrac{e}{\sqrt[3]{5}} = 3 - \ln x$

18. Solve $A = P(1 + i)^x$ for x.

19. A healing law for skin wounds states that $A = A_0 e^{-0.1t}$, where A is the number of square centimeters of unhealed skin after t days when the original area of the wound was A_0. How many days does it take for half the wound to heal?

20. In 1992, it was reported that the number of teenagers with AIDS doubles every 14 months. Find an equation to model the number of teenagers that may be infected over the next 10 years.

GROUP RESEARCH

Working in small groups is typical of most work environments, and being able to work with others to communicate specific ideas is an important skill to learn. Work with three or four other students to submit a single report based on each of the following questions.

G29. The entrance of the Aquarium of Americas in New Orleans has a gigantic building-size curve called a *logarithmic spiral*. Find out how to construct a logarithmic spiral, and write a paper about what you learned. Why do you suppose it would appear on the front of an aquarium?

Karl J. Smith

G30. If we assume that the world population grows exponentially, then it is also reasonable to assume that the use of some nonrenewable resource (such as petroleum) will also grow exponentially. In calculus it is shown that for some constant k, under these assumptions, the formula for the amount of the resource, A, consumed from time $t = 0$ to $t = T$ is given by the formula

$$A = \frac{A_0}{k}(e^{rT} - 1)$$

where r is the relative growth rate of annual consumption.

a. Solve this equation for T to find a formula for life expectancy of a particular resource.

b. According to the Energy Information Administration, the annual world production (in billions of barrels per day) of petroleum is shown in the following table:

Year:	1975	1980	1985	1990	1995	2000	2003
Quantity:	52.42	62.39	52.97	60.90	61.85	66.03	67.00

Find an exponential equation for these data.

c. If in 1998, the world petroleum reserves are 2.8 trillion barrels, estimate the life expectancy for petroleum.

INDIVIDUAL RESEARCH PROBLEMS

Learning to use sources outside your classroom and textbook is an important skill, and here are some ideas for extending some of the ideas in this chapter. You can find references to these projects in a library or at

www.mathnature.com

PROJECT 8.1 HISTORICAL QUESTION Write an essay on John Napier. Include what he is famous for today, and what he considered to be his crowning achievement. Also include a discussion of "Napier's bones."

PROJECT 8.2 Write an essay on earthquakes. In particular, discuss the Richter scale for measuring earthquakes. What is its relationship to logarithms?

PROJECT 8.3 Population analysis. See the Web pages for Section 8.3 for this individual project.

PROJECT 8.4 From your local chamber of commerce, obtain the population figures for your city for the years 1980, 1990, and 2000. Find the rate of growth for each period. Forecast the population of your city for the year 2010. Include charts and graphs. List some factors, such as new zoning laws, that could change the growth rate of your city.

PROJECT 8.5 Write an essay on carbon-14 dating. What is its relationship to logarithms?

9 The Nature of Financial Management

Finance is the art of passing money from hand to hand until it finally disappears.
—ROBERT W. SARNOFF

CONTENTS

OVERVIEW

The stated goal of this book is to strengthen your ability to solve problems—not the classroom type of problems, but those problems that you may encounter as an employee, a manager, or in everyday living. You can apply your problem-solving ability to your financial life. A goal of this chapter might well be to put some money into your bank account that you would not have had if you had not read this chapter. As a preview to this chapter, consider the question asked in Example 2 of Section 9.5:

> Suppose you are 21 years old and will make monthly deposits to a bank account paying 10% annual interest compounded monthly. Which is the better option?
> *Option I:* Pay yourself $200 per month for 5 years and then leave the balance in the bank until age 65. (Total amount of deposits is $200 × 5 × 12 = $12,000.)
> *Option II:* Wait until you are 40 years old (the age most of us start thinking seriously about retirement) and then deposit $200 per month until age 65. (Total amount of deposits is $200 × 25 × 12 = $60,000.)

Warning!! The wrong answer to this question could cost you $4,000/mo for the rest of your life!

IMPORTANT IDEAS

Future value formula for compound interest. [9.1]
Simple interest formula. [9.1]
Distinguish between ordinary and exact interest. [9.1]
Know the variables used with interest formulas. [9.1]
Calculate credit card interest. [9.2]
Distinguish between sequences and series [9.3, 9.4]
Distinguish arithmetic, geometric, and Fibonacci sequences. [9.3, Table 9.1]
Distinguish between add-on interest and amortization. [9.2, 9.6]
Distinguish among the various financial problems. [9.7]

BOOK REPORTS

Write a 500-word report on this book:
What Are Numbers?, (Glenview, IL: Scott, Foresman and Company, 1969) by Louis Auslander. Here is a problem from this book (p. 112): "Suppose a ball has the property that whenever it is dropped, it bounces up 1/2 the distance that it fell. For instance, if we drop it from 6 feet, it bounces up 3 feet. Let us hold the ball 6 feet off the ground and drop it. How far will the ball go before coming to rest?" Include the solution of this problem as part of your report.

LINKS

Individual Projects
Group Projects
Research Links

www.mathnature.com

9.1 INTEREST

Amount of Simple Interest

Certain arithmetic skills enable us to make intelligent decisions about how we spend the money we earn. One of the most fundamental mathematical concepts that consumers, as well as business people, must understand is *interest.* Simply stated, **interest** is money paid for the use of money. We receive interest when we let others use our money (when we deposit money in a savings account, for example), and we pay interest when we use the money of others (for example, when we borrow from a bank).

The amount of the deposit or loan is called the **principal** or **present value,** and the interest is stated as a percent of the principal, called the **interest rate.** The **time** is the length of time for which the money is borrowed or lent. The interest rate is usually an *annual interest rate,* and the time is stated in years unless otherwise given. These variables are related in what is known as the **simple interest formula.**

Simple Interest Formula

Remember this fundamental formula.

$$\text{INTEREST} = \text{PRESENT VALUE} \times \text{RATE} \times \text{TIME}$$

$$I = Prt$$

$I = $ AMOUNT OF INTEREST
$P = $ PRESENT VALUE (or PRINCIPAL)
$r = $ ANNUAL INTEREST RATE
$t = $ TIME (in years)

Suppose you save 20¢ per day, but only for a year. At the end of a year you will have saved $73. If you then put the money into a savings account paying 3.5% interest, how much interest will the bank pay you after one year? The present value (P) is $73, the rate ($r$) is 3.5% = 0.035, and the time (t, in years) is 1. Therefore,

$$I = Prt$$
$$= 73(0.035)(1)$$
$$= 2.555 \qquad \text{You can do this computation on a calculator:} \quad \boxed{73} \boxed{\times} \boxed{.035} \boxed{=}$$

Round money answers to the nearest cent: After one year, the interest is $2.56.

EXAMPLE 1

How much interest will you earn in three years with an initial deposit of $73?

Solution $I = Prt = 73(0.035)(3) = 7.665.$ After three years, the interest is $7.67. ◆

Future Value

There is a difference between asking for the amount of interest, as illustrated in Example 1, and asking for the **future value.** The future amount is the amount you will have after the interest is added to the principal, or present value. Let $A = $ FUTURE VALUE. Then

$$A = P + I$$

EXAMPLE 2

Suppose you see a car with a price of $12,436 that is advertised at $290 per month for 5 years. What is the amount of interest paid?

Solution   The present value is $12,436. The future value is the total amount of all the payments:

Monthly payment   Number of years
↓
$290   × 12 × 5   = $17,400
Number of payments per year

Therefore, the amount of interest is

$$I = A - P$$
$$= 17,400 - 12,436$$
$$= 4,964$$

The amount of interest is $4,964.   ◆

Interest for Part of a Year

The numbers in Example 2 were constructed to give a "nice" answer, but the length of time for an investment is not always a whole number of years. There are two ways to convert a number of days into a year:

Exact interest: 365 days per year
Ordinary interest: 360 days per year

Most applications and businesses use ordinary interest. So in this book, unless it is otherwise stated, assume ordinary interest; that is, use 360 for the number of days in a year:

Time in Ordinary Interest Formula

$$t = \frac{\text{ACTUAL NUMBER OF DAYS}}{360}$$

Calculating the time using ordinary interest often requires a calculator, as illustrated by Example 3.

EXAMPLE 3

Suppose you want to save $3,650, and put $3,000 in the bank at 8% simple interest. How long must you wait?

Solution

$$I = Prt$$
$$650 = 3,000(0.08)t \quad I = 3650 - 3000 = 650; P = 3,000; r = 0.08$$
$$650 = 240t$$
$$\frac{650}{240} = t$$

This is 2 years plus some part of a year. To change the fractional part to days, do not clear your calculator, but subtract 2 (the number of years); then multiply the fractional part by 360:

650 ÷ 240 − 2 = × 360 =   *Display:*   255

The time is 2 years 255 days.   ◆

EXAMPLE 4

Suppose that you borrow $1,200 on March 25 at 21% simple interest. How much interest accrues by September 15 (174 days later)? What is the total amount that must be repaid?

Solution We are given $P = 1{,}200$, $r = 0.21$, and

$$t = \frac{174}{360} \quad \begin{array}{l} \leftarrow \text{Actual number of days} \\ \leftarrow \text{Assume ordinary interest} \end{array}$$

$$I = Prt$$

$$= 1{,}200(0.21)\left(\frac{174}{360}\right) \quad \text{Substitute known values.}$$

$$= 121.8 \quad \text{Use a calculator to do the arithmetic.}$$

The amount of interest is $121.80. To find the amount that must be repaid, find the future value:

$$A = P + I = 1{,}200 + 121.80 = 1{,}321.80$$

The amount that must be repaid is $1,321.80. ◆

It is worthwhile to derive a formula for future value because sometimes we will not calculate the interest separately as we did in Example 4.

$$\text{FUTURE VALUE} = \text{PRESENT VALUE} + \text{INTEREST}$$

$$A = P + I$$

$$= P + Prt \quad \text{Substitute } I = Prt.$$

$$= P(1 + rt) \quad \text{Distributive property}$$

Future Value Formula (Simple Interest)

$$A = P(1 + rt)$$

STOP This is also known as the present value formula for simple interest when solved for the variable P.

EXAMPLE 5

If $10,000 is deposited in an account earning $5\frac{3}{4}\%$ simple interest, what is the future value in 5 years?

Solution We identify $P = 10{,}000$, $r = 0.0575$, and $t = 5$.

$$A = P(1 + rt)$$

$$= 10{,}000(1 + 0.0575 \times 5)$$

$$= 10{,}000(1 + 0.2875) \quad \begin{array}{l} \text{Don't forget order of} \\ \text{operations: multiplication} \\ \text{first.} \end{array}$$

$$= 10{,}000(1.2875)$$

$$= 12{,}875$$

The future value in 5 years is $12,875. ◆

EXAMPLE 6

Suppose you have decided that you will need $4,000 per month on which to live in retirement. If the rate of interest is 8%, how much must you have in the bank when you retire so that you can live on interest only?

Solution We are given $I = 4{,}000$, $r = 0.08$, and $t = \frac{1}{12}$ (one month $= \frac{1}{12}$ year):

$$I = Prt$$

$$4{,}000 = P(0.08)\left(\tfrac{1}{12}\right) \quad \text{Substitute.}$$

$$48{,}000 = (0.08)P \quad \text{Multiply both sides by 12.}$$

$$600{,}000 = P \quad \text{Divide both sides by 0.08.}$$

You must have $600,000 on deposit to earn $4,000 per month at 8%. ◆

The boy that by addition grows
And suffers no subtraction
Who multiplies the thing he
  knows
And carries every fraction
Who well divides the precious
  time
The due proportion given
To sure success aloft will climb
Interest compound receiving.

Compounding Interest

Most banks do not pay interest according to the simple interest formula; instead, after some period of time, they add the interest to the principal and then pay interest on this new, larger amount. When this is done, it is called **compound interest.**

EXAMPLE 7

Compare simple and compound interest for a $1,000 deposit at 8% interest for 3 years.

Solution   First, calculate the future value using simple interest:

$$A = P(1 + rt)$$
$$= 1,000(1 + 0.08 \times 3) \quad \text{\footnotesize Substitute known values.}$$
$$= 1,000(1.24) \quad \text{\footnotesize Order of operations: multiplication first}$$
$$= 1,240$$

With simple interest, the future value in 3 years is $1,240.

Next, assume that the interest is **compounded annually.** This means that the interest is added to the principal after 1 year has passed. This new amount then becomes the principal for the following year. Since the time period for each calculation is 1 year, we let $t = 1$ for each calculation.

First year $(t = 1)$: $A = P(1 + r)$
$$= 1,000(1 + 0.08)$$
$$= 1,080$$
$$\downarrow$$
Second year $(t - 1)$. $A = P(1 + r)$ \quad \text{\footnotesize One year's principal is previous year's balance.}
$$= 1,080(1 + 0.08)$$
$$= 1,166.40$$
$$\downarrow$$
Third year $(t = 1)$: $A = P(1 + r)$
$$= 1,166.40(1 + 0.08)$$
$$= 1,259.71$$

With interest compounded annually, the future value in 3 years is $1,259.71. The earnings from compounding are $19.71 more than from simple interest.   ◆

The problem with compound interest relates to the difficulty of calculating it. Notice that, to simplify the calculations in Example 7, the variable representing time t was given the value 1, and the process was repeated three times. Also notice that, after the future value was found, it was used as the principal in the next step. What if we wanted to compound annually for 20 years instead of for 3 years? Look at Example 7 to discover the following pattern:

Simple interest $A = P(1 + rt)$
(20 years)
$$= 1,000(1 + 0.08 \times 20)$$
$$= 1,000(1 + 1.6)$$
$$= 1,000(2.6)$$
$$= 2,600$$

> Money makes money, and the money that money makes makes more money.
>
> Benjamin Franklin

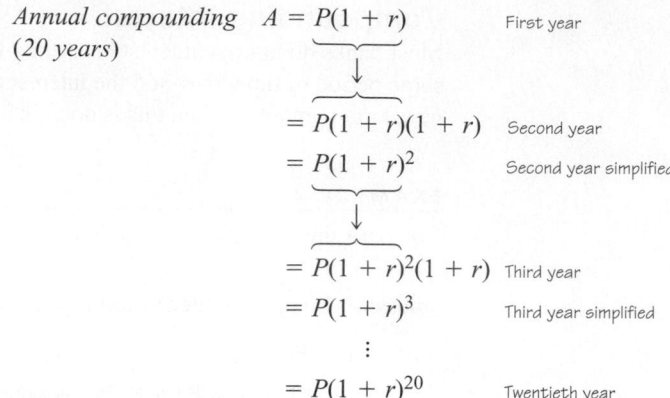

Annual compounding $A = P(1 + r)$ First year
(20 years)

$$= P(1 + r)(1 + r)$$ Second year

$$= P(1 + r)^2$$ Second year simplified

$$= P(1 + r)^2(1 + r)$$ Third year

$$= P(1 + r)^3$$ Third year simplified

$$\vdots$$

$$= P(1 + r)^{20}$$ Twentieth year

For a period of 20 years, starting with \$1,000 at 8% compounded annually, we have

$$A = 1{,}000(1.08)^{20}$$

The difficulty lies in calculating this number. For years we relied on extensive tables for obtaining numbers such as this, but the availability of calculators has made such calculations accessible to all. You will need an exponent key. These are labeled in different ways, depending on the brand of calculator. It might be $\boxed{y^x}$ or $\boxed{x^y}$ or $\boxed{\wedge}$. In this book we will show exponents by using $\boxed{y^x}$, but you should press the appropriate key on your own brand of calculator.

STOP You will need to know how to do this on your calculator to be able to successfully complete this chapter. Use this as a test problem:

$\boxed{1000}$ $\boxed{\times}$ $\boxed{1.08}$ $\boxed{y^x}$ $\boxed{20}$ $\boxed{=}$ *Display:*   4660.957144

CAUTION Note rounding rule for this chapter.

Round money answers to the nearest cent: \$4,660.96 is the future value of \$1,000 compounded annually at 8% for 20 years. This compounding yields \$2,060.96 *more* than simple interest.

Most banks compound interest more frequently than once a year. For instance, a bank may pay interest as follows:

Semiannually: twice a year or every 180 days
Quarterly: 4 times a year or every 90 days
Monthly: 12 times a year or every 30 days
Daily: 360 times a year

If we repeat the same step for more frequent intervals than annual compounding, we again begin with the simple interest formula $A = P(1 + rt)$.

Semiannually, then $t = \frac{1}{2}$: $A = P(1 + r \cdot \frac{1}{2})$

Quarterly, then $t = \frac{1}{4}$: $A = P(1 + r \cdot \frac{1}{4})$

Monthly, then $t = \frac{1}{12}$: $A = P(1 + r \cdot \frac{1}{12})$

Daily, then $t = \frac{1}{360}$: $A = P(1 + r \cdot \frac{1}{360})$

We now compound for t years, and introduce a new variable, n, as follows:

Annual compounding, $n = 1$: $A = P(1 + r)^t$

Semiannual compounding, $n = 2$: $A = P(1 + r \cdot \frac{1}{2})^{2t}$

Quarterly compounding, $n = 4$: $A = P(1 + r \cdot \frac{1}{4})^{4t}$

Monthly compounding, $n = 12$: $A = P(1 + r \cdot \frac{1}{12})^{12t}$

Daily compounding, $n = 360$: $A = P(1 + r \cdot \frac{1}{360})^{360t}$

We are now ready to state the future value formula for compound interest, which is sometimes called the **compound interest formula.** For these calculations you will need access to a calculator with an exponent key.

**Future Value Formula
(Compound Interest)**

$$A = P\left(1 + \frac{r}{n}\right)^{nt}$$

The variables we use in this formula are presented in a separate box because these variables will be used throughout this chapter. We contrast simple and compound interest in the same summary box.

**Contrast Simple and
Compound Interest**

**Variables Used with
Interest Formulas**

STOP Spend some time here; you will need to remember what these variables represent. These are the variables used in this chapter.

If interest is withdrawn from the amount in the account, use the *simple interest formula*. If the interest is deposited into the account to accrue future interest, then use the *compound interest formula*. The variables given below are used with *all* the interest formulas in this chapter.

A = FUTURE VALUE   This is the principal plus interest.

P = PRESENT VALUE   This is the same as the principal.

r = INTEREST RATE   This is the *annual* interest rate.

t = TIME   This is the time in *years*.

n = NUMBER OF COMPOUNDING PERIODS EACH YEAR

m = periodic payment (usually monthly)

EXAMPLE 8

Find the future value of $1,000 invested for 10 years at 8% interest

a. compounded annually.

b. compounded semiannually.

c. compounded quarterly.

d. compounded daily.

Solution   Identify the variables: $P = 1,000, r = 0.08, t = 10$.

a. $n = 1$: $A = \$1000(1 + 0.08)^{10} = \$2,158.92$

b. $n = 2$: $A = \$1,000\left(1 + \frac{0.08}{2}\right)^{2\cdot10} = \$2,191.12$

c. $n = 4$: $A = \$1,000\left(1 + \frac{0.08}{4}\right)^{4\cdot10} = \$2,208.04$

d. $n = 360$: $A = \$1,000\left(1 + \frac{0.08}{360}\right)^{360\cdot10} = \$2,225.34$ ◆

Continuous Compounding

A reasonable extension of the current discussion is to ask the effect of more frequent compounding. To model this situation, consider the following contrived example. Suppose $1 is invested at 100% interest for 1 year compounded at different intervals. The compound interest formula for this example is

$$A = \left(1 + \frac{1}{n}\right)^n$$

where n is the number of times of compounding in 1 year. The calculations of this formula for different values of n are shown in the following table.

Number of Periods	Formula	Amount
Annual, $n = 1$	$\left(1 + \dfrac{1}{1}\right)^1$	$2.00
Semiannual, $n = 2$	$\left(1 + \dfrac{1}{2}\right)^2$	$2.25
Quarterly, $n = 4$	$\left(1 + \dfrac{1}{4}\right)^4$	$2.44
Monthly, $n = 12$	$\left(1 + \dfrac{1}{12}\right)^{12}$	$2.61
Daily, $n = 360$	$\left(1 + \dfrac{1}{360}\right)^{360}$	$2.71

Looking only at this table, you might (incorrectly) conclude that as the number of times the investment is compounded increases, the amount of the investment increases without bound. Let us continue these calculations for even larger n:

$n = 8{,}640$ (compounding every hour):	2.718124536
$n = 518{,}400$ (every minute):	2.718279142
$n = 1{,}000{,}000$	2.718280469
$n = 10{,}000{,}000$	2.718281693
$n = 100{,}000{,}000$	2.718281815

The spreadsheet we are using for these calculations can no longer distinguish the values of $(1 + 1/n)^n$ for larger n. These values are approaching a particular number. This number, it turns out, is an irrational number, and it does not have a convenient decimal representation. (That is, its decimal representation does not terminate and does not repeat.) Mathematicians, therefore, have agreed to denote this number by using the symbol e. This number is called the **natural base** or **Euler's number.**

The Number e

Remember that $e \approx 2.72$.

STOP

> As n increases without bound, the **number e** is the irrational number that is the limiting value of the formula
>
> $$\left(1 + \frac{1}{n}\right)^n$$

In Section 4.5, we noted that the number e is irrational, and consequently does not have a terminating or repeating decimal representation. This same irrational number was used extensively in Chapter 8 as the base number in growth/decay applications as well as in evaluating natural logarithms. Nevertheless, you must wait until you discuss the concept of limit in calculus for a formal definition of e, but the preceding discussion should be enough to convince you that

$$e \approx \mathbf{2.7183}$$

Even though you will not find a bank that compounds interest every minute, you will find banks that use this limiting value to compound **continuously.** When using this model, we assume the year has 365 days.

Future Value Formula (continuous compounding)

> The future value, A, of an investment of P, *compounded continuously* at a rate of r for t years, is found by
>
> $$A = Pe^{rt}$$

EXAMPLE 9

Find the future value of $890 invested at 21.3% for 3 years, 240 days, compounded continuously.

Solution We use the formula $A = Pe^{rt}$ where $P = 890$, $r = 0.213$. For continuous compounding, use a 365-day year, so

$$3 \text{ years, } 240 \text{ days} = 3 + \frac{240}{365} = 3.657534247 \text{ years}$$

Remember that t is in years, and also remember to use this calculator value, and not a rounded value.

$$A = 890e^{0.213t} \approx 1,939.676057$$

The future value is $1,939.68.   ◆

EXAMPLE 10

Let P dollars be invested at an annual rate of r for t years. Then the future value A depends on the number of times the money is compounded each year. How long will it take for $1,250 to grow to $2,000 if it is invested at

a. 8% compounded daily? Use exact interest; that is, let $n = 365$.

$$\text{Use the formula } A = P\left(1 + \frac{r}{365}\right)^{365t}.$$

b. 8% compounded continuously? Use the formula $A = Pe^{rt}$.

Give your answer to the nearest day (assume that one year is 365 days).

Solution

a. $P = \$1,250$, $A = \$2,000$, $r = 0.08$, and t is the unknown.

$$A = P\left(1 + \frac{r}{365}\right)^{365t} \qquad \text{Given formula}$$

$$2,000 = 1,250\left(1 + \frac{0.08}{365}\right)^{365t} \qquad \text{Substitute known values.}$$

$$1.6 = \left(1 + \frac{0.08}{365}\right)^{365t} \qquad \text{Divide both sides by 1,250.}$$

$$365t = \log_{(1+0.08/365)}1.6 \qquad \text{Definition of logarithm.}$$

$$t = \frac{\log_{(1+0.08/365)}1.6}{365} \qquad \text{Divide both sides by 365.}$$
$$\text{Evaluate as } \frac{\log 1.6}{\log(1+0.08/365)} \div 365.$$

$$\approx 5.875689182$$

Approximate solution

We see that it is almost 6 years, but we need the answer to the nearest day. The time is 5 years + 0.875689182 year. Multiply 0.875689182 by 365 to find 319.6265516. This means that on the 319th day of the 6th year, we are still a bit short of $2,000, so the time necessary is 5 years 320 days.

b. For continuous compounding, use the formula $A = Pe^{rt}$.

$$A = Pe^{rt} \qquad \text{\textit{Given}}$$
$$2,000 = 1,250e^{0.08t} \qquad \text{\textit{Substitute known values.}}$$
$$1.6 = e^{0.08t} \qquad \text{\textit{Divide both sides by 1,250.}}$$
$$0.08t = \ln 1.6 \qquad \text{\textit{Definition of logarithm}}$$
$$t = \frac{\ln 1.6}{0.08} \qquad \text{\textit{Divide both sides by 0.08; this is the exact solution.}}$$
$$\approx 5.875045366 \qquad \text{\textit{Approximate solution}}$$

To find the number of days, we once again subtract 5 and multiply by 365 to find that the time necessary is 5 years 320 days. (*Note:* 319.39 means that on the 319th day, you do not quite have the $2,000. The necessary time is 5 years 320 days.) Notice that for this example, it does not matter whether we compound daily or continuously. ◆

Inflation

Any discussion of compound interest is incomplete without a discussion of **inflation.** The same procedure we used to calculate compound interest can be used to calculate the effects of inflation. The government releases reports of monthly and annual inflation rates. In 1981 the inflation rate was nearly 9%, but in 2002 it was less than 3%. Keep in mind that inflation rates can vary tremendously and that the best we can do in this section is to assume different constant inflation rates. For our purposes in this book, we will assume continuous compounding when working inflation problems.

EXAMPLE 11

If your salary today is $35,000 per year, what would you expect your salary to be in 20 years (rounded to the nearest thousand dollars) if you assume that inflation will continue at a constant rate of 6% over that time period?

Solution Inflation is an example of continuous compounding. The problem with estimating inflation is that you must "guess" the value of future inflation, which in reality does not remain constant. However, if you look back 20 years and use an average inflation rate for the past 20 years—say, 6%—you may use this as a reasonable estimate for the next 20 years. Thus, you may use $P = 35,000$, $r = 0.06$, and $t = 20$ to find

$$A = Pe^{rt} = 35,000e^{0.06(20)}$$

Note: Be sure to use parentheses for the exponent:

$$\boxed{35,000} \times \boxed{e^\wedge} \boxed{(} \boxed{.06} \times \boxed{20} \boxed{)} \boxed{=} \qquad \textit{Display:} \quad 116204.0923$$

The answer means that, if inflation continues at a constant 6% rate, an annual salary of $116,000 will have about the same purchasing power in 20 years as a salary of $35,000 today. ◆

Present Value

Sometimes we know the future value of an investment and wish to know its present value. Such a problem is called a *present value problem*. The formula follows directly from the future value formula (by division).

Present Value Formula

$$P = A \div \left(1 + \frac{r}{n}\right)^{nt} = A\left(1 + \frac{r}{n}\right)^{-nt}$$

Monday, March 1, 1982	
6-mo. adjustable rate	**17.5%**

Tuesday, March 1, 1983	
6-mo. adjustable rate	**12.5%**

Friday, March 1, 1985	
3-mo. adjustable rate	**10.5%**

Monday, March 1, 1993	
3 mo. adjustable rate	**4.5%**

Tuesday, March 1, 1994	
3-mo. adjustable rate	**3.0%**

Friday, March 1, 1996	
3-mo. adjustable rate	**4.9%**

Wednesday, March 1, 2000	
3-mo. adjustable rate	**5.2%**

Saturday, March 1, 2003	
6-mo. adjustable rate	**1.6%**

EXAMPLE 12

Suppose that you want to take a trip to Tahiti in 5 years and you decide that you will need \$5,000. To have that much money set aside in 5 years, how much money should you deposit now into a bank account paying 6% compounded quarterly?

Solution In this problem, P is unknown and A is given: $A = 5,000$. We also have $r = 0.06$, $t = 5$, and $n = 4$. Calculate:

$$P = A\left(1 + \frac{r}{n}\right)^{-nt} = 5,000\left(1 + \frac{0.06}{4}\right)^{-20}$$
$$= \$3,712.35 \qquad \blacklozenge$$

EXAMPLE 13

An insurance agent wishes to sell you a policy that will pay you \$100,000 in 30 years. What is the value of this policy in today's dollars, if we assume a 9% inflation rate?

Solution This is a present value problem for which $A = 100,000$, $r = 0.09$, $n = 1$, and $t = 30$. To find the present value, calculate

$$P = 100,000(1 + 0.09)^{-30} = \$7,537.11$$

This means that the agent is offering you an amount comparable to \$7,537.11 in terms of today's dollars. $\qquad \blacklozenge$

EXAMPLE 14   PÓLYA'S METHOD

Your first child has just been born. You want to give her 1 million dollars when she retires at age 65. If you invest your money on the day of her birth, how much do you need to invest so that she will have \$1,000,000 on her 65th birthday?

Solution We use Pólya's problem-solving guidelines for this example.

Understand the Problem. You want to make a single deposit and let it be compounded for 65 years, so that at the end of that time there will be 1 million dollars. Neither the rate of return nor the compounding period is specified. We assume daily compounding, a constant rate of return over the time period, and need to determine whether we can find an investment to meet our goals.

Devise a Plan. With daily compounding, $n = 360$. We will use the present value formula, and experiment with different interest rates:

$$P = 1,000,000\left(1 + \frac{r}{360}\right)^{-(360 \cdot 65)}$$

Carry Out the Plan.

Interest Rate	Formula	Value of P
2%	$1,000,000\left(1 + \dfrac{0.02}{360}\right)^{-(360 \cdot 65)}$	$\approx 272,541.63$
5%	$1,000,000\left(1 + \dfrac{0.05}{360}\right)^{-(360 \cdot 65)}$	$\approx 38,782.96$
8%	$1,000,000\left(1 + \dfrac{0.08}{360}\right)^{-(360 \cdot 65)}$	$\approx 5,519.75$
12%	$1,000,000\left(1 + \dfrac{0.12}{360}\right)^{-(360 \cdot 65)}$	≈ 410.27
20%	$1,000,000\left(1 + \dfrac{0.20}{360}\right)^{-(360 \cdot 65)}$	≈ 2.27

Look Back. Look down the list of interest rates and compare the rates with the amount of deposit necessary. Generally, the greater the risk of an investment, the higher the rate. Insured savings accounts may pay lower rates, bonds may pay higher rates for long-term investments, and other investments in stamps, coins, or real estate may pay the highest rates. The amount of the investment necessary to build an estate of 1 million dollars is dramatic! ◆

Problem Set 9.1

LEVEL 1

1. **IN YOUR OWN WORDS** What is interest?

2. **IN YOUR OWN WORDS** Contrast amount of interest and interest rate.

3. **IN YOUR OWN WORDS** Compare and contrast simple and compound interest.

4. **IN YOUR OWN WORDS** Compare and contrast present value and future value.

5. **IN YOUR OWN WORDS** What is the subject of Example 14? Discuss the real-life application of this example in your own life.

Use estimation to select the best response in Problems 6–15. Do not calculate.

6. If you deposit $100 in a bank account for a year, then the amount of interest is likely to be
   A. $2   B. $5   C. $102
   D. impossible to estimate

7. If you deposit $100 in a bank account for a year, then the future value is likely to be
   A. $2   B. $5   C. $102
   D. impossible to estimate

8. If you purchase a new automobile and finance it for four years, the amount of interest you might pay is
   A. $400   B. $100   C. $4,000
   D. impossible to estimate

9. What is a reasonable monthly income when you retire?
   A. $300   B. $10,000   C. $500,000
   D. impossible to estimate

10. In order to retire and live on the interest only, what is a reasonable amount to have in the bank?
    A. $300   B. $10,000   C. $500,000
    D. impossible to estimate

11. If $I = Prt$ and $P = \$49,236.45$, $r = 10.5\%$, and $t = 2$ years, estimate I.
    A. $10,000   B. $600
    C. $50,000   D. $120,000

12. If $I = Prt$ and $I = \$398.90$, $r = 9.85\%$, and $t = 1$ year, estimate P.
    A. $400   B. $40
    C. $40,000   D. $4,000

13. If $t = 3.52895$, then the time is about 3 years and how many days?
    A. 30   B. 300
    C. 52   D. 200

14. If a loan is held for 450 days, then t is about
    A. 450   B. 3
    C. $1\frac{1}{4}$ D. 5

15. If a loan is held for 180 days, then t is about
    A. 180   B. $\frac{1}{2}$
    C. $\frac{1}{4}$ D. 3

In Problems 16–19, calculate the amount of simple interest earned.

16. $1,000 at 8% for 5 years

17. $5,000 at 10% for 3 years

18. $2,000 at 12% for 5 years

19. $1,000 at 14% for 30 years

In Problems 20–25, find the future value, using the future value formula and a calculator.

20. $350 at $4\frac{3}{4}\%$ simple interest for 2 years

21. $835 at 3.5% compounded semiannually for 6 years

22. $575 at 5.5% compounded quarterly for 5 years

23. $9,730.50 at 7.6% compounded monthly for 7 years

24. $45.67 at 3.5% compounded daily for 3 years

25. $119,400 at 7.5% compounded continuously for 30 years

In Problems 26–30, find the present value, using the present value formula and a calculator.

26. Achieve $5,000 in three years at 3.5% simple interest.

27. Achieve $2,500 in five years at 8.2% interest compounded monthly.

28. Achieve $420,000 in 30 years at 6% interest compounded monthly.

29. Achieve a million dollars in 30 years at 6% interest compounded continuously.

30. Achieve $225,500 at 8.65% compounded continuously for 8 years, 135 days.

31. If $12,000 is invested at 4.5% for 20 years, find the future value if the interest is compounded:
    a. annually
    b. semiannually
    c. quarterly
    d. monthly
    e. daily
    f. every minute ($N = 525,600$)
    g. continuously
    h. simple (not compounded)

32. If $34,500 is invested at 6.9% for 30 years, find the future value if the interest is compounded:
    a. annually
    b. semiannually
    c. quarterly
    d. monthly
    e. daily
    f. every minute ($N = 525,600$)
    g. continuously
    h. simple (not compounded)

LEVEL 2

Find the total amount that must be repaid on the notes described in Problems 33–34.

33. $1,500 borrowed at 21% simple interest. What is the total amount to be repaid 55 days later?

34. $8,553 borrowed at 16.5% simple interest. What is the total amount to be repaid 3 years 125 days later?

35. Find the cost of each item in 5 years, assuming an inflation rate of 9%.
    a. cup of coffee, $0.75
    b. Sunday paper, $1.25
    c. Big Mac, $1.95
    d. gallon of gas, $1.55
    e. TV set, $600
    f. small car, $14,000
    g. car, $28,000
    h. tuition, $16,000

36. Find the cost of each item in 10 years, assuming an inflation rate of 5%.
    a. movie admission, $5.00
    b. CD, $14.95
    c. textbook, $60.00
    d. electric bill, $65
    e. phone bill, $45
    f. pair of shoes, $85
    g. new suit, $370
    h. monthly rent, $600

†*In Problems 37–40, calculate the time necessary to achieve an investment goal. Give your answer to the nearest day. Use a 365-day year.*

37. $1,000 at 8% simple interest; deposit $750

38. $3,500 at 6% simple interest; deposit $3,000

39. $5,000 at 5% daily interest; deposit $3,500

40. $5,000 at 4.5% compounded continuously; deposit $3,500

41. How much would you have in 5 years if you purchased a $1,000 5-year savings certificate that paid 4% compounded quarterly?

42. What is the interest on $2,400 for 5 years at 12% compounded continuously?

43. What is the future value after 15 years if you deposit $1,000 for your child's education and the interest is guaranteed at 16% compounded continuously?

44. Suppose you see a car with an advertised price of $18,490 at $480 per month for 5 years. What is the amount of interest paid?

45. Suppose you see a car with an advertised price of $14,500 at $410.83 per month for 4 years. What is the amount of interest paid?

46. Suppose you buy a home and finance $285,000 at $2,293.17 per month for 30 years. What is the amount of interest paid?

47. Suppose you buy a home and finance $170,000 at $1,247.40 per month for 30 years. What is the amount of interest paid?

48. Find the cost of a home in 30 years, assuming an annual inflation rate of 10%, if the present value of the house is $125,000.

49. Find the cost of the monthly rent for a two-bedroom apartment in 30 years, assuming an annual inflation rate of 10%, if the current rent is $650.

50. Suppose that an insurance agent offers you a policy that will provide you with a yearly income of $50,000 in 30 years. What is the comparable salary today, assuming an inflation rate of 6%?

51. If a friend tells you she earned $5,075 interest for the year on a 5-year certificate of deposit paying 5% simple interest, what is the amount of the deposit?

52. If Rita receives $45.33 interest for a deposit earning 3% simple interest for 240 days, what is the amount of her deposit?

†Problems 37–40 require Chapter 8 (solving exponential equations).

53. In 2000, the U.S. national debt was 5.7 trillion dollars.
    a. If this debt is shared equally by the 300 million citizens, how much would it cost each of us?
    b. If the interest rate is 6%, what is the interest on the national debt *each second?* Assume a 365-day year.

 You can check on the current national debt at **http://www.brillig.com/debt_clock/** This link, as usual, can be accessed through **www.mathnature.com**

54. If John wants to retire with $10,000 per month,* how much principal is necessary to generate this amount of monthly income if the interest rate is 15%?

55. If Melissa wants to retire with $50,000 per month,* how much principal is necessary to generate this amount of monthly income if the interest rate is 12%?

56. If Jack wants to retire with $1,000 per month, how much principal is necessary to generate this amount of monthly income if the interest rate is 6%?

LEVEL 3

† 57. Suppose that $1,000 is invested at 7% interest compounded monthly. Use the formula
$$A = P\left(1 + \frac{r}{n}\right)^{nt}$$
    a. How long (to the nearest month) before the value is $1,250?
    b. How long (to the nearest month) before the money doubles?
    c. What is the interest rate (compounded monthly and rounded to the nearest percent) if the money doubles in 5 years?

† 58. Suppose that $1,000 is invested at 5% interest compounded continuously. Use the formula
$$A = Pe^{rt}$$
    a. How long (to the nearest day) before the value is $1,250?
    b. How long (to the nearest day) before the money doubles?
    c. What is the interest rate (compounded continuously and rounded to the nearest tenth of a percent) if the money doubles in 5 years?

---

*You might think these are exorbitant monthly incomes, but if you assume 10% average inflation for 40 years, a monthly income of $220 today will be equivalent to about $10,000 per month in 40 years.

†Problems 57 and 58 require Chapter 8 (solving exponential equations).

PROBLEM SOLVING

59. The News Clip below is typical of what you will see in a newspaper.

> **NEW, HIGHEST INTEREST RATE EVER ON INSURED SAVINGS**
> **8.33%** annual yield on
> **8%** interest compounded daily
> Annual yield based on daily compounding when funds and interest remain on deposit a year. Note: Federal regulations require a substantial interest penalty for early withdrawal of principal from Certificate Accounts.

It gives two rates, the *annual yield* or *effective rate* (8.33%) and a *nominal rate* (8%). Since banks pay interest compounded for different periods (quarterly, monthly, daily, for example), they calculate a rate for which annual compounding will yield the same amount at the end of 1 year. That is, for an 8% rate:

Nominal Rate	*Effective Rate*
8%, annual compounding	8%
8%, semiannual compounding	8.16%
8%, quarterly compounding	8.24%
8%, monthly compounding	8.30%
8%, daily compounding	8.33%

To find a formula for effective rate, we recall that the compound interest formula is
$$A = P\left(1 + \frac{r}{n}\right)^{nt}$$
and the future value formula for simple interest is
$$A = P(1 + Yt)$$
The effective rate, Y, is a rate such that, at the end of one year ($t = 1$), the future value for the simple interest is equal to the future value for the compound interest rate r with n compounding periods. That is,
$$P\left(1 + \frac{r}{n}\right)^n = P(1 + Y)$$
Find a formula for effective (annual) rate, Y, for which the future value for compound interest is equal to the future value for simple interest at the end of 1 year.

60. Find the effective yield for the following investments (see Problem 59). Round to the nearest hundredth of a percent.
    a. 6%, compounded quarterly
    b. 6%, compounded monthly
    c. 4%, compounded semiannually
    d. 4%, compounded daily

9.2 | INSTALLMENT BUYING

Consumer Loans

Two types of consumer credit allow you to make installment purchases. The first, called **closed-ended,** is the traditional installment loan. An **installment loan** is an agreement to pay off a loan or a purchase by making equal payments at regular intervals for some specific period of time. In this book, it is assumed that all installment payments are made monthly.

There are two common ways of calculating installment interest. The first uses simple interest and is called *add-on interest,* and the second uses compound interest and is called *amortization.* We discuss the simple interest application in this section, and the compound interest application in Section 9.6.

In addition to closed-ended credit, it is common to obtain a type of consumer credit called **open-ended, revolving credit,** or more commonly, a **credit card** loan. Master-Card, VISA, and Discover cards, as well as those from department stores and oil companies, are examples of open-ended loans. This type of loan allows for purchases or cash advances up to a specified maximum **line of credit** and has a flexible repayment schedule.

MasterCard is a registered trademark of MasterCard International, Inc.

Add-On Interest

The most common method for calculating interest on installment loans is by a method known as **add-on interest.** It is nothing more than an application of the simple interest formula. It is called *add-on interest* because the interest is *added to* the amount borrowed so that both interest and the amount borrowed are paid for over the length of the loan. You should be familiar with the following variables:

$$P = \text{AMOUNT TO BE FINANCED (present value)}$$
$$r = \text{ADD-ON INTEREST RATE}$$
$$t = \text{TIME (in years) TO REPAY THE LOAN}$$
$$I = \text{AMOUNT OF INTEREST}$$
$$A = \text{AMOUNT TO BE REPAID (future value)}$$
$$m = \text{AMOUNT OF THE MONTHLY PAYMENT}$$
$$N = \text{NUMBER OF PAYMENTS}$$

Installment Loan Formulas

Do you see why this is called add-on interest? Do these formulas make sense to you?

AMOUNT OF INTEREST: $I = Prt$

AMOUNT TO BE REPAID: $A = P + I$ or $A = P(1 + rt)$

NUMBER OF PAYMENTS: $N = 12t$

AMOUNT OF EACH PAYMENT $m = \dfrac{A}{N}$

EXAMPLE 1

You want to purchase a computer that has a price of $1,399, and you decide to pay for it with installments over 3 years. The store tells you that the interest rate is 15%. What is the amount of each monthly payment?

Solution You ask the clerk how the interest is calculated, and you are told that the store uses add-on interest. Thus,

$$P = 1{,}399, \quad r = 0.15, \quad t = 3, \quad N = 36$$

<table>
<tr><td align="center">***Two-Step Solution***</td><td align="center">***One-Step Solution***</td></tr>
<tr><td>$I = Prt = 1{,}399(0.15)(3) = 629.55$</td><td>$A = P(1 + rt)$</td></tr>
<tr><td>$A = P + I = 1{,}399 + 629.55$</td><td>$= 1{,}399(1 + 0.15 \cdot 3)$</td></tr>
<tr><td>$= 2{,}028.55$</td><td>$= 2{,}028.55$</td></tr>
</table>

$$m = \frac{2{,}028.55}{36} \approx 56.35$$

The amount of each monthly payment is $56.35. ◆

The most common applications of installment loans are for the purchase of a car or a home. Interest for purchasing a car is determined by the add-on method, but interest for purchasing a home is not. We will, therefore, delay our discussion of home loans until after we have discussed periodic payments with compound interest. The next example shows a calculation for a car loan.

EXAMPLE 2   PÓLYA'S METHOD

Suppose that you have decided to purchase a Saturn Ion Coupe and want to determine the monthly payment if you pay for the car in 4 years. The value of your trade-in is $4,100.

Solution We use Pólya's problem-solving guidelines for this example.

Understand the Problem. Not enough information is given, so you need to ask some questions of the car dealer:

Sticker price of the car (as posted on the window): $17,865
Dealer's preparation charges (as posted on the window): $350.00
Total asking (sticker) price: $18,215
Tax rate (determined by the state):   7%
Add-on interest rate: 8%
You need to make an offer.

Devise a Plan. The plan is to offer the dealer 5% over dealer's cost. Assuming that the dealer accepts that offer, then we will calculate the monthly payment.

Carry Out the Plan. If you are serious about getting the best price, find out the **dealer's cost**—the price the dealer paid for the car you want to buy. In this book, we will tell you the dealer's cost, but in the real world you will need to do some homework to find it (consult an April issue of *Consumer Reports*). Assume that the dealer's cost for this car is $15,993.45. You decide to offer the dealer 5% *over* this cost. We will call this a **5% offer:**

$$\$15{,}993.45(1 + 0.05) = \$16{,}793.12$$

You will notice that we ignored the sticker price and the dealer's preparation charges. Our offer is based only on the *dealer's cost.* Most car dealers will accept an offer that is between 5% and 10% over what they actually paid for the car. For this example, we will assume that the dealer accepted a price of $16,800. We also assume that we have a trade-in with a value of $4,100.

Here is a list of calculations shown on the sales contract:

Sale price of Saturn:	$16,800.00	Tax (7% rate)	1,190.00
Destination charges:	200.00	Less trade-in	4,100.00
Subtotal:	17,000.00	Amount to be financed:	14,090.00

We now calculate several key amounts:

Interest: $I = Prt = 14{,}090(0.08)(4) = 4{,}508.80$

Amount to be repaid: $A = P + I = 14{,}090 + 4{,}508.80 = 18{,}598.80$

Monthly payment: $m = \dfrac{18{,}598.80}{48} = 387.475$

Look Back. The monthly payment for the car is $387.48. ◆

Annual Percentage Rate (APR)

An important aspect of add-on interest is that the actual rate you pay exceeds the quoted add-on interest rate. The reason for this is that you do not keep the entire amount borrowed for the entire time. For the car payments calculated in Example 2, the principal used was $14,090, but you do not *owe* this entire amount for 4 years. After the first payment, you will owe *less* than this amount. In fact, after you make 47 payments, you owe only $293.32; but the calculation shown in Example 2 assumes that the principal remains constant for 4 years.

To see this a little more clearly, consider a simpler example. Suppose you borrow $2,000 for 2 years with 10% add on-interest. The amount of interest is

$$\$2{,}000 \times 0.10 \times 2 = \$400$$

Now if you pay back $2,000 + $400 at the end of two years, the annual interest rate is 10%. However, if you make a partial payment of $1,200 at the end of the first year and $1,200 at the end of the second year, your total paid back is still the same ($2,400), but you have now paid a higher annual interest rate. Why? Take a look at Figure 9.1. On the left we see the interest on $2,000 is $400. But if you make a partial payment (figure on the right), we see that $200 for the first year is the correct interest, but the remaining $200 interest piled on the remaining balance of $1,000 is 20% interest (not the stated 10%). Note that since you did not owe $2,000 for 2 years, the interest rate, *r*, necessary to give $400 interest can be calculated using $I = Prt$:

$$(2{,}000)r(1) + (1{,}000)r(1) = 400$$

$$3{,}000r = 400$$

$$r = \frac{400}{3{,}000} \approx 0.13333 \quad \text{or} \quad 13.3\%$$

This number, 13.3%, is called the *annual percentage rate*. This number is too difficult to calculate, as we have just done here, if the number of months is very large. We will, instead, use the formula given in the following box.

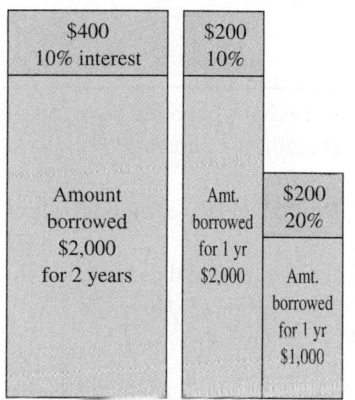

Figure 9.1 **Interest on a $2,000 two-year loan**

APR Formula

> The **annual percentage rate,** or **APR,** is the rate paid on a loan when that rate is based on the actual amount owed for the length of time that it is owed. It can be found for an add-on interest rate, *r,* with *N* payments by using the formula
>
> $$\text{APR} = \frac{2Nr}{N+1}$$

We can verify this formula for the example illustrated in Figure 9.1:

$$\text{APR} = \frac{2(2)(0.10)}{2+1} \approx 0.1333$$

In 1969, a Truth-in-Lending Act was passed by Congress; it requires all lenders to state the true annual interest rate, which is called the *annual percentage rate* (APR) and is based on the actual amount owed. Regardless of the rate quoted, when you ask a salesperson what the APR is, the law requires that you be told this rate. This regulation enables you to compare interest rates *before* you sign a contract, which must state the APR even if you haven't asked for it.

EXAMPLE 3

In Example 1, we considered the purchase of a computer with a price of $1,399, paid for in installments over 3 years at an add-on rate of 15%. What is the APR (rounded to the nearest tenth of a percent)?

Solution Knowing the amount of the purchase is not necessary when finding the APR. We need to know only N and r. Since N is the number of payments, we have $N = 12(3) = 36$, and r is given as 0.15:

$$\text{APR} = \frac{2(36)(0.15)}{36 + 1} \approx 0.292$$

The APR is 29.2%. ◆

EXAMPLE 4

Consider a Blazer with a price of $18,436 that is advertised at a monthly payment of $384.00 for 60 months. What is the APR (to the nearest tenth of a percent)?

Solution We are given $P = 18,436$, $m = 384$, and $N = 60$. The APR formula requires that we know the rate r.

The future value is the total amount to be repaid ($A = P + I$) and the amount of interest is $I = Prt$. Now, $A = 384(60) = 23,040$, so $I = A - P = 23,040 - 18,436 = 4,604$. Since $N = 12t$, we see that $t = 5$ when $N = 60$.

$I = Prt$	Interest formula
$4,604 = 18,436(r)(5)$	Substitute known values.
$920.8 = 18,436r$	Divide both sides by 5.
$0.0499457583 \approx r$	Divide both sides by 18,436.

Finally, for the APR formula, $\text{APR} = \dfrac{2Nr}{N + 1}$,

$$\text{APR} = \frac{2(60)(0.0499457583)}{61} \approx 0.098 \quad \text{Don't round until the last step.}$$

You will really need a calculator for a problem like this one, so we will show you the appropriate steps:

Find A.
$\boxed{384} \times \boxed{60} - \boxed{18436} = \div \boxed{5} \div \boxed{18436} = \times \boxed{2} \times \boxed{60} \div \boxed{61}$
Find I.   Find r.   Find APR.

$=$ *Display:* .0982539508

The APR is 9.8%. ◆

Many automobiles are being offered at 0% interest! Good deal? It seems so, but buyer beware! Consider the following example.

EXAMPLE 5

A local car dealer offered a 2002 Dodge 3/4-ton 4 x 4 pick-up truck with an MSRP of $33,290 less factory and dealer rebates of $5,489 for an "out of door no-haggle price of

$27,801." The advertisement also offered "0% APR for 60 months" and in smaller print "In lieu of rebate." Suppose you can get a 2.5% add-on rate from the credit union. Should you choose the 0% APR or the 2.5% add-on rate? What is the credit union's APR rate?

Solution   The 0% APR would finance $33,290 for 60 months;

$$\frac{\$33,290}{60} = \$554.83/\text{mo}$$

The credit union rate is 2.5% for 60 months, so

$$I = Prt = \$27,801(0.025)(5) = \$3,475.13$$
$$A = P + I = \$31,276.13$$

The monthly payment is

$$\frac{\$31,276.13}{60} = \$521.27$$

If we convert the 2.5% credit union add-on rate to an APR, we find

$$\text{APR} = \frac{2(0.025)(60)}{61} \approx 4.9\%$$

Note that the 0% APR financing is more costly than the credit union's 4.9% APR.   ◆

Open-Ended Credit

The most common type of open-ended credit used today involves credit cards issued by VISA, MasterCard, Discover, American Express, department stores, and oil companies. Because you don't have to apply for credit each time you want to charge an item, this type of credit is very convenient.

When comparing the interest rates on loans, you should use the APR. Earlier, we introduced a formula for add-on interest; but for credit cards, the stated interest rate *is* the APR. However, the APR on credit cards is often stated as a daily or a monthly rate. For credit cards, we use a 365-day year rather than a 360-day year.

EXAMPLE 6

Convert the given credit card rate to APR (rounded to the nearest tenth of a percent).

a. $1\frac{1}{2}\%$ per month   **b.** Daily rate of 0.05753%

Solution

a. Since there are 12 months per year, multiply a monthly rate by 12 to get the APR:

$$1\frac{1}{2}\% \times 12 = 18\% \text{ APR}$$

b. Multiply the daily rate by 365 to obtain the APR:

$$0.05753\% \times 365 = 20.99845\%$$

Rounded to the nearest tenth, this is equivalent to 21.0% APR.   ◆

Many credit cards charge an annual fee; some charge $1 every billing period the card is used, whereas others are free. These charges affect the APR differently, depending on how much the credit card is used during the year and on the monthly balance. If you always pay your credit card bill in full as soon as you receive it, the card with no yearly fee would obviously be the best for you. On the other hand, if you use your credit card to stretch out your payments, the APR is more important than the flat fee. For our purposes, we won't use the yearly fee in our calculations of APR on credit cards.

BANKAMERICARD

5412
4000 1234 3456 7890
VALID THRU 11/97
K J SMITH   **VISA**

VISA is a registered trademark of the Bank of America, N.T. and S.A.

Like annual fees, the interest rates or APRs for credit cards vary greatly. Because VISA and MasterCard are issued by many different banks, the terms can vary greatly even in one locality.

Credit Card Interest

An interest charge added to a consumer account is often called a **finance charge.** The finance charges can vary greatly even on credit cards that show the *same* APR, depending on the way the interest is calculated. There are three generally accepted methods for calculating these charges: *previous balance, adjusted balance,* and *average daily balance.*

Methods of Calculating Interest on Credit Cards

Pay attention! You may have credit card offers and you have a choice of which card you choose.

For credit card interest, use the simple interest formula, $I = Prt$.

Previous balance method: Interest is calculated on the previous month's balance. With this method, P = previous balance, r = annual rate, and $t = \frac{1}{12}$.

Adjusted balance method: Interest is calculated on the previous month's balance *less* credits and payments. With this method, P = adjusted balance, r = annual rate, and $t = \frac{1}{12}$.

Average daily balance method: Add the outstanding balances for *each day* in the billing period, and then divide by the number of days in the billing period to find what is called the *average daily balance.* With this method, P = average daily balance, r = annual rate, and t = number of days in the billing period divided by 365.

In Example 7, we compare the finance charges on a $1,000 credit card purchase, using these three different methods.

EXAMPLE 7

Calculate the interest on a $1,000 credit card bill that shows an 18% APR, assuming that $50 is sent on January 3 and is recorded on January 10. Contrast the three methods for calculating the interest.

Consumer Reports magazine states that, according to one accounting study, the interest costs for average daily balance and previous balance methods are about 16% higher than those for the adjusted balance method.

Solution The three methods are *previous balance method, adjusted balance method,* and *average daily balance method.* All three methods use the formula $I = Prt$.

Method:	*Previous Balance*	*Adjusted Balance*	*Average Daily Balance*
P:	$1,000	$1,000 − $50 = $950	Balance is $1,000 for 10 days of 31-day month; balance is $950 for 21 days of 31-day month; $\dfrac{10 \times \$1,000 + 21 \times \$950}{31}$ = $966.13
r:	0.18	0.18	0.18
t:	$\frac{1}{12}$	$\frac{1}{12}$	$\frac{31}{365}$
$I = Prt$	$1,000(0.18)$(\frac{1}{12})$ = $15.00	$950(0.18)$(\frac{1}{12})$ = $14.25	$966.13(0.18)$(\frac{31}{365})$ = $14.77

◆

You can sometimes make good use of credit cards by taking advantage of the period during which no finance charges are levied. Many credit cards charge no interest if you pay in full within a certain period of time (usually 20 or 30 days). This is called the **grace period.** On the other hand, if you borrow cash on your credit card, you should know that many credit cards have an additional charge for cash advances—and these can be as high as 4%. This 4% is *in addition to* the normal finance charges.

Problem Set 9.2

LEVEL 1

1. **IN YOUR OWN WORDS** What is add-on interest?

2. **IN YOUR OWN WORDS** What is APR?

3. **IN YOUR OWN WORDS** Compare and contrast open-ended and closed-ended credit.

4. **IN YOUR OWN WORDS** Discuss the methods of calculating credit card interest.

5. **IN YOUR OWN WORDS** If you have a credit card, describe the method of calculating interest on your card. Name the bank issuing the credit card. If you do not have a credit card, contact a bank and obtain an application to answer this question. Name the bank.

6. **IN YOUR OWN WORDS** Describe a good procedure for saving money with the purchase of an automobile.

Use estimation to select the best response in Problems 7–24. Do not calculate.

7. If you purchase a $2,400 item and pay for it with monthly installments for 2 years, the monthly payment is
   A. $100 per month
   B. more than $100 per month
   C. less than $100 per month

8. If you purchase a $595.95 item and pay for it with monthly installments for 1 year, the monthly payment is
   A. about $50   B. more than $50
   C. less than $50

9. If I do not pay off my credit card each month, the most important cost factor is
   A. the annual fee   B. the APR
   C. the grace period

10. If I pay off my credit card balance each month, the most important cost factor is
    A. the annual fee
    B. the APR
    C. the grace period

11. The method of calculation most advantageous to the consumer is the
    A. previous balance method
    B. adjusted balance method
    C. average daily balance method

12. If you purchase an item for $1,295 at an interest rate of 9.8%, and you finance it for 1 year, then the amount of add-on interest is about
    A. $13.00   B. $500   C. $130

13. If you purchase an item for $1,295 at an interest rate of 9.8%, and you finance it for 4 years, then the amount of add-on interest is about
    A. $13.00   B. $500   C. $130

14. If you purchase a new car for $10,000 and finance it for 4 years, the amount of interest you would expect to pay is about
    A. $4,000   B. $400   C. $24,000

15. A reasonable APR to pay for a 3-year installment loan is
    A. 1%   B. 12%   C. 32%

16. A reasonable APR to pay for a 3-year automobile loan is
    A. 6%   B. 40%   C. $2,000

17. If you wish to purchase a car with a sticker price of $10,000, a reasonable offer to make to the dealer is:
    A. $10,000   B. $9,000   C. $11,000

18. A reasonable APR for a credit card is
    A. 1%   B. 30%   C. 12%

19. In an application of the average daily balance method for the month of August, t is
    A. $\frac{1}{12}$ B. $\frac{30}{365}$ C. $\frac{31}{365}$

20. When using the average daily balance method for the month of September, t is
    A. $\frac{1}{12}$ B. $\frac{30}{365}$ C. $\frac{31}{365}$

21. If your credit card balance is $650 and the interest rate is 12% APR, then the credit card interest charge is
    A. $6.50   B. $65   C. $8.25

22. If your credit card balance is $952, you make a $50 payment, the APR is 12%, and the interest is calculated according to the previous balance method, then the finance charge is
    A. $9.52   B. $9.02   C. $9.06

23. If your credit card balance is $952, you make a $50 payment, the APR is 12%, and the interest is calculated according to the adjusted balance method, then the finance charge is
    A. $9.52   B. $9.02   C. $9.06

24. If your credit card balance is $952, you make a $50 payment, the APR is 12%, and the interest is calculated according to the average daily balance method, then the finance charge is
    A. $9.52
    B. $9.02
    C. $9.06

*Convert each credit card rate in Problems 25–30 to the APR.**

25. Oregon, $1\frac{1}{4}$% per month

26. Arizona, $1\frac{1}{3}$% per month

27. New York, $1\frac{1}{2}$% per month

28. Tennessee, 0.02740% daily rate

29. Ohio, 0.02192% daily rate

30. Nebraska, 0.03014% daily rate

Calculate the monthly finance charge for each credit card transaction in Problems 31–34. Assume that it takes 10 days for a payment to be received and recorded, and that the month is 30 days long.

31. $300 balance, 18%, $50 payment
    a. previous balance method
    b. adjusted balance method
    c. average daily balance method

32. $300 balance, 18%, $250 payment
    a. previous balance method
    b. adjusted balance method
    c. average daily balance method

33. $3,000 balance, 15%, $50 payment
    a. previous balance method
    b. adjusted balance method
    c. average daily balance method

34. $3,000 balance, 15%, $2,500 payment
    a. previous balance method
    b. adjusted balance method
    c. average daily balance method

LEVEL 2

Round your answers in Problems 35–38 to the nearest dollar.

35. Make a 6% offer on a Chevrolet Corsica that has a sticker price of $14,385 and a dealer cost of $13,378.

36. Make a 5% offer on a Ford Escort that has a sticker price of $13,205 and a dealer cost of $12,412.70.

37. Make a 10% offer on a Saturn that has a sticker price of $19,895 and a dealer cost of $17,250.

38. Make a 10% offer on a Nissan Pathfinder that has a sticker price of $32,129 and a dealer cost of $28,916.

Find the APR (rounded to the nearest tenth of a percent) for each of the loans described in Problems 39–42.

39. Purchase a living room set for $3,600 at 12% add-on interest for 3 years.

40. Purchase a stereo for $2,500 at 13% add-on interest for 2 years.

41. Purchase an oven for $650 at 11% add-on interest for 2 years.

42. Purchase a refrigerator for $2,100 at 14% add-on interest for 3 years.

Assume the cars in Problems 43–46 can be purchased for 0% down for 60 months (in lieu of rebate).
a. *Find the monthly payment if financed for 60 months at 0% APR.*
b. *Find the monthly payment if financed at 2.5% add-on interest for 60 months.*
c. *Find the APR for part b.*
d. *State whether the 0% APR or the 2.5% add-on rate should be preferred.*

43. A Dodge Ram that has a sticker price of $20,650 with factory and dealer rebates of $2,000

44. A BMW that has a sticker price of $62,650 with factory and dealer rebates of $6,000

45. A car with a sticker price of $42,700 with factory and dealer rebates of $5,100

46. A car with a sticker price of $36,500 with factory and dealer rebates of $4,200

LEVEL 3

For each of the car loans described in Problems 47–52, give the following information.
a. *Amount to be paid*
b. *Amount of interest*
c. *Interest rate*
d. *APR (rounded to the nearest tenth percent)*

47. A newspaper advertisement offers a $9,000 car for nothing down and 36 easy monthly payments of $317.50.

48. A newspaper advertisement offers a $4,000 used car for nothing down and 36 easy monthly payments of $141.62.

49. A newspaper advertisement offers a $14,350 car for nothing down and 48 easy monthly payments of $488.40.

50. A car dealer will sell you the $16,450 car of your dreams for $3,290 down and payments of $339.97 per month for 48 months.

51. A car dealer will sell you a used car for $6,798 with $798 down and payments of $168.51 per month for 48 months.

52. A car dealer will sell you the $30,450 car of your dreams for $6,000 down and payments of $662.06 per month for 60 months.

53. A car dealer carries out the following calculations:

List price	$5,368.00
Options	$1,625.00
Destination charges	$200.00
Subtotal	$7,193.00
Tax	$431.58
Less trade-in	$2,932.00
Amount to be financed	$4,692.58
8% interest for 48 months	$1,501.63
Total	$6,194.21
MONTHLY PAYMENT	$129.05

What is the annual percentage rate?

54. A car dealer carries out the following calculations:

List price	$15,428.00
Options	$3,625.00
Destination charges	$350.00
Subtotal	$19,403.00
Tax	$1,164.18
Less trade-in	$7,950.00
Amount to be financed	$12,617.18
5% interest for 48 months	$2,523.44
Total	$15,140.62
MONTHLY PAYMENT	$315.43

What is the annual percentage rate?

55. A car dealer carries out the following calculations:

List price	$9,450.00
Options	$1,125.00
Destination charges	$300.00
Subtotal	$10,875.00
Tax	$652.50
Less trade-in	$.00
Amount to be financed	$11,527.50
11% interest for 48 months	$5,072.10
Total	$16,599.60
MONTHLY PAYMENT	$345.83

What is the annual percentage rate?

PROBLEM SOLVING

56. The finance charge statement on a Sears Revolving Charge Card statement is shown here. Why do you suppose that the limitation on the 50¢ finance charge is for amounts less than $28.50?

SEARS ROEBUCK AND CO.
SEARSCHARGE SECURITY AGREEMENT

4. FINANCE CHARGE. If I do not pay the entire New Balance within 30 days (28 days for February statements) of the monthly billing date, a FINANCE CHARGE will be added to the account for the current monthly billing period. **THE FINANCE CHARGE** will be either a minimum of $0.50 if the Average Daily Balance is $28.50 or less, or a periodic rate of 1.75% per month (**ANNUAL PERCENTAGE RATE** of **21%**) on the Average Daily Balance.

---

57. Marsha needs to have a surgical procedure done and does not have the $3,000 cash necessary for the operation. Upon talking to an administrator at the hospital, she finds that it will accept MasterCard, VISA, and Discover credit cards. All of these credit cards have an APR of 18%, so she figures that it does not matter which card she uses, even though she plans to take a year to pay off the loan. Assume that Marsha makes a payment of $300 and then receives a bill. Show the interest from credit cards of 18% APR according to the previous balance, adjusted balance, and average daily balance methods. Assume that the month has 31 days and that it takes 14 days for Marsha's payment to be mailed and recorded.

58. Karen and Wayne need to buy a refrigerator because theirs just broke. Unfortunately, their savings account is depleted, and they will need to borrow money in order to buy a new one. The bank offers them a personal loan at 21% (APR), and Sears offers them an installment loan at 15% (add-on rate). Suppose that the refrigerator at Sears costs $1,598 plus 5% sales tax, and Karen and Wayne plan to pay for the refrigerator for 3 years. Should they finance it with the bank or with Sears?

59. Karen and Wayne need to buy a refrigerator because theirs just broke. Unfortunately, their savings account is depleted, and they will need to borrow money in order to buy a new one. Sears offers them an installment loan at 15% (add-on rate). If the refrigerator at Sears costs $1,598 plus 5% sales tax, and Karen and Wayne plan to pay for the refrigerator for 3 years, what is the monthly payment?

60. **Rule of 78** With a typical installment loan, you are asked to sign a contract stating the terms of repayment. If you pay off the loan early, you are entitled to an interest rebate. For example, if you finance $500 and are charged $90 interest (APR 8.46%), the total to be repaid is $590 with 24 monthly payments of $24.59. After 1 year, you decide to pay off the loan, so you figure that the rebate should be $45 (half of the interest for 2 years), but instead you are told the interest rebate is only $23.40. What happened? Look at the fine print on the contract. It says interest will be refunded according to the Rule of 78. The formula for the rebate is as follows:

$$\text{INTEREST REBATE} = \frac{k(k+1)}{n(n+1)} \times \text{FINANCE CHARGE}$$

where k is the number of payments remaining and n is the total number of payments. Determine the interest rebate on the following:

a. $1,026 interest on an 18-month loan; pay off loan after 12 months.

b. $350 interest on a 2-year loan with 10 payments remaining.

c. $10,200 borrowed at 11% on a 4-year loan with 36 months remaining.

d. $51,000 borrowed at 10% on a 5-year loan with 18 payments remaining.

9.3 SEQUENCES

Sequences or Progressions

Patterns are sometimes used as part of an IQ test. It used to be thought that an IQ test measured "innate intelligence" and that a person's IQ score was fairly constant. Today, it is known that this is not the case. IQ test scores can be significantly changed by studying the types of questions asked. Even if you have never taken an IQ test, you have taken (or will take) tests that ask pattern-type questions.

An example of an IQ test is shown below. It is a so-called "quickie" test, but it illustrates a few mathematical patterns (see Problems 1–6 on this IQ test). The purpose of this section is to look at some simple patterns to become more proficient in recognizing them, and then to apply them to develop some important financial formulas.

This test was distributed by MENSA, the high-IQ society. Here is the scoring for this test:

Give yourself 1 point for each correct answer. If you completed the test in 15 minutes or less, give yourself an additional 4 points.

1. M 2. 15 3. 8 4. 6 5. 5
6. 22 7. 2 8. 3 9. 2 10. 4
11. 4 12. 4 13. 2 14. 3
15. 3

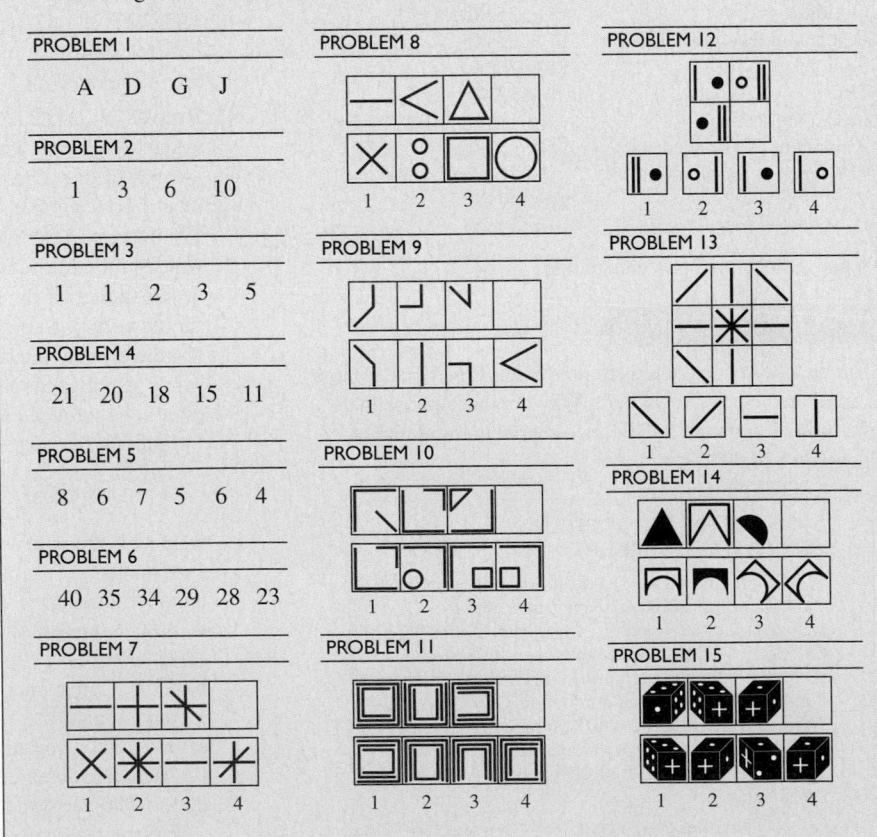

Are You A Genius?

Each problem is a series of some sort—that is, a succession of either letters, numbers or drawings—with the last item in the series missing. Each series is arranged according to a different rule and, in order to identify the missing item, you must figure out what that rule is.

Now, it's your turn to play. Give yourself a maximum of 20 minutes to answer the 15 questions. If you haven't finished in that time, stop anyway. In the test problems done with drawings, it is always the top row or group that needs to be completed by choosing one drawing from the bottom row.

PROBLEM 1

A D G J

PROBLEM 2

1 3 6 10

PROBLEM 3

1 1 2 3 5

PROBLEM 4

21 20 18 15 11

PROBLEM 5

8 6 7 5 6 4

PROBLEM 6

40 35 34 29 28 23

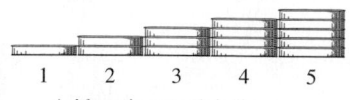

1   2   3   4   5

Arithmetic growth is linear.

Arithmetic Sequences

Perhaps the simplest pattern is the counting numbers themselves: 1, 2, 3, 4, 5, 6, ... A list of numbers having a first number, a second number, a third number, and so on, is called a **sequence.** The numbers in a sequence are called the **terms** of the sequence. The sequence of counting numbers is formed by adding 1 to each term to obtain the next term. Your math assignments may well have been identified by some sequence: "Do the multiples of 3 from 3 to 30," that is, 3, 6, 9, . . . , 27, 30. This sequence is formed by adding 3 to each term. Sequences obtained by adding the same number to each term to obtain the next term are called *arithmetic sequences* or *arithmetic progressions.*

Arithmetic Sequence

> An **arithmetic sequence** is a sequence whose consecutive terms differ by the same real number, called the **common difference.**

EXAMPLE 1

Show that each sequence is arithmetic, and find the missing term.

a. 1, 4, 7, 10, 13, ____ , . . .

b. 20, 14, 8, 2, −4, −10, ____ , . . .

c. $a_1, a_1 + d, a_1 + 2d, a_1 + 3d, a_1 + 4d,$ ____ , . . .

Solution

a. Look for a common difference by subtracting each term from the succeeding term:

$$4 - 1 = \mathbf{3}, \quad 7 - 4 = \mathbf{3}, \quad 10 - 7 = \mathbf{3}, \quad 13 - 10 = \mathbf{3}$$

If the difference between each pair of consecutive terms of the sequence is the same number, then that number is the common difference; in this case it is 3. To find the missing term, simply add the common difference. The next term is

$$13 + \mathbf{3} = 16$$

b. The common difference is −6. The next term is found by adding the common difference:

$$-10 + (-\mathbf{6}) = -16$$

c. The common difference is d, so the next term is

$$(a_1 + 4d) + \mathbf{d} = a_1 + 5d \qquad \blacklozenge$$

Example 1c leads us to a formula for arithmetic sequences:

a_1 is the first term of an arithmetic sequence.

a_2 is the second term of an arithmetic sequence, and

$$a_2 = a_1 + d$$

a_3 is the third term of an arithmetic sequence, and

$$a_3 = a_2 + d = (a_1 + d) + d = a_1 + 2d$$

a_4 is the fourth term of an arithmetic sequence, and

$$a_4 = a_3 + d = (a_1 + 2d) + d = a_1 + 3d$$
$$\vdots$$

a_{43} is the 43rd term of an arithmetic sequence, and

$$a_{43} = a_1 + \underset{\text{one less than the term number}}{\underline{42}}d$$
$$\vdots$$

CAUTION

Do not check only the first difference. All the differences must be the same.

This pattern leads to the following definition.

General Term of an Arithmetic Sequence

The *general term* of an arithmetic sequence $a_1, a_2, a_3, \ldots, a_n$, with common difference d is

$$a_n = a_1 + (n - 1)d$$

EXAMPLE 2

If $a_n = 26 - 6n$, list the sequence.

Solution

$$a_1 = 26 - 6(\mathbf{1}) = 20 \quad \text{\footnotesize a_1 means evaluate } 26 - 6n \text{ for } n = 1.$$
$$a_2 = 26 - 6(\mathbf{2}) = 14$$
$$a_3 = 26 - 6(\mathbf{3}) = 8$$
$$a_4 = 26 - 6(\mathbf{4}) = 2$$

The sequence is 20, 14, 8, 2, ...   ◆

Geometric Sequences

A second type of sequence is the *geometric sequence* or *geometric progression*. If each term is *multiplied* by the same number (instead of *added* to the same number) to obtain successive terms, a geometric sequence is formed.

Geometric Sequence

A **geometric sequence** is a sequence whose consecutive terms have the same quotient, called the **common ratio.**

If the sequence is geometric, the number obtained by dividing any term into the following term of that sequence will be the same nonzero number. No term of a geometric sequence may be zero.

EXAMPLE 3

Show that each sequence is geometric, and find the common ratio.

a. 2, 4, 8, 16, 32, ———, . . .

b. $10, 5, \frac{5}{2}, \frac{5}{4}, \frac{5}{8},$ ———, . . .

c. $g_1, g_1 r, g_1 r^2, g_1 r^3, g_1 r^4, \ldots$

Solution

a. First, verify that there is a common ratio:

$$\frac{4}{2} = \mathbf{2}, \quad \frac{8}{4} = \mathbf{2}, \quad \frac{16}{8} = \mathbf{2}, \quad \frac{32}{16} = \mathbf{2}$$

The common ratio is 2, so to find the next term, multiply the common ratio by the preceding term. The next term is

$$32(\mathbf{2}) = 64$$

b. The common ratio is $\frac{1}{2}$ (be sure to check *each* ratio). The next term is found by multiplication:

$$\frac{5}{8}\left(\frac{\mathbf{1}}{\mathbf{2}}\right) = \frac{5}{16}$$

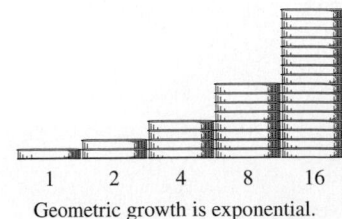

1   2   4   8   16
Geometric growth is exponential.

c. The common ratio is r, and the next term is

$$g_1 r^4(r) = g_1 r^5$$ ◆

As with arithmetic sequences, we denote the terms of a geometric sequence by using a special notation. Let $g_1, g_2, g_3, \ldots, g_n, \ldots$ be the terms of a *geometric* sequence. Example 3c leads us to a formula for geometric sequences.

$$g_2 = rg_1$$
$$g_3 = rg_2 = r(rg_1) = r^2 g_1$$
$$g_4 = rg_3 = r(r^2 g_1) = r^3 g_1$$
$$\vdots$$

one less than the term number

$$g_{92} = r^{\overbrace{91}} g_1$$
$$\vdots$$

Look for patterns to find the following formula.

General Term of a Geometric Sequence

For a geometric sequence $g_1, g_2, g_3, \ldots, g_n, \ldots$, with common ratio r, the *general term* is

$$g_n = g_1 r^{n-1}$$

EXAMPLE 4

List the sequence generated by $g_n = 50(2)^{n-1}$.

Solution

$$g_1 = 50(2)^{1-1} = 50$$
$$g_2 = 50(2)^{2-1} = 100$$
$$g_3 = 50(2)^{3-1} = 200$$
$$g_4 = 50(2)^{4-1} = 400$$

The sequence is 50, 100, 200, 400, ... ◆

Fibonacci-Type Sequences

Even though our attention is focused on arithmetic and geometric sequences, it is important to realize there can be other types of sequences.

The next type of sequence came about, oddly enough, by looking at the birth patterns of rabbits. In the 13th century, Leonardo Fibonacci wrote a book, *Liber Abaci,* in which he discussed the advantages of the Hindu–Arabic numerals over Roman numerals. In this book, one problem was to find the number of rabbits alive after a given number of generations. Let us consider what he did with this problem.

EXAMPLE 5 PÓLYA'S METHOD

Suppose a pair of rabbits will produce a new pair of rabbits in their second month, and thereafter will produce a new pair every month. The new rabbits will do exactly the same. Start with one pair. How many pairs will there be in 10 months?

Solution We use Pólya's problem-solving guidelines for this example, since it does not seem to match the previous problems we have encountered.

Understand the Problem. We can begin to understand the problem by looking at the following chart:

Number of Months	Number of Pairs	Pairs of Rabbits (the pairs shown in color are ready to reproduce in the next month)
Start	1	
1	1	
2	2	
3	3	
4	5	
5	8	
⋮	⋮	Same pair (rabbits never die)

FIBONACCI NUMBERS in NATURE

1 2 3 5 8 13 21 34 55 89 144 …

Dale Seymour Publications

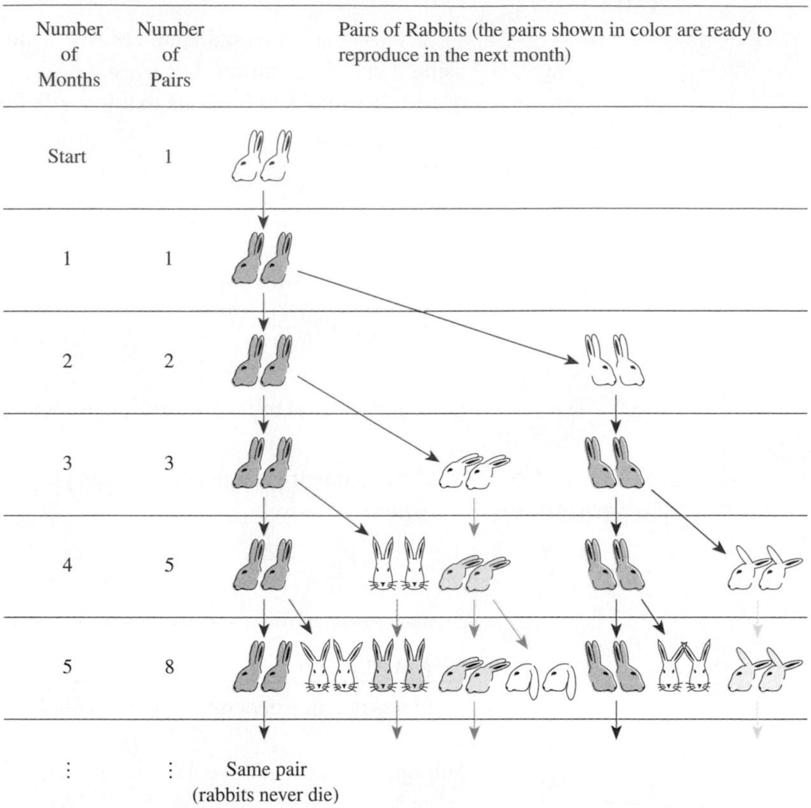

Devise a Plan. We look for a pattern with the sequence 1, 1, 2, 3, 5, 8, . . . ; it is not arithmetic and it is not geometric. It looks as if (after the first two months) each new number can be found by adding the two previous terms.

Carry Out the Plan.

$$1 + 1 = 2$$
$$1 + 2 = 3$$
$$2 + 3 = 5$$
$$3 + 5 = 8$$
$$5 + 8 = ?$$

Do you see the pattern? The sequence is 1, 1, 2, 3, 5, 8, 13, 21, 34, 55, 89, ...

Look Back. Using this pattern, Fibonacci was able to compute the number of pairs of rabbits alive after 10 months (it is the tenth term after the first 1): 89. He could also compute the number of pairs of rabbits after the first year or any other interval. Without a pattern, the problem would indeed be a difficult one. ◆

General Terms of a Fibonacci-Type Sequence

Fibonacci Sequence

> A **Fibonacci-type sequence** is a sequence in which the *general term* is given by the formula
>
> $$s_n = s_{n-1} + s_{n-2}$$
>
> where s_1 and s_2 are given. *The* **Fibonacci sequence** is that sequence for which $s_1 = s_2 = 1$.

In other words, since s_n represents the nth term, s_{n-1} and s_{n-2} are the two previous terms. A Fibonacci-type sequence is one in which terms are found by adding the two previous terms. The first two terms of the sequence must be given. If it is *the* Fibonacci sequence, then the first two terms are 1.

EXAMPLE 6

a. If $s_1 = 5$ and $s_2 = 2$, list the first five terms of this Fibonacci-type sequence.

b. If s_1 and s_2 represent any first numbers, list the first eight terms of this Fibonacci-type sequence.

Solution

a. 5, 2, 7, 9, 16

with 5+2 under 7, 2+7 over 9, 7+9 under 16.

b. s_1 and s_2 are given.

$$s_3 = s_2 + s_1$$
$$s_4 = s_3 + s_2 = (s_1 + s_2) + s_2 = s_1 + 2s_2$$
$$s_5 = s_4 + s_3 = (s_1 + 2s_2) + (s_1 + s_2) = 2s_1 + 3s_2$$
$$s_6 = s_5 + s_4 = (2s_1 + 3s_2) + (s_1 + 2s_2) = 3s_1 + 5s_2$$
$$s_7 = s_6 + s_5 = (3s_1 + 5s_2) + (2s_1 + 3s_2) = 5s_1 + 8s_2$$
$$s_8 = s_7 + s_6 = (5s_1 + 8s_2) + (3s_1 + 5s_2) = 8s_1 + 13s_2$$
$$\vdots$$

Look at the coefficients in the algebraic simplification at the right and notice that the Fibonacci sequence 1, 1, 2, 3, 5, 8, . . . is part of the construction of any Fibonacci-type sequence regardless of the terms s_1 and s_2. ◆

Historically, there has been much interest in the Fibonacci sequence. It is used in botany, zoology, business, economics, statistics, operations research, archeology, architecture, education, and sociology. There is even an official Fibonacci Association. An example of Fibonacci numbers occurring in nature is illustrated by a sunflower. The seeds are arranged in spiral curves as shown in Figure 9.2. If we count the number of counterclockwise spirals (13 and 21 in this example), they are successive terms in the Fibonacci sequence. This is true of all sunflowers and, indeed, of the seed head of any composite flower such as the daisy or aster.

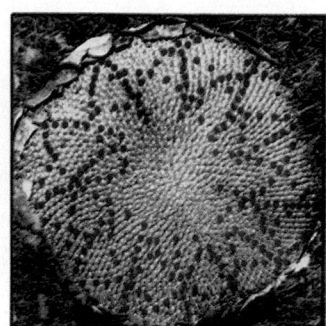

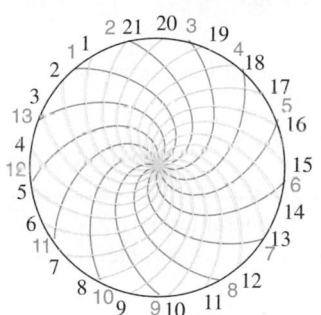

Figure 9.2 **The arrangement of the pods (phyllotaxy) of a sunflower illustrates a Fibonacci sequence.**

EXAMPLE 7

Classify the given sequences as arithmetic, geometric, Fibonacci-type, or none of the above. Find the next term for each sequence and give its general term if it is arithmetic, geometric, or Fibonacci-type.

a. 15, 30, 60, 120, . . .   **b.** 15, 30, 45, 60, . . .

c. 15, 30, 45, 75, . . .   **d.** 15, 20, 26, 33, . . .

e. 3, 3, 3, 3, . . .   **f.** 15, 30, 90, 360, . . .

Solution

a. 15, 30, 60, 120, . . . does not have a common difference, but a common ratio of 2, so this is a geometric sequence. The next term is 120(2) = 240. The general term is $g_n = 15(2)^{n-1}$.

b. 15, 30, 45, 60, . . . has a common difference of 15, so this is an arithmetic sequence. The next term of the sequence is $60 + 15 = 75$. The general term is given by $a_n = 15 + (n - 1)15 = 15 + 15n - 15 = 15n$.

c. 15, 30, 45, 75, . . . does not have a common difference or a common ratio. Next, we check for a Fibonacci-type sequence by adding successive terms: $15 + 30 = 45$; $30 + 45 = 75$, so this is a Fibonacci-type sequence. The next term is $45 + 75 = 120$. The general term is $s_n = s_{n-1} + s_{n-2}$ where $s_1 = 15$ and $s_2 = 30$.

d. 15, 20, 26, 33, . . . does not have a common difference or a common ratio. Since $15 + 20 \neq 26$, we see it is not a Fibonacci-type sequence. We do see a pattern, however, when looking at the differences:

$$20 - 15 = 5; \quad 26 - 20 = 6; \quad 33 - 26 = 7$$

The next difference is 8. Thus, the next term is $33 + 8 = 41$.

e. 3, 3, 3, 3, . . . has a common difference of 0 and the common ratio is 1, so it is both arithmetic and geometric. The next term is 3. The general term is $a_n = 3 + (n - 1)0 = 3$ or $g_n = 3(1)^{n-1} = 3$.

f. 15, 30, 90, 360, . . . does not have a common difference or a common ratio. We see a pattern when looking at the ratios:

$$\frac{30}{15} = 2; \quad \frac{90}{30} = 3; \quad \frac{360}{90} = 4$$

The next ratio is 5. Thus, the next term is $360(5) = 1{,}800$. ◆

General terms can be given for sequences that are not arithmetic, geometric, or Fibonacci-type. Consider the following example.

EXAMPLE 8

Find the first four terms for the sequences with the given general terms. If the general term defines an arithmetic, geometric, or Fibonacci-type sequence, so state.

a. $s_n = n^2$

b. $s_n = (-1)^n - 5n$

c. $s_n = s_{n-1} + s_{n-2}$, where $s_1 = -4$ and $s_2 = 6$

d. $s_n = 2n$

e. $s_n = 2n + (n - 1)(n - 2)(n - 3)(n - 4)$

Solution

a. Since $s_n = n^2$, we find $s_1 = (1)^2 = 1$; $s_2 = (2)^2 = 4$, $s_3 = (3)^2$, ...
The sequence is 1, 4, 9, 16, ...

b. Since $s_n = (-1)^n - 5n$, we find

$$
\begin{aligned}
s_1 &= (-1)^1 - 5(1) = -6 \\
s_2 &= (-1)^2 - 5(2) = -9 \\
s_3 &= (-1)^3 - 5(3) = -16 \\
s_4 &= (-1)^4 - 5(4) = -19 \\
&\vdots
\end{aligned}
$$

The sequence is $-6, -9, -16, -19, \ldots$.

c. $s_n = s_{n-1} + s_{n-2}$ is the form of a Fibonacci-type sequence. Using the two given terms, we find:

$$s_1 = -4 \quad \text{\small{Given}}$$
$$s_2 = 6 \quad \text{\small{Given}}$$
$$s_3 = s_2 + s_1 = 6 + (-4) = 2$$
$$s_4 = s_3 + s_2 = 2 + 6 = 8$$

The sequence is $-4, 6, 2, 8, \ldots$.

d. $s_n = 2n;\ s_1 = 2(1) = 2;\ s_2 = 2(2) = 4;\ s_3 = 2(3) = 6;\ \ldots$

The sequence is $2, 4, 6, 8, \ldots$; this is an arithmetic sequence.

e. $s_n = 2n + (n - 1)(n - 2)(n - 3)(n - 4)$

$$s_1 = 2(1) + (1 - 1)(1 - 2)(1 - 3)(1 - 4) = 2$$
$$s_2 = 2(2) + (2 - 1)(2 - 2)(2 - 3)(2 - 4) = 4$$
$$s_3 = 2(3) + (3 - 1)(3 - 2)(3 - 3)(3 - 4) = 6$$
$$s_4 = 2(4) + (4 - 1)(4 - 2)(4 - 3)(4 - 4) = 8$$

The sequence is $2, 4, 6, 8, \ldots$. ◆

Examples 8d and 8e show that if only a finite number of successive terms are known and no general term is given, then a *unique* general term cannot be given. That is, if we are given the sequence

$$2, 4, 6, 8, \ldots$$

the next term is probably 10 (if we are thinking of the general term of Example 8d), but it *may* be something different. In Example 8e,

$$s_5 = 2(5) + (5 - 1)(5 - 2)(5 - 3)(5 - 4) - 34$$

This gives the unlikely sequence $2, 4, 6, 8, 34, \ldots$. In general, you are looking for the simplest general term; nevertheless, you must remember that answers are not unique *unless the general term is given.*

Procedure for Classifying Sequences

To classify the sequence $s_1, s_2, \ldots, s_n$:

1. Check to see if it is *arithmetic.* Find the successive differences:

$$s_2 - s_1 = d_1$$
$$s_3 - s_2 = d_2$$
$$s_4 - s_3 = d_3$$
$$\vdots$$

If all these differences
$$d_1 = d_2 = d_3 = \cdots$$
are the same, then it is an arithmetic sequence.

2. Check to see if it is *geometric.* Find successive ratios:

$$s_2/s_1 = r_1; \quad s_3/s_2 = r_2; \quad s_4/s_3 = r_3, \ldots$$

If all these ratios
$$r_1 = r_2 = r_3 = \cdots$$
are the same, then it is a geometric sequence.

(continued)

Procedure for Classifying Sequences
continued

3. Check to see if it is *Fibonacci-type.* Check to see if

$$s_3 = s_2 + s_1$$
$$s_4 = s_3 + s_2$$
$$\vdots$$

If these sums check, then it is a Fibonacci-type sequence.

Computational Window

If you have access to a spreadsheet program, it is easy to generate the terms of a sequence.
The spreadsheet below shows the first 19 terms for the sequences given in Example 8.

Spreadsheet Application

	A	B	C	D	E	F	G	H	I	J	K	L
1	Term	a	b	c	d	e						
2	1	1	-6	-4	2	2						
3	2	4	-9	6	4	4						
4	3	9	-16	2	6	6						
5	4	16	-19	8	8	8						
6	5	25	-26	10	10	34						
7	6	36	-29	18	12	132						
8	7	49	-36	28	14	374						
9	8	64	-39	46	16	856						
10	9	81	-46	74	18	1698						
11	10	100	-49	120	20	3044						
12	11	121	-56	194	22	5062						
13	12	144	-59	314	24	7944						
14	13	169	-66	508	26	11906						
15	14	196	-69	822	28	17188						
16	15	225	-76	1330	30	24054						
17	16	256	-79	2152	32	32792						
18	17	289	-86	3482	34	43714						
19	18	324	-89	5634	36	57156						
20	19	361	-96	9116	38	73478						

The formulas for the cells correspond to the formulas in Example 8.

Term: 1; +A2+1; replicate

a: +A2^2; +A3^2; replicate

b: +(−1)^A2 − 5∗A2; +(−1)^A3 − 5∗A3; replicate

c: −4; 6; +D2+D3; replicate

d: +2∗A2; +2∗A3; replicate

e: +2∗A2+(A2 − 1)∗(A2 − 2)∗(A2 − 3)∗(A2 − 4); replicate

Problem Set 9.3

LEVEL 1

1. **IN YOUR OWN WORDS** What is a sequence?
2. **IN YOUR OWN WORDS** What do we mean by a general term?
3. **IN YOUR OWN WORDS** What is an arithmetic sequence?
4. **IN YOUR OWN WORDS** What is a geometric sequence?
5. **IN YOUR OWN WORDS** What is a Fibonacci sequence?

In Problems 6–31,
a. *Classify the sequences as arithmetic, geometric, Fibonacci, or none of these.*
b. *If arithmetic, give d; if geometric, give r; if Fibonacci, give the first two terms; and if none of these, state a pattern using your own words.*
c. *Supply the next term.*

6. $2, 4, 6, 8,$ ——— $, \ldots$
7. $2, 4, 8, 16,$ ——— $, \ldots$
8. $2, 4, 6, 10,$ ——— $, \ldots$
9. $5, 15, 25,$ ——— $, \ldots$
10. $5, 15, 45,$ ——— $, \ldots$
11. $5, 15, 20,$ ——— $, \ldots$
12. $1, 5, 25,$ ——— $, \ldots$
13. $25, 5, 1,$ ——— $, \ldots$
14. $9, 3, 1,$ ——— $, \ldots$
15. $1, 3, 9,$ ——— $, \ldots$
16. $21, 20, 18, 15, 11,$ ——— $, \ldots$
17. $8, 6, 7, 5, 6, 4,$ ——— $, \ldots$
18. $2, 5, 8, 11, 14,$ ——— $, \ldots$
19. $3, 6, 12, 24, 48,$ ——— $, \ldots$
20. $5, -15, 45, -135, 405,$ ——— $, \ldots$
21. $10, 10, 10,$ ——— $, \ldots$
22. $2, 5, 7, 12,$ ——— $, \ldots$
23. $3, 6, 9, 15,$ ——— $, \ldots$
24. $1, 8, 27, 64, 125,$ ——— $, \ldots$
25. $8, 12, 18, 27,$ ——— $, \ldots$
26. $3^2, 3^5, 3^8, 3^{11},$ ——— $, \ldots$
27. $4^5, 4^4, 4^3, 4^2,$ ——— $, \ldots$
28. $\frac{1}{2}, \frac{1}{3}, \frac{2}{3}, \frac{1}{4}, \frac{3}{4}, \frac{1}{5}, \frac{2}{5}, \frac{3}{5}, \frac{4}{5}, \frac{1}{6},$ ——— $, \ldots$
29. $\frac{1}{10}, \frac{1}{5}, \frac{3}{10}, \frac{2}{5}, \frac{1}{2},$ ——— $, \ldots$
30. $\frac{4}{3}, 2, 3, 4\frac{1}{2},$ ——— $, \ldots$
31. $\frac{7}{12}, \frac{2}{3}, \frac{3}{4}, \frac{5}{6},$ ——— $, \ldots$

LEVEL 2

In Problems 32–47,
a. *Find the first three terms of the sequences whose nth terms are given.*
b. *Classify the sequence as arithmetic (give d), geometric (give r), both, or neither.*

32. $s_n = 4n - 3$
33. $s_n = -3 + 3n$
34. $s_n = 10n$
35. $s_n = 2 - n$
36. $s_n = 7 - 3n$
37. $s_n = 10 - 10n$
38. $s_n = \dfrac{2}{n}$
39. $s_n = 1 - \dfrac{1}{n}$
40. $s_n = \dfrac{n-1}{n+1}$
41. $s_n = \dfrac{1}{2}n(n+1)$
42. $s_n = \dfrac{1}{4}n^2(n+1)^2$
43. $s_n = (-1)^n$
44. $s_n = -5$
45. $s_n = \dfrac{2}{3}$
46. $s_n = (-1)^{n+1}$
47. $s_n = (-1)^n(n+1)$

Find the requested terms in Problems 48–55.

48. Find the 15th term of the sequence
$$s_n = 4n - 3$$
49. Find the 69th term of the sequence
$$s_n = 7 - 3n$$
50. Find the 20th term of the sequence
$$s_n = (-1)^n(n+1)$$
51. Find the 3rd term of the sequence
$$s_n = (-1)^{n+1}5^{n+1}$$
52. Find the first five terms of the sequence where
$$s_1 = 2 \text{ and } s_n = 3s_{n-1}, n \geq 2$$
53. Find the first five terms of the sequence where
$$s_1 = 3 \text{ and } s_n = \tfrac{1}{3}s_{n-1}, n \geq 2$$
54. Find the first five terms of the sequence where
$$s_1 = 1, s_2 = 1, \text{ and } s_n = s_{n-1} + s_{n-2}, n \geq 3$$
55. Find the first five terms of the sequence where
$$s_1 = 1, s_2 = 2, \text{ and } s_n = s_{n-1} + s_{n-2}, n \geq 3$$

PROBLEM SOLVING

56. Is the following sequence a Fibonacci sequence?
a_n is one more than the nth term of the Fibonacci sequence.
57. Is the following sequence a Fibonacci sequence?
$1, 1, 2, 3, 5, 8, \ldots, a_n$, where a_n is the integer nearest
to $\dfrac{1}{\sqrt{5}}\left[\dfrac{1+\sqrt{5}}{2}\right]^n$

58. Apartment blocks of n floors are to be painted blue and yellow, with the rule that no two adjacent floors can be blue. (They can, however, be yellow.) Let a_n be the number of ways to paint a block with n floors.*

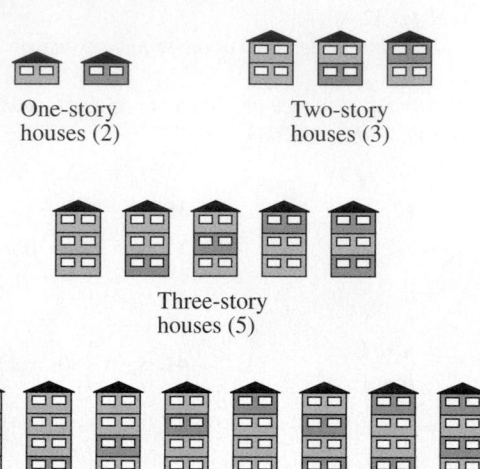

One-story houses (2)

Two-story houses (3)

Three-story houses (5)

Four-story houses (8)

Does the sequence $a_1, a_2, a_3, \ldots$ form a Fibonacci sequence?

59. Consider the following magic trick. The magician asks the audience for any two numbers (you can limit it to the counting numbers between 1 and 10 to keep the arithmetic manageable). Add these two numbers to obtain a third number. Add the second and third numbers to obtain a fourth. Continue until ten numbers are obtained. Then ask the audience to add the ten numbers, while the magician instantly gives the sum.

Consider the example shown here:

First number:	5
Second number:	9
(3) Add:	14
(4)	23
(5)	37
(6)	60
(7)	97
(8)	157
(9)	254
(10)	411
Add column:	1,067

The trick depends on the magician's ability to multiply quickly and mentally by 11. Consider the following pattern of multiplication by 11:

$$11 \times 10 = 110$$
$$11 \times 11 = 121$$
$$11 \times 12 = 132$$

$$11 \times 13 = 1\ 3 = 143$$

(Sum of the two digits; original two digits)

$$11 \times 52 = 5\ 2$$

(sum of the two digits)

$$= 572$$

$$11 \times 74 = 7\ 4$$

$$= 7\ \boxed{11}\ 4 = 814 \quad \text{Do the usual carry.}$$

Consider 11 times the 7th number in the example in the pattern:

$$11 \times 97 = 9\ \boxed{16}\ 7 = 1{,}067$$

Note that this is the sum of the ten numbers. Explain why this magician's trick works, and why it is called Fibonacci's magic trick.

60. Fill in the blanks so that

___ , 8, ___ , ___ , 27, ___ , . . . is

a. an arithmetic sequence.
b. a geometric sequence.
c. a sequence that is neither arithmetic nor geometric, for which you are able to write a general term.

9.4 | SERIES

If the terms of a sequence are added, the expression is called a **series.** We first consider a finite sequence along with its associated sum. Note that a capital letter is used to indicate the sum.

Finite Series

CAUTION

Write a lowercase s on your paper; now write a capital S. Do they look different? You need to distinguish between the lowercase s and capital S in your own written work.

The indicated sum of the terms of a finite sequence

$$s_1, s_2, s_3, \ldots, s_n$$

is called a **finite series** and is denoted by

$$S_n = s_1 + s_2 + s_3 + \cdots + s_n$$

EXAMPLE 1

a. Find S_4 where $s_n = 26 - 6n$.   **b.** Find S_3 where $s_n = (-1)^n n^2$.

Solution

a. $S_4 = s_1 + s_2 + s_3 + s_4$

$$= \overbrace{[26 - 6(1)]}^{s_1} + \overbrace{[26 - 6(2)]}^{s_2} + \overbrace{[26 - 6(3)]}^{s_3} + \overbrace{[26 - 6(4)]}^{s_4}$$

$$= 20 + 14 + 8 + 2$$

$$= 44$$

b. $S_3 = s_1 + s_2 + s_3$

$$= \overbrace{[(-1)^1(1)^2]}^{s_1} + \overbrace{[(-1)^2(2)^2]}^{s_2} + \overbrace{[(-1)^3(3)^2]}^{s_3}$$

$$= -1 + 4 + (-9)$$

$$= -6 \qquad \qquad \blacklozenge$$

The terms of the sequence in Example 1b alternate in sign: $-1, 4, -9, 16, \ldots$. A factor of $(-1)^n$ or $(-1)^{n+1}$ in the general term will cause the sign of the terms to alternate, creating a series called an **alternating series.**

Summation Notation

Before we continue to discuss finding the sum of the terms of a sequence, we need a handy notation, called **summation notation.** In Example 1a we wrote

$$S_4 = s_1 + s_2 + s_3 + s_4$$

Using summation notation (or, as it is sometimes called, **sigma notation**), we could write this sum using the Greek letter Σ:

$$S_4 = \sum_{k=1}^{4} s_k = s_1 + s_2 + s_3 + s_4$$

The sigma notation evaluates the expression (s_k) immediately following the sigma (Σ) sign, first for $k = 1$, then for $k = 2$, then for $k = 3$, and finally for $k = 4$, and then adds these numbers. That is, the expression is evaluated for *consecutive counting numbers* starting with the value of k listed at the bottom of the sigma ($k = 1$) and ending with the value of k listed at the top of the sigma ($k = 4$). For example, consider $s_k = 2k$ with $k = 1, 2, 3, 4, 5, 6, 7, 8, 9, 10$. Then

This is the last natural number in the domain. It is called the *upper limit.*
↓
$$\sum_{k=1}^{10} 2k \}$$ ←This is the function being evaluated. It is called the *general term.*
↑
This is the first natural number in the domain. It is called the *lower limit.*

Thus, $\displaystyle\sum_{k=1}^{10} 2k = 2(1) + 2(2) + 2(3) + 2(4) + 2(5) + 2(6) + 2(7) + 2(8) + 2(9) + 2(10) = 110$

The words **evaluate** and **expand** are both used to mean write out an expression in summation notation, and then sum the resulting terms, if possible.

EXAMPLE 2

a. Evaluate $\sum\limits_{k=3}^{6}(2k+1)$ **b.** Expand $\sum\limits_{k=3}^{n}\dfrac{1}{2^k}$

Solution

a. $\sum\limits_{k=3}^{6}(2k+1) = \underbrace{(2\cdot\mathbf{3}+1)}_{\substack{\text{Evaluate the expression } 2k+1 \text{ for } k=3.}} + \overbrace{(2\cdot\mathbf{4}+1)}^{k=4} + \underbrace{(2\cdot\mathbf{5}+1)}_{k=5} + \underbrace{(2\cdot\mathbf{6}+1)}_{k=6}$

$$= 7 + 9 + 11 + 13 = 40$$

b. $\sum\limits_{k=3}^{n}\dfrac{1}{2^k} = \dfrac{1}{2^3} + \dfrac{1}{2^4} + \dfrac{1}{2^5} + \dfrac{1}{2^6} + \cdots + \dfrac{1}{2^{n-1}} + \dfrac{1}{2^n}$ ◆

Arithmetic Series

An **arithmetic series** is the sum of the terms of an arithmetic sequence. Let us consider a rather simple-minded example. How many blocks are shown in the stack in Figure 9.3?

We can answer this question by simply counting the blocks: there are 34 blocks. Somehow it does not seem like this is what we have in mind with this question. Suppose we ask a better question. How many blocks are in a similar building with n rows? We notice that the number of blocks (counting from the top) in each row forms an arithmetic series:

$$1 + 6 + 11 + 16 + \cdots$$

Look for a pattern:

Denote one row by $A_1 = 1$ block

two rows: $A_2 = 1 + 6 = 7$

three rows: $A_3 = 1 + 6 + 11 = 18$

four rows: $A_4 = 1 + 6 + 11 + 16 = 34$ (shown in Figure 9.3)

$$\vdots$$

What about 10 rows?

$$A_{10} = 1 + 6 + 11 + 16 + 21 + 26 + 31 + 36 + 41 + 46$$

Instead of adding all these numbers directly, let us try an easier way. Write down A_{10} twice, once counting from the top and once counting from the bottom:

$$A_{10} = 1 + 6 + 11 + 16 + 21 + 26 + 31 + 36 + 41 + 46$$
$$\updownarrow \ \updownarrow \ \updownarrow \ \updownarrow \ \updownarrow \ \updownarrow \ \updownarrow \ \updownarrow \ \updownarrow \ \updownarrow$$
$$A_{10} = 46 + 41 + 36 + 31 + 26 + 21 + 16 + 11 + 6 + 1$$

Add these equations:

$$2A_{10} = \underset{\text{number of terms}}{47 + 47 + 47 + 47 + 47 + 47 + 47 + 47 + 47 + 47}$$

$$2A_{10} = \overbrace{10}^{\text{number of terms}}\underbrace{(47)}_{\text{sum of 1st and last terms}} = 470$$

$$A_{10} = 10\underbrace{\left(\dfrac{47}{2}\right)}_{\text{average of 1st and last terms}} = 235 \qquad \text{Divide both sides by 2.}$$

Row 1: 1
Row 2: 6
Row 3: 11
Row 4: 16

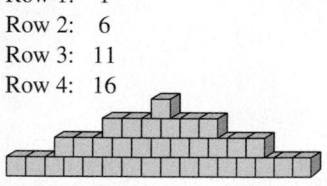

Figure 9.3 **How many blocks?**

This pattern leads us to a formula for n terms. We note that the number of blocks is an arithmetic sequence with $a_1 = 1$, $d = 5$; thus since $a_n = a_1 + (n-1)d$, we have for the blocks in Figure 9.3 a stack starting with 1 and ending (in the nth row) with

$$a_n = 1 + (n-1)5 = 1 + 5n - 5 = 5n - 4$$

Thus,

$$\underset{\substack{\uparrow \\ \text{number of blocks in } n \text{ rows}}}{A_n} = \underset{\substack{\uparrow \\ \text{number of rows}}}{n} \underbrace{\left[\frac{1 + (5n - 4)}{2}\right]}_{\text{average of 1st and last terms}}$$

$$= n\left[\frac{5n - 3}{2}\right]$$

$$= \tfrac{1}{2}(5n^2 - 3n)$$

This formula can be used for the number of blocks for any number of rows. Looking back, we see

$$n = 1: \quad A_1 = \tfrac{1}{2}[5(1)^2 - 3(1)] = 1$$
$$n = 4: \quad A_4 = \tfrac{1}{2}[5(4)^2 - 3(4)] = 34 \quad \text{(Figure 9.3)}$$
$$n = 10: \quad A_{10} = \tfrac{1}{2}[5(10)^2 - 3(10)] = 235$$

If we carry out these same steps for A_n where $a_n = a_1 + (n-1)d$, we derive the following formula for the sum of the terms of an arithmetic sequence.

Arithmetic Series Formula

The sum of the terms of an arithmetic sequence $a_1, a_2, a_3, \ldots, a_n$ with common difference d is

$$A_n = \sum_{k=1}^{n} a_k = n\left(\frac{a_1 + a_n}{2}\right) \quad \text{or} \quad A_n = \frac{n}{2}[2a_1 + (n-1)d]$$

In other words, the sum of n terms of an arithmetic sequence is n times the average of the first and last terms.

The last part of the formula for A_n is used when the last term is not explicitly stated or known. To derive this formula we know $a_n = a_1 + (n-1)d$ so

$$A_n = n\left(\frac{a_1 + a_n}{2}\right)$$

$$= n\left(\frac{a_1 + [a_1 + (n-1)d]}{2}\right)$$

$$= \frac{n}{2}[a_1 + a_1 + (n-1)d]$$

$$= \frac{n}{2}[2a_1 + (n-1)d]$$

EXAMPLE 3

In a classroom of 35 students, each student "counts off" by threes (i.e., 3, 6, 9, 12, . . .). What is the sum of the students' numbers?

Solution We recognize the sequence 3, 6, 9, 12, . . . as an arithmetic sequence with the first term $a_1 = 3$ and the common difference $d = 3$. The sum of these numbers is denoted by A_{35} since there are 35 students "counting off":

$$A_{35} = \frac{35}{2}[2(3) + (35 - 1)3] = 1,890 \qquad \blacklozenge$$

PEANUTS © 1967. Reprinted by permission of United Feature Syndicate, Inc.

Geometric Series

A **geometric series** is the sum of the terms of a geometric sequence. To motivate a formula for a geometric series, we once again consider an example. Suppose Charlie Brown receives a chain letter, and he is to copy this letter and send it to six of his friends.

You may have heard that chain letters "do not work." Why not? Consider the number of people who could become involved with this chain letter if we assume that everyone carries out their task and does not break the chain. The first mailing would consist of six letters. The second mailing involves 42 letters since the second mailing of 36 letters is added to the total:

$$6 + 36 = 42$$

1st mailing:

2nd mailing:

The number of letters in each successive mailing is a number in the geometric sequence

$$6, 36, 216, 1296, \ldots \quad \text{or} \quad 6, 6^2, 6^3, 6^4, \ldots$$

How many people receive letters with 11 mailings, assuming that no person receives a letter more than once? To answer this question, consider the series associated with a geometric sequence. We begin with a pattern:

Denote one mailing by $G_1 = 6$

two mailings: $G_2 = 6 + 6^2$

three mailings: $G_3 = 6 + 6^2 + 6^3$

$\vdots$

eleven mailings: $G_{11} = 6 + 6^2 + \cdots + 6^{11}$

We could probably use a calculator to find this sum, but we are looking for a formula, so we try something different. In fact, this time we will work out the general formula. Let

$$G_n = g_1 + g_2 + g_3 + \cdots + g_n$$
$$G_n = g_1 + g_1 r + g_1 r^2 + \cdots + g_1 r^{n-1}$$

Multiply both sides of this latter equation by r:

$$rG_n = g_1 r + g_1 r^2 + g_1 r^3 + \cdots + g_1 r^n$$

Notice that, except for the first and last terms, all the terms in the expansions for G_n and rG_n are the same, so that if we subtract one equation from the other, we have

$$G_n - rG_n = g_1 - g_1 r^n$$

We now solve for G_n:

$$(1 - r)G_n = g_1(1 - r^n)$$

$$G_n = \frac{g_1(1 - r^n)}{1 - r} \quad \text{if } r \neq 1$$

For Charlie Brown's chain letter problem, $g_1 = 6$, $n = 11$, and $r = 6$ so we find

$$G_{11} = \frac{6(1 - 6^{11})}{1 - 6} = \frac{6}{5}(6^{11} - 1) = 435,356,466$$

This is more than the number of people in the United States! The number of letters in only two more mailings would exceed the number of men, women, and children in the whole world.

Geometric Series Formula

> The sum of the terms of a geometric sequence $g_1, g_2, g_3, \ldots, g_n$ with common ratio r $(r \neq 1)$ is
>
> $$G_n = \frac{g_1(1 - r^n)}{1 - r}$$

EXAMPLE 4

Suppose some eccentric millionaire offered to hire you for a month (say, 31 days) and offered you the following salary choice. She will pay you $500,000 per day or else will pay you 1¢ for the first day, 2¢ for the second day, 4¢ for the third day, and so on for the 31 days. Which salary should you accept?

Solution   If you are paid $500,000 per day, your salary for the 31 days is

$$\$500,000(31) = \$15,500,000$$

Now, if you are paid using the doubling scheme, your salary is (in cents)

$$\overbrace{1 + 2 + 4 + 8 + \cdots + \text{last day}}^{\text{31st day}} \quad \text{or} \quad 2^0 + 2^1 + 2^2 + 2^3 + \cdots + 2^{30}$$

We see that this is the sum of the geometric sequence where $g_1 = 1$ and $r = 2$. We are looking for G_{31}:

$$G_{31} = \frac{1(1 - 2^{31})}{1 - 2} = -(1 - 2^{31}) = 2^{31} - 1$$

Using a calculator, we find this to be 2,147,483,647 cents or $21,474,836.47. You should certainly accept the doubling scheme (starting with 1¢, but do not ask for any days off). ◆

EXAMPLE 5   PÓLYA'S METHOD

The NCAA men's basketball tournament has 64 teams. How many games are necessary to determine a champion?

Solution   We use Pólya's problem-solving guidelines for this example.

Understand the Problem. Most tournaments are formed by drawing an elimination schedule similar to the one shown in Figure 9.4. This is sometimes called a *two-team elimination tournament.*

Devise a Plan. We could obtain the answer to the question by direct counting, but instead we will find a general solution, working backward. We know there will be

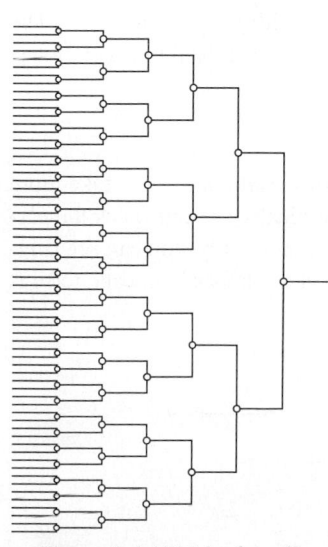

Figure 9.4 **NCAA playoffs**

The infinite! No other question has ever moved so profoundly the spirit of man [or woman]; no other idea has so fruitfully stimulated his [her] intellect; yet no other concept stands in greater need of clarification than that of the infinite . . . .

David Hilbert

1 championship game, and 2 semifinal games; continuing to work backward, there are 4 quarter-final games, . . . :

$$1 + 2 + 2^2 + 2^3 + \cdots$$

We recognize this as a geometric series. To reach 64 teams we note that since there are 64 teams, the first round has $64/2 = 32$ games. Also, we know that $2^5 = 32$, so we must find

$$1 + 2 + 2^2 + 2^3 + 2^4 + 2^5$$

Carry Out the Plan. We note that $g_1 = 1$, $r = 2$, and $n = 6$:

$$G_6 = \frac{1(1 - 2^6)}{1 - 2} = 2^6 - 1 = 63$$

Thus, the NCAA playoffs will require 63 games.

Look Back. We not only see that the NCAA tournament has 63 games for a playoff tournament, but, in general, if there are 2^n teams in a tournament, there will be $2^n - 1$ games. However, do not forget to check by estimation or by using common sense. Each game eliminates one team, and so 63 teams must be eliminated to crown a champion. ◆

Infinite Geometric Series

We have just found a formula for the sum of the first n terms of a geometric sequence. Sometimes it is also possible to find the sum of an entire infinite geometric series. Suppose an infinite geometric series

$$g_1 + g_2 + g_3 + g_4 + \cdots$$

is denoted by G. This is an **infinite series.** The **partial sums** are defined by

$$G_1 = g_1; \quad G_2 = g_1 + g_2; \quad G_3 = g_1 + g_2 + g_3; \quad \cdots$$

Consider the partial sums for an infinite geometric series with $g_1 = \frac{1}{2}$ and $r = \frac{1}{2}$. The geometric sequence is $\frac{1}{2}, \frac{1}{4}, \frac{1}{8}, \frac{1}{16}, \ldots$. The first few partial sums can be found as follows:

$$G_1 = \frac{1}{2}; \quad G_2 = \frac{1}{2} + \frac{1}{4} = \frac{3}{4}; \quad G_3 = \frac{1}{2} + \frac{1}{4} + \frac{1}{8} = \frac{7}{8}$$

Does this series have a sum if you add *all* its terms? It does seem that as you take more terms of the series, the sum is closer and closer to 1. Analytically, we can check the partial sums on a calculator or spreadsheet (see margin). Geometrically, you can see (Figure 9.5) that if the terms are laid end-to-end as lengths on a number line, each term is half the remaining distance to 1.

Spreadsheet Application _ □ ✕

It is easy to use a spreadsheet (review Section 5.3) to look at the partial sums of a sequence. For the sequence $\frac{1}{2}, \frac{1}{4}, \frac{1}{8}, \ldots$, define the cells as shown:

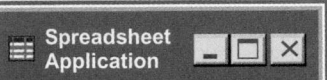

	A	B	C
1	n	term	partial sum
2	1	0.5	+B2
3	+A2+1	+B2*.5	+C2+B3
4	+A3+1	+B3*.5	+C3+B4

Replicate subsequent rows. The output is:

n	term	partial sum
1	.5	.5
2	.25	.75
3	.125	.875
4	.0625	.9375
5	.03125	.96875
6	.015625	.984375
7	.0078125	.9921875
8	.00390625	.99609375
9	.00195313	.99804688
10	.00097656	.99902344
11	.00048828	.99951172
12	.00024414	.99975586
13	.00012207	.99987793
14	.00006104	.99993896
15	.00003052	.99996948
16	.00001526	.99998474
17	.00000763	.99999237
18	.00000381	.99999619
19	.00000191	.99999809
20	.00000095	.99999905

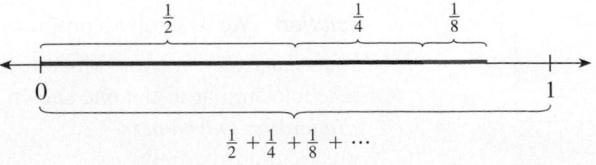

Figure 9.5 Series $\frac{1}{2} + \frac{1}{4} + \frac{1}{8} + \cdots$

It appears that the partial sums are getting closer to 1 as n becomes larger. We *can* find the sum of an infinite geometric sequence. Consider

$$G_n = \frac{g_1(1 - r^n)}{1 - r} = \frac{g_1 - g_1 r^n}{1 - r} = \frac{g_1}{1 - r} - \frac{g_1}{1 - r} r^n$$

Now, g_1, r, and $1 - r$ are fixed numbers. If $|r| < 1$, then r^n approaches 0 as n grows, and thus G_n approaches $\dfrac{g_1}{1 - r}$.

Infinite Geometric Series Formula

If $g_1, g_2, g_3, \ldots, g_n, \ldots$ is an infinite geometric sequence with a common ratio r such that $|r| < 1$, then its sum is denoted by G and is found by

$$G = \frac{g_1}{1 - r}$$

If $|r| \geq 1$, the infinite geometric series has no sum.

EXAMPLE 6

The path of each swing of a pendulum is 0.85 as long as the path of the previous swing (after the first). If the path of the tip of the first swing is 36 in. long, how far does the tip of the pendulum travel before it eventually comes to rest?

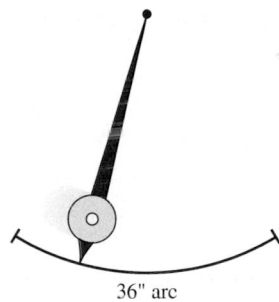

36" arc

Solution

$$\text{TOTAL DISTANCE} = 36 + 36(0.85) + 36(0.85)^2 + \cdots \quad \text{Infinite geometric series;}$$
$$g_1 = 36; r = 0.85$$
$$= \frac{36}{1 - 0.85}$$
$$= 240$$

The tip of the pendulum travels 240 in.   ◆

Summary of Sequence and Series Formulas

We conclude this section by repeating the important formulas related to sequences and series, as shown in Table 9.1 (page 468).

Procedure for Distinguishing a Series from a Sequence

Given a *sequence* of numbers, a *series* arises by considering the *sum* of the terms of the sequence. The formulas for the *general term of a sequence* and for the *sum* (of terms of the sequence; that is, a *series*) are given in Table 9.1.

Table 9.1

Type	Definition	Notation	Formula
Sequences	A list of numbers having a first term, a second term, . . .	s_n	
Arithmetic	Sequence whose terms differ by a constant called the common difference, d	a_n	$a_n = a_1 + (n-1)d$
Geometric	Sequence whose terms differ by a constant called the common ratio, r	g_n	$g_n = g_1 r^{n-1}$
Fibonacci	Sequence whose first two terms are given, and subsequent terms are found by adding the two previous terms.		$s_n = s_{n-1} + s_{n-2}$ $n \geq 3$
Series	The indicated sum of terms of a sequence	S_n	
Arithmetic	Sum of terms of an arithmetic sequence $$A_n = \sum_{k=1}^{n} a_k = a_1 + a_2 + a_3 + \cdots + a_n$$	A_n	$A_n = n\left(\dfrac{a_1 + a_n}{2}\right)$ Use when first and last terms are known. $A_n = \dfrac{n}{2}[2a_1 + (n-1)d]$ Use when first term and common difference are known.
Geometric	Sum of terms of a geometric sequence $$G_n = \sum_{k=1}^{n} g_k = g_1 + g_2 + g_3 + \cdots + g_n$$	G_n	$G_n = \dfrac{g_1(1 - r^n)}{1 - r} \quad (r \neq 1)$
	Sum of terms of an infinite geometric sequence $$G = g_1 + g_2 + g_3 + \cdots$$	G	$G = \dfrac{g_1}{1 - r} \quad |r| < 1$

Problem Set 9.4

LEVEL 1

1. **IN YOUR OWN WORDS** Distinguish a sequence and a series.

2. **IN YOUR OWN WORDS** Explain summation notation.

3. **IN YOUR OWN WORDS** What is a partial sum?

4. **IN YOUR OWN WORDS** Distinguish a geometric series and an infinite geometric series.

Find the requested values in Problems 5–10.

5. S_5 when $s_n = 15 - 3n$
6. S_8 when $s_n = 5n$
7. S_4 when $s_n = 5 \cdot 2^n$
8. S_6 when $s_n = (-1)^n$
9. S_7 when $s_n = (-1)^n$
10. S_3 when $s_n = 8 \cdot 5^n$

Evaluate the expressions in Problems 11–18.

11. $\sum_{k=3}^{5} k$
12. $\sum_{k=1}^{4} k^2$
13. $\sum_{k=2}^{6} k^2$
14. $\sum_{k=2}^{5} (100 - 5k)$
15. $\sum_{k=1}^{10} [1^k + (-1)^k]$
16. $\sum_{k=1}^{5} (-2)^{k-1}$
17. $\sum_{k=0}^{4} 3(-2)^k$
18. $\sum_{k=1}^{3} (-1)^k (k^2 + 1)$

LEVEL 2

If possible, find the sum of the infinite geometric series in Problems 19–24.

19. $1 + \frac{1}{2} + \frac{1}{4} + \cdots$
20. $1 + \frac{3}{2} + \frac{9}{4} + \cdots$
21. $1 + \frac{1}{3} + \frac{1}{9} + \cdots$
22. $100 + 50 + 25 + \cdots$
23. $-20 + 10 - 5 + \cdots$
24. $-100 + 50 - 25 + \cdots$

25. Find the sum of the first 5 odd positive integers.
26. Find the sum of the first 5 even positive integers.
27. Find the sum of the first 5 positive integers.
28. Find the sum of the first 10 odd positive integers.
29. Find the sum of the first 10 even positive integers.
30. Find the sum of the first 10 positive integers.

31. Find the sum of the first 100 odd positive integers.

32. Find the sum of the first 100 even positive integers.

33. Find the sum of the first 100 positive integers.

34. Find the sum of the first n odd positive integers.

35. Find the sum of the first n even positive integers.

36. Find the sum of the first n positive integers.

37. Find the sum of the first 20 terms of the arithmetic sequence whose first term is 100 and whose common difference is 50.

38. Find the sum of the first 50 terms of the arithmetic sequence whose first term is -15 and whose common difference is 5.

39. Find the sum of the even integers between 41 and 99.

40. Find the sum of the odd integers between 48 and 136.

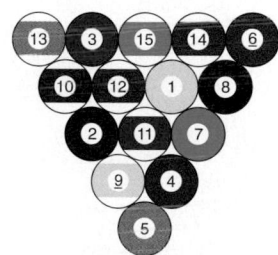

Figure 9.6 Pool balls

The game of pool uses 15 *balls numbered from* 1 *to* 15 *(see Figure 9.6). In the game of rotation, a player attempts to "sink" a ball in a pocket of the table and receives the number of points on the ball. Answer the questions in Problems* 41–44.

41. How many points would a player who "runs the table" receive? (To "run the table" means to sink all the balls.)

42. Suppose Missy sinks balls 1 through 8, and Shannon sinks balls 9 through 15. What are their respective scores?

43. Suppose Missy sinks the even-numbered balls and Shannon sinks the odd-numbered balls. What are their respective scores?

44. Suppose we consider a game of "super pool," which has 30 consecutively numbered balls on the table. How many points would a player receive to "run the table"?

45. The *Peanuts* cartoon (p. 464) expresses a common feeling regarding chain letters. Consider the total number of letters sent after a particular mailing:

1st mailing:	6
2nd mailing:	$6 + 36 = 42$
3rd mailing:	$6 + 36 + 216 = 258$

Determine the total number of letters sent in five mailings of the chain letter.

46. How many blocks would be needed to build a stack like the one shown in Figure 9.7 if the bottom row has 28 blocks?

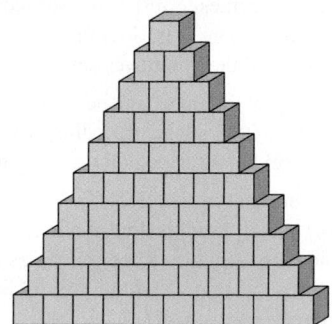

Figure 9.7 How many blocks?

47. Repeat Problem 46 if the bottom row has 87 blocks.

48. Repeat Problem 46 if the bottom row has 100 blocks.

49. A culture of bacteria increases by 100% every 24 hours. If the original culture contains 1 million bacteria ($a_0 = 1$ million), find the number of bacteria present after 10 days.

50. Use Problem 49 to find a formula for the number of bacteria present after d days.

51. How many games are necessary for a two-team elimination tournament with 32 teams?

52. Games like "Wheel of Fortune" and "Jeopardy" have one winner and two losers. A three-team game tournament is illustrated by Figure 9.8. If "Jeopardy" has a Tournament of Champions consisting of 27 players, what is the necessary number of games?

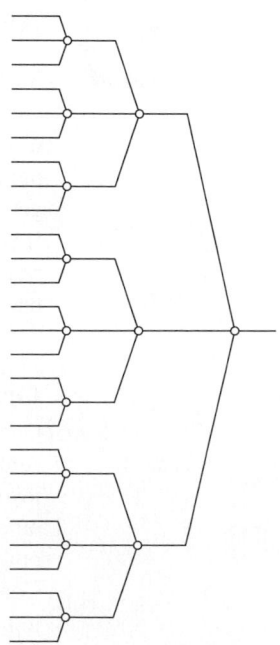

Figure 9.8 3-team tournament

53. How many games are necessary for a three-team elimination tournament with 729 teams?

54. A pendulum is swung 20 cm and allowed to swing freely until it eventually comes to rest. Each subsequent swing of the bob of the pendulum is 90% as far as the preceding swing. How far will the bob travel before coming to rest?

55. The initial swing of the tip of a pendulum is 25 cm. If each swing of the tip is 75% of the preceding swing, how far does the tip travel before eventually coming to rest?

56. A flywheel is brought to a speed of 375 revolutions per minute (rpm) and allowed to slow and eventually come to rest. If, in slowing, it rotates three-fourths as fast each subsequent minute, how many revolutions will the wheel make before returning to rest?

57. A rotating flywheel is allowed to slow to a stop from a speed of 500 rpm. While slowing, each minute it rotates two-thirds as many times as in the preceding minute. How many revolutions will the wheel make before coming to rest?

58. Advertisements say that a new type of superball will rebound to 9/10 of its original height. If it is dropped from a height of 10 ft, how far, based on the advertisements, will the ball travel before coming to rest?

59. A tennis ball is dropped from a height of 10 ft. If the ball rebounds 2/3 of its height on each bounce, how far will the ball travel before coming to rest?

PROBLEM SOLVING

60. **a.** How many blocks are there in the solid figure shown in Figure 9.9?

Figure 9.9 How many blocks?

b. How many blocks are there in a similar figure with 50 layers?

9.5 ANNUITIES

Sir Isaac Newton, one of the greatest mathematicians of all time, worked at the mint, but apparently did not like to apply mathematics to money. In a book by Rev. J. Spence in 1858, it is written, "Sir Isaac Newton, though so deep in algebra and fluxions, could not readily make up a common account: and, when he was Master of the Mint, used to get somebody else to make up his accounts for him." There is an important lesson here. If you do not like to work with money, hire someone to help you. When Albert Einstein was asked what was the most amazing formula he knew, you might think he would have given his famous formula $E = mc^2$, but he did not. He thought the most amazing formula was the compound interest formula.

Ordinary Annuities

One of the most fundamental mathematical concepts for business people and consumers is the idea of interest. We have considered present- and future-value problems involving a fixed amount, so we refer to such problems as **lump-sum problems.** On the other hand, it is far more common to encounter financial problems based on monthly or other periodic payments, so we refer to such problems as **periodic payment problems.** Consider the situation in which the *monthly payment* is known and the future value is to be determined. A sequence of payments into or out of an interest-bearing account is called an **annuity.** If the payments are made into an interest-bearing account at the end

of each time period, and if the frequency of payments is the same as the frequency of compounding, the annuity is called an *ordinary annuity*. The amount of an annuity is the sum of all payments made plus all accumulated interest. In this book we will assume that all annuities are ordinary annuities.

The best way to understand what we mean by an annuity is to consider an example. Suppose you decide to give up smoking and save the $2 per day you spend on cigarettes. How much will you save in 5 years? If you save the money without earning any interest, you will have

$$\$2 \times 365 \times 5 = \$3,650$$

However, let us assume that you save $2 per day, and at the end of each month you deposit the $60 (assume that all months are 30 days; that is, assume ordinary interest) into an account earning 12% interest compounded monthly. Now, how much will you have in 5 years? This is an example of an annuity. We could solve the problem using a spreadsheet to simulate our bank statement. Part of such a spreadsheet is shown here. Notice that even though the interest on $60 for one month is $0.60, the total monthly increase in interest is not linear, because the interest is compounded.

	A	B	C	D
	Time	Amt saved	Interest	Total in acct
1	Time	Amt	Interest	Total
2		saved		in acct
3	start	$0	$0	$0
4	1 mo	$60	$0	$60
5	2 mo	$60	$0.60	$120.00
6	3 mo	$60	$1.21	$181.81
7	4 mo	$60	$1.82	$243.63
8	5 mo	$60	$2.44	$306.07
9	6 mo	$60	$3.06	$369.13
10				

To derive a formula for annuities, let us consider this problem for a period of 6 months, and calculate the amounts plus interest for *each* deposit separately. We need a new variable to represent the amount of periodic deposit. Since this is usually a monthly payment, we let m = periodic payment. The **monthly payment** is a periodic payment that is made monthly.

We are given $m = 60$, $r = 0.12$, and $n = 12$ (monthly deposit means monthly compounding for an ordinary annuity). Let $i = r/n$. For this example, $i = 0.12/12 = 0.01$. The time, t, varies for each deposit. Remember that the deposit comes at the *end* of the month.

First deposit will earn 5 months' interest:	$60(1 + 0.01)^5 = 63.06$
Second deposit will earn 4 months' interest:	$60(1 + 0.01)^4 = 62.44$
Third deposit will earn 3 months' interest:	$60(1 + 0.01)^3 = 61.82$
Fourth deposit will earn 2 months' interest:	$60(1 + 0.01)^2 = 61.21$
Fifth deposit will earn 1 month's interest:	$60(1 + 0.01)^1 = 60.60$
Sixth deposit will earn no interest:	60.00
TOTAL IN THE ACCOUNT	369.13

This leads us to the following pattern (using variables). The total after 6 months:

$$A = m + m(1 + i)^1 + m(1 + i)^2 + m(1 + i)^3 + m(1 + i)^4 + m(1 + i)^5$$

$$= \sum_{k=1}^{6} m(1 + i)^{k-1}$$

For the total after nt periods, we recognize this as a geometric series. That is, it is the sum of the terms of a geometric sequence with $g_1 = m$ and common ratio $(1 + i)$. The

We are using the following formula for the sum of the terms of a geometric sequence:

$$G_n = \dfrac{\overset{m}{\overset{\downarrow}{g_1}}(1 - r^n)}{\underset{\underset{1+i}{\uparrow}}{1 - r}}$$

$\underset{\underset{A}{\uparrow}}{}$

sum, G_n, is the future value A. Thus (from the formula for the sum of a geometric sequence), we have

$$A = \frac{m[1 - (1 + i)^{nt}]}{1 - (1 + i)}$$

$$= \frac{m[1 - (1 + i)^{nt}]}{-i}$$

$$= \frac{m[(1 + i)^{nt} - 1]}{i}$$

Ordinary Annuity Formula

 STOP

This is one of the most useful formulas for you to use in your personal financial planning.

The future value, A, of an annuity is found with the formula

$$A = m \left[\frac{\left(1 + \dfrac{r}{n}\right)^{nt} - 1}{\dfrac{r}{n}} \right]$$

where r is the annual rate, m the periodic payment, t the time (in years), and n is the number of payments per year.

EXAMPLE 1

How much do you save in 5 years if you deposit $60 at the end of each month into an account paying 12% compounded monthly?

Solution Begin by identifying the problem type as well as the variables: **annuity;** $t = 5, m = \$60, r = 0.12,$ and $n = 12$. Next, evaluate the formula:

$$A = m \left[\frac{\left(1 + \dfrac{r}{n}\right)^{nt} - 1}{\dfrac{r}{n}} \right] = 60 \left[\frac{(1 + 0.01)^{60} - 1}{0.01} \right] \approx 4{,}900.180191$$

The future value is $4,900.18. A graph is shown in Figure 9.10.   ◆

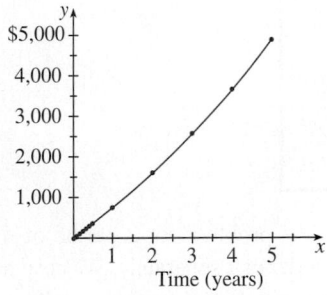

Figure 9.10 **Growth of an account with a $60 monthly deposit. The graph shows both deposits and interest compounded monthly at 12% annual rate.**

EXAMPLE 2   PÓLYA'S METHOD

Suppose you are 21 years old and will make monthly deposits to a bank account paying 10% annual interest compounded monthly.

> *Option I:*   Pay yourself $200 per month for 5 years and then leave the balance in the bank until age 65. (Total amount of deposits is $200 × 5 × 12 = $12,000.)

> *Option II:*   Wait until you are 40 years old (the age most of us start thinking seriously about retirement) and then deposit $200 per month until age 65. (Total amount of deposits is $200 × 25 × 12 = $60,000.)

Compare the amounts you would have from each of these options.

Solution We use Pólya's problem-solving guidelines for this example.

Understand the Problem. When most of us are 21 years old we do not think about retirement. However, if we do, the results can be dramatic. With this example, we investigate the differences if we save early (for 5 years), or later (for 25 years).

Devise a Plan. We calculate the value of the annuity for the 5 years of the first option, and then calculate the effect of leaving the value at the end of 5 years (the annuity) in a savings account until retirement. This part of the problem is a future value problem because it becomes a lump sum problem when deposits are no longer made (after 5 years). For the second option, we calculate the value of the annuity for 25 years.

Carry Out the Plan.

Option I: $200 per month for 5 years at 10% annual interest is an annuity with $m = 200$, $r = 0.10$, $t = 5$, and $n = 12$; this is an **annuity.**

$$A = 200\left[\frac{\left(1 + \frac{0.1}{12}\right)^{12 \cdot 5} - 1}{\frac{0.1}{12}}\right] \approx 15{,}487.41443$$

At the end of 5 years the amount in the account is $15,487.41. This money is left in the account, so it is now a future value problem for 39 years (ages 26 to 65); this part of the problem is a **future value problem.** $P = 15{,}487.41$, $r = 0.1$, $t = 39$, and $n = 12$, so that

$$A = P\left(1 + \frac{r}{n}\right)^{nt} = 15{,}487.41\left(1 + \frac{0.1}{12}\right)^{12 \cdot 39} \approx 752{,}850.86$$

(It is 752,851.08 if you use the calculator value for A from the first part of the problem.) The amount you would have at age 65 from 5 years of payments to yourself starting at age 21 is $752,850.86.

Option II: If you wait until you are 40 years old, this is an **annuity** problem with $m = 200$, $r = 0.10$, $t = 25$, and $n = 12$:

$$A = 200\left[\frac{\left(1 + \frac{0.1}{12}\right)^{12 \cdot 25} - 1}{\frac{0.1}{12}}\right] \approx 265{,}366.6806$$

The amount you would have at age 65 from 25 years of payments to yourself starting at age 40 is $265,366.68.

Look Back: Retirement Options I and II are compared graphically in Figure 9.11. Clearly, it is to your advantage to choose Option I.   ◆

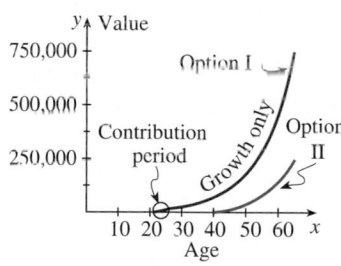

Figure 9.11 **Comparison of Options I and II**

EXAMPLE 3

The previous example gave you a choice:

Option I: Pay yourself $200 per month for 5 years beginning when you are 21 years old (this is comparable to the car payments you might make).

Option II: Wait until age 40, then make payments of $200 per month to yourself for the next 25 years.

If you were to select one of these options, what effect would it have on your retirement monthly income assuming you decide to live on the interest only and that the interest rate is 10%?

Solution   *Option I* gives you a retirement fund of $752,850.86. This will provide a monthly income as determined by the simple interest formula (because the interest is not left to accumulate):

$$I = Prt \qquad \text{Simple interest formula}$$

$$= 752{,}850.86(0.10)\left(\frac{1}{12}\right) \qquad \text{Substitute values.}$$

$$\approx 6{,}273.76 \qquad \text{Simplify.}$$

Option II gives you a retirement fund of $265,366.68. This will provide a monthly income found as follows:

$$I = Prt$$

$$= 265{,}366.68(0.10)\left(\frac{1}{12}\right) \qquad \text{\small Repeat the steps from Option I.}$$

$$\approx 2{,}211.39$$

The better choice (Option I) would mean that you would have about $4,000 *more* each and every month that you live beyond age 65. Did you ever ask the question "Why should I take math?" ◆

Sinking Funds

We now consider the situation in which we need (or want) to have a lump sum of money (a future value) in a certain period of time. The present value formula will tell us how much we need to have today, but we frequently do not have that amount available. Suppose your goal is $10,000 in 5 years. You can obtain 8% compounded monthly, so the present value formula yields

$$P = A\left(1 + \frac{r}{n}\right)^{-nt}$$

$$= \$10{,}000\left(1 + \frac{0.08}{12}\right)^{-5\cdot12}$$

$$\approx \$6{,}712.10$$

However, this is more than you can afford to put into the bank now. The next choice is to make a series of small equal investments to accumulate at 8% compounded monthly, so that the end result is the same—namely, $10,000 in 5 years. The account you set up to receive those investments is called a **sinking fund.**

To find a formula for a sinking fund, we begin with the formula for an ordinary annuity, and solve for *m* (which is the unknown for a sinking fund). Once again, we let $i = r/n$:

$$A = m\left[\frac{(1 + i)^{nt} - 1}{i}\right]$$

$$Ai = m[(1 + i)^{nt} - 1] \qquad \text{\small Multiply both sides by } i.$$

$$\frac{Ai}{(1 + i)^{nt} - 1} = m \qquad \text{\small Divide both sides by } [(1 + i)^{nt} - 1].$$

Sinking Fund Formula

CAUTION

Spend a few minutes with this idea; in your own words, can you explain when you would use this formula?

If the future value (*A*) is known, and you wish to find the amount of the periodic payment (*m*), use the *sinking fund formula*

$$m = \frac{A\left(\dfrac{r}{n}\right)}{\left(1 + \dfrac{r}{n}\right)^{nt} - 1}$$

where *r* is the annual rate, *t* is the time (in years), and *n* is the number of times per year the payments are made.

EXAMPLE 4

Suppose you want to have $10,000 in 5 years and decide to make monthly payments into an account paying 8% compounded monthly. What is the amount of each monthly payment?

Solution   You might begin by estimation. If you are paid *no* interest, then since you are making 60 equal deposits, the amount of each deposit would be

$$\frac{10,000}{60} \approx \$166.67$$

Therefore, since you are paid interest, you would expect the amount of each deposit to be somewhat less than this estimation. This is a **sinking fund problem.** We identify the known values: $A = \$10,000$ (future value), $r = 0.08$, $t = 5$, and $n = 12$. Finally, use the appropriate formula:

$$m = \frac{A\left(\dfrac{r}{n}\right)}{\left(1 + \dfrac{r}{n}\right)^{nt} - 1} = \frac{10,000\left(\dfrac{0.08}{12}\right)}{\left(1 + \dfrac{0.08}{12}\right)^{12 \cdot 5} - 1} \approx 136.0972762$$

The necessary monthly payment to the account is $136.10.   ◆

Calculator Window

Most graphing calculators have the ability to accept programs. Modern calculators require no special ability to write a program. We input the values of the variables into the memory locations as shown on the following chart:

Variable	*Input to Storage*
P	STO P
A	STO A
m	STO M
r	STO R
n	STO N

When using a calculator, generally the only difficulty is the proper use of parentheses:

	Formula	*Calculator Notation*
For future value:	$A = P\left(1 + \dfrac{r}{n}\right)^{nt}$	P*(1+R/N)^(N*T)
For present value:	$P = A\left(1 + \dfrac{r}{n}\right)^{-nt}$	A*(1+R/N)^(−N*T)
For an annuity:	$A = m\left[\dfrac{\left(1 + \dfrac{r}{n}\right)^{nt} - 1}{\dfrac{r}{n}}\right]$	M*(((1+R/N)^(N*T)−1)/(R/N))
For a sinking fund:	$m = \dfrac{A\left(\dfrac{r}{n}\right)}{\left(1 + \dfrac{r}{n}\right)^{nt} - 1}$	(A*R/N)/((1+R/N)^(N*T)−1)

To write a program, press the PRGM key and then input the formula. Most calculator programs also require some statement that forces the output of an answer. Check with your owner's manual, but this is usually a one-step process. To use your calculator for one of these formulas, you simply need to identify the type of financial problem, input the values for the appropriate variables, and then run the program containing the formula you wish to evaluate.

Problem Set 9.5

LEVEL 1

1. **IN YOUR OWN WORDS** What do we mean by a lump-sum problem?

2. **IN YOUR OWN WORDS** Why should we call an annuity a periodic payment problem?

3. **IN YOUR OWN WORDS** What is an annuity?

4. **IN YOUR OWN WORDS** What is a sinking fund?

5. **IN YOUR OWN WORDS** Distinguish an annuity from a sinking fund.

6. **IN YOUR OWN WORDS** Describe Example 3 and comment on its possible relevance.

*Use a calculator to evaluate an **ordinary annuity formula.***

$$A = m \left[\frac{\left(1 + \frac{r}{n}\right)^{nt} - 1}{\frac{r}{n}} \right]$$

for m, r, and t (respectively) given in Problems 7–22. Assume monthly payments.

7. $50; 5%; 3 yr	8. $50; 6%; 3 yr
9. $50; 8%; 3 yr	10. $50; 12%; 3 yr
11. $50; 5%; 30 yr	12. $50; 6%; 30 yr
13. $50; 8%; 30 yr	14. $50; 12%; 30 yr
15. $100; 5%; 10 yr	16. $100; 6%: 10 yr
17. $100; 8%; 10 yr	18. $100; 12%; 10 yr
19. $150; 5%; 35 yr	20. $150; 6%; 35 yr
21. $650; 5%; 30 yr	22. $650; 6%; 30 yr

In Problems 23–34, find the value of each annuity at the end of the indicated number of years. Assume that the interest is compounded with the same frequency as the deposits.

	Amount of Deposit m	Frequency Compounded n	Rate r	Time (in yr) t
23.	$500	annually	8%	30
24.	$500	annually	6%	30
25.	$250	semiannually	8%	30
26.	$600	semiannually	2%	10
27.	$300	quarterly	6%	30
28.	$100	monthly	4%	5
29.	$200	quarterly	8%	20
30.	$400	quarterly	11%	20
31.	$30	monthly	8%	5
32.	$5,000	annually	4%	10
33.	$2,500	semiannually	8.5%	20
34.	$1,250	quarterly	3%	20

Find the amount of periodic payment necessary for each deposit to a sinking fund in Problems 35–46.

	Amount Needed A	Frequency Compounded n	Rate r	Time (in yr) t
35.	$7,000	annually	8%	5
36.	$25,000	annually	11%	5
37.	$25,000	semiannually	12%	5
38.	$50,000	semiannually	14%	10
39.	$165,000	semiannually	2%	10
40.	$3,000,000	semiannually	3%	20
41.	$500,000	quarterly	8%	10
42.	$55,000	quarterly	10%	5
43.	$100,000	quarterly	8%	8
44.	$35,000	quarterly	8%	12
45.	$45,000	monthly	8%	6
46.	$120,000	monthly	7%	30

LEVEL 2

47. Self-employed persons can make contributions for their retirement into a special tax-deferred account called a *Keogh account.* Suppose you are able to contribute $20,000 into this account at the end of each year. How much will you have at the end of 20 years if the account pays 8% annual interest?

48. The owner of Sebastopol Tree Farm deposits $650 at the end of each quarter into an account paying 8% compounded quarterly. What is the value of the account at the end of 5 years?

49. The owner of Oak Hill Squirrel Farm deposits $1,000 at the end of each quarter into an account paying 8% compounded quarterly. What is the value at the end of 5 years, 6 months?

50. Clearlake Optical has a $50,000 note that comes due in 4 years. The owners wish to create a sinking fund to pay this note. If the fund earns 8% compounded semiannually, how much must each semiannual deposit be?

51. A business must raise $70,000 in 5 years. What should be the size of the owners' quarterly payment to a sinking fund paying 8% compounded quarterly?

52. A lottery offers a $1,000,000 prize to be paid in 20 equal installments of $50,000 at the end of each year. What is the value of this annuity if the current annual rate is 5%?

PROBLEM SOLVING

53. A lottery offers a $1,000,000 prize to be paid in 29 equal installments of $20,000 with a 30th final payment of $420,000. What is the total value of this annuity if the current annual rate is 5%?

54. Clearlake Optical has developed a new lens. The owners plan to issue a $4,000,000 30-year bond with a contract rate of 5.5% paid annually to raise capital to market this new lens. This means that Clearlake will be required to pay 5.5% interest each year for 30 years. To pay off the debt, they will also set up a sinking fund paying 8% interest compounded annually. What size annual payment is necessary for interest and sinking fund combined?

55. The owners of Bardoza Greeting Cards wish to introduce a new line of cards but need to raise $200,000 to do it. They decide to issue 10-year bonds with a contract rate of 6% paid semiannually. This means Bardoza must make interest payments to the bond holders each 6 months for 10 years. They also set up a sinking fund paying 8% interest compounded semiannually. How much money will they need to make the semiannual interest payments as well as payments to the sinking fund?

56. John and Rosamond want to retire in 5 years and can save $150 every three months. They plan to deposit the money at the end of each quarter into an account paying 6.72% compounded quarterly. How much will they have at the end of 5 years?

57. You want to retire at age 65. You decide to make a deposit to yourself at the end of each year into an account paying 13%, compounded annually. Assuming you are now 25 and can spare $1,200 per year, how much will you have when you retire at age 65?

58. In 2002 the maximum Social Security deposit was $5,263.80. Suppose you are 25 and make a deposit of this amount into an account at the end of each year. How much would you have (to the nearest dollar) when you retire if the account pays 4% compounded annually and you retire at age 65?

59. Repeat Problem 57 using your own age.

60. Repeat Problem 58 using your own age.

9.6 | AMORTIZATION

Present Value of an Annuity

In the previous sections we considered the lump-sum formulas for present value and for future value. We also considered the periodic payment formula, which provided the future value of periodic payments into an interest-bearing account. With the annuity formula, we assumed that we knew the amount of the periodic payment.

In the real world, if you need or want to purchase an item and you do not have the entire amount necessary, you can generally save for the item, or purchase the item and pay for it with monthly payments. Let us consider an example. Suppose you can afford $275 per month for an automobile. One possibility is to put the $275 per month into a bank account until you have enough money to buy a car. To find the future value of this monthly payment, you would need to know the interest rate (say, 5.2%) and the length of time you will make these deposits (say, 48 months or monthly payments for 4 years). The future value in the account is found using the annuity formula:

$$A = m\left[\frac{\left(1 + \frac{r}{n}\right)^{nt} - 1}{\frac{r}{n}}\right] = 275\left[\frac{\left(1 + \frac{0.052}{12}\right)^{48} - 1}{\frac{0.052}{12}}\right] \approx 14{,}638.04004$$

This means that if you make these payments to yourself, in 4 years you will have saved $275 \times 48 = $13,200$, but because of the interest paid on this account, the total amount in the account is $14,638.04.

However, it is more common to purchase a car and agree to make monthly payments to pay for it. Given the annuity calculation above, can we conclude that we can look for a car that costs $14,638? No, because if we borrow to purchase the car we need to *pay* interest charges rather than *receive* the interest as we did with the annuity. Well, then, how much can we borrow when we assume that we know the monthly payment, the interest rate, and the length of time we will pay? This type of financial problem is called **present value of an annuity.**

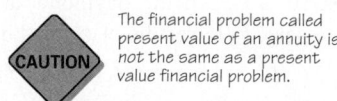

The financial problem called *present value of an annuity* is *not* the same as a *present value financial problem.*

Continuing with this car payment example, we need to ask, "What is the present value of 14,638.04004?" This is found using the formula $P = A(1 + \frac{r}{n})^{-nt}$:

$$P = 14{,}638.04004\left(1 + \frac{0.052}{12}\right)^{-48} \approx 11{,}894.46293$$

This means we could finance a loan in the amount of $11,894.46 and pay it off in 48 months with payments of $275, provided the interest rate is 5.2% compounded monthly.

We will now derive a formula for the present value of an annuity. For convenience, we let $i = r/n$.

Since $A = P(1 + i)^{nt}$ and $A = m\left[\dfrac{(1 + i)^{nt} - 1}{i}\right]$, we see

$$P(1 + i)^{nt} = m\left[\frac{(1 + i)^{nt} - 1}{i}\right] \qquad \text{Substitution}$$

$$P = \frac{m}{i}\left[\frac{(1 + i)^{nt} - 1}{(1 + i)^{nt}}\right] \qquad \text{Divide both sides by } (1 + i)^{nt}.$$

$$= \frac{m}{i}\left[1 - \frac{1}{(1 + i)^{nt}}\right]$$

$$= \frac{m}{i}\left[1 - (1 + i)^{-nt}\right]$$

Present Value of an Annuity Formula

If the periodic payment is known (m) and you wish to find the present value of those periodic payments, use the *present value of an annuity* formula:

$$P = m\left[\frac{1 - \left(1 + \dfrac{r}{n}\right)^{-nt}}{\dfrac{r}{n}}\right]$$

where P is the present value of the annuity, r is the annual interest rate, and n is the number of payments per year.

EXAMPLE 1

You look at your budget and decide that you can afford $250 per month for a car. What is the *maximum loan* you can afford if the interest rate is 13% and you want to repay the loan in 4 years?

Solution   This requires the **present value of an annuity** formula. First, determine the given variables:

$$m = 250, r = 0.13, t = 4, n = 12; \text{ thus,}$$

$$P = m\left[\frac{1 - \left(1 + \dfrac{r}{n}\right)^{-nt}}{\dfrac{r}{n}}\right] = 250\left[\frac{1 - \left(1 + \dfrac{0.13}{12}\right)^{-12 \cdot 4}}{\dfrac{0.13}{12}}\right] \approx 9{,}318.797438$$

This means that you can afford to finance about $9,319. To keep the payments at $250 (or less), a down payment should be made, if necessary, to lower the price of the car to the finance amount. ◆

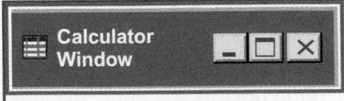

Calculator Window ▭▭✕

A formula for the present value of an annuity would be input into a calculator program as follows:
M*((1 − (1+R/N)^
(−N*T))/(R/N))

EXAMPLE 2   PÓLYA'S METHOD

Suppose your budget (or your banker) tells you that you can afford $1,575 per month for a house payment. If the current interest rate for a home loan is 10.5% and you will finance the home for 30 years, what is the expected price of the home you can afford? Assume that you have saved the money to make the required 20% down payment.

Solution   We use Pólya's problem-solving guidelines for this example.

Understand the Problem. Most textbook problems give you a price for a home, and ask for the monthly payment. This problem takes the real-life situation in which you know the monthly payment and wish to know the price of the house you can afford.

Devise a Plan. We first find the amount of the loan, and then find the down payment by subtracting the amount of the loan from the price of the home we wish to purchase.

Carry Out the Plan. We first find the amount of the loan; this is the **present value of an annuity.** We know $m = 1575$, $r = 0.105$, $t = 30$, $n = 12$; thus,

$$P = m\left[\frac{1 - \left(1 + \frac{r}{n}\right)^{-nt}}{i}\right] = 1,575\left[\frac{1 - \left(1 + \frac{0.105}{12}\right)^{-12\cdot30}}{\frac{0.105}{12}}\right] \approx 172,180.2058$$

The loan that you can afford is about $172,180.

What about the down payment? If there is a 20% down payment, then the amount of the loan is 80% of the price of the home. We know the amount of the loan, so the problem is now to determine the home price such that 80% of the price is equal to the amount of the loan. In symbols,

$$0.80x = 172,180.2058$$

which means that you simply divide the calculator output for the amount of the loan by 0.8 to find 215,225.2573.

Look Back. Given the conditions of the problem, we can look back to check our work (rounded to the nearest dollar for the check):

    Purchase price of home: $215,225
    Less 20% down payment: $ 43,045
    Amount of loan: $172,180   ◆

EXAMPLE 3

Theresa's Social Security benefit is $450 per month if she retires at age 62 instead of age 65. What is the present value of an annuity that would pay $450 per month for 3 years if the current interest rate is 10% compounded monthly?

Solution   Here, $m = 450$, $r = 0.10$, $t = 3$, and $n = 12$; then,

$$P = m\left[\frac{1 - (1 + i)^{-N}}{i}\right] = 450\left[\frac{1 - \left(1 + \frac{0.10}{12}\right)^{-36}}{\frac{0.10}{12}}\right] \approx 13,946.05601$$

The present value of this decision is $13,946.06. This represents the value of the additional 3 years of Social Security payments.   ◆

Monthly Payments

The process of paying off a debt by systematically making partial payments until the debt (principal) and the interest are repaid is called **amortization.** If the loan is paid off in regular equal installments, then we use the formula for the present value of an annuity to find the monthly payments by algebraically solving for m, where we let $i = r/n$:

$$P = m\left[\frac{1 - (1 + i)^{-nt}}{i}\right]$$

$$Pi = m[1 - (1 + i)^{-nt}] \quad \text{Multiply both sides by } i.$$

$$\frac{Pi}{1 - (1 + i)^{-nt}} = m \quad \text{Divide both sides by } 1 - (1 + i)^{-nt}.$$

Amortization Formula

If the amount of the loan is known (P), and you wish to find the amount of the periodic payment (m), use the formula

$$m = \frac{P\left(\dfrac{r}{n}\right)}{1 - \left(1 + \dfrac{r}{n}\right)^{-nt}}$$

where r is the annual rate, t is the time (in years), and n is the number of payments per year.

EXAMPLE 4

In 1990, the average price of a new home was $162,000 and the interest rate was 11%. If this amount is financed for 30 years at 11% interest, what is the monthly payment?

Solution This is an **amortization** problem. Given $P = 162,000$, $r = 0.11$, $t = 30$, and $n = 12$; then

$$m = \frac{P\left(\dfrac{r}{n}\right)}{1 - \left(1 + \dfrac{r}{n}\right)^{-nt}} = \frac{162,000\left(\dfrac{0.11}{12}\right)}{1 - \left(1 + \dfrac{0.11}{12}\right)^{-12 \cdot 30}} \approx 1,542.763901$$

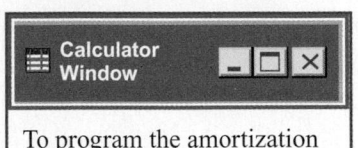

To program the amortization formula into a calculator input it as
$(P*R/N)/(1 - (1+R/N)^\wedge (-N*T))$

The monthly payment is $1,542.76. This is the payment for interest and principal to pay off a 30-year 11% loan of $162,000. ◆

The schedule of payments on a loan showing how much of each payment goes to pay the principal and how much goes to repay the interest is called an **amortization schedule.** For a home loan this schedule is rather long, so we show the input information for the loan in Example 4 in the following Spreadsheet Application. If you have a computer and spreadsheet program (review page 232), you can output the entire amortization schedule.

It is noteworthy to see how much interest is paid for the home loan of Example 4. There are 360 payments of $1,542.76, so the total amount repaid is

$$360(1,542.76) = 555,393.60$$

Since the loan was for $162,000, the interest is

$$\$555,393.60 - \$162,000 = \$393,393.60$$

Spreadsheet Application

The following spreadsheet was written to give an amortization schedule for Example 4.

	A	B	C	D	E	F	G	H	I	J
1	**Amortization Schedule**									
2	End of				Outstanding					
3	Period	Payment	Interest	Principal	balance					
4	0				162000.00					
5	+A4+1	1542.76	+E4*(.11/12)	+B5 - C5	+E4 - B5+C5					
6	replicate	replicate	replicate	replicate	replicate					

Using these spreadsheet entries for payments 1 to 360, we find the following entries:*

End of Period	Payment	Interest	Principal	Outstanding Balance	End of Period	Payment	Interest	Principal	Outstanding Balance
0				$162,000.00	348	$1,542.76	$172.66	$1,370.10	$17,465.46
1	$1,542.76	$1,485.00	$57.76	$161,942.24	349	$1,542.76	$160.10	$1,382.66	$16,082.80
2	$1,542.76	$1,484.47	$58.29	$161,883.95	350	$1,542.76	$147.43	$1,395.33	$14,687.47
3	$1,542.76	$1,483.94	$58.82	$161,825.13	351	$1,542.76	$134.64	$1,408.12	$13,279.34
4	$1,542.76	$1,483.40	$59.36	$161,765.76	352	$1,542.76	$121.73	$1,421.03	$11,858.31
5	$1,542.76	$1,482.85	$59.91	$161,705.86	353	$1,542.76	$108.70	$1,434.06	$10,424.25
6	$1,542.76	$1,482.30	$60.46	$161,645.40	354	$1,542.76	$95.56	$1,447.20	$8,977.05
7	$1,542.76	$1,481.75	$61.01	$161,584.39	355	$1,542.76	$82.29	$1,460.47	$7,516.58
8	$1,542.76	$1,481.19	$61.57	$161,522.82	356	$1,542.76	$68.90	$1,473.86	$6,042.72
9	$1,542.76	$1,480.63	$62.13	$161,460.69	357	$1,542.76	$55.39	$1,487.37	$4,555.35
10	$1,542.76	$1,480.06	$62.70	$161,397.98	358	$1,542.76	$41.76	$1,501.00	$3,054.35
11	$1,542.76	$1,479.48	$63.28	$161,334.70	359	$1,542.76	$28.00	$1,514.76	$1,539.59
12	$1,542.76	$1,478.90	$63.86	$161,270.85	360	$1,553.70	$14.11	$1,539.59	$0.00

*Disregard any small differences for the last payment on an amortization schedule.

The amount of interest paid can be reduced by making a larger down payment. For Example 4, if a 20% down payment (which is fairly standard) was made, then $162,000(0.80) = $129,600 is the amount to be financed. If this amount is amortized over 20 years (instead of 30), the monthly payments are $1,337.72. The total amount paid on *this* loan is

$$\underbrace{240(\overbrace{1,337.72}^{\text{amount of each payment}}) + \underbrace{32,400}_{\text{down payment}} = 353,452.80}_{}$$

number of payments

The savings yielded over the term of the loan by making a 20% down payment and financing the balance for 20 years rather than financing the entire amount for 30 years is $555,393.60 − $353,452.80 = $201,940.80.

It is interesting to see how much is saved in Example 4 by lowering the interest rates. In late 1997 the average fixed interest rate was 6.45%. If $P = \$162,000$, $r = 0.0645$, $t = 30$, $n = 12$, then (using the amortization formula)

$$m = \frac{P\left(\dfrac{r}{n}\right)}{1 - \left(1 + \dfrac{r}{n}\right)^{-nt}} = \frac{162,000\left(\dfrac{0.0645}{12}\right)}{1 - \left(1 + \dfrac{0.0645}{12}\right)^{-12 \cdot 30}} \approx 1,018.629046$$

The interest on the 30-year 11% loan in Example 4 was $393,393.60; now at 6.45% interest for 30 years the total interest is

$$360(\$1,018.63) - \$162,000 = \$204,706.80$$

a savings of almost $190,000.

Most of the time when you obtain a home loan, you will need to make a choice about which loan offer is best for your particular situation.

EXAMPLE 5

Suppose you are considering a $400,000 30-year home loan and there are two possibilities:

Loan A: 7.25% + 0 pts

Loan B: 6.875% + 2.375 pts

Which is the better loan if the home is sold in 10 years?

Solution   You might think that the lower rate is always the best choice, but you need to take into consideration the loan fees. One point on a loan is 1% of the amount of the loan. This means that 2.375 points represents a loan charge of

$$0.02375(\$400,000) = \$9,500$$

First, use the amortization formula for each loan:

$$\text{Loan A:}\quad m = \frac{\$400,000\left(\dfrac{0.0725}{12}\right)}{1 - \left(1 + \dfrac{0.0725}{12}\right)^{-(12)(30)}} \approx \$2,728.71$$

$$\text{Loan B:}\quad m = \frac{\$400,000\left(\dfrac{0.06875}{12}\right)}{1 - \left(1 + \dfrac{0.06875}{12}\right)^{-(12)(30)}} \approx \$2,627.72$$

The difference in payments is $100.99/mo. Now, find the present value of an annuity with $m = \$100.99$ using the 6.875% rate for 10 years:

$$A = \$100.99\left[\frac{1 - \left(1 + \dfrac{0.06875}{12}\right)^{-(12)(30)}}{\dfrac{0.06875}{12}}\right] \approx 8,746.36$$

Since the fees ($9,500) are greater than the present value of this annuity, the better choice is loan A, even though it has the higher rate.   ◆

We should note that even though both amortization and add-on interest calculate the monthly (or periodic) payment, you should pay attention to which procedure or formula to use.

Distinguish Add-On Simple Interest and Amortization

To calculate the monthly (or periodic) payment, consider one of the following methods:

Simple interest: Use this with installment loans or loans that use the words "add-on interest." These loans calculate the interest, add it to the amount of the loan ($A = P + I$), and then divide by the number of payments (N):

$$m = \frac{A}{N}$$

The interest rate, r, used for the add-on rate is not the same as the annual interest rate (APR).

Compound interest: Use this with long-term loans, such as when purchasing real estate. These loans are *amortized,* which means the interest rate, r, is the same as the annual interest rate, and the payments pay off both the principal and the interest. Use the *amortization formula*

$$m = \frac{P\left(\dfrac{r}{n}\right)}{1 - \left(1 + \dfrac{r}{n}\right)^{-nt}}$$

Problem Set 9.6

LEVEL 1

1. **IN YOUR OWN WORDS** What does amortization mean?

2. **IN YOUR OWN WORDS** Describe when you would use the present value of an annuity formula.

3. The variables m, n, r, t, A, and P are used in the various financial formulas. Tell what each of these variables represents.

*Use a calculator to evaluate the **present value of an annuity** formula*

$$P = m\left[\frac{1 - \left(1 + \dfrac{r}{n}\right)^{-nt}}{\dfrac{r}{n}}\right]$$

for the values of the variables m, r, and t (respectively) given in Problems 4–11. Assume n = 12.

4. $50; 5%; 5 yr
5. $50; 6%; 5 yr
6. $50; 8%; 5 yr
7. $150; 5%; 30 yr
8. $150; 6%; 30 yr
9. $150; 8%; 30 yr
10. $1,050; 5%; 30 yr
11. $1,050; 6%; 30 yr

*Use a calculator to evaluate the **amortization** formula*

$$m = \frac{P\left(\dfrac{r}{n}\right)}{1 - \left(1 + \dfrac{r}{n}\right)^{-nt}}$$

for the values of the variables P, r, and t (respectively) given in Problems 12–19. Assume n = 12.

12. $14,000; 5%; 5 yr
13. $14,000; 10%; 5 yr
14. $14,000; 19%; 5 yr
15. $150,000; 8%; 30 yr
16. $150,000; 9%; 30 yr
17. $150,000; 10%; 30 yr
18. $260,000; 12%; 30 yr
19. $260,000; 9%; 30 yr

Find the present value of the ordinary annuities in Problems 20–31.

	Amount of Deposit m	Frequency Compounded n	Rate r	Time (in yr) t
20.	$500	annually	8%	30
21.	$500	annually	6%	30
22.	$250	semiannually	8%	30
23.	$600	semiannually	2%	10

Amount of Deposit m	Frequency Compounded n	Rate r	Time (in yr) t
24. $300	quarterly	6%	30
25. $100	monthly	4%	5
26. $200	quarterly	8%	20
27. $400	quarterly	11%	20
28. $30	monthly	8%	5
29. $75	monthly	4%	10
30. $50	monthly	8.5%	20
31. $100	quarterly	3%	40

LEVEL 2

Find the monthly payment for the loans in Problems 32–43.

32. $500 loan for 12 months at 12%

33. $100 loan for 18 months at 18%

34. $4,560 loan for 20 months at 21%

35. $3,520 loan for 30 months at 19%

36. Used-car financing of $2,300 for 24 months at 15%

37. New-car financing of 2.9% on a 30-month $12,450 loan

38. Furniture financed at $3,456 for 36 months at 23%

39. A refrigerator financed for $985 at 17% for 15 months

40. A $112,000 home bought with a 20% down payment and the balance financed for 30 years at 11.5%

41. A $108,000 condominium bought with a 30% down payment and the balance financed for 30 years at 12.05%

42. Finance $450,000 for a warehouse with a 12.5% 30-year loan

43. Finance $859,000 for an apartment complex with a 13.2% 20-year loan

44. How much interest (to the nearest dollar) would be saved in Problem 40 if the home were financed for 15 rather than 30 years?

45. How much interest (to the nearest dollar) would be saved in Problem 41 if the condominium were financed for 15 rather than 30 years?

46. Melissa agrees to contribute $500 to the alumni fund at the end of each year for the next 5 years. Shannon wants to match Melissa's gift, but he wants to make a lump-sum contribution. If the current interest rate is 12.5% compounded annually, how much should Shannon contribute to equal Melissa's gift?

47. A $1,000,000 lottery prize pays $50,000 per year for the next 20 years. If the current rate of return is 12.25%, what is the present value of this prize?

48. Suppose you have an annuity from an insurance policy and you have the option of being paid $250 per month for 20 years or having a lump-sum payment of $25,000.

Which has more value if the current rate of return is 10%, compounded monthly?

49. An insurance policy offers you the option of being paid $750 per month for 20 years or a lump sum of $50,000. Which has more value if the current rate of return is 9%, compounded monthly, and you expect to live for 20 years?

50. You look at your budget and decide that you can afford $250 per month for a car. What is the maximum loan you can afford if the interest rate is 13% and you want to repay the loan in 4 years?

PROBLEM SOLVING

51. I recently found a real-life advertisement in the newspaper:

(Only the phone number has been changed.) Suppose that you have won a $20,000,000 lottery, paid in 20 annual installments. How much would be a fair price to be paid today for the assignment of this prize? Assume the interest rate is 5%.

52. The bottom notice (not circled) states that $6,000 is needed for 90 days, and that the advertiser is willing to pay 20% interest. How much would you expect to receive in 90 days if you lent this party the $6,000?

53. Suppose your gross monthly income is $5,500 and your current monthly payments are $625. If the bank will allow you to pay up to 36% of your gross monthly income (less current monthly payments) for a monthly house payment, what is the maximum loan you can obtain if the rate for a 30-year mortgage is 9.65%?

54. Suppose your gross monthly income is $4,550 and your spouse's gross monthly salary is $3,980. Your monthly bills are $1,235. The home you wish to purchase costs $355,000 and the loan is an 11.85% 30-year loan. How much down payment (rounded to the nearest hundred dollars) is necessary to be able to afford this home? This down payment is what percent of the cost of the home? Assume the bank will allow you to pay up to 36% of your gross monthly income (less current monthly payments) for house payments.

55. Suppose you want to purchase a home for $225,000 with a 30-year mortgage at 10.24% interest. Suppose also that you can put down 25%. What are the monthly payments? What is the total amount paid for principal and interest? What is the amount saved if this home is financed for 15 years instead of for 30 years?

56. The McBertys have $30,000 in savings to use as a down payment on a new home. They also have determined that they can afford between $1,500 and $1,800 per month for mortgage payments. If the mortgage rates are 11% per year compounded monthly, what is the price range for houses they should consider for a 30-year loan?

57. Rework Problem 56 for a 20-year loan instead of a 30-year loan.

58. Rework Problem 56 if interest rates go down to 10.2%.

59. As the interest rate increases, determine whether each of the given amounts increases or decreases. Assume that all other variables remain constant.
a. future value
b. present value
c. annuity

60. As the interest rate increases, determine whether each of the given amounts increases or decreases. Assume that all other variables remain constant.
a. monthly value for a sinking fund
b. monthly value for amortization
c. present value of an annuity

9.7 SUMMARY OF FINANCIAL FORMULAS

For many students, the most difficult part of working with finances (other than the problem of lack of funds) is in determining *which* formula to use. If we wish to free ourselves from the problem of lack of funds, we must use our knowledge in a variety of situations that occur outside the classroom.

We will now outline a procedure for using the financial formulas we have introduced in this chapter. First, review the meaning of the variables, as shown in the margin.

When classifying financial formulas, the first question to ask is:

Definitions of variables:
P = present value (principal)
A = future value
I = amount of interest
r = annual interest rate
t = number of years
n = periods; number of times compounded each year

Is the interest compounded?

If the interest is *removed* from the account, then use the *simple interest formula*

$I = Prt$ Use this for amount of interest; any one of the variables I, P, r, or t, may be unknown.

> *Examples: amount of interest; add-on interest; car payments; use the r from the simple interest formula when calculating the APR; finding the APR when given the purchase price and the monthly payment; credit card interest*

If the interest is *added* to the account, then use a *compound interest formula*. In such a case there are two remaining questions to ask when classifying a compound interest problem.

Is it a lump sum problem? If it is, then what is the unknown?

If FUTURE VALUE is the unknown, then it is a *future value* problem.

$$A = P\left(1 + \frac{r}{n}\right)^{nt} \quad \text{or} \quad A = Pe^{rt}$$

Examples: future value; inflation; retirement planning

If PRESENT VALUE is the unknown, then it is a *present value* problem.

$$P = A\left(1 + \frac{r}{n}\right)^{-nt}$$

Examples: present value; retirement planning

Is it a periodic payment problem? If it is, then is the periodic payment known?

PERIODIC PAYMENT KNOWN

Find the future value: ORDINARY ANNUITY

$$A = m\left[\frac{\left(1 + \frac{r}{n}\right)^{nt} - 1}{\frac{r}{n}}\right]$$

Examples: future value of periodic payments made to an interest-bearing account; retirement planning

Find the present value: PRESENT VALUE OF AN ANNUITY

$$P = m\left[\frac{1 - \left(1 + \frac{r}{n}\right)^{-nt}}{\frac{r}{n}}\right]$$

Examples: if the monthly payment, time, and rate are known, how much loan is possible? the current value of a known annuity (as in selling off the proceeds from an insurance policy or winning the lottery)

PERIODIC PAYMENT UNKNOWN

Future value known: SINKING FUND

$$m = \frac{A\left(\frac{r}{n}\right)}{\left(1 + \frac{r}{n}\right)^{nt} - 1}$$

Example: the periodic payment necessary to make to an interest-bearing account in order to reach some known financial goal

Present value known: AMORTIZATION

$$m = \frac{P\left(\frac{r}{n}\right)}{1 - \left(1 + \frac{r}{n}\right)^{-nt}}$$

Example: the monthly payment for an amortized loan

Problem Set 9.7

LEVEL 1

1. **IN YOUR OWN WORDS** What is the formula for the present value of an annuity? What is the unknown?

2. **IN YOUR OWN WORDS** What is the amortization formula? What is the unknown?

3. **IN YOUR OWN WORDS** What are a reasonable down payment and monthly payment for a home in your area?

4. **IN YOUR OWN WORDS** What are a reasonable down payment and monthly payment for a new automobile?

5. **IN YOUR OWN WORDS** Outline a procedure for identifying the type of financial formula for a given applied problem.

Classify the type of financial formula for the information given in Problems 6–11.

Lump sum problems

P	A
6. known	unknown
7. unknown	known

Periodic payment problems

	P	A	m
8.	unknown		known
9.		unknown	known
10.		known	unknown
11.	known		unknown

12. Match each formula in Column A with the type of financial problem in Column B.

Column A	Column B

$$A = m\left[\frac{\left(1 + \frac{r}{n}\right)^{nt} - 1}{\frac{r}{n}}\right]$$ Annuity

$$P = m\left[\frac{1 - \left(1 + \frac{r}{n}\right)^{-nt}}{\frac{r}{n}}\right]$$ Amortization

$$m = \frac{A\left(\frac{r}{n}\right)}{\left(1 + \frac{r}{n}\right)^{nt} - 1}$$ Present value of an annuity

$$m = \frac{P\left(\frac{r}{n}\right)}{1 - \left(1 + \frac{r}{n}\right)^{-nt}}$$ Sinking fund

Classify the financial problems in Problems 13–16, *and then answer each question by assuming a* 12% *interest rate compounded annually.*

13. Find the value of a $1,000 certificate in 3 years.

14. Deposit $300 at the end of each year. What is the total in the account in 10 years?

15. An insurance policy pays $10,000 in 5 years. What lump-sum deposit today will yield $10,000 in 5 years?

16. What annual deposit is necessary to give $10,000 in 5 years?

LEVEL 2

In Problems 17-50:

a. *State the type; and* b. *Answer the question.*

17. Find the value of a $1,000 certificate in $2\frac{1}{2}$ years if the interest rate is 12% compounded monthly.

18. You deposit $300 at the end of each year into an account paying 12% compounded annually. How much is in the account in 10 years?

19. An insurance policy pays $10,000 in 5 years. What lump-sum deposit today will yield $10,000 in 5 years if it is deposited at 12% compounded quarterly?

20. A 5-year term insurance policy has an annual premium of $300, and at the end of 5 years, all payments and interest are refunded. What lump-sum deposit is necessary to equal this amount if you assume an interest rate of 10% compounded annually?

21. What annual deposit is necessary to have $10,000 in 5 years if all the money is deposited at 9% interest compounded annually?

22. A $5,000,000 apartment complex loan is to be paid off in 10 years by making 10 equal annual payments. How much is each payment if the interest rate is 14% compounded annually?

23. The price of automobiles has increased at 6.25% per year. How much would you expect a $20,000 automobile to cost in 5 years if you assume annual compounding?

24. The amount to be financed on a new car is $9,500. The terms are 7% for 4 years. What is the monthly payment?

25. What deposit today is equal to 33 annual deposits of $500 into an account paying 8% compounded annually?

26. If you can afford $875 for your house payments, what is the loan you can afford if the interest rate is 6.5% and the monthly payments are made for 30 years?

27. What is the monthly payment for a home costing $125,000 with a 20% down payment and the balance financed for 30 years at 12%?

28. Ricon Bowling Alley will need $80,000 in 4 years to resurface the lanes. What lump sum would be necessary today if the owner of the business can deposit it in an account that pays 9% compounded semiannually?

29. Rita wants to save for a trip to Tahiti, so she puts $2.00 per day into a jar. After 1 year she has saved $730 and puts the money into a bank account paying 10% compounded annually. She continues to save in this manner and makes her annual $730 deposit for 15 years. How much does she have at the end of that time period?

30. Karen receives a $12,500 inheritance that she wants to save until she retires in 20 years. If she deposits the money in a fixed 11% account, compounded daily ($n = 365$), how much will she have when she retires?

31. You want to give your child a million dollars when he retires at age 65. How much money do you need to deposit into an account paying 9% compounded monthly if your child is now 10 years old?

32. An accounting firm agrees to purchase a computer for $150,000 (cash on delivery) and the delivery date is in 270 days. How much do the owners need to deposit in an account paying 18% compounded quarterly so that they will have $150,000 in 270 days?

33. For 5 years, Thompson Cleaners deposits $900 at the end of each quarter into an account paying 8% compounded quarterly. What is the value of the account at the end of 5 years?

34. What is the necessary amount of monthly payments to an account paying 18% compounded monthly in order to have $100,000 in $8\frac{1}{3}$ years if the deposits are made at the end of each month?

35. Thomas' Grocery Store is going to be remodeled in 5 years, and the remodeling will cost $300,000. How much should be deposited now in order to pay for this remodeling if the account pays 12% compounded monthly?

36. If an apartment complex will need painting in $3\frac{1}{2}$ years and the job will cost $45,000, what amount needs to be deposited into an account now in order to have the necessary funds? The account pays 12% interest compounded semiannually.

37. Teal and Associates needs to borrow $45,000. The best loan they can find is one at 12% that must be repaid in monthly installments over the next $3\frac{1}{2}$ years. How much are the monthly payments?

38. Certain Concrete Company deposits $4,000 at the end of each quarter into an account paying 10% interest compounded quarterly. What is the value of the account at the end of $7\frac{1}{2}$ years?

39. Major Magic Corporation deposits $1,000 at the end of each month into an account paying 18% interest compounded monthly. What is the value of the account at the end of $8\frac{1}{3}$ years?

40. What is the future value of $112,000 invested for 5 years at 14% compounded monthly?

41. What is the future value of $800 invested for 1 year at 10% compounded daily?

42. What is the future value of $9,000 invested for 4 years at 20% compounded monthly?

43. If $5,000 is compounded annually at 5.5% for 12 years, what is the future value?

44. If $10,000 is compounded annually at 8% for 18 years, what is the future value?

45. You owe $5,000 due in 3 years, but you would like to pay the debt today. If the present interest rate is compounded annually at 11%, how much should you pay today so that the present value is equivalent to the $5,000 payment in 3 years?

46. Sebastopol Movie Theater will need $20,000 in 5 years to replace the seats. What deposit should be made today in an account that pays 9% compounded semiannually?

47. The Fair View Market must be remodeled in 3 years. It is estimated that remodeling will cost $200,000. How much should be deposited now (to the nearest dollar) to pay for this remodeling if the account pays 10% compounded monthly?

48. A laundromat will need seven new washing machines in $2\frac{1}{2}$ years for a total cost of $2,900. How much money (to the nearest dollar) should be deposited at the present time to pay for these machines? The interest rate is 11% compounded semiannually.

49. A computerized checkout system is planned for Able's Grocery Store. The system will be delivered in 18 months at a cost of $560,000. How much should be deposited today (to the nearest dollar) into an account paying 7.5% compounded daily?

50. A lottery offers you a choice of $1,000,000 per year for 5 years or a lump-sum payment. What lump-sum payment (rounded to the nearest dollar) would equal the annual payments if the current interest rate is 14% compounded annually?

PROBLEM SOLVING

51. A contest offers the winner $50,000 now or $10,000 now and $45,000 in one year. Which is the better choice if the current interest rate is 10% compounded monthly and the winner does not intend to use any of the money for one year?

Problems 52–55 are based on a 30-year fixed-rate home loan of $185,500 with an interest rate of 7.75%.

52. What is the monthly payment?

53. What is the total amount of interest paid?

54. Use a computer to print out an amortization schedule.

55. Suppose you reduce the term to 20 years. What is the total amount of interest paid, and what is the savings over the 30-year loan?

Problems 56–59 are based on a 30-year fixed-rate home loan of $418,500 with an interest rate of 8.375%.

56. What is the monthly payment?

57. What is the total amount of interest paid?

58. Use a computer to print out an amortization schedule.

59. Suppose you reduce the term to 22 years. What is the total amount of interest paid, and what is the savings over the 30-year loan?

60. In 1982 the inflation rate hit 16%. Suppose that the average cost of a textbook in 1982 was $15. What is the expected cost in the year 2002 if we project this rate of inflation on the cost? (Assume annual compounding.) If the average cost of a textbook in 2002 is $80, what is the actual inflation rate (rounded to the nearest tenth percent)?*

*This problem requires Chapter 8 (solving exponential equations).

Fields Medal

The Fields Medal is often referred to as the "Nobel Prize of Mathematics." The existence of the award is related to the fact that Alfred Nobel chose not to include mathematics in his areas of recognition. There is no documentary evidence to explain this exclusion, but the general gossip attributes it to a personal conflict between Nobel and Mittag-Leffler. John Charles Fields (1863–1932) was disturbed by the lack of such a mathematical award, so he worked toward the establishment of these awards from 1922 until his death in 1932. It was Fields' last will and testament that provided the necessary funds for the establishment of the award, which was finalized on January 4, 1934. The first Fields Medals were awarded in 1936 to L. V. Ahlfors (Harvard) and Jesse Douglas (MIT). Because of World War II the next award was not until 1950; it has been awarded every 4 years at the International Congress of Mathematics.

My thanks to Henry S. Tropp, Humboldt State University, for this information. Professor Tropp was my first mathematics professor and was instrumental in my interest in mathematics. I thank him not only for his research on the Fields Medal, but also for his influence in my life.

The Fields Institute and The University of Toronto

CHAPTER SUMMARY

"To get money is difficult, to keep it is more difficult, but to spend it wisely is most difficult of all."

Anonymous

IMPORTANT TERMS

Add-on interest [9.2]
Adjusted balance method [9.2]
Alternating series [9.4]
Amortization [9.6]
Amortization schedule [9.6]
Annual compounding [9.1]
Annual percentage rate [9.2]
Annuity [9.5]
APR [9.2]
Arithmetic sequence [9.3]
Arithmetic series [9.4]
Average daily balance method [9.2]
Closed-ended loan [9.2]
Common difference [9.3]
Common ratio [9.3]
Compound interest [9.1]
Compound interest formula [9.1]
Continuous compounding [9.1]

Credit card [9.2]
Daily compounding [9.1]
Dealer's cost [9.2]
e [9.1]
Euler's number [9.1]
Evaluate a summation [9.4]
Exact interest [9.1]
Expand a summation [9.4]
Fibonacci sequence [9.3]
Fibonacci-type sequence [9.3]
Finance charge [9.2]
Finite series [9.4]
Five-percent offer [9.2]
Future value [9.1]
Future value formula [9.1]
Geometric sequence [9.3]
Geometric series [9.4]
Grace period [9.2]

Infinite series [9.4]
Inflation [9.1]
Installment loan [9.2]
Interest [9.1]
Interest rate [9.1]
Line of credit [9.2]
Lump-sum problem [9.5]
Monthly compounding [9.1]
Monthly payment [9.5]
Natural base [9.1]
Open-ended loan [9.2]
Ordinary interest [9.1]
Partial sum [9.4]
Periodic payment problem [9.5]
Present value [9.1]
Present value formula [9.1]
Present value of an annuity [9.6]
Previous balance method [9.2]

Principal [9.1]
Quarterly compounding [9.1]
Revolving credit [9.2]
Semiannual compounding [9.1]
Sequence [9.3]

Series [9.4]
Sigma notation [9.4]
Simple interest formula [9.1]
Sinking fund [9.5]

Sticker price [9.2]
Summation notation [9.4]
Term of a sequence [9.3]
Time [9.1]

TYPES OF PROBLEMS

Be able to estimate reasonable answers to financial problems. [9.1–9.6]

Find the amount of interest if you are given the purchase price, length of loan, and the monthly payment. [9.1]

Find the total amount to be repaid for a simple interest loan. [9.1]

Calculate the future value. [9.1]

Calculate the present value. [9.1]

Compare the amount of interest and future value for simple and compound interest. [9.1]

Be able to calculate the future value due to inflation. [9.1]

Be able to calculate the present value due to inflation. [9.1]

Calculate the time necessary to achieve an investment goal. [9.1]

How much must be deposited into a bank account to provide a given monthly income? [9.1]

Be able to calculate the monthly payment using add-on interest. [9.2]

Make an appropriate offer on a new car. [9.2]

Calculate the APR for installment loans or for credit cards. [9.2]

Be able to calculate the amount of credit card interest by using the previous balance method, the adjusted balance method, and the average daily balance method; which is better from the consumer point of view? [9.2]

Find the amount of interest, the monthly payment, and the APR for consumer transactions. [9.2]

Determine the monthly payment for an automobile using add-on interest. [9.2]

Find the APR if you are given the price, length of time, and the add-on interest rate. [9.2]

Calculate the APR given the amount financed, the monthly payment, the length of time financed. [9.2]

Classify a given sequence as arithmetic, geometric, or Fibonacci; be able to find the next term. [9.3]

Write out the terms of a sequence when given the general term; find a specific term. [9.3]

Evaluate (expand) summation expressions. [9.4]

Find sums by classifying and using the series formulas. [9.4]

Distinguish between sequences and series in application problems. [9.4]

Calculate the amount you can save by making a periodic payment into an interest-bearing account (annuity). [9.5]

Calculate the amount you need to deposit into an interest-bearing account to have a given amount at some time in the future (sinking fund). [9.5]

Find the present value of an annuity. [9.6]

Calculate the amount of money you can borrow, given a monthly payment, interest rate, and length of time (present value of an annuity). [9.6]

Find the monthly payment for an amortized loan. [9.6]

Calculate the down payment for a home, given the price; given the amount of the down payment, calculate the price. [9.6]

Given a monthly payment, interest rate, and length of time, find the amount of loan. [9.6]

Work applied financial problems. [9.1–9.7]

CHAPTER 9 REVIEW QUESTIONS

1. **IN YOUR OWN WORDS** Explain the difference between a sequence and a series. Include in your discussion how to distinguish arithmetic, geometric, and Fibonacci-type sequences, and give the formulas for their general terms. What are the relevant formulas for arithmetic and geometric series?

2. **IN YOUR OWN WORDS** Outline a procedure for identifying financial formulas. Include in your discussion future value, present value, ordinary annuity, present value of an annuity, amortization, and sinking fund classifications.

3. Classify each sequence as arithmetic, geometric, Fibonacci-type, or none of these. Find an expression for the general term if it is one of these types; otherwise, give the next two terms.
   a. 5, 10, 15, 20, . . .   **b.** 5, 10, 20, 40, . . .   **c.** 5, 10, 15, 25, . . .
   d. 5, 10, 20, 35, . . .   **e.** 5, 50, 500, 5,000, . . .   **f.** 5, 50, 5, 50, . . .

4. **a.** Evaluate $\sum_{k=1}^{3}(k^2 - 2k + 1)$ **b.** Expand $\sum_{k=1}^{4}\frac{k-1}{k+1}$

5. Suppose someone tells you she has traced her family tree back 10 generations. What is the minimum number of people on her family tree if there were no intermarriages?

6. A certain bacterium divides into two bacteria every 20 minutes. If there are 1,024 bacteria in the culture now, how many will there be in 24 hours, assuming that no bacteria die? Leave your answer in exponential form.

7. Suppose that the car you wish to purchase has a sticker price of $22,730 with a dealer cost of $18,579. Make a 5% offer for this car (rounded to the nearest hundred dollars).

8. Suppose that the amount to be financed for a car purchase is $13,500 at an add-on interest rate of 2.9% for 2 years.
   a. What are the monthly installment and the amount of interest that you will pay?
   b. Find the APR for this loan.

9. If a car with a cash price of $11,450 is offered for nothing down with 48 monthly payments of $353.04, what is the APR?

10. From the consumer's point of view, which method of calculating interest on a credit card is most advantageous? Illustrate the three types of calculating interest for a purchase of $525 with 9% APR for a 31-day month in which it takes 7 days for your $100 payment to be received and recorded.

11. Suppose that you want to have $1,000,000 in 50 years. To achieve this goal, how much do you need to deposit today if you can earn 9% interest compounded monthly?

12. **a.** What is the monthly payment for a home loan of $154,000 if the rate is 8% and the time is 20 years?
    b. If you paid a 20% down payment, what is the price of the home?
    c. What would you expect this home to be worth in 20 years if you assume an inflation rate of 4%?

13. Suppose that you expect to receive a $100,000 inheritance when you reach 21 in three years and four months. What is the present value of your inheritance if the current interest rate is 6.4% compounded monthly?

Use the advertisement shown in Figure 9.12 as a basis for answering Problems 14–20. Assume the current interest rate is 5%.

14. What is the present value of the $420,000 payment to be received in 30 years?

15. What is the actual value of the annuity portion of the prize?

16. What is the present value of the annuity?

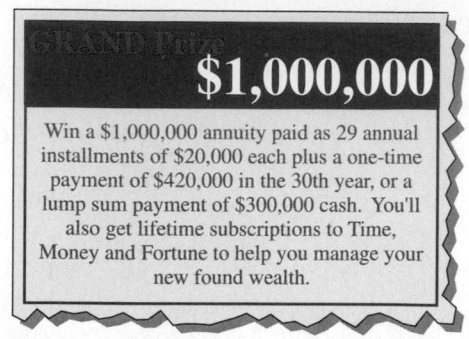

Figure 9.12 Contest prize notice

17. Suppose you take the $300,000 cash payment, and still want to have $420,000 in 30 years. How much of the $300,000 do you need to set aside now to have $420,000 in 30 years?

18. Suppose you take the $300,000 cash payment, and set aside $100,000 for savings. You plan on using the interest on the remaining $200,000. What are your annual earnings from interest?

19. Which option should you take and why?

20. If the current interest rate is 10%, which option should you take and why?

GROUP RESEARCH

Working in small groups is typical of most work environments, and learning to work with others to communicate specific ideas is an important skill. Work with three or four other students to submit a single report based on each of the following questions.

G31. Suppose you have just inherited $30,000 and need to decide what to do with the money. Write a paper discussing your options and the financial implications of those options. The paper you turn in should offer several alternatives and then the members of your group should reach a consensus of the best course of action.

G32. Write a short paper about Fibonacci numbers. You might check *The Fibonacci Quarterly,* particularly "A Primer on the Fibonacci Sequence," Parts I and II, in the February and April 1963 issues. The articles, written by Verner Hogatt and S. L. Basin, are considered classic articles on the subject. One member of your group should investigate the relationship of the Fibonacci numbers to nature, another the algebraic properties of the sequence, and another the history of the sequence.

G33. Suppose you were hired for a job paying $21,000 per year and were given the following options:

OPTION A:	Annual salary increase of $1,200
OPTION B:	Semiannual salary increase of $300
OPTION C:	Quarterly salary increase of $75
OPTION D:	Monthly salary increase of $10

Each person should write the arithmetic series for the total amount of money earned in 10 years under a different option.

Your group should reach a consensus as to which is the best option. Give reasons and show your calculations in the paper that your group submits.

G34. It is not uncommon for the owner of a home to receive a letter similar to the one shown below. Write a paper based on this letter. Different members of your group can work on different parts of the question, but you should submit one paper from your group.

a. What is the letter about?

```
Dear Customer:

Previously we provided you an enrollment package for our BiWeekly
Advantage Plan.  We have had an excellent response from our mortgagors,
yet we have not received your signed enrollment form and program fee to
start your program.

How effective is the BiWeekly Advantage Plan?  Reviewing your loan as of
the week of this letter we estimate:

                MORTGAGE INTEREST SAVINGS OF $99,287.56
                LOAN TERM REDUCED BY 7 YEARS.

Many financial advisors agree it is a sound investment practice to pay off
a mortgage early, particularly in light of growing economic uncertainty.

Act today!  Don't let procrastination or indecision deny you and your
family the opportunity to obtain mortgage-free homeownership sooner than
you ever expected.

Please call us toll-free at 1-800 555-6060 if you would like more
information about our BiWeekly Advantage Plan and a free personalized
mortgage analysis.  Our telephone representatives are available to
serve you from 9 am to 5 pm (EST).

                                            Sincerely,
```

b. A computer printout (shown at the right) was included with the letter. Assuming that these calculations are correct, discuss the advantages or disadvantages of accepting this offer.

c. The plan as described in the letter costs $375 to sign up. I called the company and asked what their plan would do that I could not do myself by simply making 13 payments a year to my mortgage holder. The answer I received was that the plan would do nothing more, but the reason people do sign up is because they do not have the self-discipline to make the midmonthly payments to themselves. Why is a biweekly payment equivalent to 13 annual payments instead of equivalent to a monthly payment?

Interest Rate:	8.3750
Current Balance:	$ 416,640.54
Monthly Payment:	**$ 3,180.91**

Homeownership: Remaining Terms		
	Years	Months
Current Payment Plan	29	5
BiWeekly Plan	22	5
REDUCTION[3]	7	0

Total Remaining Principal & Interest Payments:	
Current Payment Plan	$ 1,122,849.24
BiWeekly Plan	$ 923,561.68
INTEREST SAVINGS[3]	$ 199,287.56

d. The representative of the company told me that more than 250,000 people have signed up. How much income has the company received from this offer?

e. You calculated the income the company has received from this offer in part **d,** but that is not all it receives. It acts as a bonded and secure "holding company" for your funds (because the mortgage company does not accept "two-week" payments). This means that the company receives the use (interest value) on your money for two weeks out of every month. This is equivalent to half the year. Let's assume that the average monthly payment is $1,000 and that the company has 250,000 payments that they hold for half the year. If the interest rate is 5% (a secure guaranteed rate), how much potential interest can be received by this company?

INDIVIDUAL RESEARCH PROBLEMS

Learning to use sources outside your classroom and textbook is an important skill, and here are some ideas for extending some of the ideas in this chapter. You can find references to these projects in a library or at

 **www.mathnature.com**

PROJECT 9.1 Conduct a survey of banks, savings and loan companies, and credit unions in your area. Prepare a report on the different types of savings accounts available and the interest rates they pay. Include methods of payment as well as interest rates.

PROJECT 9.2 Do you expect to live long enough to be a millionaire? Suppose that your annual salary today is $39,000. If inflation continues at 6%, how long will it be before $39,000 increases to an annual salary of a million dollars?

PROJECT 9.3 Consult an almanac or some government source, and then write a report on the current inflation rate. Project some of these results to the year of your own expected retirement.

PROJECT 9.4 Karen says that she has heard something about APR rates but doesn't really know what the term means. Wayne says he thinks it has something to do with the prime rate, but he isn't sure what. Write a short paper explaining APR to Karen and Wayne.

PROJECT 9.5 Some savings and loan companies advertise that they pay interest *continuously.* Do some research to explain what this means.

PROJECT 9.6 Select a car of your choice, find the list price, and calculate 5% and 10% price offers. Check out available money sources in your community, and prepare a report showing the different costs for the same car. Back up your figures with data.

PROJECT 9.7 Outline a program for your own retirement. In the process of writing this paper, answer the following questions. You will need to state your assumptions about interest and inflation rates.

    a. What monthly amount of money today would provide you a comfortable living?

    b. Using the answer to part **a,** project that amount to your own retirement, calculating the effects of inflation. Use your own age and assume that you will retire at age 65.

    c. How much money would you need to have accumulated to provide the amount you found in part **b,** if you decide to live on the interest only?

    d. If you set up a sinking fund to provide the amount you found in part **c,** how much would you need to deposit each month?

    e. Offer some alternatives to the sinking fund you considered in part **d.**

    f. Draw some conclusions about your retirement.

10 The Nature of Set Theory and Counting

The first time I met eminent proof theorist Gaisi Takeuti I asked him what set theory was really about. "We are trying to get exact description of thoughts of infinite mind," he said. And then he laughed, as if filled with happiness by this impossible task.

RUDY RUCKER

CONTENTS

OVERVIEW

Sets are considered to be one of the most fundamental building blocks of mathematics. In fact, most mathematics books from basic arithmetic to calculus must introduce the concept early in the book. (Notice that this book first introduced sets in Section 1.4.) Small children learn to categorize the concept of sets when they learn numbers, colors, shapes, and sizes. The PBS show *Sesame Street* teaches the concept of set building with their song, "One of these things is not like the others." A quick search of the Internet will show you that the ideas of set theory can be as elementary as counting and as complex as logic, calculus, and abstract algebra. In this chapter we will consider the basic ideas of set theory and of counting.

IMPORTANT IDEAS

Find the union, intersection, or complement of given sets. [10.1]
Draw a Venn diagram for union, intersection, and complement. [10.1]
Distinguish among the symbols "⊆", "⊂", and "∈." [10.1]
Find the number of elements in a union. [10.1]
Carry out combined set operations. [10.2]
Prove set relationships using Venn diagrams. [10.2]
Find the number of elements in parts of a survey problem. [10.2]
Fundamental counting principle. [10.3]
Multiplication property of factorials. [10.3]
Formula for permutation. [10.3]
Formula for distinguishable permutations. [10.3]
Formula for combination. [10.4]
Classify and identify counting techniques. [10.5]
Expand binomials. [10.6]
Find a binomial coefficient. [10.6]
Work applied problems using the binomial theorem. [10.6]
Applications of counting (Rubik's Cube and Instant Insanity). [10.7]

BOOK REPORTS

Write a 500-word report on this book:
Innumeracy: Mathematical Illiteracy and Its Consequences. John Allen Paulos (New York: Hill and Wang, 1988).

LINKS

Individual Projects

Group Projects

Research Links

www.mathnature.com

10.1 SETS, SUBSETS, AND VENN DIAGRAMS

The following problem appeared on the 1987 Examination of the California Assessment Program as an open-ended problem.

> *James knows that half the students from his school are accepted at the public university nearby. Also, half are accepted at the local private college. James thinks that this adds up to 100%, so he will surely be accepted at one or the other institution. Explain why James may be wrong. If possible, use a diagram in your explanation.*

A Question of Thinking: A First Look at Students' Performance on Open-ended Questions in Mathematics.

We introduced some of the terminology and notation of sets in Section 1.4, but the study of sets, called **set theory,** permeates a great deal of mathematics. Georg Cantor (1845–1918) was the originator of set theory and of the study of transfinite numbers. In this chapter we see how set theory can help us answer this examination question. We will also use set theory extensively when discussing counting in this chapter and probability in the next chapter.

Venn Diagrams

A **set** is a collection and a useful way to depict sets is to draw a circle or an oval as a representation for the set. Recall (from Section 1.4) that the objects in the set are called its *members* or *elements.* The elements are depicted inside the circle, and objects not in the set are shown outside the circle. The **universal set** contains all the elements under consideration in a given discussion and is depicted as a rectangle. This representation of a set is called a **Venn diagram,** after John Venn (1834–1923). The Swiss mathematician Leonhard Euler (1707–1783) also used circles to illustrate principles of logic, so sometimes these diagrams are called **Euler circles.** However, Venn was the first to use them in a general way.

EXAMPLE 1

Let the universal set be all of the cards in a deck of cards. Draw a Venn diagram for the set of hearts.

Solution It is customary to represent the universal set as a rectangle (labeled *U*) and the set of hearts (labeled *H*) as a circle. The sets involved are too large to list all of the elements in either *H* or *U,* but we can say that the two of hearts (labeled ♥2) is a member of *H,* whereas the two of diamonds (labeled ♦2) is not a member of *H.* We write ♥2 ∈ *H,* whereas ♦2 ∉ *H.*

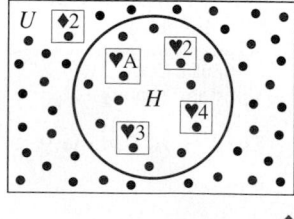

The customary notation for listing the elements in a set is to use braces. For Example 1,

$$H = \{♥A, ♥2, ♥3, ♥4, ♥5, ♥6, ♥7, ♥8, ♥9, ♥10, ♥J, ♥Q, ♥K\}$$

The set of elements that are not in *H* is referred to as the **complement** of *H,* and this is written using an overbar: $\overline{H} = \{\text{spades, diamonds, clubs}\}$. A Venn diagram for complement is shown in Figure 10.1.

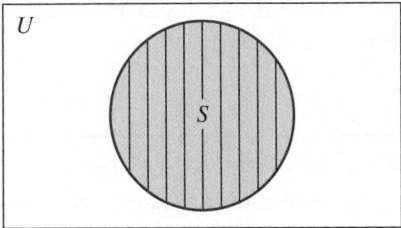
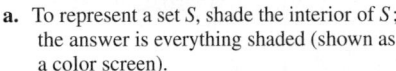
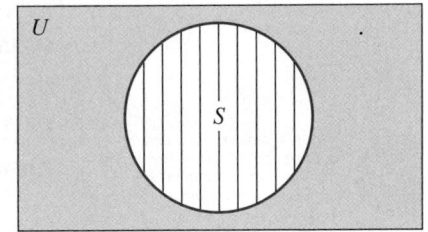

a. To represent a set S, shade the interior of S; the answer is everything shaded (shown as a color screen).

b. To represent a set $\overline{S}$, shade S; the answer is everything not shaded (shown as a color screen).

Figure 10.1 **Venn diagrams for a set and for its complement**

Notice that any set S divides the universe into two regions as shown in Figure 10.2.

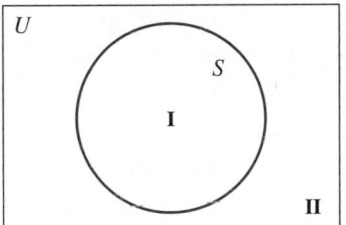

Figure 10.2 **General representation of a set S**

In Section 1.4, we talked about the *cardinality of a set*. It is the number of elements in the set. For example, the cardinality of U in Example 1 is 52, and the cardinality of H is 13. In Chapter 1, we symbolized the cardinality of a set as follows:

$$|H| = 13 \quad \text{and} \quad |U| = 52$$

> **CAUTION** Note that $H \neq |H|$. In words, a set is **not** the same as its cardinality.

Subsets and Proper Subsets

Most applications will involve more than one set, so we begin by considering the relationships between two sets A and B. The various possible relationships are shown in Figure 10.3. We say that A is a **subset** of B, which in set theory is written $A \subseteq B$. Recall, this means that every element of A is also an element of B (see Figure 10.3**a**). Similarly, $B \subseteq A$, if every element of B is also an element of A (Figure 10.3**b**). Two sets A and B are **equal,** written $A = B$, if the sets have exactly the same elements (Figure 10.3**c**). Finally, A and B are **disjoint** if they have no elements in common (Figure 10.3**d**).

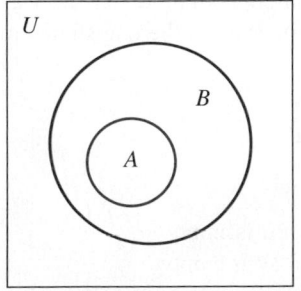
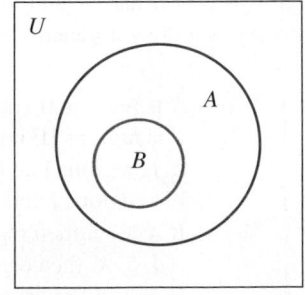
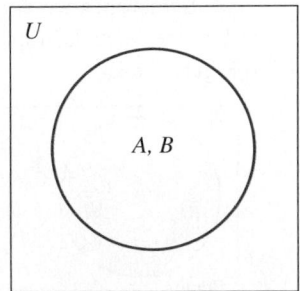
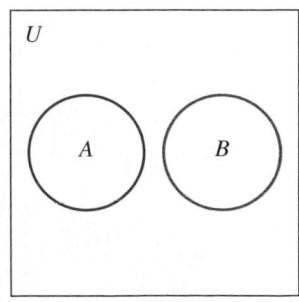

a. $A \subseteq B$ **b.** $B \subseteq A$ **c.** $A = B$ **d.** A and B are disjoint

Figure 10.3 **Relationships between two sets A and B**

If A is a subset of B, and also $A \neq B$, then we say that A is a *proper subset* of B. In set theory we write this as $A \subset B$.

STOP *This might seem like a simple example, but it shows some technical differences in the symbolism of set theory.*

EXAMPLE 2

Answer each of the given true–false questions.

$A = \{$American-manufactured automobile$\}$

$C = \{$Chevrolet, Camaro, Corvette]

$R = \{$Colors in the rainbow$\}$

$S = \{$red, orange, yellow, green, blue, indigo, violet$\}$

a. $A \subseteq C$ **b.** $C \subseteq A$

c. Chevrolet $\subseteq A$ **d.** red $\in R$

e. $\{$red$\} \in R$ **f.** $R \subseteq S$

g. $R \subset S$ **h.** $\varnothing \subseteq A$

i. $\{\ \ \} \subset C$ **j.** $\{\varnothing\} \subset R$

Solution

a. "A is a subset of C" is *false* because there is at least one American manufactured automobile, say the Cadillac, that is not in listed in the set C.

b. "C is a subset of A" is *true* because each element of C is also an element of A.

c. "Chevrolet is a subset of A" is *false* since "Chevrolet" is an element, not a set.

d. "Red is an element of the set R" is *true*. Contrast the notation in parts **c** and **d**. If part **c** were $\{$Chevrolet$\} \subseteq A$, it would have been true.

e. "The set consisting of the color red is an element of the set R" is *false* since we see "$\{$red$\}$" listed in the set S. Contrast the notation in parts **d** and **e**.

f. "R is a subset of S" is *true* since every color of the rainbow is listed in the set S.

g. "R is a proper subset of S" is *false* because $R = S$. To be a proper subset, there must be some element of S that is not in the set R.

h. "The empty set is a subset of A" is *true,* since the empty set is a subset of every set.

i. "The empty set is a proper subset of C" is *true*. The empty set is a proper subset of every *nonempty* set. $\varnothing \subset \varnothing$ is false, but $\varnothing \subseteq \varnothing$ is true.

j. "The set containing the empty set is a subset of R" is *false* since "$\varnothing$" is not listed inside the set R.   ◆

Sometimes we are given two sets X and Y, and we know nothing about the way they are related. In this situation, we draw a general figure, such as the one shown in Figure 10.4.

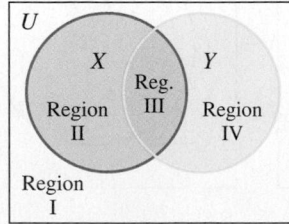

X is regions II and III.

Y is regions III and IV.

$\overline{X}$ is regions I and IV.

$\overline{Y}$ is regions I and II.

If $X \subseteq Y$, then region II is empty.

If $Y \subseteq X$, then region IV is empty.

If $X = Y$, then regions II and IV are empty.

If X and Y are disjoint, then region III is empty.

Figure 10.4 **General Venn diagram for two sets**

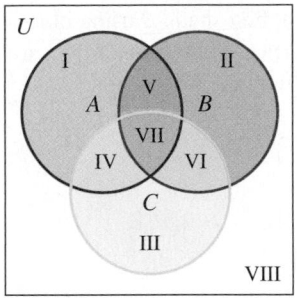

Figure 10.5 General Venn diagram for three sets

We can generalize for more sets. The general Venn diagram for three sets divides the universe into eight regions, as shown in Figure 10.5.

EXAMPLE 3

Name the regions in Figure 10.5 described by each of the following.

a. A **b.** C **c.** $\overline{A}$ **d.** $\overline{B}$ **e.** $A \subseteq B$ **f.** A and C are disjoint

Solution

a. A is regions I, IV, V, and VII.

b. C is regions III, IV, VI, and VII.

c. $\overline{A}$ is regions II, III, VI, and VIII.

d. $\overline{B}$ is regions I, III, IV, and VIII.

e. $A \subseteq B$ means that regions I and IV are empty.

f. A and C are disjoint means that regions IV and VII are empty.   ◆

Operations with Sets

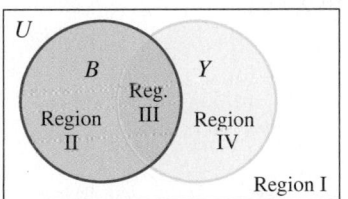

Figure 10.6 Venn diagram showing intersection and union

Suppose we consider two general sets, B and Y, as shown in Figure 10.6. If we show the set Y using a yellow highlighter, and the set B using a blue highlighter, it is easy to visualize two operations. The *intersection* of the sets is the region shown in green (the parts that are *both* yellow and blue). We see this is region III, and we describe this using the word **"and."** The *union* of the sets is the part shown in *any* color (the parts that are yellow or blue or green). We see this is regions II, III, and IV, and we describe this using the word **"or."**

Operations on Sets: Intersection and Union

Intersection (∩) is translated as **and,** and union (∪) as **or.**

> The **intersection** of sets A and B, denoted by $A \cap B$, is the set consisting of all elements common to A and B.
>
> The **union** of sets A and B, denoted by $A \cup B$, is the set consisting of all elements of A or B or both.

EXAMPLE 4

Draw Venn Diagrams for union and intersection

a. $B \cap Y$ **b.** $B \cup Y$

Solution   Highlighter pens work well when drawing Venn diagrams.

a. "$B \cap Y$" is the intersection of the sets B and Y. Draw two circles as shown in Figure 10.6. Shade in one of the circles using a blue highlighter and the other with a yellow highlighter. What do you see? After the ink dries you should see three colors. The blue and yellow circles combine in the intersection to form a green portion. (See Figure 10.6.) this is the intersection. In practice, we use a highlighter to show only the intersection portion. In Figure 10.7, first shade B using horizontal lines, and then shade the second set, Y, using vertical lines. The *intersection* is all parts that are shaded twice (both horizontal and vertical), as shown with the highlighter.

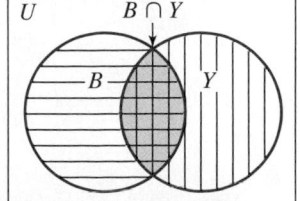

Figure 10.7 $B \cap Y$

b. "$B \cup Y$" is the union of the set B and Y. In Figure 10.8, first shade B using horizontal lines and then shade the second set, Y, using vertical lines. The *union* is all parts that are shaded (either once or twice), as shown with the highlighter.

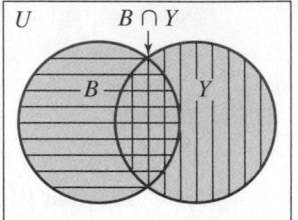

Figure 10.8 $B \cup Y$ ◆

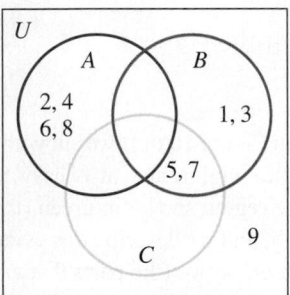

Figure 10.9

EXAMPLE 5

Let $U = \{1, 2, 3, 4, 5, 6, 7, 8, 9\}$, $A = \{2, 4, 6, 8\}$, $B = \{1, 3, 5, 7\}$, and $C = \{5, 7\}$. A Venn diagram showing these sets is shown in Figure 10.9. Find:
a. $A \cup C$ **b.** $B \cup C$ **c.** $B \cap C$ **d.** $A \cap C$

Solution

a. $A \cup C = \{2, 4, 6, 8\} \cup \{5, 7\}$
$\qquad = \{2, 4, 5, 6, 7, 8\}$

Notice that the union consists of all elements in A or in C or in both. Also note that the order in which the elements are listed is not important.

b. $B \cup C = \{1, 3, 5, 7\} \cup \{5, 7\}$
$\qquad = \{1, 3, 5, 7\}$

Notice that, even though the elements 5 and 7 appear in both sets, they are listed only once. That is, the sets $\{1, 3, 5, 7\}$ and $\{1, 3, 5, 5, 7, 7\}$ are equal (exactly the same).

c. $B \cap C = \{1, 3, 5, 7\} \cap \{5, 7\} = \{5, 7\}$

The intersection contains the elements common to both sets. Notice that the resulting set has a name (it is called C), so we write

$\qquad B \cap C = C$

d. $A \cap C = \{2, 4, 6, 8\} \cap \{5, 7\} = \{\ \}$

These sets have no elements in common, so we write $\{\ \}$ or $\varnothing$.   ◆

Suppose we consider the cardinality of the various sets in Example 5:

$$|U| = 9, \quad |A| = 4, \quad |B| = 4, \quad \text{and} \quad |C| = 2$$
$$|A \cup C| = 6 \quad \text{(part a)}$$
$$|B \cup C| = 4 \quad \text{(part b)}$$
$$|B \cap C| = 2 \quad \text{(part c)}$$
$$|A \cap C| = 0 \quad \text{(part d)}$$
$$|\varnothing| = 0 \quad \text{(part d)}$$

Consider the set S formed from the sets in Example 5:

$$S = \{U, A, B, C, \varnothing\}$$

This is a set of sets; there are five sets in S, so $|S| = 5$. Furthermore, if we remove sets from S, one-by-one, we find:

$$T = \{A, B, C, \varnothing\}, \quad \text{so } |T| = 4$$
$$U = \{B, C, \varnothing\}, \quad \text{so } |U| = 3$$
$$V = \{C, \varnothing\}, \quad \text{so } |V| = 2$$

Finally,

$$W = \{\varnothing\}, \quad \text{so } |W| = 1$$

Thus $|\{\varnothing\}| = 1$, but $|\varnothing| = 0$, so $\{\varnothing\} \neq \varnothing$.

Cardinality of Intersections and Unions

The *cardinality of an intersection* is easy; it is found by looking at the number of elements in the intersection.

The *cardinality of a union* is a bit more difficult. For sets with small cardinalities, we can find the cardinality of the unions by direct counting, but if the sets have large cardinalities, it might not be easy to find the union and then the cardinality by direct counting. Some students might want to find $|B \cup C|$ by adding $|B|$ and $|C|$, but you can see from Example 5 that $|B \cup C| \neq |B| + |C|$. However, if you look at the Venn diagram for the number of elements in the union of two sets, the situation becomes quite clear, as shown in Figure 10.10.

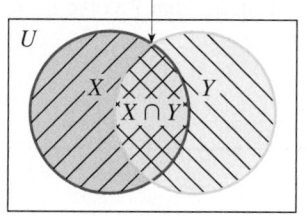

$|X| + |Y|$ adds this region twice

Figure 10.10 **Venn diagram for the number of elements in the union of two sets**

Formula for the Cardinality of the Union of Two Sets

STOP

This formula will be used later in the book.

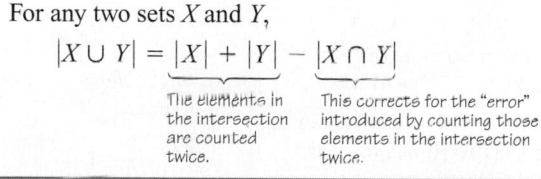

For any two sets X and Y,
$$|X \cup Y| = \underbrace{|X| + |Y|}_{\substack{\text{The elements in} \\ \text{the intersection} \\ \text{are counted} \\ \text{twice.}}} - \underbrace{|X \cap Y|}_{\substack{\text{This corrects for the "error"} \\ \text{introduced by counting those} \\ \text{elements in the intersection} \\ \text{twice.}}}$$

EXAMPLE 6

Suppose a survey indicates that 45 students are taking mathematics and 41 are taking English. How many students are taking math or English?

Solution   At first, it might seem that all you do is add 41 and 45, but such is not the case. Let $M = \{$persons taking math$\}$ and $E = \{$persons taking English$\}$.

To find out how many students are taking math and English, we need to know the number in this intersection.

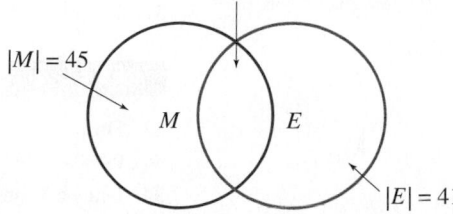

$|M| = 45$ $|E| = 41$

As you can see, you need further information. Problem solving requires that you not only recognize what known information is needed when answering a question, but also recognize when additional information is needed. Suppose 12 students are taking both math and English. In this case we see that:

By formula:

$$|M \cup E| = |M| + |E| - |M \cap E|$$
$$= 45 + 41 - 12$$
$$= 74$$

By diagram:
First, fill in 12 in $M \cap E$. Then,
$|M| = 45$ $\qquad$ $|E| = 41$
Fill in 33 $\qquad$ Fill in 29
$(45 - 12 = 33)$ $\quad$ $(41 - 12 = 29)$
The total number is $33 + 12 + 29 = 74$. ◆

Example 6 looks very much like the open-ended examination question we posed at the beginning of this section. You will find that open-ended question in the problem set. In the next section, we will consider survey questions that involve more than two sets.

Problem Set 10.1

LEVEL 1

1. IN YOUR OWN WORDS What do we mean by the operations of *union, intersection,* and *complementation?*

2. IN YOUR OWN WORDS Give an example of a set with cardinality 0.

3. IN YOUR OWN WORDS Give an example of a set with cardinality greater than 1 million.

4. IN YOUR OWN WORDS This section began with an open-ended question from the 1987 Examination of the California Assessment Program:

James knows that half the students from his school are accepted at the public university nearby. Also, half are accepted at the local private college. James thinks that this adds up to 100%, so he will surely be accepted at one or the other institution. Explain why James may be wrong. If possible, use a diagram in your explanation.

Perform the given set operations in Problems 5–14. Let
$U = \{1, 2, 3, 4, 5, 6, 7, 8, 9, 10\}.$

5. $\{2, 6, 8\} \cup \{6, 8, 10\}$

6. $\{2, 6, 8\} \cap \{6, 8, 10\}$

7. $\{2, 5, 8\} \cup \{3, 6, 9\}$

8. $\{2, 5, 8\} \cap \{3, 6, 9\}$

9. $\{1, 2, 3, 4, 5\} \cap \{3, 4, 5, 6, 7\}$

10. $\overline{\{2, 8, 9\}}$

11. $\{1, 2, 3, 4, 5\} \cup \{3, 4, 5, 6, 7\}$

12. $\overline{\{1, 2, 5, 7, 9\}}$

13. $\overline{U}$ $\qquad\qquad\qquad$ **14.** $\overline{\varnothing}$

Let $U = \{1, 2, 3, 4, 5, 6, 7\}, A = \{1, 2, 3, 4\},$
$B = \{1, 2, 5, 6\},$ *and* $C = \{3, 5, 7\}.$ *List all the members of each of the sets in Problems* 15–26.

15. $A \cup B$ $\qquad\qquad$ **16.** $A \cap B$

17. $A \cup C$

18. $A \cap C$

19. $B \cap C$

20. $B \cup C$

21. $\overline{A}$

22. $\overline{B}$

23. $\overline{C}$

24. $\overline{U}$

25. $\varnothing \cup A$

26. $\varnothing \cap B$

In Problems 27–28, use set notation to identify the shaded region.

27. $\qquad\qquad\qquad\qquad$ **28.**

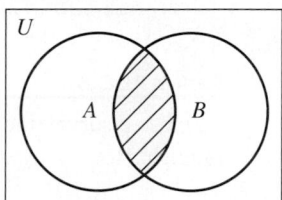 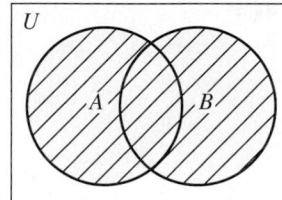

Draw Venn diagrams for each of the relationships in Problems 29–32.

29. $X \cup Y$ $\qquad\qquad$ **30.** $X \cap Z$

31. $\overline{Y}$ $\qquad\qquad\qquad$ **32.** $\overline{Z}$

LEVEL 2

33. Draw a Venn diagram showing people who are over 30, people who are 30 or under, and people who drive a car.

34. Draw a Venn diagram showing males, females, and those people who ride bicycles.

35. Draw a Venn diagram showing that all Chevrolets are automobiles.

36. Draw a Venn diagram showing that all cell phones are communication devices.

Decide whether each statement in Problem 37–53 is true or false. Give reasons for your answers.

37. {*m, a, t, h*} ⊆ {*m, a, t, h, e, i, c, s*}

38. *m* ∈ {*m, a, t, h*}   **39.** {*m*} ∈ {*m, a, t, h*}

40. {*m, a, t, h*} ⊆ {*h, t, a, m*}

41. {*m, a, t, h*} ⊂ {*h, t, a, m*}

42. {math} ∈ {*m, a, t, h*}

43. {math, history} ⊂ {high school subjects}

44. {   } ⊆ {Jeff, Maureen, Terry}

45. {blue} ∈ {colors of the rainbow}

46. 1 ∈ {1, 2, 3, 4, 5}   **47.** {1} ∈ {1, 2, 3, 4, 5}

48. 1 ∈ {{1}, {2}, {3}, {4}}

49. {1} ∈ {{1}, {2}, {3}, {4}}

50. {1} ⊂ {{1}, {2}, {3}, {4}}

51. 0 = {   }   **52.** ∅ = {   }   **53.** {∅} = {   }

54. Santa Rosa Junior College enrolled 29,000 students in the fall of 1999. It was reported that of that number, 58% were female and 42% were male. In addition, 62% were over the age of 25. How many students were there in each category if 40% of those over the age of 25 were male? Draw a Venn diagram showing these relationships.

55. Montgomery College has a 50-piece band and a 36-piece orchestra. If 14 people are members of both the band and the orchestra, can the band and orchestra travel in two 40-passenger buses?

56. From a survey of 100 college students, a marketing research company found that 75 students owned stereos, 45 owned cars, and 35 owned both cars and stereos.
   a. How many students owned either a car or a stereo (but not both)?
   b. How many students do not own either a car or a stereo?

57. In a survey of a TriDelt chapter with 50 members, 18 were taking mathematics, 35 were taking English, and 6 were taking both. How many were not taking either of these subjects?

58. In a senior class at Rancho Cotati High School, there were 25 football players and 16 basketball players. If 7 persons played both sports, how many different people played in these sports?

59. The fire department wants to send booklets on fire hazards to all teachers and homeowners in town. How many booklets does it need, using these statistics?

   50,000 homeowners
   4,000 teachers
   3,000 teachers who own their own homes

PROBLEM SOLVING

60. Each of the circles in Figure 10.11 is identified by a letter, each having a number value from 1 to 9. Where the circles overlap, the number is the sum of the values of the letters in the overlapping circles. What is the number value for each letter?*

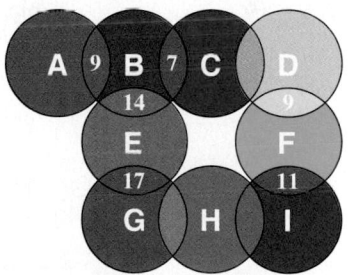

Figure 10.11 **Circle intersection puzzle**

*From "Perception Puzzles," by Jean Moyer, *Sky,* January 1995, p. 120. *Math Puzzles and Logic Problems,* © 1995 Dell Magazines, a division of Penny Marketing Limited Partnership, reprinted by permission of Dell Magazines.

10.2   COMBINED OPERATIONS WITH SETS

Operations with Sets

In the last section, we introduced three operations: *intersection, union,* and *complement.* In this section, we consider mixed operations with more than two sets. For sets, we perform operations from left to right; however, if there are parentheses, operations within them are performed first.

EXAMPLE 1

Illustrate using Venn diagrams: **a.** $\overline{A} \cup \overline{B}$ **b.** $\overline{A \cup B}$

Solution

a. This is a combined operation that should be read from left to right. Find the complements of A and B and *then* find the union. This is called a *union of complements*.

Step 1. Shade $\overline{A}$ (vertical lines).   Then shade $\overline{B}$ (horizontal lines).

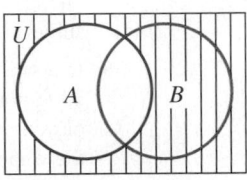   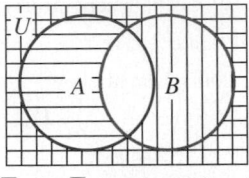

$\overline{A}$ (vertical lines) $\overline{A}$ with $\overline{B}$ (horizontal lines)

Step 2. $\overline{A} \cup \overline{B}$ is every portion that is shaded with horizontal or vertical lines. We show that here using a color highlighter.

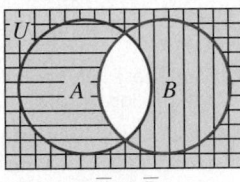

$\overline{A} \cup \overline{B}$

b. This is a combined operation that should be interpreted to mean $(\overline{A \cup B})$, which is the complement of the union. *First* find $A \cup B$ (vertical lines), and *then* find the complement (color highlighter). This is called the *complement of a union*. Compare this with part **a.**

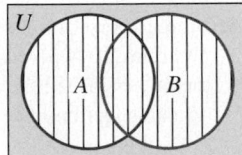

◆

Notice that $\overline{A \cup B} \neq \overline{A} \cup \overline{B}$. If they were equal, the final highlighted color portions of the Venn diagrams from Example 1 would be the same.

EXAMPLE 2   **PÓLYA'S METHOD**

Prove $\overline{A \cup B} = \overline{A} \cap \overline{B}$.

Solution   We use Pólya's problem-solving guidelines for this example.

Understand the Problem. We wish to prove the given statement is true *for all* sets A and B, so we cannot work with a *particular* example.

Devise a Plan. The procedure is to draw separate Venn diagrams for the left and the right sides, and then to compare them to see if they are identical.

Carry Out the Plan.

Step 1. Draw a diagram for the expression on the left side of the equal sign. The final result is shown with color highlighter. (See Example 1 for details.)

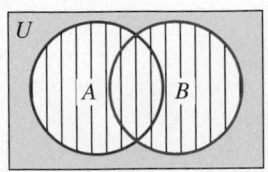

$\overline{A \cup B}$

Step 2. Draw a diagram for the expression on the right side of the equal sign.

$\overline{A}$ (vertical lines)

$\overline{B}$ (horizontal lines)

The final result, $\overline{A} \cap \overline{B}$, is the part with both vertical and horizontal lines (as shown with the color highlighter).

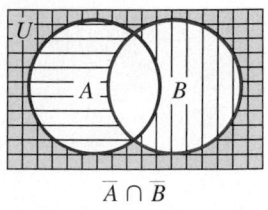

$\overline{A} \cap \overline{B}$

Step 3. Compare the portions shaded by the color highlighter in the two Venn diagrams. They are the same, so we have proved $\overline{A \cup B} = \overline{A} \cap \overline{B}$.   ◆

The result proved in Example 2 is called **De Morgan's law.** In the problem set you are asked to prove the second part of De Morgan's law.

De Morgan's Laws

For any sets X and Y:

$$\overline{X \cup Y} = \overline{X} \cap \overline{Y} \qquad \overline{X \cap Y} = \overline{X} \cup \overline{Y}$$

Next, we find the regions described by some combined operations for three sets.

EXAMPLE 4

Using the eight regions labeled at the right, describe each of the following sets:

a. $A \cup B$

b. $A \cap C$

c. $B \cap C$

d. $\overline{A}$

e. $\overline{A \cup B}$

f. $A \cap B \cap C$

g. $A \cup (B \cap C)$

h. $\overline{A \cup B} \cap C$

Solution

a. A (vertical lines); B (horizontal lines) $A \cup B$ is everything shaded and is highlighted.

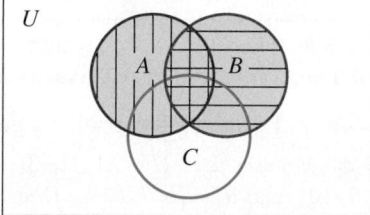

Regions I, II, IV, V, VI, and VII

b. A (vertical lines); C (horizontal lines) $A \cap C$ is everything shaded twice and is highlighted.

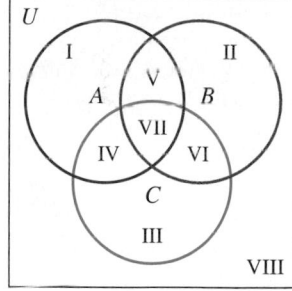

Regions IV and VII

c. B (vertical lines); C (horizontal lines); $B \cap C$ is everything shaded twice, as highlighted.

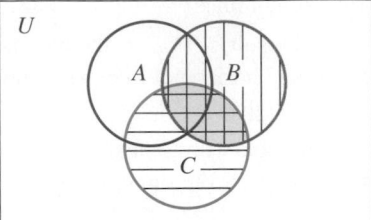

Regions VI and VII

d. A (vertical lines); $\overline{A}$ is everything not shaded, as highlighted.

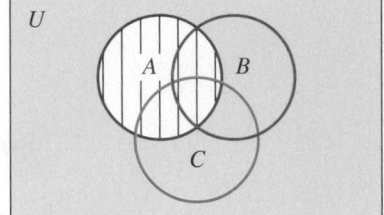

Regions II, III, VI, and VIII

e. $A \cup B$ (vertical lines); $\overline{A \cup B}$ is everything not shaded, as highlighted.

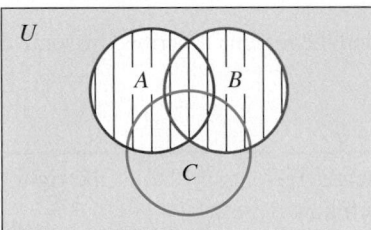

Regions III and VIII

f. $A \cap B \cap C$ is $(A \cap B) \cap C$, so we show $A \cap B$ (vertical lines) and C (horizontal lines); highlight regions shaded twice.

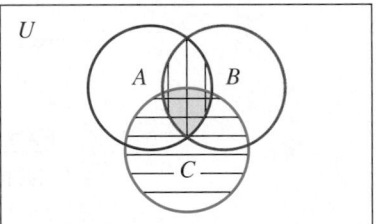

Region VII

g. Parentheses first, $B \cap C$ (vertical lines); A (horizontal lines); $A \cup (B \cap C)$ is everything shaded, as highlighted.

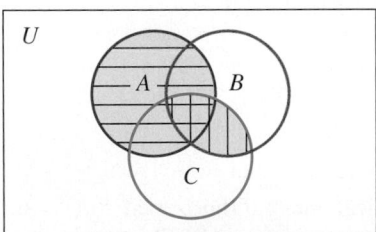

Regions I, IV, V, VI, and VII

h. $\overline{A \cup B}$ (vertical lines); and C (horizontal lines); everything shaded twice is highlighted.

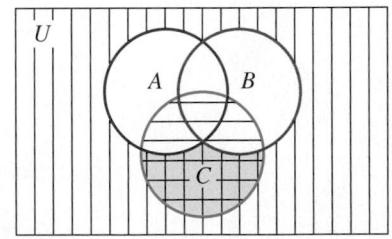

Region III ◆

EXAMPLE 4   PÓLYA'S METHOD

If $(P \cup Q) \cup R = P \cup (Q \cup R)$, we say that the operation of union is *associative*. Is the operation of union for sets an associative operation?

Solution   We use Pólya's problem-solving guidelines for this example.

Understand the Problem. Let $U = \{1, 2, 3, 4, 5, 6, 7, 8, 9, 10\}$, $P = \{1, 4, 7\}$, $Q = \{2, 4, 9, 10\}$, and $R = \{6, 7, 8, 9\}$. Does $(P \cup Q) \cup R = P \cup (Q \cup R)$?

$$(P \cup Q) \cup R = \{1, 2, 4, 7, 9, 10\} \cup \{6, 7, 8, 9\} = \{1, 2, 4, 6, 7, 8, 9, 10\}$$
$$P \cup (Q \cup R) = \{1, 4, 7\} \cup \{2, 4, 6, 7, 8, 9, 10\} = \{1, 2, 4, 6, 7, 8, 9, 10\}$$

For this example, the operation of union for sets is associative. If we had observed $(P \cup Q) \cup R \neq P \cup (Q \cup R)$, then we would have had a counterexample. Although they are equal, we cannot say that the property is true for all possibilities. However, all is not lost because it did help us to understand the question.

Devise a Plan. Use Venn diagrams.

Carry Out the Plan. Recall that the union is the entire shaded area.

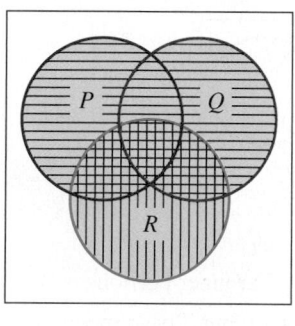 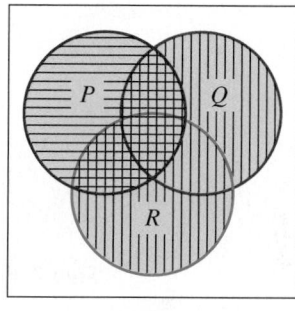

$$(P \cup Q) \cup R \qquad\qquad P \cup (Q \cup R)$$

Look Back. The operation of union for sets is an associative operation since the parts shaded in yellow are the same for both diagrams.   ◆

Survey Problems

We used the cardinality of a union property in the last section to find the number of elements in various regions formed by two sets. For three sets, the situation is a little more involved. There is a formula for the number of elements, but it is easier to use Venn diagrams, as illustrated by Example 5. Remember, the usual procedure is to fill in the number in the innermost region first and work your way outward through the Venn diagram using subtraction.

EXAMPLE 5   **PÓLYA'S METHOD**

A survey of 100 randomly selected students gave the following information:

 45 students are taking mathematics.
 41 students are taking English.
 40 students are taking history.
 15 students are taking math and English.
 18 students are taking math and history.
 17 students are taking English and history.
 7 students are taking all three.

a. How many are taking only mathematics?
b. How many are taking only English?
c. How many are taking only history?
d. How many are not taking any of these courses?

Solution We use Pólya's problem-solving guidelines for this example.

Understand the Problem. We are considering students who are members of one or more of three sets. If U represents the universe, then $|U| = 100$. We also define the three sets:

$M = \{\text{students taking mathematics}\}$
$E = \{\text{students taking English}\}$
$H = \{\text{students taking history}\}$

Devise a Plan. The plan is to draw a Venn diagram, and then to fill in the various regions. We fill in the innermost region first, and then work our way outward (using subtraction) until the number of elements of the eight regions formed by the three sets is known.

Carry Out the Plan.

Step 1. We note $|M \cap E \cap H| = 7$.

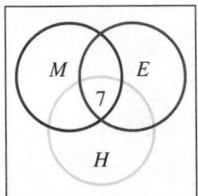

Step 2. Fill in the other inner portions.

$|E \cap H| = 17$, but 7 have previously been accounted for, so an additional 10 members ($17 - 7 = 10$) are added to the Venn diagram.

$|M \cap H| = 18$; fill in $18 - 7 = 11$.

$|M \cap E| = 15$; fill in $15 - 7 = 8$.

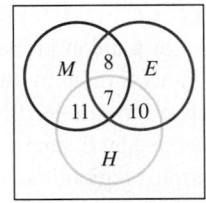

Step 3. Fill in the other regions.

$|H| = 40$, but 28 have previously been accounted for in the set H, so there are an additional 12 members ($40 - 11 - 7 - 10 = 12$).

$|E| = 41$; fill in $41 - 8 - 7 - 10 = 16$.

$|M| = 45$; fill in $45 - 11 - 7 - 8 = 19$.

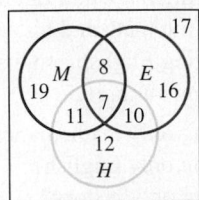

Step 4. Add all the numbers listed in the sets of the Venn diagram to see that 83 students have been accounted for. Since 100 students were surveyed, we see that 17 are not taking any of the three courses. We now have the answers to the questions directly from the Venn diagram: **a.** 19 **b.** 16 **c.** 12 **d.** 17

Look Back. Does our answer make sense? Add all the numbers in the Venn diagram as a check to see that we have accounted for the 100 students. ◆

Problem Set 10.2

LEVEL 1

1. **IN YOUR OWN WORDS** What do we mean by *De Morgan's laws?*

2. **IN YOUR OWN WORDS** What is the general procedure for drawing a Venn diagram for a survey problems?

Let $U = \{1, 2, 3, 4, 5, 6, 7\}, A = \{1, 2, 3, 4\},$
$B = \{1, 2, 5, 6\},$ *and* $C = \{3, 5, 7\}.$ *List all the members of each of the sets in Problems 3–10.*

3. $(A \cup B) \cap C$

4. $A \cup (B \cap C)$

5. $\overline{A \cup B} \cap C$

6. $A \cup \overline{B \cap C}$

7. $\overline{A} \cup (B \cap C)$

8. $(A \cup B) \cap \overline{C}$

9. $\overline{(A \cup B) \cap C}$

10. $\overline{A} \cup (\overline{B} \cap \overline{C})$

Consider the sets X and Y. Write each of the statements in Problems 11–18 in symbols.

11. a union of complements

12. a complement of a union

13. a complement of an intersection

14. an intersection of complements

15. the complement of the union of X and Y

16. the union of the complements of X and Y

17. the intersection of the complements of X and Y

18. the complement of the intersection of X and Y

Draw Venn diagrams for each of the relationships in Problems 19–34.

19. $\overline{A} \cup B$

20. $A \cap \overline{B}$

21. $\overline{A \cap B}$

22. $\overline{A \cup B}$

23. $A \cap (B \cup C)$

24. $A \cup (B \cup C)$

25. $A \cap \overline{B} \cup C$

26. $\overline{A \cup B \cup C}$

27. $\overline{A \cup B} \cap C$

28. $A \cup \overline{B \cap C}$

29. $\overline{A} \cup (B \cap C)$

30. $(A \cup B) \cap \overline{C}$

31. $\overline{(A \cup B) \cap C}$

32. $\overline{A} \cup (\overline{B} \cap \overline{C})$

33. $\overline{(A \cup B) \cup C}$

34. $(A \cap B) \cup (A \cap C)$

LEVEL 2

35. Draw a Venn diagram showing the relationship among cats, dogs, and animals.

36. Draw a Venn diagram showing the relationship among trucks, buses, and cars.

37. Draw a Venn diagram showing people in a classroom wearing some black, people wearing some blue, and people wearing some brown.

38. Draw a Venn diagram showing birds, bees, and living creatures.

39. In 1995 the United States population was approximately 263 million. It was reported that of that number, 80% were white, 12% were black, and 9% were Hispanic. If $\frac{1}{2}$% have one black and one white parent, 2% have one black and one Hispanic parent, and 1% have one white and one Hispanic parent, how many people are there in each category? Draw a Venn diagram showing these relationships.

40. Three bills were voted on in the U.S. Senate in the 107th Congress, second session (2002). Here are the results of the 8 senators voting:

Senator	*Bill A* *S.* *2514*	*Bill B* *H.R.* *2356*	*Bill C* *Amen.* *3017*
Clinton (D)	yea	yea	yea
Kennedy (D)	yea	yea	yea
Bayh (D)	nay	yea	nay
Inouye (D)	yea	yea	nay
Campbell (R)	yea	nay	nay
Helms (R)	yea	nay	nay
Thurmond (R)	yea	nay	nay
Lott (R)	yea	nay	nay

Place each senator in the appropriate region of Figure 10.12. A "yea" vote is recorded as supporting the bill and consequently places the senator inside the circle representing that particular bill.

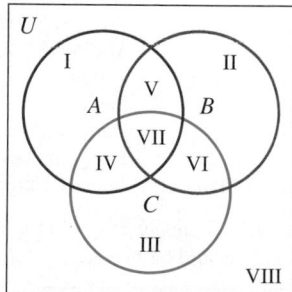

Figure 10.12 Senatorial votes

In Problems 41–44, use set notation to identify the shaded region.

41.

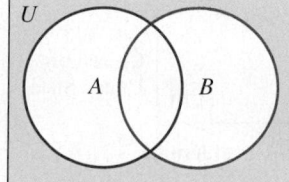

42.

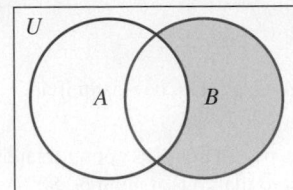

43.

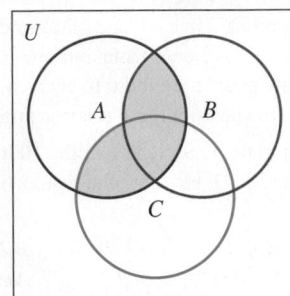

44.

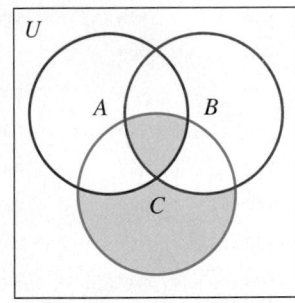

45. Listed below are five female and five male Wimbledon tennis champions, along with their country of citizenship and handedness.

Female	*Male*
Steffi Graf, Germany, right	Michael Stich, Germany, right
Martina Navratilova, U.S., left	
Chris Evert Lloyd, U.S., right	Stefan Edberg, Sweden, right
Evonne Goolagong, Australia, right	Boris Becker, Germany, right
	Pat Cash, Australia, right
Virginia Wade, Britain, right	John McEnroe, U.S., left

Using Figure 10.13, indicate in which region each of the above individuals would be placed.

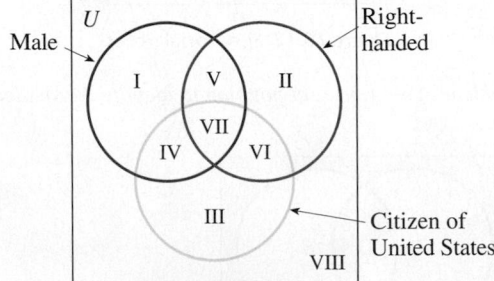

Figure 10.13 **Problem 45**

In Problems 46–51, use Venn diagrams to prove or disprove each statement. Remember to draw a diagram for the left side of the equation and another for the right side. If the final shaded portions are the same, then you have proved the result. If the final shaded portions are not identical, then you have disproved the result.

46. $\overline{A \cup B} = \overline{A} \cup \overline{B}$

47. $\overline{A \cap B} = \overline{A} \cup \overline{B}$

48. $(A \cup B) \cup C = A \cup (B \cup C)$

49. $A \cup (B \cup C) = (A \cup B) \cup (A \cup C)$

50. $A \cap (B \cup C) = (A \cap B) \cap (A \cap C)$

51. De Morgan's law: $\overline{X \cap Y} = \overline{X} \cup \overline{Y}$

52. In a recent survey of 100 persons, the following information was gathered:

59 use shampoo A.
51 use shampoo B.
35 use shampoo C.
24 use shampoos A and B.
19 use shampoos A and C.
13 use shampoos B and C.
11 use all three.

Let

$A = \{$persons who use shampoo A$\}$
$B = \{$persons who use shampoo B$\}$
$C = \{$persons who use shampoo C$\}$

Use a Venn diagram to show how many persons are in each of the eight possible categories.

53. Matt E. Matic was applying for a job. To determine whether he could handle the job, the personnel manager sent him out to poll 100 people about their favorite types of TV shows. His data were as follows:

59 preferred comedies.
38 preferred variety shows.
42 preferred serious drama.
18 preferred comedies and variety programs.
12 preferred variety and serious drama.
16 preferred comedies and serious drama.
7 preferred all types.
2 did not like any of these TV show types.

If you were the personnel manager, would you hire Matt on the basis of this survey?

54. A poll was taken of 100 students at a commuter campus to find out how they got to campus. The results were:

42 said they drove alone.
28 rode in a carpool.
31 rode public transportation.
9 used both carpools and public transportation.
10 used both a carpool and sometimes their own cars.
6 used buses as well as their own cars.
4 used all three methods.

How many used none of the above-mentioned means of transportation?

55. In an interview of 50 students,

12 liked Proposition 8 and Proposition 13.
18 liked Proposition 8, but not Proposition 5.
4 liked Proposition 8, Proposition 13, and Proposition 5.
25 liked Proposition 8.
15 liked Proposition 13.
10 liked Proposition 5, but not Proposition 8 or
  Proposition 13.
1 liked Proposition 13 and Proposition 5, but not
  Proposition 8.

a. Show the completed Venn diagram.
b. Of those surveyed, how many did not like any of the
three propositions?
c. How many like Proposition 8 and Proposition 5?

PROBLEM SOLVING
---

56. On the *NBC Nightly News* on Thursday, May 25, 1995,
Tom Brokaw read a brief report on computer use in the
United States. The story compared computer users by ethnic background, and Brokaw reported that 14% of blacks
and 13% of Hispanics use computers, but 27% of whites
use computers. Brokaw then commented that computer
use by whites was equal to that of blacks and Hispanics
combined.
a. Draw a Venn diagram showing the white, black, Hispanic, Asian, and Native American populations of the
United States, along with U.S. computer users.
b. Use the Venn diagram from part **a** to show that percentages cannot be added as was done by Brokaw.

57. Draw a Venn diagram for four sets, and label the 16
regions.

58. The Venn diagram in Figure 10.14 shows five sets. It was
drawn by Allen J. Schwenk of the U.S. Naval Academy. As

you can see, there are 32 separate regions. Describe the
following regions:

a. 1   **b.** 11   **c.** 21   **d.** 31   **e.** 16

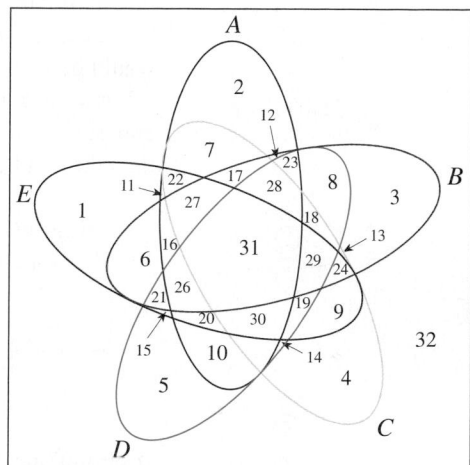

Figure 10.14 **General Venn diagram for 5 sets**

59. Using the Venn diagram in Figure 10.14, specify which
region is described by:
a. $A \cup B \cup C \cup D \cup E$ **b.** $A \cap \overline{B \cup C \cup D \cup E}$

60. Human blood is typed Rh^+ (positive blood) or Rh^- (negative blood). This Rh factor is called an *antigen*. There are
two other antigens known as A and B types. Blood lacking
both A and B types is called type O. Sketch a Venn diagram showing the three types of antigens A, B, and Rh,
and label each of the eight regions. For example, O^+ is
inside the Rh set (anything in Rh is positive), but outside
both A and B. On the other hand, O^- is that region outside
all three circles.

10.3 PERMUTATIONS

We now turn to an investigation of the nature of counting. It may seem like a simple
topic for a college math book, since everyone knows how to count in the usual "one,
two, three, . . ." method. However, there are many times we need to know "How
many?" but can't find out by direct counting.

Election Problem

Consider a club with five members?

$A = \{$Alfie, Bogie, Calvin, Doug, Ernie$\}$

In how many ways could they elect a president and secretary? We call this the *election
problem*.

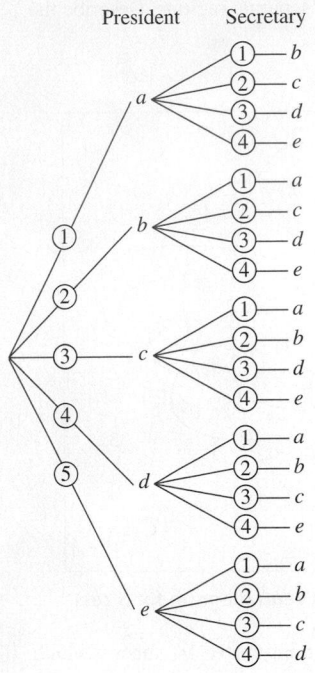

Figure 10.15

There are several ways to solve this problem. The first, and perhaps the easiest, is by making a picture listing all possibilities, known as a **tree diagram** (see Figure 10.15). We see there are 20 possibilities. This method is effective for "small" tree diagrams, but the technique quickly gets out of hand. For example, if we wished to see how many ways the club could elect a president, secretary, and treasurer, this technique would be very lengthy.

A second method of solution is by using boxes or "pigeon holes" representing each choice separately. Here we determine the number of ways of choosing a president, and then the number of ways of choosing a secretary.

Ways of choosing a president	*Ways of choosing a secretary*
5	**4**

↑
Since we have chosen a president,
only 4 remain.

If we multiply the numbers in the pigeon holes,

$$5 \cdot 4 = 20$$

and we see that the result is the same as from the tree diagram.

A third method is to use the fundamental counting principle (from Section 1.4). We repeat this principle here for review.

Fundamental Counting Principle

You can guess that something with this title is important!

STOP

> The **fundamental counting principle** gives the number of ways of performing two or more tasks. If task A can be performed in m ways, and if, after task A is performed, a second task, B, can be performed in n ways, then task A followed by task B can be performed in $m \cdot n$ ways.

EXAMPLE 1

In how many ways could the given club choose a president, secretary, and treasurer?

Solution

President   Secretary   Treasurer

The answer is $5 \times 4 \times 3 = 60$. ◆

EXAMPLE 2

In how many ways could this club choose a president, vice president, secretary, and treasurer?

Solution

$$5 \times 4 \times 3 \times 2 = 120$$ ◆

When set symbols { } are used, the order in which the elements are listed is not important. Suppose now that you wish to select elements from $A = \{a, b, c, d, e\}$ by picking them in a certain order. The selected elements are enclosed in parentheses to signify order and are called an **arrangement** of elements of A. For example, if the elements a and b are selected from A, then there are two pairs, or arrangements

(a, b) and (b, a)

These arrangements are called **ordered pairs.** Remember, when parentheses are used, the order in which the elements are listed *is* important. This example shows ordered pairs, but you could also select an **ordered triple** such as (d, c, a). These arrangements are said to be selected *without repetitions,* since a symbol cannot be used twice in the same arrangement.

If we are given some set of elements and we are to choose *without repetition* a certain number of elements from the set, we can choose them so that the order in which the choices are made is important, or so that it is not. If the *order is important,* we call our result a **permutation,** and if the order is not important it is called a **combination.** We now consider permutations, and we will consider combinations in the next section.

Suppose we consider the election problem for a larger set than A; we wish to elect a president, secretary, and treasurer from among

{Frank, George, Hans, Iris, Jane, Karl}

We could use one of the previously mentioned three methods, but we wish to generalize, so we ask, "Is it possible to count the number of arrangements without actually counting them?"

In answering this question, we note that we are selecting 3 persons from a group of 6 people. If an arbitrary finite set S has n elements and r elements are selected from S (where $r \leq n$), then an arrangement without repetitions of the r selected elements is called a *permutation.*

Permutation

> A **permutation** of r elements selected from a set S with n elements is an ordered arrangement of those r elements selected without repetitions.

EXAMPLE 3

How many permutations of two elements can be selected from a set of six elements?

Solution   Let $B = \{a, b, c, d, e, f\}$ and select two elements:

$(a, b), (a, c), (a, d), (a, e), (a, f)$

$(b, a), (b, c), (b, d), (b, e), (b, f)$

$(c, a), (c, b), (c, d), (c, e), (c, f)$

$(d, a), (d, b), (d, c), (d, e), (d, f)$

$(e, a), (e, b), (e, c), (e, d), (e, f)$

$(f, a), (f, b), (f, c), (f, d), (f, e)$

There are 30 permutations of two elements selected from a set of six elements.   ◆

> **STOP** Remember that since we are considering permutations, the order is important; that is, (a, b) is NOT the same as (b, a).

Example 3 brings up two difficulties. The first is the lack of notation for the phrase, "the number of permutations of two elements selected from a set of six elements," and the second is the inadequacy of relying on direct counting, especially if the sets are very large.

Notation for Permutations

> $_nP_r$ is a symbol used to denote the **number of permutations of** r elements selected from a set of n elements.

Example 3 can now be shortened by writing

$$_6P_2 = 30$$

Actual number of permutations is 30.
From the fundamental counting principle, we
can also write $_6P_2 = 6 \cdot 5 = 30$.

Two factors

$$_nP_r$$

Total number in set   Number we are selecting from the set

The next example leads us to a method for calculating permutations.

EXAMPLE 4

Evaluate: **a.** $_{52}P_2$ **b.** $_7P_3$ **c.** $_{10}P_4$ **d.** $_{10}P_1$ **e.** $_nP_r$

Solution   We use the fundamental counting principle:

a. $_{52}P_2 = \underbrace{52 \cdot 51}_{\text{two factors}} = 2,652$

b. $_7P_3 = \underbrace{7 \cdot 6 \cdot 5}_{\text{three factors}} = 210$

c. $_{10}P_4 = \underbrace{10 \cdot 9 \cdot 8 \cdot 7}_{\text{four factors}} = 5,040$

d. $_{10}P_1 = \underbrace{10}_{\text{one factor}}$

e. $_nP_r = \underbrace{n(n-1)(n-2) \cdot \cdots \cdot (n-r+1)}_{\text{r factors}}$

EXAMPLE 5

a. Can r be larger than n?   **b.** Can r be 0?   **c.** Can $r = n$?

Solution

a. No, since n is the total number of objects.

b. You must be careful. This example does not exactly fit our definition, so we must interpret it separately. Now, $_6P_0$ means the number of ways you can permute six objects by taking none of them. There is only one way to take no objects, and that is by not taking any. Thus, we define $_nP_0 = 1$ for any counting number n.

c. Yes; consider

$$_5P_5 = 5 \cdot 4 \cdot 3 \cdot 2 \cdot 1 = 120$$

$$_6P_6 = 6 \cdot 5 \cdot 4 \cdot 3 \cdot 2 \cdot 1 = 720$$

◆

FACTORIAL

In our work with permutations, we will frequently encounter products such as (see Example 5c)

$$6 \cdot 5 \cdot 4 \cdot 3 \cdot 2 \cdot 1 \quad \text{or} \quad 10 \cdot 9 \cdot 8 \cdot 7 \cdot 6 \cdot 5 \cdot 4 \cdot 3 \cdot 2 \cdot 1 \quad \text{or}$$
$$52 \cdot 51 \cdot 50 \cdot 49 \cdot \cdots \cdot 4 \cdot 3 \cdot 2 \cdot 1$$

Since these are rather lengthy, we use *factorial notation*.

Factorial

> For any counting number n, the **factorial of n** is defined by
>
> $$n! = n(n-1)(n-2) \cdots \cdots 3 \cdot 2 \cdot 1$$
>
> Also, $0! = 1$.

Factorial notation was first used by Christian Kramp in 1808.

EXAMPLE 6

Evaluate factorials for the numbers $0, 1, \ldots, 10$.

Solution

$0! = 1$	$1! = 1$
$2! = 2 \cdot 1 = 2$	$3! = 3 \cdot 2 \cdot 1 = 6$
$4! = 4 \cdot 3 \cdot 2 \cdot 1 = 24$	$5! = 5 \cdot 4 \cdot 3 \cdot 2 \cdot 1 = 120$
$6! = 6 \cdot 5 \cdot 4 \cdot 3 \cdot 2 \cdot 1 = 720$	$7! = 7 \cdot 6 \cdot 5 \cdot 4 \cdot 3 \cdot 2 \cdot 1 = 5{,}040$

$8! = 8 \cdot 7 \cdot 6 \cdot 5 \cdot 4 \cdot 3 \cdot 2 \cdot 1 = 40{,}320$

$9! = 9 \cdot 8 \cdot 7 \cdot 6 \cdot 5 \cdot 4 \cdot 3 \cdot 2 \cdot 1 = 362{,}880$

$10! = 10 \cdot 9 \cdot 8 \cdot 7 \cdot 6 \cdot 5 \cdot 4 \cdot 3 \cdot 2 \cdot 1 = 3{,}628{,}800$ ◆

Notice (from Example 6) that these numbers get big pretty fast. We need not wonder why the notation "!" was chosen! For example,

$$52! \approx 8.065817517 \times 10^{67}$$

If you actually carry out the calculations in Example 6, you will discover a useful property of factorials: $3! = 3 \cdot 2!$, $4! = 4 \cdot 3!$, $5! = 5 \cdot 4!$; that is, to calculate $11!$ you would not need to "start over" but simply multiply 11 and the answer found for $10!$. This property is called the **multiplication property of factorials** or the **count-down property**.

Count-down Property

> $$n! = n(n-1)!$$

EXAMPLE 7

Evaluate: **a.** $6! - 4!$ **b.** $(6 - 4)!$ **c.** $\dfrac{8!}{7!}$ **d.** $\dfrac{12!}{9!}$ **e.** $\dfrac{100!}{98!}$

Solution

a. $6! - 4! = 720 - 24 = 696$

b. $(6 - 4)! = 2! = 2$ *Compare order of operations in parts **a** and **b**.*

c. $\dfrac{8!}{7!} = \dfrac{8 \cdot 7!}{7!} = 8$ *Count-down property*

d. $\dfrac{12!}{9!} = \dfrac{12 \cdot 11 \cdot 10 \cdot 9!}{9!} = 12 \cdot 11 \cdot 10 = 1{,}320$ *Repeated use of the count-down property*

e. $\dfrac{100!}{98!} = \dfrac{100 \cdot 99 \cdot 98!}{98!} = 9{,}900$ *This example shows that you need the count-down property even if you are using a calculator.* ◆

Using factorials, we can find a formula for $_nP_r$.

$$_nP_r = n(n-1)(n-2) \cdots \cdots (n-r+1)$$

$$= n(n-1)(n-2) \cdots \cdots (n-r+1) \frac{(n-r)!}{(n-r)!}$$

$$= n(n-1)(n-2) \cdots \cdots \frac{(n-r+1)(n-r)!}{(n-r)!}$$

$$= \frac{n!}{(n-r)!}$$

This is the general formula for $_nP_r$.

Permutation Formula

$$_nP_r = \frac{n!}{(n-r)!}$$

EXAMPLE 8

Evaluate: **a.** $_{10}P_2$ **b.** $_nP_0$ **c.** $_8P_8$

Solution

a. $_{10}P_2 = \dfrac{10!}{(10-2)!} = \dfrac{10!}{8!} = \dfrac{10 \cdot 9 \cdot 8!}{8!} = 90$

b. $_nP_0 = \dfrac{n!}{(n-0)!} = \dfrac{n!}{n!} = 1$

c. $_8P_8 = \dfrac{8!}{(8-8)!} = \dfrac{8!}{0!} = 8! = 40{,}320$ ◆

EXAMPLE 9

A dispatcher is assigning five taxi drivers to eight possible cars. In how many ways can this be done?

Solution   Five of the eight cars will be assigned to a driver, and since the car one receives could make a difference, the order of selection is important, so this is a permutation problem:

$$_8P_5 = \frac{8!}{(8-5)!} = \frac{8!}{3!} = \frac{8 \cdot 7 \cdot 6 \cdot 5 \cdot 4 \cdot 3!}{3!} = 6{,}720 \qquad ◆$$

EXAMPLE 10

We return to the situation at the beginning of this section. In how many ways can we select a president, secretary, and treasurer from a club of six members?

Solution   The election is a permutation problem:

$$_6P_3 = \frac{6!}{(6-3)!} = 120 \qquad ◆$$

Distinguishable Permutations

We now consider a generalization of permutations in which one or more of the selected items are *indistinguishable* from the others.

Possibilities if H's are considered indistinguishable

H_1ATH_2	H_2ATH_1
H_1AH_2T	H_2AH_1T
H_1TAH_2	H_2TAH_1
H_1TH_2A	H_2TH_1A
H_1H_2TA	H_2H_1TA
H_1H_2AT	H_2H_1AT
AH_1H_2T	AH_2H_1T
AH_1TH_2	AH_2TH_1
ATH_1H_2	ATH_2H_1
TH_1H_2A	TH_2H_1A
TH_1AH_2	TH_2AH_1
TAH_1H_2	TAH_2H_1

EXAMPLE 11

Find the number of arrangements of letters in the given word:

a. MATH   **b.** HATH

Solution

a. This is a permutation of four objects taken four at a time:
$$_4P_4 = 4! = 24$$

b. This is different from part **a** because not all the letters in the word HATH are distinguishable; that is, the first and last letters are both H's so they are *indistinguishable*. If we make the letters distinguishable by labeling them as H_1ATH_2, we see that now the list is the same as in part **a**. Let us list them as shown in the margin. We see that there are 24 possibilities, but notice how we have arranged this listing. If we consider $H_1 = H_2$ (that is, the H's are indistinguishable), we see that the first and the second columns are the same. Since there are *two* indistinguishable letters, we divide the total by 2:
$$\frac{4!}{2} = \frac{24}{2} = 12$$ ◆

EXAMPLE 12

How many permutations are there of the letters in the word ASSIST?

Solution

There are six letters, and if you consider the letters as *distinguishable*, as in
$$AS_1S_2IS_3T$$
there are $_6P_6 = 6! = 720$ possibilities. However,
$$AS_1S_2IS_3T, \quad AS_1S_3IS_2T, \quad AS_2S_1IS_3T, \quad AS_2S_3IS_1T, \quad AS_3S_1IS_2T, \quad AS_3S_2IS_1T$$
are all indistinguishable, so you must divide the total by $3! = 6$:
$$\frac{6!}{3!} = \frac{6 \cdot 5 \cdot 4 \cdot 3!}{3!} = 120$$
There are 120 permutations of the letters in the word ASSIST. ◆

These examples can be generalized to include several categories.

EXAMPLE 13

Find the number of permutations of the letters in the word ATTRACT.

Solution   The total number of arrangements of the letters in the word is 7!; this is divided by factorials of the numbers of repeated letters in subcategories:

$$\frac{7!}{3!2!1!1!}$$

Letter T occurs three times. Letter A occurs twice. Letters R and C occur once each.

This number can now be simplified: $\dfrac{7 \cdot 6 \cdot 5 \cdot 4 \cdot 3!}{3! \cdot 2} = 420$ ◆

These examples suggest a general result.

Formula for the Number of Distinguishable Permutations

The number of **distinguishable permutations** of n objects in which n_1 are of one kind, n_2 are of another kind, . . . , and n_k are of a further kind, so that

$$n = n_1 + n_2 + \cdots + n_k$$

is denoted by $\left(n_1, n_2, \overset{n}{\ldots}, n_k\right)$ and is defined by the formula

$$\left(n_1, n_2, \overset{n}{\ldots}, n_k\right) = \frac{n!}{n_1! n_2! \cdots \cdots n_k!}$$

EXAMPLE 14

What is the number of distinguishable permutations of the letters in the words COLLEGE ALGEBRA?

Solution

$$\frac{14!}{3!3!2!2!1!1!1!1!} = \frac{\overset{7}{\cancel{14}} \cdot 13 \cdot \cancel{12} \cdot 11 \cdot \overset{5}{\cancel{10}} \cdot \overset{3}{\cancel{9}} \cdot 8!}{\cancel{3} \cdot \cancel{2} \cdot \cancel{3} \cdot \cancel{2} \cdot \cancel{2} \cdot \cancel{2}}$$ Reduce where possible.

$$= 7 \cdot 13 \cdot 11 \cdot 5 \cdot 3 \cdot 8!$$ Factored form answer

$$= 605,404,800$$ Calculator answer

If you also consider the space and where it occurs (which would be necessary if you were programming this on a computer), then the number of possibilities is found by

$$\frac{15!}{3!3!2!2!1!1!1!1!1!} = 9,081,072,000$$ ◆

Problem Set 10.3

LEVEL 1

1. IN YOUR OWN WORDS What is a permutation? What is the formula for permutations?

Evaluate each expression in Problems 2–33.

2. $_9P_1$ **3.** $_9P_2$ **4.** $_9P_3$ **5.** $_9P_4$

6. $_9P_0$ **7.** $_5P_4$ **8.** $_{52}P_3$ **9.** $_7P_2$

10. $_4P_4$ **11.** $_{100}P_1$ **12.** $_{12}P_5$ **13.** $_5P_3$

14. $_8P_4$ **15.** $_8P_0$ **16.** $_gP_h$ **17.** $_{92}P_0$

18. $_{52}P_1$ **19.** $_7P_5$ **20.** $_{16}P_3$ **21.** $_nP_4$

22 $_7P_3$ **23.** $_5P_5$ **24.** $_{50}P_4$ **25.** $_{25}P_1$

26. $_mP_3$ **27.** $_8P_3$ **28.** $_{12}P_0$ **29.** $_{10}P_2$

30. $_{11}P_4$ **31.** $_nP_5$ **32.** $_5P_r$ **33.** $_xP_y$

How many permutations are there of the words given in Problems 34–43?

34. HOLIDAY

35. ANNEX

36. ESCHEW

37. OBFUSCATION

38. MISSISSIPPI

39. CONCENTRATION

40. BOOKKEEPING

41. GRAMMATICAL

42. APOSIOPESIS

43. COMBINATORICS

LEVEL 2

44. In how many ways can a president, vice president, secretary, and treasurer be elected from a group of five people?

45. In how many ways can a president and a vice president be elected from a group of 15 people?

46. In how many ways can a president, a vice president, and a secretary be elected from a group of 10 people?

47. In how many ways can a chairperson and a vice chairperson be selected from a committee of 25 senators?

48. How many outfits consisting of a suit and a tie can a man select if he has two suits and eight ties?

49. In how many different ways can eight books be arranged on a shelf?

50. In how many ways can you select and read three books from a shelf of eight books?

51. In how many ways can a row of three contestants for a TV game show be selected from an audience of 362 people?

52. How many seven-digit telephone numbers are possible if the first two digits cannot be ones or zeros?

53. Foley's Village Inn offers the following menu in its restaurant:

Main Course	Dessert	Beverage
Prime rib	Ice cream	Coffee
Steak	Sherbet	Tea
Chicken	Cheesecake	Milk
Ham		Sanka
Shrimp		

In how many different ways can someone order a meal consisting of one choice from each category?

54. Most ATMs require that you enter a four-digit code, using the digits 0 to 9. How many four-digit codes are available?

55. Some automobiles have five button locks. How many different five-digit codes are available for these locks?

56. A museum wishes to display eight paintings next to one another on a wall. In how many ways may this be done?

57. If there are nine baseball players that need to be arranged in a batting order, in how many ways might this be done?

58. The "Pick 3" at horse racetracks requires that a person select the winning horse for three consecutive races. If the first race has nine entries, the second race eight entries, and the third race ten entries, how many different possible tickets might be purchased?

59. Tarot cards are used for telling fortunes, and in a reading, the arrangement of the cards is as important as the cards themselves. How many different readings are possible if three cards are selected from a set of seven Tarot cards?

PROBLEM SOLVING

60. Suppose you flip a coin and keep a record of the results. In how many ways could you obtain at least one head if you flip the coin six times?

10.4 COMBINATIONS

Committee Problem

Reconsider the club example given at the beginning of the previous section:

$$A = \{\text{Alfie, Bogie, Calvin, Doug, Ernie}\}$$

In how many ways could they elect a committee of two persons? We call this the *committee problem.*

One method of solution is easy (but tedious) and involves the enumeration of all possibilities:

{a, b}	{b, c}	{c, d}	{d, e}
{a, c}	{b, d}	{c, e}	
{a, d}	{b, e}		
{a, e}			

We see there are ten possible two-member committees. Do you see why we cannot use the fundamental counting principle for this committee problem in the same way we did for the election problem?

We have presented two different types of counting problems, the *election problem* and the *committee problem.* For the election problem, we found an easy numerical method of counting, using the fundamental counting principle. For the committee problem, we did not; further investigation is necessary.

Let us compare the committee and election problems. We saw in the preceding section, while considering the election problem, that the following arrangements are *different*:

	President	Secretary	Treasurer
1.	Alfie	Bogie	Calvin
2.	Bogie	Alfie	Calvin

That is, the order was important. However, in the committee problem, the following three-member committees are the *same*.

1.   Alfie   Bogie   Calvin
2.   Bogie   Alfie   Calvin

In this case, the order *is not important.* When we list objects in which the order they are listed is not important, we call the list a *combination,* and represent it as a *subset of A.*

EXAMPLE 1

Select two elements from the set $A = \{a, b, c, d, e\}$, and list all possible arrangements as well as all possible subsets.

Solution

Permutations—arrangements
Order important

$(a, b), (a, c), (a, d), (a, e)$
$(b, a), (b, c), (b, d), (b, e)$
$(c, a), (c, b), (c, d), (c, e)$
$(d, a), (d, b), (d, c), (d, e)$
$(e, a), (e, b), (e, c), (e, d)$

There are 20 permutations; note that ordered pair notation is used.

Combinations—subsets
Order not important

$\{a, b\}, \{a, c\}, \{a, d\}, \{a, e\}$
↑ $\{b, c\}, \{b, d\}, \{b, e\}$
Do not list
{b, a} since $\{c, d\}, \{c, e\}$
{a, b} = {b, a}.
                              $\{d, e\}$

There are 10 combinations; note that set notation is used.   ◆

We now state a definition of a combination.

Combination

> A **combination** of r elements of a finite set S of n elements is a subset of S that contains r distinct elements. The notations $\binom{n}{r}$ and $_nC_r$ are both used to denote the number of combinations of r elements selected from a set of n elements $(r \le n)$.

The notation $_nC_r$ is similar to the notation used for permutations, but since $\binom{n}{r}$ is more common in your later work in mathematics, we will use this notation for combinations. The notation $\binom{n}{r}$ is read as "*n* choose *r*."

The formula for the number of permutations leads directly to a formula for the number of combinations since each subset of r elements has $r!$ permutations of its members. Thus, by the fundamental counting principle,

$$\binom{n}{r} \cdot r! = {}_nP_r \quad \text{so} \quad \binom{n}{r} = \frac{{}_nP_r}{r!} = \frac{n!}{r!(n-r)!}$$

Combination Formula

$$\binom{n}{r} = \frac{n!}{r!(n-r)!}$$

EXAMPLE 2

Evaluate: **a.** $\binom{10}{3}$ **b.** $\binom{n}{0}$ **c.** $\binom{m-1}{2}$

Solution

a. $\begin{pmatrix} 10 \\ 3 \end{pmatrix} = \dfrac{10!}{3!7!} = \dfrac{10 \cdot 9 \cdot 8 \cdot 7!}{3 \cdot 2 \cdot 7!} = 120$

b. $\begin{pmatrix} n \\ 0 \end{pmatrix} = \dfrac{n!}{0!(n-0)!} = 1$

c. $\begin{pmatrix} m-1 \\ 2 \end{pmatrix} = \dfrac{(m-1)!}{2!(m-1-2)!}$

$\qquad\qquad = \dfrac{(m-1)(m-2)(m-3)!}{2 \cdot 1 \cdot (m-3)!}$

$\qquad\qquad = \dfrac{(m-1)(m-2)}{2}$ ◆

EXAMPLE 3

In how many ways can a club of five members select a three-person committee?

Solution In an election, the order of selection is important, but in choosing a committee, the order of selection is not important, so this is a combination. Three members are selected from five, so we have "5 choose 3":

$$\begin{pmatrix} 5 \\ 3 \end{pmatrix} = \dfrac{5!}{3!2!} = \dfrac{5 \cdot 4}{2!} = 10$$ ◆

Many applications deal with an ordinary **deck of cards,** as shown in Figure 10.16.

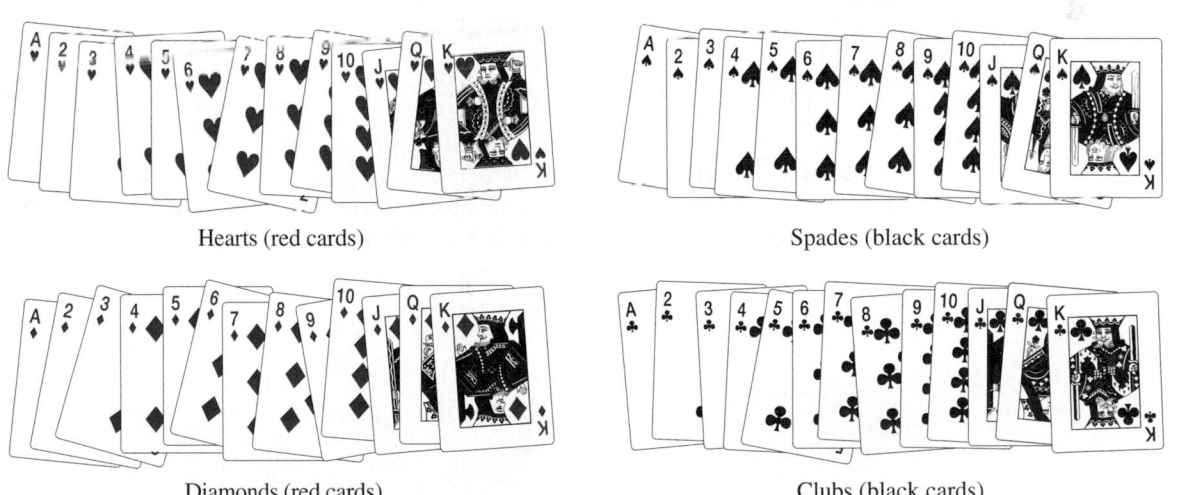

Hearts (red cards)   Spades (black cards)

Diamonds (red cards)   Clubs (black cards)

Figure 10.16 **A deck of cards**

EXAMPLE 4

Find the number of five-card hands that can be drawn from an ordinary deck of cards.

Solution

$$\begin{pmatrix} 52 \\ 5 \end{pmatrix} = \dfrac{52!}{5!47!} = \dfrac{52 \cdot 51 \cdot 50 \cdot 49 \cdot 48 \cdot 47!}{5 \cdot 4 \cdot 3 \cdot 2 \cdot 1 \cdot 47!} = 2{,}598{,}960$$ ◆

EXAMPLE 5

In how many ways can a heart flush be drawn in poker? (A heart flush is any hand of five hearts.)

Solution We need to determine the number of combinations of 13 objects (hearts) taken 5 at a time. Thus,

$$\binom{13}{5} = \frac{13!}{5!8!} = \frac{13 \cdot 12 \cdot 11 \cdot 10 \cdot 9 \cdot 8!}{5 \cdot 4 \cdot 3 \cdot 2 \cdot 1 \cdot 8!} = 1{,}287$$ ◆

EXAMPLE 6

In how many ways can a flush be drawn in poker? (A flush is five cards of one suit.)

Solution There are four possible suits, so we use the fundamental counting principle, along with the result from Example 5.

$$\underbrace{4}_{\text{Number of suits}} \cdot \underbrace{1{,}287}_{\text{Number of ways for a particular suit}} = 5{,}148$$ ◆

EXAMPLE 7

In how many ways can a full house of three tens and two queens be dealt?

Solution Begin with the fundamental counting principle:

$$\boxed{\text{Ways of obtaining two queens}} \cdot \boxed{\text{Ways of obtaining three tens}}$$

Each of the numbers in each pigeonhole is a combination (since the order in which the cards are dealt is not important):

$$\underbrace{\binom{4}{2}}_{\text{Queens}} \cdot \underbrace{\binom{4}{3}}_{\text{Tens}} = \frac{4!}{2!2!} \cdot \frac{4!}{3!1!} = \frac{4 \cdot 3 \cdot 4}{2} = 24$$ ◆

Pascal's Triangle

For many applications, if n and r are relatively small (which will be most of the problems for this course), you should notice the following relationship:

$$_0C_0 = 1$$
$$_1C_0 = 1 \qquad _1C_1 = 1$$
$$_2C_0 = 1 \qquad _2C_1 = 2 \qquad _2C_2 = 1$$
$$_3C_0 = 1 \qquad _3C_1 = 3 \qquad _3C_2 = 3 \qquad _3C_3 = 1$$
$$_4C_0 = 1 \qquad _4C_1 = 4 \qquad _4C_2 = 6 \qquad _4C_3 = 4 \qquad _4C_4 = 1$$
$$\vdots$$

That is, $_nC_r$ is the number in the nth row, rth diagonal of Pascal's triangle. We repeat the triangle (from Section 1.1) at the top of the next page for easy reference.

EXAMPLE 8

Find **a.** $_8C_3$ **b.** $_6C_5$ **c.** $_{14}C_6$ **d.** $_{12}C_{10}$

Solution

a. $_8C_3 = 56$ row 8, diagonal 3   **b.** $_6C_5 = 6$ row 6, diagonal 5

c. $_{14}C_6 = 3{,}003$ row 14, diagonal 6   **d.** $_{12}C_{10} = 66$ row 12, diagonal 10 ◆

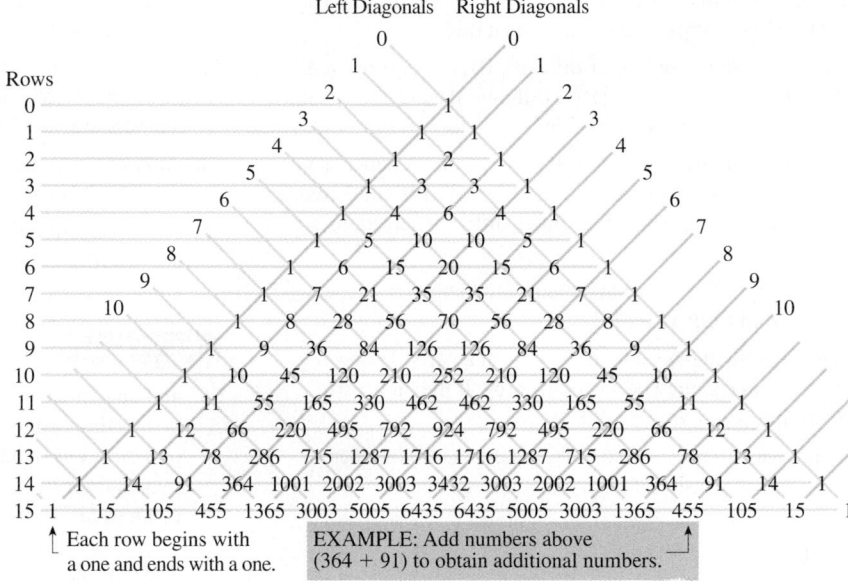

Left Diagonals   Right Diagonals

↑ Each row begins with a one and ends with a one.

EXAMPLE: Add numbers above (364 + 91) to obtain additional numbers.

Problem Set 10.4

Evaluate each expression in Problems 1–26.

1. $\binom{9}{1}$ **2.** $\binom{9}{2}$ **3.** $\binom{9}{3}$ **4.** $\binom{9}{4}$

5. $\binom{9}{0}$ **6.** $\binom{5}{4}$ **7.** $\binom{52}{3}$ **8.** $\binom{7}{2}$

9. $\binom{4}{4}$ **10.** $\binom{100}{1}$ **11.** $\binom{7}{3}$ **12.** $\binom{5}{5}$

13. $\binom{50}{48}$ **14.** $\binom{25}{1}$ **15.** $\binom{g}{h}$ **16.** $\binom{n}{4}$

17. $_kP_4$ **18.** $_nC_5$ **19.** $_mC_n$ **20.** $_mP_n$

21. $\binom{10}{4, 3, 3}$ **22.** $\binom{7}{2, 2, 3}$ **23.** $\binom{9}{2, 3, 4}$

24. $\binom{9}{1}\binom{8}{4}\binom{4}{4}$ **25.** $\binom{7}{1}\binom{6}{4}\binom{2}{2}$ **26.** $\binom{10}{6}\binom{4}{2}\binom{2}{1}$

27. A bag contains 12 pieces of candy. In how many ways can five pieces be selected?

28. If the Senate is to form a new committee of five members, in how many different ways can the committee be chosen if all 100 senators are available to serve on this committee?

29. In how many ways can three aces be drawn from a deck of cards?

30. In how many ways can two kings be drawn from a deck of cards?

31. In how many ways can four aces be drawn from a deck of cards?

32. In how many ways can a club flush be obtained?

33. In how many ways can a heart flush be obtained?

34. In how many ways can a full house of three aces and two kings be obtained?

35. In how many ways can a full house of three jacks and a pair of twos be obtained?

36. In how many ways can a committee of five be formed from a group of 15 people?

In Problems 37–55, decide whether you would use a permutation, a combination, or neither. Next, write the solution using permutation notation or combination notation, if possible, and finally, answer the question asked.

37. A club with 30 members is to select a committee of five persons. In how many ways can this be done?

38. A club with 30 members is to select five officers (president, vice-president, secretary, treasurer, and historian). In how many ways can this be done?

39. Tom is leading a workshop for 20 people. He wants each person to meet every other person (shake hands once). How many handshakes occur?

40. Rita is asked to answer three of five essay questions on an exam. How many choices does Rita have?

41. An ice cream parlor has 31 different flavors. A super sundae allows a person to select 3 different flavors. How many different selections are possible?

42. An ice cream parlor has 31 different flavors. A super sundae allows a person to select 3 different flavors and 3 (out of 8) different toppings. how many different selections are possible?

43. How many different arrangements are there of the letters in the word CORRECT?

44. There are three boys and three girls at a party. In how many ways can they be seated in a row if they must sit alternating boys and girls?

45. If there are 10 people in a club, in how many ways can they choose a dishwasher and a bouncer?

46. In how many ways can you be dealt two cards from an ordinary deck of 52 cards?

47. In how many ways can five taxi drivers be assigned to six cars?

48. How many arrangements are there of the letters in the word GAMBLE?

49. A student is asked to answer 10 out of 12 questions on an exam. In how many ways can the student select the questions to be answered?

50. In how many ways can a group of seven choose a committee of four?

51. In how many ways can seven books be arranged on a bookshelf?

52. In how many ways can you choose two books to read from a bookshelf containing seven books?

53. A certain mathematics test consists of 10 questions. In how many ways can the test be answered if the possisble answers are "true" and "false"?

54. Answer the question in Problem 53 if the possible answers are "true," "false," and "maybe."

55. Answer the question in Problem 53 if the possible answers are (a), (b), (c), (d), and (e); that is, the test is multiple choice.

LEVEL 3

56. a. Draw three points on a circle. How many triangles can you draw connecting the points?

   b. Draw four points on a circle. How many triangles can you draw by connecting any three of these points?

   c. Draw six points on a circle. How many triangles can you draw by connecting any three of these points?

   d. Draw n points on a circle ($n \geq 3$). How many triangles can you draw by connecting any three of these points?

57. Draw n points on a circle ($n \geq 5$). How many pentagons can you draw by connecting any five of these points?

58. A club consists of 16 men and 19 women. In how many ways can they choose a president, vice president, treasurer, and secretary, along with an advisory committee of six people?

59. In Problem 58, how many ways can the selection be made if exactly two of the officers must be women?

60. Prove: $_nC_r = {_nC_{n-r}}$

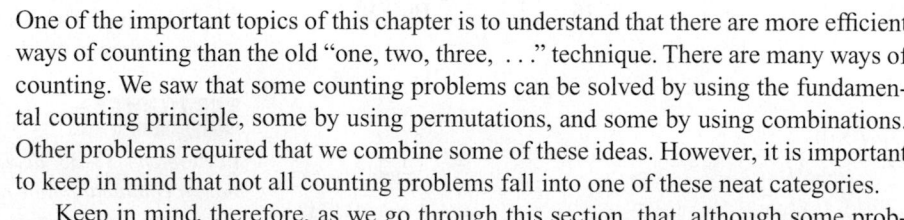

10.5 COUNTING WITHOUT COUNTING

One of the important topics of this chapter is to understand that there are more efficient ways of counting than the old "one, two, three, . . ." technique. There are many ways of counting. We saw that some counting problems can be solved by using the fundamental counting principle, some by using permutations, and some by using combinations. Other problems required that we combine some of these ideas. However, it is important to keep in mind that not all counting problems fall into one of these neat categories.

Keep in mind, therefore, as we go through this section, that, although some problems may be permutation problems and some may be combination problems, there are many counting problems that are neither.

In practice, we are usually required to decide whether a given counting problem is a permutation or a combination before we can find a solution. For the sake of review, recall the difference between the election and committee problems of the previous sections. The election problem is a permutation problem, and the committee problem is a combination problem. We summarize the definition of permutation and combination.

"Five trillion, four hundred eighty billion, five hundred twenty-three million, two hundred ninety-seven thousand, one hundred and sixty-two . . ."

Cartoon by Doug. Reprinted by permission from The Saturday Evening Post © 1976.

Permutations and Combinations

STOP

Remember:
Permutation: Order is important.
Combination: Order is not important.

> A *permutation* of a set of objects is an arrangement of certain of these objects in a *specific order*. A *combination* of a set of objects is an arrangement of certain of these objects *without regard to their order*.

EXAMPLE 1

Classify the following as permutations, combinations, or neither.

a. The number of three-letter "words" that can be formed using the letters $\{m, a, t, h\}$.

b. The number of ways you can change a $1 bill with 5 nickels, 12 dimes, and 6 quarters.

c. The number of ways a five-card hand can be drawn from a deck of cards.

d. The number of different five-numeral combinations on a combination lock.

e. The number of license plates possible in Florida.

Solution

a. Permutation, since *mat* is different from *tam*.

b. Combination, since "2 quarters and 5 dimes" is the same as "5 dimes and 2 quarters."

c. Combination, since the order in which you receive the cards is unimportant.

d. Permutation, since "5 to the L, 6 to the R, 3 to the L, . . ." is different from "6 to the R, 4 to the L, 3 to the L, . . ." We should not be misled by everyday usage of the word *combination*. We made a strict distinction between combination and permutation—one that is not made in everyday terminology. (The correct terminology would require that we call these "permutation locks.")

e. Neither; even though the *order* in which the elements are arranged is important, this does not actually fit the definition of a permutation because the objects are separated into two categories (pigeonholes). The arrangement of letters is a permutation and the arrangement of numerals is a permutation, but to count the actual number of arrangements for this problem would require permutations *and* the fundamental counting principle. ◆

License Plate Problem

States issue license plates, and as the population increases, new plates are designed with more numerals and letters. For example, the state of California has some plates consisting of three letters followed by three digits. When they ran out of these possibilities, they began making license plates with three digits followed by three letters. Most recently they have issued plates with one digit followed by three letters in turn followed by three more digits.

EXAMPLE 2

How many possibilities are there for each of the following license plate schemes?

a. Three letters followed by three digits?

b. Three digits followed by three letters?

c. One digit followed by three letters followed by three digits?

Solution

a. $26 \times 26 \times 26 \times 10 \times 10 \times 10 = 17,576,000$ *Fundamental counting principle*

b. This is the same as part **a.** However, if both schemes are in operation at the same time, then the effective number of possible plates is

$$2 \times 17{,}576{,}000 = 35{,}152{,}000$$

c. The addition of one digit is ten times the number in part **a:**

$$10 \times 17{,}576{,}000 = 175{,}760{,}000$$

Note: A state, such as California, that decides to leave the old plates in use as they move through parts **a, b,** and **c** could have the following number of license plates:

$$17{,}576{,}000 + 17{,}576{,}000 + 175{,}670{,}000 = 210{,}912{,}000$$

Notice that the fundamental counting principle is not used for all counting problems. ◆

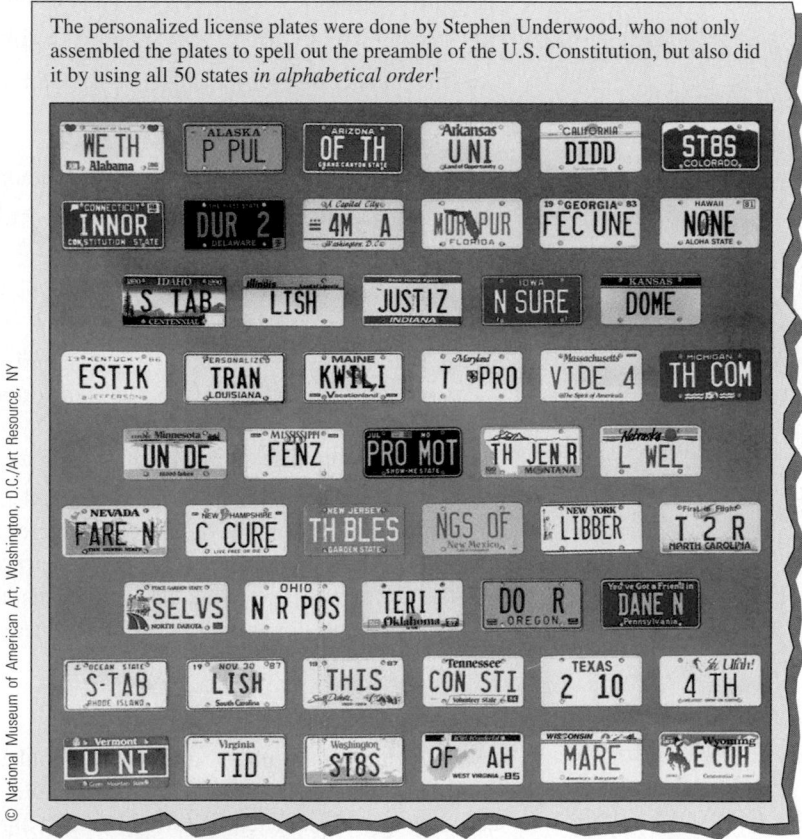

The personalized license plates were done by Stephen Underwood, who not only assembled the plates to spell out the preamble of the U.S. Constitution, but also did it by using all 50 states *in alphabetical order*!

EXAMPLE 3

How many license plates can be formed if repetitions of letters or digits is not allowed, and the state uses the scheme of three numerals followed by three letters?

Solution

To make sure you understand the question, which of the following plates would be considered a success (no repetitions)?

123ABC Success
122ABC Failure; repeated digit
456AAB Failure; repeated letter
111AAA Failure; repeated letter and repeated numeral
890XYZ Success

We can use the fundamental counting principle to count the number of license plates that *do not* have a repetition:

number of digits   letters in the alphabet

↓   ↓

$10 \times 9 \times 8 \times 26 \times 25 \times 24 = 11{,}232{,}000$

↑   ↑

digits left   letters left after the first one (no repetitions)

◆

Which Method?

We have now looked at several counting schemes: tree diagrams, pigeonholes, the fundamental counting principle, permutations, distinguishable permutations, and combinations. In practice, you will generally not be told what type of counting problem you are dealing with—you will need to decide. Table 10.1 should help with that decision. Remember, tree diagrams and pigeonholes are applications of the fundamental counting principle, so they are not listed separately in the table.

Table 10.1 Counting Methods

Fundamental Counting Principle	Permutations	Combinations	Distinguishable Permutations
Counts the total number of separate tasks	Number of ways of selecting r items out of n items		n elements divided into k categories
Repetitions are allowed.	Repetitions are not allowed.		Repetitions are not allowed.
If tasks $1, 2, 3, \ldots, k$ can be performed in $n_1, n_2, n_3, \ldots, n_k$ ways, respectively, then the total number of ways the tasks can be done is $n_1 \cdot n_2 \cdot n_3 \cdots n_k$	Order important— *arrangements* $$_nP_r = \frac{n!}{(n-r)!}$$	Order not important— *subsets* $$\binom{n}{r} = \frac{n!}{r!(n-r)!}$$	Order of categories is important. $$\binom{n}{n_1, n_2, \ldots, n_k} = \frac{n!}{n_1!n_2! \cdots n_k!}$$ where $n = n_1 + n_2 + \cdots + n_k$

EXAMPLE 4

What is the number of license plates possible if each license plate consists of three letters followed by three digits, and we add the condition that repetition of letters or digits is not permitted?

Solution This is a permutation problem since the *order* in which the elements are arranged is important, and the choice is without repetition. The number is found by using permutations along with the fundamental counting principle:

$$_{26}P_3 \cdot {}_{10}P_3 = 26 \cdot 25 \cdot 24 \cdot 10 \cdot 9 \cdot 8 = 11{,}232{,}000$$

◆

By comparing the solutions to Examples 3 and 4, you can see that alternate approaches still provide the same result.

EXAMPLE 5

A quartet is to be selected from a choir. There is to be one soprano selected from a group of six sopranos, two tenors selected from a group of five tenors, and a bass selected from three basses.

a. In how many ways can the quartet be formed?

b. In how many ways can the quartet be formed if one of the tenors is designated lead tenor?

Solution

a. Begin with the fundamental counting principle:

$$\overset{\text{soprano}}{\binom{6}{1}} \cdot \overset{\text{two tenors}}{\binom{5}{2}} \cdot \overset{\text{bass}}{\binom{3}{1}} = 6 \cdot 10 \cdot 3 = 180$$

b. Since the order of selecting the tenors is important, the middle factor from part **a** is replaced by

$$_5P_2 = \frac{5!}{3!} = 20$$

The number of ways of selecting the quartet is $6 \cdot 20 \cdot 3 = 360$. ◆

EXAMPLE 6

Suppose that the Sharp Investment Company has 15 sales representatives who are to be reassigned to three geographical areas as follows: four in the North, five in the South, and six in the West. In how many ways could the sales representatives be assigned to the geographical areas?

Solution Begin with the fundamental counting principle to fill three pigeonholes with combinations (order not important):

$$\overset{\text{North}}{\binom{15}{4}} \cdot \overset{\text{South}}{\binom{11}{5}} \cdot \overset{\text{West}}{\binom{6}{6}} = \frac{15!}{4!11!} \cdot \frac{11!}{5!6!} \cdot \frac{6!}{6!0!} = \frac{15!}{4!5!6!} = 630,630$$

Note the connection between combinations and distinguishable permutations. We could also model this problem by noting that the geographical areas divide up all the sales representatives ($4 + 5 + 6 = 15$):

$$\binom{15}{4, 5, 6} = \frac{15!}{4!5!6!} = 630,630$$ ◆

EXAMPLE 7

A club with 42 members wants to elect a president, a vice president, and a treasurer. From the remaining members, an advisory committee of five people is to be selected. In how many ways can this be done?

Solution This is both a permutation and a combination problem, with the final result calculated by using the fundamental counting principle.

$$\overset{\text{officers}}{_{42}P_3} \cdot \overset{\text{committee}}{\binom{39}{5}} = \frac{42!}{39!} \cdot \frac{39!}{5!34!} = 39,658,142,160$$ ◆

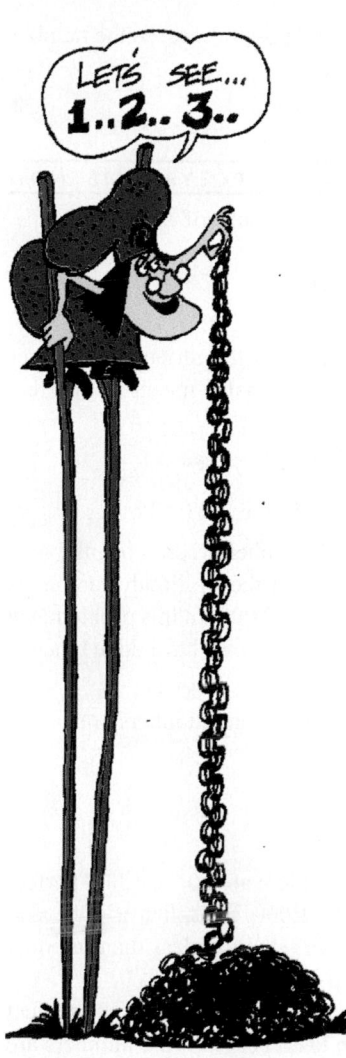

Sometimes it is easier to count what we are not counting than to enumerate all those items with which we are concerned. For example, suppose we wish to know the number of four-member committees that can be appointed from the club:

{Alfie, Bogie, Calvin, Doug, Ernie}

1. We could count directly:
    a. Alfie, Bogie, Calvin, Doug
    b. Alfie, Bogie, Calvin, Ernie
    c. Alfie, Bogie, Doug, Ernie
2. We could use the formula

$$_5C_4 = \frac{_5P_4}{4!} = \cdots$$

3. We could count those we are not counting: For each four-member committee, there is one person left out. We can leave out Alfie, Bogie, Calvin, Doug, or Ernie. Therefore, there are five four-member committees.

EXAMPLE 8

Suppose you flip a coin three times and keep a record of the result. In how many ways can you obtain at least one head?

Solution   Consider all possibilities:

H   H   H
H   H   T
H   T   H
H   T   T
T   H   H
T   H   T
T   T   H
T   T   T

You could count directly to obtain the answer, but you could also count what you are not counting by noticing that in only one out of eight possibilities do you obtain no heads. Thus, there are

$$8 - 1 = 7$$

possibilities in which you obtain at least one head.   ◆

The principle of counting without counting is particularly useful when the results become more complicated, as illustrated by the following example.

EXAMPLE 9

Find the number of ways of obtaining at least one diamond when drawing five cards from an ordinary deck of cards.

Solution   This is very difficult if we proceed directly, but we can compute the number of ways of not drawing a diamond:

$$_{39}C_5 = \frac{39 \cdot 38 \cdot 37 \cdot 36 \cdot 35}{5 \cdot 4 \cdot 3 \cdot 2 \cdot 1} = 575{,}757$$

According to the *Guinness Book of World Records*, the longest recorded paper-link chain was 6,077 ft long and was made by the first- and second-grade children at Rose City Elementary School, Indiana.

From Example 4, Section 10.4, there is a total of 2,598,960 possibilities, so the number of ways of drawing at least one diamond is

$$2{,}598{,}960 - 575{,}757 = 2{,}023{,}203$$ ◆

EXAMPLE 10   PÓLYA'S METHOD

What is the millionth positive integer that is not the square or cube of an integer?

Solution We use Pólya's method for solving this problem.

Understand the Problem. Imagine a long list of the numbers: 1, 2, 3, . . . , 999,999, 1,000,000, 1,000,001, 1,000,002. Suppose we asked for the millionth positive integer on this list—that's easy. It is 1,000,000. Do you understand what is meant by "perfect squares" and "perfect cubes"?

Perfect squares: $1^2 = 1; 2^2 = 4, 3^2 = 9; 4^2 = 16; 25; 36; 49; 64; 81; . . .$

Perfect cubes: $1^3 = 1; 2^3 = 8; 3^3 = 27; 4^3 = 64; 125; 216; 343; 512; . . .$

Now, suppose we cross out the perfect squares; for each number crossed off, the millionth number in the list changes. That is, cross off 1 and the millionth number is 1,000,001; cross off 1 and 4 and the millionth number is 1,000,002. In this problem, we need to cross out all the perfect squares *and* perfect cubes. *After* we have done this, we look at the list and find the millionth number.

Devise a Plan. It should be clear that we need to know how many numbers are crossed out.

$$U = \{1, 2, 3, . . . , 1{,}000{,}000\}$$

$$S = \{\text{perfect squares}\} \quad \text{and} \quad C = \{\text{perfect cubes}\}$$

We wish to find $|S|$: We know that $1{,}000{,}000 = (10^3)^2$ so there are $10^3 = 1{,}000$ perfect squares less than or equal to 1,000,000. Therefore, $|S| = 1{,}000$. Next, find $|C|$: We also know that $1{,}000{,}000 = (10^2)^3$ so there are $10^2 = 100$ perfect cubes less than or equal to 1,000,000. Therefore, $|C| = 100$.

If you think about crossing out all of the perfect squares and then all of the perfect cubes, you must notice that some numbers are on both lists. That is, what numbers are in the set $S \cap C$? These are the sixth powers:

$$S \cap C: \quad 1^6 = 1; 2^6 = 64; 3^6 = 729; 4^6 = 4{,}096; . . . , 10^6 = 1{,}000{,}000$$

There are 10 perfect sixth powers, so $|S \cap C| = 10$.

Carry Out the Plan. We use the formula for the cardinality of a union from Section 10.1 to find $|S \cup C|$:

$$|S \cup C| = |S| + |C| - |S \cap C| = 1{,}000 + 100 - 10 = 1{,}090$$

If you cross out 1,090 numbers, then the millionth number not crossed out is 1,001,090.

Look Back. The answer 1,001,090 is not necessarily correct because we crossed out perfect squares, cubes, and sixth powers up to 1,000,000. But each time we crossed out a number, the end number was extended by 1. Thus we need to know whether there are any perfect squares, cubes, or sixth powers between 1,000,000 and 1,001,090 (the target range). Let's check the next number on each of our lists:

Perfect squares: $1{,}001^2 = 1{,}002{,}001$, so it is not in the target range;

Perfect cubes: $101^3 = 1{,}030{,}301$, so it is not in the target range;

Perfect sixth powers: $11^6 = 1{,}771{,}561$, so it also is not in the target range.

The millionth number that is not a perfect square or perfect cube is 1,001,090. ◆

Problem Set 10.5

LEVEL 1

1. **IN YOUR OWN WORDS** State the fundamental counting principle, and explain it in your own words.

2. **IN YOUR OWN WORDS** Explain the difference between a permutation and a combination.

3. How many skirt–blouse outfits can a woman wear if she has three skirts and five blouses?

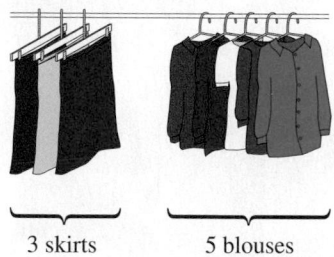

3 skirts   5 blouses

4. If a state issued license plates using the scheme of one letter followed by five digits, how many plates could it issue?

5. Repeat Problem 4 if repetitions are not allowed.

6. In how many ways can a group of 15 people elect a president, vice president, and secretary?

7. How many two-member committees can be formed from a group of seven people?

8. New York license plates consist of three letters followed by three numerals, and 245 letter arrangements are not allowed. How many plates can New York issue?

9. Boats often relay messages by using flags. How many messages can be made using five flags out of a package of 30 different flags?

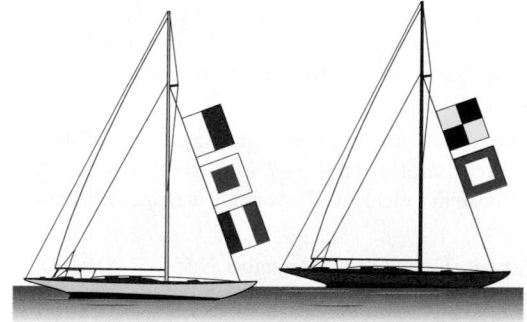

10. If $U = \{1, 2, 3, \ldots 100\}$, $A = \{$multiples of 2$\}$, and $B = \{$multiples of 3$\}$, find
    a. $|A|$ **b.** $|B|$ **c.** $|A \cap B|$ **d.** $|A \cup B|$

11. A certain lock has four tumblers, and each tumbler can assume six positions. How many different possibilities are there?

12. Suppose you flip a coin and keep a record of the results. In how many ways could you obtain at least one head if you flip the coin four times?

13. You flip a coin five times and keep a record of the results. In how many ways could you obtain at least one tail?

14. You flip a coin six times and keep a record of the results. In how many ways could you obtain at least one tail?

15. You flip a coin seven times and keep a record of the results. In how many ways could you obtain at least one tail?

16. You flip a coin n times and keep a record of the results. In how many ways could you obtain at least one tail?

17. In how many ways could a club of 15 members choose a president, vice president, and treasurer?

18. In how many ways could a club of 30 members choose a president, vice president, and treasurer?

19.

Many states offer personalized license plates. The state of California, for example, allows personalized plates with seven spaces for numerals or letters, or one of the following four symbols:

What is the total number of license plates possible using this counting scheme? (Assume that each available space is occupied by a numeral, letter, symbol, or space.)

20. How many different ways (boy–girl patterns) can a family have five children?

21. Suppose a restaurant offers the following prix fixe menu:

    Main course: prime rib, steak, chicken, filet of sole, shrimp
    Side dish: soup, salad, crab cakes
    Dessert: cheesecake, chocolate delight, ice cream
    Beverage: coffee, tea, milk

    In how many ways can someone order a meal consisting of one choice from each category?

22. In how many ways could a club of 15 appoint a committee of 4 people?

23. In how many ways could a club of 30 appoint a committee of 4 people?

In Problems 24–39, classify each as a permutation, a combination, or neither, and then aswer the question asked.

24. At Mr. Furry's Dance Studio, every man must dance the last dance. If there are five men and eight women, in how many ways can dance couples be formed for the last dance?

25. Martin's Ice Cream Store sells sundaes with chocolate, strawberry, butterscotch, or marshmallow toppings, nuts, and whipped cream. If you can choose exactly three of these, how many possible sundaes are there?

26. Five people are to dine together at a rectangular table, but the hostess cannot decide on a seating arrangement. In how many ways can the guests be seated?

27. In how many ways can three hearts be drawn from a deck of cards?

28. A shipment of 100 TV sets is received. Six sets are to be chosen at random and tested for defects. In how many ways can six sets be chosen?

29. A night watchman visits 125 offices every night. To prevent others from knowing when he will be at a particular office, he varies the order of his visits. In how many ways can this be done?

30. A certain manufacturing process calls for the mixing of six chemicals. One liquid is to be poured into the vat, and then the others are to be added in turn. All possibilities must be tested to see which gives the best results. How many tests are required?

31. There are three boys and three girls at a party. In how many ways can they be seated if they can sit down four at a time?

32. The number of arrangements in the letters of the word KARL.

33. The receiving line at a wedding consists of the bride and groom, parents of the bride, parents of the groom, the best man, and the maid of honor (all in that order). In addition, there are five groomsmen (in any order) followed by five bridesmaids (in any order). In how many ways can the receiving line be arranged?

34. There are ten people in a club; in how many ways can they choose a dishwasher and a bouncer?

35. How many three-digit numbers can be formed using the numerals {3, 4, 6, 7} without repetition?

36. In how many ways can you be dealt two cards from an ordinary deck of cards?

37. In how many ways can five taxi drivers be assigned to six cars?

38. How many arrangements are there of the letters in the word GAMBLE?

39. A student is asked to answer 10 out of 12 questions on an exam. In how many ways can she select the questions to be answered?

LEVEL 2

40. The advertisement shown here claims the following 12 mix-and-match items give 122 different outfits.

Nothing to Wear?
This Simple Wardrobe Will
Solve All Your Problems

12 Mix & Match Items
Give You 122 Outfits

The wardrobe consists of 4 blouses, 2 slacks, 2 skirts, 1 sweater, 2 jackets, and 1 scarf. Assume that the model must choose one top and one bottom item of clothing. She may or may not choose a sweater or jacket or a scarf. How many different outfits are possible?

41. Consider selecting two elements, say, *a* and *b,* from the set $A = \{a, b, c, d, e\}$. List all possible subsets of those two elements, as well as all possible arrangements.

42. Consider selecting three elements, say, *c, d,* and *e,* from the set $A = \{a, b, c, d, e\}$. List all possible subsets of those three elements. How many arrangements are there?

43. Consider selecting four elements, say, *a, b, c,* and *d,* from the set $A = \{a, b, c, d, e\}$. List all possible subsets of those four elements. How many arrangements are there?

44. Consider selecting five elements, say, *a, b, c, d,* and *e,* from the set $A = \{a, b, c, d, e\}$. List all possible subsets of those five elements. How many arrangements are there?

45. A typical social security number is 555-47-5593. How many social security numbers are possible if the first two digits cannot be 0.

46. A club consists of four men and three women. In how many ways can this club elect a president, vice president, and secretary, in that order, if the president must be a woman, and the other two officers must be men?

47. A history teacher gives a 20-question T–F exam. In how many ways can the test be answered if the possible

answers are T or F or possibly to leave the answer blank?

48. A JVC advertisement appeared in several national periodicals. It claimed that the SEA graphic equalizer system can create 371,293 different sounds from the thirteen possible zone levels on each of five different controls. Can this claim be substantiated from the advertisement and the knowledge of this chapter? If so, explain how; if not, tell why this number is impossible.

49. In how many ways can 26 books be arranged on a bookshelf?

PROBLEM SOLVING

50. A die can be held so that one, two, or three of its faces can be seen at any one time.

One face showing a 5

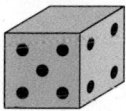

Two faces showing a total of 9

Three faces showing a total of 10

Is it possible to hold a die in different ways so that at different times the visible numbers add up to every number from 1 through 15?

51.

"I did it without counting. There are – – – beans in the jar."

Outline a procedure that would allow the fellow in the cartoon to proclaim that he "did it without counting." Estimate the number assuming that the container is a cylinder 15 in. tall and 10 in. across, and that it is filled with kidney beans.

52. The old lady jumping rope in the cartoon at the beginning of this section is counting one at a time. Assume that she jumps the rope 50 times per minute and that she jumps 8 hours a day, 5 days a week, 50 weeks a year. Estimate the length of time necessary for her to jump the rope the number of times indicated in the cartoon.

53. A marginal note in this section describes the longest paper-link chain. If there are 12 paper links per foot, and it takes 1 minute to construct each link, estimate the length of time necessary to build this paper-link chain.

54. What numerical property is exhibited by the arrangement of billiard balls shown in Figure 10.17?

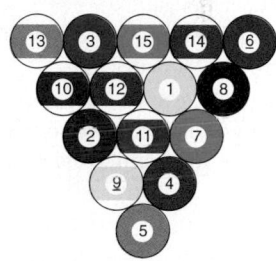

Figure 10.17 Billiard balls

55. The News Clip below presents an argument proving that at least two New Yorkers must have *exactly* the same number of hairs on their heads! For this problem, just *suppose* you knew the number of hairs on your head, as well as the *exact* number of hairs on anyone else's head. Now multiply the number of hairs on your head by the number of hairs on your neighbor's head. Take this result and multiply by the number of hairs on the heads of each person in your town or city. Continue this process until you have done this for everyone in the entire world! Make a guess (you can use scientific notation, if you like) about the size of this answer. (*Hint:* The author has worked this out and claims to know the exact answer.)

An example about reaching a conclusion about the number of objects in a set is given by M. Cohen and E. Nagel in *An Introduction to Logic*. They conclude that there are at least two people in New York City who have the same number of hairs on their heads. This conclusion was reached not through counting the hairs on the heads of 8 million inhabitants of the city, but through studies revealing that: (1) the maximum number of hairs on the human scalp could never be as many as 5,000 per square centimeter, and (2) the maximum area of the human scalp could never reach 1,000 square centimeters. We can now conclude that no human head could contain even
$$5,000 \times 1,000 = 5,000,000$$
hairs. Since this number is less than the population of New York City, it follows that at least two New Yorkers must have the same number of hairs on their heads!

The following television advertisement is the basis for Problems 56–59.

LITTLE CAESAR'S PIZZA PIZZA!!!

Customer: So what's this new deal?
Pizza chef: Two pizzas
Customer [Toward four-year-old boy]:
  Two pizzas. Write that down.
Pizza chef: And on the two pizzas choose any toppings—
  up to five [*from a list of* 11 *toppings*].
Older boy: Do you . . .
Pizza chef: . . . have to pick the same toppings on each
  pizza? NO!
Four-year-old-math whiz: Then the possibilities are
endless.
Customer: What do you mean? Five plus five are ten.
Math whiz: Actually, there are 1,048,576 possibilities.
Customer: Ten was just a ballpark figure.
Old man: You got that right

First aired on national television on November 8, 1993.

56. If we accept the facts of the advertisement and order one pizza per day, how long will we be able to order different pizzas before we are forced to order a pizza previously ordered?

57. How many different pizza orders are possible?

58. Jean Sherrod of Little Caesar's Enterprises, Inc., explains that an order where the first pizza is pepperoni and the second pizza is ham would be considered different from an order in which the first pizza is ham and the second pizza is pepperoni. Using this scheme, how many pizza orders are possible?

59. Double toppings are not included in the calculation in the advertisement. However, little Caesar's does allow double toppings. If double toppings are possible, how many different pizzas orders are possible?

60. In the text, we found

$$|A \cup B| = |A| + |B| - |A \cap B|$$

Show that

$$|A \cup B \cup C| = |A| + |B| + |C|$$
$$- |A \cap B| - |A \cap C| - |B \cap C|$$
$$+ |A \cap B \cap C|$$

10.6 BINOMIAL THEOREM

Combinations can be used to help us form a pattern for the expansion of a binomial $(a + b)^n$. We first introduced the binomial theorem in Section 5.1. This problem occurs frequently in mathematics, and direct calculation is too tedious for a very large exponent n. Also, we sometimes need to find only one term in the expansion, and a pattern will help us find only that term without the necessity of finding all the others.

Consider the powers of $(a + b)$ listed here, which are found by direct multiplication:

$$(a + b)^0 = 1$$
$$(a + b)^1 = 1 \cdot a + 1 \cdot b$$
$$(a + b)^2 = 1 \cdot a^2 + 2 \cdot ab + 1 \cdot b^2$$
$$(a + b)^3 = 1 \cdot a^3 + 3 \cdot a^2 b + 3 \cdot ab^2 + 1 \cdot b^3$$
$$(a + b)^4 = 1 \cdot a^4 + 4 \cdot a^3 b + 6 \cdot a^2 b^2 + 4 \cdot ab^3 + 1 \cdot b^4$$
$$(a + b)^5 = 1 \cdot a^5 + 5 \cdot a^4 b + 10 \cdot a^3 b^2 + 10 \cdot a^2 b^3 + 5 \cdot ab^4 + 1 \cdot b^5$$
$$\vdots \qquad \qquad \vdots$$

First, ignore the coefficients (shown in color) and focus on the variables:

$(a + b)^1$:	a	b			
$(a + b)^2$:	a^2	ab	b^2		
$(a + b)^3$:	a^3	$a^2 b$	ab^2	b^3	
$(a + b)^4$:	a^4	$a^3 b$	$a^2 b^2$	ab^3	b^4
$(a + b)^5$:	a^5	$a^4 b$	$a^3 b^2$	$a^2 b^3$	ab^4	b^5

$$\vdots \qquad \qquad \qquad \vdots$$

Do you see a pattern? As you read from left to right, the powers of a decrease and the powers of b increase. Note that the sum of the exponents for each term is the same as the original exponent:

$$(a + b)^n: \quad a^n b^0 \quad a^{n-1} b^1 \quad a^{n-2} b^2 \cdots a^{n-r} b^r \cdots a^2 b^{n-2} \quad a^1 b^{n-1} \quad a^0 b^n$$

Next, consider the numerical coefficients (shown in color):

$(a + b)^0$:   1

$(a + b)^1$:   1   1

$(a + b)^2$:   1   2   1

$(a + b)^3$:   1   3   3   1

$(a + b)^4$:   1   4   6   4   1

$(a + b)^5$:   1   5   10   10   5   1

Do you see the pattern? Recall Pascal's triangle from Section 1.1, and notice that we represent each term in this pattern by $\begin{pmatrix} n \\ k \end{pmatrix}$ as shown here:

$$\begin{pmatrix} 0 \\ 0 \end{pmatrix} = 1$$

$$\begin{pmatrix} 1 \\ 0 \end{pmatrix} = 1 \quad \begin{pmatrix} 1 \\ 1 \end{pmatrix} = 1$$

$$\begin{pmatrix} 2 \\ 0 \end{pmatrix} = 1 \quad \begin{pmatrix} 2 \\ 1 \end{pmatrix} = 2 \quad \begin{pmatrix} 2 \\ 2 \end{pmatrix} - 1$$

$$\begin{pmatrix} 3 \\ 0 \end{pmatrix} = 1 \quad \begin{pmatrix} 3 \\ 1 \end{pmatrix} = 3 \quad \begin{pmatrix} 3 \\ 2 \end{pmatrix} = 3 \quad \begin{pmatrix} 3 \\ 3 \end{pmatrix} = 1$$

$$\begin{pmatrix} 4 \\ 0 \end{pmatrix} = 1 \quad \begin{pmatrix} 4 \\ 1 \end{pmatrix} = 4 \quad \begin{pmatrix} 4 \\ 2 \end{pmatrix} = 6 \quad \begin{pmatrix} 4 \\ 3 \end{pmatrix} = 4 \quad \begin{pmatrix} 4 \\ 4 \end{pmatrix} = 1$$

$$\begin{pmatrix} 5 \\ 0 \end{pmatrix} = 1 \quad \begin{pmatrix} 5 \\ 1 \end{pmatrix} = 5 \quad \begin{pmatrix} 5 \\ 2 \end{pmatrix} = 10 \quad \begin{pmatrix} 5 \\ 3 \end{pmatrix} = 10 \quad \begin{pmatrix} 5 \\ 4 \end{pmatrix} = 5 \quad \begin{pmatrix} 5 \\ 5 \end{pmatrix} = 1$$

We now state a very important theorem in mathematics, call the **binomial theorem.**

Binomial Theorem

CAUTION — This looks worse than it really is! Relax, take your time, and you will be able to understand this theorem.

For any positive integer n,

$$(a + b)^n = \begin{pmatrix} n \\ 0 \end{pmatrix} a^n + \begin{pmatrix} n \\ 1 \end{pmatrix} a^{n-1} b + \begin{pmatrix} n \\ 2 \end{pmatrix} a^{n-2} b^2 + \cdots + \begin{pmatrix} n \\ r \end{pmatrix} a^{n-r} b^r$$

$$+ \cdots + \begin{pmatrix} n \\ n - 2 \end{pmatrix} a^2 b^{n-2} + \begin{pmatrix} n \\ n - 1 \end{pmatrix} ab^{n-1} + \begin{pmatrix} n \\ n \end{pmatrix} b^n$$

Pascal's triangle is efficient for finding the numerical coefficients for exponents that are relatively small. However, for larger exponents we use the combination formula. Recall that

$$\begin{pmatrix} n \\ k \end{pmatrix} = \frac{n!}{k!(n - k)!}$$

EXAMPLE 1

Find: $(x + y)^8$

Solution Use Pascal's triangle (row 8) to obtain the coefficients.

$(x + y)^8 = x^8 + 8x^7y + 28x^6y^2 + 56x^5y^3 + 70x^4y^4 + 56x^3y^5 + 28x^2y^6 + 8xy^7 + y^8$ ◆

EXAMPLE 2

Find: $(x - 2y)^4$

Solution In this example, let $a = x$ and $b = -2y$, and look at row 4 of Pascal's triangle for the coefficients.

$$(a + b)^4 = a^4 + 4a^3b + 6a^2b^2 + 4ab^3 + b^4 \quad \text{\small Binomial theorem, } n = 4$$
$$(x - 2y)^4 = x^4 + 4x^3(-2y) + 6x^2(-2y)^2 + 4x(-2y)^3 + (-2y)^4$$
$$= x^4 - 8x^3y + 24x^2y^2 - 32xy^3 + 16y^4 \qquad ◆$$

EXAMPLE 3

Find: $(x + y)^{15}$

Solution The power is rather large, so use the theorem and the combination formula for the coefficients

$$(x + y)^{15} = \binom{15}{0}x^{15} + \binom{15}{1}x^{14}y + \binom{15}{2}x^{13}y^2 + \cdots + \binom{15}{14}xy^{14} + \binom{15}{15}y^{15}$$

$$= \frac{15!}{0!15!}x^{15} + \frac{15!}{1!14!}x^{14}y + \frac{15!}{2!13!}x^{13}y^2 + \cdots + \frac{15!}{14!1!}xy^{14} + \frac{15!}{15!0!}y^{15}$$

$$= x^{15} + 15x^{14}y + 105x^{13}y^2 + \cdots + 15xy^{14} + y^{15} \qquad ◆$$

EXAMPLE 4

Find the coefficient of the term x^2y^{10} in the expansion of $(x + 2y)^{12}$.

Solution We note that $n = 12$, $r = 10$, $a = x$, and $b = 2y$; we then look at the rth term in the binomial expansion:

$$\binom{12}{10}x^2(2y)^{10} = \frac{12!}{10!2!}(2)^{10}x^2y^{10} = 66(1,024)x^2y^{10} = 67,584x^2y^{10}$$

The coefficient is 67,584. ◆

EXAMPLE 5

If a family has five children, how many different birth orders could the parents have if there are three boys and two girls?

Solution Part of understanding this problem might involve estimation. For example, if a family has one child, there are two ways (boy and girl). If a family has two children, there are four ways (BB, BG, GB, GG); for three children, eight ways; for four children, 16 ways; and for five children, a total of 32 ways. This means that an answer of 140 ways, for example, is an unreasonable answer, since for five children the number of all possibilities is 32.

We might try direct enumeration: BBBGG, BBGBG, BBGGB, ... This seems to be too tedious.

Look for a pattern:

One child: B ← One way   *Two children:*   BB ← One way

               G ← One way   BG ⎤

                                                   GB ⎦ ← Two ways

Three children: BBB ← One way   GG ← One way

               BBG ⎤

               BGB ⎬ ← Three ways

               GBB ⎦

               BGG ⎤

               GBG ⎬ ← Three ways

               GGB ⎦

               GGG ← One way

If we write this as follows, it should look familiar:

One child: 1   1

Two children: 1   2   1

Three children: 1   3   3   1

If we write $(B + G)^3 = B^3 + 3B^2G + 3BG^2 + G^3$, we see that the exponents correspond to the genders of the children and the coefficients give the related number of ways these gender combinations can occur. For this question, we find

$$(B + G)^5 = B^5 + 5B^4G + 10B^3G^2 + 10B^2G^3 + 5BG^4 + G^5$$

This says that the family could have three boys and two girls (look at the coefficient of B^3G^2) a total of 10 ways.   ◆

EXAMPLE 6

How many subsets of $\{1, 2, 3, 4, 5\}$ are there?

Solution   There are $\binom{5}{5}$ subsets of 5 elements, $\binom{5}{4}$ subsets of 4 elements, and so forth. Thus, the *total* number of subsets is

$$\binom{5}{0} + \binom{5}{1} + \binom{5}{2} + \binom{5}{3} + \binom{5}{4} + \binom{5}{5}$$

Instead of carrying out all of the arithmetic in evaluating these combinations, we note that

$$(a + b)^5 = \binom{5}{0}a^5 + \binom{5}{1}a^4b + \binom{5}{2}a^3b^2 + \binom{5}{3}a^2b^3 + \binom{5}{4}ab^4 + \binom{5}{5}b^5$$

so that if we let $a = 1$ and $b = 1$, we have

$$(1 + 1)^5 = \binom{5}{0}1^5 + \binom{5}{1}1^41 + \binom{5}{2}1^31^2 + \binom{5}{3}1^21^3 + \binom{5}{4}1 \cdot 1^4 + \binom{5}{5}1^5$$

$$2^5 = \binom{5}{0} + \binom{5}{1} + \binom{5}{2} + \binom{5}{3} + \binom{5}{4} + \binom{5}{5}$$

We see that there are $2^5 = 32$ subsets of a set containing five elements.   ◆

Number of Subsets   A set of n distinct elements has 2^n subsets.

EXAMPLE 7

How many subsets are there for a set containing 10 elements?

Solution   There are $2^{10} = 1,024$ subsets.   ◆

Problem Set 10.6

1. **IN YOUR OWN WORDS** State the binomial theorem.
2. **IN YOUR OWN WORDS** How many subsets are there for a set with t elements?

LEVEL 1

Evaluate the expressions in Problems 3–14 using both Pascal's triangle and the combination formula.

3. $\binom{8}{1}$ 4. $\binom{5}{4}$ 5. $\binom{8}{2}$

6. $\binom{7}{5}$ 7. $\binom{8}{3}$ 8. $\binom{9}{5}$

9. $\binom{12}{1}$ 10. $\binom{15}{0}$ 11. $\binom{20}{20}$

12. $\binom{32}{31}$ 13. $\binom{18}{2}$ 14. $\binom{46}{2}$

In Problems 15–28, expand using the binomial theorem.

15. $(x + 1)^4$ 16. $(x + 1)^8$
17. $(x + 4)^4$ 18. $(a + b)^6$
19. $(a + b)^7$ 20. $(x - 1)^5$
21. $(x - 1)^9$ 22. $(x - y)^6$
23. $(x - y)^5$ 24. $(x + 2)^5$
25. $(x - 2)^6$ 26. $(x - 3)^5$
27. $(2x + 3y)^4$ 28. $(x - 2y)^8$

LEVEL 2

Find the coefficient of the given term in the expansion of the given binomial in Problems 29–38.

29. a^5b^6 in $(a - b)^{11}$ 30. a^4b^7 in $(a + b)^{11}$
31. $x^{10}y^4$ in $(x + y)^{14}$ 32. $x^{10}y^5$ in $(x - y)^{15}$
33. x^{12} in $(x - 1)^{16}$ 34. y^8 in $(y + 1)^{12}$
35. r^5 in $(r + 2)^9$ 36. s^5 in $(s - 2)^{10}$
37. a^7b in $(a - 2b)^8$ 38. a^4b^4 in $(a + 2b)^8$

Decide whether each statement in Problems 39–42 is true or false. Explain your reasoning.

39. $(a + b)^3 = a^3 + b^3$ 40. $(2x - 3y)^5$ has six terms.

41. $(H + T)^5$ can be used to find the number of ways of obtaining three heads and two tails by looking at the coefficient of HT.

42. The number of subsets of a set with five elements is $2 \cdot 5$.

Find the first four terms in the expansion of the given binomial in Problems 43–50.

43. $(x - y)^{15}$
44. $(x + 2y)^{16}$
45. $(x - 2y)^{12}$
46. $(x - 3y)^{10}$
47. $(ab - 2b)^{15}$
48. $(rs - 3t)^{13}$
49. $(z^2 + 5k)^{11}$
50. $(z^3 - k^2)^7$

51. How many different subsets can be chosen from a set of seven elements?

52. How many different subsets can be chosen from the U.S. Senate?

53. If a family has five children, in how many ways could the parents have two boys and three girls?

54. If a family has six children, in how many ways could the parents have three boys and three girls?

55. If a family has seven children, in how many ways could the parents have four boys and three girls?

56. Suppose a coin is tossed eight times. How many outcomes of five heads and three tails are possible?

57. Suppose a coin is tossed 10 times. How many outcomes of four heads and six tails are possible?

58. Suppose a coin is tossed nine times. How many different outcomes of two heads and seven tails are possible?

PROBLEM SOLVING

59. Show that the sum of the entries of the nth row of Pascal's triangle is 2^n.

60. Show that $\binom{n - 1}{r - 1} + \binom{n - 1}{r} = \binom{n}{r}$.

10.7 RUBIK'S CUBE AND INSTANT INSANITY*

Rubik's Cube

In the summer of 1974, a Hungarian architect invented a three-dimensional object that could rotate about *all three axes* (sounds impossible, doesn't it?). He wrote up the details of the cube and obtained a patent in 1975. The cube is now known worldwide as **Rubik's cube.** In case you have not seen one of these cubes, it is shown in Figure 10.18.

There is an interesting site that allows you to "play" with a Rubik's cube online: *www.pitcom.com/rubik.html*

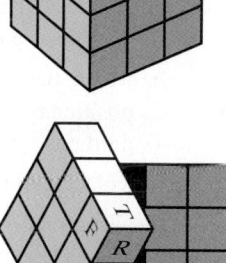

Watch this cube

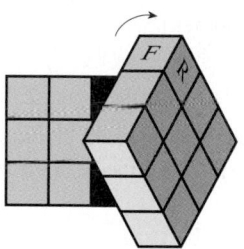

a. Rotate the top right corner with the right face.

b. Rotate the top right corner with the front face.

c. Rotate the top right corner with the top face.

Figure 10.18 **Rubik's cube**

HISTORICAL NOTE

The puzzle known as Rubik's cube was designed by Ernö Rubik, an architect and teacher in Budapest, Hungary. It was also invented independently by Terutoshi Ishige, an engineer in Japan. Both applied for patents in the mid-1970s. The cubes were first manufactured in Hungary. In 1978, a Hungarian mathematics professor brought several with him to the International Congress of Mathematics in Helsinki, Finland. This formally introduced the cube to Europe and the rest of the world. They have been widely available in the United States since 1980. They are often used today in mathematics classes to illustrate mathematical ideas, particularly in group theory.

When you purchase the cube, it is arranged so that each face is showing a different color, but after a few turns it seems next to impossible to return to the start. In fact, the manufacturer claims there are 8.86×10^{22} possible arrangements, of which there is only one correct solution. This claim is incorrect; there are 2,048 possible solutions among the 8.86×10^{22} claimed arrangements (actually, $8.85801027 \times 10^{22}$). This means there is one solution for each 4.3×10^{19} (actually 43,252,003,274,489,856,000) arrangements.

EXAMPLE 1

If it took you one-half second for each arrangement, how long would it take you to move the cube into all arrangements?

Solution

4.3×10^{19}	Arrangements
2.15×10^{19}	Divide by 2 for the number of seconds.
3.58×10^{17}	Divide by 60 for the number of minutes.
5.97×10^{15}	Divide by 60 for the number of hours.
2.49×10^{14}	Divide by 24 for the number of days.
6.81×10^{11}	Divide by 365.25 for the number of years.
6.81×10^{9}	Divide by 100 for the number of centuries.
6.81×10^{8}	Divide by 10 for the number of millennia.

*The term "Rubik's Cube" is a trademark of Ideal Toy Corporation, Hollis, New York. The term "Rubik's cube" as used in this book means the cube puzzle sold under any trademark. "Instant Insanity" is a trademark of Parker Brothers, Inc.

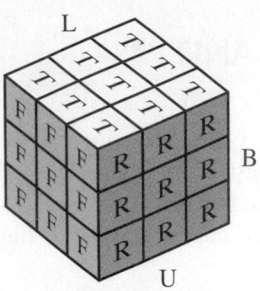

Figure 10.19 Standard-position Rubik's cube

The Bettmann Archive. © CORBIS

Our problem ends here because it would require over

> 681,000,000 *millennia!*

If a high-speed computer listed 1,000,000 arrangements per minute, it would take over 80,000 millennia to print out the possibilities. Remember, it has only been 2 millennia since Christ! ◆

Let's consider what we call the *standard-position cube,* as shown in Figure 10.19. Label the faces Front (F), Right (R), Left (L), Back (B), Top (T), and Under (U), as shown. Hold the cube in your left hand with T up and F toward you so that L is against your left palm. Now describe the results of the moves in Example 2.

EXAMPLE 2

a. Rotate the right face 90° clockwise; denote this move by R. Return the cube to standard position; we denote this move by R^{-1}.

b. Rotate the right face 180° clockwise; denote this by R^2. Return the cube to standard position by doing another R^2. Notice that $R^2R^2 = R^4$, which returns the cube to standard position.

c. Rotate the top 90° clockwise; call this T. Return the cube by doing T^{-1}.

d. *TR* means rotate the top face 90° clockwise, *then* rotate the right face 90° clockwise. Describe the steps necessary to return the cube to standard position.

Solution

a.

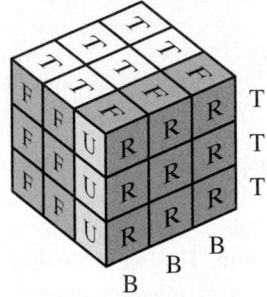

b.

c.

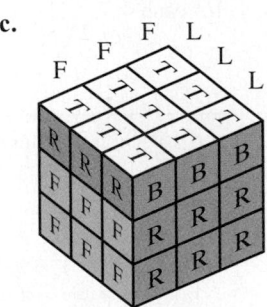

d.

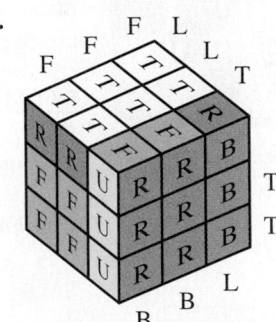

To return this cube to standard position, you need $R^{-1}T^{-1}$. ◆

Now, a rearrangement of the 54 colored faces of a small cube is a *permutation.* We discussed permutations of a square in Problem 60, Section 4.6, at which time we called them *symmetries of a square.*

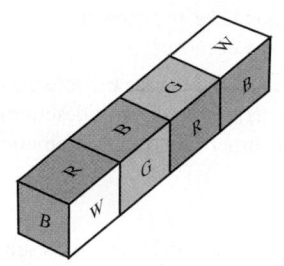

Figure 10.20 Instant Insanity

Instant Insanity

An older puzzle, simpler than Rubik's cube, is called **Instant Insanity.** It provides four cubes colored red, white, blue, and green, as indicated in Figure 10.20. The puzzle is to assemble them into a $1 \times 1 \times 4$ block so that all four colors appear on each side of the block.

EXAMPLE 3

In how many ways can this Instant Insanity puzzle be arranged?

Solution   A common mistake is to assume that a cube can have 6 different arrangements (because of our experience with dice). In the case of Instant Insanity, we are interested not only in the top, but also in the other sides. There are 6 possible faces for the top and *then* 4 possible faces for the front. Thus, by the fundamental counting principle, there are

$$6 \cdot 4 = 24 \text{ arrangements for one cube}$$

Now, for four cubes, again use the fundamental counting principle to find

$$24 \cdot 24 \cdot 24 \cdot 24 = 331{,}776$$

This is *not* the number of *different* possibilities. You are asked to find that number in the problem set.   ◆

Problem Set 10.7

LEVEL 1

1. **IN YOUR OWN WORDS** Example 3 shows that Instant Insanity has 331,776 possibilities (not all different). If you could make one different move per second, how long do you think it would take you to go through all these arrangements? Estimate your answer first, then calculate the correct answer.

2. **IN YOUR OWN WORDS** Suppose a computer could print out 1,000 arrangements in Problem 1 every minute. How long would it take for this computer to print out all possibilities?

Show the result of the moves on Rubik's cube indicated in Problems 3–15. Remember that R, F, L, B, T, and U mean rotate 90° clockwise the right, front, left, back, top, and under faces, respectively. Use the standard Rubik's cube shown in Figure 10.19 as your starting point. Consider each clockwise rotation by first turning the cube so that the side you are rotating is facing you.

3. **a.** F **b.** B **c.** U **d.** L

4. **a.** B^{-1} **b.** F^{-1} **c.** L^{-1} **d.** T^{-1}

5. **a.** F^2 **b.** T^2 **c.** L^2 **d.** B^2

6. **a.** F^3 **b.** R^3 **c.** T^3 **d.** U^3

7. RL 8. TU 9. FB^{-1}

10. $F^{-1}B$ 11. RT 12. FT

13. FT^{-1} 14. FR^{-1} 15. $F^{-1}T^{-1}$

LEVEL 2

Name the move or moves that will return the cube to standard position after making the move shown in Problems 16–23.

16. F 17. T^2

18. F^3 19. TU

20. FB^{-1} 21. L^{-1}

22. FT^{-1} 23. $F^{-1}B^{-1}$

LEVEL 3

24. Are two consecutive moves on Rubik's cube commutative?

25. Are three consecutive moves on Rubik's cube associative?

26. We saw in Example 2 that $R^{-1}T^{-1}$ reversed the move TR. Would $T^{-1}R^{-1}$ also reverse the move TR? Why or why not?

PROBLEM SOLVING

27. Find the number of *different* arrangements for the Instant Insanity blocks.

28. The most difficult part of understanding a solution to Rubik's cube is understanding the notation used by the author in stating the solution. In one solution, the Ledbetter–Nering algorithm (see the references in the Group Research Projects), the authors describe a sequence

of moves that they call THE MAD DOG. This series of moves can be described using our notation by

$$(RTF)^5$$

Start with the standard-position cube and show the result after carrying out THE MAD DOG.

29. David Singmaster, in his book *Notes on Rubik's "Magic Cube"* (see the references in the Group Research Projects), describes a sequence of moves to put the upper corners in place (after certain other moves). Using our notation, this set of moves is

$$FU^2F^{-1}U^2R^{-1}UR$$

Start with the standard-position cube and show the result after carrying out this sequence of moves.

30. In *The Simple Solution to Rubik's Cube* (see the references in the Group Research Projects), James Nourse describes a process to orient the bottom corner cubes. Using our notation, this set of moves is

$$R^{-1}U^{-1}RU^{-1}R^{-1}U^2RU^2$$

Start with the standard-position cube and show the result after carrying out this sequence of moves.

CHAPTER SUMMARY

"There is no problem in all mathematics that cannot be solved by direct counting. But with the present implements of mathematics many operations can be performed in a few minutes which without mathematical methods would take a lifetime."

Ernst Mach

IMPORTANT TERMS

And [10.1]
Arrangement [10.3]
Binomial theorem [10.6]
Cards, deck of [10.4]
Combination [10.3, 10.4]
Complement [10.1]
Count-down property [10.3]
De Morgan's laws [10.2]
Disjoints sets [10.1]
Distinguishable permutation [10.3]

Equal sets [10.1]
Euler circles [10.1]
Factorial [10.3]
Fundamental counting principle [10.3]
Instant Insanity [10.7]
Intersection [10.1]
Multiplication property of factorials [10.3]
Or [10.1]

Ordered pair [10.3]
Ordered triple [10.3]
Permutation [10.3]
Rubik's cube [10.7]
Set [10.1]
Set theory [10.1]
Subset [10.1]
Tree diagram [10.3]
Union [10.1]
Venn diagram [10.1]

TYPES OF PROBLEMS

Find unions, intersections, and complements of given sets. [10.1]
Draw Venn diagrams for complement, union, and intersection. [10.1]
Know the formula for the cardinality of the union of two sets. [10.1]
Carry out combined operations with sets. [10.2]
Answer questions involving surveys and the cardinality of sets. [10.2]
Prove statements involving sets, including De Morgan's laws. [10.2]
Apply the fundamental counting principle. [10.3]
Simplify expressions involving factorials. [10.3]
Know the permutation formula. [10.3]
Evaluate permutations and distinguishable permutations. [10.3]
Know the combination formula. [10.4]
Evaluate combinations. [10.4]
Distinguish among a permutation, a combination, a distinguishable permutation, and the fundamental counting principle. [10.5]

Answer applied counting questions. [10.3–10.6]
Expand binomials using the binomial theorem. [10.6]
Use the notation involving Rubik's cube. [10.7]
Explain the Instant Insanity problem. [10.7]

CHAPTER 10 REVIEW QUESTIONS

Let $U = \{1, 2, 3, 4, 5, 6, 7, 8, 9, 10\}$, $A = \{1, 3, 5, 7, 9\}$, $B = \{2, 4, 6, 9, 10\}$ *in Problems 1–6. Find:*

1. a. $A \cup B$ **b.** $A \cap B$ **2. a.** $\overline{B}$ **b.** $|\varnothing|$

3. a. $|U|$ **b.** $|A|$ **4. a.** $\overline{A \cap B}$ **b.** $\overline{A} \cup \overline{B}$

5. $\overline{A} \cap (B \cup A)$ **6.** $\overline{(A \cup B) \cap A}$

7. If $U = \{1, 2, 3, \ldots, 49, 50\}$, $N = \{\text{odd number}\}$, $P = \{\text{primes}\}$, find:

   a. $|N|$ **b.** $|P|$ **c.** $|N \cap P|$ **d.** $|N \cup P|$

8. Draw Venn diagrams for

   a. $X \cap Y$ **b.** $X \cup Y$ **c.** $\overline{X}$

9. Draw Venn diagrams for

   a. $\overline{X} \cap Y$ **b.** $\overline{X \cup Y}$

10. Simplify each of the given expressions:

   a. $8! - 3!$ **b.** $8 - 3!$ **c.** $(8 - 3)!$ **d.** $\left(\dfrac{8}{2}\right)!$ **e.** $\dbinom{8}{2}$

11. Find the numerical value of each expression.

   a. $_5C_3$ **b.** $_8P_3$ **c.** $_{12}P_0$ **d.** $_{14}C_4$ **e.** $_{100}P_3$

12. In how many ways can a three-member committee be chosen from a group of 12 people?

13. In how many ways can five people line up at a bank teller's window?

14. How many permutations are there of the letters of the words HAPPY and COLLEGE?

15. A jar contains four red and six white balls. Three balls are drawn at random. In how many ways can at least one red ball be drawn?

16. Prove or disprove:

   a. $A \cup (B \cap C) = (A \cup B) \cap C$ **b.** $(A \cup B) \cap C = (A \cap C) \cup (B \cap C)$

17. A survey of 70 college students showed the following data:

   42 had a car; 50 had a TV; 30 had a bicycle; 17 had a car and a bicycle; 35 had a car and a TV; 25 had a TV and a bicycle; 15 had all three

   How many students had none of the three items?

18. Advertisements for Wendy's Old Fashioned Hamburgers claim that you can have your hamburgers 256 ways. If they offer catsup, onion, mustard, pickles, lettuce, tomato, mayonnaise, and relish, is their claim correct? Explain why or why not.

19. The Wednesday Luncheon Club, consisting of ten members, decided to celebrate its first anniversary by having lunch at a fancy restaurant. When the members arrived and were ready to take their seats, they could not decide where to sit.

Just when they were ready to leave because of their embarrassment at being unable to decide how to sit around the rectangular table, the manager came to the rescue and told them to sit down just where they were standing. Then the secretary of the club was to write down where they were sitting, and they were to return the following day and sit in a different order. If they continued this until they had tried all arrangements, the manager would give them anything on the menu, free of charge.

It sounded like a very good deal, so they agreed. However, the day of the free lunch never came. Can you explain why?

20. One variation of Instant Insanity is a puzzle with five blocks instead of four.
    a. How many arrangements are possible?
    b. If you carry out one arrangement every second, how much time is required to show the number of arrangements in part **a**?

GROUP RESEARCH PROBLEMS

Working in small groups is typical of most work environments, and being able to work with others to communicate specific ideas is an important skill to learn. Work with three or four other students to submit a single report based on each of the following questions.

G35. What is the millionth positive integer that is not a square, cube, or fifth power?

G36. A teacher assigned five problems, A, B, C, D, and E. Not all students turned in answers to all of the problems. Here is a tally of the percentage of students turning in each problem:

A: 46%	A, B: 25%	A, B, C: 13%	A, B, C, D: 7%	A, B, C, D, E: 4%
B: 40%	B, C: 26%	A, B, E: 19%	A, B, C, E: 8%	
C: 43%	C, D: 26%	A, D, E: 16%	A, B, D, E: 9%	
D: 38%	D, E: 22%	B, C, D: 12%	A, C, D, E: 6%	
E: 41%	A, E: 30%	C, D, E: 14%	B, C, D, E: 6%	

What percent of the students did not turn in any problems? Assume that no students turned in combinations not listed.

G37. A famous mathematician, Bertrand Russell, created a whole series of paradoxes by considering situations such as the following *barber's rule:* "Suppose in the small California town of Ferndale it is the practice of many of the men to be shaved by the barber. Now, the barber has a rule that has come to be known as the barber's rule: *He shaves those men and only those men who do not shave themselves.* The question is: Does the barber shave himself?" If he does shave himself, then according to the barber's rule, he does not shave himself. On the other hand, if he does not shave himself, then, according to the barber's rule, he shaves himself. We can only conclude that there can be no such barber's rule. But why not? Write a paper explaining what is meant by a *paradox.* Use the Historical Note below for some suggestions about mathematicians who have done work in this area. You might begin with the links for this problem at **www.mathnature.com**

HISTORICAL NOTE

About the time that Cantor's work began to gain acceptance, certain inconsistencies began to appear. One of these inconsistencies, called **Russell's paradox,** is what Problem G37 is about. Other famous paradoxes in set theory have been studied by many famous mathematicians, including Zermelo, Frankel, von Neumann, Bernays, and Poincaré. These studies have given rise to three main schools of thought concerning the foundation of mathematics.

G38. Prepare a strip of paper as shown in Figure 10.21a. Turn it over and mark the other side as shown in Figure 10.21b.

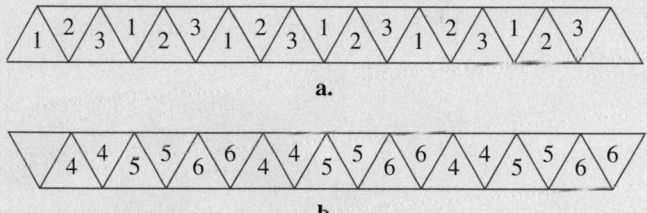

Figure 10.21 **Strips for constructing a hexahexaflexagon. Make sure that each of the numbered triangles is equilateral.**

Starting from the left of Figure 10.21b, fold the 4 onto the 4, the 5 onto the 5, 6 onto 6, 4 onto 4, and so on until your paper looks like the one shown in Figures 10.21c and 10.21d.

c. Front   **d.** Back

Figure 10.21 **Hexahexaflexagon after the first fold**

Continue by folding 1 onto the 1 from the front, by folding the 1 onto the 1 from the back, and finally by bringing the 1 up from the bottom so that it rests on top of the 1 on the top. Your paper should look like the one shown in Figures 10.21e and 10.21f.

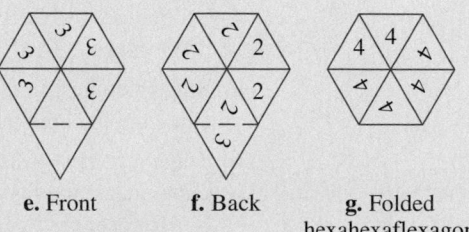

e. Front   **f.** Back   **g.** Folded hexahexaflexagon

Figure 10.21 **Hexahexaflexagon after the second fold**

Paste the blank onto the blank, and the result is called a **hexahexaflexagon,** as shown in Figure 10.21g. With a little practice you'll be able to "flex" your hexahexaflexagon (see Figure 10.22) so that you can obtain a side with all 1s, another with all 2s, . . . , and another with all 6s. After you have become fairly proficient at "flexing," count the number of flexes required to obtain all six "sides." What do you think is the fewest number of flexes necessary to obtain all six sides?

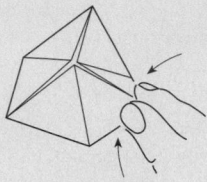 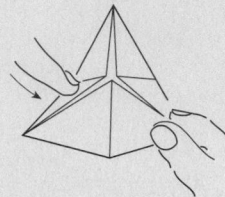 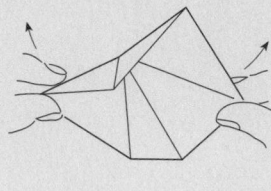

Figure 10.22 **To "flex" your hexahexaflexagon, pinch together two of the triangles (left two figures). The inner edge may then be opened with the other hand (rightmost picture). If the hexahexaflexagon cannot be opened, an adjacent pair of triangles is pinched. If it opens, turn it inside out, finding a side that was not visible before. Be careful not to tear the hexahexaflexagon by forcing the flex.**

G39. A puzzle sold under the name *The Avenger* is pictured in Figure 10.23.

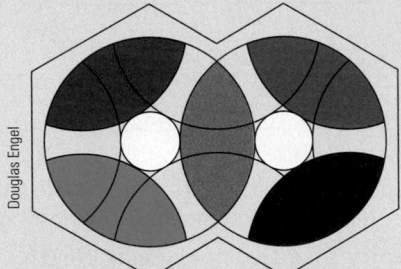 

Douglas Engel

Figure 10.23 **The Avenger Puzzle**

There are four problems posed in the article shown in the reference. Write a report on this article.

> **Reference** "Group Theory, Rubik's Cube and The Avenger," *Games,* June/July 1987, pp. 44–45.

G40. Consult one of the references and learn to solve Rubik's cube. Demonstrate your skill to the class. Nourse names the following categories:

20 minutes:	WHIZ
10 minutes:	SPEED DEMON
5 minutes:	EXPERT
3 minutes:	MASTER OF THE CUBE

a. Stage a contest in front of the class to see which members of your group can complete one face of a Rubik's cube.

b. Stage a contest to see which member of your group can solve the Rubik's cube puzzle the fastest. Report the results to the class.

> **References** Ledbetter and Nering, *The Solution to Rubik's Cube* (Rohnert Park, CA, Noah's Ark Enterprises, 1980).
> James G. Nourse, *The Simple Solution to Rubik's Cube* (New York: Bantam Books, 1981).
> David Singmaster, *Notes on Rubik's "Magic Cube,"* 5th ed. (Hillside, NJ: Enslow Publishers, 1980).

INDIVIDUAL RESEARCH PROBLEMS

Learning to use sources outside your classroom and textbook is an important skill, and here are some ideas for extending some of the ideas in this chapter. You can find references to these projects in a library or at

www.mathnature.com

PROJECT 10.1 Write a paper on the famous *Tower of Hanoi* problem.

PROJECT 10.2 Symbolically name the 32 regions formed by a Venn diagram with five sets.

PROJECT 10.3 Write a report discussing the creation of colors using additive color mixing and subtractive color mixing.

PROJECT 10.4 How can all the constructions of Euclidean geometry be done by folding paper? What assumptions are made when paper is folded to construct geometric figures? What is a hexaflexagon? What is origami?

PROJECT 10.5 Find a solution for the Instant Insanity puzzle.

11 The Nature of Probability

It is a truth very certain that, when it is not in our power to determine what is true, we ought to follow what is most probable.
René Descartes

CONTENTS

OVERVIEW

We see examples of probability every day. Weather forecasts, stock market analyses, contests, children's games, political polls, game shows, and gambling all involve ideas of probability. Probability is the mathematics of uncertainty.

If you have ever played the lottery (hundreds of billions are wagered legally each year) or bought a life insurance policy, you have indirectly used probability. Animals and plants are bred to enhance certain traits to be passed through generations, and these breeding techniques are directed by the probability of genetics. Research in human inherited diseases such as cystic fibrosis, sickle-cell anemia, and Huntington's disease all involve probability.

In this chapter, we'll investigate the definition of probability and some of the procedures for dealing with probability.

IMPORTANT IDEAS

Probability by counting [11.1]
Expectation with and without a cost for playing [11.2]
Finding odds (given probability) [11.3]
Finding probability (given odds) [11.3]
Procedure for using tree diagrams [11.3]
Probabilities of unions and intersections [11.4]
Drawing with and without replacement [11.4]
Tree diagrams to find probabilities [11.4]
Find probabilities of binomial experiments. [11.5]

BOOK REPORTS

Write a 500-word report on one of the following books:
How to Take a Chance. Darrell Huff and Irving Geis (New York: Norton, 1959).
Beat the Dealer. Edward O. Thorp (New York: Vintage Books, 1966).
The Power of Logical Thinking. Marilyn vos Savant (New York: St. Martin's Press, 1997). This book deals with people's perception of probability and statistics.

LINKS
www.mathnature.com

Individual Projects

Group Projects

Research Links

11.1 INTRODUCTION TO PROBABILITY

Terminology

An **experiment** is an observation of any physical occurrence. The **sample space** of an experiment is the set of all its possible outcomes. An **event** is a subset of the sample space. If an event is the empty set, it is called the **impossible event;** and if it has only one element, it is called a **simple event.**

EXAMPLE 1

Suppose a researcher wishes to study how the color of a child's eyes are related to the parents' eye color.
a. List the sample space.
b. Give an example of a simple event.
c. Give an example of an impossible event.

Solution
a. The sample space is a listing of all possible eye colors:

{green, blue, brown, hazel, other}

b. A simple event is an event with only one element, say,

{blue eyes}

c. An impossible event is one that is empty, say,

{purple eyes}   ◆

The next example involves a **die,** which is a cube showing six faces marked 1, 2, 3, 4, 5, and 6. If each outcome of an experiment has the same chance of occurring as any other outcome, then they are said to be **equally likely outcomes.**

EXAMPLE 2

a. What is the sample space for the experiment of tossing a coin and then rolling a die?
b. List the following events for the sample space in part **a:**
   E = {rolling an even number on the die}
   H = {tossing a head}
   X = {rolling a six and tossing a tail}
   Which of these (if any) are simple events?

Solution   A coin is considered *fair* if the outcomes of head and tail are *equally likely.* We also note that a fair die is one for which the outcomes from rolling it are *equally likely.* A die for which one outcome is more likely than the others is called a *loaded die.* In this book, we will assume fair dice and fair coins unless otherwise noted.

a. You can visualize the sample space as shown in Figure 11.1. If the sample space is very large, it is sometimes worthwhile to make a direct listing, as shown in

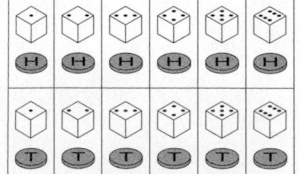

Figure 11.1 **Sample space for tossing a coin and rolling a die**

Figure 11.1. Sometimes, it is helpful to build sample spaces by using what is called a *tree diagram:*

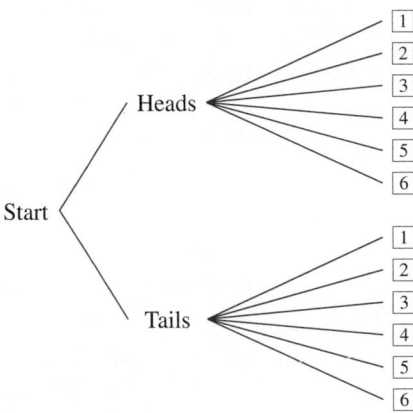

We see that *S* (the sample space) is

$$S = \{H1, H2, H3, H4, H5, H6, T1, T2, T3, T4, T5, T6\}$$

b. An event must be a *subset* of the sample space. Notice that *E, H,* and *X* are all subsets of *S*. Therefore,

$$E = \{H2, H4, H6, T2, T4, T6\}$$
$$H = \{H1, H2, H3, H4, H5, H6\}$$
$$X = \{T6\}$$

X is a simple event, because it has only one element.   ◆

Two events *E* and *F* are said to be **mutually exclusive** if $E \cap F = \varnothing$, that is, they cannot both occur simultaneously.

EXAMPLE 3

Suppose that you perform an experiment of rolling a die. Find the sample space, and then let $E = \{1, 3, 5\}$, $F = \{2, 4, 6\}$, $G = \{1, 3, 6\}$, and $H = \{2, 4\}$. Which of these are mutually exclusive?

Solution   The sample space for a single die, as shown in Figure 11.2, is

$$S = \{1, 2, 3, 4, 5, 6\} \quad \text{We look only at the spots on top.}$$

Evidently,   *E* and *F* are mutually exclusive, since $E \cap F = \varnothing$.

G and *H* are mutually exclusive, since $G \cap H = \varnothing$.

E and *H* are mutually exclusive, since $E \cap H = \varnothing$.

But   *F* and *H* are *not* mutually exclusive.

E and *G* are *not* mutually exclusive.

F and *G* are *not* mutually exclusive.   ◆

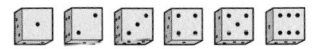

Figure 11.2 **Sample space for a single die**

Life is a school of probability.

Walter Bagebot

Probability

A **probabilistic model** deals with situations that are random in character and attempts to predict the outcomes of events with a certain stated or known degree of accuracy.

For example, if we toss a coin, it is impossible to predict in advance whether the outcome will be a head or a tail. Our intuition tells us that it is equally likely to be a head or a tail, and somehow we sense that if we repeat the experiment of tossing a coin a large number of times, head will occur "about half the time." To check this out, I recently flipped a coin 1,000 times and obtained 460 heads and 540 tails. The percentage of heads is $\frac{460}{1,000} = 0.46 = 46\%$, which is called the *relative frequency.* If an experiment is repeated n times and an event occurs m times, then

$\frac{m}{n}$ is called the **relative frequency** of the event.

Our task is to create a model that will assign a number p, called the *probability of an event,* which will predict the relative frequency. This means that *for a sufficiently large number of repetitions* of an experiment,

$$p \approx \frac{m}{n}$$

Probabilities can be obtained in one of three ways:

1. **Empirical probabilities** (also called *a posteriori* models) are obtained from experimental data. For example, an assembly line producing brake assemblies for General Motors produces 1,500 items per day. The probability of a defective brake can be obtained by experimentation. Suppose the 1,500 brakes are tested and 3 are found to be defective. Then the empirical probability is the relative frequency of occurrence, namely

$$\frac{3}{1,500} = 0.002 \text{ or } 0.2\%$$

2. **Theoretical probabilities** (also called *a priori* models) are obtained by logical reasoning according to stated definitions. For example, the probability of rolling a die and obtaining a 3 is $\frac{1}{6}$, because there are six possible outcomes, each with an equal chance of occurring, so a 3 should appear $\frac{1}{6} \approx 17\%$ of the time.

3. **Subjective probabilities** are obtained by experience and indicate a measure of "certainty" on the part of the speaker. These probabilities are not necessarily arrived at through experimentation or theory. For example, a TV reporter studies the satellite maps and then issues a prediction about tomorrow's weather based on experience: 80% chance of rain tomorrow.

Our focus in this chapter is on theoretical probabilities, but we should keep in mind that our theoretical model should be predictive of the results obtained by experimentation (empirical probabilities); if they are not consistent, and we have been careful about our record-keeping in arriving at an empirical probability, we would conclude our theoretical model to be faulty. The reason for this conclusion is called the **law of large numbers.**

Law of Large Numbers

As an experiment is repeated more and more times, the empirical probability (that is, the proportion of outcomes favorable to any particular event) will tend to come closer and closer to the theoretical probability of that event.

This law of large numbers keeps us "honest" in maintaining our records and in setting up models for finding the theoretical probability.

The simplest probability model makes certain assumptions about the sample space. If the sample space can be divided into *mutually exclusive* and *equally likely* outcomes,

Heads, I win; tails, you lose.

—*17th century English saying*

www

www.mathnature.com
has a link to an interactive
coin-flipping site.

we can define the probability of an event. Let's consider the experiment of tossing a single coin. A suitable sample space is

$$S = \{\text{heads, tails}\}$$

Suppose we wish to consider the event of obtaining heads; we'll call this event *A:*

$$A = \{\text{heads}\}$$

and this is a simple event.

We wish to define the probability of event *A,* which we denote by $P(A)$. Notice that the outcomes in the sample space are mutually exclusive; that is, if one occurs, the other cannot occur. If we flip a coin, there are two possible outcomes, and *one and only one* outcome can occur on a toss. If each outcome in the sample space is equally likely, we define the probability of *A* as

$$P(A) = \frac{\text{NUMBER OF SUCCESSFUL RESULTS}}{\text{NUMBER OF POSSIBLE RESULTS}}$$

A "successful" result is a result that corresponds to the event whose probability we are seeking—in this case, {heads}. Since we can obtain a head (success) in only one way, and the total number of possible outcomes is two, the probability of heads is given by this definition as

$$P(\text{heads}) = P(A) = \frac{1}{2}$$

This must correspond to the empirical results you would obtain if you repeated the experiment a large number of times. In Problem 48 you are asked to repeat this experiment 100 times.

Definition of Probability

STOP

This is the basic definition for this chapter.

> If an experiment can result in any of n mutually exclusive and equally likely outcomes, and if s of these outcomes are considered favorable, then the **probability** of an event E, denoted $P(E)$, is
>
> $$P(E) = \frac{s}{n} = \frac{\text{NUMBER OF OUTCOMES FAVORABLE TO } E}{\text{NUMBER OF ALL POSSIBLE OUTCOMES}}$$

EXAMPLE 4

Use the definition of probability to find first the probability of white and second the probability of black, using the spinner shown and assuming that the arrow will never lie on a border line.

Solution Looking at the spinner, we note that it is divided into three areas of the same size. We assume that the spinner is equally likely to land in any of these three areas.

$$P(\text{white}) = \frac{2}{3} \qquad \leftarrow \text{Two sections are white.}$$
$$\leftarrow \text{Three sections all together}$$

$$P(\text{black}) = \frac{1}{3} \qquad \leftarrow \text{One section is black.}$$
$$\leftarrow \text{Three sections all together}$$

These are theoretical probabilities.   ◆

EXAMPLE 5

Consider a jar that contains marbles as shown. Suppose that each marble has an equal chance of being picked from the jar. Find:

a. $P(\text{black})$ **b.** $P(\text{gray})$ **c.** $P(\text{white})$

Solution

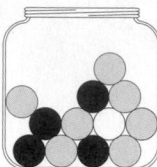

$$P(\text{black}) = \frac{4}{12} \quad \leftarrow 4 \text{ black marbles in jar}$$
$$\qquad\qquad\quad \leftarrow 12 \text{ marbles in jar}$$

$$= \frac{1}{3} \quad \text{Reduce fractions.}$$

$$P(\text{gray}) = \frac{7}{12}$$

$$P(\text{white}) = \frac{1}{12}$$

This is a theoretical probability; also note that the probabilities of all the simple events sum to 1:

$$\frac{1}{3} + \frac{7}{12} + \frac{1}{12} = 1 \qquad \blacklozenge$$

Reduced fractions are used to state probabilities when the fractions are fairly simple. If, however, the fractions are not simple, and you have a calculator, it is acceptable to state the probabilities as decimals, as shown in Example 6.

EXAMPLE 6

Suppose that, in a certain study, 46 out of 155 people showed a certain kind of behavior. Assign a probability to this behavior.

Solution

$$P(\text{behavior shown}) = \frac{46}{155} \approx 0.30$$

This is an empirical probability because it was arrived at through experimentation. $\blacklozenge$

EXAMPLE 7

Consider two spinners as shown. You and an opponent are to spin a spinner simultaneously, and the one with the higher number wins. Which spinner should you choose, and why?

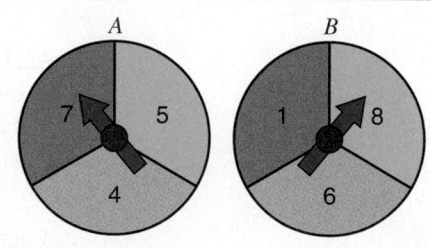

Solution We begin by listing the sample space:

B \ A	1	6	8
4	(4, 1)	(4, 6)	(4, 8)
5	(5, 1)	(5, 6)	(5, 8)
7	(7, 1)	(7, 6)	(7, 8)

The times that A wins are enclosed in boxes.

$P(A \text{ wins}) = \dfrac{4}{9}$; $P(B \text{ wins}) = \dfrac{5}{9}$

We would choose spinner B because it has a greater probability of winning. $\blacklozenge$

EXAMPLE 8

Suppose that a single card is selected from an ordinary deck of 52 cards. Find:

a. $P(\text{ace})$ **b.** $P(\text{heart})$ **c.** $P(\text{face card})$

Solution   The sample space for a *deck of cards* is shown in Figure 10.16, page 523.

a. An ace is a card with one spot. $P(\text{ace}) = \dfrac{4}{52} = \dfrac{1}{13}$

b. $P(\text{heart}) = \dfrac{13}{52} = \dfrac{1}{4}$

c. $P(\text{face card}) = \dfrac{12}{52}$ ← A face card is a jack, queen, or king.

← Number of cards in the sample space

$\qquad\qquad\qquad = \dfrac{3}{13}$ ◆

The following example uses two **dice** (plural form of **die,** which is a common randomizing device).

EXAMPLE 9

Suppose you are just beginning a game of Monopoly™. You roll a pair of dice. What is the probability that you land on a railroad on the first roll of the dice?

Solution   You need to know the locations of the railroads on a Monopoly board. There are four railroads, which are positioned so that only one can be reached on one roll of a pair of dice. The required number to roll is a 5. We begin by listing the sample space.

You might try {2, 3, 4, 5, 6, 7, 8, 9, 10 11, 12}, but these possible outcomes are not equally likely, which you can see by considering a tree diagram. The roll of the first die has 6 possibilities, and then *each* of these in turn can combine with any of 6 possibilities for a total of 36 possibilities. The sample space is summarized in Figure 11.3.

Irving Hertzel, a professor from Iowa State, used a computer to find the most probable squares on which you can land in the game of Monopoly™.

1. Illinois Avenue
2. Go
3. B.&O. Railroad
4. Free Parking

See Group Research Project G41 for some questions based on Monopoly.

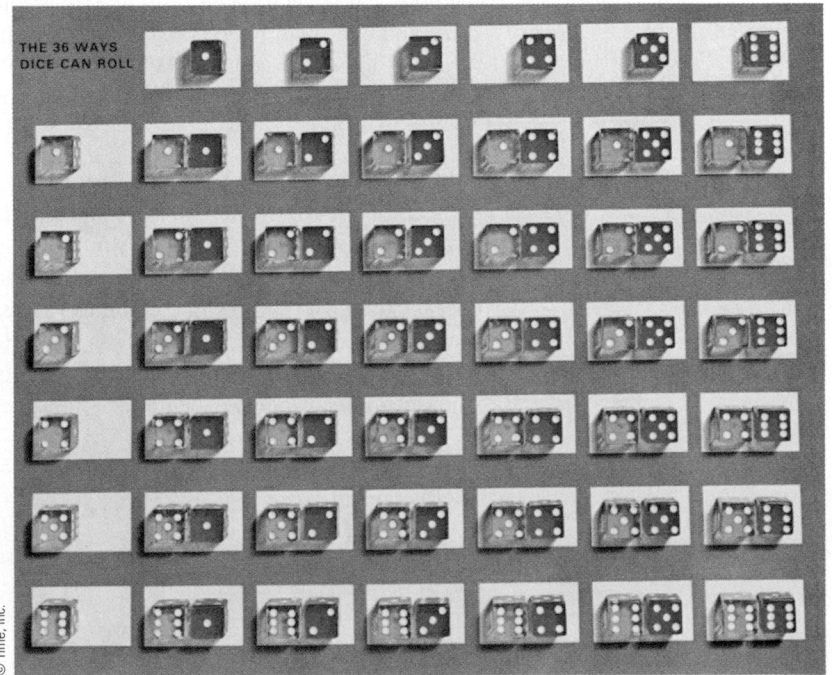

Figure 11.3 **Sample space for a pair of dice**

© Time, Inc.

Thus, $n = 36$ for the definition of probability. We need to look at Figure 11.3 to see how many possibilities there are for obtaining a 5. We find (1, 4), (2, 3), (3, 2), and (4, 1), so $s = 4$. Then

$$P(\text{five}) = \frac{4}{36} \quad \leftarrow \text{4 ways to obtain a 5}$$
$$\phantom{P(\text{five})} \quad \leftarrow \text{36 ways to roll a pair of dice}$$
$$= \frac{1}{9}$$

◆

You should use Figure 11.3 when working probability problems dealing with rolling a pair of dice. You will be asked to perform an experiment (Problem 49) in which you roll a pair of dice 100 times and then compare the results you obtain with the probabilities you calculate by using Figure 11.3.

When using the definition of probability, you must be careful not to overlook the restrictions on s and n. The requirement that the listing of the elements of the sample space are *mutually exclusive* simply means that each outcome of the experiment will be counted in only one category (that is easy). The importance of the second part of the restriction, that the n possible outcomes are *equally likely,* is evident in Example 9. If we roll a pair of dice and list the outcomes as 2, 3, 4, 5, 6, 7, 8, 9, 10, 11, and 12, we note these are *not equally likely* possibilities and do not lead to the correct theoretical probabilities. If we use the sample space in Figure 11.3 for a pair of dice, we note these *are equally likely possibilities,* so these 36 possibilities are used for n when rolling a pair of dice. The following example illustrates the importance of looking at the correct sample space.

EXAMPLE 10

A contest game show has a large prize and a small joke prize, but the contestant does not know which one is concealed behind a door. The emcee shows the contestant another large prize and places it behind the door. Thus, there are now two prizes behind the door and the contestant does not know whether it is two large prizes or one large prize and one small prize. The emcee then shows one of the prizes at random and it turns out to be a large prize. What is the probability that the contestant obtains both large prizes?

There is a similar, but different problem, known as the "Monty Hall Dilemma." An interactive link for this variation can be found at **www.mathnature.com**

Solution It is an *error* to reason as follows: *The probability of receiving both large prizes is 50% because there are two prizes left, one large and one small, so the sample space has two possibilities, one of which is success.* If you reason this way, you are not considering equally likely possibilities in the sample space. (Problem 54 gives an empirical experiment to illustrate this fact.) Here is a correct arrangement of the sample space. The two possible original prizes are B (for big prize) and S (for small prize). The sample space is B or S. When a new third prize N (which is big) is added, then the possibilities for the two prizes behind the door are:

Shown prize:	N	B	N	S̸
Remaining prize:	B	N	S	N̸

⎧ Can't be this one
⎨ because the shown
⎩ prize is big.

We have three equally likely choices for the remaining prize—B, N, or S—and since two of these are successes we have

$$P(\text{big prize}) = \frac{2}{3}$$

◆

0   0.5   1
impossible even chance certain

$P(\emptyset) = 0$ $P(S) = 1$

The probability of the empty set is 0, which means that the event cannot occur. In Problem Set 11.1, you will be asked to show that the probability of an event that *must* occur is 1. These are the two extremes. All other probabilities fall somewhere in between. The closer a probability is to 1, the more likely the event is to occur; the closer a probability is to 0, the less likely the event is to occur.

If the probability you are finding has a sample space that is easy to count, and if all simple events in the sample space are equally likely, then there is a simple procedure that we can use to find probabilities.

Probability by Counting

STOP *Pay attention to this procedure; it will suffice for much of what you will do in this chapter.*

> If the probability you are finding has a sample space that is easy to count, and if all simple events in the sample space are equally likely, then there is a simple procedure that you can use to find probabilities.
>
> 1. Describe and identify the sample space, S. The number of elements in S is n.
> 2. Count the number of occurrences that satisfy the event of interest (we call this success); denote it by s.
> 3. Compute the probability of the event using the formula
>
> $$P(E) = \frac{s}{n}$$

Keep in mind that this procedure does not apply to every situation. If it doesn't, you need a more complicated model, or else you must proceed experimentally.

Probabilities of Unions and Intersections

The word *or* is translated as $\cup$ (union), and the word *and* is translated as $\cap$ (intersection). We will find the probabilities of compound events involving the words *or* and *and* by finding unions and intersections of events.

EXAMPLE 11

Suppose that a single card is selected from an ordinary deck of cards.

a. What is the probability that it is a two or a king?

b. What is the probability that it is a two or a heart?

c. What is the probability that it is a two and a heart?

d. What is the probability that it is a two and a king?

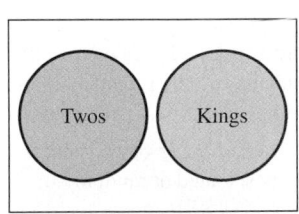

Mutually exclusive

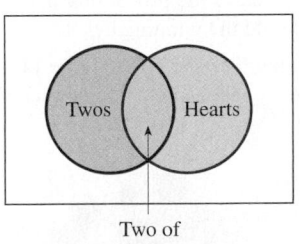

Two of hearts

Not mutually exclusive

Solution

a. $P(\text{two or a king}) = P(\text{two} \cup \text{king})$

Look at Figure 10.16 if you are not familiar with a deck of cards.

two = {two of hearts, two of spades, two of diamonds, two of clubs}
king = {king of hearts, king of spades, king of diamonds, king of clubs}

two $\cup$ king = {two of hearts, two of spades, two of diamonds, two of clubs, king of hearts, king of spades, king of diamonds, king of clubs}

There are 8 possibilities for success. It is usually not necessary to list all of these possibilities to know that there are 8 possibilities (four twos and four kings):

$$P(\text{two} \cup \text{king}) = \frac{8}{52} = \frac{2}{13}$$

b. This seems to be very similar to part **a,** but there is one important difference. Look at the sample space and notice that although there are 4 twos and 13 hearts, the total number of successes is *not* 4 + 13 = 17, *but rather* 16.

> two = {**two of hearts,** two of spades, two of diamonds, two of clubs}
> heart = {ace of hearts, **two of hearts,** three of hearts, . . . , king of hearts}
>
> two ∪ heart = {**two of hearts,** two of spades, two of diamonds, two of clubs, ace of hearts, three of hearts, four of hearts, five of hearts, six of hearts, seven of hearts, eight of hearts, nine of hearts, ten of hearts, jack of hearts, queen of hearts, king of hearts}.

In the previous chapter, we found |*two ∪ hearts*|:

$$|T \cup H| = |T| + |H| - |T \cap H|$$
$$= 4 + 13 - 1$$
$$= 16$$

It is not necessary to list these possibilities. The purpose of doing so in this case was to reinforce the fact that there are *actually* 16 (not 17) possibilities.

$$P(\text{two} \cup \text{heart}) = \frac{16}{52} = \frac{4}{13}$$

c. two ∩ heart = {two of hearts} so there is one element in the intersection:

$$P(\text{two} \cap \text{heart}) = \frac{1}{52}$$

d. two ∩ king = ∅, so there are no elements in the intersection:

$$P(\text{two} \cap \text{king}) = \frac{0}{52} = 0$$

◆

Problem Set 11.1

LEVEL 1

1. **IN YOUR OWN WORDS** What is the difference between empirical and theoretical probabilities?

2. **IN YOUR OWN WORDS** Define probability.

3. **IN YOUR OWN WORDS** Write a simple argument to convince someone why all probabilities must be between 0 and 1 (including 0 and 1).

For the spinners in Problems 4–7, assume that the pointer can never lie on a border line.

4. **a.** $P(A)$ 5. **a.** $P(D)$ 6. **a.** $P(G)$ 7. **a.** $P(J)$
   b. $P(B)$ **b.** $P(E)$ **b.** $P(H)$ **b.** $P(G \text{ or } H)$
   c. $P(C)$ **c.** $P(F)$ **c.** $P(I)$ **c.** $P(I \text{ or } J)$

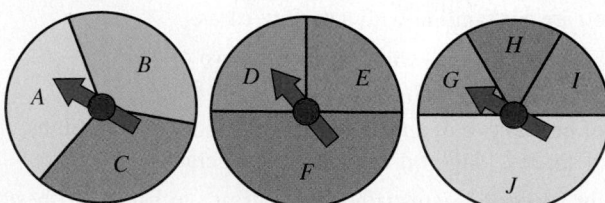

Give the probabilities in Problems 8–11 in decimal form (correct to two decimal places). A calculator may he helpful with these problems.

8. Last year, 1,485 calculators were returned to the manufacturer. If 85,000 were produced, assign a number to specify the probability that a particular calculator would be returned.

9. Last semester, a certain professor gave 13 A's out of 285 grades. If one of the 285 students is to be selected at random, what is the probability that his or her grade is an A?

10. Last year in Ferndale, CA, it rained on 75 days. What is the probability of rain on a day selected at random?

11. The campus vets club is having a raffle and is selling 1,500 tickets. If the people on your floor of the dorm bought 285 of those tickets, what is the probability that someone on your floor will hold the winning ticket?

12. Consider the jar containing marbles shown in Figure 11.4. Suppose each marble has an equal chance of being picked from the jar. Find:
   a. $P(\text{white})$
   b. $P(\text{black})$
   c. $P(\text{gray})$

Figure 11.4 A jar of marbles

Table 11.1 Poker Hands*

Hand		Cards
Royal flush 4 hands		
Other straight flush 36 hands		
Four of a kind 624 hands		
Full house 3,744 hands		
Flush 5,108 hands		
Straight 10,200 hands		
Three of a kind 54,912 hands		
Two pair 123,552 hands		
One pair 1,098,240 hands		
Other hands 1,302,540 hands		

Poker is a common game in which players are dealt five cards from a deck of cards. It can be shown that there are 2,598,960 different possible poker hands. The winning hands (from highest to lowest) are shown in Table 11.1. Find the requested probabilities in Problems 13–22. Use a calculator, and show your answers to whatever accuracy possible on your calculator.

13. P(royal flush)   **14.** P(straight flush)
15. P(four of a kind)   **16.** P(full house)
17. P(flush)   **18.** P(straight)
19. P(three of a kind)   **20.** P(two pair)
21. P(one pair)   **22.** P(no pair or better)

*All of these possibilities are mutually exclusive. That is, the 36 straight flushes do not include the 4 royal flushes, and the 5,108 flushes do not include the better flush hands of straight or royal flushes.

LEVEL 2

A single card is selected from an ordinary deck of cards. The sample space is shown in Figure 10.16. Find the probabilities in Problems 23–26.

23. a. P(five of clubs)   **b.** P(five)   **c.** P(club)
24. a. P(jack)   **b.** P(spade)   **c.** P(jack of spades)
25. a. P(five and a jack)   **b.** P(five or a jack)
26. a. P(heart and a jack)   **b.** P(heart or a jack)

Suppose that you toss a coin and roll a die in Problems 27–30. The sample space is shown in Figure 11.1.

27. What is the probability of obtaining:
   a. Tails and a five?   **b.** Tails or a five?
   c. Heads and a two?

28. What is the probability of obtaining:
   a. Tails?   **b.** Heads or a two?
   c. One, two, three, or four?

29. What is the probability of obtaining:
   a. Heads and an odd number?
   b. Heads or an odd number?

30. What is the probability of obtaining:
   a. Heads and a five?   **b.** Heads or a five?

Use the sample space shown in Figure 11.3 to find the probabilities in Problems 31–39 for the experiment of rolling a pair of dice.

31. P(five)   **32.** P(six)   **33.** P(seven)
34. P(eight)   **35.** P(nine)   **36.** P(two)
37. P(four or five) **38.** P(even)   **39.** P(eight or ten)

Suppose you and an opponent each pick one of the spinners shown here. A "win" means spinning a higher number. Construct a sample space to answer each question, and tell which of the two spinners given in Problems 40–47 you would choose in each case.

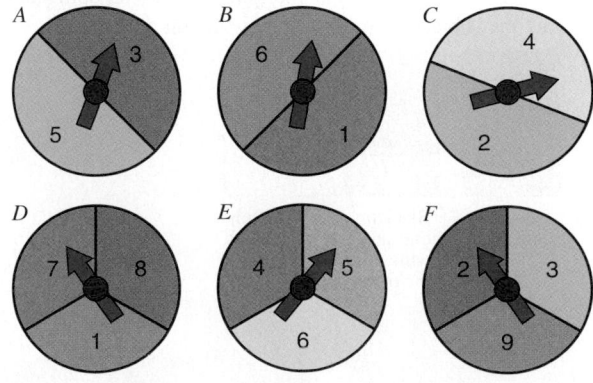

40. A plays B **41.** A plays C **42.** B plays C
43. D plays E **44.** E plays F **45.** D plays F
46. B plays D **47.** F plays C

LEVEL 3

Perform the experiments in Problems 48–51, tally your results, and calculate the probabilities (to the nearest hundredth).

48. Toss a coin 100 times. Make sure that, each time the coin is flipped, it rotates several times in the air and lands on a table or on the floor. Keep a record of the results of this experiment. Based on your experiment, what is $P(\text{heads})$?

49. Roll a pair of dice 100 times. Keep a record of the results. Based on your experiment, find: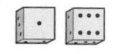

   a. $P(\text{two})$ b. $P(\text{three})$ c. $P(\text{four})$
   d. $P(\text{five})$ e. $P(\text{six})$ f. $P(\text{seven})$
   g. $P(\text{eight})$ h. $P(\text{nine})$ i. $P(\text{ten})$
   j. $P(\text{eleven})$ k. $P(\text{twelve})$

50. Flip three coins simultaneously 100 times, and note the results. The possible outcomes are:
   a. three heads
   b. two heads and one tail
   c. two tails and one head
   d. three tails

   Based on your experiment, find the probabilities of each of these events. Do these appear to be equally likely?

51. Simultaneously toss a coin and roll a die 100 times, and note the outcome of each trail. The possible outcomes are H1, H2, H3, H4, H5, H6, T1, T2, T3, T4, T5, and T6. Do these appear to be equally likely outcomes?

52. Suppose it is certain that an earthquake will occur someday. What is the probability (to the nearest percent) that it will occur while you are at work? Assume you are at work 8 hours per day, 240 days per year.

53. Suppose it is certain that an earthquake will occur some day. What is the probability (to the nearest percent) that it will occur while you are at school? Assume you are at school 5 hours per day, 174 days per year.

54. Prepare three cards that are identical except for the color. One card is black on both sides, one is white on both sides, and one is black on one side and white on the other.

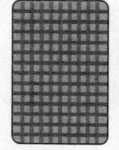

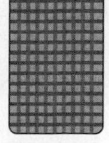

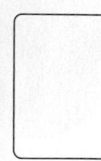

Black on both sides   Black on one side, white on the other   White on both sides

One side of one card is selected at random and placed flat on the table. You will see either a black or a white card; record the color of the face (the top). This is *not* the event with which we are concerned; rather, we are interested in finding the probability of the *other* side (the side you cannot see) being black or white. Record the underside, as shown in the following table. Repeat the experiment 50 times and find the probability of occurrence with respect to the known color. Do these appear to be equally likely outcomes?

Color of face	Frequency	Outcome (color of the underside)	Frequency	Probability
White		White		
		Black		
Black		White		
		Black		

B.C. By permission of Johnny Hart and Creators Syndicate, Inc.

55. Dice is a popular game in gambling casinos. Two dice are tossed, and various amounts are paid according to the outcome. If a seven or eleven occurs on the first roll, the player wins.
   a. What is the probability of winning on the first roll?
   b. The player loses if the outcome of the first roll is a two, three, or twelve. What is the probability of losing on the first roll?

56. In dice, a pair of ones is called *snake eyes*. What is the probability of losing a dice game by rolling snake eyes?

57. Consider a die with only four sides, marked one, two, three, and four.
  a. Write out a sample space similar to the one in Figure 11.3 for rolling a pair of these dice.
  Assuming equally likely outcomes, find the probability that the sum of the dice is the given number.
  b. P(two)   **c.** P(three)   **d.** P(four)
  e. P(five)   **f.** P(six)   **g.** P(seven)

58. The game of Dungeons and Dragons uses nonstandard dice. Consider a die with eight sides marked one, two, three, four, five, six, seven, and eight. Write a sample space similar to the one in Figure 11.3 for rolling a pair of these dice.

> **PROBLEM SOLVING**

59. A mad scientist has captured you and is showing you around his foul-smelling laboratory. He motions to an opaque, formalin-filled jar. "This jar contains one organ, either a kidney or a brain," he cheerily informs you. His voice seems an octave too high as he gives you a twisted leer. You watch as the madman grabs a brain lying on his worktable and drops it into the jar as well. He then shakes

the jar and quickly withdraws a single organ. It proves to be a brain. He turns to you and says, "What is now the chance of removing another brain?" Fearing that the scientist might remove *your* brain in his next ghoulish experiment, you want to give him the right answer. What is your response?*

60. This question is to test your intuition. *Read* the following exercises (do *not* carry out the experiment).
  A. Open a phone book to any page in the white pages. Select 100 consecutive phone numbers and tally the number of times each of the digits 0, 1, 2, 3, 4, 5, 6, 7, 8, 9 occurs as a first digit.
  B. Repeat A for the third digit.
  C. Repeat A for the last digit.
  a. Do you think the occurrence of each digit will be about the same for each of these experiments?
  b. Do you think the results of all three experiments will be about the same?
  c. *Now,* after answering parts **a** and **b,** carry out the experiments described in A, B, and C.

*Problem by Clifford A. Pickover from *Discover*®, published by Walt Disney Magazine Publishing Group Inc., March 1997, p. 94. Reprinted with permission.

11.2 MATHEMATICAL EXPECTATION

Expected Value

Smiles toothpaste is giving away $10,000. All you must do to have a chance to win is send a postcard with your name on it (the fine print says you do not need to buy a tube of toothpaste). Is it worthwhile to enter?

Suppose the contest receives 1 million postcards (a conservative estimate). We wish to compute the **expected value** (or your **expectation**) of winning this contest. We find the expectation for this contest by multiplying the amount to win by the probability of winning:

$$\text{EXPECTATION} = (\text{AMOUNT TO WIN}) \times (\text{PROBABILITY OF WINNING})$$

$$= \$10,000 \times \frac{1}{1,000,000}$$

$$= \$0.01$$

What does this expected value mean? It means that if you were to play this "game" a large number of times, you would expect your *average winnings per game* to be $0.01.

Fair Game

> A game is said to be **fair** if the expected value is 0. If the expected value is positive, then the game is in your favor; and if the expected value is negative, then the game is not in your favor.

Is the Smiles toothpaste giveaway game fair? If the toothpaste company charges you 1¢ to play the game, then it is fair. But how much does the postcard cost? (A cost is a negative amount to win.) We see that this is not a fair game.

EXAMPLE 1

Suppose that you draw a card from a deck of cards and are paid $10 if it is an ace. What is the expected value?

Solution

$$\text{EXPECTATION} = \$10 \times \frac{4}{52} \approx \$0.77$$

Since expected value involves a dollar amount, round your answer to the nearest cent. ◆

Sometimes there is more than one payoff, and we define the expected value (or expectation) as the sum of the expected values from each separate payoff.

EXAMPLE 2

A recent contest offered one grand prize worth $10,000, two second prizes worth $5,000 each, and ten third prizes worth $1,000 each. What is the expected value if you assume that there are 1 million entries?

Solution

$$P(\text{1st prize}) = \frac{1}{1,000,000}; \quad P(\text{2nd prize}) = \frac{2}{1,000,000}; \quad P(\text{3rd prize}) = \frac{10}{1,000,000}$$

$$\text{EXPECTATION} = \underbrace{\$10,000}_{\text{amount of 1st prize}} \times \underbrace{\frac{1}{1,000,000}}_{P(\text{1st prize})} + \$5,000 \times \underbrace{\frac{2}{1,000,000}}_{\text{2nd prize}} + \$1,000 \times \underbrace{\frac{10}{1,000,000}}_{\text{3rd prize}}$$

$$= \$0.01 + \$0.01 + \$0.01$$

$$= \$0.03$$ ◆

You can use mathematical expectation to help you make a decision, as suggested by the following example.

> Let the king prohibit gambling and betting in his kingdom, for these are vices that destroy the kingdoms of princes.
>
> The *Code of Manu*, ca. 100 A.D.

EXAMPLE 3

You are offered two games:

Game A: Two dice are rolled. You will be paid $3.60 if you roll two ones, and will not receive anything for any other outcome.

Game B: Two dice are rolled. You will be paid $36.00 if you roll any pair, but you must pay $3.60 for any other outcome.

Which game should you play?

Solution You might say, "I'll play the first game because, if I play that game, I cannot lose anything." This strategy involves *minimizing your losses*. On the other hand, you can use a strategy that *maximizes your winnings*. In this book, we will base our decisions on maximizing the winnings. That is, we wish to select the game that provides the larger expectation.

Game A: EXPECTATION = $\$3.60 \times \dfrac{1}{36} = \$0.10$

Game B: When calculating the expected value with a charge (a loss), write that charge as a negative number (a negative payoff is a loss).

$$\text{EXPECTATION} = \$36.00 \times \frac{6}{36} + (-\$3.60) \times \frac{30}{36}$$

$$= \$6.00 + (-\$3.00)$$

$$= \$3.00$$

This means that, if you were to play each game 100 times, you would expect your winnings for Game A to be 100($0.10) or about $10 and those from playing Game B to be 100($3.00) or about $300. You should choose to play Game B. ◆

Now we give a formal definition of expectation.

Mathematical Expectation

STOP

Study this formula to be sure you understand what it is saying.

> If an event E has several possible outcomes with probabilities $p_1, p_2, p_3, \ldots, p_n$, and if for each of these outcomes the amount that can be won is $a_1, a_2, a_3, \ldots, a_n$, respectively, then the **mathematical expectation** (or expected value) of E is
>
> $$\text{EXPECTATION} = a_1p_1 + a_2p_2 + a_3p_3 + \cdots + a_np_n$$

Since we know that $p_1 + p_2 + \cdots + p_n = 1$, we note that in many examples some of the probabilities may be zero. For Example 1, we might have said that there are two payoffs, $10 if you draw an ace (probability $\frac{4}{52}$) and $0 otherwise (probability $\frac{48}{52}$) so that

$$\text{EXPECTATION} = \$10\left(\frac{4}{52}\right) + \$0\left(\frac{48}{52}\right) \approx \$0.77$$

EXAMPLE 4

A contest offered the prizes shown here. What is the expected value for this contest?

Solution
We note the following values:

$a_1 = \$15,000; p_1 = 0.000008$

$a_2 = \$1,000; p_2 = 0.000016$

$a_3 = \$625; p_3 = 0.000016$

$a_4 = \$525; p_4 = 0.000016$

$a_5 = \$390; p_5 = 0.000032$

$a_6 = \$250; p_6 = 0.000032$

Prize	Value	Probability
Grand Prize Trip	$15,000	0.000008
Samsonite Luggage	$1,000	0.000016
Magic Chef Range	$625.00	0.000016
Murray Bicycle	$525.00	0.000016
Lawn Boy Mower	$390.00	0.000032
Weber Kettle	$250.00	0.000032

EXPECTATION $= \$15,000(0.000008) + \$1,000(0.000016) + \$625(0.000016)$

$+ \$525(0.000016) + \$390(0.000032) + \$250(0.000032)$

$\approx \$0.17$ ◆

It is not necessary that the expectation or expected value be reported in terms of dollars and cents. Consider the following example.

EXAMPLE 5

Suppose a family has three children. What is the expected number of girls?

Solution The sample space for this problem can be found using a tree diagram: {GGG, GGB, GBG, GBB, BGG, BGB, BBG, BBB}. All possible outcomes and their probabilities are listed in the following table:

Number of girls	*Probability*	*Product*
0	$\frac{1}{8}$	0
1	$\frac{3}{8}$	$\frac{3}{8}$
2	$\frac{3}{8}$	$\frac{6}{8}$
3	$\frac{1}{8}$	$\frac{3}{8}$

$$\text{EXPECTED VALUE} = 0 + \frac{3}{8} + \frac{6}{8} + \frac{3}{8} = \frac{12}{8} = 1.5$$

Notice from Example 5 that the expected value may be a number that can never occur; it is obvious that 1.5 girls will never occur. An expected value of 1.5 simply means that if we record the numbers of girls in a large number of different three-child families, the *average* number of girls for all these families will be 1.5. We will discuss averages in the next chapter.

Expectation with a Cost of Playing

Many games charge you a fee to play. If you must pay to play, this cost of playing should be taken into consideration when you calculate the expected value. Remember, if the expected value is 0, it is a fair game; if the expected value is positive, you should play, but if it is negative, you should not.

Expectations with a Cost for Playing

If there is a cost of playing a game, the cost of playing must be subtracted from the expectation.

$$\text{EXPECTATION} = \text{AMT. TO WIN}(\text{PROB. OF WINNING}) - \text{COST OF PLAYING}$$

EXAMPLE 6

Consider a game that consists of drawing a card from a deck of cards. If it is a face card, you win $20. Should you play the game if it costs $5 to play?

Solution

$$\text{EXPECTATION} = \underbrace{\$20}_{\text{Winnings}} \underbrace{\left(\frac{12}{52}\right)}_{\text{Prob. of winning}} \underbrace{- \ \$5}_{\text{Amt. to play}} \approx -\$0.38$$

You should not play this game, because it has a negative expectation.

EXAMPLE 7

Eva, who is a realtor, knows that if she takes a listing to sell a house, it will cost her $1,000. However, if she sells the house, she will receive 6% of the selling price. If another realtor sells the house, Eva will receive 3% of the selling price. If the house remains unsold after 3 months, she will lose the listing and receive nothing. Suppose that the probabilities for selling a particular $200,000 house are as follows: The probability that Eva will sell the house is 0.4; the probability that another agent will sell the house is 0.2; and the probability that the house will remain unsold is 0.4. What is Eva's expectation if she takes this listing?

Solution   First, we must decide whether this is an "entrance fee" problem. Is Eva required to pay the $1,000 before the "game" of selling the house is played? The answer is yes, so the $1,000 must be subtracted from the expectation.

Now let's calculate the payoffs:

$$6\% \text{ of } \$200,000 = 0.06(\$200,000) = \$12,000$$

$$3\% \text{ of } \$200,000 = 0.03(\$200,000) = \$6,000$$

$$\underset{\text{Eva sells the house.}}{\text{EXPECTATION} = (\$12,000)(0.4)} + \underset{\text{Another agent sells.}}{(\$6,000)(0.2)} - \underset{\text{Cost of playing}}{(\$1,000)}$$

$$= \$5,000$$

Eva's expectation is $5,000.   ◆

Sometimes it is easier to subtract the cost of playing as you go. For example, you could calculate the expectation of Example 7 as follows:

$$\text{EXPECTATION} = \underset{\text{Eva sells the house.}}{(\$12,000 - \$1,000)(0.4)} + \underset{\text{Another agent sells.}}{(\$6,000 - \$1,000)(0.2)} + \underset{\text{House doesn't sell.}}{(-\$1,000)(0.4)}$$

$$= \$11,000(0.4) + \$5,000(0.2) - \$1,000(0.4)$$

$$= \$4,400 + \$1,000 - \$400$$

$$= \$5,000$$

You must understand the nature of the game to know whether to subtract the cost of playing, as we did in Example 7. If you surrender your money to play, then it must be subtracted, but if you "leave it on the table," then you do not subtract it. Consider the following example of a U.S. roulette game. In this game, your bet is placed on the table but is not collected until after the play of the game and it is determined that you lost.

EXAMPLE 8

What is the expectation for playing roulette if you bet $1 on number 5?

Solution   A U.S. roulette wheel has 38 numbered slots (1–36, 0, and 00), as shown in Figure 11.5 on page 568. Some of the more common bets and payoffs are shown. If the payoff is listed as 6 to 1, you would receive $6 for each $1 bet. In addition, you would keep the $1 you originally wagered. One play consists of having the dealer spin the wheel and a little ball in opposite directions. As the ball slows to a stop, it lands in one of the 38 numbered slots, which are colored black, red, or green. A single number bet has a payoff of 35 to 1. The $1 you bet is collected only if you lose.

Now, you can calculate the expected value:

$$\text{EXPECTATION} = \overset{\text{Payoff for a win}}{35} \underset{\text{Probability of winning}}{\left(\frac{1}{38}\right)} + \overset{\text{Payoff for a loss}}{(-1)} \underset{\text{Probability of losing}}{\left(\frac{37}{38}\right)}$$

$$\approx -\$0.05$$

The expected loss is about 5¢ per play.   ◆

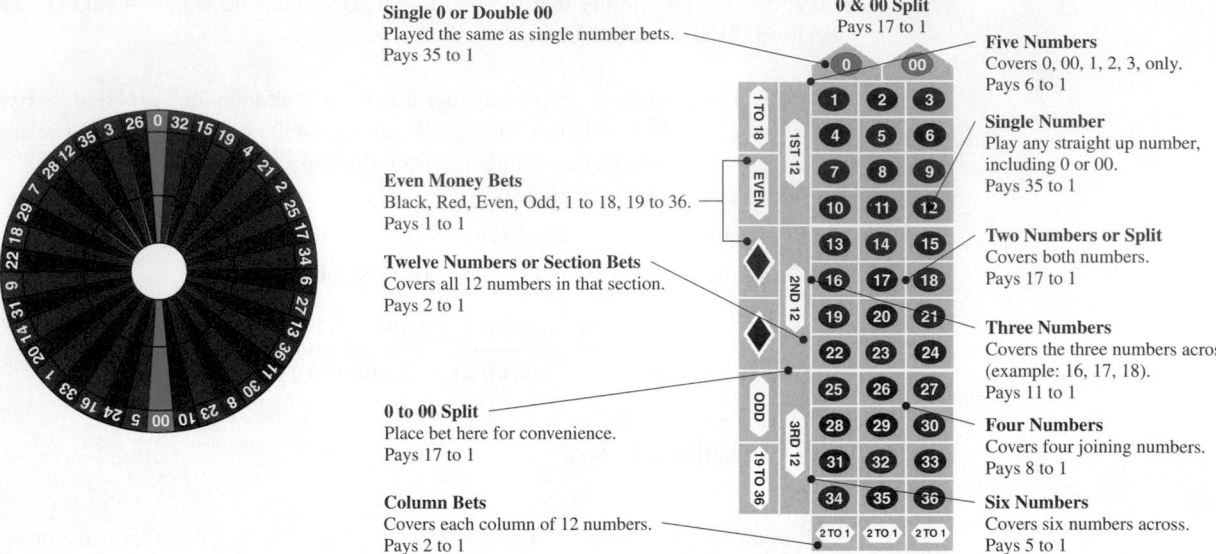

Figure 11.5 **U.S. roulette wheel and board**

Problem Set 11.2

LEVEL 1

1. **IN YOUR OWN WORDS** *True or false?* In roulette, if you bet on black, the probability of winning is $\frac{1}{2}$ because there are the same number of black and red spots. Explain.

2. **IN YOUR OWN WORDS** *True or false?* An expected value of $5 means that you should expect to win $5 each time you play the game. Explain.

3. **IN YOUR OWN WORDS** *True or false?* If the expected value of a game is positive, then it is a game you should play. Explain.

4. **IN YOUR OWN WORDS** If you were asked to choose between a sure $10,000, or an 80% chance of winning $15,000 and a 20% chance of winning nothing, which would you take?

   Game A:

   $E = \$10,000(1) = \$10,000$

   This is a sure thing.

   Game B:

   $E = \$15,000(0.8) + \$0(0.2) = \$12,000$

   True or false? The better choice, according to expected value, is to take the 80% chance of winning $15,000. Explain.

5. **IN YOUR OWN WORDS** Suppose that you buy a lottery ticket for $1. The payoff is $50,000, with a probability of winning 1/1,000,000. Therefore, the expected value is

   $$E = \$50,000\left(\frac{1}{1,000,000}\right) = \$0.05$$

   Is this statement true or false? Explain.

Use estimation to select the best response in Problems 6–11. Do not calculate.

6. The expectation from playing a game in which you win $950 by correctly calling heads or tails when you flip a coin is about
   A. $500   B. $50   C. $950

7. The expectation from playing a game in which you win $950 by correctly calling heads or tails on each of five flips of a coin is about
   A. $500   B. $50   C. $950

8. If the expected value of playing a $1 game of blackjack is $0.04, then after playing the game 100 times you should have netted about
   A. $104   B. −$4   C. $4

9. If your expected value when playing a $1 game of roulette is −$0.05, then after playing the game 100 times you should have netted about
   A. −$105   B. −$5   C. $5

10. The probability of correctly guessing a telephone number is about
   A. 1 out of 100
   B. 1 out of 1,000
   C. 1 out of 1,000,000

11. Winning the grand prize in a state lottery is about as probable as
   A. having a car accident
   B. having an item fall out of the sky into your yard
   C. being a contestant on *Jeopardy*

12. Suppose that you roll two dice. You will be paid $5 if you roll a double. You will not receive anything for any other outcome. How much should you be willing to pay for the privilege of rolling the dice?

13. A magazine subscription service is having a contest in which the prize is $80,000. If the company receives 1 million entries, what is the expectation of the contest?

14. A box contains one each of $1, $5, $10, $20, and $100 bills. You reach in and withdraw one bill. What is the expected value?

15. A box contains one each of $1, $5, $10, $20, and $100 bills. It costs $20 to reach in and withdraw one bill. What is the expected value?

16. Suppose that you have 5 quarters, 5 dimes, 10 nickels, and 5 pennies in your pocket. You reach in and choose a coin at random so that you can tip your barber. What is the barber's expectation? What tip is the barber most likely to receive?

17. A game involves tossing two coins and receiving 50¢ if they are both heads. What is a fair price to pay for the privilege of playing?

18. Krinkles potato chips is having a "Lucky Seven Sweepstakes." The one grand prize is $70,000; 7 second prizes each pay $7,000; 77 third prizes each pay $700; and 777 fourth prizes each pay $70. What is the expectation of this contest, if there are 10 million entries?

19. A punch-out card contains 100 spaces. One space pays $100, five spaces pay $10, and the others pay nothing. How much should you pay to punch out one space?

LEVEL 2

20. In old gangster movies on TV, you often hear of "number runners" or the "numbers racket." This numbers game, which is still played today, involves betting $1 on the last three digits of the number of stocks sold on a particular day in the future as reported in *The Wall Street Journal*. If the payoff is $500, what is the expectation for this numbers game?

21. In a TV game show, four prizes are hidden on a game board which contains 20 spaces. One prize is worth $10,000, two prizes are worth $5,000, and the other prize

is worth $1,000. The remaining spaces contain no prizes. The game show host offers a sure prize of $1,000 not to play this game. Should the contestant choose the sure prize or play the game?

22. Suppose it costs $2,400 to advertise and list a $350,000 house for sale. The listing agent will earn 5% of the selling price if the listing agent sells the property, but only 2.5% if the house is sold by another agent. If the house is unsold after 4 months, the listing (and the cost of advertising) will be lost. Suppose that the probabilities for seling the house are as follows:

Event	Probability
Sells the house alone	0.40
Sells through another agent	0.30
Not sell in 4 months	0.30

What is the expected profit from listing this house?

23. A realtor who takes the listing on a house to be sold knows that she will spend $800 trying to sell the house. If she sells it herself, she will earn 6% of the selling price. If another realtor sells a house from her list, the first realtor will earn only 3% of the price. If the house remains unsold after 6 months, she will lose the listing. Suppose that probabilities are as follows:

Event	Probability
Sell by herself	0.50
Sell by another realtor	0.30
Not sell in 6 months	0.20

What is the expected profit from listing a $185,000 house?

24. An oil-drilling company knows that it costs $25,000 to sink a test well. If oil is hit, the income for the drilling company will be $425,000. If only natural gas is hit, the income will be $125,000. If nothing is hit, there will be no income. If the probability of hitting oil is 1/40 and if the probability of hitting gas is 1/20, what is the expectation for the drilling company? Should the company sink the test well?

25. In Problem 24, suppose that the income for hitting oil is changed to $825,000 and the income for gas to $225,000. Now what is the expectation for the drilling company? Should the company sink the test well?

26. Consider the following game in which a player rolls a single die. If a prime (2, 3, or 5) is rolled, the player wins $2. If a square (1 or 4) is rolled, the player wins $1. However, if the player rolls a perfect number (6), it costs the player $11. Is this a good deal for the player or not?

27. A game involves drawing a single card from an ordinary deck. If an ace is drawn, you receive 50¢; if a face card is drawn, you receive 25¢; if the two of spades is drawn, you receive $1. If the cost of playing is 10¢, should you play?

28. A company held a contest, and the following information was included in the fine print:

Prize	Number of Prizes	Probability of Winning Indicated Prize
$10,000	13	0.000005
$1,000	52	0.00002
$100	520	0.0002
$10	28,900	0.010886
TOTAL	29,485	0.011111

Read this information carefully, and calculate the expectation (to the nearest cent) for this contest.

29. A company held a bingo contest for which the following chances of winning were given:

Playing One Card, Your Chances of Winning Are at Least:			
	1 Time	7 Times	13 Times
$25 prize	1 in 21,252	1 in 3,036	1 in 1,635
$3 prize	1 in 2,125	1 in 304	1 in 163
$1 prize	1 in 886	1 in 127	1 in 68
Any prize	1 in 609	1 in 87	1 in 47

What is the expectation (to the nearest cent) from playing one card 13 times?

30. Heights (in inches) obtained by a group of people in a random survey is reported in the following table:

Heights	Probability
55	0.001
60	0.022
65	0.136
70	0.341
75	0.341
80	0.136
85	0.022
90	0.001

What is the expected height (in inches)?

31. In a certain school, the probabilities of the number of students who are reported tardy is shown on the following table:

Number tardy:	0	1	2	3	4
Probability	0.15	0.25	0.31	0.21	0.08

What is the expected number of tardies (rounded to two decimal places)?

32. Calculate the expectation (to the nearest cent) for the *Reader's Digest* sweepstakes described. Assume there are 197,000,000 entries.

$10,500,000.00
SWEEPSTAKES ENTRY DOCUMENT

Official Disclosure of Dates
To be eligible to win the *FIVE MILLION DOLLAR* Grand Prize, you must return the attached Sweepstakes Entry Document by August 19, 1991. Failure to respond by that date will result in the forfeiture of Grand Prize eligibility. To be eligible to win any of 58,567 other prizes (but not the Grand Prize), return your Sweepstakes Entry Document by March 2, 1992.

Grand Prize Distribution Information
If you are chosen Grand Prize winner, you will receive *FIVE MILLION DOLLARS* in your choice of payment options: Either 30 equal yearly payments of $167,000.00 each *OR* 360 equal monthly payments of $14,000.00 each. You must specify your choice of payment option now by detaching the

appropriate seal at left and affixing it onto the box provided on your Sweepstakes Entry Document.

OFFICIAL PRIZE LIST
1 First Prize....................$100,000.00
2 Second Prizes.................$50,000.00
3 Third Prizes...................$20,000.00
4 Fourth Prizes...................$5,000.00
10 Fifth Prizes........................$500.00
400 Sixth Prizes.....................$100.00
58,147 Seventh Prizes......... ...Winner's choice of a "Special Edition" Men's or Women's Wristwatch, an $89.00 approximate retail value.

Sponsor *J.P. Burr*

SWEEPSTAKES DIRECTOR

Courtesy of Reader's Digest, Inc.

33. A merchant is considering a purchase of autographed sports memorabilia for resale to collectors. The probabilities of various possible incomes from reselling these collectables are estimated in the following table:

Income	*Probability*
$1,000	0.12
$800	0.38
$500	0.45
$200	0.05

The merchant believes that, in order to make this purchase worthwhile, the expected profit must be at least $200. Should the collectables be purchased if the cost is $500?

What is the expectation for the $1 bets in Problems 34–43 on a U.S. roulette wheel? See Figure 11.5 on page 568.

34. Black

35. Odd

36. Single-number bet

37. Double-number bet

38. Three-number bet

39. Four-number bet

40. Five-number bet

41. Six-number bet

42. Twelve-number bet

43. Column bet

Consider the spinners in Problems 44–47. Determine which represent fair games. Assume that the cost to spin the wheel once is $5.00 and that you will receive the amount shown on the spinner after it stops.

44.   **45.**

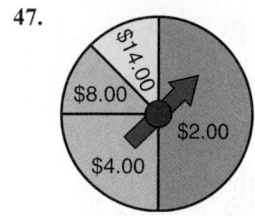

46.   **47.**

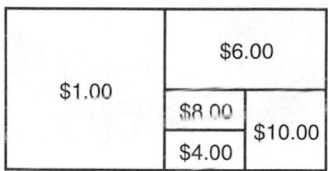

48. Assume that a dart is randomly thrown at the dart board shown here and strikes the board every time. The payoffs are listed on the board. How much should you be willing to pay for the opportunity to play this game?

	$6.00
$1.00	
	$8.00
	$10.00
	$4.00

49. Assume that a dart is randomly thrown at the dart board shown here and strikes the board every time. The payoffs are listed on the board. How much should you be willing to pay for the opportunity to play this game?

$16	
	−$5
$8 $2 $1	
$4	

50. Consider a state lottery that has a weekly television show. On this show, a contestant receives the opportunity to win $1 million. The contestant picks from four hidden windows. Behind each is one of the following: $150,000, $200,000, $1 million, or a "stopper." Before beginning, the contestant is offered $100,000 to stop. Mathematically speaking, should the contestant take the $100,000?

51. Consider a state lottery that has a weekly television show. On this show, a contestant receives the opportunity to win $1 million. The contestant picks from four hidden windows. Behind each is one of the following: $150,000, $200,000, $1 million, or a "stopper." If the contestant picks the window containing $150,000 or $200,000, the contestant is asked whether he or she wishes to quit or continue. Suppose a certain contestant has picked both the $150,000 and $200,000 windows. Mathematically speaking, should the contestant pick "one more time?"

52. **IN YOUR OWN WORDS** Read Problem 51. Discuss what you would do. Consider this situation: Suppose you owned a $350,000 home free and clear; would you gamble it on a 50/50 chance to win $1,000,000? According to the mathematical expectation, what should you do? Discuss.

53. **St. Petersburg Paradox** Suppose you toss a coin and will win $1 if it comes up heads. If it comes up tails, you toss again. This time you will receive $2 if it comes up heads. If it comes up tails, toss again. This time you will receive $4 if it is heads and nothing if it comes up tails. What is the mathematical expectation for this game?

54. **St. Petersburg Paradox** Suppose you toss a coin and will win $1 if it comes up heads. If it comes up tails, you toss again. This time you will receive $2 if it comes up heads. If it comes up tails, toss again. This time you will receive $4 if it is heads. Continue in this fashion for a total of 10 flips of the coin, after which you receive nothing if it comes up tails. What is the mathematical expectation for this game?

55. **St. Petersburg Paradox** Suppose you toss a coin and will win $1 if it comes up heads. If it comes up tails, you toss again. This time you will receive $2 if it comes up heads. If it comes up tails, toss again. This time you will receive $4 if it is heads. Continue in this fashion for a total of 1,000 flips of the coin, after which you receive nothing if it comes up tails. What is the mathematical expectation for this game?

56. **IN YOUR OWN WORDS** **St. Petersburg Paradox** Suppose you toss a coin and will win $1 if it comes up heads. If it comes up tails, you toss again. This time you will receive $2 if it comes up heads. If it comes up tails, toss again. This time you will receive $4 if it is heads. You continue in this fashion until you finally toss a head. Would you pay $100 for the privilege of playing this game? What is the mathematical expectation for this game?

57. **IN YOUR OWN WORDS** Suppose you are in class and your instructor makes you the following legitimate offer. Take out a piece of paper and, without communicating with your classmates, write one of the following messages under your name:

☐ I share the wealth and will receive $1,000 if *everyone* in the class checks this box.

☐ I will not share the wealth and want a certain $100.

If *everyone* checks the first box, then all will receive $1,000. If *one* person checks the second box, then only those who check the second box will receive $100. Which box would you check, and why?

58. IN YOUR OWN WORDS Repeat Problem 57 except change the stakes to $110 and $100, respectively.

59. IN YOUR OWN WORDS Repeat Problem 57 except change the stakes to $10,000 and $10, respectively.

60. IN YOUR OWN WORDS Repeat Problem 57, except change the stakes to be an A grade in the class if you *all* check the first box, but if anyone checks the second box, the following will occur: Those who check it will have a 50% chance of an A and a 50% chance of an F, but those who do not check the second box will be given an F in this course.

11.3 PROBABILITY MODELS

Engraving by Darcis: *Le Trente-et-un*

© The Bettmann Archive/CORBIS

HISTORICAL NOTE

The engraving depicts gambling in 18th-century France. The mathematical theory of probability arose in France in the 17th century when a gambler, Chevalier de Méré, became interested in adjusting the stakes so that he could win more often than he lost. In 1654 he wrote to Blaise Pascal, who in turn sent his questions to Pierre de Fermat. Together they developed the first theory of probability.

Complementary Probabilities

In Section 11.1 we looked at the probability of an event E that consists of n mutually exclusive and equally likely outcomes where s of these outcomes are considered favorable. Now we wish to expand our discussion. Let

$s =$ NUMBER OF GOOD OUTCOMES (successes)

$f =$ NUMBER OF BAD OUTCOMES (failures)

$n =$ TOTAL NUMBER OF POSSIBLE OUTCOMES ($s + f = n$)

Then the probability that event E occurs is

$$P(E) = \frac{s}{n}$$

The probability that event E does not occur is

$$P(\overline{E}) = \frac{f}{n}$$

An important property is found by adding these probabilities:

$$P(E) + P(\overline{E}) = \frac{s}{n} + \frac{f}{n} = \frac{s+f}{n} = \frac{n}{n} = 1$$

Property of Complements

$$P(E) = 1 - P(\overline{E}) \quad \text{or} \quad P(\overline{E}) = 1 - P(E)$$

The probabilities $P(E)$ and $P(\overline{E})$ are called **complementary probabilities.** In other words, the property of complements says that probabilities whose sum is 1 are complementary.

EXAMPLE 1

Use Table 11.1 (page 561) to find the probability of not obtaining one pair with a poker hand.

Solution

From Table 11.1 we see that $P(\text{pair}) = \dfrac{1,098,240}{2,598,960} \approx 0.42$, so that

$$P(\text{no pair}) = P(\overline{\text{pair}}) = 1 - P(\text{pair}) \approx 1 - 0.42 = 0.58 \qquad \blacklozenge$$

EXAMPLE 2

What is the probability of obtaining at least one head in three flips of a coin?

Solution

Let $F = \{\text{obtain at least one head in three flips of a coin}\}$.

Method I: Work directly; use a tree diagram to find the possibilities.

	First	*Second*	*Third*		*First*	*Second*	*Third*		*Success*
					H	H	H		yes
					H	H	T		yes
					H	T	H		yes
Start					H	T	T		yes
					T	H	H		yes
					T	H	T		yes
					T	T	H		yes
					T	T	T		no

$$P(F) = \frac{7}{8}$$

Method II: For one coin, there are 2 outcomes (heads and tails); for three coins we have

$$2 \times 2 \times 2 = 8 \text{ possibilities} \qquad \text{\small Fundamental counting principle}$$

We answer the question by finding the complement; $\overline{F}$ is the event of receiving no heads (that is, of obtaining all tails). *Without* drawing the tree diagram, we note that there is only one way of obtaining all tails (TTT). Thus

$$P(F) = 1 - P(\overline{F}) = 1 - \frac{1}{8} = \frac{7}{8}$$ ◆

Odds

Related to probability is the notion of odds. Instead of forming ratios

$$P(E) = \frac{s}{n} \quad \text{and} \quad P(\overline{E}) = \frac{f}{n}$$

we form the following ratios:

Odds in favor of an event E: $\dfrac{s}{f}$ (ratio of success to failure)

Odds against an event E: $\dfrac{f}{s}$ (ratio of failure to success)

Recall:

s = NUMBER OF SUCCESSES

f = NUMBER OF FAILURES

n = NUMBER OF POSSIBILITIES

EXAMPLE 3

If a jar has 2 quarters, 200 dimes, and 800 pennies, and a coin is to be chosen at random, what are the odds against picking a quarter?

Solution We are interested in picking a quarter, so let $s = 2$; a failure is not obtaining a quarter so $f = 1{,}000$. (Find f by adding 200 and 800.)

Odds in favor of obtaining a quarter: $\dfrac{s}{f} = \dfrac{2}{1{,}000} = \dfrac{1}{500}$

Odds against obtaining a quarter: $\dfrac{f}{s} = \dfrac{1{,}000}{2} = \dfrac{500}{1}$

Do not write $\frac{500}{1}$ as 500.

The odds against picking a quarter when a coin is selected at random are 500 to 1. ◆

EXAMPLE 4   PÓLYA'S METHOD

The odds against winning a lottery are 50,000,000 to 1. Make up an example to help visualize these odds.

This example ties together some of the ideas of this chapter with ideas from Chapter 1. It is here to get you to think twice before purchasing a lottery ticket.

Solution We use Pólya's problem-solving guidelines for this example.

Understand the Problem. It is difficult for the human mind to comprehend numbers like 50 million to 1, so the point of this problem is to help us "visualize" its magnitude.

Devise a Plan. We will fill a house with 50 million ping pong balls, and then paint 1 of them red.

Carry Out the Plan. Imagine one red ping pong ball and 50,000,000 white ping pong balls. Winning a lottery is equivalent to reaching into a container containing these 50,000,001 ping pong balls and obtaining the red one. To visualize this, consider the size container you would need. Assume that a ping pong ball takes up a volume of 1 in.^{3}. A home of 1,200 ft^2 with an 8-ft ceiling has a volume of

$$1{,}200 \text{ ft}^2 \times 8 \text{ ft} = 9{,}600 \text{ ft}^3$$

$$= 9{,}600 \times (12 \text{ in.} \times 12 \text{ in.} \times 12 \text{ in.}) = 16{,}588{,}800 \text{ in.}^3$$

Look Back. Even though the number 16,588,800 is still not directly comprehensible, you should be able to imagine a house filled with ping pong balls. However, we can now say we would need to *fill* about 3 homes with ping pong balls and imagine that one ball is painted red. We do not even know which house contains the red ball, but we can imagine picking one of the three houses at random, and then reaching into that one house and choosing one ball. If the red one is chosen we win the lottery!   ◆

Sometimes you know the probability and want to find the odds, or you may know the odds and want to find the probability. These relationships are easy if you remember:

$$s + f = n$$

Finding the Odds Given the Probability

Suppose that you *know* $P(E)$.

Odds in favor of an event E: $\dfrac{P(E)}{P(\overline{E})}$

Odds against an event E: $\dfrac{P(\overline{E})}{P(E)}$

You can show these formulas to be true:

$$\frac{P(E)}{P(\overline{E})} = \frac{\dfrac{s}{n}}{\dfrac{f}{n}} = \frac{s}{n} \cdot \frac{n}{f} = \frac{s}{f} = \text{odds in favor}$$

The verification of the second formula is left for the problem set.

Finding the Probability Given Odds in Favor or Odds Against an Event

Suppose that you *know* the odds in favor of an event E:

s to f

or the odds against an event E:

f to s.

Then,

$$P(E) = \frac{s}{s+f} \quad \text{and} \quad P(\overline{E}) = \frac{f}{s+f}$$

EXAMPLE 5

If the probability of an event is 0.45, what are the odds in favor of the event?

Solution $P(E) = 0.45 = \dfrac{45}{100} = \dfrac{9}{20}$ This is given.

$$P(\overline{E}) = 1 - \frac{9}{20} = \frac{11}{20}$$

Then the odds in favor of E are $\dfrac{P(E)}{P(\overline{E})} = \dfrac{\frac{9}{20}}{\frac{11}{20}} = \dfrac{9}{20} \cdot \dfrac{20}{11} = \dfrac{9}{11}$.

The odds in favor are 9 to 11. ◆

EXAMPLE 6

If the odds against you are 20 to 1, what is the probability of the event?

Solution Odds against an event are f to s, so $f = 20$ and $s = 1$; thus,

$$P(E) = \frac{1}{20 + 1} = \frac{1}{21} \approx 0.048$$

◆

EXAMPLE 7

If the odds in favor of some event are 2 to 5, what is the probability of the event?

Solution Odds in favor are s to f, so $s = 2$ and $f = 5$; thus,

$$P(E) = \frac{2}{2 + 5} = \frac{2}{7} \approx 0.286$$

◆

EXAMPLE 8

What is the probability that a two-child family will have children of opposite sex?

Second Child

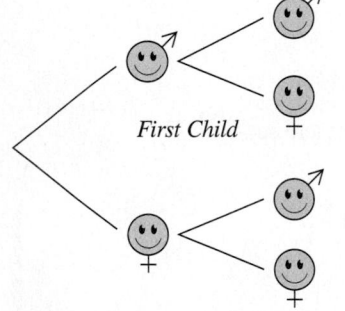

First Child

Solution The fundamental counting principle tells us that, since there are 2 ways of having a child (B or G) for 2 children, there is a total of

$$2 \cdot 2 = 4 \text{ ways}$$

We verify this by looking at a tree diagram (shown in the margin). There are 4 equally likely outcomes: BB, BG, GB, and GG. Thus, the probability of having a boy and a girl in a family of two children is

$$\frac{\text{NUMBER OF SUCCESSFUL OUTCOMES}}{\text{TOTAL NUMBER OF ALL POSSIBLE OUTCOMES}} = \frac{2}{4} = \frac{1}{2}$$

◆

The fundamental counting principle can be used repeatedly for more than two possibilities.

EXAMPLE 9

What is the probability that a four-child family will have two boys and two girls?

Solution The fundamental counting principle tells us the number of possibilities is

$$2 \times 2 \times 2 \times 2 = 16$$

Thus, $n = 16$. To find s we list the events that bring success:

$$\{\text{BBGG, BGBG, BGGB, GBGB, GBBG, GGBB}\}$$

Since there are 6 elements in this set, we see that $s = 6$. Thus, the desired probability is

$$\frac{6}{16} = \frac{3}{8}$$

◆

Conditional Probability

Frequently, we wish to compute the probability of an event but we have additional information that will alter the sample space. For example, suppose that a family has two children. What is the probability that the family has two boys?

$$P(2 \text{ boys}) = \frac{1}{4}$$ Sample space: BB, BG, GB, GG; 1 success out of 4 possibilities

Now, let's complicate the problem a little. Suppose that we know that the older child is a boy. We have altered the sample space as follows:

Original sample space: BB, BG, GB, GG; but we need to cross out the last two possibilities because we *know* that the older child is a boy:

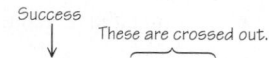

Altered sample space: BB, BG, ~~GB, GG~~ ; therefore,

Altered sample space has two elements.

$$P(2 \text{ boys given the older is a boy}) = \frac{1}{2}$$

This is a problem involving a **conditional probability**—namely, a *probability of an event **given** that another event F has occurred.* We denote this by

$$P(E \mid F) \qquad \text{Read this as: "probability of E given F."}$$

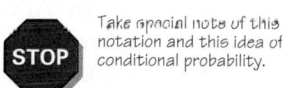

EXAMPLE 10

Suppose that you toss two coins (or a single coin twice). What is the probability that two heads are obtained if you know that at least one head is obtained?

Solution Consider an altered sample space: HH, HT, TH, ~~TT~~. The probability is $\frac{1}{3}$.   ◆

EXAMPLE 11

Suppose that you draw two jellybeans from a jar containing 50 green, 30 white, and 20 red jellybeans. Find the following probabilities. Let $R = \{\text{the second jellybean drawn is red}\}$.

a. $P(R)$

b. $P(R \mid \text{a red jellybean is drawn on the first draw})$

c. $P(R \mid \text{a red jellybean is not drawn on the first draw})$

Solution

a. $P(R) = \dfrac{20}{100} = 0.2$

Remember, it does not matter what happened on the first draw because we do not know what happened on that draw. The second jellybean "does not remember" what is drawn on the first draw.

b. This time we know what happened on the first draw, so $n = 99$ and $s = 19$ (because a red jellybean was drawn on the first draw):

$$P(R \mid \text{a red jellybean is drawn on the first draw}) = \frac{19}{99} \approx 0.192$$

c. We still have $n = 99$, but this time $s = 20$:

$$P(R \mid \text{a red jellybean is not drawn on the first draw}) = \frac{20}{99} \approx 0.202 \qquad \blacklozenge$$

EXAMPLE 12   PÓLYA'S METHOD

In a life science experiment, it is necessary to examine fruit flies and to determine their sex and whether they have mutated after exposure to a certain dose of radiation. Suppose a single fruit fly is selected at random from the radiated fruit flies. Find the following probabilities (correct to the nearest hundredth).

a. It is male.   **b.** It is a normal male.

c. It is normal, given that it is a male.   **d.** It is a male, given that it is normal.

e. It is mutated, given that it is a male.

Solution   We use Pólya's problem-solving guidelines for this example.

Understand the Problem. We obviously need more information to find the desired probabilities. We need to know the number of fruit flies that are part of the experiment. Assume that the total number is 1,000. We also need some data. Upon examination, we find that for 1,000 fruit flies examined, there were 643 females and 357 males. Also, 403 of the females were normal and 240 were mutated; of the males, 190 were normal and 167 were mutated.

Devise a Plan. We will display the data in table form to make it easy to calculate the desired probabilities.

Carry Out the Plan.

	Mutated	Normal	Total
Male	167	190	357
Female	240	403	643
Total	407	593	1,000

a. $P(\text{male}) = \dfrac{357}{1,000} \approx 0.36$ Round answers to the nearest hundredth.

b. $P(\text{normal male}) = \dfrac{190}{1,000} = 0.19$

c. $P(\text{normal} \mid \text{male}) = \dfrac{190}{357} \approx 0.53$ Note the altered sample space.

d. $P(\text{male} \mid \text{normal}) = \dfrac{190}{593} \approx 0.32$ Yet another altered sample space

e. $P(\text{mutated} \mid \text{male}) = \dfrac{167}{357} \approx 0.47$

Look Back. All probabilities seem reasonable, and all are between 0 and 1. $\blacklozenge$

CAUTION   This is a crucial example. Spend some time working through all the parts.

EXAMPLE 13

Two cards are drawn from a deck of cards. Find the probability of the given events.

a. The first card drawn is a heart.

b. The first card drawn is not a heart.

c. The second card drawn is a heart if the first card drawn was a heart.

d. The second card drawn is not a heart if the first card drawn was a heart.

e. The second card drawn is a heart if the first card drawn was not a heart.

f. The second card drawn is a not a heart if the first card drawn was not a heart.

g. The second card drawn is a heart.

h. Use a tree diagram to represent the indicated probabilities.

Solution

Let $H_1 = \{$first card drawn is a heart$\}$ and $H_2 = \{$second card drawn is a heart$\}$.

a. $P(H_1) = \dfrac{13}{52} = \dfrac{1}{4}$

There are 13 hearts in a deck of 52 cards.

b. $P(\overline{H_1}) = \dfrac{39}{52} = \dfrac{3}{4}$

There are 39 nonhearts in a deck of 52 cards.

c. $P(H_2 \mid H_1) = \dfrac{12}{51} = \dfrac{4}{17}$

If a heart is obtained on the first draw, then for the second draw there are 12 hearts in the deck of 51 remaining cards.

d. $P(\overline{H_2} \mid H_1) = \dfrac{39}{51} = \dfrac{13}{17}$

If a heart is obtained on the first draw, then for the second draw there are 39 nonhearts in a deck of 51 remaining cards.

e. $P(H_2 \mid \overline{H_1}) = \dfrac{13}{51}$

If a heart is not obtained on the first draw, then for the second draw there are 13 hearts in a deck of 51 remaining cards.

f. $P(\overline{H_2} \mid \overline{H_1}) = \dfrac{38}{51}$

If a nonheart is obtained on the first draw, then for the second draw there are 38 nonhearts in a deck of 51 cards.

g. $P(H_2) = \dfrac{1}{4}$

If we do not know what happened on the first draw, then this is not a conditional probability, so we consider this probability to be the same as $P(H_1)$.

h. We use a tree diagram to illustrate this situation.

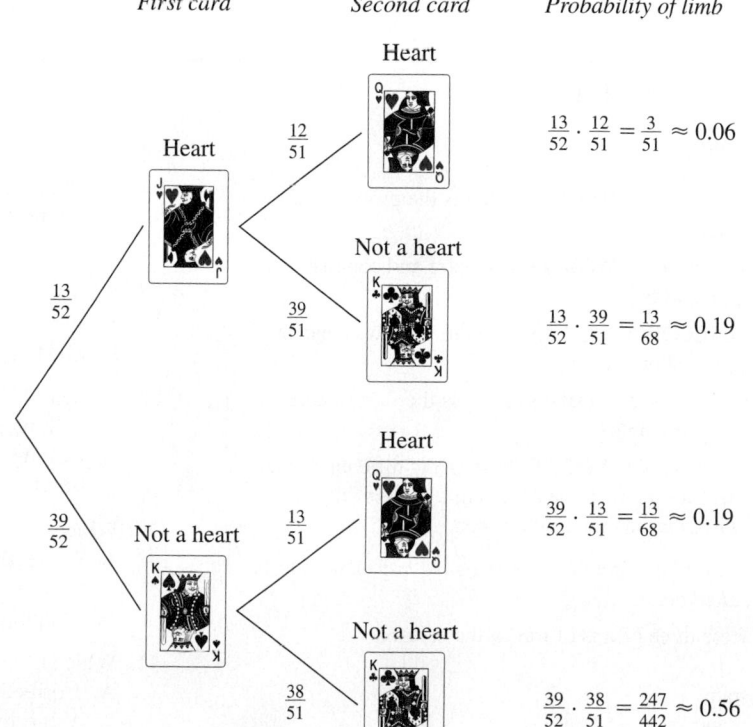

First card   Second card   Probability of limb

Heart $\dfrac{13}{52} \cdot \dfrac{12}{51} = \dfrac{3}{51} \approx 0.06$

Not a heart $\dfrac{13}{52} \cdot \dfrac{39}{51} = \dfrac{13}{68} \approx 0.19$

Heart $\dfrac{39}{52} \cdot \dfrac{13}{51} = \dfrac{13}{68} \approx 0.19$

Not a heart $\dfrac{39}{52} \cdot \dfrac{38}{51} = \dfrac{247}{442} \approx 0.56$

Note that we use the word *limb* to mean a sequence of branches that starts at the beginning. Also note that we have filled in the probabilities on each branch. To find the probabilities, we multiply as we move horizontally across a limb, and we add as we move vertically from limb to limb. Verify that the probabilities for the first card add to 1:

$$P(H_1) + P(\overline{H}_1) = \frac{13}{52} + \frac{39}{52} = \frac{52}{52} = 1$$

Also, for the second card, the probabilities for each branch add to 1. For $P(H_2)$, we can confirm our reasoning in part **g** by considering the first and third limbs of the tree diagram:

$$P(H_2) = \frac{13}{52} \cdot \frac{12}{51} + \frac{39}{52} \cdot \frac{13}{51} = \frac{663}{2,652} = \frac{1}{4}$$

Notice that conditional probabilities are found in the tree diagram by beginning at their condition.

Verify that all of the information in parts **a–g** can be found directly from the tree diagram. For this reason, we often use tree diagrams to assist us in finding probabilities and conditional probabilities. ◆

Procedure for Using Tree Diagrams

CAUTION You will find using tree diagrams a very helpful procedure for finding probabilities.

Multiply when moving horizontally across a limb.

Add when moving vertically from limb to limb.

Conditional probabilities start at their condition; unconditional probabilities start at the beginning of the tree.

Problem Set 11.3

LEVEL 1

1. **IN YOUR OWN WORDS** What is the property of complements?

2. **IN YOUR OWN WORDS** Compare and contrast odds and probability.

3. **IN YOUR OWN WORDS** Explain what we mean by "conditional probability."

4. **IN YOUR OWN WORDS** What is the fundamental counting principle?

5. **IN YOUR OWN WORDS** How many different sequences are there for the first five moves in a game of tick-tack-toe? Give reasons for your answer.

Use estimation to select the best response in Problems 6–11. Do not calculate.

6. Of these three means of travel, the safest is
   A. car
   B. train
   C. plane

7. Which of the following is most probable?
   A. Winning the grand prize in a state lottery
   B. Being struck by lightning
   C. Appearing on the *Tonight Show*

8. Which of the following is more probable?
   A. Flipping a coin 3 times and obtaining at least 2 heads
   B. Flipping a coin 4 times and obtaining at least 2 heads

9. Which of the following is more probable?
   A. Rolling a die 3 times and obtaining a six 2 times
   B. Rolling a die 3 times and obtaining a six at least 2 times

10. Which of the following is more probable?
    A. Correctly guessing all the answers on a 20-question true-false examination
    B. Flipping a coin 20 times and obtaining all heads

11. Which of the following is more probable?
    A. Correctly guessing all the answers on a 10-question 5-part multiple-choice test

B. Your living room is filled with white ping pong balls. There is also one red ping pong ball in the room. You reach in and select a ping pong ball at random and select the red ping pong ball.

Find the requested probabilities in Problems 12–15.

12. $P(\overline{A})$ if $P(A) = 0.6$

13. $P(\overline{B})$ if $P(B) = \frac{4}{5}$

14. $P(C)$ if $P(\overline{C}) = \frac{9}{13}$

15. $P(D)$ if $P(\overline{D}) = 0.005$

16. A card is selected from an ordinary deck. What is the probability that it is not a face card?

17. Choose a natural number between 1 and 100, inclusive. What is the probability that the number chosen is not a multiple of 5?

18. Three fair coins are tossed. What is the probability that at least one is a head?

19. Find the probability of obtaining at least one head in four flips of a coin.

20. What are the odds in favor of drawing an ace from an ordinary deck of cards?

21. What are a four-child family's odds against having four boys?

22. The probability of drawing a heart from a deck of cards is $\frac{1}{4}$; what are the odds in favor of drawing a heart?

23. Suppose the probability of an event is 0.82. What are the odds in favor of this event?

24. The odds against an event are ten to one. What is the probability of this event?

25. Racetracks quote the approximate odds for each race on a large display board called a *tote board*. Here's what it might say for a particular race:

Horse Number	Odds
1	2 to 1
2	15 to 1
3	3 to 2
4	7 to 5
5	1 to 1

What would be the probability of winning for each of these horses? *Note:* The odds stated are for the horse's *losing*. Thus,

$$P(\text{horse 1 losing}) = \frac{2}{2+1} = \frac{2}{3}$$

so

$$P(\text{horse 1 winning}) = 1 - \frac{2}{3} = \frac{1}{3}$$

26. Suppose the odds in favor are 9 to 1 that a man will be bald by the time he is 60. State this as a probability.

27. Suppose the odds are 33 to 1 that someone will lie to you at least once in the next seven days. State this as a probability.

28. Suppose that a family wants to have four children.
a. What is the sample space?
b. What is the probability of 4 girls? 4 boys?
c. What is the probability of 1 girl and 3 boys? 1 boy and 3 girls?
d. What is the probability of 2 boys and 2 girls?
e. What is the sum of your answers in parts **b** through **d?**

29. What is the probability of obtaining exactly three heads in four flips of a coin, given that at least two are heads?

30. The Emory Harrison family of Tennessee had 13 boys.

© AP/Wide World Photos

a. What is the probability of a 13-child family having 13 boys?
b. What is the probability that the next child of the Harrison family will be a boy (giving them 14 boys in a row)?

A single card is drawn from a standard deck of cards. Find the probabilities if the given information is known about the chosen card in Problems 31–36. A face card is a jack, queen, or king.

31. $P(\text{face card} \mid \text{jack})$

32. $P(\text{jack} \mid \text{face card})$

33. $P(\text{heart} \mid \text{not a spade})$

34. $P(\text{two} \mid \text{not a face card})$

35. $P(\text{black} \mid \text{jack})$

36. $P(\text{jack} \mid \text{black})$

Two cards are drawn from a standard deck of cards, and one of the two cards is noted and removed. Find the probabilities of the second card, given the information about the removed card provided in Problems 37–42.

37. $P(\text{ace} \mid \text{two})$

38. $P(\text{king} \mid \text{king})$

39. $P(\text{heart} \mid \text{heart})$

40. $P(\text{heart} \mid \text{spade})$

41. $P(\text{black} \mid \text{red})$

42. $P(\text{black} \mid \text{black})$

43. What is the probability of getting a license plate that has a repeated letter or digit if you live in a state that has two letters followed by four numerals?

44. What is the probability of getting a license plate that has a repeated letter or digit if you live in a state that has one numeral followed by three letters followed by three numerals?

45. What is the probability of getting a license plate that has a repeated letter or digit if you live in a state where the scheme is three numerals followed by three letters?

46. Consider this letter to Dear Abby.*

> **DEAR ABBY:** My husband and I just had our eighth child. Another girl, and I am really one disappointed woman. I suppose I should thank God she was healthy, but, Abby, this one was supposed to have been a boy. Even the doctor told me the law of averages was in our favor 100 to one. What is the probability of our next child being a boy?

What is the family's probability of having eight girls in a row? What are the odds against?

47. Consider the following table showing the results of a survey of TV network executives.

Opinion of current programming

Executive	satisfied, S	Not satisfied, $\overline{S}$	Total
NBC, N	18	7	25
CBS, C	21	9	30
ABC, A	15	10	25
Total	54	26	80

Suppose one network executive is selected at random. Find the indicated probabilities.
a. What is the probability that it is an NBC executive?
b. What is the probability that the selected person is satisfied?
c. What is the probability the selected person is from CBS if we know the person is satisfied with current programming?
d. What is the probability the selected person is satisfied, if we know the person is from CBS?

48. Suppose a single die is rolled. Find the probabilities.
a. 6, given that an odd number was rolled
b. 5, given that an odd number was rolled

*Adapted from DEAR ABBY column by Abigail Van Buren. © 1974. Dist. by Universal Press Syndicate. Reprinted by permission. All rights reserved.

49. Suppose a single die is rolled. Find the probabilities.
a. odd, given that the rolled number was a 6
b. odd, given that the rolled number was a 5

LEVEL 3

50. Suppose a pair of dice are rolled. Consider the sum of the numbers on the top of the dice and find the probabilities.
a. 7, given that the sum is odd
b. odd, given that a 7 was rolled
c. 7, given that at least one die came up 2

51. Suppose a pair of dice are rolled. Consider the sum of the numbers on the top of the dice and find the probabilities.
a. 5, given that exactly one die came up 2
b. 3, given that exactly one die came up 2
c. 2, given that exactly one die came up 2

52. Suppose a pair of dice are rolled. Consider the sum of the numbers on the top of the dice and find the probabilities.
a. 8, given that a double was rolled
b. a double, given that an 8 was rolled

53. Show that the odds against an event E can be found by computing $P(\overline{E})/P(E)$.

54. Two cards are drawn from a deck of cards. Find the requested probabilities.
a. The first card drawn is a club.
b. The first card drawn is not a club.
c. The second card drawn is a club if the first card drawn was a club.
d. The second card drawn is not a club if the first card drawn was a club.
e. The second card drawn is a club if the first card drawn was not a club.
f. The second card drawn is not a club if the first card drawn was not a club.
g. The second card drawn is a club.
h. Use a tree diagram to represent the indicated probabilities.

55. A sorority has 35 members, 25 of whom are full members and 10 are pledges. Two persons are selected at random from the membership list of the sorority. Find the requested probabilities.
a. The first person selected is a pledge.
b. The first person selected is not a pledge.
c. The second person selected is a pledge if the first person selected was also a pledge.
d. The second person selected was a full member if the first person selected was a pledge.
e. The second person selected is a pledge if the first person selected was a full member.
f. The second person selected is a full member if the first person selected was also a full member.
g. The second person selected was a pledge.
h. Use a tree diagram to represent the indicated probabilities.

56. The odds against winning a certain lottery are a million to one. Make up an example to help visualize these odds.

57. The odds against winning a certain lottery are ten million to one. Make up an example to help visualize these odds.

58. The odds against winning a certain "Power Ball" lotto are 150 million to 1. Make up an example to help visualize these odds.

PROBLEM SOLVING

59. One roulette system is to bet $1 on black. If black comes up on the first spin of the wheel, you win $1 and the game is over. If black does not come up, double your bet ($2). If you win, the game is over and your net winnings for two spins is still $1 (show this). If black does not come up, double your bet again ($4). If you win, the game is over and your net winnings for three spins is still $1 (show this). Continue in this doubling procedure until you eventually win; in every case your net winnings amount to $1 (show this). What is the fallacy with this betting system?

60. "Last week I won a free Big Mac at McDonald's. I sure was lucky!" exclaimed Charlie. "Do you go there often?" asked Pat. "Only twenty or thirty times a month. And the odds of winning a Big Mac were only 20 to 1." "Don't you mean 1 to 20?" queried Pat. "Don't confuse me with details. I don't even understand odds at the racetrack, and I go all the time." Is Charlie or Pat correct about the odds?

11.4 CALCULATED PROBABILITIES

In this section, we formulate some probability problems using tree diagrams, combinations, permutations, and the fundamental counting principle. We will also consider some additional models that will allow us to break up more complicated probability problems into some simpler types.

Keno Games

EXAMPLE 1

A very popular lotto game in several states, as well as in most casinos, is a game called **Keno.** The game consists of a player trying to guess in advance which numbers will be selected from a pot containing 80 numbers. Twenty numbers are then selected at random. The player may choose from 1 to 15 spots and gets paid according to Table 11.2 on page 584. Suppose a person picks one number and is paid $3.00 if the number picked is among the 20 chosen for the game. The cost for playing this game (which is collected in advance) is $1.00. Is this a fair game?

YOU CAN WIN $50,000
KENO

Solution In this problem we are picking 1 out of 80 and we will win if the one number we pick is among the 20 numbers chosen from the pot. Since the order in which the 20 numbers are picked from the pot does not matter, we see this is a combination.

$$P(\text{picking one number}) = \frac{\text{NUMBER OF WAYS OF CHOOSING 1 NO. FROM 20}}{\text{NUMBER OF WAYS OF CHOOSING 1 NO. FROM 80}}$$

$$= \frac{{}_{20}C_1}{{}_{80}C_1} = \frac{20}{80} = \frac{1}{4}$$

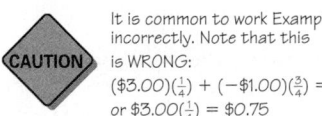

It is common to work Example 1 incorrectly. Note that this is WRONG:
$(\$3.00)(\frac{1}{4}) + (-\$1.00)(\frac{3}{4}) = 0$
or $\$3.00(\frac{1}{4}) = \$0.75$
On the other hand,
$(\$2.00)(\frac{1}{4}) - \$1.00(\frac{3}{4}) = -0.25$ is acceptable.

We now calculate the mathematical expectation:

$$\text{EXPECTATION} = \underbrace{\$3.00(\tfrac{1}{4})}_{} - \$1.00 = -\$0.25$$

Subtract $1.00 because you pay first

No, it is not a fair game, since the expectation is negative. ◆

The game of Keno becomes more complicated when picking more numbers.

Table 11.2 Keno Payoffs

MARK 1 SPOT

Catch	Play $1.00	Play $3.00	Play $5.00
Win	3.00	9.00	15.00

MARK 2 SPOTS

Catch	Play $1.00	Play $3.00	Play $5.00
2 Win	12.00	36.00	60.00

MARK 3 SPOTS

Catch	Play $1.00	Play $3.00	Play $5.00
2 Win	1.00	3.00	5.00
3 Win	40.00	120.00	200.00

MARK 4 SPOTS

Catch	Play $1.00	Play $3.00	Play $5.00
2 Win	1.00	3.00	5.00
3 Win	3.00	9.00	15.00
4 Win	113.00	339.00	565.00

MARK 5 SPOTS

Catch	Play $1.00	Play $3.00	Play $5.00
3 Win	1.00	3.00	5.00
4 Win	10.00	30.00	50.00
5 Win	750.00	2,250.00	3,750.00

MARK 6 SPOTS

Catch	Play $1.00	Play $3.00	Play $5.00
3 Win	1.00	3.00	5.00
4 Win	3.00	9.00	15.00
5 Win	90.00	270.00	450.00
6 Win	1,480.00	4,400.00	7,400.00

MARK 7 SPOTS

Catch	Play $1.00	Play $3.00	Play $5.00
4 Win	1.00	3.00	5.00
5 Win	18.00	54.00	90.00
6 Win	400.00	1,200.00	2,000.00
7 Win	8,000.00	24,000.00	40,000.00

MARK 8 SPOTS

Catch	Play $1.00	Play $3.00	Play $5.00
5 Win	8.00	24.00	40.00
6 Win	100.00	300.00	500.00
7 Win	1,480.00	4,440.00	7,400.00
8 Win	17,000.00	**50,000.00**	50,000.00

MARK 9 SPOTS

Catch	Play $1.00	Play $3.00	Play $5.00
5 Win	3.00	9.00	15.00
6 Win	42.00	126.00	210.00
7 Win	350.00	1,050.00	1,750.00
8 Win	4,200.00	12,600.00	21,000.00
9 Win	18,000.00	**50,000.00**	50,000.00

MARK 10 SPOTS

Catch	Play $1.00	Play $3.00	Play $5.00
5 Win	2.00	6.00	10.00
6 Win	20.00	60.00	100.00
7 Win	130.00	390.00	650.00
8 Win	900.00	2,700.00	4,500.00
9 Win	4,500.00	13,500.00	22,500.00
10 Win	20,000.00	**50,000.00**	50,000.00

MARK 11 SPOTS

Catch	Play $1.00	Play $3.00	Play $5.00
6 Win	9.00	27.00	45.00
7 Win	75.00	225.00	375.00
8 Win	380.00	1,140.00	1,900.00
9 Win	2,000.00	6,000.00	10,000.00
10 Win	12,500.00	37,500.00	**50,000.00**
11 Win	21,000.00	**50,000.00**	50,000.00

MARK 12 SPOTS

Catch	Play $1.00	Play $3.00	Play $5.00
6 Win	5.00	15.00	25.00
7 Win	28.00	84.00	140.00
8 Win	200.00	600.00	1,000.00
9 Win	850.00	2,550.00	4,250.00
10 Win	2,400.00	7,200.00	12,000.00
11 Win	13,000.00	39,000.00	**50,000.00**
12 Win	25,000.00	**50,000.00**	50,000.00

MARK 13 SPOTS ($2.00 minimum)

Catch	Play $2.00	Play $3.00	Play $5.00
6 Win	6.00	9.00	15.00
7 Win	24.00	36.00	60.00
8 Win	150.00	225.00	375.00
9 Win	1,400.00	2,100.00	3,500.00
10 Win	4,000.00	6,000.00	10,000.00
11 Win	18,000.00	27,000.00	45,000.00
12 Win	28,000.00	42,000.00	**50,000.00**
13 Win	**50,000.00**	**50,000.00**	50,000.00

MARK 14 SPOTS ($2.00 minimum)

Catch	Play $2.00	Play $3.00	Play $5.00
6 Win	4.00	6.00	10.00
7 Win	16.00	24.00	40.00
8 Win	64.00	96.00	160.00
9 Win	600.00	900.00	1,500.00
10 Win	1,600.00	2,400.00	4,000.00
11 Win	5,000.00	7,500.00	12,500.00
12 Win	24,000.00	36,000.00	**50,000.00**
13 Win	36,000.00	**50,000.00**	50,000.00
14 Win	**50,000.00**	**50,000.00**	50,000.00

MARK 15 SPOTS ($2.00 minimum)

Catch	Play $2.00	Play $3.00	Play $5.00
6 Win	2.00	3.00	5.00
7 Win	14.00	21.00	35.00
8 Win	42.00	63.00	105.00
9 Win	200.00	300.00	500.00
10 Win	800.00	1,200.00	2,000.00
11 Win	4,000.00	6,000.00	10,000.00
12 Win	16,000.00	24,000.00	40,000.00
13 Win	24,000.00	36,000.00	**50,000.00**
14 Win	**50,000.00**	**50,000.00**	50,000.00
15 Win	**50,000.00**	**50,000.00**	50,000.00

EXAMPLE 2

What is the mathematical expectation for playing a three-spot $1.00 Keno ticket?

There are four possibilities:

Pick 0 numbers, win $0.

Pick 1 numbers, win $0.

Pick 2 numbers, win $1.00.

Pick 3 numbers, win $40.00.

Solution $P(\text{pick } 3) = \dfrac{s}{n}$ Where s is the number of ways of picking 3 numbers from the 20 winning numbers, and n is the number of ways of picking 3 numbers from the 80 possible numbers.

$$= \frac{_{20}C_3}{_{80}C_3} = \frac{20 \cdot 19 \cdot 18}{80 \cdot 79 \cdot 78} \approx 0.0138753651$$

$P(\text{pick } 2) = \dfrac{s}{n}$ Where s is the number of ways of picking 2 numbers from the 20 winning numbers and also picking one number from the 60 nonwinning numbers, and n is the number of ways of picking 3 numbers from the 80 possible numbers.

$$= \frac{_{20}C_2 \cdot {_{60}C_1}}{_{80}C_3} = \frac{20 \cdot 19 \cdot 60 \cdot 3}{80 \cdot 79 \cdot 78} \approx 0.1387536514$$

Similarly,

$$P(\text{pick 1}) = \frac{_{20}C_1 \cdot {}_{60}C_2}{_{80}C_3} \approx 0.4308666018$$

$$P(\text{pick 0}) = \frac{_{60}C_3}{_{80}C_3} \approx 0.4165043817$$

These Keno payoffs are found in Table 11.2.

We see that the expected value of one $1.00 play is:

EXPECTED VALUE = AMT. TO WIN • PROB. OF WINNING − AMT. TO PLAY

$$= \$40 \,(\text{PROB. OF PICKING 3}) + \$1(\text{PROB. OF PICKING 2}) - \$1$$

$$\approx \$40(0.0138753651) + \$1(0.1387536514) - \$1$$

$$\approx -0.3062$$

An alternate method is to subtract the amount you pay from each payoff as you go.

The loss is about $0.31 for each play.   ◆

Independent Events

Consider the following problem dealing with four cards. Suppose four cards are taken from an ordinary deck of cards, and we form two stacks—one with an ace (one) and a deuce (two) and the other with an ace and a jack. If a card is drawn at random from each pile, what is the probability that a blackjack will occur (an ace and a jack)?

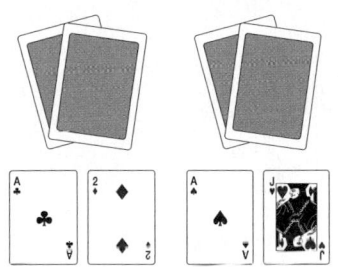

The tree diagram illustrates the possibilities. Notice that the probability of obtaining an ace from the first stack is 1/2 and the probability of obtaining a jack from the second stack is also 1/2.

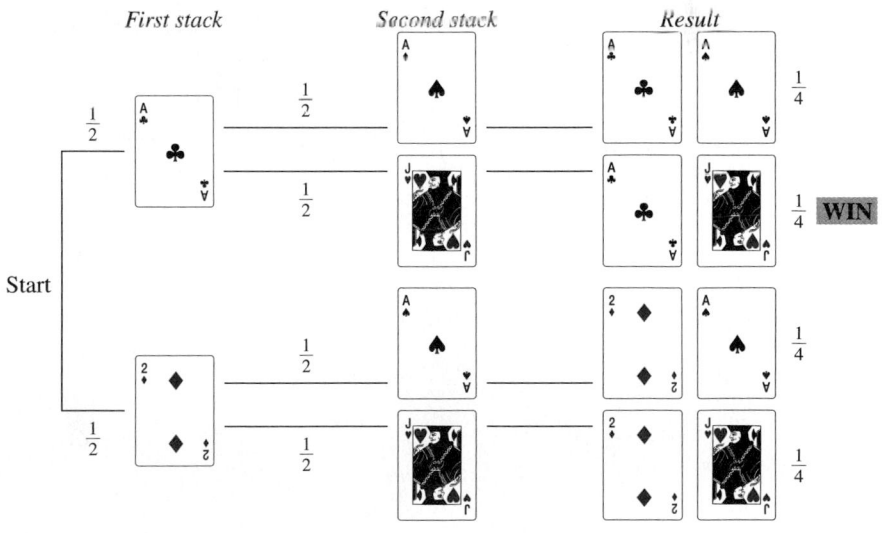

The probability of a blackjack is $\frac{1}{4}$. Notice *for this example* that

$$\text{PROBABILITY OF ACE FROM FIRST STACK} = \frac{1}{2}$$

$$\text{PROBABILITY OF JACK FROM SECOND STACK} = \frac{1}{2}$$

$$\text{PROBABILITY OF BLACKJACK} = \frac{1}{2} \times \frac{1}{2} = \frac{1}{4}$$

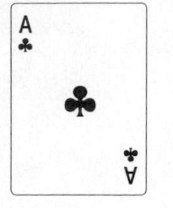

If one event (draw from the first stack) has no effect on the outcome of the second event (draw from the second stack), then we say that the events are **independent.**

To understand how independence is determined, consider the following alternative problem. Suppose that all four cards are put together into one pile. This is an entirely different situation. There are four possibilities for the first draw, and three possibilities for the second draw, as illustrated here.

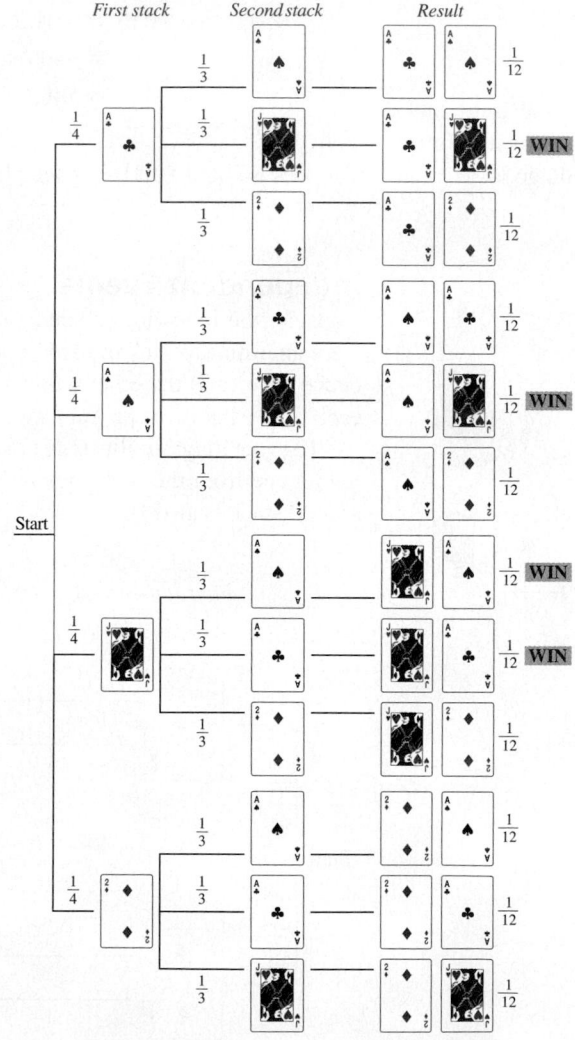

The probability of a blackjack is $\frac{4}{12} = \frac{1}{3}$. We find this by looking at the sample space of equally likely possibilities. We *cannot* find this probability by multiplying the component parts for the first and second draws as we did when considering independent stacks.

For this second situation we see that the first and second draws are **dependent** because the probabilities of selecting a particular second card are influenced by the draw on the first card.

Probability of an Intersection

If two events are independent, then we can find the probability of an intersection by multiplication. This property is called the **multiplication property of probability** or the *probability of an intersection*.

Multiplication Property of Probability

If events E and F are independent events, then we can find the probability of an intersection as follows:

$$P(E \text{ and } F) = P(E \cap F) = P(E \mid F) \cdot P(F)$$
$$= P(E) \cdot P(F)$$

EXAMPLE 3

What is the probability of black occurring on two successive plays on a U.S. roulette wheel?

Solution Recall (Figure 11.5) that a U.S. roulette wheel has 38 compartments, 18 of which are black. Since the results of the spin of the wheel for one game are independent of the results of another spin, the plays are independent. Let $B_1 = \{$black on first spin$\}$ and $B_2 = \{$black on second spin$\}$. Then

$$P(B_1 \cap B_2) = P(B_1) \cdot P(B_2) = \frac{18}{38} \cdot \frac{18}{38} \approx 0.2243767313 \qquad \blacklozenge$$

The idea of independence is sometimes a difficult idea to communicate outside of the classroom. On July 8, 1995, in a preliminary hearing in the O. J. Simpson murder trial, an objection was raised that an adequate foundation had not been laid to establish the independence (and consequently the use of the multiplication property of probability) of blood-type factors. It was a perfect place for the mathematical principle of independence to be introduced, but instead the response focused on how many previous times the multiplication principle had been allowed by the court.

Another example involving a misuse of the multiplication principle involves gamblers. If you have ever been at a casino and watched players bet on black or red on a roulette game, you have seen that if one color comes up several times in a row, players start betting on the other color because they think it is "due." The spins of a roulette wheel are independent and it is a fallacy to think that previous spins have any effect on any future spins. For example, let's continue with Example 3 and ask, what is the probability of black on the *next* or third spin:

$$P(B_3) = \frac{18}{38} \approx 0.4736842105 \quad \text{where } B_3 = \{\text{black on third spin}\}$$

But if we ask, what is the probability of obtaining black on the *next* three consecutive spins, it is

$$P(B_1 \cap B_2 \cap B_3) = P(B_1) \cdot P(B_2) \cdot P(B_3) = \left(\frac{18}{38}\right)^3 \approx 0.1062837148$$

You might also note that the multiplication property can be used for two *or more* independent events.

Independence and the multiplication property are often used with the idea of complementary events. We now consider the mathematics associated with an experiment called the **birthday problem.**

EXAMPLE 4 PÓLYA'S METHOD

See www.mathnature.com for links to an interactive site exploring this problem.

WWW

What is the probability that at least 2 of 4 unrelated persons share the same birthday? This example is considered in more detail as a Group Research Project at the end of this chapter.

Solution We use Pólya's problem-solving guidelines for this example.

Understand the Problem. We assume that the birthdays of the individuals are independent, and we ignore leap years. We also are not considering the birthday year. We want to know, for example, whether any 2 (or more) of the 4 persons have the same birthday—say, August 3.

Devise a Plan. The first person's birthday can be any day (365 possibilities out of 365 days). We will find the probability that the second person's birthday is *different* from the first person's birthday. This is

$$1 - \tfrac{1}{365} = \tfrac{364}{365}$$

The probability that the third person's birthday is different from the first two birthdays is $\tfrac{363}{365}$ and the probability that the fourth person's birthday is different is $\tfrac{362}{365}$. Since these are independent events, we use the multiplication property of probability.

Carry Out the Plan.

$$P(\text{match}) = 1 - P(\text{no match}) = 1 - \frac{365}{365} \times \frac{364}{365} \times \frac{363}{365} \times \frac{362}{365} \approx 0.0163559125$$

Look Back. There are about 2 chances out of 100 there will be a birthday match with 4 persons in the group. It is worth noting that if we consider this for 23 persons, we have

$$1 - \frac{{}_{365}P_{23}}{365^{23}} \approx 0.5072972343$$

This means that if you look at a group of more than 23 persons, it is more likely than not that there will be a birthday match. ◆

Probability of a Union

The multiplication property of probability is used to find the probability of an intersection (E and F); now let's turn to the *probability of a union* (E or F).

$$P(E \text{ or } F) = P(E \cup F) = \frac{|E \cup F|}{n} \qquad \text{Definition of probability}$$

$$= \frac{|E| + |F| - |E \cap F|}{n} \qquad \text{Cardinality of a union}$$

$$= \frac{|E|}{n} + \frac{|F|}{n} - \frac{|E \cap F|}{n}$$

$$= P(E) + P(F) - P(E \cap F) \qquad \text{Definition of probability}$$

We have derived a property called the **addition property of probability.**

Addition Property of Probability

For any events E and F, the probability of their union can be found by

$$P(E \cup F) = P(E \text{ or } F) = P(E) + P(F) - P(E \cap F)$$

EXAMPLE 5

Suppose a coin is tossed and a die is simultaneously rolled. What is the probability of tossing a tail or rolling a 4? Is this event more or less likely than flipping a coin and obtaining a head?

Solution   Let $T = \{$tail is tossed$\}$ and $F = \{4$ is rolled$\}$.

$$
\begin{aligned}
P(T \cup F) &= P(T) + P(F) - P(T \cap F) && \text{Addition property}\\
&= P(T) + P(F) - P(T) \cdot P(F) && \text{Multiplication property; } T \text{ and } F \text{ are independent.}\\
&= \tfrac{1}{2} + \tfrac{1}{6} - \tfrac{1}{2} \cdot \tfrac{1}{6}\\
&= \tfrac{7}{12}
\end{aligned}
$$

Since $\frac{7}{12} \approx 0.58$, we see that this event is more likely than flipping a coin and obtaining a head.   ◆

EXAMPLE 6

If A, B, and C are independent events so that $P(A) = 0.2$, $P(B) = 0.6$, and $P(C) = 0.3$, find $P[\overline{(A \cup B) \cap C}]$.

Solution   We need to do some algebra before we evaluate.

$$
\begin{aligned}
P[\overline{(A \cup B) \cap C}] &= 1 - P[(A \cup B) \cap C)] && \text{Complementary probabilities}\\
&= 1 - P(A \cup B) \cdot P(C) && \text{Multiplication property of probability}\\
&= 1 - [P(A) + P(B) - P(A \cap B)] \cdot P(C) && \text{Addition property of probability}\\
&= 1 - [P(A) + P(B) - P(A) \cdot P(B)] \cdot P(C) && \text{Multiplication property of probability}\\
&= 1 - P(A) \cdot P(C) - P(B) \cdot P(C) + P(A) \cdot P(B) \cdot P(C) && \text{Distributive property (watch signs)}
\end{aligned}
$$

We now evaluate:

$$
\begin{aligned}
P[\overline{(A \cup B) \cap C}] &= 1 - (0.2)(0.3) - (0.6)(0.3) + (0.2)(0.6)(0.3)\\
&= 0.796
\end{aligned}
$$ ◆

Drawing With and Without Replacement

To highlight the difference between a combination model and a permutation model, we focus on the idea of *replacement.* Drawing **with replacement** means choosing the first item, noting the result, and then replacing the item back into the sample space before selecting the second item. Drawing **without replacement** means selecting the first item, noting the result, and then selecting a second item *without* replacing the first item.

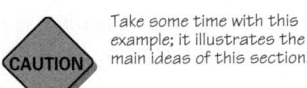
Take some time with this example; it illustrates the main ideas of this section.

EXAMPLE 7

Find the probability of each event when drawing two cards from an ordinary deck.

a. Drawing a spade on the first draw and a heart on the second draw with replacement

b. Drawing a spade on the first draw or a heart on the second draw with replacement

c. Drawing two hearts with replacement

d. Drawing a spade on the first draw and a heart on the second draw without replacement

e. Drawing a spade on the first draw or a heart on the second draw without replacement

f. Drawing two hearts without replacement

Rank the events from the most probable to the least probable.

Solution We state the decimal approximations for each probability for ease of comparison. Let $S_1 = \{$draw a spade on the first draw$\}$ and $H_2 = \{$draw a heart on the second draw$\}$.

With replacement

a. With replacement, the events are independent. Thus
$$P(S_1 \cap H_2) = P(S_1) \cdot P(H_2) = \tfrac{1}{4} \cdot \tfrac{1}{4} = \tfrac{1}{16} = 0.0625$$

b. $P(S_1 \cup H_2) = P(S_1) + P(H_2) - P(S_1 \cap H_2)$
$$= \tfrac{1}{4} + \tfrac{1}{4} - \tfrac{1}{16} \quad \text{\small From part } \mathbf{a}$$
$$= \tfrac{7}{16} = 0.4375$$

c. With replacement, the events are independent. Then
$$P(\text{two hearts}) = \tfrac{1}{4} \cdot \tfrac{1}{4} = \tfrac{1}{16} = 0.0625$$

Without replacement

d. We use the permutation model since the order is important. The number of possibilities is $_{52}P_2 = 52 \cdot 51$; the number of successes (fundamental counting principle) is $13 \cdot 13$. Thus,
$$P(S_1 \cap H_2) = \frac{13 \cdot 13}{_{52}P_2} = \frac{13 \cdot 13}{52 \cdot 51} = \frac{13}{204} \approx 0.0637254902$$

e. $P(S_1 \cup H_2) = P(S_1) + P(H_2) - P(S_1 \cap H_2)$
$$= \tfrac{1}{4} + \tfrac{1}{4} - \tfrac{13}{204} \quad \text{\small From part } \mathbf{d}$$
$$= \tfrac{89}{204} \approx 0.4362745098$$

f. Without replacement, draw both cards at once. We use combinations since the order is not important.
$$P(\text{two hearts}) = \frac{_{13}C_2}{_{52}C_2} = \frac{\dfrac{13 \cdot 12}{2}}{\dfrac{52 \cdot 51}{2}} = \frac{1}{17} \approx 0.0588235294$$

The ranking is: (1) part **b**, (2) part **e** (probabilities of parts **b** and **e** are about the same), (3) part **d**, (4) parts **a** and **c** (tie), (5) part **f**. The probabilities of parts **a, c, d,** and **f** are about the same.   ◆

Note the first three words of this subsection.

Tree Diagrams

A powerful tool in handling probability problems is a device we have frequently used —a tree diagram. If the events are independent, we find the probabilities using the multiplication property of probability.

EXAMPLE 8

Consider a game consisting of at most three cuts with a deck of cards. You win and the game is over if a face card turns up on any of the cuts, but you lose if a face card does not turn up. (A face card is a jack, queen, or king.) If you stand to win or lose $1 on this game, should you play?

Solution

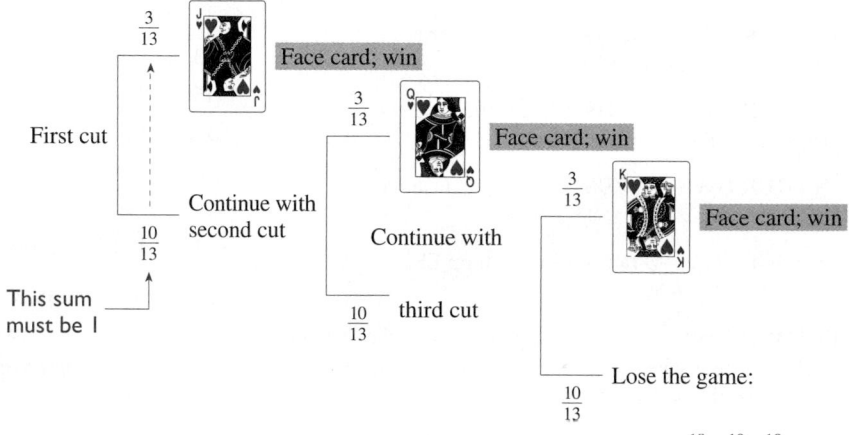

$$P(\text{lose the game}) = \tfrac{10}{13} \cdot \tfrac{10}{13} \cdot \tfrac{10}{13} \approx 0.455$$

$$P(\text{win}) = 1 - P(\text{lose}) \approx 1 - 0.455 = 0.545$$

$$\text{EXPECTATION} = \text{AMOUNT TO WIN} \cdot P(\text{win}) + \text{AMOUNT TO LOSE} \cdot P(\text{lose})$$

$$\approx \$1 \cdot (0.545) + (-\$1) \cdot (0.455)$$

$$= \$0.09$$

Since the mathematical expectation is positive, the recommendation is to play the game. If you played this game 1,000 times, you could expect to be ahead about 1,000($0.09) = $90.00. ◆

Guest Essay: Extrasensory Perception

Another presumed kind of extrasensory perception is the predictive dream. Everyone has an Aunt Matilda who had a vivid dream of a fiery car crash the night before Uncle Mortimer wrapped his Ford around a utility pole. I'm my own Aunt Matilda: When I was a kid I once dreamed of hitting a grand-slam home run and two days later I hit a bases-loaded triple. (Even believers in precognitive experiences don't expect an exact correspondence.) When one has such a dream and the predicted event happens, it's hard not to believe in precognition. But as the following derivation shows, such experiences are more rationally accounted for by coincidence.

Assume the probability to be one out of 10,000 that a particular dream matches a few vivid details of some sequence of events in real life. This is a pretty unlikely occurrence, and means that the chances of a nonpredictive dream are an overwhelming 9,999 out of 10,000. Also assume that whether or not a dream matches experience one day is independent of whether or not some other dream matches experience some other day. Thus, the probability of having two successive nonmatching dreams is, by the multiplication principle for probability, the product of 9,999/10,000 and 9,999/10,000. Likewise, the probability of having N straight nights of nonmatching dreams is $(9{,}999/10{,}000)^N$; for a year's worth of nonmatching or nonpredictive dreams, the probability is $(9{,}999/10{,}000)^{365}$.

Since $(9{,}999/10{,}000)^{365}$ is about .964, we can conclude that about 96.4 percent of the people who dream every night will have only nonmatching dreams during a one-year span. But that means that about 3.6 percent of the people who dream every night will have a predictive dream. 3.6 percent is not such a small fraction; it translates into millions of apparently precognitive dreams every year. Even if we change the probability to one in a million for such a predictive dream, we'll still get huge numbers of them by chance alone in a country the size of the United States. There's no need to invoke any special parapsychological abilities; the ordinariness of apparently predictive dreams does not need any explaining. What would need explaining would be the nonoccurrence of such dreams.

John Paulos, *Innumeracy*, pp. 54–55

Problem Set 11.4

1. **IN YOUR OWN WORDS** What do we mean by independent events?

2. **IN YOUR OWN WORDS** What is the formula for the probability of an intersection?

3. **IN YOUR OWN WORDS** What is the formula for the probability of a union?

4. **IN YOUR OWN WORDS** What is the birthday problem?

5. **IN YOUR OWN WORDS** Comment on ESP as described in the guest essay by John Paulos.

6. **IN YOUR OWN WORDS** Compare/contrast the following two betting schemes for roulette that are sometimes used by gamblers.

   Double on each loss Bet $1 on black. If black comes up, you win $1, and the game is over. If black does not come up, double your bet ($2). If black now comes up, your net winnings are $1 (lose $1, then win $2), and the game is over. If black does not come up, double your bet ($4). If black now comes up, your net winnings are $1 (lose $1, lose $2, then win $4). Continue with the doubling procedure until you eventually win; in every case, your net winnings are $1.

   Double on each win Bet $1 on black. If black does not come up, you lose $1, and the game is over. If black comes up, double your bet ($2), and play again. If black does not come up on the next spin, you lose $1 (win $1, then lose $2), and the game is over. If black comes up, double your bet ($4) and play again. If black does not come up on the next spin, you lose $1 ($+1 + 2 - 4$), and the game is over. If black comes up, double your bet ($8), and play again. If black does not come up on the next spin, you lose $1 ($+1 + 2 + 4 - 8$), and the game is over. If black does come up, double your bet ($16) one last time. If black does not come up on the next spin, you lose $1 ($+1 + 2 + 4 + 8 - 16$), and the game is over. If black does come up, the game is over and you win $31 ($+1 + 2 + 4 + 8 + 16$). Using this scheme, you are risking a maximum loss of $1 for a chance at winning $31.

Suppose events A, B, and C are independent and

$$P(A) = \tfrac{1}{2} \qquad P(B) = \tfrac{1}{3} \qquad P(C) = \tfrac{1}{6}$$

Find the probabilities in Problems 7–24.

7. $P(\overline{A})$
8. $P(\overline{B})$
9. $P(\overline{C})$
10. $P(A \cap B)$
11. $P(A \cap C)$
12. $P(B \cap C)$
13. $P(A \cup B)$
14. $P(A \cup C)$
15. $P(B \cup C)$
16. $P(\overline{A \cap B})$
17. $P(\overline{A \cap C})$
18. $P(\overline{B \cap C})$
19. $P(\overline{A \cup B})$
20. $P(\overline{A \cup C})$
21. $P(\overline{B \cup C})$
22. $P(A \cap B \cap C)$
23. $P(\overline{A \cap B \cap C})$
24. $P[(A \cup B) \cap C]$

In Problems 25–36, suppose a die is rolled twice and let

$$A = \{first\ toss\ is\ a\ prime\}$$
$$B = \{first\ toss\ is\ a\ 3\}$$
$$C = \{second\ toss\ is\ a\ 2\}$$
$$D = \{second\ toss\ is\ a\ 3\}$$

Answer the questions or find the requested probabilities.

25. Are A and B independent?
26. Are A and C independent?
27. Are A and D independent?
28. Are B and C independent?
29. Are B and D independent?
30. Are C and D independent?
31. $P(A \cap B)$
32. $P(A \cap C)$
33. $P(A \cup B)$
34. $P(A \cup C)$
35. $P(B \cup D)$
36. $P(C \cup D)$

37. A certain slot machine has three identical independent wheels, each with 13 symbols as follows: 1 bar, 2 lemons, 2 bells, 3 plums, 2 cherries, and 3 oranges. Suppose you spin the wheels and one of the 13 symbols on each wheel is selected. Find the probabilities and leave your answers in decimal form.
    a. $P(3\ bars)$
    b. $P(3\ oranges)$
    c. $P(3\ plums)$
    d. $P(cherries\ on\ the\ first\ wheel)$
    e. $P(cherries\ on\ the\ first\ 2\ wheels)$

38. Suppose a slot machine has three independent wheels as shown in Figure 11.6. Find the probabilities.

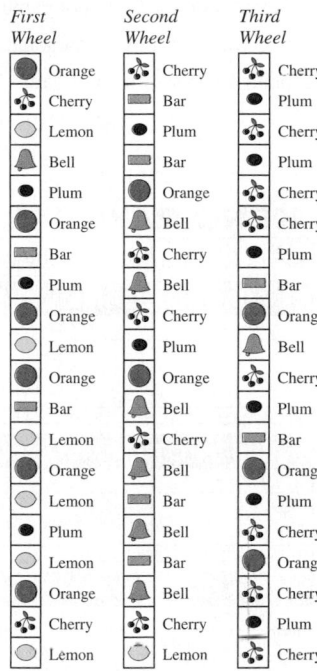

First Wheel	Second Wheel	Third Wheel
Orange	Cherry	Cherry
Cherry	Bar	Plum
Lemon	Plum	Cherry
Bell	Bar	Plum
Plum	Orange	Cherry
Orange	Bell	Cherry
Bar	Cherry	Plum
Plum	Bell	Bar
Orange	Cherry	Orange
Lemon	Plum	Bell
Orange	Orange	Cherry
Bar	Bell	Plum
Lemon	Cherry	Bar
Orange	Bell	Orange
Lemon	Bar	Plum
Plum	Bell	Cherry
Lemon	Bar	Orange
Orange	Bell	Cherry
Cherry	Cherry	Plum
Lemon	Lemon	Cherry

Figure 11.6 **Slot machine wheels**

a. P(3 bars)
b. P(3 bells)
c. P(3 cherries)
d. P(cherries on the first wheel)
e. P(cherries on the first 2 wheels)

39. The payoffs for the slot machine shown in Figure 11.6 are as follows:

First one cherry	2 coins
First two cherries	5 coins
First two wheels are cherries and the third wheel a bar	10 coins
Three cherries	10 coins
Three oranges	14 coins
Three plums	14 coins
Three bells	20 coins
Three bars (jackpot)	50 coins

What is the mathematical expectation for playing the game with the coin being a quarter dollar? Assume the three wheels are independent.

40. A "high rollers" Keno is offered in which you pay $749 to play a one-spot ticket. If you catch your number, you are paid $2,247. What is your expectation for this game?

41. A special "catch all" Keno ticket allows you to play a six-spot ticket that pays only if you pick all 6 numbers. It costs $5 to play and pays $27,777 if you win. What is your expectation for this game?

42. What is the expectation for playing Keno by picking two numbers and paying $1.00 to play? (See Table 11.2 for payoffs.)

43. What is the expectation for playing a three-spot Keno and paying $3.00 to play? (See Table 11.2 for payoffs.)

44. What is the expectation for playing a four-spot Keno and paying $5.00 to play? (See Table 11.2 for payoffs.)

45. What is the probability that at least 2 of 5 unrelated persons share the same birthday?

46. What is the probability that at least two people in your class (assume a class of 30 students) have the same birthday?

47. What is the probability of obtaining five tails when a coin is flipped five times?

48. What is the probability of obtaining at least one tail when a coin is flipped five times?

49. Assume a jar has five red marbles and three black marbles. Draw out 2 marbles with and without replacement. Find the requested probabilities.
    a. P(two red marbles)
    b. P(two black marbles)
    c. P(one red and one black marble)
    d. P(red on the first draw and black on the second draw)

50. Suppose that in an assortment of 20 calculators, there are 5 with defective switches. Draw with and without replacement.
    a. If one machine is selected at random, what is the probability it has a defective switch?
    b. If two machines are selected at random, what is the probability that both have defective switches?
    c. If three machines are selected at random, what is the probability that all three have defective switches?

51. a. A game consists of at most three cuts with a deck of 52 cards. You win $1 and the game is over if a heart turns up, but lose $1 otherwise. Should you play?
    b. Repeat this game, but remove the card that turns up on the cut.

52. a. A game consists of removing a card from a deck of 52 cards. If the card is a face card, you win $1 and the game is over. If it is not a face card, remove another card. If that one is a face card, you win $1. Repeat the process again, and then again (for a total of 4 times). If you have not removed any face card, then you lose the game, and must pay $1. Should you play this game?
    b. Repeat this game, but replace the card before doing the following draw.

PROBLEM SOLVING

53. We know that in a game of U.S. roulette, the probability that the ball drops into any one slot is 1 out of 38. Suppose that there are two balls spinning at once, and that it is physically possible for two balls to fit into the same slot. What is the probability that *both* balls would drop into the same slot?

54. Chevalier de Méré used to bet that he could get at least one 6 in four rolls of a die. He also bet that, in 24 tosses of

a pair of dice, he would get at least one 12. He found that he won more often than he lost with the first bet, but not with the second. He did not know why, so he wrote to Pascal seeking the probabilities of these events. What are the probabilities for winning these two games?

55. In a book by John Fisher, *Never Give a Sucker an Even Break* (Pantheon Books, 1976), we find the following problem:

Take a small opaque bottle and seven olives, two of which are green, five black. The green ones are considered the "unlucky" ones. Place all seven olives in the bottle, the neck of which should be of such a size that it will allow only one olive to pass through at a time. Ask the sucker to shake them and then wager that he will not be able to roll out three olives without getting an unlucky green one amongst them. If a green olive shows, he loses.

What is the probability that you win? Remember, the sucker bets that he can roll out three black olives without getting one of the green olives.

Many states conduct lotteries; a typical lottery (Keno) payoff ticket is shown here. Assume there are 20 numbers chosen from a set of 80 possible numbers. Use this information for Problems 56–60.

56. What is the expectation for a three-spot game?

57. What is the expectation for a four-spot game?

58. What is the expectation for a five-spot game?

59. What is the expectation for a six-spot game?

60. Which game illustrated on this Keno payoff ticket has the best expectation?

KENO PAYOUTS

NUMBER OF SPOTS MATCHED	PLAY $1.00...WIN:

6 SPOT GAME

MATCH	PRIZE
6	$1,000
5	$25
4	$4
3	$1

Overall odds of winning a prize in this game — 1:6.2

5 SPOT GAME

MATCH	PRIZE
5	$250
4	$10
3	$2

Overall odds of winning a prize in this game — 1:10.3

4 SPOT GAME

MATCH	PRIZE
4	$50
3	$4
2	$1

Overall odds of winning a prize in this game — 1:3.9

3 SPOT GAME

MATCH	PRIZE
3	$20
2	$2

Overall odds of winning a prize in this game — 1:6.6

2 SPOT GAME

MATCH	PRIZE
2	$8

Overall odds of winning a prize in this game — 1:16.5

1 SPOT GAME

MATCH	PRIZE
1	$2

Overall odds of winning a prize in this game — 1:4.0

Source: The California Lottery.

THE BINOMIAL DISTRIBUTION

11.5

Random Variables

The management at Daimler-Chryster want to introduce a "zero defects" advertising campaign on a new line of automobiles, so they decide to do extensive testing of parts coming off the assembly line. Suppose 150 parts come off the assembly line every day, and inspectors keep track of the number of items with any type of defect. If X is the number of defective items, then X can obviously assume any of the values 0, 1, 2, 3, . . . , 149, 150. This variable X is an example of a *random variable*. A **random variable** X associated with the sample space S of a probability is an assignment of a real number to each simple event in S. If S has a finite number of outcomes, then X is called a *finite* (or *discrete*) *random variable*; if X can assume any real value on an interval, then it is called a *continuous random variable*.

Consider now a common type of experiment—one with only two outcomes, A and $\bar{A}$. Suppose that $P(A) = p$ and $P(\bar{A}) = q = 1 - p$. We are interested in n repetitions of the experiment. If $P(A)$ remains the same for each repetition and we let X represent the number of times that event A has occurred, then we call X a **binomial random variable.**

Toss a coin four times and keep track of the results. The sample place is shown below.

Number of heads	Outcomes				
4	HHHH				
3	HHHT	HHTH	HTHH	THHH	
2	HHTT	HTHT	HTTH	THTH	THHT	TTHH
1	TTTH	TTHT	THTT	HTTT	
0	TTTT				

Let the random variable X represent the number of heads that have occurred. That is, $X = 0$ if we obtain no heads; $X = 1$ means one head is obtained; $X = 2$ means two heads are obtained; $X = 3$, $X = 4$ mean three and four heads are obtained, respectively. Since there are 16 possibilities,

$$P(X = 4) = \frac{1}{16}$$

$$P(X = 3) = \frac{4}{16} = \frac{1}{4}$$

$$P(X = 2) = \frac{6}{16} = \frac{3}{8}$$

$$P(X = 1) = \frac{4}{16} = \frac{1}{4}$$

$$P(X = 0) = \frac{1}{16}$$

Note that the sum of these probabilities is one:

$$\frac{1}{16} + \frac{4}{16} + \frac{6}{16} + \frac{4}{16} + \frac{1}{16} = 1$$

This introduction illustrates a common probability model called the **binomial distribution,** which is a list of outcomes and probabilities for what we call a *binomial experiment.*

Binomial Experiments

Binomial Experiment

A **binomial experiment** is an experiment that satisfies four conditions:

1. There must be a fixed number of trials. Denote this number by n.

2. There must be two possible mutually exclusive outcomes for each trial. Call them *success* and *failure*.

3. Each trial must be independent. That is, the outcome of a particular trial is not affected by the outcome of any other trial.

4. The probabilities of success and failure must remain constant for each trial.

Consider the manufacture of computer chips. Suppose three items are chosen at random from a day's production and are classified as defective (F) or nondefective (S). We are interested in the number of successes obtained. Suppose that an item has a probability of 0.1 of being defective and therefore a probability of 0.9 of being nondefective. We will assume that these probabilities remain the same throughout the experiment and that the classification of any particular item is independent of the classification of any other item. The sample space and the probabilities are shown in the margin.

If we let $X =$ the number of successes obtained, then we can calculate the probabilities using the information from the margin:

Sample space	*Associated probabilities*
1. FFF	$(0.1)(0.1)(0.1) = (0.1)^3$
2. FFS	$(0.1)(0.1)(0.9) = (0.9)(0.1)^2$
3. FSF	$(0.1)(0.9)(0.1) = (0.9)(0.1)^2$
4. SFF	$(0.9)(0.1)(0.1) = (0.9)(0.1)^2$
5. FSS	$(0.1)(0.9)(0.9) = (0.9)^2(0.1)$
6. SFS	$(0.9)(0.1)(0.9) = (0.9)^2(0.1)$
7. SSF	$(0.9)(0.9)(0.1) = (0.9)^2(0.1)$
8. SSS	$(0.9)(0.9)(0.9) = (0.9)^3$

S = success and F = failure. These are independent, so the probability of each of the three repetitions is the product of the individual probabilities.

$$P(X = 0) = (0.1)^3 \qquad \text{Line 1 of the sample space}$$

$$P(X = 1) = 3(0.9)(0.1)^2 \qquad \begin{array}{l}\text{This is found from lines 2, 3, and 4:}\\ \quad \text{FFS} \quad (0.9)(0.1)^2\\ \quad \text{FSF} \quad (0.9)(0.1)^2\\ \quad \text{SFF} \quad (0.9)(0.1)^2\end{array}$$

$$P(X = 2) = 3(0.9)^2(0.1) \qquad \begin{array}{l}\text{This is found from lines 5, 6, and 7:}\\ \quad \text{FSS} \quad (0.9)^2(0.1)\\ \quad \text{SFS} \quad (0.9)^2(0.1)\\ \quad \text{SSF} \quad (0.9)^2(0.1)\end{array}$$

$$P(X = 3) = (0.9)^3 \qquad \text{Line 8 of the sample space}$$

Note that this same result can be achieved by simply considering the following binomial expansion:*

$$(0.1 + 0.9)^3 = (0.1)^3 + 3(0.1)^2(0.9) + 3(0.1)(0.9)^2 + (0.9)^3$$

This leads us to the **binomial distribution theorem.**

Binomial Distribution Theorem

Let X be a random variable for the number of successes in n independent and identical repetitions of an experiment with two possible outcomes, success and failure. If p is the probability of success, then

$$P(X = k) = \binom{n}{k} p^k (1 - p)^{n-k}$$

where $k = 0, 1, \ldots, n$.

EXAMPLE 1

Suppose a sociology teacher always gives true–false tests with 10 questions.

a. What is the probability of getting exactly 70% correct by guessing?

b. What is the probability of getting 70% correct or better by guessing?

c. If a student can be sure of getting five questions correct, but must guess at the others, what is the probability of getting 70% correct or better?

*You might wish to review the binomial theorem in Section 10.6.

Solution

For this problem $P(\text{success}) = \frac{1}{2}$ and $P(\text{failure}) = \frac{1}{2}$.

a.
$$P(X = 7) = \binom{10}{7}\left(\frac{1}{2}\right)^7\left(\frac{1}{2}\right)^3$$

$$= \frac{10!}{7!3!} \cdot \frac{1}{2^{10}}$$

$$= \frac{120}{1,024}$$

$$= \frac{15}{128} \approx 0.117$$

b.
$$P(X = 8) = \binom{10}{8}\left(\frac{1}{2}\right)^8\left(\frac{1}{2}\right)^2$$

$$= \frac{10!}{8!3!} \cdot \frac{1}{2^{10}}$$

$$= \frac{45}{1,024} \approx 0.044$$

$$P(X = 9) = \binom{10}{9}\left(\frac{1}{2}\right)^9\left(\frac{1}{2}\right)$$

$$= \frac{10!}{9!1!} \cdot \frac{1}{2^{10}}$$

$$= \frac{5}{512} \approx 0.098$$

$$P(X = 10) = \binom{10}{10}\left(\frac{1}{2}\right)^{10}$$

$$= 1 \cdot \frac{1}{2^{10}}$$

$$= \frac{1}{1,024} \approx 0.001$$

$$P(X \geq 7) = \frac{120}{1,024} + \frac{45}{1,024} + \frac{10}{1,024} + \frac{1}{1,024}$$

$$= \frac{176}{1,024}$$

$$= \frac{11}{64} \approx 0.172$$

c. $P(2 \leq X \leq 5) = \binom{5}{2}\left(\frac{1}{2}\right)^5 + \binom{5}{3}\left(\frac{1}{2}\right)^5 + \binom{5}{4}\left(\frac{1}{2}\right)^5 + \binom{5}{5}\left(\frac{1}{2}\right)^5$

$$= 10 \cdot \frac{1}{32} + 10 \cdot \frac{1}{32} + 5 \cdot \frac{1}{32} + 1 \cdot \frac{1}{32}$$

$$= \frac{13}{16} = 0.8125$$

◆

Binomial experiments are also sometimes called **Bernoulli trials,** after Jacob Bernoulli (1654–1705).

EXAMPLE 2

If the probabilities for the two outcomes of n Bernoulli trials are p and $q = 1 - p$, and S denotes success and F denotes failure, find a formula for k successes.

Solution

Sometimes Bernoulli trials are cast in terms of randomly drawing balls from an urn filled with r red balls and b black balls. The proportion of

$$\text{red balls is } p = \frac{r}{r + b} \quad \text{and} \quad \text{black balls is } q = \frac{b}{r + b}$$

If each ball is replaced after drawing (and the urn is stirred up to make sure the next ball is drawn at random), then p and q will remain constant for each drawing. If the balls are not replaced, then p and q will change after each drawing. For example, if balls are drawn 8 times with replacement and the sequence is

RBRBBBBB

the probability of this sequence is

$$pqpqqqqq = p^2q^6$$

On the other hand, if the order is not important and we want the probability of obtaining 2 red and 6 black balls, then each sequence such as

RBRBBBBB or RBBBBBBR or . . . .

occurs with probability p^2q^6. Further, the number of sequences in which 2 red balls and 6 black balls are drawn is given as the coefficient of p^2q^6 in the binomial expansion of $p + q$. This number is

$$\binom{8}{2} \quad \text{so that the probability is} \quad \binom{8}{2}p^2q^6$$

The probability of k successes in n trials of a Bernoulli experiment is

$$P(k \text{ successes}) = \binom{n}{k}p^kq^{n-k}$$

for $k = 0, 1, 2, \ldots, n$. ◆

EXAMPLE 3

A missile has a probability of 0.1 of penetrating enemy defenses and reaching its target. If five missiles are aimed at the same target, what is the probability that exactly one will hit its target? What is the probability that at least one will hit its target?

Solution

$$P(X = 1) = \binom{5}{1}\left(\frac{1}{10}\right)^1\left(\frac{9}{10}\right)^4$$

$$= 0.32805 \qquad \text{By calculator}$$

$$P(X \geq 1) = 1 - P(X = 0) \qquad \text{Complementary probabilities}$$

$$= 1 - \binom{5}{0}\left(\frac{1}{10}\right)^0\left(\frac{9}{10}\right)^5$$

$$= 0.40951$$

The probability that one missile will hit its target is 0.33 (or about 1/3 of the time), and the probability that at least one will hit its target is 0.41 (or about 40% of the time). ◆

EXAMPLE 4

Suppose the probability that a page of a term paper is properly prepared is 0.25. What is the probability that there will be exactly one page with at least one error in a term paper of five pages?

Solution

The probability of at least one error on a page is $1 - 0.25 = 0.75$. The solution is given by

$$P(X = 1) = \binom{5}{1}(0.75)^1(0.25)^4$$

$$\approx 0.0146484375 \quad \text{By calculator}$$

The probability that there is exactly one page with at least one error is about 1.5%. ◆

Problem Set 11.5

LEVEL 1

1. **IN YOUR OWN WORDS** What is a binomial random variable?

2. **IN YOUR OWN WORDS** What is a Bernoulli trial?

Find the binomial probabilities in Problems 3–10.

3. $n = 5, X = 3, p = 0.30$

4. $n = 4, X = 3, p = 0.25$

5. $n = 12, X = 6, p = 0.65$

6. $n = 10, X = 4, p = 0.80$

7. $n = 6, X = 6, p = 0.50$

8. $n = 8, X = 8, p = 0.75$

9. $n = 7, X = 5, p = 0.10$

10. $n = 15, X = 13, p = 0.40$

11. Find the probability of obtaining exactly three heads on five tosses of a fair coin.

12. Find the probability of obtaining exactly four heads on five tosses of a fair coin.

13. Find the probability of obtaining exactly three heads on six tosses of a fair coin.

14. Find the probability of obtaining exactly five tails on six tosses of a fair coin.

15. Find the probability of obtaining exactly two threes on five rolls of a fair die.

16. Find the probability of obtaining exactly three twos on five rolls of a fair die.

17. A couple plans to have four children. Find the probability that they will have more than two girls.

18. A couple plans to have five children. Find the probability that they will have more than three girls.

Suppose you are taking a true–false test with ten questions. If you guess at the answers on this test, find the probabilities in Problems 19–22.

19. Exactly eight correct answers

20. Exactly nine correct answers

21. Fewer than two correct answers

22. More than eight correct answers

Suppose a jar contains 3 pens: 1 red, 1 blue, and 1 green. Three people sign a document, one at a time. Each person selects a pen and signs the document and then replaces the pen before the next person selects a pen at random. Find the probabilities in Problems 23–26.

23. The red pen is never selected.

24. Exactly one person uses the red pen.

25. Exactly two people use the red pen.

26. All three people select the red pen.

All the cows in a certain herd are white-faced. The probability that a white-faced calf will be born by mating with a certain bull is 0.9. Suppose four cows are bred to the same bull. Find the probabilities in Problems 27–32.

27. Four white-faced calves

28. Three white-faced calves

29. No white-faced calves

30. One white-faced calf

31. At least three white-faced calves

32. No more than two white-faced calves

A researcher chooses three leaves from a target environment and classifies each sample as fungus-free or contaminated. Suppose that a leaf has a probability of 0.2 of being infected. In Problems 33–37, find the requested probabilities.

33. No infected leaves are found.

34. All three samples are infected.

35. Two are found to be infected.

36. One is infected.

37. One or more is infected.

38. Numerically, how are your answers for Problems 33–37 related?

LEVEL 2

39. The probability that children in a specific family will have blond hair is $\frac{3}{4}$. If there are four children in the family, what is the probability that two of them are blond?

40. The probability that the children in a specific family will have brown eyes is $\frac{3}{8}$. If there are four children in the family, what is the probability that at least two of them have brown eyes?

41. What is the probability that if a pair of dice is rolled six times, exactly one sum of seven is rolled?

42. What is the probability that if a pair of dice is rolled twelve times, exactly two sums of seven are rolled?

43. If Marcy can make 70% of free throws in a basketball game, what is the probability that she makes five of the next six free throws?

44. If Bart can make only 40% of free throws in a basketball game, what is the probability that he will make one of the next three free throws?

45. Lymnozyme cures most infections in Koi fish caused by bacteria; in fact, it has been shown to be 96% effective if used according to the directions. If 5 Koi with a bacterial infection are treated, what is the probability that 4 of the fish are cured?

46. Lymnozyme cures most infections in Koi fish caused by bacteria; in fact, it has been shown to be 96% effective if used according to the directions. If 15 Koi with a bacterial infection are treated, what is the probability that 14 of the fish are cured?

47. It is known that 85% of the graduates of Foley's School of Motel Management are placed in a job within 6 months of graduation. If a class has 20 graduates, what is the probability that 15 will be placed within 6 months?

48. If a graduating class from Problem 47 has 10 graduates, what is the probability that at least 9 will be placed within 6 months?

49. Suppose the National League team has a probability of $\frac{3}{5}$ of winning a World Series game and the American League team has a probability of $\frac{2}{5}$. The series is over as soon as one team wins four games. Find the probability that the National League team wins the series in seven games.

50. Suppose the National League team has a probability of $\frac{3}{5}$ of winning the World Series and the American League team has a probability of $\frac{2}{5}$. The series is over as soon as one team wins four games. Find the probability that the series is over in seven games.

51. The batting average of the star baseball player Mike Thompson is 0.300 (that is, $P(\text{hit}) = 0.3$). What is the probability that Thompson will get at least three hits in four times at bat?

52. Eighty percent of the widgets of the Ampex Widget Company meet the specifications of its customers. If a sample of six widgets is tested, what is the probability that three or more of them would fail to meet the specifications?

LEVEL 3

53. Suppose that research has shown that the probability that a missile penetrates enemy defenses and reaches its target is 0.1. Find the smallest number of identical missiles that are necessary in order to be 80% certain of hitting the target at least once.

54. Suppose a person claims to have ESP and says she can read mental images 75% of the time. We set up an experiment where she must determine whether we are looking at a picture of a circle or a square. We agree that if she can correctly identify the mental image at least five out of six times we will grant her claim.
    a. What is the probability of granting her claim if she is in fact only guessing?
    b. What is the probability of denying her claim if she really does have the ability she claims?

55. Repeat Problem 54 where the person claims to read mental images 90% of the time.

56. Fire frequency has a great impact on national parks. If the fire frequency is too severe, generally species that occur in later successional habitats will be excluded. Likewise, the absence of fire will lead to a paucity of species that require early successional habitats. If, on average, fire occurs in a site forest once every 10 years, then the probability is 10% that this site will burn in a given year. If there are 100 such site forests, what is the probability that fire will occur in at least one site over the next year? Assume that all sites are independent of each other; that is, fire does not spread into neighboring patches.

PROBLEM SOLVING

57. In a certain office, three men determine who will pay for coffee by each flipping a coin. If one of them has an outcome that is different from the other two, he must pay for the coffee. What is the probability that in any play of the game there will be an "odd man out"?

58. In the game of Problem 57, what is the probability that there will be an odd man out in a particular play if there are four players?

59. In the game of Problem 57, what is the probability that there will be an odd man out in a particular play if there are five players?

60. Generalize Problem 57 for *n* people playing the game of "odd man out."

CHAPTER SUMMARY

"The mathematics curriculum should include the continued study of probability so that all students can use experimental or theoretical probability, as appropriate, to represent and solve problems involving uncertainty."

NCTM Standards

IMPORTANT TERMS

Addition property of probability [11.4]
Bernoulli trial [11.5]
Binomial distribution theorem [11.5]
Binomial experiment [11.5]
Binomial random variable [11.5]
Birthday problem [11.4]
Complementary probabilities [11.3]
Conditional probability [11.3]
Dependent events [11.4]
Dice [11.1]
Die [11.1]
Empirical probability [11.1]

Equally likely outcomes [11.1]
Event [11.1]
Expectation [11.2]
Expected value [11.2]
Experiment [11.1]
Fair game [11.2]
Impossible event [11.1]
Independent events [11.4]
Law of large numbers [11.1]
Keno [11.4]
Mathematical expectation [11.2]
Multiplication property of probability [11.4]
Mutually exclusive [11.1]

Odds against [11.3]
Odds in favor [11.3]
Probabilistic model [11.1]
Probability [11.1]
Property of complements [11.3]
Random variable [11.5]
Relative frequency [11.1]
Sample space [11.1]
Simple event [11.1]
Subjective probability [11.1]
Theoretical probability [11.1]
With replacement [11.4]
Without replacement [11.4]

TYPES OF PROBLEMS

Find empirical probabilities. [11.1]
Find probabilities by looking at the sample space. [11.1]
Decide which of two events is more probable. [11.1]
Find the expectation with a cost of playing. [11.2]
Find the mathematical expectation for a simple event. [11.2]
Find the mathematical expectation for a compound event. [11.2]
Make decisions based on mathematical expectation. [11.2]
Find expectations for roulette. [11.2]
Know the procedure for using tree diagrams to find probabilities. [11.3]
Find the probability of a complement. [11.3]
Find the odds, given the probability. [11.3]
Find the probability, given the odds. [11.3]
Find conditional probabilities. [11.3]
Find the probabilities of unions, intersections, and complements. [11.4]

Find slot machine probabilities. [11.4]
Calculate Keno probabilities. [11.4]
Use tree diagrams to find probabilities. [11.4]
Find binomial probabilities. [11.5]
Answer questions involving applied probabilities. [11.1–11.5]

CHAPTER 11 REVIEW QUESTIONS

1. If a sample from an assembly line reveals four defective items out of 1,000 sampled items, what is the probability that any one item is defective?
2. What is the probability of obtaining a prime in a single roll of a die?
3. A pair of dice is rolled. What is the probability that the resulting sum is eight?
4. A card is selected from an ordinary deck of cards. What is the probability that it is a jack or better? (A jack or better is a jack, queen, king, or ace.)
5. A single card is drawn from a standard deck of cards. What is the probability that it is an ace, if you know that one ace has already been removed from the deck?
6. For a roll of a pair of dice, find the following.
   a. P(5 on at least one of the dice)
   b. P(5 on one die or 4 on the other)
   c. P(5 on one die and 4 on the other)
7. If the probability of dropping an egg is 0.01, what is the probability of not dropping the egg?
8. If $P(E) = 0.9$, what are the odds in favor of E?
9. If the odds against an event are 1,000 to 1, what is the probability of the event?
10. A game consists of rolling a die and receiving $12 if a one is rolled and nothing otherwise. What is the mathematical expectation?
11. A box contains three orange balls and two purple balls, and two are selected at random. What is the probability of the second ball being orange if we know that the first draw was orange and we draw the second ball:
    a. after replacing the first ball?   b. without replacing the first ball?
12. If A and B are independent events with $P(A) = \frac{2}{3}$ and $P(B) = \frac{3}{5}$, what is $P(A \cup B)$?
13. A manufacturer of water test kits knows that about 0.1% of the test strips in the kits are defective. Write the binomial probability formula that would be used to determine the probability that exactly x test strips in a kit of n strips are defective.

The weather channel makes accurate predictions about 85% of the time. Use this information to answer the questions in Problems 14–15.

14. What is the probability that they are accurate on exactly 3 of the next 5 days?
15. What is the probability that they are accurate exactly 4 of the next 4 days?
16. A friend says she has two new baby cats to show you, but she doesn't know whether they're both male, both female, or a pair. You tell her that you want only a male, and she telephones the vet and asks, "Is at least one a male?" The vet answers, "Yes!" What is the probability that, if you select one kitten, then it is a male? Assume that a baby cat is equally likely to be female or male.
17. The probability that an individual who is selected at random from a population has blue eyes has been estimated to be 0.35. Find the probability that at least one of four people selected at random has blue eyes.

18. Consider the set of octahedral dice shown in Figure 11.17. Suppose your opponent chooses die *D*. Which die would you choose, and why, if the game consists of each player rolling their chosen die once and the higher number on top is the winner?

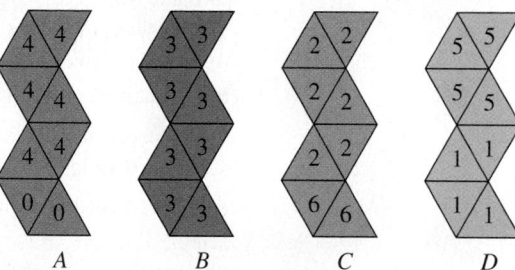

A   B   C   D

Figure 11.7 **Octahedral dice**

19. The following problem appeared in an "Ask Marilyn" column, by Marilyn Vos Savant:

> *There are five cars on display as prizes, and their five ignition keys are in a box. You get to pick one key out of the box and try it in the ignition of one car. If it fits, you win the car. What are your chances of winning a car?*

Answer this question.

20. Answer the question in Problem 19 assuming that you are in the audience and want the probability that the contestant will win the car you have selected.

GROUP RESEARCH

Working in small groups is typical of most work environments, and learning to work with others to communicate specific ideas is an important skill. Work with three or four other students to submit a single report based on each of the following questions.

G41. a. In how many possible ways can you land on jail (just visiting) on your first turn when playing a Monopoly game?

b. Is it possible to make it from GO to Park Place on your first roll of the dice in a Monopoly game? If so, what is the probability that not only would that happen, but also that you would obtain a 2 on your next roll to complete a set (a monopoly)?

G42. Birthday problem:

a. Experiment: Consider the birthdates of some famous mathematicians:

Abel	August 5, 1802
Cardano	September 24, 1501
Descartes	March 31, 1596
Euler	April 15, 1707
Fermat	August 17, 1601
Galois	October 25, 1811
Gauss	April 30, 1777
Newton	December 25, 1642
Pascal	June 19, 1623
Riemann	September 17, 1826

See the links for this problem at
www.mathnature.com

Add to this list the birthdates of the members of your class. *But before you compile this list, guess the probability that at least two people in this group will have exactly the same birthday (not counting the year).* Be sure to make your guess *before* finding out the birthdates of your classmates. The answer, of course, depends on the number of people on the list. Ten mathematicians are listed and you may have 20 people in your class, giving 30 names on the list.

b. Find the probability of at least one birthday match among 3 randomly selected people. (See Example 4, Section 11.4.)

c. Find the probability of at least one birthday match among 23 randomly selected people. Have each person in your group pick 23 names at random from a biographical dictionary or a *Who's Who,* and verify empirically the probability you calculated.

d. Draw a graph showing the probability of a birthday match given a group of n people. How many people are necessary for the probability actually to reach 1?

e. In the previous parts of this problem we interpreted two people having the same birthday as meaning at least 2 have the same birthday (see Example 4, Section 11.4). We now refine this idea. Find the following probabilities for a group of 5 randomly selected people:

Exactly 2 of the 5 have the same birthday.
Exactly 3 have the same birthday.
Exactly 4 have the same birthday.
All 5 have the same birthday.
There are exactly two pairs sharing (a different) birthday.
There is a full house of birthdays (that is, three share one birthday, and two share another).

Show that the questions of this problem account for all the possibilities; that is, show that the sum of the probabilities for all of these possibilities is the same as for the original birthday problem involving 5 persons: What is the probability of a birthday match among 5 randomly selected people?

G43. Consider the following classroom activity. Suppose the floor consists of square tiles 9 in. on each side. The players will toss a circular disk onto the floor. If the disk comes to rest on the edge of any tile, the player loses $1. Otherwise, the player wins $1. What is the probability of winning if the disk is:

a. a dime **b.** a quarter **c.** a disk with a diameter of 4 in.

d. Now, the real question: What size should the disk be so that the probability that the player wins is 0.45?

INDIVIDUAL RESEARCH PROBLEMS

Learning to use sources outside your classroom and textbook is an important skill, and here are some ideas for extending some of the ideas in this chapter. You can find references to these projects in a library or at

www.mathnature.com

PROJECT 11.1 What do the following people have in common?
Paul Painlevé, former President of France
Omar Khayyám, author of *The Rubaiyat*
Emanual Lasker, world chess champion
James Moriarty, Sherlock Holmes's nemesis, author of *The Dynamics of an Asteroid*

PROJECT 11.2 Devise a fair scheme for eliminating coins in this country.

PROJECT 11.3 Find the probability that the 13th day of a randomly chosen month will be a Friday.

PROJECT 11.4 You have five alternatives from which to choose. List your preferences for the alternatives from best to worst.

1. Sure win of $5 and no chance of loss

2. 6.92% chance to win $20 and 93.08% chance to win $3.98

3. 27.52% chance to win $20 and 72.48% chance to lose 69 cents

4. 61.85% chance to win $20 and 38.15% chance to lose $19.31

5. 90.46% chance to win $20 and 9.54% chance to lose $137.20

   a. Answer the question based on your own feelings.

   b. Answer the question using mathematical expectation as a basis for selecting your answer.

   c. Conduct a survey of at least 10 people and summarize your results.

   d. What are the conclusions of the study?

PROJECT 11.5

1. Choose between A and B:

   A. A sure gain of $240

   B. 25% chance to gain $1,000 and a 75% chance to gain $0

2. Choose between C and D:

   C. A sure loss of $700

   D. 75% chance to lose $1,000 and 25% chance to lose nothing

3. Choose between E and F:

   E. Imagine that you have decided to see a concert and have paid the admission price of $10. As you enter the concert hall, you discover that you have lost your ticket. Would you pay $10 for another ticket?

    F. Imagine that you have decided to see a concert where the admission is $10. As you begin to enter the concert ticket line, you discover that you have lost one of your $10 bills. Would you still pay $10 for a ticket to the concert?

    a. Answer each of the questions* based on your own feelings.

    b. Answer questions (1) and (2) using mathematical expectation as a basis for selecting your answers.

    c. Conduct a survey of at least 10 people and summarize your results.

PROJECT 11.6 Do some research on Keno probabilities. Write a paper on playing Keno.

---

*These questions are from "A Bird in the Hand," by Carolyn Richbart and Lynn Richbart, *The Mathematics Teacher,* November 1996, pp. 674–676.

12 The Nature of Statistics

There are three kinds of lies: lies, damned lies, and statistics.
G. B. HALSTED

CONTENTS

OVERVIEW

Undoubtedly, you have some idea about what is meant by the term *statistics.* For example, we hear about:

1. Statistics on population growth
2. Information about the depletion of the rain forests or the ozone layer
3. The latest statistics on the cost of living
4. The Gallup Poll's use of statistics to predict election outcomes
5. The Nielsen ratings, which indicate that one show has 30% more viewers than another
6. Baseball or sports statistics

We could go on and on, but you can see from these examples that there are two main uses for the word **statistics.** First, we use the term to mean a mass of data, including charts and tables. This is the everyday, nontechnical use of the word. Second, the word refers to a methodology for collecting, analyzing, and interpreting data. In this chapter, we'll examine some of these statistical methods. We do not intend to present a statistics course, but rather to prepare you to use statistics in your everyday life, as well as possibly preparing you to take a college-level statistics course.

IMPORTANT IDEAS

Read and interpret bar graphs, line graphs, circle graphs, and pictographs [12.1]
Recognize and describe misuses of graphs [12.1]
Decide on an appropriate measure of central tendency [12.2]
Finding the mean, median, and mode [12.2]
Formula for weighted mean [12.2]
Procedure for finding the standard deviation [12.2]
Percentages for the standard deviations on a normal curve [12.3]
Find and use z-scores [12.3]
Meanings associated with a normal curve [12.3]
Formula for the linear correlation coefficient [12.4]
Formulas for the slope and y-intercept of the least squares (or regression) line [12.4]
Difference between descriptive and inferential statistics [12.5]
Difference between a sample and a population [12.5]
Type I and Type II sampling error [12.5]

BOOK REPORTS

Write a 500-word report on the following book:
How to Lie with Statistics, Darrell Huff (New York: Norton, 1954).

LINKS
Individual Projects
Group Projects
Research Links

www.mathnature.com

12.1 FREQUENCY DISTRIBUTIONS AND GRAPHS

Frequency Distributions

Computers, spreadsheets, and simulation programs have done a lot to help us deal with hundreds or thousands of pieces of information at the same time. We can deal with large batches of data by organizing them into groups, or **classes.** The difference between the lower limit of one class and the lower limit of the next class is called the **interval** of the class. After determining the number of values within a class, termed the **frequency,** you can use this information to summarize the data. The end result of this classification and tabulation is called a **frequency distribution.** For example, suppose that you roll a pair of dice 50 times and obtain these outcomes:

3, 2, 6, 5, 3, 8, 8, 7, 10, 9, 7, 5, 12, 9, 6, 11, 8, 11, 11, 8, 7, 7, 7, 10, 11, 6, 4, 8, 8, 7, 6, 4, 10, 7, 9, 7, 9, 6, 6, 9, 4, 4, 6, 3, 4, 10, 6, 9, 6, 11

We can organize these data in a convenient way by using a frequency distribution, as shown in Table 12.1.

Table 12.1

Frequency Distribution for 50 Rolls of a Pair of Dice		
Outcome	Tally	Frequency
2	|	1
3	|||	3
4	⊞⊞	5
5	||	2
6	⊞⊞ ||||	9
7	⊞⊞ |||	8
8	⊞⊞ |	6
9	⊞⊞ |	6
10	||||	4
11	⊞⊞	5
12	|	1

EXAMPLE 1

Make a frequency distribution for the information in Table 12.2.

Solution Make three columns. First, list the sales tax categories; next tally; and finally, count the tallies to determine the frequency of each (see margin). To account for irregularities (such as 2.9%), the categories are often grouped. For this example, the data are divided into 8 categories (or groups). The number of categories is usually arbitrary and chosen for convenience. This is called a **grouped frequency distribution.**

Table 12.2 State Sales Tax Rates*

Alabama	4%	Hawaii	4%	Massachusetts	5%	New Mexico	5%	South Dakota	4%
Alaska	0%	Idaho	5%	Michigan	6%	New York	4%	Tennessee	6%
Arizona	5.6%	Illinois	$6\frac{1}{4}$%	Minnesota	$6\frac{1}{2}$%	North Carolina	4.5%	Texas	6.25%
Arkansas	5.125%	Indiana	5%	Mississippi	7%	North Dakota	5%	Utah	4.75%
California	$7\frac{1}{4}$%	Iowa	5%	Missouri	4.225%	Ohio	5%	Vermont	5%
Colorado	2.9%	Kansas	5.3%	Montana	0%	Oklahoma	4.5%	Virginia	4.5%
Connecticut	6%	Kentucky	6%	Nebraska	5%	Oregon	0%	Washington	$6\frac{1}{2}$%
Delaware	0%	Louisiana	4%	Nevada	6.5%	Pennsylvania	6%	West Virginia	6%
Florida	6%	Maine	5%	New Hampshire	0%	Rhode Island	7%	Wisconsin	5%
Georgia	4%	Maryland	5%	New Jersey	6%	South Carolina	5%	Wyoming	4%
				District of Columbia, 5.75%					

*Does not include local sales taxes.

Sales tax	Tally	Frequency
0%	⊞⊞	5
2.9%	|	1
4%	⊞⊞ ||	7
4.225%	|	1
$4\frac{1}{2}$%	|||	3
4.75%	|	1
5%	⊞⊞ ⊞⊞ |||	13
5.125%	|	1
5.3%	|	1
5.6%	|	1
5.75%	|	1
6%	⊞⊞ |||	8

Sales tax	Tally	Frequency
$6\frac{1}{4}$%	||	2
$6\frac{1}{2}$%	|||	3
7%	||	2
$7\frac{1}{4}$%	|	1

Grouped Frequency Distribution

Sales Tax	Tally		Frequency
0–1%	⊞⊞		5
1^+–2%			0
2^+–3%	|		1
3^+–4%	⊞⊞ ||		7
4^+–5%	⊞⊞ ⊞⊞ ⊞⊞ |||		18
5^+–6%	⊞⊞ ⊞⊞ ||		12
6^+–7%	⊞⊞ ||		7
Over 7%	|		1

◆

A procedure that is useful for organizing large sets of data is called a **stem-and-leaf plot.** Consider the data shown in Table 12.3.*

Table 12.3 Best Actors, 1928–2002

Year	Actor	Movie	Age	Year	Actor	Movie	Age
1928	Emil Jannings	*The Way of All Flesh*	44	1965	Lee Marvin	*Cat Ballou*	41
1929	Warner Baxter	*In Old Arizona*	38	1966	Paul Scofield	*A Man for All Seasons*	44
1930	George Arliss	*Disraeli*	46	1967	Rod Steiger	*In the Heat of the Night*	42
1931	Lionel Barrymore	*A Free Soul*	53	1968	Cliff Robertson	*Charly*	43
1932	Fredric March ⎫ tie	*Dr. Jekyll and Mr. Hyde*	35	1969	John Wayne	*True Grit*	62
1932	Wallace Berry ⎭	*The Champ*	47	1970	George C. Scott	*Patton*	43
1933	Charles Laughton	*The Private Life of Henry VIII*	34	1971	Gene Hackman	*The French Connection*	40
1934	Clark Gable	*It Happened One Night*	33	1972	Marlon Brando	*The Godfather*	48
1935	Victor McLaglen	*The Informer*	49	1973	Jack Lemmon	*Save the Tiger*	48
1936	Paul Muni	*The Story of Louis Pasteur*	41	1974	Art Carney	*Harry and Tonto*	56
1937	Spencer Tracy	*Captains Courageous*	37	1975	Jack Nicholson	*One Flew over the Cuckoo's Nest*	38
1938	Spencer Tracy	*Boys' Town*	38	1976	Peter Finch	*Network*	60
1939	Robert Donat	*Goodbye Mr. Chips*	34	1977	Richard Dreyfuss	*The Goodbye Girl*	32
1940	James Stewart	*The Philadelphia Story*	32	1978	Jon Voight	*Coming Home*	40
1941	Gary Cooper	*Sergeant York*	40	1979	Dustin Hoffman	*Kramer vs. Kramer*	42
1942	James Cagney	*Yankee Doodle Dandy*	43	1980	Robert De Niro	*Raging Bull*	37
1943	Paul Lukas	*On the Rhine*	48	1981	Henry Fonda	*On Golden Pond*	76
1944	Bing Crosby	*Going My Way*	43	1982	Ben Kingsley	*Gandhi*	39
1945	Ray Milland	*The Lost Weekend*	40	1983	Robert Duvall	*Tender Mercies*	55
1946	Fredric March	*The Best Years of Our Lives*	49	1984	F. Murray Abraham	*Amadeus*	45
1947	Ronald Colman	*A Double Life*	56	1985	William Hurt	*Kiss of the Spider Woman*	35
1948	Laurence Olivier	*Hamlet*	41	1986	Paul Newman	*Color of Money*	61
1949	Broderick Crawford	*All the King's Men*	38	1987	Michael Douglas	*Wall Street*	33
1950	Jose Ferrer	*Cyrano de Bergerac*	38	1988	Dustin Hoffman	*Rainman*	51
1951	Humphrey Bogart	*The African Queen*	52	1989	Daniel Day-Lewis	*My Left Foot*	31
1952	Gary Cooper	*High Noon*	51	1990	Jeremy Irons	*Reversal of Fortune*	42
1953	William Holden	*Stalag 17*	35	1991	Anthony Hopkins	*Silence of the Lambs*	55
1954	Marlon Brando	*On the Waterfront*	30	1992	Al Pacino	*Scent of a Woman*	52
1955	Ernest Borgnine	*Marty*	38	1993	Tom Hanks	*Philadelphia*	37
1956	Yul Brynner	*The King and I*	41	1994	Tom Hanks	*Forrest Gump*	38
1957	Alec Guinness	*The Bridge on the River Kwai*	43	1995	Nicholas Cage	*Leaving Las Vegas*	31
1958	David Niven	*Separate Tables*	49	1996	Geoffrey Rush	*Shine*	45
1959	Charlton Heston	*Ben Hur*	35	1997	Jack Nicholson	*As Good As It Gets*	59
1960	Burt Lancaster	*Elmer Gantry*	47	1998	Roberto Benigni	*Life Is Beautiful*	46
1961	Maximilian Schell	*Judgment at Nuremburg*	31	1999	Kevin Spacey	*American Beauty*	41
1962	Gregory Peck	*To Kill a Mockingbird*	46	2000	Russell Crowe	*Gladiator*	36
1963	Sidney Poitier	*Lilies of the Field*	39	2001	Denzel Washington	*Training Day*	47
1964	Rex Harrison	*My Fair Lady*	56	2002	Adrien Brody	*The Pianist*	29

For these data we form stems representing decades of the ages of the actors. For example, Denzel Washington, who won the best actor award in 2001 for his role in *Training Day,* was 47 years of age when he won the award. We would say that the stem for this age is 4 (for 40), and the leaf is 7.

*The idea for this example came from "Ages of Oscar-winning Best Actors and Actresses" by Richard Brown and Gretchen Davis, *The Mathematics Teacher,* February 1990, pp. 96–102.

EXAMPLE 2

Construct a stem-and-leaf plot for the best actor ages in Table 12.3.

Solution

Stem-and-Leaf Plot of Ages of Best Actor, 1928–2002

2	9
3	0 1 1 1 2 2 3 3 4 4 5 5 5 5 6 7 7 7 8 8 8 8 8 8 9 9
4	0 0 0 0 1 1 1 1 1 2 2 2 3 3 3 3 3 4 4 5 5 6 6 6 7 7 7 8 8 8 9 9 9
5	1 1 2 2 3 5 5 6 6 6 9
6	0 1 2
7	6

This plot is useful because it is easy to see that most best actor winners received the award in their forties. ◆

The data sets {32, 56, 47, 30, 41} and {3.2, 5.6, 4.7, 3.0, 4.1} have the same stem-and-leaf diagrams with the first set having a leaf unit of 1 and the second set having a leaf unit of 0.1.

To help us understand the relationship between and among variables, we use a diagram called a **graph.** In this section, we consider *bar graphs, line graphs, circle graphs,* and *pictographs.*

Bar Graphs

A **bar graph** compares several related pieces of data using horizontal or vertical bars of uniform width. There must be some sort of scale or measurement on both the horizontal and vertical axes. An example of a bar graph is shown in Figure 12.1, which shows the data from Table 12.4.

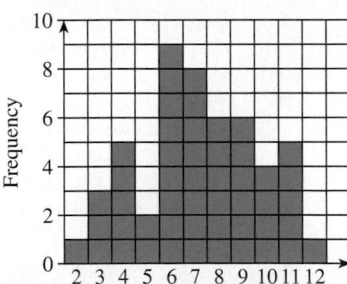

Figure 12.1 **Outcomes of experiment of rolling a pair of dice**

Table 12.4

Frequency Distribution for 50 Rolls of a Pair of Dice		
Outcome	Tally	Frequency
2	|	1
3	|||	3
4	⊞	5
5	||	2
6	⊞ ||||	9
7	⊞ |||	8
8	⊞ |	6
9	⊞ |	6
10	||||	4
11	⊞	5
12	|	1

EXAMPLE 3

Construct a bar graph for the data given in Example 1. Use the grouped categories.

Solution To construct a bar graph, draw and label the horizontal and vertical axes, as shown in part **a** of Figure 12.2. It is helpful (although not necessary) to use graph paper. Next, draw marks indicating the frequency, as shown in part **b** of Figure 12.2. Finally, complete the bars and shade them as shown in part **c.**

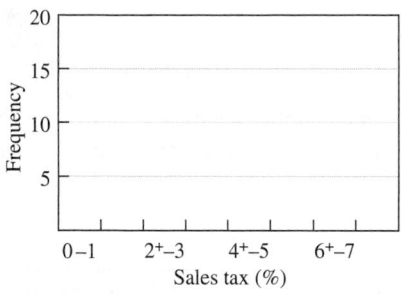

a. Drawing and labeling the axes.

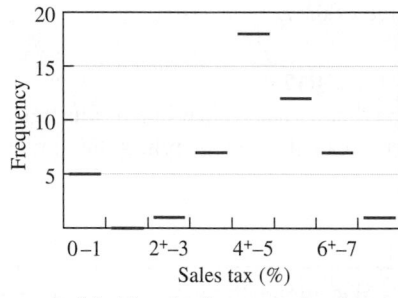

b. Marking the frequency levels.

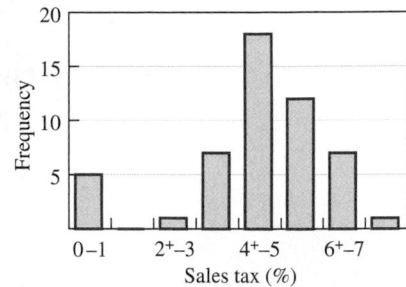

c. Completing and shading the bars.

Figure 12.2 Bar graph for sales tax data

◆

You will frequently need to look at and interpret bar graphs in which bars of different lengths are used for comparison purposes.

EXAMPLE 4

Refer to Figure 12.3 to answer the following questions.

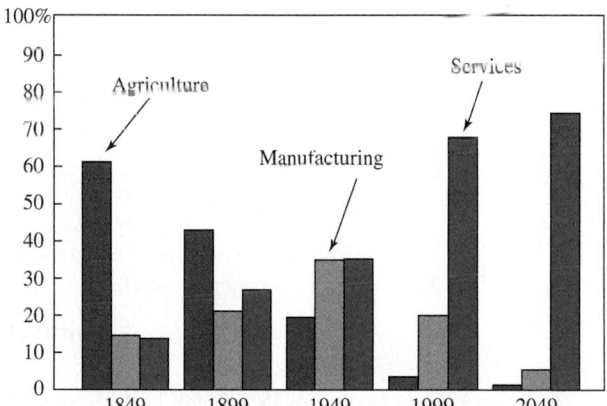

Figure 12.3 Share of U.S. employment in agriculture, manufacturing, and services from 1849 to 2049

Source: U.S. Bureau of Labor Statistics and Historical Statistics of the United States, p. 139; Paine Webber estimates.

a. What was the share of U.S. employment in manufacturing in 1999?

b. In which year(s) was U.S. employment in services approximately equal to employment in manufacturing?

c. From the graph, form a conclusion about U.S. employment in agriculture for the period 1849–2049.

Solution

a. We see the bar representing manufacturing in 1999 is approximately 20%. (Use a straightedge for help.)

b. Employments in services and manufacturing were approximately equal in 1949.

c. Conclusions, of course, might vary, but one obvious conclusion is that there has been a dramatic decline of employment in agriculture in the United States over this period of time.   ◆

Line Graphs

A graph that uses a broken line to illustrate how one quantity changes with respect to another is called a **line graph.** A line graph is one of the most widely used kinds of graph.

EXAMPLE 5

Draw a line graph for the data given in Example 1. Use the previously grouped categories.

Solution   The line graph uses points instead of bars to designate the locations of the frequencies. These points are then connected by line segments, as shown in Figure 12.4. To plot the points, use the frequency distribution to find the category (0^+–1%, for example); then plot a point showing the frequency (5, in this example). This step is shown in part **a** of Figure 12.4. The last step is to connect the dots with line segments, as shown in part **b.**

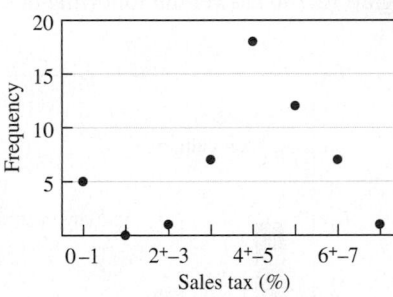
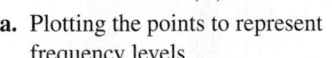
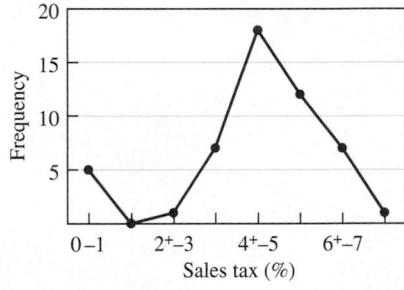

a. Plotting the points to represent frequency levels

b. Connecting the dots with line segments

***Figure 12.4* Constructing a line graph for sales tax data**   ◆

Just as with bar graphs, you need to be able to read and interpret line graphs.

EXAMPLE 6

Refer to Figure 12.5 to answer the following questions.

a. In which year was the voter turnout in a presidential election the greatest?

b. During the period 1932–1996, were there more periods of economic growth or economic decline?

c. What was the voter turnout in 1996? Did a Republican or Democrat win that election? In which year was the voter turnout the closest to that in 1996? Did a Republican or a Democrat win in that election?

Solution

a. Voter turnout was the greatest (63.1%) in 1960.

b. Growth is shown as a white region and decline as a shaded region; there are 12 periods of each, but the growth periods are of greater duration.

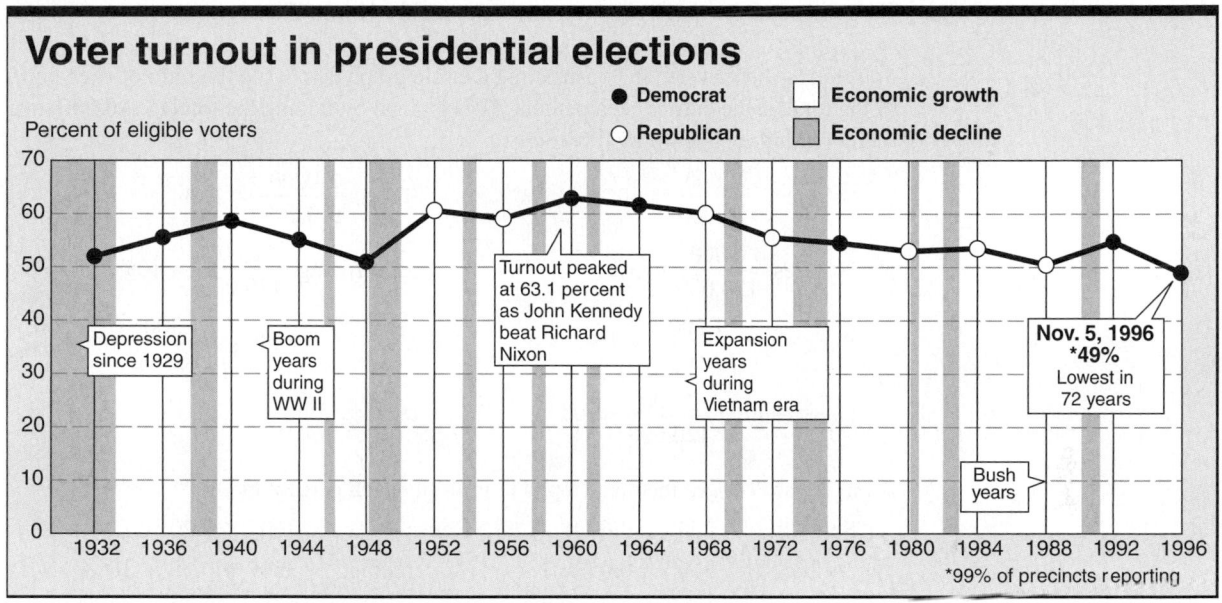

Figure 12.5 Voter turnout in presidential elections, 1932–1996

c. In 1996 the voter turnout was 49% and a Democrat won the election. Voter turnout had not been that low since 1988 (about 51%), when the winner was a Republican, or since 1948 (about 52%) when the winner was a Democrat. ◆

Circle Graphs

Another type of commonly used graph is the **circle graph,** also known as a **pie chart.** This graph is particularly useful in illustrating how a whole quantity is divided into parts—for example, income or expenses in a budget.

To create a circle graph, first express the number in each category as a percentage of the total. Then convert this percentage to an angle in a circle. Remember that a circle is divided into 360°, so we multiply the percent by 360 to find the number of degrees for each category. You can use a protractor to construct a circle graph, as shown in Example 7.

EXAMPLE 7

The 2002 expenses for Karlin Enterprises are shown in Figure 12.6.

KE KARLIN ENTERPRISES	**EXPENSE REPORT** FY 2002
Salaries	$ 72,000
Rents, taxes, insurance	$ 24,000
Utilities	$ 6,000
Advertising	$ 12,000
Shrinkage	$ 1,200
Materials and supplies	$ 1,200
Depreciation	$ 3,600
TOTAL	**$120,000**

Figure 12.6 Karlin Enterprises expenses

Construct a circle graph showing the expenses for Karlin Enterprises.

Solution The first step in constructing a circle graph is to write the ratio of each entry to the total of the entries, as a percent. This is done by finding the total ($120,000) and then dividing each entry by that total:

Salaries: $\dfrac{72,000}{120,000} = 60\%$ Rents: $\dfrac{24,000}{120,000} = 20\%$

Utilities: $\dfrac{6,000}{120,000} = 5\%$ Advertising: $\dfrac{12,000}{120,000} = 10\%$

Shrinkage: $\dfrac{1,200}{120,000} = 1\%$ Materials/supplies: $\dfrac{1,200}{120,000} = 1\%$

Depreciation: $\dfrac{3,600}{120,000} = 3\%$

A circle has 360°, so the next step is to multiply each percent by 360°:

Salaries: $360° \times 0.60 = 216°$ Rents: $360° \times 0.20 = 72°$

Utilities: $360° \times 0.05 = 18°$ Advertising: $360° \times 0.10 = 36°$

Shrinkage: $360° \times 0.01 = 3.6°$ Depreciation: $360° \times 0.03 = 10.8°$

Materials/supplies: $360° \times 0.01 = 3.6°$

Finally, use a protractor to construct the circle graph as shown in Figure 12.7.

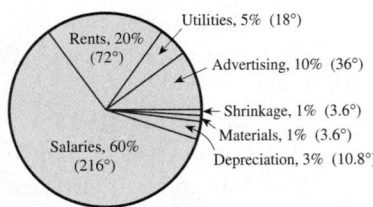

Figure 12.7 **Circle graph showing the expenses for Karlin Enterprises**   ◆

Pictographs

A **pictograph** is a representation of data that uses pictures to show quantity. Consider the raw data shown in Table 12.5, and the bar graph of those data shown in Figure 12.8.

Table 12.5 Marital Status of Persons Age 65 and Older* (in millions)

	Married	Widowed	Divorced	Never Married
Women	5.4	7.2	0.4	0.8
Men	7.5	1.4	0.3	0.6

*Figures are rounded to the nearest 100,000.

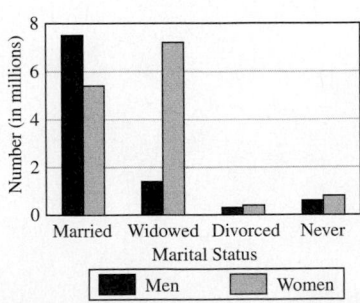

Figure 12.8 **Bar graph showing the marital status of persons age 65 and older**

A pictograph uses a picture to illustrate data; it is normally used only in popular publications, rather than for scientific applications. For these data, suppose that we draw pictures of a woman and a man so that each picture represents 1 million persons, as shown in Figure 12.9.

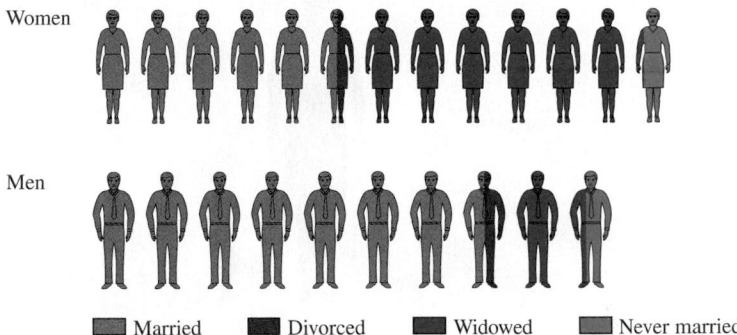

Women

Men

■ Married   ■ Divorced   ■ Widowed   ■ Never married

Figure 12.9 **Pictograph showing the marital status of persons age 65 and older**

Misuses of Graphs

The scales on the axes of either bar or line graphs are frequently chosen to exaggerate or diminish real difference. Even worse, graphs are often presented with no scale whatsoever. For instance, Figure 12.10b shows a graphical "comparison" between Anacin and "regular strength aspirin."

The most misused type of graph is the pictograph. Consider the data from Table 12.5. Such data can be used to determine the height of a three-dimensional object, as in part **a** of Figure 12.11. When an object (such as a person) is viewed as three-dimensional, differences seem much larger than they actually are. Look at

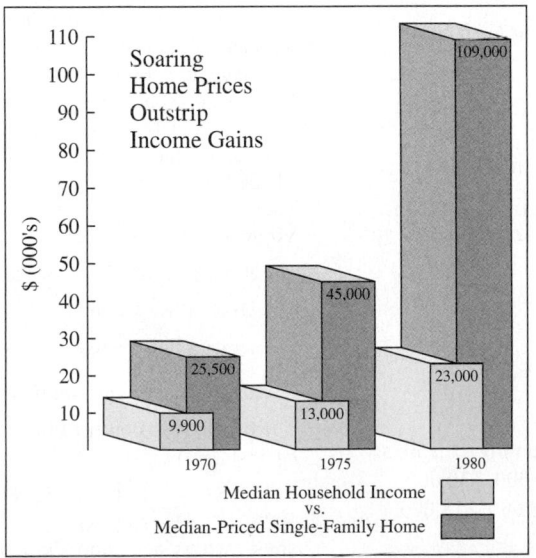

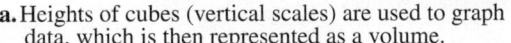

a. Heights of cubes (vertical scales) are used to graph data, which is then represented as a volume.

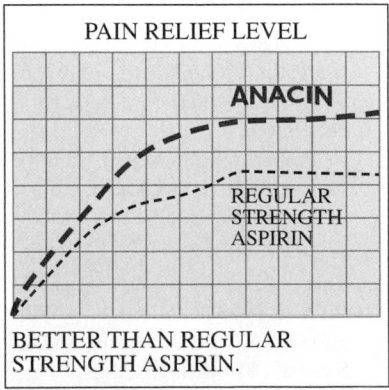

b. No scale is shown in this Anacin advertisement.

Figure 12.10 **Misuse of graphs**

part **b** of Figure 12.11 and notice that, as the height and width are doubled, the volume is actually increased eightfold.

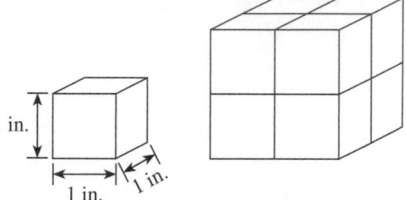

a. Pictograph showing the number of widowed persons of age 65 and older

b. Change in volume as a result of changes in length and width

Figure 12.11 **Examples of misuses in pictographs**

Problem Set 12.1

LEVEL 1

1. **IN YOUR OWN WORDS** Distinguish between a frequency distribution and a stem-and-leaf plot.

2. **IN YOUR OWN WORDS** Write down what you think the following advertisement means:
Nine out of ten dentists recommend Trident for their patients who chew gum.

3. **IN YOUR OWN WORDS** Write down what you think the following advertisement means:
Eight out of ten owners said their cats prefer Whiskas.

4. The heights of 30 students are as follows (rounded to the nearest inch): 66, 68, 65, 70, 67, 67, 68, 64, 64, 66, 64, 70, 72, 71, 69, 64, 63, 70, 71, 63, 68, 67, 67, 65, 69, 65, 67, 66, 69, 69.
   a. Prepare a frequency distribution.
   b. Draw a stem-and-leaf plot.
   c. Draw a bar graph.
   d. Draw a line graph.

5. The wages of employees of a small accounting firm are as follows: $20,000, $25,000, $25,000, $25,000, $30,000, $30,000, $35,000, $50,000, $60,000, $18,000, $18,000, $16,000, $14,000.
   a. Prepare a frequency distribution.
   b. Draw a stem-and-leaf plot.
   c. Draw a bar graph.
   d. Draw a line graph.

Use the following information in Problems 6–8.

The waiting times, in days, for a marriage license in the 50 states are as follows:

Alabama, 0; Alaska, 3; Arizona, 0; Arkansas, 0; California, 0; Colorado, 0; Connecticut, 4; Delaware, 0; Florida, 3; Georgia, 3; Hawaii, 0; Idaho, 0; Illinois, 0; Indiana, 3; Iowa, 3; Kansas, 3; Kentucky, 0; Louisiana, 3; Maine, 3; Maryland, 2; Massachusetts, 3; Michigan, 3; Minnesota, 5; Mississippi, 3; Missouri, 0; Montana, 0; Nebraska, 0; Nevada, 0; New Hampshire, 3; New Jersey, 3; New Mexico, 0; New York, 0; North Carolina, 0; North Dakota, 0; Ohio, 5; Oklahoma, 0; Oregon, 3; Pennsylvania, 3; Rhode Island, 0; South Carolina, 1; South Dakota, 0; Tennessee, 3; Texas, 0; Utah, 0; Vermont, 3; Virginia, 0; Washington, 3; West Virginia, 3; Wisconsin, 5; Wyoming, 0.

6. Prepare a frequency distribution.

7. Draw a bar graph.

8. Draw a line graph.

Use the following information in Problems 9–10.

The purchasing power of the dollar (1983 = $1.00) is (to the nearest cent):

1955	$3.74	1960	$3.41	1965	$3.20
1970	$2.64	1975	$1.92	1980	$1.28
1985	$0.95	1990	$0.79	1995	$0.67
2000	$0.59				

9. Draw a line graph.

10. Draw a bar graph.

Use the information in Table 12.6 in Problems 11–14.

Table 12.6 U.S. Minimum Wage

Year	Minimum wage	Value in 1983 dollars	Value in 2000 dollars
1956	$1.00	$3.44	$6.50
1961	$1.15	$3.60	$6.82
1963	$1.25	$3.82	$7.23
1967	$1.40	$3.93	$7.43
1968	$1.60	$4.32	$8.18
1974	$2.00	$3.92	$7.42
1975	$2.10	$3.81	$7.20
1976	$2.30	$3.94	$7.46
1978	$2.65	$4.01	$7.58
1979	$2.90	$3.99	$7.54
1980	$3.10	$3.83	$7.25
1981	$3.35	$3.79	$7.16
1990	$3.80	$2.96	$5.60
1991	$4.25	$3.18	$6.01
1996	$4.75	$2.85	$5.39
1997	$5.15	$2.75	$5.20
2000	$5.15	$2.72	$5.15

Current minimum wage in 2002 is $5.15 and is established by the Fair Labor Standards Act.

11. Draw a bar graph for the minimum wage using year 2000 dollar values.

12. Draw a line graph of the minimum wage using the 1983 dollar values and on the same set of axes graph the purchasing power of the dollar as given in Problems 9 and 10.

13. In what year will the lines in Problem 12 cross?

14. The actual minimum wage is best represented as a "step graph." To see what this looks like, draw a line graph with the minimum wage remaining constant at $1.00 from 1956 to 1961 with a jump in 1961 to $1.15. Continue in the same fashion until you reach the current minimum wage of $5.15 in 2000.

15. Figure 12.8 shows a bar graph for the information contained in Table 12.5. Draw a line graph to represent these data.

16. Figure 12.12 shows a line graph.

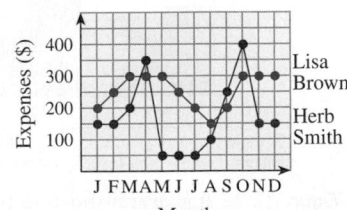

Figure 12.12 Expenses for two salespeople of the Leadwell Pencil Company

a. During which month did Herb incur the most expenses?

b. During which month did Lisa incur the least expenses?

17. Figure 12.13 shows a line graph.

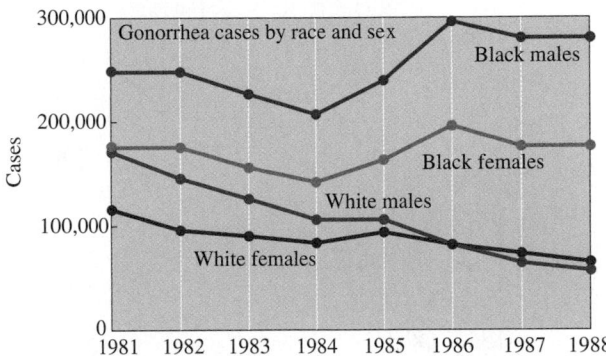

Figure 12.13 Gonorrhea cases by race and by sex

a. What are the categories being graphed?

b. In what year were the number of cases in two of the categories the same?

c. During which period was the number of cases increasing for black females?

d. How many cases of gonorrhea were there among white males in 1985?

18. In Section 12.4 we will discuss scatter diagrams, which are used to study the relationship between two variables, such as student final exam score and final grade in Smith's Math 10 class (see Figure 12.14).

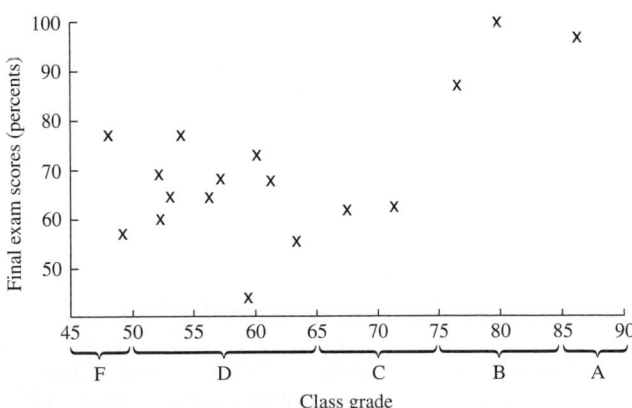

Figure 12.14 Grades and final exam scores

Approximately what is the lowest final exam score received by a student awarded a final grade of C or better?

A. 50%

B. 55%

C. 60%

D. 65%

19. The scatter diagram in Figure 12.15 relates the incomes from crops and livestock for selected states in the United States.

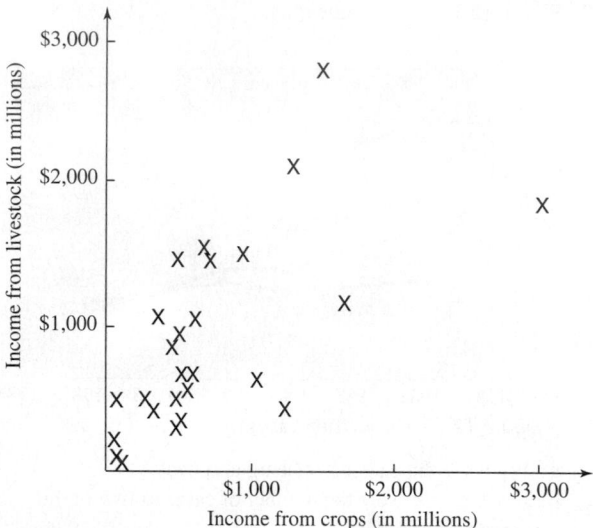

Figure 12.15 **Scatter diagram of state income between crops and livestock**

For states with income from crops greater than a billion dollars, the number of states with income from livestock greater than a billion dollars is about:

A. 16   B. 5   C. 10   D. 22

Use the line graph in Figure 12.16 *to answer the questions in Problems 20–25.*

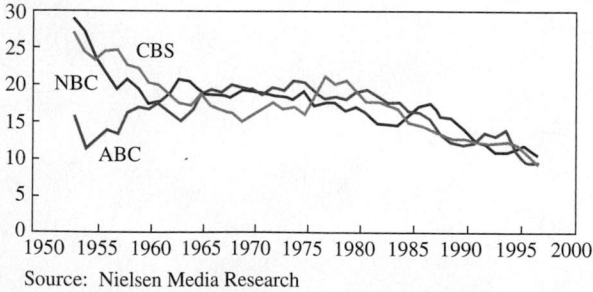

Source: Nielsen Media Research

Figure 12.16 **Major TV network ratings**

20. How many networks are tracked? Can you think of any other networks that might have been included? If so, why do you think they have not been included?

21. In which year(s) did all three tracked networks have the same percent share?

22. Which network had the best market share in 1997?

23. What is the maximum rating share by any of the tracked networks? What was the percent share and what was the network?

24. What is the minimum rating share by any of the tracked networks? What was the percent share and what was the network?

25. By looking at the line graph, form a conclusion.

Use the bar graph in Figure 12.17 *to answer the questions in Problems 26–29.*

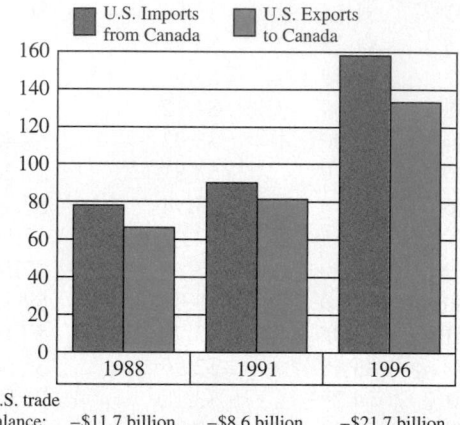

Figure 12.17 **Bar graph showing U.S. trade with Canada**

26. What is being illustrated in this graph?

27. Did the United States have more imports or exports for the years 1988–1996?

28. What was the approximate balance of trade with Canada in 1991?

29. What was the approximate dollar amount of imports from Canada in 1991?

Use the bar graph in Figure 12.18 *to answer the questions in Problems 30–33.*

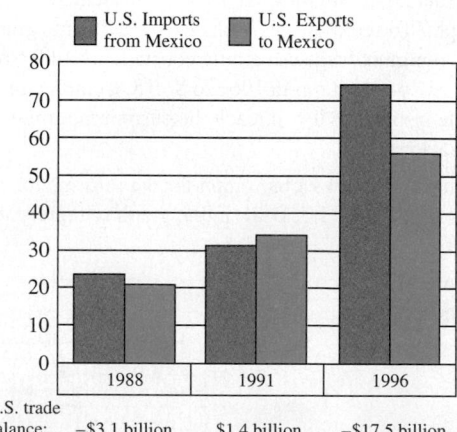

Figure 12.18 **Bar graph showing U.S. trade with Mexico**

30. What is being illustrated in this graph?

31. Did the United States have more imports or exports for the years 1988–1996?

32. What was the approximate balance of trade with Mexico in 1991?

33. What was the approximate dollar amount of exports to Mexico in 1996?

LEVEL 2

When a person in California renews the registration for an automobile, the bar graph shown in Figure 12.19 is included with the bill. Use this bar graph to answer the questions in Problems 34–39. The following statement is included with the graph:

> *There is no safe way to drive after drinking. These charts show that a few drinks can make you an unsafe driver. They show that drinking affects your BLOOD ALCOHOL CONCENTRATION (BAC). The BAC zones for various numbers of drinks and time periods are printed in white, gray, and red. HOW TO USE THESE CHARTS: First, find the chart that includes your weight. For example, if you weigh 160 lbs., use the "150 to 169" chart. Then look under "Total Drinks" at the "2" on this "150 to 169" chart. Now look below the "2" drinks, in the row for 1 hour. You'll see your BAC is in the gray shaded zone. This means that if you drive after 2 drinks in 1 hour, you could be arrested. In the gray zone, your chances of having an accident are 5 times higher than if you had no drinks. But if you had 4 drinks in 1 hour, your BAC would be in the red shaded area . . . and your chances of having an accident 25 times higher.*

BAC Zones:	90 to 109 lbs.								110 to 129 lbs.								130 to 149 lbs.								150 to 169 lbs.							
TIME FROM 1st DRINK	TOTAL DRINKS								TOTAL DRINKS								TOTAL DRINKS								TOTAL DRINKS							
	1	2	3	4	5	6	7	8	1	2	3	4	5	6	7	8	1	2	3	4	5	6	7	8	1	2	3	4	5	6	7	8
1 hr																																
2 hrs																																
3 hrs																																
4 hrs																																

BAC Zones:	170 to 189 lbs.								190 to 209 lbs.								210 to 229 lbs.								230 lbs. & up							
TIME FROM 1st DRINK	TOTAL DRINKS								TOTAL DRINKS								TOTAL DRINKS								TOTAL DRINKS							
	1	2	3	4	5	6	7	8	1	2	3	4	5	6	7	8	1	2	3	4	5	6	7	8	1	2	3	4	5	6	7	8
1 hr																																
2 hrs																																
3 hrs																																
4 hrs																																

SHADINGS IN THE CHARTS ABOVE MEAN:

☐ (.01% – .04%) Seldom illegal   ▓ (.05% – .09%) May be illegal   ■ (.10% Up) Definitely illegal

▓ (.05% – .09%) Illegal if under 18 yrs. old

Prepared by the Department of Motor Vehicles in cooperation with the California Highway Patrol.

Figure 12.19 **Blood Alcohol Concentration (BAC) charts**

34. Suppose that you weigh 115 pounds and that you have two drinks in 2 hours. If you then drive, how much more likely are you to have an accident than if you had refrained from drinking?

35. Suppose that you weigh 115 pounds and that you have four drinks in 3 hours. If you then drive, how much more likely are you to have an accident than if you had refrained from drinking?

36. Suppose that you weigh 195 pounds and have two drinks in 2 hours. According to Figure 12.19, are you seldom illegal, maybe illegal, or definitely illegal?

37. Suppose that you weigh 195 pounds and that you have four drinks in 3 hours. According to Figure 12.19, are you seldom illegal, maybe illegal, or definitely illegal?

38. If you weigh 135 pounds, how many drinks in 3 hours would you need to be definitely illegal?

39. If you weigh 185 pounds and drink a six-pack of beer during a 3-hour baseball game, can you legally drive home?

40. The financial report from a small college shows the following income:

Tuition	$1,388,000
Development	90,000
Fund-raising	119,000
Athletic Dept.	44,000
Interest	75,000
Tax subsidy	29,000
TOTAL	$1,745,000

Draw a circle graph showing this income.

41. The financial report from a small college shows the following expenses:

Salaries	$1,400,000
Academic	$68,000
Plant operations	$196,000
Student activities	$31,000
Athletic programs	$50,000
TOTAL:	$1,745,000

Draw a circle graph showing this income.

42. The amount of electricity used in a typical all-electric home is shown in the circle graph in Figure 12.20.

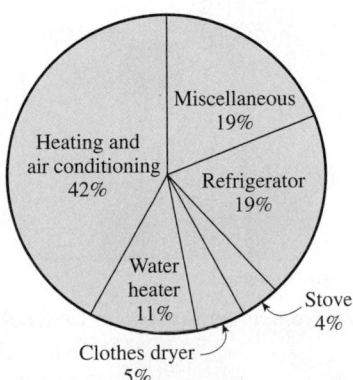

Figure 12.20 **Home electricity usage**

If, in a certain month, a home used 1,100 kwh (kilowatt-hours), find the amounts of electricity used from the graph.
a. The amount of electricity used by the water heater
b. The amount of electricity used by the stove
c. The amount of electricity used by the refrigerator
d. The amount of electricity used by the clothes dryer

43. The national debt is of growing concern in the United States. It reached $1 billion in 1916 during World War I, and climbed to $278 billion by the end of World War II. It reached its first trillion on October 1, 1981, and rose to $2 trillion on April 3, 1986. The third trillion milestone was reached on April 4, 1990. However, the magnitude of our national debt became headline news when it reached $4 trillion and became a campaign slogan for Ross Perot in 1992. On July 1, 1999, the national debt was just over 5.6 trillion, and on January 1, 2003, it was 6.3 trillion. Represent this information on a line graph.

44. **Top women on Wall Street** Draw a pictograph to represent the following data on women in big Wall Street firms. In 1996, women comprised approximately 11% of the top-tier executives.

Regina Dolan is the chief financial officer of Paine Webber. She is one of 46 women among 465 managing directors.

Robin Neustein is the chief of staff at Goldman Sachs. She is one of 9 women among 173 managing directors.

Theresa Lang is the company treasurer for Merrill Lynch. She is one of 76 women among 694 managing directors.

What is wrong, if anything, with each of the statements in Problems 45–49? *Explain your reasoning.*

45. Consider the graph shown in Figure 12.21. Clearly, Anacin is better.

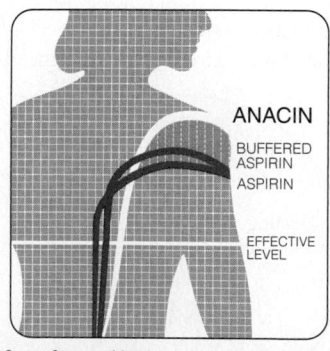

Figure 12.21 From an advertisement for a pain reliever

46. Consider the graph shown in Figure 12.22. The potential commercial forest growth as compared with the current commercial forest growth, is almost double (44.9 to 74.2).

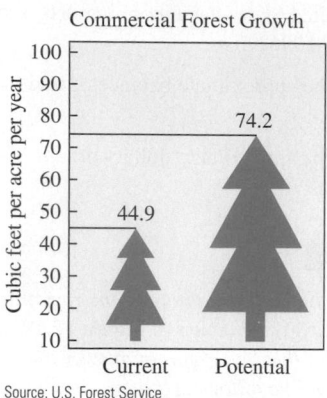

Figure 12.22 From the U.S. Forest Service

47. Consider the graph shown in Figure 12.23.

How much is "in the bank"?
Conflicting estimates of undiscovered oil and gas reserves—which can be recovered and produced.

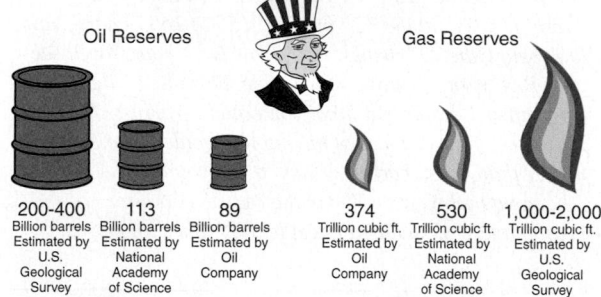

Figure 12.23 Pictograph showing conflicting estimates of oil and gas reserves

48. The following graph was circulated throughout the mathematics department of a leading college.

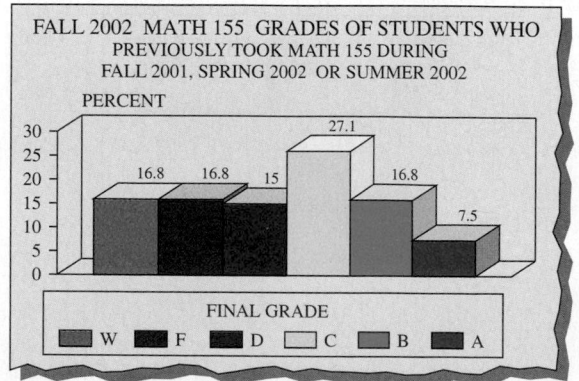

49. Consider the Saab advertisement.

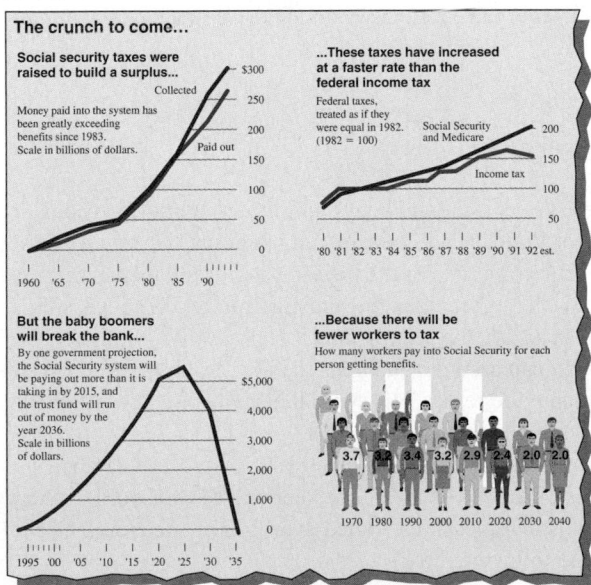

There are two cars built in Sweden.
Before you buy theirs, drive ours.

When people who know cars think about Swedish cars, they think of them as being strong and durable. And conquering some of the toughest driving conditions in the world.

But, unfortunately, when most people think about buying a Swedish car, the one they think about usually isn't ours. (Even though ours doesn't cost any more.)

Ours is the SAAB 99E. It's strong and durable. But it's a lot different from their car.

Our car has Front-Wheel Drive for better traction, stability and handling.

It has a 1.85 liter, fuel-injected, 4 cylinder, overhead cam engine as standard in every car. 4-speed transmission is standard too. Or you can get a 3-speed automatic (optional).

Our car has four-wheel disc brakes and dual-diagonal braking system so you can stop straight and fast every time.

It has a wide stance. (About 55 inches.) So it rides and handles like a sports car.

Outside, our car is smaller than a lot of "small" cars. 172" overall length, 57" overall width.

Inside, our car has bucket seats up front and a full five feet across in the back so you can easily accommodate five adults.

It has more headroom than a Rolls Royce and more room from the brake pedal to the back seat than a Mercedes 280. And it has factory air conditioning as an option.

There are a lot of other things that make our car different from their car. Like roll cage construction and a special "hot seat" for cold winter days.

So before you buy their car, stop by your nearest SAAB dealer and drive our car. The SAAB 99E. We think you'll buy it instead of theirs.

SAAB 99E

Source: © 1971 by SAAB-Scania of America, Inc.

A newspaper article discussing whether Social Security could be cut offered information in Figure 12.24. Use the information in these graphs to answer the questions in Problems 50–56.

The crunch to come...

Social security taxes were raised to build a surplus...

Money paid into the system has been greatly exceeding benefits since 1983. Scale in billions of dollars.

...These taxes have increased at a faster rate than the federal income tax

Federal taxes, treated as if they were equal in 1982. (1982 = 100)

But the baby boomers will break the bank...

By one government projection, the Social Security system will be paying out more than it is taking in by 2015, and the trust fund will run out of money by the year 2036. Scale in billions of dollars.

...Because there will be fewer workers to tax

How many workers pay into Social Security for each person getting benefits.

Figure 12.24 **Social Security Reform**

50. What is the difference between the amount of money collected from Social Security taxes and the amount paid out in 1992?

51. How many workers pay into Social Security for each person getting benefits in 1990?

52. What is a "baby boomer"?

53. When will the baby boomers break the bank?

54. How much money is projected for the Social Security system for the year 2000?

55. There seems to be an error in one of the three line graphs. What is it?

56. There seems to be an error in the pictograph. What is it?

The amount of time it takes for three leading pain relievers to reach your bloodstream is as follows: Brand A, 480 seconds; Brand B, 500 seconds; Brand C, 490 seconds. Use this information for Problems 57–59.

57. Draw a bar graph using the scale shown in Figure 12.25a.

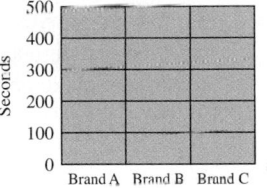
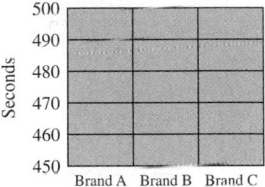

a. Scale from 0 to 500 seconds   **b.** Scale from 450 to 500 seconds

Figure 12.25 **Time for pain reliever to reach your bloodstream**

58. Draw a bar graph using the scale shown in Figure 12.25b.

59. If you were an advertiser working on a promotion campaign for Brand A, which graph from Figure 12.25 would seem to give your product a more distinct advantage?

60. The following advertisements (adapted from "Caveat Emptor" by Robert Leighton, *Games,* January 1988, p.38) show some vastly overpriced items. See whether you can read between the lines and identify the item each ad is describing.
   a. Miniature, sparkling white crystals harvested in exotic Hawaii make perfect specimens for studying the wonders of science! Each educationally decorated packet contains hundreds of refined crystals. Completely safe and non-toxic. Only **$3.95.**
   b. It's like having U.S. history in your pocket! Miniature portraits of America's best-loved presidents, Washington sites, and more. Richly detailed copper and silver engraving. Set of four, just **$4.50.**
   c. Today multimillion-dollar computers need special programs to generate random numbers. But this GEOMETRICALLY DESIGNED device ingeniously bypasses computer technology to provide random numbers quickly and easily! Exciting? You bet! Batteries not required. Only **$4.00.**

d. They said we couldn't do it, but we are! For a limited time, our LEADING HOLLYWOOD PROP HOUSE is dividing up and selling—yes, SELLING—parcels of this famous movie setting. You've seen it featured in *Lawrence of Arabia*, *Beach Blanket Bingo*, and *Ishtar!*

Our warehouse must be cleared! Own a vial of movie history—a great conversation piece! Only **$4.50.**

e. As you command water—or any liquid—to flow upward! Could you be tampering with the secrets of the Earth? Your friends will be amazed! Just **$31.99.**

PEANUTS Reprinted by permission of UFS, Inc.

Baseball, as everyone knows, is played on a field with a bat and a ball. But baseball is also played on paper and computers with numbers and decimals. The name of this game-within-a-game is statistics, and to some fans, it is more engrossing and more real than the action on the field. Most sports keep statistics, but baseball statistics are in a league by themselves.

12.2 DESCRIPTIVE STATISTICS

Descriptive statistics is concerned with the accumulation of data, measures of central tendency, and dispersion.

Measures of Central Tendency

In Section 12.1, we organized data into a frequency distribution and then discussed their presentation in graphical form. However, some properties of data can help us interpret masses of information. We will use the following *Peanuts* cartoon to introduce the notion of *average*.

Do you suppose that Violet's dad bowled better on Monday nights (185 avg) than on Thursday nights (170 avg)? Don't be too hasty to say "yes" before you look at the scores that make up these averages:

	Monday night	*Thursday night*
Game 1	175	180
Game 2	150	130
Game 3	160	161
Game 4	180	185
Game 5	160	163
Game 6	183	185
Game 7	287	186
Totals	1,295	1,190

To find the averages used by Violet in the cartoon, we divide these totals by the number of games:

Monday night   *Thursday night*

$$\frac{1,295}{7} = 185 \qquad \frac{1,190}{7} = 170$$

If we consider the averages, Violet's dad did better on Mondays; but if we consider the games separately, we see that Violet's dad typically did better on Thursday (five out of seven games). Would any other properties of the bowling scores tell us this fact?

Since we must often add up a list of numbers in statistics, as we did above, we use the symbol Σx to mean *the sum of all the values that x can assume*. Similarly, Σx^2 means to square each value that x can assume, and then add the results; $(\Sigma x)^2$ means to first add the values and then square the result. The symbol Σ is the Greek capital letter sigma (which is chosen because S reminds us of "sum").

The average used by Violet is only one kind of statistical measure that can be used. It is the measure that most of us think of when we hear someone use the word *average*. It is called the *mean*. Other statistical measures, called **averages** or **measures of central tendency,** are defined in the following box.

Measures of Central Tendency: Mean, Median, Mode

STOP

These ideas are essential for understanding many concepts.

*The **mean** is the most sensitive average. It reflects the entire distribution and is the most common average.*

*The **median** gives the middle value. It is useful when there are a few extraordinary values to distort the mean.*

*The **mode** is the average that measures "popularity." It is possible to have no mode or more than one mode.*

1. **Mean.** The number found by adding the data and then dividing by the number of data values. The mean is usually denoted by $\bar{x}$:

$$\bar{x} = \frac{\Sigma x}{n}$$

2. **Median.** The middle number when the numbers in the data values are arranged in order of size. If there are two middle numbers (in the case of an even number of data), the median is the mean of these two middle numbers.

3. **Mode.** The value that occurs most frequently. If no number occurs more than once, there is no mode. It is possible to have more than one mode.

Violet used the mean and called it the average. Let us consider other measures of central tendency for Violet's dad's bowling scores.

Median Rearrange the data values from smallest to largest when finding the median:

Monday night		Thursday night
150		130
160		161
160		163
175	← **Middle number** →	**180**
180	**is the median.**	185
183		185
287		186

Mode Look for the number that occurs most frequently:

Monday night		Thursday night	
150		130	
160 ⎱ **Most frequent**		161	
160 ⎰ **is the mode.**		163	
175		180	
180		**185** ⎱ **Most frequent**	
183		**185** ⎰ **is the mode.**	
287		186	

If we compare the three measures of central tendency for the bowling scores, we find the following:

	Monday night	Thursday night
Mean	**185**	170
Median	175	**180**
Mode	160	**185**

We are no longer convinced that Violet's dad did better on Monday nights than on Thursday nights. (See boldface for the winning night, according to each measure of central tendency—Thursday wins two out of three.)

EXAMPLE 1

Many calculators have built-in statistical functions. The method of inputting data varies from brand to brand, and you should check with your owner's manual. Once the data are entered, your calculator will find the mean, median, and sometimes the mode by pressing a single statistical button.

Find the mean, median, and mode for the following sets of numbers.

a. 3, 5, 5, 8, 9   **b.** 4, 10, 9, 8, 9, 4, 5   **c.** 6, 5, 4, 7, 1, 9

Solution

a. *Mean:* $\dfrac{\text{SUM OF TERMS}}{\text{NUMBER OF TERMS}} = \dfrac{3 + 5 + 5 + 8 + 9}{5} = \dfrac{30}{5} = 6$

Median: Arrange in order: 3, 5, 5, 8, 9. The middle term is the median: 5.

Mode: The most frequently occurring term is the mode: 5.

b. *Mean:* $\dfrac{4 + 10 + 9 + 8 + 9 + 4 + 5}{7} = \dfrac{49}{7} = 7$

Median: 4, 4, 5, 8, 9, 9, 10; the median is 8.

Mode: The data have two modes; 4 and 9. If data have two modes, we say they are **bimodal.**

c. *Mean:* $\dfrac{6 + 5 + 4 + 7 + 1 + 9}{6} = \dfrac{32}{6} \approx 5.33$

Median: 1, 4, 5, 6, 7, 9; the median is $\dfrac{11}{2} = 5.5$.

$\dfrac{5 + 6}{2} = \dfrac{11}{2}$

Mode: There is **no mode** because no term appears more than once.   ◆

A rather nice physical model illustrates the idea of the mean. Consider a seesaw that consists of a plank and a movable support (called a *fulcrum*). We assume that the plank has no weight and is marked off into units as shown in Figure 12.26.

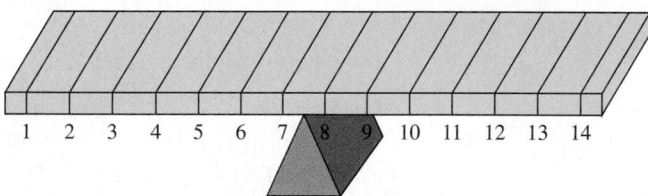

Figure 12.26 **Fulcrum and plank model for mean**

Now let's place some 1-lb weights in the positions of the numbers in some given distribution. The balance point for the plank is the mean. For example, consider the data from part **a** of Example 1: 3, 5, 5, 8, 9. If weights are placed on these locations on the plank, the balance point is 6, as shown in Figure 12.27.

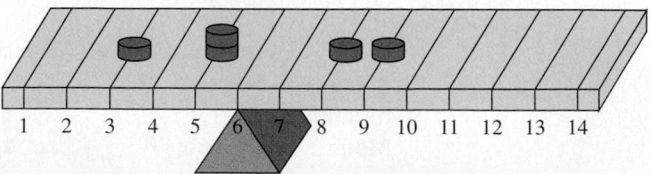

Figure 12.27 **Balance point for the data set 3, 5, 5, 8, 9 is 6.**

EXAMPLE 2

The graph below represents the distribution of students in Clark's math class. Which of the statements is true?

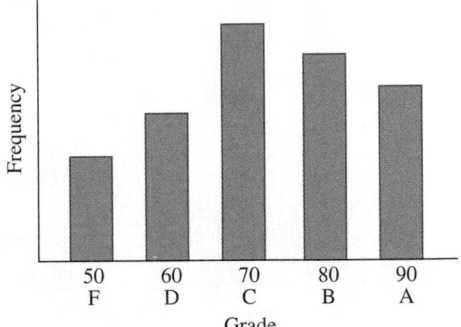

Notice that you are given data values, but are not given the corresponding frequencies or relative frequencies. Therefore, we are not able to compute the values of the mean and median.

A. The mean is the same as the mode.
B. The mean is greater than the mode.
C. The median is less than the mode.
D. The mean is less than the mode.

Solution   The only measure of central tendency we can determine exactly is the mode: *The mode is the value under the tallest bar.* We see the mode is 70. We now estimate the mean. In finding the mean, the number 70 would be added most frequently. If the value of 60 occurred the same number of times as 80, they would have a mean of 70, but we see there are more 80's than there are 60's, so the mean must be greater than 70. Similarly, we see that adding the 50's and 90's would give us a mean of more than 70. Thus, the mean is greater than 70, and the correct answer is B.   ◆

EXAMPLE 3

Consider the number of days one must wait for a marriage license in the various states in the United States.

Days' wait	Frequency
0	25
1	1
2	1
3	19
4	1
5	3
Total	50

What are the mean, the median, and the mode for these data?

Solution

Mean: To find the mean, we could, of course, add all 50 individual numbers, but instead, notice that

0 occurs 25 times, so write	0×25
1 occurs 1 time, so write	1×1
2 occurs 1 time, so write	2×1
3 occurs 19 times, so write	3×19
4 occurs 1 time, so write	4×1
5 occurs 3 times, so write	5×3

Thus, the mean is

$$\bar{x} = \frac{0 \times 25 + 1 \times 1 + 2 \times 1 + 3 \times 19 + 4 \times 1 + 5 \times 3}{50} = \frac{79}{50} = 1.58$$

Median: Since the median is the middle number, and since there are 50 values, the median is the mean of the 25th and 26th numbers (when they are arranged in order):

$$\left.\begin{array}{l} \text{25th term is } 0 \\ \text{26th term is } 1 \end{array}\right\} \quad \frac{0 + 1}{2} = \frac{1}{2}$$

Mode: The mode is the value that occurs most frequently, which is 0.   ◆

When finding the mean from a frequency distribution, you are finding what is called a *weighted mean.*

Weighted Mean

> If a list of scores $x_1, x_2, x_3, \ldots, x_n$ occurs $w_1, w_2, \ldots w_n$ times, respectively, then the **weighted mean** is
>
> $$\bar{x} = \frac{\Sigma(w \cdot x)}{\Sigma w}$$

Measures of Position

The median divides the data into two equal parts, with half the values above the median and half below the median, so the median is called a **measure of position.** Sometimes we use benchmark positions that divide the data into more than two parts. **Quartiles,** denoted by Q_1 (first quartile), Q_2 (second quartile), and Q_3 (third quartile), divide the data into four equal parts. **Deciles** are nine values that divide the data into ten equal parts, and **percentiles** are 99 values that divide the data into 100 equal parts. For example, when you take the Scholastic Assessment Test (SAT), your score is recorded as a percentile score. If you scored in the 92nd percentile, it means that you scored better than approximately 92% of those who took the test.

Measures of Dispersion

The measures we've been discussing can help us interpret information, but they do not give the entire story. For example, consider these sets of data:

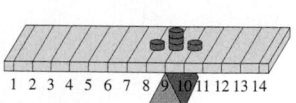

a. $A = \{8, 9, 9, 9, 10\}$

Set *A:* $\{8, 9, 9, 9, 10\}$ *Mean:* $\dfrac{8 + 9 + 9 + 9 + 10}{5} = 9$

Median: 9

Mode: 9

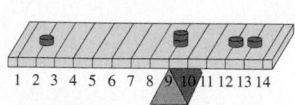

b. $B = \{2, 9, 9, 12, 13\}$

Figure 12.28 **Visualization of dispersion of sets of data**

Set *B:* $\{2, 9, 9, 12, 13\}$ *Mean:* $\dfrac{2 + 9 + 9 + 12 + 13}{5} = 9$

Median: 9

Mode: 9

Notice that, for sets *A* and *B,* the measures of central tendency do not distinguish the data. However, if you look at the data placed on planks, as shown in Figure 12.28, you will see that the data in Set *B* are relatively widely dispersed along the plank, whereas the data in Set *A* are clumped around the mean.

We'll consider three **measures of dispersion:** the *range,* the *standard deviation,* and the *variance.*

Range

> The **range** of a set of data is the difference between the largest and the smallest numbers in the set.

EXAMPLE 4

Find the ranges for the following data sets:

a. Set $A = \{8, 9, 9, 9, 10\}$ **b.** Set $B = \{2, 9, 9, 12, 13\}$

Solution Notice from Figure 12.28 that the mean for each of these sets of data is the same. The range is found by comparing the difference between the largest and smallest values in the set.

a. $10 - 8 = 2$ **b.** $13 - 2 = 11$ ◆

Notice that the range is determined by only the largest and the smallest numbers in the set; it does not give us any information about the other numbers.

The range is used, along with quartiles, to construct a statistical tool called a *box plot*. For a given set of data, a **box plot** consists of a rectangular box positioned above a numerical scale, drawn from Q_1 (the first quartile) to Q_3 (the third quartile). The median (Q_2, or second quartile) is shown as a dashed line, and a segment is extended to the left to show the distance to the minimum value; another segment is extended to the right for the maximum value. We illustrate with an example.

EXAMPLE 5

Draw a box plot for the ages of the actors who received the best actor award at the Academy Awards, as reported in Table 12.3 on page 609.

Solution We find the quartiles by first finding Q_2, the median. Since there are 76 entries, the median is the middle item. For this example, the middle is the mean of the 38th and 39th items:

$$Q_2 = \frac{37 + 42}{2} = 39.5$$

Next, we find Q_1 (first quartile), which is the median of all items below Q_2 (the mean of the 19th and 20th items):

$$Q_1 = \frac{38 + 38}{2} = 38$$

Finally, find Q_3 (third quartile), which is the median of all items above Q_2:

$$Q_3 = \frac{48 + 48}{2} = 48$$

The minimum age is 29 (Adrien Brody), and the maximum age is 76 (Henry Fonda), so we have a box plot as shown in Figure 12.29.

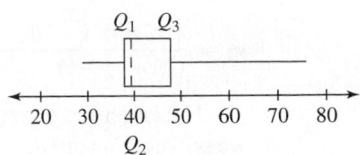

Figure 12.29 **Box plot**   ◆

Sometimes a box plot is called a *box-and-whisker plot.* Its usefulness should be clear when you look at Figure 12.29. It shows:

1. the median (a measure of central tendency);

2. the location of the middle half of the data (represented by the extent of the box);

3. the range (a measure of dispersion);

4. the skewness (the nonsymmetry of both the box and the whiskers).

The *variance* and *standard deviation* are measures that use all the numbers in the data set to give information about the dispersion. When finding the variance, we must make a distinction between the **variance of the entire population** and the **variance of a random sample** from the population. When the variance is based on a set of sample scores, it is denoted by s^2; and when it is based on all scores in a population, it is denoted by σ^2 (σ is the lowercase Greek letter sigma). We will discuss populations and samples in Section 12.5. The variance for a random sample is found by

$$s^2 = \frac{\Sigma(x - \bar{x})^2}{n - 1}$$

To understand this formula for the sample variance, we will consider an example before summarizing a procedure. Again, let's use the data sets we worked with in Example 4.

Set $A = \{8, 9, 9, 9, 10\}$ Set $B = \{2, 9, 9, 12, 13\}$
Mean is 9.   Mean is 9.

Find the deviations by subtracting the mean from each term:

$8 - 9 = -1$	$2 - 9 = -7$
$9 - 9 = 0$	$9 - 9 = 0$
$9 - 9 = 0$	$9 - 9 = 0$
$9 - 9 = 0$	$12 - 9 = 3$
$10 - 9 = 1$	$13 - 9 = 4$
↑	↑
mean	mean

If we sum these deviations (to obtain a measure of the total deviation), in each case we obtain 0, because the positive and negative differences "cancel each other out." Next we calculate the *square of each of these deviations:*

Set $A = \{8, 9, 9, 9, 10\}$ Set $B = \{2, 9, 9, 12, 13\}$

$(8 - 9)^2 = (-1)^2 = 1$	$(2 - 9)^2 = (-7)^2 = 49$
$(9 - 9)^2 = 0^2 = 0$	$(9 - 9)^2 = 0^2 = 0$
$(9 - 9)^2 = 0^2 = 0$	$(9 - 9)^2 = 0^2 = 0$
$(9 - 9)^2 = 0^2 = 0$	$(12 - 9)^2 = (3)^2 = 9$
$(10 - 9)^2 = (1)^2 = 1$	$(13 - 9)^2 = (4)^2 = 16$

Finally, we find the sum of these squares and divide by one less than the number of items to obtain the variance:

Set A: Set B:

$$s^2 = \frac{1 + 0 + 0 + 0 + 1}{5 - 1} = \frac{2}{4} = 0.5 \qquad s^2 = \frac{49 + 0 + 0 + 9 + 16}{5 - 1} = \frac{74}{4} = 18.5$$

The larger the variance, the more dispersion there is in the original data. However, we will continue to develop a true picture of the dispersion. Since we squared each difference (to eliminate the effect of positive and negative differences), it seems reasonable that we should find the square root of the variance as a more meaningful measure of dispersion. This number, called the **standard deviation,** is denoted by the lowercase Greek letter sigma (σ) when it is based on a population and by s when it is based on a sample. You will need a calculator to find square roots. For the data sets of Example 4:

$$\text{Set } A: \quad s = \sqrt{0.5} \qquad\qquad \text{Set } B: \quad s = \sqrt{18.5}$$
$$\approx 0.707 \qquad\qquad\qquad\qquad \approx 4.301$$

We summarize these steps in the following box.

Standard Deviation

Spend some time with this box; make sure you understand not only the procedure, but also the concept.

The standard deviation of a sample, denoted by s, is the square root of the variance. To find it, carry out these steps:

1. Determine the mean of the set of numbers.
2. Subtract the mean from each number in the set.
3. Square each of these differences.
4. Find the sum of the squares of the differences.
5. Divide this sum by one less than the number of pieces of data. This is the *variance* of the sample.
6. Take the square root of the variance. This is the *standard deviation* of the sample.

EXAMPLE 6

Suppose that Missy received the following test scores in a math class: 92, 85, 65, 89, 96, and 71. Find s, the standard deviation, for her test scores.

Solution   First we calculate the mean, as in

Step 1: $\bar{x} = \dfrac{92 + 85 + 65 + 89 + 96 + 71}{6} = 83$ This is the mean.

We summarize steps 2 through 4 by using a table format:

Score	(Deviation from the mean)^2
92	$(92 - 83)^2 = 9^2 = 81$
85	$(85 - 83)^2 = 2^2 = 4$
65	$(65 - 83)^2 = (-18)^2 = 324$
89	$(89 - 83)^2 = 6^2 = 36$
96	$(96 - 83)^2 = 13^2 = 169$
71	$(71 - 83)^2 = (-12)^2 = 144$

Step 5: Divide the sum by 5 (one less than the number of scores):

$$\frac{81 + 4 + 324 + 36 + 169 + 144}{6 - 1} = \frac{758}{5} = 151.6$$

We note that this number, 151.6, is called the variance. If you do not have access to a calculator, you can use the variance as a measure of dispersion. However, we assume you have a calculator and can find the standard deviation.

Step 6: $s = \sqrt{\dfrac{758}{5}} \approx 12.31$ ◆

How can we use the standard deviation? We will begin the discussion in Section 12.3 with this question. For now, however, we will give one example. Suppose that Missy obtained 65 on an examination for which the mean was 50 and the standard deviation was 15, whereas Shannon in another class scored 74 on an examination for which the mean was 80 and the standard deviation was 3. Did Shannon or Missy do better in her respective class? We see that Missy scored one standard deviation *above* the mean ($50 + 15 = 65$), whereas Shannon scored two standard deviations *below* the mean ($80 - 2 \times 3 = 74$); therefore Missy did better compared to her classmates than did Shannon.

Problem Set 12.2

LEVEL 1

1. IN YOUR OWN WORDS What do we mean by average?

2. IN YOUR OWN WORDS What is a measure of central tendency?

3. IN YOUR OWN WORDS What is a measure of dispersion?

4. IN YOUR OWN WORDS Compare and contrast mean, median, and mode.

5. IN YOUR OWN WORDS In 1989, Andy Van Slyke's batting average was better than Dave Justice's; and in 1990 Andy once again beat Dave. Does it follow that Andy's combined 1989–1990 batting average is better than Dave's?

	Andy			Dave		
	Hits	AB	Avg.	Hits	AB	Avg.
1989	113	476	0.237	12	51	0.235
1990	140	493	0.284	124	439	0.282

6. IN YOUR OWN WORDS Is the standard deviation always smaller than the variance?

7. IN YOUR OWN WORDS Suppose that a variance is zero. What can you say about the data?

8. IN YOUR OWN WORDS A professor gives five exams. Two students' scores have the same mean, although one student did better on all the tests except one. Give an example of such scores.

9. IN YOUR OWN WORDS A professor gives six exams. Two students' scores have the same mean, although one student's scores have a small standard deviation and the other student's scores have a large standard deviation. Give an example of such scores.

10. IN YOUR OWN WORDS Comment on the following quotation, which was printed in the May 28, 1989, issue of *The Community, Technical, and Junior College TIMES:* "Half the American students are below average."

11. IN YOUR OWN WORDS Comment on the following quotation: "I just received my SAT scores and I scored in the 78th percentile in English, and 50th percentile in math."

In Problems 12–21, find the three measures of central tendency (the mean, median, and mode).

12. 1, 2, 3, 4, 5   **13.** 17, 18, 19, 20, 21

14. 103, 104, 105, 106, 107   **15.** 765, 766, 767, 768, 769

16. 3, 5, 8, 13, 21   **17.** 1, 4, 9, 16, 25

18. 79, 90, 95, 95, 96   **19.** 70, 81, 95, 79, 85

20. 1, 2, 3, 3, 3, 4, 5   **21.** 0, 1, 1, 2, 3, 4, 16, 21

In Problems 22–31, find the range and the standard deviation (correct to two decimal places). If you do not have a calculator, find the range and the variance.

22. 1, 2, 3, 4, 5   **23.** 17, 18, 19, 20, 21

24. 103, 104, 105, 106, 107   **25.** 765, 766, 767, 768, 769

26. 3, 5, 8, 13, 21   **27.** 1, 4, 9, 16, 25

28. 79, 90, 95, 95, 96   **29.** 70, 81, 95, 79, 85

30. 1, 2, 3, 3, 3, 4, 5   **31.** 0, 1, 1, 2, 3, 4, 16, 21

32. By looking at Problems 12–15 and 22–25, and discovering a pattern, answer the following questions about the numbers 217,849, 217,850, 217,851, 217,852, and 217,853.
  a. What is the mean?
  b. What is the standard deviation?

33. The 2001 monthly cost for health care for an employee and two dependents is given in the following table.

Provider	Cost	Provider	Cost
Maxicare	$415.24	Kaiser	$433.50
Cigna	$424.77	Aetna U.S. Healthcare	$436.11
Health Net	$427.48	Blue Shield HMO	$442.28
Pacific Care	$428.05	Omni Healthcare	$457.86
Health Plan of the Redwoods	$431.52	Lifeguard	$457.94

Draw a box plot for the monthly costs of health care (ranked from lowest to highest).

34. Find the mean, the median, and the mode of the following salaries of employees of the Moe D. Lawn Landscaping Company:

Salary	Frequency
$25,000	4
28,000	3
30,000	2
45,000	1

35. G. Thumb, the leading salesperson for the Moe D. Lawn Landscaping Company, turned in the following summary of sales for the week of October 23–28:

Date	Number of clients contacted by G. Thumb	Date	Number of clients contacted by G. Thumb
Oct. 23	12	Oct. 26	16
Oct. 24	9	Oct. 27	10
Oct. 25	10	Oct. 28	21

Find the mean, median, and mode.

36. Find the mean, the median, and the mode of the following scores:

Test score	Frequency
90	1
80	3
70	10
60	5
50	2

37. A class obtained the following scores on a test:

Test score	Frequency
90	1
80	6
70	10
60	4
50	3
40	1

Find the mean, the median, the mode, and the range for the class.

38. A class obtained the following test scores:

Test score	Frequency
90	2
80	4
70	9
60	5
50	3
40	1
30	2
0	4

Find the mean, median, mode, and range for the class.

39. The county fair reported the following total attendance (in thousands).

Year	2000	1999	1998	1997	1996	1995
Attendance	366	391	358	373	346	364

Find the mean, median, and mode for the attendance figures (rounded to the nearest thousand).

40. The following salaries for the executives of a certain company are known:

Position	Salary
President	$170,000
1st VP	140,000
2nd VP	120,000
Supervising manager	54,000
Accounting manager	40,000
Personnel manager	40,000
Department manager	30,000
Department manager	30,000

Find the mean, the median, and the mode. Which measure seems to best describe the average executive salary for the company?

41. The graph in Figure 12.30 shows the distribution of scores on an examination.

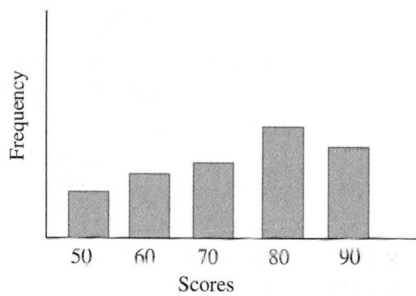

Figure 12.30 **Scores on an examination**

Which of the following statements is true about the distribution?

A. The mean is less than the mode.
B. The median is the same as the mode.
C. The median is greater than the mode.
D. The mean is the same as the mode.

42. The graph in Figure 12.31 shows the salaries of employees for a certain company.

Figure 12.31 **Company salaries**

Which of the following statements is true about the distribution?

A. The mean is less than the mode.
B. The median is the same as the mode.
C. The median is greater than the mode.
D. The mean is the same as the mode.

43. Shoe sizes for a team of football players range from size 10 to size 14. The majority of players wear size 12, and the number of players wearing size 11 is the same as the number wearing size 13. It is also true that the number of players wearing size 10 is the same as the number wearing size 14. Which statement is true about the distribution?
A. The median is less than the mode.
B. The mean is less than the mode.
C. The mean is the same as the median.
D. The median is greater than the mean.

44. In a survey, members of a math class were asked how many minutes per week they exercised. There were three possible responses: 10 minutes or less (70% checked this), between 10 and 60 minutes (20% checked this one), and 60 minutes or more (10%). Which statement is true about the distribution?
A. The mode is less than the mean.
B. The median is less than the mode.
C. The mean is less than the median.
D. The mean is the same as the mode.

Use the nutritional information about candy bars given in Table 12.7 to answer the questions in Problems 45 and 46.

45. a. What is the mean calories from fat?
b. What is the median (in grams) of total fat?
c. What is the mode (in grams) of the serving size?

46. a. Divide the serving size into quartiles.
b. Draw a box plot for the calories from fat.
c. What are the mean and standard deviation for total size in Table 12.7?

Table 12.7 Nutritional Information about Candy Bars

Company	Serving size	Total size	Calories from fat	Total fat
Almond Joy	36	180	90	10
Baby Ruth	60	280	110	12
Butterfingers	42	200	70	8
Clark Bar	50	240	90	10
5th Avenue	57	280	110	12
Heath Bar	40	210	110	13
Hershey choc. bar	43	230	120	13
Hershey choc. w/alm.	41	230	120	14
100 Grand	43	200	70	8
Kit Kat	42	220	110	12
Milky Way	61	280	100	11
Mr. Good Bar	49	280	160	18
Nestle milk choc.	41	220	110	13
Nut Rageous	45	250	140	15
Peppermint Patty	42	170	35	4
Reese's Cup	45	240	130	14
Rolo	54	230	110	12
Snickers	59	280	120	14
3 Musketeers	60	260	70	8
Twix	57	280	130	14

Find the standard deviation (rounded to the nearest unit) for the data indicated in Problems 47–52.

47. Problem 35   **48.** Problem 34

49. Problem 37   **50.** Problem 36

51. Problem 39   **52.** Problem 38

53. The number of miles driven on each of five tires was 17,000, 19,000, 19,000, 20,000, and 21,000. Find the mean, the range, and the standard deviation (rounded to the nearest unit) for these mileages.

54. Roll a single die until all six numbers occur at least once. Repeat the experiment 20 times. Find the mean, the median, the mode, and the range of the number of tosses.

55. Roll a pair of dice until all 11 possible sums occur at least once. Repeat the experiment 20 times. Find the mean, the median, the mode, and the range of the number of tosses.

56. When you take the Scholastic Assessment Test (SAT), your score is recorded as a percentile score. If you scored in the 92nd percentile, it means that you scored better than approximately 92% of those who took the test.
a. If Lisa's score was 85 and that score was the 23rd score from the top in a class of 240 scores. What is Lisa's percentile rank?
b. Lee has received a percentile rank of 85% in a class of 50 students. What is Lee's rank in the class?

LEVEL 3

57. The mean defined in the text is sometimes called the *arithmetic mean* to distinguish it from other possible means. For example, a different mean, called the *harmonic mean* (H.M.), is used to average speeds. This mean is defined to be the sum of the reciprocals of all scores divided into the number of scores. For example, the harmonic mean of the numbers 4, 5, 6, 6, 7, 8 is

$$\text{H.M.} = \frac{n}{\sum \frac{1}{x}}$$

$$= \frac{6}{\frac{1}{4} + \frac{1}{5} + \frac{1}{6} + \frac{1}{6} + \frac{1}{7} + \frac{1}{8}} \approx 5.7$$

a. Find the arithmetic mean and the harmonic mean of the numbers 2, 2, 5, 5, 7, 8, 8, 9, 9, 10.
b. A trip from San Francisco to Disneyland is approximately 460 miles. If the southbound trip averaged 52 mph and the return trip averaged 61 mph, what is the average speed for the round trip? Compare the arithmetic and harmonic means.

58. The mean defined in the text is sometimes called the *arithmetic mean* to distinguish it from other possible means. For example, a different mean, called the *geometric mean* (G.M.), is used in business and economics for finding average rates of change, average rates of growth, or average ratios. This mean is defined to be the *n*th root of the product of the numbers. For example, the geometric mean of 4, 5, and 6 is

$$\text{G.M.} = (4 \cdot 5 \cdot 6)^{1/3}$$
$$= \sqrt[3]{120} \approx 4.9$$

a. Find the arithmetic and geometric means for the numbers 2, 3, 5, 7, 7, 8, and 10.

b. The growth rates for three cities are 1.5%, 2.0%, and 0.9%. Compare the arithmetic and geometric means.

PROBLEM SOLVING

59. The example about averages of bowling scores at the beginning of this section is an instance of what is known as *Simpson's paradox*. The following example illustrates this paradox.*

Player A

	At bat	Hits	Avg.
Against right-handed pitchers	202	45	0.223
Against left-handed pitchers	250	71	0.284
Overall	452	116	0.257

Player B

	At bat	Hits	Avg.
Against right-handed pitchers	250	58	0.232
Against left-handed pitchers	108	32	0.296
Overall	358	90	0.251

*From "Instances of Simpson's Paradox," by Thomas R. Knapp, *College Mathematics Journal*, July 1985, pp. 209–211.

Notice that Player A has a better overall batting average than Player B, but yet is worse against both right-handed pitchers and left-handed pitchers. The following is an algebraic statement of Simpson's Paradox.

Consider two populations for which the overall rate r of occurrence of some phenomenon in population A is greater than the corresponding rate R in population B. Suppose that each of the two populations is composed of the same two categories C_1 and C_2, and the rates of occurrence of the phenomenon for the two categories in population A are r_1 and r_2, and in population B are R_1 and R_2. If $r_1 < R_1$ and $r_2 < R_2$, despite the fact that $r > R$, then Simpson's paradox is said to have occurred.

a. Relate the variables in this statement to the numbers in the example.

b. An example of Simpson's paradox from real life is the fact that the overall federal income tax rate increased from 1974 to 1978, but decreased for each bracket. Make up a fictitious example to show how this might be possible.

60. Let Q_1, Q_2, and Q_3 be the quartiles (from smallest to largest) for a large population of scores. True or false (with reasons):

$$Q_2 - Q_1 = Q_3 - Q_2$$

12.3 | THE NORMAL CURVE

Cumulative Distributions

Sometimes we represent frequencies in a cumulative way, especially when we want to find the position of one case relative to the performance of the group. A **cumulative frequency** is the sum of all preceding frequencies in which some order has been established.

EXAMPLE 1

Table 12.8 Number of moving violations

Number of moving violations	Percent of drivers
0	21%
1	25%
2	24%
3	13%
4	8%
5	5%
6 or more	4%

The number of moving violations, along with the percent of drivers, is shown in Table 12.8. Write a cumulative frequency for this example. Find the mean, median, and mode. What percentage of clients have at least 3 moving violations?

Solution

Number of moving violations	Percent of drivers	Cumulative percent
0	21%	21%
1	25%	46%
2	24%	70%
3	13%	83%
4	8%	91%
5	5%	96%
6 or more	4%	100%

The mean is found (using the idea of a weighted mean):

$$\bar{x} = \frac{0 \times 0.21 + 1 \times 0.25 + 2 \times 0.24 + 3 \times 0.13 + 4 \times 0.08 + 5 \times 0.05 + 6 \times 0.04}{1}$$

$$= 1.93$$

Note: The reason for dividing by 1 is that the sum of the percents is 1.

The median is the number of moving violations for which the cumulative percent first exceeds 50%; we see this is 2 moving violations.

The mode is the number of moving violations that occurs most frequently; we see this is 1 moving violation.

"At least 3" moving violations means "3 or more," so look at the percents for 3, 4, 5, and 6 or more moving violations:

$$0.13 + 0.08 + 0.05 + 0.04 = 0.30 \qquad \blacklozenge$$

Bell-Shaped Curves

Cartoon by David Pascal.

The cartoon in the margin suggests that most people do not like to think of themselves or their children as having "normal intelligence." But what do we mean by *normal* or *normal intelligence?*

Suppose we survey the results of 20 children's scores on an IQ test. The scores (rounded to the nearest 5 points) are 115, 90, 100, 95, 105, 95, 105, 105, 95, 125, 120, 110, 100, 100, 90, 110, 100, 115, 105, and 80. We can find $\bar{x} = 103$ and $s \approx 10.93$. A frequency graph of these data is shown in part **a** of Figure 12.32. If we consider 10,000 scores instead of only 20, we might obtain the frequency distribution shown in part **b** of Figure 12.32.

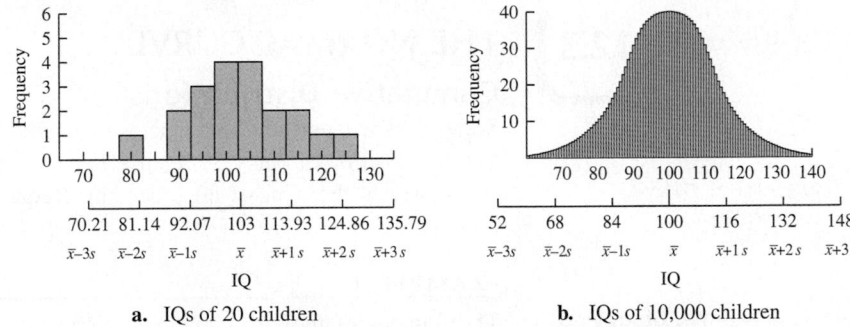

a. IQs of 20 children   **b.** IQs of 10,000 children

Figure 12.32 **Frequency distributions for IQ scores**

The data illustrated in Figure 12.32 approximate a commonly used curve called a *normal frequency curve,* or simply a **normal curve.** (See Figure 12.33.)

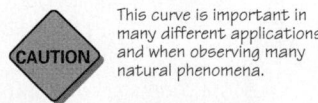

This curve is important in many different applications and when observing many natural phenomena.

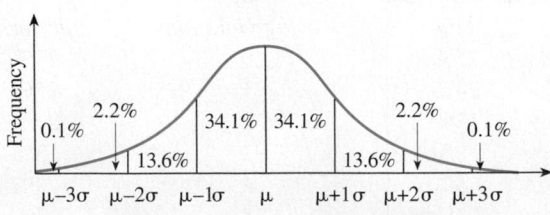

Figure 12.33 **A normal curve**

If we obtain the frequency distribution of a large number of measurements (as with IQ), the corresponding graph tends to look normal, or **bell-shaped.** The normal curve has some interesting properties. In it, the mean, the median, and the mode all have the same value, and all occur exactly at the center of the distribution; we denote this value by the Greek letter mu (μ). The standard deviation for this distribution is σ (sigma). Roughly 68% of all values lie within the region from 1 standard deviation below to 1 standard deviation above the mean. About 95% lie within 2 standard deviations on either side of the mean, and virtually all (99.8%) values lie within 3 standard deviations on either side. These percentages are the same regardless of the particular mean or standard deviation.

The normal distribution is a **continuous** (rather than a discrete) **distribution,** and it extends indefinitely in both directions, never touching the x-axis. It is symmetric about a vertical line drawn through the mean, μ. Graphs of this curve for several choices of σ are shown in Figure 12.34.

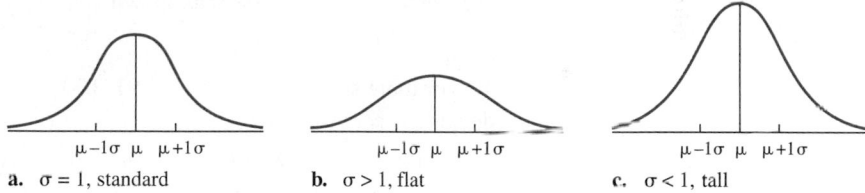

a. $\sigma = 1$, standard   **b.** $\sigma > 1$, flat   **c.** $\sigma < 1$, tall

Figure 12.34 **Variations of normal curves**

EXAMPLE 2

Predict the distribution of IQ scores of 1,000 people if we assume that IQ scores are normally distributed, with a mean of 100 and a standard deviation of 15.

Solution   First, find the breaking points around the mean. For $\mu = 100$ and $\sigma = 15$:

$$\mu + \sigma = 100 + 15 = 115 \qquad \mu - \sigma = 100 - 15 = 85$$
$$\mu + 2\sigma = 100 + 2 \cdot 15 = 130 \qquad \mu - 2\sigma = 100 - 2 \cdot 15 = 70$$
$$\mu + 3\sigma = 100 + 3 \cdot 15 = 145 \qquad \mu - 3\sigma = 100 - 3 \cdot 15 = 55$$

We use Figure 12.34 to find that 34.1% of the scores will be between 100 and 115 (i.e., between μ and $\mu + 1\sigma$):

$$0.341 \times 1,000 = 341$$

About 13.6% will be between 115 and 130 (between $\mu + 1\sigma$ and $\mu + 2\sigma$):

$$0.136 \times 1,000 = 136$$

About 2.2% will be between 130 and 145 (between $\mu + 2\sigma$ and $\mu + 3\sigma$):

$$0.022 \times 1,000 = 22$$

About 0.1% will be above 145 (more than $\mu + 3\sigma$):

$$0.001 \times 1,000 = 1$$

The distribution for intervals below the mean is identical, since the normal curve is the same to the left and to the right of the mean. The distribution is shown in the margin. ◆

Scores	%	Expected number
55 or below	0.1	1
55$^+$–70	2.2	22
70$^+$–85	13.6	136
85$^+$–100	34.1	341
100$^+$–115	34.1	341
115$^+$–130	13.6	136
130$^+$–145	2.2	22
Above 145	0.1	1
Totals	100.0	1,000

EXAMPLE 3

Suppose that an instructor "grades on a curve." Show the grading distribution on an examination of 45 students, if the scores are normally distributed with a mean of 73 and a standard deviation of 9.

Solution Grading on a curve means determining students' grades according to the percentages shown in Figure 12.35.

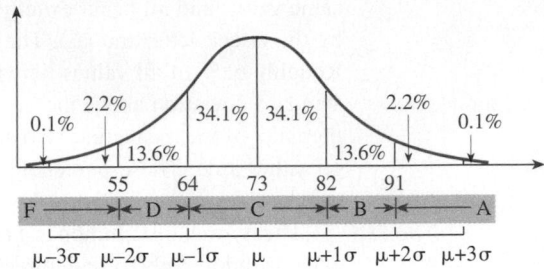

Figure 12.35 Grade distribution for a class "graded on a curve"

We calculate these numbers as shown:

Calculation	Scores	Grade	Calculation	Number
Two or more standard deviations above the mean	91–100	A	$0.023 \times 45 = 1.04$	1
$\mu + 2\sigma = 73 + 2 \cdot 9 = 91$	82–90	B	$0.136 \times 45 = 6.12$	6
$\mu + 1\sigma = 73 + 1 \cdot 9 = 82$				
Mean: $\mu = 73$	64–81	C	$0.682 \times 45 = 30.69$	31
$\mu - 1\sigma = 73 - 1 \cdot 9 = 64$				
$\mu - 2\sigma = 73 - 2 \cdot 9 = 55$	55–63	D	same as B	6
Two or more standard deviations below the mean	0–54	F	same as A	1

Grading on a curve means that the person with the top score in this class of 45 receives an A; the next 6 ranked persons (from the top) receive B grades; the bottom score in the class receives an F; the next 6 ranked persons (from the bottom) receive D grades; and finally, the remaining 31 persons receive C grades. Notice that the majority of the class (34.1% + 34.1% = 68.2%) will receive an "average" C grade. ◆

z-SCORES

Sometimes we want to know the percent of occurrence for scores that do not happen to be 1, 2, or 3 standard deviations from the mean. For Example 3, we can see from Figure 12.35 that 34.1% of the scores are between the mean and 1 standard deviation above the mean. Suppose we wish to find the percent of scores that are between the mean and 1.2 standard deviations above the mean. To find this percent, we use Table 12.9 for this purpose. First, we introduce some terminology.

We use *z-scores* (sometimes called *standard scores*) to determine how far, in terms of standard deviations, that a given score is from the mean of the distribution. For example, if $z = 1$, we can use the Table 12.9 to find the percent of scores between the mean and the value that is 1 standard deviation above the mean. (From Figure 12.35, we know that the percent is 34.1% or 0.341.) In Table 12.9, look in the row labeled (at the left) 1.0 and in the column headed 0.00: The entry is 0.3413, which is 34.13%. For $z = 1.2$, look at the entry in the row marked 1.2 and the 0.00 column: It is 0.3849.

Table 12.9 Standard Normal Distribution; z-Scores

For a particular value, this table gives the percent of scores between the mean and the *z*-value of a normally distributed random variable.

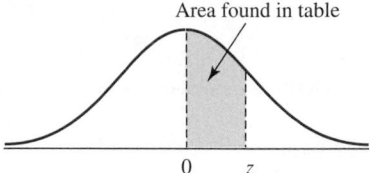

Area found in table

0   z

z	0.00	0.01	0.02	0.03	0.04	0.05	0.06	0.07	0.08	0.09
0.0	0.0000	0.0040	0.0080	0.0120	0.0160	0.0199	0.0239	0.0279	0.0319	0.0359
0.1	0.0398	0.0438	0.0478	0.0517	0.0557	0.0596	0.0636	0.0675	0.0714	0.0753
0.2	0.0793	0.0832	0.0871	0.0910	0.0948	0.0987	0.1026	0.1064	0.1103	0.1141
0.3	0.1179	0.1217	0.1255	0.1293	0.1331	0.1368	0.1406	0.1443	0.1480	0.1517
0.4	0.1554	0.1591	0.1628	0.1664	0.1700	0.1736	0.1772	0.1808	0.1844	0.1879
0.5	0.1915	0.1950	0.1985	0.2019	0.2054	0.2088	0.2123	0.2157	0.2190	0.2224
0.6	0.2257	0.2291	0.2324	0.2357	0.2389	0.2422	0.2454	0.2486	0.2517	0.2549
0.7	0.2580	0.2611	0.2642	0.2673	0.2704	0.2734	0.2764	0.2794	0.2823	0.2852
0.8	0.2881	0.2910	0.2939	0.2967	0.2995	0.3023	0.3051	0.3078	0.3106	0.3133
0.9	0.3159	0.3186	0.3212	0.3238	0.3264	0.3289	0.3315	0.3340	0.3365	0.3389
1.0	0.3413	0.3438	0.3461	0.3485	0.3508	0.3531	0.3554	0.3577	0.3599	0.3621
1.1	0.3643	0.3665	0.3686	0.3708	0.3729	0.3749	0.3770	0.3790	0.3810	0.3830
1.2	0.3849	0.3869	0.3888	0.3907	0.3925	0.3944	0.3962	0.3980	0.3997	0.4015
1.3	0.4032	0.4049	0.4066	0.4082	0.4099	0.4115	0.4131	0.4147	0.4162	0.4177
1.4	0.4192	0.4207	0.4222	0.4236	0.4251	0.4265	0.4279	0.4292	0.4306	0.4319
1.5	0.4332	0.4345	0.4357	0.4370	0.4382	0.4394	0.4406	0.4418	0.4429	0.4441
1.6	0.4452	0.4463	0.4474	0.4484	0.4495	0.4505	0.4515	0.4525	0.4535	0.4545
1.7	0.4554	0.4564	0.4573	0.4582	0.4591	0.4599	0.4608	0.4616	0.4625	0.4633
1.8	0.4641	0.4649	0.4656	0.4664	0.4671	0.4678	0.4686	0.4693	0.4699	0.4706
1.9	0.4713	0.4719	0.4726	0.4732	0.4738	0.4744	0.4750	0.4756	0.4761	0.4767
2.0	0.4772	0.4778	0.4783	0.4788	0.4793	0.4798	0.4803	0.4808	0.4812	0.4817
2.1	0.4821	0.4826	0.4830	0.4834	0.4838	0.4842	0.4846	0.4850	0.4854	0.4857
2.2	0.4861	0.4864	0.4868	0.4871	0.4875	0.4878	0.4881	0.4884	0.4887	0.4890
2.3	0.4893	0.4896	0.4898	0.4901	0.4904	0.4906	0.4909	0.4911	0.4913	0.4916
2.4	0.4918	0.4920	0.4922	0.4925	0.4927	0.4929	0.4931	0.4932	0.4934	0.4936
2.5	0.4938	0.4940	0.4941	0.4943	0.4945	0.4946	0.4948	0.4949	0.4951	0.4952
2.6	0.4953	0.4955	0.4956	0.4957	0.4959	0.4960	0.4961	0.4962	0.4963	0.4964
2.7	0.4965	0.4966	0.4967	0.4968	0.4969	0.4970	0.4971	0.4972	0.4973	0.4974
2.8	0.4974	0.4975	0.4976	0.4977	0.4977	0.4978	0.4979	0.4979	0.4980	0.4981
2.9	0.4981	0.4982	0.4982	0.4983	0.4984	0.4984	0.4985	0.4985	0.4986	0.4986
3.0	0.4987	0.4987	0.4987	0.4988	0.4988	0.4989	0.4989	0.4989	0.4990	0.4990

Note: For values of *z* above 3.09, use 0.4999.

Suppose we want to find $z = 1.68$; look in Table 12.9 at the row labeled 1.6 and the column headed 0.08 to find the entry 0.4535. This means that 45.35% of the values in a normal distribution are between the mean and 1.68 standard deviations above the mean.

We use the *z*-score to translate any normal curve into a standard normal curve (the particular normal curve with a mean of 0 and a standard deviation of 1) by using the definition.

z-SCORE

If x is a value from a normal distribution with mean μ and standard deviation σ, then its **z-score** is

$$z = \frac{x - \mu}{\sigma}$$

The z-score is used with Table 12.9 to find the percent of occurrence between the mean and the number of standard deviations above the mean specified by the z-score, as illustrated in Figure 12.36.

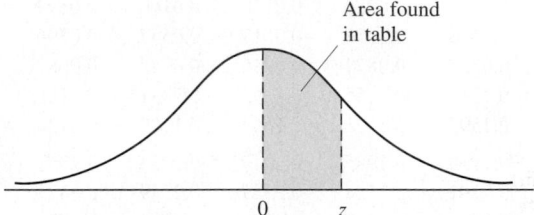

Area found in table

Figure 12.36 Percent of occurrence using a z-score

EXAMPLE 4

The Eureka Lightbulb Company tested a new line of lightbulbs and found their lifetimes to be normally distributed, with a mean life of 98 hours and a standard deviation of 13 hours.

a. What percentage of bulbs will last less than 72 hours?

b. What percentage of bulbs will last less than 100 hours?

c. What is the probability that a bulb selected at random will last longer than 111 hours?

d. What is the probability that a bulb will last between 106 and 120 hours?

Solution   Draw a normal curve with mean 98 and standard deviation 13, as shown in Figure 12.37.

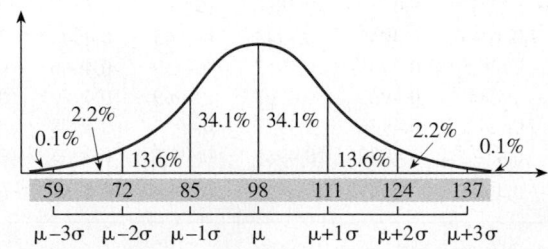

Figure 12.37 Lightbulb lifetimes are normally distributed.

For this example, we are given $\mu = 98$ and $\sigma = 13$.

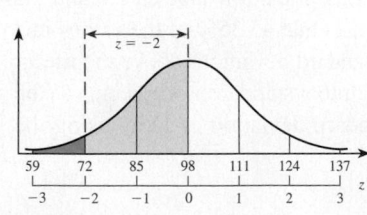

a. For $x = 72$, $z = \dfrac{72 - 98}{13} = -2$

This is 2 standard deviations below the mean (which is the same as the z-score). The percentage we seek is shown in blue:

About 2.3% (2.2% + 0.1% = 2.3%) will last less than 72 hours.

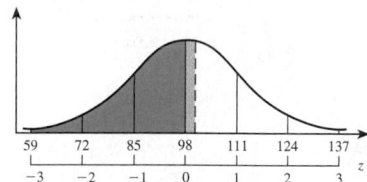

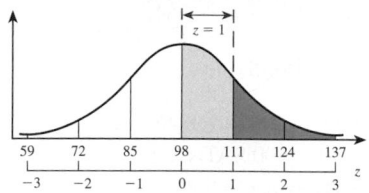

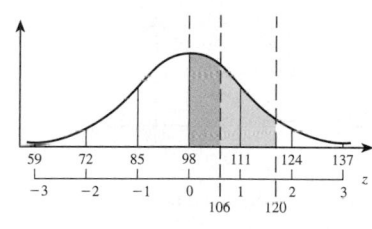

b. For $x = 100$, $z = \dfrac{100 - 98}{13} \approx 0.15$

This is 0.15 *standard deviation above the mean* (which is the same as the z-score). From Table 12.9, we find 0.0596 (this is shown in green). Since we want the percent of values less than 100, we must add 50% for the numbers below the mean (shown in blue). The percentage we seek is $0.5000 + 0.0596 = 0.5596$ or about 56.0%.

c. For $x = 111$, $z = \dfrac{111 - 98}{13} = 1$

This is one standard deviation above the mean, which is the same as the z-score. The percentage we seek is shown in blue. We know that about 15.9% ($13.6\% + 2.2\% + 0.1\% = 15.9\%$) of the bulbs will last longer than 111 hours, so

$$P(\text{bulb life} > 111 \text{ hours}) \approx 0.159$$

We can also use Table 12.9. For $z = 1.00$, the table entry is 0.3413, and we are looking for the area to the right, so we compute

$$0.5000 - 0.3413 = 0.1587$$

d. We first find the z-scores using $\mu = 98$ and $\sigma = 13$. For $x = 106$,

$$z = \frac{106 - 98}{13} \approx 0.62$$

From Table 12.9, the area between the z-score and the mean is 0.2324 (shown in green). For $x = 120$,

$$z = \frac{120 - 98}{13} \approx 1.69$$

From Table 12.9, the area between this z-score and the mean is 0.4545. The desired answer (shown in yellow) is approximately

$$0.4545 - 0.2324 = 0.2221$$

Since percent and probability are the same, we see the probability that the life of the bulb is between 106 and 120 hours is about 22.2%. ◆

Example 4 leads us to the observation that there are three equivalent ideas associated with the normal curve. These are summarized in the following box.

Meanings Associated With a Normal Curve

In a standard normal curve, the following three quantities are equivalent:

Probability of a randomly chosen item lying in an interval;

Percentage of total items that lie in an interval;

Area under a normal curve along an interval.

We will come back to the idea of the area under this curve in Section 17.4.

Sometimes data do not fall into a normal distribution, but are **skewed,** which means their distribution has more tail on one side or the other. For example, Figure 12.38a (page 640) shows that the 1941 scores on the SAT exam (when the test was first used) were normally distributed. However, by 1990, the scale had become skewed to the left, as shown in Figure 12.38b.

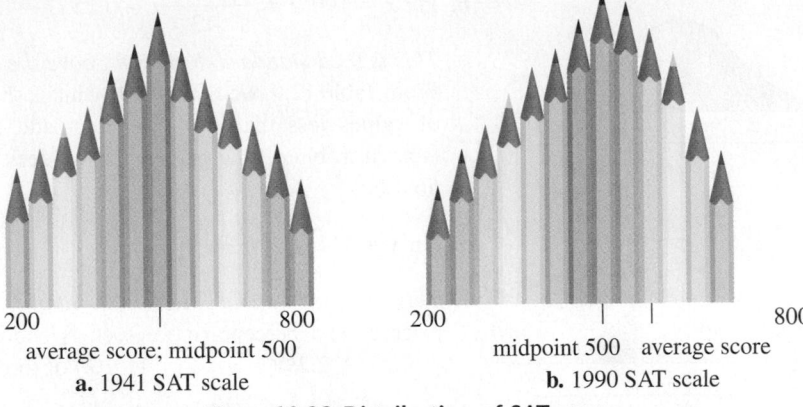

average score; midpoint 500
a. 1941 SAT scale

midpoint 500   average score
b. 1990 SAT scale

Figure 10.38 **Distribution of SAT scores**

In a normal distribution, the mean, median, and mode all have the same value, but if the distribution is skewed, the relative positions of the mean, median, and mode would be as shown in Figure 12.39.

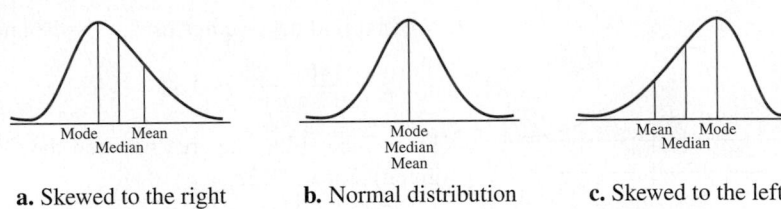

a. Skewed to the right   **b.** Normal distribution   **c.** Skewed to the left

Figure 12.39 **Comparison of three distributions**

Problem Set 12.3

LEVEL 1

1. **IN YOUR OWN WORDS** What do we mean by a cumulative frequency?

2. **IN YOUR OWN WORDS** What does it mean to "grade on a curve"?

3. **IN YOUR OWN WORDS** What is a normal curve?

4. **IN YOUR OWN WORDS** What is a distribution that is skewed to the right?

5. **IN YOUR OWN WORDS** What does the z-score, or standard score, represent?

6. **IN YOUR OWN WORDS** When will a z-score be negative?

Find the cumulative distribution, the mean, median, and mode for each of the tables in Problems 7–12.

7. The number of bedrooms of homes in a certain community is shown in the table below.

Number of bedrooms	Percent homes
0	5%
1	11%
2	29%
3	34%
4	15%
5	5%
6 or more	1%

8. The cumulative grade point average of graduating seniors at a small liberal arts college is shown in the following table.

GPA	Interval mean	Percent of class
3.25–4.00	3.625	23%
2.50–3.24	2.87	41%
1.75–2.49	2.12	19%
1.00–1.74	1.37	6%
0.00–0.99	0.495	11%

9. The table gives the distribution of the number of exemptions claimed by employees of a large university.

Number of exemptions	Proportion of employees
0	1%
1	11%
2	35%
3	21%
4	20%
5	6%
6	3%
7	2%
8	1%

10. The table shows the distribution of households, according to the number of motor vehicles per household.

Number of motor vehicles	Proportion of households
0	0.18
1	0.24
2	0.32
3	0.16
4	0.09
5 or more	0.01

11. A store orders tubes of glue in the proportions shown in the following table.

Size (oz)	Proportion
2	0.28
6	0.12
8	0.35
12	0.06
16	0.19

12. The distribution of number of years of education of employees of a large company is shown in the following table.

Number of years of school	Proportion of employees
11 or fewer	0.01
12	0.27
13	0.13
14	0.07
15	0.05
16	0.32
17	0.02
18	0.11
19 or more	0.02

What percent of the total population is found between the mean and the z-score given in Problems 13–24?

13. $z = 1.4$ 14. $z = 0.3$ 15. $z = 2.43$

16. $z = 1.86$ 17. $z = 3.25$ 18. $z = -0.6$

19. $z = -2.33$ 20. $z = -0.50$ 21. $z = -0.46$

22. $z = -1.19$ 23. $z = -2.22$ 24. $z = -3.41$

In Problems 25–29, *suppose that people's heights (in centimeters) are normally distributed, with a mean of* 170 *and a standard deviation of* 5. *We find the heights of* 50 *people.*

25. **a.** How many would you expect to be between 165 and 175 cm tall?
    b. How many would you expect to be taller than 168 cm?

26. **a.** How many would you expect to be between 170 and 180 cm?
    b. How many would you expect to be taller than 176 cm?

27. What is the probability that a person selected at random is taller than 163 cm?

28. What is the variance in heights for this experiment?

29. What is the cumulative distribution?

In Problems 30–34 *suppose that, for a certain exam, a teacher grades on a curve. It is known that the mean is* 50 *and the standard deviation is* 5. *There are* 45 *students in the class.*

30. **a.** How many students should receive a C?
    b. How many students should receive an A?

31. What score would be necessary to obtain an A?

32. If an exam paper is selected at random, what is the probability that it will be a failing paper?

33. What is the variance in scores for this exam?

34. What is the cumulative distribution?

LEVEL 2

In Problems 35–39, *suppose that, for a certain mathematics class, the scores are normally distributed with a mean of* 75 *and a standard deviation of* 8. *The teacher wishes to give As to the top* 6% *of the students and Fs to the bottom* 6%. *The next* 16% *in either direction will be given Bs and Ds, with the other students receiving Cs. Find the bottom cutoff for the grades in Problems* 35–38.

35. A   36. B   37. C   38. D

39. What is the cumulative distribution?

40. **IN YOUR OWN WORDS** In a distribution that is skewed to the right, which has the greatest value—the mean, median, or mode? Explain why this is the case.

41. **IN YOUR OWN WORDS** In a distribution that is skewed to the left, which has the greatest value—the mean, median, or mode? Explain why this is the case.

A normal distribution has a mean of 85.7 and a standard deviation of 4.85. Find data values corresponding to the values of z given in Problems 42–45.

42. $z = 0.85$

43. $z = 2.55$

44. $z = -1.25$

45. $z = -3.46$

46. Suppose that the breaking strength of a rope (in pounds) is normally distributed, with a mean of 100 pounds and a standard deviation of 16. What is the probability that a certain rope will break when subjected to a force of 130 pounds or less?

47. The diameter of an electric cable is normally distributed, with a mean of 0.9 inch and a standard deviation of 0.01 inch. What is the probability that the diameter will exceed 0.91 inch?

48. Suppose that the annual rainfall in Ferndale, California, is known to be normally distributed, with a mean of 35.5 inches and a standard deviation of 2.5 inches. About 2.3% of the time, the annual rainfall will exceed how many inches?

49. About what percent of the years will it rain more than 36 inches in Ferndale (see Problem 48)?

50. In Problem 48, what is the probability that the rainfall in a given year will exceed 30.5 inches in Ferndale?

51. The breaking strength (in pounds) of a certain new synthetic is normally distributed, with a mean of 165 and a variance of 9. The material is considered defective if the breaking strength is less than 159 pounds. What is the probability that a single, randomly selected piece of material will be defective?

52. The diameter of a pipe is normally distributed, with a mean of 0.4 inch and a variance of 0.0004. What is the probability that the diameter of a randomly selected pipe will exceed 0.44 inch?

53. The diameter of a pipe is normally distributed, with a mean of 0.4 inch and a variance of 0.0004. What is the probability that the diameter of a randomly selected pipe will exceed 0.41 inch?

54. Suppose the neck size of men is normally distributed, with a mean of 15.5 inches and a standard deviation of 0.5 inch. A shirt manufacturer is going to introduce a new line of shirts. Assume that if your neck size falls between two shirt sizes, you purchase the next larger shirt size. How many of each of the following sizes should be included in a batch of 1,000 shirts?

a. 14   **b.** 14.5   **c.** 15   **d.** 15.5

e. 16   **f.** 16.5   **g.** 17

55. A package of Toys Galore Cereal is marked "Net Wt. 12 oz." The actual weight is normally distributed, with a mean of 12 oz and a variance of 0.04.
   a. What percent of the packages will weigh less than 12 oz?
   b. What weight will be exceeded by 2.3% of the packages?

56. Instant Dinner comes in packages with weights that are normally distributed, with a standard deviation of 0.3 oz. If 2.3% of the dinners weigh more than 13.5 oz, what is the mean weight?

LEVEL 3

57. The equation for the normal curve is an exponential equation. Let $\mu = 0$ and $\sigma = 1$ to obtain the equation for the *standard normal curve:*
$$y = \frac{e^{-x^2/2}}{\sqrt{2\pi}}$$
Evaluate this formula for $x = -4, -3, \ldots, 3, 4$.

58. **IN YOUR OWN WORDS** Calculate values for the equation $y = 2^{-x^2}$ and compare with the standard normal curve in Problem 57. List some similarities and some differences.

59. The equation for the normal curve is
$$y = \frac{e^{-(x-\mu)^2/(2\sigma^2)}}{\sigma\sqrt{2\pi}}$$
Calculate values for $x = 20, 30, \ldots, 70, 80$, where $\mu = 50$ and $\sigma = 10$.

60. The graph shown in Figure 12.40 is from the February 1991 issue of *Scientific American*. If the curve in the middle is a standard normal curve, describe the upper curve (labeled **a**) and then describe the lower curve (labeled **b**).

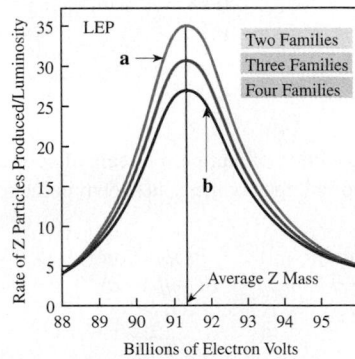

Figure 12.40 **Normal curves**

CORRELATION AND REGRESSION

Regression Analysis

In mathematical modeling, it is often necessary to deal with numerical data and to make assumptions regarding the relationship between two variables. For example, you may want to examine the relationship between

IQ and salary
Study time and grades
Age and heart disease
Runner's speed and runner's brand of shoe
Math grades in the 8th grade and amount of TV viewing
Teachers' salaries and beer consumption

All are attempts to relate two variables in some way or another. If it is established that there is a **correlation,** then the next step in the modeling process is to identify the nature of the relationship. This is called **regression analysis.** In this section we consider only linear relationships. It is assumed that you are familiar with the slope–intercept form of the equation of a line (namely, $y = mx + b$). If you need a review of graphing lines, you should look at Section 13.1.

We are interested in finding a *best-fitting line* by a technique called the **least squares method.** The derivations of the results and the formulas given in this section are, for the most part, based on calculus, and are therefore beyond the scope of this book. We will attempt to focus instead on how to use and interpret the formulas.

The first consideration is one of correlation. We want to know whether two variables are related. Let us call one variable x and the other y. These variables can be represented as ordered pairs (x, y) in a graph called a **scatter diagram.**

EXAMPLE 1

A survey of 20 students compared the grade received on an examination with the length of time the student studied. Draw a scatter diagram to represent the data in the table.

Student number	1	2	3	4	5	6	7	8	9	10	11	12	13	14	15	16	17	18	19	20
Length of study time (nearest 5 min.)	30	40	30	35	45	15	15	50	30	0	20	10	25	25	25	30	40	35	20	15
Grade (100 possible)	72	85	75	78	89	58	71	94	78	10	75	43	68	60	70	68	82	75	65	62

Solution Let x be the study time (in minutes) and let y be the grade (in points). The graph is shown in Figure 12.41.

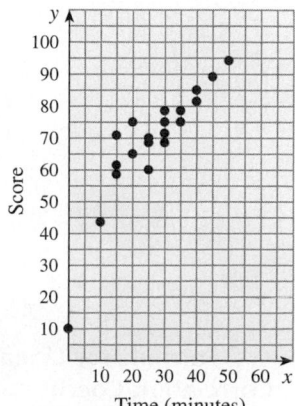

Figure 12.41 **Correlation beween study time and grade**

Correlation is a measure to determine whether there is a statistically significant relationship between two variables. Intuitively, it should assign a measure consistent with the scatter diagrams as shown in Figure 12.42.

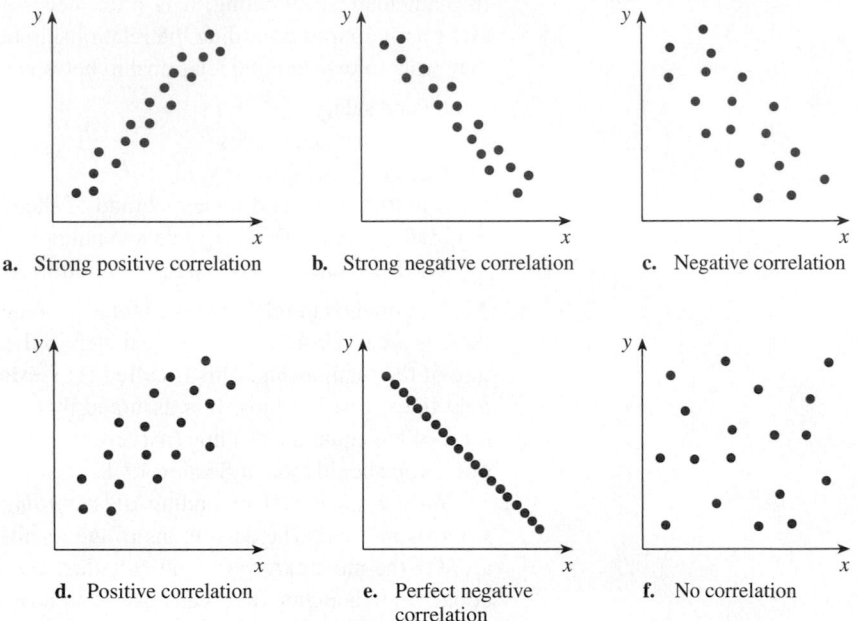

a. Strong positive correlation **b.** Strong negative correlation **c.** Negative correlation

d. Positive correlation **e.** Perfect negative correlation **f.** No correlation

Figure 12.42 **Examples of correlation**

Such a measure, called the *linear correlation coefficient, r,* is defined so that it has the following properties:

1. r measures the correlation between x and y.

2. r is between -1 and 1.

3. If r is close to 0, it means there is little correlation.

4. If r is close to 1, it means there is a strong positive correlation.

5. If r is close to -1, it means there is a strong negative correlation.

To write a formula for r, let n denote the number of pairs of data present; and as before:

Σx	denotes the sum of the x-values.
Σx^2	means square the x-values and then sum.
$(\Sigma x)^2$	means sum the x-values and then square.
Σxy	means multiply each x-value by the corresponding y-value and then sum.
$n\Sigma xy$	means multiply n times Σxy.
$(\Sigma x)(\Sigma y)$	means multiply Σx times Σy.

Formula for Linear Correlation Coefficient

The **linear correlation coefficient, r,** is

$$r = \frac{n\Sigma xy - (\Sigma x)(\Sigma y)}{\sqrt{n(\Sigma x^2) - (\Sigma x)^2}\ \sqrt{n(\Sigma y^2) - (\Sigma y)^2}}$$

EXAMPLE 2

Find r for the following data (from Example 1):

Student number	1	2	3	4	5	6	7	8	9	10	11	12	13	14	15	16	17	18	19	20
Length of study time (nearest 5 min.)	30	40	30	35	45	15	15	50	30	0	20	10	25	25	25	30	40	35	20	15
Grade (100 *possible*)	72	85	75	78	89	58	71	94	78	10	75	43	68	60	70	68	82	75	65	62

Solution

Study time, x	Score, y	xy	x^2	y^2
30	72	2,160	900	5,184
40	85	3,400	1,600	7,225
30	75	2,250	900	5,625
35	78	2,730	1,225	6,084
45	89	4,005	2,025	7,921
15	58	870	225	3,364
15	71	1,065	225	5,041
50	94	4,700	2,500	8,836
30	78	2,340	900	6,084
0	10	0	0	100
20	75	1,500	400	5,625
10	43	430	100	1,849
25	68	1,700	625	4,624
25	60	1,500	625	3,600
25	70	1,750	625	4,900
30	68	2,040	900	4,624
40	82	3,280	1,600	6,724
35	75	2,625	1,225	5,625
20	65	1,300	400	4,225
15	62	930	225	3,844
Total 535	1,378	40,575	17,225	101,104
↑	↑	↑	↑	↑
Σx	Σy	Σxy	Σx^2	Σy^2

$$r = \frac{n\Sigma xy - (\Sigma x)(\Sigma y)}{\sqrt{n(\Sigma x^2) - (\Sigma x)^2}\sqrt{n(\Sigma y^2) - (\Sigma y)^2}}$$

$$= \frac{20(40,575) - (535)(1,378)}{\sqrt{20(17,225) - (535)^2}\ \sqrt{20(101,104) - (1,378)^2}}$$

$$= \frac{74,270}{\sqrt{58,275}\ \sqrt{123,196}}$$

$$\approx 0.877 \qquad \blacklozenge$$

Table 12.10 Correlation coefficient, r

n	$\alpha = 0.05$	$\alpha = 0.01$
4	0.950	0.999
5	0.878	0.959
6	0.811	0.917
7	0.754	0.875
8	0.707	0.834
9	0.666	0.798
10	0.632	0.765
11	0.602	0.735
12	0.576	0.708
13	0.553	0.684
14	0.532	0.661
15	0.514	0.641
16	0.497	0.623
17	0.482	0.606
18	0.468	0.590
19	0.456	0.575
20	0.444	0.561
25	0.396	0.505
30	0.361	0.463
35	0.335	0.430
40	0.312	0.402
45	0.294	0.378
50	0.279	0.361
60	0.254	0.330
70	0.236	0.305
80	0.220	0.286
90	0.207	0.269
100	0.196	0.256

*The derivation of this table is beyond the scope of this course. It shows the critical values of the **Pearson correlation coefficient**.*

Example 2 shows a very strong positive correlation. But if r for this example had been 0.46, would we still have been able to conclude that there is a strong correlation? This question is a topic of major concern in statistics. The term **significance level** is used to denote the cutoff between results attributed to chance and results attributed to significant differences. Table 12.10 gives *critical values* for determining whether two variables are correlated. If $|r|$ is greater than the given table value, then you may assume a correlation exists between the variables. If you use the column labeled $\alpha = 0.05$, then we say the significance level is 5%. This means that the probability is 0.05 that you will say the variables are correlated when, in fact, the results should be attributed to chance, and similarly for a significance level of 1% ($\alpha = 0.01$). For Example 2, since $n = 20$, we see in Table 12.10 that $r = 0.877$ shows a significant linear correlation at both the 1% and 5% levels. On the other hand, if $r = 0.46$ and $n = 20$, then there is a significant linear correlation at a 5% level, but not at a 1% level.

EXAMPLE 3

Find the critical value of the linear correlation coefficient for 13 pairs of data and a significance level of 0.05.

Solution From Table 12.10, the critical value is 0.553. For $n = 13$, any value greater than $r = 0.553$ or less than -0.553 is evidence of linear correlation. ◆

EXAMPLE 4

If $r = -0.85$ and $n = 13$, are the variables correlated at a significance level of 1%?

Solution For $n = 13$ and $\alpha = 0.01$, the Table 12.10 table entry is 0.684. Since r is negative and $|r| > 0.684$, we see that there is a negative linear correlation. ◆

EXAMPLE 5

Once again, we remind you to use a calculator to carry out calculations such as the ones shown in Example 5. You should check the work shown in this example on your own calculator. Keep in mind that many calculators have built-in function keys for finding the correlation.

The following table shows a sample of some past annual mean salaries for college professors, along with the annual per capita beer consumption (in gallons) for Americans. Find the correlation coefficient.

Year	1995	1993	1991	1990	1989	1988
Mean teacher salary	$58,400	$57,400	$55,800	$53,200	$50,100	$47,200
Per capita beer consumption	22.6	22.8	23.1	24.0	25.4	24.7

Solution $n = 6$ $\qquad \Sigma x^2 = 1.738705 \times 10^{10}$

$\Sigma x = 322{,}100 \qquad \Sigma y^2 = 3{,}395.46$

$\Sigma y = 142.6 \qquad (\Sigma x)^2 = 1.0374841 \times 10^{11}$

$\Sigma xy = 7{,}632{,}720 \qquad (\Sigma y)^2 = 20{,}334.76$

$$r = \frac{n\Sigma xy - (\Sigma x)(\Sigma y)}{\sqrt{n(\Sigma x^2) - (\Sigma x)^2} \ \sqrt{n(\Sigma y^2) - (\Sigma y)^2}}$$

$$= \frac{6(7{,}632{,}720) - (322{,}100)(142.6)}{\sqrt{6(1.738705 \times 10^{10}) - 1.0374841 \times 10^{11}} \ \sqrt{6(3{,}395.46) - 20{,}334.76}}$$

$$\approx -0.9151$$

The number $r \approx -0.9151$ is statistically significant at the 5% level, but not at the 1% level. This means that there is a negative correlation between these two phenomena. ◆

The significance of the negative correlation in Example 5 implies that professors' salaries and beer drinking are related, but be careful! In any event, the techniques in this chapter can be used only to establish a *statistical* linear relationship. *We cannot establish the existence or absence of any inherent cause-and-effect relationship on the basis of a correlation analysis.*

Best-Fitting Line

The final step in our discussion of correlation is to find the best-fitting line.* That is, we want to find a line $y' = mx + b$ so that the sum of the distances of the data points from this line will be as small as possible. (We use y' instead of y to distinguish between the actual second component, y, and the predicted y-value, y'.) Since some of these dis-

---

*If you need a review of graphing lines, see Section 13.1.

tances may be positive and some negative, and since we do not want large opposites to "cancel each other out," we minimize the sum of the *squares* of these distances. Therefore, the regression line is sometimes called the *least squares line*.

Least Squares Line

> The **least squares** (or *regression*) **line** is $y' = mx + b$, where
> $$m = \frac{n(\Sigma xy) - (\Sigma x)(\Sigma y)}{n(\Sigma x^2) - (\Sigma x)^2} \qquad b = \frac{\Sigma y - m(\Sigma x)}{n}$$
> This is the line of *best fit*.

EXAMPLE 6

Find the best-fitting line for the data in Example 1:

Student number	1	2	3	4	5	6	7	8	9	10	11	12	13	14	15	16	17	18	19	20
Length of study time (nearest 5 min.)	30	40	30	35	45	15	15	50	30	0	20	10	25	25	25	30	40	35	20	15
Grade (100 possible)	72	85	75	78	89	58	71	94	78	10	75	43	68	60	70	68	82	75	65	62

Solution   From Example 2, $n = 20$ and

$$
\begin{array}{ccccc}
535 & 1{,}378 & 40{,}575 & 17{,}225 & 101{,}104 \\
\uparrow & \uparrow & \uparrow & \uparrow & \uparrow \\
\Sigma x & \Sigma y & \Sigma xy & \Sigma x^2 & \Sigma y^2
\end{array}
$$

$$m = \frac{n(\Sigma xy) - (\Sigma x)(\Sigma y)}{n(\Sigma x^2) - (\Sigma x)^2} = \frac{20(40{,}575) - (535)(1{,}378)}{20(17{,}225) - (535)^2} \approx 1.27447$$

$$b = \frac{\Sigma y - m(\Sigma x)}{n} = \frac{1{,}378 - 1.27447(535)}{20} \approx 34.8078$$

Then we approximate the best-fitting line as the line with equation $y' = 1.3x + 35$. This line (along with the data points) is shown in Figure 12.43.

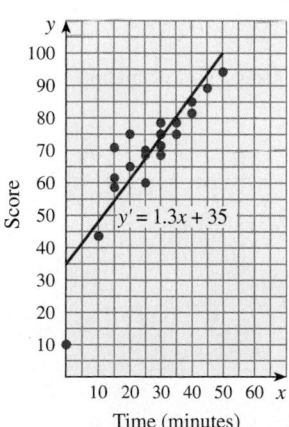

Figure 12.43 **Regression line**   ◆

EXAMPLE 7

Use the regression line in Example 6 to predict the score of a person who studied $\frac{1}{2}$ hour.

Solution $x = 30$ minutes, so $y' = 1.3x + 35 = 1.3(30) + 35 = 74$.   ◆

A Final Word of Caution. You should use the regression line only if r indicates that there is a significant linear correlation, as shown in Table 12.10.

Problem Set 12.4

LEVEL 1

1. **IN YOUR OWN WORDS** What do we mean by correlation?

2. **IN YOUR OWN WORDS** How do you find a linear correlation coefficient?

3. **IN YOUR OWN WORDS** How do you determine whether there is a linear correlation between two variables x and y?

4. **IN YOUR OWN WORDS** What is a least squares line?

5. **IN YOUR OWN WORDS** Discuss the correlation shown by the following chart.

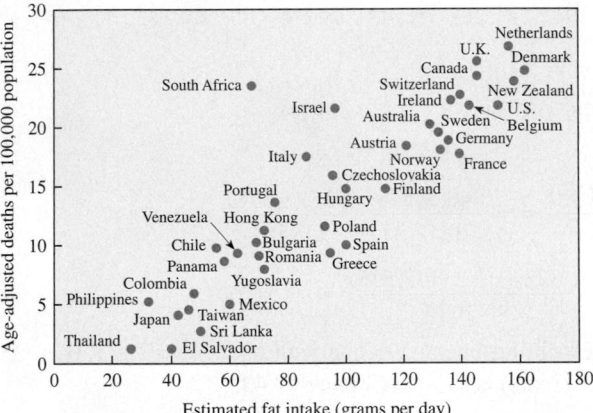

Source: Graph by Slim Films, from "Diet and Cancer," by Leonard A. Cohen, *Scientific American*, November 1987, p. 44. © 1987 by Scientific American, Inc. All rights reserved.

6. **IN YOUR OWN WORDS** Discuss the correlation shown by the following chart.

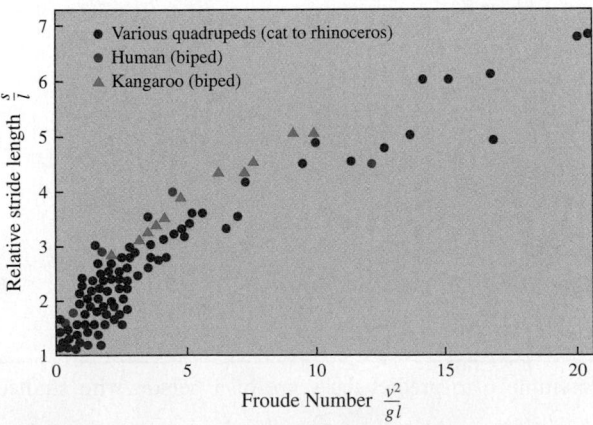

Source: Graph by Patricia J. Wynne, from "How Dinosaurs Ran," by R. McNeill Alexander, *Scientific American*, April 1991, p. 132. © 1991 by Scientific American, Inc. All rights reserved.

In Problems 7–18, a sample of paired data gives a linear correlation coefficient r. In each case, use Table 12.10 to determine whether there is a significant linear correlation.

7. $n = 10, r = 0.7$; 1% level

8. $n = 30, r = 0.4$; 1% level

9. $n = 30, r = 0.4$; 5% level

10. $n = 15, r = -0.732$; 5% level

11. $n = 35, r = -0.413$; 1% level

12. $n = 50, r = -0.3416$; 1% level

13. $n = 25, r = 0.521$; 1% level

14. $n = 100, r = -0.4109$; 1% level

15. $n = 40, r = 0.416$; 5% level

16. $n = 20, r = -0.214$; 5% level

17. $n = 10, r = -0.56$; 1% level

18. $n = 10, r = 0.7$; 5% level

Draw a scatter diagram and find r for the data shown in each table in Problems 19–24.

19. x	y	20. x	y	21. x	y
4	0	1	1	1	30
5	−10	2	5	3	22
10	−10	3	8	3	19
10	−20	4	13	5	15
				8	10

22. x	y	23. x	y	24. x	y
0	25	85	80	10	20
1	19	90	40	20	48
2	16	100	30	30	60
3	12	102	28	30	58
4	10	105	25	50	70
				60	75

Find the regression line for the data points in Problems 25–30.

25. x	y	26. x	y	27. x	y
4	0	1	1	1	30
5	−10	2	5	3	22
10	−10	3	8	3	19
10	−20	4	13	5	15
				8	10

28. x	y	29. x	y	30. x	y
0	25	85	80	10	20
1	19	90	40	20	48
2	16	100	30	30	60
3	12	102	28	30	58
4	10	105	25	50	70
				60	75

Match the equation and correlation in Problems 31–36 with a graph.

31. $y = 0.6x + 2; r = 0.9$

32. $y = 0.5x + 2; r = 0.7$

33. $y = 0.4x + 2; r = 0.5$ **34.** $y = -0.4x + 2; r = -0.4$

35. $y = -0.5x + 2; r = -0.6$ **36.** $y = -0.7x + 2; r = -0.8$

A.
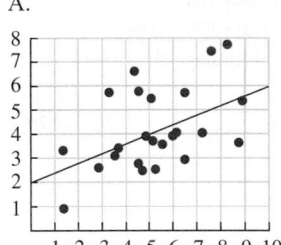

B.
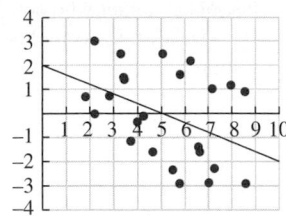

C.
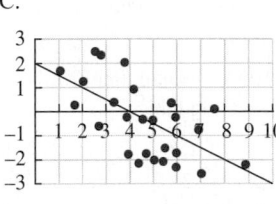

D.
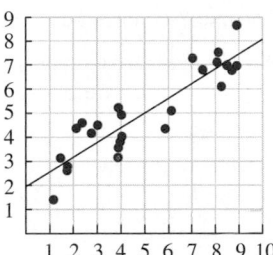

E.
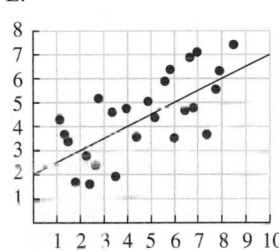

F.
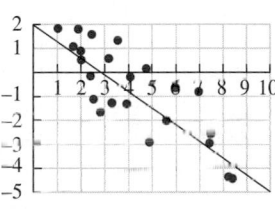

LEVEL 2

Given the information in Problems 37–40, find the equation for the least squares line as well the correlation coefficient.

37. $n = 8, \Sigma x = 40, \Sigma y = 48, \Sigma xy = 240,$
$\Sigma x^2 = 210, \Sigma y^2 = 288$

38. $n = 10, \Sigma x = 30, \Sigma y = 20, \Sigma xy = 60,$
$\Sigma x^2 = 90, \Sigma y^2 = 40$

39. $n = 6, \Sigma x = 36, \Sigma y = 48, \Sigma xy = 322,$
$\Sigma x^2 = 250, \Sigma y^2 = 418$

40. $n = 10, \Sigma x = 57, \Sigma y = 1, \Sigma xy = -13,$
$\Sigma x^2 = 353, \Sigma y^2 = 45$

41. U.S. wine consumption over the last few years (in gallons per person per year) is shown in the following table.

Year	1988	1989	1990	1991	1993	1995
Wine consumption	2.24	2.09	2.05	1.85	1.74	1.80

Compare these numbers with the U.S. beer consumption (found in Example 5, page 646) for those years in which both numbers are reported. Are beer drinking and wine drinking correlated? What is the correlation coefficient?

42. A group of ten people were selected and given a standard IQ test. These scores were then compared with their high school grades.

IQ	117	105	111	96	135	81	103	99	107	109
Grade (GPA)	3.1	2.8	2.5	2.8	3.4	1.9	2.1	3.2	2.9	2.3

Find r and determine whether it is statistically significant at the 1% level.

43. A new computer circuit was tested and the times (in nanoseconds) required to carry out different subroutines were recorded.

Difficulty	1	2	2	3	4	5	5	5
Time	10	11	13	8	15	18	21	19

Find r and determine whether it is statistically significant at the 1% level.

44. Find the regression line for the data in Problem 42. Assume x is the IQ and y is the GPA.

45. Find the regression line for the data in Problem 43. Assume x is the difficulty level and y is the time.

46. The following data are measurements of temperature ($x = °F$) and chirping frequency ($y = $ chirps per second) for the striped ground cricket. Is there a correlation between temperature and chirping, and if so, is it significant at the 5% or the 1% significance level?

Temperature	31.4	22.0	34.1	29.1	27.0	24.0	20.9
Frequency	20.0	16.0	19.8	18.4	17.1	15.5	14.7

Temperature	27.8	20.8	28.5	26.4	28.1	27.0	28.6	24.6
Frequency	17.1	15.4	16.2	15.0	17.2	16.0	17.0	14.4

47. The following data are the number of years of full-time education (x) and the annual salary in thousands of dollars (y) for 15 persons. Is there a correlation between education and salary, and if so, is it significant at the 5% or the 1% significance level?

Education	20	27	28	18	13	18	9	16
Salary	35.2	24.6	23.7	33.3	24.4	33.4	11.2	32.3

Education	16	12	12	19	16	14	13
Salary	25.1	22.1	18.9	37.8	25.9	28.4	29.6

48. A researcher chooses and interviews a group of 15 male workers in an automobile plant. The researcher then gives a score (x) ranging from 1 to 20 based on a scale of patriotism—the higher the score, the more patriotic the person appears to be. Each person is then given a written test and is scored (y) on their patriotism. Is there a correlation between the researcher score and the test score, and if so, is it significant at the 5% or the 1% significance level?

Researcher	10	14	15	17	17	18	18	19
Test	15	12	19	8	9	16	17	6

Researcher	16	18	20	12	14	9	17
Test	11	14	12	11	10	12	6

49. A bank records the number of mortgage applications and its own prevailing interest rate (at the first of the month) for each of 16 consecutive months. Is there a correlation between the interest rate (x) and the number of applicants (y), and if so, is it significant at the 5% or the 1% significance level?

Interest	9.5	9.9	10.0	10.5	11.0	11.5	11.0	12.0	12.0
Number	27	29	25	25	19	20	17	13	15

Interest	12.5	13.0	13.5	13.0	12.5	11.5	11.5
Number	10	10	6	5	5	11	14

50. Find the best-fitting line for the data in Problem 46.

51. Find the best-fitting line for the data in Problem 47.

52. Find the best-fitting line for the data in Problem 48.

53. Find the best-fitting line for the data in Problem 49.

Problems 54–59 are based on a 100-mile footrace, called the Hardrock 100, held in Colorado each year. *

54. The winner of the 1994 race was Scott Hirst. Here are his arrival times at the first 6 aid stations:

Time[†] (in hours)	2:57	5:03	7:22	9:26	10:10	11:16
Distance (in miles)	12.2	19.0	28.0	33.4	36.6	43.8

Predict his arrival time at the next aid station at 51.7 miles.

55. IN YOUR OWN WORDS The winner of the 1994 race was Scott Hirst. Here are his arrival times at the first 6 aid stations:

Time[†] (in hours)	2:57	5:03	7:22	9:26	10:10	11:16
Distance (in miles)	12.2	19.0	28.0	33.4	36.6	43.8

Predict his arrival time at the finish of the race (100.1 miles). His actual time of arrival was 32:00. What is the difference between the predicted and actual arrival times? Comment on the results.

56. The last-place runner who completed the 1994 race was John DeWalt. Here are his arrival times at the first 6 aid stations:

Time[†] (in hours)	3:53	6:56	10:36	14:12	15:26	17:50
Distance (in miles)	12.2	19.0	28.0	33.4	36.6	43.8

Predict his arrival time at the next aid station at 51.7 miles.

LEVEL 3

57. IN YOUR OWN WORDS The Hardrock 100 is a 100-mile footrace held in Colorado each year.* The last-place runner who completed the 1994 race was John DeWalt. Here are his arrival times at the first 6 aid stations:

Time (in hours)	3:53	6:56	10:36	14:12	15:26	17:50
Distance (in miles)	12.2	19.0	28.0	33.4	36.6	43.8

Predict his arrival time at the finish of the race (100.1 miles). His actual time of arrival was 47:50. What is the difference between the predicted and actual arrival times? Comment on the results.

58. The ages of the runners in the 1994 race and their finishing positions are given in the following table.

Finish	1	2	3	4	5	6
Age	33	46	32	31	42	38
Finish	7	8	9	10	11	12
Age	43	39	43	44	35	38
Finish	13	14	15	16	17	18
Age	42	27	49	49	40	40
Finish	19	20	21	22	23	24
Age	43	47	41	36	59	48
Finish	25	26	27	28	29	30
Age	51	36	31	33	51	43
Finish	31	32	33	34	36	37
Age	34	56	54	46	35	58

Is there a significant correlation between age and running times?[‡]

59. The ages of the runners in the 1994 race and their finishing times are given in the following table.[§]

Age	33	46	32	31	42	38
Time	32:00	32:20	32:29	33:57	34:36	35:29
Age	43	39	43	44	35	38
Time	35:52	36:54	38:04	38:04	38:20	38:43
Age	42	27	49	49	40	40
Time	39:18	39:21	41:50	42:06	42:06	42:06
Age	43	47	41	36	59	48
Time	42:59	43:41	43:41	43:41	44:46	44:46
Age	51	36	31	33	51	43
Time	44:46	45:21	46:19	46:26	46:49	47:08
Age	34	56	54	46	35	58
Time	47:21	47:27	47:43	47:46	47:50	47:50

Is there a significant correlation between age and running times?

60. IN YOUR OWN WORDS A study was made to see whether a correlation existed between students' grades in a course and students' rankings of the professor's teaching.[*] The results of this survey are shown in Table 12.11. Use these results to formulate a conclusion. Remove the last data point from the table (class no. 47). This was a class who rated the professor's teaching low, and the students' grades were also very low. Does this change your conclusion? Elaborate on your answer.

[*]"The Correlation Coefficient and Influential Data Points," by Donald J. Dessart, *The Mathematics Teacher*, March 1997, pp. 242–246.

Table 12.11 Mean Student Academic Grades and Mean Teacher Ratings for Forty-seven Classes

Class No.	Grade	Rating	Class No.	Grade	Rating
1	3.19	1.00	24	3.00	2.10
2	3.58	1.06	25	2.68	2.22
3	3.89	1.17	26	2.36	2.25
4	2.50	1.30	27	3.35	1.00
5	3.11	1.38	28	3.27	1.00
6	2.81	1.39	29	3.11	1.13
7	3.73	1.40	30	3.18	1.27
8	3.06	1.43	31	3.16	1.31
9	3.55	1.44	32	2.83	1.33
10	3.63	1.50	33	3.05	1.50
11	3.71	1.55	34	3.61	1.50
12	2.00	1.56	35	2.92	1.53
13	3.19	1.60	36	3.17	1.58
14	3.23	1.70	37	2.18	1.58
15	3.38	1.70	38	3.41	1.73
16	3.21	1.79	39	3.55	1.80
17	3.25	1.82	40	2.69	1.82
18	2.63	1.83	41	2.55	1.90
19	3.27	1.85	42	3.08	1.91
20	2.58	2.00	43	3.00	1.92
21	3.13	2.00	44	3.67	2.20
22	3.19	2.00	45	3.67	2.27
23	3.46	2.07	46	2.95	2.60
			47	1.68	2.80

12.5   SAMPLING

The first three sections of this chapter dealt with what is called *descriptive statistics,* which is concerned with the accumulation of data, measures of central tendency, and dispersion. A second branch of statistics is **inferential statistics,** which is concerned with making generalizations or predictions about a population based on a sample from that population.

A **sample** is a group of items chosen to represent a larger group. The larger group is called a **population.** Thus, the sample is a proper subset of the population. The population to be considered for a statistical application is called the **target population.** The sample is analyzed; then based on this analysis, some conclusion about the entire population is made. Sampling necessarily involves some error, because the sample and the population are not identical. A great deal of effort in statistics is devoted to the methodology of sampling. Here are some common sampling procedures:

Simple random sampling: This sample is obtained in a way that allows every member of the target population to have the same chance of being chosen. This is the type of sampling we assumed so far in Chapters 11–12.

Systematic sampling: The sample is obtained by drawing every *k*th item on a list or production line. The first item is determined by using a random number.

Cluster sampling: This procedure is applied on a geographical basis. The result is sometimes known as an *area sample.*

Stratified sampling: The entire population is divided into parts, called *strata,* according to some factor (such as sex, age, or income). When a population has varied characteristics, it is desirable to separate the population into homogeneous strata, and then take a random sample from each stratum.

To choose the most appropriate procedure for selecting an unbiased sample from a target population, we must take into account two considerations:

1. *Is the procedure random?* For example, asking people to voluntarily mail or phone in their preferences is not random, since the respondents select themselves (and there is nothing to prevent someone from sending in more than one response). Likewise, selecting names from an alphabetical listing may introduce an ethnic bias and is not random.

2. *Does the procedure take into account the intended target population?* For example, if you wish to know how many students at a certain college are interested in expanding the hours of operation at the college library, then it would not make sense to send the survey forms to the entire population of the city. Any procedure that would be appropriate in this case would ensure that only college students were surveyed.

EXAMPLE 1

A long-distance phone provider wants to sell a new "unlimited minutes" plan, and wishes to conduct a survey to find the degree of interest its customers have in such a plan. Which of the following methods would be most appropriate for obtaining an unbiased sample?

A. Survey the current customers whose names are chosen from an alphabetical listing of all customers on the west coast.

B. Survey a random selection of customers whose names are chosen from an alphabetical listing of all customers.

C. Survey the first 1,000 customers whose names are chosen from an alphabetical listing of all the customers.

D. Survey the first 1,000 customers whose names are chosen from a list purchased from a competing phone company.

E. Survey 1,000 persons whose names are chosen from the local telephone directory.

Solution

The solution depends on identifying the target population; it is the current customers of the long-distance phone company.

Choices D and E are not appropriate, because they do not reach the target population. Choice C hits the target population, but is not sufficiently random. Choice A hits only part of the target population. Therefore, choice B is the most appropriate since it addresses the target population in a random manner. ◆

The inference drawn from a poll can, of course, be wrong, so statistics is also concerned with estimating the error involved in predictions based on samples. In 1936, the *Literary Digest* predicted that Alfred Landon would defeat Franklin D. Roosevelt—who was subsequently reelected president by a landslide. (The magazine ceased publication the following year.) In 1948, the *Chicago Daily Tribune* drew an incorrect conclusion from its polls and declared in a headline that Thomas Dewey had just been

© CORBIS

elected president over Harry S. Truman. And in 1976, the *Milwaukee Sentinel* printed the erroneous headline shown here about the Wisconsin Democratic primary, again based on the result of its polls.

In an attempt to minimize error in their predictions, statisticians follow very careful procedures:

Step 1. Propose some hypothesis about a population.

Step 2. Gather a sample from the population.

Step 3. Analyze the data.

Step 4. Accept or reject the hypothesis.

Suppose that you want to decide whether a certain coin is a "fair" coin. You decide to test the hypothesis, "This is a fair coin," by flipping the coin 100 times. This provides a *sample*. Suppose the result is

> Heads: 55   Tails: 45

Do you accept or reject the hypothesis that "This coin is fair"? The expected number of heads is 50, but certainly a fair coin might well produce the results obtained.

As you can readily see, two types of errors are possible:

Type I: Rejection of the hypotheses when it is true

Type II: Acceptance of the hypothesis when it is false

How can we minimize the possibility of making either error? Let's carry this example further, and repeat the experiment of flipping the coin 100 times:

Trial number	1	2	3	4	5	6	. . .
Number of heads	55	52	54	57	59	55	. . .

If the coin is fair and we repeat the experiment a large number of times, it can be shown mathematically that the distribution should be normal, with a mean of 50 and a standard deviation of 5, as shown in Figure 12.44. The question is whether to accept the *unknown* test coin as fair.

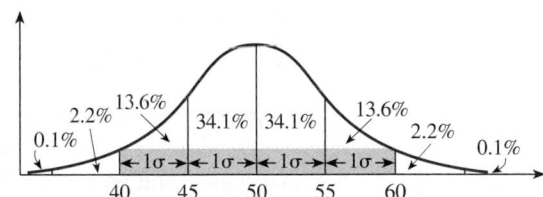

Figure 12.44 Given normal distribution for the number of heads upon flipping a fair coin 100 times

Suppose that you are willing to accept the coin as fair only if the number of heads falls between 45 and 55 (that is, within 1 standard deviation of the expected number). If you adopt this standard, you know you will be correct 68% of the time if the coin is fair. How do you know this? Look at Figure 12.44, and note that 34.1% of the results are within $+1\sigma$ and 34.1% are within -1σ of the mean; the total is $34.1\% + 34.1\% = 68.2\%$.

But a friend says, "Yes, you will be correct 68% of the time, but you will also be rejecting a lot of fair coins!" You respond, "But suppose that a coin really is a bad coin (it really favors heads), with a mean number of heads of 60 and a standard deviation of 5. If I adopted the same standard ($\pm 1\sigma$), I'd be accepting all the coins in the shaded region of Figure 12.45" (page 654).

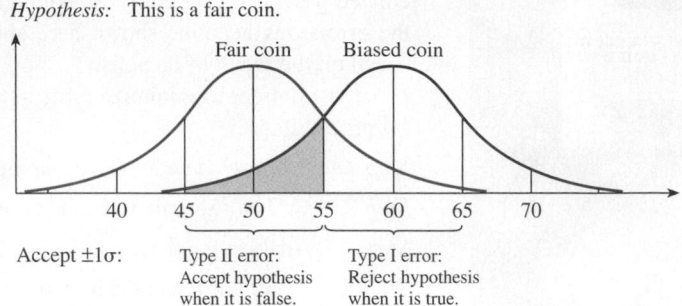

Hypothesis: This is a fair coin.

Figure 12.45 **Comparison of Type I and Type II errors**

EXAMPLE 2

Consider the hypothesis that the mean grade for a class is 75. We conduct a survey of 5 people from the class and find that their grades are 85, 65, 70, 52, and 96. There are four possibilities:

We decide to reject the conclusion that the mean grade is 75 and:
A. The actual mean is 75.
B. The actual mean is not 75.

We decide to accept the conclusion that the mean grade is 75 and:
C. The actual mean is 75.
D. The actual mean is not 75.

For each of the four possibilities, state whether the conclusion is true or which type of error has been made.

Solution

A. reject a true conclusion; Type I error

B. correct decision

C. correct decision

D. accept a false conclusion; Type II error   ◆

As you can see, decreasing the Type I error probability increases the Type II error probability, and vice versa. Deciding which type of error to minimize depends on the stakes involved and on some statistical calculations that go beyond the scope of this course.

Consider a company that produces two types of valves. The first type is used in jet aircraft, and the failure of this valve might cause many deaths. A sample of the valves is taken and tested, and the company must accept or reject the entire shipment on the basis of these test results. Under these circumstances, the company would rather reject many good valves than accept a bad one. On the other hand, the second valve is used in toy airplanes; the failure of this valve would merely cause the crash of the model. In this case, the company wouldn't want to reject too many good valves, so they would minimize the probability of rejecting the good valves.

Many times we read of a poll in the newspaper or hear of it on the evening news, and a percentage is given. For example, "The candidate was favored by 48% of those sampled." What is not often stated (or is stated only in small print) is that there is a margin of error and a confidence level. For example, "The margin of error is 4 percent at a confidence level of 95%." This means that we are 95% confident that the actual percent of people who favor the candidate is between 44% and 52%.

> Poll-taker to boss:
>   "Our latest opinion poll showed that 90% of the people aren't interested in the opinions of others."

EXAMPLE 3

My mother, who is 78, has been a smoker since she was 18 years old. Even though she has emphysema, she claims that she will live well into her eighties, which is proof that the smoking warning claims do not mean a thing. What is wrong with this reasoning?

Solution

It is known that the average nonsmoker lives about $5\frac{1}{2}$ years longer than the average smoker. Consider the normal curves shown in Figure 12.46.

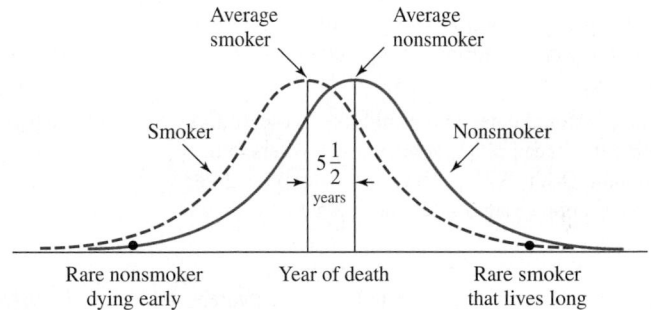

Figure 12.46 Life expectancies for smokers and nonsmokers

As you can see, sometimes the nonsmoker will die early (analogous to a Type II error) and sometimes the smoker will live long (analogous to a Type I error).   ◆

Many times you may hear reasoning or forming a conclusion by looking at one particular case, which may be an exception. It occurs so often, in fact, that we give this type of reasoning a name— we call it the **fallacy of exceptions.**

Problem Set 12.5

LEVEL 1

1. **IN YOUR OWN WORDS** When you hear the word *statistics,* what comes to mind? Look up the word in a dictionary, and see if you wish to supplement or change any part of your answer.

2. **IN YOUR OWN WORDS** Explain the difference between descriptive and inferential statistics.

3. **IN YOUR OWN WORDS** Explain the fallacy of the exception.

4. An ice cream store wants to use a survey to determine which newspapers its customers regularly read. Which of the following procedures would be most appropriate for obtaining a statistically unbiased sample?
   A. Survey a randomly selected sample of people living in the city's limits.
   B. Survey the first 100 customers on an alphabetical listing of all the ice cream store's customers.
   C. Survey 100 customers whose names are randomly chosen from an alphabetical listing of all customers of the ice cream store.
   D. Survey 100 people whose names are randomly chosen from the local telephone directory.

5. In a certain geographical market, Sprint wishes to increase its market share by taking a survey of customers of its competitor, Pacific Bell, in an effort to find out which parts of its service are least satisfactory. Which of the following procedures would be most appropriate for obtaining a statistically unbiased sample?
   A. Survey a random selection of people chosen from a list of Sprint's customers.
   B. Survey a random selection of people chosen from a list of Pacific Bell's customers.
   C. Place a newspaper advertisement asking for survey participants, and then randomly choose 100 people from those who respond.
   D. Call local phone numbers and ask the person who answers if they are satisfied with their phone service.

6. A movie theater would like to use a survey to determine which factors are most important to those who come to their theater. Which of the following procedures would be most appropriate for obtaining a statistically unbiased sample?
   A. Survey a random selection of people passing by the theater in the mall in front of the theater.
   B. Survey a random selection of shoppers in the mall.
   C. Survey a random selection of people who are leaving the theater after watching a movie.
   D. Induce people to answer survey questions by offering discount coupons for various city events, including the movie theater.

7. A local congressional candidate would like to use a survey to determine the issues of importance to the voters in her congressional district. Which of the following procedures would be most appropriate for obtaining a statistically unbiased sample?
   A. Survey a selection of people whose names are randomly chosen from a local list of registered voters who are party members.
   B. Survey a selection of people whose names are randomly chosen from a local list of all registered voters.
   C. Survey a selection of people whose names are randomly chosen from the telephone directory.
   D. Survey a selection of people whose names are randomly chosen from a national list of party members.

8. A school board wishes to determine opinions of parents regarding the assigning of homework in mathematics classes. Which of the following procedures would be most appropriate for obtaining a statistically unbiased sample?
   A. Survey a selection of parents from the official school roster of parents.
   B. Survey a selection of people whose names are randomly chosen from the telephone directory.
   C. Survey people chosen from the first 1,000 names on an alphabetical listing of parents of mathematics students.
   D. Survey a selection of people whose names are randomly chosen from a listing of parents of mathematics students.

9. An environmental group wants to use a survey to determine the extent to which its members would be willing to protest a recent decision of the governor of a particular state. Which of the following procedures would be most appropriate for obtaining a statistically unbiased sample?
   A. Survey a selection of people whose names are randomly chosen from the telephone directory.
   B. Have sign-ups for workers to gather signatures at local malls throughout the state.
   C. Survey members attending the annual convention of the environmental group.
   D. Survey a selection of people whose names are randomly chosen from a list of members of the environmental group.

10. Cal Trans wants to use a survey to determine the extent of public approval for a project to construct a new freeway through a particular neighborhood. Which of the following procedures would be most appropriate for obtaining a statistically unbiased sample?
    A. Go door-to-door in the affected neighborhood, soliciting opinions.
    B. Hold a public forum in the affected neighborhood, and distribute questionnaires there.
    C. Place advertisements in newspapers in the affected neighborhood, asking readers to mail in their opinions.
    D. Survey a selection of people whose names are randomly chosen from the city telephone directory.

11. Betty's Koi Emporium has decided to conduct a survey to determine the number of its customers who order supplies on the Internet. Which of the following procedures would be most appropriate for obtaining a statistically unbiased sample?
    A. Survey a random selection of customers chosen from a list of all people who have purchased an item from Betty's Koi Emporium.
    B. Distribute questionnaires to people who make a purchase in the next 30 days.
    C. Post notices soliciting the necessary information on bulletin boards in the Koi Emporium.
    D. Survey people coming out of a nearby shopping mall.

12. Krispy Kreme is considering building a franchise in a particular city. It will use a survey to determine the extent of interest among residents of the city. Which of the following procedures would be most appropriate for obtaining a statistically unbiased sample?
    A. Survey a random selection of people whose names are obtained from national sales records of Krispy Kreme customers.
    B. Survey a selection of people who are listed on the residential roll sheets of the city.
    C. Survey a selection of people whose names are randomly chosen from the telephone directory of the target city.
    D. Survey a random selection of customers coming out of other donut shops in the city.

13. An espresso coffee franchise is considering expanding its services to a new location in the city, but prior to doing so it will use a survey to determine the extent to which its customers are interested in a second location. Which of the following procedures would be most appropriate for obtaining a statistically unbiased sample?
    A. Survey a selection of people whose names are randomly chosen from the telephone directory for the city.
    B. Ask customers who drive through to voluntarily phone in their preferences.
    C. Ask customers who drive through to voluntarily mail in their preferences.
    D. Survey a selection of people whose names are randomly chosen from a list of all customers.

In Problems 14–23, *decide on a reasonable means for conducting the survey to obtain the desired information.*

14. A newsstand wishes to use a survey to determine which out-of-town newspapers it should regularly stock.

15. Sprint would like to take customers from AT&T, one of its competitors in a certain geographical area. Sprint would like to survey AT&T's customers in order to find out which elements of its service are least satisfactory.

16. A pet store would like to use a survey to determine which factors are most important to cat owners in determining the brand of cat food that they purchase.

17. A local political party would like to use a survey to determine the level of support among party members for the party's candidates in the upcoming primary election.

18. A health and fitness club would like to survey its members in order to determine which new equipment they would prefer.

19. A union wants to use a survey to determine the extent to which its members approve of a newly negotiated collective bargaining agreement.

20. The city council wants to use a survey to determine the extent of public approval for a project to construct a new playground in a certain residential neighborhood.

21. The college student government will use a survey to determine the level of support among students for a proposal to lower the student fees.

22. A retailer is considering offering extended warranty policies on video recorders. It will use a survey to determine the extent of interest among owners of the video recorders.

23. A supermarket is considering expanding its services by adding a florist, but prior to doing so it will use a survey to determine the extent to which its customers are interested in such a service.

24. Suppose you were interested in knowing the mean score for a certain test in a mathematics class with 20 students. You sample four members of the class and find their test scores to be 49, 56, 72, and 96. The instructor reports that the mean was 72, but you are not sure that is truthful. List the four possibilities and categorize the conclusions including the types of possible errors.

25. Suppose you were interested in knowing the mean score for a certain test in a mathematics class with 20 students. You sample four members of the class and find their test scores to be 52, 56, 96, and 99. The instructor reports that the mean was 72, but you are not sure that is truthful. List the four possibilities and categorize the conclusions including the types of possible errors.

26. Suppose that of 80 workers randomly selected and interviewed, 55 were opposed to an increase in Social Security taxes. Would you accept or reject the hypothesis that the majority of workers are opposed to the increase in taxes? List the four possibilities and categorize the conclusion including the types of possible errors.

27. Conduct a survey asking the following questions.
    a. Do you doodle?
    b. If you doodle, which of the following do you doodle?
       (1) circles, curves, or spirals
       (2) squares, rectangles, or other straight-line designs
       (3) people or human features
       (4) symbols, such as stars, arrows, or other objects
       (5) words, letters, or numerals
       (6) other
    c. When do you most often doodle?
       (1) while at a meeting or class
       (2) while on the phone at home
       (3) when solving a problem
       (4) when you have nothing else to do
       (5) when writing a letter
       (6) at meals

28. Conduct a survey to determine the major worry of college students.

29. Conduct a survey to determine whether there is a significant correlation between math scores in the 8th grade and amount of TV viewing.

30. Suppose that John hands you a coin to flip and wants to bet on the outcome. Now, John has tried this sort of thing before, and you suspect that the coin is "rigged." You decide to test this hypothesis by taking a sample. You flip the coin twice, and it is heads both times. You say, "Aha, I knew it was rigged!" John replies, "Don't be silly. Any coin can come up heads twice in a row." The following scheme was devised by mathematician John von Neumann to allow fair results even if the coin is somewhat biased. The coin is flipped twice. If it comes up heads both times or tails both times, it is flipped twice again. If it comes up heads–tails, this will decide the outcome in favor of the first party; and if it comes up tails–heads, this will decide the outcome in favor of the second party. Show that this will result in a fair toss even if the coins are biased.

CHAPTER SUMMARY

"If you want to understand nature, you must be conversant with the language in which nature speaks to us."

Richard Feynman, Everybody Counts: A Report to the Nation

IMPORTANT TERMS

Average [12.2]
Bar graph [12.1]
Bell-shaped curve [12.3]
Bimodal [12.2]
Box plot [12.2]
Circle graph [12.1]
Classes [12.1]
Continuous distribution [12.3]
Correlation [12.4]
Cumulative frequency [12.3]
Deciles [12.2]
Descriptive statistics [12.2]
Fallacy of exceptions [12.5]
Frequency [12.1]
Frequency distribution [12.1]
Graph [12.1]
Grouped frequency distribution [12.1]
Inferential statistics [12.5]

Interval [12.1]
Least squares line [12.4]
Least squares method [12.4]
Line graph [12.1]
Linear correlation coefficient [12.4]
Mean [12.2]
Measures of central tendency [12.2]
Measures of dispersion [12.2]
Measures of position [12.2]
Median [12.2]
Mode [12.2]
Normal curve [12.3]
Pearson correlation coefficient [12.4]
Percentile [12.2]
Pictograph [12.1]
Pie chart [12.1]
Population [12.5]

Quartile [12.2]
Range [12.2]
Regression analysis [12.4]
Sample [12.5]
Scatter diagram [12.4]
Significance level [12.4]
Skewed distribution [12.3]
Standard deviation [12.2]
Statistics [Overview]
Stem-and-leaf plot [12.1]
Target population [12.5]
Type I error [12.5]
Type II error [12.5]
Variance of a population [12.2]
Variance of a random sample [12.2]
Weighted mean [12.2]
***z*-score** [12.3]

TYPES OF PROBLEMS

Prepare a frequency distribution. [12.1]
Draw a bar graph. [12.1]
Draw a line graph. [12.1]
Draw a stem-and-leaf plot. [12.1]
Draw a circle graph. [12.1]
Draw a pictograph. [12.1]
Read and interpret relationships presented in graphical form. [12.1]
Recognize misuses of graphs. [12.1]
Find the mean, median, and mode for a set of data. [12.2]
Find the range, standard deviation, and variance for a set of data. [12.2]
Find a cumulative distribution. [12.3]
Interpret information given in table form. [12.3]
Find the expected numbers for ranges of a normally distributed set of data. [12.3]
Determine the probability of falling within a certain range of a normally distributed set of data. [12.3]
Find and use z-scores. [12.3]
Draw a scatter diagram for a data set. [12.4]
Decide whether there is a significant linear correlation between two given variables. [12.4]
Find a regression line for a data set. [12.4]
Discuss the type of correlation for a given data set. [12.5]
Determine whether there is a significant linear correlation, given the number of items and the correlation coefficient. [12.4]

Find the correlation coefficient for a given set of data. [12.4]
Choose an appropriate procedure for selecting an unbiased sample. [12.5]
Classify Type I and Type II errors. [12.5]
Make an inference about a population by taking a sample. [12.5]

CHAPTER 12 REVIEW QUESTIONS

1. **a.** Make a frequency table for the following results of tossing a coin 40 times:
    HTTTT   HHTHH   TTHHT   THTHT   HHTTT   THHTH   HHTHT   TTTTT
   b. Draw a bar graph for the number of heads and tails given in part **a.**

2. The table below shows the U.S. government's expenditures for defense.

Year	Expenditures (billions)
1993	$279
1994	$269
1995	$260
1996	$253

   a. Draw a bar graph to represent these data.
   b. Suppose that you had the following viewpoint: *Defense spending down only* $0.3 *trillion from* 1993 *to* 1996. Draw a graph that shows very little decrease in the expenditures.
   c. Suppose you had the following viewpoint: *Defense expenditures drastically cut* $26,000,000,000 *from* 1993 *to* 1996. Draw a graph that shows a tremendous decrease in the expenditures.
   d. In view of parts **b** and **c**, discuss the possibilities of using statistics to mislead or support different views.

Use the data set {5, 21, 21, 25, 30, 40} *in Problems 3–7 to find the requested value.*

3. mean   4. median   5. mode   6. range   7. standard deviation

8. A small grocery store stocked several sizes of Copycat cola last year. The sales figures are shown in the table. Find the mean, the median, and the mode. If the store manager decides to cut back the variety and stock only one size, which measure of central tendency will be most useful in making this decision?

Size	No. of cases sold
6 oz	5
10 oz	10
12 oz	35

9. A student's scores in a certain math class are 72, 73, 74, 85, and 91. Find the mean, the median, and the mode. Which measure of central tendency is most representative of the student's scores?

10. The 1993–1994 school superintendents' salaries in Sonoma County are shown:

Rohnert Park, $90,000	Santa Rosa, $86,958	Sonoma Valley, $86,004
Cloverdale, $84,665	Analy, $84,522	Winsor, $83,500
SCOE, $83,473	Mark West, $82,543	Rincon Valley, $81,000
Petaluma, $80,814	Forestville, $76,944	Roseland, $74,370
Gravenstein, $73,378	Bellevue, $72,057	Twin Hills, $71,564
Old Adobe, $71,000	Guerneville, $69,965	Sebastopol, $69,123
Geyersville, $69,000	Wright, $67,691	Healdsburg, $67,562
Piner-Olivet, $67,300	Bennett Valley, $65,600	Harmony, $64,009
Waugh, $60,000	Two Rock, $57,760	Wilmar, $57,000
Alexander Valley, $54,528	Cinnabar, $54,500	Monte Rio, $49,016
Liberty, $47,412	West Side, $44,474	Oak Grove, $43,625
Fort Ross, $42,515	Horicon, $41,800	Montgomery, $39,000
Kenwood, $28,550	Dunham, $16,950	

Find the mean, the median, and the mode. Which measure of central tendency is most representative of the superintendents' salaries?

11. The blockbuster movie *Titanic* brought about a renewed interest in that disaster. Use Table 12.12 to answer the following questions.

Table 12.12 Passengers and Survivors on the *Titanic*

Passenger category	Number of passengers	Number of survivors
Children, 1st class	6	6
Children, 2nd class	24	24
Children, 3rd class	79	27
Women, 1st class	144	140
Women, 2nd class	93	80
Women, 3rd class	165	76
Men, 1st class	175	57
Men, 2nd class	168	14
Men, 3rd class	462	75

    a. Was a difference in the survival rates related to the class of passenger? Draw a graph of survival rates of first-, second-, and third-class passengers.

    b. Was a difference in the survival rates related to the gender or age of the passenger? Draw a graph of survival rates of men, women, and children.

12. Consider the line graph showing the median family income in 1990 dollars.

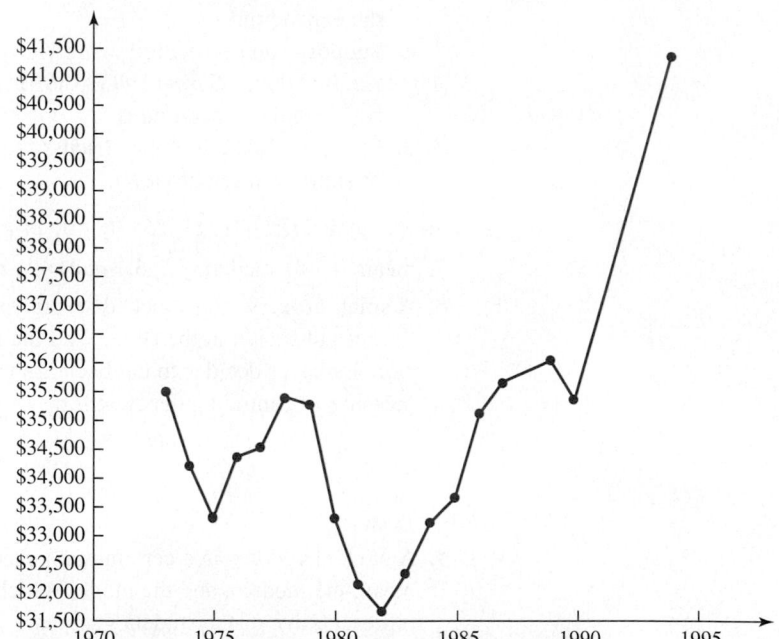

    a. In which year (1973–1994) was the median family income the least?

    b. In which year (1973–1994) was the median family income the greatest?

    c. What was the approximate median family income in 1994?

13. The following table compares age with blood pressure.

Age (in years)	20	25	30	40	50	35	68	55
Blood pressure	85	91	84	93	100	86	94	92

    a. Draw a scatter diagram and find the regression line for this set of data.

    b. Find the linear correlation coefficient and determine whether the variables are significantly correlated at either the 1% or 5% level.

14. Read the *Dear Abby* column. Abby's answer was consoling and gracious, but not very statistical. If pregnancy durations have a normal distribution with a mean of 266 and a standard deviation of 16 days, what is the probability of having a 314-day pregnancy?

15. The fastest growing cities in California, and their populations are:

Cities	Population
Coalinga	15,200
Brisbane	4,060
Brentwood	23,100
Dublin	32,500
Rio Vista	4,850
Cupertino	52,900
La Quinta	24,250
Rocklin	35,250
Lincoln	9,675
Temecula	53,800

   a. Draw a bar graph showing the population of each of these cities.
   b. What are the mean, median, and mode for these populations?

GROUP RESEARCH

Working in small groups is typical of most work environments, and learning to work with others to communicate specific ideas is an important skill. Work with three or four other students to submit a single report based on each of the following questions.

G44.a. Consider a class in which the following test scores were obtained: 96, 92, 92, 89, 88, 87, 87, 87, 87, 80, 79, 79, 78, 76, 76, 76, 76, 74, 73, 72, 72, 71, 71, 71, 70, 66, 66, 60, 53, and 20. Find the mean, median, mode, and standard deviation. If this class is graded "on a curve," prepare a grading scale.

   b. Find the mean, median, mode, range, and standard deviation of the ages of the actors winning the best actor award at the Academy Awards. See Table 12.3 on page 609.

G45. Toss a toothpick onto a hardwood floor 1,000 times as described in Figure 12.47, or toss 1,000 toothpicks, one at a time, onto the floor. Let ℓ be the length of the toothpick and d be the distance between the parallel lines determined by the floorboards.

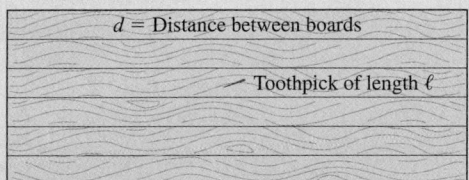

Figure 12.47 Buffon's needle problem

Equipment needed: A box of toothpicks (of uniform length) and a large sheet of paper with equidistant parallel lines. A hardwood floor works very well instead of using a sheet of paper. The length of a toothpick should be less than the perpendicular distance between the parallel lines.

a. Guess the probability p that a toothpick will cross a line. *Do this before you begin the experiment.* The members of your group should reach a consensus before continuing.

b. Perform the experiment and find p empirically. That is, to find p, divide the number of toothpicks crossing a line by the number of toothpicks tossed (1,000 in this case).

c. By direct measurement, find ℓ and d.

d. Calculate 2ℓ and pd, and $\dfrac{2\ell}{pd}$.

e. Formulate a conclusion. This is an experiment known as Buffon's needle problem.

G46. You are interested in knowing the number and ages of children (0–18 years) in a part (or all) of your community. You will need to sample 50 families, finding the number of children in each family and the age of each child. It is important that you select the 50 families at random. How to do this is the subject of a course in statistics. For this problem, however, follow these steps:

Step 1. Determine the geographic boundaries of the area with which you are concerned.

Step 2. Consider various methods for selecting the families at random. For example, could you:

  (i) select the first 50 homes at which someone is at home when you call?

  (ii) select 50 numbers from a phone book that covers the same geographic boundaries as those described in step 1?

  (iii) Using (i) or (ii) could result in a biased sample. Can you guess why this might be true? In a statistics course, you might explore other ways of selecting the homes. For this problem, use one of these methods.

Step 3. Consider different ways of asking the question. Can the way the family is approached affect the response?

Step 4. Gather your data.

Step 5. Organize your data. Construct a frequency distribution for the children, with integral values from 0 to 18.

Step 6. Find out the number of families who actually live in the area you've selected. If you can't do this, assume that the area has 1,000 families.

a. What is the average number of children per family?

b. What percent of the children are in the first grade (age 6)?

c. If all the children aged 12–15 are in junior high, how many are in junior high for the geographic area you are considering?

d. See if you can actually find out the answers to parts **b** and **c**, and compare these answers with your projections.

e. What other inferences can you make from your data?

INDIVIDUAL RESEARCH PROBLEMS

Learning to use sources outside your classroom and textbook is an important skill, and here are some ideas for extending some of the ideas in this chapter. You can find references to these projects in a library or at

 **www.mathnature.com**

PROJECT 12.1 Collect examples of good statistical graphs and examples of misleading graphs. Use some of the leading newspapers and national magazines, or Web sites.

PROJECT 12.2 Carry out the following experiment: *A cat has two bowls of food; one bowl contains Whiskas and the other contains some other brand. The cat eats Whiskas and leaves the other untouched. Make a list of possible reasons why the cat ignored the second bowl.* Describe the circumstances under which you think the advertiser could claim: *Eight out of ten owners said their cat preferred Whiskas.*

PROJECT 12.3 Roll a pair of dice 36 times, and draw a bar graph of the outcomes.

    a. Find the mean, the variance, and the standard deviation for this model.

    b. Repeat the experiment.

    c. Compare the results of parts **a** and **b**.

PROJECT 12.4 Prepare a report or exhibit showing how statistics are used in baseball.

PROJECT 12.5 Prepare a report or exhibit showing how statistics are used in educational testing.

PROJECT 12.6 Prepare a report or exhibit showing how statistics are used in psychology.

PROJECT 12.7 Prepare a report or exhibit showing how statistics are used in business. Use a daily report of transactions on the New York Stock Exchange. What inferences can you make from the information reported?

PROJECT 12.8 Investigate the work of Adolph Quetelet, Francis Galton, Karl Pearson, R. A. Fisher, and Florence Nightingale. Prepare a report or an exhibit of their work in statistics.

PROJECT 12.9 "We need privacy and a consistent wind," said Wilbur. "Did you write to the Weather Bureau to find a suitable location?" "Well," replied Orville, "I received this list of possible locations and Kitty Hawk, North Carolina, looks like just what we want. Look at this . . . ." However, Orville and Wilbur spent many days waiting in frustration after they arrived in Kitty Hawk, because the winds weren't suitable. The Weather Bureau's information gave the averages, but the Wright brothers didn't realize that an acceptable average can be produced by unacceptable extremes. Write a paper explaining how it is possible to have an acceptable average produced by unacceptable extremes.

PROJECT 12.10 Select something that you think might be normally distributed (for example, the ring size of students at your college). Next, select 100 people and make the appropriate measurements (in this example, ring size). Calculate the mean and standard deviation. Illustrate your finding using a bar graph. Do your data appear to be normally distributed?

PROJECT 12.11 Five identical containers (shoe boxes, paper cups, etc.) must be prepared for this problem, with contents as follows: There are five boxes containing red and white items (such as marbles, poker chips, or colored slips of paper).

Box	Contents
#1	15 red and 15 white
#2	30 red and 0 white
#3	25 red and 5 white
#4	20 red and 10 white
#5	10 red and 20 white

Select one of the boxes at random so that you don't know its contents.

Step 1. Shake the box.

Step 2. Select one marker, note the result, and return it to the box.

Step 3. Repeat the first two steps 20 times with the same box.

    a. What do you think is inside the box you have sampled?

    b. Could you have guessed the contents by repeating the experiment five times? Ten times? Do you think you should have more than 20 observations per experiment?

13 The Nature of Graphs and Functions

Algebra is but written geometry and geometry is but figured algebra.
G. B. HALSTED

CONTENTS

OVERVIEW

One of the most revolutionary ideas in the history of human thought was the joining together of algebra and geometry. It was basically a simple idea (as are most revolutionary ideas)—namely, representing an ordered pair from algebra as a point on a flat surface, called a plane. This was summarized by the great mathematician Lagrange, who said, "As long as algebra and geometry proceeded along separate paths, their advance was slow and their application limited. But when these sciences joined company, they drew from each other fresh vitality and thenceforth marched on at rapid pace toward perfection."

In this chapter, we introduce a tie between algebra and geometry: the idea that every curve can be associated with an equation and every equation can be associated with a graph. We consider both lines and nonlinear curves in this chapter.

IMPORTANT IDEAS

Graphing a line by plotting points and by slope–intercept [13.1]
Distinguish vertical lines and horizontal lines [13.1]
Graphing half-planes [13.2]
Graphing equations by plotting points [13.3]
Graphing parabolas [13.3]
Graphing exponential curves [13.3]
Recognizing conic sections from the general-form equation [13.4]
Graph standard-form conic sections [13.4]
Using functional notation [13.5]

BOOK REPORT

Write a 500-word report on the following book:
Beyond Numeracy: Ruminations of a Numbers Man. John Paulos (New York: Random House, 1992).

LINKS

Individual Projects

Group Projects

Research Links

www.mathnature.com

13.1 CARTESIAN COORDINATES AND GRAPHING LINES

Solving Equations with Two Variables

Let's consider an equation with two variables, say, x and y. If there are two values x and y that make an equation true, then we say that the ordered pair (x, y) **satisfies** the equation and that it is a **solution** of the equation.

> As long as algebra and geometry proceeded along separate paths, their advance was slow and their application limited. But when these sciences joined company, they drew from each other fresh vitality and thenceforth marched on at rapid pace toward perfection.
>
> Lagrange

EXAMPLE 1

Tell whether each ordered pair satisfies the equation $3x - 2y = 7$.

a. $(1, -2)$ **b.** $(-2, 1)$ **c.** $(-1, -5)$

Solution   You should substitute each pair of values of x and y into the given equation to see whether the equation is true or false. If it is true, the ordered pair is a solution.

a. $(1, -2)$ means $x = 1$ and $y = -2$:

$$3x - 2y = 3(1) - 2(-2)$$
$$= 3 + 4$$
$$= 7 \quad \text{Since } 7 = 7, \text{ we see that } (1, -2) \text{ is a solution.}$$

b. $(-2, 1)$ means $x = -2$ and $y = 1$ (compare with part **a** and notice that the order is important in determining which variable takes which value).

$$3x - 2y = 3(-2) - 2(1)$$
$$= -6 - 2$$
$$= -8$$

Since $-8 \neq 7$, we see that $(-2, 1)$ is not a solution.

c. $(-1, -5)$:

$$3x - 2y = 3(-1) - 2(-5)$$
$$= -3 + 10$$
$$= 7 \quad \text{The ordered pair } (-1, -5) \text{ is a solution.} \qquad \blacklozenge$$

Cartesian Coordinate System

Have we found all the ordered pairs that satisfy the equation $3x - 2y = 7$? Can you find others? Let's represent this information by *drawing a graph*. A **Cartesian coordinate system** is formed by drawing two perpendicular number lines, called **axes.** The point of intersection is called the **origin,** and the plane is divided into four parts called **quadrants,** which are labeled as shown in Figure 13.1.

Point $P(a, b)$ in Figure 13.1 is found in the plane by counting a units from the origin in a horizontal direction (either positive or negative), and then b units (either positive or negative) in a vertical direction. The number a is sometimes called the **abscissa** or *first component* and b the **ordinate** (or *second component*) of the point P. Together, a and b are called the **coordinates** of the point P.

Suppose we plot the ordered pairs in Example 1a and 1c, along with four others, as shown in Figure 13.2a.

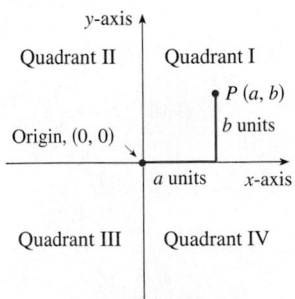

Figure 13.1 Cartesian coordinate system

Graphing a Line by Plotting Points

The process of graphing a line requires that you find ordered pairs that make an equation true. To do this, *you* must choose convenient values for x and then solve the resulting equation to find a corresponding value for y.

René Descartes
(1596–1650)
René Descartes, the person after whom we name the coordinate system, was a person of frail health. He had a lifelong habit of lying in bed until late in the morning or even the early afternoon. It is said that these hours in bed were probably his most productive. During one of these periods, it is presumed, Descartes made the discovery of the coordinate system. This discovery led to a branch of modern mathematics called **analytic geometry**, which can be described as an algebraic way of looking at geometry, or a geometric way of looking at algebra; simply put, it is the blending of algebra and geometry.

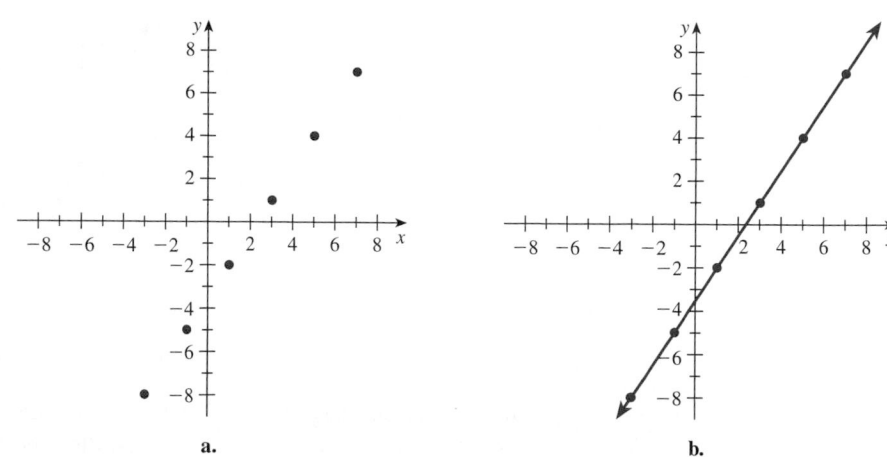

Figure 13.2 (a) Points satisfying $3x - 2y = 7$ and (b) the line passing through the points

EXAMPLE 2

Find three ordered pairs that satisfy the equation $y = -2x + 3$.

Solution *You* choose any x value—say, $x = 1$. Substitute this value into the given equation to find a corresponding y-value:

$$y = -2x + 3 \qquad \text{Given equation}$$
$$= -2(1) + 3 \qquad \text{Substitute chosen value.}$$
$$= -2 + 3$$
$$= 1$$

You choose this value.
↓
The first ordered pair is $(1, 1)$.
↑
You find this value by substitution into the equation.

Choose a second value—say, $x = 2$. Then,

$$y = -2x + 3 \qquad \text{Start with given equation.}$$
$$= -2(2) + 3 \qquad \text{Substitute.}$$
$$= -1 \qquad \text{Simplify.}$$

The second ordered pair is $(2, -1)$.

Choose a third value—say, $x = -1$. Then,

$$y = -2x + 3$$
$$= -2(-1) + 3$$
$$= 5$$

The third ordered pair is $(-1, 5)$. ◆

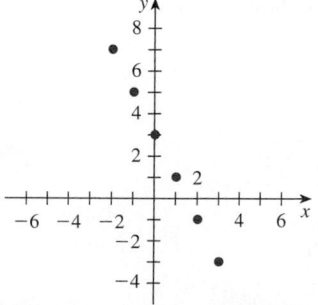

Figure 13.3 Points that satisfy $y = -2x + 3$

Have we found *all* the ordered pairs that satisfy the equation $y = -2x + 3$? Can you find others? Suppose we plot the ordered pairs in Example 2, along with three additional points, as shown in Figure 13.3. Do you notice anything about the arrangement of these points in the plane? Suppose that we draw a line passing through these points, as shown in Figure 13.4 (page 668). This line represents the set of *all* ordered pairs that satisfy the equation.

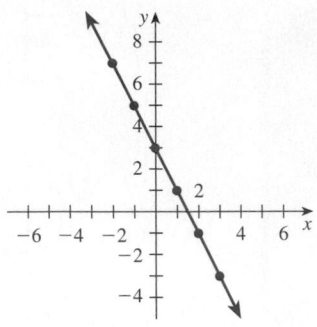

Figure 13.4 **Representation of all points that satisfy $y = -2x + 3$**

If we carry out the process of finding three ordered pairs that satisfy an equation of a line, and then we draw the line through those points, we say that we are *graphing the line,* and the final set of points we have drawn represents the **graph** of the line.

Procedure for Graphing a Line by Plotting Points

To graph a line, find two ordered pairs that lie on the line. These two points determine the line. Find a third point as a check. Draw the line (using a straight-edge) passing through these three points.

If the three points don't lie on a straight line, then you have made an error.

EXAMPLE 3

Graph $y = 2x + 2$.

Solution It is generally easier to pick x and find y.

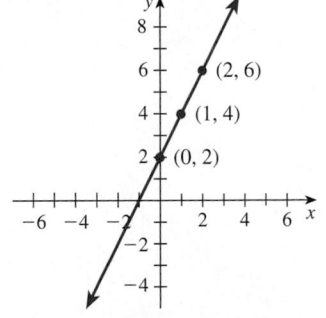

Figure 13.5 **Graph of $y = 2x + 2$**

If $x = 0$:	$y = 2x + 2$	*Given equation*
	$= 2(0) + 2$	*Substitute.*
	$= 2$	*Simplify.* Plot the point $(0, 2)$.
If $x = 1$:	$y = 2x + 2$	
	$= 2(1) + 2$	
	$= 4$	Plot $(1, 4)$.
If $x = 2$:	$y = 2x + 2$	
	$= 2(2) + 2$	
	$= 6$	Plot $(2, 6)$.

Draw the line passing through the three plotted points, as shown in Figure 13.5. ◆

Graphing a Line by Slope–Intercept

Let's introduce some terminology.

Standard Form

CAUTION

Study the form of this equation so that in the future you will be able to recognize lines by looking at the equation.

The **standard form** of the equation of a line is
$$Ax + By + C = 0$$
where A, B, and C are real numbers (A and B not both 0) and (x, y) is any point on the line. This equation is called:

a **first-degree equation** in two variables, or

a **linear equation** in two variables.

Graphing a line by plotting points is generally not the most efficient way. Consider the standard form of the equation of a line:

$$Ax + By + C = 0$$

If $B \neq 0$, we can solve for y:

$$By = -Ax - C$$

$$y = -\frac{A}{B}x - \frac{C}{B}$$

Notice that if $m = -\dfrac{A}{B}$ and $b = -\dfrac{C}{B}$ then the first-degree equation in two variables x and y (with $B \neq 0$) can be written in the form

$$y = mx + b$$

This is a very useful way of writing the equation of a line. Note that lowercase b in this form is not the same as capital B in the standard form. This form is easy to use when you *know* a value for x and want to find a value for y. We call x, the first component, the **independent variable** and calculate the value of y, the second component, called the **dependent variable.**

The points where a graph crosses the coordinate axes are usually easy to find, and they are often used to help sketch the curve. A **y-intercept** is a point where a graph crosses the y-axis and, consequently, it is a point with a first component of 0. When we say b is a y-intercept, we mean the curve crosses the y-axis at the point $(0, b)$. An **x-intercept** is a point where a graph crosses the x-axis, and it has a second component of 0. If we speak of an x-intercept of a, we mean the graph crosses the x-axis at the point $(a, 0)$.

The **slope** of a line passing through the points (x_1, y_1) and (x_2, y_2) is the steepness of that line, and is measured by rise divided by run:

$$m = \frac{\text{RISE}}{\text{RUN}} = \frac{\text{CHANGE IN THE VERTICAL DISTANCE}}{\text{CHANGE IN THE HORIZONTAL DISTANCE}} = \frac{y_2 - y_1}{x_2 - x_1}$$

Figure 13.6 shows how slope can be used to measure the steepness of a roof.

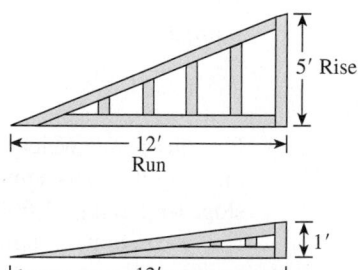

Figure 13.6 Cross section of roof gables is an example of slope; it is called the *pitch* of a roof.

Slope–Intercept Form

STOP

This is the most important form of the equation of a line.

The first-degree equation

$$y = mx + b$$

where b is the y-intercept and m is the slope is called the **slope–intercept form** of the equation of a line.

EXAMPLE 4

Graph the line passing through $(-6, 1)$ with slope $m = -\frac{2}{5}$.

Solution

Begin by plotting the point $(-6, 1)$. Think of the slope as a fraction, $m = \frac{-2}{5}$, for example, where RISE $= -2$ and RUN $= 5$. These steps are shown in Figure 13.7.

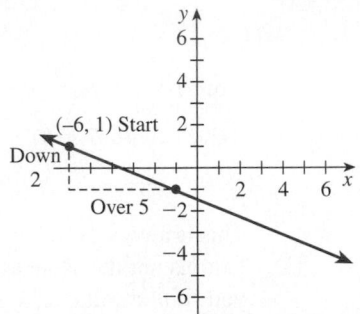

Figure 13.7 Graph of the line passing through $(-6, 1)$ with slope $-2/5$ ◆

We can use the procedure illustrated in Example 4 to graph a line when we are given the equation of a line. We use the *y*-intercept as the given point.

EXAMPLE 5

Graph the line $x + 2y = 6$.

Solution Solve for *y* to put the equation into slope–intercept form:

$$x + 2y = 6$$
$$2y = -x + 6 \qquad \text{Subtract x from both sides.}$$
$$y = -\tfrac{1}{2}x + 3 \qquad \text{Divide both sides (all terms) by 2.}$$

The *y*-intercept is $(0, 3)$; plot this point.

The slope is $m = -\frac{1}{2}$; *start* at the *y*-intercept and count down 1 unit and over (rightward) 2 units:

$$m = \frac{\text{RISE}}{\text{RUN}} = -\frac{1}{2} = \frac{-1}{2}$$

Plot this point, which we call a **slope point,** meaning that it is found by counting out the slope (rise and run) from a given point. Using this slope point, as well as the *y*-intercept, we now draw the line as shown in Figure 13.8.

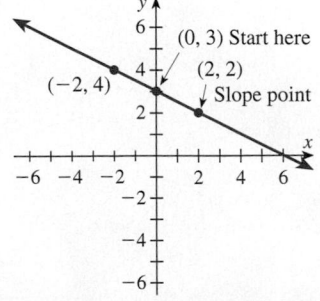

Figure 13.8 Graph of $x + 2y = 6$ ◆

A general procedure can now be stated for graphing a line using the slope–intercept method. This procedure is shown in Figure 13.9.

CAUTION Be careful when writing the slope as a fraction. $-\dfrac{2}{5}$ can be written as $\dfrac{-2}{5}$ or $\dfrac{2}{-5}$ but NOT as $\dfrac{-2}{-5}$.

STOP Study the procedure for graphing a line as described in Figure 13.9.

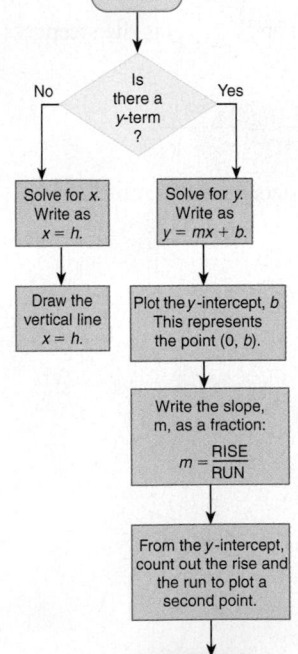

Figure 13.9 General procedure for graphing lines

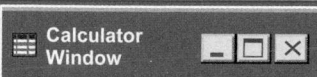

Many calculators available today do a variety of types of graphing problems. The cost is still somewhat prohibitive ($50–$70), but it is worthwhile to show how easily you can graph lines and other curves by using a graphing calculator.

There is generally a button labeled Y=. Press this button and input the equation you wish to graph. For Example 6, we wish to graph

$$y = \frac{2}{3}x + 2$$

After inputting the equation, we usually obtain the graph by pressing a button labeled GRAPH. The graph is shown.

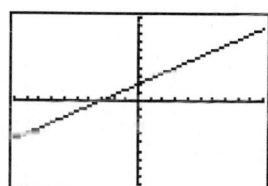

To see the coordinates of points on a graph, we can press TRACE, which displays the coordinates of the point on the curve corresponding to the location of the cursor.

EXAMPLE 6

Sketch the graph of $2x - 3y + 6 = 0$.

Solution   *For most graphs,* the most efficient method is to solve the equation for y and then find the y-intercept and the slope by inspection. This is the general procedure illustrated in Figure 13.9.

$$2x - 3y + 6 = 0$$
$$2x + 6 = 3y$$
$$y = \frac{2}{3}x + 2$$

Thus, $b = 2$; plot the y-intercept $(0, 2)$ as shown in Figure 13.10.

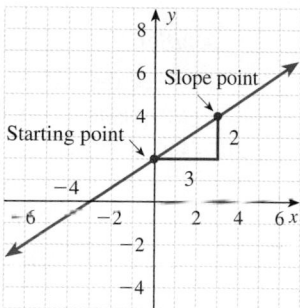

Figure 13.10 **Graph of $2x - 3y + 6 = 0$**

Since $m = \frac{2}{3}$, we start at the y-intercept and move up 2 and over (rightward) 3; plot another point (which we have called a slope point). Draw the line passing through the two plotted points.   ◆

Even though the slope–intercept form for graphing a line is the preferred method, sometimes it is not convenient to plot the y-intercept.

EXAMPLE 7

Sketch the graph of $7x + y + 28 = 0$.

Solution   Since $y = -7x - 28$, we have $m = -7$ and $b = -28$. The y-intercept is a little awkward to plot, so find some other point to use. If $x = -4$, then

$$y = -7x - 28$$
$$= -7(-4) - 28 = 28 - 28 = 0$$

and $(-4, 0)$ is a point on the line. Write the slope as $m = -7 = \frac{-7}{1}$ and find a slope point by counting down 7 and over 1. Note that another slope point can be found by using $m = -7 = \frac{7}{-1}$. The graph is shown in Figure 13.11.

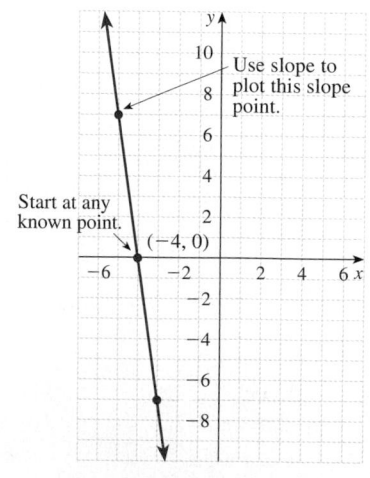

Figure 13.11 **Graph of $7x + y + 28 = 0$**   ◆

In Example 7 note that $m = -7 = \dfrac{7}{-1}$ means "up 7 and back (leftward) 1," and

$m = -7 = \dfrac{-7}{1}$ means "down 7 and over 1." A *positive slope* (as in Example 6) indicates that the line *increases* (goes uphill) from left to right, and a *negative slope* (as in Example 7) indicates that the line *decreases* (goes downhill) from left to right. A *zero slope* (as in the following example) indicates that the line is horizontal. For a vertical line, we say that the *slope does not exist.*

EXAMPLE 8

Graph   **a.** $y = 4$ **b.** $x = 3$

Solution

a. Since the only requirement is that y (the second component) equal 4, we see that there is no restriction on the choice for x. Thus, $(0, 4)$, $(1, 4)$, and $(-2, 4)$ all satisfy the equation that $y = 4$. If you plot and connect these points, you will see that the line formed is a **horizontal line.** A horizontal line has slope equal to zero.

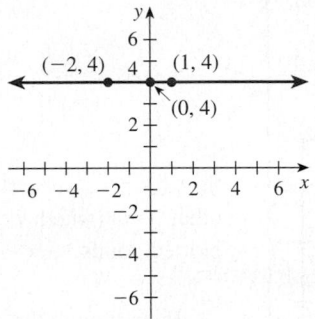

b. Since the only requirement is that x (the first component) equal 3, we see that there is no restriction on the choice for y. Thus, $(3, 2)$, $(3, -1)$, and $(3, 0)$ all satisfy the condition that $x = 3$. If you plot and connect these points, you will see that the line formed is a **vertical line.** A vertical line does not have any slope. Note that a line with no slope is vertical, but a line with zero slope is horizontal.

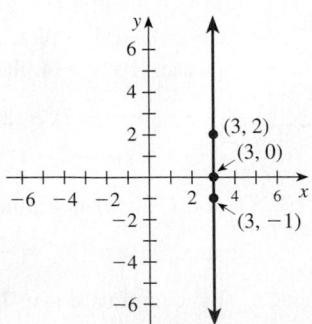

Problem Set 13.1

LEVEL 1

1. Use the map of Venus shown in Figure 13.12 to name the landmarks at the specified locations.

    a. $(90, 0)$ **b.** $(0, 65)$ **c.** $(165, 65)$

    d. $(-80, 32)$ **e.** $(-30, -50)$

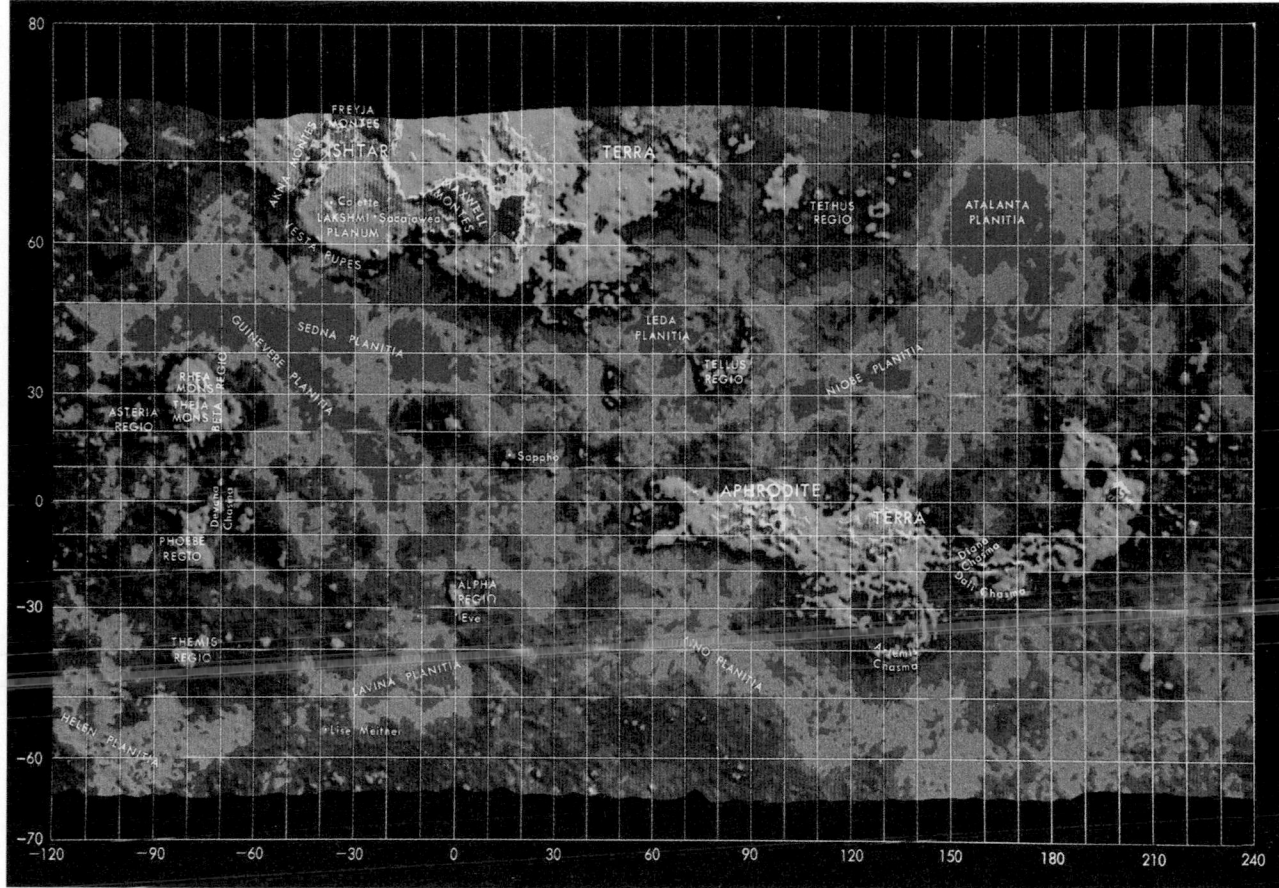

Figure 13.12 **Map of Venus**

2. **IN YOUR OWN WORDS** What does it mean to solve a linear equation in two variables?

3. **IN YOUR OWN WORDS** What is a *y*-intercept of a line? What is slope?

4. **IN YOUR OWN WORDS** Outline a procedure for graphing a line by plotting points.

5. **IN YOUR OWN WORDS** Contrast horizontal and vertical lines and their slopes.

6. **IN YOUR OWN WORDS** Outline a procedure for graphing a line using the slope–intercept method.

For each equation in Problems 7–18, find three ordered pairs that satisfy the equation, and then use this information to graph each line.

7. $y = x + 5$
8. $y = 2x - 1$

9. $y = 2x + 5$
10. $y = x - 4$

11. $y = x - 1$
12. $y = x + 1$

13. $y = -2x + 1$
14. $y = -3x + 1$

15. $y = 2x + 1$
16. $y + 3x = 1$

17. $3x + 4y = 8$
18. $x + 2y = 4$

Graph the lines through the given points and with the given slopes as indicated in Problems 19–30.

19. $(2, 5); m = \frac{1}{3}$
20. $(7, -3); m = \frac{1}{5}$

21. $(-4, -3); m = \frac{3}{4}$
22. $(5, 0); m = -\frac{2}{3}$

23. $(1, -1); m = -\frac{1}{7}$
24. $(6, 3); m = -\frac{2}{5}$

25. $(-1, 3); m = 1$
26. $(1, -1); m = 2$

27. $(1, 3); m = 3$
28. $(2, 3); m = 0$

29. $(2, 3);$ no slope
30. $(2, 3); m = 1.5$

LEVEL 2

Graph the lines in Problems 31–44.

31. $y = 2x + 3$

32. $y = 4x + 5$

33. $x + y + 4 = 0$

34. $3x + y + 2 = 0$

35. $2x + 3y + 6 = 0$

36. $x + 3y - 2 = 0$

37. $3x + 2y - 5 = 0$

38. $3x + 4y - 5 = 0$

39. $x = 5$

40. $y = -2$

41. $x + y - 100 = 0$

42. $y = 50x$

43. $y = -100x$

44. $y = -0.001x$

LEVEL 3

Problems 45–50 show scatter diagrams along with a best-fitting line. Match the best-fitting line in each with an appropriate equation named A–F.

45.

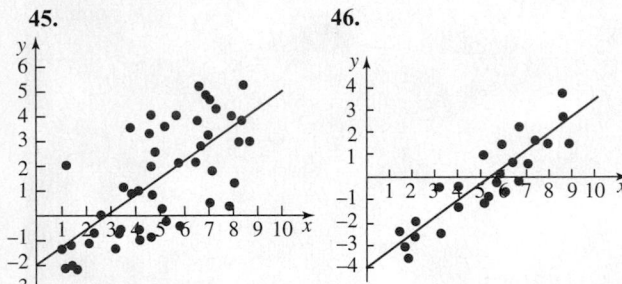

46.

47.

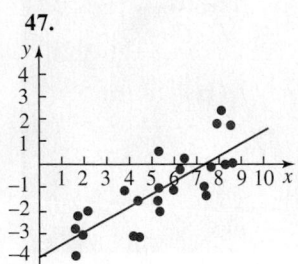

48.

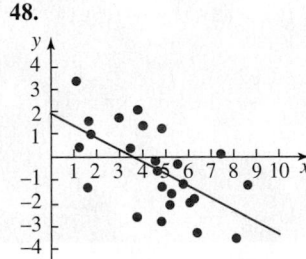

49.

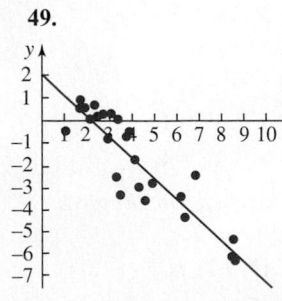

50.

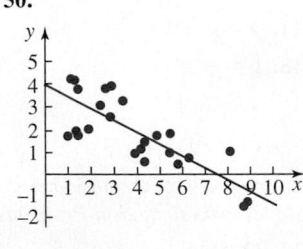

A. $y = 0.5x - 4$ B. $y = -0.5x + 2$

C. $y = 0.7x - 2$ D. $y = -0.5x + 4$

E. $y = -0.9x + 2$ F. $y = 0.9x - 2$

G. $y = -0.7x + 4$ H. $y = 0.7x - 4$

PROBLEM SOLVING

51. **IN YOUR OWN WORDS** The Brazilian Institute of Geography and Statistics reported the following fertility rates for Brazil, measured as the average number of children born per woman of childbearing age:

1970: 5.8   1980: 4.4   1991: 2.7   1994: 2.2

  a. Do you think the fertility rate is declining linearly? (That is, if you plot these points, is the graph linear?) Support your answer.

  b. Let the base year ($x = 0$) be 1970. Pick two data points and find the equation of the line between these points.

  c. Predict the fertility rate in the year 2005. Will this answer vary depending on the selected data points? Support your answer.

52. Two lines in the same plane are parallel if there is no point of intersection. Without graphing the lines, determine how you can decide whether the graphs of two equations are parallel lines.

53. If a present value P is invested at the simple interest rate r for t years, then the *future value* after t years is given by the formula $A = P(1 + rt)$. Suppose you invest \$10,000 at 8% simple interest per year.

  a. Graph the amount you will have in t years.

  b. What is the slope of the graph?

  c. What is the A-intercept of the graph?

54. A business purchasing an item for business purposes may use *straight-line depreciation* to obtain a tax deduction. The formula for the present value, P, after t years is

$$P = C - \left(\frac{C - s}{L}\right)t$$

where C is the cost and s is the scrap value after L years. The number L is called the *useful life* of the item.

  a. If a certain piece of equipment costs \$20,000 and has a scrap value of \$2,000 after 8 years, write an equation to represent the present value after t years.

  b. Graph the amount you will have in t years.

  c. What is the slope of the graph?

55. Suppose the *profit P* (in dollars) of a certain item is given by $P = 1.25x - 850$, where x is the number of items sold.

  a. Graph this profit relationship.

  b. Interpret the value of P when $x = 0$.

  c. The slope of this graph is called the *marginal profit.* Find the marginal profit, and give an interpretation.

56. Suppose the *cost C* (in dollars) of x items is given by $C = 2.25x + 550$.

  a. Graph this cost relationship.

  b. Interpret the value of C when $x = 0$.

  c. The slope of the graph is called the *marginal cost.* Find the marginal cost and give an interpretation.

57. The population of Texas was 17.0 million in 1990 and 19.8 million in 1998. Let x be the year (let 1990 be the base year; that is, $x = 0$ represents the year 1990 and $x = 8$ represents 1998) and y be the population. Use this information to write a linear equation, and then use this equation to predict the population of Texas in the year 2005.

58. Show that the equation of a line passing through (h, k) with slope m is

$$y - k = m(x - h)$$

This is called the *point–slope form.*

59. Begin with the point–slope form in Problem 58 and derive the *two-point form.* If the line passes through (x_1, y_1) and (x_2, y_2), then

$$y - y_1 = \left(\frac{y_2 - y_1}{x_2 - x_1}\right)(x - x_1)$$

60. Show that the equation of the line with x-intercept a and y-intercept b is

$$\frac{x}{a} + \frac{y}{b} = 1$$

$(a \neq 0, b \neq 0)$. This is called the *intercept form.*

13.2 GRAPHING HALF-PLANES

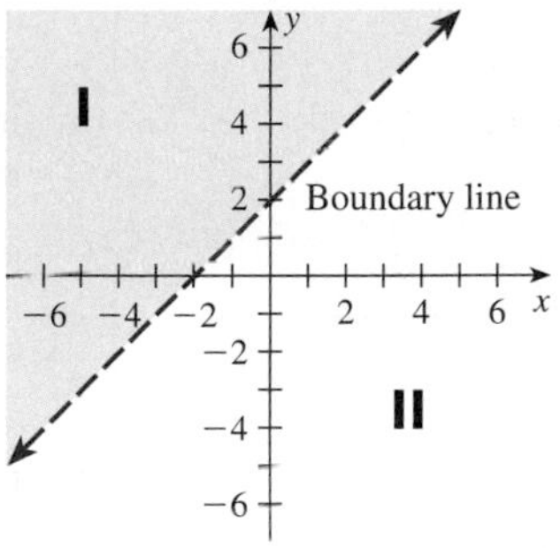

Once you know how to graph a line, you can also graph linear inequalities. We begin by noting that every line divides a plane into three parts, as shown in Figure 13.13. Two parts are labeled I and II; these are called **half-planes.** The third part is called the **boundary** and is the line separating the half-planes. The solution of a first-degree inequality in two unknowns is the set of all ordered pairs that satisfy the given inequality. This solution set is a half-plane. The following table offers some examples of first-degree inequalities with two unknowns, along with some associated terminology.

Figure 13.13 **Half-planes**

Example	Inequality *symbol*	*Boundary included*	*Term*
$3x - y > 5$	$>$	no	**open half-plane**
$3x - y < 5$	$<$	no	open half-plane
$3x - y \geq 5$	$\geq$	yes	**closed half-plane**
$3x - y \leq 5$	$\leq$	yes	closed half-plane

We can now summarize the procedure for graphing a first-degree inequality in two unknowns.

Procedure for Graphing a Linear Inequality

Step 1. **Graph the boundary.**

Replace the inequality symbol with an equality symbol and draw the resulting line. This is the boundary line.

Use a solid line when the boundary is included ($\leq$ or $\geq$).

Use a dashed line when the boundary is not included ($<$ or $>$).

Step 2. **Test a point.**

Choose any point in the plane, called a **test point,** that is not on the boundary line; the point $(0, 0)$ is usually the simplest choice.

If this test point makes the *inequality* true, shade in the half-plane that contains the test point. That is, the shaded plane is the solution set.*

If the test point makes the *inequality* false, shade in the other half-plane for the solution.

*A highlighter pen does a nice job of shading in your work.

This process sounds complicated, but if you know how to draw lines from equations, you will not find this difficult.

EXAMPLE 1

Graph $3x - y \geq 5$.

Solution Note that the inequality symbol is $\geq$, so the boundary is included.

Step 1. Graph the boundary; draw the (solid) line corresponding to

$$3x - y = 5 \qquad \text{Replace the inequality symbol with an equality symbol.}$$
$$y = 3x - 5$$

The y-intercept is -5 and the slope is 3; the boundary line is shown in Figure 13.14.

Step 2. Choose a test point; we choose $(0, 0)$.

Plot $(0, 0)$ in Figure 13.14 and note that it lies in one of the half-planes determined by the boundary line. We now check this test point with the given *inequality:*

$$3x - y \geq 5$$
$$3(0) - (0) \geq 5 \qquad \text{You can usually test this in your head.}$$
$$0 \geq 5 \quad \text{This is false.}$$

Therefore, shade the half-plane that does *not* contain $(0, 0)$, as shown in Figure 13.14. ◆

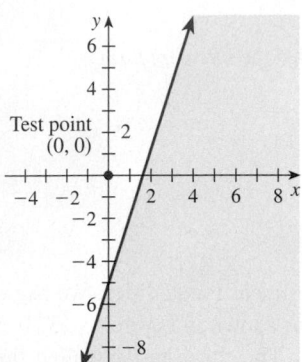

Figure 13.14 Graph of $3x - y \geq 5$

EXAMPLE 2

Graph $y < x$.

Solution Note that the inequality symbol is $<$, so the boundary line is not included.

Step 1. Draw the (dashed) boundary line, $y = x,$ as shown in Figure 13.15.

Step 2. Choose a test point. We can't pick $(0, 0)$, because $(0, 0)$ is on the boundary line. Since we must choose some point not on the boundary, we choose $(1, 0)$:

$$y < x$$
$$0 < 1 \quad \text{This is true.}$$

Therefore, shade the half-plane that contains the test point, as shown in Figure 13.15. ◆

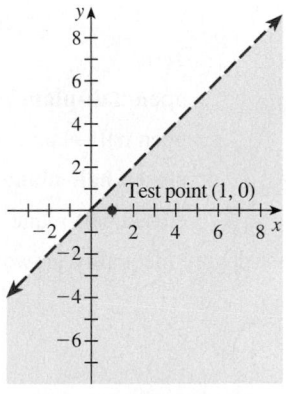

Figure 13.15 Graph of $y < x$

Problem Set 13.2

1. **IN YOUR OWN WORDS** What is a linear inequality in two variables?

2. **IN YOUR OWN WORDS** Outline a procedure for graphing first-degree inequalities in two variables.

State whether each statement in Problems 3–11 is true or false. If it is false, explain why you think that is the case.

3. The linear inequality $2x + 5y < 2$ does not have a boundary line, because the inequality symbol is $<$.

4. A good test point for the linear inequality $y \geq x$ is $(0, 0)$.

5. The test point $(0, 0)$ does not work for $y > x$ because 0 is not greater than 0.

6. The test point $(0, 0)$ satisfies the inequality $2y - 3x < 2$.

7. The test point $(0, 0)$ satisfies the inequality $3x - 2y \geq -1$.

8. The test point $(0, 0)$ satisfies the inequality $3x > 2y$.

9. The test point $(-2, 4)$ satisfies the inequality $y > 2x - 1$.

10. The test point $(-2, 4)$ satisfies the inequality $5x + 2y \leq 9$.

11. The test point $(-2, 4)$ satisfies the inequality $4x < 3y$.

LEVEL 2

Graph the first-degree inequalities in two unknowns in Problems 12–29.

12. $y \leq 2x + 1$

13. $y \geq 5x - 3$

14. $y > 5x - 3$

15. $y \geq -2x + 3$

16. $y > 3x - 3$

17. $y < -2x + 5$

18. $3x \leq 2y$

19. $2x < 3y$

20. $x \geq y$

21. $y > x$

22. $y \geq 0$

23. $x \leq 0$

24. $x - 3y \geq 9$

25. $6x - 2y < 1$

26. $3x + 2y > 1$

27. $5x - 2y > 1$

28. $x + 2y + 4 \leq 0$

29. $2x - 3y + 6 \geq 0$

PROBLEM SOLVING

30. Will a baseball player's batting average always change when the player goes up to bat? That is, under what conditions will the batting average stay the same? *Hint:* Let a be the number of times at bat, and let h be the number of hits. Then the batting average is $\dfrac{h}{a}$. If the player gets a hit, then the new batting average is

$$\frac{h + 1}{a + 1}$$

and if the player does not get a hit, then the new batting average is

$$\frac{h}{a + 1}$$

13.3 GRAPHING CURVES

Curves by Plotting Points

When cannons were introduced in the 13th century, their primary use was to demoralize the enemy. It was much later that they were used for strategic purposes. In fact, cannons existed nearly three centuries before enough was known about the behavior of projectiles to use them with any accuracy. The cannonball does not travel in a straight line, it was discovered, because of an unseen force that today we know as *gravity*. Consider Figure 13.16. It is a scale drawing (graph) of the path of a cannonball fired in a particular way.

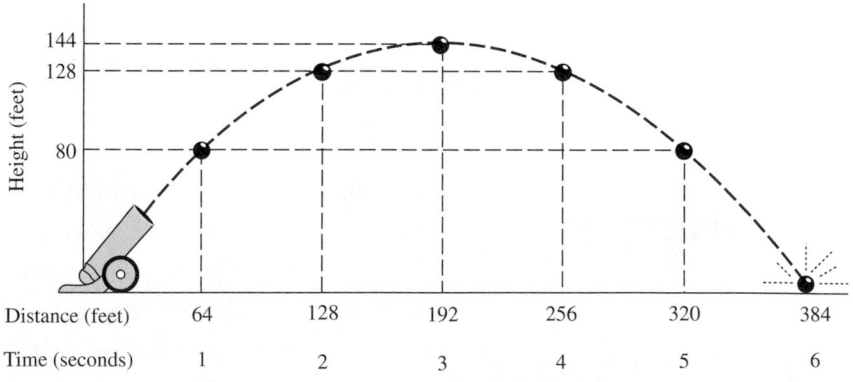

Figure 13.16 **Path of a cannonball**

The path described by a projectile is called a **parabola.** Any projectile—a ball, an arrow, a bullet, a rock from a slingshot, even water from the nozzle of a hose or sprinkler—will travel a parabolic path. Note that this parabolic curve has a maximum height and is symmetric about a vertical line through that height. In other words, the ascent and descent paths are symmetric.

EXAMPLE 1

Graph $y = x^2$.

Solution   We will choose x-values and find corresponding y-values.

Let $x = 0$: $y = 0^2 = 0$; plot $(0, 0)$.

Let $x = 1$: $y = 1^2 = 1$; plot $(1, 1)$.

Let $x = -1$: $y = (-1)^2 = 1$; plot $(-1, 1)$.

Notice in Figure 13.17 that these points do not fall in a straight line. If we find two more points, we can see the shape of the graph.

Let $x = 2$: $y = 2^2 = 4$; plot $(2, 4)$.

Let $x = -2$: $y = (-2)^2 = 4$; plot $(-2, 4)$.

Connect the points to form a smooth curve as shown in Figure 13.18. To **graph** an equation means to plot all ordered pairs that satisfy the equation.

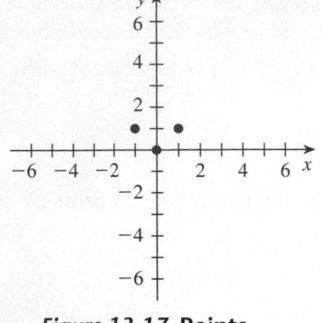

Figure 13.17 **Points satisfying the equation $y = x^2$ do not lie on a straight line.**

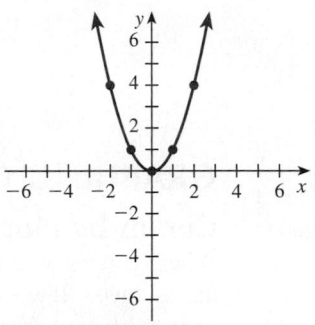

Figure 13.18 **Graph of $y = x^2$**   ◆

The curve shown in Figure 13.18 is a parabola that is said to *open upward*. The lowest point, $(0, 0)$ in Example 1, is called the **vertex.** The following example is a parabola that *opens downward*. Notice that the number a in $y = ax^2$ is negative.

EXAMPLE 2

Sketch $y = -\frac{1}{2}x^2$.

Solution

Let $x = 0$: $y = -\frac{1}{2}(0)^2 = 0$; plot $(0, 0)$.

Let $x = 1$: $y = -\frac{1}{2}(1)^2 = -\frac{1}{2}$; plot $(1, -\frac{1}{2})$.

Let $x = -1$: $y = -\frac{1}{2}(-1)^2 = -\frac{1}{2}$; plot $(-1, -\frac{1}{2})$.

Let $x = 2$: $y = -\frac{1}{2}(2)^2 = -2$; plot $(2, -2)$.

Let $x = -2$: $y = -\frac{1}{2}(-2)^2 = -2$; plot $(-2, -2)$.

Let $x = 4$: $y = -\frac{1}{2}(4)^2 = -8$; plot $(4, -8)$.

Connect these points to form a smooth curve, as shown in Figure 13.19.   ◆

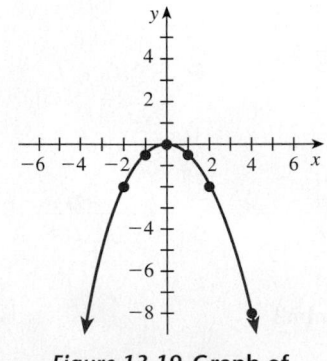

Figure 13.19 **Graph of $y = -\frac{1}{2}x^2$**

Notice from Example 2 that you can use symmetry to plot points on one side of the vertex. If the vertex is at the origin, $(0, 0)$, we call the curve a **standard parabola.**

Standard Parabolas

> The graphs of the following curves are parabolas with vertex $(0, 0)$:
>
> $y = ax^2$　　opens up
>
> $y = -ax^2$　　opens down
>
> $x = ay^2$　　opens rights
>
> $x = -ay^2$　　opens left

You can sketch many different curves by plotting points. The procedure is to decide whether you should pick x-values and find the corresponding y-values, or pick y-values and find the corresponding x-values. Find enough ordered pairs so that you can connect those points with a smooth graph. Many graphs in mathematics are not smooth, but we will not consider those in this course.

If the equation is

$$y = ax^2 + bz + c$$

then the graph is a parabola that opens upward if $a > 0$ and downward if $a < 0$. If the equation is

$$x = ay^2 + by + c$$

then the graph is a parabola that opens to the right if $a > 0$ and to the left if $a < 0$

EXAMPLE 3

Sketch the graph of the curve $x = y^2 - 6y + 4$.

Solution　We use the same procedure to find points on the graph, except that in this equation we see that it will be easier to choose y-values and find corresponding x-values.

Let $y = 0$:　$x = 0^2 - 6 \cdot 0 + 4 = 4$; plot $(4, 0)$.

Let $y = 1$:　$x = 1^2 - 6 \cdot 1 + 4 = 1 - 6 + 4 = -1$; plot $(-1, 1)$.

Let $y = 2$:　$x = 2^2 - 6 \cdot 2 + 4 = 4 - 12 + 4 = -4$; plot $(-4, 2)$.

Let $y = 3$:　$x = 3^2 - 6 \cdot 3 + 4 = 9 - 18 + 4 = -5$; plot $(-5, 3)$.

Let $y = 4$:　$x = 4^2 - 6 \cdot 4 + 4 = 16 - 24 + 4 = -4$; plot $(-4, 4)$.

Let $y = 5$:　$x = 5^2 - 6 \cdot 5 + 4 = 25 - 30 + 4 = -1$; plot $(-1, 5)$.

Let $y = 6$:　$x = 6^2 - 6 \cdot 6 + 4 = 36 - 36 + 4 = 4$; plot $(4, 6)$.

How many points should we find? For lines, we needed to find two points, with a third point as a check. For curves that are not lines, we do not have a particular number of points to check. We must plot as many points as are necessary to enable us to draw a smooth curve. We connect the points we have plotted to draw the curve shown in Figure 13.20.　◆

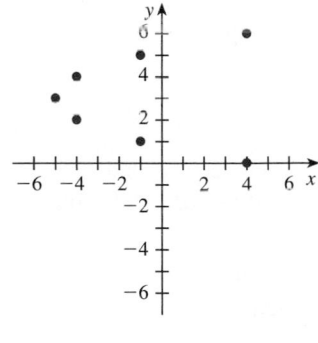

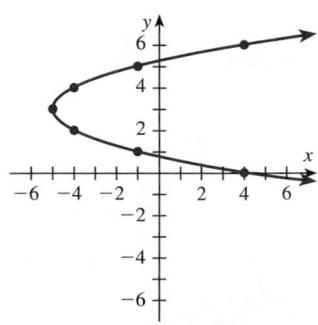

Figure 13.20 **Graph of** $x = y^2 - 6y + 4$

Exponential Curves

We considered exponential equations in Section 8.2. Recall that an *exponential equation* is one in which a variable appears as an exponent. Consider the equation $y = 2^x$. This equation represents a doubling process. The graph of such an equation is called an **exponential curve** and is graphed by plotting points.

EXAMPLE 4

Sketch the graph of $y = 2^x$.

Solution   Choose x-values and find corresponding y-values. These values form ordered pairs (x, y). Plot enough ordered pairs so that you can see the general shape of the curve, and then connect the points with a smooth curve.

Let $x = 0$: $y = 2^0 = 1$; plot $(0, 1)$.

Let $x = 1$: $y = 2^1 = 2$; plot $(1, 2)$.

Let $x = 2$: $y = 2^2 = 4$; plot $(2, 4)$.

Let $x = 3$: $y = 2^3 = 8$; plot $(3, 8)$.

Let $x = -1$: $y = 2^{-1} = 1/2$; plot $(-1, \frac{1}{2})$.

Let $x = -2$: $y = 2^{-2} = 1/4$; plot $(-2, \frac{1}{4})$.

Let $x = -3$: $y = 2^{-3} = 1/8$; plot $(-3, \frac{1}{8})$.

Connect the points with a smooth curve, as shown in Figure 13.21.

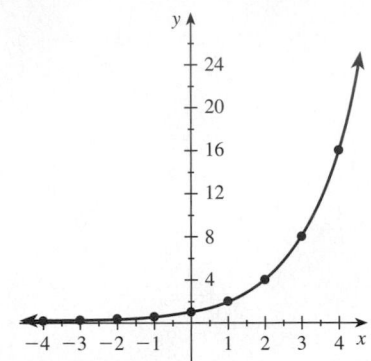

Figure 13.21 **Graph of $y = 2^x$**   ◆

EXAMPLE 5   PÓLYA'S METHOD

Marsha and Tony have owned a parcel of property within the city limits of their small town for a number of years. Now a developer is threatening to build a large housing project on the land next to their property, but to do so the developer is seeking a zoning change from the City Planning Commission. The hearing is next week, and Tony believes that the population projections presented by the developer are vague and unfounded. Tony needs a mathematical method for projecting population and needs to communicate his findings to the Planning Commission.

Solution   We use Pólya's problem-solving guidelines for this example.

Understand the Problem.   Tony calls his local Chamber of Commerce and finds that the growth rate of his town is now 5%. Also, according to the 1990 census, the population was 2,500.

Devise a Plan.   Tony decides to draw a graph showing the population between the years 1990 and 2010. The formula for population growth (from Section 8.3) is $A = A_0 e^{rt}$.

Carry Out the Plan.   For this problem,

$$A_0 = 2,500 \quad \text{and} \quad r = 5\% = 0.05$$ <small>Remember, to change a percent to a decimal, move the decimal point two places to the left.</small>

The equation of the graph is $A = A_0 e^{rt} \approx 2,500 e^{0.05t}$.

If $t = 0$: $A = 2,500 e^0 = 2,500$.   <small>Plot the point $(0, 2500)$. This is the 1990 population; 1990 is called the base year. It is also called the "present time" (even if it is now 1999). Thus, if $t = 5$, then the population corresponds to the year 1995. If $t = 10$, the population is for the year 2000.</small>

If $t = 10$: $A = 2,500 e^{0.05(10)} \approx 4,120$.

Press: [2500] [×] [e] [^] [(] [.05] [×] [10] [)] [=]

Display: 4121.803177

This means that the predicted population in the year 2000 is about 4,120. Plot the point $(10, 4120)$.

If $t = 20$: $A = 2,500 e^{0.05(20)} \approx 6,800$.

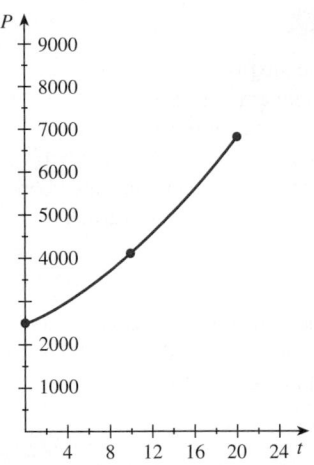

Figure 13.22 Graph showing population A for 1990 to 2010 (base year, $t = 0$, is 1990)

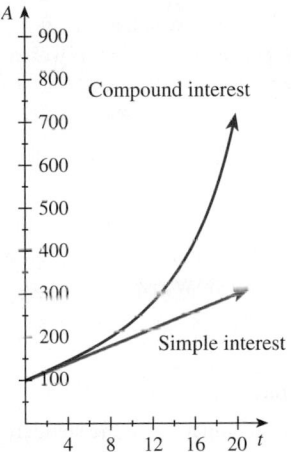

Figure 13.23 Comparison of simple interest and compound interest

This means that the predicted population for the year 2010 is 6,800.

Display: 6795.704571   Plot the point (20, 6800).

We have plotted these points in Figure 13.22.

Look Back. To do a good job with persuasion, you need facts and data to back up your argument. The visual impact of a graph can be quite impressive.   ◆

One of the most important concepts for intelligent functioning in today's world is interest and how interest works. We can now return to this idea and look at graphs showing the growth of money as the result of interest accumulation. Simple interest is represented by a line, whereas compound interest requires an exponential graph. This means that, for short periods of time, there is very little difference between simple and compound interest, but as time increases, the differences become significant.

EXAMPLE 6

Suppose that you deposit $100 at 10% interest. Graph the total amounts you will have if you invest your money at simple interest and if you invest your money at compound interest. Recall the appropriate formulas:

$$\text{Simple interest:}\quad A = P(1 + rt)$$
$$\text{Compound interest:}\quad A = P(1 + r)^t$$

Solution   For this example, $P = 100$ and $r = 0.10$; the variables are t and A. We begin by finding some ordered pairs. A calculator is necessary for calculating compound interest.

Year	Simple interest	Compound interest
$t = 0$	$A = 100(1 + 0.1 \cdot 0) = 100$ Point (0, 100)	$A = 100(1 + 0.1)^0 = 100$ Point (0, 100)
$t = 1$	$A = 100(1 + 0.1 \cdot 1) = 110$ Point (1, 110)	$A = 100(1 + 0.1)^1 = 110$ Point (1, 110)
$t = 10$	$A = 100(1 + 0.1 \cdot 10) = 200$ Point (10, 200)	$A = 100(1 + 0.1)^{10} \approx 259$ Point (10, 259)
$t = 20$	$A = 100(1 + 0.1 \cdot 20) = 300$ Point (20, 300)	$A = 100(1 + 0.1)^{20} \approx 673$ Point (20, 673)

The graphs through these points are shown in Figure 13.23.   ◆

Problem Set 13.3

LEVEL 1

1. **IN YOUR OWN WORDS** What is a parabola?

2. **IN YOUR OWN WORDS** What does the symbol e represent?

3. **IN YOUR OWN WORDS** Explain how you might make a prediction of population growth.

Sketch the graph of each equation in Problems 4–31.

4. $y = 3x^2$
5. $y = 2x^2$
6. $y = 10x^2$
7. $y = -x^2$
8. $y = -2x^2$
9. $y = -3x^2$
10. $y = -5x^2$
11. $y = 5x^2$
12. $y = \frac{1}{2}x^2$
13. $x = 2y^2$
14. $x = -3y^2$
15. $x = \frac{1}{2}y^2$
16. $y = 2x^3$
17. $y = -x^3$
18. $x = -y^3$
19. $y = \frac{1}{3}x^2$
20. $y = -\frac{2}{3}x^2$
21. $y = \frac{1}{10}x^2$
22. $y = -\frac{1}{10}x^2$
23. $y = \frac{1}{1,000}x^2$
24. $y = x^2 - 4$
25. $y = x^2 + 4$

26. $y = 9 - x^2$ **27.** $y = -3x^2 + 4$
28. $y = 2x^2 - 3$ **29.** $y = -2x^2 + 3$
30. $y = x^2 - 2x + 1$ **31.** $y = x^2 + 4x + 4$

LEVEL 2

Sketch the graphs of the equations in Problems 32–43.

32. $y = 3^x$ **33.** $y = 4^x$
34. $y = 5^x$ **35.** $y = -3^x$
36. $y = -6^x$ **37.** $y = -7^x$
38. $y = 10^x$ **39.** $y = 100 - 2^x$
40. $y = \left(\frac{1}{2}\right)^x$ **41.** $y = \left(\frac{1}{3}\right)^x$
42. $y = \left(\frac{1}{10}\right)^x$ **43.** $y = -\left(\frac{1}{10}\right)^x$

44. a. Graph $h = -16t^2 + 96t$.
  b. Does the parabola in part **a** open upward or downward?
  c. If we relate the graph in part **a** to the cannonball in Figure 13.16, can t be negative? Draw the graph of the parabola in part **a** for $0 \le t \le 6$.

45. a. Graph $d = 16t^2$.
  b. Does the parabola in part **a** open upward or downward?
  c. Graph the parabola in part **a** for $0 \le t \le 16$.

Draw the graphs in Problems 46–51.

46. $y = -2x^2 + 4x - 2$ **47.** $y = \frac{1}{4}x^2 - \frac{1}{2}x + \frac{1}{4}$
48. $y = \frac{1}{2}x^2 + x + \frac{1}{2}$ **49.** $y = x^2 - 2x + 3$
50. $x = 3y^2 + 12y + 14$ **51.** $x = 4y - y^2 - 4$

52. Change the growth rate in Example 5 to 6%, and graph the population curve.

53. Change the growth rate in Example 5 to 1.5%, and graph the population curve.

54. Rework Example 6 for a 12% interest rate.

55. Rework Example 6 for a 5% interest rate.

PROBLEM SOLVING

56. In archaeology, carbon-14 dating is a standard method of determining the age of certain artifacts. Decay rate can be measured by half-life, which is the time required for one-half of a substance to decompose. Carbon-14 has a half-life of approximately 5,600 years. If 100 grams of carbon were present originally, then the amount A of carbon present today is given by the formula

$$A = 100\left(\frac{1}{2}\right)^{t/5,600}$$

where t is the time since the artifact was alive until today. Graph this relationship by letting $t = 5,600$, $t = 11,200$, $t = 16,800 \ldots$, and adjust the scale accordingly.

57. The healing law for skin wounds states that

$$A = \frac{A_0}{e^{0.1t}}$$

where A is the number of square centimeters of unhealed skin after t days when the original area of the wound was A_0. Graph the healing curve for a wound of 50 cm^2.

58. A learning curve describes the rate at which a person learns certain specific tasks. If N is the number of words per minute typed by a student, then, for an average student

$$N = 80(1 - e^{-0.016n})$$

where n is the number of days of instruction. Graph this formula.

59. The equation for the standard normal curve (the normal curve with mean 0 and standard deviation 1) graphs as an exponential curve. Graph this curve, whose equation is

$$y = \frac{e^{-x^2/2}}{\sqrt{2\pi}}$$

by using a calculator or by plotting points.

60. Graph the curve $y = 2^{-x^2}$ and compare with the standard normal curve (Problem 59). List some similarities and some differences.

13.4 CONIC SECTIONS

In the last section we graphed parabolas by plotting points. The parabola is one of the curves studied by the Greeks, in particular Apollonius (ca. 262–190 B.C.). His methods, expounded in an eight-volume work called *Conics,* were so modern that they are sometimes considered as an analytic geometry preceding René Descartes. (See the Historical Note on page 667.)

Here is the foundation of this study by Apollonius. Consider the intersection of a cone and a plane. The possibilities are shown in Figure 13.24. The set of points that results from this intersection of two three-dimensional objects (a cone and a plane) is a two-dimensional curve known as a **conic section.**

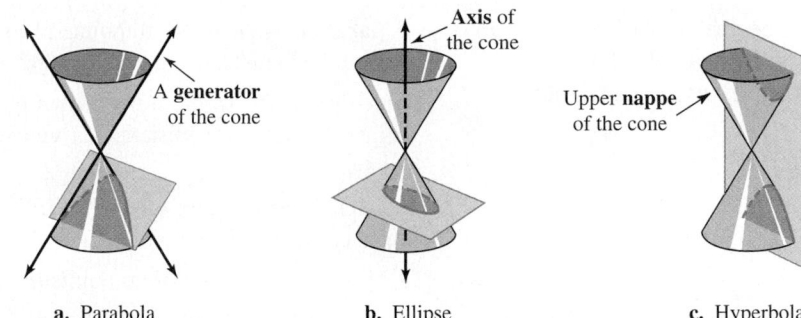

a. Parabola

The plane is parallel to one of the generators of the cone.

b. Ellipse

The plane intersects only one nappe and is not parallel to a generator.

c. Hyperbola

The plane intersects both nappes of the cone.

Figure 13.24 **Conic sections**

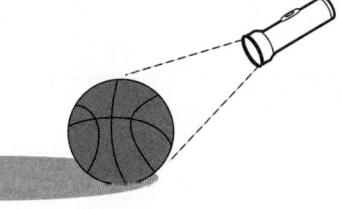

Parabola

You can see the conic sections using a flashlight and a basketball. If you hold the flashlight off to the side (as shown in the margin), you will be able to generate a shadow that is a parabola. If you adjust the angle of the flashlight as shown below the shadow's shape turns to an ellipse. If you hold the flashlight directly above the basketball, you will see the shadow of a circle. A circle is considered to be a special case of an ellipse.

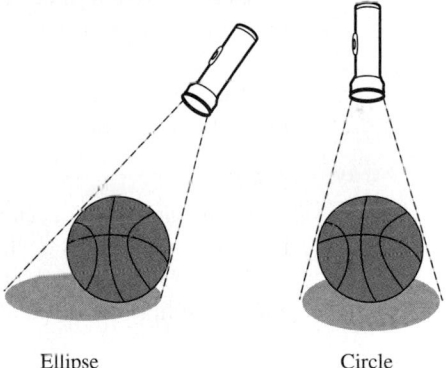

Ellipse   Circle

To picture a shadow that is a hyperbola, you need two basketballs (or you can "imagine" a second ball), as shown below.

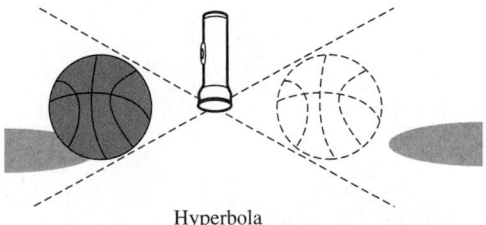

Hyperbola

Geometric Definition of Conic Sections

In the last section we graphed parabolas by plotting points, but for the other conic sections, the algebra and arithmetic necessary for graphing by plotting points are more complicated, so in this section we will consider some important characteristics of the curve that will help us easily sketch a graph.

We begin with a geometric definition for each of the conic sections.

**Geometric Definition
of a Parabola, Ellipse,
Circle, and Hyperbola**

This is a tough definition; we
will spend some time
interpreting it below.

> A **parabola** is the set of all points in a plane equidistant from a given point (called the **focus**) and a given line (called the **directrix**).
>
> An **ellipse** is the set of all points in a plane such that, for each point on the ellipse, the sum of its distances from two fixed points (called the **foci**) is a constant.
>
> A **circle** (a special case of an ellipse) is the set of all points in a plane a given distance from a given point, called the **center of the circle.**
>
> A **hyperbola** is the set of all points in a plane such that, for each point on the hyperbola, the difference of its distances from two fixed points (called the **foci**) is a constant.

To help us understand this definition, we will draw each of these curves by using the definition.

EXAMPLE 1

Draw a circle using the definition.

Solution

Consider a circle with center at point P and a given radius of r. This circle is shown in Figure 13.25. You can use a compass to help you draw a circle.   ◆

EXAMPLE 2

Draw a parabola using the definition.

Solution

Label the given point F and the given line L. (Any point or any line is acceptable, as long as the point is not on the line.) Imagine a special "parabola graph paper" as shown in Figure 13.26. It consists of circles drawn at F with radii of 1, 2, 3, ... It also consists of lines drawn parallel to L at 1, 2, 3, ...

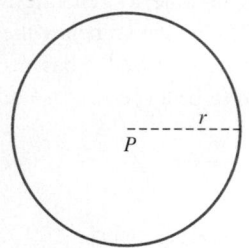

Figure 13.25 **A circle graphed from definition**

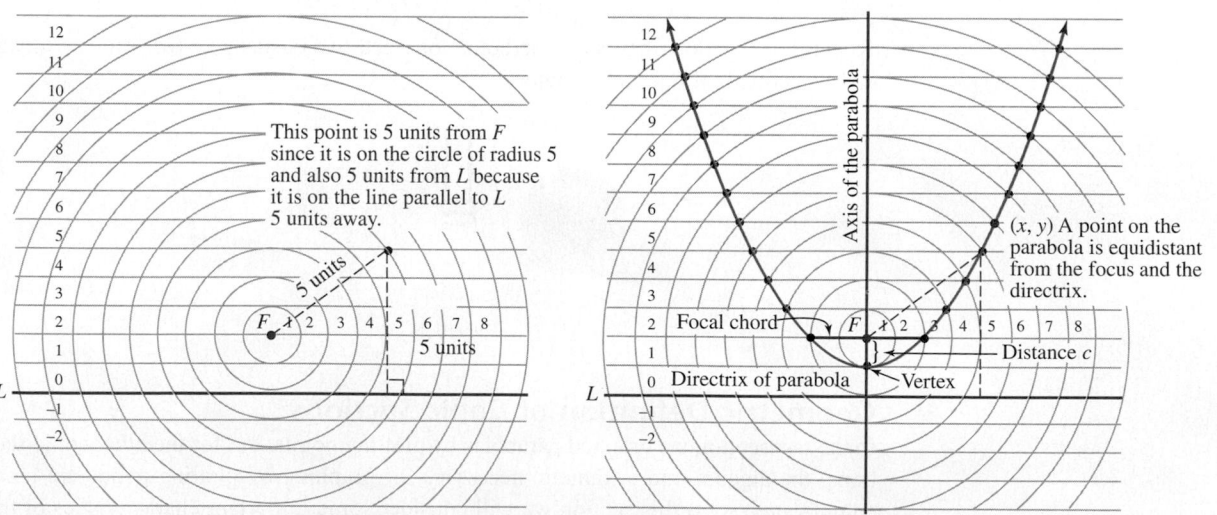

a. Parabola graph paper

b. Parabola on graph paper

Figure 13.26 **Parabola graphed from definition**

To sketch a parabola using the definition, plot points in the plane equidistant from the focus F and the directrix L. Draw a line through the focus and perpendicular to the directrix. This line is called the **axis of the parabola.** Let V be the point on this line halfway between the focus and the directrix. It is called the **vertex of the parabola.** Plot other points equidistant from F and L, as shown in Figure 13.26b. We denote the distance between the vertex and the focus of the parabola as c. ◆

EXAMPLE 3

Draw an ellipse using the definition.

Solution

Let F_1 and F_2 be any distinct points, and imagine a special type of graph paper shown in Figure 13.27a. This paper is constructed by plotting two points, labeled F_1 and F_2, and then drawing concentric circles with radii $r = 1, r = 2, r = 3, \ldots$ emanating from each focus.

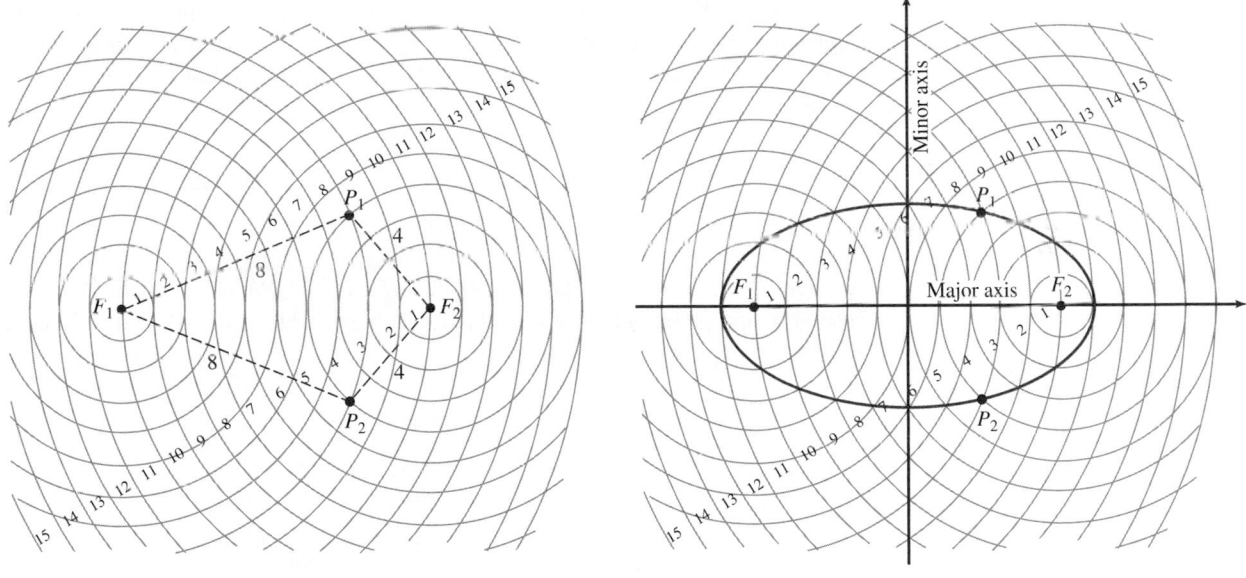

a. Ellipse graph paper

b. Ellipse on graph paper

Figure 13.27 **Graphing an ellipse**

We can use this special graph paper to graph an ellipse. Our objective is to plot all points such that the sum of the distances from two fixed points is a constant, as stated in the definition. We begin by picking a constant. This constant must be greater than the straight-line distance from F_1 to F_2. For this demonstration, we note that F_1 and F_2 in Figure 13.27 are 10 units apart, so the given constant is 12. We plot all the points so that the sum of their distances from the foci is 12. This is easy with our special graph paper. If a point is 8 units from F_1, for example, then it is 4 units from F_2, so we look at the intersection of the circle with $r = 8$ from F_1 with the circle with $r = 4$ from F_2, as shown in Figure 13.27a. We continue this process by plotting the following points:

Distance from F_1	Distance from F_2
8	4
9	3
10	2
11	1

Do you see why 11 units is the maximum integer distance from F_1?

7	5
6	6
5	7
4	8
3	9
2	10
1	11

Do you also see why 11 units is the maximum integer distance from F_2?

There are, of course, other rational and irrational choices that could be made, but connecting these points as shown in Figure 13.27 will give a good picture of an ellipse with constant 12 and a distance of 10 units between F_1 and F_2. ◆

The line passing through F_1 and F_2 is called the **major axis.** The **center of an ellipse** is the midpoint of the segment $\overline{F_1F_2}$. The line passing through the center perpendicular to the major axis is called the **minor axis.** The ellipse is symmetric with respect to both the major and minor axes. Even though we know that lines do not have length, it is common to talk about the *length of the major axis.* This is the distance between the vertices along the major axis. This distance is 2*a*. Similarly, we define the *length of the minor axis* to be 2*b*.

EXAMPLE 4

Draw a hyperbola using the definition.

Solution We are given two points; call them F_1 and F_2. We use the definition and ellipse graphing paper to draw a hyperbola. The hyperbola shown in Figure 13.28 has a given constant of 8.

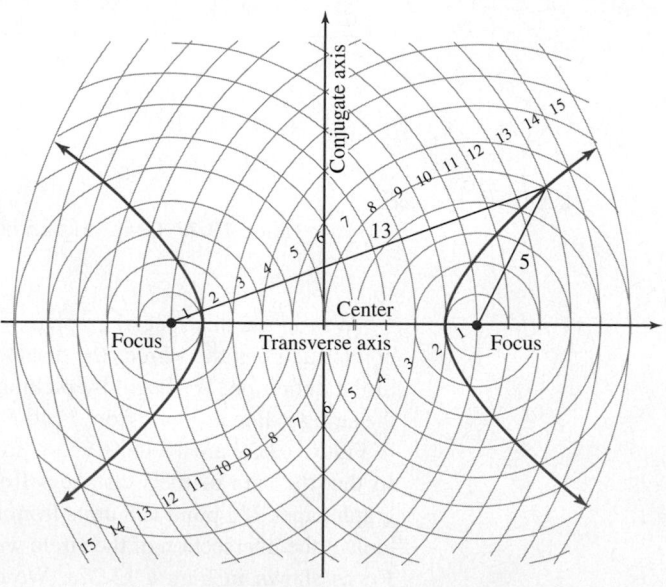

Figure 13.28 **Hyperbola on graph paper** ◆

The line passing through the foci is called the **transverse axis.** The **center of a hyperbola** is the midpoint of the segment connecting the foci. The line passing through the center perpendicular to the transverse axis is called the **conjugate axis.** The hyperbola is symmetric with respect to both the transverse and the conjugate axes.

Algebraic Representation of Conic Sections

Analytic geometry allows us to give algebraic as well as geometric representations to the conic sections.

General Form

Study the form of this equation so that in the future you will be able to recognize each of the conic sections by looking at the equation.

The **general form** of the equation of a conic section is
$$Ax^2 + Bxy + Cy^2 + Dx + Ey + F = 0$$
where $A, B, C, D, E,$ and F are real numbers and (x, y) is any point on the curve. This equation is called a:

first-degree equation if $A = B = C = 0$;

second-degree equation otherwise.

If $B = 0$, then we classify the conic sections as follows:
If $A = C = 0$, then the conic section is a **line;**
If $A = 0$ and $C \neq 0$ or if $A \neq 0$ and $C = 0$, it is a **parabola;**
If A and C have the same signs, it is an **ellipse;**
If $A = C \neq 0$, it is a **circle;** and
If A and C have opposite signs, it is a **hyperbola.**
If $B \neq 0$, then the conic is rotated.

In this course we will not consider rotated conic sections, so we assume that $B = 0$.

EXAMPLE 5

Identify each conic.

a. $4x^2 + 9y^2 = 36$ **b.** $4x^2 + 9y = 36$ **c.** $9x^2 - 16y^2 = 144$

d. $9x - 16y = 144$ **e.** $25x^2 + 25y^2 = 1$ **f.** $25x^2 + 4y^2 = 10$

g. $25x^2 - 4y^2 + 3x - 5y + 6 = 0$ **h.** $25x^2 + 4y^2 + 50x - 8y = 0$

Solution

To identify each conic, compare the given form with the form
$$Ax^2 + Cy^2 + Dx + Ey + F = 0$$

a. $4x^2 + 9y^2 = 36$; second-degree in both x and y, with $A = 4$ and $C = 9$ both positive; the graph is an ellipse.

b. $4x^2 + 9y = 36$; no y^2 term ($C = 0$), so the graph is a parabola ($A = 4$).

c. $9x^2 - 16y^2 = 144$; second-degree in both x and y with $A = 9$ and $C = -16$ opposite in sign; thus, the graph is a hyperbola.

d. $9x - 16y = 144$; no x^2 and y^2 terms ($A = 0, C = 0$); thus, the graph is a line.

e. $25x^2 + 25y^2 = 1$; $A = C = 25$, so the graph is an ellipse (in fact, it is also a circle).

f. $25x^2 + 4y^2 = 10$; A and C have the same sign, so the graph is an ellipse.

g. $25x^2 - 4y^2 + 3x - 5y + 6 = 0$; A and C have opposite signs, so the graph is a hyperbola.

h. $25x^2 + 4y^2 + 50x - 8y = 0$; A and C have the same sign, so the graph is an ellipse. ◆

Graphing Conic Sections

We noted that in this book we will not consider rotated conic sections (that is, we assume $B = 0$). We further assume that the conic sections we graph from a given equation will be centered at the origin. Such conics are said to be in **standard form.** Conics can be located anywhere in the plane, and a process called *completing the square* can be used to transform the general-form equation into one of several standard forms, but the method is beyond the scope of this text.

We have considered graphing lines and parabolas earlier in this chapter, so we conclude this section by considering graphing ellipses and hyperbolas.

Ellipses

Standard-Form Equations of an Ellipse

The standard-form **equations for an ellipse** are

$$\frac{x^2}{a^2} + \frac{y^2}{b^2} = 1 \qquad \frac{y^2}{a^2} + \frac{x^2}{b^2} = 1$$

If $a = b$, then the ellipse is a circle, and we let $a = b = r$.

The standard-form **equation for a circle** is

$$x^2 + y^2 = r^2$$

Here are the directions for sketching an ellipse from a standard-form equation. Write the equation in standard form, so that there is a 1 on the right and the coefficients of the square terms are also 1. The center is $(0, 0)$; plot the intercepts on the x-axis and the y-axis. For the x-intercepts, plot $\pm$ the square root of the number under the x^2; for the y-intercepts, plot $\pm$ the square root of the number under the y^2. Finally, draw the ellipse using these intercepts. The longer axis is called the major axis; if this major axis is horizontal, then it is called a **horizontal ellipse;** if the major axis is vertical, it is called a **vertical ellipse.**

EXAMPLE 6

Sketch $\dfrac{x^2}{9} + \dfrac{y^2}{4} = 1$.

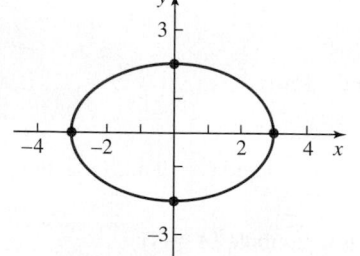

Figure 13.29
Graph of $\dfrac{x^2}{9} + \dfrac{y^2}{4} = 1$

Solution
The center of the ellipse is at $(0, 0)$. The x-intercepts are ± 3 ($\pm\sqrt{9} = \pm 3$; these are the vertices), and the y-intercepts are $\pm\sqrt{4} = \pm 2$. Sketch the ellipse using the four intercepts, as shown in Figure 13.29.

Because the horizontal distance (3 for this example) is larger than the vertical distance (2 for this example), we see this is an example of a horizontal ellipse. ◆

EXAMPLE 7

Sketch $9x^2 + 4y^2 = 36$.

Solution
Divide both sides by 36 to put this equation into standard form:

$$\frac{9x^2}{36} + \frac{4y^2}{36} = \frac{36}{36} \quad \text{or} \quad \frac{x^2}{4} + \frac{y^2}{9} = 1$$

We see $a^2 = 9$ and $b^2 = 4$, which indicates an ellipse with the major axis vertical because the larger number is under the y^2-term; thus, this is an example of a vertical ellipse. The x-intercepts are ± 2 and the y-intercepts are ± 3 (these are the vertices). The sketch is shown in Figure 13.30.

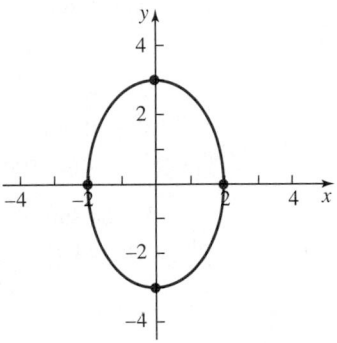

Figure 13.30 Graph of $\dfrac{x^2}{4} + \dfrac{y^2}{9} = 1$ ◆

The lengths of a and b are related to the location of the foci, as well as to the amount of "roundness" of the ellipse. For example, if $a = b$, it is clear that the ellipse is "perfectly round," but if $a = 5b$, then the ellipse is "elongated." We know (from the definition) that the foci of the ellipse are on the major axis and if the distance from the center to the foci is c (c a positive number),

$$c^2 = a^2 - b^2$$

A measure of the roundness of an ellipse is the **eccentricity,** which is defined as

$$\epsilon = \frac{c}{a} = \sqrt{1 - \frac{b^2}{a^2}}$$

Since $c < a$, we see that the eccentricity is between 0 and 1. If $a = b$, then $\epsilon = 0$ and the conic is a circle. For an ellipse, we see

$$0 \le \epsilon < 1$$

One of the most interesting applications of ellipses is their use in modeling planetary orbits. The orbit of a planet can be described by an ellipse with the sun at one focus. The orbit is commonly identified by the length of its major axis $2a$ and its eccentricity ϵ. The *aphelion* is the point where a planet is farthest from the sun, and the *perihelion* is the point where a planet is closest to the sun.

EXAMPLE 8   **PÓLYA'S METHOD**

Find an equation that models the earth's orbit around the sun.

Solution   We use Pólya's problem-solving guidelines for this example.

Understand the Problem. Before we can do this, we need additional information. An elementary astronomy book, an encyclopedia, or an almanac will tell us that the length of the major axis of the earth's orbit is 1.86×10^8 mi and the eccentricity of the earth's orbit is about 0.017.

Devise a Plan.

Using this information, we have $\epsilon = 0.017$ and $2a = 1.86 \times 10^8$ mi or $a = 9.3 \times 10^7$. Then

$$\epsilon = \frac{c}{a}$$

$$0.017 = \frac{c}{9.3 \times 10^7}$$

$$c = 1.581 \times 10^6$$

We can now find b^2 and substitute into the equation for an ellipse.

Carry Out the Plan.

$$c^2 = a^2 - b^2$$

$$b^2 = (9.3 \times 10^7)^2 - (1.581 \times 10^6)^2$$

$$= 8.646500439 \times 10^{15}$$

If we take the sun to be at one of the foci of the ellipse, we find the equation of the earth's orbit to be

$$\frac{x^2}{8.649 \times 10^{15}} + \frac{y^2}{8.647 \times 10^{15}} = 1$$

Look Back. We graph the equation that we have obtained, which is shown in Figure 11.31a. The graph is complicated by the scale, which you can simplify by making a good scale choice. For this example,

$$a^2 = 8.649 \times 10^{15} \qquad\qquad b^2 = 8.647 \times 10^{15}$$

$$a = \sqrt{8.649 \times 10^{15}} \qquad\qquad b = \sqrt{8.647 \times 10^{15}}$$

$$= \sqrt{8.649} \times \sqrt{10^{15}} \qquad\qquad = \sqrt{8.647} \times \sqrt{10^{15}}$$

Choose each unit to be $\sqrt{10^{15}}$ to obtain the equation

$$\frac{x^2}{8.649} + \frac{y^2}{8.647} = 1$$

This simpler, but equivalent, graph is shown in Figure 13.31b.

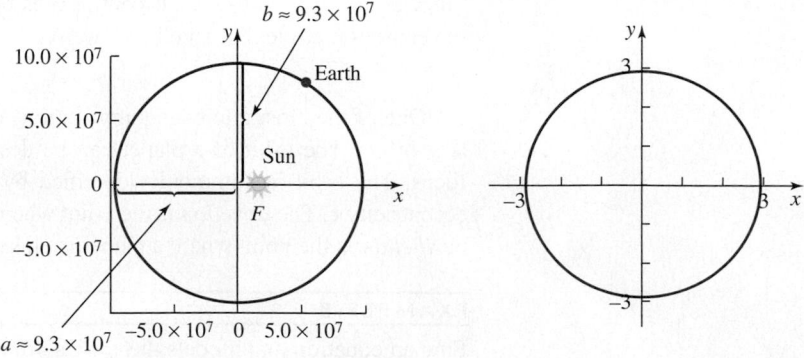

a. Graph with units in miles   **b.** Graph with units of $\sqrt{10^{15}}$

Figure 13.31 **Graph of the earth's orbit around the sun**

We see that the orbit is almost circular with the sun at the center. The fact that $\epsilon = 0.017$ tells us that the earth's orbit is almost circular. ◆

Another application of ellipses is the so-called "whispering room" phenomenon. Here is how these rooms work: One person stands at each focus of an elliptic dome. If one person whispers, the other will clearly hear what is said, but anyone *not* near a focus will not hear anything. This is especially impressive when the foci are far apart. Some famous buildings with this property are St. Paul's Cathedral in London, England, the old U.S. Capitol in Washington, D.C., and the Mormon Tabernacle in Salt Lake City, Utah. These galleries use the geometric property of ellipses that a wave (a sound wave, for example) from one focus will reflect an ellipse and be projected back to the other focus (see Figure 13.32a).

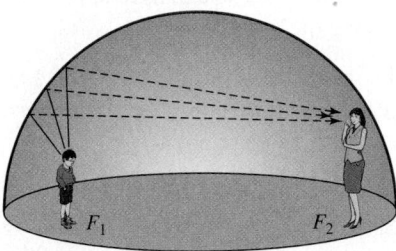 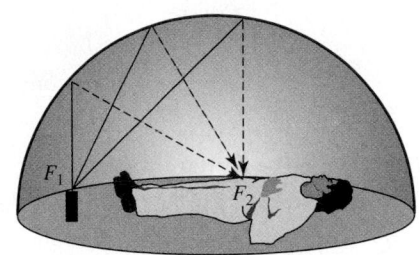

a. A whispering room: Only a person standing at focus F_2 clearly hears a sound emanating from focus F_1.

b. A pulse emanating from focus F_1 is concentrated on a kidney stone at focus F_2.

Figure 13.32 Applications of ellipses

The elliptic reflective property is also used in a procedure for disintegrating kidney stones. A patient is placed in an ellipsoidal tub of water (see Figure 13.32b—an ellipsoid is a three-dimensional elliptic figure) such that the kidney stone is at one focus. A pulse generated at the other focus is then concentrated at the kidney stone.

Hyperbolas

Standard-Form Equations of a Hyperbola

The standard-form **equations for a hyperbola** are

$$\frac{x^2}{a^2} - \frac{y^2}{b^2} = 1 \qquad \frac{y^2}{a^2} - \frac{x^2}{b^2} = 1$$

For the hyperbola, $c^2 = a^2 + b^2$.

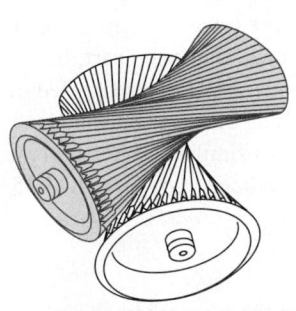

To transmit motion to a skew shaft, gears with blades form hyperbolas. Such gears are called *hyperboloidal gears*.

As with the other conic sections, we will sketch a hyperbola by inspection of the equation. The points of intersection of the hyperbola with the transverse axis are called the **vertices.** For

$$\frac{x^2}{a^2} - \frac{y^2}{b^2} = 1 \quad \text{and} \quad \frac{y^2}{a^2} - \frac{x^2}{b^2} = 1$$

notice that the vertices for the first equation occur at $(a, 0)$ and $(-a, 0)$. These are found by letting $y = 0$ and solving for x. This is the equation of a **horizontal hyperbola,** and we call the distance $2a$ the *length of the transverse axis.* The foci are located at $(c, 0)$ and $(-c, 0)$. The vertices for the second equation occur at $(0, a)$ and $(0, -a)$. This is the equation of a **vertical hyperbola** with foci located at $(0, c)$ and $(0, -c)$.

A hyperbola does not intersect the conjugate axis, but if we plot the points $(0, b)$, $(0, -b)$ and $(-b, 0)$, $(b, 0)$, respectively, we determine a segment on the conjugate axis whose length is called the *length of the conjugate axis.*

The procedure we will use to sketch a hyperbola is illustrated in the next example.

EXAMPLE 9

Sketch $\dfrac{x^2}{4} - \dfrac{y^2}{9} = 1$.

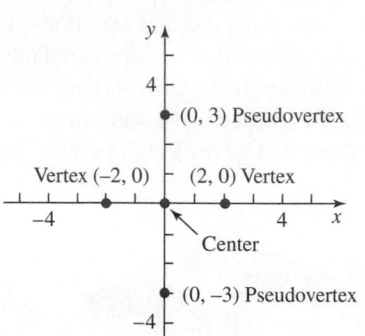

Vertex (−2, 0) (2, 0) Vertex

(0, 3) Pseudovertex

Center

(0, −3) Pseudovertex

Figure 13.33 Preliminary sketch

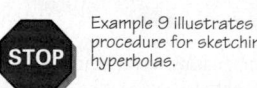

STOP

Example 9 illustrates a procedure for sketching hyperbolas.

Solution
The center of the hyperbola is at (0, 0), and $a = 2$ and $b = 3$, by inspection.

Step 1. Plot the vertices 2 units and −2 units from the center, (0, 0), at ±2, as shown in Figure 13.33. The transverse axis is along the x-axis, and the conjugate axis is along the y-axis.

Step 2. Plot the endpoints of the conjugate axis at 3 units and −3 units from the center. We call these points the *pseudovertices,* since the curve does not actually pass through these points.

Step 3. Draw lines through the vertices and pseudovertices parallel to the axes of the hyperbola. These lines form what we call the *central rectangle*. The diagonal lines passing through the corners of the central rectangle are **slant asymptotes** for the hyperbola, as shown in Figure 13.34.

Step 4. Draw the hyperbola using the slant asymptotes and the vertices.

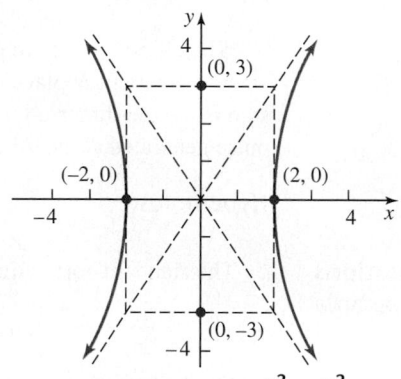

(0, 3)

(−2, 0)   (2, 0)

(0, −3)

Figure 13.34 Graph of $\dfrac{x^2}{4} - \dfrac{y^2}{9} = 1$ ◆

Hyperbolas also have a useful reflection property. To illustrate, suppose an aircraft has crashed somewhere in the desert. A device in the wreckage emits a "beep" at regular intervals. Two observers, located at listening posts a known distance apart, time a beep. It turns out that the time difference between the two listening posts multiplied by the speed of sound gives the value $2a$ for a hyperbola on which the airplane is located. A third listening post will determine two more hyperbolas in a similar fashion, and the airplane can thus be located at the intersection of these hyperbolas. See Figure 13.35.

Recall that the eccentricity ϵ of an ellipse satisfies

$$0 \le \epsilon < 1$$

with a circle having eccentricity $\epsilon = 0$. Since $\epsilon = \dfrac{c}{a}$, we see for a hyperbola that $\epsilon > 1$ (because $c > a$). The remaining possibility for the positive number ϵ is the case where $\epsilon = 1$. In this case, the conic section is a parabola.

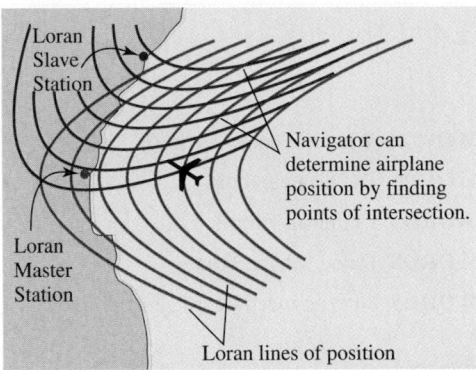

Figure 13.35 **LORAN measures the differences in the time of arrival of signals from two sets of stations. The plane's position at the intersection lines is charted on a special map based on a hyperbolic coordinate system.**

Parabolic Reflectors

Parabolic curves are used in the design of lighting systems, telescopes, and radar antennas, mainly because of the property illustrated in Figure 13.36.

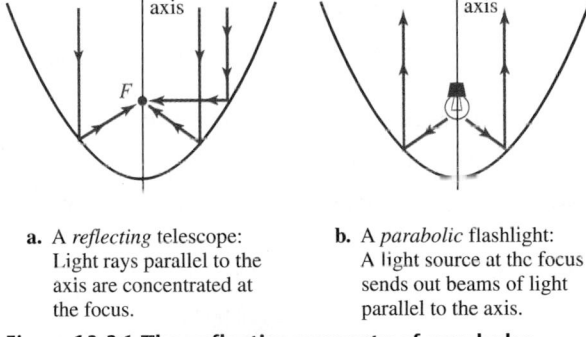

a. A *reflecting* telescope: Light rays parallel to the axis are concentrated at the focus.

b. A *parabolic* flashlight: A light source at the focus sends out beams of light parallel to the axis.

Figure 13.36 **The reflective property of parabolas**

As an illustration of the reflective property of parabolas, let us examine its application to reflecting telescopes (see Figure 13.36a). The eyepiece of such a telescope is placed at the focus of a parabolic mirror. Light enters the telescope in rays that are parallel to the axis of the parabola. It is known from physics that when light is reflected, the angle of incidence equals the angle of reflection. Hence, the parallel rays of light strike the parabolic mirror so that they all reflect through the focus, which means that all the parallel rays are concentrated at the eyepiece, thereby maximizing the light-gathering ability of the mirror.

Flashlights and automobile headlights (see Figure 13.36b) simply reverse the process: A light source is placed at the focus of a parabolic mirror, the light rays strike the mirror with an angle of incidence equal to the angle of reflection, and each ray is reflected along a path parallel to the axis, thus emitting a light beam of parallel rays.

Radar utilizes both of these properties. First, a pulse is transmitted from the focus to a parabolic surface. As with a reflecting telescope, parallel pulses are transmitted in this way. The reflected pulses then strike the parabolic surface and are sent back to be received at the focus.

Problem Set 13.4

1. **IN YOUR OWN WORDS** What is a conic section?

2. **IN YOUR OWN WORDS** Define a parabola.

3. **IN YOUR OWN WORDS** Define an ellipse.

4. **IN YOUR OWN WORDS** Define a hyperbola.

5. **IN YOUR OWN WORDS** Give a procedure for sketching an ellipse using its equation.

6. **IN YOUR OWN WORDS** Give a procedure for sketching a hyperbola using its equation.

7. **IN YOUR OWN WORDS** Explain how you recognize each of the following conics by inspecting the equation.
   a. line   **b.** parabola   **c.** ellipse   **d.** hyperbola

8. **IN YOUR OWN WORDS** What is the eccentricity for each of the given conics? Also, explain how changes in the eccentricity affect the curve.
   a. parabola   **b.** ellipse   **c.** hyperbola

If $Ax^2 + Cy^2 + Dx + Ey + F = 0$, for Problems 9–13, then list conditions on the constants to assure that the indicated graph results.

9. a hyperbola   10. a circle   11. a parabola

12. an ellipse   13. a line

Sketch the requested conic sections in Problems 14–23 using the definition.

14. A circle with radius 3.

15. A circle with radius 5.

16. A parabola with the distance between the directrix and focus 2 units.

17. A parabola with the distance between the directrix and focus 4 units.

18. A parabola with the distance between the directrix and focus 6 units.

19. A parabola with the distance between the directrix and focus 1 unit.

20. An ellipse with the distance between the foci 10 units and the sum of the distances 12 units.

21. An ellipse with the distance between the foci 10 units and the sum of the distances 14 units.

22. A hyperbola with the distance between the foci 10 units and the difference of the distances 8 units.

23. A hyperbola with the distance between the foci 10 units and the difference of the distances 6 units.

Identify the curves in Problems 24–27.

24. **a.** $2x - y - 8 = 0$ **b.** $4x^2 - 16y = 0$
    c. $(x - 1) = -2(y + 2)$

25. **a.** $2x + y - 10 = 0$
    b. $x^2 + 8(y - 12)^2 = 16$
    c. $y^2 - 4x + 10y + 13 = 0$

26. **a.** $x^2 + y^2 - 3y = 0$
    b. $y^2 + 4x - 3y + 1 = 0$
    c. $x^2 - 9y^2 - 6x + 18 = 0$

27. **a.** $2y^2 + 8y - 20x + 148 = 0$
    b. $9x^2 - 6y^2 + 18y - 23 = 0$
    c. $9x^2 + 6y^2 + 18x - 23 = 0$

Sketch the curves using the equations given in Problems 28–51.

28. $x^2 + y^2 = 1$ 29. $x^2 + y^2 = 64$

30. $x^2 + y^2 = 50$ 31. $x^2 + y^2 = 250$

32. $\dfrac{x^2}{4} + \dfrac{y^2}{9} = 1$ 33. $\dfrac{x^2}{25} + \dfrac{y^2}{36} = 1$

34. $x^2 + \dfrac{y^2}{9} = 1$ 35. $4x^2 + 9y^2 = 36$

36. $25x^2 + 16y^2 = 400$ 37. $16x^2 + 9y^2 = 144$

38. $x^2 - y^2 = 1$ 39. $y^2 - x^2 = 4$

40. $\dfrac{x^2}{9} - \dfrac{y^2}{4} = 1$ 41. $\dfrac{x^2}{4} - \dfrac{y^2}{9} = 1$

42. $\dfrac{y^2}{36} - \dfrac{x^2}{9} = 1$ 43. $\dfrac{y^2}{16} - \dfrac{x^2}{36} = 1$

44. $36y^2 - 25x^2 = 900$ 45. $3y^2 = 4x^2 + 12$

46. $3x^2 - 4y^2 = 12$ 47. $3y^2 = 4x^2 + 5$

48. $4y^2 - 4x^2 = 5$ 49. $3x^2 - 4y^2 = 5$

50. $4y^2 - x^2 = 9$ 51. $x^2 - y^2 = 9$

52. If the length of the major axis of the earth's orbit is 186,000,000 mi and its eccentricity is 0.017, how far is the earth from the sun when it is at *aphelion* and at *perihelion?*

53. If the length of the semimajor axis of the orbit of Mars is 1.4×10^8 mi and the eccentricity is about 0.093, determine the greatest and least distance of Mars from the sun.

54. If the planet Mercury is 28 million miles from the sun at *perihelion*, and the eccentricity of its orbit is about 1/5, how long is the major axis of Mercury's orbit?

55. The moon's orbit is elliptical with the earth at one focus. The point at which the moon is farthest from the earth is called the *apogee*, and the point at which it is closest is called the *perigee*. If the moon is 199,000 miles from the earth at apogee and the length of the major axis of its orbit is 378,000 miles, what is the eccentricity of the moon's orbit?

PROBLEM SOLVING

56. A parabolic archway has the dimensions shown in Figure 13.37. Find the equation of the parabolic portion.

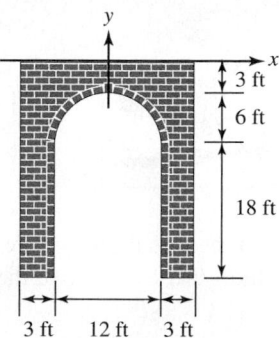

Figure 13.37 **A parabolic archway**

57. A radar antenna is constructed so that a cross-section along its axis is a parabola with the receiver at the focus. Find the focus if the antenna is 12 m across and its depth is 4 m, as shown in Figure 13.38.

58. A parabolic reflector (see Figure 13.38) is constructed so that a cross section along its axis is a parabola with the light source at the focus. Find the location of the focus if the reflector is 16 cm across and its depth is 8 cm.

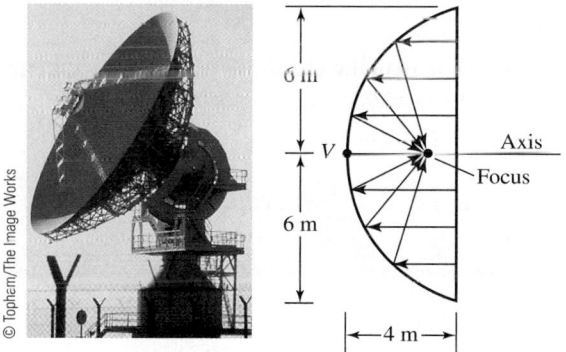

Figure 13.38 **Radar antenna**

59. Beams of light parallel to the axis of the parabolic mirror shown in Figure 13.39 strike the mirror and are reflected. Find the distance from the vertex to the point where the beams concentrate.

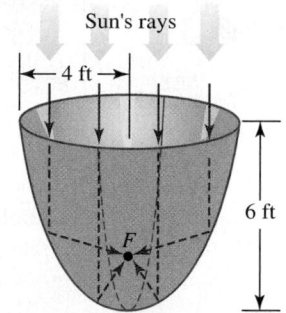

Figure 13.39 **A parabolic mirror**

60. A stone tunnel is to be constructed such that the opening is a semielliptic arch as shown in Figure 13.40. It is necessary to know the height at 4-ft intervals from the center. That is, how high is the tunnel at 4, 8, 12, 16, and 20 ft from the center? (Answer to the nearest tenth of a foot.)

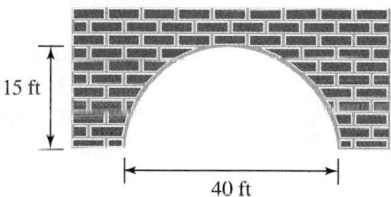

Figure 13.40 **Semielliptic arch**

13.5 | FUNCTIONS

The idea of looking at two sets of variables at the same time was introduced in Section 13.3. Sets of ordered pairs provide a very compact and useful way to represent relationships between various sets of numbers. To consider this idea, let's look at the *B.C.* cartoon at the top of page 696.

The distance an object will fall depends on (among other things) the length of time it falls. If we let the variable d be the distance the object has fallen (in feet) and the variable t be the time it has fallen (in seconds), and if we disregard air resistance, the formula is

$$d = 16t^2$$

B.C. By permission of Johnny Hart and Creators Syndicate, Inc.

Therefore, in the *B.C.* cartoon, if the well is 16 seconds deep (and we neglect the time it takes for the sound to come back up), we know that the depth of the well (in feet) is

$$d = 16(16)^2$$
$$= 16(256)$$
$$= 4,096$$

The formula $d = 16t^2$ gives rise to a set of data:

Time (in seconds)	0	1	2	3	4	. . .	15	16
Distance (in ft)	0	16	64	144	256	. . .	3,600	4,096

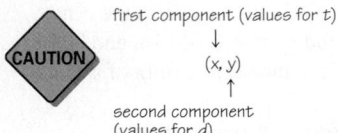

For every nonnegative value of t, there is a corresponding value of d. We can represent the data in the table as a set of ordered pairs in which the first component represents a value for t and the second component represents a corresponding value for d. For this example, we have $(0, 0)$, $(1, 16)$, $(2, 64)$, $(3, 144)$, $(4, 256)$, . . . , $(15, 3600)$, $(16, 4096)$.

Whenever we have a situation comparable to the one illustrated by this example— namely, whenever the first component of an ordered pair is associated with exactly one second component—we call the set of ordered pairs a *function*.

Function

> A **function** is a set of ordered pairs in which the first component is associated with exactly one second component.

Not all sets of ordered pairs are functions, as we can see from the following examples.

EXAMPLE 1

Which of the following sets of ordered pairs are functions?

a. $\{(0, 0), (1, 2), (2, 4), (3, 9), (4, 16)\}$ **b.** $\{(0, 0), (1, 1), (1, -1), (4, 2), (3, -2)\}$

c. $\{(1, 3), (2, 3), (3, 3), (4, 3)\}$ **d.** $\{(3, 1), (3, 2), (3, 3), (3, 4)\}$

Solution

a. We see that

$0 \rightarrow 0$

$1 \rightarrow 2$

$2 \rightarrow 4$

$3 \rightarrow 9$

$4 \rightarrow 16$

Sometimes it is helpful to think of the first component as the "picker" and the second component as the "pickee." Given an ordered pair (x, y), we find that each replacement for x "picks" a partner, or a second value. We can symbolize this by x → y.

Since each first component is associated with exactly one second component, the set is a function.

b. For this set,

$$0 \rightarrow 0$$

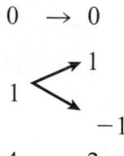

Since 1 picks two values as a partner or second component, we call the number 1 a "fickle picker." But *a function is a set of ordered pairs for which there are no fickle pickers.*

$$4 \rightarrow 2$$
$$3 \rightarrow -2$$

Since the first component can be associated with more than one second component, the set is not a function.

c. For this set,

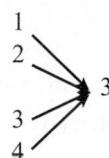

There are no fickle pickers, so it is a function.

This is an example of a function.

d. Finally,

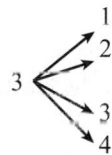

The number 3 is a fickle picker, so it is not a function.

Since the first component is associated with several second components, the set is not a function. ◆

Another way to consider functions is with the idea of a **function machine,** as shown in Figure 13.41. Think of this machine as having an input where items are entered and an output where results are obtained, much like a vending machine. If a number 2 is dropped into the input, a function machine will output a single value. If the name of the function machine is f, then the output value is called "f of 2" and is written as $f(2)$. This is called **functional notation.** The set of all possible replacements for x in $f(x)$ is called the **domain** and the set of all possible values for $f(x)$ is called the **range.**

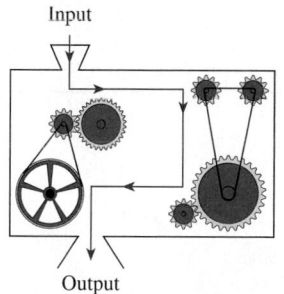

Figure 13.41 **Function machine**

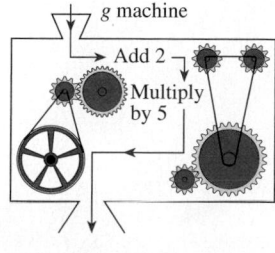

EXAMPLE 2

If you input each of the given values into a function machine named g, what is the resulting output value?

a. 4   **b.** -3 **c.** π **d.** x

Solution

a. Input is 4; output is $g(4)$, pronounced "gee of four." To calculate $g(4)$, first add 2, then multiply by 5:

Input value
↓
Output value

$$(4 + 2) \times 5 = 6 \times 5 = 30$$

We write $g(4) = 30$.

b. Input is -3; output is $g(-3)$, pronounced "gee of negative three":

$$g(-3) = (-3 + 2) \times 5 = -1 \times 5 = -5$$

c. Input is π; output is $g(\pi)$, pronounced "gee of pi":

$$g(\pi) = (\pi + 2) \times 5 = 5(\pi + 2)$$

d. Input is x; output is $g(x)$, pronounced "gee of ex":

$$g(x) = (x + 2) \times 5 = 5(x + 2)$$ ◆

EXAMPLE 3

If a function machine f squares the input value, we write $f(x) = x^2$, where x represents the input value. We usually define functions by simply saying "Let $f(x) = x^2$." Identify the value of f for the given value.

a. $f(2)$ **b.** $f(8)$ **c.** $f(-3)$ **d.** $f(t)$

Solution

a. $f(2) = 2^2 = 4$ **b.** $f(8) = 8^2 = 64$

c. $f(-3) = (-3)^2 = 9$ **d.** $f(t) = t^2$ ◆

Pay attention to the correct use of functional notation. If we write "let $f(x) = x^2$" we are defining a *function* named "f" by giving the formula that tells us how to find the second component "$f(x)$" for a given first component "x." That is, f represents the *function,* whereas $f(x)$ represents a *number.* In the next example, we use functional notation to find what is sometimes called a **difference quotient** for a given function. This difference quotient is defined to be the function

$$\frac{f(x + h) - f(x)}{h}$$

which will be used in calculus (Section 17.3).

EXAMPLE 4

If $f(x) = \frac{1}{2}x^2$, find the difference quotient.

Solution

We find the difference quotient by listing four steps (as shown in boldface).

1. $f(x) = \frac{1}{2}x^2$ List the given function.

2. $f(x + h) = \frac{1}{2}(x + h)^2$ Replace x by x + h in the given function.

3. $f(x + h) - f(x) = \frac{1}{2}(x + h)^2 - \frac{1}{2}x^2$ Subtract f(x) from f(x + h) and then simplify.

$$= \frac{1}{2}(x^2 + 2xh + h^2) - \frac{1}{2}x^2$$

$$= \frac{1}{2}x^2 + xh + \frac{1}{2}h^2 - \frac{1}{2}x^2$$

$$= xh + \frac{1}{2}h^2$$

4. $\dfrac{f(x + h) - f(x)}{h} = \dfrac{xh + \frac{1}{2}h^2}{h}$ Divide the result of step 3 by h.

$$= \frac{h(x + \frac{1}{2}h)}{h}$$ Factor (common factor of h).

$$= x + \frac{1}{2}h$$ Simplify.   ◆

If a function is defined as a graph, there is an easy test to decide whether the graph represents a function. It is called the **vertical line test,** and it states that every vertical line passes through the graph of a function in at most one point. This means that if you sweep a vertical line across a graph and it simultaneously intersects the curve at more than one point, then the curve is *not* the graph of a function.

EXAMPLE 5

Use the vertical line test to determine whether the given curve is the graph of a function. Name the probable domain and range by looking at the graph.

a.

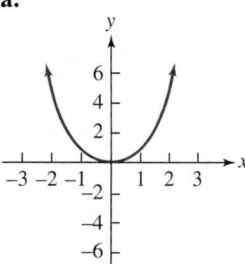

b.

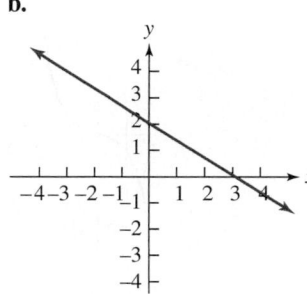

c.

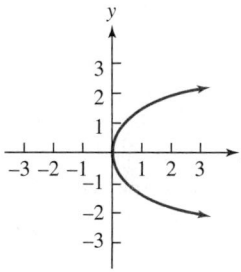

d.

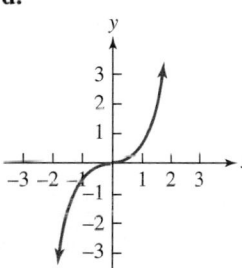

e.

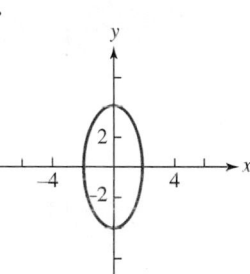

f.

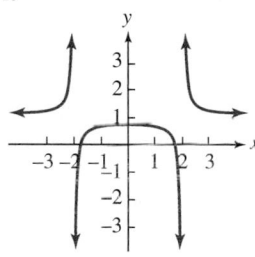

Solution

a.

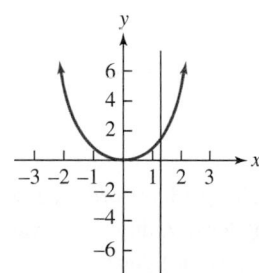

Passes the vertical line test; it is a function.
Domain: $\mathbb{R}$
Range: $y \geq 0$

b.

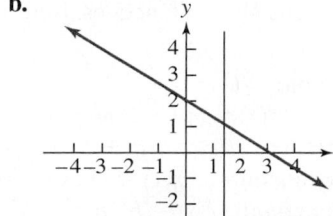

Passes the vertical line test; it is a function.
Domain: $\mathbb{R}$
Range: $\mathbb{R}$

c.

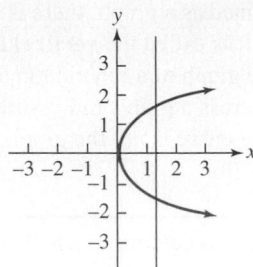

Does not pass the vertical line test; it is not a function.
Domain: $x \geq 0$
Range: $\mathbb{R}$

d.

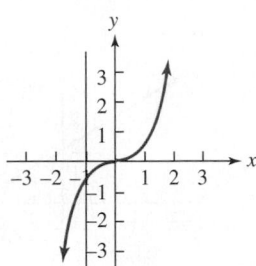

Passes the vertical line test; it is a function.
Domain: $\mathbb{R}$
Range: $\mathbb{R}$

e.

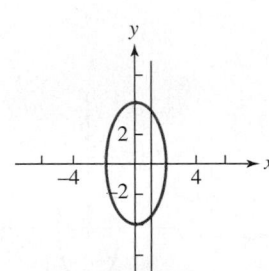

Does not pass the vertical line test; it is not a function.
Domain: $-2 \leq x \leq 2$
Range: $-4 \leq y \leq 4$

f.

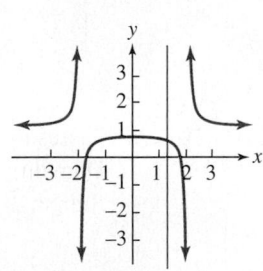

Passes the vertical line test; it is a function.
Domain: $x \neq -2, x \neq 2$
Range: $y \leq \frac{3}{4}$ or $y > 1$

◆

Many of the concepts we have previously discussed are functions, but we need to be careful with our characterizations. For example, consider the conic sections. Lines are generally functions, but a vertical line is not a function. Parabolas that open up or down are functions, but those that open right or left are not functions. Circles, ellipses, and hyperbolas are generally not functions. Here are the forms of some important functions:

constant function: $f(x) = k$
linear function: $f(x) = mx + b, m \neq 0$
quadratic function: $f(x) = ax^2 + bx + c, a \neq 0$
trigonometric functions: $f(x) = \sin x$ or $f(x) = \cos x$ or $f(x) = \tan x$
exponential function: $f(x) = b^x, b > 0, b \neq 1$
logarithmic function: $f(x) = \log_b x, b > 0, b \neq 0, x > 0$
probability function: A function P that satisfies the following properties: $0 \leq P(E) \leq 1, P(S) = 1$, and if E and F are mutually exclusive events, then $P(E \cup F) = P(E) + P(F)$.

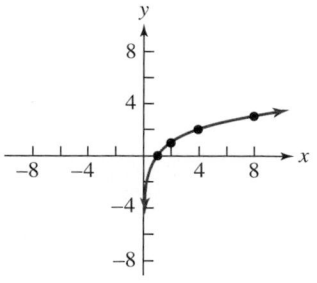

Figure 13.42 **Graph of**
$g(x) = \log_2 x$

EXAMPLE 6

Graph and then classify the function $g(x) = \log_2 x$.

Solution   We graph this function by plotting points. Recall that $\log_2 x$ is the exponent on a base 2 that gives the result x. In finding $g(x)$, we select values of x that are powers of 2:

x	$g(x)$	
1	0	$\log_2 1$ is the exponent on 2 that gives 1: $2^0 = 1$
2	1	$\log_2 2$ is the exponent on 2 that gives 2: $2^1 = 2$
4	2	$\log_2 4$ is the exponent on 2 that gives 4: $2^2 = 4$
8	3	$\log_2 8$ is the exponent on 2 that gives 8: $2^3 = 8$

This is a logarithmic function and the graph is shown in Figure 13.42.   ◆

Problem Set 13.5

LEVEL 1

1. **IN YOUR OWN WORDS** What is a function?
2. **IN YOUR OWN WORDS** Explain the notation $f(x)$.

Which of the sets in Problems 3–14 are functions?

3. $\{(1, 4), (2, 5), (4, 7), (9, 12)\}$
4. $\{(4, 1), (5, 2), (7, 4), (12, 9)\}$
5. $\{(1, 1), (2, 1), (3, 4), (4, 4), (5, 9), (6, 9)\}$
6. $\{(1, 1), (1, 2), (4, 3), (4, 4), (9, 5), (9, 6)\}$
7. $\{(4, 3), (17, 29), (18, 52), (4, 19)\}$
8. $\{(13, 4), (29, 4), (5, 4), (9, 4)\}$
9. $\{(19, 4), (52, 18), (29, 17), (3, 4)\}$
10. $\{(4, 9), (4, 4), (4, 29), (4, 19)\}$
11. $\{(5, 0)\}$
12. $\{(0, 0)\}$
13. $\{1, 2, 3, 4, 5\}$
14. $\{69, 82, 44, 37\}$

Tell what the output value is for each of the function machines in Problems 15–20 for **(a)** 4, **(b)** 6, **(c)** -8, **(d)** $\frac{1}{2}$, **(e)** x.

15.

16.

17.

18.

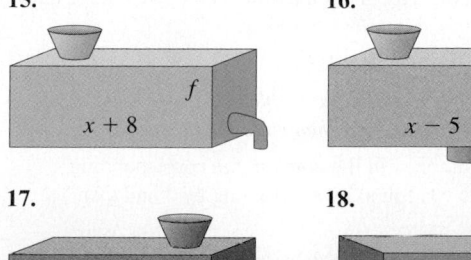

19.

20.

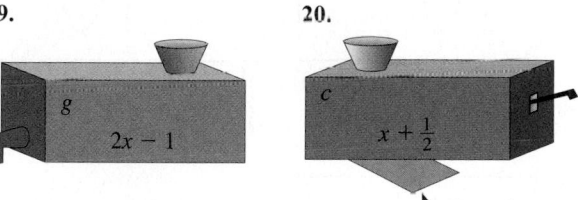

Trap door

21. If $f(x) = x + 7$, find
    a. $f(15)$ **b.** $f(-9)$ **c.** $f(p)$
22. If $g(x) = 2x$, find
    a. $g(100)$ **b.** $g(-25)$ **c.** $g(m)$
23. If $h(x) = 3x - 1$, find
    a. $h(0)$ **b.** $h(-10)$ **c.** $h(a)$
24. If $f(x) = x^2 + 1$, find
    a. $f(-3)$ **b.** $f(\frac{1}{2})$ **c.** $f(b)$
25. If $g(x) = \dfrac{x}{2}$, find
    a. $g(10)$ **b.** $g(-4)$ **c.** $g(3)$
26. If $h(x) = 0.6x$, find
    a. $h(4.1)$ **b.** $h(2.3)$ **c.** $h(\frac{5}{2})$

Use the vertical line test in Problems 27–32 to determine whether the curve is a function. Also state the probable domain and range.

27.

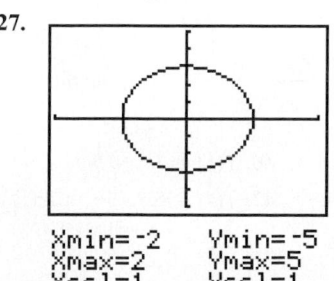

28.

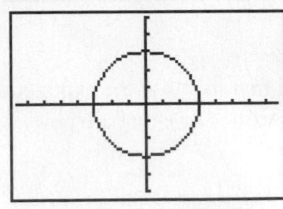

```
Xmin=-7.580645…
Xmax=7.5806451…
Xscl=1
Ymin=-5
Ymax=5
Yscl=1
```

29.

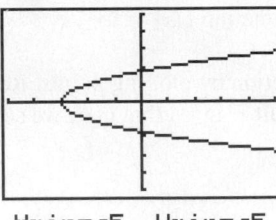

```
Xmin=-5    Ymin=-5
Xmax=5     Ymax=5
Xscl=1     Yscl=1
```

30.

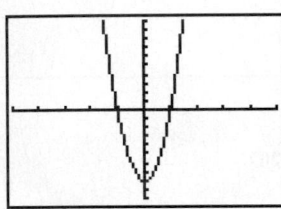

```
Xmin=-10   Ymin=-100
Xmax=10    Ymax=100
Xscl=2     Yscl=10
```

31.

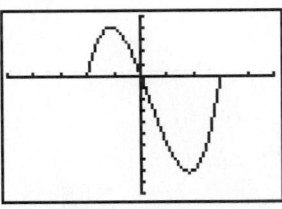

```
Xmin=-5    Ymin=-10
Xmax=5     Ymax=5
Xscl=1     Yscl=1
```

32.

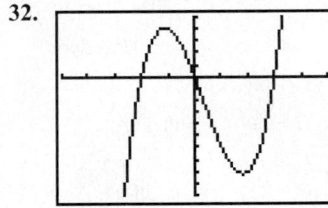

```
Xmin=-5    Ymin=-10
Xmax=5     Ymax=5
Xscl=1     Yscl=1
```

In Problems 33–38, graph each function and then classify as a linear, quadratic, exponential, logarithmic, or probability function.

33. $f(x) = 2x^2$

34. $f(x) = 2x$

35. $f(x) = \log x$

36. $f(x) = (\frac{1}{2})^x$

37. $f(x) = \dfrac{e^{-x^2/2}}{\sqrt{2\pi}}$

38. $f(x) = 5 - x$

LEVEL 2

Find the difference quotient, $\dfrac{f(x + h) - f(x)}{h}$*, for the functions given in Problems 39–44.*

39. $f(x) = 3x - 5$

40. $f(x) = 3x^2 - 5$

41. $f(x) = x^3$

42. $f(x) = 5x^3$

43. $f(x) = \dfrac{1}{x}$

44. $f(x) = \dfrac{1}{x^2}$

LEVEL 3

45.

YOUR WELL IS 16 SECONDS DEEP

The velocity v (in feet per second) of the rock dropped into the well in the *B.C.* cartoon at the beginning of this section is also related to time t (in seconds) by the formula

$$v = 32t$$

Complete the table showing the time and velocity of the rock.

Time (in sec)	0	1	2	3	4	...	8	...	16
Velocity (in ft per sec)	0	32	**a.**	**b.**	**c.**		**d.**		**e.**

46. Using the table of values in Problem 45, find the velocity of the rock at the instant it was released, after 8 seconds, and when it hit the bottom of the well. Write your answers in the form (t, v).

47. An independent distributor bought a new vending machine for $2,000. It had a probable scrap value of $100 at the end of its expected 10-year life. The value V at the end of n years is given by

$$V = 2,000 - 190n$$

Complete the table showing the year and the value of the machine.

Year	0	1	3	5	7	9	10
Value	2,000	1,810	**a.**	**b.**	**c.**	**d.**	**e.**

48. Using the table of values in Problem 47, find the value of the machine when it is purchased, when it is 5 years old, and when it is scrapped. Write your answers in the form (n, V).

49. Let $P(x)$ be the number of prime numbers less than x. Find
   a. $P(10)$ **b.** $P(-10)$ **c.** $P(100)$

50. Let $S(x)$ be the exponent on a base 2 that gives the result x. Find
   a. $S(32)$ **b.** $S(\frac{1}{8})$ **c.** $S(\sqrt{2})$

For each verbal description in Problems 51–54, write a rule in the form of an equation and then state the domain.

51. For each number x in the domain, the corresponding range value y is found by multiplying by 3 and then subtracting 5.

52. For each number x in the domain, the corresponding range value y is found by squaring and then subtracting 5 times the domain value.

53. For each number x in the domain, the corresponding range value y is found by taking the square root of the difference of the domain value subtracted from 5.

54. For each number x in the domain, the corresponding range value y is found by adding 1 to the domain value and then dividing that result into 5 added to 5 times the domain value.

PROBLEM SOLVING

55. Let A be the set of the following cities:

A = {Arm, MI; Bone, ID; Cheek, TX; Doublehead, AL; Elbow Lake, MN}

a. Suppose that these cities are connected by direct service as shown in Figure 13.43a. We define a set of ordered pairs (x, y) if and only if x and y are connected by direct service. Find the elements of this set of ordered pairs. Is this set a function?

b. Suppose that we make the connections one-way, as shown by the arrows in Figure 13.43b. We define a set of all pairs (x, y) for which there is one-way service from x to y. Find the elements of this set. Is this set a function?

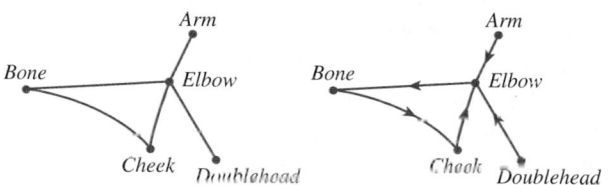

a. City network   **b.** Directed network

Figure 13.43 **Networks**

56. From a square whose side has length x, create a new square whose side is 5 in. longer. Find an expression for the difference between the areas of the two squares as a function of x. Graph this expression.

57. "Look out! That old abandoned well is very dangerous, Huck. I think that it should be capped so that nobody could fall in and kill his self." "Ah, shucks, Tom, let's climb down and see how deep it is. I'll bet it is a mile down to the bottom." If we assume that it is not possible for Tom or Huck to climb down into the well, how can they find out the depth of the well? Assume that the well is a mile deep and explain how Tom and Huck might determine this fact.

58. It is estimated that t years from now, the population of a certain suburban community will be

$$P(t) = 20 - \frac{6}{t+1}$$

thousand people.

a. What will the population of the community be 9 years from now?

b. By how much will the population increase during the ninth year?

c. What will happen to the size of the population in the "long run"?

59. Find the area of a square as a function of its perimeter.

60. Find the area of a circle as a function of its circumference.

CHAPTER SUMMARY

"Nature herself exhibits to us measurable and observable quantities in definite mathematical dependence; . . . Nearly all the 'known' functions have presented themselves in the attempt to solve geometrical, mechanical, or physical problems."

J. T. Mertz

IMPORTANT TERMS

Abscissa [13.1]
Analytic geometry [13.1]
Axes [13.1]
Axis of a parabola [13.4]
Boundary [13.2]
Cartesian coordinate system [13.1]
Center of a circle [13.4]
Center of an ellipse [13.4]
Center of a hyperbola [13.4]

Circle [13.4]
Closed half-plane [13.2]
Conic sections [13.4]
Conjugate axis [13.4]
Constant function [13.5]
Coordinates [13.1]
Dependent variable [13.1]
Difference quotient [13.5]
Directrix [13.4]

Domain [13.5]
Eccentricity [13.4]
Ellipse [13.4]
Exponential curve [13.3]
Exponential function [13.5]
First-degree equation [13.1, 13.4]
Focus (*pl.* foci) [13.4]
Function [13.5]
Function machine [13.5]

Functional notation [13.5]
General form [13.4]
Graph [13.1, 13.3]
Half-plane [13.2]
Horizontal ellipse [13.4]
Horizontal hyperbola [13.4]
Horizontal line [13.1]
Hyperbola [13.4]
Independent variable [13.1]
Line [13.4]
Linear equation [13.1]
Linear function [13.5]
Logarithmic function [13.5]
Major axis [13.4]

Minor axis [13.4]
Open half-plane [13.2]
Ordinate [13.1]
Origin [13.1]
Parabola [13.3, 13.4]
Probability function [13.5]
Quadrant [13.1]
Quadratic function [13.5]
Range [13.5]
Satisfy [13.1]
Second-degree equation [13.4]
Slant asymptotes [13.4]
Slope [13.1]

Slope–intercept form [13.1]
Slope point [13.1]
Solution [13.1]
Standard form [13.1, 13.4]
Test point [13.2]
Transverse axis [13.4]
Vertex [13.3, 13.4]
Vertical ellipse [13.4]
Vertical hyperbola [13.4]
Vertical line [13.1]
Vertical line test [13.5]
x-intercept [13.1]
y-intercept [13.1]

TYPES OF PROBLEMS

Graph lines by plotting points and by using the slope–intercept. [13.1]
Draw a line when given a point and the slope. [13.1]
Match a line with an equation. [13.1]
Solve applied problems involving lines, including future value, depreciation, profit, marginal profit, cost, and marginal cost. [13.1]
Graph first-degree inequalities with two unknowns. [13.2]
Graph parabolas and exponential curves by plotting points. [13.3]
Solve applied problems involving exponential and parabolic models. [13.3]
Identify a conic section by looking at its equation. [13.4]
Sketch a conic section using its geometric definition. [13.4]
Sketch a conic section using its standard-form equation. [13.4]
Solve applied problems involving the conic sections. [13.4]
Decide whether a given set is a function. [13.5]
Use the vertical line test to decide whether a given graph represents a function. [13.5]
Determine the output value for a function. [13.5]
Evaluate functions. [13.5]
Graph a function and then classify it as a linear, quadratic, exponential, logarithmic, or probability function. [13.5]
Find $\dfrac{f(x + h) - f(x)}{h}$ for a given function f. [13.5]

CHAPTER 13 REVIEW QUESTIONS

Graph the lines, curves, or half-planes in Problems 1–12.

1. $5x - y = 15$

2. $y = -\frac{4}{5}x - 3$

3. $2x + 3y = 15$

4. $x = -\frac{2}{3}y + 1$

5. $x = 150$

6. $x < 3y$

7. $y = 1 - x^2$

8. $y = -2^x$

9. $\dfrac{x^2}{16} - \dfrac{y^2}{9} = 1$

10. $\dfrac{x^2}{20} + \dfrac{y^2}{10} = 1$

11. $x^2 + y^2 = 1$

12. $x^2 - y^2 = 1$

13. Is $\{(4, 3), (5, -2), (6, 3)\}$ a function? Tell why or why not.

Find the value of each function in Problems 14–17.

14. $f(x) = 3x + 2$; find $f(6)$.

15. $g(x) = x^2 - 3$; find $g(0)$.

16. $F(x) = 5x + 25$; find $F(10)$.

17. $m(x) = 5$; find $m(10)$.

18. An amount of money A results from investing a sum P at a simple interest rate of 7% for 10 years, and is specified by the formula

$$A = 1.7P$$

Graph this equation, where P is the independent variable.

19. If a cannonball is fired upward with an initial velocity of 128 feet per second, its height can be calculated according to the formula

$$y = 128t - 16t^2$$

where t is the length of time (in seconds) after the cannonball is fired. Sketch this equation by letting $t = 0, 1, 2, \ldots, 7, 8$. Connect these points with part of a parabola.

20. Draw a population curve for a city whose growth rate is 1.3% and whose present population is 53,000. The equation is

$$P = P_0 e^{rt}$$

Let $t = 0, 10, \ldots, 50$ to help you find points for graphing this curve.

GROUP RESEARCH

Working in small groups is typical of most work environments, and learning to work with others to communicate specific ideas is an important skill. Work with three or four other students to submit a single report based on each of the following questions.

G47. If the path of a baseball is parabolic and is 200 ft wide at the base and 50 ft high at the vertex, write an equation that specifies the path of the baseball if the origin is the point of departure for the ball, and the form of the equation is $y - k = a(x - h)^2$, where (h, k) are the coordinates of the highest point of the baseball.

G48. According to the Centers for Disease Control, the number of AIDS-related cases is shown in Table 13.1.

a. Use the formula $A = A_0 e^{rt}$ to predict the number of cases in 1996.

b. Plot the data points and approximate by using a normal curve with mean at 1988 and standard deviation $\sqrt{5}$. Predict the number of cases in 1996.

c. Research the number of AIDS cases in 1996 and decide whether the model in part **a** or part **b** makes a better prediction.

Table 13.1 Number of AIDS-related cases

Year	No. of new cases
1982	920
1983	2,573
1984	5,237
1985	9,328
1986	14,705
1987	19,333
1988	21,978

G49. Investigate the topic of conic sections. Build models and/or find three-dimensional models for the conic sections. What did the Greeks know of the conic sections?

G50. Prepare a list of women mathematicians from the history of mathematics. Answer the question, "Why were so few mathematicians female?"

> **References** Teri Perl, *Math Equals: Biographies of Women Mathematicians plus Related Activities.* (Reading, MA: Addison-Wesley Publishing Co., 1978).
>
> Loretta Kelley, "Why Were So Few Mathematicians Female?" *The Mathematics Teacher,* October 1996.
>
> Barbara Sicherman and Carol H. Green, eds. *Notable American Women: The Modern Period.* A Biographical Dictionary. (Cambridge, MA: Belknap Press, Harvard University Press, 1980).
>
> *Outstanding Women in Mathematics and Science* (National Women's History Project, Windsor, CA 95492, 1991).

G51. Prepare a list of black mathematicians from the history of mathematics.

> **Reference** Virginia Newell et al., eds. *Black Mathematicians and Their Works* (Ardmore, PA: Dorrance & Company, 1980).

G52. Prepare a list of mathematicians with the first name of Karl.

G53. Write a news article about a historical mathematician as if you were a contemporary of the person you are writing about. Put it in newspaper style and include other newsworthy items from the period.

INDIVIDUAL RESEARCH PROBLEMS

Learning to use sources outside your classroom and textbook is an important skill, and here are some ideas for extending some of the ideas in this chapter. You can find references to these projects in a library or at

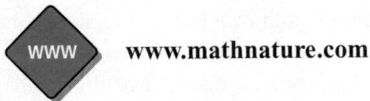

www.mathnature.com

PROJECT 13.1 This problem is a continuation of Problem 30, Section 13.2. A player's batting average really isn't simply the ratio $\frac{h}{a}$. It is the value of $\frac{h}{a}$ rounded to the nearest thousandth. It is possible that a batting average could be raised or lowered, but the reported batting average might remain the same when rounded. Write a paper on this topic.

PROJECT 13.2 The population in California was 31,910,000 in January 1995, and 32,231,000 in January 1996. Predict California's population in the year 2000. Check the Internet or an almanac to verify the 2000 population using the information of this problem.

PROJECT 13.3 The population in Sebastopol, California, was 7,475 in January 1995, and 7,525 in January 1996. Predict Sebastopol's population in the year 2005.

PROJECT 13.4 Predict the population of your city or state for the year 2005.

PROJECT 13.5 Ships at sea locate their positions using the LOng RAnge Navigation system known as LORAN. In this system, a master station sends signals that can be received by ships at sea. To fix the position of a particular ship, a secondary sending station also emits signals that can be received by the ship. Since the ship monitoring the two signals will usually be nearer one of the two stations, there will be a difference in the distances that the two signals travel. Because $d = rt$, there will be a slight time difference between the signals. If the ship follows a path so that the time difference remains constant, what is the path the ship will follow? Suppose the difference in the arrival of the time signals is 300 μsec. (*Note:* μsec is a microsecond—that is, one millionth of a second.) Also, suppose that the foci are 100 miles apart. Finally, suppose that signals travel at 980 ft/μsec.

PROJECT 13.6 Write a short paper exploring the concept of the eccentricity of an ellipse.

14

The Nature of Mathematical Systems

Now, when these studies reach the point of intercommunion and connection with one another, and come to be considered in their mutual affinities, then, I think, but not till then, will the pursuit of them have a value for our objects; otherwise there is no profit in them.

PLATO

CONTENTS

OVERVIEW

Solving systems of equations is a procedure that arises throughout mathematics. One of the fundamental building blocks of this book has been problem solving, and in most real-life situations, solving a problem requires (1) understanding the problem, (2) devising a method of solution, (3) carrying out the method, and (4) checking. Devising a method of solution may involve many quantities that are unknown, as well as many relationships between and among those quantities. Systems of equations and inequalities may facilitate the solution of a particular problem. This chapter begins by reviewing those methods that you have probably studied in a previous course, and then proceed to two methods that are particularly suitable for calculator or computer solutions—namely, Gauss–Jordan elimination and Inverse matrix methods. The chapter concludes by considering some applied problems that involve finding a maximum or minimum value subject to a set of constraints.

IMPORTANT IDEAS

Solving systems of equations by graphing, substitution, and addition [14.1]
Solve coin, rate, supply/demand, and mixture word problems [14.2]
Four elementary row operations [14.3]
Relationship between a system of equations and a corresponding matrix [14.3]
Process of pivoting [14.3]
Gauss–Jordan elimination [14.3]
Know the matrix operations of addition, multiplication by a scalar, subtraction, and multiplication [14.4]
Know the properties of matrices including commutative, associative, identity, inverse, and distributive properties [14.4]
Procedure for finding the inverse of a matrix [14.4]
Solving systems of inequalities [14.5]
Linear programming theorem [14.6]
Procedure for solving a linear programming problem [14.6]

BOOK REPORT

Write a 500-word report on this book:
On the Shoulders of Giants: New Approaches to Numeracy, Lynn Arthur Steen, Editor (Washington, DC: National Academy Press, 1990).

LINKS

Individual Projects

Group Projects

Research Links

www.mathnature.com

14.1   SYSTEMS OF LINEAR EQUATIONS

Two or more equations that are to be solved at the same time make up a **system of equations.** The **simultaneous solution** of a system of equations is the intersection of the solution sets of the individual equations. We use a brace to show that we are looking for a simultaneous solution. If all the equations in a system are linear, it is called a **linear system.**

We begin by reviewing those methods of solving linear systems that you first encountered in your previous algebra courses, and then we generalize first to matrices, and later in this chapter to systems of linear inequalities.

Graphing Method

The graph of each equation in a system of linear equations in two variables is a line; in the Cartesian plane, two lines must be related to each other in one of three ways:

1. They intersect at a single point.
2. The graphs are parallel lines. In this case, the solution set is empty, and the system is called *inconsistent.* In general, any system that has an empty solution set is referred to as an **inconsistent system.**
3. The graphs are the same line. In this case, there are infinitely many points in the solution set, and any solution of one equation is also a solution of the other. Such a system is called a **dependent system.**

In other words: Graph the equations and check for the intersection point. If the graphs do not intersect, then the equations represent an inconsistent system. If they coincide, then they represent a dependent system. When we solve a system by graphing the equations and then looking for points of intersection, we call it the **graphing method.**

Calculator Window

You can graph these lines with a graphing calculator:

Y1=(2/3)X+(8/3)

Y2=−X+6

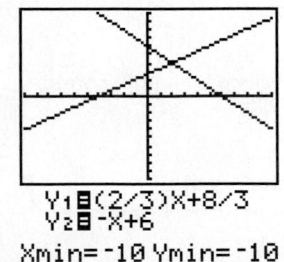

```
Y₁▤(2/3)X+8/3
Y₂▤-X+6
Xmin=-10 Ymin=-10
Xmax=10 Ymax=10
Xscl=1   Yscl=1
```

Use a [TRACE] or [ISECT] to find X=2 Y=4.

EXAMPLE 1

Solve the given systems by graphing:

a. $\begin{cases} 2x - 3y = -8 \\ x + y = 6 \end{cases}$ **b.** $\begin{cases} 2x - 3y = -8 \\ 4x - 6y = 0 \end{cases}$

c. $\begin{cases} 2x - 3y = -8 \\ y = \frac{2}{3}x + \frac{8}{3} \end{cases}$

Solution

a. Graph the line $2x - 3y = -8$:

$$3y = 2x + 8$$
$$y = \frac{2}{3}x + \frac{8}{3}$$

Graph the line $x + y = 6$:

$$y = -x + 6$$

Look at the point(s) of intersection, as shown in Figure 14.1. The solution appears to be $(2, 4)$, which can be verified by direct substitution into both the given equations.

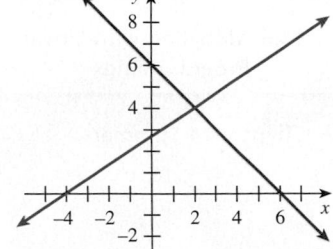

Figure 14.1 **Graph of**
$\begin{cases} 2x - 3y = -8 \\ x + y = 6 \end{cases}$

b. The graphs of the given lines are shown in Figure 14.2.

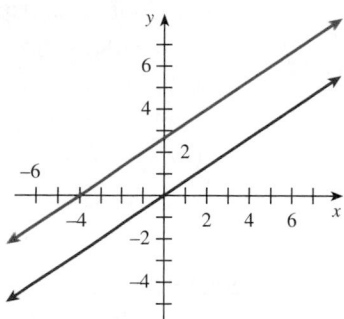

Figure 14.2 **Graph of** $\begin{cases} 2x - 3y = -8 \\ 4x - 6y = 0 \end{cases}$

Notice that these lines are parallel; you can show this analytically by noting that the slopes of the lines are the same. Since they are distinct parallel lines, there is not a point of intersection. This is an *inconsistent system.*

c. The graphs of the given lines are shown in Figure 14.3.

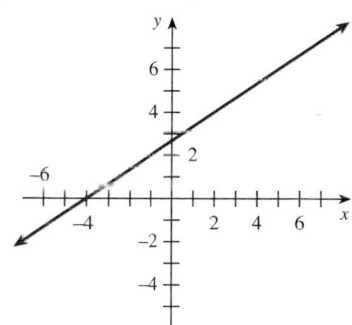

Figure 14.3 **Graph of** $\begin{cases} 2x - 3y = -8 \\ y = \frac{2}{3}x + \frac{8}{3} \end{cases}$

The equations represent the same line. This is a *dependent system.* ◆

Substitution Method

The graphing method can give solutions as accurate as the graphs you can draw, and consequently it is adequate for many applications. However, there is often a need for more exact methods.

In general, given a system, the procedure for solving is to write a simpler equivalent system. Two systems are said to be **equivalent** if they have the same solution set. In this book we limit ourselves to finding only real roots. There are several ways to go about writing equivalent systems. The first nongraphical method we consider comes from the substitution property of real numbers and leads to a **substitution method** for solving systems.

Substitution Method for Solving Systems of Equations with Two Equations and Two Unknowns

1. *Solve* one of the equations for one of the variables.

2. *Substitute* the expression that you obtain into the other equation.

3. *Solve* the resulting equation.

4. *Substitute* that solution into either of the original equations to find the value of the other variable.

5. *State* the solution.

EXAMPLE 2

Solve: $\begin{cases} 2p + 3q = 5 \\ q = -2p + 7 \end{cases}$

Solution

Since $q = -2p + 7$, substitute $-2p + 7$ for q in the other equation:

$$2p + 3q = 5$$
$$2p + 3(-2p + 7) = 5$$
$$2p - 6p + 21 = 5$$
$$-4p = -16$$
$$p = 4$$

Now, substitute 4 for p in either of the given equations:

$$q = -2p + 7$$
$$= -2(4) + 7$$
$$= -1$$

The solution is $(p, q) = (4, -1)$. If the variables are not x and y, then you must also show the variables along with the ordered pair. This establishes which variable is associated with which number—namely, $p = 4$ and $q = -1$. ◆

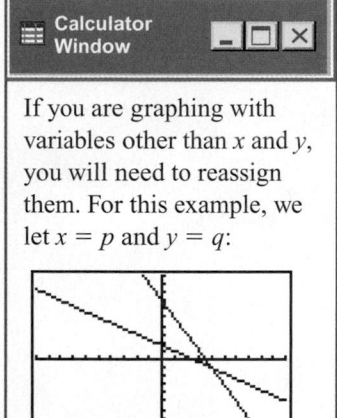

Calculator Window ⊟ ▢ ✕

If you are graphing with variables other than x and y, you will need to reassign them. For this example, we let $x = p$ and $y = q$:

Y₁⊟-(2/3)X+5/3
Y₂⊟-2X+7

Xmin=-10 Ymin=-10
Xmax=10 Ymax=10
Xscl=1   Yscl=1

Many calculators have an
[ISECT] function:
Intersection
$x = 4$ $y = -3$

Linear Combination (Addition) Method

A third method for solving systems is called the **linear combination method** (or, as it is often called, the **addition method**). It involves substitution and the idea that if equal quantities are added to equal quantities, the resulting equation is equivalent to the original system. In general, addition will not simplify the system unless the numerical coefficients of one or more terms are opposites. However, you can often force them to be opposites by multiplying one or both of the given equations by nonzero constants.

Linear Combination Method for Solving Systems of Equations

1. *Multiply* one or both of the equations by a constant or constants so that the coefficients of one of the variables become opposites.

2. *Add* corresponding members of the equations to obtain a new equation in a single variable.

3. *Solve* the derived equation for that variable.

4. *Substitute* the value of the found variable into either of the original equations, and solve for the second variable.

5. *State* the solution.

In other words: Multiply one or both of the equations by a constant or constants so that the coefficients of one of the variables are opposites, and then add the equations to eliminate the variable.

EXAMPLE 3

Solve the given system by the linear combination method.

$$\begin{cases} 3x + 5y = -2 \\ 2x + 3y = 0 \end{cases}$$

Solution Multiply **both sides** of the first equation by 2, and **both sides** of the second equation by -3. This procedure, denoted as shown below, forces the coefficients of x to be opposites:

$$\begin{array}{c} 2 \\ -3 \end{array} \begin{cases} 3x + 5y = -2 \\ 2x + 3y = 0 \end{cases}$$

This means you should add the equations.

$$+ \begin{cases} 6x + 10y = -4 \\ -6x - 9y = 0 \end{cases}$$

$$y = -4 \quad \text{Mentally, add the equations.}$$

If $y = -4$, then $2x + 3y = 0$ means $2x + 3(-4) = 0$, which implies that $x = 6$. The solution is $(6, -4)$. ◆

Problem Set 14.1

LEVEL 1

1. **IN YOUR OWN WORDS** What is a system of linear equations?

2. **IN YOUR OWN WORDS** Distinguish inconsistent and dependent systems.

3. **IN YOUR OWN WORDS** Describe the graphing method for solving a system of equations.

4. **IN YOUR OWN WORDS** Describe the substitution method for solving a system of equations.

5. **IN YOUR OWN WORDS** Describe the addition method for solving a system of equations.

Solve the systems in Problems 6–13 by graphing.

6. $\begin{cases} x - y = 2 \\ 2x + 3y = 9 \end{cases}$

7. $\begin{cases} 3x - 4y = 16 \\ -x + 2y = -6 \end{cases}$

8. $\begin{cases} y = 3x + 1 \\ x - 2y = 8 \end{cases}$

9. $\begin{cases} 2x - 3y = 12 \\ -4x + 6y = 18 \end{cases}$

10. $\begin{cases} x - 6 = y \\ 4x + y = 9 \end{cases}$

11. $\begin{cases} 6x + y = -5 \\ x + 3y = 2 \end{cases}$

12. $\begin{cases} 4x - 3y = -1 \\ -2x + 3y = -1 \end{cases}$

13. $\begin{cases} 3x + 2y = 5 \\ 4x - 3y = 1 \end{cases}$

Solve the systems in Problems 14–25 by the substitution method.

14. $\begin{cases} y = 3 - 2x \\ 3x + 2y = -17 \end{cases}$

15. $\begin{cases} 5x - 2y = -19 \\ x = 3y + 4 \end{cases}$

16. $\begin{cases} y = 5 - 3x \\ 2x + 3y = 1 \end{cases}$

17. $\begin{cases} 3x - y = -1 \\ x = 2y + 3 \end{cases}$

18. $\begin{cases} 2x - 3y = 15 \\ y = \frac{2}{3}x - 8 \end{cases}$

19. $\begin{cases} x + y = 12 \\ 0.6y = 0.5(12) \end{cases}$

20. $\begin{cases} 4y + 5x = 2 \\ y = \frac{5}{4}x + 2 \end{cases}$

21. $\begin{cases} \frac{x}{3} - y = 7 \\ x + \frac{y}{2} = 7 \end{cases}$

22. $\begin{cases} 3t_1 + 5t_2 = 1{,}541 \\ t_2 = 2t_1 + 160 \end{cases}$

23. $\begin{cases} x = -7y - 3 \\ 2x + 5y = 3 \end{cases}$

24. $\begin{cases} x = 3y - 4 \\ 5x - 4y = -9 \end{cases}$

25. $\begin{cases} x + 3y = 0 \\ x = 5y + 16 \end{cases}$

Solve the systems in Problems 26–37 by the addition method.

26. $\begin{cases} x + y = 16 \\ x - y = 10 \end{cases}$

27. $\begin{cases} x + y = 560 \\ x - y = 490 \end{cases}$

28. $\begin{cases} x + y = 6 \\ x - 2y = 12 \end{cases}$ **29.** $\begin{cases} 3x + y = 13 \\ x - 2y = 9 \end{cases}$

30. $\begin{cases} 6r - 4s = 10 \\ 2s = 3r - 5 \end{cases}$ **31.** $\begin{cases} 3u + 2v = 5 \\ 4v = 10 - 6u \end{cases}$

32. $\begin{cases} 3a_1 + 4a_2 = -9 \\ 5a_1 + 7a_2 = -14 \end{cases}$ **33.** $\begin{cases} 5s_1 + 2s_2 = 23 \\ 2s_1 + 7s_2 = 34 \end{cases}$

34. $\begin{cases} s + t = 12 \\ s - 2t = -4 \end{cases}$ **35.** $\begin{cases} 2u - 3v = 16 \\ 5u + 2v = 21 \end{cases}$

36. $\begin{cases} 2x + 5y = 7 \\ 2x + 6y = 14 \end{cases}$ **37.** $\begin{cases} 5x + 4y = 5 \\ 15x - 2y = 8 \end{cases}$

LEVEL 2

Solve the systems in Problems 38–55 for all real solutions, using any suitable method.

38. $\begin{cases} x + y = 7 \\ x - y = -1 \end{cases}$ **39.** $\begin{cases} x - y = 8 \\ x + y = 2 \end{cases}$

40. $\begin{cases} -x + 2y = 2 \\ 4x - 7y = -5 \end{cases}$ **41.** $\begin{cases} x - 6y = -3 \\ 2x + 3y = 9 \end{cases}$

42. $\begin{cases} y = 3x + 1 \\ x - 2y = 8 \end{cases}$ **43.** $\begin{cases} 2x + 3y = 9 \\ x = 5y - 2 \end{cases}$

44. $\begin{cases} x + y = 4 \\ 2x + 3y = 9 \end{cases}$ **45.** $\begin{cases} 3x + 4y = 8 \\ x + 2y = 2 \end{cases}$

46. $\begin{cases} 2x - y = 6 \\ 4x + y = 3 \end{cases}$ **47.** $\begin{cases} 6x + 9y = -4 \\ 9x + 3y = 1 \end{cases}$

48. $\begin{cases} 5x - 2y = -1 \\ 3x + y = 17 \end{cases}$ **49.** $\begin{cases} 5x + 4y = 9 \\ 9x + 3y = 12 \end{cases}$

50. $\begin{cases} 100x - y = 0 \\ 50x + y = 300 \end{cases}$ **51.** $\begin{cases} x = \frac{3}{4}y - 2 \\ 3y - 4x = 5 \end{cases}$

52. $\begin{cases} q + d = 147 \\ 0.25q + 0.10d = 24.15 \end{cases}$ **53.** $\begin{cases} x + y = 10 \\ 0.4x + 0.9y = 0.5(10) \end{cases}$

54. $\begin{cases} 12x - 5y = -39 \\ y = 2x + 9 \end{cases}$ **55.** $\begin{cases} y = 2x - 1 \\ y = -3x - 9 \end{cases}$

PROBLEM SOLVING

56. **JOURNAL PROBLEM** (From "When Does a Dog Become Older Than Its Owner?" by Anne Larson Quinn and Karen R. Larson, *The Mathematics Teacher,* December 1996, pp. 734–737.) The premise of this problem is that since dogs supposedly age seven times as quickly as humans, at some point the dog will become "older" than its owner. Karen wanted to determine exactly on which day this milestone would occur for her and her dog, Sydney, so that they could celebrate the occasion. Here are the basic facts. Karen was born on December 7, 1970, and Sydney was born on April 18, 1992. For every year that Karen aged, Syndey aged seven equivalent people-years. On what date are Karen and Sydney "the same age"?

57. Assume that Sydney in Problem 56 is a cat instead of a dog, and assume that cats age four times as quickly as humans. Using this assumption, when will Karen and Sydney be "the same age"?

58. Assume that you obtain a dog that was born today. When will this dog be older (in dog years) than you are?

59. **HISTORICAL QUESTION** The Louvre Tablet from the Babylonian civilization is dated about 1500 B.C. It shows a system equivalent to

$$\begin{cases} xy = 1 \\ x + y = a \end{cases}$$

Solve this system for x and y in terms of a. This is not a linear system, and you will need to use the quadratic formula after substituting.

60. **HISTORICAL QUESTION** The following problem was written by Leonhard Euler: "Two persons owe conjointly 29 pistoles;* they both have money, but neither of them enough to enable him, singly, to discharge this common debt." The first debtor says therefore to the second, "If you give me $\frac{2}{3}$ of your money, I can immediately pay the debt." The second answers that he also could discharge the debt, if the other would give him $\frac{3}{4}$ of his money. Required, how many pistoles each had?

*A *pistole* is a unit of money.

14.2 PROBLEM SOLVING WITH SYSTEMS

Problem solving has been a major theme in this book since it was first introduced in Chapter 1. The procedure that we described as "evolving to a single variable" is equivalent to solving a system by substitution. However, as we see in this chapter, substitution is only one of the techniques for solving systems. In this chapter, we look at systems that can be solved by a variety of techniques. This section introduces some of the usual types of word problems, and as we develop the rest of this chapter, we will introduce some of the more unusual types of problems that use systems of equations.

As we have seen, substitution is one of the primary tools used in solving word problems. Many word problems are given in a form that indicates more than two variables. There are at least three ways you can reduce the number of variables in an applied problem:

1. Substitute numbers for variables. Use this type of substitution when the value of a variable is known.
2. Substitute variables for other variables by using a known formula.
3. Substitute variables for other variables when relationships between those variables are given in the problem.

Coin Problems

The first type of problem we consider in this section is coin problems. We begin here because coin problems vividly illustrate the need to be careful in defining the variables you use. In money problems, you must distinguish between the *number of coins* and the *value of the coins* because, for example, the number of quarters you have is not the same as the value of those quarters. The formulas you will need for money problems are:

$$\text{VALUE OF QUARTERS} = 25(\text{NUMBER OF QUARTERS})$$

$$\text{VALUE OF DIMES} = 10(\text{NUMBER OF DIMES})$$

$$\text{VALUE OF NICKELS} = 5(\text{NUMBER OF NICKELS})$$

$$\text{VALUE OF PENNIES} = \text{NUMBER OF PENNIES}$$

Notice that the values in the preceding formulas are in terms of cents and not dollars.

EXAMPLE 1

A box of coins has dimes and nickels totaling $4.20. If there are three more nickels than dimes, how many of each type of coin is in the box?

Solution Restate the problem in equation form. It does not matter if you use more than two variables.

$$\begin{cases} \text{VALUE OF DIMES} + \text{VALUE OF NICKELS} = \text{TOTAL VALUE} \\ \text{NUMBER OF NICKELS} = \text{NUMBER OF DIMES} + 3 \end{cases}$$

We used five different unknowns in this statement.

Substitute known numbers:

Substitute: TOTAL VALUE $= 420$
$$\downarrow$$

$$\begin{cases} \text{VALUE OF DIMES} + \text{VALUE OF NICKELS} = \text{TOTAL VALUE} \\ \text{NUMBER OF NICKELS} = \text{NUMBER OF DIMES} + 3 \end{cases}$$

Substitute using a formula:

$$10(\text{NUMBER OF DIMES}) \qquad 5(\text{NUMBER OF NICKELS}) \qquad \textit{Substitute.}$$
$$\downarrow \qquad\qquad\qquad \downarrow$$

$$\begin{cases} \text{VALUE OF DIMES} + \text{VALUE OF NICKELS} = 420 \\ \text{NUMBER OF NICKELS} = \text{NUMBER OF DIMES} + 3 \end{cases}$$

$$\begin{cases} 10(\text{NUMBER OF DIMES}) + 5(\text{NUMBER OF NICKELS}) = 420 \\ \text{NUMBER OF NICKELS} = \text{NUMBER OF DIMES} + 3 \end{cases}$$

Now that we have only two unknowns, we choose variables. Let $d =$ NUMBER OF DIMES and $n =$ NUMBER OF NICKELS.

$$\begin{cases} 10d + 5n = 420 \\ n = d + 3 \end{cases}$$

Solve *this* system by substitution:

$$10d + 5n = 420$$
$$10d + 5(d + 3) = 420 \quad \text{\small Substitute } n = d + 3.$$
$$10d + 5d + 15 = 420$$
$$15d = 405$$
$$d = 27$$

If $d = 27$, then $n = d + 3 = 27 + 3 = 30$. There are 27 dimes and 30 nickels. *Check:* Three more nickels than dimes and the value is $27(\$0.10) + 30(\$0.05) = \$4.20$. ◆

Combining Rates

On a recent visit to an airport, I was walking from the terminal to the gate and there was a moving sidewalk. This caused me to think (of course) of a variety of algebra problems. If I walk at a rate of 100 feet per minute and the sidewalk travels at 80 ft per minute, then my rate is:

100 feet per minute if I walk to the terminal without using the moving sidewalk;
80 feet per minute if I stand on the moving sidewalk;
$100 + 80 = 180$ feet per minute if I walk on the moving sidewalk;
$100 - 80 = 20$ feet per minute if I walk against the movement of the moving sidewalk.

This illustrates a general principle. When you move in air, in water, on a treadmill, or on some other medium that is also moving, you combine rates. *If you move in the same direction, your rate is added to the rate of the medium. If you move against the movement of the medium, the rates are subtracted.* Consider the following examples.

EXAMPLE 2

A deep-sea fishing boat travels at 20 mph going out with the tide. Sometimes it comes in against the tidal current and is able to make only 15 mph. What is the boat's rate without the current, and what is the rate at which the tide moves the water?

Solution We are given:

RATE WITH THE TIDE is 20 mph.

RATE AGAINST THE TIDE is 15 mph.

This gives us the following system of equations:

$$\begin{cases} (\text{RATE OF BOAT}) + (\text{RATE OF TIDE}) = 20 \quad \text{\small Same direction: Add rates.} \\ (\text{RATE OF BOAT}) - (\text{RATE OF TIDE}) = 15 \quad \text{\small Opposite direction: Subtract rates.} \end{cases}$$

We assume that the rate of the boat is greater so that we obtain a positive distance. There are two unknowns; let $b =$ RATE OF BOAT and $t =$ RATE OF TIDE.

$$+ \begin{cases} b + t = 20 \\ b - t = 15 \end{cases}$$
$$\overline{2b = 35}$$
$$b = 17.5$$

Substitute to find t:

$$17.5 + t = 20$$
$$t = 2.5$$

The boat's speed is 17.5 mph, and the tide moves at 2.5 mph.   ◆

EXAMPLE 3

On a certain day a large bird was clocked at 80 mph flying with the wind, but could fly only 10 mph against the wind. What is the speed of the bird in still air?

Solution We are given:

RATE WITH THE WIND is 80 mph.

RATE AGAINST THE WIND is 10 mph.

This gives us the following system:

$$\begin{cases} \text{BIRD'S RATE} + \text{WIND'S RATE} = 80 \\ \text{BIRD'S RATE} - \text{WIND'S RATE} = 10 \end{cases}$$

Let $b = $ BIRD'S RATE and $w = $ WIND'S RATE:

$$+ \begin{cases} b + w = 80 \\ b - w = 10 \end{cases} \quad \text{\small Assume the bird's rate is greater than the wind's rate.}$$
$$2b = 90$$
$$b = 45$$

The bird's rate in still air is 45 mph.   ◆

> ⬦ **CAUTION** Remember to answer the question that was asked. In Example 3, we were not asked to find the rate of the wind.

Supply and Demand

We have solved word problems with linear combinations and with substitution. Sometimes it is useful to solve word problems by graphing. One such application has to do with supply and demand. If supply greatly exceeds demand, then money will be lost because of unsold items. On the other hand, if demand greatly exceeds supply, then money will be lost because of insufficient inventory. The most desirable situation is when the supply and demand are equal. The point for which the supply and demand are equal is called the **equilibrium point.** In this section we assume that supply and demand functions are linear.

EXAMPLE 4   PÓLYA'S METHOD

Suppose you have a small product that is marketable to the students on your campus. You want to know what price to charge for this product to maximize your profit.

Solution We use Pólya's problem-solving guidelines for this example.

Understand the Problem. There is not sufficient information given to answer this question. *A little market research shows that only* 200 *people would buy the product if it were priced at* $10, *but* 2,000 *would buy it at* $1. This information represents the **demand.** Let $p = $ PRICE and $n = $ NUMBER OF ITEMS. Then, because the demand, n, is determined by the price, let the price, p, be the independent variable. That is, let the ordered pairs be (p, n). From the given information, we see that the demand curve passes through the points (10, 200) and (1, 2000). If we assume that demand is linear (a straight line), we label the line passing through these points as the "demand" line.

There is still insufficient information. *We call a local shop and find that they can make the product during slack time and could supply* 300 *items at a price that allows*

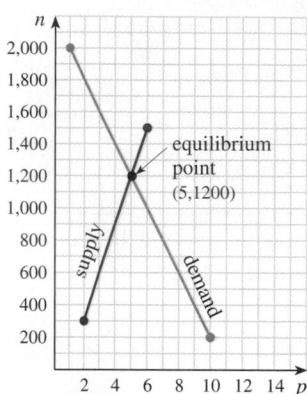

Figure 14.4 **Supply and demand**

you to sell them for $2. To supply more, the shop must use overtime. If the shop supplies 1,500 *items, you will have to charge $6 for your product.* This information represents the **supply.** Use (p, n) as defined for demand. From the given information, the supply curve passes through (2, 300) and (6, 1500). If we assume that supply is linear, we draw the line through these points and label it the "supply" line, as shown in Figure 14.4.

Devise a Plan. We now have sufficient information to solve the problem. Profit is maximized at the equilibrium point. This is the point of intersection of the supply and demand lines. We will look for the intersection point of the supply and demand lines.

Carry Out the Plan. We see from Figure 14.4 that the equilibrium point is (5, 1200). This means that the price charged should be $5. It also says that you should expect to sell 1,200 items.

Look Back. As a real modeling problem, you would need to research the parts shown in italics, but for our purposes in this book, you will be supplied this information. ◆

Age Problems

If you are comparing birthdates of individuals, then the *younger* person has the *larger* birthdate (year). For example, if you were born in 1975, and your sister is 6 years younger, then her birthdate is

years between ages
↓

$\underbrace{1975}_{\text{birthdate of older person}}$ + $\underbrace{6}$ $= \underbrace{1981}_{\text{birthdate of younger person}}$

EXAMPLE 5

Ron Howard is 9 years older than actor Tom Cruise. If the sum of their birth years is 3,915, in what year was Tom Cruise born?

Solution Let

x = Ron Howard's birth year

y = Tom Cruise's birth year

Since Ron Howard is 9 years older, you add 9 to his birth year to obtain Cruise's birth year: $x + 9 = y$. Thus, we have the following system

$$\begin{cases} x + 9 = y \\ x + y = 3{,}915 \end{cases}$$

We solve this system by substituting $x + 9$ for y in the second equation (this is easy since the first equation is already solved for y):

$$x + (x + 9) = 3{,}915$$
$$2x + 9 = 3{,}915$$
$$2x = 3{,}906$$
$$x = 1953$$

The solution to the equation may not be the answer to the question asked.

We are looking for Tom Cruise's birth year, so $y = x + 9 = 1953 + 9 = 1962$. ◆

Mixture Problems

Another common textbook application problem is the so-called *mixture problem.* The process involves combining two (or more) ingredients to obtain a mixture. Each ingredient has a certain quantity, and the quantity of the mixture is the sum of the quantities of the added ingredients. Quantities are measured in some appropriate unit; for example, consider the following spreadsheet:

	A	B	C	D	E	F
1	**INGREDIENT I**	**+**	**INGREDIENT II**	**=**	**MIXTURE**	
2 **a.**	5 lb	+	10 lb	=	15 lb	
3 **b.**	x lb	+	y lb	=	$(x + y)$ lb	
4 **c.**	18 oz	+	20 oz	=	38 oz	
5 **d.**	s oz	+	t oz	=	$(s + t)$ oz	
6 **e.**	8 L	+	p L	=	$(8 + p)$ L	
7						

Spreadsheet Application

This idea seems easy enough, but with mixture problems we have additional considerations. We begin with a simple example of mixing peanuts and cashews.

EXAMPLE 6

a. If you mix 16 lb of peanuts with 4 lb of cashews, how many pounds are in this mixture (called mixture I)?

b. If you mix 2 lb of peanuts with 8 lb of cashews, how many pounds are in this mixture (called mixture II)?

c. If you mix mixtures I and II, what are the percentage of peanuts and the percentage of cashews?

Solution

	PEANUTS	+	CASHEWS	=	MIXTURE
a.	16 lb	+	4 lb	=	20 lb (mixture I)
b.	2 lb	+	8 lb	=	10 lb (mixture II)
c.	$(16 + 2)$ lb	+	$(4 + 8)$ lb	=	$(20 + 10)$ lb (final mixture)

Notice the double check; columns add down and rows add across to determine the total amount in the final mixture. As percentages these are:

$$\frac{18}{30} = 0.6 \text{ or } 60\% \qquad \frac{12}{30} = 0.4 \text{ or } 40\%$$

◆

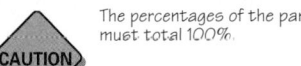
The percentages of the parts must total 100%.

CAUTION

Mixture problems are often stated using percentages. The following example is a restatement of Example 6.

EXAMPLE 7

If you mix together 20 lb of a peanut/cashew mixture consisting of 80% peanuts with 10 lb of a peanut/cashew mixture consisting of 20% peanuts to obtain 30 lb of a peanut/cashew mixture consisting of 60% peanuts, how many pounds of peanuts and cashews are present in mixture I, mixture II, and the final mixture?

Solution We must mix together

	MIXTURE I	+	MIXTURE II	=	FINAL MIXTURE
Given:	20 lb	+	10 lb	=	30 lb

We now use percentages to find the component parts of the final mixture (and to answer the question):

	MIXTURE I	+	MIXTURE II	=	FINAL MIXTURE
PEANUTS	$[0.8(20) = 16 \text{ lb}]$	+	$[0.2(10) = 2 \text{ lb}]$	=	18 lb
CASHEWS	$[0.2(20) = 4 \text{ lb}]$	+	$[0.8(10) = 8 \text{ lb}]$	=	12 lb
TOTAL	$[(16 + 4) \text{ lb}]$	+	$[(2 + 8) \text{ lb}]$	=	$(18 + 12)$ lb

Check: Final mixture should be 60% peanuts: $0.60(30) = 18$ lb.

◆

The last of this related trilogy of introductory examples repeats the information in the previous two examples in the form of a typical mixture problem. If you have difficulty with this example, you can look back at the previous two examples.

EXAMPLE 8

How much of a peanut/cashew mixture consisting of 80% peanuts must be mixed with a peanut/cashew mixture consisting of 20% peanuts to obtain 30 lb of a peanut/cashew mixture consisting of 60% peanuts?

Solution   We must mix together

	MIXTURE I	+	MIXTURE II	=	FINAL MIXTURE
Given:					30 lb

Let x = number of pounds of mixture I

y = number of pounds of mixture II

$$x \quad + \quad y \quad = \quad 30$$

These mixtures are made up of peanuts and cashews:

PEANUTS	$0.8x$	+	$0.2y$	=	$0.6(30) = 18$ lb
CASHEWS	$0.2x$	+	$0.8y$	=	$0.4(30) = 12$ lb

Notice that the percentages add to 100% (or as decimals add to 1). This gives rise to the following system:

$$\begin{cases} x + y = 30 \\ 0.8x + 0.2y = 18 \\ 0.2x + 0.8y = 12 \end{cases}$$

You notice that we have three equations with two unknowns. We will discuss such systems in general in the following section, but for a problem like this one we can use substitution. From the first equation, we have $y = 30 - x$, which we substitute into either of the other equations:

$$0.8x + 0.2y = 18$$
$$0.8x + 0.2(30 - x) = 18$$
$$0.8x + 6 - 0.2x = 18$$
$$0.6x + 6 = 18$$
$$0.6x = 12$$
$$x = 20$$

If $x = 20$, then $y = 10$; combine 20 pounds of mixture I with 10 pounds of mixture II.
Check:

$$\begin{cases} x + y = 30 & \quad 20 + 10 = 30 \quad ✔ \\ 0.8x + 0.2y = 18 & \quad 0.8(20) + 0.2(10) = 18 \quad ✔ \\ 0.2x + 0.8y = 12 & \quad 0.2(20) + 0.8(10) = 12 \quad ✔ \end{cases}$$ ◆

EXAMPLE 9

How many liters of water must be added to 3 liters of an 80% acid solution to obtain a 30% acid solution? (By an 80% acid solution we mean a solution that is 80% acid and 20% water; remember that the sum of the percentages of all ingredients must be 100%.)

Solution   The basic relationship for this mixture is

AMOUNT OF ACID = PERCENT ACID × AMOUNT OF SOLUTION

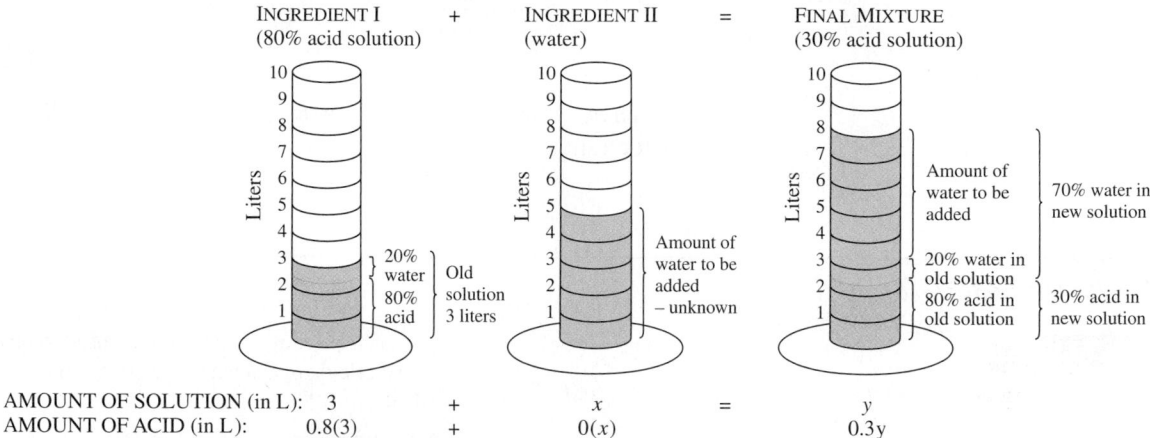

| INGREDIENT I (80% acid solution) | + | INGREDIENT II (water) | = | FINAL MIXTURE (30% acid solution) |

AMOUNT OF SOLUTION (in L): 3 + x = y
AMOUNT OF ACID (in L):   0.8(3) + 0(x)   0.3y

where x = AMOUNT OF WATER TO BE ADDED

y = AMOUNT AFTER MIXING

This leads to the system

$$\begin{cases} 3 + x = y \\ 2.4 + 0 = 0.3y \end{cases}$$

Solve the second equation for y to obtain $y = 8$, and substitute this into the first equation to obtain

$$3 + x = 8$$

$$x = 5$$

The amount of water to be added is 5 liters. ◆

EXAMPLE 10

Milk containing 10% butterfat and cream containing 80% butterfat are mixed to produce half-and-half, which is 50% butterfat. How many gallons of each must be mixed to make 140 gallons of half-and-half?

Solution

	BEFORE MIXING		AFTER MIXING
AMT. OF SOLUTION:	MILK + CREAM	=	HALF-AND-HALF
AMT. OF BUTTERFAT:	0.10(MILK) + 0.80(CREAM)	=	0.50(HALF-AND-HALF)

We see that there are two unknowns; let

m = AMOUNT OF MILK

c = AMOUNT OF CREAM

140 = AMOUNT OF HALF-AND-HALF

We have the system of equations (with variables):

AMOUNT OF SOLUTION: $\begin{cases} m + c = 140 \\ 0.10m + 0.80c = 0.50(140) \end{cases}$

AMOUNT OF BUTTERFAT:

Multiply both sides of the second equation by -10:

$$+ \begin{cases} m + c = 140 \\ -m - 8c = -700 \end{cases}$$
$$-7c = -560$$
$$c = 80$$

If $c = 80$, then $m + 80 = 140$ so that $m = 60$. You must mix 60 gallons of milk and 80 gallons of cream. ◆

Problem Set 14.2

LEVEL 1

Carefully interpret each problem, restate the information and relationships until you are able to write the equations, solve the equations, and state an answer.

Problems 1–3 are patterned after Example 1.

1. A box contains twenty-five more dimes than nickels. How many of each type of coin is there if the total value is $7.15?

2. A box contains only dimes and quarters. The number of dimes is three less than twice the number of quarters. If the total value of the coins is $22.65, how many dimes are in the box?

3. Forty-two coins have a total value of $6.90. If the coins are all nickels and quarters, how many are nickels?

Problems 4–6 are patterned after Examples 2–3.

4. A boat travels 25 mph relative to the riverbank while going downstream and only 15 mph returning upstream. What is the boat's speed in still water?

5. A plane travels 200 mph relative to the ground while flying with a strong wind and only 150 mph returning against it. What is the plane's speed in still air?

6. An airliner flies at 510 mph in the jet stream, but only 270 mph returning against it. What is the speed of the plane in still air?

Problems 7–9 are patterned after Example 4.

7. The demand for a product varies from 150,000 units at $110 per unit to 300,000 at $20 per unit. Also, 300,000 could be supplied at $90 per unit, whereas only 200,000 could be supplied for $10 each. Find the equilibrium point for the system.

8. California Instruments, a manufacturer of calculators, finds by test marketing its calculators at UCLA that 180 calculators could be sold when they were priced at $10, but only 20 calculators could be sold when they were priced at $40. On the other hand, they find that 20 calculators can be supplied at $10 each. If they were supplied at $40 each, overtime shifts could be used to raise the supply to 180 calculators. What is the optimum price for the calculators?

9. A manufacturer of lapel buttons test marketed a new item at the University of California, Davis. It was found that 900 items could be sold if they were priced at $1, but only 300 items could be sold if the price were raised to $7. On the other hand, they find that 600 items can be supplied at $1 each. If they were supplied at $9 each, overtime shifts could be used to raise the supply to 1,000 items. What is the optimum price for the items?

Problems 10–12 are patterned after Example 5.

10. Mel Brooks is seven years older than fellow funnyman Dom DeLuise. If the sum of their birth years is 3,859, when was DeLuise born?

11. The sum of the birth years of actresses Debra Winger and Meryl Streep is 3,904. If Debra is six years younger, in what year was she born?

12. Ray Charles is fifteen years older than fellow musician José Feliciano. If the sum of their years of birth is 3,875, then in what year was Feliciano born?

Problems 13–18 are patterned after Examples 6–7. The information is based on the spreadsheet shown on page 719. Suppose Ingredient I is made up of 80% micoden and 20% water, Ingredient II is made up of 30% micoden, 50% bixon, and 20% water, and these ingredients are mixed together.

	A	B	C	D	E	F
1	Ingredient I	Ingredient II	MIXTURE			
2	a. 5 lb	10 lb	15 lb			
3	b. x lb	y lb	$(x + y)$ lb			
4	c. 8 L	p L	$(8 + p)$ L			
5						
6						

13. How much micoden is in the mixture **a**?

14. How much water is in the mixture **b**?

15. How much bixon is in the mixture **c**?

16. What is the percentage of water in mixture **a**?

17. What is the percentage of bixon in mixture **a?**

18. What is the percentage of micoden in mixture **a?**

Problems 19–21 are patterned after Examples 8–10.

19. How many ounces of a base metal (no silver) must be alloyed with 100 ounces of 21% silver alloy to obtain an alloy that is 15% silver?

20. An after-shave lotion is 50% alcohol. If you have 6 fluid ounces of the lotion, how much water must be added to reduce the mixture to 20% alcohol?

21. Milk containing 20% butterfat is mixed with cream containing 60% butterfat to produce half-and-half, which is 50% butterfat. How many gallons of each must be mixed to make 180 gallons of half-and-half?

LEVEL 2

Problems 22–58 provide a variety of types of word problems. Answer each question.

22. Suppose a car rental agency gives the following choices:

    Option A: $30 per day plus 40¢ per mile
    Option B: Flat $50 per day (unlimited miles)

    At what mileage are both rates the same if you rent the car for three days?

23. Suppose a car rental agency gives the following options:

    Option A: $40 per day plus 50¢ per mile
    Option B: Flat $60 per day with unlimited mileage

    At what mileage are both rates the same if you rent the car for four days?

24. The supply for a certain commodity is linear and determined to be $p = 0.005n + 12$, whereas the demand is linear with $p = 150 - 0.01n$, where p is the price and n is the number of items. What is the equilibrium point for (p, n)?

25. A certain item has a linear supply curve $p = 0.0005n - 3$ and a linear demand $p = 8 - 0.0006n$, where p is the price and n is the number of items. What is the equilibrium point (p, n)?

26. There are six more dimes than quarters in a container. How many of each coin is there if the total value is $3.75?

27. A box contains $8.40 in quarters and dimes. The number of quarters is twice the number of dimes. How many of each type of coin is in the box?

28. A canoeist rows downstream in $1\frac{1}{2}$ hr and back upstream in 3 hr. What is the rate of the current if the canoeist rows 9 miles in each direction?

29. A plane makes a 660-mile flight with the wind in $2\frac{1}{2}$ hours. Returning against the wind takes 3 hours. Find the wind speed.

30. Charles Bronson was born nine years before another movie hardguy, Clint Eastwood. If the sum of their years of birth is 3,853, in what year was Eastwood born?

31. Kristy McNichol is just a year older than fellow actress Tatum O'Neal. The sum of their years of birth is 3,935. In what year was Kristy born?

32. You have a 24% silver alloy and some pure silver. How much of each must be mixed to obtain 100 oz of 43% silver?

33. How much water must be added to a gallon of 80% antifreeze to obtain a 60% mixture?

34. How much antifreeze must be added to a gallon of 60% antifreeze to obtain an 80% mixture?

35. The combined area of New York and California is 204,192 square miles. The area of California is 108,530 square miles more than that of New York. Find the land area of each state.

36. The area of Texas is 208,044 square miles greater than that of Florida. Their combined area is 316,224 square miles. What is the area of each state?

37. Forty-two coins have a total value of $9.50. If the coins are all nickels and quarters, how many are quarters?

38. A collection of coins has a value of $4.76. There is the same number of nickels and dimes but four fewer pennies than nickels or dimes. How many pennies are in the collection if there are 86 coins?

39. A bunch of change contains nickels, dimes, and quarters. There are the same number of dimes and quarters, and there are eight more nickels than either dimes or quarters. How many dimes are there if the value of the 98 coins is $12.40?

40. A box contains $8.40 in nickels, dimes, and pennies. How many of each type of coin is in the box if the number of dimes is six less than twice the number of pennies, and there is an equal number of dimes and nickels in the box?

41. Sherlock Holmes was called in as a consultant to solve the Great Bank Robbery. He was told that the thief had made away with a bag of money containing $5, $10, and $20 bills totaling $1,390. When the bankers were checking serial numbers to see how many of each denomination were taken, Holmes said, "It is elementary, since there were five times as many $10 bills as $5 bills and three more than twice as many $20 bills as $5 bills." How many of each denomination were taken?

42. A plane makes an 870-mile flight in $3\frac{1}{3}$ hours against a strong head wind, but returns in 50 minutes less with the wind. What is the plane's speed without the wind?

43. A plane with a tail wind makes its 945-mile flight in 3 hours. The return flight against the wind takes a half hour longer. What is the wind speed?

44. A chemist has two solutions of sulfuric acid. One is a 50% solution, and the other is a 75% solution. How many liters of each does the chemist mix to get 10 liters of a 60% solution?

45. A dairy has cream containing 23% butterfat and milk that is 3% butterfat. How much of each must be mixed to obtain 30 gallons of a richer milk containing 4% butterfat?

46. A winery has a large amount of a wine labeled "Lot I" that is a mixture of 92% Merlot wine and 8% Cabernet Sauvignon wine. It also has a second wine labeled "Lot II" that is a mixture of 88% Cabernet Sauvignon and 12% Merlot. The winemaster decides to mix together these two lots to obtain 500 gallons of a mixture that is 50% Merlot and 50% Cabernet Sauvignon. How much of each should be used to obtain this blend?

47. A pain remedy contains 12% aspirin, and a stronger formula has 25% aspirin, but is otherwise the same. A chemist mixes some of each to obtain 100 mg of a mixture with 20% aspirin. How much of each is used?

48. The combined height of the Transamerica Tower and the Bank of America Building is 1,632 ft. The Transamerica Tower is 74 ft taller. What is the height of each of the San Francisco skyscrapers?

49. The Standard Oil and the Sears buildings have a combined height of 2,590 ft. The Sears Building is 318 ft taller. What is the height of each of these Chicago towers?

50. The combined length of the Golden Gate and San Francisco Bay bridges is 6,510 ft. If the Golden Gate is 1,890 ft longer, what is its length?

51. End to end, the Verrazano-Narrows and the George Washington bridges would span 7,760 ft. If the Verrazano-Narrows is the longer of the two New York structures by 760 ft, how long is it?

52. Noxin Electronics has investigated the feasibility of introducing a new line of magnetic tape. The study shows that both supply and demand are linear. The supply can increase from 1,000 items at $2 each to 5,000 units at $4 each. The demand ranges from 1,000 items at $4 to 7,000 at $3. What is the equilibrium point of this supply-and-demand system?

53. Noxin Electronics is considering producing a small cassette line. Research shows linear demand to be from 40,000 cassettes at $2 to 100,000 at $1. Similarly, the supply goes from 20,000 cassettes at 50¢ to 80,000 at $5. What is the equilibrium point?

54. Sterling silver contains 92.5% silver. How many grams of pure silver and sterling silver must be mixed to get 100 grams of a 94% alloy?

55. The radiator of a car holds 17 quarts of liquid. If it now contains 15% antifreeze, how many quarts must be replaced by antifreeze to give the car a 60% solution in its radiator?

56. How many gallons of 24% butterfat cream must be mixed with 500 gallons of 3% butterfat milk to obtain a 4% butterfat milk?

57. The supply curve for a new software product is given by $n = 2.5p - 500$, and the demand curve for the same product is $n = 200 - 0.5p$, where n is the number of items and p is the number of dollars.
   a. At $250 for the product, how many items would be supplied? How many would be demanded?
   b. At what price would no items be supplied?
   c. At what price would no items be demanded?
   d. What is the equilibrium price for this product?
   e. How many units will be produced at the equilibrium price?

58. The supply curve for a certain commodity is $n = 2,500p - 500$, and the demand curve for the same product is $n = 31,500 - 1,500p$, where n is the number of items and p is the number of dollars.
   a. At $15 per unit of the commodity, how many items would be supplied? How many would be demanded?
   b. At what price would no items be supplied?
   c. At what price would no items be demanded?
   d. What is the equilibrium price for this product?
   e. How many units will be produced at the equilibrium price?

PROBLEM SOLVING

What is wrong, if anything, with each of the statements in Problems 59–60? *Explain your reasoning.*

59. A bottle and a cork cost $1.10. What is the cost of the cork alone, if the bottle costs a dollar more?
   Answer: The cork costs 10¢ and the bottle costs $1.

60. You have identical cups, one containing coffee and one containing cream. One teaspoon of the cream is added to the coffee and stirred in. Now a teaspoon of the coffee/cream mixture is added back to the cup of cream and stirred in. Both cups contain identical amounts of liquid, but is there more cream in the coffee or more coffee in the cream?
   Answer: There is obviously more cream in the coffee since it was added before any coffee was removed.

14.3 | MATRIX SOLUTION OF A SYSTEM OF EQUATIONS

One of the most common types of problems to which we can apply mathematics in a variety of different disciplines is to the solution of systems of equations. In fact, common real-world problems require the simultaneous solution of systems involving 3, 4,

5, or even 20 or 100 unknowns. The methods of the previous section will not suffice and, in practice, techniques that will allow computer or calculator help in solving systems are common. In this section, we introduce a way of solving large systems of equations in a general way so that we can handle the solution of a system of m equations with n unknowns.

Definition of a Matrix

You are already familiar with matrices from everyday experiences. For example, the following table shows a rental chart in the form of a matrix.

Country	Fiat	Opel	Renault
Austria	$149	$219	$289
Belgium	$130	$222	$273
Denmark	$164	$269	$408
France	$189	$206	$215
G. Britain	$160	$208	$225
Holland	$156	$213	$299
Italy	$179	$259	$353
Spain	$156	$210	$247
Sweden	$178	$246	$281

Considering the numerical entries in this chart as a 9 by 3 matrix, we would say that the price of the Renault in Belgium is found in row 2, column 3 (namely, $273). In contrast, the entry in row 3, column 2 ($269), is the cost of the Opel in Denmark.

Consider a system with unknowns x_1 and x_2. We use **subscripts** 1 and 2 to denote the unknowns, instead of using variables x and y, because we want to be able to handle n unknowns, which we can easily denote as $x_1, x_2, x_3, \ldots, x_n$; if we continued by using x, y, z, $\ldots$ for systems in general, we would soon run out of letters. Here is the way we will write a general system of two equations with two unknowns:

$$\begin{cases} a_{11}x_1 + a_{12}x_2 = b_1 \\ a_{21}x_1 + a_{22}x_2 = b_2 \end{cases}$$

The coefficients of the unknowns use **double subscripts** to denote their position in the system; a_{11} is used to denote the numerical coefficient of the first variable in the first row; a_{12} denotes the numerical coefficient of the second variable in the first row; and so on. The constants are denoted by b_1 and b_2.

We now separate the parts of this system of equations into *rectangular arrays* of numbers. An **array** of numbers is called a **matrix.** A matrix is denoted by enclosing the array in large brackets.

Let [A] be the matrix (array) of coefficients: $\begin{bmatrix} a_{11} & a_{12} \\ a_{21} & a_{22} \end{bmatrix}$

Let [X] be the matrix of unknowns: $\begin{bmatrix} x_1 \\ x_2 \end{bmatrix}$

Let [B] be the matrix of constants: $\begin{bmatrix} b_1 \\ b_2 \end{bmatrix}$

We will write the system of equations as a **matrix equation,** [A][X] = [B], but before we do this, we will do some preliminary work with matrices.

Matrices are classified by the number of (horizontal) **rows** and (vertical) **columns.** The numbers of rows and columns of a matrix need not be the same; but if they are, the matrix is called a **square matrix.**

The **order** or **dimension** of a matrix is given by an expression $m \times n$ (pronounced "m by n"), where m is the number of rows and n is the number of columns. For example, [A] (shown above) is a matrix of order 2×2, and matrices (plural for matrix) [X] and [B] have order 2×1.

Matrix Form of a System of Equations

We write a system of equations in the form of an **augmented matrix.** The matrix refers to the matrix of coefficients, and we *augment* (add to, or affix) this matrix by writing the constant terms at the right of the matrix (separated by a dashed line):

$$\begin{cases} a_{11}x_1 + a_{12}x_2 = b_1 \\ a_{21}x_1 + a_{22}x_2 = b_2 \end{cases} \text{ in matrix form is } \begin{bmatrix} a_{11} & a_{12} & \vdots & b_1 \\ a_{21} & a_{22} & \vdots & b_2 \end{bmatrix}$$

EXAMPLE 1

Give the order of each system, and then write it in augmented matrix form.

a. $\begin{cases} 2x + y = 3 \\ 3x - y = 2 \\ 4x + 3y = 7 \end{cases}$
b. $\begin{cases} 5x - 3y + z = -3 \\ 2x + 5z = 14 \end{cases}$
c. $\begin{cases} x_1 - 3x_3 + x_5 = -3 \\ x_2 + x_4 = -1 \\ x_3 + x_5 = 7 \\ x_1 + x_2 - x_3 + 4x_4 = -8 \\ x_1 + x_2 + x_3 + x_4 + x_5 = 8 \end{cases}$

Solution   Note that some coefficients are negative and some are zero.

a. $\begin{bmatrix} 2 & 1 & | & 3 \\ 3 & -1 & | & 2 \\ 4 & 3 & | & 7 \end{bmatrix}$
  Order: 3×3

b. $\begin{bmatrix} 5 & -3 & 1 & | & -3 \\ 2 & 0 & 5 & | & 14 \end{bmatrix}$
  Order: 2×4

c. $\begin{bmatrix} 1 & 0 & -3 & 0 & 1 & | & -3 \\ 0 & 1 & 0 & 1 & 0 & | & -1 \\ 0 & 0 & 1 & 0 & 1 & | & 7 \\ 1 & 1 & -1 & 4 & 0 & | & -8 \\ 1 & 1 & 1 & 1 & 1 & | & 8 \end{bmatrix}$ Order: 5×6 ◆

EXAMPLE 2

Write a system of equations (use $x_1, x_2, x_3, \ldots$) that has the given augmented matrix.

a. $\begin{bmatrix} 2 & 1 & -1 & | & -3 \\ 3 & -2 & 1 & | & 9 \\ 1 & -4 & 3 & | & 17 \end{bmatrix}$

b. $\begin{bmatrix} 1 & 0 & 0 & | & 3 \\ 0 & 1 & 0 & | & -2 \\ 0 & 0 & 1 & | & 21 \end{bmatrix}$

c. $\begin{bmatrix} 1 & 0 & 0 & | & 5 \\ 0 & 1 & 0 & | & 12 \\ 0 & 0 & 1 & | & -3 \\ 0 & 0 & 0 & | & 4 \end{bmatrix}$

d. $\begin{bmatrix} 1 & 0 & 0 & | & -7 \\ 0 & 1 & 0 & | & 3 \\ 0 & 0 & 1 & | & -1 \\ 0 & 0 & 0 & | & 0 \end{bmatrix}$

Solution

a. $\begin{cases} 2x_1 + x_2 - x_3 = -3 \\ 3x_1 - 2x_2 + x_3 = 9 \\ x_1 - 4x_2 + 3x_3 = 17 \end{cases}$

b. $\begin{cases} x_1 = 3 \\ x_2 = -2 \\ x_3 = 21 \end{cases}$

c. $\begin{cases} x_1 = 5 \\ x_2 = 12 \\ x_3 = -3 \\ 0 = 4 \end{cases}$

d. $\begin{cases} x_1 = -7 \\ x_2 = 3 \\ x_3 = -1 \\ 0 = 0 \end{cases}$ ◆

The goal of this section is to solve a system of m equations with n unknowns. We have already looked at systems of two equations with two unknowns. In previous courses you may have solved three equations with three unknowns. Now, however, we want to be able to solve problems with two equations and five unknowns, or three equations and two unknowns, or systems with any number of linear equations and unknowns. The procedure for this section—*Gauss–Jordan elimination*—is a general method for solving all these types of systems. We write the system in augmented matrix form (as in Example 1), then carry out a process that transforms the matrix until the solution is obvious. Look back at Example 2—the solution to part **b** is obvious. Part **c** shows $0 = 4$ in the last equation, so this system has no solution (0 cannot equal 4), and part **d** shows $0 = 0$ (which is true for all replacements of the variable), which means that the solution is found by looking at the other equations (namely, $x_1 = -7$, $x_2 = 3$, and $x_3 = -1$). The terms with nonzero coefficients in these examples (that is, parts **b, c,** and **d**) are arranged on a diagonal, and such a system is said to be in **diagonal form.**

Elementary Row Operations and Pivoting

What process will allow us to transform a matrix into diagonal form? We begin with some steps called **elementary row operations.** Elementary row operations change the *form* of a matrix, but the new form represents an equivalent system. Matrices that represent equivalent systems are called **equivalent matrices;** we now introduce the elementary row operations, which allow us to write equivalent matrices. Let us work with a system consisting of three equations and three unknowns (any size will work the same way).

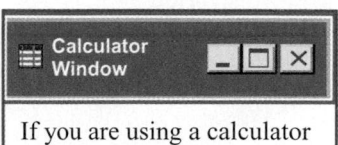

$$\begin{array}{cc} \textit{System format} & \textit{Matrix format} \\ \begin{cases} 2x - 2y + 4z = 14 \\ x - y - 2z = -9 \\ 3x + 2y + z = 16 \end{cases} & [A] = \begin{bmatrix} 2 & -2 & 4 & | & 14 \\ 1 & -1 & -2 & | & -9 \\ 3 & 2 & 1 & | & 16 \end{bmatrix} \end{array}$$

For the discussion, we call this matrix A, and denote it by [A].

Elementary Row Operation 1: RowSwap

Interchanging two equations is equivalent to interchanging two rows in the matrix format, and certainly, if we do this, the solution to the system will be the same:

$$\begin{array}{cc} \textit{System format} & \textit{Matrix format} \\ \begin{cases} x - y - 2z = -9 \\ 2x - 2y + 4z = 14 \\ 3x + 2y + z = 16 \end{cases} & \begin{bmatrix} 1 & -1 & -2 & | & -9 \\ 2 & -2 & 4 & | & 14 \\ 3 & 2 & 1 & | & 16 \end{bmatrix} \end{array}$$

In this example we interchanged the first and the second rows of the matrix. If we denote the original matrix as [A], then we indicate the operation of interchanging the first and second rows of matrix A by RowSwap([A],1,2).

Calculator Window

Check your owner's manual for the matrix operations. Most will have a [MATRIX] key and then you will need to name the matrix. Most will allow for three matrices (called [A], [B], and [C]). Then you will need to input the order (most will handle up to order 6×6). After inputting a matrix, it is a good idea to recall it to make sure it is input correctly. For this example, after a row swap of rows 1 and 2 the display shows:

[1 −1 −2 −9]
[2 −2   4   14]
[3   2   1   16]

Elementary Row Operation 2: Row+

Since adding the entries of one equation to the corresponding entries (similar terms) of another equation will not change the solution to a system of equations, the second elementary row operation is called row addition. (This is the step called linear combinations in the previous section.) In terms of matrices, we see that this operation corresponds to adding one row to another:

$$\begin{array}{cc} \textit{System format} & \textit{Matrix format} \\ \begin{cases} 2x - 2y + 4z = 14 \\ x - y - 2z = -9 \\ 3x + 2y + z = 16 \end{cases} & [A] = \begin{bmatrix} 2 & -2 & 4 & | & 14 \\ 1 & -1 & -2 & | & -9 \\ 3 & 2 & 1 & | & 16 \end{bmatrix} \end{array}$$

Add row 1 to row 3:

$$\begin{array}{cc} \textit{System format} & \textit{Matrix format} \\ \begin{cases} 2x - 2y + 4z = 14 \\ x - y - 2z = -9 \\ 5x + 5z = 30 \end{cases} & \begin{bmatrix} 2 & -2 & 4 & | & 14 \\ 1 & -1 & -2 & | & -9 \\ 5 & 0 & 5 & | & 30 \end{bmatrix} \end{array}$$

Notice that only row 3 changes; we call the row being added to (that is, the row that is being changed) the **target row.** We indicate this operation by $\text{Row}+([A],1,3)$.

↑
target row

Elementary Row Operation 3: *Row

Multiplying or dividing both sides of an equation by any nonzero number does not change the simultaneous solution; then, in matrix format, the solution will not be changed if any row is multiplied or divided by a nonzero constant. In this context we call the constant a **scalar.** For example, we can multiply both sides of the first equation of the original system by $\frac{1}{2}$. (Note that dividing both sides of an equation by 2 can be considered as multiplying both sides by $\frac{1}{2}$.)

<table>
<tr><td>System format</td><td>Matrix format</td></tr>
<tr><td>$\begin{cases} 2x - 2y + 4z = 14 \\ x - y - 2z = -9 \\ 3x + 2y + z = 16 \end{cases}$</td><td>$[A] = \begin{bmatrix} 2 & -2 & 4 & \vdots & 14 \\ 1 & -1 & -2 & \vdots & -9 \\ 3 & 2 & 1 & \vdots & 16 \end{bmatrix}$</td></tr>
</table>

Multiply the first row by $\frac{1}{2}$ to obtain:

<table>
<tr><td>System format</td><td>Matrix format</td></tr>
<tr><td>$\begin{cases} x - y + 2z = 7 \\ x - y - 2z = -9 \\ 3x + 2y + z = 16 \end{cases}$</td><td>$\begin{bmatrix} 1 & -1 & 2 & \vdots & 7 \\ 1 & -1 & -2 & \vdots & -9 \\ 3 & 2 & 1 & \vdots & 16 \end{bmatrix}$</td></tr>
</table>

We indicate this operation by $*\text{Row}(\frac{1}{2},[A],1)$.

scalar
↓
↑
target row

Elementary Row Operation 4: *Row+

When solving systems, more often than not we need to multiply both sides of an equation by a scalar before adding to make the coefficients opposites. Elementary row operation 4 combines row operations 2 and 3 so that this can be accomplished in one step. Let us return to the original system:

<table>
<tr><td>System format</td><td>Matrix format</td></tr>
<tr><td>$\begin{cases} 2x - 2y + 4z = 14 \\ x - y - 2z = -9 \\ 3x + 2y + z = 16 \end{cases}$</td><td>$[A] = \begin{bmatrix} 2 & -2 & 4 & \vdots & 14 \\ 1 & -1 & -2 & \vdots & -9 \\ 3 & 2 & 1 & \vdots & 16 \end{bmatrix}$</td></tr>
</table>

We can change this system by multiplying the second equation by -2 and adding the result to the first equation. In matrix terminology we say that we multiply the second row by -2 and add it to the first row. Denote this by

scalar
↓

$$*\text{Row}+(-2,[A],2,1)$$

↑
target row

<table>
<tr><td>System format</td><td>Matrix format</td></tr>
<tr><td>$\begin{cases} 8z = 32 \\ x - y - 2z = -9 \\ 3x + 2y + z = 16 \end{cases}$</td><td>$\begin{bmatrix} 0 & 0 & 8 & \vdots & 32 \\ 1 & -1 & -2 & \vdots & -9 \\ 3 & 2 & 1 & \vdots & 16 \end{bmatrix}$</td></tr>
</table>

Once again, multiply the second row, this time by -3, and add it to the third row. Wait! Why -3? Where did that come from? The idea is the same one we used in the linear combination method—we use a number that will give a zero coefficient to the x in the third equation.

$$
\begin{array}{cc}
\textit{System format} & \textit{Matrix format} \\
\begin{cases} 8z = 32 \\ x - y - 2z = -9 \\ 5y + 7z = 43 \end{cases} &
\begin{bmatrix} 0 & 0 & 8 & \vdots & 32 \\ 1 & -1 & -2 & \vdots & -9 \\ 0 & 5 & 7 & \vdots & 43 \end{bmatrix}
\end{array}
$$

Note that the multiplied row is not changed; instead, the changed row is the one to which the multiplied row is added. We call the original row the **pivot row.** Note also that we did not work with the original matrix [A] but rather with the previous answer, so we indicate this by

$$
\begin{array}{c}
\text{scalar} \quad\quad \text{pivot row} \\
\downarrow \quad\quad\quad \downarrow \\
*\text{Row}+(-3\ ,[\text{Ans}], 2\quad,\ 3\quad\ ) \\
\uparrow \\
\text{target row}
\end{array}
$$

There you have it! You can carry out these four elementary operations until you have a system for which the solution is obvious, as illustrated by the following example.

EXAMPLE 3

Solve using both system format and matrix format: $\begin{cases} 2x - 5y = 5 \\ x - 2y = 1 \end{cases}$

Solution

System format	Matrix format	Operation performed
$\begin{cases} 2x - 5y = 5 \\ x - 2y = 1 \end{cases}$	$[A] = \begin{bmatrix} 2 & -5 & \vdots & 5 \\ 1 & -2 & \vdots & 1 \end{bmatrix}$	RowSwap([A],1,2)
$\begin{cases} x - 2y = 1 \\ 2x - 5y = 5 \end{cases}$	$\begin{bmatrix} 1 & -2 & \vdots & 1 \\ 2 & -5 & \vdots & 5 \end{bmatrix}$	*This matrix is called* [Ans] *because it is the result of the previous matrix operation.*
$\begin{cases} x - 2y = 1 \\ -y = 3 \end{cases}$	$\begin{bmatrix} 1 & -2 & \vdots & 1 \\ 0 & -1 & \vdots & 3 \end{bmatrix}$	*Row+(−2,[Ans],1,2) *This matrix is now referred to as* [Ans].
$\begin{cases} x - 2y = 1 \\ y = -3 \end{cases}$	$\begin{bmatrix} 1 & -2 & \vdots & 1 \\ 0 & 1 & \vdots & -3 \end{bmatrix}$	*Row(−1,[Ans],2)
$\begin{cases} x = -5 \\ y = -3 \end{cases}$	$\begin{bmatrix} 1 & 0 & \vdots & -5 \\ 0 & 1 & \vdots & -3 \end{bmatrix}$	*Row+(2,[Ans],2,1)

The solution $(-5, -3)$ is now obvious. ◆

As you study Example 3, first look at how the elementary row operations led to a system equivalent to the first—but one for which the solution is obvious. Next, try to decide *why* a particular row operation was chosen when it was. Many students quickly learn the elementary row operations, but then use a series of (almost random) steps until the obvious solution results. This often works, but is not very efficient. The steps chosen in Example 3 illustrate a very efficient method of using the elementary row operations to determine a system whose solution is obvious. Let us restate the elementary row operations and the operations called *pivoting.*

Pivoting

There are four *elementary row operations* for producing equivalent matrices:

1. **RowSwap**   Interchange any two rows.

2. **Row+**   Row addition—add a row to any other row.

3. **∗Row**   Scalar multiplication—multiply (or divide) all the elements of a row by the same nonzero real number.

4. **∗Row+**   Multiply all the entries of a row (*pivot row*) by a nonzero real number and add each resulting product to the corresponding entry of another specified row (*target row*).

This operation changes only the target row.

These elementary row operations are used together in a process called **pivoting,** which means:

1. Divide all entries in the row in which the pivot appears (called the *pivot row*) by the nonzero pivot element so that the pivot entry becomes a 1. This uses elementary row operation 3.

2. Obtain zeros above and below the pivot element by using elementary row operation 4.

Gauss–Jordan Elimination

You are now ready to see the method worked out by Gauss and Jordan; it is known as **Gauss–Jordan elimination.** It efficiently uses the elementary row operations to diagonalize the matrix. That is, the first pivot is the first entry in the first row, first column; the second is the entry in the second row, second column; and so on until the solution is obvious. A **pivot** element is an element that is used to eliminate elements above and below it in a given column by using elementary row operations.

Gauss–Jordan Elimination

Step 1. Select as the first pivot the element in the first row, first column, and pivot.

Step 2. The next pivot is the element in the second row, second column; pivot.

Step 3. Repeat the process until you arrive at the last row, or until the pivot element is a zero. If it is a zero and you can interchange that row with a row below it, so that the pivot element is no longer a zero, do so and continue. If it is zero and you cannot interchange rows so that it is not a zero, continue with the next row. The final matrix is called the **row-reduced form.**

EXAMPLE 4

Solve $\begin{cases} x + 2y - z = 0 \\ 2x + 3y - 2z = 3 \\ -x - 4y + 3z = -2 \end{cases}$

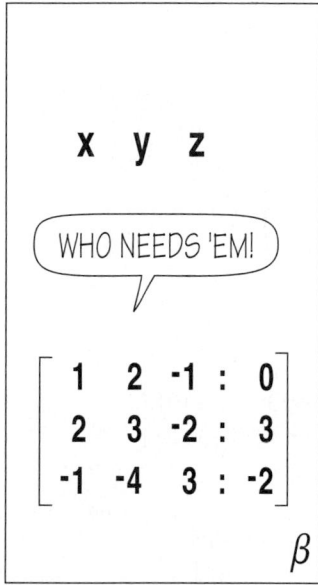

Courtesy of Patrick J. Boyle

Pay special attention to the correct interpretation of what you see with a calculator, and the correct answer.

Solution We solve this system by choosing the steps according to the Gauss–Jordan method.

$$[A] = \begin{bmatrix} 1 & 2 & -1 & | & 0 \\ 2 & 3 & -2 & | & 3 \\ -1 & -4 & 3 & | & -2 \end{bmatrix}$$

This matrix represents the given system.

First pivot (row 1, col. 1)

$$\rightarrow \begin{bmatrix} 1 & 2 & -1 & | & 0 \\ 0 & -1 & 0 & | & 3 \\ 0 & -2 & 2 & | & -2 \end{bmatrix}$$

*Row+(−2,[A],1,2)
*Row+(1,[Ans],1,3)

$$\rightarrow \begin{bmatrix} 1 & 2 & -1 & | & 0 \\ 0 & 1 & 0 & | & -3 \\ 0 & -2 & 2 & | & -2 \end{bmatrix}$$

*Row(−1,[Ans],2)

Second pivot (row 2, col. 2)

$$\rightarrow \begin{bmatrix} 1 & 0 & -1 & | & 6 \\ 0 & 1 & 0 & | & -3 \\ 0 & 0 & 2 & | & -8 \end{bmatrix}$$

*Row+(−2,[Ans],2,1)
*Row+(2,[Ans],2,3)

$$\rightarrow \begin{bmatrix} 1 & 0 & -1 & | & 6 \\ 0 & 1 & 0 & | & -3 \\ 0 & 0 & 1 & | & -4 \end{bmatrix}$$

*Row(0.5,[Ans],3)

Third pivot (row 3, col. 3)

$$\rightarrow \begin{bmatrix} 1 & 0 & 0 & | & 2 \\ 0 & 1 & 0 & | & -3 \\ 0 & 0 & 1 & | & -4 \end{bmatrix}$$

*Row+(1,[Ans],3,1)

The solution $(2, -3, -4)$ is found by inspection since this matrix represents the system

$$\begin{cases} 1x + 0y + 0z = 2 \\ 0x + 1y + 0z = -3 \\ 0x + 0y + 1z = -4 \end{cases}$$

To check this answer, substitute the x-, y-, and z-values into the original equations:

$$\begin{cases} 2 + 2(-3) - (-4) = 0 & ✔ \\ 2(2) + 3(-3) - 2(-4) = 3 & ✔ \\ -(2) - 4(-3) + 3(-4) = -2 & ✔ \end{cases}$$

◆

EXAMPLE 5

A rancher has to mix three types of feed for her cattle. The following analysis shows the amounts per bag (100 lb) of grain:

Grain	Protein	Carbohydrates	Sodium
A	7 lb	88 lb	1 lb
B	6 lb	90 lb	1 lb
C	10 lb	70 lb	2 lb

How many bags of each type of grain should she mix to provide 71 lb of protein, 854 lb of carbohydrates, and 12 lb of sodium?

Solution Let a, b, and c be the number of bags of grains A, B, and C, respectively, that are needed for the mixture. Then

Grain	Protein	Carbohydrates	Sodium
A	7a lb	88a lb	a lb
B	6b lb	90b lb	b lb
C	10c lb	70c lb	2c lb
Total needed:	71 lb	854 lb	12 lb

Thus, $\begin{cases} 7a + 6b + 10c = 71 \\ 88a + 90b + 70c = 854 \\ a + b + 2c = 12 \end{cases}$

Let $[A] = \begin{bmatrix} 7 & 6 & 10 & | & 71 \\ 88 & 90 & 70 & | & 854 \\ 1 & 1 & 2 & | & 12 \end{bmatrix}$.

We show the steps in Gauss–Jordan elimination:

$$\begin{bmatrix} 7 & 6 & 10 & | & 71 \\ 88 & 90 & 70 & | & 854 \\ 1 & 1 & 2 & | & 12 \end{bmatrix} \rightarrow \begin{bmatrix} 1 & 1 & 2 & | & 12 \\ 88 & 90 & 70 & | & 854 \\ 7 & 6 & 10 & | & 71 \end{bmatrix}$$
RowSwap([A],1,3)

$$\rightarrow \begin{bmatrix} 1 & 1 & 2 & | & 12 \\ 0 & 2 & -106 & | & -202 \\ 0 & -1 & -4 & | & -13 \end{bmatrix} \rightarrow \begin{bmatrix} 1 & 1 & 2 & | & 12 \\ 0 & 1 & -53 & | & -101 \\ 0 & -1 & -4 & | & -13 \end{bmatrix}$$
*Row+(−88,[Ans],1,2)   *Row(1/2,[Ans],2)
*Row+(−7,[Ans],1,3)

$$\rightarrow \begin{bmatrix} 1 & 0 & 55 & | & 113 \\ 0 & 1 & -53 & | & -101 \\ 0 & 0 & -57 & | & -114 \end{bmatrix} \rightarrow \begin{bmatrix} 1 & 0 & 55 & | & 113 \\ 0 & 1 & -53 & | & -101 \\ 0 & 0 & 1 & | & 2 \end{bmatrix}$$
*Row+(−1,[Ans],2,1)   *Row(−1/57,[Ans],3)
*Row+(1,[Ans],2,3)

$$\rightarrow \begin{bmatrix} 1 & 0 & 0 & | & 3 \\ 0 & 1 & 0 & | & 5 \\ 0 & 0 & 1 & | & 2 \end{bmatrix}$$
*Row+(−55,[Ans],3,1)

*Row+(53,[Ans],3,2)

Mix three bags of grain A, five bags of grain B, and two bags of grain C.   ◆

Problem Set 14.3

LEVEL 1

1. IN YOUR OWN WORDS What is a matrix?

2. IN YOUR OWN WORDS List the elementary row operations and briefly describe each.

3. IN YOUR OWN WORDS What is Gauss–Jordan elimination?

In Problems 4–13, decide whether the statement is true or false. If it is false, tell what is wrong.

4. Entry a_{34} is in row 3, column 4.

5. The matrix for the system $\begin{cases} x_1 = 1 \\ x_2 = 2 \\ x_3 = 0 \\ x_4 = 5 \end{cases}$ has order 4×4.

6. In *Row+ notation, the first number listed is the target row.

7. In *Row and *Row+ notation, the target row is the last number listed.

8. If row 3 of a matrix [A] is added to row 5 of [A], then the correct notation is *Row+([A],3,5).

9. If row 7 of a matrix [B] is multiplied by −2 and then added to row 6 of [B], then the correct notation is *Row+(−2,[B],7,6).

10. The notation *Row+(2, [ANS], 6, 3) means multiply row 6 of the previous matrix by 2 and add the corresponding entries to row 3.

11. The notation *Row+(3^{-1},[C],4,2) means multiply row 2 of matrix [C] by $\frac{1}{3}$ and add the corresponding entries to the entries in row 4.

12. In the notation *Row+(3,[A],4,5), the target row is 3.

13. For the matrix $[A] = \begin{bmatrix} 0 & 7 & 8 & | & 3 \\ 1 & 2 & 3 & | & 4 \\ 0 & 1 & 3 & | & 4 \end{bmatrix}$

the first step in Gauss–Jordan elimination is RowSwap([A],1,3).

14. Write each system in augmented matrix form.

a. $\begin{cases} 4x + 5y = -16 \\ 3x + 2y = 5 \end{cases}$
b. $\begin{cases} x + y + z = 4 \\ 3x + 2y + z = 7 \\ x - 3y + 2z = 0 \end{cases}$

c. $\begin{cases} x_1 + 2x_2 - 5x_3 + x_4 = 5 \\ x_1 - 3x_3 + 6x_4 = 0 \\ x_3 - 3x_4 = -15 \\ x_2 - 5x_3 + 5x_4 = 2 \end{cases}$

15. Write a system of equations that has the given augmented matrix.

a. $\begin{bmatrix} 6 & 7 & 8 & | & 3 \\ 1 & 2 & 3 & | & 4 \\ 0 & 1 & 3 & | & 4 \end{bmatrix}$

b. $\begin{bmatrix} 1 & 0 & 0 & | & 3 \\ 0 & 1 & 2 & | & 4 \end{bmatrix}$

c. $\begin{bmatrix} 1 & 0 & 0 & | & 32 \\ 0 & 1 & 0 & | & 27 \\ 0 & 0 & 1 & | & -5 \\ 0 & 0 & 0 & | & 3 \end{bmatrix}$

Given the matrices in Problems 16–19, perform elementary row operations to obtain a 1 in the row 1, column 1 position. Answers may vary.

16. $[A] = \begin{bmatrix} 3 & 1 & 2 & | & 1 \\ 0 & 2 & 4 & | & 5 \\ 1 & 3 & -4 & | & 9 \end{bmatrix}$

17. $[B] = \begin{bmatrix} -2 & 3 & 5 & | & 9 \\ 1 & 0 & 2 & | & -8 \\ 0 & 1 & 0 & | & 5 \end{bmatrix}$

18. $[C] = \begin{bmatrix} 2 & 4 & 10 & | & -12 \\ 6 & 3 & 4 & | & 6 \\ 10 & -1 & 0 & | & 1 \end{bmatrix}$

19. $[A] = \begin{bmatrix} 5 & 20 & 15 & | & 6 \\ 7 & -5 & 3 & | & 2 \\ 12 & 0 & 1 & | & 4 \end{bmatrix}$

Given the matrices in Problems 20–23, perform elementary row operations to obtain zeros under the 1 in the first column. Answers may vary.

20. $[A] = \begin{bmatrix} 1 & 2 & -3 & | & 0 \\ 0 & 3 & 1 & | & 4 \\ 2 & 5 & 1 & | & 6 \end{bmatrix}$

21. $[B] = \begin{bmatrix} 1 & 3 & -5 & | & 6 \\ -3 & 4 & 1 & | & 2 \\ 0 & 5 & 1 & | & 3 \end{bmatrix}$

22. $[C] = \begin{bmatrix} 1 & 2 & 4 & | & 1 \\ -2 & 5 & 0 & | & 2 \\ -4 & 5 & 1 & | & 3 \end{bmatrix}$

23. $[A] = \begin{bmatrix} 1 & 5 & 3 & | & 2 \\ 2 & 3 & -1 & | & 4 \\ 3 & 2 & 1 & | & 0 \end{bmatrix}$

Given the matrices in Problems 24–27, perform elementary row operations to obtain a 1 in the second row, second column without changing the entries in the first column. Answers may vary.

24. $[A] = \begin{bmatrix} 1 & 3 & 5 & | & 2 \\ 0 & 2 & 6 & | & -8 \\ 0 & 3 & 4 & | & 1 \end{bmatrix}$

25. $[B] = \begin{bmatrix} 1 & 5 & -3 & | & 5 \\ 0 & 3 & 9 & | & -15 \\ 0 & 2 & 1 & | & 5 \end{bmatrix}$

26. $[C] = \begin{bmatrix} 1 & 4 & -1 & | & 6 \\ 0 & 5 & 1 & | & 3 \\ 0 & 4 & 6 & | & 5 \end{bmatrix}$

27. $[A] = \begin{bmatrix} 1 & 3 & -2 & | & 0 \\ 0 & 4 & 2 & | & 9 \\ 0 & 3 & 6 & | & 1 \end{bmatrix}$

Given the matrices in Problems 28–31, perform elementary row operations to obtain a zero (or zeros) above and below the 1 in the second column without changing the entries in the first column. Answers may vary.

28. $[A] = \begin{bmatrix} 1 & 5 & -3 & | & 2 \\ 0 & 1 & 4 & | & 5 \\ 0 & 3 & 4 & | & 2 \end{bmatrix}$

29. $[B] = \begin{bmatrix} 1 & 3 & 6 & | & 12 \\ 0 & 1 & -2 & | & -5 \\ 0 & -2 & 2 & | & 6 \end{bmatrix}$

30. $[C] = \begin{bmatrix} 1 & 6 & -3 & 4 & | & 1 \\ 0 & 1 & 7 & 3 & | & 0 \\ 0 & 3 & 4 & 0 & | & -2 \\ 0 & -2 & 3 & 1 & | & 0 \end{bmatrix}$

31. $[D] = \begin{bmatrix} 1 & 5 & -1 & 2 & | & 8 \\ 0 & 1 & 5 & 2 & | & 0 \\ 0 & 1 & 4 & 0 & | & 5 \\ 0 & 2 & -3 & 1 & | & 7 \end{bmatrix}$

Given the matrices in Problems 32–35, perform elementary row operations to obtain a 1 in the third row, third column without changing the entries in the first two columns. Answers may vary.

32. $[A] = \begin{bmatrix} 1 & 0 & 4 & | & 5 \\ 0 & 1 & -3 & | & 6 \\ 0 & 0 & 5 & | & 10 \end{bmatrix}$

33. $[B] = \begin{bmatrix} 1 & 0 & 4 & | & -5 \\ 0 & 1 & 3 & | & 6 \\ 0 & 0 & 8 & | & 12 \end{bmatrix}$

34. $[C] = \begin{bmatrix} 1 & 0 & -2 & 1 & | & 1 \\ 0 & 1 & 6 & 2 & | & 0 \\ 0 & 0 & 4 & 0 & | & -2 \\ 0 & 0 & 2 & 1 & | & 0 \end{bmatrix}$

35. $[D] = \begin{bmatrix} 1 & 0 & -8 & 2 & | & 8 \\ 0 & 1 & -1 & 3 & | & 2 \\ 0 & 0 & 2 & 0 & | & 10 \\ 0 & 0 & -2 & 1 & | & 6 \end{bmatrix}$

Given the matrices in Problems 36–39, perform elementary row operations to obtain zeros above and below the 1 in the third column without changing the entries in the first or second columns. Answers may vary.

36. $[A] = \begin{bmatrix} 1 & 0 & -1 & | & 5 \\ 0 & 1 & 2 & | & 6 \\ 0 & 0 & 1 & | & 4 \end{bmatrix}$

37. $[B] = \begin{bmatrix} 1 & 0 & -3 & | & -2 \\ 0 & 1 & 4 & | & 5 \\ 0 & 0 & 1 & | & 3 \end{bmatrix}$

38. $[C] = \begin{bmatrix} 1 & 0 & -1 & 4 & | & 1 \\ 0 & 1 & 4 & 3 & | & 0 \\ 0 & 0 & 1 & 0 & | & -2 \\ 0 & 0 & -3 & 1 & | & 0 \end{bmatrix}$

39. $[D] = \begin{bmatrix} 1 & 0 & -8 & 2 & | & 8 \\ 0 & 1 & 4 & 2 & | & 0 \\ 0 & 0 & 1 & 0 & | & 2 \\ 0 & 0 & -1 & 1 & | & 7 \end{bmatrix}$

LEVEL 2

Solve the systems in Problems 40–57 by the Gauss–Jordan method.

40. $\begin{cases} x + y = 7 \\ x - y = -1 \end{cases}$

41. $\begin{cases} x - y = 8 \\ x + y = 2 \end{cases}$

42. $\begin{cases} -x + 2y = 2 \\ 4x - 7y = -5 \end{cases}$

43. $\begin{cases} x - 6y = -3 \\ 2x + 3y = 9 \end{cases}$

44. $\begin{cases} 4x - 3y = 1 \\ 5x + 2y = 7 \end{cases}$

45. $\begin{cases} 3x + 7y = 5 \\ 4x + 9y = 7 \end{cases}$

46. $\begin{cases} x - y = 2 \\ 2x + 3y = 9 \end{cases}$

47. $\begin{cases} 3x - 4y = 16 \\ -x + 2y = -6 \end{cases}$

48. $\begin{cases} x - y = 1 \\ x + z = 1 \\ y - z = 1 \end{cases}$

49. $\begin{cases} x + y = 2 \\ x - z = 1 \\ -y + z = 1 \end{cases}$

50. $\begin{cases} x + 5z = 9 \\ y + 2z = 2 \\ 2x + 3z = 4 \end{cases}$

51. $\begin{cases} x + 2z = 13 \\ 2x + y = 8 \\ -2y + 9z = 41 \end{cases}$

52. $\begin{cases} 4x + y = -2 \\ 3x + 2z = -9 \\ 2y + 3z = -5 \end{cases}$

53. $\begin{cases} 5x + z = 9 \\ x - 5z = 7 \\ x + y - z = 0 \end{cases}$

54. $\begin{cases} x + y = -2 \\ y + z = 2 \\ x - y - z = -1 \end{cases}$

55. $\begin{cases} x + y = -1 \\ y + z = -1 \\ x + y + z = 1 \end{cases}$

56. $\begin{cases} x + 2z = 9 \\ 2x + y = 13 \\ 2y + z = 8 \end{cases}$

57. $\begin{cases} x + 2z = 0 \\ 3x - y + 2z = 0 \\ 4x + y = 6 \end{cases}$

PROBLEM SOLVING

58. To control a certain type of crop disease, it is necessary to use 23 gal of chemical A and 34 gal of chemical B. The dealer can order commercial spray I, each container of which holds 5 gal of chemical A and 2 gal of chemical B, and commercial spray II, each container of which holds 2 gal of chemical A and 7 gal of chemical B. How many containers of each type of commercial spray should be used to obtain exactly the right proportion of chemicals needed?

59. A candy maker mixes chocolate, milk, and mint extract to produce three kinds of candy (I, II, and III) with the following proportions:

   I: 7 lb chocolate, 5 gal milk, 1 oz mint extract
  II: 3 lb chocolate, 2 gal milk, 2 oz mint extract
 III: 4 lb chocolate, 3 gal milk, 3 oz mint extract

If 67 lb of chocolate, 48 gal of milk, and 32 oz of mint extract are available, how much of each kind of candy can be produced?

60. Using the data from Problem 59, how much of each type of candy can be produced with 62 lb of chocolate, 44 gal of milk, and 32 oz of mint extract?

14.4 INVERSE MATRICES

The availability of computer software and calculators that can carry out matrix operations has considerably increased the importance of a matrix solution to systems of equations, which involves the inverse of a matrix. If we let [A] be the matrix of coefficients of a system of equations, [X] the matrix of unknowns, and [B] the matrix of constants, we can then represent the system of equations by the **matrix equation**

$$[A][X] = [B]$$

Note that $[A]^{-1}$ does not mean

$$\frac{1}{[A]}$$

but rather means the inverse of matrix [A].

If we can define an inverse matrix, denoted by $[A]^{-1}$, we should be able to solve the *system* by finding

$$[X] = [A]^{-1}[B]$$

To understand this simple process, we need to understand what it means for matrices to be inverses and develop a basic algebra for matrices. Even though these matrix operations are difficult with a pencil and paper, they are easy with the aid of computer software or a calculator that does matrix operations.

Matrix Operations

The next box gives a definition of matrix equality along with the fundamental matrix operations.

Matrix Operations

Equality
[M] = [N] if and only if matrices [M] and [N] are the same order and the corresponding entries are the same.

Addition
[M] + [N] = [S] if and only if [M] and [N] are the same order and the entries of [S] are found by adding the corresponding entries of [M] and [N].

Multiplication by a scalar
$c[M] = [M]c$ is the matrix in which each entry of [M] is multiplied by the scalar (real number) c.

Subtraction
[M] − [N] = [D] if and only if [M] and [N] are the same order and the entries of [D] are found by subtracting the entries of [N] from the corresponding entries of [M].

Multiplication
Let [M] be an $m \times r$ matrix and [N] an $r \times n$ matrix. The product matrix [M][N] = [P] is an $m \times n$ matrix. The entry in the ith row and jth column of [M][N] is *the sum of the products formed by multiplying each entry of the ith row of* [M] *by the corresponding element in the jth column of* [N].

All of these definitions, except multiplication, are straightforward, so we will consider multiplication separately after Example 1. If an addition or multiplication cannot be performed because of the order of the given matrices, the matrices are said to be **nonconformable.**

EXAMPLE 1

Let $[A] = [5 \quad 2 \quad 1]$, $[B] = [4 \quad 8 \quad -5]$, $[C] = \begin{bmatrix} 7 & 3 & 2 \\ 5 & -4 & -3 \end{bmatrix}$,

$[D] = \begin{bmatrix} 4 & -2 & 1 \\ -3 & 3 & -1 \\ 2 & 4 & -1 \end{bmatrix}$, $[E] = \begin{bmatrix} 3 & 4 & -1 \\ 2 & 0 & 5 \\ -4 & 2 & 3 \end{bmatrix}$. Find:

a. [A] + [B]   **b.** [A] + [C]   **c.** [D] + [E]   **d.** (−5)[C]   **e.** 2[A] − 3[B]

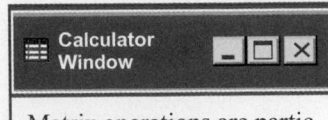

Matrix operations are particularly easy using a calculator that handles matrices. For this example:

$[A] + [B] = [9 \ 10 \ -4]$

$[A] + [C]$ error

$[E] + [D] =$
$\begin{bmatrix} 7 & 2 & 0 \end{bmatrix}$
$\begin{bmatrix} -1 & 3 & 4 \end{bmatrix}$
$\begin{bmatrix} -2 & 6 & 2 \end{bmatrix}$

$-5*[C] =$
$\begin{bmatrix} -35 & -15 & -10 \end{bmatrix}$
$\begin{bmatrix} -25 & 20 & 15 \end{bmatrix}$

$2[A] - 3[B] =$
$\begin{bmatrix} -2 & -20 & 17 \end{bmatrix}$

Solution

a. $[A] + [B] = [5 \ \ 2 \ \ 1] + [4 \ \ 8 \ \ -5]$
$= [5 + 4 \ \ 2 + 8 \ \ 1 + (-5)]$
$= [9 \ \ 10 \ \ -4]$

b. $[A] + [C]$ is not defined because $[A]$ and $[C]$ are nonconformable.

c. $[D] + [E] = \begin{bmatrix} 4 & -2 & 1 \\ -3 & 3 & -1 \\ 2 & 4 & -1 \end{bmatrix} + \begin{bmatrix} 3 & 4 & -1 \\ 2 & 0 & 5 \\ -4 & 2 & 3 \end{bmatrix} = \begin{bmatrix} 7 & 2 & 0 \\ -1 & 3 & 4 \\ -2 & 6 & 2 \end{bmatrix}$

Add entry by entry.

d. $(-5)[C] = (-5)\begin{bmatrix} 7 & 3 & 2 \\ 5 & -4 & -3 \end{bmatrix} = \begin{bmatrix} -35 & -15 & -10 \\ -25 & 20 & 15 \end{bmatrix}$

Multiply each entry by -5.

e. $2[A] - 3[B] = 2[5 \ \ 2 \ \ 1] + (-3)[4 \ \ 8 \ \ -5]$
$= [10 \ \ 4 \ \ 2] + [-12 \ \ -24 \ \ 15]$
$= [-2 \ \ -20 \ \ 17]$

◆

EXAMPLE 2

We illustrate matrix multiplication for several different examples.

Solution

a. $\underbrace{[2 \ \ 3 \ \ 4 \ \ 5]}_{1 \times 4 \text{ matrix}} \underbrace{\begin{bmatrix} a \\ b \\ c \\ d \end{bmatrix}}_{4 \times 1 \text{ matrix}} = \underbrace{[2a + 3b + 4c + 5d]}_{1 \times 1 \text{ matrix answer}}$

To be conformable, these numbers must be the same.

b. $\underbrace{[2 \ \ 3 \ \ 4 \ \ 5]}_{1 \times 4 \text{ matrix}} \underbrace{\begin{bmatrix} 1 \\ -3 \\ 0 \\ 2 \end{bmatrix}}_{4 \times 1 \text{ matrix}} = \underbrace{[2(1) + 3(-3) + 4(0) + 5(2)]}_{1 \times 1 \text{ matrix}} = [3]$

Same

c. $\underbrace{[5 \ \ 3 \ \ 2]}_{1 \times 3 \text{ matrix}} \underbrace{\begin{bmatrix} 1 \\ 2 \\ 3 \\ 4 \end{bmatrix}}_{4 \times 1 \text{ matrix}}$ not defined

Not the same

d. $\underbrace{[2 \ \ 3 \ \ 4 \ \ 5]}_{1 \times 4 \text{ matrix}} \underbrace{\begin{bmatrix} 1 & 2 \\ -3 & 1 \\ 0 & -2 \\ 2 & 3 \end{bmatrix}}_{4 \times 2 \text{ matrix}} = [\underbrace{2(1) + 3(-3) + 4(0) + 5(2)}_{(\text{row 1, column 1}) \text{ entry}} \qquad]$

Same: Answer is a 1×2 matrix

$[2 \ \ 3 \ \ 4 \ \ 5]\begin{bmatrix} 1 & 2 \\ -3 & 1 \\ 0 & -2 \\ 2 & 3 \end{bmatrix} = [3 \qquad \underbrace{2(2) + 3(1) + 4(-2) + 5(3)}_{(\text{row 1, column 2}) \text{ entry}}]$

$= [3 \ \ 14]$

e. $[2 \quad -1 \quad 2] \begin{bmatrix} 4 & 2 & -1 \\ 1 & 0 & 2 \\ 3 & -1 & 3 \end{bmatrix} = [\underbrace{2(4) + (-1)(1) + 2(3)}_{\text{(row 1, column 1) entry}} \qquad]$

$[2 \quad -1 \quad 2] \begin{bmatrix} 4 & 2 & -1 \\ 1 & 0 & 2 \\ 3 & -1 & 3 \end{bmatrix} = [13 \qquad \underbrace{2(2) + (-1)(0) + 2(-1)}_{\text{(row 1, column 2) entry}} \qquad]$

$[2 \quad -1 \quad 2] \begin{bmatrix} 4 & 2 & -1 \\ 1 & 0 & 2 \\ 3 & -1 & 3 \end{bmatrix} = [13 \quad 2 \quad \underbrace{2(-1) + (-1)(2) + 2(3)]}_{\text{(row 1, column 3) entry}}$

$= [13 \quad 2 \quad 2]$

This example is written out to clearly demonstrate what is happening. Your work, how-ever, should look like this:

$$[2 \quad -1 \quad 2] \begin{bmatrix} 4 & 2 & -1 \\ 1 & 0 & 2 \\ 3 & -1 & 3 \end{bmatrix} = [8 - 1 + 6 \quad 4 + 0 - 2 \quad -2 - 2 + 6]$$

$$= [13 \quad 2 \quad 2] \qquad \blacklozenge$$

▦ Calculator Window   ⎯ ⬜ ✕

One of the principal advantages of using matrix multiplication to solve systems of equations is the ease with which matrix multiplication can be done using a matrix calculator. You should check the following calculations using a calculator.

EXAMPLE 3

Find [A][B] and [B][A].

a. Let $[A] = \begin{bmatrix} 1 & 2 & 3 & 4 \\ 5 & 6 & 7 & 8 \end{bmatrix}$ and $[B] = \begin{bmatrix} -3 & 1 & -2 \\ 0 & -1 & 5 \\ -4 & 3 & -1 \\ 2 & 3 & -2 \end{bmatrix}$.

b. Let $[A] = \begin{bmatrix} 3 & -1 & 4 \\ 2 & 1 & 0 \\ -1 & 3 & 2 \end{bmatrix}$ and $[B] = \begin{bmatrix} 5 & 1 & -1 \\ 2 & 3 & -2 \\ 0 & 3 & 4 \end{bmatrix}$.

Solution

a. $[A][B] = \begin{bmatrix} 1 & 2 & 3 & 4 \\ 5 & 6 & 7 & 8 \end{bmatrix} \begin{bmatrix} -3 & 1 & -2 \\ 0 & -1 & 5 \\ -4 & 3 & -1 \\ 2 & 3 & -2 \end{bmatrix}$

$= \begin{bmatrix} 1(-3) + 2(0) + 3(-4) + 4(2) & 1(1) + 2(-1) + 3(3) + 4(3) & 1(-2) + 2(5) + 3(-1) + 4(-2) \\ 5(-3) + 6(0) + 7(-4) + 8(2) & 5(1) + 6(-1) + 7(3) + 8(3) & 5(-2) + 6(5) + 7(-1) + 8(-2) \end{bmatrix}$

$= \begin{bmatrix} -7 & 20 & -3 \\ -27 & 44 & -3 \end{bmatrix}$

$$[B][A] = \begin{bmatrix} -3 & 1 & -2 \\ 0 & -1 & 5 \\ -4 & 3 & -1 \\ 2 & 3 & -2 \end{bmatrix} \begin{bmatrix} 1 & 2 & 3 & 4 \\ 5 & 6 & 7 & 8 \end{bmatrix}$$

[B] and [A] are not conformable. Note that $[A][B] \neq [B][A]$.

b. $[A][B] = \begin{bmatrix} 3 & -1 & 4 \\ 2 & 1 & 0 \\ -1 & 3 & 2 \end{bmatrix} \begin{bmatrix} 5 & 1 & -1 \\ 2 & 3 & -2 \\ 0 & 3 & 4 \end{bmatrix}$

$$= \begin{bmatrix} 3(5) + (-1)(2) + 4(0) & 3(1) + (-1)(3) + 4(3) & 3(-1) + (-1)(-2) + 4(4) \\ 2(5) + (1)(2) + 0(0) & 2(1) + (1)(3) + 0(3) & 2(-1) + (1)(-2) + 0(4) \\ (-1)(5) + 3(2) + 2(0) & (-1)(1) + 3(3) + 2(3) & (-1)(-1) + 3(-2) + 2(4) \end{bmatrix}$$

$$= \begin{bmatrix} 13 & 12 & 15 \\ 12 & 5 & -4 \\ 1 & 14 & 3 \end{bmatrix}$$

$$[B][A] = \begin{bmatrix} 5 & 1 & -1 \\ 2 & 3 & -2 \\ 0 & 3 & 4 \end{bmatrix} \begin{bmatrix} 3 & -1 & 4 \\ 2 & 1 & 0 \\ -1 & 3 & 2 \end{bmatrix} = \begin{bmatrix} 18 & -7 & 18 \\ 14 & -5 & 4 \\ 2 & 15 & 8 \end{bmatrix}$$

Once again, note that $[A][B] \neq [B][A]$. ◆

EXAMPLE 4

Let $[A] = \begin{bmatrix} 1 & 2 & 3 \\ 4 & -1 & 5 \\ 3 & 2 & -1 \end{bmatrix}$, $[X] = \begin{bmatrix} x \\ y \\ z \end{bmatrix}$, and $[B] = \begin{bmatrix} 3 \\ 16 \\ 5 \end{bmatrix}$.

What is $[A][X] = [B]$?

Solution $[A][X] = \begin{bmatrix} x + 2y + 3z \\ 4x - y + 5z \\ 3x + 2y - z \end{bmatrix}$ so $[A][X] = [B]$ is a matrix equation

representing the system $\begin{cases} x + 2y + 3z = 3 \\ 4x - y + 5z = 16 \\ 3x + 2y - z = 5 \end{cases}$ ◆

Communication Matrices

A **communication matrix** is a square matrix in which the entries symbolize the occurrence of some facet or event with a 1 and the nonoccurrence with a 0. Consider the following example.

EXAMPLE 5   **PÓLYA'S METHOD**

Use matrix notation to summarize the following information: The United States has diplomatic relations with Russia and with Mexico, but not with Cuba. Mexico has diplomatic relations with the United States and Russia, but not with Cuba. Russia has diplomatic relations with the United States, Mexico, and Cuba. Finally, Cuba has diplomatic relations with Russia, but not with the United States and not with Mexico. Note that a country is not considered to have diplomatic relations with itself.

a. Write a matrix showing direct communication possibilities.

b. Write a matrix showing the channels of communication that are open to the various countries if they are willing to speak through an intermediary.

Solution We use Pólya's problem-solving guidelines for this example.

Understand the Problem. Part **a** is fairly easy to understand, so we will do this first:

$$[A] = \begin{array}{c} \\ \text{U.S.} \\ \text{Russia} \\ \text{Cuba} \\ \text{Mexico} \end{array} \begin{array}{cccc} \text{U.S.} & \text{Russia} & \text{Cuba} & \text{Mexico} \\ \begin{bmatrix} 0 & 1 & 0 & 1 \\ 1 & 0 & 1 & 1 \\ 0 & 1 & 0 & 0 \\ 1 & 1 & 0 & 0 \end{bmatrix} \end{array}$$

For part **b,** we want to find a matrix that will tell us the number of ways the countries can communicate through an intermediary. For example, the United States can talk to Cuba, by talking through Russia. A country can also communicate with itself (to test security, perhaps) if it does so through an intermediary.

Devise a Plan. The plan could list all possibilities by considering one possibility at a time, but instead we will use matrix multiplication to find $[A]^2$; we will then verify that this is the desired matrix.

Carry Out the Plan.

$$[A]^2 = \begin{bmatrix} 0 & 1 & 0 & 1 \\ 1 & 0 & 1 & 1 \\ 0 & 1 & 0 & 0 \\ 1 & 1 & 0 & 0 \end{bmatrix}\begin{bmatrix} 0 & 1 & 0 & 1 \\ 1 & 0 & 1 & 1 \\ 0 & 1 & 0 & 0 \\ 1 & 1 & 0 & 0 \end{bmatrix}$$

$$= \begin{bmatrix} 0+1+0+1 & 0+0+0+1 & 0+1+0+0 & 0+1+0+0 \\ 0+0+0+1 & 1+0+1+1 & 0+0+0+0 & 1+0+0+0 \\ 0+1+0+0 & 0+0+0+0 & 0+1+0+0 & 0+1+0+0 \\ 0+1+0+0 & 1+0+0+0 & 0+1+0+0 & 1+1+0+0 \end{bmatrix}$$

$$= \begin{bmatrix} 2 & 1 & 1 & 1 \\ 1 & 3 & 0 & 1 \\ 1 & 0 & 1 & 1 \\ 1 & 1 & 1 & 2 \end{bmatrix}$$

Look Back. We see that the United States can communicate with Cuba (entry in the first row, third column) through an intermediary. The zeros indicate that Russia and Cuba cannot communicate through intermediaries (they can communicate only directly). Matrix $[A]^3$ tells us how many ways the countries can communicate if they use two intermediaries. ◆

Algebraic Properties of Matrices

Properties for an algebra of matrices can also be developed. The $m \times n$ **zero matrix,** denoted by $[0]$, is the matrix with m rows and n columns in which each entry is 0. The *identity matrix for multiplication,* denoted by $[I_n]$, is the square matrix with n rows and n columns consisting of a 1 in each position on the **main diagonal** (entries m_{11}, m_{22}, $m_{33}, \ldots, m_{nn}$) and zeros elsewhere:

$$[I_2] = \begin{bmatrix} 1 & 0 \\ 0 & 1 \end{bmatrix}, \quad [I_3] = \begin{bmatrix} 1 & 0 & 0 \\ 0 & 1 & 0 \\ 0 & 0 & 1 \end{bmatrix}, \quad [I_4] = \begin{bmatrix} 1 & 0 & 0 & 0 \\ 0 & 1 & 0 & 0 \\ 0 & 0 & 1 & 0 \\ 0 & 0 & 0 & 1 \end{bmatrix}$$

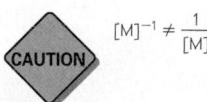
$[M]^{-1} \neq \dfrac{1}{[M]}$

The **additive inverse** of a matrix $[M]$ is denoted by $[-M]$ and is defined by $(-1)[M]$; the **multiplicative inverse** of a matrix $[M]$ is denoted by $[M]^{-1}$ if it exists. Table 14.1 summarizes the properties of matrices. Assume that $[M]$, $[N]$, and $[P]$ all have order $n \times n$, which forces them to be conformable for the given operations. If the context makes the order of the identity obvious, we sometimes write [I] to denote the identity matrix; that is, we assume that [I] means [I_n].

Table 14.1 Properties of Matrices

Property	Addition	Multiplication
Commutative	$[M] + [N] = [N] + [M]$	$[M][N] \neq [N][M]$
Associative	$([M] + [N]) + [P] = [M] + ([N] + [P])$	$([M][N])[P] = [M]([N][P])$
Identity	$[M] + [0] = [0] + [M]$	$[I][M] = [M][I] = [M]$
Inverse	$[M] + [-M] = [-M] + [M] = [0]$	$[M][M]^{-1} = [M]^{-1}[M] = [I]$
Distributive		$[M]([N] + [P]) = [M][N] + [M][P]$
	and	$([N] + [P])[M] = [N][M] + [P][M]$

Inverse Property

The property from this list that is particularly important for us in solving systems of equations is the inverse property. There are two unanswered questions about the inverse property. Given a *square* matrix [M], when does $[M]^{-1}$ exist? And if it exists, how do you find it?

Inverse of a Matrix

If [A] is a square matrix and if there exists a matrix $[A]^{-1}$ such that

$$[A]^{-1}[A] = [A][A]^{-1} = [I]$$

where [I] is the identity matrix for multiplication, then $[A]^{-1}$ is called the **inverse** of [A] for multiplication.

Usually, in the context of matrices, when we talk simply of the inverse of [A] we mean the inverse of [A] for multiplication, and when we talk of the identity matrix we mean the identity matrix for multiplication.

EXAMPLE 6

Show that [A] and [B] are inverses, given:

a. $[A] = \begin{bmatrix} 2 & 1 \\ 3 & 2 \end{bmatrix}$; $[B] = \begin{bmatrix} 2 & -1 \\ -3 & 2 \end{bmatrix}$

b. $[A] = \begin{bmatrix} 0 & 1 & 2 \\ -1 & 1 & 2 \\ 1 & -2 & -5 \end{bmatrix}$; $[B] = \begin{bmatrix} 1 & -1 & 0 \\ 3 & 2 & 2 \\ -1 & -1 & -1 \end{bmatrix}$

Solution

a. We must show that $[A][B] = [I]$ and $[B][A] = [I]$.

$$[A][B] = \begin{bmatrix} 2 & 1 \\ 3 & 2 \end{bmatrix}\begin{bmatrix} 2 & -1 \\ -3 & 2 \end{bmatrix} = \begin{bmatrix} 4-3 & -2+2 \\ 6-6 & -3+4 \end{bmatrix} = \begin{bmatrix} 1 & 0 \\ 0 & 1 \end{bmatrix} = [I]$$

$$[B][A] = \begin{bmatrix} 2 & -1 \\ -3 & 2 \end{bmatrix}\begin{bmatrix} 2 & 1 \\ 3 & 2 \end{bmatrix} = \begin{bmatrix} 4-3 & 2-2 \\ -6+6 & -3+4 \end{bmatrix} = \begin{bmatrix} 1 & 0 \\ 0 & 1 \end{bmatrix} = [I]$$

Since $[A][B] = [B][A] = [I]$, we see that $[B] = [A]^{-1}$.

b.

$$[A][B] = \begin{bmatrix} 0 & 1 & 2 \\ -1 & 1 & 2 \\ 1 & -2 & -5 \end{bmatrix} \begin{bmatrix} 1 & -1 & 0 \\ 3 & 2 & 2 \\ -1 & -1 & -1 \end{bmatrix}$$

$$= \begin{bmatrix} 0+3-2 & 0+2-2 & 0+2-2 \\ -1+3-2 & 1+2-2 & 0+2-2 \\ 1-6+5 & -1-4+5 & 0-4+5 \end{bmatrix}$$

$$= \begin{bmatrix} 1 & 0 & 0 \\ 0 & 1 & 0 \\ 0 & 0 & 1 \end{bmatrix} = [I]$$

$$[B][A] = \begin{bmatrix} 1 & -1 & 0 \\ 3 & 2 & 2 \\ -1 & -1 & -1 \end{bmatrix} \begin{bmatrix} 0 & 1 & 2 \\ -1 & 1 & 2 \\ 1 & -2 & -5 \end{bmatrix}$$

$$= \begin{bmatrix} 0+1+0 & 1-1+0 & 2-2+0 \\ 0-2+2 & 3+2-4 & 6+4-10 \\ 0+1-1 & -1-1+2 & -2-2+5 \end{bmatrix}$$

$$= \begin{bmatrix} 1 & 0 & 0 \\ 0 & 1 & 0 \\ 0 & 0 & 1 \end{bmatrix} = [I]$$

Since $[A][B] = [I] = [B][A]$, then $[B] = [A]^{-1}$. ◆

If a given matrix has an inverse, we say it is **nonsingular.** The unanswered question, however, is how to *find* an inverse matrix.

▦ Calculator Window   _ ▢ ✕

If you have a calculator that does matrix operations, use a MATRIX function and then press $[A]^{-1}$ to find the inverse of a matrix $[A]$. We will show this step in the margin as we find the inverse matrix in the next two examples.

EXAMPLE 7

Find the inverse of $[A] = \begin{bmatrix} 1 & 2 \\ 1 & 4 \end{bmatrix}$.

Calculator Window _ ▢ ✕

If you input MATRIX $[A]$ and then press $[A]^{-1}$, the output will look like:

$$\begin{bmatrix} 2 & -1 \\ -0.5 & 0.5 \end{bmatrix}$$

Solution Find a matrix $[B]$, if it exists, so that $[A][B] = [I]$; since we do not know $[B]$, let its entries be variables:

$$[B] = \begin{bmatrix} x_1 & x_2 \\ y_1 & y_2 \end{bmatrix}$$

Then, $[A][B] = \begin{bmatrix} 1 & 2 \\ 1 & 4 \end{bmatrix} \begin{bmatrix} x_1 & x_2 \\ y_1 & y_2 \end{bmatrix}$

$$= \begin{bmatrix} x_1 + 2y_1 & x_2 + 2y_2 \\ x_1 + 4y_1 & x_2 + 4y_2 \end{bmatrix}$$

$$= \begin{bmatrix} 1 & 0 \\ 0 & 1 \end{bmatrix}$$

By definition of equality of matrices, we see that

$$\begin{cases} x_1 + 2y_1 = 1 \\ x_1 + 4y_1 = 0 \end{cases} \quad \text{and} \quad \begin{cases} x_2 + 2y_2 = 0 \\ x_2 + 4y_2 = 1 \end{cases}$$

Solve each of these systems simultaneously to find $x_1 = 2$, $x_2 = -1$, $y_1 = -\frac{1}{2}$, $y_2 = \frac{1}{2}$. Thus, the inverse is

$$[B] = \begin{bmatrix} x_1 & x_2 \\ y_1 & y_2 \end{bmatrix} = \begin{bmatrix} 2 & -1 \\ -\frac{1}{2} & \frac{1}{2} \end{bmatrix}$$

◆

EXAMPLE 8

Find the inverse for $[A] = \begin{bmatrix} 1 & -1 & 0 \\ 3 & 2 & 2 \\ -1 & -1 & -1 \end{bmatrix}$.

Solution We need to find a matrix $\begin{bmatrix} x_1 & x_2 & x_3 \\ y_1 & y_2 & y_3 \\ z_1 & z_2 & z_3 \end{bmatrix}$ so that

$$\begin{bmatrix} 1 & -1 & 0 \\ 3 & 2 & 2 \\ -1 & -1 & -1 \end{bmatrix} \begin{bmatrix} x_1 & x_2 & x_3 \\ y_1 & y_2 & y_3 \\ z_1 & z_2 & z_3 \end{bmatrix} = \begin{bmatrix} 1 & 0 & 0 \\ 0 & 1 & 0 \\ 0 & 0 & 1 \end{bmatrix}$$

The definition of equality of matrices gives rise to three systems of equations.

$$\begin{cases} x_1 - y_1 + 0z_1 = 1 \\ 3x_1 + 2y_1 + 2z_1 = 0 \\ -x_1 - y_1 - z_1 = 0 \end{cases} \quad \begin{cases} x_2 - y_2 + 0z_2 = 0 \\ 3x_2 + 2y_2 + 2z_2 = 1 \\ -x_2 - y_2 - z_2 = 0 \end{cases} \quad \begin{cases} x_3 - y_3 + 0z_3 = 0 \\ 3x_3 + 2y_3 + 2z_3 = 0 \\ -x_3 - y_3 - z_3 = 1 \end{cases}$$

We could solve these as three separate systems using Gauss–Jordan elimination; however, all the steps would be identical since the coefficients are the same in each system. Therefore, suppose we augment the matrix of the coefficients by the *three* columns of constants and do all three at once. Write the augmented matrix as $[B] = [A \mid I]$.

$$\begin{bmatrix} 1 & -1 & 0 & \vdots & 1 & 0 & 0 \\ 3 & 2 & 2 & \vdots & 0 & 1 & 0 \\ -1 & -1 & -1 & \vdots & 0 & 0 & 1 \end{bmatrix} \rightarrow \begin{bmatrix} 1 & -1 & 0 & \vdots & 1 & 0 & 0 \\ 0 & 5 & 2 & \vdots & -3 & 1 & 0 \\ 0 & -2 & -1 & \vdots & 1 & 0 & 1 \end{bmatrix}$$

*Row$+(-3,[B],1,2)$
*Row$+(1,[Ans],1,3)$

$$\rightarrow \begin{bmatrix} 1 & -1 & 0 & \vdots & 1 & 0 & 0 \\ 0 & 1 & 0 & \vdots & -1 & 1 & 2 \\ 0 & -2 & -1 & \vdots & 1 & 0 & 1 \end{bmatrix} \rightarrow \begin{bmatrix} 1 & 0 & 0 & \vdots & 0 & 1 & 2 \\ 0 & 1 & 0 & \vdots & -1 & 1 & 2 \\ 0 & 0 & -1 & \vdots & -1 & 2 & 5 \end{bmatrix}$$

*Row$+(2,[Ans],3,2)$ *Row$+(1,[Ans],2,1)$
                        *Row$+(2,[Ans],2,3)$

$$\rightarrow \begin{bmatrix} 1 & 0 & 0 & \vdots & 0 & 1 & 2 \\ 0 & 1 & 0 & \vdots & -1 & 1 & 2 \\ 0 & 0 & 1 & \vdots & 1 & -2 & -5 \end{bmatrix}$$

*Row$(-1,[Ans],3)$

Now, if we relate this to the original three systems, we see that the inverse matrix is

$$\begin{bmatrix} 0 & 1 & 2 \\ -1 & 1 & 2 \\ 1 & -2 & -5 \end{bmatrix}$$

◆

By studying Example 8, we are led to a procedure for finding the inverse of a nonsingular matrix.

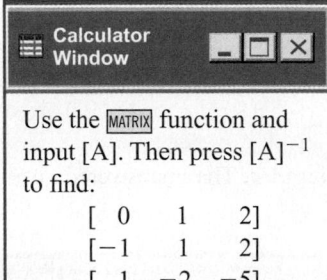

Calculator Window

Use the MATRIX function and input $[A]$. Then press $[A]^{-1}$ to find:

$$\begin{bmatrix} 0 & 1 & 2 \\ -1 & 1 & 2 \\ 1 & -2 & -5 \end{bmatrix}$$

Procedure for Finding the Inverse of a Matrix

To find the inverse of a *square* matrix *A*:

1. Augment [A] with [I]; that is, write [A ⋮ I], where [I] is the identity matrix of the same order as [A].

2. Perform elementary row operations using Gauss–Jordan elimination to change the matrix *A* into the identity matrix (if possible).

3. If at any time you obtain all zeros in a row or column to the left of the dashed line, then there will be no inverse.

4. If steps 1 and 2 can be performed, the result in the augmented part is the inverse of [A].

EXAMPLE 9

Find the inverse, if possible, of $[A] = \begin{bmatrix} 1 & 2 \\ 0 & 0 \end{bmatrix}$.

Calculator Window ▬ ▢ ✕

If you try to find the inverse of a **singular** matrix (one that does not have an inverse), you will obtain an error message.

Solution   Write the augmented matrix [A ⋮ I]:

$$\begin{bmatrix} 1 & 2 & \vdots & 1 & 0 \\ 0 & 0 & \vdots & 0 & 1 \end{bmatrix}$$

We want to make the left-hand side look like the corresponding identity matrix. This is impossible since there are no elementary row operations that will put it into the required form. Thus, there is no inverse.   ◆

EXAMPLE 10

Use matrix methods to find the inverse, if possible, of $[A] = \begin{bmatrix} 1 & 2 \\ 1 & 4 \end{bmatrix}$. Compare with Example 7.

Solution   Enter the augmented matrix [B] = [A ⋮ I].

$$\begin{bmatrix} 1 & 2 & \vdots & 1 & 0 \\ 1 & 4 & \vdots & 0 & 1 \end{bmatrix} \rightarrow \begin{bmatrix} 1 & 2 & \vdots & 1 & 0 \\ 0 & 2 & \vdots & -1 & 1 \end{bmatrix} \rightarrow \begin{bmatrix} 1 & 2 & \vdots & 1 & 0 \\ 0 & 1 & \vdots & -\frac{1}{2} & \frac{1}{2} \end{bmatrix}$$

    *Row+(−1,[B],1,2)   *Row(.5,[Ans],2)

$$\rightarrow \begin{bmatrix} 1 & 0 & \vdots & 2 & -1 \\ 0 & 1 & \vdots & -\frac{1}{2} & \frac{1}{2} \end{bmatrix}$$

    *Row+(−2,[Ans],2,1)

The inverse is on the right of the dashed line: $\begin{bmatrix} 2 & -1 \\ -\frac{1}{2} & \frac{1}{2} \end{bmatrix}$; this result agrees with Example 7.

When a matrix has fractional entries, it is often rewritten with a fractional coefficient and integer entries to simplify the arithmetic. For example, we could rewrite this inverse matrix as

$$[A]^{-1} = \frac{1}{2} \begin{bmatrix} 4 & -2 \\ -1 & 1 \end{bmatrix}$$ ◆

EXAMPLE 11

Find the inverse, if possible, of $[A] = \begin{bmatrix} 0 & 1 & 2 \\ 2 & -1 & 1 \\ -1 & 1 & 0 \end{bmatrix}$.

Solution Write the augmented matrix $[B] = [A \mid I]$ and make the left-hand side look like the corresponding identity matrix (if possible):

$$
\begin{bmatrix} 0 & 1 & 2 & \vdots & 1 & 0 & 0 \\ 2 & -1 & 1 & \vdots & 0 & 1 & 0 \\ -1 & 1 & 0 & \vdots & 0 & 0 & 1 \end{bmatrix} \rightarrow
\begin{bmatrix} -1 & 1 & 0 & \vdots & 0 & 0 & 1 \\ 2 & -1 & 1 & \vdots & 0 & 1 & 0 \\ 0 & 1 & 2 & \vdots & 1 & 0 & 0 \end{bmatrix}
$$
$$\text{RowSwap}([B],1,3)$$

$$
\rightarrow \begin{bmatrix} 1 & -1 & 0 & \vdots & 0 & 0 & -1 \\ 2 & -1 & 1 & \vdots & 0 & 1 & 0 \\ 0 & 1 & 2 & \vdots & 1 & 0 & 0 \end{bmatrix} \rightarrow
\begin{bmatrix} 1 & -1 & 0 & \vdots & 0 & 0 & -1 \\ 0 & 1 & 1 & \vdots & 0 & 1 & 2 \\ 0 & 1 & 2 & \vdots & 1 & 0 & 0 \end{bmatrix}
$$
$$*\text{Row}(-1,[\text{Ans}],1) \qquad *\text{Row}+(-2,[\text{Ans}],1,2)$$

$$
\rightarrow \begin{bmatrix} 1 & 0 & 1 & \vdots & 0 & 1 & 1 \\ 0 & 1 & 1 & \vdots & 0 & 1 & 2 \\ 0 & 0 & 1 & \vdots & 1 & -1 & -2 \end{bmatrix} \rightarrow
\begin{bmatrix} 1 & 0 & 0 & \vdots & -1 & 2 & 3 \\ 0 & 1 & 0 & \vdots & -1 & 2 & 4 \\ 0 & 0 & 1 & \vdots & 1 & -1 & -2 \end{bmatrix}
$$
$$*\text{Row}+(1,[\text{Ans}],2,1) \qquad *\text{Row}+(-1,[\text{Ans}],3,1)$$
$$*\text{Row}+(-1,[\text{Ans}],2,3) \qquad *\text{Row}+(-1,[\text{Ans}],3,2)$$

Thus $[A]^{-1} = \begin{bmatrix} -1 & 2 & 3 \\ -1 & 2 & 4 \\ 1 & -1 & -2 \end{bmatrix}$

🖿 Calculator Window _ □ ✕

The answer to Example 11 was "nice," but most matrices will have "ugly" entries. If you are using a calculator, those fractions will be shown in decimal form. For example, the inverse of

$$[A] = \begin{bmatrix} 3 & -1 & 4 \\ 2 & 1 & 0 \\ -1 & 3 & 2 \end{bmatrix} \text{ is found by calculator to be (approximately)}$$

$$[A]^{-1} = \begin{bmatrix} .0526315789 & .3684210526 & -.1052631579 \\ -.1052631579 & .2631578947 & .2105263158 \\ .1842105263 & -.2105263158 & .1315789474 \end{bmatrix}$$

Most real-life applications will involve inverses such as the one shown here.

Systems of Equations

In Example 4 we saw how a system of equations can be written in matrix form. We can now see how to solve a system of linear equations by using the inverse. Consider a system of n linear equations with n unknowns whose matrix of coefficients $[A]$ has an inverse $[A]^{-1}$:

$$[A][X] = [B] \qquad \text{Given system}$$
$$[A]^{-1}[A][X] = [A]^{-1}[B] \qquad \text{Multiply both sides by } [A]^{-1}.$$
$$([A]^{-1}[A])[X] = [A]^{-1}[B] \qquad \text{Associative property}$$
$$[I][X] = [A]^{-1}[B] \qquad \text{Inverse property}$$
$$[X] = [A]^{-1}[B] \qquad \text{Identity property}$$

In other words: To solve a system of equations, look to see whether the number of equations is the same as the number of unknowns. If so, find the inverse of the matrix of the coefficients (if it exists) and multiply it (on the right) by the matrix of the constants.

EXAMPLE 12

Solve: $\begin{cases} y + 2z = 0 \\ 2x - y + z = -1 \\ y - x = 1 \end{cases}$

Solution:

Write in matrix form: $[A] = \begin{bmatrix} 0 & 1 & 2 \\ 2 & -1 & 1 \\ -1 & 1 & 0 \end{bmatrix}$, $[X] = \begin{bmatrix} x \\ y \\ z \end{bmatrix}$, $[B] = \begin{bmatrix} 0 \\ -1 \\ 1 \end{bmatrix}$.

From Example 11, $[A]^{-1} = \begin{bmatrix} -1 & 2 & 3 \\ -1 & 2 & 4 \\ 1 & -1 & -2 \end{bmatrix}$. Thus,

$$[X] = [A]^{-1}[B] = \begin{bmatrix} -1 & 2 & 3 \\ -1 & 2 & 4 \\ 1 & -1 & -2 \end{bmatrix} \begin{bmatrix} 0 \\ -1 \\ 1 \end{bmatrix} = \begin{bmatrix} 1 \\ 2 \\ -1 \end{bmatrix}$$

Therefore, the solution to the system is $(x, y, z) = (1, 2, -1)$.   ◆

☰ Calculator Window

The calculator, of course, is what really makes this inverse method worthwhile. It makes the solution of any system *that has the same number of equations and variables* almost trivial. For Example 12, simply enter [A] and [B] and then press

$$[A]^{-1}[B]$$

The output is:

$$\begin{bmatrix} 1 \end{bmatrix}$$
$$\begin{bmatrix} 2 \end{bmatrix}$$
$$\begin{bmatrix} -1 \end{bmatrix}$$

That is all there is! Does this make the method worthwhile?

The method of solving a system by using the inverse matrix is very efficient if you know the inverse. Unfortunately, *finding* the inverse for one system is usually more work than using another method to solve the system. However, there are certain applications that yield the same system over and over, and the only thing to change is the constants. In this case you should think about the inverse method. And, finally, computers and calculators can find approximations for inverse matrices quite easily, so this method becomes the method of choice if you have access to this technology.

Problem Set 14.4

LEVEL 1

1. **IN YOUR OWN WORDS** When are two matrices equal?

2. **IN YOUR OWN WORDS** Describe the process for adding matrices.

3. **IN YOUR OWN WORDS** Describe the process for multiplying matrices.

4. **IN YOUR OWN WORDS** What is the inverse of a matrix?

5. **IN YOUR OWN WORDS** What is a nonsingular matrix?

6. **IN YOUR OWN WORDS** Describe a procedure for finding the inverse of a square matrix.

7. **IN YOUR OWN WORDS** Describe a procedure for using the inverse of a matrix to solve a system of equations.

8. IN YOUR OWN WORDS What is a communication matrix?

In Problems 9–16, find the indicated matrices if possible.

$$[A] = \begin{bmatrix} 1 & 2 \\ 4 & 0 \\ -1 & 3 \\ 2 & 1 \end{bmatrix} \qquad [B] = \begin{bmatrix} 4 & 2 \\ -1 & 3 \end{bmatrix}$$

$$[C] = \begin{bmatrix} 1 & 0 & 0 & 0 \\ 0 & 1 & 0 & 0 \\ 0 & 0 & 1 & 0 \\ 0 & 0 & 0 & 1 \end{bmatrix}$$

$$[D] = \begin{bmatrix} 4 & 1 & 3 & 6 \\ -1 & 0 & -2 & 3 \end{bmatrix} \qquad [E] = \begin{bmatrix} 1 & 0 & 2 \\ 3 & -1 & 2 \\ 4 & 1 & 0 \end{bmatrix}$$

$$[F] = \begin{bmatrix} 1 & 4 & 0 \\ 3 & -1 & 2 \\ -2 & 1 & 5 \end{bmatrix} \qquad [G] = \begin{bmatrix} 8 & 1 & 6 \\ 3 & 5 & 7 \\ 4 & 9 & 2 \end{bmatrix}$$

9. a. $[E] + [F]$ **b.** $2[E] - [G]$
10. a. $[E][F]$ **b.** $[E][G]$
11. a. $[F][G]$ **b.** $[G][F]$
12. a. $[A][B]$ **b.** $[B][D]$
13. a. $[B]^2$ **b.** $[C]^3$
14. a. $[E]([F] + [G])$ **b.** $[E][F] + [E][G]$
15. a. $([B] + [C])[A]$ **b.** $[B][A] + [C][A]$
16. a. $([E][F])[G]$ **b.** $[E]([F][G])$

Write $[A][X] = [B]$, if possible, for the matrices given in Problems 17–18.

17. $[A] = \begin{bmatrix} 1 & 2 & 4 \\ -3 & 2 & 1 \\ 2 & 0 & 1 \end{bmatrix},$

$[X] = \begin{bmatrix} x \\ y \\ z \end{bmatrix}, \quad \text{and} \quad [B] = \begin{bmatrix} 13 \\ 11 \\ 0 \end{bmatrix}.$

18. $[A] = \begin{bmatrix} 4 & 1 & 0 \\ 3 & -1 & 2 \\ 2 & 3 & 1 \end{bmatrix},$

$[X] = \begin{bmatrix} x \\ y \\ z \end{bmatrix}, \quad \text{and} \quad [B] = \begin{bmatrix} 2 \\ 11 \\ -1 \end{bmatrix}.$

Show that the given matrices are inverses in Problems 19–20.

19. $[A] = \begin{bmatrix} 2 & 7 \\ 1 & 4 \end{bmatrix}; \quad [B] = \begin{bmatrix} 4 & -7 \\ -1 & 2 \end{bmatrix}$

20. $[A] = \begin{bmatrix} -16 & -2 & 7 \\ 7 & 1 & -3 \\ -3 & 0 & 1 \end{bmatrix}; \quad [B] = \begin{bmatrix} 1 & 2 & -1 \\ 2 & 5 & 1 \\ 3 & 6 & -2 \end{bmatrix}$

LEVEL 2

Find the inverse of each matrix in Problems 21–26, if it exists.

21. $\begin{bmatrix} 4 & -7 \\ -1 & 2 \end{bmatrix}$ **22.** $\begin{bmatrix} 8 & 6 \\ -2 & 4 \end{bmatrix}$

23. $\begin{bmatrix} 1 & 0 & 2 \\ 2 & 1 & 0 \\ 0 & -2 & 9 \end{bmatrix}$ **24.** $\begin{bmatrix} 6 & 1 & 20 \\ 1 & -1 & 0 \\ 0 & 1 & 3 \end{bmatrix}$

25. $\begin{bmatrix} 1 & 0 & 0 & 1 \\ 0 & 2 & 0 & 0 \\ 0 & 0 & 0 & 1 \\ 2 & 0 & 1 & 0 \end{bmatrix}$ **26.** $\begin{bmatrix} 0 & 1 & 2 & 0 \\ 0 & 0 & 0 & 1 \\ 1 & 1 & 3 & 0 \\ 2 & 4 & 0 & 0 \end{bmatrix}$

Solve the systems in Problems 27–54 by solving the corresponding matrix equation with an inverse, if possible.

Problems 27–32 use the inverse found in Problem 21.

27. $\begin{cases} 4x - 7y = -2 \\ -x + 2y = 1 \end{cases}$ **28.** $\begin{cases} 4x - 7y = -65 \\ -x + 2y = 18 \end{cases}$

29. $\begin{cases} 4x - 7y = 48 \\ -x + 2y = -13 \end{cases}$ **30.** $\begin{cases} 4x - 7y = 2 \\ -x + 2y = 3 \end{cases}$

31. $\begin{cases} 4x - 7y = 5 \\ -x + 2y = 4 \end{cases}$ **32.** $\begin{cases} 4x - 7y = -3 \\ -x + 2y = 8 \end{cases}$

Problems 33–38 use the inverse found in Problem 22.

33. $\begin{cases} 8x + 6y = 12 \\ -2x + 4y = -14 \end{cases}$ **34.** $\begin{cases} 8x + 6y = 16 \\ -2x + 4y = 18 \end{cases}$

35. $\begin{cases} 8x + 6y = -6 \\ -2x + 4y = -26 \end{cases}$ **36.** $\begin{cases} 8x + 6y = -28 \\ -2x + 4y = 18 \end{cases}$

37. $\begin{cases} 8x + 6y = -26 \\ -2x + 4y = 12 \end{cases}$ **38.** $\begin{cases} 8x + 6y = -36 \\ -2x + 4y = -2 \end{cases}$

Problems 39–44 all use the same inverse.

39. $\begin{cases} 2x + 3y = 9 \\ x - 6y = -3 \end{cases}$ **40.** $\begin{cases} 2x + 3y = 2 \\ x - 6y = 16 \end{cases}$

41. $\begin{cases} 2x + 3y = 2 \\ x - 6y = -14 \end{cases}$ **42.** $\begin{cases} 2x + 3y = 9 \\ x - 6y = 42 \end{cases}$

43. $\begin{cases} 2x + 3y = -22 \\ x - 6y = 49 \end{cases}$ **44.** $\begin{cases} 2x + 3y = 12 \\ x - 6y = -24 \end{cases}$

Problems 45–50 use the inverse found in Problem 23.

45. $\begin{cases} x + 2z = 7 \\ 2x + y = 16 \\ -2y + 9z = -3 \end{cases}$ **46.** $\begin{cases} x + 2z = 4 \\ 2x + y = 0 \\ -2y + 9z = 19 \end{cases}$

47. $\begin{cases} x + 2z = 4 \\ 2x + y = 0 \\ -2y + 9z = 31 \end{cases}$ **48.** $\begin{cases} x + 2z = 7 \\ 2x + y = 1 \\ -2y + 9z = 28 \end{cases}$

49. $\begin{cases} x + 2z = 12 \\ 2x + y = 0 \\ -2y + 9z = 10 \end{cases}$ **50.** $\begin{cases} x + 2z = 5 \\ 2x + y = 8 \\ -2y + 9z = 9 \end{cases}$

Problems 51–54 use the inverse found in Problem 24.

51. $\begin{cases} 6x + y + 20z = 27 \\ x - y = 0 \\ y + 3z = 4 \end{cases}$ **52.** $\begin{cases} 6x + y + 20z = 14 \\ x - y = 1 \\ y + 3z = 1 \end{cases}$

53. $\begin{cases} 6x + y + 20z = -12 \\ x - y = 6 \\ y + 3z = -7 \end{cases}$ **54.** $\begin{cases} 6x + y + 20z = 57 \\ x - y = 1 \\ y + 3z = 5 \end{cases}$

LEVEL 3

55. If we take the first few rows of Pascal's triangle and arrange them into a lower triangular matrix, we form what is called a *Pascal matrix:*

$$[P] = \begin{bmatrix} 1 & 0 & 0 & 0 & 0 \\ 1 & 1 & 0 & 0 & 0 \\ 1 & 2 & 1 & 0 & 0 \\ 1 & 3 & 3 & 1 & 0 \\ 1 & 4 & 6 & 4 & 1 \end{bmatrix}$$

  a. Find the inverse of this matrix.
  b. Make a general statement about the inverse of any order Pascal's matrix.

56. If we add Pascal's matrix (Problem 55) and the identity matrix, we find $[P] + [I]$. For the order shown in Problem 55,

$$[P] + [I] = \begin{bmatrix} 2 & 0 & 0 & 0 & 0 \\ 1 & 2 & 0 & 0 & 0 \\ 1 & 2 & 2 & 0 & 0 \\ 1 & 3 & 3 & 2 & 0 \\ 1 & 4 & 6 & 4 & 2 \end{bmatrix}$$

  Find the inverse of this matrix.

57. How many channels of communication are open to each country in Example 5 if the countries are willing to speak to each other through two intermediaries?

58. Consider the map of airline routes shown in Figure 14.5.

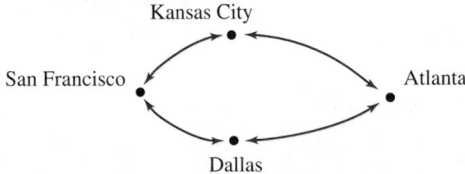

Kansas City

San Francisco   Atlanta

Dallas

Figure 14.5 **Travel routes for four cities**

  a. Fill in the blanks in the following communication matrix representing the airline routes:

$$[T] = \begin{array}{c} \\ \text{SF} \\ \text{D} \\ \text{A} \\ \text{KC} \end{array} \begin{bmatrix} & & & \\ & & & \\ & & & \\ & & & \\ & & & \end{bmatrix} \begin{array}{cccc} \text{SF} & \text{D} & \text{A} & \text{KC} \end{array}$$

  b. Write the communication matrix showing the number of routes among these cities if you make exactly one intermediate stop. Use this matrix to state in how many ways you can travel from Kansas City to San Francisco making exactly one intermediate stop.
  c. Write the communication matrix showing the number of routes among these cities if you make exactly two intermediate stops. Use this matrix to state in how many ways you can travel from San Francisco to Kansas City making two intermediate stops.

PROBLEM SOLVING

59. The Seedy Vin Company produces Riesling, Charbono, and Rosé wines. There are three procedures for producing each wine (the procedures for production affect the cost of the final product). One procedure allows an outside company to bottle the wine; a second allows the wine to be produced and bottled at the winery; a third allows the wine to be estate bottled. The amount of wine produced by the Seedy Vin Company by each method is shown by the matrix

$$[W] = \begin{array}{c} \\ \\ \\ \\ \end{array} \begin{array}{ccc} \text{Riesling} & \text{Charbono} & \text{Rosé} \end{array}$$
$$[W] = \begin{bmatrix} 2 & 1 & 3 \\ 4 & 3 & 6 \\ 1 & 2 & 4 \end{bmatrix} \begin{array}{l} \text{outside bottling} \\ \text{produced and bottled} \\ \text{at winery} \\ \text{estate bottled} \end{array}$$

  Suppose the cost for each method of production is given by the matrix

$$[C] = [\text{outside} \quad \text{winery} \quad \text{estate}]$$
$$= [1 \quad 4 \quad 6]$$

  Suppose the production cost of a unit of each type of wine is given by the matrix

$$[D] = \begin{bmatrix} 40 \\ 60 \\ 30 \end{bmatrix} \begin{array}{l} \text{Riesling} \\ \text{Charbono} \\ \text{Rosé} \end{array}$$

  a. Find the cost of producing each of the three types of wine. *Hint:* This is $[C][W]$.
  b. Find the dollar amount for producing each unit of the different types of wine for these methods of production. *Hint:* This is $[W][D]$.
  c. Find $([C][W])[D]$ and $[C]([W][D])$. Give a verbal interpretation for these products.

60. Sociologists often study the dominance of one group over another. Suppose in a certain society there are four classes which we will call: abigweel, aweel, upancomer, and pon. *Hint:* These are pronounced "A big wheel," "A wheel," "Up and comer," and "pea-on."

  Abigweel dominate aweel, upancomer, and pon.
  Aweel dominate upancomer and pon.
  Upancomer dominate pon.

a. Write a communication matrix, [D], representing these dominance relationships.

b. We say that an individual has two-stage dominance over another if 1 appears in the matrix $[D]^2$ for those individuals. Write the two-stage dominance matrix.

c. We define the *power* of an individual as the sum of the entries in the appropriate row of the matrix

$$[P] = [D] + [D]^2$$

Determine the power of abigweel, aweel, upancomer, and pon.

d. Rank the abigweel, aweel, upancomer, and pon.

14.5 SYSTEMS OF INEQUALITIES

In previous sections, we have discussed the simultaneous solution of a system of equations. In this section, we discuss the graphical solution of a simultaneous **system of inequalities.** The solution of a *system of inequalities* refers to the intersection of the solutions of the individual inequalities in the system. The procedure for graphing the solution of a system of inequalities is to graph the solution of the individual inequalities and then shade in the intersection of all the graphs.

EXAMPLE 1

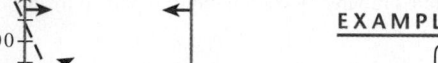

Solve:
$$\begin{cases} y > -20x + 110 \\ x \geq 0 \\ y \geq 0 \\ x \leq 8 \end{cases}$$

Solution Graph each half-plane, but instead of shading, mark the appropriate regions with arrows. For this example, the two inequalities

$$x \geq 0$$
$$y \geq 0$$

give the first quadrant, which is marked with arrows as shown in the margin. Next, graph the equation $y = -20x + 110$ and use a test point, say, $(0, 0)$, to determine the half-plane. Finally draw $x = 8$ and mark it with arrows to show that $x \leq 8$, as shown in the margin.

The last step is to shade the intersection, as shown in Figure 14.6. ◆

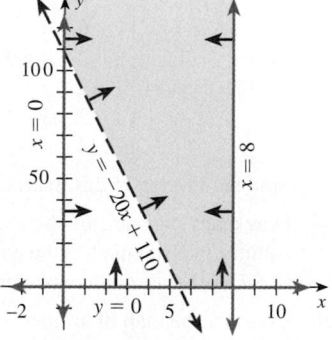

Figure 14.6 **Graph of a system of inequalities**

EXAMPLE 2

Graph the solution of the system:
$$\begin{cases} 2x + y \leq 3 \\ x - y > 5 \\ x \geq 0 \\ y \geq -10 \end{cases}$$

Solution The graph of the individual inequalities and their intersection is shown in Figure 14.7. Note the use of arrows to show the solutions of the individual inequalities. This device replaces the use of a lot of shading, which can be confusing if there are many inequalities in the system. In this book we show the intersection (solution) in color. You can show the intersection in your work in a variety of different ways, but a color highlighter works well. ◆

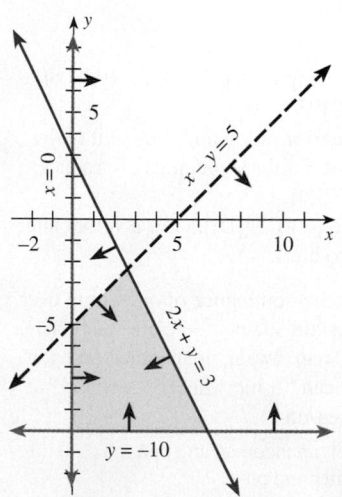

Figure 14.7 **System of inequalities**

Problem Set 14.5

1. **IN YOUR OWN WORDS** What is a system of linear inequalities?

2. **IN YOUR OWN WORDS** Describe a procedure for solving a system of linear inequalities.

Graph the solution of each system given in Problems 3–16.

3. **a.** $\begin{cases} x \geq 0 \\ y \leq 0 \end{cases}$ **b.** $\begin{cases} x \geq 0 \\ y \geq 0 \end{cases}$

4. **a.** $\begin{cases} x \leq 0 \\ y \leq 0 \end{cases}$ **b.** $\begin{cases} x \leq 0 \\ y \geq 0 \end{cases}$

5. $\begin{cases} x \geq 0 \\ y \geq 0 \\ x < 5 \\ y < 6 \end{cases}$ 6. $\begin{cases} x \geq 0 \\ y \geq 0 \\ x < 500 \\ y < 1{,}000 \end{cases}$

7. $\begin{cases} -10 < x < 6 \\ 2 \leq y \leq 5 \end{cases}$ 8. $\begin{cases} -4 \leq x \leq -2 \\ -5 \leq y \leq 9 \end{cases}$

9. $\begin{cases} 2x + y > 3 \\ 3x - y < 2 \end{cases}$ 10. $\begin{cases} y \leq 3x - 4 \\ y \geq -2x + 5 \end{cases}$

11. $\begin{cases} 3x - 2y \geq 6 \\ 2x + 3y \leq 6 \end{cases}$ 12. $\begin{cases} y - 5 \leq 0 \\ y \geq 0 \end{cases}$

13. $\begin{cases} x - 10 \leq 0 \\ x \geq 0 \end{cases}$ 14. $\begin{cases} y - 25 \leq 0 \\ y \geq 0 \end{cases}$

15. $\begin{cases} x - y \geq 0 \\ y \leq 0 \end{cases}$ 16. $\begin{cases} x + 5 \geq 0 \\ x \leq 0 \end{cases}$

Graph the solution of each system given in Problems 17–30.

17. $\begin{cases} -10 \leq x \\ x \leq 6 \\ -3 < y \\ y < 8 \end{cases}$ 18. $\begin{cases} -5 < x \\ 3 \geq x \\ 5 > y \\ 2 \leq y \end{cases}$

19. $\begin{cases} y \geq \frac{3}{4}x - 4 \\ y \leq -\frac{3}{4}x + 11 \\ x \geq 6 \end{cases}$ 20. $\begin{cases} y \geq \frac{3}{2}x + 3 \\ y \leq \frac{3}{2}x + 6 \\ 3 \leq y \leq 6 \end{cases}$

21. $\begin{cases} 5x + 2y \leq 30 \\ 5x + 2y \geq 20 \\ x \geq 0 \\ y \geq 0 \end{cases}$ 22. $\begin{cases} 5x - 2y + 30 \geq 0 \\ 5x - 2y + 20 \leq 0 \\ x \leq 0 \\ y \geq 0 \end{cases}$

23. $\begin{cases} 8x + 3y \leq 9 \\ y - 4 \geq -\frac{8}{3}(x + 2) \\ -5 \leq y \leq 3 \end{cases}$ 24. $\begin{cases} x + y - 9 \leq 0 \\ x + y + 3 \geq 0 \\ x - y \leq 7 \\ y - x \leq 5 \end{cases}$

25. $\begin{cases} x \geq 0 \\ y \geq 0 \\ x + y \leq 9 \\ 2x - 3y \geq -6 \\ x - y \leq 3 \end{cases}$ 26. $\begin{cases} x \geq 0 \\ y \geq 0 \\ x + y \leq 8 \\ y \leq 4 \\ x \leq 6 \end{cases}$

27. $\begin{cases} 2x + y \leq 8 \\ y \leq 5 \\ x - y \leq 2 \\ 3x - y > 5 \end{cases}$ 28. $\begin{cases} 2x + 3y \leq 30 \\ 3x + 2y \geq 20 \\ x \geq 0 \\ y \geq 0 \end{cases}$

29. $\begin{cases} 2x - 3y + 30 \geq 0 \\ 3x - 2y + 20 \leq 0 \\ x \leq 0 \\ y \geq 0 \end{cases}$ 30. $\begin{cases} x + y - 10 \leq 0 \\ x + y + 4 \geq 0 \\ x - y \leq 6 \\ y - x \leq 4 \end{cases}$

14.6 MODELING WITH LINEAR PROGRAMMING

This section gives us the opportunity to use mathematical modeling to solve an important type of real-life problem. The problem of interest is to maximize or minimize a linear expression subject to a set of limitations. This type of problem was first analyzed during World War II and developed into a modeling procedure known as **linear programming.** It is used to maximize profits, minimize costs, find efficient shipping schedules, minimize waste, secure the proper mix of ingredients, control inventories, and efficiently assign tasks to personnel.

We begin with some terminology. A **convex set** is a set that contains the line segment joining any two of its points. The expression to be maximized or minimized is called the **objective function,** and the limitations are called **constraints.** In linear programming, these constraints are specified by a system of linear inequalities whose solution forms a convex set S. Each point in the set S is called a **feasible solution,** and a point at which the objective function takes on a maximum or a minimum value is called an **optimum solution.**

Linear Programming Theorem

A linear expression in two variables

$$c_1 x + c_2 y$$

defined over a convex set S whose sides are line segments takes on its maximum value at a corner point of S and its minimum value at a corner point of S. If S is unbounded, there may or may not be an optimum value, but if there is, then it must occur at a corner point.

Procedure for Solving a Linear Programming Problem

To solve a linear programming problem:

Step 1. Find the objective function (the quantity to be maximized or minimized).

Step 2. Graph the constraints defined by a system of linear inequalities; this graph is called the set S.

Step 3. Find the corners of S; for each corner, this may require the solution of a system of two equations with two unknowns.

Step 4. Find the value of the objective function for the coordinates of each corner point. The largest value is the maximum; the smallest value is the minimum.

EXAMPLE 1

Maximize $T = 4x + 5y$ subject to: $\begin{cases} 2x + 5y \le 25 \\ 6x + 5y \le 45 \\ x \ge 0 \\ y \ge 0 \end{cases}$

Solution   The objective function is the function to be maximized or minimized; for this example, it is $T = 4x + 5y$.

Next, graph the set of constraints. This is the feasible region and is shown in Figure 14.8. The goal of this linear programming problem is to find that point in the feasible set that gives the largest possible value of T. Consider the following list of possible solutions:

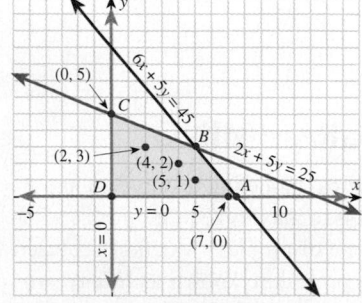

Figure 14.8 Graph of feasible solutions

	$T = 4x + 5y$
A point in the feasible set	*Value of objective function*
$(2, 3)$	$T = 4(2) + 5(3) = 23$
$(4, 2)$	$T = 4(4) + 5(2) = 26$
$(5, 1)$	$T = 4(5) + 5(1) = 25$
$(7, 0)$	$T = 4(7) + 5(0) = 28$
$(0, 5)$	$T = 4(0) + 5(5) = 25$

The point from the table that makes the objective function the largest is $(7, 0)$. But is this the largest for all feasible solutions? How about $(6, 1)$ or $(5, 3)$? The linear programming theorem tells us that the maximum value occurs at a corner point. The corner points are labeled *A, B, C,* and *D* on Figure 14.8. Some corner points can be found by inspection—for example, $D = (0, 0)$ and $C = (0, 5)$. Other corner points may require some work with boundary lines.

Use the equations (not the inequalities) for the boundaries.

Point *A*:   Solve the system $\begin{cases} y = 0 \\ 6x + 5y = 45 \end{cases}$

Solve by substitution to find $x = \frac{15}{2}$ and $y = 0$.

Point *B*:   Solve the system $\begin{cases} 2x + 5y = 25 \\ 6x + 5y = 45 \end{cases}$

Solve by adding to find $x = 5$ and $y = 3$.

We now have the coordinates of all the corner points. This is the *entire* list of all the points we need to check from the feasible set.

$$T = 4x + 5y$$

Corner point	Objective function
A: $(\frac{15}{2}, 0)$	$T = 4(\frac{15}{2}) + 5(0) = 30$
B: $(5, 3)$	$T = 4(5) + 5(3) = 35$
C: $(0, 5)$	$T = 4(0) + 5(5) = 25$
D: $(0, 0)$	$T = 4(0) + 5(0) = 0$

Look for the maximum value of *T* on *this* list; it is 35 so the maximum value of *T* is 35.   ◆

**Narendra Karmarkar
(1955 –   )**

In 1984, a 28-year-old mathematician at Lucent Technologies made a startling theoretical breakthrough in the solving of systems of equations, which often grow too vast and complex for even the powerful computers. Dr. Narendra Karmarkar devised a radically new procedure that avoids the inevitable computer slowdowns with linear programming problems. "This is a path-breaking result," said Dr. Ronald L. Graham, director of mathematical sciences for Lucent Technologies in Murray Hill, N.J. "Science has its moments of great progress, and this may well be one of them."

Real-life linear programming problems can involve about 800,000 variables, and Karmarkar's algorithm can solve such a problem in about 10 hours of computer time. This same problem would have taken weeks to solve if the usual process, known as the *simplex method*, were used.

EXAMPLE 2   **PÓLYA'S METHOD**

A farmer has 100 acres on which to plant two crops, corn and wheat, and the problem is to maximize the profit. There are several considerations; the first is the expense:

	Cost per acre	
	Corn	Wheat
seed	$ 12	$ 40
fertilizer	$ 58	$ 80
planting/care/harvesting	$ 50	$ 90
TOTAL	$120	$210

After the harvest, the farmer must store the crops while awaiting proper market conditions. Each acre yields an average of 110 bushels of corn or 30 bushels of wheat. The limitations of resources are as follows:

Available capital: $15,000
Available storage facilities: 4,000 bushels

If the net profit (after all expenses have been subtracted) per bushel of corn is $1.30 and for wheat is $2.00, how should the farmer plant the 100 acres to maximize the profits?

Solution   We use Pólya's problem-solving guidelines for this example.

Understand the Problem. First, you might try to solve this problem by using your intuition.

Plant 100 acres in wheat

Production: 100 acres @ $30 bu/acre = 3,000 bushels

Net profit: 3,000 bu × $2.00/bu = $6,000

Costs: 100 acres @ $210/acre = $21,000

These costs are more than the $15,000 available, so the farmer cannot plant 100 acres in wheat.

Plant 100 acres in corn

Production: 100 acres @ 110 bu/acre = 11,000 bushels

Net profit: 11,000 bu × $1.30/bu = $14,300

Costs: 100 acres @ $120/acre = $12,000

These costs can be met with the available $15,000, but the production of 11,000 bushels exceeds the available storage capacity of 4,000 bu.

Clearly, some mix of wheat and corn is necessary. Let us build a mathematical model.

Devise a Plan.

Let x = number of acres to be planted in corn;

y = number of acres to be planted in wheat.

With a mathematical model, we must make certain assumptions. For this example, we assume:

These first two assumptions (constraints) will apply in almost every linear programming model.

$x \geq 0$ The number of acres of corn cannot be negative.

$y \geq 0$ The number of acres of wheat cannot be negative.

$x + y \leq 100$ The amount of available land is 100 acres. We do not assume that $x + y = 100$, because it might be more profitable to leave some land unplanted.

EXPENSES ≤ 15,000

The total expenses cannot exceed $15,000. We also know

EXPENSES = EXPENSE FOR CORN + EXPENSE FOR WHEAT

= $120x + 210y$

Thus, this constraint (in terms of x and y) is

$120x + 210y \leq 15,000$

TOTAL YIELD ≤ 4,000

The total yield cannot exceed the storage capacity of 4,000 bushels. We also know:

TOTAL YIELD = YIELD FOR CORN + YIELD FOR WHEAT

= $110x + 30y$

Thus, this constraint (in terms of x and y) is

$110x + 30y \leq 4,000$

The farmer wants to maximize the profit. Let P = TOTAL PROFIT.

$$\text{TOTAL PROFIT} = \underbrace{\text{PROFIT FROM CORN}}_{\downarrow} + \underbrace{\text{PROFIT FROM WHEAT}}_{\downarrow}$$

$$= \underbrace{\text{CORN VALUE} \times \text{CORN AMOUNT}}_{} + \underbrace{\text{WHEAT VALUE} \times \text{WHEAT AMOUNT}}_{}$$

$$= 1.30 \times 110x \qquad\qquad + \qquad\qquad 2.00 \times 30y$$

$$= 143x + 60y$$

We have formulated the following linear programming problem:

Maximize: $P = 143x + 60y$

Subject to: $\begin{cases} x \geq 0 \\ y \geq 0 \\ x + y \leq 100 \\ 120x + 210y \leq 15{,}000 \\ 110x + 30y \leq 4{,}000 \end{cases}$

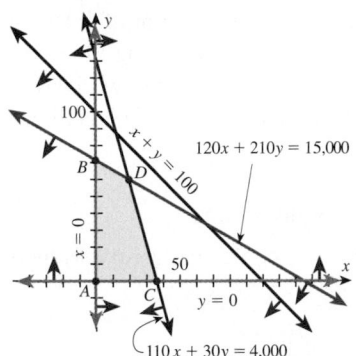

Figure 14.9 **Farmer problem**

Carry Out the Plan.
To solve this problem, we must graph the constraints (see Figure 14.9). Note that the corner points are labeled *A, B, C,* and *D.* Find the coordinates of these points:

$A(0, 0)$ By inspection

$B(0, \frac{500}{7})$ Solve the system $\begin{cases} 120x + 210y = 15{,}000 \\ x = 0 \end{cases}$

$C(\frac{400}{11}, 0)$ Solve the system $\begin{cases} 110x + 30y = 4{,}000 \\ y = 0 \end{cases}$

$D(20, 60)$ Solve the system $\begin{cases} 110x + 30y = 4{,}000 \\ 120x + 210y = 15{,}000 \end{cases}$

Use the linear programming theorem to check these corner points:

Corner point	Objective function, $P = 143x + 60y$
$A(0, 0)$	$P = 143(0) + 60(0) = 0$
$B(0, \frac{500}{7})$	$P = 143(0) + 60(\frac{500}{7}) \approx 4{,}286$
$C(\frac{400}{11}, 0)$	$P = 143(\frac{400}{11}) + 60(0) = 5{,}200$
$D(20, 60)$	$P = 143(20) + 60(60) = 6{,}460$

Look Back. The maximum value of P is at $(20, 60)$. This means that, to maximize the profit subject to the constraints, the farmer should plant 20 acres in corn, plant 60 acres in wheat, and leave 20 acres unplanted. ◆

Notice from the graph in Example 2 that some of the constraints could be eliminated from the problem and everything else would remain unchanged. For example, the boundary $x + y = 100$ was not necessary in finding the maximum value of $P.$ Such a condition is said to be a **superfluous constraint.** It is not uncommon to have superfluous constraints in a linear programming problem. Suppose, however, that the farmer in Example 2 contracted to have the grains stored at a neighboring farm and now the contract calls for *at least* 4,000 bushels to be stored. This change from $110x + 30y \leq 4{,}000$ to $110x + 30y \geq 4{,}000$ *now* makes the condition $x + y \leq 100$ important to the solution of the problem (see Problem 27). Therefore, you must be careful about superfluous constraints even though they do not affect the solution at the present time.

The next example is solved more succinctly to show you the way your work will probably look.

EXAMPLE 3   PÓLYA'S METHOD

The Sticky Widget Company makes two types of widgets: regular and deluxe. Each widget is produced at a station consisting of a machine and a person who finishes the widgets by hand. The regular widget requires 2 hr of machine time and 1 hr of finishing time. The deluxe widget requires 3 hr of machine time and 5 hr of finishing time. The profit on the regular widget is $25; on the deluxe widget it is $30. If the workday is 8 hr, how many of each type of widget should be produced at each station to maximize the profit?*

Solution   We use Pólya's problem-solving guidelines for this example.

Understand the Problem. We are trying to find the number of regular and deluxe widgets that should be produced to maximize the profits.

Devise a Plan.

Let x = NUMBER OF REGULAR WIDGETS PRODUCED

           y = NUMBER OF DELUXE WIDGETS PRODUCED

Maximize profit: $P = 25x + 30y$

Subject to: $\begin{cases} x \geq 0 \\ y \geq 0 \\ 2x + 3y \leq 8 \\ x + 5y \leq 8 \end{cases}$ *The workday is no more than 8 hr.*

The set of feasible solutions is found by graphing this system of inequalities, as shown in Figure 14.10. The corner points are found by considering the intersection of the boundary lines.

Carry Out the Plan. The linear programming theorem requires the values summarized with the following table.

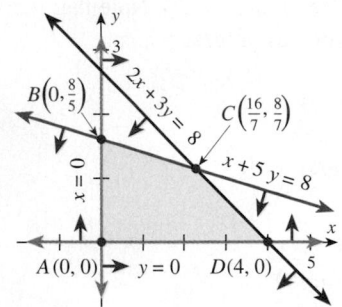

Figure 14.10 **Maximizing profit**

Corner point	System to solve	Solution	Objective function $25x + 30y$	Comment
A	$\begin{cases} x = 0 \\ y = 0 \end{cases}$	$(0, 0)$	$25(0) + 30(0) = 0$	**minimum**
B	$\begin{cases} x = 0 \\ x + 5y = 8 \end{cases}$	$(0, \frac{8}{5})$	$25(0) + 30(\frac{8}{5}) = 48$	
C	$\begin{cases} 2x + 3y = 8 \\ x + 5y = 8 \end{cases}$	$(\frac{16}{7}, \frac{8}{7})$	$25(\frac{16}{7}) + 30(\frac{8}{7}) \approx 91.43$	
D	$\begin{cases} y = 0 \\ 2x + 3y = 8 \end{cases}$	$(4, 0)$	$25(4) + 30(0) = 100$	**maximum**

*It is difficult to set up textbook problems to model real-life problems, but our goal in this book is to teach you a *process* of problem solving. Remember that, in real life, you cannot be sure of what information to use and what information is not necessary. This problem, for example, would be stated in real life as: Maximize the profit. All of the other information would be part of the assumptions you would gather for the solution of the problem. Deciding on the assumptions often requires a great deal of work when solving *real-life* problems.

Look Back. Profits are maximized if only regular widgets are produced. The company should produce four regular widgets per day at each station.   ◆

The method discussed here can be generalized to higher dimensions by means of more sophisticated methods for solving these linear programming problems. The general method for solving a linear programming problem is called the *simplex method* and is usually discussed in a course called linear programming or finite mathematics.

Problem Set 14.6

LEVEL 1

1. IN YOUR OWN WORDS What is a linear programming problem?

2. IN YOUR OWN WORDS What is an objective function? What do we mean by constraints?

3. Decide whether the given point is a feasible solution for the constraints

$$\begin{cases} x \geq 0 \\ y \geq 0 \\ 3x + 2y \leq 10 \\ 2x + 4y \leq 8 \end{cases}$$

a. (1, 3)   **b.** (?, 1)   **c.** (1, 2)
d. (−1, 4)   **e.** (2, 2)   **f.** (0, 4)

4. Decide whether the given point is a corner point for the constraints

$$\begin{cases} x \geq 0 \\ y \geq 0 \\ 2x + 3y \geq 120 \\ 2x + y \geq 80 \end{cases}$$

a. (0, 0)   **b.** (0, 80)   **c.** (80, 0)
d. (60, 0)   **e.** (30, 20)   **f.** (20, 30)

LEVEL 2

Find the corner points for the set of feasible solutions for the constraints given in Problems 5–16.

5. $\begin{cases} x \geq 0 \\ y \geq 0 \\ 2x + y \leq 12 \\ x + 2y \leq 9 \end{cases}$

6. $\begin{cases} x \geq 0 \\ y \geq 0 \\ 2x + 5y \leq 20 \\ 2x + y \leq 12 \end{cases}$

7. $\begin{cases} x \geq 0 \\ y \geq 0 \\ 3x + 2y \leq 12 \\ x + 2y \leq 8 \end{cases}$

8. $\begin{cases} x \geq 0 \\ y \geq 0 \\ x \leq 10 \\ y \leq 8 \\ 3x + 2y \geq 12 \end{cases}$

9. $\begin{cases} x \geq 0 \\ y \geq 0 \\ x + y \leq 8 \\ y \leq 4 \\ x \leq 6 \end{cases}$

10. $\begin{cases} x \geq 0 \\ y \geq 0 \\ x + y \geq 6 \\ -2x + y \geq -16 \\ y \leq 9 \end{cases}$

11. $\begin{cases} x \geq 0 \\ y \geq 0 \\ 3x + 2y \leq 8 \\ x + 5y \leq 8 \end{cases}$

12. $\begin{cases} x \geq 0 \\ y \geq 0 \\ x \leq 8 \\ y \geq 2 \\ x + y \leq 10 \\ x \leq 3y \end{cases}$

13. $\begin{cases} x \geq 0 \\ y \geq 0 \\ 10x + 5y \geq 200 \\ 2x + 5y \geq 100 \\ 3x + 4y \geq 120 \end{cases}$

14. $\begin{cases} x \geq 0 \\ y \geq 0 \\ x + y \leq 9 \\ 2x - 3y \geq -6 \\ x - y \leq 3 \end{cases}$

15. $\begin{cases} x \geq 0 \\ y \geq 0 \\ 2x + y \geq 8 \\ y \leq 5 \\ x - y \leq 2 \\ 3x - 2y \geq 5 \end{cases}$

16. $\begin{cases} x \geq 0 \\ y \geq 0 \\ 2x + y \geq 8 \\ x - 2y \leq 7 \\ x - y \geq -3 \\ x \leq 9 \end{cases}$

Find the optimum value for each objective function given in Problems 17–21.

17. Maximize $W = 30x + 20y$ subject to the constraints of Problem 5.

18. Maximize $T = 100x + 10y$ subject to the constraints of Problem 5.

19. Maximize $P = 100x + 100y$ subject to the constraints of Problem 7.

20. Minimize $K = 6x + 18y$ subject to the constraints of Problem 8.

21. Minimize $A = 2x - 3y$ subject to the constraints of Problem 9.

LEVEL 3

Write a linear programming model, including the objective function and the set of constraints, for Problems 22–25. DO NOT SOLVE, but be sure to define all your variables.

22. The Wadsworth Widget Company manufactures two types of widgets: regular and deluxe. Each widget is produced at a station consisting of a machine and a person who finishes each widget by hand. The regular widget requires 3 hr of machine time and 2 hr of finishing time. The deluxe widget requires 2 hr of machine time and 4 hr of finishing time. The profit on the regular widget is $25; on the deluxe widget it is $30. If the workday is 8 hours, how many of each type of widget should be produced at each station per day to maximize the profit?

23. A convalescent hospital wishes to provide, at a minimum cost, a diet that has a minimum of 200 g of carbohydrates, 100 g of protein, and 20 g of fat per day. These requirements can be met with two foods, A and B:

Food	Carbohydrates	Protein	Fats
A	10 g	2 g	3 g
B	5 g	5 g	4 g

If food A costs $0.29 per unit and food B costs $0.15 per unit, how many units of each food should be purchased for each patient per day to meet the minimum requirements at the lowest cost?

24. Karlin Enterprises manufactures two games. Standing orders require that at least 24,000 space-battle games and 5,000 football games be produced per month. The Gainesville plant can produce 600 space-battle games and 100 football games per day; the Sacramento plant can produce 300 space-battle games and 100 football games per day. If the Gainesville plant costs $20,000 per day to operate and the Sacramento factory costs $15,000 per day, find the number of days per month each factory should operate to minimize the cost. (Assume each month is 30 days.)

25. Brown Bros., Inc., is an investment company doing an analysis of the pension fund for a certain company. The fund has a maximum of $10 million to invest in two places: no more than $8 million in stocks yielding 12%, and at least $2 million in long-term bonds yielding 8%. The stock-to-bond investment ratio cannot be more than 3 to 1. How should Brown Bros. advise its client so that the investments yield the maximum yearly return?

PROBLEM SOLVING

Solve the linear programming problems in Problems 26–30.

26. Suppose the net profit per bushel of corn in Example 2 increased to $2.00 and the net profit per bushel of wheat dropped to $1.50. Maximize the profit if the other conditions in the example remain the same.

27. Suppose the farmer in Example 2 contracted to have the grain stored at a neighboring farm and the contract calls for at least 4,000 bushels to be stored. How many acres should be planted in corn and how many in wheat to maximize profit if the other conditions in Example 2 remain the same?

28. The Thompson Company manufactures two industrial products, standard ($45 profit per item) and economy ($30 profit per item). These items are built using machine time and manual labor. The standard product requires 3 hr of machine time and 2 hr of manual labor. The economy model requires 3 hr of machine time and no manual labor. If the week's supply of manual labor is limited to 800 hr and machine time to 15,000 hr, how much of each type of product should be produced each week to maximize the profit?

29. The following carbohydrate information is given on the side of the respective cereal boxes (for 1 oz of cereal with $\frac{1}{2}$ cup of whole milk):

	Starch and related carbohydrates	Sucrose and other sugars
Kellogg's Corn Flakes	23 g	7 g
Post Honeycombs	14 g	17 g

What is the minimum cost to receive at least 322 g starch and 119 g sucrose by consuming these two cereals if Corn Flakes cost $0.07 per ounce and Honeycombs cost $0.19 per ounce?

30. Your broker tells you of two investments she thinks are worthwhile. She advises a new issue of Pertec stock, which should yield 20% over the next year, and then to balance your account she advises Campbell Municipal Bonds with a 10% yearly yield. The bond-to-stock ratio should not be greater than 3 to 1. If you have no more than $100,000 to invest and do not want to invest more than $70,000 in Pertec or less than $20,000 in bonds, how much should be invested in each to maximize your return?

CHAPTER SUMMARY

"How can students compete in a mathematical society when they leave school knowing so little mathematics?"

Lester Thurow, *in Everybody Counts*

IMPORTANT TERMS

Addition of matrices [14.4]
Addition method [14.1]
Additive inverse [14.4]
Array [14.3]
Associative property [14.4]
Augmented matrix [14.3]
Column [14.3]
Communication matrix [14.4]
Commutative property [14.4]
Constraint [14.6]
Convex set [14.6]
Demand [14.2]
Dependent system [14.1]
Diagonal form [14.3]
Dimension [14.3]
Distributive property [14.4]
Double subscripts [14.3]
Elementary row operations [14.3]
Equal matrices [14.4]
Equilibrium point [14.2]
Equivalent matrices [14.3]
Equivalent systems [14.1]

Feasible solution [14.6]
Gauss–Jordan elimination [14.3]
Graphing method [14.1]
Identity matrix [14.4]
Inconsistent system [14.1]
Inverse matrix [14.4]
Inverse property [14.4]
Linear combination method [14.1]
Linear programming [14.6]
Linear system [14.1]
Main diagonal [14.4]
Matrix [14.3]
Matrix equation [14.3, 14.4]
Multiplication of matrices [14.4]
Multiplicative inverse [14.4]
Nonconformable matrices [14.4]
Nonsingular matrix [14.4]
Objective function [14.6]
Optimum solution [14.6]
Order [14.3]
Pivot [14.3]
Pivot row [14.3]

Pivoting [14.3]
Row [14.3]
Row+ [14.3]
Row-reduced form [14.3]
RowSwap [14.3]
Scalar [14.3]
Scalar multiplication [14.4]
Simultaneous solution [14.1]
Singular matrix [14.4]
Square matrix [14.3]
Subscript [14.3]
Substitution method [14.1]
Subtraction of matrices [14.4]
Superfluous constraint [14.6]
Supply [14.2]
System of equations [14.1]
System of inequalities [14.5]
Target row [14.3]
*Row [14.3]
*Row+ [14.3]
Zero matrix [14.4]

TYPES OF PROBLEMS

Solve systems of equations by graphing, substitution, or addition, as directed. [14.1]
Solve systems of equations by selecting the most appropriate method. [14.1]
Solve applied problems, including coin problems, combining rates, supply and demand, and mixture problems. [14.2]
Perform elementary row operations on a given matrix. [14.3]
Solve systems of equations by the Gauss–Jordan method. [14.3]
Carry out matrix operations, including finding the inverse of a given matrix. [14.4]
Solve a system of equations using the inverse matrix method. [14.4]
Solve a system of inequalities. [14.5]
Decide whether a given point is a feasible solution for a set of constraints. [14.6]
Decide whether a given point is a corner point for a set of constraints. [14.6]
Find the corner points for a set of feasible solutions. [14.6]
Maximize or minimize an objective function subject to a set of constraints. [14.6]
Solve applied problems using a linear programming model. [14.6]

CHAPTER 12 REVIEW QUESTIONS

In Problems 1–4, perform the indicated matrix operations, if possible, where

$$[A] = \begin{bmatrix} 1 & 0 \\ 2 & -1 \end{bmatrix} \quad [B] = \begin{bmatrix} 2 & -1 & 0 \\ 1 & 0 & 1 \end{bmatrix} \quad [C] = \begin{bmatrix} 2 & 0 \\ 1 & 2 \\ -1 & 1 \end{bmatrix} \quad [D] = \begin{bmatrix} 0 & 1 & 0 \\ -1 & 0 & 0 \\ 0 & 1 & -1 \end{bmatrix}$$

1. $[C][B] - 3[D]$ **2.** $[A][B][C]$ **3.** $[B][A][D]$ **4.** $[C][A][B]$

Find the inverse of each matrix in Problems 5–6.

5. $\begin{bmatrix} 2 & 1 \\ -\frac{3}{2} & -\frac{1}{2} \end{bmatrix}$ **6.** $\begin{bmatrix} 1 & 3 & 3 \\ 1 & 4 & 3 \\ 1 & 3 & 4 \end{bmatrix}$

Solve the systems in Problems 7–11 by the indicated method.

7. By graphing: $\begin{cases} 2x - y = 2 \\ 3x - 2y = 1 \end{cases}$ **8.** By addition: $\begin{cases} x + 3y = 3 \\ 4x - 6y = -6 \end{cases}$

9. By substitution: $\begin{cases} y = 1 - 2x \\ 5x + 2y = 1 \end{cases}$ **10.** By Gauss–Jordan: $\begin{cases} x + y + z = 2 \\ x + 2y - 2z = 1 \\ x + y + 3z = 4 \end{cases}$

11. By using an inverse matrix: $\begin{cases} 2x + y = 13 \\ -\frac{3}{2}x - \frac{1}{2}y = 10 \end{cases}$

12. Graph the solution of the system: $\begin{cases} 2x - y + 2 \le 0 \\ 2x - y + 12 \ge 0 \\ 3x + 2y + 1 > 0 \\ 3x + 2y - 18 < 0 \end{cases}$

13. To manufacture a certain alloy, it is necessary to use 33 oz of metal A and 56 oz of metal B. It is cheaper for the manufacturer if she buys and mixes two products that come as metal bars: Product I, each bar of which contains 3 oz of metal A and 5 oz of metal B; Product II, each bar of which contains 4 oz of metal A and 7 oz of metal B. How many bars of each product should she use to produce the desired alloy?

14. A manufacturer of auto accessories uses three basic parts, A, B, and C, in its three products, in the following proportions:

	A	B	C
Product I	2	1	1
Product II	2	2	1
Product III	3	2	2

The inventory shows 1,250 of part A, 900 of part B, and 750 of part C on hand. How many of each product may be manufactured using all the inventory on hand?

15. A farmer has 500 acres on which to plant two crops: corn and wheat. To produce these crops, there are certain expenses:

| | Cost per acre | |
	Corn	Wheat
Seed	$ 12	$ 10
Fertilizer	$ 58	$ 20
Planting/care/harvesting	$ 50	$ 30
TOTAL	$120	$ 60

After the harvest, the farmer must store the crops while awaiting proper market conditions. Each acre yields an average of 100 bushels of corn or 40 bushels of wheat. The farmer has available capital of $24,000 and is not able to store more than 18,000 bushels. If the net profit (after all expenses have been subtracted) per bushel of corn is $2.10 and for wheat is $2.50, how should the farmer plant the 500 acres to maximize the profit, and what is the maximum profit?

GROUP RESEARCH

Working in small groups is typical of most work environments, and learning to work with others to communicate specific ideas is an important skill. Work with three or four other students to submit a single report based on each of the following questions.

G54. Team A beats F, and ties C; team B beats A, C, and F; team C beats E and F; team D ties A and beats F; team E beats A and F; team F beats C and ties D. Rank these teams. If two teams tie, enter 0.5 in the communication matrix instead of 1.

G55. Suppose your group conducts an experiment at a local department store. You walk up a rising escalator and you take one step per second to reach the top in 20 seconds. Next, you walk up the same rising escalator at the rate of two steps per second and this time it takes 32 steps. How many steps would be required to reach the top on a stopped escalator?

G56. Two ranchers sold a herd of cattle and received as many dollars for each animal as there were cattle in the herd. With the money they bought a flock of sheep at $10 a head and then a lamb with the rest of the money (less than $10). Finally, they divided the animals between them, with one rancher obtaining an extra sheep and the other the lamb. The rancher who got the lamb was given his friend's new watch as compensation. What is the value of the watch?

G57. Suppose your group has just been hired by a company called Alco. You are asked to analyze its operations and make some recommendations about how it can comply at a minimum cost with recent orders of the Environmental Protection Agency (EPA). To prepare your report, you study the operation and obtain the following information:

- Alco Cement Company produces cement.

- The EPA has ordered Alco to reduce the amount of emissions released into the atmosphere during production.

- Alco wants to comply, but wants to do so at the least possible cost.

- Present production is 2.5 million barrels of cement, and 2 pounds of dust are emitted for every barrel of cement produced.

- The cement is produced in kilns that are presently equipped with mechanical collectors.

- To reduce the emissions to the required level, the mechanical collectors must be replaced either by four-field electrostatic precipitators, which would reduce emission to 0.5 pound of dust per barrel of cement, or by five-field precipitators, which would reduce emission to 0.2 pound per barrel.

- The capital and operating costs for the four-field precipitator are 14¢ per barrel of cement produced; for the five-field precipitator, costs are 18¢ per barrel.

- To comply with the EPA, Alco must reduce particulate emission by at least 4.2 million pounds.*

Use mathematical modeling to write your paper. Mathematical modeling involves creating equations and procedures to make predictions about the real world. Typical textbook problems focus on limited, specific skills, but in the real world you need to sift through the given information to decide what information you need and what information you do not need. You may need to do some research to gather data not provided.

G58. Knot Theory.   Get a piece of string with two free ends, and tie those ends together with a knot. Some knots that you can tie will hold the ends of the string together and other knots will not (no pun intended!). In mathematics, there is a branch of mathematics known as *knot theory*. Mathematically, a knot is defined as a *closed piecewise linear curve* in $\mathbb{R}^3$. Two or more knots together is known as a *link*. Knots can be cataloged according to the number of crossings (ignoring mirror reflections). There is only one knot with crossing number three (called the *cloverleaf knot*), one knot with crossing number four, two with crossing number five, and three with crossing number six.

a. How many knots are possible with crossing number seven?

b. How many knots are possible with crossing number eight?

Write a paper on knot theory.

 See **www.mathnature.com** (Section 14.6, reference links) for an excellent site with links.

---

* This research project is adapted from R. E. Kohn, "A Mathematical Programming Model for Air Pollution Control," *Science and Mathematics,* June 1969, pp. 487–499.

INDIVIDUAL RESEARCH PROBLEM

Learning to use sources outside your classroom and textbook is an important skill, and here are some ideas for extending some of the ideas in this chapter. You can find references to these projects in the library or at

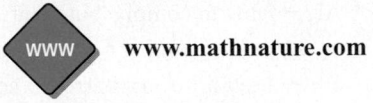 **www.mathnature.com**

PROJECT 14.1 HISTORICAL QUESTION In the Boston Museum of Fine Arts is a display of carefully made stone cubes found in the ruins of Mohenjo-Daro of the Indus. The stones are a set of weights that exhibit the binary pattern, 1, 2, 4, 8, 16, ... The fundamental unit displayed is just a bit lighter than the ounce in the U.S. measurement system. The old European standard of 16 oz for 1 pound may be a relic of the same idea. Write a paper showing how a set of such stones can successfully be used to measure any reasonable given weight of more than one unit.

15 The Nature of Networks and Graph Theory

There is nothing in the world except empty space. Geometry bent one way here describes gravitation. Rippled another way somewhere else it manifests all the qualities of an electromagnetic wave. Excited at still another plane, the magic material that is space shows itself as a particle. There is nothing that is foreign or "physical" immersed in space. Everything that is, is constructed out of geometry.

JOHN A. WHEELER

CONTENTS

IMPORTANT IDEAS

Königsberg bridge problem [15.1]
Euler's circuit theorem [15.1]
Hamiltonian cycles and the traveling salesperson problem (TSP) [15.1]
Kruskal's algorithm [15.2]
Topologically equivalent figures, four-color problem, and fractals [15.3]

OVERVIEW

When we use the word "network" today we probably think of a computer network, but we use networks every time we make a phone call, send an e-mail or "snail mail," drive on an interstate, or fly in an airplane.

In this chapter we look at mathematical ideas called *circuits, cycles,* and *trees,* which are part of geometry known was *graph theory.* When used in the context of graph theory, the word *graph* means something different from the way we used graph in referring to the coordinate plane or the way we used graph to represent statistical data.

Even though these topics sound abstract, they have many interesting and useful applications such as finding the minimum cost of traveling to a number of locations (known as the *traveling salesperson problem*), installing an irrigation system, using a search engine on the Internet, or even coloring maps or making realistic-looking original landscapes in movies.

Many of the applications of this chapter are part of a new branch of mathematics (new in the sense that it began in the late 1930s) and grew out of mathematical problems associated with World War II. This branch of mathematics, called **operations research** (or *operational research*), deals with the application of scientific methods to management decision making, especially for the allocation of resources. Some examples include forecasting water pollution or predicting the scope of the AIDS epidemic.

BOOK REPORTS

Write a 500-word report on one of the following books:
The Traveling Salesman Problem: A Guided Tour of Combinatorial Optimization, E. L. Lawler, J. K. Lenstra, A. H. G. Rinnooy Kan, D. B. Shmoys (New York: John Wiley and Sons, 1987).
Graph Theory and Its Applications, Jonathan Gross and Jay Yellen (Boca Raton: CRC Press, LLC, 1998).

LINKS

Individual Projects

Group Projects

Research Links

www.mathnature.com

15.1 EULER CIRCUITS AND HAMILTONIAN CYCLES

Euler Circuits

In the 18th century, in the German town of Königsberg (now a Russian city), a popular pastime was to walk along the bank of the Pregel River and cross over some of the seven bridges that connected two islands, as shown in Figure 15.1.

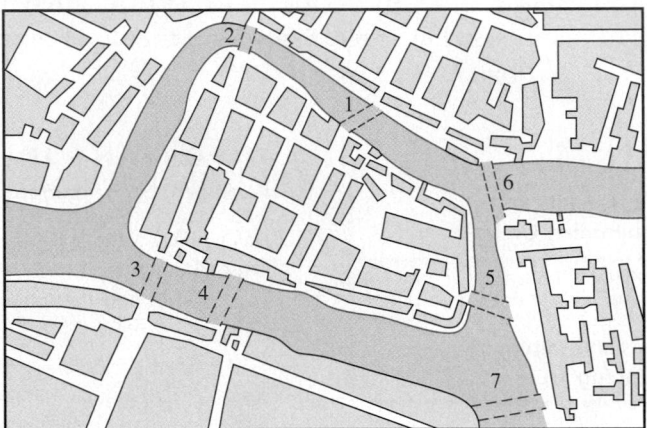

Figure 15.1 **Königsberg bridges**

One day a native asked a neighbor this question, "How can you take a walk so that you cross each of our seven bridges once and only once and end up where you started?" The problem intrigued the neighbor, and soon caught the interest of many other people of Königsberg as well. Whenever people tried it, they ended up either not crossing a bridge at all or else crossing one bridge twice. This problem was brought to the attention of the Swiss mathematician Leonhard Euler, who was serving at the court of the Russian empress Catherine the Great in St. Petersburg. The method of solution we discuss here was first developed by Euler, and led to the development of two major topics in geometry. The first is *networks,* which we discuss in this section, and the second is *topology,* which we discuss in Section 15.3.

We will use Pólya's problem-solving method for the Königsberg bridge problem.

Understand the Problem. To understand the problem, Euler began by drawing a diagram for the problem, as shown in Figure 15.2a.

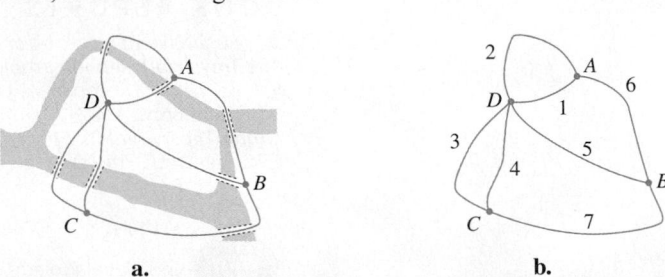

Figure 15.2 **Königsberg bridge problem**

Next, Euler used one of the great problem-solving procedures—namely, to change the conceptual mode. That is, he **let the land area be represented as points (sometimes called *vertices* or *nodes*), and let the bridges be represented by arcs or line**

segments (sometimes called *edges*) connecting the given points. As part of understanding the problem, we can do what Euler did—we can begin by tracing a diagram
like the one shown in Figure 15.2b.

Devise a Plan. To solve the bridge problem, we need to draw the figure without lifting
the pencil from the paper. Figures similar to the one in Figure 15.2b are called **networks** or **graphs.** In a network, the points where the line segments meet (or cross) are
called **vertices,** and the lines representing bridges are called **edges** or **arcs.** Each separated part of the plane formed by a network is called a **region.**

EXAMPLE 1

Complete the table for each of the given networks.

a.    b.    c.

d.    e.    f.

Solution

Graph	Edges (E)	Vertices (V)	Regions (R)	$V + R - 2$
a.	3	3	2	3
b.	4	3	3	4
c.	5	3	4	5
d.	4	4	2	4
e.	5	4	3	5
f.	6	4	4	6

Note that $V + R - 2 = E$. Do you think this will always be true?   ◆

   A network is said to be **traversable** if it can be traced in one sweep without lifting
the pencil from the paper and without tracing the same edge more than once. Vertices
may be passed through more than once. The **degree** of a vertex is the number of edges
that meet at that vertex.

EXAMPLE 2

List the number of edges and the degree of each vertex shown in Example 1. Find the
sum of the degrees of the vertices, and tell whether each network is traversable.

Solution

	Number of edges	*Degree of each vertex*	*Sum*	*Traversable*
a.	3	2; 2; 2	6	yes
b.	4	3; 2; 3	8	yes
c.	5	4; 3; 3	10	yes
d.	4	2; 2; 2; 2	8	yes
e.	5	2; 3; 2; 3	10	yes
f.	6	3; 3; 3; 3	12	no
◆

First, note that *the sum of the degrees of the vertices in Example 2 equals twice the number of edges.* Do you see why this must always be true? Consider any graph. Each edge must be connected at both ends, so the sum of all of those ends must be twice the number of vertices.

Now, consider a second observation regarding traversability. It is assumed that you worked Example 2 by actually tracing out the networks. However, a more complicated network, such as the Königsberg bridge problem, will require some analysis. The goal is to begin at some vertex, travel on each edge exactly once, and then return to the starting vertex. Such a path is called an **Euler circuit.** We can now rephrase the Königsberg bridge problem: "Does the network in Figure 15.2 have an Euler circuit?"

To answer this question, we will follow Euler's lead and classify vertices. Vertex A in Figure 15.2 is degree 3, so the vertex A is called an **odd vertex.** In the same way, D is an odd vertex, because it is 5th degree. A vertex with even degree is called an **even vertex.** Euler discovered that only a certain number of odd vertices can exist in any network if you are to travel it in one journey without retracing any edge. You may start at any vertex and end at any other vertex, as long as you travel the entire network. Also the network must connect each point (this is called a **connected network**).

Let's examine networks more carefully and look for a pattern, as shown in Table 15.1.

Table 15.1 Arrivals and departures for networks

Number of arcs at a vertex	Description	Possibilities
1	1 departure (starting point) 1 arrival (ending point)	
2	1 arrival (arrive then depart) and 1 departure (depart then arrive)	
3	1 arrival, 2 departures: (depart–arrive–depart; starting point) 2 arrivals, 1 departure: (arrive–depart–arrive; ending point)	
4	two arrivals, 2 departures	
5	2 arrivals, 3 departures (starting point) 3 arrivals, 2 departures (ending point)	

We see that, if the vertex is odd, then it must be a starting point or an ending point. What is the largest number of starting and ending points in any network? [*Answer:* Two—one starting point and one ending point.] This discussion leads us to **devise a plan,** which we now state without proof.

Count the number of odd vertices:

If there are no odd vertices, the network is traversable and any point may be a starting point. The point selected will also be the ending point.

If there is one odd vertex, the network is not traversable. A network cannot have only one starting or ending point without the other.

If there are two odd vertices, the network is traversable; one odd vertex must be a starting point and the other odd vertex must be the ending point.

If there are more than two odd vertices, the network is not traversable. A network cannot have more than one starting point and one ending point.

Carry Out the Plan.   Classify the vertices; there are four odd vertices, so the network is not traversable.

Look Back. We have solved the Königsberg bridge problem, but you should note that saying it cannot be done is not the same thing as saying "I can't do the problem." We can do the problem, and the solution is certain.

We summarize this investigation.

Euler's Circuit Theorem

Every vertex on a graph with an Euler circuit has an even degree, and, conversely, if in a connected graph every vertex has an even degree, then the graph has an Euler circuit.

EXAMPLE 3

Which of the following networks have an Euler circuit? Do not answer by trial and error, but by analyzing the number of odd vertices.

a.

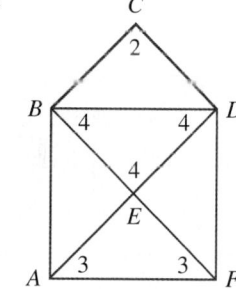

b.

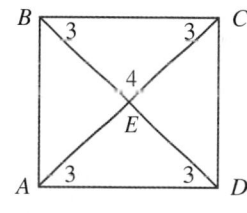

Solution

a. I first saw this network as a child's puzzle in elementary school. It has four even vertices (*B, C, D,* and *E*) and two odd vertices (*A* and *F*), and it is therefore traversable. To traverse it, you must start at *A* or *F* (that is, at an odd vertex). The path is shown. Note that we do not end at the beginning point, so it does not have an Euler circuit (even though it is traversable).

b. This network has one even vertex and four odd vertices, so it is not traversable, and does not have an Euler circuit. ◆

Applications of Euler Circuits

Euler circuits have a wide variety of applications. We will mention a few.

- *Supermarket problem*   Set up the shelves in a market or convenience store so that it is possible to enter the store at the door and travel in each aisle exactly once (once and only once) and leave by the same door.

- *Police-patrol problem*   Suppose a police car needs to patrol a gated subdivision, and would like to enter the gate, cruise all the streets exactly once, and then leave by the same gate.

- *Floor-plan problem*   Suppose you have a floor plan of a school (or any other building) and want to deliver a message to each classroom by going through each classroom doorway exactly once.

- *Water-pipe problem*   Suppose you have a network of water pipes, and you wish to inspect the pipeline. Can you pass your hand over each pipe exactly once without lifting your hand from a pipe, and without going over some pipe a second time?

We will examine one of these applications, and leave the others for the problem set. Let's look at the *floor-plan problem*. This problem, which is related to the Königsberg bridge problem, involves taking a trip through all the rooms and passing through each door only once. There is, however, one important difference between these two problems. The Königsberg bridge problem requires an Euler circuit, but the floor-plan problem does not. In other words, with the bridges we must end up where we started, but the floor-plan problem seeks only traversability. Let's label the rooms in Figure 15.3a as *A, B, C, D, E,* and *F.* Rooms *A, C, E,* and *F* have two doors, and rooms *B* and *D* have three doors; in Figure 15.3b, it looks as if there are five rooms, but since there are doors that lead to the "outside," we must count the outside as a room. So this figure also has six rooms labeled *A, B, C, D, E,* and *F.* Rooms *A, B,* and *C* each have 5 doors, rooms *D* and *E* each have 4 doors, and room *F* has 9 doors.

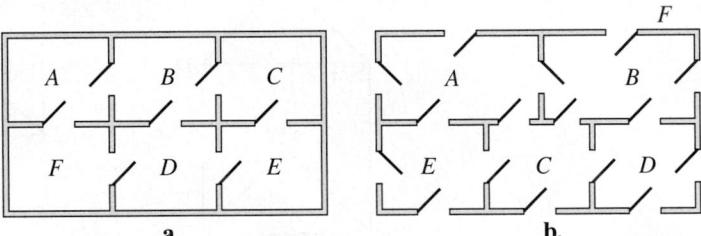

Figure 15.3 **Floor-plan problem**

Make a conjecture about the solution to this problem. If there are no rooms with an odd number of doors, then it will be traversable. If there are two rooms with an odd number of doors, then it will be traversable: Start in one of these rooms and end up in the other.

EXAMPLE 4   **PÓLYA'S METHOD**

Solve the floor-plan problems:

a.   **b.**

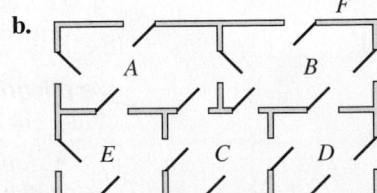

Solution   We use Pólya's problem-solving guidelines for this example.

Understand the Problem.   The floor-plan problem asks the question: "Can we travel into each room and pass through every door once?"

Devise a Plan.   Classify each room as even or odd, according to the number of doors in that room. It will be possible if there are no rooms with an odd number of doors, or if there are exactly two rooms with an odd number of doors.

Carry Out the Plan.

a. There are six rooms, and rooms B and D are odd, so this floor plan can be traversed. The solution requires that we begin in either room B or room D, and finish in the other.

b. There are six rooms, and rooms A, B, C, and F are odd (with D and E even). Since there are more than two odd rooms, this floor plan cannot be traversed. If one of the doors connecting two of the odd rooms is blocked, then the floor plan could be traversed.

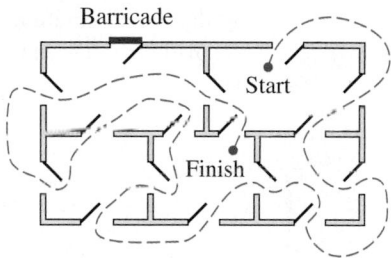

Look Back.   We can check our work by actually drawing possible routes, as shown for part **b** above.   ◆

Hamiltonian Cycles

One application that cannot be solved using Euler circuits is the so-called **traveling salesperson problem:** A salesperson starts at home and wants to visit several cities without going through any city more than once, and then return to the starting city. This problem is so famous with so many people working on its solution that it is often referred to in the literature as **TSP.** The salesperson would like to do this in the most efficient way (that is, least distance, least time, smallest cost, . . .). To answer this question, we reverse the roles of the vertices and edges of an Euler circuit. Now, we ask whether we can visit each vertex exactly once and end at the original vertex. Such a path is called a **Hamiltonian cycle.**

EXAMPLE 5

Find a Hamiltonian cycle for the given network.

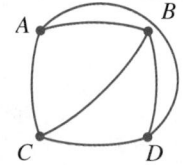

Solution   Note that this network is the one given in Example 1f. We found that there was *not* an Euler circuit for this network. On the other hand, it is easy to find a Hamiltonian cycle:

$$A \to C \to B \to D \to A$$

◆

It seems as if the problem of deciding whether a network has a Hamiltonian cycle should have a solution similar to that of the Euler circuit problem, but such is not the

See www.mathnature.com for links for solving the TSP.

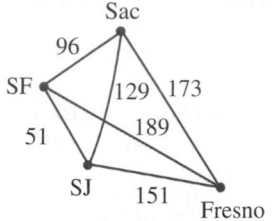

Figure 15.4 TSP for four cities

case. In fact, no solution is known at this time, and it is one of the great unsolved problems of mathematics. In this book, the best we will be able to do is a trial-and-error solution. If you are interested in seeing some of the different attempts at finding a solution to this problem, you can check the Web address shown in the margin.

EXAMPLE 6   **PÓLYA'S METHOD**

A salesman wants to visit four California cities, San Francisco, Sacramento, San Jose, and Fresno. Driving distances are shown in Figure 15.4. What is the shortest trip starting and ending in San Francisco that visits each of these cities?

Solution   Since the best we can do is to offer some possible methods of attack, we will use this example to help build your problem-solving skills. We use Pólya's problem-solving guidelines.

Understand the Problem. Part of understanding the problem is to decide what we mean by the "best" solution. For this problem, let us assume it is the least miles traveled. We also note that, in terms of miles traveled, each route and its reverse are equivalent. That is,

$$SF \rightarrow San\ Jose \rightarrow Fresno \rightarrow Sacramento \rightarrow SF$$

is the same as

$$SF \rightarrow Sacramento \rightarrow Fresno \rightarrow San\ Jose \rightarrow SF$$

Devise a Plan. There are several possible methods of attack for this problem: *brute force* (listing all possible routes); *nearest neighbor* (at each city, go to the nearest neighbor next); sometimes the nearest-neighbor plan will form a loop without going to some city so we repair this problem using a method called the *sorted-edge* method. In the sorted-edge method, we sort the choices by selecting the nearest neighbor that does not form a loop.

Carry Out the Plan.
Brute force:

$$SF \xrightarrow{96} S \xrightarrow{173} F \xrightarrow{151} SJ \xrightarrow{51} SF \qquad Total: \quad 471\ miles$$
$$SF \xrightarrow{96} S \xrightarrow{129} SJ \xrightarrow{151} F \xrightarrow{189} SF \qquad Total: \quad 565\ miles$$
$$SF \xrightarrow{51} SJ \xrightarrow{129} S \xrightarrow{173} F \xrightarrow{189} SF \qquad Total: \quad 542\ miles$$

Here are the reverse trips (so we don't need to calculate these):

$$SF \rightarrow SJ \rightarrow F \rightarrow S \rightarrow SF$$
$$SF \rightarrow F \rightarrow SJ \rightarrow S \rightarrow SF$$
$$SF \rightarrow F \rightarrow S \rightarrow SJ \rightarrow SF$$

We see that 471 is the minimum number of miles.

Nearest neighbor:

$$SF \xrightarrow{51} SJ \xrightarrow{129} S \xrightarrow{96} SF \qquad$$ A loop is formed; Fresno is not included because it is never the nearest neighbor if we start in San Francisco.

Sorted edge:
For this method, we sort the distances (edges of the graph) from smallest to largest: 51, 96, 129, 151, 173, and 189. This gives the following trip (skipping 96 and 151 because these choices would form a loop):

$$SF \xrightarrow{51} SJ \xrightarrow{129} S \xrightarrow{173} F \xrightarrow{189} SF \qquad Total: \quad 542\ miles$$

Look Back. With this simple problem, it is easy to see that the best overall solution is a trip with 471 miles, but as you can imagine, for a larger number of cities the solution may not at all be obvious.   ◆

We summarize the sorted-edge method for finding an approximate solution to a traveling salesperson problem.

Sorted-edge Method

Draw a graph showing the cities and the distances; identify the starting vertex.

Step 1: Choose the edge attached to the starting vertex that has the shortest distance or the lowest cost. Travel along this edge to the next vertex.

Step 2: At the second vertex, travel along the edge with the shortest distance or lowest cost. Do not choose a vertex that would lead to a vertex already visited.

Step 3: Continue until all vertices are visited until arriving back at the original vertex.

The sorted-edge method may not produce the optimal solution, so you should also check other methods. Since the brute-force method requires that we check all the routes, it is worthwhile to find a formula that tells us the number of routes we need to check. Note in Example 6 we found three possible routes (along with three reversals). Consider the next example, which generalizes the number of routes we found by brute force in Example 6.

EXAMPLE 7

a. How many routes are there for four cities, say, San Francisco, Sacramento, San Jose, and Fresno?

b. How many routes are there for n cities?

Solution

a. If we start in San Francisco, there are 3 cities to which we can travel. Then, by the fundamental counting principle, we have

$$3 \cdot 2 \cdot 1 = 6 \text{ routes}$$

Since half the routes are reversals of the others, we have

$$\frac{3 \cdot 2 \cdot 1}{2} = 3 \text{ routes}$$

b. Following the steps in part **a,** we note that from the first city there are $n - 1$ cities to visit, so (from the fundamental counting principle) there are

$$(n - 1)(n - 2)(n - 3) \cdots 3 \cdot 2 \cdot 1 \text{ routes}$$

and if we disregard reversals there are

$$\frac{(n - 1)(n - 2)(n - 3) \cdots 3 \cdot 2 \cdot 1}{2} \text{ routes}$$ ◆

CAUTION

This formula will be used later in the text.

Problem Set 15.1

1. **IN YOUR OWN WORDS** Describe the Königsberg bridge problem.

2. **IN YOUR OWN WORDS** Describe the floor-plan problem.

3. **IN YOUR OWN WORDS** Describe the solution to the Königsberg bridge problem.

4. **IN YOUR OWN WORDS** Describe the traveling salesperson problem.

5. **IN YOUR OWN WORDS** Contrast Euler circuits and Hamiltonian cycles.

Which of the networks in Problems 6–11 are Euler circuits? If a network can be traversed, show how.

6.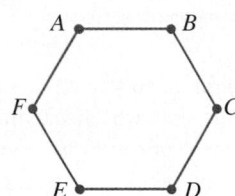

7.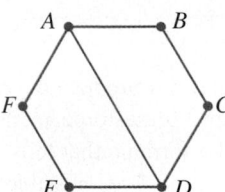

8.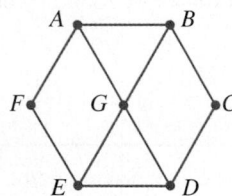

9.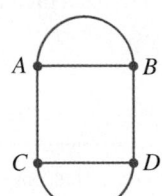

10.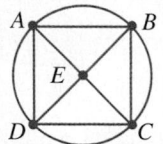

11.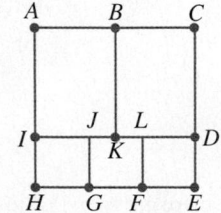

Which of the networks in Problems 12–17 have Hamiltonian cycles? If a network has one, describe it.

12.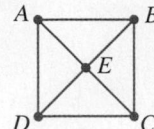

13.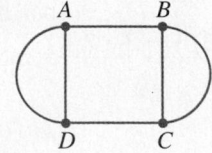

14.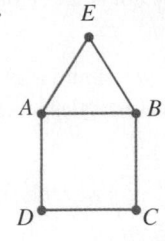

15.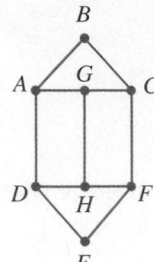

16.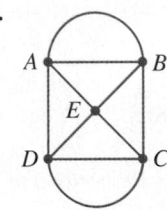

17.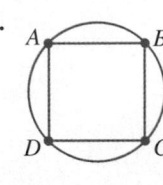

For which of the floor plans in Problems 18–23 can you pass through all the rooms while going through each door exactly once? If it is possible, show how it might be done.

18.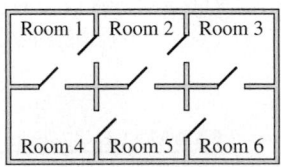

19.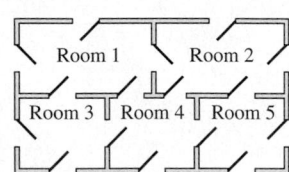

20.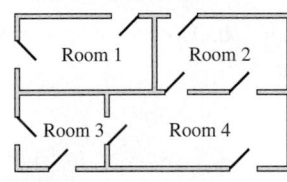

21.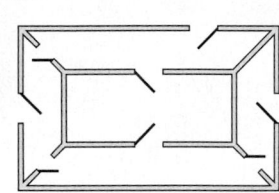

22.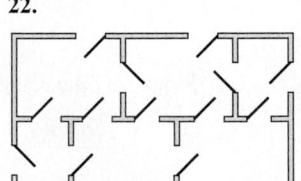

23.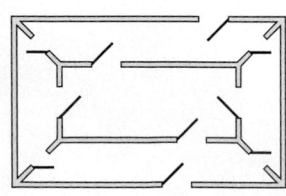

Which of the networks in Problems 24–27 are Euler circuits? If a network can be traversed, show how. Note these are the same networks as those given in Problems 28–31.

24.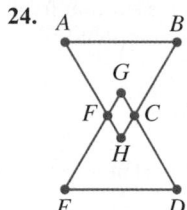

25.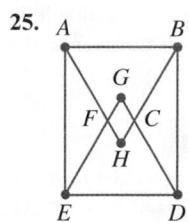

26.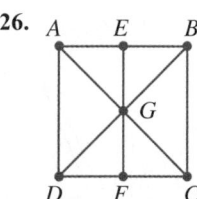

27.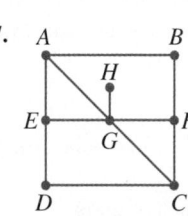

Which of the networks in Problems 28–31 have Hamiltonian cycles? If a network has one, describe it. Notice these are the same networks as those given in Problems 24–27.

28.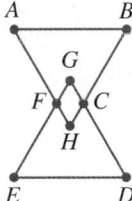

29.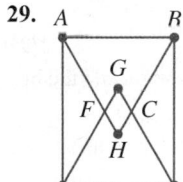

30.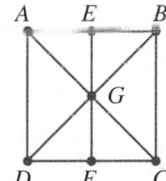

31.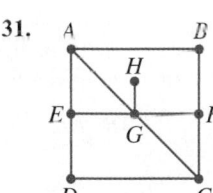

32. After Euler solved the Königsberg bridge problem, an eighth bridge was built as shown in Figure 15.5. Is this network traversable? If so, show how.

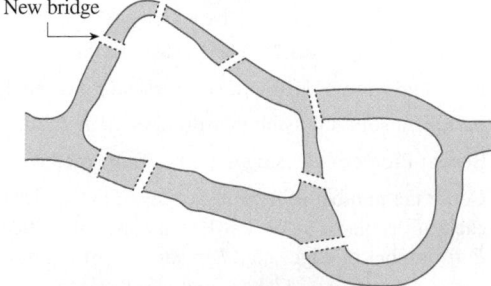

Figure 15.5 Königsberg with eight bridges

33. Traveler's Dodecahedron This problem was sold in the last half of the 19th century as a puzzle known as the "Traveler's Dodecahedron" or "A Voyage 'Round the World." It consisted of 20 pegs (called *cities*), and the point of the puzzle was to use string to connect each peg only once arriving back to the same peg you started from. Find a route (starting at Brussels—labeled 1) that visits each of the 20 cities on the dodecahedron shown in Figure 15.6.

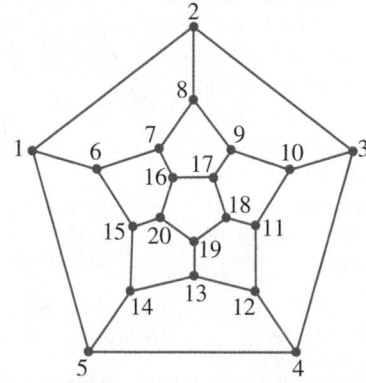

Figure 15.6 Hamilton's "Traveler's Dodecahedron"

34. Is there an Euler circuit for the Traveler's Dodecahedron shown in Figure 15.6? If so, show it.

35. Is there an Euler circuit for the graph shown in Figure 15.7?

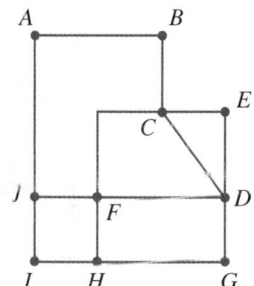

Figure 15.7
Network problem

36. Suppose you want to get from point *A* to point *B* in New York City, and also suppose you wish to cross over each of the six bridges shown in the following photograph. Is it possible? If so, show one such path.

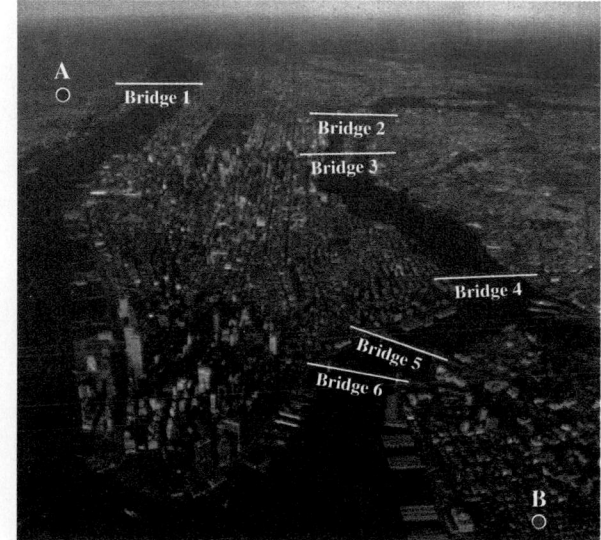

37. A simplified map of New York City, showing the subway connections between Manhattan and The Bronx, Queens, and Brooklyn, is shown in Figure 15.8. Is it possible to travel on the New York subway system and use each subway exactly once? You can visit each borough (The Bronx, Queens, Brooklyn, or Manhattan) as many times as you wish.

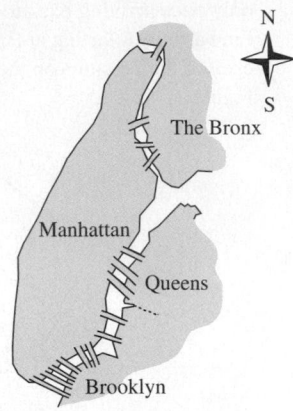

Figure 15.8 New York City subways

38. A portion of London's Underground transit system is shown in Figure 15.9. Is it possible to travel the entire system and visit each station while taking each route exactly once?

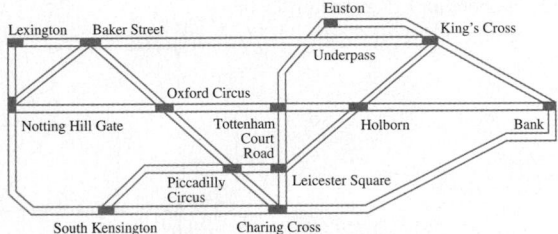

Figure 15.9 London Underground

39. In Massachusetts there is a re-creation of an 1830s New England village called Old Sturbridge Village. A map is shown in Figure 15.10. Is it possible to stroll the streets marked in color? Give reasons for your answer.

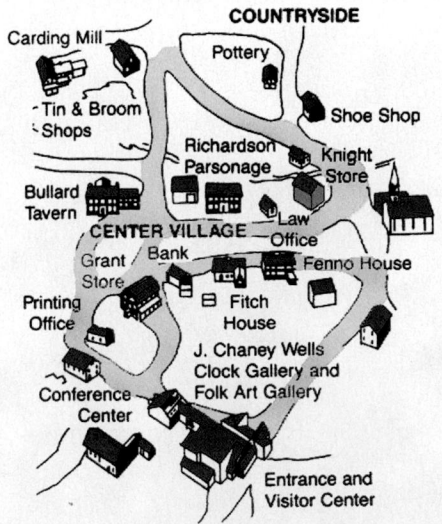

Figure 15.10 Old Sturbridge Village problem

40. Reconsider the question in Problem 39 if the street in front of the church across from the Knight Store is opened (colored).

LEVEL 3

41. The edges of a cube form a three-dimensional network. Are the edges of a cube traversable?

42. A saleswoman wants to visit eastern cities, New York City, Boston, Cleveland, and Washington, D.C. Driving distances are as shown in Figure 15.11. What is the shortest trip starting in New York that visits each of these cities?

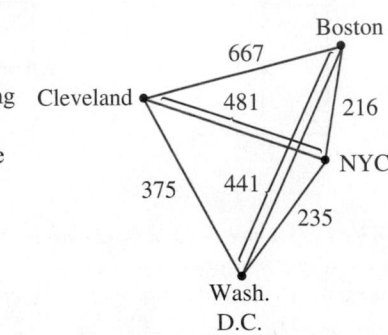

Figure 15.11 TSP for four cities

Find a solution using the brute-force method.

43. Repeat Problem 42 using the indicated method.
   a. Find a solution using the nearest-neighbor method.
   b. Find a solution using the sorted-edge method.

44. A salesperson wants to visit each of the cities Denver, St. Louis, Los Angeles, and New Orleans. Driving distances are as shown in Figure 15.12. What is the shortest trip starting in Denver that visits each of these cities?

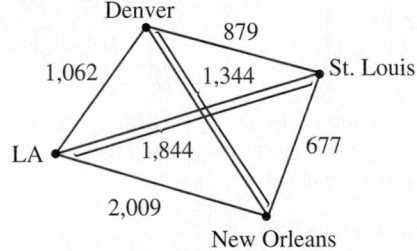

Figure 15.12 TSP for four cities

   a. Find a solution using the nearest-neighbor method.
   b. Find a solution using the sorted-edge method.

45. Repeat Problem 44 using the brute-force method.

46. Count the number of vertices, edges (arcs), and regions for each of Problems 6–17. Let V = number of vertices, E = number of edges, and R = number of regions. Compare $V + R$ with E. Make a conjecture relating V, R, and E. This relationship is called *Euler's formula for networks*.

PROBLEM SOLVING

47. The saleswoman in Problem 42 needs to add Atlanta to her itinerary. Driving distances are shown. What is the shortest trip starting in New York that visits each of these cities?

	A	B	C	NYC	D.C.
A	—	1,115	780	887	634
B	1,115	—	667	216	441
C	780	667	—	481	375
NYC	887	216	481	—	235
D.C.	634	441	375	235	—

48. A quality control inspector must visit franchises in Atlanta, Boston, Chicago, Dallas, and Minneapolis. Since this inspection must be monthly, the inspector, who lives in Chicago, would like to find the most efficient route (in terms of distances). Driving distances are shown. What is the most efficient route?

	A	B	C	D	M
A	—	1,115	717	691	1,131
B	1,115	—	1,013	1,845	1,619
C	717	1,013	—	937	420
D	691	1,845	937	—	963
M	1,131	1,619	420	963	—

49. What is the sum of the measures of the angles of a tetrahedron? *Hint:* Consider the sum of the measures of the face angles of a cube. A cube has six square faces, and since each face has four right angles, the sum of the measures of the angles on each face is 360°; hence, the sum of the measures of the face angles of a cube is 6(360°) = 2,160°.

50. What is the sum of the measures of the angles of a pentagonal prism? (See Figure 15.13.)

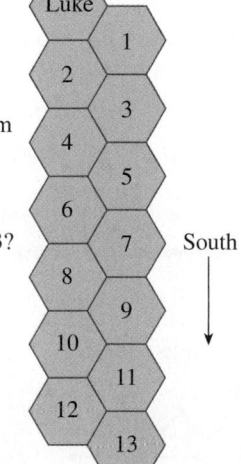

Figure 15.13 **Pentagonal prism**

51. On a planet far, far away, Luke finds himself in a strange building with hexagon-shaped rooms as shown in the figure. In his search for the princess, he always moves to an adjacent room and always in a southerly direction.
    a. How many paths are there to room 1? to room 2? to room 3? to room 4?
    b. How many paths are there to room 10?
    c. How many paths are there to room 13?

52. How many paths are there to room *n* in Problem 51?

53. Emil Torday told the story of seeing some African children playing with sand:

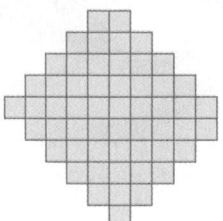

*The children were drawing, and I was at once asked to perform certain impossible tasks; great was their joy when the white man failed to accomplish them.**

One task was to trace the figure in the sand with one continuous sweep of the finger.
    a. What is the children's secret for successfully drawing this pattern?
    b. Draw this figure; why is it difficult to do this without knowing something about networks?

54. About a century ago, August Möbius made the discovery that, if you take a strip of paper, give it a single half-twist, and paste the ends together, you will have a piece of paper with only one side! Construct a Möbius strip, and verify that it has only one side. How many edges does it have?

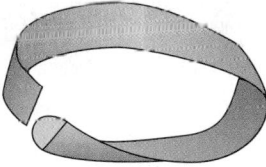

55. Construct a Möbius strip (see Problem 54). Cut the strip in half down the center. Describe the result.

56. Construct a Möbius strip (see Problem 54). Cut the strip in half down the center. Cut it in half again. Describe the result.

57. Construct a Möbius strip (see Problem 54). Cut the strip along a path that is one-third the distance from the edge. Describe the result.

58. Construct a Möbius strip (see Problem 54). Mark a point *A* on the strip. Draw an arc from *A* around the strip until you return to the point *A*. Do you think you could connect *any* two points on the sheet of paper without lifting your pencil?

59. Take a strip of paper 11 in. by 1 in., and give it three half-twists; join the ends together. How many edges and sides does this band have? What happens if you cut down the center of this piece?

* Quoted by Claudia Zaslavsky in *Africa Counts* (Boston: Prindle, Weber, & Schmidt, 1973) from *On the Trail of the Bushongo* by Emil Torday.

60. What is a Klein bottle? Examine the picture.

Klein bottle

Can you build or construct a physical model? You can use the limerick as a hint.

> A mathematician named Klein
> Thought the Möbius strip was divine.
> Said he, "If you glue the edges of two,
> You'll get a weird bottle like mine."

15.2 TREES AND MINIMUM SPANNING TREES

Trees

A *circuit* could be defined as a path/route that begins and ends at the same vertex.

In the last section, we considered graphs with circuits: an Euler circuit (which is a round-trip path traveling all the edges), and a Hamiltonian cycle (a path that visits each vertex exactly once). In this section, we consider another kind of graph, called a *tree*, which does not have a circuit. Let us begin with an example.

EXAMPLE 1

Suppose you wish to draw a family tree showing yourself, your parents, and your maternal and paternal grandparents.

Solution One possibility for showing this family tree is shown in Figure 15.14.

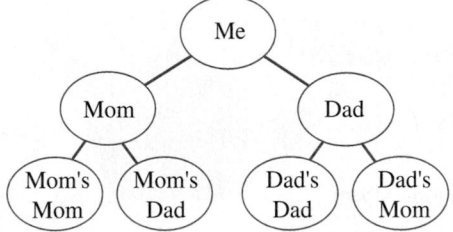

Figure 15.14 Personal family tree ◆

Simplified family tree

The family tree shown in Example 1 has two obvious properties. It is a **connected graph** because there is at least one path between each pair of vertices, and there are no circuits in this family tree. A simplified tree diagram for Example 1 is shown in the margin.

Tree

> A **tree** is a graph that is connected and has no circuits.

EXAMPLE 2

Determine which of the given graphs are trees.

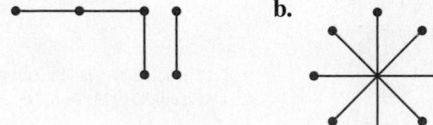

d. 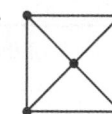 **e.**

Solution

a. This is not a tree because it is not connected.

b. This is a tree.

c. This is a tree.

d. This is not a tree because there is at least one circuit.

e. There is a circuit, so it is not a tree. ◆

In a tree, there is always exactly one path from each vertex in the graph to any other vertex in the graph. We illustrate this property of trees with the following example.

EXAMPLE 3

Ben wishes to install a sprinkler system to water the areas shown in Figure 15.15. Show how this might be done.

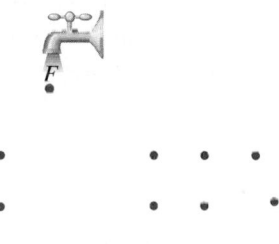

Figure 15.15 Locations of a faucet and sprinkler heads

Solution

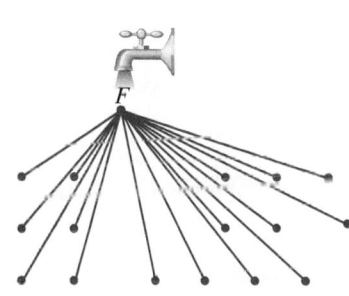

We know there is at least one way to build a tree from each vertex (in this case, the faucet, labeled F). We show one such way in the margin. ◆

The solution shown for Example 3 may or may not be an efficient solution to the sprinkler system problem. Suppose we connect the vertices in Example 3 without regard to whether the graph is a tree, as shown in Figure 15.16a. Next, we remove edges until the resulting graph is a tree. A tree that is created from another graph by removing edges but keeping a path to each vertex is called a **spanning tree.** Can you form a spanning tree for the graph in Figure 15.16a? If you think about it for a moment, you will see that any connected graph will have a spanning tree, and that if the original graph has at least one circuit, then it will have several different spanning trees. Figure 15.16b shows a spanning tree for the sprinkler problem of Example 3.

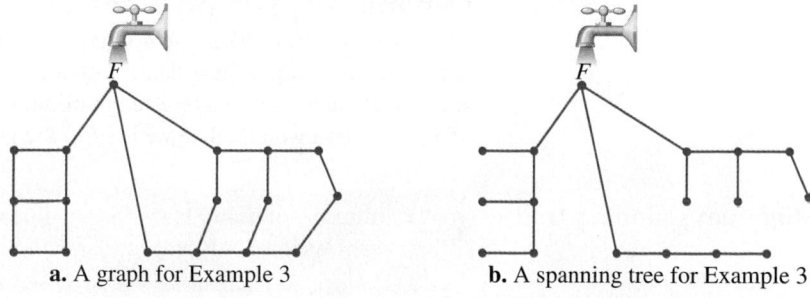

a. A graph for Example 3   **b.** A spanning tree for Example 3

Figure 15.16 Comparison of a graph and a spanning tree

EXAMPLE 4

Find two different spanning trees for each of the given graphs.

a.

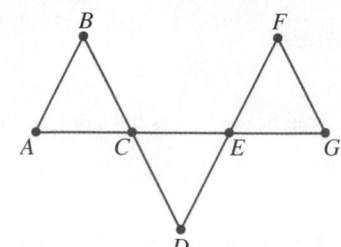

b.

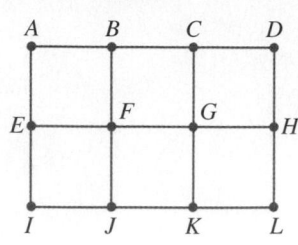

Solution

Since a spanning tree must have a path connecting all vertices, but cannot have any circuits, we remove edges (one at a time) without moving any of the vertices and without creating a disconnected graph. We show two different possibilities for each of the given graphs, while noting that others are possible.

a. This graph has three circuits: $A \to B \to C \to A$, $C \to D \to E \to C$, and $E \to F \to G \to E$. To obtain a spanning tree, we must break up each of these circuits, but at the same time not disconnect the graph. There are many ways we could do this. In the first tree we remove edges *BC*, *CE*, and *EF*. In the second tree, we remove edges *AB*, *CD*, and *FG*.

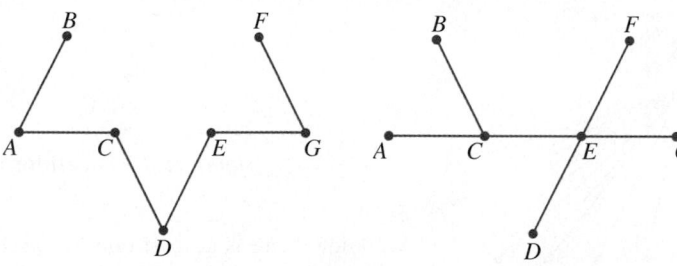

b. There are many circuits, and we show two possible spanning trees here.

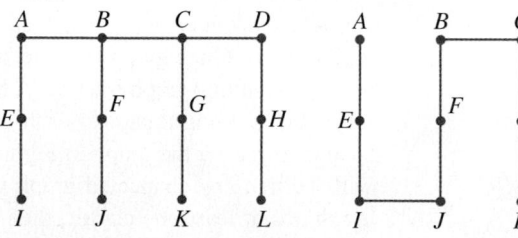

Minimum Spanning Trees

As we can see from Example 4, there may be several spanning trees. Sometimes the length of each edge is associated with a cost or a length, called the edge's **weight.** In such cases we are often interested in minimizing the cost or the distance. If the edges of a graph have weight, then we refer to the graph as a **weighted graph.**

Minimum spanning tree

A **minimum spanning tree** is a spanning tree for which the sum of the numbers associated with the edges is a minimum.

EXAMPLE 5   PÓLYA'S METHOD

A portion of the Santa Rosa Junior College campus, along with some walkways (lengths shown in feet) connecting the buildings, is shown in Figure 15.17.

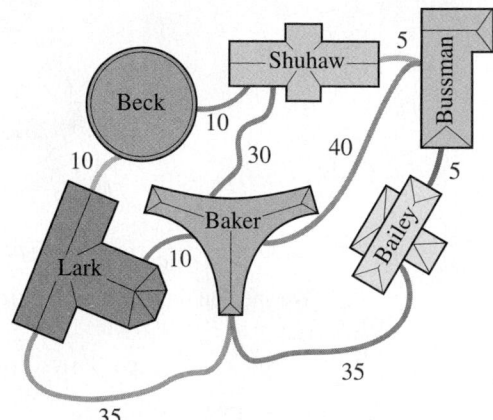

Figure 15.17 **Portion of campus**

Suppose the decision is made to connect each building with a brick walkway, and the only requirement is that there be one brick walkway connecting each of the buildings. We assume that the cost of installing a brick walkway is $100/ft. What is the minimum cost for this project?

Solution

Let's use Pólya's problem-solving guidelines.

Understand the Problem. To make sure we understand the problem, we consider a simpler problem. Consider the simple graph shown in the margin with costs shown in color. We consider the vertices to be buildings since the only requirement is that there be one brick walkway connecting each of the buildings. To find the best way to construct the walkways, we consider minimum spanning trees. Since this is a circuit, we can break this circuit in one of three ways: eliminate one of the sides, *AB, BC,* or *AC.*

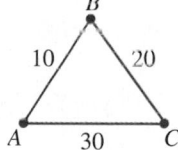

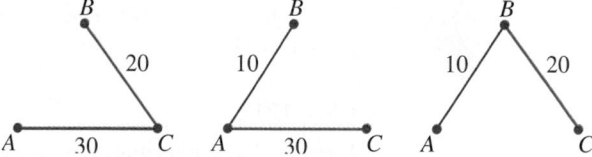

We see that the cost associated with each of these trees is:

$$20 + 30 = 50 \qquad 10 + 30 = 40 \qquad 10 + 20 = 30$$

The minimal cost for this simplified problem is 30.

Devise a Plan. We will carry out the steps for the Santa Rosa campus as follows: Look at Figure 15.17 and find the side with the smallest weight (because we wish to keep the smaller weights). We see there are two sides labeled 5; select either of these. Next, select a side with the smallest remaining weight (it is also 5). Continue by each time selecting the smallest remaining weight until every vertex is connected, but *do not select any edge that creates a circuit.*

Carry Out the Plan. Following this procedure, we select both of the edges labeled 5, as well as the three labeled 10. The resulting pathways are shown in Figure 15.18 on page 778.

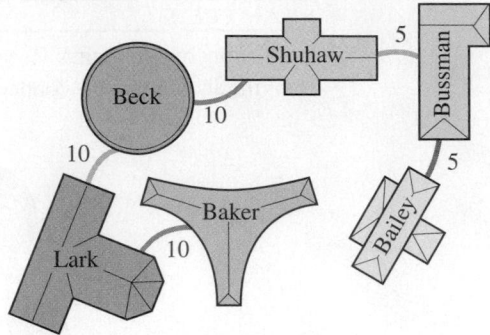

Figure 15.18 **Minimum spanning tree**

We see that the graph in Figure 15.18 is a minimum spanning tree, so the total distance is

$$5 + 5 + 10 + 10 + 10 = 40$$

with a total cost of

$$40 \times \$100 = \$4,000$$

Look Back. We can try other possible routes connecting all of the buildings, but in each case, the cost is more than $4,000. ◆

The process used in Example 5 illustrates a procedure called **Kruskal's algorithm.**

Kruskal's Algorithm

To construct a minimum spanning tree from a weighted graph:

1. Select any edge with minimum weight.

2. Select the next edge with minimum weight among those not yet selected.

3. Continue to choose edges of minimum weight from those not yet selected, but make sure not to select any edge that forms a circuit.

4. Repeat the process until the tree connects all of the vertices of the original graph.

WWW

There are many Web sites that illustrate Kruskal's algorithm. You might wish to explore this idea by checking the links at **www.mathnature.com**.

EXAMPLE 6

Use Kruskal's algorithm to find the minimum spanning tree for the weighted graph in Figure 15.19. The numbers represent hundreds of dollars.

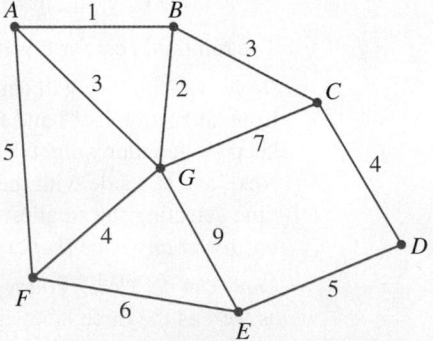

Figure 15.19 **Find the minimum spanning tree**

Solution

Step 1:   Choose the side with weight 1:

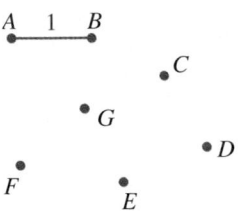

Step 2:   Choose the side with weight 2:

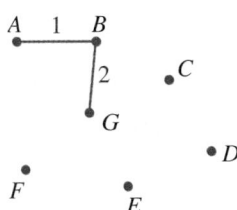

Step 3:   Choose the side with weight 3. Do not connect AG because that would form a circuit.

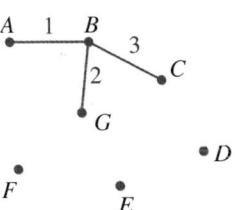

Continue with this process to select the two sides with weights of 4.

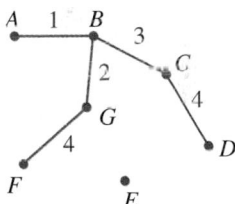

Step 4:   When we connect the side with the next lowest weight, namely, ED with weight 5, we know we are finished because we now have a tree with all of the vertices connected.

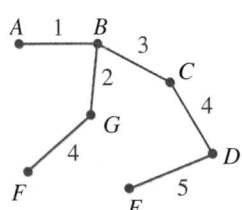

We can now calculate the weight of this tree:

$$1 + 2 + 3 + 4 + 4 + 5 = 19$$

Since these weights are in hundreds of dollars, the weight of the minimum spanning tree is $1,900.   ◆

The next example is adapted from a standardized test given in the United Kingdom in 1995.

EXAMPLE 7

A company is considering building a gas pipeline network to connect seven wells (*A, B, C, D, E, F, G*) to a processing plant *H*. The possible pipelines that they can construct and their costs (in hundreds of thousands of dollars) are listed in the following table.

Pipeline	AB	AE	AD	BC	BE	BF	CG	DE	DF	EH	FG	FH
Cost	23	17	19	15	30	27	10	14	20	28	11	35

What pipelines do you suggest be built and what is the total cost of your suggested pipeline network?

Solution

Begin by drawing a graph to represent the data. This graph is shown in Figure 15.20.

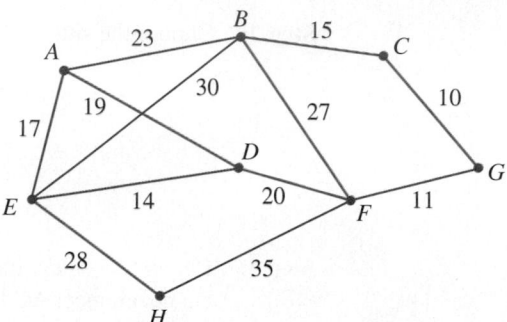

Figure 15.20 Building a gas pipeline

We now apply Kruskal's algorithm (in table form).

	Link	**Cost**	**Decision**
Step 1:	CG	10	Smallest value; add to tree
Step 2:	FG	11	Next smallest value; add to tree
Step 3:	DE	14	Add to tree; note that graph does not need to be connected at this step
	BC	15	Add to tree
	AE	17	Add to tree
	AD	19	Reject; it forms a circuit *ADE*
	DF	20	Add to tree
	AB	23	Reject; it forms a circuit *ABCGFDE*
	BF	27	Reject; it forms a circuit *BFGC*
Step 4:	EH	28	Add to tree; stop because all vertices are now included.

The completed minimum spanning tree is shown in Figure 15.21.

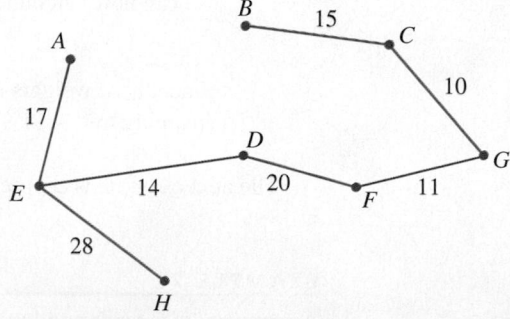

Figure 15.21 Minimum spanning tree for pipeline problem

The minimal cost is

$$10 + 11 + 14 + 15 + 17 + 20 + 28 = 115$$

so, the cost of the pipeline is \$115,000. ◆

Note in Example 7 that there were eight given vertices, and that there were seven links added to form the minimal spanning tree. This is a general result.

Number-of-vertices-and-edges-in-a-tree Theorem

If a graph is a tree with n vertices, then the number of edges is $n - 1$.

You are asked to explain why this seems plausible in Problem 59. There is another related property that says the converse of this property holds for connected graphs. If the number of edges is one less than the number of vertices in a connected graph, then the graph is a tree.

Problem Set 15.2

LEVEL 1

1. **IN YOUR OWN WORDS** What do we mean by a tree?

2. **IN YOUR OWN WORDS** What do we mean by a spanning tree?

3. **IN YOUR OWN WORDS** State Kruskal's algorithm. When would you use this algorithm?

Determine whether each of the graphs in Problems 4–11 is a tree. If it is not, explain why.

4.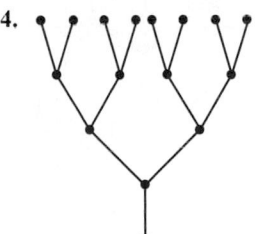

5.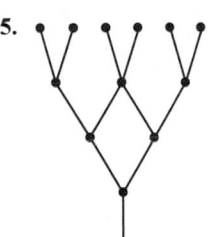

6.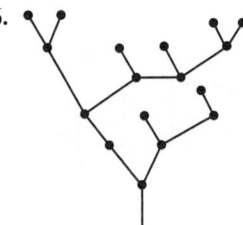

7.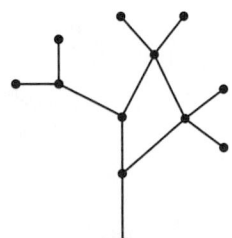

8.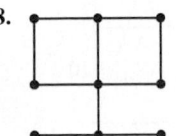

9.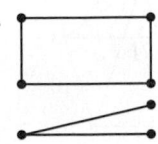

10.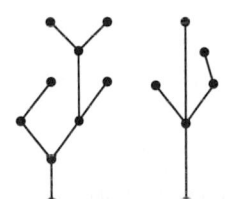

11.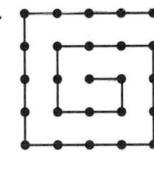

Find two different spanning trees for each graph in Problems 12–19.

12.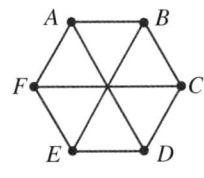

13.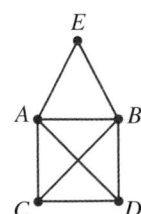

14.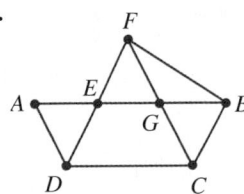

15.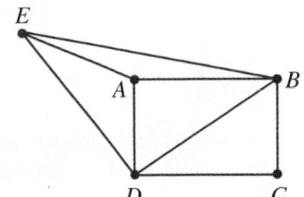

16.

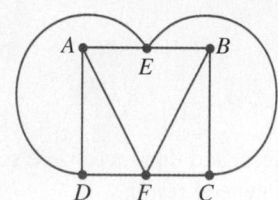

17.

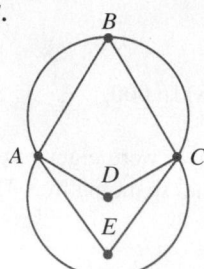

18.

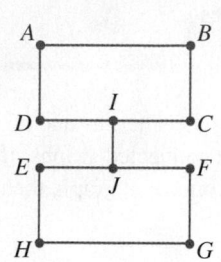

19.

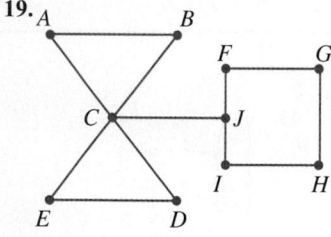

Find all the spanning trees for the graphs in Problems 20–25.

20.

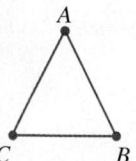

21.

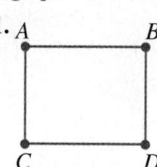

22.

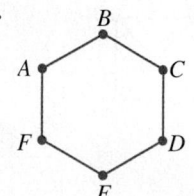

23.

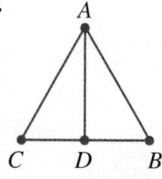

24.

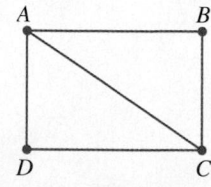

25.

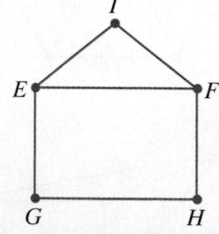

In chemistry, molecules are represented as ball-and-stick models or by bond-line diagrams. For example, the ball-and-stick model for ethane is shown in Figure 15.22a and the bond-line diagram is shown in Figure 15.22b. A corresponding tree diagram is shown in Figure 15.22c.

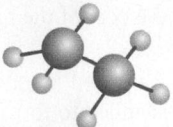
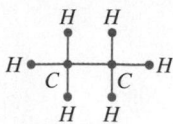

a. Ball-and-stick model **b.** Bond-line drawing **c.** Tree diagram

Figure 15.22 **Ethane molecule**

Draw graphs for each of the molecules in Problems 26–31. If the graph does not form a tree, tell why.

26. Methane

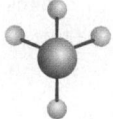

27. Propane

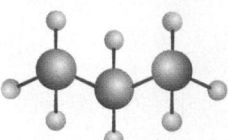

28. Butane

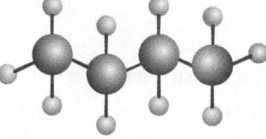

29. Isobutane

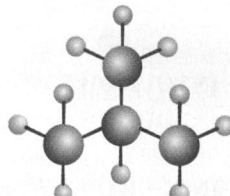

30. Cyclopropane

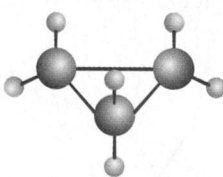

31. Cyclohexane

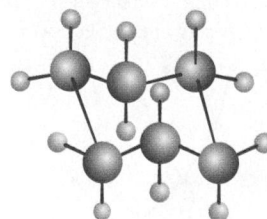

Find the minimum spanning tree for each of the graphs in Problems 32–41.

32.

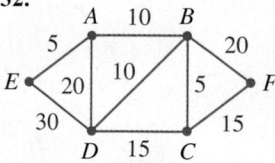

33.

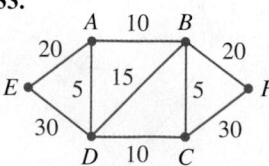

34.

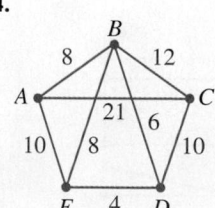

35.

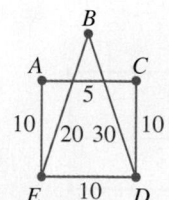

36.

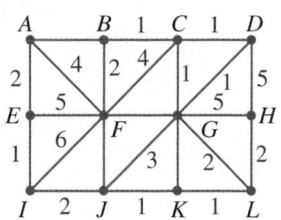

37.

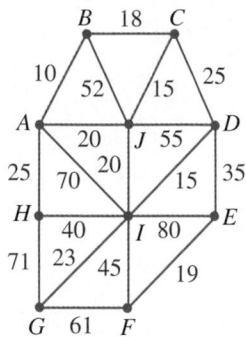

38.

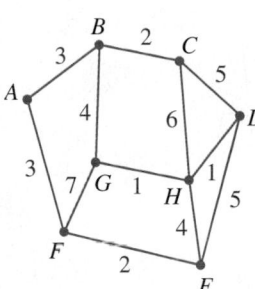

39.

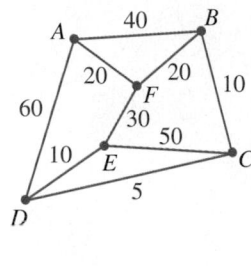

40.

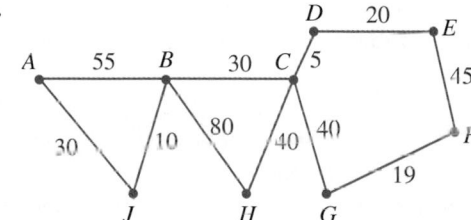

41.

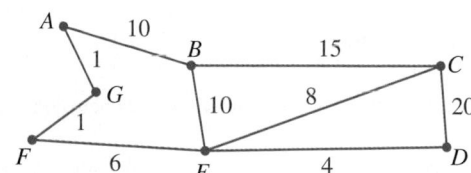

LEVEL 2

42. A chemist is studying a chemical compound with a tree-like structure that contains 39 atoms. How many chemical bonds are there in this molecule?

43. A chemist is studying a chemical compound with a tree-like structure that contains 65 atoms. How many chemical bonds are there in this molecule?

44. A Web site called "The Oracle of Bacon" is hosted by the University of Virginia. (Go to **www.mathnature.com** for a link to this site, if you wish.) It links other actors to Kevin Bacon by the movies in which the actors appeared. For example, I entered the movie star from *Gone with the Wind,* Clark Gable, and the Oracle told me that Clark

Gable had a Bacon number of two because Clark Gable was in the 1953 movie *Mogambo* with Donald Siden, and Donald Siden was in the 1995 movie *Balto* with Kevin Bacon. I tried it again with Will Smith, and the Oracle said that he also had a Bacon number of two since he was in the 2000 movie *The Legend of Bagger Vance* with Charlize Theron, who in turn was in the 2002 movie *Trapped* with Kevin Bacon. Robert De Niro has a Bacon number of one because he was in the 1996 movie *Sleepers* with Kevin Bacon. This trivia game was started by Albright College students Craig Fass, Brian Turtle, and Mike Ginelli, who hypothesized that all actors—living or dead—have a Bacon number of 6 or less. If you could draw a graph of the relationships between Kevin Bacon and other actors, would the result be a tree?

45. Suppose you use the *Yahoo* search engine to do research for a term paper. You start on the *Yahoo* page and follow the links. When you do this you decide to keep a record of the sites you visit. If you keep the record as a graph, will it form a tree?

46. Use a tree to show the following family tree. You have two children, Shannon and Melissa. Shannon has three children, Soren, Thoren, and Floren. Melissa has two children, Hannah and Banana.

47. Use a tree to show the following management relationships for a college. The positions are a college president; an academic vice president, who reports directly to the president and who supervises six academic deans, each of whom supervises three departments; a vice president for business services, who reports directly to the president and who supervises the personnel office, scholarships and grants, as well as the bookstore and food services; and finally, a vice president of operations, who is in charge of supervising facilities, grounds, and certified staff and who also reports directly to the president.

48. How many edges are there in a tree with 15 vertices?

49. How many vertices are there in a tree with 48 edges?

50. Suppose you wish to install a drip sprinkler system and need to run a drip water line to five areas, as shown in the given graph. The numbers show the distances in feet. What is the smallest number of feet of drip hose necessary to install?

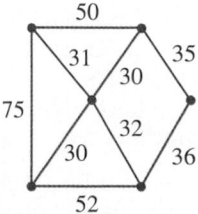

51. Suppose your college is planning on installing some covered walkways connecting six buildings as shown in the following map. The plan is to allow a person to walk to any building under cover, and the numbers shown on the

map represent distances measured in feet. If the covered walkway costs $350/ft, what is the minimum cost for this project?

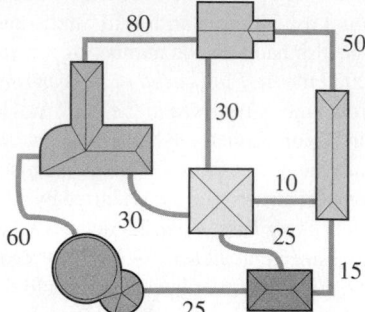

52. A mutual water system obtained estimates for installing water pipes among their respective properties (labeled *A, B, C, D,* and *E*). These amounts (in dollars) are shown in color in Figure 15.23. Which lines should they install to minimize the cost?

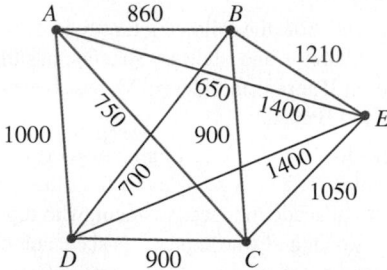

Figure 15.23 Water line cost estimates

53. The map in Figure 15.24 shows driving distances and times between California and Nevada cities. Use Kruskal's algorithm to find the minimum spanning tree for the following cities: Santa Rosa, San Francisco, Oakland, Manteca, Yosemite Village, Merced, Fresno, and San Jose.

54. Use the map in Figure 15.24 and Kruskal's algorithm to construct the minimum spanning tree for the cities of Reno, Carson City, Lee Vining, Fallon, Austin, Tonopah, Bishop, Beatty, Death Valley, and Lone Pine.

LEVEL 3

55. Suppose XYZ drilling has four oil wells that must be connected via pipelines to a storage tank. The cost of each pipeline (in millions of dollars) is shown in the following table:

From/To	#1	#2	#3	#4	Tank
#1	—	1	4	2	3
#2	1	—	3	1	2
#3	4	3	—	1	4
#4	2	1	1	—	1
Tank	3	2	4	1	—

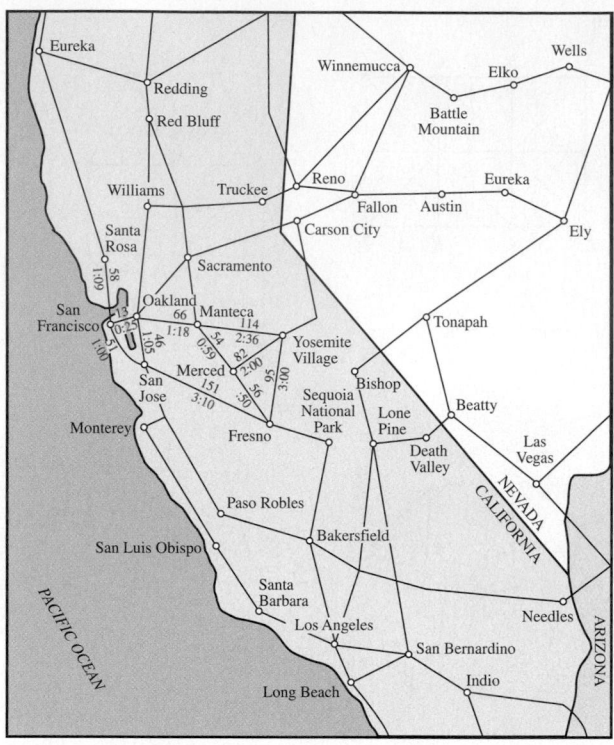

Source: Adapted from California–Nevada Driving Distances map © 2002 AAA.

Figure 15.24 California–Nevada driving distances

a. Represent this information with a weighted graph.
b. Use Kruskal's algorithm to find a minimum spanning tree.
c. What is the minimum cost that links together all of the wells and the tank?

56. Consider the following table showing costs between vertices.

From/To	#1	#2	#3	#4	Tank
#1	—	3	7	3	2
#2	3	—	6	8	5
#3	7	6	—	10	9
#4	3	8	10	—	4
#5	2	5	9	4	—

a. Represent this information with a weighted graph.
b. Use Kruskal's algorithm to find a minimum spanning tree.
c. What is the minimum cost that links together all of the vertices?

57. Suppose a network is to be built connecting the Florida cities of Tallahassee, Jacksonville, St. Petersburg, Orlando, and Miami. The given numbers show the miles between cities.

| | Tallahassee | St. Petersburg | Miami | | |
		Jacksonville		Orlando	
Tallahassee	—	172	249	253	476
Jacksonville	172	—	219	160	410
St. Petersburg	249	219	—	105	237
Orlando	253	160	105	—	252
Miami	476	410	237	252	—

a. Represent this information with a weighted graph.

b. Use Kruskal's algorithm to find a minimum spanning tree.

c. What is the minimum cost that links together all of the cities if the cost is \$85/mi?

58. Suppose a network is to be built connecting the cities of Norfolk, Raleigh, Charlotte, Atlanta, and Savannah. The given numbers show the miles between cities.

| | Norfolk | | Charlotte | | Savannah |
		Raleigh		Atlanta	
Norfolk	—	170	333	597	519
Raleigh	170	—	175	427	358
Charlotte	333	175	—	250	249
Atlanta	597	427	250	—	256
Savannah	519	358	249	256	—

a. Represent this information with a weighted graph.

b. Use Kruskal's algorithm to find a minimum spanning tree.

c. What is the minimum cost that links together all of the cities if the cost is \$205/mi?

59. *Number-of-vertices-and-edges-in-a-tree property*

a. State the number-of-vertices-and-edges-in-a-tree theorem.

b. Consider a tree with one vertex. What is the number of edges? Does the property hold in this case?

c. Consider a tree with two vertices. How many edges can you have and still have a tree? Explain why you cannot have two or more edges.

d. Consider a tree with three vertices. How many edges can you have and still have a tree? Explain why you cannot have three or more edges.

60. In 1889, Arthur Cayley (see Historical Note on p. 738) proved that a complete graph with n vertices has n^{n-2} spanning trees.

a. How many spanning trees are there for a complete graph with 3 vertices, according to Cayley's theorem? Verify this number by drawing a complete graph with 3 vertices and then by finding all the spanning trees.

b. How many spanning trees are there for a complete graph with 4 vertices, according to Cayley's theorem? Verify this number by drawing a complete graph with 4 vertices and then by finding all the spanning trees.

c. How many spanning trees are there for a complete graph with 5 vertices?

d. How many spanning trees are there for a complete graph with 6 vertices?

15.3 TOPOLOGY AND FRACTALS

> The golden age of mathematics— that was not the age of Euclid, it is ours.
>
> C. J. Keyser

A brief look at the history of geometry illustrates, in a very graphical way, the historical evolution of many mathematical ideas and the nature of changes in mathematical thought. The geometry of the Greeks included very concrete notions of space and geometry. They considered space to be a locus in which objects could move freely about, and their geometry was a geometry of congruence. This geometry is known as Euclidean geometry. In this section we investigate two very different branches of geometry that question, or alter, the way we think of space and dimension.

Topology

In the 17th century, space came to be conceptualized as a set of points, and, with the non-Euclidean geometries of the 19th century, mathematicians gave up the notion that geometry had to describe the physical universe. The existence of multiple geometries was accepted, but space was still thought of as a geometry of congruence. The emphasis shifted to sets, and geometry was studied as a mathematical system. Space could be conceived as a set of points together with an abstract set of relations in which these points are involved. The time was right for geometry to be considered as the theory of

HISTORICAL NOTE

The set theory of Cantor (see the Historical Note on page 498) provided a basis for topology, which was presented for the first time by Jules-Henri Poincaré (1854–1912) in *Analysis Situs*. A second branch of topology was added in 1914 by Felix Hausdorff (1868–1942) in *Basic Features of Set Theory*. Earlier mathematicians, including Euler, Möbius, and Klein, had touched on some of the ideas we study in topology, but the field was given its major impetus by L. E. J. Brouwer (1882–1966). Today much research is being done in topology, which has practical applications in astronomy, chemistry, economics, and electrical circuitry.

such a space, and in 1895 Jules-Henri Poincaré published a book using this notion of space and geometry in a systematic development. This book was called *Vorstudien zur Topologie (Introductory Studies in Topology)*. However, topology was not the invention of any one person, and the names of Cantor, Euler, Fréchet, Hausdorff, Möbius, and Riemann are associated with the origins of **topology.** Today it is a broad and fundamental branch of mathematics.

To obtain an idea about the nature of topology, consider a device called a *geoboard*. You may have used a geoboard in elementary school. Suppose we stretch one rubber band over the pegs to form a square and another to form a triangle, as shown in Figure 15.25.

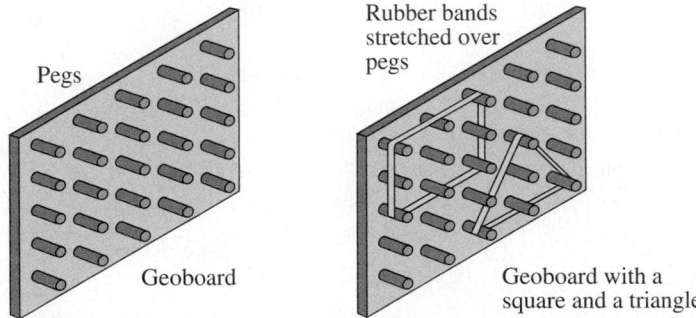

Figure 15.25 **Creating geometric figures with a geoboard**

In high-school geometry, the square and the triangle in Figure 15.25 would be seen as different. However, in topology, these figures are viewed as the same object. Topology is concerned with discovering and analyzing the essential similarities and differences between sets and figures. One important idea is called *elastic motion,* which includes bending, stretching, shrinking, or distorting the figure in any way that allows the points to remain distinct. It does not include cutting a figure unless we "sew up" the cut *exactly* as it was before.

Topologically Equivalent Figures

> Two geometric figures are said to be **topologically equivalent** if one figure can be elastically twisted, stretched, bent, shrunk, or straightened into the same shape as the other. One can cut the figure, provided at some point the cut edges are "glued" back together again to be exactly the same as before.

Rubber bands can be stretched into a wide variety of shapes. All forms in Figure 15.26 are topologically equivalent. We say that a curve is **planar** if it lies flat in a plane.

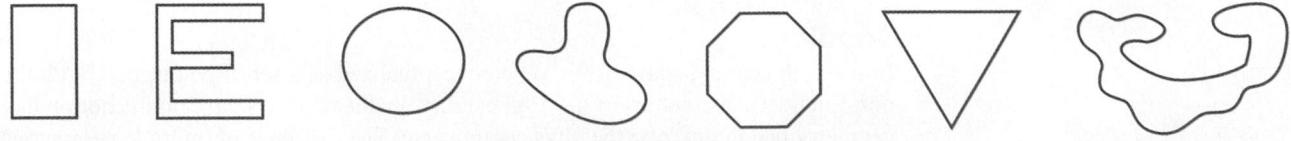

Figure 15.26 **Topologically equivalent curves**

All of the curves in Figure 15.26 are *planar simple closed curves.* A curve is **closed** if it divides the plane into three disjoint subsets: the set of points on the curve itself, the

The children and their distorted images are topologically equivalent.

set of point *interior* to the curve, and the set of points *exterior* to the curve. It is said to be **simple** if it has only one interior. Sometimes a simple closed curve is called a **Jordan curve.** Notice that, to pass from a point in the interior to a point in the exterior, it is necessary to cross over the given curve an odd number of times. This property remains the same for any distortion, and is therefore called an *invariant* property.

Two-dimensional surfaces in a three-dimensional space are classified according to the number of cuts possible without slicing the object into two pieces. The number of cuts that can be made without cutting the figure into two pieces is called its **genus.** The genus of an object is the number of holes in the object. (See Figure 15.27.)

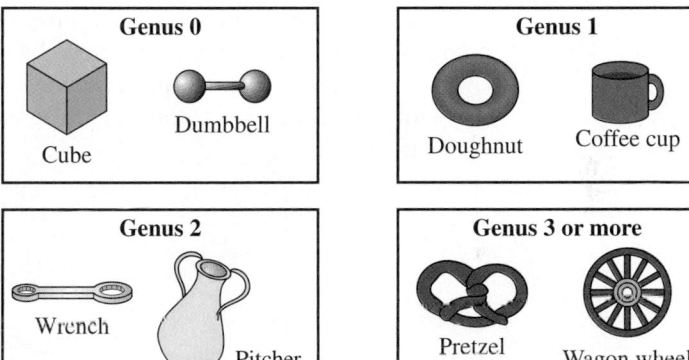

Figure 15.27 Genus of the surfaces of some everyday objects. Look at the number of holes in the object.

For example, no cut can be made through a sphere without cutting it into two pieces, so its genus is 0. In three dimensions, you can generally classify the genus of an object by looking at the number of holes the object has. A doughnut, for example, has genus 1 since it has 1 hole. In mathematical terms, we say it has genus 1 since only one closed cut can be made without dividing it into two pieces. All figures with the same genus are topologically equivalent. Figure 15.28 shows that a doughnut and a coffee cup are topologically equivalent, and Figure 15.29 shows objects of genus 0, genus 1, and genus 2.

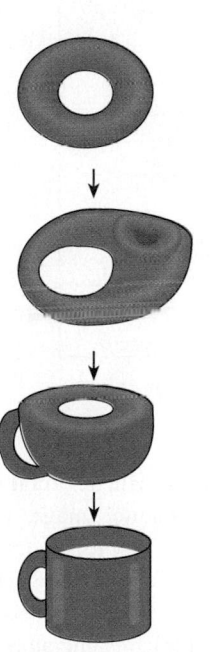

Figure 15.28 A doughnut is topologically equivalent to a coffee cup.

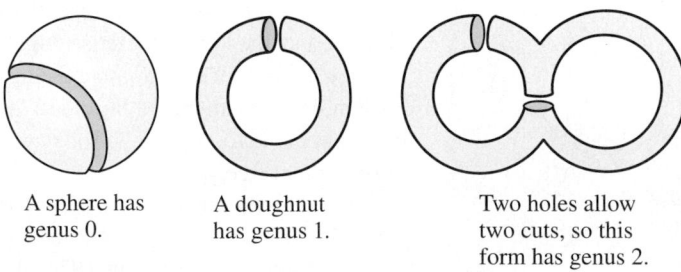

A sphere has genus 0.

A doughnut has genus 1.

Two holes allow two cuts, so this form has genus 2.

Figure 15.29 Genus of a sphere, a doughnut, and a two-holed doughnut

Four-Color Problem

One of the earliest and most famous problems in topology is the four-color problem. It was first stated in 1850 by the English mathematician Francis Guthrie. It states that any map on a plane or a sphere can be colored with at most four colors so that any two countries that share a common boundary are colored differently. (See Figure 15.30, for example.) All attempts to prove this conjecture had failed until Kenneth Appel and

Wolfgang Haken of the University of Illinois announced their proof in 1976. The university honored their discovery using the illustrated postmark.

FOUR COLORS

SUFFICE

Source: University of Illinois

Since the theorem was first stated, many unsuccessful attempts have been made to prove it. The first published "incorrect proof" is due to Kempe, who enumerated four cases and disposed of each. However, in 1990, an error was found in one of those cases, which it turned out was subcategorized as 1,930 different cases. Appel and Haken reduced the map to a graph as Euler did with the Königsberg bridge problem. They reduced each country to a point and used computers to check every possible arrangement of four colors for each case, requiring more than 1,200 hours of computer time to verify the proof.

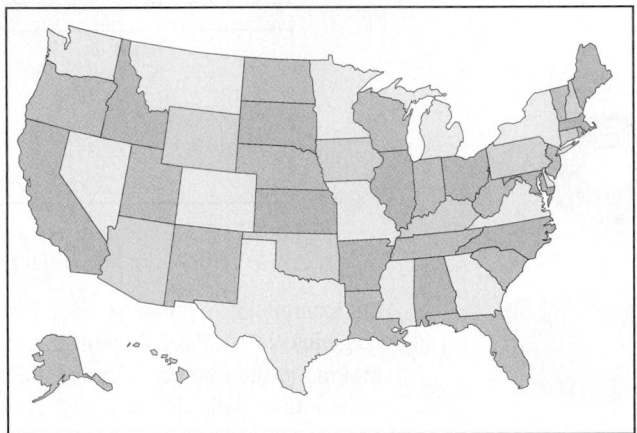

Figure 15.30 **Every map can be colored with four colors.**

Fractal Geometry

One of the newest and most exciting branches of mathematics is called **fractal geometry.** Fractals have been used recently to produce realistic computer images in many movies, and the new supercrisp high-definition television (HDTV) uses fractals to squeeze the HDTV signal into existing broadcast channels. In February 1989, Iterated System, Inc., began marketing a $32,500 software package for creating models of biological systems from fractals. Today you can find hundreds of fractal generators online, most of them free.

Fractals were invented by Benoit B. Mandelbrot over 30 years ago, but have become important only in the last few years because of computers. Mandelbrot's first book on fractals appeared in 1975; in it he used computer graphics to illustrate the fractals. The book inspired Richard Voss, a physicist at IBM, to create stunning landscapes, earthly and otherworldly (see Figure 15.31). "Without computer graphics, this work could have been completely disregarded," Mandelbrot acknowledges.

What exactly is a fractal? We are used to describing the dimension of an object without having a precise definition: A point has 0 dimension; a line, 1 dimension; a plane, 2 dimensions; and the world around us, 3 dimensions. We can even stretch our imagination to believe that Einstein used a four-dimensional model. However, what about a dimension of 1.5? Fractals allow us to define objects with noninteger dimension. For example, a jagged line is given a fractional dimension between 1 and 2, and the exact value is determined by the line's "jaggedness."

We will illustrate this concept by constructing the most famous fractal curve, the so-called "snowflake curve." Start with a line segment $\overline{AB}$:

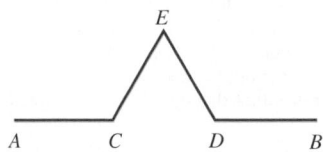

Divide this segment into thirds, by marking locations C and D:

$N = 3$ segments

$r = \dfrac{1}{3}$ is the length of each segment

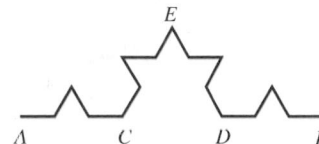

Now construct an equilateral triangle CED on the middle segment and then remove the middle segment $\overline{CD}$:

$N = 4$ segments

$r = \dfrac{1}{3}$ is the length of each segment

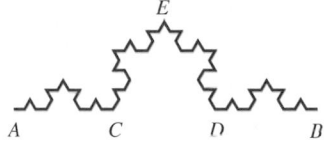

Now, repeat the above steps for each segment:

$N = 16$ segments

$r = \dfrac{1}{9}$ is the length of each segment

Again, repeat the process:

If you repeat this process (until you reach any desired level of complexity), you have a fractal curve with dimension between 1 and 2.

The actual description of the dimension is more difficult to understand. Mandelbrot defined the dimension as follows:

$N = 3; r = \dfrac{1}{3}; D = 1$

$N = 4; r = \dfrac{1}{3}; D \approx 1.26$

$N = 16; r = \dfrac{1}{9}; D \approx 1.26$

$N = 64; r = \dfrac{1}{27}; D \approx 1.26$

$$D = \frac{\log N}{\log \frac{1}{r}}$$

Logarithms are discussed in Chapter 8.

where N is any integer and r is the length of each segment.* For the illustrations above, the dimension is calculated in the margin. Some fractal images are shown in Figure 15.31.

Source: *Fractal Vision: Put Fractals to Work for You.* © 1992 SAMS Publishing, a division of Prentice-Hall Computer Publishing.

Figure 15.31 **Fractal images of a mountainscape (generated by Richard Voss) and fern**

*Mandelbrot defined r as the ratio $\dfrac{L}{N}$, where L is the sum of the lengths of the N line segments.

Guest Essay: WHAT GOOD ARE FRACTALS?

Okay, fractals can make sense out of chaos (see the Guest Essay at the end of this section), but what can you do with them? It is a question currently being asked by physicists and other scientists at many professional meetings, says Alan Norton, a former associate of Mandelbrot who is now working on computer architectures at IBM. For a young idea still being translated into the dialects of each scientific discipline, the answer, Norton says, is: Quite a bit. The fractal dimension may give scientists a way to describe a complex phenomenon with a single number.

Harold Hastings, professor of mathematics at Hofstra University on Long Island, is enthusiastic about modeling the Okefenokee Swamp in Georgia with fractals. From aerial photographs, he has studied vegetation patterns and found that some key tree groups, such as cypress, are patchier and show a larger fractal dimension than others.

Shaun Lovejoy, a meteorologist who works at Météorologie Nationale, the French national weather service in Paris, confirmed that clouds follow fractal patterns. Again, by analyzing satellite photographs, he found similarities in the shapes of many cloud types that formed over the Indian Ocean. From tiny puff-like clouds to an enormous mass that extended from Central Africa to Southern India, all exhibited the same fractal dimension. Prior to Mandelbrot's discovery of fractals, cloud shapes had not been candidates for mathematical analysis and meteorologists who theorize about the origin of weather ignored them. Lovejoy's work suggests that the atmosphere on a small-scale weather pattern near the Earth's surface resembles that on a large-scale weather pattern extending many miles away, an idea that runs counter to current theories.

The occurrence of earthquakes. The surfaces of metal fractures. The path a computer program takes when it scurries through its memory. The way our own neurons fire when we go searching through *our* memories. The wish list for fractal description grows. Time will tell whether the fractal dimension becomes invaluable to scientists interested in building mathematical models of the world's workings.

From "Geometrical Forms Known as Fractals Find Sense in Chaos," by Jeanne McDermott, *Smithsonian*, December 1983, p.116.

Figure 15.32 Escher print: Circle Limit I

Tessellations

The construction of the snowflake curve reminds us of another interesting mathematical construction, called a **tessellation.** By skillfully altering a basic polygon (such as a triangle, rectangle, or hexagon), the artist Escher was able to produce artistic tessellations such as that shown in Figure 15.32. We can describe a procedure for reproducing a simple tessellation based on the Escher print in Figure 15.32.

Step 1. Start with an equilateral triangle *ABC*. Mark off the same curve on sides *AB* and *AC* as shown in Figure 15.33. Mark off another curve on side *BC* that is symmetric about the midpoint *P*. If you choose the curves carefully, as Escher did, an interesting figure suitable for tessellating will be formed.

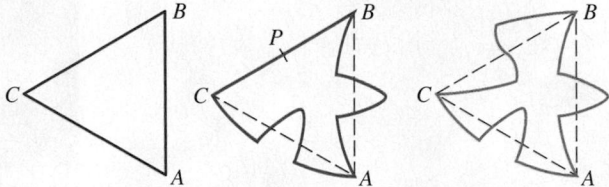

Figure 15.33 Tessellation pattern

Step 2. Six of these figures accurately fit together around a point, forming a hexagonal array. If you trace and cut one of these basic figures, you can continue the tessellation over as large an area as you wish.

Problem Set 15.3

1. **IN YOUR OWN WORDS** What do we mean by topology?

2. **IN YOUR OWN WORDS** What is the four-color problem?

3. Group the figures into classes so that all the elements within each class are topologically equivalent, and no elements from different classes are topologically equivalent.

A. 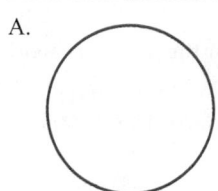 B.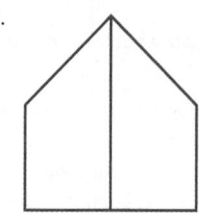

C. 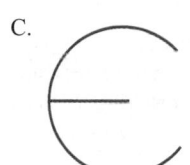 D.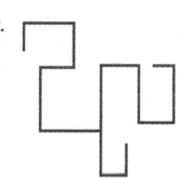

E.  F.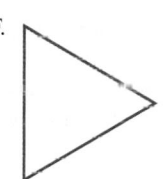

G. 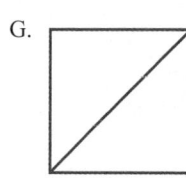 H.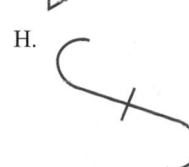

4. Group the figures into classes so that all the elements within each class are topologically equivalent, and no elements from different classes are topologically equivalent.

A. 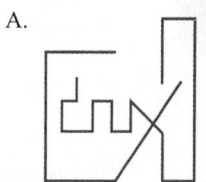 B.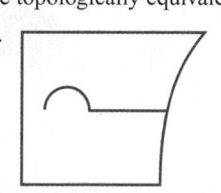

C. 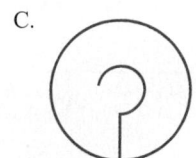 D.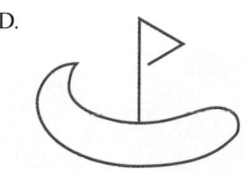

E. 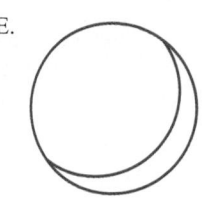 F.

G.  H.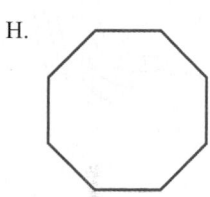

5. Which of the figures in Problem 3 are simple closed curves?

6. Which of the figures in Problem 4 are simple closed curves?

7. Group the letters of the alphabet into classes so that all the elements within each class are topologically equivalent and no elements from different classes are topologically equivalent.

A   B   C   D   E   F   G   H   I   J   K   L   M

N   O   P   Q   R   S   T   U   V   W   X   Y   Z

8. Group the objects into classes so that all the elements within each class are topologically equivalent and no elements from different classes are topologically equivalent.
   A. a glass
   B. a bowling ball
   C. a sheet of typing paper
   D. a sphere
   E. a ruler
   F. a banana
   G. a sheet of two-ring-binder paper

9. Group the objects into classes so that all the elements within each class are topologically equivalent and no elements from different classes are topologically equivalent.
   A. a bolt
   B. a straw
   C. a horseshoe
   D. a sewing needle
   E. a brick
   F. a pencil
   G. a funnel with a handle

In Problems 10–14, determine whether each of the points A, B, and C is inside or outside of the simple closed curve.

10.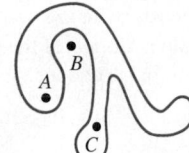

11.

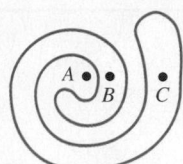

12.

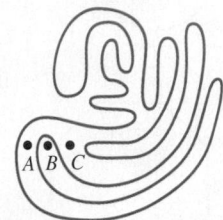

13.

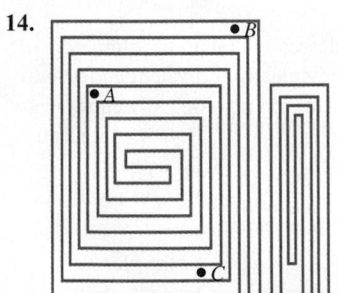

14.

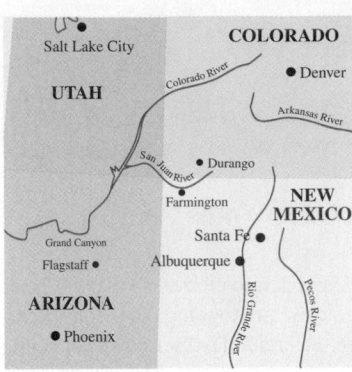

How many colors are needed for the 4-corners area?

17. If a map (on a plane or on the surface of a sphere) is partitioned into two or more regions with each vertex of even degree, then the resulting map can be colored with exactly two colors. Draw a map illustrating this fact.

18. If a map (on a plane or on the surface of a sphere) is partitioned into regions, each with an even number of edges, and if each vertex is of degree 3, the resulting map can be colored with exactly three colors. Draw a map illustrating this fact.

19. If a map (on a plane or on the surface of a sphere) is partitioned into at least five regions, each sharing its borders with exactly three neighboring regions, the resulting map can be colored in three colors. Draw a map illustrating this fact.

LEVEL 3

20. Find the length of the following snowflake curve:

Snowflake curve

Begin with a line segment of length 1 unit, divided into thirds:

15. a. Let X be a point obviously outside the figure given in Problem 12. Draw $\overline{AX}$. How many times does it cross the curve? Repeat for $\overline{BX}$ and $\overline{CX}$.
  b. Repeat part **a** for the figure given in Problem 13.
  c. Repeat part **a** for the figure given in Problem 14.
  d. Make a conjecture based on parts **a–c**. This conjecture involves a theorem called the *Jordan curve theorem*.

16. One of the simplest map-coloring rules of topology involves a map of "countries" with straight lines as boundaries. How many colors would be necessary for the 4-corners area on a U.S. map? (See the map at the top of the next column.) Note that a common point is not considered a common boundary.

a. What is the length of the snowflake curve after the first step?

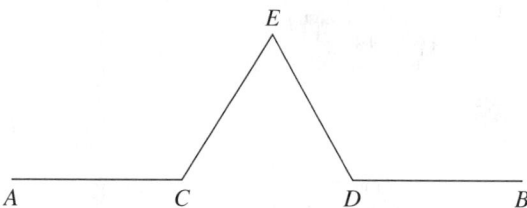

b. What is the length of the snowflake curve after the second step?

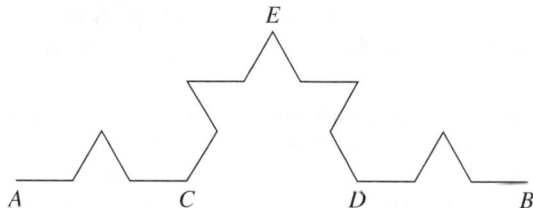

c. What is the length of the snowflake curve after the third step?

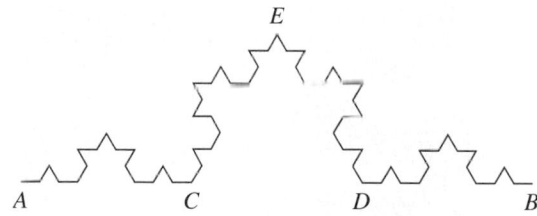

21. Construct a fractal curve by forming squares (rather than triangles as shown in the text). For example, the first step is:

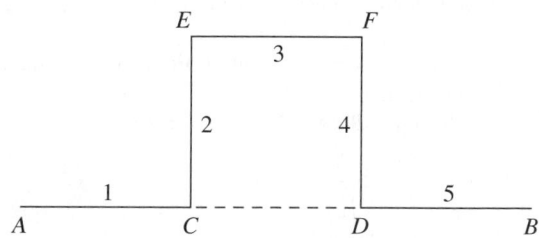

22. Construct a tessellation using triangles.

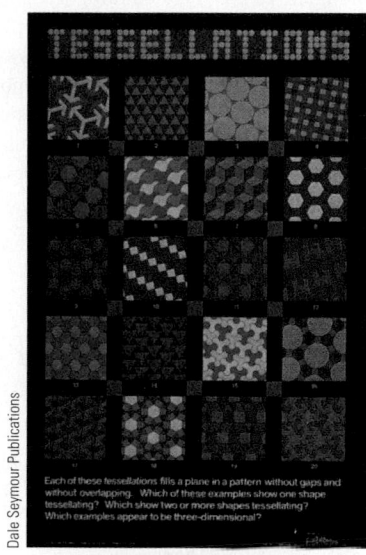

Each of these tessellations fills a plane in a pattern without gaps and without overlapping. Which of these examples show one shape tessellating? Which show two or more shapes tessellating? Which examples appear to be three-dimensional?

Dale Seymour Publications

23. Construct a tessellation using rectangles.

24. Two identical squares joined so that they have a common edge is called a *domino,* as shown in Figure 15.34. Three identical squares can be joined together to form a *tromino,* and they come in two different shapes. *Pentominos,* composed of five squares, come in 12 different shapes.

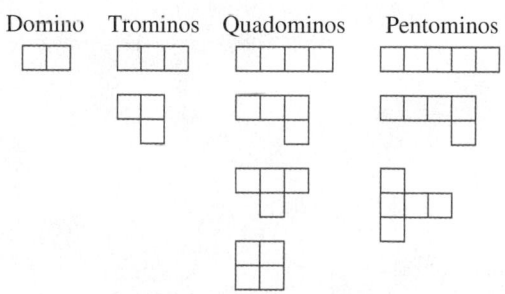

Figure 15.34 **Mosaic tiles**

Complete Figure 15.34 by showing the other 9 pentomino shapes.

25. Forming a mosaic pattern using polygons, like those shown in Figure 15.34, is called *tiling*. A tiling and a tessellation are really the same idea, although the word *tessellation* is usually used when the pattern is more complicated than a polygon. For example, most classroom floors are tiled with simple squares, a very simple pattern with dominos. An early example of a tromino pattern is found in an 18th century painting. This pattern is shown in Figure 15.35.

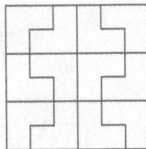

Figure 15.35 **A tiling pattern for a 6 × 6 square using trominos**

Use pentominos to tile a rectangle of the given size.
a. 3 × 20   **b.** 4 × 15

26. Design a mosaic using a pentagon.

27. Design a mosaic using a hexagon.

28.

This is an example of an 18th century painting showing two ladies and a servant, serenaded by a musician. The process of painting a distorted image on canvas, which becomes complete when viewed as a reflection, is called *anamorphosis*. What topological relation does the picture on canvas have with the image in the reflecting glass?

PROBLEM SOLVING

29. Answer the questions after reading the poem in the News Clip.

> He killed the noble Mudjokivis.
> Of the skin he made him mittens,
> Made them with the fur side inside.
> Made them with the skin side
>   outside.
> He, to get the warm inside,
> Put the skin side outside;
> He, to get the cold side outside,
> Put the warm side fur side inside.
> That's why he put the fur side
>   inside.
> Why he put the skin side outside,
> Why he turned them inside outside.

a. If a right-handed mitten is turned inside out, as is suggested in the poem, will it still fit a right hand?

b. Is a right-handed mitten topologically equivalent to a left-handed mitten?

30. Some mathematicians were reluctant to accept the proof of the four-color problem because of the necessity of computer verification. The proof was not "elegant" in the sense that it required the computer analysis of a large number of cases. Study the map in Figure 15.36 and determine for yourself whether it is the *first five-color map*, providing a counterexample for the computerized "proof."

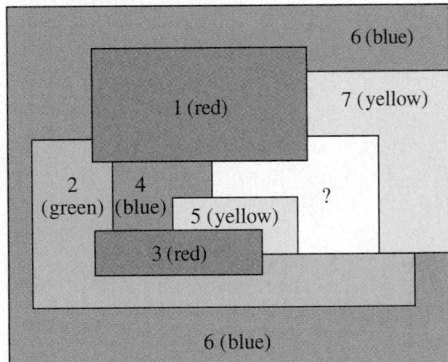

Figure 15.36 **Is this the world's first five-color map?**

What is the color of the white region? Consider the following table:

Region	Blue	Yellow	Green	Red
1	x	x	x	
2	x	x		x
3	x	x	x	
4		x	x	x
5	x		x	x
6		x	x	x
7	x		x	x
?	x	x	x	x

The x's indicate the colors that share a boundary with the given region. As you can see, the white region is bounded by all four colors, so therefore requires a fifth color.

Tibor Hirsch

Guest Essay: CHAOS
Jack Wadhams, Golden West College

An exciting new topic in mathematics and an attempt to bring order to the universe has been labeled **chaos theory**, which provides refreshing insight into how very simple beginnings can yield structures of incredible complexity and of enchanting beauty. Let's trace a chaos path to simple beginnings.

Water flowing through a pipe offers one of the simplest physical models of chaos. Pressure is applied to the end of the pipe and the water flows in straight lines. More pressure increases the speed of the laminar flow until the pressure reaches a critical value, and a radically new situation evolves—turbulence. A simple laminar

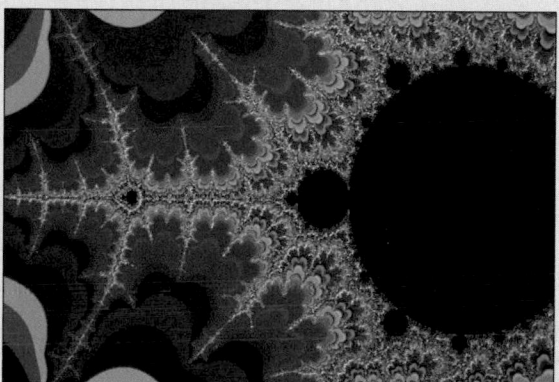

flow suddenly changes to a flow of beautiful complexity consisting of swirls within swirls. Before turbulence, the path of any particle was quite predictable. After a minute change in the pressure, turbulence occurred and predictability was lost. **Chaos** is concerned with systems in which minute changes suddenly transform predictability into unpredictability.

The simple quadratic equation $z_{n+1} = z_n^2 - c$ (where z is a complex number, a number you considered in algebra, but is not in the domain of real numbers we are assuming in this book) offers a fascinating mathematical example of chaos. With this equation and a computer, a graph of unimaginable complexity and surprising beauty can be constructed. To create a graph we plot a grid of points (often more than a million) from a 4×4 square region R centered at the origin of the complex plane. We let $z_0 = 0$ and $c = a + bi$ be a point in the region R. Substitute z_1 and c into the right-hand side of the equation to obtain a second number z_2. After many such iterations, we obtain the sequence of numbers z_0, $z_1 ..., z_i, ...$, which approaches a fixed value or it does not. If it does approach a fixed value, we paint a black dot on the computer screen at $a + bi$; otherwise, we paint a white dot. This process of selecting a point from region R, done millions of times, produces regions of black and white dots. If the boundary between the convergent and divergent regions is connected, a bumpy curve called the Mandelbrot set is formed. The graph is painted black on the interior of the Mandelbrot set and white on the exterior. We find that points selected far from the Mandelbrot set (the boundary) behave in a very predictable fashion (like the laminar flow in the pipe). However, when we choose points near the Mandelbrot set (the boundary), the behavior is quite unpredictable. We have encountered chaos. Because of the complexity of the Mandelbrot set, it is difficult to determine whether a given point is outside or inside the set.

This path to chaos has led to the Mandelbrot set, the most complicated set in mathematics, and it turns out to be fractal in nature, yet it is generated from a simple quadratic equation. That is, from a very simple quadratic equation, a set of surprising complexity and beauty emerges. Regions of chaotic behavior, under magnification, reveal a self-similar fractal. Our mathematical analogy awakens the possibility that the universe may also originate from a few simple relations that evolve into profoundly complex structures, like our brain, capable of understanding its own origins and still asking about its origins. This chaos path to simple beginnings seems to be fractal.

CHAPTER 15 SUMMARY

"... *Mathematics is an aspect of culture as well as a collection of algorithms*"

Carl Boyer

IMPORTANT TERMS

Arc [15.1]
Closed curve [15.3]
Connected graph [15.2]
Connected network [15.1]
Degree (vertex) [15.1]
Edge [15.1]
Euler circuit [15.1]
Euler's circuit theorem [15.1]
Even vertex [15.1]
Four-color problem [15.3]
Fractal geometry [15.3]
Genus [15.3]
Graph [15.1]

Hamiltonian cycle [15.1]
Jordan curve [15.3]
Kruskal's algorithm [15.2]
Minimum spanning tree [15.2]
Network [15.1]
Number-of-edges theorem [15.2]
Odd vertex [15.1]
Operations research [overview]
Planar curve [15.3]
Region [15.1]
Simple curve [15.3]
Sorted-edge method [15.1]

Spanning tree [15.2]
Tessellation [15.3]
Topologically equivalent [15.3]
Topology [15.3]
Traveling salesperson problem (TSP) [15.1]
Traversable network [15.1]
Tree [15.2]
Vertex [15.1]
Weight [15.2]
Weighted graph [15.2]

TYPES OF PROBLEMS

Decide whether a network is an Euler circuit. [15.1]
Work floor-plan problems. [15.1]
Solve applied problems involving traversable networks. [15.1]
Determine whether a given graph is a tree. [15.2]
Find spanning trees for a given graph. [15.2]
Given a weighted graph, find the minimum spanning tree. [15.2]
Sort figures into topologically equivalent classes. [15.3]
Decide whether a given point is an interior or exterior point. [15.3]
Solve applied problems involving the four-color theorem. [15.3]
Design mosaics (tessellations). [15.3]

CHAPTER 15 REVIEW QUESTIONS

In Problems 1–3, tell whether the network is traversable. If the network is traversable, show how.

1. 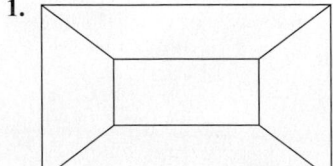 **2.**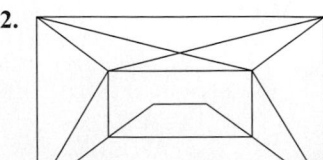

3.

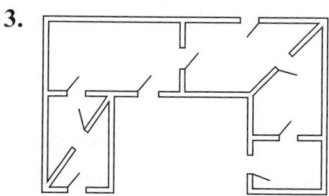

4. Draw a map with seven regions such that the indicated number of colors is required so that no two bordering regions have the same color.
   a. Two colors   **b.** Three colors   **c.** Four colors   **d.** Five colors

5. The *San Francisco Chronicle* reported that two Stanford graduates, Dave Kaval and Brad Null, set a goal to see a game in every major league stadium. (*Note:* There are 16 National League and 14 American League teams.) They began in San Francisco and selected the route shown in Figure 15.37.

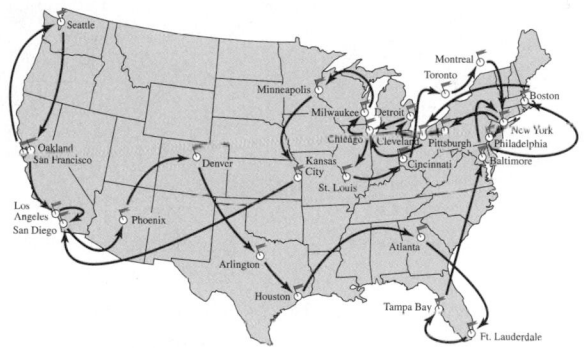

Figure 15.37 Tour of major league stadiums (TSP)

   a. In how many ways could Kaval and Null start in San Francisco and visit the cities of Minneapolis, Milwaukee, Chicago, St. Louis, Detroit, Cleveland, Cincinnati, and Pittsburgh?

   b. If we use the brute-force method for solving the TSP of Kaval and Null, the number of possible routes for the 30 major league cities is the astronomical number 4.4×10^{30}. Use the brute-force method for the simplified problem of finding the best way to begin in San Francisco and visit Los Angeles, San Diego, and Phoenix. The mileage chart is shown below. What is the mileage using the brute-force method?

From/To	SF	LA	SD	P
SF	—	369	502	755
LA	369	—	133	372
SD	502	133	—	353
P	755	372	353	—

6. **a.** Use the nearest-neighbor method to approximate the optimal route for the mileage chart shown in Problem 5b. What is the mileage when using this route?

   b. Show a complete, weighted graph for these cities. Is there a minimum spanning tree? If so, what is the mileage using this method?

7. The Big 10 football conference consists of the following schools:

Ohio State	Illinois
Penn State	Northwestern (Ill.)
Michigan	Indiana
Michigan State	Purdue (Indiana)
Wisconsin	Minnesota
Iowa	

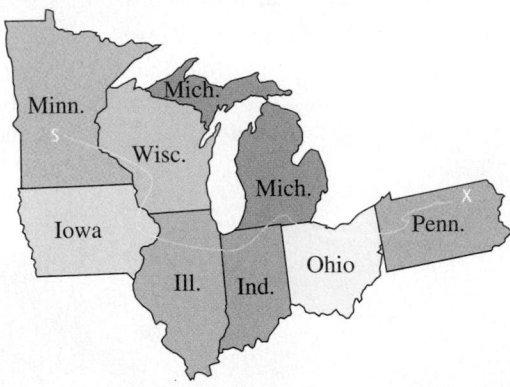

a. Is it possible to visit each of these schools by crossing each common state border exactly once? If so, find an Euler path.
b. Is it possible to find an Euler circuit? That is, is it possible to start the trip in any given state and end the trip in the state in which you started?

8. Consider the graph in Figure 15.38.
   a. Find the cost of the nearest-neighbor tour, starting at A.
   b. Find the cost of the nearest neighbor tour, starting at M.
   c. How many tours starting at K would be necessary to find the most efficient solution by using the brute-force method?
   d. Use Kruskal's algorithm to find the cost of the minimum spanning tree.

Figure 15.38 Weighted graph

9. a. Is there an Euler circuit for the graph in Figure 15.39?
   b. Find a Hamiltonian circuit for the graph in Figure 15.39.
   c. Find a spanning tree for the graph in Figure 15.39.

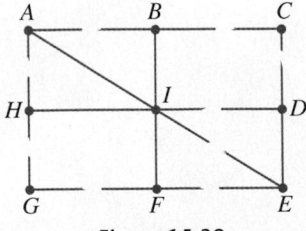

Figure 15.39

10. Take a strip of paper 11 in. by 1 in. and give it four half-twists; join the edges together. How many edges and sides does this band have? Cut the band down the center. What is the result?

GROUP RESEARCH

Working in small groups is typical of most work environments, and learning to work with other students to communicate specific ideas is an important skill. Work with three or four other students to submit a single report based on each of the following questions.

G59. What exactly are fractals?

To get you started on your paper, we ask the following question that relates the ideas of series and fractals using the *snowflake curve*. Cut an equilateral triangle of side a out of paper, as shown in Figure 15.40a. Next, three equilateral triangles, each of side $a/3$, are cut out and placed in the middle of each side of the first triangle, as shown in Figure 15.40b. Then 12 equilateral triangles, each of side $a/9$, are placed halfway along each of the sides of this figure, as shown in Figure 15.40c. Figure 15.40d shows the result of adding 48 equilateral triangles, each of side $a/27$, to the previous figure. As part of the work on this paper, find the perimeter and the area of the snowflake curve formed if you continue this process indefinitely.

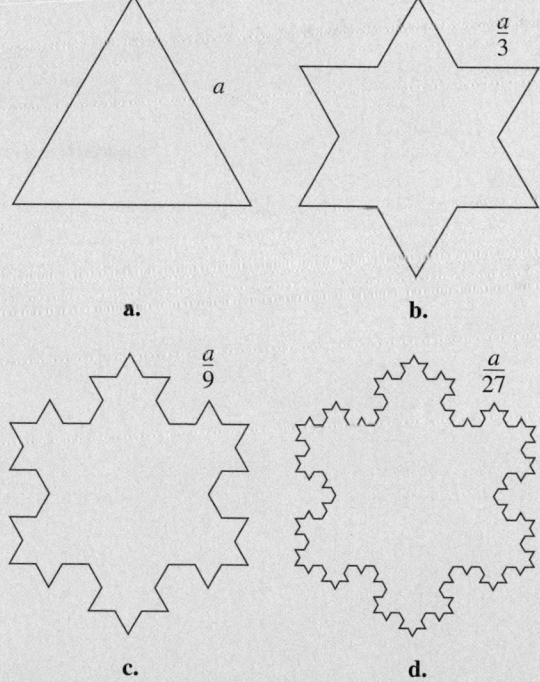

Figure 15.40 Construction of a snowflake curve

References Anthony Barcellos, "The Fractal Geometry of Mandelbrot," *The College Mathematics Journal,* March 1984, pp. 98–114.

"Interview, Benoit B. Mandelbrot," *OMNI,* February 1984, pp. 65–66.

Benoit Mandelbrot, *Fractals: Form, Chance, and Dimension* (San Francisco: W. H. Freeman, 1977).

Benoit Mandelbrot, *The Fractal Geometry of Nature* (San Francisco: W. H. Freeman, 1982).

INDIVIDUAL RESEARCH PROBLEMS

Learning to use sources outside your classroom and textbook is an important skill, and here are some ideas for extending some of the ideas in this chapter. You can find references to these projects in a library or at

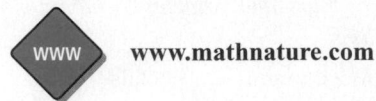

www.mathnature.com

PROJECT 15.1 Write a report on Ramsey theory.

PROJECT 15.2 Write a report on the geometry of the Garden Houses of the second-century city of Ostia.

PROJECT 15.3 The German artist Albrecht Dürer (1471–1528) was not only a Renaissance artist, but also somewhat of a mathematician. Do some research on the mathematics of Dürer.

PROJECT 15.4 Prepare a classroom demonstration of topology by drawing geometric figures on a piece of rubber inner tube, and illustrate the ways these figures can be distorted.

PROJECT 15.5

> Your aid I want, nine trees to plant
> In rows just half a score;
> And let there be in each row three.
> Solve this: I ask no more.

The problem shown in the News Clip was first published by John Jackson in 1821. Without the poetry, the puzzle can be stated as follows: **Arrange nine trees so they occur in ten rows of three trees each.** Find a solution.

16 The Nature of Voting and Apportionment

I think that society must choose among a number of alternative policies. These policies may be thought of as quite comprehensive, covering a number of aspects: foreign policy, budgetary policy, or whatever. Now, each individual member of the society has a preference, or a set of preferences, over these alternatives. I guess that you can say one alternative is better than another.

KENNETH J. ARROW

CONTENTS

OVERVIEW

This chapter investigates the question of how groups can best arrive at decisions. We refer to the study of the decision-making process as **social choice theory.** This study considers the possibilities by which individual preferences are translated into a single group choice. Social choice theory arose to help explain the processes of **voting,** or indicating an individual's preferences. You have no doubt participated as a voter in many contexts, but you may not realize that the voting method used can significantly affect the outcome of the election.

Historically, certain voting inconsistencies were discussed by Jean-Charles de Borda (1733–1799) and Marie Jean Antoine Nicolas Caritat, the Marquis of Condorcet (1743–1794). It was not until 1951 that the mathematician Kenneth Arrow (1921–   ) proved that all attempts to arrive at a suitable technique for voting are doomed to failure!

Since all voting methods have inherent fallacies, in this chapter, we will investigate different voting methods, as well as some reasons why we need to carefully consider the election process. We also look at the process of selecting a representative government using the apportionment of votes.

IMPORTANT IDEAS

Selecting a winning candidate using a system of voting [16.1]

Summary of voting methods, Table 16.2 [16.1]

Fair voting principles: majority criterion, Condorcet criterion, monotonicity criterion, irrelevant alternatives criterion [16.2]

Arrow's impossibility theorem [16.2]

Apportion legislative seats or other resources [16.3]

Recognize flaws or inconsistencies in the apportionment process [16.4]

BOOK REPORTS

Write a 500-word report on one of the following books:
Voting Procedures, Michael Dummett (Oxford, UK: Oxford University Press, 1984).
Considerations on Representative Government, John Stuart Mill (New York: Harper and Brothers, 1862).

LINKS

Individual Projects

Group Projects

Research Links

www.mathnature.com

16.1 VOTING

The process of selection can be accomplished in many different ways. If one person alone makes the decision, we call it a **dictatorship.** If the decision is made by a group, it is called a **vote**. The different methods of selection using a vote, such as voting on a proposal, resolution, law, or a choice between candidates is an area of study called *social choice theory.* In this section we will discuss several ways of counting the votes to declare a winner.

Majority Rule

One common method of voting is by *majority rule.* By **majority rule,** we mean voting to find an alternative that receives more than 50% of the vote. Be careful how you interpret this. If there are 11 voters with two alternatives A and B, then 6 votes for A is a majority. If there are 12 voters and A receives 6 votes, then A does not have a majority. Finally, if there are more than two alternatives, say, A, B, and C, with 12 voters and A receives 5 votes, B receives 3 votes, and C receives 4 votes, then no alternative has a majority. We summarize with the following box.

Majority Rule

> If the number of votes is n and n is even, then a majority is $\frac{n}{2} + 1$.
>
> If the number of votes is odd, then a majority is $\frac{n+1}{2}$.

©AFP/CORBIS

2000 Republican National Convention in Philadelphia

EXAMPLE 1

Consider an election with three alternatives. These might be Republican, Democrat, and Green party candidates. To keep our notation simple, we will designate the candidates as A, B, and C. There are 6 possible rankings, regardless of the number of voters. We list these using the following notation:

  Choices: (ABC)(ACB)(BAC)(BCA)(CAB)(CBA)

The symbol "(ACB)" means that the ranking of a voter is candidate A for first place, C for second place, and B for third place. The six sets of three letters here indicate all possible ways for a voter to rank three candidates. Now, suppose there are 12 voters and we list the rankings of these voters using the following notation:

Choices:	(ABC)	(ACB)	(BAC)	(BCA)	(CAB)	(CBA)
No. of votes:	5	0	2	1	0	4

This means that 5 of the 12 voters ranked the candidates ABC, whereas none of the voters ranked the candidates in the order ACB. Notice that we have accounted for all 6 possibilities even though in some cases some possibilities have no voters. The sum of the number of votes for all possibilities equals the number of voters—12 for this example. Use the majority rule to find a winner.

Solution For the election we look only at first choices (even though the voters ranked all the candidates); we see A received 5 votes ($5 + 0 = 5$), B received 3 votes ($2 + 1 = 3$), and C received 4 votes ($0 + 4 = 4$). If we use the majority rule, $\frac{12}{2} + 1 = 7$ votes, we see that there is no winner. ◆

Notice from Example 1 that with three candidates, we listed 6 possibilities. In social choice theory, the principle that asserts that any set of individual rankings is possible is called the *principle of unrestricted domain.* If there are *n* candidates, then there are *n* first choices, *n* − 1 second choices, and so on. So, by the fundamental counting principle, the total number of choices is

$$n(n - 1)(n - 2) \cdots 3 \cdot 2 \cdot 1$$

We denote this product by writing *n*!. That is, $5! = 5 \cdot 4 \cdot 3 \cdot 2 \cdot 1$.

The majority rule satisfies a principle called *symmetry.* This principle ensures that if one voter prefers choice A to choice B and another prefers choice B to A, then their votes should cancel each other out.

Plurality Method

In the case of no winner by majority rule, we often want to have an alternative way to select a winner. By the **plurality method** we mean the winner is the candidate with the highest number of votes.

Plurality Method

> Each voter votes for one candidate. The candidate receiving the most votes is declared the winner.

EXAMPLE 2

Reconsider the Republican, Democrat, and Green Party vote from Example 1.

Choices:	(ABC)	(ACB)	(BAC)	(BCA)	(CAB)	(CBA)
No. of votes:	5	0	2	1	0	4

Who is the winner using the plurality method?

Solution   A: $5 + 0 = 5$ votes; B: $2 + 1 = 3$ votes; C: $0 + 4 = 4$ votes. The winner is the candidate with the most votes, namely, candidate A.   ◆

EXAMPLE 3

Consider the following voting situation:

Choices:	(ABC)	(ACB)	(BAC)	(BCA)	(CAB)	(CBA)
No. of votes:	3	2	2	0	1	4

Who is the winner using the plurality method?

Solution   We see the outcome is A: 5 votes ($3 + 2 = 5$), B: 2 votes ($2 + 0 = 2$), and C: 5 votes ($1 + 4 = 5$). There is no majority winner, and there is no winner using the plurality method.   ◆

Somehow we cannot be satisfied with the result of Example 3. The very nature of voting seems to imply that we want a winner, so we need some set of tie-breaking rules. In situations governed by *Robert's Rules of Order,* the chairperson is allowed to vote to make or break a tie. Social choice theory calls this the *principle of decisiveness.*

Borda Count

A common way of determining a winner when there is no majority is to assign a point value to each voter's ranking. The last-place candidate is given 1 point, each next-to-the-last candidate is given 2 points, and so on. This counting scheme, called a **Borda count,** is defined in the following box.

Borda Count

> Each voter ranks the candidates. If there are n candidates, then n points are assigned to the first choice for each voter, with $n - 1$ points for the next choice, and so on. The points for each candidate are added and if one has more votes, that candidate is declared the winner.

We illustrate this process in Example 4.

EXAMPLE 4

Consider the following election with four candidates, A, B, C, and D. There are $4! = 4 \times 3 \times 2 \times 1 = 24$ possible rankings for these four candidates, so instead of listing all those possibilities as we did in the previous examples of this section, we ask the voters to rank their choices.

Voter	Ranking
#1	B, D, C, A
#2	D, C, A, B
#3	B, A, C, D
#4	B, A, D, C
#5	D, A, B, C
#6	A, B, C, D

Note the results: A: 1 vote; B: 3 votes; C: 0 votes; D: 2 votes. There are a total of 6 voters, so a majority would require $6/2 + 1 = 4$ votes, so there is no majority. Declare a winner using a Borda count.

Solution   A Borda count with 4 candidates awards 4 points for each first-place ranking, 3 points for each second-place ranking, followed by 2 points and 1 point. We summarize the voters' rankings in table form:

				Voter			
Candidate	#1	#2	#3	#4	#5	#6	*Total*
A	1	2	3	3	3	4	16
B	4	1	4	4	2	3	18
C	2	3	2	1	1	2	11
D	3	4	1	2	4	1	15

The winner is B, since this candidate has the highest total.   ◆

An example of voting using the Borda method with which you may be familiar is the annual voting for the Heisman Trophy in collegiate football. In 2001, the winner Eric Crouch of the University of Nebraska was selected after 871 ballots were mailed to media personnel across the country, 53 Heisman winners, and one Suzuki fan, for a total of 925 electors. Each elector votes for three choices and a point total is reached by a system of three points for a first place vote, two for a second, and one for a third. It was reported that Crouch received 770 points, with the runner-up Rex Grossman of the University of Florida garnering 708 points.

The principle of decisiveness forces us to consider a structure for selecting a winner when the method we use does not produce a winner. One such method is to hold a runoff election. A runoff election is an attempt to obtain a majority vote by eliminating one or more alternatives and voting again on the remaining choices.

Hare Method

The first runoff method we will consider was proposed in 1861 by Thomas Hare (1806–1891). In this method, votes are transferred from eliminated candidates to remaining candidates. We summarize this method in the following box.

Hare Method

> Each voter votes for one candidate. If a candidate receives a majority of the votes, that candidate is declared to be a *first-round winner*. If no candidate receives a majority of the votes, then with the **Hare method,** also known as the *plurality with an elimination runoff method,* then the candidate(s) with the fewest number of first-place votes is (are) eliminated.
>
> Each voter votes for one candidate in the second round. If a candidate receives a majority, that candidate is declared to be a *second-round winner.* If no candidate receives a majority of the second-round votes, then eliminate the candidate(s) with the fewest number of first-place votes. Repeat this process until a candidate receives a majority.

EXAMPLE 5

Reconsider Example 3:

Choices:	(ABC)(ACB)(BAC)(BCA)(CAB)(CBA)
No. of votes:	3 2 2 0 1 4

There is neither a majority winner nor a plurality winner. Find a solution using the Hare method.

Solution   We see that A receives 5 (3 + 2 = 5) first-round votes; B receives 2 votes; and C receives 5 votes. We hold a runoff election by eliminating the alternative with the fewest votes; this is choice B. For convenience, in this book we assume that a voter's order of preference will remain the same for subsequent rounds of voting. Thus, we now have the following possibilities, where we have crossed out candidate B:

Choices:	(AB̸C)(ACB̸)(B̸AC)(B̸CA)(CAB̸)(CB̸A)
No. of votes:	3 2 2 0 1 4
	↓ ↓ ↓ ↓ ↓ ↓

Choices:	(AC)	(CA)
No. of votes:	3 + 2 + 2 = 7	0 + 1 + 4 = 5

We now declare a second-round winner, A, using the *majority rule*.   ◆

The assumption that we made in Example 5 about consistent voting has a name in social choice theory. It is called the *principle of independence of irrelevant alternatives.* In other words, we assume consistent voting, which means that if a voter prefers A to B with C a possible choice, then the voter still prefers A to B when C is not a possible choice.

In countries with many political parties, such as France, the Hare method is used for electing their president.

Pairwise Comparison Method

Runoff elections are not always appropriate. It seems reasonable that if everyone in a group of voters prefers candidate X over candidate Y, then under its voting method, the group should prefer X to Y. Social choice theory calls this the *Pareto principle*. Thus, it is desirable that the pairwise methods we consider satisfy this principle. The characteristic property of these pairwise methods is that they pair up the competitors, two at a time. Such methods are called **binary voting.** We begin with the most important of these methods.

Pairwise Comparison Method

> In the **pairwise comparison method** of voting, the voters rank the candidates by making a series of comparisons in which each candidate is compared to each of the other candidates. If choice A is preferred to choice B, then A receives 1 point. If B is preferred to A, then B receives 1 point. If the candidates tie, each receives $\frac{1}{2}$ point. The candidate with the most points is declared the winner.

EXAMPLE 6

Consider the election described by Example 1:

Choices:	(ABC)(ACB)(BAC)(BCA)(CAB)(CBA)
No. of votes:	5 0 2 1 0 4

Find the winner using the pairwise comparison method.

Solution   A over B: $5 + 0 + 0 = 5$; B over A: $2 + 1 + 4 = 7$
B obtains 1 point.

A over C: $5 + 0 + 2 = 7$; C over A: $1 + 0 + 4 = 5$
A obtains 1 point.

B over C: $5 + 2 + 1 = 8$; C over B: $0 + 0 + 4 = 4$
B obtains 1 point.

Totals:

     A: 1 point

     B: $1 + 1 = 2$ points

     C: 0 points

Candidate B wins the election by the pairwise comparison method.   ◆

Tournament Method

Another form of binary voting is a *king-of-the-hill* situation in which the competitors are paired, the winner of one pairing taking on the next competitor. These types of runoff elections are sometimes called **sequential voting.**

     One of the most common examples of sequential voting is called the **tournament method.** With this method, candidates are teamed head-to-head, with the winner of one pairing facing a new opponent for the next election. Tennis matches and other sporting events are often decided in this fashion.

EXAMPLE 7

Consider the following election:

Choices:	(ABC)(ACB)(BAC)(BCA)(CAB)(CBA)
No. of votes:	5 0 3 0 0 4

Find the winner using the tournament method.

Solution

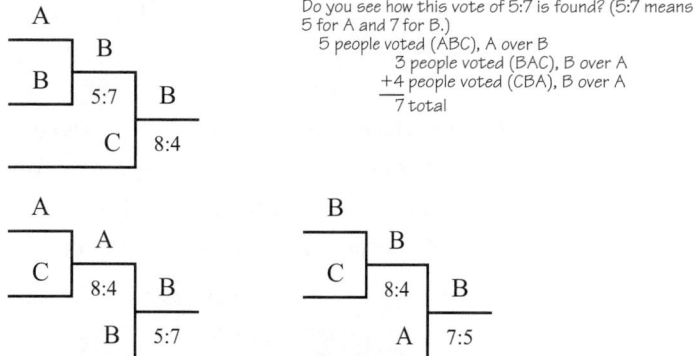

Do you see how this vote of 5:7 is found? (5:7 means 5 for A and 7 for B.)
5 people voted (ABC), A over B
  3 people voted (BAC), B over A
 +4 people voted (CBA), B over A
  7 total

The tournament pairing finds B to be the winner.   ◆

In Example 7 there were three candidates, and the number of comparisons charts was three (A with B, A with C, and B with C). In general, the number of necessary comparisons for *n* candidates is

$$\frac{n(n-1)}{2}$$

In this book we will not have examples with more than 5 candidates, so Table 16.1 shows the number of choices for *n* between 3 and 6 (inclusive).

Even though we can use some tie-breaking voting procedures, ties may still exist. Decisiveness requires that we specify some method for breaking ties. Sometimes *breaking a tie* can be accomplished by using another voting method, by choosing the candidate with the most first-place votes, by voting by the presiding officer, or even by flipping a coin. If the voting process is to be fair, the tie-breaking procedure should be specified before the vote.

Table 16.1 Number of Pairwise Comparisons

n	Number
3	3
4	6
5	10
6	15

EXAMPLE 8

Consider the following voting situation:

Choices: (ABC)(ACB)(BAC)(BCA)(CAB)(CBA)

No. of votes: 2   0   0   2   1   0

a. Is there a majority winner?

b. Is there a plurality winner?

c. If there is a tie, break it by deleting the candidate with the fewest first-place votes.

d. If there is a tie, break it by deleting the candidate with the most last-place votes.

Solution

a. A majority of the 5 votes is 3 votes; there is no majority.

b. There is no plurality.

c. There is a tie; we drop the candidate with the *fewest first-place votes*. (Recall that this is called the Hare method.)

 Number of first-place votes:

   A: 2 + 0 = 2

   B: 0 + 2 = 2

   C: 1 + 0 = 1

Choice C has the fewest first-place votes; compress the ratings:

Choices:	(AB)	(BA)
No. of votes:	$2 + 0 + 1 = 3$	$0 + 2 + 0 = 2$

The winner is A.

d. There is a tie; we drop the candidate with the *most last-place votes*.

A is in last place for $2 + 0 = 2$ of the votes.

B is in last place for $0 + 1 = 1$ of the votes.

C is in last place for $2 + 0 = 2$ of the votes.

There is a tie, so we use this rule again; drop the candidate with the *most last-place votes* or *next-to-last-place votes*.

A has $0 + 1 = 1$ next-to-the-last-place votes and 2 last-place votes, for a total of 3 votes.

B has $2 + 0 = 2$ next-to-the-last-place votes and 1 last-place vote, for a total of 3 votes.

C has $0 + 2 = 2$ next-to-the-last-place votes and 2 last-place votes, for a total of 4 votes.

Delete C and compress the ratings:

(AB)	(BA)
$2 + 0 + 1 = 3$	$0 + 2 + 0 = 2$

The winner is A.   ◆

Approval Voting

Historically, the most recent voting method replaces the "one person, one vote" method with which we are familiar in the United States with a system that allows a voter to cast one vote for each of the candidates. There is no limit on the number of candidates an individual can vote for.

Approval Voting Method

> The **approval voting method** allows each voter to cast one vote for each candidate that meets with his or her approval. The candidate with the most votes is declared the winner.

This is the method used to select the secretary general of the United Nations and is popular in those countries in which there are many candidates. It was designed, in part, to prevent the election of minority candidates in multicandidate contests.

EXAMPLE 9

The Milwaukee Booster Club (7 members) is meeting to decide ways to bring people to the downtown area, and they are brainstorming ideas. Here is a list of their suggestions:

A: Hold an art show.

B: Hire a big-time consultant.

C: Hold a contest.

R: Advertise on the radio.

N: Advertise in the newspaper.

Here are the members' rankings and the number of choices within each ranking that they intend to vote for.

(ABCRN)	(RNCAB)	(NRACB)	(CANRB)
1	2	1	3

intent: vote for 5   vote for 3   vote for 5   vote for 2

a. What is the maximum number of votes that can be cast?

b. What is the outcome if each voter votes for his or her top three?

c. What is the outcome if each voter votes the number shown by his or her intent.

Solution

a. Since there are 7 members and 5 choices, the maximum number of possible votes is $7 \times 5 = 35$.

b.

					Total
A:	1		1	3	**5**
B:	1				**1**
C:	1	2		3	**6**
R:		2	1		**3**
N:		2	1	3	**6**

We see that advertising in the newspaper (N) and holding a contest (C) tie with 6 votes each.

c.

					Total
A:	1		1	3	**5**
B:	1		1		**2**
C:	1	2	1	3	**7**
R:	1	2	1		**4**
N:	1	2	1		**4**

We see that holding a contest (C) wins with 7 votes.   ◆

We conclude with an example comparing the voting methods introduced in this section. These methods are summarized in Table 16.2 on page 810.

EXAMPLE 10

The town of Ferndale has four candidates running for mayor: the town barber, Darrell; the fire chief, Clough; the grocer, Abel; and a housewife, Belle. A poll of 1,000 voters shows the following results:

(D, A, B, C)	(C, A, B, D)	(C, A, D, B)	(B, A, D, C)
225	190	210	375

a. How many different votes are possible (4 are shown)? What is the vote for the possibilities not shown?

b. Is there a majority winner?

c. Is there a winner using the plurality method?

d. What is the Borda count, and is there a winner using this method?

Table 16.2 Summary of Voting Methods

Method	Description
Majority method	Each voter votes for one candidate. If the number of voters is n and n is even, then the candidate with $\frac{n}{2} + 1$ or more votes is the winner. If the number n is odd, then the candidate with $\frac{n+1}{2}$ or more votes is the winner.
Plurality method	Each voter votes for one candidate. The candidate receiving the most votes is the winner.
Borda count method	Each voter ranks the candidates. Each last-place candidate is given 1 point, each next-to-the-last candidate is given 2 points, and so on. The candidate with the highest number of points is the winner.
Hare method	Each voter votes for one candidate. If a candidate receives a majority of the votes, that candidate is the winner. If no candidate receives a majority, eliminate the candidate with the fewest first-place votes and repeat the process until there is a majority candidate, who is the winner.
Pairwise comparison method	Each voter ranks the candidates. Each candidate is compared to each of the other candidates. If choice A is preferred to choice B, then A receives 1 point. If B is preferred to A, then B receives 1 point. If the candidates tie, then each receives $\frac{1}{2}$ point. The candidate with the most points is the winner.
Tournament method	This method compares the entire slate of candidates two at a time, in a predetermined order. The first and second candidates are compared, the candidate with the fewer votes is eliminated, and the winner is then compared with the third candidate. These pairwise comparisons continue until the final pairing, which selects the winner.
Approval method	Each voter casts one vote for all candidates that meet with his or her approval. The candidate with the most votes is declared the winner.

Solution

a. There are $4! = 24$ possibilities; there are no votes for the 20 possibilities not shown.

b. There is no majority winner: A: 0 votes; B: 375 votes; C: $190 + 210 = 400$ votes; D: 225 votes.

c. The plurality winner is C (because it has the most votes).

d. We show the Borda count in the following table:

	A	B	C	D	Total A	B	C	D
225:	3	2	1	4	675	450	225	900
190:	3	2	4	1	570	380	760	190
210:	3	1	4	2	630	210	840	420
375:	3	4	1	2	1,125	1,500	375	750
TOTAL:					3,000	2,540	2,200	2,260

The Borda count declares A the winner. ◆

Problem Set 16.1

LEVEL 1

1. **IN YOUR OWN WORDS** Which of the following are examples of a dictatorship?
   a. Dad comes home with a surprise for the family—he just bought tickets for the family to vacation in Hawaii.
   b. The choice of selecting the champagne for the toast is left to the cellar master at the restaurant.
   c. The chairperson makes a decision to poll the members of the committee regarding their opinion about the impeachment vote.
   d. The department chair decides that the meeting will be held at 4:00 P.M. on Thursday.

2. **IN YOUR OWN WORDS** Describe and discuss the plurality voting method.

3. **IN YOUR OWN WORDS** Describe and discuss the Borda count voting method.

4. **IN YOUR OWN WORDS** Describe and discuss the Hare voting method.

5. **IN YOUR OWN WORDS** Describe and discuss the pairwise comparison method.

6. **IN YOUR OWN WORDS** Describe and discuss the tournament method.

7. **IN YOUR OWN WORDS** Describe and discuss the approval voting method.

8. **IN YOUR OWN WORDS** Give one example in which you have participated in voting where the count was tabulated by the plurality voting method. Your example can be made up or factual, but you should be specific.

9. **IN YOUR OWN WORDS** Give one example in which you have participated in voting where the count was tabulated by the Borda count method. Your example can be made up or factual, but you should be specific.

10. **IN YOUR OWN WORDS** Give one example in which you have participated in voting where the count was tabulated by the Hare voting method. Your example can be made up or factual, but you should be specific.

11. **IN YOUR OWN WORDS** Give one example in which you have participated in voting where the count was tabulated by using the tournament method. Your example can be made up or factual, but you should be specific.

12. **IN YOUR OWN WORDS** Give one example in which you have participated in voting where the count was tabulated by the approval voting method. Your example can be made up or factual, but you should be specific.

In voting between three candidates the outcomes are reported as:

(ABC)(ACB)(BAC)(BCA)(CAB)(CBA)
  8   4   3   0   2   5

Use this information to answer the questions in Problems 13–18.

13. What is the total number of votes?

14. How many votes would be necessary for a majority?

15. a. What does the notation (CAB) mean?
    b. What does the "5" under (CBA) mean?

16. a. What does the notation (ACB) mean?
    b. What does the "4" under (ACB) mean?

17. a. If a person ranks A as their first choice, B as their second choice, and C last, how would this be written?
    b. How many voters have this preference?

18. a. What does the notation (BCA) mean?
    b. What does the "0" under (BCA) mean?

In voting between four candidates the outcomes are reported as:

(ADBC)(DACB)(BADC)(BDCA)(DCAB)
   8   5   3   1   2

Use this information to answer the questions in Problems 19–24.

19. What is the total number of votes?

20. How many votes would be necessary for a majority?

21. a. What does the notation (DACB) mean?
    b. What does the "5" under (DACB) mean?

22. a. What does the notation (ADBC) mean?
    b. What does the "8" under (ADBC) mean?

23. a. If a person ranks B as their first choice, C as their second choice, D as their third choice, and A last, how would this be written?
    b. How many voters have this preference?

24. a. How many possibilities are there for voter preferences with four candidates?
    b. Name those possibilities that have 0 voter preferences.

25. a. How many different ways can a voter rank 3 candidates (no ties are allowed)?
    b. How many different ways can a voter rank 4 candidates (no ties are allowed)?
    c. How many different ways can a voter rank 5 candidates (no ties are allowed)?

LEVEL 2

26. a. If there are 10 voters and 5 candidates, how many total points would there be in a Borda count?
    b. If there are 20 voters and 4 candidates, how many total points would there be in a Borda count?
    c. If there are 200 voters and 3 candidates, how many total points would there be in a Borda count?

27. How many different ways can a voter rank n candidates (no ties are allowed)?

28. It can be shown that

$$1 + 2 + 3 + \cdots + m = \frac{m(m + 1)}{2}$$

Use this formula to determine how many total points there would be in a Borda count with n voters and m candidates.

In voting between three candidates the outcomes are reported as:

(ABC)(ACB)(BAC)(BCA)(CAB)(CBA)

  8   4   3   0   2   5

Determine the winner, if any, using the voting methods in Problems 29–34.

29. Majority rule   **30.** Plurality method

31. Borda count method   **32.** Hare method

33. Pairwise comparison method

34. Tournament method

35. Twelve board members are voting on after-meeting activities, and they are asked to check any that they might like. The outcome of their choices is shown here:

	1	2	3	4	5	6	7	8	9	10	11	12
Snacks	x	x	x	x	x	x		x	x	x		x
Drinks	x	x		x	x	x		x		x		x
Travel slides						x				x	x	
Guest speaker	x	x		x	x		x	x	x		x	x

What is the outcome using approval voting?

36. Twelve board members are voting on admitting two new board members. They interview 5 candidates and vote "x" for an acceptable candidate and no vote for an unacceptable candidate. The outcome of their choices is shown here:

	1	2	3	4	5	6	7	8	9	10	11	12
A	x	x	x	x	x	x		x	x	x	x	
B	x	x	x	x		x	x	x		x		x
C	x	x	x	x	x	x	x		x	x		x
D	x	x			x		x	x	x	x		x
E	x	x	x			x	x	x			x	x

What is the outcome using approval voting?

37. In the 2000 race for the governor of Vermont, the state vote was as follows:

Marilyn Christian	1,054
Howard Dean	148,059
Ruth Dwyer	111,359
Richard Gottlieb	337
Hardy Macia	785
Anthony Pollina	28,116
Phil Stannard	2,148
Joel Williams	1,359
Others	255
TOTAL	293,472

Was there a majority winner in this election, and if so, who was it?

38. In the 2000 race for the 36th Congressional District of California, the vote was as follows:

Jane Harmon	106,975
John Konopka	3,297
Steven Kuykendall	103,142
Matt Ornati	2,078
Daniel Sherman	5,615
TOTAL	221,107

Was there a majority winner in this election, and if so, who was it?

Twelve people serve on a board and are considering three alternatives: A, B, and C. Here are the choices followed by vote:

(ABC)(ACB)(BAC)(BCA)(CAB)(CBA)

  2   4   2   1   2   1

Determine the winner, if any, using the voting methods in Problems 39–44.

39. Majority rule

40. Plurality method

41. Borda count method

42. Hare method

43. Pairwise comparison method

44. Tournament method

Seventeen people serve on a board and are considering three alternatives: A, B, and C. Here are the choices followed by vote:

(ABC)(ACB)(BAC)(BCA)(CAB)(CBA)

  1   3   4   3   5   1

Determine the winner, if any, using the voting methods in Problems 45–50.

45. Majority rule

46. Plurality method

47. Borda count method

48. Hare method

49. Pairwise comparison method

50. Tournament method

LEVEL 3

Suppose your college transcripts show the following distribution of grades:

  A: 2

  B: 6

  C: 5

  D: 1

  F: 0

If all of these grades are in three-unit classes, use this information to answer the questions in Problems 51–52.

51. a. Which grade is the most common?
   b. Which voting method describes how you answered part **a**?

52. a. What is your GPA?
   b. Which voting method describes how you answered part **a**?

Suppose your college transcripts show the following distribution of grades:

A: 14

B: 21

C: 35

D: 5

F: 2

If all of these grades are in three-unit classes, use this information to answer the questions in Problems 53–54.

53. a. Which grade is the most common?
   b. Which voting method describes how you answered part **a**?

54. a. What is your GPA?
   b. Which voting method describes how you answered part **a**?

In Problems 55–59, consider the following situation. A political party holds a national convention with 1,100 delegates. At the convention, five persons (which we will call A, B, C, D,

and E) have been nominated as the party's presidential candidate. After the speeches and hoopla, the delegates are asked to rank all five candidates according to his or her choice. However, before the vote, caucuses have narrowed the choices down to six different possibilities. The results of the first ballot are shown (choices, followed by the number of votes):

(ADEBC)(BEDCA)(CBEDA)

   360   240   200

(DCEBA)(EBDCA)(ECDBA)

   180   80   40

Answer questions and give reasons.

55. How many possible rankings are there?

56. a. Is there a majority winner? If so, who?
   b. Is there a plurality winner? If so, who?

57. Who would win in a runoff election using the principle of eliminating the candidate with the fewest first-place votes?

58. Who would win in a runoff election using the principle of eliminating the candidate with the most last-place votes?

59. What is the Borda count for each person? Who wins the Borda count?

60. IN YOUR OWN WORDS By looking at your answers to Problems 56–59, who would you say should be declared the winner? Look at the title of the next section. Relate your answer to this question for the need to study the next section.

16.2   VOTING DILEMMAS

Principles in social choice theory do not behave in the same fashion as principles of mathematics. In mathematics, a correctly stated principle has no exceptions. On the other hand, we frequently find exceptions to voting principles. In the last section, we considered four reasonable and often-used voting methods. In this section, we will consider four voting principles that most would agree are desirable, and then we will show that all of the voting methods will fail one or more of the principles. We will call these **fair voting principles:** *majority criterion, Condorcet criterion, monotonicity criterion,* and *irrelevant alternatives criterion.* Let us consider these voting principles, one at a time.

Majority Criterion

The first and most obvious criterion is called the **majority criterion.**

Majority Criterion

> If a candidate receives a majority of the first-place votes, then that candidate should be declared the winner.

Only a Borda count method can violate this criterion. Consider the following example.

EXAMPLE 1

The South Davis Faculty Association is using the Borda count method to vote for their collective bargaining representative. Their choices are the All Faculty Association (A), American Federation of Teachers (B), and California Teachers Association (C). Here are the results of the voting:

$$(ABC)(ACB)(BAC)(BCA)(CAB)(CBA)$$

Votes: 16   0   0   8   0   7

a. Which organization is selected for collective bargaining?

b. Does the choice in part **a** violate the majority criterion?

Solution

a. Here is a tally of the Borda count:

				Totals		
	A	B	C	A	B	C
(ABC) 16:	3	2	1	48	32	16
(BCA) 8:	1	3	2	8	24	16
(CBA) 7:	1	2	3	7	14	21
Total:				63	70	53

The highest Borda count number goes to choice B, the American Federation of Teachers.

b. The majority votes are A: 16 votes, which is a majority of the 31 votes that were cast, so the Borda count violates the majority criterion.   ◆

Example 1 shows that the Borda count method can violate the majority criterion. However, all the other methods must satisfy this criterion. Suppose that a candidate X is the first choice for more than half the voters. It follows that X will have more first-place votes than any other single candidate and must win by the plurality method. If the Hare method is used, then X would always have at least the votes that it started with, and since that is more than half the votes, X could never be eliminated and would wind up the winner. And finally, since X has the majority of the votes in each of its pairwise matchups, X would win in each of those matchups. No other candidate can win as many pairs as X does, so X wins the election.

Thus, we conclude that the Borda count method presents a dilemma. Although it takes into account voters' preferences by having all candidates ranked, a candidate with a majority of first-place votes can lose an election!

Condorcet Criterion

About a decade after Borda proposed his counting procedure, the mathematician Marquis de Condorcet became interested in some of the apparent dilemmas raised by the Borda count methods. He proposed a head-to-head election to rank the candidates. The candidate who wins all the one-to-one matchups is the **Condorcet candidate.** The **Condorcet criterion** asserts that the Condorcet candidate should win the election. Some elections do not yield a Condorcet candidate because none of the candidates can win over *all* the others.

Condorcet Criterion

> If a candidate is favored when compared one-on-one with every other candidate, then that candidate should be declared the winner.

Before most major elections in the United States, we hear the results of polls for each of the political parties pairing candidates and telling preferences in a one-on-one election. Is this a valid way of considering the candidates? Consider the following example.

EXAMPLE 2

The seniors at Weseltown High School are voting on where to go for their senior trip. They are deciding on Angel Falls (A), Bend Canyon (B), Cedar Lake (C), or Danger Gap (D). The results of the preferences are:

(DABC)	(ACBD)	(BCAD)	(CBDA)	(CBAD)
120	100	90	80	45

a. Who is the Condorcet candidate?

b. Is there a majority winner? If not, is there a plurality winner? Does this violate the Condorcet criterion?

c. Who wins the Borda count? Does this violate the Condorcet criterion?

d. Who wins using the Hare method? Does this violate the Condorcet criterion?

e. Who wins using the pairwise comparison method? Does this violate the Condorcet criterion?

Solution

a. The best way to examine the one-on-one matchups is to construct a table with all possibilities listed as the row and column headings.

	A	B	C	D
A	—	*		
B	*	—		
C			—	
D				—

* This is for A with B; this means that A and B are matched.

$$\underbrace{\text{(DABC)(ACBD)}}_{\text{A wins}} \quad \underbrace{\text{(BCAD)(CBDA)(CBAD)}}_{\text{B wins}}$$

For the numbers, look at the line right under the preferences:

$$120 + 100 = 220 \qquad 90 + 80 + 45 = 215; \qquad \text{A wins.}$$

	A	B	C	D
A	—	**A**		
B	**A**	—		
C			—	
D				—

We similarly fill in the rest of the table by comparing the items, one-on-one:

A with C: A: $120 + 100 = 220$; C: $90 + 80 + 45 = 215$; A wins.

A with D: A: $100 + 90 + 45 = 235$; D: $120 + 80 = 200$; A wins.

B with C: B: $120 + 90 = 210$; C: $100 + 80 + 45 = 225$; C wins.

B with D: B: $100 + 90 + 80 + 45 = 315$; D: 120; B wins.

C with D: C: $100 + 90 + 80 + 45 = 315$; D: 120; C wins.

We complete the table as shown:

	A	B	C	D
A	—	A	A	A
B	A	—	C	B
C	A	C	—	C
D	A	B	C	—

We see that the Condorcet choice is A (Angel Falls) since the column headed A and the row headed A both have all entries of A.

b. The first place votes are:

A: 100

B: 90

C: $80 + 45 = 125$

D: 120

Since there were 435 votes cast, a majority would be

$$\frac{435 + 1}{2} = 218 \text{ votes}$$

There is no majority. The winner of the plurality vote is C, Cedar Lake. This example shows that the plurality method can violate the Condorcet criterion.

c. The Borda count is shown in the following table:

					Total			
	A	B	C	D	A	B	C	D
120:	3	2	1	4	360	240	120	480
100:	4	2	3	1	400	200	300	100
90:	2	4	3	1	180	360	270	90
80:	1	3	4	2	80	240	320	160
45:	2	3	4	1	90	135	180	45
TOTALS:					1110	1175	1190	875

Location C, Cedar Falls, wins the Borda count. This example shows that the Borda count can violate the Condorcet criterion.

d. Using the Hare method, there is no majority, so we eliminate the candidate with the fewest first-place votes; this is choice B. The remaining tally is

A: 100   C: $90 + 80 + 45 = 215$ D: 120

Since a majority vote is 216, there is still no majority, so we now eliminate A. The result now is:

C: $100 + 90 + 80 + 45 = 315$ D: 120

The declared winner is C, Cedar Falls. This example demonstrates that the Hare method can violate the Condorcet criterion.

e. For the pairwise comparison method, we can use the table of pairings we had in part **a** to award the points:

A is favored over B, C, and D, giving A 3 points.

B is favored over D, giving B 1 point.

C is favored over B and D, giving C 2 points.

D is not favored.

The choice is A, Angel Falls. Note that if a certain choice is favored over all other candidates, then this candidate will have the largest point value. Thus, the pairwise comparison method can never violate the Condorcet criterion. ◆

Monotonicity Criterion

Another property of voting has to do with elections that are held more than once. Historically, there have been many pairings of the same two candidates, and at a personal level, we are often part of a process in which a nonbinding vote is taken before all the discussion takes place. Such a vote is known as a **straw vote.** It would seem obvious that if a winning candidate in the first election gained strength before the second election, then that candidate should win the second election. A statement of this property is called the **monotonicity criterion.**

Monotonicity Criterion

> A candidate who wins a first election and then gains additional support, without losing any of the original support, should also win a second election.

As obvious as this criterion may seem, the following example introduces another voting dilemma by showing it is possible for the winner of the first election to gain additional support before a second election, and then lose that second election.

EXAMPLE 3

In 1995 the 105th International Olympic Committee met in Budapest to select the 2002 Winter Olympic site. The cities in the running were Québec (Q), Salt Lake City (L), Östersund (T), and Sion (S). Consider the following fictional account of how the voting might have been conducted. The voting takes place over a two-day period using the Hare method. The first day the 87 members of the IOC take a nonbinding vote, and then on the second day they take a binding vote.

a. On the first day, the rankings of the IOC members were

    (TLSQ)(LQTS)(QSTL)(TQSL)(TSLQ)
      21   24   30   6   6

What are the results of the election using the Hare method for the first nonbinding day of voting?

b. On the evening of the first day of voting, representatives from Salt Lake City offered bribes to the 12 members with the bottom votes. They were able to convince these IOC members to move Québec to the top of their list, because, after all, Québec won the day's straw votes anyway. Now for the second day, the rankings of the IOC committee were:

    (TLSQ)(LQTS)(QSTL)(QTSL)(QLTS)
      21   24   30   6   6

What are the results of the election using the Hare method for the second binding day of voting?

Solution

a. A majority is $\dfrac{87 + 1}{2} = 44$ votes. On the first day,

Round 1: T: 21 + 6 + 6 = 33

L: 24

Q: 30

S: 0

No city has a majority (45) of votes, so Sion is eliminated from the voting.

Round 2: T: 21 + 6 + 6 = 33

L: 24

Q: 30

No city has a majority, so now Salt Lake City is eliminated from the voting.

Round 3: T: 21 + 6 + 6 = 33

Q: 24 + 30 = 54

Québec has a majority of the votes and is the winner from the first nonbinding day of voting using the Hare method.

b. On the second day,

Round 1: T: 21

L: 24

Q: 30 + 6 + 6 = 42

S: 0

No city has a majority (44) of votes, so Sion is eliminated from the voting.

Round 2: T: 21

L: 24

Q: 30 + 6 + 6 = 42

No city has a majority, so now Ostersund (T) is eliminated from the voting.

Round 3: L: 21 + 24 = 45 votes

Q: 30 + 6 + 6 = 42

Salt Lake City has a majority of the votes and is the winner using the Hare method.

◆

As you can see from this remarkable example, it is possible for the winning candidate on a first vote (Québec) to receive more votes and end up losing the election! We see (from Example 3) the Hare method can violate the monotonicity criterion. It is also possible to find examples showing that Borda's and the pairwise comparison methods can also violate the monotonicity criterion. The plurality method cannot violate the monotonicity criterion.

Irrelevant Alternatives Criterion

In the controversial 2000 presidential election, there was much talk about the final vote of the election. Although there is no such thing as an official final figure, the numbers in Table 16.3 are an aggregate of state numbers that appear to be final. Suppose the president were selected by popular vote (rather than the Electoral College). As close as the election was, if we use the numbers in Table 16.3, we see that the winner would have been Al Gore. Now, suppose that another election were held, and this time Ralph Nader dropped out before the vote. Since Nader really had no chance of winning, we

Table 16.3 Summary of Popular and Electoral Vote in the 2000 U.S. Presidential Election

Candidate	Party	Vote	Percentage	Electoral College Vote
Harry Browne	Libertarian	386,024	00.37	0
Pat Buchanan	Reform	448,750	00.42	0
George W. Bush	Republican	50,456,167	47.88	271
Al Gore	Democrat	50,996,277	48.39	267
Ralph Nader	Green	2,864,810	02.72	0
14 others		238,300	00.23	0
TOTAL		105,390,115		538

might conclude this action should not have any effect on the outcome. But as you can see from these numbers, such is not the case. The Nader voters could have swung the election either way. We might consider this a voting dilemma because it would violate the following criterion, called the **irrelevant alternatives criterion.**

Irrelevant Alternatives Criterion

If a candidate is declared the winner of an election, and in a second election one or more of the other candidates is removed, then the previous winner should still be declared the winner.

EXAMPLE 1

The mathematics department (22 members) has interviewed five candidates for a new instructor position. They are Alicia (A), Benito (B), Carmelia (C), Doug (D), and Erin (E). There are $5! = 120$ possible rankings, but the voting process has narrowed the voting to seven possible rankings:

(BDCEA)(BDEAC)(EDACB)(ACEBD)(DECBA)(CBDEA)(CEDBA)

  6   4   4   4   2   1   1

Just before the voting, Carmelia withdraws from the hiring process.

a. Is there a majority winner? If not, who is the plurality winner?

b. Who is the Borda count winner?

c. Who wins using the Hare method?

d. Use the pairwise comparison method to determine the winner.

e. Suppose one of the department members convinces the others that voting should take place without excluding Carmelia. Who wins the pairwise comparison method?

Solution   If we delete Carmelia from the process, the vote becomes

(BDEA)(EDAB)(AEBD)(DEBA)(EDBA)

  11   4   4   2   1

Note, with C deleted there are three of these voting preferences that are the same, so we added $6 + 4 + 1 = 11$.

a. The first place votes are

A: 4

B: 11

D: 2

E: 4 + 1 = 5

A majority is $\frac{22}{2} + 1 = 12$, so there is no majority winner. We see that Benito is the plurality winner.

b. The Borda count numbers are

	A	B	D	E		Total		
					A	B	D	E
11:	1	4	3	2	11	44	33	22
4:	2	1	3	4	8	4	12	16
4:	4	2	1	3	16	8	4	12
2:	1	2	4	3	2	4	8	6
1:	1	2	3	4	1	2	3	4
TOTAL:					38	62	60	60

Benito is the Borda count winner.

c. The fewest first-place votes are for Doug, so when he is eliminated for the second round, the results of this round are

A: 4

B: 11

E: 4 + 2 + 1 = 7

There is still no majority, so on this round Alicia is eliminated. The third-round results are

B: 11

E: 4 + 4 + 2 + 1 = 11

There is a tie. The chairperson could break the tie, or another method could be chosen to break the tie.

d. A over B: 4 + 4 = 8; and B over A: 11 + 2 + 1 = 14
B wins 1 point.

A over D: 4; and D over A: 11 + 4 + 2 + 1 = 18
D wins 1 point.

A over E: 4; and E over A: 11 + 4 + 2 + 1 = 18
E wins 1 point.

B over D: 11 + 4 = 15; and D over B: 4 + 2 + 1 = 7
B wins 1 point.

B over E: 11; and E over B: 4 + 4 + 2 + 1 = 11
Tie; B wins 1/2 point and E wins 1/2 point.

D over E: 11 + 2 = 13; and E over D: 4 + 4 + 1 = 9
D wins 1 point.

The tally of points is

> A: 0
>
> B: $2\frac{1}{2}$ points
>
> D: 2 points
>
> E: $1\frac{1}{2}$ points

Benito is the winner using the pairwise comparison method.

e. From the solutions to parts **a–d,** it certainly appears that Benito should be the victor. But what effect did Carmelia's withdrawal from the process have on the outcome? The irrelevant alternatives criterion says it should have no effect. However, this example shows us otherwise. Consider the original rankings, and calculate the winner using the pairwise comparison method.

A over B: $4 + 4 = 8$; and B over A: $6 + 4 + 2 + 1 + 1 = 14$
B wins 1 point.

A over C: $4 + 4 + 4 = 12$; and C over A: $6 + 2 + 1 + 1 = 10$
A wins 1 point.

A over D: 4; and D over A: $6 + 4 + 4 + 2 + 1 + 1 = 18$
D wins 1 point.

A over E: 4; and E over A: $6 + 4 + 4 + 2 + 1 + 1 = 18$
E wins 1 point.

B over C: $6 + 4 = 10$; and C over B: $4 + 4 + 2 + 1 + 1 = 12$
C wins 1 point.

B over D: $6 + 4 + 4 + 1 = 15$; and D over B: $4 + 2 + 1 = 7$
B wins 1 point.

B over E: $6 + 4 + 1 = 11$; and E over B: $4 + 4 + 2 + 1 = 11$
Tie; B wins 1/2 point and E wins 1/2 point.

C over D: $4 + 1 + 1 = 6$; and D over C: $6 + 4 + 4 + 2 = 16$
D wins 1 point.

C over E: $6 + 4 + 1 + 1 = 12$; and E over C: $4 + 4 + 2 = 10$
C wins 1 point.

D over E: $6 + 4 + 2 + 1 = 13$; and E over D: $4 + 4 + 1 = 9$
D wins 1 point.

The tally of points is

> A: 1
>
> B: $2\frac{1}{2}$ points
>
> C: 2 points
>
> D: 3 points
>
> E: $1\frac{1}{2}$ points

Doug is the winner using the pairwise comparison method, and this violates the irrelevant alternatives criterion. ◆

An outstanding factual example illustrating this dilemma occurred in the 1991 Louisiana gubernatorial race. The candidates were the incumbent Governor "Buddy" Roemer and his challengers, former governor Edwin Edwards and David Duke. Now, David Duke was a former leader of the Ku Klux Klan, and the former governor was indicted for corruption, so it is reasonable to assume that Roemer would have beaten either of his opponents in a one-on-one race, but instead he came in last.

Arrow's Impossibility Theorem

We have now considered four criteria that would seem to be desirable properties of any voting system. We refer to these four criteria as the **fairness criteria.**

Fairness Criteria

Majority criterion
If a candidate receives a majority of the first-place votes, then that candidate should be declared the winner.
Condorcet criterion
If a candidate is favored when compared one on one with every other candidate, then that candidate should be declared the winner.
Monotonicity criterion
A candidate who wins a first election and then gains additional support, without losing any of the original support, should also win a second election.
Irrelevant alternatives criterion
If a candidate is declared the winner of an election, and in a second election one or more of the other candidates is removed, then the previous winner should still be declared the winner.

We compare these criteria with the voting methods we have considered in Table 16.4.

Table 16.4 Comparison of Voting Methods and Fairness Criteria

Voting Method:	Plurality	Hare	Borda Count	Pairwise Comparison
Criterion: **Majority**	Satisfied	Satisfied	Not satisfied (Example 1)	Satisfied
Condorcet	Not satisfied (Example 2b)	Not satisfied (Example 2d)	Not satisfied (Example 2c)	Satisfied
Monotonicity	Satisfied	Not satisfied (Example 3)	Not satisfied	Not satisfied
Irrelevant Alternatives	Not satisfied	Not satisfied	Not satisfied	Not satisfied (Example 4)

In 1951, the economist Kenneth Arrow (1921–   ) proved that there is exactly one method for voting that satisfies all four of these principles, and this method is a *dictatorship.* Stated in another way, it is known as *Arrow's paradox:* Perfect democratic voting is, not just in practice but in principle, impossible. Essentially this says that there is no perfect voting method.

EXAMPLE 5

In the voting for the 2004 Olympics site, five cities were in the running: Athens (A), Buenos Aires (B), Cape Town (C), Rome (R), and Stockholm (S). The first round votes of the 107 IOC members were

A: 32

B: 16

C: 16

R: 23

S: 20

Who wins by the

a. majority/plurality methods?

b. Hare method?

c. Pairwise comparisons method?
(Note: The voters' rankings are not available, so we can't do the Borda count.)

Solution

a. A majority would require $\dfrac{107 + 1}{2} = 54$ votes. There is no majority. The plurality vote goes to Athens.

b. The Hare method eliminates the choice with the smallest number of first-place votes. There is a tie (16), so a run-off election is held between Buenos Aires and Cape Town. Suppose the result of this election is B: 45, C: 62, so Buenos Aires is eliminated from Round 2.

Here are round 2 votes:

A: 38

C: 22

R: 28

S: 19

There is no majority, so Stockholm is eliminated.
The results from Round 3 are

A: 52

C: 20

R: 35

There is still no majority, so Cape Town is eliminated.
The results from Round 4 voting are:

A: 66

R: 41

The majority choice is Athens.

c. For the pairwise matchings:
A wins 4 points because it has more votes than any of the other choices. No other choice can beat 4 points, so Athens is the choice.   ◆

The previous example illustrates that it is not *necessary* to have a vote with contradictions. It is possible to satisfy all of the fairness criteria. However, the following example illustrates quite the reverse. Any of the candidates A, B, or C could be declared the winner using the tournament method!

EXAMPLE 6

Consider the following election:

Choices: (ABC)(ACB)(BAC)(BCA)(CAB)(CBA)

No. of votes: 1   0   0   1   1   0

Determine the winner using the tournament method.

Solution The pairing AB gives A one point; the pairing AC gives C one point; and the pairing BC gives B one point. All three are tied in points.

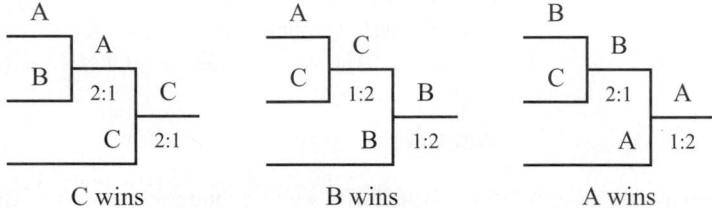

C wins   B wins   A wins

However, if we play this as a tournament, we see that any of A, B, or C could win, depending on the initial pairing. This shows that there is tremendous power in the hands of the tournament director or committee chair who has the opportunity to set the agenda, if the group choice is to be made using a tournament (pairwise) voting method. *Different agendas may produce different winners.* This is called the *agenda effect.* ◆

Insincere Voting

A variation of the agenda effect is insincere voting, or the offering of amendments with the purpose of changing an election. Consider the following example.

Assume that Tom, Ann, and Linda each have the choice of voting on whether to lower the drinking age to 18. The current law sets the drinking age at 21. Voting against the new law (age 18) means that the old law (age 21) will prevail. Let's assume that Tom and Ann are in favor of the new law, but Linda is against it. Here is a table of their preferences:

	First choice	*Second choice*
Tom:	new law (age 18)	old law (age 21)
Ann:	new law (age 18)	old law (age 21)
Linda:	old law (age 21)	new law (age 18)

Let's also assume that this law will pass or fail, depending on the outcome of the votes of these three people. If the vote is taken now, all three persons know the vote will be 2 for the new law and 1 against the new law. However, Linda decides to defeat the new law by insincere voting, and she introduces an amendment that she knows Tom would like most of all, but Ann likes least of all. Suppose that Linda knows Tom would like to have no law regarding drinking, but that Ann would find that offensive. Linda offers an amendment *pretending* to prefer the old law (age 21) over the amendment (no age) and the amendment (no age) over the new law (age 18). Here are the choices:

	First choice	*Second choice*	*Third choice*
Tom:	amendment (no age)	new law (age 18)	old law (age 21)
Ann:	new law (age 18)	old law (age 21)	amendment (no age)
Linda:	old law (age 21)	amendment (no age)	new law (age 18)

The vote is taken first for the amendment, and it passes because Tom votes for it and Linda votes insincerely by voting for the amendment, so the amendment carries with a vote of 2 to 1. Now, the vote on the floor is for no age limitation or for the old law. Tom votes for no limitation and Linda and Ann vote for the old law, which carries by a vote of 2 to 1.

What does this say? If you are sitting through a meeting conducted by *Robert's Rules of Order,* it may be better to enter the more preferred outcomes at a later stage of the discussion. The chances of success are better when there are fewer remaining votes.

There is a curious possibility that seems to violate the transitive law in mathematics. The **transitive law** states:

If A beats B, and B beats C, then A should beat C.

Example 6 violates this law and leads to a paradox. Notice that A beats B by a vote of 2 to 1; B beats C by a vote of 2 to 1. The transitive law says that A should beat C, but that is NOT the case! C beats A by a vote of 2 to 1. This paradox was first described by Marquis de Condorcet. He wrote a treatise, *Essay on the Application of Analysis to the Probability of Majority Decisions,* in 1785, and he described this paradox, which today is known as **Condorcet's paradox** or *the paradox of voting.*

We now state the result known as **Arrow's impossibility theorem,** which he proved in 1951.

Arrow's Impossibility Theorem

> No social choice rule satisfies all six of the following conditions.
>
> 1. **Unrestricted domain**   Any set of rankings is possible; if there are n candidates, then there are $n!$ possible rankings.
> 2. **Decisiveness**   Given any set of individual rankings, the method produces a winner.
> 3. **Symmetry and transitivity**   The voting system should be symmetric and transitive over the set of all outcomes.
> 4. **Independence of irrelevant alternatives**   If a voter prefers A to B with C as a possible choice, then the voter still prefers A to B when C is not a possible choice.
> 5. **Pareto principle**   If each voter prefers A over B, then the group chooses A over B.
> 6. There should be **no dictator.**

We conclude with an example that I call the Rodney King example; it forces "everyone to get along" because everyone wins!

EXAMPLE 7
_____

Example 5 used the historical vote of the International Olympic Committee to select the site for the 2004 Olympics. Remember, there were five cities in the running: Athens (A), Buenos Aires (B), Cape Town (C), Rome (R), and Stockholm (S). In this example, we replace the actual vote with a fictitious preference schedule:

(ARSCB)	(BSRCA)	(CBSRA)	(RCSBA)	(SBRCA)	(SCRBA)
36	24	20	18	8	4

a. How many preference schedules received 0 votes?

b. Who is the majority/plurality winner?

c. What is the Borda count, and who is the winner from this method?

d. Who is the winner using the Hare method?

e. Who wins from the pairwise comparison method?

f. Suppose there is a runoff in which the top two contenders of the plurality method face each other in a runoff. (What, you say . . . we have not previously considered this method! You are right, but we didn't want one of the cities to feel left out.)

Solution

a. Since there are 5 cities, there are $5! = 5 \times 4 \times 3 \times 2 \times 1 = 120$ possible preference schedules. Since we have shown 6 of them, it follows that there are 114 that received 0 votes.

b. There are 110 votes so a majority is $110/2 + 1 = 56$ votes; there is no majority. The plurality winner is Athens (A) with 36 votes.

c. We set up a table for the Borda count:

	A	B	C	R	S	Total A	B	C	R	S
36:	5	1	2	4	3	180	36	72	144	108
24:	1	5	2	3	4	24	120	48	72	96
20:	1	4	5	2	3	20	80	100	40	60
18:	1	2	4	5	3	18	36	72	90	54
8:	1	4	2	3	5	8	32	16	24	40
4:	1	2	4	3	5	4	8	16	12	20
TOTALS:						254	312	324	382	378

The Borda count winner is Rome (R).

d. There is no majority winner; the lowest first-place votes are for Stockholm, so the second-round vote is

Athens:	36
Buenos Aires:	$24 + 8 = 32$
Cape Town:	$20 + 4 = 24$
Rome:	18

The least number of votes is for Rome, so that city is eliminated. The third-round vote is:

Athens:	36
Buenos Aires:	$24 + 8 = 32$
Cape Town:	$20 + 18 + 4 = 42$

The least number of votes for this round is for Buenos Aires, so the final-round votes are

Athens:	36
Cape Town:	$24 + 20 + 18 + 8 + 4 = 74$

The Hare method winner is Cape Town (C).

e. For the pairwise comparison method, there are $\dfrac{(5)(4)}{2} = 10$ matchups. The vote is (the details are left for you)

A: 0 points

B: 1 points

C: 2 points

R: 3 points

S: 4 points

Stockholm (S) is the winner from the pairwise comparison method.

f. A faces off against B:

A: 36

B: $24 + 20 + 18 + 8 + 4 = 74$

Buenos Aires (B) wins the election.   ◆

Admittedly, the numbers used in Example 7 were contrived to make a point, but nevertheless this example shows that we used five different common voting procedures to come up with five different winners. You can see that those with the power to select the voting method may have the power to determine the outcome of the election.

PROBLEM SET 16.2

LEVEL 1

1. **IN YOUR OWN WORDS** What is the majority criterion?
2. **IN YOUR OWN WORDS** What is the Condorcet criterion?
3. **IN YOUR OWN WORDS** What is the monotonicity criterion?
4. **IN YOUR OWN WORDS** What is the irrelevant alternatives criterion?
5. **IN YOUR OWN WORDS** What are the fairness criteria?
6. **IN YOUR OWN WORDS** If you could have only one of the fairness criteria, which one would you choose and why?
7. **IN YOUR OWN WORDS** If you could have only two of the fairness criteria, which ones would you choose and why?
8. **IN YOUR OWN WORDS** What is insincere voting?
9. **IN YOUR OWN WORDS** Why is Arrow's impossibility theorem important?
10. **IN YOUR OWN WORDS** What is the Pareto principle?
11. The South Davis Faculty Association is using the Hare method to vote for their collective bargaining representative. Their choices are the All Faculty Association (A), American Federation of Teachers (B), and California Teachers Association (C). Here are the results of the voting:

    (ABC)(ACB)(BAC)(BCA)(CAB)(CBA)

    Votes: 11   1   3   6   3   7

    a. Which organization is selected for collective bargaining using the Hare method?
    b. Does the choice in part **a** violate the majority criterion?
12. An election with three candidates has the following rankings:

    (ABC)(BCA)(CBA)

       5   4   3

    a. Is there a majority? If not, who wins the plurality vote?

b. Who wins using the Borda count method?
c. Does the Borda method violate the majority criterion?

13. An election with three candidates has the following rankings:

    (BAC)(ACB)(CAB)

       5   6   3

    a. Is there a majority? If not, who wins the plurality vote?
    b. Who wins using the Borda count method?
    c. Does the Borda method violate the majority criterion?

14. The South Davis Faculty Association is using the Borda count method to vote for their collective bargaining unit. Their choices are the All Faculty Association (A), American Federation of Teachers (B), and California Teachers Association (C). Here are the results of the voting:

    (ABC)(ACB)(BAC)(BCA)(CAB)(CBA)

    Votes: 16   0   0   5   0   9

    a. Is there a majority? If not, who wins the plurality vote?
    b. Who wins using the Borda count method?
    c. Does the Borda method violate the majority criterion?

15. Consider the following voting situation:

    (ABC)(ACB)(BAC)(BCA)(CAB)(CBA)

    Votes: 2   0   0   2   1   0

    Notice that there is no winner using the majority or plurality rules.

    a. Who would win in a runoff election by dropping the choice with the fewest first-place votes?
    b. Who would win if B withdraws before the election?
    c. Does this violate any of the fairness criteria?

16. Consider the following voting situation:

    (ABC)(ACB)(BAC)(BCA)(CAB)(CBA)

    Votes: 9   0   9   0   0   7

Notice that there is no winner using the majority or plurality rules.

a. Who would win in a runoff election by dropping the choice of the fewest first-place votes?

b. Who would win if C withdraws before the election?

c. Does this violate any of the fairness criteria?

17. The philosophy department is selecting a chairperson, and the candidates are Andersen (A), Bailey (B), and Clark (C). Here are the preferences of the 27 department members:

(ABC)(BCA)(BAC)(CAB)
  8   5   8   6

a. Who is the Condorcet candidate, if there is one?

b. Is there a majority winner? If not, who wins the plurality vote? Does this violate the Condorcet criterion?

18. The philosophy department is selecting a chairperson, and the candidates are Andersen (A), Bailey (B), and Clark (C). Here are the preferences of the 27 department members:

(ABC)(BCA)(BAC)(CAB)
  8   5   8   6

a. Who is the Condorcet candidate, if there is one?

b. Who wins according the Borda count method? Does this violate the Condorcet criterion?

19. The Adobe School District is hiring a vice principal and has interviewed four candidates: Andrew (A), Bono (B), Carol (C), and Davy (D). The hiring committee has indicated their preferences:

(ACDB)(CBAD)(BCDA)(DBCA)
   7   5   3   2

a. Who is the winner using the plurality method?

b. Suppose that Carol drops out of the running before the vote is taken. Who is the winner using the plurality method?

c. Do the results of parts a and b violate the irrelevant alternatives criterion?

LEVEL 2

The seniors at Weseltown High School are voting for where to go for their senior trip. They are deciding on Angel Falls (A), Bend Canyon (B), Cedar Lake (C), or Danger Gap (D). The results of the preferences are

(DABC)(ACBD)(BCAD)(CBDA)(CBAD)
   30   25   22   20   11

Use this information for Problems 20–24.

20. Who is the Condorcet candidate, if there is one?

21. Is there a majority winner? If not, is there a plurality winner? Does this violate the Condorcet criterion?

22. Who wins the Borda count? Does this violate the Condorcet criterion?

23. Who wins using the Hare method? Does this violate the Condorcet criterion?

24. Who wins using the pairwise comparison method? Does this violate the Condorcet criterion?

The seniors at Weseltown High School are voting for where to go for their senior trip. They are deciding on Angel Falls (A), Bend Canyon (B), Cedar Lake (C), or Danger Gap (D). The results of the preferences are

(DABC)(ACBD)(BCAD)(CBDA)(CBAD)
   80   45   30   10   50

Use this information for Problems 25–29.

25. Who is the Condorcet candidate, if there is one?

26. Is there a majority winner? If not, is there a plurality winner? Does this violate the Condorcet criterion?

27. Who wins the Borda count? Does this violate the Condorcet criterion?

28. Who wins using the Hare method? Does this violate the Condorcet criterion?

29. Who wins using the pairwise comparison method? Does this violate the Condorcet criterion?

A focus group of 33 people for ABC TV were asked to rank the government spending priorities of education (E), military spending (M), health care (H), immigration (I), and lowering taxes (T). Here are the preferences:

(EIHTM)(MIEHT)(HMETI)(TMEIH)
   15   6   6   6

Use this information to answer Problems 30–35.

30. Who is the winner using the pairwise comparison method?

31. Who is the winner using a Borda count?

32. Suppose that the losing issues of health care and lowering taxes are removed from the table. Now, who is the winner using the pairwise comparison method? Does this violate the irrelevant alternatives criterion?

33. Suppose that the losing issues of health care and lowering taxes are removed from the table. Now, who is the winner using the Borda count method? Does this violate the irrelevant alternatives criterion?

34. Suppose that the losing issues of health care, lowering taxes, and immigration are removed from the table. Now, who is the winner using the pairwise comparison method? Does the pairwise comparison method violate the irrelevant alternatives criterion?

35. Suppose that the losing issues of health care, lowering taxes, and immigration are removed from the table. Now, who is the winner using the Borda count method? Does the Borda count method violate the irrelevant alternatives criterion?

In 1988 *the 94th International Olympic Committee (IOC) met in Seoul to select the* 1994 *Winter Olympics site. The cities in the running were Anchorage (A), Lillehammer (L), Ostersund (T) and Sofia (S). Consider the following fictional account of how the voting might have been conducted. The voting takes place over a two-day period using the Hare method. The first day the 93 members of the IOC take a nonbinding vote, and then on the second day they take a binding vote. Use this information in Problems 36 and 37.*

36. On the first day, the rankings of the IOC members were:

(TLSA)(LATS)(ASLT)(ASTL)(STLA)

   31   35   20   4   3

What are the results of the election using the Hare method for the first (nonbinding) day of voting?

37. On the evening of the first day of voting (see Problem 36), representatives from Ostersund wine, dine, and sweet talk the 7 members with the bottom votes. They were able to convince seven of those IOC members to move Lillehammer to the top of their list. After all, they argued, Lillehammer won the straw vote anyway—everyone likes to vote for the winner. Also, one of the (TLSA) voters changed to (LTSA). On the second day, the rankings of the IOC committee were:

(TLSA)(LTSA)(LATS)(ASLT)(LAST)(LSTA)

   30   1   35   20   4   3

What are the results of the election using the Hare method for the second (binding) day of voting? Does this vote count violate the monotonicity criterion?

In 1993 *the 101st International Olympic Committee met in Monaco to select the* 2000 *Winter Olympics site. The cities in the running were Beijing (B), Berlin (L), Istanbul (I), Manchester (M), and Sydney (S). Suppose we look at their voting preferences:*

(BLIMS)(LBSIM)(IBLSM)(MSBLI)

   32   3   5   8

(LSBIM)(SBLMI)(IMSBL)(MBSLI)

   6   30   2   3

Use this information to answer the questions in Problems 38 and 39.

38. a. Is there a majority winner? If not, which country wins the plurality vote?
   b. Find the results of the election using the Hare method. Just after the third vote, one of the committee members voting for Manchester was accused of cheating and was disqualified. Because of that scandal, one member admitted she was voting insincerely, and changed her vote from Manchester to Sydney. What is the result of using the Hare method?
   c. Do the results of parts **a** and **b** violate any of the fairness criteria?

39. a. Find the result of the election using a Borda count.
   b. Using the result of part **a,** determine whether any of the fairness criteria have been violated.

40. The U.S. president is elected with a vote of the Electoral College. However, if the vote were conducted using the Hare method, what is the outcome? Use the data in Table 16.3, and assume that the second choice of the Browne voters is Gore, and the second choice of the Buchanan voters is Bush. Assume that the second choice of 80% of the Nader voters is Gore and for 20% of them the second choice is Bush. Finally, assume that the other voters split the second choice 50%–50% between Gore and Bush. Who is the winner of this election?

41. Article 7 of the French constitution states, "The President of the Republic is elected by an absolute majority of votes cast. If this is not obtained on the first ballot, a second round of voting must be held, to take place two Sundays later. Only two candidates may stand for election on the second ballot, these being the two that obtained the greatest number of votes in the first round."

The 2002 French presidential election incumbent President Jacques Chirac (center-right political party) and Prime Minister Lionel Jospin (Socialist party) were shoo-ins for the second round. Here are the results of the first-round voting (rounded to the nearest percent):

Jacques Chirac:	20%
Jean-Marie Le Pen (extreme right)	17%
Lionel Jospin:	16%
Others:	47%

These percents are of the votes cast. It was estimated that 28% of the voters abstained.

   a. Who are the two candidates in the runoff election?
   b. Jean-Marie Le Pen is described as a racist. Here is a statement from *A la française forum* (translation by LKL):
   "Today, the strongest feeling I have is SHAME. For the first time in my life, I'm ashamed to be French. All of the values that I believe in (culture, tolerance, integration . . .) have been scorned and denounced by 17% of my country's voters." Comment on this quotation in light of the fairness criteria.
   c. The vote in the second round of the election was:

Jacques Chirac:	82%
Jean-Marie Le Pen:	18%

It was estimated that 19% of the voters abstained for this ballot. Give at least one possible change in the voting preferences to account for both the first and the second votes.

A group of fun-loving people have decided to play a practical joke on one of their friends, but they can't decide which friend, Alice (A), Betty (B), or Connie (C). Their preferences are

(ABC)(CBA)(BCA)

   6   5   4

Use this information to answer the questions in Problems 42–45.

42. a. Is there a Condorcet candidate?
   b. Is there a majority? If not, who wins the plurality vote? Does this violate the Condorcet criterion?

43. Who wins using the Borda count method? Does this violate the Condorcet criterion?

44. Who wins the election using the Hare method? Does this violate the Condorcet criterion?

45. Who wins the election using the pairwise comparison method? Does this violate any of the conditions in Arrow's impossibility theorem?

The fraternity $\Sigma\Delta\Gamma$ is electing a national president and there are four candidates: Alberto (A), Bate (B), Carl (C), and Dave (D). The voter preferences are:

(BDCA)(BDAC)(CDAB)(ADCB)
   100   120   130   150

Use this information to answer the questions in Problems 46–49.

46. a. How many votes were cast?
   b. Is there a majority? If not, who wins the plurality vote?
   c. Is there a Condorcet candidate?

47. Who wins using the Borda count method? Does this violate any of the fairness criteria?

48. Who wins the election using the Hare method? Does this violate any of the fairness criteria?

49. Who is the winner by using the pairwise comparison method? Does this violate any of the fairness criteria?

50. Consider an election with three candidates with the following results:

(ABC)(BCA)(CBA)
   5   3   3

   a. Is there a majority winner? If not, who is the plurality winner?
   b. Who wins using the pairwise comparison method?
   c. Is the ordering for the choices for candidates in part **b** transitive?

51. Consider an election with four candidates with the following results:

(ABCD)(ABDC)(CDAB)(CDBA)(DACB)
   10   9   8   7   6

   a. Is there a winner using the pairwise comparison method?
   b. Is there a winner using the tournament method?
   c. Do either of these methods violate any conditions of Arrow's impossibility theorem?

52. Repeat Problem 51 where there are 10 votes for each listed possibility.

53. Consider an election with four candidates with the following results:

(ABCD)(BCAD)(CABD)
   20   20   10

   a. Who wins the election using a Borda count method?
   b. Does the Borda count method violate the irrelevant alternative criterion?

54. Consider an election with three candidates with the following results:

(ACB)(BAC)(CAB)
   4   2   5

   a. Is there a majority winner? If not, who is the plurality winner?
   b. If a majority is required for election, there must be a runoff between the second and third choices. Who will win that runoff?
   c. How can the voters who support C vote insincerely to enable C to win the election?

55. Consider an election with three candidates with the following results:

(ACB)(BAC)(CBA)(CAB)
   2   5   4   2

   a. Is there a majority winner? If not, who is the plurality winner?
   b. Who wins the election using the Borda count method?
   c. Who wins if they first eliminate the one with the most last-place votes and then have a runoff between the other two?
   d. Could the two voters with preference (CAB) change the outcome of the election in part **c** if they voted insincerely and pretended to have the preference (CBA)?

56. Suppose that 100 Senators must vote on an appropriation: a new bridge in Alabama (A), a new freeway interchange in California (C), or a grain subsidy for Iowa (I). The Senate Whip estimates the preferences of the senators is

(ACI)(CIA)(ICA)
   10   38   52

   a. Which project wins using the Borda count method?
   b. Iowa argues that they should win because they have a majority vote. However, since the vote is to be by Borda count, those who favor California believe that Iowa could win and thus they vote insincerely for Alabama as their second choice. How would this affect the Borda count?
   c. If those who favor Iowa believe that the insincere voting in part **b** might take place, should they still vote so as to cause the funds to go to Iowa?

LEVEL 3

57. Suppose that Jane, Linda, Ann, and Melissa are members of a committee of the Tuesday Afternoon Club and Jane, Linda, and Ann all prefer a new rule that says the meeting time will change to the evenings. Jane proposed this new rule because she absolutely cannot come if it is not in the evening, and Linda prefers it because Jane picks her up for the meetings, but she could drive herself if Jane were not available. Ann, on the other hand, says she will go along with anything that does not interfere with her Wednesday morning golf lessons. Melissa argues that if they adopt the new rule they will need to change the name of their club, which will require a by-law change, so she is opposed to the new rule. Can you suggest an amendment that Melissa could offer to force the vote in her favor? Note that for a new rule to pass, three votes are required, and on a tie vote of two to two, the old rule stands.

58. Make up an example of a vote that is not transitive.

59. Suppose there are nine serious, but almost indistinguishable, candidates for a U.S. presidential primary. Also suppose that at the last possible minute a very radical candidate enters the race. He is so radical, in fact, that 90% of the voters rank this candidate at the bottom. Show how it might be possible for voters to elect this candidate.

PROBLEM SOLVING

60. **The game of WIN**   Construct a set of nonstandard dice as shown in Figure 16.1.

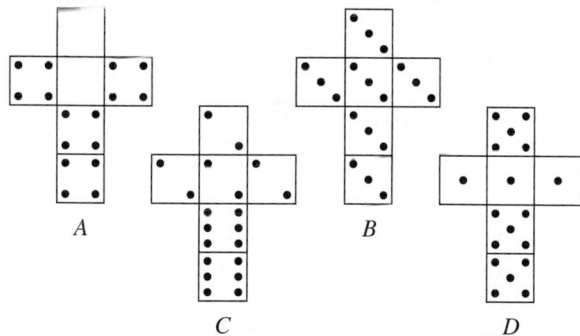

Figure 16.1 **Faces on the dice for a game of WIN**

Suppose that one player picks die A and that the other picks die B. Then we can enumerate the sample space as shown here.

A:	0	0	4	4	4	4
B:						
3	(3, 0)	(3, 0)	(3, 4)	(3, 4)	(3, 4)	(3, 4)
3	(3, 0)	(3, 0)	(3, 4)	(3, 4)	(3, 4)	(3, 4)
3	(3, 0)	(3, 0)	(3, 4)	(3, 4)	(3, 4)	(3, 4)
3	(3, 0)	(3, 0)	(3, 4)	(3, 4)	(3, 4)	(3, 4)
3	(3, 0)	(3, 0)	(3, 4)	(3, 4)	(3, 4)	(3, 4)
3	(3, 0)	(3, 0)	(3, 4)	(3, 4)	(3, 4)	(3, 4)

                        B wins   *A wins*

We see that A's probability of winning is $\frac{24}{36}$, or $\frac{2}{3}$. There are, of course, many other possible choices for the dice played. If you were to play the game of WIN, would you choose your die first or second? What is your probability of winning at WIN? This problem reminds you of which concept introduced in this section?

16.3 APPORTIONMENT

The framing of the United States Constitution during the Constitutional Convention in 1787 has been the subject of books, movies, and plays. Many issues of grave importance were introduced and debated, but one of the most heated and important debates concerned how the states would be represented in the new legislature. The large states wanted proportional representation based on population, and the smaller states wanted representation by state. From this debate came the Great Compromise, which led to the formation of two sides of the legislative branch of government. The compromise allowed the Senate to have two representatives per state (advantageous for the smaller states) and the House of Representatives to determine the number of representatives for a particular state by the size of the population (advantageous for the larger states). The historical note at the left gives the exact wording from the United States Constitution, and if you read it you will notice that it does not specify *how* to determine the number of representatives for each state. The process of making this decision is called **apportionment.** To *apportion* means to divide or share out according to a plan. It usually refers to dividing representatives in Congress or taxes to the states, but it can refer to judicial decisions or to the assignment of goods or people to different jurisdictions.

In this section, we will consider five apportionment plans: *Adams' plan, Jefferson's plan, Hamilton's plan, Webster's plan, and Huntington–Hill's plan (HH plan).* You recognize, no doubt, some of these names from American history.

We begin with a simple example to introduce us to some of the terminology used with apportionment. Table 16.5 shows the population over the years for five New York boroughs.

Table 16.5 Populations of New York Boroughs

Year	Total	Manhattan	Bronx	Brooklyn	Queens	Staten Island
				Population (in thousands)		
1790	49	32	2	5	6	4
1800	81	61	2	6	7	5
1840	697	516	8	139	19	15
1900	3,438	1,850	201	1,167	153	67
1940	7,454	1,890	1,395	2,698	1,297	174
1990	7,324	1,488	1,204	2,301	1,952	379
2000	8,007	1,537	1,333	2,465	2,229	443

One of the reoccurring problems with apportionment is working with approximate data and the problems caused by rounding. For example, if you look at the total population for the five New York Boroughs in the year 2000, you will find the total to be 8,008,000. If you add the numbers in Table 16.5, you will find the total to be 8,007,000. The discrepancy is caused by rounding, and we should not just sweep it under the rug and ignore such discrepancies. How we handle rounding is a large part of distinguishing apportionment problems. Consider the following example.

EXAMPLE 1
_____

In 1790 the population in New York City was 49,000. Suppose the city council at that time consisted of 8 members.

a. How many people did each city council member represent?

b. If the assignment of council seats is proportional to the borough's population, use Table 16.5 to allocate the council seats.

Solution

a. Since the population was 49,000, each council member should represent

$$\frac{49,000}{8} = 6,125$$

This number is called the *standard divisor*.

b. To find the appropriate representation, we need to divide the population of each borough by the standard divisor; this number is known as the *standard quota*. Round your results to the nearest hundredth. We show rounding the standard quota to the nearest whole number, and we also show the result of rounding up and of rounding down. The result of rounding up is called the **upper quota** and the result of rounding down is called the **lower quota**.

	Standard quota	*Nearest*	*Up* *(Upper quota)*	*Down* *(Lower quota)*
Manhattan:	$\frac{32,000}{6,125} \approx 5.22$	5	6	5
Bronx:	$\frac{2,000}{6,125} \approx 0.33$	0	1	0
Brooklyn:	$\frac{5,000}{6,125} \approx 0.82$	1	1	0
Queens:	$\frac{6,000}{6,125} \approx 0.98$	1	1	0
Staten Island:	$\frac{4,000}{6,125} \approx 0.65$	1	1	0
TOTAL:		8	10	5

Now, seats on a city council must be whole numbers, so we should use the numbers from one of the columns of rounded numbers. But which column? Since we need to fill 8 seats, we see for this example if we round to the nearest unit, we will fill the 8 seats. ◆

Example 1 raises some questions. First, how would you like it if you lived in the Bronx? You would have no representation. On the other hand, if you rounded each of the numbers up, the number of seats would increase to 10, and certainly rounding down for this example would not satisfy anyone (except perhaps those from Manhattan). There are other difficulties caused by the rounding in Example 1, but before we take a closer look, we need the following definition.

Standard Divisor

$$\text{STANDARD DIVISOR} = \frac{\text{TOTAL POPULATION}}{\text{NUMBER OF SHARES}}$$

Standard Quota

$$\text{STANDARD QUOTA} = \frac{\text{TOTAL POPULATION}}{\text{STANDARD DIVISOR}}$$

EXAMPLE 2

Use the data for New York City in 1790 to find the standard divisor and standard quota for each borough for the number of seats requested. Round these standard quotas to the nearest integer, up (upper quota), and down (lower quota).

a. 12 seats

b. 13 seats

Solution The total population is 49,000. Let d be the standard divisor and q be the standard quota for each borough.

a. 12 seats

Standard divisor: $\quad d = \dfrac{49{,}000}{12} \approx 4083.33$

	Standard quota, q	Nearest	Upper quota	Lower quota
Manhattan:	$\dfrac{32{,}000}{4{,}083.33} \approx 7.84$	8	8	7
Bronx:	$\dfrac{2{,}000}{4{,}083.33} \approx 0.49$	0	1	0
Brooklyn:	$\dfrac{5{,}000}{4{,}083.33} \approx 1.22$	1	2	1
Queens:	$\dfrac{6{,}000}{4{,}083.33} \approx 1.47$	1	2	1
Staten Island:	$\dfrac{4{,}000}{4{,}083.33} \approx 0.98$	1	1	0
TOTAL:		11	14	9

b. 13 seat $\quad d = \dfrac{49{,}000}{13} \approx 3{,}769.23$

		Nearest	Up	Down
Manhattan:	$\dfrac{32{,}000}{3{,}769.23} \approx 8.49$	8	9	8
Bronx:	$\dfrac{2{,}000}{3{,}769.23} \approx 0.53$	1	1	0
Brooklyn:	$\dfrac{5{,}000}{3{,}769.23} \approx 1.33$	1	2	1
Queens:	$\dfrac{6{,}000}{3{,}769.23} \approx 1.59$	2	2	1
Staten Island:	$\dfrac{4{,}000}{3{,}769.23} \approx 1.06$	1	2	1
TOTAL:		13	16	11

Take a look at the results of Example 2, and you will see if the required number of seats is 12, none of the rounding methods shown will work. Also note that if there are fewer than 13 seats, there is no representation for the Bronx. You may also have noticed that the apportionment we seek must be either the upper quota or the lower quota. This is known as the **quota rule.**

The Quota Rule

> The number assigned to each represented unit must be either the standard quota rounded down to the nearest integer or the standard quota rounded up to the nearest integer.

Table 16.6 U. S. Population in the 1790 Census*

State	Population
Connecticut	237,655
Delaware	59,096
Georgia	82,548
Kentucky	73,677
Maryland	319,728
Mass.	475,199
New Hamp.	141,899
New Jersey	184,139
New York	340,241
N. Carolina	395,005
Pennsylvania	433,611
Rhode Island	69,112
S. Carolina	249,073
Vermont	85,341
Virginia	747,550
TOTAL	3,893,874

*We are using demographic information from the Consortium for Political and Social Research, Study 00003. The population numbers actually used for the 1794 apportionment were slightly different from these, so the historical record does not exactly match these academic examples. Among the reasons for this discrepancy is that prior to 1870, the population base included the total free population of the states and three-fifths of the number of slaves, and it excluded American Indians not taxed.

Before we decide on a rounding scheme, we will consider one more historical example from the United States Congress. The first congressional apportionment was to occur after the 1790 census. (It actually occurred in 1794.) The results of the census of 1790 are shown in Table 16.6.

EXAMPLE 3

Use the results of Table 16.6, and the fact that the number of seats in the House of Representatives was to be raised from 65 to 105. Find the standard divisor and the standard quotas for each state. Round each of the standard quotas to the nearest number, as well as to give the lower and upper quotas.

Solution   Standard divisor: $d = \dfrac{3,893,874}{105} \approx 37,084.51$

	Standard Quota, q	Nearest	Lower Quota	Upper Quota
Connecticut:	$\dfrac{237,655}{d} \approx 6.41$	6	6	7
Delaware:	$\dfrac{59,096}{d} \approx 1.59$	2	1	2
Georgia:	$\dfrac{82,548}{d} \approx 2.23$	2	2	3
Kentucky:	$\dfrac{73,677}{d} \approx 1.99$	2	1	2
Maryland:	$\dfrac{319,728}{d} \approx 8.62$	9	8	9
Massachusetts:	$\dfrac{475,199}{d} \approx 12.81$	13	12	13
New Hampshire:	$\dfrac{141,899}{d} \approx 3.83$	4	3	4
New Jersey:	$\dfrac{184,139}{d} \approx 4.97$	5	4	5
New York:	$\dfrac{340,241}{d} \approx 9.17$	9	9	10
North Carolina:	$\dfrac{395,005}{d} \approx 10.65$	11	10	11
Pennsylvania:	$\dfrac{433,611}{d} \approx 11.69$	12	11	12
Rhode Island:	$\dfrac{69,112}{d} \approx 1.86$	2	1	2
South Carolina:	$\dfrac{249,073}{d} \approx 6.72$	7	6	7
Vermont:	$\dfrac{85,341}{d} \approx 2.30$	2	2	3
Virginia:	$\dfrac{747,550}{d} \approx 20.16$	20	20	21
TOTAL:		106	96	111

The quota rule tells us that the actual representation for each state must be the lower quota or the upper quota.   ◆

The First Congress of the United States needed to decide how to round the standard quotas, q, of Example 3. There were three plans proposed initially, and we will consider them one at a time. One of them rounded up, one rounded down, and a third rounded down with some additional conditions.

Adams' Plan

The first plan we will consider was proposed by the sixth President of the United States, John Quincy Adams, so it is known as **Adams' plan.**

Adams' Plan

Round up!

CAUTION

> Any standard quota with a decimal portion must be *rounded up* to the next whole number. To give the appropriate number of seats, use the following procedure.
> 1. Compute the standard divisor, d.
> 2. Compute the standard quota, q, for each state.
> 3. Round the standard quotas up and add these quotas to find the total number of seats. It will be correct, or it will be too large. Raise the standard divisor in small increments until the rounded quotas provide the appropriate total; this divisor is called a **modified divisor,** D. The number Q for each state is called the **modified quota.**

HISTORICAL NOTE

John Quincy Adams
(1767–1848)

John Quincy Adams was the sixth president of the United States. His best-known achievement was the formation of the Monroe Doctrine in 1823. He was the son of the second president, as well as the minister to the Netherlands, Prussia, and Russia. He was also a U.S. Senator and, after his presidency, served in the House of Representatives. He proposed a method of apportionment, but it was never adopted for use.

Back in 1790 the process of finding the modified quota was quite a task, but with the help of spreadsheets today it is not very difficult. Look at Example 3 and note that the upper quotas total 111 and we are looking for a total of 105 seats. Look at the spreadsheet in Figure 16.2. Note that $d = 37,084.51$ (from Example 3) and if we raise this to $D = 38,000$ the total number of seats is 108, so we raised d too little. Next, we raise d to $D = 40,000$ and the total number of seats is now 103, so we raised d too much. (Remember, the goal is 105.) Finally, you will see if we choose $D = 39,600$ we obtain the target number of seats, which is 105. The number of seats for each of the original states according to Adams' apportionment plan is shown in Figure 16.2.

EXAMPLE 4

Advanced mathematics is taught at five high schools in the Santa Rosa Unified School District. The district has just received a grant of 200 TI-Voyage 200 calculators that are to be distributed to the five high schools. If these calculators are distributed according to Adams' plan, how should they be apportioned to the schools? Use the data in Table 16.7.

Table 16.7 School Statistics for Santa Rosa Unified School District High Schools

School	# Students	# Advanced math students
Elsie Allen	1,524	72
Maria Carrillo	1,687	131
Montgomery	1,755	243
Piner	1,519	95
Santa Rosa	1,797	71
TOTAL	8,282	612

	A	B	C	D	E	F	G	H	I	J	K
		Spreadsheet Application									
1		CN	DE	GA	KY	MD	MS	NH	NJ	NY	
2	Population	237,655	59,096	82,548	73,677	319,728	475,199	141,899	184,139	340,241	
3	q	6.41	1.59	2.23	1.99	8.62	12.81	3.83	4.97	9.17	
4	Nearest	6	2	2	2	9	13	4	5	9	
5	Round up	7	2	3	2	9	13	4	5	10	
6	Round dn	6	1	2	1	8	12	3	4	9	
7	Trial 1	7	2	3	2	9	13	4	5	9	
8	Q	6.25	1.56	2.17	1.94	8.41	12.51	3.73	4.85	8.95	
9	Trial 2	6	2	3	2	8	12	4	5	9	
10	Q	5.94	1.48	2.06	1.84	7.99	11.88	3.55	4.60	8.51	
11	Adams	6	2	3	2	9	13	4	5	9	
12	Q	6.00	1.49	2.06	1.86	8.07	12.00	3.58	4.65	8.59	
13											
14											
15		NC	PA	RI	SC	VT	VA	TOTAL	No. of Seats	d	
16	Population	395,005	433,611	69,112	249,073	85,341	747,550	3,893,874	105	37,084.51	
17	q	10.65	11.69	1.86	6.72	2.30	20.16	105			
18	Nearest	11	12	2	7	2	20	106			
19	Round up	11	12	2	7	3	21	111			
20	Round dn	10	11	1	6	2	20	96		D	
21	Trial 1	11	11	2	7	3	20	106	Raise d to:	38,000	too little
22	Q	10.39	11.41	1.82	6.55	2.25	19.67				
23	Trial 2	10	11	2	7	3	19	103	Raise d to:	40,000	too much
24	Q	9.88	10.84	1.73	6.23	2.13	18.69				
25	Adams	10	11	2	7	3	19	105	Raise d to:	39,600	just right
26	Q	9.97	10.95	1.75	6.29	2.16	18.88				

Figure 16.2 Apportionment calculations for Adams' apportionment plan

Solution The standard divisor is $d = \dfrac{612}{200} = 3.06$.

The standard and upper quotas are

	Population	Standard quota, q	Upper quota
Elsie Allen	72	$72/d \approx 23.53$	24
Maria Carrillo	131	$131/d \approx 42.81$	43
Montgomery	243	$243/d \approx 79.41$	80
Piner	95	$95/d \approx 31.04$	32
Santa Rosa	71	$71/d \approx 23.20$	23
TOTAL			202

We now find the modified divisor, D, by raising d until the sum of the upper quotas give us 200. Suppose we raise $d = 3.06$ to $D = 3.1$. Now, the modified quotas are

	Modified quota, Q	Adams' plan
Elsie Allen	$72/D \approx 23.23$	24
Maria Carrillo	$131/D \approx 42.26$	43
Montgomery	$243/D \approx 78.39$	79
Piner	$95/D \approx 30.65$	31
Santa Rosa	$71/D \approx 22.90$	23
TOTAL		200

♦

Jefferson's Plan

Since the second plan was proposed by the third president of the United States, Thomas Jefferson, it is known as **Jefferson's plan.** It is the same as Adams' plan, except you round down instead of rounding up.

Jefferson's Plan

Round down!

Any standard quota with a decimal portion must be *rounded down* to the next lower whole number. To give the appropriate number of seats, use the following procedure.

1. Compute the standard divisor, d.

2. Compute the standard quota, q, for each state.

3. Round the standard quotas down and find the total number of seats. It will be correct, or it will be too small. Lower the standard divisor in small increments to find a modified divisor that will give modified quotas that provide the appropriate total.

We again turn to Example 3 and note that the lower quotas total 96 and we are looking for a total of 105 seats. Look at the spreadsheet in Figure 16.3. This time we look at the lower quotas. Note that $d = 37,084.51$ (from Example 3) and if we lower this to $D = 36,000$ the total number of seats is 104, so we lowered d too little. (Remember, the

	A	B	C	D	E	F	G	H	I	J	
1		CN	DE	GA	KY	MD	MS	NH	NJ	NY	
2	Population	237,655	59,096	82,548	73,677	319,728	475,199	141,899	184,139	340,241	
3	q	6.41	1.59	2.23	1.99	8.62	12.81	3.83	4.97	9.17	
4	Nearest	6	2	2	2	9	13	4	5	9	
5	Round up	7	2	3	2	9	13	4	5	10	
6	Round dn	6	1	2	1	8	12	3	4	9	
7	Trial 1	6	1	2	2	8	13	3	5	9	
8	Q	6.80	1.64	2.29	2.05	8.88	13.20	3.94	5.11	9.45	
9	Trial 2	7	1	2	2	9	14	4	5	10	
10	Q	7.20	1.79	2.50	2.23	9.69	14.40	4.30	5.58	10.31	
11	Jefferson	6	1	2	2	9	13	4	5	9	
12	Q	6.79	1.69	2.36	2.11	9.14	13.58	4.05	5.26	9.72	
13											
14											
15		NC	PA	RI	SC	VT	VA	TOTAL	No. of Seats	d	
16	Population	395,005	433,611	69,112	249,073	85,341	747,550	3,893,874	105	37,084.51	
17	q	10.65	11.69	1.86	6.72	2.30	20.16	105			
18	Nearest	11	12	2	7	2	20	106			
19	Round up	11	12	2	7	3	21	111			
20	Round dn	10	11	1	6	2	20	96		D	
21	Trial 1	10	12	1	6	2	20	100	Lower d to:	36,000	too little
22	Q	10.97	12.04	1.92	6.92	2.37	20.77				
23	Trial 2	11	13	2	7	2	22	111	Lower d to:	33,000	too much
24	Q	11.97	13.14	2.09	7.55	2.59	22.65				
25	Jefferson	11	12	1	7	2	21	105	Lower d to:	35,000	just right
26	Q	11.29	12.39	1.97	7.12	2.44	21.36				

Figure 16.3 **Apportionment calculations for Jefferson's apportionment plan**

goal is 105.) Next, we lower d to $D = 33,000$ and the total number of seats is now 111, so we lowered d too much. Finally, if we choose $D = 35,000$, we obtain the target number of seats, which is 105. The number of seats for each of the original states according to Jefferson's apportionment plan is shown in Figure 16.3.

EXAMPLE 5

Use Table 16.7 and Jefferson's plan to distribute the 200 calculators to the five high schools based on the number of advanced math students.

Solution The standard divisor is $d = \dfrac{612}{200} = 3.06$.

The standard and upper quotas are:

	Population	Standard quota, q	Lower quota
Elsie Allen	72	$72/d \approx 23.53$	23
Maria Carrillo	131	$131/d \approx 42.81$	42
Montgomery	243	$243/d \approx 79.41$	79
Piner	95	$95/d \approx 31.04$	31
Santa Rosa	71	$71/d \approx 23.20$	23
TOTAL			198

Suppose we lower $d = 3.06$ to $D = 3.01$ (this is the modified divisor). Now, the standard and lower quotas are

	Modified quota, Q	Jefferson's plan
Elsie Allen	$72/D \approx 23.92$	23
Maria Carrillo	$131/D \approx 43.52$	43
Montgomery	$243/D \approx 80.73$	80
Piner	$95/D \approx 31.56$	31
Santa Rosa	$71/D \approx 23.59$	23
TOTAL		200

◆

Hamilton's Plan

As you might have guessed by now, both Adams' and Jefferson's plans are a bit complicated. Historically, many complained that using this "modified divisor" to "tweak" the numbers is a bit "like magic." Alexander Hamilton, secretary of the treasury under George Washington, proposed the next plan we will consider, which is called, of course, **Hamilton's plan.** It is easier and more straightforward than the previous two plans we have considered.

Hamilton's Plan

Round down!

1. Compute the standard divisor, d.
2. Compute the standard quota, q, for each state.
3. Round the standard quotas down to the nearest integer, but *each must be at least one.*
4. Give any additional seats one at a time (until no seats are left) to the states with the largest fractional parts of their standard quotas.

Note that Hamilton took into account the problem we discovered in Example 1. It just does not seem right that a district should have no representation, so if Hamilton's plan is used, all voters will have *some* representation.

Once again, look at Example 3 and note that the lower quotas total 96 and we are looking for a total of 105 seats. In the spreadsheet in Figure 16.4, we show you all three methods for easy comparison. To carry out Hamilton's plan, start by looking at the value of q in the second row. Look across and locate the one with the greatest decimal portion; it is Kentucky (1.99, but look just at the decimal portion; 0.99); add one seat to Kentucky (see #1). Next, pick New Jersey (0.97); add one seat to New Jersey (see #2). Next, pick Rhode Island because it has the next largest decimal portion (0.86); add one seat to Rhode Island (see #3). Continue to add seats in order (see #4 to #9). Note that now the total is 105 seats, so the process is complete.

Spreadsheet Application

	A	B	C	D	E	F	G	H	I	J
1		CN	DE	GA	KY	MD	MS	NH	NJ	NY
2	Population	237,655	59,096	82,548	73,677	319,728	475,199	141,899	184,139	340,241
3	q	6.41	1.59	2.23	1.99	8.62	12.81	3.83	4.97	9.17
4	Nearest	6	2	2	2	9	13	4	5	9
5	Round up	7	2	3	2	9	13	4	5	10
6	Round dn	6	1	2	1	8	12	3	4	9
7	Adams	6	2	3	2	9	13	4	5	9
8	Jefferson	6	1	2	2	9	13	4	5	9
9	Hamilton	6	1	2	2	9	13	4	5	9
10					#1	#9	#5	#4	#2	
11										
12		NC	PA	RI	SC	VT	VA	TOTAL	No. of Seats	d
13	Population	395,005	433,611	69,112	249,073	85,341	747,550	3,893,874	105	37,084.51
14	q	10.65	11.69	1.86	6.72	2.30	20.16	105		
15	Nearest	11	12	2	7	2	20	106		
16	Round up	11	12	2	7	3	21	111		
17	Round dn	10	11	1	6	2	20	96		
18	Adams	10	11	2	7	3	19	105		
19	Jefferson	11	12	1	7	2	21	105		
20	Hamilton	11	12	2	7	2	20	105		
21		#8	#7	#3	#6					

Figure 16.4 Apportionment calculations for Adams', Jefferson's, and Hamilton's apportionment plans

EXAMPLE 6

Use Table 16.7 and Hamilton's plan to distribute the 200 calculators to the five high schools based on the number of advanced mathematics students.

Solution The standard divisor is $d = \dfrac{612}{200} = 3.06$.

The standard and lower quotas, along with Hamilton's plan, are shown:

	Standard quota, q	Lower quota	Hamilton's plan	
Elsie Allen	$72/d \approx 23.53$	23	24	#2 (0.53)
Maria Carrillo	$131/d \approx 42.81$	42	43	#1 (0.81)
Montgomery	$243/d \approx 79.41$	79	79	
Piner	$95/d \approx 31.04$	31	31	
Santa Rosa	$71/d \approx 23.20$	23	23	
TOTAL		198	200	◆

The following example projects these mathematical methods of apportionment into the political process.

EXAMPLE 7

Take a good look at Figure 16.4. to answer the following equations.

a. If you were from a small state, which plan would you probably favor?

b. If you were from a large state, which plan would you probably favor?

c. Which plan do you think was the first plan to be adopted by the First Congress?

Solution

a. There are seven states (out of 15) for which the representation changes depending on the adopted plan. Adams' plan favors Delaware and Georgia, but hurts North Carolina and Pennsylvania. Jefferson's plan hurts Rhode Island. It seems as if *Adams' plan favors the smaller states.*

b. By the analysis in part **a,** it seems that the larger states would favor Jefferson's plan. It is not a coincidence that *Jefferson's plan favors the larger states* and that Jefferson was from Virginia.

c. It would seem that the plan to compromise the positions should be Hamilton's plan.   ◆

By considering Example 7, we can understand why the first apportionment plan to pass was Hamilton's plan. When the bill that would have adopted Hamilton's plan reached President Washington's desk, it became the first presidential veto in the history of our country. Washington's objection was to the fourth step in Hamilton's plan. Congress was not able to override the veto, so with a second bill Congress adopted Jefferson's plan, which was used until 1840 when it was replaced because flaws in Hamilton's plan showed up after the 1820 and 1830 censuses. We will discuss these flaws in the next section.

Webster's Plan

Daniel Webster, a senator from Massachusetts, ran for president on the Whig party, and was appointed secretary of state by President William H. Harrison. When the reapportionment based on the 1830 census was done, New York had a standard quota of 38.59 but was awarded 40 seats using Jefferson's plan. Webster argued that this was unconstitutional (violated the *quota rule*) and suggested a compromise, which became known as **Webster's plan.** His plan is similar to both Adams' and Jefferson's plans. Instead of rounding up or down from the modified quota, Webster proposed that any quota with a decimal portion must be rounded to the *nearest* whole number. This method is based on the *arithmetic mean;* round down if the standard quotient is less than the *arithmetic mean* and round up otherwise.

Webster's Plan

Use the arithmetic mean to round.

Any standard quota with a decimal portion must be *rounded to the nearest whole number.* To give the appropriate number of seats, use the following procedure.

1. Compute the standard divisor, *d.*

2. Compute the standard quota, *q,* for each state.

3. Round the standard quotas down if the fractional part is less than 0.5 and up if the fractional part is greater than or equal to 0.5. It will be correct, or it will not. Lower or raise the standard divisor in small increments to find a modified divisor that will give modified quotas that provide the appropriate total.

EXAMPLE 8

The city of St. Louis, MO, passed a ballot measure to provide and pay for 130 surveillance cameras for high crime areas. The city council mandated that these cameras be apportioned among the five highest crime areas, based on the 2001 crime statistics, summarized in Table 16.8. Use Webster's plan.

Table 16.8 Serious Crimes in St. Louis, MO in 2001

Precinct	*Number of Violent Crimes/100 Residents*
Downtown	24.45
Fairground	10.04
Columbus Square	9.75
Downtown West	9.43
Peabody	9.01
TOTAL	62.68

Solution In this case, the "population" is the total number of crimes/hundred residents. This is 62.68, and the number of items is 130. Thus, the

$$d = \text{STANDARD DIVISOR} = \frac{62.68}{130} \approx 0.48$$

The standard quotas are

		Rounded
Downtown:	$\dfrac{24.45}{d} \approx 50.71$	51
Fairground:	$\dfrac{10.04}{d} \approx 20.82$	21
Columbus Square:	$\dfrac{9.75}{d} \approx 20.22$	20
Downtown West:	$\dfrac{9.43}{d} \approx 19.56$	20
Peabody:	$\dfrac{9.01}{d} \approx 18.69$	19
TOTAL:		131

The total of the standard quotas is too large, so we should raise the standard divisor to form a modified divisor. By trial and error we find that three decimal place rounding is necessary in order to give the correct number of cameras.

Modified divisor: 0.484

The modified quotas are

		Rounded
Downtown:	$\dfrac{24.45}{0.484} \approx 50.52$	51
Fairground:	$\dfrac{10.04}{0.484} \approx 20.74$	21
Columbus Square:	$\dfrac{9.75}{0.484} \approx 20.14$	20
Downtown West:	$\dfrac{9.43}{0.484} \approx 19.48$	19
Peabody:	$\dfrac{9.01}{0.484} \approx 18.62$	19
TOTAL:		130

◆

At the time that Webster's plan was adopted, no one suspected that it had the same flaw as Jefferson's plan. It was used after the 1840 census and from 1900 to 1941 when it was replaced by Huntington–Hill's plan. Oddly enough from 1850 to 1900 Hamilton's plan was used, so the only method never actually used in Congress was Adams' plan.

Huntington–Hill's Plan

Edward Huntington, a professor of mechanics and mathematics at Harvard University, and Joseph Hill, chief statistician for the Bureau of the Census, devised a rounding method currently used by the U.S. legislature. From 1850 to 1911, the size of the House of Representatives (number of seats) changed as states were added. In 1911, the House size was fixed at 433, with a provision for the addition of one seat each for Arizona and New Mexico. The House size has remained at 435 since, except for a temporary increase to 437 at the time of admission of Alaska and Hawaii. The seats went back to 435 after the subsequent census. In 1910, a plan called the *Huntington–Hill's plan*, which we will abbreviate as the *HH plan*, was adopted. It is the same as Webster's plan except it rounds using the geometric mean, rather than the arithmetic mean used by Webster's plan. First, we present an example reviewing the concept of geometric mean.

EXAMPLE 9

The following modified quotas are given. Find the upper and lower quotas, and the arithmetic and geometric means of the upper and lower quotas. Then round the modified quota by comparing it to the arithmetic mean and then to the geometric mean.

a. 9.49   **b.** 1.42   **c.** 4.53   **d.** 6.42

Solution   If a and b are two numbers, then

$$\textbf{arithmetic mean} = \frac{a + b}{2} \qquad \textbf{geometric mean} = \sqrt{ab}$$

	Q	Round down	Round up	A.M.	Round compared to A.M.	G.M.	Round compared to G.M.
a.	9.49	9	10	9.5	9	$\sqrt{9 \times 10} \approx 9.486$	10
b.	1.42	1	2	1.5	1	$\sqrt{1 \times 2} \approx 1.414$	2
c.	4.53	4	5	4.5	5	$\sqrt{4 \times 5} \approx 4.472$	5
d.	6.42	6	7	6.5	6	$\sqrt{6 \times 7} \approx 6.48$	6 ◆

Huntington–Hill's (HH) Plan

Round to the geometric mean!

To give the appropriate number of seats using **Huntington–Hill's plan,** follow this procedure.

1. Compute the standard divisor, d.

2. Compute the standard quota, q, for each state. Let a = lower quota and b = upper quota for each state.

3. Round the standard quotas down if the fractional part is less than $\sqrt{ab}$ and up if the fractional part is greater than or equal to $\sqrt{ab}$. Add these rounded numbers; it will be correct, or it will not. Lower or raise the standard divisor in small increments to find a modified divisor that will give modified quotas that provide the appropriate total.

EXAMPLE 10

Disney World has several resorts with lakes:

Grand Floridian, 904 rooms

Contemporary Resort, 1,041 rooms

Polynesian Resort, 853 rooms

Yacht Club Resort, 630 rooms

Beach Club Resort, 584 rooms

If they purchase 400 paddle boats, how many boats would each resort be assigned if they were assigned according to the number of rooms using HH's plan?

Solution   There are 4,012 rooms

$$d = \text{STANDARD DIVISOR} = \frac{4,012}{400} \approx 10.03$$

The standard quotas are

		Geometric mean	*Rounded*
Contemporary:	$\dfrac{1,041}{d} \approx 103.79$	$\sqrt{103 \times 104} \approx 103.499$	104
Polynesian:	$\dfrac{853}{d} \approx 85.04$	$\sqrt{85 \times 86} \approx 85.499$	85
Grand:	$\dfrac{904}{d} \approx 90.13$	$\sqrt{90 \times 91} \approx 90.499$	90
Yacht Club:	$\dfrac{630}{d} \approx 62.81$	$\sqrt{62 \times 63} \approx 62.498$	63
Beach Club:	$\dfrac{584}{d} \approx 58.23$	$\sqrt{58 \times 59} \approx 58.498$	58
TOTAL:			400 ◆

We conclude this section with an example that allows us to compare and contrast the apportionment methods introduced in this section. We have summarized these in Table 16.9.

Table 16.9 Summary of Apportionment Methods

Method	*Divisor*	*Apportionment*
Adams' plan	Round **up; raise** the standard divisor to find the modified divisor.	Round the standard quotas up. Apportion to each group its modified upper quota. It favors the smaller states.
Jefferson's plan	Round **down; lower** the standard divisor to find the modified divisor.	Round the standard quotas down. Apportion to each group its modified lower quota. It favors the larger states.
Hamilton's plan	Use the standard divisor. Round down.	Round the standard quotas down. Distribute additional seats one at a time until all items are distributed.
Webster's plan	Use modified divisors. May round up or down.	Round by comparing with the **arithmetic mean** of the upper and lower quotas.
HH's plan	Use modified divisors. May round up or down.	Round by comparing with the **geometric mean** of the upper and lower quotas.

EXAMPLE 11   PÓLYA'S METHOD

Consider College Town with district populations as follows:

   North: 8,600   South: 5,400   East: 7,200   West: 3,800

Historically, suppose the population of College Town was 24,000 and there are exactly 10 council seats. Apportion the seats according to the indicated method.*

a. Adams' plan

b. Jefferson's plan

c. Hamilton's plan

d. Webster's plan

e. HH's plan

Solution
The standard divisor is $d = 2,400$.
The standard quota is

$$q = \frac{24,000}{10} = 2,400$$

Over time, populations change, and the ten seats must be divided fairly for the four districts.

Understand the Problem. Let's begin by doing some simple arithmetic. The present population is 25,000 and the number of seats is 10, so we begin by calculating the standard divisor:

$$\text{STANDARD DIVISOR} = \frac{25,000}{10} = 2,500$$

Next, calculate the standard quotas for the revised population:

North: $\dfrac{8,600}{2,500} = 3.44$ South: $\dfrac{5,400}{2,500} = 2.16$

East: $\dfrac{7,200}{2,500} = 2.88$ West: $\dfrac{3,800}{2,500} = 1.52$

Devise a Plan. Historically, there were five plans devised, and this example asks us to use each of these plans.

Carry Out the Plan. The results using the standard quota for each of these five plans are shown in the following table.

Results of City Council Apportionment

District	Population	Std Quota	Adams	Jefferson	Webster	Hamilton	HH
North	8,600	3.44	4	3	3	3	3
South	5,400	2.16	3	2	2	2	2
East	7,200	2.88	3	2	3	3	3
West	3,800	1.52	2	1	2	2	2
TOTAL	25,000		12	8	10	10	10

---

*This problem and solution are adapted from the article "Decimals, Rounding, and Apportionment," by Kay I. Meeks, *The Mathematics Teacher,* October 1992, pp. 523–525.

We see that the results from Adams' and Jefferson's plans require that we consider modified quotas.

a. Adams' plan: The modified divisor should be greater than the standard divisor. Try $D = 3000$. The modified quotas are shown in the following table.

b. Jefferson's plan: The modified divisor should be less than the standard divisor. Try $D = 2000$. The modified quotas are shown in the following table.

Additional Results of City Council Apportionment

District	Population	Std Quota $d = 2,400$	Modified Quotas Adams $D = 3000$	Jefferson $D = 2000$	Adams	Jefferson
North	8,600	3.44	2.87	4.30	3	4
South	5,400	2.16	1.80	2.70	2	2
East	7,200	2.88	2.40	3.60	3	3
West	3,800	1.52	1.27	1.90	2	1
TOTAL	25,000				10	10

c. Hamilton's plan:

Rank the original standard quotas:

North: $\dfrac{8,600}{25,000} \cdot 10 \approx 3.44$ Decimal rank 3

South: $\dfrac{5,400}{25,000} \cdot 10 \approx 2.16$ Decimal rank 4

East: $\dfrac{7,200}{25,000} \cdot 10 \approx 2.88$ Decimal rank 1

West: $\dfrac{3,800}{25,000} \cdot 10 \approx 1.52$ Decimal rank 2

Hamilton's plan adds one seat to the East (highest decimal, so it is rank 1) and one to the West (second highest decimal, so it is rank 2).

d. Webster's plan: The results of rounding to the nearest whole number (that is, rounding to the arithmetic mean) provided the correct apportionment.

e. Huntington–Hill's plan:

Consider the calculations shown in the table:

	Geometric mean (G.M.)	Compare standard quota with G.M.	Round	Seats
North:	G.M. $= \sqrt{3 \cdot 4} \approx 3.46$	less	down	3
South:	G.M. $= \sqrt{2 \cdot 3} \approx 2.45$	less	down	2
East:	G.M. $= \sqrt{2 \cdot 3} \approx 2.45$	greater	up	3
West:	G.M. $= \sqrt{1 \cdot 2} \approx 1.41$	greater	up	2
			TOTAL:	10

Notice that the total equals the required number of city council seats.

Look Back. Do all the methods provide the appropriate number of seats? Do any violate the quota rule?   ◆

In the next section we will see that since Adams', Jefferson's, Webster's, and HH's methods all use modified quotas, they can, under certain circumstances, violate the quota rule. We also see in the next section that every apportionment process will have some sort of anomaly under certain conditions.

PROBLEM SET 16.3

LEVEL 1

1. **IN YOUR OWN WORDS** Discuss the idea of apportionment. Describe some different apportionment schemes.

2. **IN YOUR OWN WORDS** Discuss the ideas of arithmetic mean and geometric mean.

3. **IN YOUR OWN WORDS** Which apportionment schemes for the U.S. Congress favor the smaller states? Which ones favor the larger states? Are any neutral as far as state size is concerned?

4. **IN YOUR OWN WORDS** What is the quota rule? Does this rule make sense to you? Discuss.

5. **IN YOUR OWN WORDS** If you round the standard quotas down, how do you need to change the standard divisor to find the modified quotas?

6. **IN YOUR OWN WORDS** If you round the standard quotas up, how do you need to change the standard divisor to find the modified quotas?

Modified quotas are given in Problems 7–14. Round your answers to two decimal places.
a. *Find the lower and upper quotas.*
b. *Find the arithmetic mean of the lower and upper quotas.*
c. *Find the geometric mean of the lower and upper quotas.*
d. *Round the given modified quota by comparing it with first the arithmetic mean, and then with the geometric mean.*

7. 3.81	**8.** 1.24
9. 1.46	**10.** 3.48
11. 2.49	**12.** 2.51
13. 1,695.4	**14.** 1,695.6

Find the standard divisor (to two decimal places) for the given populations and number of representative seats in Problems 15–22.

	Population	# seats		Population	# seats
15.	52,000	8	**16.**	135,000	8
17.	630	5	**18.**	540	7
19.	1,450,000	12	**20.**	8,920,000	12
21.	23,000,000	125	**22.**	62,300,000	225

For the given year, find the standard quotas for the New York City boroughs given in Table 16.5 in Problems 23–28. Assume there are eight council seats.

23. 1800		**24.** 1840	
25. 1900		**26.** 1940	
27. 1990		**28.** 2000	

Consider the populations given in Problems 29–32.
a. *Find the standard divisor.*
b. *Find the standard quota for each precinct.*
c. *Total, rounding the standard quotas down.*
d. *Find a modified divisor that will give modified quotas to produce the desired number of seats.*

29. 10 seats

Population	
1st Precinct	35,000
2nd Precinct	21,000
3rd Precinct	12,000
4th Precinct	48,000
TOTAL	116,000

30. 12 seats

Population	
1st Precinct	35,000
2nd Precinct	21,000
3rd Precinct	12,000
4th Precinct	48,000
TOTAL	116,000

31. 10 seats

Population	
1st Precinct	135,000
2nd Precinct	231,000
3rd Precinct	118,000
4th Precinct	316,000
TOTAL	800,000

32. 12 seats

Population	
1st Precinct	135,000
2nd Precinct	231,000
3rd Precinct	118,000
4th Precinct	316,000
TOTAL	800,000

Consider the populations given in Problems 33–36.
a. *Find the standard divisor.*
b. *Find the standard quota for each precinct.*
c. *Total, rounding the standard quotas up.*
d. *Find a modified divisor that will give modified quotas to produce the desired number of seats.*

33. 10 seats

Population	
1st Precinct	35,000
2nd Precinct	21,000
3rd Precinct	12,000
4th Precinct	48,000
TOTAL	116,000

34. 12 seats

Population	
1st Precinct	35,000
2nd Precinct	21,000
3rd Precinct	12,000
4th Precinct	48,000
TOTAL	116,000

35. 10 seats

Population	
1st Precinct	135,000
2nd Precinct	231,000
3rd Precinct	118,000
4th Precinct	316,000
TOTAL	800,000

36. 12 seats

Population	
1st Precinct	135,000
2nd Precinct	231,000
3rd Precinct	118,000
4th Precinct	316,000
TOTAL	800,000

LEVEL 2

In the apportionment of the House of Representatives based on the 1790 *Census (Example 3), there are* 15 *states. At that time Maine was still considered part of Massachusetts. If Maine had been a separate state, it would have shared in the distribution of the seats in the House. In the* 1790 *census, Maine's population was* 96,643. *Subtract this number from Massachusetts' population and then answer the questions in Problems* 37–40 *based on* 16 *states.*

37. What are the standard quotas for the 16 states?

38. What is the total of the quotas rounded to the nearest number?

39. What is the sum of the lower quotas?

40. What is the sum of the upper quotas?

Consider the following apportionment problem for College Town:

North:	8,700
South:	5,600
East:	7,200
West:	3,500

Suppose each council member is to represent approximately 2,500 *citizens. Use the apportionment plan requested in Problems* 41–45 *assuming there must be* 10 *representatives.*

41. Adams' plan

42. Jefferson's plan

43. Hamilton's plan

44. Webster's plan

45. HH plan

Consider the following apportionment problem:

North:	18,200
South:	12,900
East:	17,600
West:	13,300

Use the apportionment plan requested in Problems 46–50 *assuming there must be* 26 *representatives.*

46. Adams' plan

47. Jefferson's plan

48. Hamilton's plan

49. Webster's plan

50. HH plan

Consider the following apportionment problem:

North:	18,200
South:	12,900
East:	17,600
West:	13,300

Use the apportionment plan requested in Problems 51–55 *assuming there must be* 16 *representatives.*

51. Adams' plan

52. Jefferson's plan

53. Hamilton's plan

54. Webster's plan

55. HH plan

LEVEL 3

Consider the following apportionment problem:

North:	1,820,000
Northeast:	2,950,000
East:	1,760,000
Southeast:	1,980,000
South:	1,200,000
Southwest:	2,480,000
West:	3,300,000
Northwest:	1,140,000

If there are to be 475 representatives, use the apportionment plan requested in Problems 56–60.

56. Adams' plan

57. Jefferson's plan

58. Hamilton's plan

59. Webster's plan

60. HH plan

16.4 APPORTIONMENT PARADOXES

As we saw in the last section, the U.S. Constitution mandated that representation in the House of Representatives be based on proportional representation by state. The first three apportionment plans, Adams', Jefferson's, and Hamilton's plans, were debated in Congress in regard to implementing this mandate using the population numbers from the 1790 census. The easiest to apply was Hamilton's plan, which, as we saw, was vetoed by Washington. This left two plans that relied on modified divisors. Remember, the quota rule states that the actual number of seats assigned to each state must be the lower or the upper quota. Consider the following historical example.

EXAMPLE 1

In the 1830 election, New York's population was 1,918,578 and the U.S. population was 11,931,000. At that time, there were 240 seats in the House of Representatives. Determine the standard, upper, and lower quotas for New York in the 1830 election.

Solution $d = \dfrac{11,931,000}{240} = 49,712.5$

The standard quota was $\dfrac{1,918,578}{d} \approx 38.59$

The lower quota is 38 and the upper quota is 39. ◆

Remember, in 1830 Jefferson's plan was in place and the modified division used for that election produced a number with a standard quota that rounded down to 40 (see Example 1; this violated the quota rule). Daniel Webster was outraged, and he argued that the result was unconstitutional. Unfortunately, he proposed a method that had the same flaw (namely, it violated the quota rule). Any plan that uses a modified divisor can violate the quota rule.

EXAMPLE 2

Packard-Hue manufactures testing equipment at four locations and has just hired 200 new employees. Those employees are to be apportioned by production levels at the four locations using Jefferson's plan. The locations and production levels are

Atlanta:	12,520
Buffalo:	4,555
Carson City:	812
Denver:	947

a. What is the standard divisor?

b. What are the standard quotas?

c. What are the lower and upper quotas?

d. Find a modified divisor that will produce an appropriate modified quota totaling the 200 new employees.

e. Show that this violates the quota rule.

Solution

a. The total production is the sum of the production levels/wk at each of the four plants; 18,834.

The standard divisor is $d = \dfrac{18,834}{200} \approx 94.17$.

b. We calculate the production quotas:

Factory	Production/wk	q	Lower quota	Upper quota
A:	12,520	132.95	132	133
B:	4,555	48.37	48	49
C:	812	8.62	8	9
D:	947	10.06	10	11
TOTALS:			198	202

c. The lower and upper quotas are found by rounding, as shown in the table in part **b.**

d. We need to *lower d* so that the values of *q increase* to give us a total of 200 when we round them down. Let $D = 93$. We calculate the modified quotients.

Factory	Production/wk	Q	Round down
A:	12,520	134.62	134
B:	4,555	48.98	48
C:	812	8.73	8
D:	947	10.18	10
TOTAL:			200

e. Notice that factory A's lower quota is 132 and its upper quota is 133, but the modified quota produces a rounded-down value that is higher than the upper quota. Thus, we say this example violates the quota rule. ◆

The same kind of inconsistencies can happen with any of the methods that use modified divisors and modified quotas to form the sum. Only Hamilton's plan is immune to this behavior, and all the other plans may violate the quota rule. You might ask, Why, then, don't we use Hamilton's plan? We have another paradox to consider.

Alabama Paradox

A serious inconsistency in Hamilton's plan was discovered in the 1880 census. Before the number of seats in the House of Representatives was fixed at 435, there was a debate on whether to have 299 or 300 seats in the House. Using Hamilton's method with 299 members, Alabama was to receive eight seats. But if the total number of representatives were *increased* to 300, Alabama would receive only seven seats!

Alabama Paradox

> A reapportionment in which an increase in the total number of seats results in a loss in the seats for some state is known as the **Alabama paradox.**

Example 6 of the previous section provides a good example of this paradox.

EXAMPLE 3

In Example 6 of Section 16.3 we used Hamilton's plan to distribute 200 calculators as shown in this table:

The standard divisor is $d = \dfrac{612}{200} = 3.06$.

	Standard quota, q	Lower quota	Hamilton's plan	
Elsie Allen	$72/d \approx 23.53$	23	24	#2 (0.53)
Maria Carrillo	$131/d \approx 42.81$	42	43	#1 (0.81)
Montgomery	$243/d \approx 79.41$	79	79	
Piner	$95/d \approx 31.04$	31	31	
Santa Rosa	$71/d \approx 23.20$	23	23	
TOTAL		198	200	

Now, suppose that a parent at Maria Carrillo High School heard of the donation and contributed one additional calculator to raise the grant to 201 calculators. What is the apportionment according to Hamilton's plan?

Solution

The standard divisor is $d = \dfrac{612}{201} \approx 3.045$.

	Standard quota, q	Lower quota	Hamilton's plan	
Elsie Allen	$72/d \approx 23.65$	23	24	#2 (0.65)
Maria Carrillo	$131/d \approx 42.02$	42	42	
Montgomery	$243/d \approx 79.81$	79	80	#1 (0.81)
Piner	$95/d \approx 31.20$	31	31	#4 (0.20)
Santa Rosa	$71/d \approx 23.32$	23	24	#3 (0.32)
TOTAL		198	201	

As you can see, the extra calculator from the Maria Carrillo parent caused that school to receive one less calculator!   ◆

If you look at the preceding example, it is easy to see how this paradox can occur. Raising the number can cause a decimal part of one number to increase faster than the decimal part of another, changing their "rank" in their decimal parts.

Population Paradox

Shortly after the discovery of the Alabama paradox, another paradox was discovered around 1900 while Congress was still using Hamilton's plan. It is possible for the population of one state to be growing at a faster rate than another state, but still lose seats to the slower-growing state.

Population Paradox

> When there is a fixed number of seats, a reapportionment that causes a state to lose a seat to another state even though the percent increase in the population of the state that loses the seat is larger than the percent increase of the state that wins the seat is called the **population paradox.**

For an example of the population paradox, consider the following example.

EXAMPLE 4

Suppose there are 100 new teachers to be apportioned to the three boroughs according to their 1990 population using Hamilton's plan. The population of each borough is

	1990	2000
Anderson Valley	3,755	3,800
Bennett Valley	10,250	10,350
Central Valley	36,100	36,150
TOTAL	50,105	50,300

Show that this example illustrates the population paradox.

Solution   In 1990, $d = \dfrac{50,105}{100} = 501.05$.

The standard quotas are (in thousands)

	Population	q	Lower quota	Hamilton's plan
A:	3,755	7.49	7	8 (#1)
B:	10,250	20.46	20	20
C:	36,100	72.05	72	72
TOTAL:	50,105		99	100

In 2000, $d = \dfrac{50,300}{100} = 503$.

	Population	q	Lower quota	Hamilton's plan
A:	3,800	7.55	7	7
B:	10,350	20.58	20	21 (#2)
C:	36,150	71.87	71	72 (#1)
TOTAL:	50,105		98	100

Now, here are the percent increases:

	1990	2000	Increase	% increase
A:	3,755	3,800	45	$45/3755 \approx 1.20\%$
B:	10,250	10,350	100	$100/10,250 \approx 0.98\%$
C:	36,100	36,150	50	$50/36,100 \approx 0.14\%$

Anderson Valley had the largest percent increase, but lost one vote to Bennett Valley. This is an example of the population paradox.   ◆

New States Paradox

The last paradox we will discuss was discovered in 1907 when Oklahoma joined the Union. If a new state is added to the existing states and the number of seats being apportioned is increased to prevent decreasing the existing apportionment, addition may cause a shift in some of the original allocations.

New States Paradox

> A reapportionment in which an increase in the total number of seats causes a shift in the apportionments of the existing states is called the **new states paradox.**

The following example is a simplified version of Example 3.

EXAMPLE 5

Suppose a school received a grant of 20 calculators to distribute to two schools:

Elsie Allen	71 advanced math students
Maria Carrillo	119 advanced math students
TOTAL	190 advanced math students

The standard divisor is $d = \dfrac{190}{20} = 9.5$

That is, there is one calculator for each 9 or 10 students. The calculators are to be apportioned according to Hamilton's plan.

		q	Round down	Hamilton's plan
Elsie Allen	71	7.47	7	7
Maria Carrillo	119	12.53	12	13 (#1)
TOTAL			19	20

Now, suppose a parent decides to donate 5 more calculators for Ridgeway High, with a student population of 51 advanced math students. Shouldn't that be just right to accommodate the new school? Show that this example illustrates the new states paradox by calculating the new apportionment using this added contribution.

Solution

Elsie Allen	71 advanced math students
Maria Carrillo	119 advanced math students
Ridgeway	51 advanced math students
TOTAL	241 advanced math students

The standard divisor is $d = \dfrac{241}{25} = 9.64$.

The calculators are to be apportioned according to Hamilton's plan.

		q	Round down	Hamilton's plan
Elsie Allen	71	7.37	7	8 (#1)
Maria Carrillo	119	12.34	12	12
Ridgeway	51	5.29	5	5
TOTAL			24	25

Notice that even though the five new calculators went to Ridgeway (as intended) that addition (new state) caused an adjustment on the prior apportionment. This is an example of the new states paradox. ◆

Balinski and Young's Impossibility Theorem

In 1980 two mathematicians proved that if any apportionment plan satisfies the quota rule, then it must permit the possibility of some other paradox. Here is a statement of the **Balinski and Young's impossibility theorem.**

Balinski and Young's Impossibility Theorem

> Any apportionment plan that does not violate the quota rule must produce paradoxes. And any apportionment plan that does not produce paradoxes must violate the quota rule.

There is a footnote, however, to this theorem. It must be qualified that the second part of this theorem is true only when the number of seats to be apportioned is fixed up front, as it is fixed with the U.S. House of Representatives to be 435.

We conclude with a summary of the paradoxes of this section, along with the quota rule, in Table 16.10.

Table 16.10 Comparison of Apportionment Paradoxes

Paradoxes	Adams'	Jefferson's	Apportionment Plans Hamilton's	Webster's	HH
Quota rule Apportionment must be either the lower or upper quota.	Yes	Yes	No	Yes	Yes
Alabama paradox An increase in the total number may result in a loss for a state.	No	No	Yes	No	No
Population paradox One state may lose seats to another even though its percent increase is greater.	No	No	Yes	No	No
New states paradox The addition of a new state may change the apportionment of another group.	No	No	Yes	No	No

Note: "Yes" means that the apportionment method may violate the stated paradox. "No" means that the apportionment method may not violate the stated paradox.

PROBLEM SET 16.4

LEVEL 1

1. **IN YOUR OWN WORDS** What is the Alabama paradox?
2. **IN YOUR OWN WORDS** What is the population paradox?
3. **IN YOUR OWN WORDS** What is the new states paradox?
4. **IN YOUR OWN WORDS** What does the Balinski and Young's impossibility theorem say?

Use Adams' plan in Problems 5–6. Show that the quota rule is violated.

5. State: A   B   C   D
   Population: 68,500   34,700   16,000   9,500
   Number of seats: 100

6. State: A   B   C   D
   Population: 685   347   160   95
   Number of seats: 100

Use Jefferson's plan in Problems 7–10. Show that the quota rule is violated.

7. State: A   B   C   D
Population: 68,500   34,700   14,800   9,500
Number of seats: 100

8. State: A   B   C   D
Population: 17,179   7,500   49,400   5,824
Number of seats: 132

9. State: A   B   C   D   E
Population: 1,100   1,100   1,515   4,590   2,010
Number of seats: 200

10. State: A   B   C   D   E
Population: 1,700   3,300   7,000   24,190   8,810
Number of seats: 150

In Problems 11–14, use Hamilton's plan to apportion the new seats to the existing states. Then increase the number of seats by one and decide whether the Alabama paradox occurs. Assume that the populations are in thousands.

11. State: A   B   C   D
Population: 181   246   812   1,485
Number of seats: 246

12. State: A   B   C   D
Population: 235   318   564   938
Number of seats: 45

13. State: A   B   C   D   E
Population: 300   301   340   630   505
Number of seats: 50

14. State: A   B   C   D   E
Population: 300   700   800   800   701
Number of seats: 82

In Problems 15–18, apportion the indicated number of representatives to three states A, B, and C using Hamilton's plan. Next, use the revised populations to reapportion the representatives. Decide whether the population paradox occurs.

15. State: A   B   C
Population: 55,200   124,900   190,000
Revised population: 61,100   148,100   215,000
Number of seats: 11

16. State: A   B   C
Population: 90,000   124,800   226,000
Revised population: 98,000   144,900   247,100
Number of seats: 13

17. State: A   B   C
Population: 89,950   124,800   226,000
Revised population: 97,950   144,900   247,100
Number of seats: 13

18. State: A   B   C
Population: 7,510   20,500   72,000
Revised population: 7,650   20,800   72,200
Number of seats: 100

In Problems 19–22, apportion the indicated number of representatives to two states, A and B, using Hamilton's plan. Next, recalculate the apportionment using Hamilton's plan for the three states, C and the original two states. Decide whether the new states paradox occurs.

19. State: A   B   C
Population: 144,899   59,096   38,240
Number of original seats: 12
Additional seats: 2

20. State: A   B   C
Population: 394,990   753,950   138,550
Number of original seats: 16
Additional seats: 1

21. State: A   B   C
Population: 7,000,500   9,290,500   1,450,000
Number of original seats: 50
Additional seats: 4

22. State: A   B   C
Population: 265,000   104,000   69,000
Number of original seats: 16
Additional seats: 2

LEVEL 2

23. Packard-Hue manufactures testing equipment at four locations, and has just hired 300 new employees. Those employees are to be apportioned using production levels at the four locations according to Jefferson's plan. The locations and production are as follows:

   Atlanta: 12,520
   Buffalo: 4,555
   Carson City: 812
   Denver: 947

 a. What is the standard divisor?
 b. What are the standard quotas?
 c. What are the lower and upper quotas?
 d. Find a modified divisor that will produce appropriate modified quotas totaling the 300 new employees.
 e. Does this violate the quota rule?

24. The English enrollments at five high schools in the Santa Rosa Unified School District are as follows:

   Elsie Allen   154
   Maria Carrillo   142
   Montgomery   165
   Piner   307
   Santa Rosa   231
   TOTAL   999

Suppose that 45 copies of an important instructional video are to be allocated to the schools by using Hamilton's plan and the school population.

a. What is the standard divisor?
b. What are the standard quotas?
c. What are the lower and upper quotas?
d. What are the allocations for the 45 videos?
e. What are the allocations for 46 videos?
f. Does this illustrate the Alabama paradox?

25. The township of Bella Rosa is divided into two districts, uptown (pop. 16,980) and downtown (pop. 3,350), and is governed by 100 council members.
    a. What is the standard divisor?
    b. What are the standard quotas?
    c. How should the seats be apportioned using Hamilton's plan?
    d. Suppose the township annexes a third district (pop. 2,500). Using the standard divisor from part **a**, it was agreed that the new district should add 12 new seats. Carry out this new apportionment for the township using Hamilton's plan.
    e. Does this example illustrate the new states paradox?

26. Suppose the annual salaries of three people are

    Employee #1: $43,100

    Employee #2: $42,150

    Employee #3: $20,000 (half-time)

    a. What are their salaries if they are given a 5% raise, and then the result is rounded to the nearest $1,000 using Hamilton's plan with a cap on the total salaries of $111,000?
    b. Suppose the salary increase is to be 6% with a cap of $111,000. What are the salaries if they are rounded to the nearest $1,000 using Hamilton's plan?
    c. If you compare parts **a** and **b**, are any of the paradoxes illustrated?

PROBLEM SOLVING

27. A fair apportionment of dividing a leftover piece of cake between two children is to let child #1 cut the cake into two pieces and then to let child #2 pick which piece he or she wants. Consider the following apportionment of dividing the leftover piece of cake among three children. Let the first child cut the cake into two pieces. Then the second child is permitted to cut one of those pieces into two parts. Child #3 can select any of the pieces, followed by child #2 selecting one of the remaining pieces, followed by child #1 who gets the remaining piece. Is this allocation process fair, if each child's goal is to maximize the size of his or her own piece of cake?

28. An elderly rancher died and left her estate to her three children. She bequeathed her 17 prize horses in the following manner: 1/2 to the eldest, 1/3 to the second child, and 1/9 to the youngest. How would you divide this estate?

29. The children (see Problem 28) decided to call in a very wise judge to help in the distribution of the rancher's estate. The judge arrived with a horse of his own. He put his horse in with the 17 belonging to the estate, and then told each child to pick from among the 18 in the proportions stipulated by the will (but be careful, he warned, not to pick his horse). The first child took nine horses, the second child took six, and the third child, two. The 17 horses were thus divided among the children. The wise judge took his horse from the corral, took a fair sum for his services, and rode off into the sunset.

    The youngest son complained that the oldest son received 9 horses (but was entitled to only $17/2 = 8.5$ horses). The judge was asked about this, and he faxed the children the following message: "You all received more than you deserved. The eldest received 1/2 of an 'extra' horse, the middle child received 1/3 more, and the youngest, 1/9 of a horse 'extra.'" Apportion the horses according to Adams', Jefferson's, and Webster's plans. Which plan gives the appropriate distribution of horses?

30. The children (see Problem 28) decided to call in a very wise judge to help in the distribution of the rancher's estate. They informed the judge that the 17 horses were not of equal value. The children agreed on a ranking of the 17 horses (#1 being the best and #17 being a real dog of a horse). They asked the judge to divide the estate fairly so that each child would receive not only the correct number of horses but horses whose average rank would also be the same. For example, if a child received horses 1 and 17, the number of horses is two and the average value is

$$\frac{1 + 17}{2} = 9.$$ How did the judge apportion the horses?

CHAPTER SUMMARY

"Representatives and direct Taxes shall be apportioned among the several States which may be included in this Union, according to their respective numbers . . ."

Article I, Section 2 of the *Constitution of the United States*

IMPORTANT TERMS

Adams' plan [16.3]
Alabama paradox [16.4]
Apportionment [16.3]
Approval voting method [16.1]
Arithmetic mean [16.3]
Arrow's impossibility theorem [16.2]
Balinski and Young's impossibility theorem [16.4]
Binary voting [16.1]
Borda count [16.1]
Condorcet candidate [16.2]
Condorcet criterion [16.2]
Condorcet's paradox [16.2]
Decisiveness (principle of) [16.1]
Dictatorship [16.1]
Fairness criteria [16.2]
Fair voting principles [16.2]

Geometric mean [16.3]
Hamilton's plan [16.3]
Hare method [16.1]
HH plan [16.3]
Huntington–Hill's plan [16.3]
Independence of irrelevant alternatives [16.1]
Irrelevant alternatives criterion [16.2]
Jefferson's plan [16.3]
Lower quota [16.3]
Majority criterion [16.2]
Majority rule [16.1]
Modified divisor [16.3]
Modified quota [16.3]
Monotonicity criterion [16.2]
New states paradox [16.4]
Pairwise comparison method [16.1]

Pareto principle [16.1]
Plurality method [16.1]
Population paradox [16.4]
Quota rule [16.3]
Runoff election [16.1]
Sequential voting [16.1]
Social choice theory [16.1]
Standard divisor [16.3]
Standard quota [16.3]
Straw vote [16.2]
Symmetry [16.1]
Tournament method [16.1]
Transitive law [16.2]
Unrestricted domain [16.2]
Upper quota [16.3]
Vote [16.1]
Webster's plan [16.3]

TYPES OF PROBLEMS

Understand and be able to use voting notation. [16.1]
Decide how many different voting orders are possible for a set of candidates. [16.1]
Use the majority rule. [16.1]
Calculate the winner of an election using the plurality method. [16.1]
Calculate the winner of an election using the Borda count method. [16.1]
Calculate the winner of an election using the pairwise comparison method. [16.1]
Calculate the winner of an election using the tournament method. [16.1]
Calculate the winner of an election using the Hare method. [16.1]
Be able to discuss and recognize the majority criterion. [16.2]
Be able to discuss and recognize the Condorcet criterion. [16.2]
Be able to discuss and use the monotonicity criterion. [16.2]
Be able to discuss and use the irrelevant alternatives criterion. [16.2]
What are the fairness criteria? [16.2]
Be able to find the standard quota and divisor. [16.3]
Apportion using Adams' plan. [16.3]
Apportion using Jefferson's plan. [16.3]
Apportion using Hamilton's plan. [16.3]
Apportion using Webster's plan. [16.3]
Apportion using Huntington–Hill's plan. [16.3]
Be able to discuss and recognize the Alabama paradox. [16.4]
Be able to discuss and recognize the population paradox. [16.4]
Be able to discuss and recognize the new states paradox. [16.4]

CHAPTER 16 REVIEW QUESTIONS

A taste test is conducted on the Atlantic City Boardwalk. People are given samples of Coke, Pepsi, and Safeway brands of cola in unmarked cups, and are then asked to rank them in order of preference. The first cup is labeled A, the second, B, and the third, C. Here are the results of the voting:

(CBA)	(ABC)	(BAC)
18	15	12

Use this information in Problems 1–5.

1. Does any item have a majority? If not, how about a plurality?
2. How many rankings of the three colas received no votes?
3. Who wins using the Hare method?
4. Who wins a pairwise comparison?
5. Are any of the fairness criteria violated?

In an election with three candidates A, B, and C, we find the following results of the voting:

(ABC)	(ACB)	(BAC)	(BCA)	(CBA)	(CAB)
22%	23%	15%	29%	7%	4%

Use this information in Problems 6–7.

6. Does anyone receive a plurality?
7. Is there a Condorcet winner?
8. In 2001 the voting for the Heisman Trophy involved 925 ballots for three college football players. The results (in alphabetical order) were as follows:

   David Carr (Fresno State): 34, 60, 58

   Eric Crouch (Nebraska): 162, 98, 88

   Ken Dorsey (Miami): 109, 122, 67

   Dwight Freeney (Syracuse): 2, 6, 24

   Rex Grossman (Florida): 137, 105, 87

   Joey Harrington (Oregon): 54, 68, 66

   Bryant McKinnie (Miami): 26, 12, 14

   Julius Peppers (North Carolina): 2, 10, 15

   Antwaan Randle El (Indiana): 46, 39, 51

   Roy Williams (Oklahoma): 13, 36, 35

   What was the Borda count for each player? Who was the winner according to the Borda count?

Consider a vote of four candidates with the following results:

(ADBC)	(CABD)	(BCDA)	(DBAC)
7	5	4	1

Use this information for Problems 9–11. If there is a tie, break the tie by having a runoff of the tied candidates.

9. Who wins by the Hare method?
10. Who wins by the pairwise comparison method?

11. If B pulls out before the election, who wins? Does this violate the irrelevant alternatives criterion?

Chemistry is taught at five high schools in the Santa Rosa Unified School District. The district has just received a grant of 100 microscopes, which are to be apportioned to the five high schools based on each school's chemistry population. Use the data in Table 16.11 for Problems 12–19.

Table 16.11 School Statistics for Santa Rosa Unified School District High Schools

School	# Students	# Chemistry students
Elsie Allen	1,524	90
Maria Carrillo	1,687	215
Montgomery	1,755	268
Piner	1,519	133
Santa Rosa	1,797	84
TOTAL	8,282	790

12. Find the standard divisor.
13. What are the standard, lower, and upper quotas?
14. Apportion the microscopes using Adams' plan.
15. Apportion the microscopes using Jefferson's plan.
16. Apportion the microscopes using Hamilton's plan.
17. Apportion the microscopes using Webster's plan.
18. Apportion the microscopes using HH's plan.
19. Can you point out any apportionment paradoxes?
20. The city of St. Louis, MO, passed a ballot measure to provide and pay for 180 surveillance cameras for high crime areas. The city council mandated that these cameras be apportioned among the five highest crime areas, based on the 2001 crime statistics, summarized in Table 16.12. Use Webster's plan.

Table 16.12 Serious Crimes in St. Louis, MO, in 2001

Precinct	Number of Violent Crimes/100 Residents
Downtown	24.45
Fairground	10.04
Columbus Square	9.75
Downtown West	9.43
Peabody	9.01

GROUP RESEARCH PROBLEMS

Working in small groups is typical of most work environments, and learning to work with others to communicate specific ideas is an important skill. Work with three or four other students to submit a single report based on each of the following questions.

G60. Research the 2000 U.S. presidential election, but do not focus on the "hanging chads." Rather, investigate the effect that third-party candidate Ralph Nader had on the outcome of the election.

G61. Write a history of apportionment in the United States House of Representatives. Pay particular attention to the paradoxes of apportionment.

G62. Investigate some item of interest to your group. It might be to predict the outcome of an upcoming election, or your favorite song or movie. Your group should make up a list of 5 or 6 choices; for example, you might be researching what is the best of the *Star Wars* movies. Make up a written ballot and ask at least 50 people to rank the items on your list. Summarize the outcome of your poll. Was there a majority winner? A plurality winner? Who wins according to the Borda count or the Hare methods? What about the pairwise comparison method? Present a summary of your results.

INDIVIDUAL RESEARCH PROBLEMS

Learning to use sources outside your classroom and textbook is an important skill, and here are some ideas for extending some of the ideas in this chapter. You can find references to these projects in a library or at

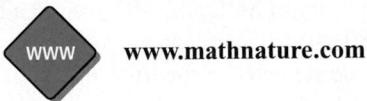 **www.mathnature.com**

PROJECT 16.1 Research how voting is conducted for the following events. Use the terminology of this chapter, not the terminology used in the original sources.

    a. Heisman Trophy

    b. Selecting an Olympic host city

    c. The Academy Awards

    d. The Nobel Prizes

    e. The Pulitzer Prize

PROJECT 16.2 Compare and contrast the voting paradoxes. Which one do you find the most disturbing, and why? Which do you find the least disturbing, and why?

PROJECT 16.3 Compare and contrast the different apportionment plans. Which method do you think is best? Support your position with examples and facts.

PROJECT 16.4 Compare and contrast the apportionment paradoxes. Which one of these do you find the most disturbing, and why? Which one of these do you find the least disturbing, and why?

17 The Nature of Calculus

The language of calculus has spread to all scientific fields; the insight it conveys about the nature of change is something that no educated person can afford to be without.

EVERYBODY COUNTS, NATIONAL RESEARCH COUNCIL

CONTENTS

OVERVIEW

In this chapter, we move from elementary mathematics to presenting an overview of the underpinnings of college mathematics. Some of the basic building blocks for advanced mathematics are functions, functional notation, and limits. These ideas are, in turn, the building blocks for a discussion of the nature of calculus. It is not the intent of this chapter to teach calculus, but to discuss the *nature* of calculus in a context that will enable you to see why its invention was not only necessary, but inevitable.

IMPORTANT IDEAS

Discuss the nature of calculus [17.1]
Mathematical modeling [17.1]
Limit of a sequence [17.2]
Limits to infinity [17.2]
Derivative [17.3]
Derivative of e^x [17.3]
Area function [17.4]
Antiderivative [17.4]
Antiderivative of e^x [17.4]

BOOK REPORT

Write a 500-word report on the following book:
To Infinity and Beyond: A Cultural History of the Infinite. Eli Maor (Boston: Birkhäuser, 1987).

LINKS

Individual Projects

Group Projects

Research Links

www.mathnature.com

17.1 | WHAT IS CALCULUS?

If there is an event that marked the coming of age of mathematics in Western culture, it must surely be the essentially simultaneous development of the calculus by Newton and Leibniz in the 17th century. Before this remarkable synthesis, mathematics had often been viewed as merely a strange but harmless pursuit, indulged in by those with an excess of leisure time. After the calculus, mathematics became virtually the only acceptable language for describing the physical universe. This view of mathematics and its association with the scientific method has come to dominate the Western view of how the world ought to be explained.

What distinguishes calculus from algebra, geometry, and trigonometry is the transition from static or discrete applications (see Figure 17.1) to those that are dynamic or continuous (see Figure 17.2). For example, in elementary mathematics we consider the slope of a line, but in calculus we define the (nonconstant) slope of a nonlinear curve. In elementary mathematics we find average values of position and velocity, but in calculus we can find instantaneous values of changes of velocity and acceleration. In elementary mathematics we find the average of a finite collection of numbers, but in calculus we can find the average value of a function with infinitely many values over an interval.

Calculus is the mathematics of motion and change, which is why calculus is a prerequisite for many courses. Whenever we move from the static to the dynamic, we would consider using calculus.

The development of calculus in the 17th century by Newton and Leibniz was the result of their attempt to answer some fundamental questions about the world and the way things work. These investigations led to two fundamental concepts of calculus—namely, the idea of a *derivative* and that of an *integral*. The breakthrough in the development of these concepts was the formulation of a mathematical tool called a *limit*.

1. **Limit:**   The limit is a mathematical tool for studying the *tendency* of a function as its variable *approaches* some value. Calculus is based on the concept of limit. We shall introduce the limit of a function informally in the next section.

2. **Derivative:**   The derivative is defined as a certain type of limit, and it is used initially to compute rates of change and slopes of tangent lines to curves. The study of derivatives is called *differential calculus*. Derivatives can be used in sketching graphs and in finding the extreme (largest and smallest) values of functions.

3. **Integral:**   The integral is found by taking a special limit of a sum of terms, and the study of this process is called *integral calculus*. Area, volume, arc length, work, and hydrostatic force are a few of the many quantities that can be expressed as integrals.

Let us take an intuitive look at each of these three essential ideas of calculus.

The Limit: Zeno's Paradox

In Chapter 5 we first introduced Zeno's paradox. Zeno (ca. 500 B.C.) was a Greek philosopher known primarily for his famous paradoxes. One of those concerns a race between Achilles, a legendary Greek hero, and a tortoise. When the race begins, the (slower) tortoise is given a head start, as shown in Figure 17.3.

ELEMENTARY MATHEMATICS

1. Slope of a line

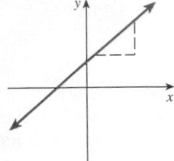

2. Tangent line to a circle

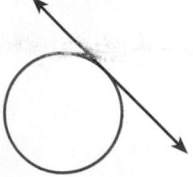

3. Area of a region bounded by line segments

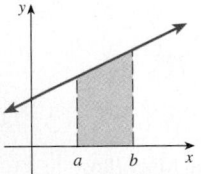

4. Average position and velocity

5. Average of a finite collection of numbers

Figure 17.1 **Topics from elementary mathematics**

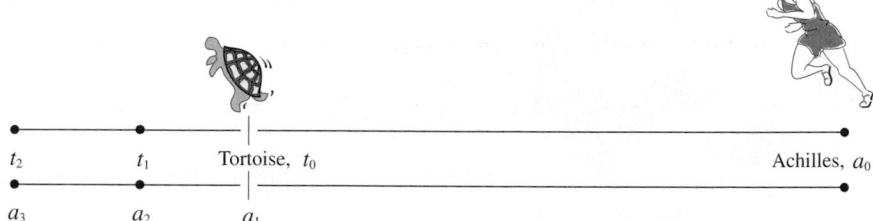

Figure 17.3 Achilles and the tortoise

CALCULUS

1. Slope of a curve

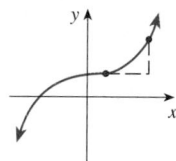

2. Tangent line to a general curve

3. Area of a region bounded by curves

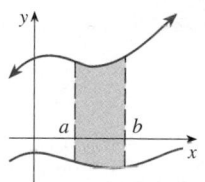

4. Instantaneous changes in position and velocity

5. Average of an infinite collection of numbers

Figure 17.2 Topics from calculus

Is it possible for Achilles to overtake the tortoise? Zeno pointed out that by the time Achilles reaches the tortoise's starting point, $a_1 = t_0$, the tortoise will have moved ahead to a new point t_1. When Achilles gets to this next point, a_2, the tortoise will be at a new point, t_2. The tortoise, even though much slower than Achilles, keeps moving forward. Although the distance between Achilles and the tortoise is getting smaller and smaller, the tortoise will apparently always be ahead.

Of course, common sense tells us that Achilles will overtake the slow tortoise, but where is the error in reasoning? The error is in the assumption that an infinite amount of time is required to cover a distance divided into an infinite number of segments. This discussion is getting at an essential idea in calculus—namely, the notion of a limit.

Consider the successive positions for both Achilles and the tortoise:

Starting position
↓

Achilles: $a_0, a_1, a_2, a_3, a_4, \ldots$

Tortoise: $t_0, t_1, t_2, t_3, t_4, \ldots$

After the start, the positions for Achilles, as well as those for the tortoise, form sets of positions that are ordered with positive integers. Such ordered listings are called *sequences* (see Section 9.3).

For Achilles and the tortoise we have two sequences $\{a_1, a_2, a_3, \ldots, a_n, \ldots\}$ and $\{t_1, t_2, t_3, \ldots, t_n, \ldots\}$, where $a_n < t_n$ for all values of n. Both the sequence for Achilles' position and the sequence for the tortoise's position have limits, and it is precisely at that limit point that Achilles overtakes the tortoise. The idea of limit will be discussed in the next section, and it is this limit idea that allows us to define the other two basic concepts of calculus: the derivative and the integral. Even if the solution to Zeno's paradox using limits seems unnatural at first, do not be discouraged. It took over 2000 years to refine the ideas of Zeno and provide conclusive answers to those questions about limits. The following example will provide an intuitive preview of a limit.

EXAMPLE 1

The sequence

$$\tfrac{1}{2}, \tfrac{2}{3}, \tfrac{3}{4}, \tfrac{4}{5}, \cdots$$

can be described by writing a *general term:* $\dfrac{n}{n+1}$, where $n = 1, 2, 3, 4, \ldots$.* Can you guess the limit, L, of this sequence? We will say that L is the number that the

---

* You might wish to review sequences and general terms in Section 9.3.

sequence with general term $\dfrac{n}{n + 1}$ tends toward as n becomes large without bound. We will define a notation to summarize this idea:

$$L = \lim_{n \to \infty} \frac{n}{n + 1}$$

Solution As you consider larger and larger values for n, you find a sequence of fractions:

$$\frac{1}{2}, \frac{2}{3}, \frac{3}{4}, \ldots, \frac{1{,}000}{1{,}001}, \frac{1{,}001}{1{,}002}, \ldots, \frac{9{,}999{,}999}{10{,}000{,}000}, \ldots$$

It is reasonable to guess that the sequence of fractions is approaching the number 1. ◆

The Derivative: The Tangent Problem

A tangent line (or, if the context is clear, simply say "tangent") to a circle at a given point P is a line that intersects the circle at P and only at P (see Figure 17.4a). This characterization does not apply for curves in general, as you can see by looking at Figure 17.4b.

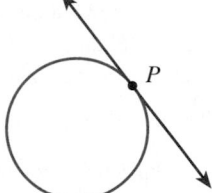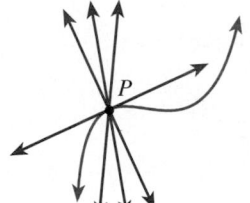

a. At each point P on a circle, there is one line that intersects the circle exactly once.

b. At a point P on a curve, there may be several lines that intersect that curve only once.

Figure 17.4 Tangent line

To find a tangent line, begin by considering a line that passes through two points on the curve, as shown in Figure 17.5a. This line is called a **secant line.**

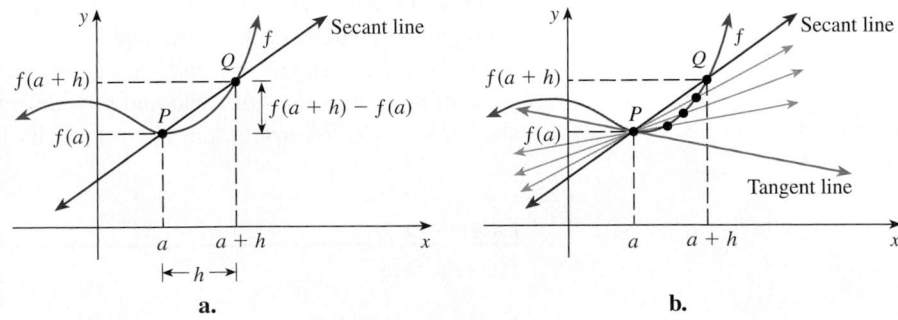

a. **b.**

Figure 17.5 Secant line

The coordinates of the two points P and Q are $P(a, f(a))$ and $Q(a + h, f(a + h))$. The slope of the secant line is

$$m = \frac{\text{RISE}}{\text{RUN}} = \frac{f(a + h) - f(a)}{h}$$

Check
www.mathnature.com
for some links to interactive
sites illustrating the
concept that a sequence
of secant lines approaches
a tangent line.

Now imagine that Q moves along the curve toward P, as shown in Figure 17.5b. You can see that the secant line approaches a limiting position as h approaches zero. We define this limiting position to be the **tangent line.** The slope of the tangent line is defined as a limit of the sequence of slopes of a set of secant lines. Once again, we can use limit notation to summarize this idea: We say that the slopes of the secant lines, as h becomes small, tend toward a number that we call the slope of the tangent line. We will define the following notation to summarize this idea:

$$\lim_{h \to 0} \frac{f(a + h) - f(a)}{h}$$

This limit forms the definition of derivative and forms the foundation for what is called **differential calculus.**

The Integral: The Area Problem

You probably know the formula for the area of a circle with radius r:

$$A = \pi r^2$$

The Egyptians were the first to use this formula over 5,000 years ago, but the Greek Archimedes (ca. 300 B.C.) showed how to derive the formula for the area of a circle by using a limiting process. Consider the areas of inscribed polygons, as shown in Figure 17.6.

Check
www.mathnature.com
for a link amplifying this idea.

 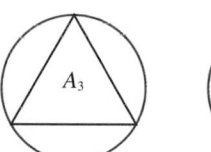 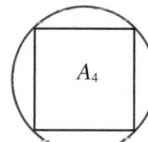 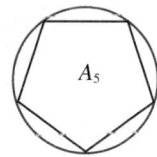 

Figure 17.6 Approximating the area of a circle

Even though Archimedes did not use the following notation, here is the essence of what he did, using a method called "exhaustion":

Let A_3 be the area of the inscribed equilateral triangle;
A_4 be the area of the inscribed square;
A_5 be the area of the inscribed regular pentagon.

How can we find the area of this circle? As you can see from Figure 17.6, if we consider the area of A_3, then A_4, then $A_5, \ldots$, we should have a sequence of areas such that each successive area more closely approximates that of the circle. We write this idea as a limit statement:

$$A = \lim_{n \to \infty} A_n$$

In this course we will use limits in yet a different way to find the areas of regions enclosed by curves. For example, consider the area shown in color in Figure 17.7. We can approximate the area by using rectangles. If A_n is the area of the nth rectangle, then the total area can be approximated by finding the sum

$$A_1 + A_2 + A_3 + \cdots + A_{n-1} + A_n$$

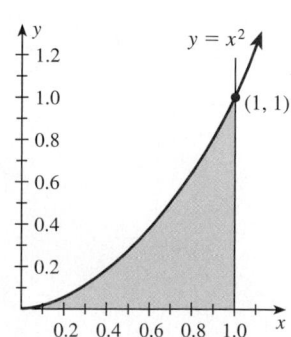

Figure 17.7 Area under a curve

This process is shown in Figure 17.8.

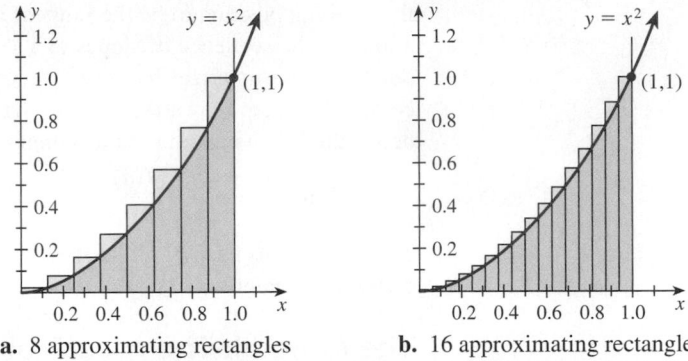

a. 8 approximating rectangles   **b.** 16 approximating rectangles

Figure 17.8 **Approximating the area using rectangles**

The area problem leads to a process called *integration,* and the study of integration forms what is called **integral calculus.** Similar reasoning allows us to calculate such quantities as volume, the length of a curve, the average value, or amount of work required for a particular task.

Mathematical Modeling

A real-life situation is usually far too complicated to be precisely and mathematically defined. When confronted with a problem in the real world, therefore, it is usually necessary to develop a mathematical framework based on certain assumptions about the real world. This framework can then be used to find a solution to the real-world problem. The process of developing this body of mathematics is referred to as **mathematical modeling.** Most mathematical models are dynamic (not static), but are continually being revised (modified) as additional relevant information becomes known.

Some mathematical models are quite accurate, particularly those used in the physical sciences. For example, one of the first models we will consider in calculus is a model for the path of a projectile. Other rather precise models predict such things as the time of sunrise and sunset, or the speed at which an object falls in a vacuum. Some mathematical models, however, are less accurate, especially those that involve examples from the life sciences and social sciences. Only recently has modeling in these disciplines become precise enough to be expressed in terms of calculus.

What, precisely, is a mathematical model? Sometimes, mathematical modeling can mean nothing more than a textbook word problem. But mathematical modeling can also mean choosing appropriate mathematics to solve a problem that has previously been unsolved. In this book, we use the term *mathematical modeling* to mean something between these two extremes. That is, it is a process we will apply to some real-life problem that does not have an obvious solution. It usually cannot be solved by applying a single formula.

The first step of what we call mathematical modeling involves *abstraction.*

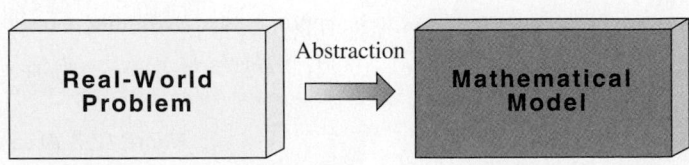

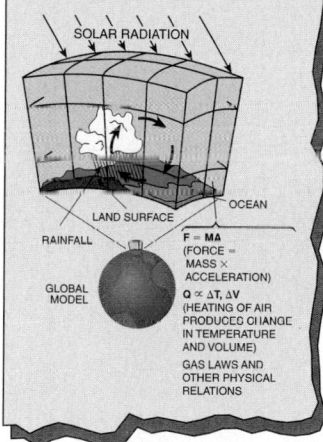

With the method of abstraction, certain assumptions about the real world are made, variables are defined, and appropriate mathematics is developed. The next step is to simplify the mathematics or derive related mathematical facts from the mathematical model.

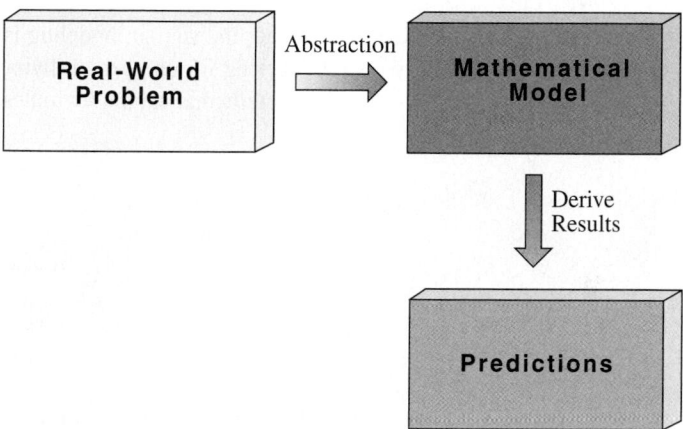

The results derived from the mathematical model should lead us to some predictions about the real world. The next step is to gather data from the situation being modeled, and then to compare those data with the predictions. If the two do not agree, then the gathered data are used to modify the assumptions used in the model.

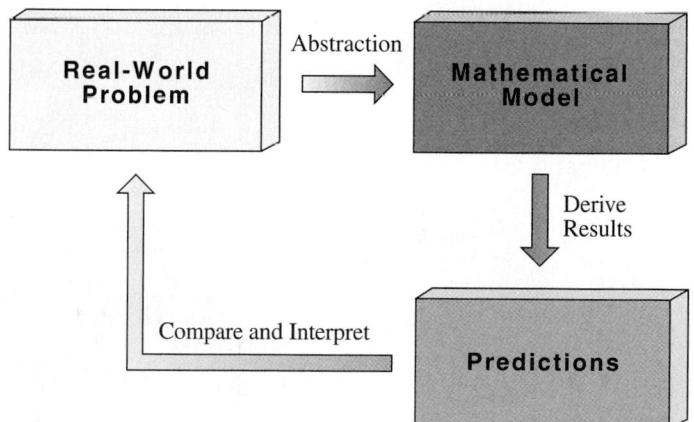

Mathematical modeling is an ongoing process. As long as the predictions match the real world, the assumptions made about the real world are regarded as correct, as are the defined variables. On the other hand, as discrepancies are noticed, it is necessary to construct a closer and a more dependable mathematical model. You might wish to read the article in *Scientific American* described in the News Clip.

EXAMPLE 2

Develop a monthly budget for your personal expenses. You can assume any family situation (for example, number of people and amount of income and expenses), but this budget requires that you are self-sufficient and that income and expenses must always be equal. *This example is not about budgets, but about the process of modeling.*

Solution   The first step in modeling is to understand the problem. Let us assume that we are budgeting for one person living at home with his or her parents. We begin by setting up a mathematical model to describe the income and expenses.

MONTHLY BUDGET		
INCOME		$275.00
EXPENSES		
Car payments	$125.00	
Savings	$ 50.00	
Movies	$ 40.00	
Gasoline	$ 60.00	
		$275.00

This budget satisfies the condition that income equals expenses. However, this model does not reflect the real world. We need to compare and interpret the results of this budget with reality. What about taxes and FICA? Let us make a second attempt at abstraction.

MONTHLY BUDGET			
INCOME			
Monthly Income before taxes	$640.00		
From parents	$530.00		
TOTAL MONTHLY INCOME	$1,170.00		
EXPENSES			
Fixed		Variable	
Rent/mortgage		Food	
Car payments	$125.00	Gasoline	$70.00
Income tax withholding	$170.00		
FICA	$85.00	Utilities	
Retirement	$110.00	Electricity	
Contributions	$65.00	Gas	
		Water/Sewer	
Installments		Telephone	$35.00
Wells Fargo	$35.00	Cable TV	
Sears	$25.00	Other	
VISA	$25.00		
		Entertainment	$30.00
Savings			
Wells Fargo		Other	
	$400.00		
Other			
Total Fixed Expenses	$1,040.00	Total Variable Expenses	$135.00
Total Variable Expenses	$135.00		
TOTAL MONTHLY EXPENSE	$1,175.00		

We can see that the budget is a rough estimate of actual real-world spending, but is not quite balanced (which it needs to be) and does not include "hidden" income and expenses, such as value for the room and board at home. "But my parents pay for that!" is the response. A model must reflect the true income and expenses for the individual, so we must go back and create a more realistic model. Even though "cash" is not exchanged for room and board, "value" is given and must be taken into account. What about income or expenses that do not occur monthly? How about clothes, car repairs, or a vacation? Once again, the modeling process compares and interprets, and then refines with another abstraction.

MONTHLY BUDGET

INCOME

Monthly Income before taxes	$640.00	Other nonmonthly income	$500.00
From parents	$530.00		$215.00
Credit cards	$125.00		
Total other/12	$59.58	Total other	$715.00
TOTAL MONTHLY INCOME	$1,354.58		

EXPENSES	Fixed		Variable		Nonmonthly
Rent/mortgage	$350.00	Food	$280.00	Car repairs	$290.00
Car payments	$125.00	Gasoline	$70.00	Home repairs	
Income tax withholding	$170.00			Car license	$85.00
FICA	$85.00	Utilities		Medical	
Retirement	$110.00	Electricity	$25.00	Dental	
Contributions	$65.00	Gas	$15.00	Clothing	$350.00
Insurance	$105.00	Water/Sewer		Gifts	$400.00
Installments		Telephone	$35.00	Contributions	
Wells Fargo	$35.00	Cable TV	$20.00	Vacation	$800.00
Sears	$25.00	Other		Professional	
VISA	$25.00				
		Entertainment	$30.00	Insurance:	
Savings					
Wells Fargo	$65.00	Other			
				Other:	
Other					
Total Fixed Expenses	$1,160.00	Total Variable Expense	$475.00	Total Nonmonthly	$1,925.00
Total Variable Expenses	$475.00				
1/12 Nonmonthly	$160.42				
TOTAL MONTHLY EXPENSE	$1,795.42				

Now this iteration of the budget is much more realistic, but it does not "balance." Each successive iteration should more closely model the real world, and to complicate the process, the real-world situation of income and expenses is always changing, which may require further revisions of the model shown below. As we can see, the process of modeling is ongoing and may never be "complete."

MONTHLY BUDGET

INCOME

Monthly Income before taxes	$640.00	Other nonmonthly income	$500.00
From parents	$580.00		$215.00
Credit cards	$125.00		
Second job	$240.00		
Total other/12	$59.58	Total other	$715.00
TOTAL MONTHLY INCOME	$1,644.58		

EXPENSES	Fixed		Variable		Nonmonthly
Rent/mortgage	$350.00	Food	$250.00	Car repairs	$290.00
Car payments	$125.00	Gasoline	$70.00	Home repairs	
Income tax withholding	$170.00			Car license	$85.00
FICA	$85.00	Utilities		Medical	
Retirement	$110.00	Electricity	$25.00	Dental	
Contributions	$65.00	Gas	$15.00	Clothing	$350.00
Insurance	$105.00	Water/Sewer		Gifts	$300.00
Installments		Telephone	$35.00	Contributions	
Wells Fargo	$35.00	Cable TV	$20.00	Vacation	
Sears	$25.00	Other		Professional	
VISA	$25.00				
		Entertainment	$30.00	Insurance:	
Savings					
Wells Fargo	$19.16				
Other		Other		Other	
Total Fixed Expenses	$1,114.16				
Total Variable Expenses	$445.00	Total Variable Expenses	$445.00	Total Nonmonthly	$1,025.00
1/12 Nonmonthly	$85.42				
TOTAL MONTHLY EXPENSE	$1,644.58				

Problem Set 17.1

LEVEL 1

1. **IN YOUR OWN WORDS** What are the three main topics of calculus?

2. **IN YOUR OWN WORDS** What is a mathematical model? Why are mathematical models necessary or useful?

3. **IN YOUR OWN WORDS** An analogy to Zeno's tortoise paradox can be made as follows.

   A woman standing in a room cannot walk to a wall. To do so, she would first have to go half the distance, then half the remaining distance, and then again half of what still remains. This process can always be continued and can never be ended.

   Draw an appropriate figure for this problem and then present an argument using sequences to show that the woman will, indeed, reach the wall.

4. **IN YOUR OWN WORDS** Zeno's paradoxes remind us of an argument that might lead to an absurd conclusion:

   Suppose I am playing baseball and decide to steal second base. To run from first to second base, I must first go half the distance, then half the remaining distance, and then again half of what remains. This process is continued so that I never reach second base. Therefore it is pointless to steal base.

   Draw an appropriate figure for this problem and then present a mathematical argument using sequences to show that the conclusion is absurd.

See **www.mathnature.com** for a link to a site on Zeno's paradox.

5. Consider the sequence 0.3, 0.33, 0.333, 0.3333, ... What do you think is the appropriate limit of this sequence?

6. Consider the sequence 6, 6.6, 6.66, 6.666, ... What do you think is the appropriate limit of this sequence?

7. Consider the sequence 0.9, 0.99, 0.999, 0.9999, ... What do you think is the appropriate limit of this sequence?

8. Consider the sequence 0.2, 0.27, 0.272, 0.2727, ... What do you think is the appropriate limit of this sequence?

9. Consider the sequence 3, 3.1, 3.14, 3.141, 3.1415, 3.14159, 3.141592, ... What do you think is the appropriate limit of this sequence?

Copy the figures in Problems 10–15 *on your paper. Draw what you think is an appropriate tangent line for each curve at the point P.*

10.
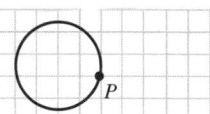

11.
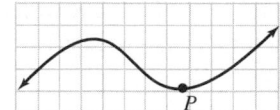

12. 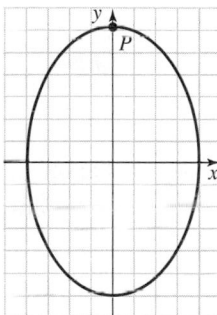   13.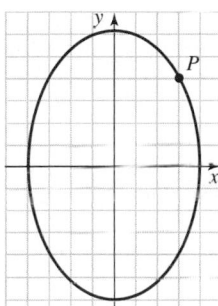

14.
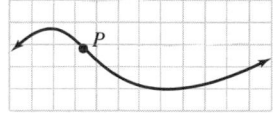

15.
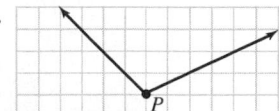

LEVEL 2

In Problems 16–21, ***guess*** *the requested limits.*

16. $\lim\limits_{n \to \infty} \dfrac{2n}{n+4}$ 17. $\lim\limits_{n \to \infty} \dfrac{2n}{3n+1}$

18. $\lim\limits_{n \to \infty} \dfrac{n+1}{n+2}$ 19. $\lim\limits_{n \to \infty} \dfrac{n+1}{2n}$

20. $\lim\limits_{n \to \infty} \dfrac{3n}{n^2+2}$ 21. $\lim\limits_{n \to \infty} \dfrac{3n^2+1}{2n^2-1}$

Estimate the area in each figure shown in Problems 22–27.

22.   23.
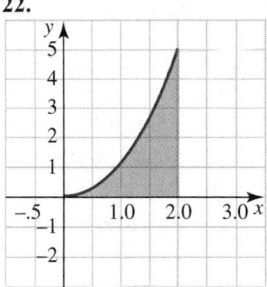 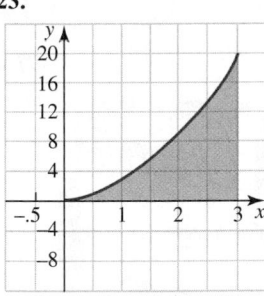

24.   25.
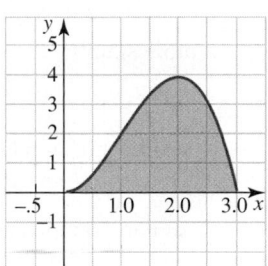 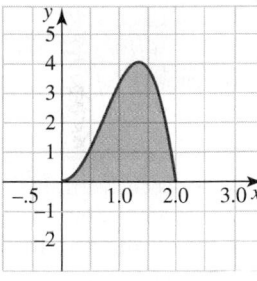

26.
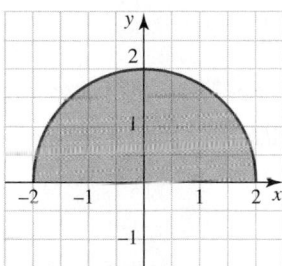

27.
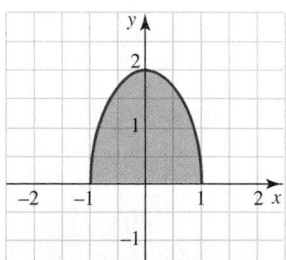

PROBLEM SOLVING

28. **a.** Calculate the sum of the areas of the rectangles shown in Figure 17.8a.
     b. Calculate the sum of the areas of the rectangles shown in Figure 17.8b.
     c. Make a guess about the shaded area under the curve.

29. Prepare a personal budget.

30. Prepare a budget for a family of four with an annual income of $100,000.

These consecutive frames from a 16-mm movie film show a racehorse. If this film were projected at a rate of 24 frames per second, the viewer would have the illusion of seeing the racehorse in action as he makes his way to the finish line.

17.2 | LIMITS

The making of a motion picture is a complex process, and editing all the film into a movie requires that all the frames of the action be labeled in chronological order. For example, R21-435 might signify the 435th frame of the 21st reel. A mathematician might refer to the movie editor's labeling procedure by saying the frames are arranged in a *sequence*. We introduced sequences in Chapter 9, but let's briefly review that idea. A **sequence** is a succession of numbers that are listed according to a given prescription or rule. Specifically, if n is a positive integer, the sequence whose nth term is the number a_n can be written as

$$a_1, a_2, \ldots, a_n, \ldots$$

or more simply

$$\{a_n\}$$

The number a_n is called the **general term** of the sequence. We will deal only with infinite sequences, so each term a_n has a **successor** a_{n+1} and for $n > 1$, a **predecessor** a_{n-1}. For example, by associating each positive integer n with its reciprocal $\dfrac{1}{n}$, we obtain the sequence denoted by

$$\left\{\frac{1}{n}\right\}, \quad \text{which represents the succession of numbers } 1, \frac{1}{2}, \frac{1}{3}, \ldots, \frac{1}{n}, \ldots$$

The general term is denoted by $a_n = \dfrac{1}{n}$.

The following examples illustrate the notation and terminology used in connection with sequences.

EXAMPLE 1

Find the 1st, 2nd, and 15th terms of the sequence $\{a_n\}$ where the general term is

$$a_n = \left(\frac{1}{2}\right)^{n-1}$$

Solution If $n = 1$, then $a_1 = \left(\frac{1}{2}\right)^{1-1} = 1$. Similarly,

$$a_2 = \left(\frac{1}{2}\right)^{2-1} = \frac{1}{2}$$

$$a_{15} = \left(\frac{1}{2}\right)^{15-1} = \left(\frac{1}{2}\right)^{14} = 2^{-14} \qquad \blacklozenge$$

The Limit of a Sequence

It is often desirable to examine the behavior of a given sequence $\{a_n\}$ as n gets arbitrarily large. For example, consider the sequence

$$a_n = \frac{n}{n+1}$$

Even though we write a_n, this is a function, $a(n) = \dfrac{n}{n+1}$, for which the domain is the set of nonnegative integers.

Because $a_1 = \frac{1}{2}, a_2 = \frac{2}{3}, a_3 = \frac{3}{4}, \ldots$, we can plot the terms of this sequence on a number line as shown in Figure 17.9a, or the sequence can be plotted in two dimensions as shown in Figure 17.9b.

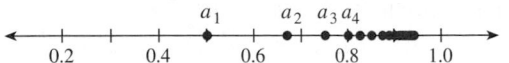

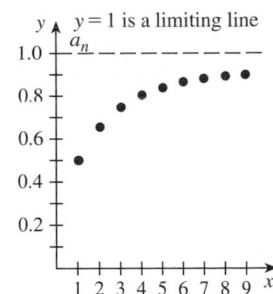

a. Graphing a sequence in one dimension

b. Graphing a sequence in two dimensions

Figure 17.9 **Graphing the sequence** $a_n = \dfrac{n}{n+1}$

Even though we know these tracks to be parallel (they do not actually approach each other), they appear to meet at the horizon. This concept of convergence at a point is used in calculus to give specific values for otherwise unmeasurable quantities.

Limits to Infinity

By looking at either graph in Figure 17.9, we can see that the terms of the sequence are approaching 1. We say that the *limit of this sequence* is $L = 1$.

In general, we say that the sequence **converges** to the limit L and write

$$\lim_{n \to \infty} a_n = L$$

if L is a *unique* number that the value of the expression a_n is moving ever and ever closer to as $n \to \infty$. The number L is called the **limit of the sequence.** The value L does not need to exist. If a limit does not exist, we say the sequence **diverges.**

EXAMPLE 2

Find the limit, if it exists.

a. $\lim_{n \to \infty} \dfrac{n}{n+1}$ 　　**b.** $\lim_{n \to \infty} n$ 　　**c.** $\lim_{n \to \infty} \dfrac{10{,}000}{n}$ 　　**d.** $\lim_{n \to \infty} \dfrac{2n-1}{n}$

Solution

a. $\lim_{n \to \infty} \dfrac{n}{n+1}$ 　This sequence is shown in Figure 17.9. We see

$$\lim_{n \to \infty} \dfrac{n}{n+1} = 1$$

b. $\lim_{n \to \infty} n$ 　This sequence is $1, 2, 3, 4, \ldots$, and we see that the values are not approaching any limit, so we say this limit does not exist.

c. $\lim_{n \to \infty} \dfrac{10{,}000}{n}$

The values of this expression are: $10{,}000$ ($n = 1$); $5{,}000$ ($n = 2$); $3{,}333.33$ ($n = 3$); $2{,}500$ ($n = 4$); $2{,}000$ ($n = 5$); $\ldots$. For large n we see that the value of the sequence is close to 0:

$$n = 10{,}000: \quad \dfrac{10{,}000}{n} = \dfrac{10{,}000}{10{,}000} = 1$$
$$\vdots$$
$$n = 10{,}000{,}000: \quad \dfrac{10{,}000}{n} = \dfrac{10{,}000}{10{,}000{,}000} = 0.001$$
$$\vdots$$

It appears that $\lim_{n \to \infty} \dfrac{10{,}000}{n} = 0$.

d. $\displaystyle\lim_{n\to\infty} \frac{2n-1}{n}$

This sequence can be shown with the following table for selected values for n:

n	1	10	100	1,000	10,000	100,000
$\dfrac{2n-1}{n}$	1	1.9	1.99	1.999	1.9999	1.99999

In this case it appears that $\displaystyle\lim_{n\to\infty} \frac{2n-1}{n} = 2.$ ◆

Example 2 leads to some conclusions.

Limits to Infinity

$$\lim_{n\to\infty} \frac{1}{n} = 0$$

Furthermore, for any A, and k a positive integer,

$$\lim_{n\to\infty} \frac{A}{n^k} = 0$$

EXAMPLE 3

Find the limit of each of these convergent sequences:

a. $\left\{\dfrac{100}{n}\right\}$ **b.** $\left\{\dfrac{2n^2+5n-7}{n^3}\right\}$ **c.** $\left\{\dfrac{3n^4+n-1}{5n^4+2n^2+1}\right\}$

Solution

a. As n grows arbitrarily large, $\dfrac{100}{n}$ gets smaller and smaller. Thus,

$$\lim_{n\to\infty} \frac{100}{n} = 0$$

A graphical representation is shown in Figure 17.10.

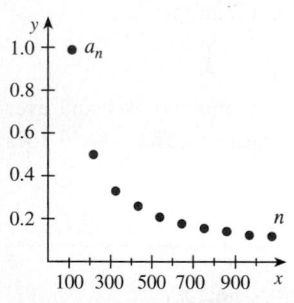

Figure 17.10 **Graphical representation of $a_n = 100/n$**

b. We begin by algebraically rewriting the expression:

$$\frac{2n^2+5n-7}{n^3} = \frac{2}{n} + \frac{5}{n^2} - \frac{7}{n^3}$$

We now use the limit to infinity property:

$$\lim_{n\to\infty} \frac{2n^2+5n-7}{n^3} = \lim_{n\to\infty} \left[\frac{2n^2}{n^3} + \frac{5n}{n^3} + \frac{-7}{n^3}\right]$$

$$= \lim_{n\to\infty} \left[\frac{2}{n} + \frac{5}{n^2} + \frac{-7}{n^3}\right]$$

If we assume that the limit of a sum is the sum of the limits, then we have

$$= 0 + 0 + 0 = 0$$

A graph is shown in Figure 17.11.

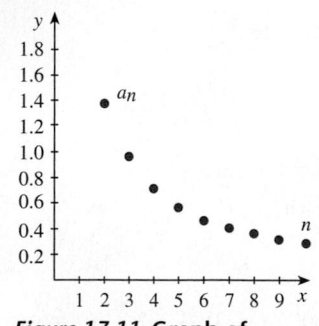

Figure 17.11 **Graph of** $a_n = \dfrac{2n^2+5n-7}{n^3}$

c. Divide the numerator and denominator by n^4 to obtain

$$\lim_{n\to\infty} \frac{3n^4+n-1}{5n^4+2n^2+1} = \lim_{n\to\infty} \frac{\dfrac{3n^4}{n^4} + \dfrac{n}{n^4} - \dfrac{1}{n^4}}{\dfrac{5n^4}{n^4} + \dfrac{2n^2}{n^4} + \dfrac{1}{n^4}}$$ Divide each term by n^4.

$$= \lim_{n\to\infty} \frac{3 + \dfrac{1}{n^3} - \dfrac{1}{n^4}}{5 + \dfrac{2}{n^2} + \dfrac{1}{n^4}} = \frac{3}{5}$$ ◆

Infinite Limits

EXAMPLE 4

Find the limit (if it exists).

a. $\{(-1)^n\}$ **b.** $0.9, 0.99, 0.999, \ldots$

c. $\lim\limits_{n\to\infty} \dfrac{n^5 + n^3 + 2}{7n^4 + n^2 + 3}$

Solution

a. The sequence defined by $\{(-1)^n\}$ is $-1, 1, -1, 1, \ldots$, and this sequence diverges by oscillation because the nth term is always either 1 or -1. Thus a_n cannot approach one specific number L as n grows large. The graph is shown in Figure 17.12a.

b. The given sequence can be written as

$$\frac{9}{10}, \frac{99}{100}, \frac{999}{1,000}, \ldots \quad \text{or} \quad \lim_{n\to\infty}\frac{n}{n+1} = \lim_{n\to\infty}\frac{1}{1+\dfrac{1}{n}} = 1$$

c. $\lim\limits_{n\to\infty} \dfrac{n^5 + n^3 + 2}{7n^4 + n^2 + 3} = \lim\limits_{n\to\infty} \dfrac{1 + \dfrac{1}{n^2} + \dfrac{2}{n^5}}{\dfrac{7}{n} + \dfrac{1}{n^3} + \dfrac{3}{n^5}}$

The numerator tends toward 1 as $n \to \infty$, and the denominator approaches 0. Hence the quotient gets ever larger and larger, passing by all numbers, so it cannot approach a specific number L; thus the sequence must diverge. The graph is shown in Figure 17.12b.

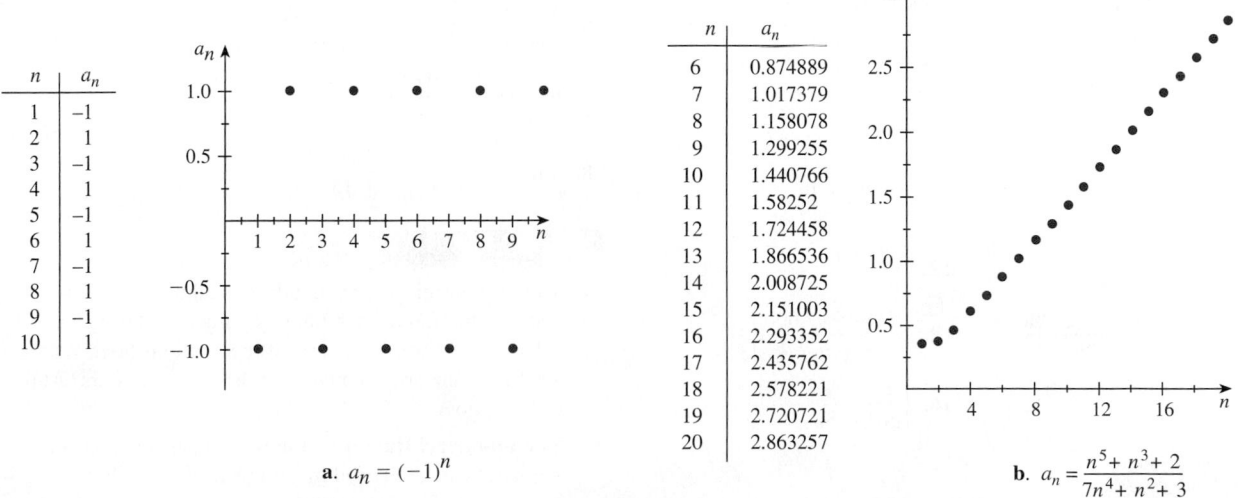

n	a_n
1	-1
2	1
3	-1
4	1
5	-1
6	1
7	-1
8	1
9	-1
10	1

n	a_n
6	0.874889
7	1.017379
8	1.158078
9	1.299255
10	1.440766
11	1.58252
12	1.724458
13	1.866536
14	2.008725
15	2.151003
16	2.293352
17	2.435762
18	2.578221
19	2.720721
20	2.863257

a. $a_n = (-1)^n$

b. $a_n = \dfrac{n^5 + n^3 + 2}{7n^4 + n^2 + 3}$

Figure 17.12 **Graphs and values for two divergent sequences**   ◆

If $\lim\limits_{n\to\infty} a_n$ does not exist because the numbers a_n become arbitrarily large as $n \to \infty$, we write $\lim\limits_{n\to\infty} a_n = \infty$. We summarize this more precisely in the following box.

Limit Notation

$\lim\limits_{n\to\infty} a_n = \infty$ means that for any real number A, we have $a_n > A$ for all sufficiently large n.

$\lim\limits_{n\to\infty} b_n = -\infty$ means that for any real number B, we have $b_n < B$ for all sufficiently large n.

The answer to Example 4c can be rewritten in this notation as

$$\lim_{n\to\infty} \frac{n^5 + n^3 + 2}{7n^4 + n^2 + 3} = \infty$$

Also notice that $\lim\limits_{n\to\infty}(-5n) = -\infty$ (that is, exceeds all bounds), whereas $\lim\limits_{n\to\infty} (-1)^n$ does not exist. Thus, the answer to Example 4a is *not* ∞ or $-\infty$.

Problem Set 17.2

LEVEL 1

1. **IN YOUR OWN WORDS** What do we mean by the limit of a sequence?

2. **IN YOUR OWN WORDS** Outline a procedure for finding the limit of a sequence.

Write out the first five terms (beginning with $n = 1$) of the sequences given in Problems 3–6.

3. $\{1 + (-1)^n\}$

4. $\left\{\left(\dfrac{-1}{2}\right)^{n+2}\right\}$

5. $\left\{\dfrac{3n+1}{n+2}\right\}$

6. $\left\{\dfrac{n^2-n}{n^2+n}\right\}$

Find each limit in Problems 7–10, if it exists.

7. $\lim\limits_{n\to\infty} \dfrac{8,000n}{n+1}$

8. $\lim\limits_{n\to\infty} \dfrac{8,000}{n-1}$

9. $\lim\limits_{n\to\infty} \dfrac{2n+1}{3n-4}$

10. $\lim\limits_{n\to\infty} \dfrac{4-7n}{8+n}$

Compute the limit of the convergent sequences in Problems 11–16.

11. $\left\{\dfrac{5n+8}{n}\right\}$

12. $\left\{\dfrac{5n}{n+7}\right\}$

13. $\left\{\dfrac{8n^2 + 800n + 5,000}{2n^2 - 1,000n + 2}\right\}$

14. $\left\{\dfrac{100n + 7,000}{n^2 - n - 1}\right\}$

15. $\left\{\dfrac{8n^2 + 6n + 4,000}{n^3 + 1}\right\}$

16. $\left\{\dfrac{n^3 - 6n^2 + 85}{2n^3 - 5n + 170}\right\}$

LEVEL 2

Find the limit (if it exists) as $n \to \infty$ for each of the sequences in Problems 17–28.

17. $\left\{\dfrac{1}{3n}\right\}$

18. $\left\{\dfrac{1}{2^n}\right\}$

19. $0.69, 0.699, 0.6999, 0.69999, \ldots$

20. $5, 5\frac{1}{2}, 5\frac{2}{3}, 5\frac{3}{4}, 5\frac{4}{5}, \ldots$

21. $\lim\limits_{n\to\infty} \dfrac{3n^2 - 7n + 2}{5n^4 + 9n^2}$

22. $\lim\limits_{n\to\infty} \dfrac{4n^4 + 10n - 1}{9n^3 - 2n^2 - 7n + 3}$

23. $\lim\limits_{n\to\infty} \dfrac{2n^4 + 5n^2 - 6}{3n + 8}$

24. $\lim\limits_{n\to\infty} \dfrac{10n^3 + 13}{7n^2 - n + 2}$

25. $\lim\limits_{n\to\infty} \dfrac{15 + 9n - 6n^2}{2n^2 - 4n + 1}$

26. $\lim\limits_{n\to\infty} \dfrac{n^4 + 5n^3 + 8n^2 - 4n + 12}{2n^4 - 7n^2 + 3n}$

27. $\lim\limits_{n\to\infty} \dfrac{-21n^3 + 52}{-7n^3 + n^2 + 20n - 9}$

28. $\lim\limits_{n\to\infty} \dfrac{12n^5 + 7n^4 - 3n^2 + 2n}{n^5 + 8n^3 + 14}$

PROBLEM SOLVING

29. Twenty-four milligrams of a drug is administered into the body. At the end of each hour, the amount of drug present is half what it was at the end of the previous hour. What amount of the drug is present at the end of 5 hours? At the end of n hours?

30. **The Fibonacci Rabbit Problem** The general term in a sequence can be defined in a number of ways. In this problem, we consider a sequence whose nth term is defined by a *recursion formula*—that is, a formula in which the nth term is given in terms of previous terms in the sequence. The problem was originally examined by Leonardo Pisano (also called Fibonacci) in the 13th century, and was first introduced in Chapter 9. Recall that we suppose rabbits

breed in such a way that each pair of adult rabbits produces a pair of baby rabbits each month.

Number of Months	Number of Pairs	Pairs of Rabbits (the pairs shown in color are ready to reproduce in the next month)
Start	1	
1	1	
2	2	
3	3	
4	5	
5	8	
⋮	⋮	Same pair (rabbits never die)

Further assume that young rabbits become adults after two months and produce another pair of offspring at that time. A rabbit breeder begins with one adult pair. Let a_n denote the number of adult pairs of rabbits in this "colony" at the end of n months.

a. Explain why $a_1 = 1$, $a_2 = 1$, $a_3 = 2$, $a_4 = 3$, and in general

$$a_{n+1} = a_{n-1} + a_n \quad \text{for } n = 2, 3, 4, \ldots$$

b. The *growth rate* of the colony during the $(n+1)$st month is $r_n = \dfrac{a_{n+1}}{a_n}$. Compute r_n for $n = 1, 2, 3, \ldots, 10$.

c. Assume that the growth rate sequence $\{r_n\}$ defined in part **b** converges, and let $L = \lim\limits_{n \to \infty} r_n$. Use the recursion formula in part **a** to show that

$$\frac{a_{n+1}}{a_n} = 1 + \frac{a_{n-1}}{a_n}$$

and conclude that L must satisfy the equation

$$L = 1 + \frac{1}{L}$$

Use this information to compute L.

17.3 DERIVATIVES

Average Rate of Change

Elementary mathematics focuses on formulas and relationships among variables, but does not include the analysis of quantities that are in a state of constant change. For example, the following problem might be found in elementary mathematics. If you drive at 55 miles per hour (mph) for a total of 3 hours, how far did you travel? This example uses a formula, $d = rt$ (distance = rate · time), and the answer is $d = 55(3) = 165$. However, this model does not adequately describe the situation in the real world. You could drive for 3 hours at an *average rate* of 55 mph, but you probably could not drive for 3 hours at a *constant rate* of 55 mph—in reality, the rate would be in a state of constant change. Other applications in which rates may not remain constant quickly come to mind:

Profits changing with sales
Population changing with the growth rate
Property taxes changing with the tax rate
Tumor sizes changing with chemotherapy
The speed of falling objects changing over time
The slope of a nonlinear curve

The rate at which one quantity changes relative to another is mathematically described by using a concept called a *derivative*. We will see that the rate of change of one quantity relative to another is mathematically determined by finding the slope of a line drawn tangent to a curve.

We begin with a simple example. Consider the speed of a moving object, say, a car. By *speed* we mean the rate at which the distance traveled varies with time. It has magnitude, but no direction. We can measure speed in two ways: *average speed* and *instantaneous speed.* We begin with the average speed of a commuter driving from home to the office.

To find the average speed, we use the formula $d = rt$ or $r = \dfrac{d}{t}$; that is, we divide the distance traveled by the elapsed time:

$$\text{AVERAGE SPEED} = \frac{\text{DISTANCE TRAVELED}}{\text{ELAPSED TIME}}$$

Table 17.1 Distance and Time for a Commuter Car

Time	Distance from home
6:09 A.M.	0 miles
6:25	16
6:30	21
6:34	25
6:36	26.7
6:37	27.7
6:39	28.5
7:01	34
7:03	35
7:09	38
7:15	43
7:28	50

EXAMPLE 1

A commuter left home one morning and set the odometer to zero. The times and distances were noted as shown in Table 17.1. Find the average speed for the requested time intervals.

a. 6:09 A.M. to 6:39 A.M. (0–30 min)

b. 6:39 A.M. to 7:09 A.M. (30–60 min)

c. 7:01 A.M. to 7:28 A.M. (52–79 min)

Solution

a. 0–30 min: AVERAGE SPEED $= \dfrac{28.5 - 0}{\frac{1}{2} - 0} = 57$ mph

b. 30–60 min: AVERAGE SPEED $= \dfrac{38 - 28.5}{1 - \frac{1}{2}} = 19$ mph

c. 52–79 min: AVERAGE SPEED $= \dfrac{50 - 34}{\frac{79}{60} - \frac{52}{60}} \approx 35.6$ mph   ◆

We can generalize the work done in Example 1 by writing the following formula:

$$\text{AVERAGE SPEED} = \frac{d_2 - d_1}{t_2 - t_1} \text{ mph}$$

where the car travels d_1 miles in t_1 hours and d_2 miles in t_2 hours. The relationship is shown graphically in Figure 17.13, where the average speed is the slope of the line joining (d_1, t_1) and (d_2, t_2).

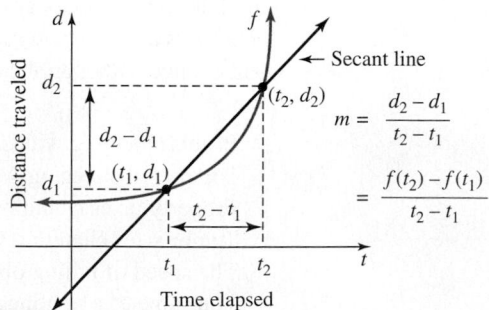

Figure 17.13 **Geometrical interpretation of average speed**

The line connecting (d_1, t_1) and (d_2, t_2) in Figure 17.13 is sometimes called the **secant line.** We see that the average rate of change is

$$\frac{f(t_2) - f(t_1)}{t_2 - t_1}$$

It is worthwhile to state this formula in terms of the usual variable x. Let $x = t_1$ and let $h = t_2 - t_1$, which is the length of the time interval over which we are finding an average. From these substitutions it follows that

$$h = t_2 - x$$
$$h + x = t_2$$

This leads to a general statement of the average rate of change.

Average Rate of Change

This is the difference quotient.

> The **average rate of change** of a function f with respect to x over an interval from x to $x + h$ is given by the formula
>
> $$\frac{f(x + h) - f(x)}{h}$$

EXAMPLE 2

Find the average rate of change of f with respect to x between $x = 3$ and $x = 5$ if $f(x) = x^2 - 4x + 7$.

Solution
Given $x = 3$ and $h = 5 - 3 = 2$:

$$f(x + h) = f(3 + 2) = f(5) = 5^2 - 4(5) + 7 = 12$$
$$f(x) = f(3) = 3^2 - 4(3) + 7 = 4$$

$$\text{AVERAGE RATE OF CHANGE} = \frac{f(x + h) - f(x)}{h}$$
$$= \frac{12 - 4}{2}$$
$$= 4$$

The slope of the secant line is 4, which is confirmed geometrically by looking at Figure 17.14. ◆

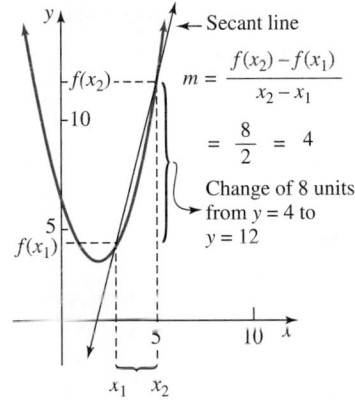

Figure 17.14 Average rate of change of f, where $f(x) = x^2 - 4x + 7$, between $x = 3$ and $x = 5$

We consider an example from the social sciences.

EXAMPLE 3

Table 17.2 shows the number of divorces for selected years from 1987 to 1996. Find the average divorce rate for each of the time periods given.

a. 1987–1992   **b.** 1987–1996

c. 1992–1996   **d.** 1995–1996

Solution
The average rate of change of f is defined by
$$\frac{f(x + h) - f(x)}{h}$$

a. $x = 1987$ and $h = 5$: $\quad \dfrac{f(1992) - f(1987)}{5} = \dfrac{1.215 - 1.157}{5}$
$$= 0.116$$

The average divorce rate (rate of change in number of divorces) between 1987 and 1992 is an average increase of 11,600 per year. *Note: 0.116 million is 11,600.*

Table 17.2 Number of U.S. Divorces

Year	Number (in millions)
1987	1.157
1992	1.215
1995	1.169
1996	1.150

b. $x = 1987$ and $h = 9$: $\dfrac{f(1996) - f(1987)}{9} = \dfrac{1.150 - 1.157}{9}$

$$= -0.000\overline{7}$$

A negative rate of change indicates a decrease. The average divorce rate between 1987 and 1996 is a decrease of 778 divorces per year.

c. $x = 1992$ and $h = 4$: $\dfrac{f(1996) - f(1992)}{4} = \dfrac{1.150 - 1.215}{4}$

$$= -0.01625$$

The average divorce rate between 1992 and 1996 is an average decrease of 16,250 divorces per year.

d. $x = 1995$ and $h = 1$: $\dfrac{f(1996) - f(1995)}{1} = 1.150 - 1.169$

$$= -0.019$$

Between 1995 and 1996, the number of divorces decreased by 19,000, which is about 2%.   ◆

Instantaneous Rate of Change

Suppose in Example 3 we are interested in the divorce rate in 1996. To estimate this rate, we can use the slopes of the secant lines (labeled **b, c,** and **d**), as shown in Figure 17.15. Which one would you use as a more accurate estimate?

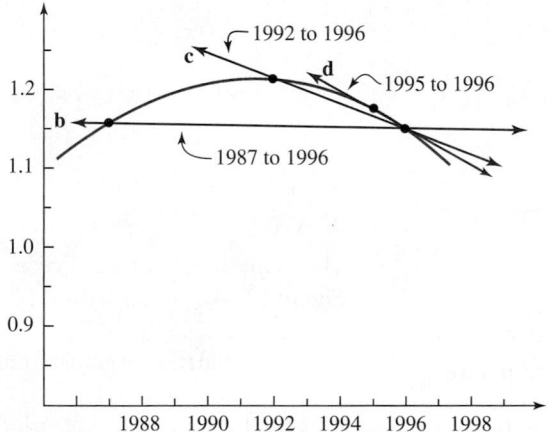

Figure 17.15 **U.S. divorces and secant lines at 1996**

To find the rate *at a particular time,* imagine that a curve called f is defined by a roller-coaster track. Consider a fixed location on this track; call it $(x, f(x))$. Now imagine the front car of the roller coaster at some point on the track a horizontal distance of h units from x. The slope of the secant line between these points is the average rate of change of the function f between the two points.

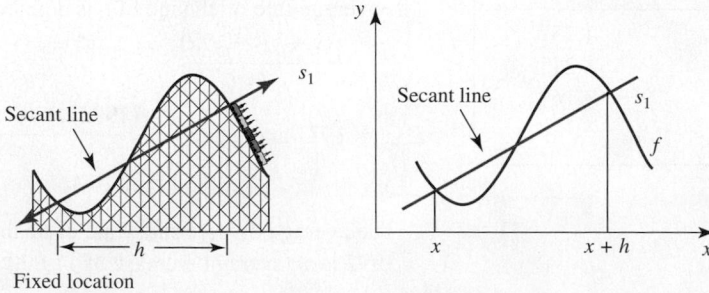

Also imagine the roller coaster car moving along the track toward the fixed location. That is, let $h \to 0$. As it moves along the track, we obtain a sequence of secant lanes. The roller coaster is shown at the left and the curve with the secant lines is shown at the right.

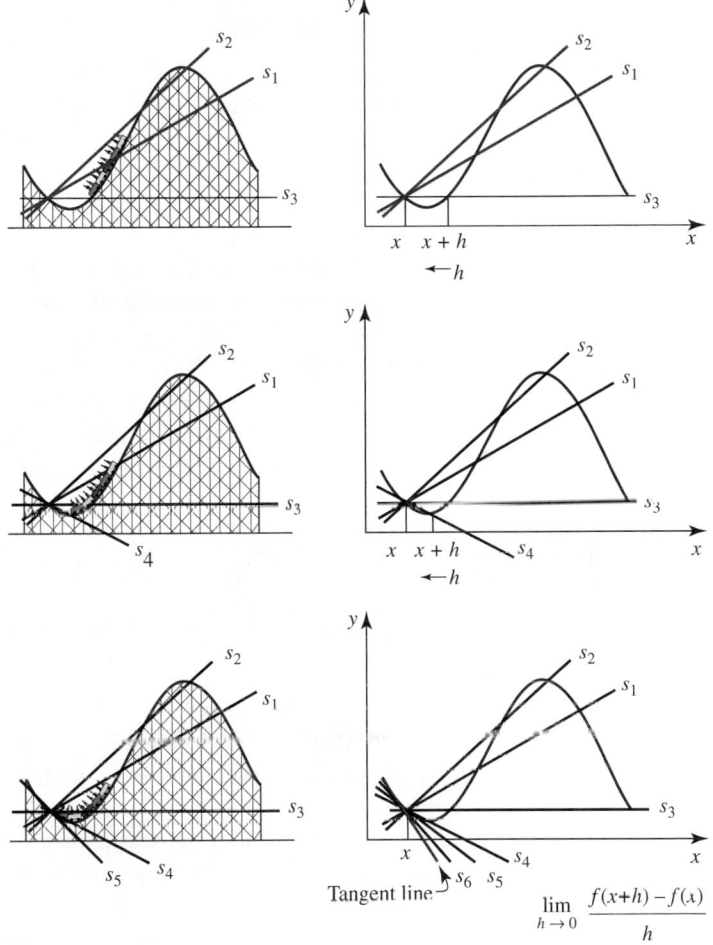

As the car gets close to the fixed location, the secant lines approach a limiting line (shown in color). This limiting line for which $h \to 0$ is called the *tangent line.*

Instantaneous Rate of Change

Given a function f and the graph of $y = f(x)$, the **tangent line** at the point $(x, f(x))$ is the line that passes through this point with slope

$$\lim_{h \to 0} \frac{f(x + h) - f(x)}{h}$$

if this limit exists. The slope of the tangent line is also referred to as the **instantaneous rate of change** of the function f with respect to x.

If you look at Figure 17.15, you can measure the slope of the tangent line to approximate the instantaneous rate of change in 1996 to be -0.02 or about -2%.

EXAMPLE 4
───
Find the instantaneous rate of change of the function $f(x) = 5x^2$ with respect to x.

Solution Find $\lim\limits_{h\to 0} \dfrac{f(x + h) - f(x)}{h}$.

We evaluate this expression in small steps.

1. $f(x) = 5x^2$ is given.

2. $f(x + h) = 5(x + h)^2 = 5x^2 + 10xh + 5h^2$

3. $f(x + h) - f(x) = (5x^2 + 10xh + 5h^2) - 5x^2 = 10xh + 5h^2$

4. $\dfrac{f(x + h) - f(x)}{h} = \dfrac{10xh + 5h^2}{h} = 10x + 5h \quad (h \neq 0)$

5. $\lim\limits_{h\to 0} \dfrac{f(x + h) - f(x)}{h} = \lim\limits_{h\to 0}(10x + 5h) = 10x$

We will refer to the small steps (shown in boldface) as *the five-step process* for finding the instantaneous rate of change. ◆

Derivative

The concept of the derivative is a very powerful mathematical idea, and the variety of applications is almost unlimited. The instantaneous rate of change of a function f per unit of change in x is only one of many applications using that specific limiting idea.

Derivative

🛑 STOP

This is one of the most important definitions in all of mathematics.

> For a given function f, we define the **derivative of f at x,** denoted by $f'(x)$, to be
>
> $$f'(x) = \lim_{h\to 0} \frac{f(x + h) - f(x)}{h}$$
>
> provided this limit exists. If the limit exists, we say f is a **differentiable function** of x.

If the limit does not exist, we say that f is *not differentiable* at x. We have now developed all of the techniques necessary to apply the definition of derivative to a variety of functions. Also note that as $h \to 0$, $h \neq 0$, so we assume $h \neq 0$ without stating so when using the five-step process for finding the derivative.

EXAMPLE 5

Use the definition to find the derivative of $y = 2x^2$.

Solution
We carry out the five-step process:

1. $f(x) = 2x^2$ is given.

2. $f(x + h) = 2(x + h)^2$ Evaluate f at $x + h$.

$\qquad\qquad = 2x^2 + 4xh + 2h^2$

3. $f(x + h) - f(x) = (2x^2 + 4xh + 2h^2) - 2x^2$ Subtract $f(x)$ from $f(x + h)$.

$\qquad\qquad\qquad = 4xh + 2h^2$

4. $\dfrac{f(x + h) - f(x)}{h} = \dfrac{4xh + 2h^2}{h}$ Now, divide by h.

$\qquad\qquad\qquad = \dfrac{h(4x + 2h)}{h}$

$\qquad\qquad\qquad = 4x + 2h$

5. $\lim\limits_{h \to 0} \dfrac{f(x + h) - f(x)}{h} = \lim\limits_{h \to 0} (4x + 2h)$ *Finally, take the limit.*

$$= 4x$$

The derivative of $y = 2x^2$ is $4x$. Sometimes we write $y' = 4x$ or $f'(x) = 4x$.   ◆

In more advanced courses, it is shown that the exponential function has a very important property involving derivatives. *The exponential function is its own derivative.*

Derivative of e^x

> If $f(x) = e^x$, then $f'(x) = e^x$, and if $f(x) = e^{ax}$, then $f'(x) = ae^{ax}$.

EXAMPLE 6

Find the derivative of the given functions:

a. $y = e^{3x}$ **b.** $y = e^{-5x}$

Solution

a. If $y = e^{3x}$, then $y' = 3e^{3x}$.   **b.** If $y = e^{-5x}$, then $y' = -5e^{-5x}$.   ◆

Tangent Line

EXAMPLE 7

Find the standard-form equation of the line tangent to the graph of $y = 2x^2$ at the point where $x = -1$.

Solution   Recall (Problem 58, Section 13.1) that the equation of a line with slope m passing through the point (h, k) is

$$y - k = m(x - h)$$

First, we need to find the point (h, k). We are given the first component $(x = -1)$, so we see that $h = -1$. To find the second component, evaluate the given equation:

$$x = -1, \text{ then } y = 2(-1)^2 = 2$$

We see $k = 2$. Now, we are looking for the tangent line passing through the point $(-1, 2)$ with slope equal to the derivative of the function $y = 2x^2$. Remember, the slope of a curve at a point is the value of the derivative at that point.

y' or f' means derivative and does not represent an exponent.

From Example 5, $y' = 4x$, so the slope of the tangent line point of the given curve is $4x$. Thus, at $x = -1$,

$$f'(-1) = 4(-1) = -4$$

Now we can use the point–slope form:

$$y - k = m(x - h) \qquad \text{Point–slope form}$$
$$y - 2 = -4(x + 1) \qquad \text{Substitute given values.}$$
$$y = -4x - 2 \qquad \text{Slope–intercept form}$$

or, in standard form,

$$4x + y + 2 = 0$$ ◆

It is worth noting that if the derivative of a function is positive, then the curve is rising at that point (since the slope of the tangent line is positive); and if the derivative is negative at a point, then the curve is falling at that point. If the derivative is 0, then the tangent line is horizontal. In Example 7, we found the equation of the tangent line of the function $y = 2x^2$ at $(-1, 2)$. This tangent line is shown in Figure 17.16 (page 884).

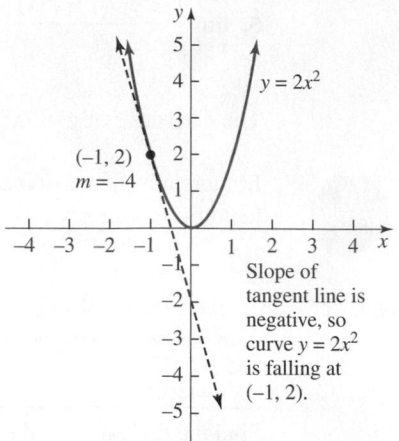

Figure 17.16 Graph of $y = 2x^2$
with tangent line at $(-1, 2)$

Velocity

A final application of derivative we will consider is the **velocity,** which is defined as an instantaneous rate of change. Consider the following example.

EXAMPLE 8

Consider a "free-fall" ride at an amusement park that involves falling 100 feet in 2.5 seconds. As you are falling, you pass the 16-ft mark at one second and the 64-ft mark at two seconds. What is the velocity at the *instant* the ride passes the 100-ft mark (as measured from the top)?

Solution

Let $s(t)$ be the distance in feet from the top t seconds after release. Also assume that $s(t) = 16t^2$.*

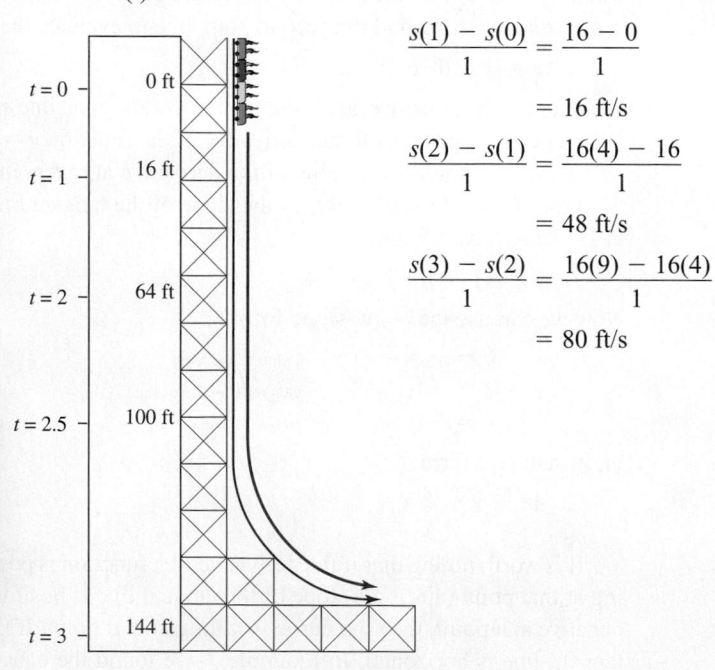

$$\frac{s(1) - s(0)}{1} = \frac{16 - 0}{1}$$
$$= 16 \text{ ft/s}$$
$$\frac{s(2) - s(1)}{1} = \frac{16(4) - 16}{1}$$
$$= 48 \text{ ft/s}$$
$$\frac{s(3) - s(2)}{1} = \frac{16(9) - 16(4)}{1}$$
$$= 80 \text{ ft/s}$$

*This is the formula for free fall in a vacuum; it is sufficiently accurate for our purposes in this problem.

In general, the average velocity from time $t = t_1$ to $t = t_1 + h$ is given by the following formula (where $h \neq 0$).

$$\text{AVERAGE VELOCITY} = \frac{\text{CHANGE IN POSITION}}{\text{CHANGE IN TIME}}$$

$$= \frac{s(t_1 + h) - s(t_1)}{h}$$

$$= \frac{16(t_1 + h)^2 - 16t_1^2}{h}$$

$$= \frac{16t_1^2 + 32t_1 h + 16h^2 - 16t_1^2}{h}$$

$$= \frac{32t_1 h + 16h^2}{h}$$

$$= 32t_1 + 16h$$

Now, to find the velocity at a particular instant in time, we simply need to consider the limit as $h \to 0$:

$$\text{INSTANTANEOUS VELOCITY} = \lim_{h \to \infty} \frac{s(t_1 + h) - s(t_1)}{h}$$

$$= \lim_{h \to \infty} (32t_1 + 16h)$$

$$= 32t_1$$

We can now use these formulas to find various average and instantaneous velocities. (We do this to show you how to apply the formula we have just derived as practice before answering the question asked in this example.)

Time	Average velocity $32t + 16h$	Instantaneous velocity $32t$
$t = 0$		$32(0) = 0$
$h = 1$	$32(0) + 16(1) = 16$	
From $t = 0$ to $t = 1$		
$t = 1$		$32(1) = 32$
$h = 2$	$32(1) + 16(2) = 64$	
From $t = 0$ to $t = 2$		
$h = 1$	$32(1) + 16(1) = 48$	
From $t = 1$ to $t = 2$		
$t = 2$		$32(2) = 64$
$h = 3$	$32(2) + 16(3) = 112$	
From $t = 0$ to $t = 3$		
$h = 2$	$32(2) + 16(2) = 96$	
From $t = 0$ to $t = 2$		
$h = 1$	$32(2) + 16(1) = 80$	
From $t = 1$ to $t = 2$		

$$\vdots$$

We are given that the ride passes 100 ft at 2.5 seconds, so the instantaneous velocity at that instant is

$$32(2.5) = 80 \text{ ft/s}$$

(By the way, 80 ft/s is approximately 55 mph.)

◆

Problem Set 17.3

1. The graph in Figure 17.17 shows the height h of a projectile after t seconds. Find the average rate of change of height (in feet) with respect to the requested changes in time t (in seconds).

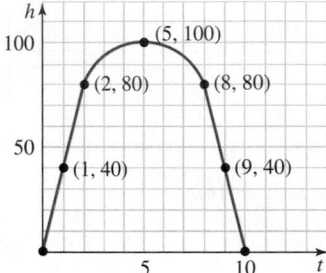

Figure 17.17 Height of a projectile in feet t seconds after fired

a. 1 to 8
b. 1 to 5
c. 1 to 2
d. What do you think the rate of change is at $t = 1$?

2. The graph in Figure 17.18 shows company output as a function of the number of workers. Find the average rate of change of output for the given change in the number of workers.

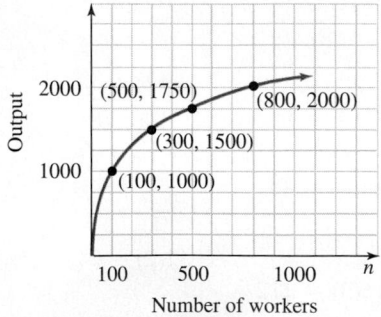

Figure 17.18 Output in production relative to the number of employees at Kampbell Construction

a. 100 to 800
b. 100 to 500
c. 100 to 300
d. What do you think the rate of change is at $n = 100$?

3. Table 17.1 gives some distances and commute times for a typical daily commute. Find the average speed for the requested time intervals.
a. 6:09 to 6:36
b. 6:36 to 7:03
c. 7:03 to 7:28
d. 6:09 to 7:28

4. The SAT scores of entering first-year college students are shown in Figure 17.19. Find the average yearly rate of change of the scores for the requested periods.

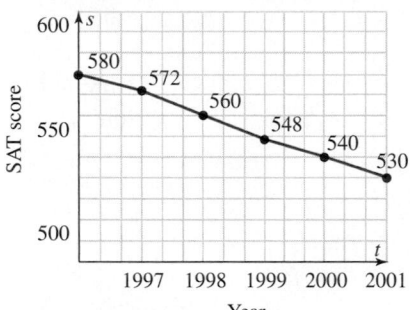

Figure 17.19 SAT scores at Riveria College

a. 1997 to 2001
b. 1998 to 2001
c. 1999 to 2001
d. 2000 to 2001

5. Table 17.3 shows the Gross National Product (GNP) in trillions of dollars for the years 1960–1996. Find the average yearly rate of change of the GNP for the requested years.

Table 17.3 Gross National Product

Year	Dollars (in trillions)
1960	0.5153
1970	1.0155
1980	2.7320
1990	5.5461
1995	7.2654
1996	7.6360

a. 1960 to 1996
b. 1970 to 1996
c. 1980 to 1996
d. 1990 to 1996
e. 1995 to 1996
f. How fast do you think the GNP is changing in 1996?

Trace the curves in Problems 6–11 onto your own paper and draw the secant line passing through P and Q. Next, imagine $h \to 0$ and draw the tangent line at P assuming that Q moves *along the curve to the point P. Finally, estimate the slope of the curve at P using the slope of the tangent line you have drawn.*

6.

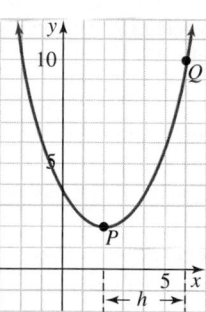

7.

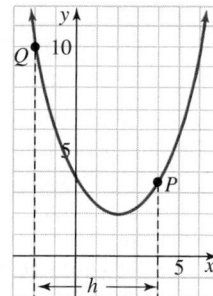

8.

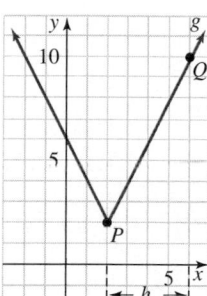

9.

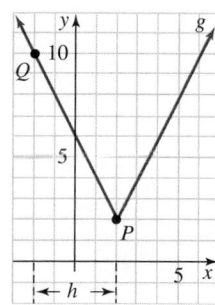

10.

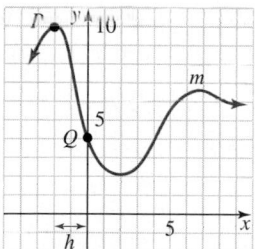

11.

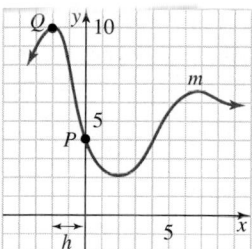

LEVEL 2

*Find the average (part **a**) and instantaneous rates of change (part **b**) of the functions in Problems 12–15.*

12. $f(x) = 4 - 3x$
   a. for $x = -3$ to $x = 2$ **b.** at $x = -3$

13. $f(x) = 5$
   a. for $x = -3$ to $x = 3$ **b.** at $x = -3$

14. $f(x) = 3x^2$
   a. for $x = 1$ to $x = 3$ **b.** at $x = 1$

15. $f(x) = -2x^2 + x + 4$
   a. for $x = 4$ to $x = 9$ **b.** at $x = 4$

Find the derivative, $f'(x)$, of each of the functions in Problems 16–21 by using the derivative definition or the derivative of the exponential function.

16. $f(x) = \frac{1}{2}x^2$

17. $f(x) = \frac{1}{3}x^3$

18. $f(x) = e^{1.2x}$

19. $y = e^{-6x}$

20. $y = 25 - 250x$

21. $f(x) = 3 + 2x - 3x^2$

Find an equation of the line tangent to the curves in Problems 22–25 at the given point.

22. $y = 5x^2$ at $x = 3$

23. $y = 2x^2$ at $x = 4$

24. $y = 4 - 5x$ at $x = -2$

25. $y = 3x^2 + 4x$ at $x = 0$

26. A freely falling body experiencing no air resistance falls $s(t) = 16t^2$ feet in t seconds. Express the body's velocity at time $t = 2$ as a limit. Evaluate this limit.

27. If you toss a ball from the top of the Tower of Pisa directly upward with an initial speed of 96 ft/s, the height h at time t is given by

$$h(t) = -16t^2 + 96t + 176$$

Figure 17.20 shows h and the velocity, v, at various times, t.
   a. What is the height of the tower?
   b. Find the velocity as a function of time.

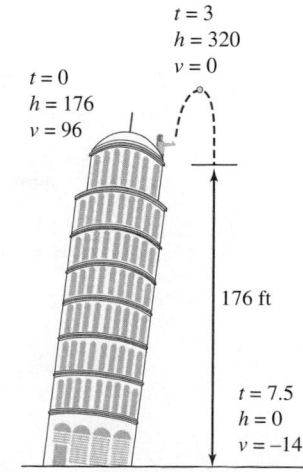

Figure 17.20 **Tower of Pisa**

28. Suppose the profit, P, measured in thousands of dollars, for a manufacturer is a function of the number of units produced, x, and behaves according to the model

$$P(x) = 50x - x^2$$

Also suppose that the present production is 20 units.
  a. What is the per-unit increase in profit if production is increased from 20 to 30 units?
  b. Repeat for 20 to 25 units.
  c. Repeat for 20 to 21 units.
  d. What is the rate of change at $x = 20$?

29. The cost, C, in dollars, for producing x items is given by

$$C(x) = 30x^2 - 100x$$

  a. Find the average rate of change of cost as x increases from 100 to 200 items.
  b. Repeat for 100 to 110 items.

  c. Repeat for 100 to 101 items.
  d. Repeat for 100 to $(100 + h)$ items.

LEVEL 3

30. Suppose the number (in millions) of bacteria present in a culture at time t is given by the formula

$$N(t) = 2t^2 - 200t + 1,000$$

  a. Derive a formula for the instantaneous rate of change of the number of bacteria with respect to time.
  b. Find the instantaneous rate of change of the number of bacteria with respect to time at time $t = 3$.
  c. Find the instantaneous rate of change of the number of bacteria with respect to time at the beginning of this experiment.

17.4 INTEGRALS

We considered several simple area formulas in Chapter 7. We found those formulas by filling regions with square units, and then counting them in some convenient manner. Sir Isaac Newton (see page 864) described area by using calculus. We will use his ideas, along with modern notation, to give you a glimpse of what we mean by *integral calculus*.

Area Function

We begin by defining an *area function*.

Area Function

STOP

Study this example until you understand this function.

> The **area function,** $A(t)$, is the area bounded below by the x-axis, above by a function $y = f(x)$, on the left by the y-axis, and on the right by the vertical line $y = t$.

The area function is shown in Figure 17.21.

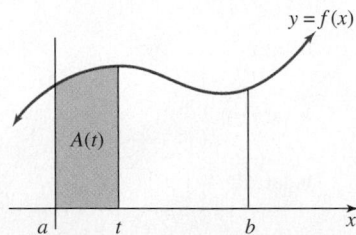

Figure 17.21 **Area function**

EXAMPLE 1

Let $y = 4$. Find: **a.** $A(5)$ **b.** $A(6)$ **c.** $A(50)$ **d.** $A(x)$

Solution

a. We draw the region as shown in Figure 17.22. $A(5)$ is the area below the line $y = 4$, above the x-axis, to the right of the y-axis, and to the left of the line $x = 5$. This is a rectangle, so we know

$$A(5) = 4(5) = 20$$

b. $A(6) = 4(6) = 24$

c. $A(50) = 4(50) = 200$

b. $A(x) = 4x$

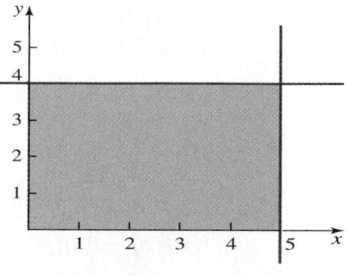

Figure 17.22 **Area function**   ◆

The task before us is to find $A(x)$ for any function $y = f(x)$.

EXAMPLE 2

Let $y = 4x$. Find: **a.** $A(5)$ **b.** $A(10)$ **c.** $A(50)$ **d.** $A(x)$

Solution

Begin by drawing the region as shown in Figure 17.23.

a. Figure 17.23 shows the area function for $A(5)$. We see the area we seek is a triangle:

$A = \frac{1}{2}bh$ Area of a triangle

$= \frac{1}{2}(5)(20)$ We can see the base is 5, so $b = 5$. To find the height we use $y = 4x$ for $x = 5$; the height is $4(5) = 20$.

$= 50$

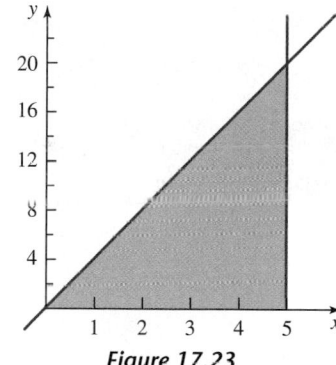

Figure 17.23

b. For $A(10)$, $x = 10$, so $b = 10$ and $h = 4(10) = 40$ and

$$A = \frac{1}{2}bh = \frac{1}{2}(10)(40) = 200$$

c. $A(50) = \frac{1}{2}(50)(200) = 5,000$

d. $A(x) = \frac{1}{2}(x)(4x) = 2x^2$ ◆

The answer to Example 2d, namely, $A(x) = 2x^2$, is called the *antiderivative* of the original function $f(x) = 4x$ because if we take the derivative of $A(x) = 2x^2$ we obtain $4x$ (see Example 5 of Section 17.3). An antiderivative of the function defined by $f(x)$ is written as $\int f(x)\, dx$.

Antiderivative

If the derivative of a function defined by $y = F(x)$ is $f(x)$, then the **antiderivative** of $f(x)$ is

$$\int f(x)\, dx = F(x)$$

This means that $F'(x) = f(x)$.

We have not yet discussed a procedure for *finding* an antiderivative, but we can check to see whether a particular function *is* an antiderivative as shown by the following example.

EXAMPLE 3

Show that $x^3 + 8$ and $x^3 - 2$ are antiderivatives of $3x^2$.

Solution
We carry out the five-step process for both functions.

1. $f(x)$:
$$x^3 + 8 \qquad\qquad\qquad x^3 - 2$$

2. $f(x + h)$:
$$(x + h)^3 + 8 \qquad\qquad (x + h)^3 - 2$$

3. $f(x + h) - f(x)$:
$$(x + h)^3 + 8 - (x^3 + 8) \qquad\qquad (x + h)^3 - 2 - (x^3 - 2)$$
$$= x^3 + 3x^2h + 3xh^2 + h^3 + 8 - x^3 - 8 \qquad = x^3 + 3x^2h + 3xh^2 + h^3 - 2 - x^3 + 2$$
$$= 3x^2h + 3xh^2 + h^3 \qquad\qquad = 3x^2h + 3xh^2 + h^3$$

4. $\dfrac{f(x + h) - f(x)}{h}$
$$\dfrac{3x^2h + 3xh^2 + h^3}{h} = 3x^2 + 3xh + h^2 \qquad \dfrac{3x^2h + 3xh^2 + h^3}{h} = 3x^2 + 3xh + h^2$$

5. $\lim\limits_{h \to 0} \dfrac{f(x + h) - f(x)}{h}$
$$\lim_{h \to 0} (3x^2 + 3xh + h^2) = 3x^2 \qquad \lim_{h \to 0} (3x^2 + 3xh + h^2) = 3x^2$$

We see that both of the given functions are antiderivatives of $3x^2$. We write this as

$$\int 3x^2 \, dx = x^3 + 8 \quad \text{and} \quad \int 3x^2 \, dx = x^3 - 2 \qquad\qquad \blacklozenge$$

Notice from Example 3 that antiderivatives are not unique. If we study the steps of this example, we see that *antiderivatives are the same except for a constant that is added or subtracted.* If we represent this constant as C, we write

$$\int 3x^2 \, dx = x^3 + C$$

where C is any constant. This says if you select any constant value for C and find the derivative of $f(x) = x^3 + C$, the result will be $3x^2$.

The exponential function $f(x) = e^x$ is the function that is equal to its own derivative, so it follows that it is a function that is its own antiderivative. We state this result in formula form.

Antiderivative of e^x

If $f(x) = e^x$, then
$$\int e^x \, dx = e^x + C \quad \text{and} \quad \int e^{ax} \, dx = \frac{e^{ax}}{a} + C$$

More advanced courses derive some properties of antiderivatives, which we summarize in the following box.

Properties of Antiderivative

Constant multiple	$\int af(x) \, dx = a \int f(x) \, dx$
Antiderivative of a sum	$\int [f(x) + g(x)] \, dx = \int f(x) \, dx + \int g(x) \, dx$

Newton found that the derivative of the area function $A(x)$ is $f(x)$. In other words, the area function is an antiderivative of f; we write

$$\int f(x)\, dx = A(x)$$

Antiderivatives as Areas

We can find an antiderivative by using areas, as illustrated by the following example.

EXAMPLE 4

Find $\int (2x + 3)\, dx$.

Solution

Let $f(x) = 2x + 3$; consider the region bounded by the x-axis, the y-axis, the line $y = 2x + 3$, and the vertical line x units to the right of the origin, as shown in Figure 17.24. We recognize this as a trapezoid, so we can write

$$\int (2x + 3)\, dx = \tfrac{1}{2}h(b + B)$$
$$= \tfrac{1}{2}x[3 + (2x + 3)]$$
$$= \tfrac{1}{2}x(2x + 6)$$
$$= x^2 + 3x$$

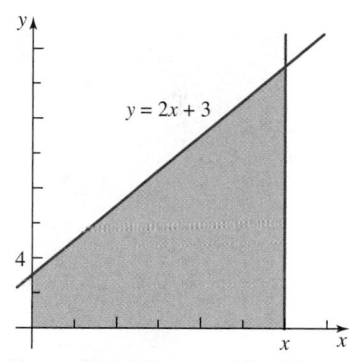

Figure 17.24 Trapezoid with $h = x$, $h = 3$, $B = 2x + 3$

Since we know that antiderivatives are not unique, but differ only by a constant, we write

$$\int (2x + 3)\, dx = x^2 + 3x + C$$

We can also use properties of an antiderivative, as shown here:

$$\int (2x + 3)\, dx = \int 2x\, dx + \int 3\, dx \qquad \text{Antiderivative of a sum}$$

$$= 2\int x\, dx + 3\int dx \qquad \text{Constant multiple}$$

$$= 2\,\frac{x^2}{2} + 3x \qquad \text{Antiderivative of } x \text{ is } \tfrac{1}{2}x^2 \text{ (Example 2d)}$$
$$\qquad\qquad\qquad \text{and antiderivative of } dx \text{ is } x \text{ (Example 1d).} \qquad \blacklozenge$$

The Definite Integral

We note that the antiderivative of a function is a function. We now consider a concept called *the definite integral,* which, although it can be related to an antiderivative, is an important concept in its own right. We introduce the definite integral with an example.

Suppose that between 1990 and 1995, the rate of oil consumption (in billions of barrels) is given by the formula

$$R(t) = 78e^{-0.04t}$$

where t is measured in years after 1990. If we use this consumption rate, we can predict the amount of oil that will be consumed from 1995 to 2000. Since the derivative is the rate of change of a function, we can say that the total consumption since 1995 is given by a function we will call T with the property that $T'(t) = R(t)$. Then

$$T(t) = \int T'(t)\, dt = \int R(t)\, dt$$

$$= \int 78e^{-0.04t}\, dt$$

$$= 78 \int e^{-0.04t}\, dt \qquad \text{Constant multiple property}$$

$$= 78 \frac{e^{-0.04t}}{-0.04} + C \qquad \text{From the formula for antiderivative of } e^{ax}$$

$$= -1{,}950e^{-0.04t} + C$$

We can find C because if $t = 0$, then oil consumption is also 0, so

$$T(0) = -1{,}950e^{-0.04(0)} + C = 0$$

$$C = 1{,}950$$

Thus, $T(t) = -1{,}950e^{-0.04t} + 1{,}950$ is the total consumption after 1990. This means that $T(5)$ is the total consumption from 1990 to 1995, and $T(10)$ is the total consumption from 1990 to 2000.

$$T(5) = -1{,}950e^{-0.04(5)} + 1{,}950 \approx 353.48$$

$$T(10) = -1{,}950e^{-0.04(10)} + 1{,}950 \approx 642.88$$

We can find the total consumption from 1995 to 2000 by subtraction:

$$T(10) - T(5) \approx 642.88 - 353.48 = 289.40$$

We would predict approximately 289 billion barrels to be consumed, assuming the rate of consumption does not change.

Let us take a closer look at what we have done. Since $T(t)$ is an antiderivative of $R(t)$ over the interval from 1990 to 2000, we see that $T(10) - T(5)$ is the *net change* of the function T over the interval. In general, if $F(x)$ is any function and a and b are real numbers with $a < b$, then the *net change of $F(x)$ over the interval* $[a, b]$ is the number

$$F(b) - F(a)$$

The quantity $F(b) - F(a)$ is often abbreviated by $F(x)\Big|_a^b$. This discussion leads to the following definition.

Definite Integral

> Let f be a function defined over the interval $[a, b]$. Then the **definite integral** of f over the interval is denoted by
>
> $$\int_a^b f(x)\, dx$$
>
> and is the net change of an antiderivative of f over that interval. Thus, if $F(x)$ is an antiderivative of $f(x)$, then
>
> $$\int_a^b f(x)\, dx = F(x)\Big|_a^b = F(b) - F(a)$$

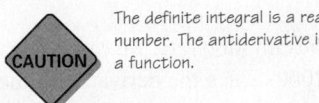

The definite integral is a real number. The antiderivative is a function.

The function f is called the **integrand** and the constants a and b are called the **limits of integration**. The letter x is called a **dummy variable** because the definite integral is a fixed number and not a function of x. To contrast a definite integral with its corresponding antiderivative, an antiderivative is sometimes called an **indefinite integral.**

Areas as Antiderivatives

In Example 4 we found an antiderivative by using a known area formula, but the real power of calculus is to use an antiderivative to find an area.

Area Under a Curve

> The **area under a curve** is defined to be the area of the region bounded by the graph of a function f and the x-axis and the vertical lines $x = a$ and $x = b$, and is
>
> $$A = \int_a^b f(x)\, dx$$

EXAMPLE 5

Find the area under the curve $y = x^2$ on $[1, 5]$.

Solution

We draw a graph of the desired region, as shown in Figure 17.25.

We now have two ways for finding this area; the first is to use an antiderivative, and the second is to approximate the actual area using known area formulas.

(1) Using an antiderivative

Suppose you know that the derivative of $\frac{1}{3}x^3$ is x^2 (for example, from Problem 17, Problem Set 17.3). Then

$$\int_1^5 x^2\, dx = \frac{x^3}{3}\bigg|_1^5 = \frac{125}{3} - \frac{1}{3} = \frac{124}{3} \approx 41.3$$

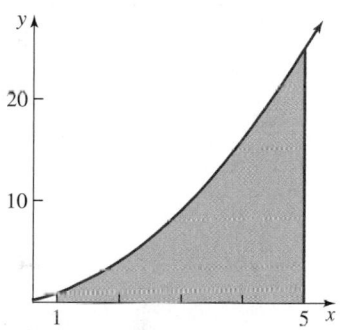

Figure 17.25 Area bounded by $y = x^2$, the x-axis, and the lines $x = 1$ and $x = 5$

(2) Using areas

We do not have a known area formula for the region shown in Figure 17.25, but we can approximate this area using rectangles.

If we draw one rectangle determined by the left endpoint of the region (namely $x = 1$), we obtain an approximation we call A_1:

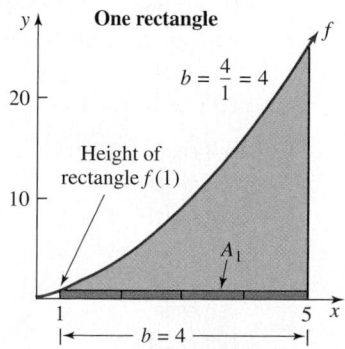

This rectangle has area $A = bh$, where $b = 4$ and $h = f(1) = 1$:

$$\int_1^5 x^2\, dx \approx A_1 = 4 \cdot f(1) = 4 \cdot 1^2 = 4$$

If we draw two rectangles determined by the left endpoints we obtain:

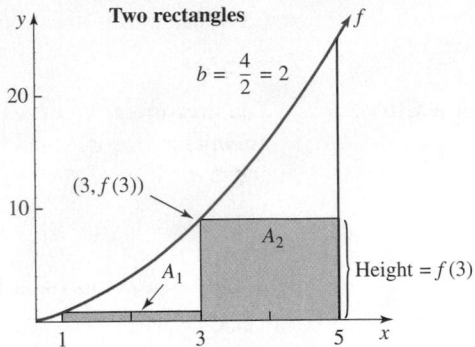

$$\int_1^5 x^2 \, dx \approx A_1 + A_2$$

$$= b \cdot f(1) + b \cdot f(3) = 2 \cdot 1^2 + 2 \cdot 3^2 = 20$$

We continue the process; for example, draw four rectangles:

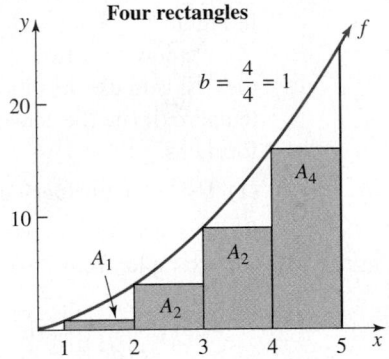

$$\int_1^5 x^2 \, dx \approx A_1 + A_2 + A_3 + A_4$$
$$= b \cdot f(1) + b \cdot f(2) + b \cdot f(3) + b \cdot f(4)$$
$$= 1 \cdot 1^2 + 1 \cdot 2^2 + 1 \cdot 3^2 + 1 \cdot 4^2$$
$$= 30$$

If we continue for 8 rectangles, we would approximate the area as 35.5. In fact, using a computer, we could generate the following table of values:

TYPE OF ESTIMATE	# OF SUB-INTERVALS	ESTIMATE OVER [1, 5]
Left endpt	4	30.0000000000
Left endpt	8	35.5000000000
Left endpt	16	38.3750000000
Left endpt	32	39.8437500000
Left endpt	64	40.5859375000
Left endpt	128	40.9589843750
Left endpt	256	41.1459960938
Left endpt	512	41.2396240234
Left endpt	1024	41.2864685059

(continued)

TYPE OF ESTIMATE	# OF SUB-INTERVALS	ESTIMATE OVER [1, 5]
Left endpt	2048	41.3098983765
Left endpt	4096	41.3216152191
Left endpt	10000	41.3285334400

If we take the limit as the number of rectangles increases, we would find this number approaches $41.\overline{3} = \frac{124}{3}$. ◆

Problem Set 17.4

LEVEL 1

Evaluate the area function for the functions given in Problems 1–6.

1. Let $y = 5$; find $A(8)$.

2. Let $y = 8.3$; find $A(x)$.

3. Let $y = 3x$; find $A(4)$.

4. Let $y = 2x$; find $A(x)$.

5. Let $y = 3x + 2$; find $A(3)$.

6. Let $y = x + 5$; find $A(x)$.

Find the antiderivative by using areas in Problems 7–12.

7. $\int 6\, dx$

8. $\int 3\, dx$

9. $\int (x + 5)\, dx$

10. $\int (x + 6)\, dx$

11. $\int (3x + 4)\, dx$

12. $\int (2x + 5)\, dx$

LEVEL 2

13. **IN YOUR OWN WORDS** What is a derivative?

14. **IN YOUR OWN WORDS** What is an integral?

15. **IN YOUR OWN WORDS** What is the area function?

16. **IN YOUR OWN WORDS** Contrast a definite integral and an indefinite integral.

17. Show that $\frac{1}{2}x^2$ is an antiderivative of x.

18. Show that $\frac{1}{2}x^2 - 5$ is an antiderivative of x.

19. Show that $2x^3$ is an antiderivative of $6x^2$.

20. Show that $2x^3 - 3$ is an antiderivative of $6x^2$.

Find the area under the curves in Problems 21–24 on the given intervals.

21. $y = x^2$ on $[1, 9]$.

22. $y = x^2$ on $[2, 5]$.

23. $y = 3x$ on $[3, 8]$.

24. $y = 2x + 1$ on $[1, 4]$

Evaluate the integrals given in Problems 25–28.

25. $\int_0^2 e^{0.5x}\, dx$

26. $\int_{-1}^3 (9 - x^2)\, dx$

27. $\int_1^5 (1 + 6x)\, dx$

28. $\int_0^2 (e^x + e^2)\, dx$

LEVEL 3

29. **IN YOUR OWN WORDS** A culture is growing at an hourly rate of

$$R'(t) = 200e^{0.5t}$$

for $0 \le t \le 10$. Find the area between the graph of this equation and the t-axis. What do you think this area represents?

PROBLEM SOLVING

30. **IN YOUR OWN WORDS** Suppose the accumulated cost of a piece of equipment is $C(t)$ and the accumulated revenue is $R(t)$, where both of these are measured in thousands of dollars and t is the number of years since the piece of equipment was installed. If it is known that

$$C'(t) = 18 \quad \text{and} \quad R'(t) = 21e^{-0.01t}$$

find the area (to the nearest unit) between the graphs of C' and R'. Do not forget that $t \ge 0$. What do you think this area represents?

CHAPTER SUMMARY

"The development of the calculus represents one of the great intellectual accomplishments in human history."

NCTM *STANDARDS*

IMPORTANT TERMS

Antiderivative [17.4]
Antiderivative of a sum [17.4]
Antiderivative of e^x [17.4]
Area function [17.4]
Area under a curve [17.4]
Average rate of change [17.3]
Calculus [17.1]
Constant multiple [17.4]
Converge [17.2]
Definite integral [17.4]
Derivative [17.1; 17.3]

Derivative of e^x [17.3]
Differentiable function [17.1]
Differential calculus [17.1]
Diverge [17.2]
Dummy variable [17.4]
General term [17.2]
Indefinite integral [17.4]
Instantaneous rate of change [17.3]
Integral [17.1]
Integral calculus [17.1]
Integral of a sum [17.4]

Integrand [17.4]
Limit [17.1]
Limit of a sequence [17.1]
Limits of integration [17.4]
Mathematical modeling [17.1]
Predecessor [17.2]
Secant line [17.1; 17.3]
Sequence [17.2]
Successor [17.2]
Tangent line [17.1; 17.3]
Velocity [17.3]

TYPES OF PROBLEMS

Describe the limit process, including Zeno's paradox. [17.1]
Describe the derivative, including the tangent problem. [17.1]
Describe the integral, including the area problem. [17.1]
Be able to guess the limit of a sequence. [17.1]
Be able to draw the line tangent to a curve at a specified point. [17.1]
Approximate an area by using rectangles. [17.1]
Explain the process of mathematical modeling. [17.1]
Write out the first 5 terms of a sequence when given a general term. [17.2]
Find the limit of a sequence. [17.2]
Solve applied problems involving limits of sequences. [17.2]
Estimate a rate of change by looking at a graph. [17.3]
Estimate the slope of a tangent line by looking at a graph. [17.3]
Find an average rate of change for a given function over an interval. [17.3]
Find an instantaneous rate of change for a given function at a particular point. [17.3]
Find the derivative by using the definition of derivative. [17.3]
Find the equation of a tangent line. [17.3]
Evaluate an area function. [17.4]
Find an antiderivative. [17.4]
Find the area under a curve. [17.4]
Approximate the value of a definite integral by using areas. [17.4]

Inches tall

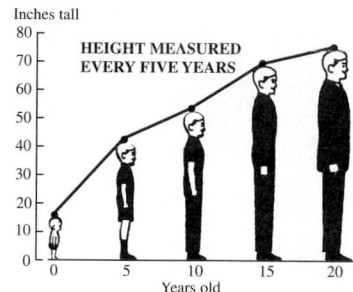

A rough estimate of a man's growth from childhood to maturity is obtained by measuring his height every five years. The straight lines between heads show the rate of growth.

Inches tall

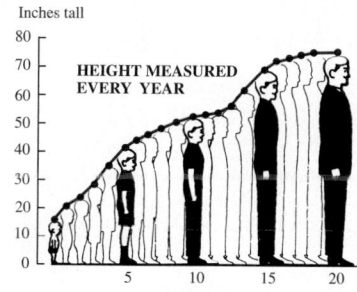

A more accurate approximation of growth is obtained by taking yearly measurements. The straight lines connecting the heads now begin to blend into a continuous curve.

Inches tall

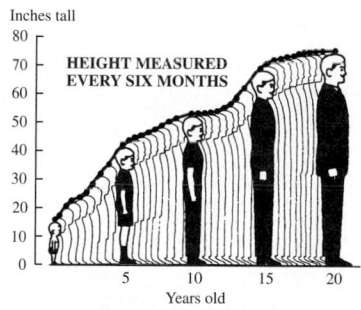

Measurements taken every six months give a still more accurate rate-of-growth line. With increasingly smaller intervals, the lines merge into an ever-smoother curve.

CHAPTER 17 REVIEW QUESTIONS

Evaluate the limits in Problems 1–4.

1. $\lim\limits_{n \to \infty} \dfrac{1}{n}$

2. $\lim\limits_{n \to \infty} \dfrac{3n^4 + 20}{7n^4}$

3. $\lim\limits_{n \to \infty} (2n + 3)$

4. $\lim\limits_{n \to \infty} \left(1 + \dfrac{1}{n}\right)^n$

5. **IN YOUR OWN WORDS** There are three pictures at the left that can be used to find the growth rate of the boy. Explain what is being illustrated with this sequence of drawings. Why do you think we might title this sequence of illustrations "instantaneous growth rate"?

6. What are the main ideas of calculus? Briefly describe each of these main ideas.

7. Use the definition of derivative to find $f'(x)$ where $f(x) = 6 - 4x^2$.

8. Evaluate $\displaystyle\int_{-1}^{0} (x - 4x^3)\, dx$.

9. Evaluate $\displaystyle\int_{1}^{2} e^x\, dx$ correct to two decimal places.

10. The rate of consumption (in billions of barrels per year) for oil conforms to the formula

$$R(t) = 32.4e^{6t/125}$$

for t years after 2000. If the total oil still left in the earth is estimated to be 670 billion barrels, estimate the length of time before all available oil is consumed if the rate does not change and no new oil reserves are discovered.

EPILOGUE—Why Not Math?
Mathematics in the Natural Sciences,
Social Sciences, and Humanities

MATHEMATICS

Since we began this book with a prologue that asked the question, "Why study math?" it seems appropriate that we end the book with an epilogue asking, "Why not study math?" Mathematics is the foundation and lifeblood of nearly all human endeavors. The German philosopher John Frederick Herbart (1776–1841) summarizes this idea:

All quantitative determinations are in the hands of mathematics, and it at once follows from this that all speculation which is heedless of mathematics, which does not enter into partnership with it, which does not seek its aid in distinguishing between the manifold modifications that must of necessity arise by a change of quantitative determinations, is either an empty play of thoughts, or at most a fruitless effort.

Historically, mathematics has always been at the core of a liberal arts education. This is a book of mathematics for the liberal arts. Karl Gauss, one of the greatest mathematicians of all time, called mathematics the "Queen of the Sciences," but mathematics goes beyond the sciences. Bertrand Russell claimed, "Mathematics, rightly viewed, possesses not only truth, but supreme beauty . . . ." And finally, Maxine Bôcher concludes, "I like to look at mathematics almost more as an art than as a science . . . ."

In fact, the mathematics degree at UCLA is classified as an art, not a science. Mathematics seems to be part of the structure of our minds, more akin to memory than to a learnable discipline. Enjoyment and use of mathematics are not dependent on "book learning," and even a casual perusal of the topics in this book will clearly illustrate that mathematics is many things to many people.

Like music, mathematics resists definition. Bertrand Russell had this to say about mathematics: "Mathematics may be defined as the subject in which we never know what we are talking about, nor whether what we are saying is true." Einstein, with his customary mildness, tells us, "so far as the theorems of mathematics are about reality, they are not certain; so far as they are certain, they are not about reality." Aristotle, who was as sure of everything as anyone can be of anything, thought mathematics to be the study of quantity, whereas Russell, in a less playful mood, thinks of it as the "class of all propositions of the type 'p implies q,'" which seems to have little to do with quantity." Willard Gibbs thought of mathematics as a language; Hilbert thought of it as a game. Hardy stressed its uselessness, Hogben its practicality. Mill thought it an empirical science, whereas to Sullivan it was an art, and to the wonderful J. J. Sylvester, it was "the music of reason."

This ambiguity should be consoling. It suggests that mathematics has so many mansions that there is room for everyone. In this epilogue we will discuss the mathematics in the natural sciences, in the social sciences, and in the humanities.

Mathematics in the Natural Sciences

When most of us think about applications of mathematics, we think of those in the natural sciences, or those sciences that deal with matter, energy, and the interrelations and transformations. Some of the major categories include (alphabetically) astronomy, biology, chemistry (and biochemistry), computer science, ecology, geology, medicine, meteorology, physics (and biophysics), statistics, and zoology.

We began this text by looking at reasoning, both inductive and deductive reasoning (Section 1.2). Inductive reasoning in the natural sciences is called the *scientific method.* With this type of reasoning, we are trying to recognize and formulate hypotheses, and then collect data through observation and experimentation in an effort to test our hypotheses to make further conclusions or conjectures based on the obtained data.

Using mathematics to make real-life predictions involves a process known as **mathematical modeling.** Since most models are dynamic, a feedback and reevaluation process is part of the formulation of a good model. We discussed this process in Section 17.1 (pp. 866–870). It involves *abstraction, deriving results, interpretation,* and then *verification.* An illustration of what we mean is shown in Figure E.1.

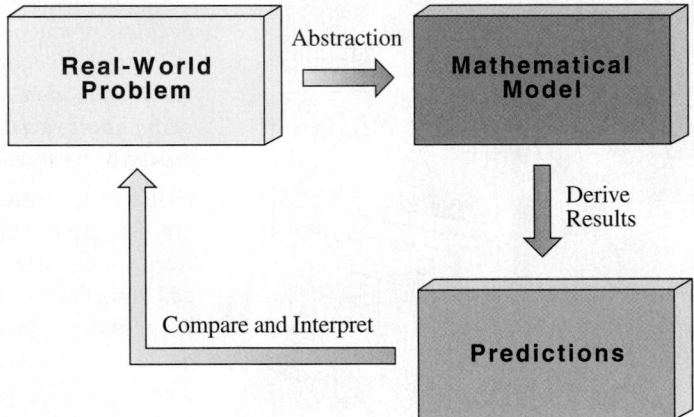

Figure E.1 **Mathematical modeling**

A good example of mathematics modeling in the natural sciences, but with implications to history and the social sciences (not to mention golf!), is modeling the path of a cannonball. (See Section 13.3, p. 677.) The *real-world* problem is to hit a target.

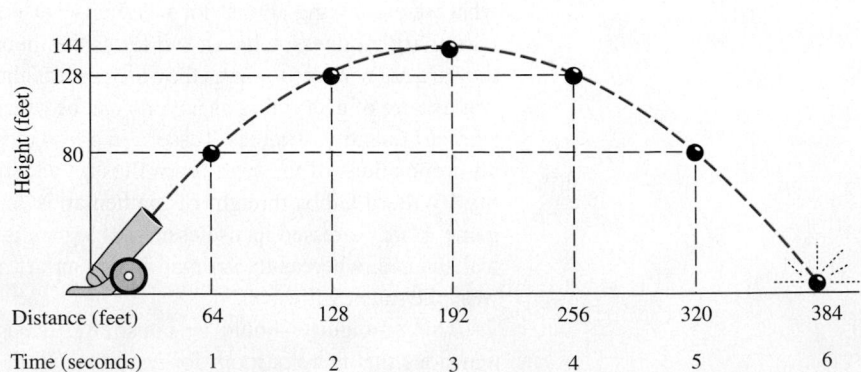

Fire a cannonball at 63 m/s at a 57° angle. We might begin by using trial and error. One possible trial is shown in Figure E.2.

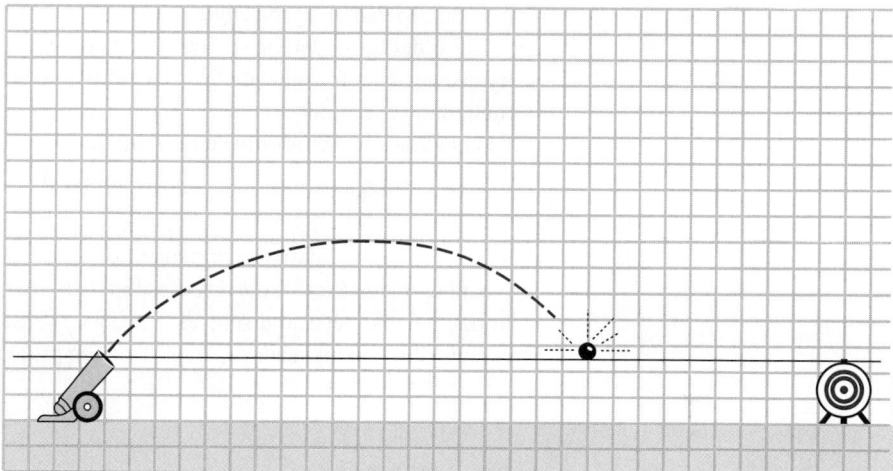

Figure E.2 **Trying to hit a target by trial and error**

The first use of a cannon was primarily to demoralize the enemy by its overwhelming power with little hope of actually hitting a specific target. In World War I, the trajectory of a cannon was trial and error, and we all remember old World War II movies where a "spotter" would phone directions to a gunner to adjust the angle of the cannon. We now need to make some assumptions; this is called *abstraction*. We assume that we are on earth (the earth's gravitational acceleration is 32 ft/s^2 or 9.8 m/s^2; we also assume that the density of the cannonball is constant, and that wind and friction are negligible). When we look at the path of a cannonball, we recognize it as having a parabolic shape, so our first attempt at modeling the path is to guess that it has an equation like $y = ax^2$ for an appropriate a. We also know that it opens downward, so a is negative. We now need to *derive some results* for the modeling process. We first guess that the equation has a vertex at (192, 144), so the equation we seek has the form

$$(x - 192)^2 = -4c(y - 144)$$

We also see that the curve passes through (0, 0), so

$$(0 - 192)^2 = -4c(0 - 144)$$
$$36,864 = 576c$$
$$c = 64$$

We guess that the equation for the path of the cannonball can be described by the equation

$$(x - 192)^2 = -256(y - 144)$$

The graph is shown in Figure E.3.

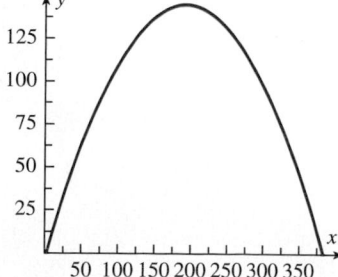

Figure E.3 **Path of a cannonball**

Modeling is an iterative process, so we now must *interpret and compare*. Even though the path of the cannonball seems correct (it passes through all of the appropriate points), it does not correctly model the given information. The path should somehow be dependent on the angle of elevation as well as the speed of the cannonball.

In order to model the path as a function of the angle of trajectory and initial velocity, we need some concepts from calculus, specifically Newton's laws of motion. Even though the derivation of these equations is beyond the scope of this course, we can understand these equations once they are stated as follows:

$$x = (v_0 \cos \theta)t \qquad y = h_0 + (v_0 \sin \theta)t - 16t^2$$

where h_0 is the initial height off the ground, v_0 is the initial velocity of the projectile in the direction of θ with the horizontal ($0° \leq \theta \leq 180°$). You might wish to review the meaning of $\cos \theta$ and $\sin \theta$ discussed in Section 6.5. The variables x and y represent the horizontal and vertical distances, respectively, measured in feet, and the variable t represents the time, in seconds, after firing the projectile.

EXAMPLE 1

Graph the equations

$$x = (v_0 \cos \theta)t \qquad y = h_0 + (v_0 \sin \theta)t - 4.9t^2$$

where $\theta = 57°$, $v_0 = 63$ ft/s, and $h_0 = 0$.

Solution
The desired equations are

$$x = (63 \cos 57°)t \qquad y = (63 \sin 57°)t - 4.9t^2$$

We can set up a table of values (as shown in the margin) or use a graphing calculator to sketch the curve as shown in Figure E.4.

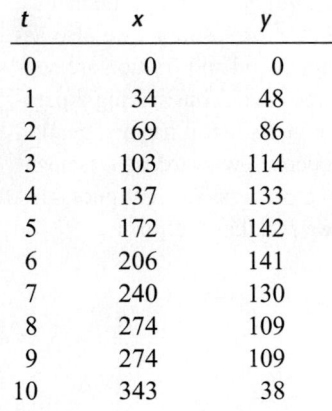

t	x	y
0	0	0
1	34	48
2	69	86
3	103	114
4	137	133
5	172	142
6	206	141
7	240	130
8	274	109
9	274	109
10	343	38

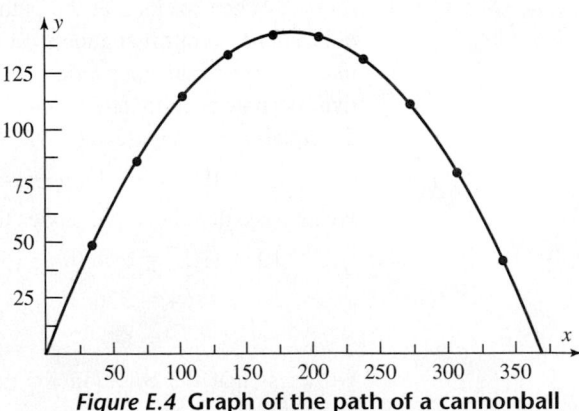

Figure E.4 **Graph of the path of a cannonball** ◆

We compare Figure E.4 with the real-world problem and it seems to correctly model the situation. In modeling, it may be necessary to go back and modify the assumptions several times before obtaining a result that appropriately models what we seek to find.

Astronomy is the study of objects and matter outside the earth's atmosphere, their motions and paths, as well as their physical and chemical properties. Science often deals with the very small (such as cells, atoms, and electrons) or the very large (as with planets and galaxies). In Section 1.3, we defined scientific notation and developed the laws of exponents to help us deal with large and small numbers.

EXAMPLE 2

In Example 3 of Section 1.3, we found that one light-year is approximately $5,870,000,000,000 = 5.87 \times 10^{12}$ miles. The closest star is Alpha Centauri, a distance of 4.35 light-years from the sun. How far is this in miles?

Solution

We use the properties of exponents:

$$4.35 \text{ light-years} = 4.35 \times (1 \text{ light-year})$$
$$= 4.35 \times (5.87 \times 10^{12} \text{ miles})$$
$$= 25.5345 \times 10^{12} \text{ miles}$$
$$= 2.55 \times 10^{13} \text{ miles}$$

Note: Because we are working with approximate numbers, all of these equal signs would generally be "≈," meaning "approximately equal to," but common usage is to write equal signs.   ◆

The subsection on "Comprehending Large Numbers," Section 1.3, is particularly important in astronomy. These large numbers are necessary in Example 8, pp. 689–690, Section 13.4, which finds the equation for the earth's orbit around the sun. This example is revisited in the following example.

EXAMPLE 3

Describe the meaning of the equation

$$\frac{x^2}{8.649 \times 10^{15}} + \frac{y^2}{8.647 \times 10^{15}} = 1$$

which models the path of the earth's orbit around the sun. The given distances are in miles. Find the eccentricity. The closer the eccentricity is to zero, the more circular the orbit.

Solution

We compare this with the equation of a *standard-form ellipse* centered at the origin, $\frac{x^2}{a^2} + \frac{y^2}{b^2} = 1$. We note that $a = \sqrt{8.649 \times 10^{15}} \approx 93,000,000$ and $b = \sqrt{8.647 \times 10^{15}} \approx 92,990,000$. Since $a \approx b$, we note that the elliptic orbit is almost circular. The measure of the circularity of the orbit is the *eccentricity*, which is defined as

$$\epsilon = \frac{c}{a} = \sqrt{1 - \frac{b^2}{a^2}} = \sqrt{1 - \frac{8.647 \times 10^{15}}{8.649 \times 10^{15}}} \approx 0.015$$

(*Note:* In reality, ϵ for the earth is about 0.016678, which we could obtain with better accuracy on the measurement of the lengths of the sides.) The *aphelion* (the greatest distance from the sun) is $a \approx 93,000,000$ miles and the *perihelion* (the closest distance from the sun) is $b \approx 92,990,000$.   ◆

Biology is defined as that branch of knowledge that deals with living organisms and vital processes, and includes the study of the plant and animal life of a region or environment. In Section 5.3, pp. 230–231, we considered a significant example from biology, specifically an application from genetics.

EXAMPLE 4

Suppose a certain population has two eye color genes: B (brown eyes, dominant) and b (blue eyes, recessive). Suppose we have an isolated population in which 60% of the genes in the gene pool are dominant B, and the other 40% are recessive b. What fraction of the population has each genotype? What percent of the population has each phenotype?

Solution

Let $p = 0.6$ and $q = 0.4$. Since p and q give us 100% of all the genes in the gene pool, we see that $p + q = 1$. Since

$$(p + q)^2 = p^2 + 2pq + q^2$$

we can find the percents:

genotype *BB:* $p^2 = (0.6)^2 = 0.36$,   so 36% have *BB* genotype

genotype *bB* or *Bb:* $2pq = 2(0.6)(0.4) = 0.48$,   so 48% have this genotype

genotype *bb:* $q^2 = (0.4)^2 = 0.16$,   so 16% have *bb* genotype

Check genotypes: $0.36 + 0.48 + 0.16 = 1.00$

As for the phenotypes, we look only at outward appearances, and since brown is dominant, *BB, bB,* and *Bb* all have brown eyes; this accounts for 84%, leaving 16% with blue eyes.   ◆

Another area of biology that relates closely with mathematics is the issue of *scale*. For example, we say that to predict a child's height as an adult, use a *scaling factor* of 2, which means if a two-year-old is 31 in. tall, then we predict that this child will be 62 in. tall as an adult. If we scale up a linear measure by a factor of two, then a two-dimensional measure will increase by a factor of $2^2 = 4$, and a three-dimensional measure will increase by a factor of $2^3 = 8$.

EXAMPLE 5

Consider a tissue sample as shown in Figure E.5.

Figure E.5 **Tissue sample**

If this sample is scaled up by a factor of 2, how does the area of the new sample compare with the area of the original?

Solution

Since the area function is a square ($A = s^2$), we know that the area of the sample is scaled up by a factor of $2^2 = 4$.   ◆

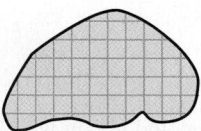

If the length and width are doubled, then the area is increased by four. If the length and width are tripled, then the area is increased by nine.

Chemistry is the science that deals with the composition, structure, and properties of substances and with the transformations that they undergo. A popular and exciting natural science called **biochemistry** deals with the chemical compounds and processes occurring in organisms.

Computer science is the study of computers, their design, and programming and includes artificial intelligence, networking, computer graphics, and computer languages. In this book Section 3.5 discusses the history of computers, and Section 3.4 the binary numeration system, which is used with computers.

Geology is the science that deals with the history of the earth and its life, especially as recorded in rocks. **Ecology** is the science that is concerned with the interrelationship of organisms and their environments. **Medicine** is the science and art dealing with the maintenance of health and the prevention, alleviation, or cure of disease. **Meteorology** is the science that deals with the atmosphere and its phenomena, and especially with weather and weather forecasting.

Physics is the science that deals with matter and energy and their interactions. **Biophysics** deals with the application of physical principles and methods to biological systems and problems.

Statistics is a branch of mathematics dealing with the collection, analysis, interpretation, and presentation of masses of numerical data. Chapter 12 of this text introduces some topics in statistics—namely, frequency distributions and graphs; measures of central tendency, position, and dispersion; normal curves; and correlation and regression analysis.

And last, but not least, on our list of major categories of the natural sciences is **zoology,** which is concerned with classification and properties of animals including their structure, function, growth, origin, evolution, and distribution. Zoology is often considered to be a branch of biology.

Here is a partial list of applications from the natural sciences that have been included in other parts of this text. They are listed in the order of presentation in the book.

Velocity of light in a vacuum, Problem 16, p. 36, Section 1.3
Distance between Earth and Mars, Problem 18, p. 36, Section 1.3
Volume of a neuron, Problem 22, p. 37, Section 1.3
Circuit design, pp. 90–94, Section 2.6
Genetics, pp. 230–231, Section 5.3; Problems 52–55; 58–59, p. 238, Section 5.3;
    Problem 19, p. 292, Chapter 5 Review
Blood gene pool, Problem 60, p. 239, Section 5.3
Planetary distances, Problem 60, p. 333, Section 6.5
Studies of the human body, pp. 337–338, Section 6.6
Alkalinity, Problem 56, p. 413, Section 8.2
Richter scale, Problem 57, p. 413, Section 8.2
Radioactive decay, Examples 5 and 6, pp. 415–416; Problems 17–20, 55–56, pp. 420–422,
    Section 8.3
Earthquakes, Problems 21–24; 31, p. 420, Section 8.3
Power of sound, Problems 25–29, p. 420, Section 8.3
Bacterial cultures, Problem 32, p. 420, Section 8.3
Newton's law of cooling, Problems 48–51, p. 421, Section 8.3
Atmospheric pressure, Problems 53–54, p. 421, Section 8.3
Power supply in a satellite, Problems 55–56, p. 421, Section 8.3
Challenger disaster, Problem 59, p. 421, Section 8.3
Arrhenius function to measure friction, Problem 60, p. 422, Section 8.3
Healing law for skin, Problem 19, p. 424, Chapter 8 Review;
    Problem 57, p. 682, Section 13.3
Life science fruit fly experiment, Example 14, p. 578, Section 11.3
Nutritional information about candy bars, Table 12.7, p. 632, Section 12.2
Mapping Venus, Problem 1, p. 673, Section 13.1
Disintegration of kidney stones, p. 691, Section 13.4
LORAN navigational system, p. 693, Section 13.4
Planetary orbits, Example 8, p. 689; Problems 52–55, p. 694, Section 13.4
Parabolic reflection property for antennas, p. 693, Section 13.4;
    Problems 57–59, p. 695, Section 13.4
Path of a cannonball, Problem 19, p. 705, Chapter 13 Review
Administration of a drug into the body, Problem 29, p. 876, Section 17.2
Rate of a growing culture, Problem 29, p. 895, Section 17.4

An excellent discussion of mathematics in modeling a projectile's path can be found in the following article in the classic book, *World of Mathematics* (New York, Simon and Schuster, 1956):

"Mathematics of Motion," by Galileo Galilei

You will find other references to mathematics in the natural sciences on the World Wide Web at

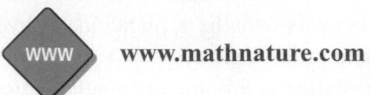 **www.mathnature.com**

Mathematics in the Social Sciences

The social sciences are the study of human society and of individual relationships in and to society. It also refers to a scholarly or scientific discipline that deals with such study. Some of the major categories include (arranged alphabetically) anthropology, archaeology, civics, history, languages, political science, psychology, and sociology.

Anthropology refers to the study of human beings in relation to distribution, origin, classification, and relationship of races, physical character, environmental and social relations, and culture. In its broadest sense, it also includes theology.

EXAMPLE 6

Discuss the invention of the zero symbol.

Solution

The term zero, written 0, is defined to be the *additive identity* (Section 4.6), which is that number with the property that $x + 0 = 0 + x$ for any number x. The Babylonians (Section 3.1) used written symbols for thousands of years before they invented a symbol for 0, which was initially introduced as a position marker to differentiate between numbers such as 123 and 1230. The first documented use of the *number* 0 is found around the first century A.D. in the Mayan numeration system. The Hindus customarily wrote numbers in columns, and used a zero symbol to represent a blank column, which is necessary for our present place-value numeration system.   ◆

Another application of mathematics to anthropology is carbon dating of artifacts using the decay formula. Review Examples 4–6 (pp. 415–416) of Section 8.3, as well as Problems 17–20 of Problem Set 8.3. In particular, Problem 57 of Problem Set 8.3 asked you to use the gathered data to set a probable date for the Shroud of Turin. If you worked this problem, you found that it is not dated from the time of Christ.

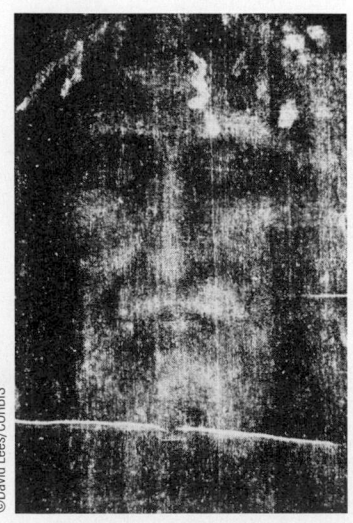

EXAMPLE 7

In 1988, a small sample of the Shroud of Turin was taken and scientists from Oxford University, the University of Arizona, and the Swiss Federal Institute of Technology were permitted to test it. Suppose the cloth contained 90.7% of the original amount of carbon. According to this information, how old is the Shroud?

Solution

We use the decay formula $A = A_0 e^{rt}$, where $A/A_0 = 0.907$ and $r = -1.209680943\text{E} - 4$ (from Example 5, pp. 415–416, Section 8.3)

$$A = A_0 e^{rt} \qquad \text{Decay formula}$$

$$\frac{A}{A_0} = e^{rt} \qquad \text{Divide both sides by } A_0.$$

$$0.907 = e^{rt} \qquad \text{Remember, } t \text{ is the unknown; } r \text{ is known.}$$

$$rt = \ln 0.907 \qquad \text{Definition of logarithm}$$

$$t = \frac{\ln 0.907}{r} \qquad r = -1.209680943E - 4$$

$$\approx 806.9 \qquad \text{This is the most probable age (in years).}$$

The probable date for the Shroud is about A.D 1200. Since the first recorded evidence of the Shroud is about 1389, we see that A.D 1200 is not only possible, but plausible. ◆

"Human sacrifices and human compassion. Greek artists and American cannibals. The birth of language and the death of civilizations. Mongol hordes and ancient toys. Missing aviators and sunken ships. Egyptian politicians and modern saints. The first human and the latest war. The past hides the key to the future. Find it in . . ." This advertisement for the periodical *Discovering Archaeology* does a good job of describing **archaeology.**

Civics is the subject that deals with the rights and duties of citizens. **History** refers to a chronological record of significant events, often including an explanation of their causes. The prologue of this book focuses on mathematical history. **Linguistics** is the study of human speech including the units, nature, structure, and modification of language, whereas **language** refers to the actual knowledge of the words and vocabulary used to communicate ideas and feelings.

Political science is a branch of social science that is the study of the description and analysis of political and especially governmental institutions and processes. Related to this idea is the problem of representation in government, namely, the issue of voting and apportionment. In Sections 16.1 and 16.2 we examined different forms of representative government, including dictatorship, majority, and plurality rules. What effect does a run-off election or a third party have on the results of an election?

Apportionment (Sections 16.3 and 16.4) was an important issue for the framers of the U.S. Constitution, and historically there were four plans that are important to government planning today. These plans were advanced by Thomas Jefferson, John Quincy Adams, Daniel Webster, and Alexander Hamilton.

Psychology, broadly defined, refers to the science of mind and behavior. Mathematical tools used in psychology include sets, statistics, and the analysis of data, which we discuss at length in this text in Chapter 10. The organization of data and surveys using Venn diagrams is introduced in this book in Section 1.4, Examples 4 and 5.

Sociology is the study of society, social institutions, and social relationships, specifically the systematic study of the development, structure, interaction, and collective behavior of organized groups of human beings. Throughout the book, we encountered a secret sect of intellectuals called the Pythagoreans. Every evening each member of the Pythagorean Society had to reflect on three questions:

1. What good have I done today?

2. What have I failed at today?

3. What have I not done today that I should have done?

EXAMPLE 8

The Pythagoreans studied numbers that were called *perfect numbers.* A perfect number is a counting number that is equal to the sum of all its divisors that are less than the number itself. Show that 28 is a perfect number.

Solution

We first find all divisors of 28 that are less than 28: 1, 2, 4, 7, and 14. Their sum is

$$1 + 2 + 4 + 7 + 14 = 28$$

So we see that 28 is perfect!   ◆

Understanding population growth is an essential part of understanding the underlying principles used in the social sciences. Population growth, introduced in Section 8.3, along with a working knowledge of the number *e,* are used to predict population sizes for different growth rates. Recall the growth formula:

$$A = A_0 e^{rt}$$

which gives the future population A after t years for an initial population A_0 and a growth rate of r. You might wish to reconsider Examples 1 and 2 of Section 8.3 (pp. 413–415).

The principal use of mathematics in the social sciences is in its heavy use of statistics (Chapter 12). The topics of surveys and sampling are important to social scientists (see Section 12.5) and proceeding in an appropriate manner is very important. To choose an unbiased sample from a target population, we must ask two questions: *Is the procedure random?* and *Does the procedure take into account the target population?* Take a look at Example 1 (p. 652) and Problems 4–26 (pp. 655–657) of Section 12.5.

Matrices, an important topic in mathematics, are used to organize and manipulate data in the social sciences. This topic is introduced and discussed in Sections 14.3 and 14.4 (pp. 724–748).

Social Choice

The study of the decision-making process by which individual preferences are translated into a single group choice is known as *social choice theory*. It is an important topic in mathematics (see Chapter 16) and culminates with the surprising mathematical result that it is impossible to find a suitable technique for voting.

Related to voting is the process by which a representative government is chosen, called *apportionment* (Sections 16.3 and 16.4).

Stable Marriages

Suppose there are n graduates of a well-known medical school, and those graduates will be matched to n medical centers to serve their internship programs. Each graduate must select his or her preferences for medical school, and in turn, each medical school must make a list of preferences for graduates. We will call this pairing a **marriage.** A pairing is called **unstable** if there is a graduate and a school who have not picked each other but who would prefer to be married to each other rather than to their current selection. Otherwise, the pairing is said to be **stable.**

EXAMPLE 9

Suppose there are three graduates, *a, b, c,* and three schools *A, B, C,* with choices as follows:

> *a(ABC)* means that *a* ranks the schools as *A,* first choice; *B,* second choice; and *C,* third choice.
> *b(ACB)* means that *b* ranks the schools *A, C,* and *B,* respectively.
> *c(ABC)* means that *c* ranks the schools *A, B,* and then *C.*

On the other hand, the schools' rankings of candidates are:

> *A(cab),*   *B(cba),*   and   *C(bca)*

How many possible marriages are there, and how many of these are stable?

Solution

We begin by representing the choices for this example in matrix form (we considered matrices in Section 14.3):

$$
\begin{array}{c@{\qquad}ccc}
 & A & B & C \\
a & (1,2) & (2,3) & (3,3) \\
b & (1,3) & (3,2) & (2,1) \\
c & (1,1) & (2,1) & (3,2)
\end{array}
$$

The entry (1, 2) for (*a*, *A*) means that *a* picks *A* as its 1st choice, and *A* picks *a* as its 2nd choice.

To answer the first question, we use the fundamental counting principle from Section 1.4. Person *a* can be paired with any of three schools, and then (since the matching is one school to one student) there are two schools left for person *b,* and finally one for person *c:*

> Number of possibilities: $3 \cdot 2 \cdot 1 = 6$

We show these choices using a tree diagram in Figure E.6. We need to consider each of these six possibilities, one at a time. The first one listed is *aA, bB,* and *cC.* We repeat the above matrix, this time with these pairings circled:

$$
\begin{array}{c@{\qquad}ccc}
 & A & B & C \\
a & \boxed{(1,2)} & (2,3) & (3,3) \\
b & (1,3) & \boxed{(3,2)} & (2,1) \\
c & (1,1) & (2,1) & \boxed{(3,2)}
\end{array}
$$

We look for a pairing in which both partners would rather be paired with someone else. This will not occur for a row in which the pairing shows a first choice, so we do not look at row *a* to find dissatisfaction; instead we look at row *b.* The pairing (3, 2) tells us that *b* is associated with his or her third choice. We are looking for another pair *in the same row* where the second component is smaller than its corresponding choice in *its column.* Note that in row *b* the entry (2, 1) has a second component of 1, which means that *C* would rather be paired with *b* than with its current pairing of *c.* We also see that person *b* would also prefer to be married to *c* (second choice) than with its current pairing, so this is *unstable.*

Let's move to the second pairing shown in Figure E.6. It is *aA, bC,* and *cB.* We show this pairing:

$$
\begin{array}{c@{\qquad}ccc}
 & A & B & C \\
a & \boxed{(1,2)} & (2,3) & (3,3) \\
b & (1,3) & (3,2) & \boxed{(2,1)} \\
c & (1,1) & \boxed{(2,1)} & (3,2)
\end{array}
$$

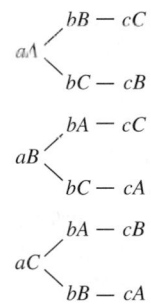

$aA \Big\langle \begin{array}{l} bB - cC \\ bC - cB \end{array}$

$aB \Big\langle \begin{array}{l} bA - cC \\ bC - cA \end{array}$

$aC \Big\langle \begin{array}{l} bA - cB \\ bB - cA \end{array}$

Figure E.6 **Possible pairings**

Row *a* is OK because *a* is paired with his or her first choice.

Row *b* is OK because *b* is paired with his or her second choice, but *C* is paired with a first choice.

Row *c* shows that *c* is paired with his or her second choice, but would rather be with *A*. *A* would also rather be with *C*, so this marriage is unstable.

The other possibilities are listed here:

$$
\begin{array}{c} & A & B & C \\ a \\ b \\ c \end{array}
\begin{bmatrix} (1,2) & \boxed{(2,3)} & (3,3) \\ \boxed{(1,3)} & (3,2) & (2,1) \\ (1,1) & (2,1) & \boxed{(3,2)} \end{bmatrix}
$$

Not stable; *a* prefers *A*, and *A* prefers *a* to *b*.

$$
\begin{array}{c} & A & B & C \\ a \\ b \\ c \end{array}
\begin{bmatrix} (1,2) & \boxed{(2,3)} & (3,3) \\ (1,3) & (3,2) & \boxed{(2,1)} \\ \boxed{(1,1)} & (2,1) & (3,2) \end{bmatrix}
$$

a prefers *A*, but *A* is paired with first choice.

b prefers *A*, but *A* is paired with first choice.

Stable

$$
\begin{array}{c} & A & B & C \\ a \\ b \\ c \end{array}
\begin{bmatrix} (1,2) & (2,3) & \boxed{(3,3)} \\ \boxed{(1,3)} & (3,2) & (2,1) \\ (1,1) & \boxed{(2,1)} & (3,2) \end{bmatrix}
$$

c prefers *A*, and *A* prefers *c*, so this is unstable.

$$
\begin{array}{c} & A & B & C \\ a \\ b \\ c \end{array}
\begin{bmatrix} (1,2) & (2,3) & \boxed{(3,3)} \\ (1,3) & \boxed{(3,2)} & (2,1) \\ \boxed{(1,1)} & (2,1) & (3,2) \end{bmatrix}
$$

b prefers *C* and *C* prefers *b*, so this is unstable.

◆

Example 9 has five unstable pairings and one stable one. It can be proved that every marriage has at least one stable pairing. In the problem set you are asked to consider the situation where the number of candidates is greater than the number of schools.

Here is a partial list of applications from the social sciences that have been included in other parts of this text. We list them in the order of presentation in the book.

An excellent discussion of mathematics and the social sciences can be found in the following list of articles in the classic book, *World of Mathematics* (New York, Simon and Schuster, 1956):

"Gustav Theodor Fechner," a commentary of the founder of psychophysics, by Edwin G. Boring. His contribution was to introduce measurement as a tool.

"Mathematics of Population and Food," by Thomas Robert Malthus, explores the thesis that all animated life tends to increase beyond the nourishment prepared for it.

"A Mathematical Approach to Ethics," by George Birkhoff

You will find other references to mathematics in the social sciences on the World Wide Web at

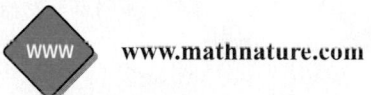

www.mathnature.com

Mathematics in the Humanities

By humanities we mean art, music, and literature. One of the major themes of this book has been the analysis of shapes and patterns, and shapes and patterns are the basis for a great deal of art and music.

There is an interdependence of mathematics and art (see Chapter 6). The shape that seems to be the most appealing to human intellect is the so-called golden rectangle, which was used in the design of the Parthenon in Athens and in *La Parade* by the French impressionist Georges Seurat.

EXAMPLE 10

An unfinished canvas by Leonardo da Vinci entitled *St. Jerome* was painted about 1481. Find a golden rectangle that fits neatly around a prominent part of this painting. Measure the length and width of the rectangle and then find the ratio of the width to length.

Solution
A portion of the work of art is shown below (with a rectangle superimposed). The width and length of the rectangle will vary with the size of the reproduction, but the ratio of width to length is about 0.61.

St. Jerome by Leonardo da Vinci, 1481

◆

In order to draw realistic-looking drawings, mathematical ideas from projective geometry are necessary (see Section 6.7). Early attempts at three-dimensional art failed miserably because they did not use projective geometry. We cite as examples Duccio's *Last Supper* on p. 340, Section 6.7, or the Egyptian art shown in Figure E.7.

Figure E.7 Judgment of the Dead-Pai on papyrus ca. 14th century B.C.

Projective geometry was an early mathematical attempt to represent three-dimensional objects on a canvas (see Dürer's *Designer of the Lying Woman* on page 341). Finally, a plan for perspective was developed, as shown in Figure E.8.

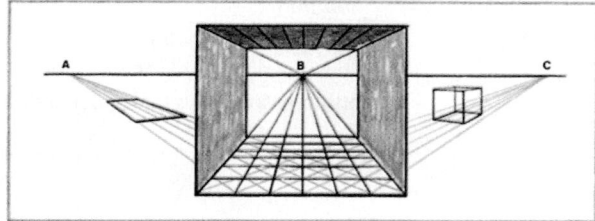

From *Mathematics* by David Bergamini. © Time, Inc. Reprinted by permission.

Figure E.8 **Perspective in art**

The rule of optical perspective begins with the horizon (line *AC*). One point *B* on line *AC* is selected to be the vanishing point, and all other lines recede directly from the viewer and converge at the vanishing point. All other lines (except verticals and parallels to the horizon) have their own vanishing points, governed by their particular angle to the plane of the picture. The vanishing point for the rectangle in Figure E.8 is point *A,* and for the cube at the right, it is point *C.*

Music is very mathematical and the Greeks included music when studying arithmetic, geometry, and astronomy. Today with a great deal of music created on a synthesizer, it is not uncommon for a computer programmer to be a musician and for a musician to be a computer programmer. The modern musical scale is divided into 13 tones, as shown in Figure E.9.

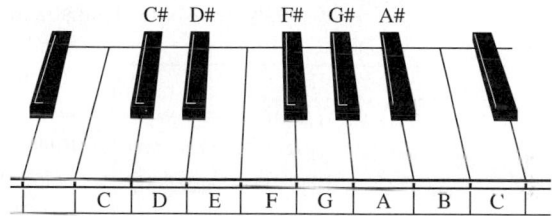

Figure E.9 **Piano keyboard showing one octave**

The Greeks plucked strings and found what types of string vibrations make pleasing sounds. They found that the sound was pleasing if it was plucked at a location that was in the ratio of 1 to 2 (which we call an *octave*), 2 to 3 (a *fifth*), 3 to 4 (a *fourth*), and so on. You might notice that the piano keyboard is divided into 8 white keys (diatonic scale) and 5 black keys (pentatonic scale). These numbers 5 and 8 are two consecutive Fibonacci numbers (introduced in Section 9.3). The middle C is tuned so the string vibrates 264 times per second, whereas the note A above middle C vibrates 440 times per second. Note that the ratio of 440 to 264 reduces to 5 to 3, two other Fibonacci numbers.

A mathematical concept known as the sine function (see Section 6.5) can be used to record the motion of the vibrations over time. These vibrations (known as sound) are shown in the following example.

EXAMPLE 11

A tuning fork vibrating at 264 Hz [frequency $f = 264$ and Hz is an abbreviation for Hertz, meaning "cycles per unit of time" after the physicist Heinrich Hertz (1857–1894)]. This sound produces middle C on the musical scale and can be described by an equation of the form

$$y = 0.0050 \sin(528 \cdot 180x)$$

Use a graphing calculator to look at this curve.

Solution

The graph is shown in Figure E.10, along with the necessary input values. Technology does not do a good job of graphing an equation such as this, and you may obtain many different looking graphs, depending on the window values.

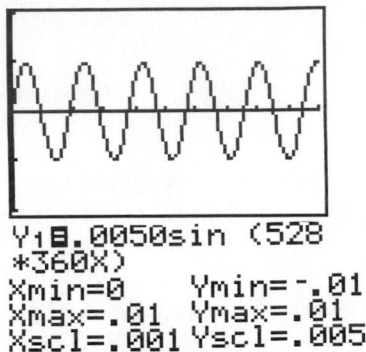

Figure E.10 **Graph of a musical note described by** $y = 0.0050 \sin(528\pi x)$ ◆

A composition by Mozart is interesting not only for its music, but also because it illustrates the mathematical idea of an inversion transformation we introduced in Section 6.1. This unique piece for violin, reproduced in Figure E.11, can be played simultaneously by two musicians facing each other and reading the music, laid flat between them, in opposite directions. The two parts, though different, will be in perfect harmony, without violating a single rule of classical composition. Bach's *Art of the Fugue* made use of mirror reflections in passages between the upper treble and lower bass figures.

Here is a partial list of applications from the humanities that have been included in other parts of this text.

Tiling patterns, Problems 24–27, pp. 793–794, Section 15.3
Anamorphosic art, Problem 28, p. 794, Section 15.3
Fractal art (chaos), Guest Essay, p. 795, Section 15.3

An excellent discussion of mathematics and the humanities can be found in the following list of articles from the classic book, *World of Mathematics* (New York, Simon and Schuster, 1956):

"Mathematics of Aesthetics," by George David Birkhoff
"Mathematics as an Art," by John William Navin Sullivan
"Mathematics and Music," by James Jeans
"Geometry in the South Pacific," by Sylvia Townsend Warner. The story is an example of mathematics in literature.

You will find other references to mathematics in the humanities on the World Wide Web at

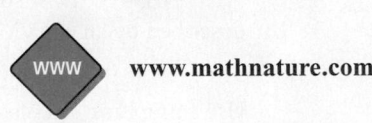

www.mathnature.com

The Selmer Bandwagon, 1961, 9(2).

Figure E.11 **Mozart composition**

Mathematics in Business and Economics

The focus of Chapter 9 in the text forms the foundations for business and economics. A fundamental, and essential, idea in business and economics is the concept of **interest.** Interest is an amount paid for the use of someone else's money. It can be compounded once, or it can be compounded at regular intervals. Related to interest are annuities (future value of interest and deposits), sinking funds (monthly payment when the future value is known), present value of an annuity (the amount you can borrow, given the monthly payment), and amortization (the monthly payment when the present value is known). All of these concepts are essential ideas for understanding the processes of business, borrowing, and economics, and they are all important topics of mathematics.

Here is a partial list of applications from business and economics that have been included in other parts of this text. We list them in the order of presentation in the book.

Gross domestic product, Problem 19, p. 36, Section 1.3
Salary calculation, Problems 26–28, p. 37, Section 1.3
Inflation, Example 11, p. 436, Section 9.1
Giving your child $1,000,000 for under $500 (for real!), Example 14, p. 437, Section 9.1
Size of the national debt, Problem 53, p. 440, Section 9.1
Saving for retirement, Problems 54–56, p. 440, Section 9.1
Effective annual yield, Problems 59–60, p. 440, Section 9.1
Consumer loans and installment buying, Examples 1–6, pp. 441–445, Section 9.2
Credit card interest, Examples 5–6, pp. 446–447; Problems 21–34, pp. 447–448, Section 9.2
Buying an automobile, Problems 35–38, 50–55, pp. 448–449, Section 9.2
Rule of 78, Problem 60, p. 449, Section 9.2
Annuities, Examples 1–3, pp. 472–474, Section 9.5
Would you rather retire on $2,000/mo or $6,000/mo for life? Which do you think costs you less? Examples 2–3, pp. 472–473, Section 9.5
Sinking funds, Example 4, p. 475, Section 9.5
Keogh account, Problem 47, p. 476, Section 9.5
Present value of an annuity, Examples 1–3, pp. 478–479, Section 9.6
Retirement on Social Security, Example 3, p. 479, Section 9.6
Amortization, Example 4, pp. 480–482, Section 9.6
Monthly payments, Examples 4–5, pp. 480–482, Section 9.6
Straight-line depreciation, Problem 54, p. 674, Section 13.1
Marginal profit, Problem 55, p. 674, Section 13.1
Marginal cost, Problem 56, p. 674, Section 13.1
Supply and demand, Example 4, pp. 717–718; Problems 7–9, p. 722; Problems 57–58, p. 724, Section 14.2
Maximization and minimization using linear programming, pp. 749–755, Section 14.6
Accumulated cost of a piece of equipment, Problem 30, p. 895, Section 17.4

An excellent discussion of mathematics in business and economics can be found in the following list of articles from the classic book, *World of Mathematics* (New York, Simon and Schuster, 1956):

"Mathematics of Value and Demand," by Augustin Cournot

"Theory of Political Economy," by William Stanley Jevons. This essay begins with the declaration, "It is clear that economics, if it is to be a science at all, must be a mathematical science. . . . Wherever the things treated are capable of being *greater or less,* there the laws and relations must be mathematical in nature."

"The Theory of Economic Behavior," by Leonid Hurwicz, introduces the important application of *Theory of Games and Economic Behavior.* Game theory is primarily concerned with the logic of strategy and was first envisioned by the great mathematician Gottfried Leibniz (1646–1716). However, the theory of games as we know it today was developed in the 1920s by John von Neumann and Emile

Borel. It gained wide acceptance in 1944 in a book entitled *Theory of Games and Economic Behavior* by von Neumann and Oskar Morgenstern.

"Theory of Games," by S. Vajda gives you further study in game theory.

Epilogue Problem Set

LEVEL 1

1. **IN YOUR OWN WORDS** Describe some similarities between the scientific method and mathematical modeling.

2. **IN YOUR OWN WORDS** Pick one of the areas that is of most interest to you: natural sciences, social sciences, humanities, or business and economics. Write a paper discussing at least one application of this area to mathematics that was not discussed in this appendix.

3. Classify each of the disciplines as humanities, a natural science, or a social science.
   a. psychology
   b. physics
   c. meteorology
   d. music
   e. history

4. Classify each of the disciplines as humanities, a natural science, or a social science.
   a. chemistry
   b. art
   c. French
   d. geology
   e. computer programming

5. Classify each of the disciplines as humanities, a natural science, or a social science.
   a. civics
   b. medicine
   c. anthropology
   d. political science
   e. biology
   f. ecology

6. Classify each of the disciplines as humanities, a natural science, or a social science.
   a. dance
   b. theology
   c. women's studies
   d. social work
   e. linguistics
   f. conflict studies

7. Classify each of the disciplines as humanities, a natural science, or a social science.
   a. plant pathology
   b. geography
   c. horticulture
   d. food science
   e. entomology
   f. forestry

8. Classify each of the disciplines as humanities, a natural science, or a social science.
   a. literature
   b. Jewish studies
   c. human nutrition
   d. agriculture
   e. gender studies
   f. kinesiology

9. If a star is located at a distance of 858,000,000,000,000,000,000 miles, what is this distance in light-years?

10. If a star is located at a distance of 5.61 light-years, what is this distance in miles?

11. The eccentricities of the planets are given in the following table:

   Mercury, 0.21   Venus, 0.01   Earth, 0.02
   Mars, 0.09   Jupiter, 0.04   Saturn, 0.06
   Uranus, 0.05   Neptune, 0.01   Pluto, 0.24

   Which planet has the most circular orbit?

12. Each person has 23 pairs of chromosomes. If we inherit one of each pair from each parent, what are the number of possibilities?

13. An average chicken egg is 2 in. from top to bottom, and an ostrich egg is 6 in. from top to bottom. What is the scaling factor? If a chicken egg weighs 2 oz, what is the expected weight of an ostrich egg?

14. Eggs are graded for both quantity and size. Jumbo eggs must have a minimum weight of 56 lb per 30-doz case and small eggs must have a minimum weight of 34 lb per 30-doz case. What is the scaling factor for the weight of one small egg compared to one jumbo egg?

15. Suppose a pea has two skin characteristics: S (smooth, dominant) and w (wrinkled, recessive). Suppose we have an isolated population in which 55% of the genes in the gene pool are dominant S and the other 45% are recessive w. What fraction of the population has each genotype?

16. Suppose a pea has two skin characteristics: S (smooth, dominant) and w (wrinkled, recessive). Suppose we have an isolated population in which 48% of the genes in the gene pool are dominant S and the other 52% are recessive w. What fraction of the population has each genotype?

17. The area of a cell magnified 2,000 times is about 20 cm^2. What is the area of the unmagnified cell?

LEVEL 2

18. a. If you change the angle in Example 1 to 45°, would you expect the cannonball to go farther or shorter than shown in the example?
   b. Use a calculator to sketch the path of a cannonball with initial velocity 63 ft/s at a 45° angle. Was your conjecture in part **a** correct?

19. What is the path of a cannonball fired at 64 ft/s at an angle of 60°?

20. Describe the meaning of the equation

$$\frac{x^2}{4.5837 \times 10^{15}} + \frac{y^2}{4.5835 \times 10^{15}} = 1$$

which models the path of Venus' orbit around the sun. The given distances are in miles.
    a. What is the eccentricity?
    b. Graph this equation.

21. Describe the meaning of the equation

$$\frac{x^2}{2.015 \times 10^{16}} + \frac{y^2}{1.995 \times 10^{16}} = 1$$

which models the path of Mars' orbit around the sun. The given distances are in miles.
    a. What is the eccentricity?
    b. Graph this equation.

22. Four professors, A, B, C, and D, are each going to hire a student. Six students have applied. How many possible marriages are there?

23. Each of four women is going to marry one of four men. How many possible marriages are there?

24. Four professors, A, B, C, and D, are each going to hire a student. Six students have applied. Here are the preferences:

$A(a, f, b, c, d, e)$ $a(C, B, D, A)$
$B(a, b, f, e, d, c)$ $b(C, D, B, A)$
$C(b, a, e, c, d, f)$ $c(B, A, C, D)$
$D(a, b, e, f, c, d)$ $d(A, B, D, C)$
                              $e(A, B, C, D)$
                              $f(B, D, A, C)$

    a. Show the matrix of their choices.
    b. List one pairing and state whether it is stable or unstable.

25. Four women are going to marry four men. Here are their preferences:

$a(A, D, B, C)$ $A(d, a, b, c)$
$b(D, A, C, B)$ $B(a, d, c, b)$
$c(D, A, B, C)$ $C(a, b, d, c)$
$d(D, B, A, C)$ $D(b, a, c, d)$

    a. Show the matrix of their choices.
    b. List one pairing and state whether it is stable or unstable.

26. An equation for middle C with frequency $f = 261.626$ and amplitude 8 has the equation

$$y = 8 \sin(360fx)$$

Use a calculator to graph this note.

27. Graph the sound wave with equation

$$y = 8\sin(360fx) + 4\sin(720fx)$$

for $f = 261.626$.

LEVEL 3

28. Suppose there are three graduates, a, b, c, and three schools, A, B, C, with choices as follows:

$a(CAB)$ $b(CAB)$ $c(CBA)$
$A(abc)$ $B(bac)$ $C(acb)$

    a. How many possible marriages are there?
    b. How many of these are stable?

29. Suppose there are three graduates, a, b, c, and three schools, A, B, C, with choices as follows:

$a(BCA)$ $b(CAB)$ $c(ABC)$
$A(abc)$ $B(bca)$ $C(cab)$

    a. How many possible marriages are there?
    b. How many of these are stable?

PROBLEM SOLVING

30. The moon's orbit is elliptical with the earth at one focus. The point at which the moon is farthest from the earth is called the *apogee*, and the point at which it is closest is called the *perigee*. If the moon is 199,000 miles from the earth at apogee and the length of the major axis of its orbit is 378,000 miles, what is the eccentricity of the moon's orbit?

A Glossary

Abelian group   A group that is also commutative.

Abscissa   The horizontal coordinate in a two-dimensional system of rectangular coordinates, usually denoted by x.

Absolute value   The absolute value of a number is the distance of that number from the origin. Symbolically,

$$|n| = \begin{cases} n & \text{if } n \geq 0 \\ -n & \text{if } n < 0 \end{cases}$$

Accuracy   One speaks of an *accurate statement* in the sense that it is true and correct, or of an *accurate computation* in the sense that it contains no numerical error. *Accurate to a certain decimal place* means that all digits preceding and including the given one are correct.

Acre   A unit commonly used in the United States system for measuring land. It contains 43,560 ft^2.

Acute angle   An angle whose measure is smaller than a right angle.

Acute triangle   A triangle with three acute angles.

Adams' apportionment plan   An apportionment plan in which the representation of a geographical area is determined by finding the quotient of the number of people in that area divided by the total number of people and then the result is rounded as follows: Any quotient with a decimal portion must be rounded up to the next whole number.

Addition   One of the fundamental undefined operations applied to the set of counting numbers.

Addition law of exponents   To multiply two numbers with like bases, add the exponents; that is, $b^m \cdot b^n = b^{m+n}$.

Addition law of logarithms   The log of the product of two numbers is the sum of the logs of those numbers. In symbols,

$$\log_b(AB) = \log_b A + \log_b B$$

Addition method   The method of solution of a system of equations in which the coefficients of one of the variables are opposites so that when the equations are added, one of the variables is eliminated.

Addition of integers   If the integers to be added have the same sign, the answer will also have that same sign and will have a magnitude equal to the sum of the absolute values of the given integers. If the integers to be added have opposite signs, the answer will have the sign of the integer with the larger absolute value, and will have a magnitude equal to the difference of the absolute values. Finally, if one or both of the given integers is 0, use the property that $n + 0 = n$ for any integer n.

Addition of matrices $[M] + [N] = [S]$ if and only if $[M]$ and $[N]$ are the same order and the entries of $[S]$ are found by adding the corresponding entries of $[M]$ and $[N]$.

Addition of rational numbers

$$\frac{a}{b} + \frac{c}{d} = \frac{ad}{bd} + \frac{bc}{bd} = \frac{ad + bc}{bd}$$

Addition principle   A method of representing numbers by repeating a symbol in a numeration system and repeatedly adding the symbol's value. For example, $\cap\cap\cap \,||||$ means $10 + 10 + 10 + 1 + 1 + 1 + 1$.

Addition property (of equations)   The solution of an equation is unchanged by adding the same number to both sides of the equation.

Addition property of inequality   The solution of an inequality is unchanged if you add the same number to both sides of the inequality.

Addition property of probabilities   For any events E and F, the probability of their union can be found by

$$P(E \cup F) = P(E \text{ or } F) = P(E) + P(F) - P(E \cap F)$$

Additive identity   The number 0, which has the property that $a + 0 = a$ for any number a.

Additive inverse   See *Opposites*. The additive inverse of a matrix $[M]$ is denoted by $[-M]$ and is defined by $(-1)[M]$.

Add-on interest   It is a method of calculating interest and installments on a loan. The amount of interest is calculated according to the formula $I = Prt$ and is then added to the amount of the loan. This sum, divided by the number of payments, is the amount of monthly payment.

Address   Designation of the location of data within internal memory or on a magnetic disk or tape.

Adjacent angles   Two angles are adjacent if they share a common side.

Adjacent side   In a right triangle, an acute angle is made up of two sides; one of those sides is the hypotenuse, and the other side is called the adjacent side.

Adjusted balance method   A method of calculating credit card interest using the formula $I = Prt$ in which P is the balance owed after the current payment is subtracted.

Alabama paradox   An increase in the total numbers of items to be apportioned resulting in a loss for a group is called the Alabama paradox.

Algebra   A generalization of arithmetic. Letters called variables are used to denote numbers, which are related by laws that hold (or are assumed) for any of the numbers in the set. The four main processes of algebra are (1) simplify, (2) evaluate, (3) factor, and (4) solve.

Algebraic expression Any meaningful combination of numbers, variables, and signs of operation.

Alternate exterior angles Two *alternate angles* are angles on opposite sides of a transversal cutting two parallel lines, each having one of the lines for one of its sides. They are *alternate exterior angles* if neither lies between the two lines cut by the transversal.

Alternate interior angles Two *alternate angles* are angles on opposite sides of a transversal cutting two parallel lines, each having one of the lines for one of its sides. They are *alternate interior angles* if both lie between the two lines cut by the transversal.

Alternating series A series that alternates in sign.

Amortization The process of paying off a debt by systematically making partial payments until the debt (principal) and interest are repaid.

Amortization schedule A table showing the schedule of payments of a loan detailing the amount of each payment that goes to repay the principal and how much goes to pay interest.

Amortized loan A loan that is fully paid off with the last periodic payment.

Analytic geometry The geometry in which position is represented analytically (or by coordinates) and algebraic methods of reasoning are used for the most part.

And See *Conjunction*. In everyday usage, it is used to join together elements that are connected in two sets simultaneously.

AND-gate An electrical circuit that simulates conjunction; that is, the circuit is on when two switches are on.

Angle Two rays or segments with a common endpoint.

Angle of depression The angle between the line of sight to an object below measured from a horizontal.

Angle of elevation The angle between the line of sight to an object above measured from a horizontal.

Annual compounding In the compound interest formula, it is when $n = 1$.

Annual percentage rate The percentage rate charged on a loan based on the actual amount owed and the actual time it is owed. The approximation formula for annual percentage rate (APR) is

$$APR = 2Nr/(N + 1)$$

Annuity A sequence of payments into or out of an interest-bearing account. If the payments are made into an interest-bearing account at the end of each time period, and if the frequency of payments is the same as the frequency of compounding, the annuity is called an *ordinary annuity*.

Antecedent See *Conditional*.

Antiderivative If the derivative of a function defined by $y = F(x)$ is $f(x)$, then the **antiderivative** of $f(x)$ is

$$\int f(x)\, dx = F(x)$$

This means that $F'(x) = f(x)$.

Apportionment The process of dividing the representation in a legislative body according to some plan.

Approval voting The approval voting method allows each voter to cast one vote for each candidate who meets with his or her approval. The candidate with the most votes is declared the winner.

APR Abbreviation for annual percentage rate. See *Compound interest formula*.

Arc (1) Part of the circumference of a circle. (2) In networks, it is a connection between two vertices.

Area A number describing the two-dimensional content of a set. Specifically, it is the number of square units enclosed in a plane figure.

Area formulas Square, s^2; rectangle, ℓw; parallelogram, bh; triangle, $\frac{1}{2}bh$; circle, πr^2; trapezoid, $\frac{1}{2}h(b_1 + b_2)$.

Area function The function that is the area bounded below by the x-axis, above by a function $y = f(x)$, on the left by the y-axis, and on the right by the vertical line $x = t$.

Area under a curve The area of the region bounded by the graph of a function f, the x-axis, and the vertical lines $x = a$ and $x = b$ is given by

$$A = \int_a^b f(x)\, dx$$

Argument (1) The statements and conclusion as a form of logical reasoning. (2) In a logarithmic expression $\log_b N$, it is the number N.

Arithmetic mean The arithmetic mean of the numbers a and b is $\dfrac{a + b}{2}$.

Arithmetic sequence A sequence, each term of which is equal to the sum of the preceding term and a constant, written $a_1, a_2 = a_1 + d, a_3 = a_1 + 2d, \ldots$; the nth term of an arithmetic sequence is $a_1 + (n - 1)d$, where a_1 is the first term and d is the *common difference*. Also called an *arithmetic progression*. See also *Sequence*.

Arithmetic series The indicated sum of the terms of an arithmetic sequence. The sum of n terms is denoted by A_n and

$$A_n = \frac{n}{2}(a_1 + a_n) \text{ or } A_n = \frac{n}{2}[2a_1 + (n - 1)d]$$

Arrangement Same as *permutation*.

Array An arrangement of items into rows and columns. See *Matrix*.

Arrow's impossibility theorem No social choice rule satisfies all six of the following conditions.

1. **Unrestricted domain** Any set of rankings is possible; if there are n candidates, then there are $n!$ possible rankings.

2. **Decisiveness** Given any set of individual rankings, the method produces a winner.

3. **Symmetry and transitive** The voting system should be symmetric and transitive over the set of all outcomes.

4. **Independence of irrelevant alternatives** If a voter prefers A to B with C as a possible choice, then the voter still prefers A to B when C is not a possible choice.

5. **Pareto principle** If each voter prefers A over B, then the group chooses A over B.
6. There should be **no dictator.**

Artificial intelligence A field of study devoted to computer simulation of human intelligence.

ASCII code A standard computer code used to facilitate the interchange of information among various types of computer equipment.

Assignment A computer term for setting the value of one variable to match the value of another.

Associative property A property of grouping that applies to certain operations (addition and multiplication, for example, but not to subtraction or division): If *a, b,* and *c* are real numbers, then

$$(a + b) + c = a + (b + c) \text{ and } (ab)c = a(bc)$$

Assuming the antecedent Same as *direct reasoning.*

Assuming the consequent A logical fallacy; same as the *fallacy of the converse.*

Augmented matrix A matrix that results after affixing an additional column to an existing matrix.

Average A single number that is used to typify or represent a set of numbers. In this book, it refers to the *mean, median,* or *mode.*

Average daily balance method A method of calculating credit card interest using the formula $I = Prt$ in which P is the average daily balance owed for a current month, and t is the number of days in the month divided by 365.

Average rate of change The average rate of change of a function f from x to $x + h$ is

$$\frac{f(x + h) - f(x)}{h}$$

Axes The intersecting lines of a Cartesian coordinate system. The horizontal axis is called the *x*-axis, and the vertical axis is called the *y*-axis. The axes divide the plane into four parts called *quadrants.*

Axiom A statement that is accepted without proof.

Axis of a parabola The line through the focus of a parabola drawn perpendicular to the directrix.

Axis of symmetry A curve is symmetric with respect to a line, called the axis of symmetry, if for any point P on the curve, there is a point Q also on the curve such that the axis of symmetry is the perpendicular bisector of the line segment $\overline{PQ}$.

Balinski and Young's impossibility theorem Any apportionment plan that does not violate the quota rule must produce paradoxes. And any apportionment method that does not produce paradoxes must violate the quota rule.

Balloon payment A single, larger payment made at the end of the time period of an installment loan that is not amortized.

Bar graph See *Graph.*

Base (1) See *Exponent.* (2) In a percent problem, it is the whole quantity.

Base angles In an isosceles triangle, the angles formed by the base with each of the equal sides.

Base b numeration system A numeration system with b elements.

Base of an exponential In $y = b^x$, the *base* is b ($b \neq 1$).

Base of a triangle In an isosceles triangle, the side that is not the same length as the sides whose lengths are the same.

Because A logical operator for p because q, which is defined to mean

$$(p \wedge q) \wedge (q \to p)$$

Bell-shaped curve See *Normal curve.*

Belong to a set To be an element of a set.

Bernoulli trail A Bernoulli trial is an experiment that meets four conditions:

1. There must be a fixed number of trials. Denote this number by n.
2. There must be two possible mutually exclusive outcomes for each trial. Call them *success* and *failure.*
3. Each trial must be independent. That is, the outcome of a particular trial is not affected by the outcome of any other trial.
4. The probability of success and failure must remain constant for each trial.

Biconditional A logical operator for simple statements p and q which is true when p and q have the same truth values, and is false when p and q have different truth values.

Billion A name for $10^9 = 1,000,000,000$.

Bimodal A distribution with two modes.

Binary numeration system A numeration system with two numerals, 0 and 1.

Binary voting Selecting a winner by pairing two competitors and then pairing up the winners to select the final victor.

Binomial A polynomial with exactly two terms.

Binomial distribution theorem Let X be a random variable for the number of successes in n independent and identical repetitions of an experiment with two possible outcomes, success and failure. If p is the probability of success, then

$$P(X = k) = \binom{n}{k} p^k (1 - p)^{n-k}$$

where $k = 0, 1, \ldots, n$.

Binomial experiment A *binomial experiment* is an experiment that meets four conditions:

1. There must be a fixed number of trials. Denote this number by n.
2. There must be two possible mutually exclusive outcomes for each trial. Call them *success* and *failure.*
3. Each trial must be independent. That is, the outcome of a particular trial is not affected by the outcome of any other trial.
4. The probability of success and failure must remain constant for each trial.

Binomial random variable   A random variable associated with an experiment having exactly two outcomes.

Binomial theorem   For any positive integer n,

$$(a + b)^n = \sum_{k=0}^{n} \binom{n}{k} a^{n-k} b^k$$

Birthday problem   The probability that two unrelated people will have their birthdays on the same day (not counting the year of birth).

Bisect   To divide into two equal or congruent parts.

Bit   Binary digit, the smallest unit of data storage with a value of either 0 or 1, thereby representing whether a circuit is open or closed.

Bond   An interest-bearing certificate issued by a government or business, promising to pay the holder a specified amount (usually $1,000) on a certain date.

Boot   To start a computer. To do this the computer must have access to an operating system on a floppy disk or hard disk.

Borda count   A common way of determining a winner when there is no majority is to assign a point value to each voter's ranking. If there are n candidates, then n points are assigned to the first choice for each voter, with $n - 1$ points for the next choice, and so on. The points for each candidate are added and if one has more votes, that candidate is declared the winner.

Boundary   See *Half-plane.*

Box plot   A rectangular box positioned above a numerical scale associated with a given set of data. It shows the maximum value, the minimum value, and the quartiles for the data.

Braces   See *Grouping symbols.*

Brackets   See *Grouping symbols.*

Bug   An error in the design or makeup of a computer program (software bug) or a hardware component of the system (hardware bug).

Bulletin board   An online means of communicating with others on a particular topic.

Byte   The fundamental block of data that can be processed by a computer. In most microcomputers, a byte is a group of eight adjacent bits and the rough equivalent of one alphanumeric symbol.

Calculus   The field of mathematics that deals with differentiation and integration of functions, and related concepts and applications.

Canceling   The process of reducing a fraction by dividing the same number into both the numerator and the denominator.

Canonical form   When a given number is written as a product of prime factors in ascending order, it is said to be in canonical form.

Capacity   A measurement for the amount of liquid a container holds.

Cardinal number   A number that designates the manyness of a set; the number of units, but not the order in which they are arranged.

Cardinality   The number of elements in a set.

Cards   A deck of 52 matching objects that are identical on one side and on the other side are divided into four suits (hearts, diamonds, spades, and clubs). The objects, called cards, are labeled A, 2, . . . , 9, 10, J, Q, and K in each suit.

Cartesian coordinate system   Two intersecting lines, called *axes,* used to locate points in a plane called a *Cartesian plane.* If the intersecting lines are perpendicular, the system is called a *rectangular coordinate system.*

Cartesian plane   See *Cartesian coordinate system.*

CD-ROM   A form of mass storage. It is a cheap read-only device, which means that you can only use the data stored on it when it was created, but it can store a massive amount of material, such as an entire encyclopedia.

Cell   A specific location on a spreadsheet. It is designated using a letter (column heading) followed by a numeral (row heading). A cell can contain a letter, word, sentence, number, or formula.

Celsius   A metric measurement for temperature for which the freezing point of water is 0° and the boiling point of water is 100°.

Center of a circle   See *Circle.*

Center of an ellipse   The midpoint of the line segment connecting the foci of the ellipse.

Center of a hyperbola   The midpoint of the line segment connecting the foci of the hyperbola.

Centi-   A prefix that means 1/100.

Centigram   One hundredth of a gram.

Centiliter   One hundredth of a liter.

Centimeter   One hundredth of a meter.

Change of base theorem

$$\log_a x = \frac{\log_b x}{\log_b a}$$

Ciphertext   A secret or coded message.

Circle   The set of points in a plane that are a given distance from a given point. The given point is called the *center,* and the given distance is called the *radius.* The diameter is twice the radius. The *unit circle* is the circle with center at $(0, 0)$ and $r = 1$.

Circle graph   See *Graph.*

Circuit   A complete path of electrical current including a power source and a switch. It may include an indicator light to show when the circuit is complete.

Circular definition   A definition that relies on the use of the word being defined, or other words that rely on the word being defined.

Circumference   The distance around a circle. The formula for finding the circumference is $C = \pi D$ or $C = 2\pi r$.

Classes   One of the groupings when organizing data. The difference between the lower limit of one class and the lower limit of the next class is called the *interval* of the class. The number of values within a class is called the *frequency.*

Closed   See *Closure property.*

Closed curve A curve that has no endpoints.

Closed-ended loan An installment loan.

Closed half-plane See *Half-plane.*

Closed network A network that connects each point.

Closed set A set that satisfies the closure property for some operation.

Closing The process of settlement on a real estate loan.

Closing costs Costs paid at the closing of a real estate loan.

Closure property A set S is *closed* for an operation $\circ$ if $a \circ b$ is an element of S for all elements a and b in S. This property is called the *closure property.*

Coefficient Any factor of a term is said to be the coefficient of the remaining factors. The *numerical coefficient* is the numerical part of the term, usually written before the variable part. In $3x$, it is the number 3, in $9x^2y^3$ it is the number 9. Generally, the word *coefficient* is taken to be the numerical coefficient of the variable factors.

Column A vertical arrangement of numbers or entries of a matrix. It is denoted by letters $A, B, C, \ldots$ on a spreadsheet.

Combination It is a selection of objects from a given set without regard to the order in which they are selected. Sometimes it refers to the number of ways this selection can be done and is denoted by ${}_nC_r$ or $\binom{n}{r}$ and is pronounced "n choose r." The formula for finding it is

$$\binom{n}{r} = \frac{n!}{r!(n-r)!}$$

Common denominator For two or more fractions, a common multiple of the denominators.

Common difference The difference between successive terms of an arithmetic sequence.

Common factor A factor that two or more terms of a polynomial have in common.

Common fraction Fractions written in the form of one integer divided by a whole number are common fractions. For example, 1/10 is common fraction representation and 0.1 is the decimal representation of the same number.

Common logarithm A logarithm to the base 10; written $\log N$.

Common ratio The ratio between successive terms of a geometric sequence.

Communication matrix A square matrix in which the entries symbolize the occurrence of some facet or event with a 1 and the nonoccurrence with a 0.

Communications package A program that allows one computer to communicate with another computer.

Commutative group A group that also satisfies the property that

$$a \circ b = b \circ a$$

for some operation $\circ$ and elements a and b in the set.

Commutative property A property of order that applies to certain operations (addition and multiplication, for example, but not to subtraction and division). If a and b are real numbers, then $a + b = b + a$ and $ab = ba$.

Comparison property For any two numbers x and y, exactly one of the following is true: (1) $x = y$; x is equal to y (the same as) (2) $x > y$; x is greater than y (bigger than) (3) $x < y$; x is less than y (smaller than). This is sometimes known as the *trichotomy property.*

Comparison rate for home loans A formula for comparing terms of a home loan. The formula is

$$\text{APR} + 0.125\left(\text{POINTS} + \frac{\text{ORIGINATION FEE}}{\text{AMOUNT OF LOAN}}\right)$$

Compass An instrument for scribing circles or for measuring distances between two points.

Compiler An operating system program that converts an entire program written in a higher-level language into machine language before the program is executed.

Complement (1) Two numbers less than 1 are called complements if their sum is 1. (2) The complement of a set is everything not in the set relative to a given universe.

Complementary angles Two angles are complementary if the sum of their measures is 90°.

Complementary probabilities

$$P(E) = 1 - P(\overline{E})$$

Two probabilities are *complementary* if

$$P(E) + P(\overline{E}) = 1$$

Completely factored An expression is completely factored if it is a product and there are no common factors and no difference of squares—that is, if no further factoring is possible.

Complex decimal A form that mixes decimal and fractional form, such as $0.12\frac{1}{2}$.

Complex fraction A rational expression a/b where a or b (or both) have fractional form.

Components See *Ordered pair.*

Composite number Sometimes simply referred to as a *composite;* it is a positive integer that has more than two divisors.

Compound interest A method of calculating interest by adding the interest to the principal at the end of the compounding period so that this sum is used in the interest calculation for the next period.

Compound interest formula $A = P(1 + i)^N$, where A = future value; P = present value (or principal); r = annual interest rate (APR); t = number of years; n = number of times compounded per year; $i = \frac{r}{n}$; and $N = nt$.

Compound statement A statement formed by combining simple statements with one or more operators.

Compounding The process of adding interest to the principal so that in the next time period the interest is calculated on this sum.

Computer A device which, under the direction of a program, can process data, alter its own program instructions, and perform computations and logical operations without human intervention.

Computer abuse A misuse of a computer.

Computer program A set of step-by-step directions that instruct a computer how to carry out a certain task.

Conclusion The statement that follows (or is to be proved to follow) as a consequence of the hypothesis of the theorem. Also called a *logical conclusion*.

Conditional The statement "if *p*, then *q*," symbolized by $p \rightarrow q$. The statement *p* is called the *antecedent* and *q* is called the *consequent*.

Conditional equation See *Equation*.

Conditional inequality See *Inequality*.

Conditional probability A probability that is found on the condition that a certain event has occurred. The notation $P(E \mid F)$ is the probability of event *E* on the condition that event *F* has occurred.

Condorcet candidate A candidate who wins all the one-to-one matchups.

Condorcet criterion If a candidate is favored when compared one on one with every other candidate, then that candidate should be declared the winner.

Condorcet's paradox There are three citizens, *A*, *B*, and *C*, and each citizen ranks three different policies as shown:

$$A: x > y > z \quad B: y > z > x \quad C: z > x > y$$

Then two citizens prefer *x* to *y*, two prefer *y* to *z*, and two prefer *z* to *x*, which is not transitive. Each voter is consistent, but the social choice is inconsistent. This is known as Condorcet's paradox.

Cone A solid with a circle for its base and a curved surface tapering evenly to an apex so that any point on this surface is in a straight line between the circumference of the base and the apex.

Congruent Of the same size and shape; if one is placed on top of the other, the two figures will coincide exactly in all their parts.

Congruent angles Two angles that have the same measure.

Congruent modulo *m* Two real numbers *a* and *b* are congruent modulo *m*, written $a \equiv b \pmod{m}$, if *a* and *b* differ by a multiple of *m*.

Congruent triangles Two triangles that have the same size and shape.

Conic sections The set of points that results from the intersection of two three-dimensional objects (a cone and a plane) is a two-dimensional curve known as a conic section. The conic sections include the parabola, ellipse (special case, circle), and hyperbola.

Conjecture A guess or prediction based on incomplete or uncertain evidence.

Conjugate axis The line passing through the center perpendicular to the transverse axis of a hyperbola.

Conjunction The conjunction of two simple statements *p* and *q* is true whenever both *p* and *q* are true, and is false otherwise. The common translation of conjunction is "and."

Connected network A network that connects each point.

Connective A rule that operates on one or two simple statements. Some examples of connectives are *and, or, not, unless, ...*

Consecutive integers Integers that differ by 1.

Consequent See *Conditional*.

Consistent system If a system of equations has at least one solution, it is consistent; otherwise it is said to be *inconsistent*.

Constant Symbol with exactly one possible value.

Constant function A function of the form $f(x) = c$.

Constant multiple It is the following property of integrals:

$$\int af(x)\,dx = a \int f(x)\,dx$$

Constraint A limitation placed on an objective function. See *Linear programming*.

Construct The process of drawing a figure that will satisfy certain given conditions.

Contained in a set An element is contained in a set if it is a member of the set.

Continuous compounding If we let the number of compounding periods in a year increase without limit, the result is called continuous compounding. The formula is

$$A = Pe^{rt}$$

for a present value of *P*, future value *A* at a rate of *r* for *t* years.

Continuous distribution A probability distribution that includes all *x*-values (as opposed to a discrete distribution, which allows a finite number of *x*-values).

Contradiction An open equation for which the solution set is empty.

Contrapositive For the implication $p \rightarrow q$, the contrapositive is $\sim q \rightarrow \sim p$.

Converge To draw near to. A series is said to converge when the sum of the first *n* terms approaches a limit as *n* increases without bound. We say that the sequence converges to a limit *L* if the values of the successive terms of the sequence get closer and closer to the number *L* as $n \rightarrow \infty$.

Converse For the implication $p \rightarrow q$, the converse is $q \rightarrow p$.

Convex set A set that contains the line segment joining any two of its points.

Coordinate plane See *Cartesian coordinate system*.

Coordinates A numerical description for a point. Also see *Ordered pair*.

Correlation The interdependence between two sets of numbers. It is a relationship between two quantities, such that when one changes the other does (simultaneous increasing or decreasing is called *positive correlation;* and one increasing, the other decreasing, *negative correlation*).

Corresponding angles Angles in different triangles that are similarly related to the rest of the triangle.

Corresponding parts Points, angles, lines, etc., in different figures, similarly related to the rest of the figures.

Corresponding sides Sides of different triangles that are similarly related to the rest of the triangle.

Cosine In a right triangle *ABC* with right angle *C*,

$$\cos A = \frac{\text{LENGTH OF ADJACENT SIDE OF } A}{\text{LENGTH OF HYPOTENUSE}}$$

Countable set   A set with cardinality $\aleph_0$. That is, a set that can be placed into a one-to-one correspondence with the set of counting numbers.

Count-down property $n! = n(n - 1)!$

Counterclockwise   In the direction of rotation opposite to that in which the hands move around the dial of a clock.

Counterexample   An example that is used to disprove a proposition.

Counting numbers   See *Natural numbers*.

CPU   <u>C</u>entral <u>P</u>rocessing <u>U</u>nit, the primary section of the computer that contains the memory, logic, and arithmetic procedures necessary to process data and perform computations. The CPU also controls the functions performed by the input, output, and memory devices.

Credit card   A card signifying that the person or business issued the card has been approved for open-ended credit. It can be used at certain restaurants, airlines, and stores accepting that card.

Cryptography   The writing or deciphering of messages in code.

Cube   (1) A solid with six equal square sides. (2) In an expression such as x^3, which is pronounced "x cubed," it means xxx.

Cube root   See *Root of a number*.

Cubed   See *Cube*.

Cubic unit   A three-dimensional unit. It is the result of cubing a unit of measurement.

Cumulative frequency   A frequency that is the total number of cases having any given score or a score that is lower.

Cup   A unit of measurement in the United States measurement system that is equivalent to 8 fluid ounces.

Cursor   Indicator (often flashing) on a computer or calculator display to designate where the next character input will be placed.

Cylinder   Suppose we are given two parallel planes and two simple closed curves C_1 and C_2 in these planes for which lines joining corresponding points of C_1 and C_2 are parallel to a given line L. A cylinder is a closed surface consisting of two bases that are plane regions bounded by such curves C_1 and C_2 and a lateral surface that is the union of all line segments joining corresponding points of C_1 and C_2.

Daily compounding   In the compound interest formula, it is when $n = 365$ (exact interest) or when $n = 360$ (ordinary interest). In this book, use ordinary interest unless otherwise indicated.

Data processing   The recording and handling of information by means of mechanical or electronic equipment.

Database   A collection of information. A database manager is a program that is in charge of the information stored in a database.

Database manager   A computer program that allows a user to interface with a database.

Dealer's cost   The actual amount that a dealer pays for the goods sold.

Debug   The organized process of testing for, locating, and correcting errors within a program.

Decagon   A polygon having ten sides.

Decay formula   Refers to exponential decay. It is described by the equation

$$A = A_0 e^{rt}$$

where r is the annual decay rate (and consequently is negative), t is the time (in years), A_0 is the amount present initially (present value), and A is the future value. If r is positive, this formula models growth, and if r is negative, the formula models decay.

Deci-   A prefix that means 1/10.

Decibel   A unit of measurement for measuring sounds. It is defined as a ratio of the intensity of one sound, I, and another sound $I_0 \approx 10^{-16}$ watt/cm², the intensity of a barely audible sound for a person with normal hearing.

Deciles   Nine values that divide a data set into ten equal parts.

Decimal   Any number written in decimal notation. The digits represent powers of ten with whole numbers and fractions being separated by a period, called a *decimal point*. Sometimes called a Hindu–Arabic numeral.

Decimal fraction   A number in decimal notation that has fractional parts, such as 23.25. If a common fraction p/q is written as a decimal fraction, the result will either be a *terminating decimal* as with $\frac{1}{4} = 0.25$ or a *repeating decimal* as with $\frac{2}{3} = 0.6666 \ldots$.

Decimal notation   The representation of a number using the decimal number system. See *Decimal*.

Decimal numeration system   A numeration system with base 10.

Decimal point   See *Decimal*.

Decisiveness   Given any set of individual rankings, the method produces a winner when voting.

Decoding key   A key that allows one to unscramble a coded message.

Deductive reasoning   A formal structure based on a set of axioms and a set of undefined terms. New terms are defined in terms of the given undefined terms and new statements, or *theorems*, are derived from the axioms by proof.

Definite integral   Let f be a function defined over the interval $[a, b]$. Then the definite integral of f over the interval is denoted by

$$\int_a^b f(x)\, dx$$

and is the net change of an antiderivative of f over that interval. Thus, if $F(x)$ is an antiderivative of $f(x)$, then

$$\int_a^b f(x)\, dx = F(x)\, \Big|_a^b = F(b) - F(a)$$

Degree   (1) The degree of a term in one variable is the exponent of the variable, or it is the sum of the exponents of the variables if there are more than one. The degree of a polynomial is the degree of its highest-degree term. (2) A unit of measurement of an angle that is equal to 1/360 of a revolution.

Deka-   A prefix that means 10.

Deleted point   A single point that is excluded from the domain.

De Morgan's laws   For sets X and Y,

$$\overline{X \cup Y} = \overline{X} \cap \overline{Y} \quad \text{and} \quad \overline{X \cap Y} = \overline{X} \cup \overline{Y}$$

Demand   The number of items that can be sold at a given price.

Denominator   See *Rational number*.

Dense set   A set of numbers with the property that between any two points of the set, there exists another point in the set that is between the two given points.

Denying the antecedent   A logical fallacy; same as the *fallacy of the inverse*.

Denying the consequent   Same as *indirect reasoning*.

Dependent events   Two events are dependent if the occurrence of one influences the occurrence of the other.

Dependent system   If *every* ordered pair satisfying one equation in a system of equations also satisfies every other equation of the given system, then we describe the system as dependent.

Dependent variable   The variable associated with the second component of an ordered pair.

Derivative   One of the fundamental operations of calculus; it is the instantaneous rate of change of a function with respect to the variable. Formally, for a given function f, we define the derivative of f at x, denoted by $f'(x)$, to be

$$f'(x) = \lim_{h \to 0} \frac{f(x + h) - f(x)}{h}$$

provided this limit exists. If the limit exists, we say f is a differentiable function of x.

Description method   A method of defining a set by describing the set (as opposed to listing its elements).

Descriptive statistics   Statistics that is concerned with the accumulation of data, measures of central tendency, and dispersion.

Diagonal form   A matrix with the terms arranged on a diagonal, from upper left to lower right and zeros elsewhere.

Diameter   See *Circle*.

Dice   Plural for the word *die*, which is a small, marked cube used in games of chance.

Dictatorship   A selection process where one person alone makes a decision.

Die   See *Dice*.

Difference   The result of a subtraction.

Difference quotient   If f is a function, then the *difference quotient* is defined to be the function

$$\frac{f(x + h) - f(x)}{h}$$

Difference of squares   A mathematical expression in the form $a^2 - b^2$.

Differentiable function   See *Derivative*.

Differential calculus   That branch of calculus that deals with the derivative and applications of the derivative.

Dimension   A configuration having length only is said to be of one dimension; area and not volume, two dimensions; volume, three dimensions. In reference to matrices, it is the numbers of rows and columns.

Direct reasoning   One of the principal forms of logical reasoning. It is an argument of the form $[(p \to q) \wedge p] \to q$.

Directrix   See *Parabola*.

Discount   A reduction from a usual or list price.

Discrete mathematics   That part of mathematics that deals with sets of objects that can be counted or processes that consist of a sequence of individual steps.

Disjoint sets   Sets that have no elements in common.

Disjunction   The disjunction of two simple statements p and q is false whenever both p and q are false, and is true otherwise. The common translation of conjunction is "or."

Disk drive   A mechanical device that uses the rotating surface of a magnetic disk for the high-speed transfer and storage of data.

Distinguishable permutation   The number of distinguishable permutations of n objects in which n_1 are of one kind, n_2 are of another kind, . . . , and n_k are of a further kind, so that $n = n_1 + n_2 + \cdots + n_k$ is denoted by

$$\binom{n}{n_1, n_2, \ldots, n_k}$$

and is defined by the formula

$$\binom{n}{n_1, n_2, \ldots, n_k} = \frac{n!}{n_1! n_2! \cdot \cdots \cdot n_k!}$$

Distributive law of exponents

$$(1)\ (ab)^m = a^m b^m; \quad (2)\ \left(\frac{a}{b}\right)^m = \frac{a^m}{b^m}$$

Distributive property (for multiplication over addition)   If a, b, and c are real numbers, then $a(b + c) = ab + ac$ and $(a + b)c = ac + bc$ for the basic operations. That is, the number outside the parentheses indicating a sum or difference is distributed to each of the numbers inside the parentheses.

Diverge   A sequence that does not converge is said to diverge.

Dividend   The number or quantity to be divided. In a/b, the dividend is a.

Divides   See *Divisibility*.

Divine proportion   If two lengths h and w satisfy the proportion

$$\frac{h}{w} = \frac{w}{h + w}$$

then the lengths are said to be in a *divine proportion*.

Divisibility   If m and d are counting numbers, and if there is a counting number k so that $m = d \cdot k$, we say that d is a divisor of m, d is a factor of m, d divides m, and m is a multiple of d.

Division $a/b = x$ is $a \div b = x$ and means $a = bx$.

Division by zero   In the definition of division, $b \neq 0$, because if $b = 0$, then $bx = 0$, regardless of the value of x.

If $a \neq 0$, then there is no such number. On the other hand, if $a = 0$, then $0/0 = 1$ checks from the definition, and so also does $0/0 = 2$, which means that $1 = 2$, another contradiction. Thus, division by 0 is excluded.

Division of integers   The quotient of two integers is the quotient of the absolute values, and is positive if the given integers have the same sign, and negative if the given numbers have opposite signs. Furthermore, division by zero is not possible and division into 0 gives the answer 0.

Division of rational numbers
$$\frac{a}{b} \div \frac{c}{d} = \frac{ad}{bc} \quad (c \neq 0)$$

Division property (of equations)   The solution of an equation is unchanged by dividing both sides of the equation by the same nonzero number.

Division property of inequality   See *Multiplication property of inequality.*

Divisor   The quantity by which the dividend is to be divided. In a/b, b is the divisor.

Dodecagon   A polygon with 12 sides.

Domain   The *domain* of a variable is the set of replacements for the variable. The *domain* of a graph of an equation with two variables x and y is the set of permissible real-number replacements for x.

Double negative $-(-a) = a$

Double subscripts   Two subscripts on a variable as in a_{12}. (Do not read this as "twelve.")

Down payment   An amount paid at the time a product is financed. The purchase price minus the down payment is equal to the amount financed.

Download   The process of copying a program from the network to your computer.

Dummy variable   A variable in a mathematical expression whose only function is as a placeholder.

e   It is Euler's number and is defined by
$$e = \lim_{n \to \infty} \left(1 + \frac{1}{n}\right)^n$$

Eccentricity   For a conic section, it is defined as the ratio
$$\epsilon = \frac{c}{a}$$
For the ellipse, $0 \leq \epsilon < 1$, where ϵ measures the amount of roundness. If $\epsilon = 0$, then the conic is a circle. For the parabola, $\epsilon = 1$; and for the hyperbola, $\epsilon > 1$.

Edge   A line or a line segment that is the intersection of two plane faces of a geometric figure, or that is in the boundary of a plane figure.

Either . . . or   A logical operator for "either p or q," which is defined to mean
$$(p \lor q) \land \sim (q \land p)$$

Element   One of the individual objects that belong to a set.

Elementary operations   Refers to the operations of addition, subtraction, multiplication, and division.

Elementary row operations   There are four elementary row operations for producing equivalent matrices: (1) *RowSwap:*

Interchange any two rows. (2) *Row+:* Row addition—add a row to any other row. (3) **Row:* Scalar multiplication—multiply (or divide) all the elements of a row by the same nonzero real number. (4) **Row+:* Multiply all the entries of a row (*pivot row*) by a nonzero real number and add each resulting product to the corresponding entry of another specified row (*target row*).

Ellipse   The set of all points in a plane such that, for each point on the ellipse, the sum of its distances from two fixed points (called the **foci**) is a constant.

Elliptic geometry   A non-Euclidean geometry in which a Saccheri quadrilateral is constructed with summit angles obtuse.

e-mail   Electronic mail sent from one computer to another.

Empirical probability   A probability obtained empirically by experimentation.

Empty set   See *Set.*

Encoding key   A key that allows one to scramble, or encode a message.

Encrypt   To scramble a message so that it cannot be read by an unwanted person.

Equal angles   Two angles that have the same measure.

Equal matrices   Two matrices are equal if they are the same order (dimension) and also the corresponding elements are the same (equal).

Equal sets   Sets that contain the same elements.

Equal to   Two numbers are equal if they represent the same quantity or are identical. In mathematics, a relationship that satisfies the axioms of equality.

Equality, axioms of   For $a, b, c \in \mathbb{R}$,

Reflexive: $a = a$
Symmetric: If $a = b$, then $b = a$.
Transitive: If $a = b$ and $b = c$, then $a = c$.
Substitution: If $a = b$, then a may be replaced throughout by b (or b by a) in any statement without changing the truth or falsity of the statement.

Equally likely outcomes   Outcomes whose probabilities of occurring are the same.

Equation   A statement of equality. If always true, an equation is called an *identity;* if always false, it is called a *contradiction.* If it is sometimes true and sometimes false, it is called a *conditional equation.* Values that make an equation true are said to *satisfy* the equation and are called *solutions* or *roots* of the equation. Equations with the same solutions are called *equivalent equations.*

Equation of a graph   Every point on the graph has coordinates that satisfy the equation, and every ordered pair that satisfies the equation has coordinates that lie on the graph.

Equation properties   There are four equation properties: (1) *Addition property:* Adding the same number to both sides of an equation results in an equivalent equation. (2) *Subtraction property:* Subtracting the same number from both sides of an equation results in an equivalent equation. (3) *Multiplication property:* Multiplying both sides of a given equation by the same nonzero number results in an equivalent equation. (4) *Division property:* Dividing both sides of a given equation by the same nonzero number results in an equivalent equation.

Equilateral triangle A triangle whose three sides all have the same length.

Equilibrium point A point for which the supply and demand are equal.

Equivalent equations See *Equation.*

Equivalent matrices Matrices that represent equivalent systems.

Equivalent sets Sets that have the same cardinality.

Equivalent systems Systems that have the same solution set.

Estimate An approximation (usually mental) of size or value used to form an opinion.

Euclidean geometry The study of geometry based on the assumptions of Euclid. These basic assumptions are called Euclid's postulates.

Euclid's postulates 1. A straight line can be drawn from any point to any other point. 2. A straight line extends infinitely in either direction. 3. A circle can be described with any point as center and with a radius equal to any finite straight line drawn from the center. 4. All right angles are equal to each other. 5. Given a straight line and any point not on this line, there is one and only one line through that point that is parallel to the given line.

Euler circles The representation of sets using interlocking circles.

Euler circuit Begin at some vertex of a graph, travel on each edge exactly once, and return to the starting vertex. The path that is a trace of the tip is called an *Euler circuit.*

Euler's circuit theorem Every vertex on a graph with an Euler circuit has an even degree, and conversely, if in a connected graph every vertex has an even degree, then the graph has an Euler circuit.

Euler's number It is the number *e*.

Evaluate To *evaluate* an expression means to replace the variables by given numerical values and then simplify the resulting numerical expression. To *evaluate* a trigonometric ratio means to find its approximate numerical value. To *evaluate* a summation means to find its value.

Even vertex In a network, a vertex with even degree—that is, with an even number of arcs or line segments connected at that vertex.

Event A subset of a sample space.

Exact interest The calculation of interest assuming that there are 365 days in a year.

Exact solution The simplified value of a logarithmic expression before approximation by calculator.

Exclusive or A translation of *p or q* which includes *p* or *q*, but not both. In this book we translate the *exclusive or* as "either *p* or *q*."

Expand To simplify by carrying out the given operations.

Expand a summation To write out a summation notation showing the individual terms without a sigma.

Expanded notation A way of writing a number that lists the meaning of each grouping symbol and the number of items in that group. For example, 382.5 written in expanded notation is

$$3 \times 10^2 + 8 \times 10^1 + 2 \times 10^0 + 5 \times 10^{-1}$$

Expectation See *Mathematical expectation.*

Expected value See *Mathematical expectation.*

Experiment An observation of any physical occurrence.

Exponent Where *b* is any nonzero real number and *n* is any natural number, exponent is defined as follows:

$$b^n = \underbrace{b \cdot b \cdot \cdots \cdot b}_{n\ \text{factors}} \quad b^0 = 1 \quad b^{-n} = \frac{1}{b^n}$$

b is called the *base,* *n* is called the *exponent,* and b^n is called a *power* or *exponential.*

Exponential See *Exponent.*

Exponential curve The graph of an exponential equation. It indicates an increasingly steep rise, and passes through the point (0, 1).

Exponential equation An equation of the form $y = b^x$, where *b* is positive and not equal to 1.

Exponential function A function that can be written as $f(x) = b^x$, where $b > 0$, $b \neq 1$.

Exponential notation A notation involving exponents.

Exponentiation The process of raising a number to some power. See *Exponent.*

Expression Numbers, variables, functions, and their arguments that can be evaluated to obtain a single result.

Extended order of operations 1. First, perform any operations enclosed in parentheses. 2. Next, perform any operations that involve raising to a power. 3. Perform multiplications and divisions as they occur by working from left to right. 4. Finally, perform additions and subtractions as they occur by working from left to right.

Exterior angle An exterior angle of a triangle is the angle on the other side of an extension on one side of the triangle.

Extraneous root A number obtained in the process of solving an equation that is not a root of the equation to be solved.

Extremes See *Proportion.*

Factor (*noun*) Each of the numbers multiplied to form a product is called a factor of the product. (*verb*) To write a given number as a product.

Factor tree The representation of a composite number showing the steps of successive factoring by writing each new pair of factors under the composite.

Factorial For a natural number *n*, the product of all the positive integers less than or equal to *n*. It is denoted by *n*! and is defined by

$$n! = n(n - 1)(n - 2) \cdot \cdots \cdot 4 \cdot 3 \cdot 2 \cdot 1$$

Also, 0! = 1.

Factoring The process of determining the factors of a product.

Factorization The result of factoring a number or an expression.

Fahrenheit A unit of measurement in the United States system for measuring temperature based on a system where the freezing point of water is 32° and the boiling point of water is 212°.

Fair coin A coin for which heads and tails are equally likely.

Fair game A game for which the mathematical expectation is zero.

Fair voing principles *See* Fairness criteria.

Fairness criteria Properties that would seem to be desirable in any voting system.

> *Majority criterion:* If a candidate receives a majority of the first-place votes, then that candidate should be declared the winner.
> *Condorcet criterion:* If a candidate is favored when compared one on one with every other candidate, then that candidate should be declared the winner.
> *Monotonicity criterion:* A candidate who wins a first election and then gains additional support, without losing any of the original support, should also win a second election.
> *Irrelevant alternatives criterion:* If a candidate is declared the winner of an election, and in a second election one or more of the other candidates is removed, then the previous winner should still be declared the winner.

Fallacy An invalid form of reasoning.

Fallacy of exceptions Reasoning or forming a conclusion by looking at one particular case, which may be an exception.

Fallacy of the converse An invalid form of reasoning that has the form $[(p \rightarrow q) \wedge q]$ and reaches the incorrect conclusion p.

Fallacy of the inverse An invalid form of reasoning that has the form $[(p \rightarrow q) \wedge (\sim p)]$ and reaches the incorrect conclusion $\sim q$.

False chain pattern An invalid form of reasoning that has the form $[(p \rightarrow q) \wedge (p \rightarrow r)]$ and reaches the incorrect conclusion $q \rightarrow r$.

Feasible solution A set of values that satisfies the set of constraints in a linear programming problem.

Fibonacci sequence The sequence 1, 1, 2, 3, 5, 8, 13, 21, ... The general term is $s_n = s_{n-1} + s_{n-2}$, for any given s_1 and s_2.

Fibonacci-type sequence A sequence with general term $s_n = s_{n-1} + s_{n-2}$, for any given s_1 and s_2. The Fibonacci sequence has first terms 1, 1, . . . , but *a* Fibonacci-type sequence can have any two first terms.

Field A set with two operations satisfying the closure, commutative, associative, identity, and inverse properties for both operations. A field also satisfies a distributive property combining both operations.

Finance charge A charge made for the use of someone else's money.

Finite series A series with n terms, where n is a counting number.

Finite set *See Set.*

First component *See Ordered pair.*

First-degree equation With one variable, an equation of the form $ax + b = 0$; with two variables, an equation of the form $y = mx + b$.

Five-percent offer An offer made that is 105% of the price paid by the dealer. That is, it is an offer that is 5% over the cost.

Fixed-point form The usual decimal representation of a number. It is usually used in the context of writing numbers in scientific notation or in floating-point form. See *Floating–point form.*

Floating-point form It is a calculator or computer variation of scientific notation in which a number is written as a number between one and ten times a power of ten where the power of ten is understood. For example, 2.678×10^{11} is scientific notation and 2.67800000000E11 or 2.67800000000 +11 are floating-point representations. The fixed-point representation is the usual decimal representation of 267,800,000,000.

Floor-plan problem Given a floor plan of some building, you wish to find a path from room to room that will proceed through all of the rooms exactly once.

Floppy disk Storage medium that is a flexible platter ($3\frac{1}{2}$ or $5\frac{1}{4}$ inches in diameter) of mylar plastic coated with a magnetic material. Data are represented on the disk by electrical impulses.

Foci Plural for *focus.*

Focus See *Parabola, Ellipse,* and *Hyperbola.*

FOIL (1) A method for multiplying binomials that requires First terms, Outer terms + Inner terms, Last terms:

$$(a + b)(c + d) = ac + (ad + bc) + bd$$

(2) A method for factoring a trinomial into the product of two binomials.

Foot A unit of linear measure in the United States system that is equal to 12 inches.

Foreclose If the scheduled payments are not made, the lender takes the right to redeem the mortgage and keeps the collateral property.

Formula A general answer, rule, or principle stated in mathematical notation.

Fractal A family of shapes involving chance whose irregularities are statistical in nature. They are shapes used, for example, to model coastlines, growth, and boundaries of clouds. Fractals model curves as well as surfaces. The term *fractal set* is also used in place of the world *fractal.*

Fractal geometry The branch of geometry that studies the properties of fractals.

Fraction See *Rational number.*

Frequency See *Classes.*

Frequency distribution For a collection of data, the tabulation of the number of elements in each class.

Function A rule that assigns to each element in the domain a single (unique) element.

Function machine A device used to help us understand the nature of functions. It is the representation of a function as a machine in which some number is input and then after being "processed" through the machine, outputs a single value.

Functional notation The representation of a function f using the notation $f(x)$.

Fundamental counting principle If one task can be performed in m ways and a second task can be performed in n ways, then the number of ways that the tasks can be performed one after the other is mn.

Fundamental operators In symbolic logic, the fundamental operators are the connectives *and, or,* and *not*.

Fundamental property of equations If P and Q are algebraic expressions, and k is a real number, then each of the following is equivalent to $P = Q$:

Addition	$P + k = Q + k$
Subtraction	$P - k = Q - k$
Nonzero multiplication	$kP = kQ, k \neq 0$
Nonzero division	$\dfrac{P}{k} = \dfrac{Q}{k}, k \neq 0$

Fundamental property of fractions If both the numerator and denominator are multiplied by the same nonzero number, the resulting fraction will be the same. That is,

$$\frac{PK}{QK} = \frac{P}{Q} \ (Q, K \neq 0).$$

Fundamental property of inequalities If P and Q are algebraic expressions, and k is a real number, then each of the following is equivalent to $P < Q$:

Addition	$P + k < Q + k$
Subtraction	$P - k < Q - k$
Positive multiplication	$kP < kQ, k > 0$
Positive division	$\dfrac{P}{k} < \dfrac{Q}{k}, k > 0$
Negative multiplication	$kP > kQ, k < 0$
Negative division	$\dfrac{P}{k} > \dfrac{Q}{k}, k < 0$

This property also applies for $\leq$, $>$, and $\geq$.

Fundamental theorem of arithmetic Every counting number greater than 1 is either a prime or a product of primes, and the prime factorization is unique (except for the order in which the factors appear).

Future value See *Compound interest formula.*

Future value formula For simple interest: $A = P(1 + rt)$; for compound interest: $A = P(1 + i)^N$.

Fuzzy logic A relatively new branch of logic used in computer programming that does not use the law of the excluded middle.

Gallon A measure of capacity in the United States system that is equal to 4 quarts.

Gates In circuit logic, it is a symbolic representation of a particular circuit.

Gauss–Jordan elimination A method for solving a system of equations that uses the following steps. *Step 1:* Select as the first pivot the element in the first row, first column, and pivot. *Step 2:* The next pivot is the element in the second row, second column; pivot. *Step 3:* Repeat the process until you arrive at the last row, or until the pivot element is a zero. If it is a zero and you can interchange that row with a row below it, so that the pivot element is no longer a zero, do so and continue. If it is zero and you cannot interchange rows so that it is not a zero, continue with the next row. The final matrix is called the **row-reduced form.**

g.c.f. An abbreviation for *greatest common factor.*

General form In relation to second-degree equations (or conic sections) it refers to the form

$$Ax^2 + Bxy + Cy^2 + Dx + Ey + F = 0$$

where A, B, C, D, E, and F are real numbers and (x, y) is any point on the curve.

General term The nth term of a sequence or series.

Genus The number of cuts that can be made without cutting a figure into two pieces. The genus is equivalent to the number of holes in the object.

Geometric mean The geometric mean of the numbers a and b is $\sqrt{ab}$.

Geometric sequence A sequence for which the ratio of each term to the preceding term is a constant, written $g_1, g_2, g_3, \ldots$. The nth term of a geometric sequence is $g_n = g_1 r^{n-1}$, where g_1 is the first term and r is the *common ratio*. It is also called a *geometric progression*.

Geometric series The indicated sum of the terms of a geometric sequence. The sum of n terms is denoted by G_n and

$$G_n = \frac{g_1(1 - r^n)}{1 - r}, \quad r \neq 1$$

If $|r| < 1$, then $G = \dfrac{g_1}{1 - r}$, where G is the sum of the infinite geometric series. If $|r| \geq 1$, the infinite geometric series has no sum.

Geometry The branch of mathematics that treats the shape and size of things. Technically, it is the study of invariant properties of given elements under specified groups of transformations.

GIGO Garbage In, Garbage Out, an old axiom regarding the use of computers.

Golden ratio The division of a line segment AB by an interior point P so that

$$\frac{AB}{AP} = \frac{AP}{PB}$$

It follows that this ratio is a root of the equation $x^2 - x - 1 = 0$, or $x = \frac{1}{2}(1 + \sqrt{5})$. This ratio is called the golden ratio and is said to be considered pleasing to the eye.

Golden rectangle A rectangle R with the property that it can be divided into a square and a rectangle similar to R; a rectangle whose sides form a golden ratio.

Googol The number with 1 followed by 100 zeros—that is, 10,000,000,000,000,000,000,000,000,000,000,-000,000,000,000,000,000,000,000,000,000,000,-000,000,000,000,000,000,000,000,000

Grace period   A period of time between when an item is purchased and when it is paid during which no interest is charged.

Gram   A unit of weight in the metric system. It is equal to the weight of one cubic centimeter of water at 4°C.

Grant's tomb properties   Two fundamental properties of logarithms:

1. $\log_b b^x = x$

2. $b^{\log_b x} = x,\quad x > 0$

Graph   (1) In statistics, it is a drawing that shows the relation between certain sets of numbers. Common forms are bar graphs, line graphs, pictographs, and pie charts (circle graphs). (2) A drawing that shows the relation between certain sets of numbers. It may be one-dimensional ($\mathbb{R}$), two-dimensional ($\mathbb{R}^2$), or three-dimensional ($\mathbb{R}^3$). (3) A set of vertices connected by arcs or line segments.

Graph of an equation   See *Equation of a graph.*

Graphing method   A method of solving a system of equations that finds the solution by looking at the intersection of the individual graphs. It is an approximate method of solving a system of equations, and depends on the accuracy of the graph that is drawn.

Great circle   A circle on a sphere that has its diameter equal to that of the sphere.

Greater than   If a lies to the right of b on a number line, then a is greater than b, $a > b$. Formally, $a > b$ if and only if $a - b$ is positive.

Greater than or equal to   Written $a \geq b$, means $a > b$ or $a = b$.

Greatest common factor   The largest divisor common to a given set of numbers.

Group   A set with one defined operation that satisfies the closure, associative, identity, and inverse properties.

Grouped frequency distribution   If the data are grouped before they are tallied, then the resulting distribution is called a *grouped frequency distribution.*

Grouping symbols   Parentheses ( ), brackets [ ], and braces { } indicate the order of operations and are also sometimes used to indicate multiplication, as in (2)(3) = 6. Also called *symbols of inclusion.*

Growth formula   Refers to exponential growth. It is described by the equation

$$A = A_0 e^{rt}$$

where r is the annual growth rate (and consequently is positive), t is the time (in years), A_0 is the amount present initially (present value), and A is the future value. If r is positive, this formula models growth, and if r is negative, the formula models decay.

Half-life   The time that it takes for a particular radioactive substance to decay to half of its original amount.

Half-line   A ray, with or without its endpoint. The half-line is said to be closed if it includes the endpoint, and open if it does not include the endpoint.

Half-plane   The part of a plane that lies on one side of a line in the plane. It is a *closed* half-plane if the line is included. It is an *open* half-plane if the line is not included. The line is the *boundary* of the half-plane in either case.

Hamiltonian cycle   A path that begins at some vertex and then visits each vertex exactly once, ending up at the original vertex.

Hamilton's apportionment plan   An apportionment plan in which the representation of a geographical area is determined by finding the quotient of the number of people in that area divided by the total number of people and then the result is rounded as follows: Allocate the remainder, one at a time, to districts on the basis of the decreasing order of the decimal portion of the quotients. This method is sometimes called the *method of the largest fractions.*

Hare method   Each voter votes for one candidate. If a candidate receives a majority of the votes, that candidate is declared to be a *first-round winner*. If no candidate receives a majority of the votes, then with the *Hare method*, also known as the *plurality with an elimination runoff method*, then the candidate(s) with the fewest number of first place votes is (are) eliminated. Each voter votes for one candidate in the second round. If a candidate receives a majority, that candidate is declared to be a *second-round winner*. If no candidate receives a majority of the second-round votes, then the candidate(s) with the fewest number of votes is(are) eliminated. Repeat this process until a candidate receives a majority.

Hecto-   A prefix meaning 100.

Heptagon   A polygon having seven sides.

Hexagon   A polygon having six sides.

HH method   See *Huntington–Hill's plan.*

Higher-level language   A computer programming language (e.g., BASIC, PASCAL, LOGO) that approaches the syntax of English and is easier both to use and to learn than machine language. It is also not system-dependent.

Hindu–Arabic numerals   Same as the usual decimal numeration system that is in everyday use.

Horizontal ellipse   An ellipse whose major axis is horizontal.

Horizontal hyperbola   A hyperbola whose transverse axis is horizontal.

Horizontal line   A line with zero slope. Its equation has the form $y = $ constant.

Hundred   Ten 10s.

Huntington–Hill's plan   An apportionment plan currently in use by the U.S. legislature. It is based on the geometric mean of two numbers, a and b, where a is the value of the exact ratio rounded down and b is the value of the exact ratio rounded up. It rounds down if the exact quota is less than the geometric mean and rounds up if it is greater than the geometric mean.

Hyperbola   The set of all points in a plane such that, for each point on the hyperbola, the difference of its distances from two fixed points (called the **foci**) is a constant.

Hyperbolic geometry   A non-Euclidean geometry in which a Saccheri quadrilateral is constructed with summit angles acute.

Hypotenuse   The longest side in a right triangle.

Hypothesis   An assumed proposition used as a premise in proving something else.

Identity   (1) A statement of equality that is true for all values of the variable. It also refers to a number I so that for some operation $\circ$, $I \circ a = a \circ I = a$ for every number a in a given set. (2) An open equation that is true for all replacements of the variable.

Identity matrix   A matrix satisfying the identity property. It is a square matrix consisting of ones along the main diagonal and zeros elsewhere.

If-then   In symbolic logic, it is a connective also called *implication*.

Implication   A statement that follows from other statements. It is also a proposition formed from two given propositions by connecting them with an "if . . . , then . . ." form. It is symbolized by $p \rightarrow q$.

Impossible event   An event for which the probability is zero—that is, an event that cannot happen.

Improper fraction   A fraction for which the numerator is greater than the denominator.

Improper subset   See *Subset*.

Inch   A linear measurement in the United States system equal in length to the following segment:

––––––––––––––

Inclusive or   This is the same as *disjunction*. The compound statement "*p* or *q*" is called the *inclusive or*.

Inconsistent system   A system for which no replacements of the variable make the equations true simultaneously.

Indefinite integral   An antiderivative.

Independence of irrelevant alternatives   If a voter prefers A to B with C as a possible choice, we assume that the voter still prefers A to B when C is not a possible choice.

Independent events   Events E and F are *independent* if the occurrence of one in no way affects the occurrence of the other.

Independent system   A system of equations such that no one of them is necessarily satisfied by a set of values of the variables that satisfy all the others.

Independent variable   The variable associated with the first component of an ordered pair.

Indirect reasoning   One of the principal forms of logical reasoning. It is an argument of the form
$[(p \rightarrow q) \wedge \sim q] \rightarrow \sim p$.

Inductive reasoning   A type of reasoning accomplished by first observing patterns and then predicting answers for more complicated similar problems.

Inequality   A statement of order. If always true, an inequality is called an *absolute inequality;* if always false, an inequality is called a *contradiction*. If sometimes true and sometimes false, it is called a *conditional inequality*. Values that

make the statement true are said to satisfy the inequality. A *string of inequalities* may be used to show the order of three or more quantities.

Inequality symbols   The symbols $>$, $\geq$, $<$, and $\leq$. Also called *order symbols*.

Inferential statistics   Statistics that is concerned with making generalizations or predictions about a population based on a sample from that population.

Infinite series   The indicated sum of an infinite sequence.

Infinite set   See *Set*.

Infinity symbol ∞

Inflation   An increase in the amount of money in circulation, resulting in a fall in its value and a rise in prices. In this book, we assume annual compounding with the future value formula; that is, use $A = P(1 + r)^n$, where r is the projected annual inflation rate, n is the number of years, and P is the present value.

Information retrieval   The locating and displaying of specific material from a description of its content.

Input   A method of putting information into a computer. *Input* includes downloading a program, typing on a keyboard, pressing on a pressure-sensitive screen.

Input device   Component of a system that allows the entry of data on a program into a computer's memory.

Installment loan   A financial problem in which an item is paid for over a period of time. It is calculated using add-on interest or compound interest.

Installments   Part of a debt paid at regular intervals over a period of time.

Instant Insanity   A puzzle game consisting of four blocks with different colors on the faces. The object of the game is to arrange the four blocks in a row so that no color is repeated on one side as the four blocks are rotated through $360°$.

Instantaneous rate of change   The instantaneous rate of change of a function f from x to $x + h$ is

$$\lim_{h \to 0} \frac{f(x + h) - f(x)}{h}$$

Integers $\mathbb{Z} = \{\ldots, -3, -2, -1, 0, 1, 2, 3, \ldots\}$, composed of the natural numbers, their opposites, and 0.

Integral   A fundamental concept of calculus, which involves finding the area bounded by a curve, the x-axis, and two vertical lines. Let f be a function defined over the interval $[a, b]$. Then the definite integral of f over the interval is denoted by

$$\int_a^b f(x)\, dx$$

and is the net change of an antiderivative of f over that interval. The numbers a and b are called the *limits of integration*.

Integral calculus   That branch of calculus that involves applications of the integral.

Integral of a sum   It is the following property of integrals:

$$\int [f(x) + g(x)]\, dx = \int f(x)\, dx + \int g(x)\, dx$$

Integrand   In an integral, it is the function to be integrated.

Integrated circuit   The plastic or ceramic body that contains a chip and the leads connecting it to other components.

Interactive   Software that allows continuous two-way communication between the user and the program.

Intercept form of the equation of a line   The form

$$\frac{x}{a} + \frac{y}{b} = 1$$

of a linear equation where the x-intercept is a and the y-intercept is b.

Intercepts   The point or points where a line or a curve crosses a coordinate axis. The x-intercepts are sometimes called the *zeros* of the equation.

Interest   An amount of money paid for the use of another's money. See *Compound interest.*

Interest-only loan   A loan in which periodic payments are for interest only so that the principal amount of the loan remains the same.

Interest rate   The percentage rate paid on financial problems. In this book it is denoted by r and is assumed to be an annual rate unless otherwise stated.

Interface   The electronics necessary for a computer to communicate with a peripheral.

Internet   A network of computers from all over the world that are connected together. It can be accessed through institutions or servers such as America Online, Compuserve, or Prodigy.

Intersection   The *intersection* of sets A and B, denoted by $A \cap B$, is the set consisting of elements in *both* A and B.

Interval   See *Classes.*

Invalid argument   An argument that is not valid.

Inverse   (1) In symbolic logic, for an implication $p \to q$, the inverse is the statement $\sim p \to \sim q$. (2) For addition, see *Opposites.* For multiplication, see *Reciprocal.* (3) For matrices, if $[A]$ is a square matrix, and if there exists a matrix $[A]^{-1}$ such that

$$[A]^{-1}[A] = [A][A]^{-1} = [I]$$

where $[I]$ is the identity matrix for multiplication, then $[A]^{-1}$ is called the inverse of $[A]$ for multiplication.

Inverse cosine   See *Inverse trigonometric ratios.*

Inverse property   For each $a \in \mathbb{R}$, there is a unique number $(-a) \in \mathbb{R}$, called the *opposite* (or *additive inverse*) of a, so that

$$a + (-a) = -a + a = 0$$

Inverse sine   See *Inverse trigonometric ratios.*

Inverse tangent   See *Inverse trigonometric ratios.*

Inverse trigonometric ratios   The inverse sine, inverse cosine, and inverse tangent are the *inverse trigonometric ratios.* For θ an acute angle in a right triangle,

$$\sin^{-1}\left(\frac{\text{OPP}}{\text{HYP}}\right) = \theta; \qquad \cos^{-1}\left(\frac{\text{ADJ}}{\text{HYP}}\right) = \theta$$

$$\tan^{-1}\left(\frac{\text{OPP}}{\text{ADJ}}\right) = \theta$$

Invert   In relation to the fraction a/b, it means to interchange the numerator and the denominator to obtain the fraction b/a.

Irrational number   A number that can be expressed as a nonrepeating, nonterminating decimal; the set of irrational numbers is denoted by $\mathbb{Q}'$.

Irrelevant alternatives criterion   If a candidate is declared the winner of an election, and in a second election one or more of the other candidates is removed, then the previous winner should still be declared the winner.

Isosceles triangle   A triangle with two sides the same length.

Isosceles triangle property   If two sides of a triangle have the same length, then angles opposite them are equal.

Jefferson's apportionment plan   An apportionment plan in which the representation of a geographical area is determined by finding the quotient of the number of people in that area divided by the total number of people and then the result is rounded as follows: Any quotient with a decimal portion must be rounded down to the previous whole number.

Jordan curve   Also called a *simple closed curve.* For example, a curve such as a circle or an ellipse or a rectangle that is closed and does not intersect itself.

Juxtaposition   When two variables, a number and a variable, or a symbol and a parenthesis are written next to each other with no operation symbol, as in xy, $2x$, or $3(x + y)$. Juxtaposition is used to indicate multiplication.

K   The symbol represents 1,024 (or 2^{10}). For example, 48K bytes of memory is the same as $48 \times 1,024$ or 49,152 bytes. It is sometimes used as an approximation for 1,000.

Keno   A lottery game that consists of a player trying to guess in advance which numbers will be selected from a pot containing n numbers. A certain number, say, m, where $m < n$, of selections is randomly made from the pot of n numbers. The player then gets paid according to how many of the m numbers were selected.

Keyboard   Typewriter-like device that allows the user to input data and commands into a computer.

Kilo-   A prefix that means 1,000.

Kilogram   1,000 grams.

Kiloliter   1,000 liters.

Kilometer   1,000 meters.

Kruskal's algorithm   To construct a minimum spanning tree from a weighted graph: 1. Select any edge with minimum weight. 2. Select the next edge with minimum weight among those not yet selected. 3. Continue to choose edges of minimum weight from those not yet selected, but make sure not to select any edge that forms a circuit. 4. Repeat this process until the tree connects all of the vertices of the original graph.

Laptop   A small portable computer.

Law of contraposition   A conditional may always be replaced by its contrapositive without having its truth value affected.

Law of detachment   Same as *direct reasoning*.

Law of double negation $\sim(\sim p) \Leftrightarrow p$

Law of the excluded middle   Every simple statement is either true or false.

Laws of exponents   There are 5 laws of exponents.

Addition law	$b^m \cdot b^n = b^{m+n}$
Multiplication law	$(b^n)^m = b^{mn}$
Subtraction law	$\dfrac{b^m}{b^n} = b^{m-n}$
Distributive laws	$(ab)^m = a^m b^m$
	$\left(\dfrac{a}{b}\right)^m = \dfrac{a^m}{b^m}$

Laws of logarithms   If A, B, and b are positive numbers, p is any real number, and $b \neq 1$, then

Addition law	$\log_b(AB) = \log_b A + \log_b B$
Subtraction law	$\log_b \dfrac{A}{B} = \log_b A - \log_b B$
Multiplication law	$\log_b A^p = p \log_b A$

Law of square roots   There are 4 laws of square roots.

(1) $\sqrt{0} = 0$ (2) $\sqrt{a^2} = a$

(3) $\sqrt{ab} = \sqrt{a}\sqrt{b}$ (4) $\sqrt{\dfrac{a}{b}} = \dfrac{\sqrt{a}}{\sqrt{b}}$

l.c.d.   An abbreviation for least common denominator.

l.c.m.   An abbreviation for least common multiple.

Least common denominator (l.c.d.)   The smallest number that is exactly divisible by each of the given numbers.

Least common multiple (l.c.m.)   The smallest number that each of a given set of numbers divides into.

Least-squares line   A line $y = mx + b$ so that the sum of the squares of the vertical distances of the data points from this line will be as small as possible.

Least-squares method   A method based on the principle that the best prediction of a quantity that can be deduced from a set of measurements or observations is that for which the sum of the squares of the deviations of the observed values (from predictions) is a minimum.

Leg of a triangle   One of the two sides of a right triangle that are not the hypotenuse.

Length   A measurement of an object from end to end.

Less than   If a is to the left of b on a number line, then a is less than b, $a < b$. Formally, $a < b$ if and only if $b > a$.

Less than or equal to   Written $a \leq b$, means $a < b$ or $a = b$.

Like terms   Terms that differ only in their numerical coefficients. Also called *similar terms*.

Limit   The formal definition of a limit is beyond the scope of this course. Intuitively, it is the tendency of a function to approach some value as its variable approaches a given value.

Limit of a sequence   The formal definition of a limit of a sequence is beyond the scope of this course. Intuitively, it is an accumulation point such that there are an infinite number of terms of the sequence arbitrarily close to the accumulation point.

Limits of integration   See *Integral*.

Line   In mathematics, it is an undefined term. It is a curve that is straight, so it is sometimes referred to as a *straight line*. It extends in both directions, is considered one-dimensional, so it has no thickness.

Line graph   See *Graph*.

Line of credit   A preapproved credit limit on a credit account. The maximum amount of credit to be extended to a borrower. That is, it is a promise by a lender to extend credit up to some predetermined amount.

Line segment   A part of a line between two points on the line.

Line of symmetry   A line with the property that for a given curve, any point P on the curve has a corresponding point Q (called the reflection point of P) so that the perpendicular bisector of $\overline{PQ}$ is on the line of symmetry.

Linear   (1) A first-degree polynomial. (2) Pertaining to a line. In two variables, a set of points satisfying the equation $Ax + By + C = 0$.

Linear combination method   See *Addition method*.

Linear correlation coefficient   A measure to determine whether there is a statistically significant linear relationship between two variables.

Linear equation   An equation of the form $ax + b = 0$ (one variable) or $Ax + By + C = 0$ (two variables). A first-degree equation with one or two variables. For example, $x + 5 = 0$ and $x + y + 5 = 0$ are linear. An equation is linear in a certain variable if it is first-degree in that variable. For example, $x + y^2 = 0$ is linear in x, but not y.

Linear function   A function whose equation can be written in the form $f(x) = mx + b$.

Linear inequality   A first-degree inequality with one or two variables.

Linear polynomial   A first-degree polynomial.

Linear programming   A type of problem that seeks to maximize or minimize a function called the *objective function* subject to a set of *restrictions* (linear inequalities) called *constraints*.

Linear programming theorem   A linear expression in two variables, $c_1 x + c_2 y$, defined over a convex set S whose sides are line segments, takes on its maximum value at a corner point of S and its minimum value at a corner point of S. If S is unbounded, there may or may not be an optimum value, but if there is, then it must occur at a corner point.

Linear system   A system of equations, each of which is first-degree.

Liter   The basic unit of capacity in the metric system. It is the capacity of 1 cubic decimeter.

Literal equation   An equation with more than one variable.

Logarithm   For $A > 0$, $b > 0$, $b \neq 1$,

$$x = \log_b A \quad \text{means} \quad b^x = A$$

x is the called the logarithm and A is called the argument.

Logarithmic equation   An equation for which there is a logarithm on one or both sides.

Logarithmic function

$$f(x) = \log_b x, \quad b > 0, \quad x > 0$$

Logarithmic scale   A scale in which logarithms are used to make data more manageable by expanding small variations and compressing large ones.

Logic   The science of correct reasoning.

Logical conclusion   The statement that follows logically as a consequence of the hypotheses of a theorem.

Logical equivalence   Two statements are *logically equivalent* if they have the same truth values.

Logical fallacy   An invalid form of reasoning.

Log of both sides theorem   If A, B, and b are positive real numbers with $b \neq 1$, then

$$\log_b A = \log_b B \quad \text{is equivalent to} \quad A = B$$

Lower quota   In apportionment, the result of a quota found by rounding down.

Lowest common denominator   For two or more fractions, the smallest common multiple of the denominators. It is the same as the *lowest common multiple*.

Lump-sum problem   A financial problem that deals with a single sum of money, called a *lump sum*. Contrast with a *periodic payment* problem.

Main diagonal   The entries $a_{11}, a_{22}, a_{33}, \ldots$ in a matrix.

Major axis   In an ellipse, the line passing though the foci.

Majority   Voting to find an alternative that receives more than 50% of the vote.

Majority criterion   If a candidate receives a majority of the first-place votes, then that candidate should be declared the winner.

Marriage   A pairing of one couple, each partner taken from a separate group. A marriage is *stable* if each partner is satisfied with the pairing and *unstable* if one or the other (or both) would prefer to be paired with another.

Mass   In this course, it is the amount of matter an object comprises. Formally, it is a measure of the tendency of a body to oppose changes in its velocity.

Mathematical expectation   A calculation defined as the product of an amount to be won and the probability that it is won. If there is more than one amount to be won, it is the sum of the expectations of all the prizes. It is also called the *expected value* or *expectation*.

Mathematical modeling   An iterative procedure that makes assumptions about real-world problems to formulate the problem in mathematical terms. After the mathematical problem is solved, it is tested for accuracy in the real world, and revised for the next step in the iterative process.

Mathematical system   A set with at least one defined operation and some developed properties.

Matrix   A rectangular array of terms called *elements*.

Matrix equation   An equation whose elements are matrices.

Maximum loan   In this book, it refers to the maximum amount of loan that can be obtained for a home with a given amount of income and a given amount of debt. To find this amount, use the present value of an annuity formula.

Mean   The number found by adding the data and dividing by the number of values in the data set. The sample mean is usually denoted by $\bar{x}$.

Means   See *Proportion*.

Measure   Comparison to some unit recognized as standard.

Measures of central tendency   Refers to the averages of mean, median, and mode.

Measures of dispersion   Refers to the measures of range, standard deviation, and variance.

Measures of position   Measures that divide a data set by position, which include median, quartiles, deciles, and percentiles.

Median   The middle number when the numbers in the data are arranged in order of size. If there are two middle numbers (in the case of an even number of data values), the median is the mean of these two middle numbers.

Member   See *Set*.

Meter   The basic unit for measuring length in the metric system.

Metric system   A decimal system of weights and measures in which the gram, the meter, and the liter are the basic units of mass, length, and capacity, respectively. One gram is the mass of one cm^3 of water and one liter is the same as 1,000 cm^3. In this book, the metric system refers to the SI metric system as revised in 1960.

Micrometer   One millionth of a meter, denoted by μm.

Mile   A unit of linear measurement in the United States system that is equal to 5,280 ft.

Milli-   A prefix that means 1/1,000.

Milligram   1/1,000 of a gram.

Milliliter   1/1,000 of a liter.

Millimeter   1/1,000 of a meter.

Million   A name for $10^6 = 1,000,000$.

Minicomputer   An everyday name for a personal computer.

Minimum spanning tree   A spanning tree for which the sum of the numbers with the edges is a minimum.

Minor axis   In an ellipse, the axis perpendicular to the major axis passing though the center of the ellipse.

Minus   Refers to the operation of subtraction. The symbol "$-$" means minus only when it appears between two numbers, two variables, or between numbers and variables.

Mixed number   A number that has both a counting number part and a proper fraction part; for example, $3\frac{1}{2}$.

Mode   The value in a data set that occurs most frequently. If no number occurs more than once, there is no mode. It is possible to have more than one mode.

Modem   A device connected to a computer that allows the computer to communicate with other computers using electric cables or phone lines.

Modified quotient   Adjust the standard divisor so that the desired number of seats is used. This adjusted number is known as the modified quotient.

Modular codes   A code based on modular arithmetic.

Modulo 5 A mathematical system consisting of five elements having the property that every number is equivalent to one of these five elements if they have the same remainder when divided by 5.

Modulo n A mathematical system consisting of n elements having the property that every number is equivalent to one of these n elements if they have the same remainder when divided by n.

Modus ponens Same as *direct reasoning.*

Modus tollens Same as *indirect reasoning.*

Monitor An output device for communicating with a computer. It is similar to a television screen.

Monomial A polynomial with one and only one term.

Monotonicity criterion A candidate who wins a first election and then gains additional support, without losing any of the original support, should also win a second election.

Monthly compounding In the compound interest formula, it is when $n = 12$.

Monthly payment In an installment application, it is a periodic payment that is made once every month.

Mortgage An agreement, or loan contract, in which a borrower pledges a home or other real estate as security.

Mouse A small plastic "box" usually with two buttons on top and a ball on the bottom so that it can be rolled around on a pad. It is attached to the computer by a long cord and is used to take over some of the keyboard functions.

Multiple See *Divisibility.*

Multiplication For $b \neq 0$, $a \times b$ means

$$\underbrace{b + b + b + \cdots + b}_{a \text{ addends}}$$

If $a = 0$, then $0 \times b = 0$.

Multiplication law of equality If $a = b$, then $ac = bc$. Also called the *multiplication property of equality* or a *fundamental property of equations.*

Multiplication law of exponents To raise a power to a power, multiply the exponents. That is, $(b^n)^m = b^{mn}$.

Multiplication law of inequality
1. If $a > b$ and $c > 0$, then $ac > bc$.
2. If $a > b$ and $c < 0$, then $ac < bc$.

Multiplication law of logarithms The log of the pth power of a number is p times the log of that number. In symbols,

$$\log_b A^p = p \log_b A$$

Multiplication of integers If the integers to be multiplied both have the same sign, the result is positive and the magnitude of the answer is the product of the absolute values of the integers. If the integers to be multiplied have opposite signs, the product is negative and has magnitude equal to the product of the absolute values of the given integers. Finally, if one or both of the given integers is 0, the product is 0.

Multiplication of matrices Let [M] be an $m \times r$ matrix and [N] an $r \times n$ matrix. The product matrix [M][N] = [P] is an $m \times n$ matrix. The entry in the ith row and jth column of [M][N] is the sum of the products formed by multiplying each entry of the ith row of [M] by the corresponding element in the jth column of [N].

Multiplication of rational numbers

$$\frac{a}{b} \times \frac{c}{d} = \frac{ac}{bd}$$

Multiplication principle In a numeration system, multiplication of the value of a symbol by some number. Also, see *Fundamental counting principle.*

Multiplication property (of equations) Both sides of an equation may be multiplied or divided by any nonzero number to obtain an equivalent equation.

Multiplication property of factorials

$$n! = n(n - 1)!$$

Multiplication property of inequality Both sides of an inequality may be multiplied or divided by a positive number, and the order of the inequality will remain unchanged. The order is reversed if both sides are multiplied or divided by a negative number. That is, if $a < b$ then $ac < bc$ if $c > 0$ and $ac > bc$ if $c < 0$. This also applies to $\leq$, $>$, and $\geq$.

Multiplication property of probability If events E and F are independent events, then we can find the probability of an intersection as follows:

$$P(E \cap F) = P(E \text{ and } F) = P(E) \cdot P(F)$$

Multiplicative identity The number 1, with the property that $1 \cdot a = a$ for any real number a.

Multiplicative inverse (1) See *Reciprocal.* (2) If [A] is a square matrix and if there exists a matrix $[A]^{-1}$ such that

$$[A]^{-1}[A] = [A][A]^{-1} = [I]$$

where [I] is the identity matrix for multiplication, then $[A]^{-1}$ is called the inverse of [A] for multiplication.

Multiplicity If a root for an equation appears more than once, it is called a *root of multiplicity.* For example,

$$(x - 1)(x - 1)(x - 1)(x - 2)(x - 2)(x - 3) = 0$$

has roots 1, 2, and 3. The root 1 has multiplicity three and root 2 has multiplicity two.

Mutually exclusive Events are *mutually exclusive* if their intersection is empty.

Natural base The natural base is e; it refers to an exponential with a base e.

Natural logarithm A logarithm to the base e, written $\ln N$.

Natural numbers $\mathbb{N} = \{1, 2, 3, 4, 5, \ldots\}$, the positive integers, also called the *counting numbers.*

Negation A logical connective that changes the truth value of a given statement. The negation of p is symbolized by $\sim p$.

Negative of a conditional The negative of a conditional, $p \rightarrow q$ is found by

$$\sim(p \rightarrow q) \Leftrightarrow p \wedge \sim q$$

Negative number A number less than zero.

Negative sign   The symbol "−" when used in front of a number, as in −5. Do not confuse with the same symbol used for subtraction.

Neither . . . nor   A logical operator for "neither p nor q," which is defined to mean $\sim(p \lor q)$.

Network   (1) A linking together of computers. (2) A set of points connected by an arc or by line segments.

New states paradox   When a reapportionment of an increased number of seats causes a shift in the apportionment of the existing states, it is known as the *new states paradox*.

***n*-gon**   A polygon with n sides.

No *p* is *q*   A logical operator for "no p is q," which is defined to mean $p \rightarrow \sim q$.

Nonagon   A polygon with 9 sides.

Nonconformable matrices   Matrices that cannot be added or multiplied because their dimensions are not compatible.

Non-Euclidean geometry   A geometry that results when Euclid's fifth postulate is not accepted.

Nonrepeating decimal   A decimal representation of a number that does not repeat.

Nonsingular matrix   A matrix that has an inverse.

Nonterminating decimal   A decimal representation of a number that does not terminate.

Normal curve   A graphical representation of a normal distribution. Its high point occurs at the mean, it is symmetric with respect to this mean, and each side of the mean has an area that includes 34.1% of the population within one standard deviation, 13.6% from one to two standard deviations, and about 2.3% of the population more than two standard deviations from the mean.

Not   A common translation for the connective of negation.

NOT-gate   A logical gate that changes the truth value of a given statement.

Null set   See *Set.*

Number   A *number* represents a given quantity, as opposed to a *numeral,* which is the symbol for the number. In mathematics, it generally refers to a specific set of numbers—for example, counting numbers, whole numbers, integers, rationals, or real numbers. If the set is not specified, the assumed usage is to the set of real numbers.

Number line   A line used to display a set of numbers graphically (the axis for a one-dimensional graph).

Numeral   Symbol used to denote a number.

Numeration system   A system of symbols with rules of combination for representing all numbers.

Numerator   See *Rational number.*

Numerical coefficient   See *Coefficient.*

Objective function   The function to be maximized or minimized in a linear programming problem.

Obtuse angle   An angle that is greater than a right angle and smaller than a straight angle.

Obtuse triangle   A triangle with one obtuse angle.

Octagon   A polygon with 8 sides.

Octal numeration system   A numeration system with 8 symbols.

Odd vertex   In a network, a vertex of odd degree—that is, with an odd number of arcs or line segments connected at that vertex.

Odds   If $s + f = n$, where s is the number of outcomes considered favorable to an event E and n is the total number of possibilities, then the *odds in favor* of E is s/f and the *odds against* E is f/s.

One   The first counting number; it is also called the *identity element for multiplication;* that is, it satisfies the property that

$$x \cdot 1 = 1 \cdot x = x$$

for all numbers x.

One-dimensional coordinate system   A real number line.

One-to-one correspondence   Between two sets A and B, this means each element of A can be matched with exactly one element of B and also each element of B can be matched with exactly one element of A.

Online   To be connected to a computer network or to the Internet.

Open-ended loan   A preapproved line of credit that the borrower can access as long as timely payments are made and the credit line is not exceeded. It is usually known as a credit card loan.

Open half-plane   See *Half-plane.*

Open equation   An equation that has at least one variable.

Operator   A rule, such as negation, that modifies the value of a simple statement, or a rule that combines two simple statements, such as conjunction or disjunction.

Opposite side   In a right triangle, an acute angle is made up of two sides. The opposite side of the angle refers to the third side that is not used to make up the sides of the angle.

Opposites   Opposites x and $-x$ are the same distance from 0 on the number line but in opposite directions; $-x$ is also called the *additive inverse* of x. Do not confuse the symbol "−" meaning opposite with the same symbol as used to mean subtraction or negative.

Optimum solution   The maximum or minimum value in a linear programming problem.

Or   A common translation for the connective of disjunction.

OR-gate   An electrical circuit that simulates disjunction. That is, the circuit is on when either of two switches is on.

Order   Refers to the direction that an inequality symbol points. In reference to matrices, it refers to the number of rows and columns in a matrix. When used in relation to a matrix, it is the same as the *dimension* of the matrix.

Order of an inequality   Refers to a $>$, $\geq$, $<$, or $\leq$ relationship.

Order of operations   If no grouping symbols are used in a numerical expression, first perform all multiplications and divisions from left to right, and then perform all additions and subtractions from left to right.

Order symbols   Refers to $>$, $\geq$, $<$, $\leq$ in an inequality. Also called inequality symbols.

Ordered pair   A pair of numbers, written (x, y), in which the order of naming is important. The numbers x and y are sometimes called the *first* and *second components* of the pair and are called the *coordinates* of the point designated by (x, y).

Ordered triple   Three numbers, written (x, y, z), in which the order of the components is important.

Ordinary annuity   See *Annuity*.

Ordinary interest   The calculation of interest assuming a year has 360 days. In this book, we assume ordinary interest unless otherwise stated.

Ordinate   The vertical coordinate in a two-dimensional system of rectangular coordinates, usually denoted by y.

Origin   The point designating 0 on a number line. In two dimensions, the point of intersection of the coordinate axes; the coordinates are $(0, 0)$.

Origination fee   A fee paid to obtain a real estate loan.

Ounce   (1) A unit of capacity in the United States system that is equal to 1/128 of a gallon. (2) A unit of mass in the United States system that is equal to 1/16 of a pound.

Output   A method of getting information out of a computer.

Output device   Component of a system that allows the output of data. The most common output device is a printer.

Overlapping sets   Sets whose intersection is not empty.

Pairwise comparision method   In the *pairwise comparison method* of voting, the voters rank the candidates. The method consists of a series of comparisons in which each candidate is compared to each of the other candidates. If choice A is preferred to choice B, then A receives 1 point. If B is preferred to A, then B receives 1 point. If the candidates tie, each receives $\frac{1}{2}$ point. The candidate with the most points is the winner.

Parabola   A set of points in the plane equidistant from a given point (called the *focus*) and a given line (called the *directrix*). It is the path of a projectile. The *axis of symmetry* is the axis of the parabola. The point where the axis cuts the parabola is the *vertex*.

Parallel circuit   Two switches connected together so that if either of the two switches is turned on, the circuit is on.

Parallel lines   Two nonintersecting straight lines in the same plane.

Parallelepiped   A polyhedron, all of whose faces are parallelograms.

Parallelogram   A quadrilateral with its opposite sides parallel.

Parentheses   See *Grouping symbols*.

Pareto principle   If each voter prefers A over B, then the group chooses A over B.

Partial sum   If $s_1, s_2, s_3, \ldots$ is a sequence, then the partial sums are $S_1 = s_1$, $S_2 = s_1 + s_2$, $S_3 = s_1 + s_2 + s_3, \ldots$.

Pascal's triangle   A triangular array of numbers that is bordered by ones and the sum of two adjacent numbers in one row is equal to the number in the next row between the two numbers.

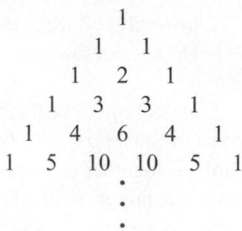

Password   A word or set of symbols that allows access to a computer account.

Pattern recognition   A computer function that entails the automatic identification and classification of shapes, forms, or relationships.

Pearson correlation coefficient   A number between -1 and $+1$ that indicates the degree of linear relationship between two sets of numbers. In the text we call this the *linear correlation coefficient*.

Pentagon   A polygon with 5 sides.

Percent   The ratio of a given number to 100; hundredths; denoted by %; that is, 5% means 5/100.

Percent markdown   The percent of an original price used to find the amount of discount.

Percent problem A is $P\%$ of W is formulated as a proportion:

$$\frac{P}{100} = \frac{A}{W}$$

Percentage   The given amount in a percent problem.

Percentile   Ninety-nine values that divide a data set into one hundred equal parts.

Perfect number   An integer that is equal to the sum of all of its factors except the number itself. For example, 28 is a perfect number since $28 = 1 + 2 + 4 + 7 + 14$.

Perfect square $1^2 = 1, 2^2 = 4, 3^2 = 9, \ldots$, so the perfect squares are $1, 4, 9, 16, 25, 36, 49, \ldots$.

Perimeter   The distance around a polygon.

Periodic payment problem   A financial problem that involves monthly or other periodic payments.

Peripheral   A device, such as a printer, that is connected to and operated by a computer.

Permutation   A selection of objects from a given set with regard to the order in which they are selected. Sometimes it refers to the number of ways this selection can be done and is denoted by $_nP_r$. The formula for finding it is

$$_nP_r = \frac{n!}{(n - r)!}$$

Perpendicular lines   Two lines are perpendicular if they meet at right angles.

Personal computer   A computer kept for and used by an individual.

Pi (π)   A number that is defined at the ratio of the circumference to the diameter of a circle. It cannot be represented exactly as a decimal, but it is a number between 3.1415 and 3.1416.

Pictograph   See *Graph*.

Pie chart   See *Graph*.

Pirating   Stealing software by copying it illegally for the use of someone other than the person who paid for it.

Pivot   A process that uses elementary row operations to carry out the following steps: 1. Divide all entries in the row in which the pivot appears (called the *pivot row*) by the nonzero pivot element so that the pivot entry becomes a 1. This uses elementary row operation 3. 2. Obtain zeros above and below the pivot element by using elementary row operation 4.

Pivot row   In an elementary row operation, it is the row that is multiplied by a constant. See *Elementary row operations*.

Pivoting   See *Pivot*.

Pixel   Any of the thousands (or millions) of tiny dots that make up a computer or calculator image.

Place-value names   Trillions, hundred billions, ten billions, billions, hundred millions, ten millions, millions, hundred thousands, ten thousands, thousands, hundreds, tens, units, tenths, hundredths, thousandths, ten-thousandths, hundred-thousandths, and millionths (from large to small).

Planar curve   A curve completely contained in a plane.

Plane   In mathematics, it is an undefined term. It is flat and level and extends infinitely in horizontal and vertical directions. It is considered two-dimensional.

Plot a point   To mark the position of a point.

Plurality rule   The winner of an election is the candidate with the highest number of votes.

Point   (1) In the decimal representation of a number, it is a mark that divides the whole number part of a number from its fractional part. (2) In relation to a home loan, it represents 1% of the value of a loan, so that 3 points would be a fee paid to a lender equal to 3% of the amount of the loan. (3) In geometry, it is an undefined word that signifies a position, but that has no dimension or size.

Point–slope form   An algebraic form of an equation of a line that is given in terms of a point (x_1, y_1) and slope m of a given line: $y - y_1 = m(x - x_1)$.

Police patrol problem   Suppose a police car needs to patrol a gated community, and would like to enter the gate, cruise all the streets exactly once, and then leave by the same gate.

Polygon   A geometric figure that has three or more straight sides that all lie in a plane so that the starting point and the ending point are the same.

Polynomial   An algebraic expression that may be written as a sum (or difference) of terms. Each *term* of a polynomial contains multiplication only.

Population   The total set of items (actual or potential) defined by some characteristic of the items.

Population growth   The population, P, at some future time can be predicted if you know the population P_0 at some time, and the annual growth rate, r. The predicted population t years after the given time is $P = P_0 e^{rt}$.

Population paradox   When there is a fixed number of seats, a reapportionment may cause a state to lose a seat to another state, even though the percent increase in the population of the state that loses the seat is larger than the percent increase of the state that wins the seat. When this occurs, it is known as the *population paradox*.

Positional system   A numeration system in which the position of a symbol in the representation of a number determines the meaning of that symbol.

Positive number   A number greater than 0.

Positive sign   The symbol "+" when used in front of a number or an expression.

Positive square root   The symbol $\sqrt{x}$ is the positive number that, when multiplied by itself, gives the number x. The symbol "$\sqrt{}$" is always positive.

Postulate   A statement that is accepted without proof.

Pound   A unit of measurement for mass in the United States system. It is equal to 16 oz.

Power   See *Exponent*.

Precision   The accuracy of the measurement; for example, a measurement is taken to the nearest inch, nearest foot, or nearest mile. It is not to be confused with accuracy that applies to the calculation.

Predecessor   In a sequence, the predecessor of an element a_n is the preceding element, a_{n-1}.

Premise   A previous statement or assertion that serves as the basis for an argument.

Present value   See *Compound interest formula*.

Present value formula   The present value, P, of a known future value A invested at an annual interest rate of r for t years compounded n times per year is found by the formula

$$P = A\left(1 + \frac{r}{n}\right)^{-nt}$$

Present value of an annuity   A financial formula that seeks the present value from periodic payments over a period of time. The formula is

$$P = m\left[\frac{1 - \left(1 + \frac{r}{n}\right)^{-nt}}{\frac{r}{n}}\right]$$

Previous balance method   A method of calculating credit card interest using the formula $I = Prt$ in which P is the balance owed before the current payment is subtracted.

Prime factorization   The factorization of a number so that all of the factors are primes and so that their product is equal to the given number.

Prime number $P = \{2, 3, 5, 7, 11, 13, 17, 19, 23, \ldots\}$; a number with exactly two factors: 1 and the number itself.

Principal   See *Compound interest formula*.

Printer   An output device for a computer.

Prism   In this book, it refers to a right prism, which is also called a parallelepiped or more commonly a box.

Probabilistic model   A model that deals with situations that are random in character and attempts to predict the outcomes of events with a certain stated or known degree of accuracy.

Probability   If an experiment can result in any of n ($n \geq 1$) mutually exclusive and equally likely outcomes, and if s of these are considered favorable to event E, then $P(E) = s/n$.

Probability function   A function P that satisfies the following properties: $0 \leq P(E) \leq 1$, $P(S) = 1$, and if E and F are mutually exclusive events, then $P(E \cup F) = P(E) + P(F)$.

Problem-solving procedure   1. *Read the problem.* Note what it is all about. Focus on processes rather than numbers. You can't work a problem you don't understand. 2. *Restate the problem.* Write a verbal description of the problem using operation signs and an equal sign. Look for equality. If you can't find equal quantities, you will never formulate an equation. 3. *Choose a variable.* If there is a single unknown, choose a variable. 4. *Substitute.* Replace the verbal phrases by known numbers and by the variable. 5. *Solve the equation.* This is the easy step. Be sure your answer makes sense by checking it with the original question in the problem. Use estimation to eliminate unreasonable answers. 6. *State the answer.* There were no variables defined when you started, so $x = 3$ is not an answer. Pay attention to units of measure and other details of the problem. Remember to answer the question that was asked.

Product   The result of a multiplication.

Profit formula $P = S - C$, where P represents the profit, S represents the selling price (or revenue), and C the cost (or overhead).

Program   A set of step-by-step instructions that instruct a computer what to do in a specified situation.

Progression   See *Sequence.*

Projective geometry   The study of those properties of geometric configurations that are invariant under projection. It was developed to satisfy the need for depth in works of art.

Prompt   In a computer program, a prompt is a direction that causes the program to print some message to help the user understand what is happening at a particular time.

Proper divisor   A divisor of a number that is less than the number itself.

Proper fraction   A fraction for which the numerator is less than the denominator.

Proper subset   See *Subset.*

Proof   A logical argument that establishes the truth of a statement.

Property of complements   See *Complementary probabilities.*

Property of proportions   If the product of the means equals the product of the extremes, then the ratios form a proportion. Also, if the ratios form a proportion, then the product of the means equals the product of the extremes.

Property of rational expressions   Let P, Q, R, S, and K be any polynomials such that all values of the variable that cause division by zero are excluded from the domain.

Equality $\dfrac{P}{Q} = \dfrac{R}{S}$ if and only if $PS = QR$.

Fundamental property $\dfrac{PK}{QK} = \dfrac{P}{Q}$

Addition $\dfrac{P}{Q} + \dfrac{R}{S} = \dfrac{PS + QR}{QS}$

Subtraction $\dfrac{P}{Q} - \dfrac{R}{S} = \dfrac{PS - QR}{QS}$

Multiplication $\dfrac{P}{Q} \cdot \dfrac{R}{S} = \dfrac{PR}{QS}$

Division $\dfrac{P}{Q} \div \dfrac{R}{S} = \dfrac{PS}{QR}$

Property of zero $AB = 0$ if and only if $A = 0$ or $B = 0$ (or both). Also called the *zero-product rule.*

Proportion   A statement of equality between two ratios. For example,

$$\frac{a}{b} = \frac{c}{d}$$

For this proportion, a and d are called the *extremes; b* and c are called the *means.*

Protractor   A device used to measure angles.

Pseudosphere   The surface of revolution of a tractrix about its asymptote. It is sometimes called a "four-dimensional sphere."

Pyramid   A solid figure having a polygon as a base, the sides of which form the bases of triangular surfaces meeting at a common vertex.

Pythagorean theorem   If a triangle with legs a and b and hypotenuse c is a right triangle, then $a^2 + b^2 = c^2$.

Quadrant   See *Axes.*

Quadratic   A second-degree polynomial.

Quadratic equation   An equation of the form $ax^2 + bx + c = 0$, $a \neq 0$.

Quadratic formula   If $ax^2 + bx + c = 0$ and $a \neq 0$, then $x = \dfrac{-b \pm \sqrt{b^2 - 4ac}}{2a}$.
The radicand $b^2 - 4ac$ is called the *discriminant* of the quadratic.

Quadratic function

$$f(x) = ax^2 + bx + c, \quad a \neq 0$$

Quadrilateral   A polygon having four sides.

Quart   A measure of capacity in the United States system equal to 1/4 of a gallon.

Quarterly compounding   In the compound interest formula, it is when $n = 4$.

Quartile   Three values that divide a data set into four equal parts.

Quota rule   The number assigned to each represented unit must be either the standard quota rounded down to the nearest integer, or the standard quota rounded up to the nearest integer.

Quotient   The result of a division.

Radical form   The $\sqrt{}$ symbol in an expression such as $\sqrt{2}$. The number 2 is called the *radicand* and an expression involving a radical is called a radical expression.

Radicand   See *Radical form.*

Radius   The distance of a point on a circle from the center of the same circle.

RAM Random-Access Memory, or memory where each location is uniformly accessible, often used for the storage of a program and the data being processed.

Random variable A *random variable X* associated with the sample space *S* of a probability is a function that assigns a real number to each simple event in *S*.

Range (1) In statistics, it is the difference between the largest and the smallest numbers in the data set. (2) The *range* of a graph of an equation with two variables *x* and *y* is the set of permissible real-number replacements for *y*.

Rate (1) In percent problems, it is the percent. (2) In tax problems, it is the level of taxation, written as a percent. (3) In financial problems, it refers to the APR.

Ratio The quotient of two numbers or expressions.

Rational equation An equation that has at least one variable in the denominator.

Rational number A number belonging to the set $\mathbb{Q}$ defined by

$$\mathbb{Q} = \left\{ \frac{a}{b} \middle|\ a \text{ is an integer, } b \text{ is a nonzero integer} \right\}$$

a is called the *numerator* and *b* is called the *denominator*. A rational number is also called a *fraction*.

Ray If *P* is a point on a line, then a ray from the point *P* is all points on the line on one side of *P*.

Real number line A line on which points are associated with real numbers in a one-to-one fashion.

Real numbers The set of all rational and irrational numbers, denoted by $\mathbb{R}$.

Reciprocal The reciprocal of *n* is $\frac{1}{n}$, also called the *multiplicative inverse of n*.

Rectangle A quadrilateral whose angles are all right angles.

Rectangular coordinate system See *Cartesian coordinate system*.

Rectangular coordinates See *Ordered pair*.

Rectangular parallelepiped In this book, it refers to a box all of whose angles are right angles.

Reduced fraction A fraction so that the numerator and denominator have no common divisors (other than 1).

Reducing fractions The process by which we make sure that there are no common factors (other than 1) for the numerator and denominator of a fraction.

Reflection Given a line *L* and a point *P*, we call the point *P'* the *reflection* about the line *L* if *PP'* is perpendicular to *L* and is also bisected by *L*.

Region In a network, a separate part of the plane.

Regression analysis The analysis used to determine the relationship or a correlation between two variables.

Regular polygon A polygon with all sides the same length.

Relation A set of ordered pairs.

Relative frequency If an experiment is repeated *n* times and an event occurs *m* times, then the relative frequency is the ratio *m/n*.

Relatively prime Two integers are relatively prime if they have no common factors other than ±1; two polynomials are relatively prime if they have no common factors except constants.

Remainder When an integer *m* is divided by a positive integer *n*, and a quotient *q* is obtained for which $m = nq + r$ with $0 \le r < n$, then *r* is the remainder.

Repeating decimal See *Decimal fraction*.

Repetitive system A property of a numeration system for which a single symbol is repeated to represent a given number. For example, ∩∩∩ in the Babylonian system means $10 + 10 + 10 = 30$.

Replication On a spreadsheet, the operation of copying a formula from one place to another.

Resolution The number of dots (or pixels) determines the clarity, or resolution, of the image on the monitor.

Revolving credit It is the same as open-end or credit card credit.

Rhombus A parallelogram with adjacent sides equal.

Richter number A number used to denote the magnitude or size of an earthquake.

Richter scale Same as *Richter number*.

Right angle An angle of 90°.

Right circular cone A cone with a circular base for which the base is perpendicular to its axis.

Right circular cylinder A cylinder with a circular base for which the base is perpendicular to its axis.

Right prism A prism whose base is perpendicular to the lateral edges.

Right triangle A triangle with one right angle.

Rise See *Slope*.

ROM Read-Only Memory, or memory that cannot be altered either by the user or a loss of power. In microcomputers, the ROM usually contains the operating system and system programs.

Root of a number An *n*th root (*n* is a natural number) of a number *b* is *a* only if $a^n = b$. If $n = 2$, then the root is called a *square root;* if $n = 3$, it is called a *cube root*.

Root of an equation See *Solution*.

Roster method A method of defining a set by listing its members.

Rounding a number Dropping decimals after a certain significant place. The procedure for rounding is: 1. Locate the rounding place digit; 2. Determine the rounding place digit: It stays the same if the first digit to its right is a 0, 1, 2, 3, or 4; it increases by 1 if the digit to the right is a 5, 6, 7, 8, or 9. 3. Change digits: All digits to the left of the rounding digit remain the same (unless there is a carry) and all digits to the right of the rounding digit are changed to zeros. 4. Drop zeros: If the rounding place digit is to the left of the decimal point, drop all trailing zeros; if the rounding place digit is to the right of the decimal point, drop all trailing zeros to the right of the rounding place digit.

Row   A horizontal arrangement of numbers or entries of a matrix. It is denoted by numerals 1, 2, 3, . . . on a spreadsheet.

Row+   An elementary row transformation that causes one row of a matrix (called the *pivot row*) to be added to another row (called the *target row*). The answer to this addition replaces the entries in the target row, entry-by-entry.

Row-reduced form   The final matrix after the process of Gauss–Jordan elimination.

RowSwap   An elementary row operation that causes two rows of a matrix to be switched, entry-by-entry.

Rubik's cube   A three-dimensional cube that can rotate about all three axes. It is a puzzle that has the object of returning the faces to a single color position.

Rules of divisibility   A number N is divisible by:

1
2 if the last digit is divisible by 2.
3 if the sum of the digits is divisible by 3.
4 if the number formed by the last two digits is divisible by 4.
5 if the last digit is 0 or 5.
6 if the number is divisible by 2 and by 3.
8 if the number formed by the last three digits is divisible by 8.
9 if the sum of the digits is divisible by 9.
10 if the last digit is 0.
12 if the number is divisible by 3 and by 4.

Run   See *Slope.*

Runoff election   An attempt to obtain a majority vote by eliminating one or more alternatives and voting again on the remaining choices.

Saccheri quadrilateral   A rectangle with base angles A and B right angles, and with sides $\overline{AC}$ and $\overline{BD}$ the same length.

Sales price   A reduced price usually offered to stimulate sales. It can be found by subtracting the discount from the original price, or by multiplying the original price by the complement of the markdown.

Sales tax   A tax levied by government bodies that is based on the sale price of an item.

Sample   A finite portion of a population.

Sample space   The set of possible outcomes for an experiment.

Satisfy   See *Equation* or *Inequality.*

Scalar   A real number.

Scalar multiplication   The multiplication of a real number and a matrix.

Scalene triangle   A triangle with no two sides having the same length.

Scatter diagram   A diagram showing the frequencies with which joint values of variables are observed. One variable is indicated along the x-axis and the other along the y-axis.

Scientific notation   Writing a number as the product of a number between 1 and 10 and a power of 10: For any real number n, $n = m \cdot 10^c$, $1 \leq m < 10$, and c is an integer. Calculators often switch to scientific notation to represent large or small numbers. The usual notation is 8.234 05, where the space separates the number from the power; thus 8.234 05 means 8.234×10^5.

Secant line   A line passing through two points of a given curve.

Second component   See *Ordered pair.*

Second-degree equation   With one variable, an equation of the form $ax^2 + bx + c = 0$; with two variables, an equation of the form

$$Ax^2 + Bxy + Cy^2 + Dx + Ey + F = 0$$

Semiannual compounding   In the compound interest formula, it is when $n = 2$.

Semicircle   Half a circle.

Sequence   An *infinite sequence* is a function whose domain is the set of counting numbers. It is sometimes called a *progression.* A *finite sequence* with n terms is a function whose domain is the set of numbers $\{1, 2, 3, . . . , n\}$.

Sequential voting   A run-off election procedure that has one vote followed by another.

Series   The indicated sum of a finite or an infinite sequence of terms.

Series circuit   Two switches connected together so that the circuit is on only if both switches are on.

Set   A collection of particular things, called the *members* or *elements* of the set. A set with no elements is called the *null set* or *empty set* and is denoted by the symbol $\varnothing$. All elements of a *finite set* may be listed, whereas the elements of an *infinite set* continue without end.

Set-builder notation   A technical notation for defining a set. For example,

$$\{a \mid a \in \mathbb{J}, 5 < a < 100\}$$

means "the set of all elements a such that a is an integer between 5 and 100."

Set theory   That branch of mathematics that studies sets.

SI system   See *Metric system.*

Sieve of Eratosthenes   A method for determining a set of primes less than some counting number n. Write out the consecutive numbers from 1 to n. Cross out 1, since it is not classified as a prime number. Draw a circle around 2, the smallest prime number. Then cross out every following multiple of 2, since each is divisible by 2 and thus is not prime. Draw a circle around 3, the next prime number. Then cross out each succeeding multiple of 3. Some of these numbers, such as 6 and 12, will already have been crossed out because they are also multiples of 2. Circle the next open prime, 5, and cross out all subsequent multiples of 5. The next prime number is 7; circle 7 and cross out multiples of 7. Continue this process until you have crossed out the primes up to $\sqrt{n}$. All of the remaining numbers on the list are prime.

Sigma notation   Sigma, the Greek letter corresponding to S, is written Σ. It is used to indicate the process of summing the first to the nth terms of a set of numbers s_1, s_2,

$s_3, \ldots, s_n$, which is written as

$$\sum_{k=1}^{n} s_k$$

This notation is also called *summation notation.*

Signed number   An integer.

Significance level   Deviations between hypothesis and observations that are so improbable under the hypothesis as not to be due merely to sampling errors or fluctuations are said to be *statistically significant.* The significance level is set at an acceptable level for a deviation to be statistically significant.

Similar figures   Two geometric figures are similar if they have the same shape, but not necessarily the same size.

Similar terms   Terms that differ only in their numerical coefficients.

Similar triangle theorem   Two triangles are similar if two angles of one triangle are equal to two angles of the other triangle. If the triangles are similar, then their corresponding sides are proportional.

Similar triangles   Triangles that have the same shape.

Similarity   Two geometric figures are *similar* if they have the same shape.

Simple curve   A curve that does not intersect itself.

Simple event   An event for which the sample space has only one element.

Simple grouping system   A numeration system is a grouping system if the position of the symbols is not important, and each symbol larger than 1 represents a group of other symbols.

Simple interest formula $I = Prt$

Simple statement   A statement that does not contain a connective.

Simplify   (1) A *polynomial:* Combine similar terms and write terms in order of descending degree. (2) A fraction (a rational expression): Simplify numerator and denominator, factor if possible, and eliminate all common factors. (3) A square root: The *radicand* (the number under the radical sign) has no factor with an exponent larger than 1 when it is written in factored form; the radicand is not written as a fraction or by using negative exponents; there are no square root symbols used in the denominators of fractions.

Simulation   Use of a computer program to simulate some real-world situation.

Simultaneous solution   The solution of a simultaneous system of equations.

Sine   In a right triangle ABC with right angle C,

$$\sin A = \frac{\text{LENGTH OF OPPOSITE SIDE OF } A}{\text{LENGTH OF HYPOTENUSE}}$$

Singular matrix   A matrix that does not have an inverse.

Sinking fund   A financial problem in which the monthly payment must be found to obtain a known future value. The formula is

$$m = \frac{A\left(\dfrac{r}{n}\right)}{\left(1 + \dfrac{r}{n}\right)^{nt} - 1}$$

Skewed distribution   A statistical distribution that is not symmetric, but favors the occurrence on one side of the mean or the other.

Slant asymptotes   In graphing a hyperbola, the diagonal lines passing through the corners of the central rectangle.

Slope   The slope of a line passing through (x_1, y_1) and (x_2, y_2) is denoted by m, and is found by

$$m = \frac{y_2 - y_1}{x_2 - x_1} = \frac{\text{VERTICAL CHANGE}}{\text{HORIZONTAL CHANGE}} = \frac{\text{RISE}}{\text{RUN}}$$

Slope–intercept form $y = mx + b$

Slope point   A point that is found after counting out the rise and the run from the y-intercept.

Software   The routines, programs, and associated documentation in a computer system.

Software package   A commercially available computer program that is written to carry out a specific purpose, for example, a database program or a word-processing program.

Solution   The values or ordered pairs of values for which an equation, a system of equations, inequality, or system of inequalities is true. Also called *roots.*

Solution set   The set of all solutions to an equation.

Solve a proportion   To find the missing term of a proportion. Procedure: First, find the product of the means or the product of the extremes, whichever does not contain the unknown term; next, divide this product by the number that is opposite the unknown term.

Solve an equation   To find the values of the variable that make the equation true.

Solve an inequality   To find the values of the variable that make the inequality true.

Some   A word used to mean *at least one.*

Spanning tree   A tree that is created from another graph by removing edges while keeping a path to each vertex.

Sphere   The set of all points in space that are a given distance from a given point.

Spreadsheet   A rectangular grid used to collect and perform calculations on data. *Rows* are horizontal and are labeled with numbers, and *columns* are vertical and are labeled with letters to designate *cells* such as A4, P604. Each cell can contain text, numbers, or formulas.

Square   (1) A quadrilateral with all sides the same length and all angles right angles. (2) In an expression such as x^2, which is pronounced "x-squared," it means xx.

Square matrix   A matrix with the same number of rows and columns.

Square number   Numbers that are squares of the counting numbers: 1, 4, 9, 16, 25, 36, 49, 64, 81, 100, 121, 144, 169, ...

Square root   See *Root of a number.*

Square unit   A two-dimensional unit. It is the result of squaring a unit of measurement.

Stable marriage   A pairing in which both partners are satisfied.

Standard deviation   It is a measure of the variation from a trend. In particular, it is the square root of the mean of the squares of the deviations from the mean.

Standard divisor

$$\text{STANDARD DIVISOR} = \frac{\text{TOTAL POPULATION}}{\text{NUMBER OF SHARES}}$$

Standard form   The standard form of the equation of a line is $Ax + By + C = 0$.

Standard quota

$$\text{STANDARD QUOTA} = \frac{\text{TOTAL POPULATION}}{\text{STANDARD DIVISOR}}$$

Statement   A declarative sentence that is either true or false, but not both true and false.

Statistics   Methods of obtaining and analyzing quantitative data.

Stem-and-leaf plot   A procedure for organizing data that can be divided into two categories. The first category is listed at the left, and the second category at the right.

Sticker price   In this book, it refers to the manufacturer's total price of a new automobile as listed on the window of the car.

Straight angle   An angle whose rays point in opposite directions; an angle whose measure is $180°$.

Straightedge   A device used as an aid in drawing a straight line segment.

Straw vote   A nonbinding vote taken before all the discussion has taken place. It precedes the actual vote.

Street problem   A problem that asks the number of possible routes from one location to another along some city's streets. The assumptions are that we always move in the correct direction and that we do not cut through the middle of a block, but rather stay on the streets or alleys.

Subjective probability   A probability obtained by experience and used to indicate a measure of "certainty" on the part of the speaker. These probabilities are not necessarily arrived at through experimentation or theory.

Subscript   A small number or letter written below and to the right or left of a letter as a mark of distinction.

Subset   A set contained within a set. There are 2^n subsets of a set with n distinct elements. A subset is *improper* if it is equivalent to the given set; otherwise, it is *proper.*

Substitution method   The method of solution of a system of equations in which one of the equations is solved for one of the variables and substituted into another equation.

Substitution property   The process of replacing one quantity or unknown by another quantity. That is, if $a = b$, then a may be substituted for b in any mathematical statement without affecting the truth or falsity of the given mathematical statement.

Subtraction   The operation of subtraction is defined by:

$$a - b = x \quad \text{means} \quad a = b + x$$

Subtraction law of exponents   To divide two numbers with the same base, subtract the exponents. That is,

$$\frac{b^m}{b^n} = b^{m-n}$$

Subtraction law of logarithms   The log of the quotient of two numbers is the log of the numerator minus the log of the denominator. In symbols,

$$\log_b\left(\frac{A}{B}\right) = \log_b A - \log_b B$$

Subtraction of integers

$$a - b = a + (-b)$$

Subtraction of matrices $[M] - [N] = [S]$ if and only if $[M]$ and $[N]$ are the same order and the entries of $[S]$ are found by subtracting the corresponding entries of $[M]$ and $[N]$.

Subtraction of rational numbers

$$\frac{a}{b} - \frac{c}{d} = \frac{ad}{bd} - \frac{bc}{bd} = \frac{ad - bc}{bd}$$

Subtraction principle   In reference to numeration systems, it is subtracting the value of some symbol from the value of the other symbols. For example, in the Roman numeration system IX uses the subtraction principle because the position of the I in front of the X indicates that the value of I (which is 1) is to be subtracted from the value of X (which is 10). IX $= 9$.

Subtraction property of equations   The solution of an equation is unchanged by subtracting the same number from both sides of the equation.

Subtraction property of inequality   See *Addition property of inequality.*

Successor   In a sequence, the successor of an element a_n is the following element, a_{n+1}.

Sum   The result of an addition.

Summation notation   See *Sigma notation.*

Supercomputer   A large, very fast mainframe computer used especially for scientific computations.

Superfluous constraint   In a linear programming problem, a constraint that does not change the outcome if it is deleted.

Supermarket problem   Set up the shelves in a market or convenience store so that it is possible to enter the store at one door and travel each aisle once (and only once) and leave by the same door.

Supplementary angles   Two angles whose sum is $180°$.

Supply   The number of items that can be supplied at a given price.

Surface   In mathematics, it is an undefined term. It is the outer face or exterior of an object; it has an extent or magnitude having length and breadth, but no thickness.

Surface area   The area of the outside faces of a solid.

Syllogism   A logical argument that involves three propositions, usually two premises and a conclusion, the conclusion necessarily being true if the premises are true.

Symbols of inclusion   See *Grouping symbols.*
Symmetric property of equality

If $a = b$, then $b = a$.

Symmetry   (1) In geometry, a graph or picture is *symmetric with respect to a line* if the graph is a mirror reflection along the line. (2) In voting, it means that if one voter prefers *A* to *B* and another *B* to *A*, then the votes should cancel each other out.
Syntax error   The breaking of a rule governing the structure of the programming language being used.
System of equations   A set of equations that are to be solved *simultaneously*. A brace symbol is used to show the equations belonging to the system.
System of inequalities   A set of inequalities that are to be solved simultaneously. The solution is the set of all ordered pairs (x, y) that satisfy all the given inequalities. It is found by finding the intersection of the half-planes defined by each inequality.
Tangent   In a right triangle *ABC* with right angle *C*,

$$\tan A = \frac{\text{LENGTH OF OPPOSITE SIDE OF } A}{\text{LENGTH OF ADJACENT SIDE OF } A}$$

Tangent line   A tangent line to a circle is a line that contains exactly one point of the circle. The tangent line to a curve at a point *P* is the limiting position, if this exists, of the secant line through a fixed point *P* on the curve and a variable point *P'* on the curve so that *P'* approaches *P* along the curve.
Target population   The population to be considered for a statistical application.
Target row   In an elementary row operation, it is the row that is changed. See *Elementary row operations.*
Tautology   A compound statement is a tautology if all values on its truth table are true.
Temperature   The degree of hotness or coldness.
Ten   A representation for "ΔΔΔΔΔΔΔΔΔΔ" objects.
Term   A number, a variable, or a product of numbers and variables. See *Polynomial*. A *term of a sequence* is one of the elements of that sequence.
Terminating decimal   See *Decimal fraction.*
Tessellation   A mosaic, repetitive pattern.
Test point   A point that is chosen to find the appropriate half-plane when graphing a linear inequality in two variables.
Theorem   A statement that has been proved. See *Deductive reasoning.*
Theoretical probability   A probability obtained by logical reasoning according to stated definitions.
Time   In a financial problem, the length of time (in years) from the present value to the future value.
*Row   An elementary row operation that multiplies each entry of a row of a matrix (called the *target row*) by some number, called a *scalar*. The elements of the row are replaced term-by-term by the products. It is denoted by *Row.

*Row+   An elementary row operation that multiplies each entry of a row of a matrix (called the *pivot row*) by some number (called a *scalar*), and then adds that product, term-by-term, to the numbers in another row (called the *target row*). The results replace the entries in the target row, term-by-term. It is denoted by *Row+.
Ton   A measurement of mass in the United States system; it is equal to 2,000 lb.
Topologically equivalent   Two geometric figures are said to be *topologically equivalent* if one figure can be elastically twisted, stretched, bent, shrunk, or straightened into the same shape as the other. One can cut the figure, provided at some point the cut edges are "glued" back together again to be exactly the same as before.
Topology   That branch of geometry that deals with the *topological properties* of figures. If one figure can be transformed into another by stretching or contracting, then the figures are said to be *topologically equivalent.*
Tournament method   A method of selecting a winner by pairing candidates head to head with the winner of one facing a new opponent for the next election.
Trailing zeros   Sometimes zeros are placed after the decimal point or after the last digit to the right of the decimal point, and if these zeros do not change the value of the number, they are called *trailing zeros.*
Transformation   A passage from one figure or expression to another, such as a reflection, translation, rotation, contraction, or dilation.
Transformational geometry   The geometry that studies transformations.
Transitive law   If A beats B, and B beats C, then A should beat C. In symbols, equality: If $a = b$ and $b = c$, then $a = c$; inequality: If $a > b$ and $b > c$, then $a > c$. Also holds for $\geq$, $<$, and $\leq$.
Transitive reasoning   If $a = b$ and $b = c$, then $a = c$.
Translating symbols   The process of writing an English sentence in mathematical symbols.
Transversal   A line that intersects two parallel lines.
Transverse axis   The line passing through the foci is called the *transverse axis.*
Trapezoid   A quadrilateral that has two parallel sides.
Traveling salesperson problem (TSP)   A salesperson starts at home and wants to visit several cities without going through any city more than once, and then returning to the starting city.
Traversable network   A network is said to be *traversable* if it can be traced in one sweep without lifting the pencil from the paper and without tracing the same edge more than once. Vertices may be passed through more than once.
Tree   A graph which is connected and has no circuits.
Tree diagram   A device used to list all the possibilities for an experiment.
Triangle   A polygon with three sides.
Trichotomy   Exactly one of the following is true, for any real numbers *a* and *b*: $a < b$, $a > b$, or $a = b$.

Trigonometric functions The same as the *trigonometric ratios.*

Trigonometric ratios The sine, cosine, and tangent ratios are known as the *trigonometric ratios.*

Trillion A name for $10^{12} = 1,000,000,000,000$.

Trinomial A polynomial with exactly three terms.

Truth set The set of values that makes a given statement true.

Truth table A table that shows the truth values of all possibilities for compound statements.

Truth value The truth value of a simple statement is true or false. The truth value of a compound statement is true or false and depends only on the truth values of its simple component parts. It is determined by using the rules for connecting those parts with well-defined operators.

Two-point form The equation of a line passing through (x_1, y_1) and (x_2, y_2) is

$$ y - y_1 = \left(\frac{y_2 - y_1}{x_2 - x_1} \right)(x - x_1) $$

Type I error Rejection of a hypothesis based on sampling when, in fact, the hypothesis is true.

Type II error Acceptance of a hypothesis based on sampling when, in fact, it is false.

Undefined terms To avoid circular definitions, it is necessary to include certain terms without specific mathematical definition.

Union The union of sets A and B, denoted by $A \cup B$, is the set consisting of elements in A or in B or in both A and B.

Unit circle A circle with radius 1 centered at the origin.

Unit distance The distance between 0 and 1 on a number line.

Unit scale The distance between the points marked 0 and 1 on a number line.

United States system The measurement system used in the United States.

Universal set The set that contains all of the elements under consideration for a problem or a set of problems.

Unless A logical operator for "*p* unless *q*" that is defined to mean $\sim q \rightarrow p$.

Unrestricted domain Any set of rankings is possible; if there are n candidates, then there are $n!$ possible rankings.

Unstable marriage A pairing in which one (or both) of the partners would prefer to be paired with another partner.

Upper quota In apportionment, the result of a quota found by rounding up.

Upload The process of copying a program from your computer to the network.

User-friendly A term used to describe software that is easy to use. It includes built-in safeguards to keep the user from changing important parts of a program.

Valid argument In logic, refers to a correctly inferred logical argument.

Variable A symbol that represents unspecified elements of a given set. On a calculator, it refers to the name given to a location in the memory that can be assigned a value.

Variable expression An expression that contains at least one variable.

Variance The square of the standard deviation. When the variance is based on a set of sample scores, it is called the *variance of a random sample* and is denoted by s^2; when it is based on the entire population, it is called the *variance of the population* and is denoted by σ^2. The formulas for variance are

$$ s^2 = \frac{\Sigma(x - \bar{x})^2}{n - 1} $$

$$ \sigma^2 = \frac{\Sigma(x - \mu)^2}{n} $$

Velocity An instantaneous rate of change; a directed speed.

Venn diagram A diagram used to illustrate relationships among sets.

Vertex (1) A *vertex* of a polygon is a corner point, or a point of intersection of two sides. (2) A *vertex* of a parabola is the lowest point for a parabola that opens upward; the highest point for one that opens downward; the leftmost point for one that opens to the right; and the rightmost point for one that opens to the left.

Vertex angle The angle included between the legs of an isosceles triangle.

Vertical angles Two angles such that each side of one is a prolongation through the vertex of a side of the other.

Vertical ellipse An ellipse whose major axis is vertical.

Vertical hyperbola A hyperbola whose transverse axis is vertical.

Vertical line A line with undefined slope. Its equation has the form x = constant.

Vertical line test Every vertical line passes through the graph of a function in at most one point. This means that if you sweep a vertical line across a graph and it simultaneously intersects the curve at more than one point, then the curve is not the graph of a function.

Volume A number describing three-dimensional content of a set. Specifically, it is the number of cubic units enclosed in a solid figure.

Vote A decision by a group on a proposal, resolution, law, or a choice between candidates for office.

Water-pipe problem Consider a network of water pipes to be inspected. Is it possible to pass a hand over each pipe exactly once without lifting it from a pipe, and without going over the same pipe more than once?

Webster's apportionment plan An apportionment plan in which the representation of a geographical area is determined by finding the quotient of the number of people in that area divided by the total number of people and then the result is rounded as follows: Any quotient with a decimal portion must be rounded to the nearest whole number.

Weight (1) In everyday usage, the heaviness of an object. In scientific usage, the gravitational pull on a body. (2) In a network or graph, a cost associated with an edge.

Weighted graph   A graph for which all its edges have weight.

Weighted mean   If the scores $x_1, x_2, x_3, \ldots, x_n$ occur $w_1, w_2, \ldots, w_n$ times, respectively, then the *weighted mean* is

$$\bar{x} = \frac{\Sigma(w \cdot x)}{\Sigma w}$$

Well-defined set   A set for which there is no doubt about whether a particular element is included in the given set.

Whole numbers   The positive integers and zero; $\mathbb{W} = \{0, 1, 2, 3, \ldots\}$.

Windows   A graphical environment for IBM format computers.

With replacement   If there is more than one step for an experiment, to perform the experiment with replacement means that the object chosen on the first step is replaced before the next steps are completed.

Without replacement   If there is more than one step for an experiment, to perform the experiment without replacement means that the object chosen on the first step is not replaced before the next steps are completed.

Word processing   The process of creating, modifying, deleting, and formatting textual materials.

World Wide Web   A network that connects together computers from all over the world. It is abbreviated by www.

WYSIWYG: An acronym for <u>W</u>hat <u>Y</u>ou <u>S</u>ee <u>I</u>s <u>W</u>hat <u>Y</u>ou <u>G</u>et.

***x*-axis**   The horizontal axis in a Cartesian coordinate system.

***x*-intercept**   The place where a graph passes through the *x*-axis.

***y*-axis**   The vertical axis in a Cartesian coordinate system.

***y*-intercept**   The place where a graph passes through the *y*-axis. For a line $y = mx + b$, it is the point b.

Yard   A linear measure in the United States system; it has the same length as 3 ft.

Zero   The number that divides the positive and negative numbers; it is also called the *identity element* for addition; that is, it satisfies the property that

$$x + 0 = 0 + x = x \text{ for all numbers } x.$$

Zero matrix   A matrix with all entries equal to 0.

Zero multiplication   If a is any real number, then $a \cdot 0 = 0 \cdot a = 0$.

Zero-product rule   If $a \cdot b = 0$, then either $a = 0$ or $b = 0$.

***z*-score**   A measure to determine the distance (in terms of standard deviations) that a given score is from the mean of a distribution.

B | Selected Answers

11. Must draw at least five cards. **13.** 1 **15.** It is impossible to trace out this circuit. **17.** Nothing can be said about Deidre's age as compared to Chelsea's age. **19.** April 2011 **21.** 119 **23.** 945 **25.** $\frac{3}{8}$ **27.** 2 **29. a.** tetrahedron **b.** cube (hexahedron) **c.** octahedron **d.** icosahedron **e.** dodecahedron **31.** 1,499 **33.** More than a million trees must be cut down to print a trillion one-dollar bills. **35.** The area is approximately 13.7 in.^{2}. **37.** $\frac{1}{3}$ **39.** She had $22 at the start of the first day. **41.** The cost is $390 if 100 make the trip. **43.** Let x be the number of weeks enrolled, and y be the total cost. Then $y = 45(10 - x)$. The graph is shown.

45. The rate at which alcohol is changing with respect to time is $\frac{dC}{dt} = 0.15e^{-t/2}$.

47. There are 33 paths.
49. It will take until March 2057. **51.** $b = \frac{2}{3}\ln 2$
53. Answers vary; there are at least three cubes that can be seen. **55.** This pattern of numbers has the property that the sum of the numbers in any row, column, or diagonal is the same (namely, 15). **57.** The population in 2000 was 153,000. The graph is shown.

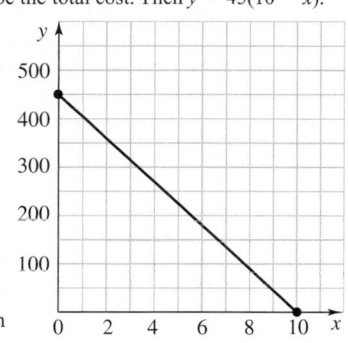

59. $1,073,741,823

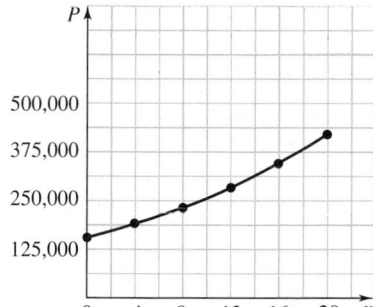

CHAPTER 1 THE NATURE OF PROBLEM SOLVING

1.1 Problem Solving, page 12

7. first diagonal **9.** yes; yes; triangle is symmetric **11.** 20 **13.** 56 **15.** 2 up, 3 over; 10 paths **17.** 4 up, 6 over; 210 paths **19.** 4 down, 3 over; 35 paths **21.** 2 up, 4 over; 15 paths **23.** Answers vary. **25.** There is a total of 27 boxes. **27. a.** 2 **b.** 4 **c.** 8 **d.** 16 **29.** 37 paths **31.** 11 paths **37.** The fly flies 20 miles. **39.** 8 miles or 12 miles **41.** 1,089; does not work for palindromes **43.** 6, 8, 10, 14, and 15 **45.** They are the same. **47.** 66 sec. **49.** The coins are a quarter and a nickel. **51.** 54 **53.** Answers vary. Take goose across; return and pick up the fox; deliver the fox and take the goose back to the first side; drop off the goose, pick up the grain and leave it with the fox; return one last time to fetch the goose. **55.** Pour the contents of glass 5 into glass 2.

57. The number of routes from the start to the finish minus the number of routes passing through the barricade. **a.** 26 **b.** 23 **c.** 27 **59.** There are 26 cards in each pile. The hidden card is 7 cards down in the pile. Three cards are removed from the second pile, leaving 23 cards. Look for a pattern:

If the 3 cards are tens, then $23 + 7 = 30$ cards down to the hidden card.
If the 3 cards total 29, then $22 + 7 = 29$ cards down.
If the 3 cards total 28, then $21 + 7 = 28$ cards down.

.
.
.

If the 3 cards total x, then $23 - (30 - x) + 7 = x$ cards down.

1.2 Inductive and Deductive Reasoning, page 21

7. a. 32 **b.** 12 **9. a.** 7 **b.** 11 **11. a.** 505 **b.** 100 **13. a.** 29 **b.** 35 **15. a.** 23 **b.** 28 **17. a.** 13 **b.** 6 **19.** deductive reasoning; answers vary **21.** 4, 8, 3, 7, 2, 6, 1, 5, 9, 4, 8, 3, ... **23.** 6, 3, 9, 6, 3, 9, 6, 3, 9, . . . **25.** $50^2 = 2,500$ **27. a.** $3 + 2 \times 4$ **b.** $3(2 + 4)$ **29. a.** $8 \times 9 + 10$ **b.** $8(9 + 10)$ **31. a.** $3^2 + 2^3$ **b.** $3^3 - 2^2$ **33. a.** $4^2 + 9^2$ **b.** $(4 + 9)^2$ **35. a.** $3(n + 4) = 16$ **b.** $5(n + 1) = 5n + 5$ **37. a.** $n^2 + 8n = 6$ **b.** $n^2 + 8n = n(8 + n)$ **39.** $A = \frac{1}{2}bh$ **41.** $A = \frac{1}{2}h(a + b)$ **43.** $V = \ell wh$ **45.** $V = \pi r^2 h$ **47.** $9 \times 54,321 - 1$; 488,888 **49.** 98,888,888,888 **51.** 1,111,111,101,000 **53.** 91 squares **55.** $\frac{3}{8}$ **57.** Balance 3 with 3; if it balances, the heavier one is in the three not weighed; if it does not balance, then the pan balance will show which set of 3 contains the heavier one. In any case, the first weighing narrows it down to a set of 3 coins, one of which is the heavy coin. Take 2 from this set of heavy 3 and balance them. If they balance, the heavy coin is the one not weighed; if they do not balance, then the pan balance will show the heavier one. In any case, it takes two weighings. **59.** Common sum is 4,440,084,513.

1.3 Scientific Notation and Estimation, page 36

9. a. 2.379×10^1; 2.379 01 **b.** 1×10^{-6}; 1 −06 **c.** 3.5×10^{10}; 3.5 10 **11. a.** 10^{100}; 1 100 **b.** 1.2003×10^6; 1.2003 06 **c.** 1.23×10^{-6}; 1.23 −06 **13. a.** 64 **b.** 0.0021 **c.** 40,700 **15. a.** 0.027 **b.** 32,170,000 **c.** 0.00000 00000 889 **17.** 8.6×10^1; 6.1×10^3 **19.** 6×10^{12}; 1.86×10^5 **21.** 906 **23.** 38,000,000,000,000,000,000,000,000 *Estimates in Problems 25–31 may vary.* **25.** Estimate $2,000 \times 500 = 1,000,000$; $1,850 \times 487 = 900,950$ **27.** Estimate $1,500 \times 12 = 1,500 \times 10 + 1,500 \times 2 = 15,000 + 3,000 = 18,000$; $$1,543 \times 12 = $18,516$ **29.** Estimate 1,000,000 people ÷ 10,000 people/yr = 100 yr; $1,000,000 \div 13,688 \approx 73$, so it will take her about 73 years. **31.** Estimate $300 \div 10 = 30$; $280 \div 10.2 = 27.451$ **33. a.** 10^4 **b.** 1.4×10^{-1} **35. a.** x^2 **b.** $8xy^4$ **37. a.** 3×10^9 **b.** 4×10^{-11} **39.** Answers vary; 30 miles **41.** Estimate 15 persons/cm^2, so about 180 people **43.** Estimate 8/cm^2, so about 300 penguins **45.** 5.157×10^{15} **47.** A million dollars would form a stack about 119 yd high. **49.** 1.356×10^{17} grains **51.** 281,474,976,710,656 **53.** Estimate 40–50 days; calculate 41 days, 7 hours, and 40 minutes. **55.** 48 minutes, 13 seconds **57.** 192 zeros **59.** D

1.4 Finite and Infinite, page 46

7. well defined   **9.** well defined   **11.** not well defined; answers vary; for example, change to, "The set of people with two ears."
13. not well defined; change to, "Bets over \$100 on the next race at Hialeah."   **15.** {m, a, t, h, e, i, c, s}   **17.** {1, 3, 5, 7, 9}
19. {p, i, e}   **21.** {6, 8, 10, 12, 14}   **23.** The set of counting numbers less than 10.   **25.** The multiples of 10 between 0 and 101.
27. The set of odd numbers between 100 and 170.   **29.** The set of all x such that x is an odd counting number; {1, 3, 5, 7, ...}
31. The set of all x such that x is a natural number (except 8); {1, 2, 3, 4, 5, 6, 7, 9, 10, ...}   **33.** The set of all x such that x is a whole number less than 8; {0, 1, 2, 3, 4, 5, 6, 7}
35. {(c, w), (c, x), (d, w), (d, x), (f, w), (f, x)}   **37.** {(1, a), (1, b), (1, c), (2, a), (2, b), (2, c), (3, a), (3, b), (3, c), (4, a), (4, b), (4, c), (5, a), (5, b), (5, c)}   **39.** $|A \times B| = 26 \times 10 = 260$
41. $|C \times D| = 100 \times 50 = 5{,}000$ **43. a.** $|A| = 3; |B| = 1;$ $|C| = 3; |D| = 1; |E| = 1; |F| = 1$ **b.** $A \leftrightarrow C; B \leftrightarrow D \leftrightarrow E \leftrightarrow F$
c. $A = C; D = E = F$
45. {m, a, t}; {m, a, t}; there are others.

$$\uparrow \uparrow \uparrow \quad \uparrow \uparrow \uparrow$$
{1, 2, 3}; {3, 2, 1}

47. { 1, 2, 3,   ..., n,   n + 1, ...,   353, 354, 355, 356, ..., 586, 587}

$$\uparrow \uparrow \uparrow \quad \uparrow \quad \uparrow \quad \uparrow \uparrow$$
{550, 551, 552, ..., n + 549, n + 550, ..., 902, 903}

49. a. finite   **b.** infinite   **c.** finite   **d.** finite
51. { 1, 2, 3, ...,   n, ...}

$$\uparrow \uparrow \uparrow \quad \uparrow$$
{−1, −2, −3, ..., −n, ...}

Since these sets can be put into a one-to-one correspondence, they have the same cardinality; namely, $\aleph_0$.
53. {1, 2, 3, ..., n − 1, n, ...}

$$\uparrow \uparrow \uparrow \quad \uparrow \quad \uparrow$$
{1, 2, 4, ..., 2^n, 2^{n+1}, ...}

Since these sets can be put into a one-to-one correspondence, they have the same cardinality; namely, $\aleph_0$.
55. $\mathbb{W} = \{0, 1, 2, 3, ...,   n, ...\}$

$$\uparrow \uparrow \uparrow \uparrow \quad \uparrow$$
{1, 2, 3, 4, ..., n + 1, ...}

57. {12, 14, 16, ...,   n, ...}

$$\uparrow \uparrow \uparrow \quad \uparrow$$
{14, 16, 18, ..., n + 2, ...}

59. Answers vary; the set, E, of even integers has cardinality $\aleph_0$; the set, O, of odd integers has cardinality $\aleph_0$. If we add the cardinality of the even integers to the cardinality of the odd integers, we have $\aleph_0 + \aleph_0$. However, if we put the even integers together with the odd integers, we have the set of counting numbers, which has cardinality $\aleph_0$. Thus, you might say, $\aleph_0 + \aleph_0 = \aleph_0$.

Chapter 1 Review Questions, page 48

1. Understand the problem, devise a plan, carry out the plan, and then look back.   **2.** Use Pascal's triangle; look at 5 blocks down and 4 blocks over to find 126.   **3.** Turn the chessboard so that it forms a triangle with the rook at the top. To get the appropriate square, look at 7 blocks down and 4 blocks over using Pascal's triangle to find 330 paths.   **4.** By patterns: $1 \times 1 = 1; 11 \times 11 = 121; 111 \times 111 = 12{,}321; ...; 111{,}111{,}111^2 = 12{,}345{,}678{,}987{,}654{,}321$ **5. a.** Order of operations: (1) First, perform any operations enclosed in parentheses.

(2) Next, perform multiplications and divisions as they occur by working from left to right. (3) Finally, perform additions and subtractions as they occur by working from left to right.   **6.** The scientific notation for a number is that number written as a power of 10 times another number x, such that $1 \le x < 10$.   **7.** Inductive reasoning; the answer was found by looking at the pattern of questions.
8. Answers vary.   **a.** ⬛2⬛ ⬛y^x⬛ ⬛63⬛ ⬛=⬛ or ⬛2⬛ ⬛^⬛ ⬛63⬛ ⬛ENTER⬛ 9.223372037E18
b. ⬛9.22⬛ ⬛EE⬛ ⬛18⬛ ⬛÷⬛ ⬛6.34⬛ ⬛EE⬛ ⬛6⬛ ⬛=⬛ 1.454258675E12
9. A. \$50/person times 263 million is about \$13 billion; not even close.   B. \$1 × 60 sec × 60 min × 24 hr × 365.25 days × 1,000 years is about \$3.1 billion; not even close.   C. 1 million people × \$80,000 + 1 billion people × 200 ≈ 280 billion; choice C comes closest.
10. $C = \dfrac{\ell w}{15}$ where C is the capacity of the boat, ℓ and w are the length and width of the boat.
11. $\frac{3}{4}$ hr × 365 = 273.75 hr. This is approximately $2\frac{3}{4}$ hr/book. Could not possibly be a complete transcription of each book.
12. Answers vary. An ice cube is about 1 in.³ and a cubic foot has 12^3 in.³. A classroom 30 ft × 50 ft × 10 ft would hold about 2.592×10^7 ice cubes. $\dfrac{7 \times 10^{16}}{2.592 \times 10^7} \approx 2.7 \times 10^9$ classrooms; this is not meaningful, but is shown here because it is typical of first attempts. It is important to look for a meaningful comparison. Let's try again. Lake Mead is about 35,154,000 m³. (**www.infoplease.com**). To convert this to "ice cubes," we look up a cubic meter (in a dictionary) to find it is 35.314667 ft³. Thus, we have

$$(3.5154 \times 10^7)(35.314667)(12^3) \approx 2.145 \times 10^{12} \text{ in.}^3$$

Finally, we divide the reported size of the iceberg by the capacity of Lake Mead to find

$$(7 \times 10^{16}) \div (2.145 \times 10^{12}) \approx 32{,}630$$

It would take more than 32,600 dams the size of the Hoover Dam (which forms Lake Mead) to hold the capacity of the iceberg. This is, no doubt, larger than the capacity of all of the lakes behind all of the world's dams, even the world's largest one to be completed in 2009 on the Yangzi River in China.
13. $6.3 \times 10^{12}; (6.3 \times 10^{12}) \div (2.9 \times 10^8) \approx 2.1724 \times 10^4$; each person's share is about \$21,724.
14. Arrange the cards as $\begin{bmatrix} 2 & 7 & 6 \\ 9 & 5 & 1 \\ 4 & 3 & 8 \end{bmatrix}$.
15. Measure a dollar bill (2.5 in. by 6 in.). The floor is 240 in. by 360 in. This is 96 bills by 60 bills, or 5,760 bills on the floor. The ceiling is 10 ft = 120 in. with 233/in., so the total number of bills in the room is

$$5{,}760 \times 120 \times 233 \approx 161{,}049{,}600$$

The national debt is \$6,300,000,000,000 so

$$\frac{63 \times 10^{12}}{161{,}049{,}600} \approx 39{,}118 \text{ classrooms}$$

This number of classrooms is still hard to comprehend, so we will fill them with \$100 bills instead. Now we can estimate it will take 392 classrooms. How many classrooms are in your city? Imagine them *filled* with \$100 bills.
16. a. The set of rational numbers is the set of all numbers of the form $\frac{a}{b}$ such that a is an integer and b is a counting number.
b. $\frac{2}{3}$; answers vary.

17. $\{5, 10, 15, \ldots, 5n, \ldots\}$
$\quad\updownarrow\ \updownarrow\ \updownarrow\qquad\updownarrow$
$\{10, 20, 30, \ldots, 10n, \ldots\}$

Since the second set is a proper subset of the first set, we see that the set F is infinite.

18. a. $\{n \mid (-n) \in \mathbb{N}\}$; answers vary.
b. $\{\ 1,\quad 2,\quad 3, \ldots,\quad n, \ldots\}$
$\quad\ \ \updownarrow\ \ \updownarrow\ \ \updownarrow\qquad\ \updownarrow$
$\{-1, -2, -3, \ldots, -n, \ldots\}$

Since the first set is the set of counting numbers, it has cardinality $\aleph_0$; so the given set also has cardinality $\aleph_0$ since it can be put into a one-to-one correspondence with the set of counting numbers.

19. Answers vary; 20 is unhappy because $2^2 + 0^2 = 4$, which is unhappy. 100 is happy because $1^2 + 0^2 + 0^2 = 1$, which is happy.

20. By patterns: $7^1 = 7$; $7^2 = 49$ ends in 9; $7^3 = 343$ ends in 3; $7^4 = 2{,}401$ ends in 1; 7^5 ends in 7; 7^6 ends in 9; 7^7 ends in 3; 7^8 ends in a 1, ... Looking ahead, we see that 7^{1000} must end in 1.

CHAPTER 2 THE NATURE OF LOGIC

2.1 Deductive Reasoning, page 60

7. a, b, and d are statements **9.** a and b are statements
11. Some dogs do not have fleas. **13.** All people pay taxes.
15. Some triangles are squares. **17.** Some counting numbers are not divisible by 1. **19.** All integers are odd.
21. a. T **b.** F **c.** T **d.** F **23. a.** Prices or taxes will rise.
b. Prices will not rise and taxes will rise. **c.** Prices will rise or taxes will not rise. **d.** Prices will not rise or taxes will not rise.
25. a. F **b.** T **c.** T **d.** F **27.** *Wording may vary.*
a. Paul is peculiar and likes to read math texts. **b.** Paul is not peculiar and likes to read math texts. **c.** Paul is peculiar or does not like to read math texts. **d.** Paul is not peculiar or does not like to read math texts. **29. a.** T **b.** F **c.** T **d.** F **31. a.** T **b.** F
33. a. F **b.** T **35. a.** T **b.** T **37. a.** T **b.** F
39. $e \wedge d \wedge t$ **41.** $\sim t \wedge \sim m$ **43.** $(j \vee i) \wedge \sim p$
45. $d \vee e$ **47. a.** $(s \wedge t) \vee c$ **b.** $s \wedge (t \vee c)$ **49.** T
51. T **53.** F **55.** F **57.** T **59.** Melissa did change her mind.

2.2 Truth Tables and the Conditional, page 67

5.

p	q	$\sim p$	$\sim p \vee q$
T	T	F	T
T	F	F	F
F	T	T	T
F	F	T	T

7.

p	q	$p \wedge q$	$\sim(p \wedge q)$
T	T	T	F
T	F	F	T
F	T	F	T
F	F	F	T

9.

r	$\sim r$	$\sim(\sim r)$
T	F	T
F	T	F

11.

p	q	$\sim q$	$p \wedge \sim q$
T	T	F	F
T	F	T	T
F	T	F	F
F	F	T	F

13.

p	q	$\sim p$	$\sim p \wedge q$	$\sim q$	$(\sim p \wedge q) \vee \sim q$
T	T	F	F	F	F
T	F	F	F	T	T
F	T	T	T	F	T
F	F	T	F	T	T

15.

p	q	$p \to q$	$p \vee (p \to q)$
T	T	T	T
T	F	F	T
F	T	T	T
F	F	T	T

17.

p	q	$p \vee q$	$p \wedge (p \vee q)$	$[p \wedge (p \vee q)] \to p$
T	T	T	T	T
T	F	T	T	T
F	T	T	F	T
F	F	F	F	T

19.

p	q	r	$p \vee q$	$(p \vee q) \vee r$
T	T	T	T	T
T	T	F	T	T
T	F	T	T	T
T	F	F	T	T
F	T	T	T	T
F	T	F	T	T
F	F	T	F	T
F	F	F	F	F

21.

p	q	r	$p \vee q$	$\sim r$	$(p \vee q) \wedge (\sim r)$	$[(p \vee q) \wedge \sim r] \wedge r$
T	T	T	T	F	F	F
T	T	F	T	T	T	F
T	F	T	T	F	F	F
T	F	F	T	T	T	F
F	T	T	T	F	F	F
F	T	F	T	T	T	F
F	F	T	F	F	F	F
F	F	F	F	T	F	F

23. Statement: $\sim p \to \sim q$. Converse: $\sim q \to \sim p$.
Inverse: $p \to q$. Contrapositive: $q \to p$.
25. Statement: $\sim t \to \sim s$. Converse: $\sim s \to \sim t$.
Inverse: $t \to s$. Contrapositive: $s \to t$.
27. Statement: If I get paid, then I will go Saturday.
Converse: If I go Saturday, then I will get paid.
Inverse: If I do not get paid, then I will not go Saturday.
Contrapositive: If I do not go Saturday, then I do not get paid.
29. If it is a triangle, then it is a polygon. **31.** If you are a good person, then you will go to heaven. **33.** If we make a proper use of those means which the God of Nature has placed in our power, then we are not weak. **35.** If it is work, then it is noble. **37.** true
39. true **41. a.** T **b.** F **c.** F **43. a.** F **b.** T **c.** T
45. $(q \wedge \sim d) \to n$, where q: The qualifying person is a child; d: This child is your dependent; n: You enter your child's name.
47. $(b \vee c) \to n$, where b: The amount on line 32 is $86,025; c: The amount on line 32 is less than $86,025; n: You multiply the number of exemptions by $2,500. **49.** $(m \vee d) \to s$, where m: You are a student; d: You are a disabled person; s: You see line 6 of instructions.
51. Assume d, c, and b are true, and w is false: $(\sim T \wedge T) \to (F \wedge T)$ is true. **53. a.** $a \to (e \vee f)$ **b.** $(a \wedge e) \to q$ **c.** $(a \wedge f) \to q$
55. $[(m \vee t \vee w \vee h) \wedge s] \to q$ **57.** $t \to (m \wedge s \wedge p)$
59. This is not a statement, since it gives rise to a paradox and is neither true nor false.

2.3 Operators and Laws of Logic, page 74

7. no   **9.** no   **11.** yes   **13.** no

15.

p	q	$p \vee q$	$\sim(p \vee q)$
T	T	T	F
T	F	T	F
F	T	T	F
F	F	F	T

17.

p	q	$\sim q$	$p \rightarrow \sim q$
T	T	F	F
T	F	T	T
F	T	F	T
F	F	T	T

19. Let h: I will buy a new house; p: All provisions of the sale are clearly understood. Not h unless p: $\sim p \rightarrow \sim h$.   **21.** Let r: I am obligated to pay the rent; s: I signed the contract. r because s: $(r \wedge s) \wedge (s \rightarrow r)$ **23.** Let m: It is a man; i: It is an island. No m is i: $m \rightarrow \sim i$ **25.** Let n: You are nice to people on your way up; m: You will meet people on your way down. n because m: $(n \wedge m) \wedge (m \rightarrow n)$.   **27.** Let f: The majority, by mere force of numbers, deprives a minority of a clearly written constitutional right; r: Revolution is justified. If f then r: $f \rightarrow r$. *Answers to Problems 29–33 may vary.*   **29.** The cherries have not turned red or they are ready to be picked.   **31.** If Melissa watches Jay Leno, then she watches the NBC late-night orchestra.
33. If the money is not available, then I will not take my vacation.
35.

p	$\sim p$	$\sim(\sim p)$	$p \leftrightarrow \sim(\sim p)$
T	F	T	T
F	T	F	T

Thus, $p \Leftrightarrow \sim(\sim p)$.

37.

p	q	$p \vee q$	$\sim(p \vee q)$	$\sim p$	$\sim q$	$\sim p \wedge \sim q$	$\sim(p \vee q) \leftrightarrow (\sim p \wedge \sim q)$
T	T	T	F	F	F	F	T
T	F	T	F	F	T	F	T
F	T	T	F	T	F	F	T
F	F	F	T	T	T	T	T

Thus, $\sim(p \vee q) \Leftrightarrow \sim p \wedge \sim q$.

39.

p	q	$p \rightarrow q$	$\sim p$	$\sim p \vee q$	$(p \rightarrow q) \leftrightarrow (\sim p \vee q)$
T	T	T	F	T	T
T	F	F	F	F	T
F	T	T	T	T	T
F	F	T	T	T	T

Thus, $(p \rightarrow q) \Leftrightarrow (\sim p \vee q)$.

41. $p \wedge q$ **43.** $\sim p \wedge q$ **45.** Jane did not go to the basketball game and she did not go to the soccer game.   **47.** Sally is on time or she did not miss the boat.   **49.** You are out of Schlitz and you have beer.   **51.** $x - 5 = 4$ and $x \neq 1$.   **53.** $x = 1$ and $y = 2$, and $2x + 3y \neq 8$.   **55.** Let d: You purchase your ticket between January 5 and February 15; f: You fly round trip between February 20 and May 3; m: You depart on Monday; t: You depart on Tuesday; w: You depart on Wednesday; h: You return on Tuesday; i: You return on Wednesday; j: You return on Thursday; s: You stay over a Saturday night; e: You obtain 40% off regular fare. Symbolic statement: $[d \wedge f \wedge (m \vee t \vee w) \wedge (h \vee i \vee j) \wedge s] \rightarrow e$
57. Let ℓ: The tenant lets the premises; m: The tenant lets a portion of the premises; s: The tenant sublets the premises; t: The tenant sublets a portion of the premises; p: Permission is obtained. Symbolic statement: $\sim p \rightarrow [\sim(\ell \vee m) \vee \sim(s \vee t)]$
59. Let a: Alfie is afraid to go; b: Bogie lied; c: Clyde lied. Symbolic statement: $[(\sim a) \vee (b \wedge c)] \wedge \sim[(\sim a) \wedge (b \wedge c)]$.

2.4 The Nature of Proof, page 81

7. We show this is a fallacy by constructing a truth table:

							$[(p \rightarrow q) \wedge (p \rightarrow r)] \rightarrow (q \rightarrow r)$	
p	q	r	$p \rightarrow q$	$p \rightarrow r$	$q \rightarrow r$	$(p \rightarrow q) \wedge (p \rightarrow r)$		
T	T	T	T	T	T	T	T
T	T	F	T	F	F	F	T
T	F	T	F	T	T	F	T
T	F	F	F	F	T	F	T
F	T	T	T	T	T	T	T
F	T	F	T	T	F	T	F
F	F	T	T	T	T	T	T
F	F	F	T	T	T	T	T

↑
Not all Ts

Since the result does not show all Ts, it is a fallacy. This is the false chain pattern.

9.

Problem-solving language	Detective language
Understand the problem.	*Understand the case.*
What is the unknown?	What are you looking for?
Devise a plan.	*Investigate the case.*
Do you know a related problem?	Have you solved a similar case?
Can you simplify the problem?	What are the facts?
Carry out the plan.	*Analyze the facts/data.*
What information is important?	What information is important?
What information is not important?	What information is not important?
What pieces of information fit together logically?	What pieces of information do not seem to fit together logically?
Which information is consistent with the given information?	Which data are inconsistent with the given information?
Look back.	*Reexamine the facts.*
Examine the solution obtained.	Do the facts support the solution?
Does it make sense?	Can we obtain a conviction?

11. a. valid; by truth table (law of the excluded middle)
b. valid; law of the excluded middle   **13. a.** valid; indirect
b. valid; direct   **15.** valid; indirect   **17.** valid; indirect
19. valid; direct   **21.** invalid; fallacy of the inverse
23. invalid; fallacy of the inverse   **25.** valid; indirect and the law of double negation   **27.** invalid; fallacy of the converse
29. valid; direct   **31.** invalid; a statement is not equivalent to its converse.   **33.** valid; indirect   **35.** valid; indirect   **37.** If you learn mathematics, then you understand human nature. *(transitive)*
39. We do not go to the concert. *(indirect)*   **41.** $b = 0$ *(excluded middle)*   **43.** We do not interfere with the publication of false information. *(indirect)*   **45.** I will not eat that piece of pie. *(indirect)*
47. We will go to Europe. *(transitive, direct)*   **49.** Babies cannot manage crocodiles. *(transitive)*   **51.** None of my poultry are officers. *(transitive, contrapositive)*   **53.** Airsecond Aircraft Company suffers financial setbacks. *(direct, indirect)*   **55.** The janitor could not have taken the elevator because the building fuses were blown.
57. Guinea pigs do not appreciate Beethoven.   **59.** This is not an easy problem.

2.5 Problem Solving Using Logic, page 89

1. Moe sees two hands and two black hats and concludes his hat must be black. If Moe's hat were white, then Harry and Larry would each

have a solution because they would each see one white hat and one black hat with two hands raised. Thus, Moe knows that his hat must be black.   **3.** White, since this could happen only at the North Pole.
5. There can be only one yellow flower, so there must be 49 red flowers.   **7.** Curly committed the murder.   **9.** Only one question is necessary: Ask Connie (who falsely claimed to have the mixed bag) to pull out one fruit. Suppose she pulls out a peach; this means she has the bag containing two peaches. Then, Alice, who falsely claimed two peaches, must have two plums. This leaves Betty with the mixed bag. Suppose she pulls out a plum; this means she has the bag containing two plums. Then, Betty, who falsely claimed two plums, must have two peaches. This leaves Alice with the mixed bag.

2.6 Logic Circuits, page 93

5. $p \wedge q$

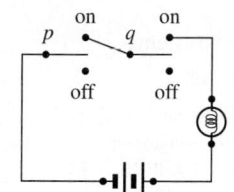

7. $\sim p \wedge q$

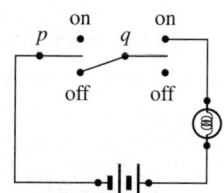

9. $\sim(p \vee q)$

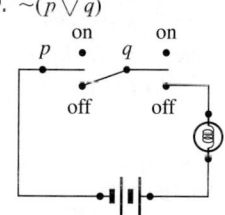

11. $p \rightarrow q$

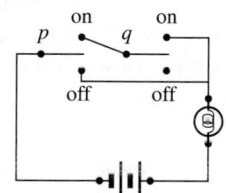

13. $p \rightarrow \sim q$

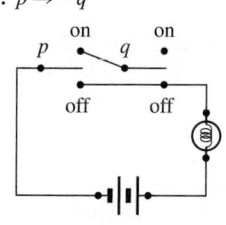

15. $p \leftrightarrow \sim q$

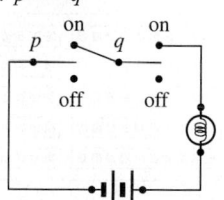

17.

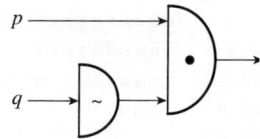

19.

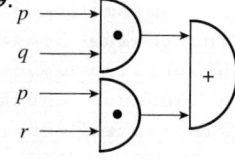

21.

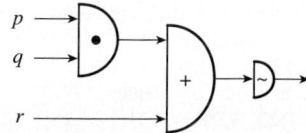

23.

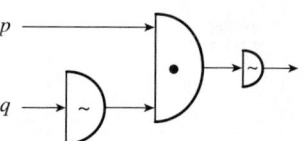

25. Notice $\sim p \rightarrow q \Leftrightarrow p \vee q$

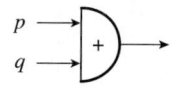

27. $\sim q \rightarrow \sim p \Leftrightarrow \sim p \vee q$

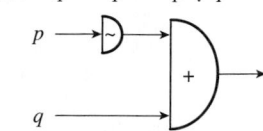

29. Let the committee members be a, b, and c, respectively. Light *on* represents a majority. The circuit is shown at the right.

Chapter 2 Review Questions, page 95

1. a. A *logical statement* is a declarative sentence that is either true or false.
b. A *tautology* is a logical statement in which the conclusion is equivalent to its premise.
c. $(p \rightarrow q) \leftrightarrow (\sim q \rightarrow \sim p)$ or, in words: A conditional may always be replaced by its contrapositive without having its truth value affected.

2.

p	q	$\sim p$	$p \wedge q$	$p \vee q$	$p \rightarrow q$	$p \leftrightarrow q$
T	T	F	T	T	T	T
T	F	F	F	T	F	F
F	T	T	F	T	T	F
F	F	T	F	F	T	T

3.

p	q	$p \wedge q$	$\sim(p \wedge q)$
T	T	T	F
T	F	F	T
F	T	F	T
F	F	F	T

4.

p	q	$\sim q$	$p \vee \sim q$	$\sim p$	$(p \vee \sim q) \wedge \sim p$	$[(p \vee \sim q) \wedge \sim p] \rightarrow \sim q$
T	T	F	T	F	F	T
T	F	T	T	F	F	T
F	T	F	F	T	F	T
F	F	T	T	T	T	T

5.

p	q	r	$p \wedge q$	$(p \wedge q) \wedge r$	$[(p \wedge q) \wedge r] \rightarrow p$
T	T	T	T	T	T
T	T	F	T	F	T
T	F	T	F	F	T
T	F	F	F	F	T
F	T	T	F	F	T
F	T	F	F	F	T
F	F	T	F	F	T
F	F	F	F	F	T

6.

p	q	$\sim p$	$\sim q$	$p \wedge q$	$\sim(p \wedge q)$	$\sim p \vee \sim q$	$\sim(p \wedge q) \leftrightarrow (\sim p \vee \sim q)$
T	T	F	F	T	F	F	T
T	F	F	T	F	T	T	T
F	T	T	F	F	T	T	T
F	F	T	T	F	T	T	T

7. $p \to q$

$$\dfrac{p}{\therefore q}$$

p	q	$p \to q$	$(p \to q) \wedge p$	$[(p \to q) \wedge p] \to q$
T	T	T	T	T
T	F	F	F	T
F	T	T	F	T
F	F	T	F	T

8. Answers vary. If you study hard, then you will get an A. You do not get an A. Therefore, you did not study hard. **9.** Answers vary; fallacy of the converse, fallacy of the inverse, or false chain pattern are the possibilities. **10.** Yes; it is indirect reasoning.

11. a. Some birds do not have feathers. **b.** No apples are rotten. **c.** Some cars have two wheels. **d.** All smart people attend college. **e.** You go on Tuesday and you can win the lottery.

12. a. T **b.** T **c.** T **d.** T **e.** T **13. a.** If P is a prime number, then $P + 2$ is a prime number. **b.** Either P or $P + 2$ is a prime number. **14. a.** Let p: This machine is a computer; q: This machine is capable of self-direction. $p \to \sim q$ **b.** Contrapositive: $\sim(\sim q) \to \sim p$; $q \to \sim p$. If this machine is capable of self-direction, then it is not a computer. **15.** Let p: There are a finite number of primes; q: There is some natural number, greater than 1, that is not divisible by any prime. Then the argument in symbolic form is: $p \to q$

$$\dfrac{\sim q}{\therefore \sim p}$$

Conclusion: There are infinitely many primes (assuming that "not a finite number" is the same as "infinitely many"). This is indirect reasoning.

16. a.

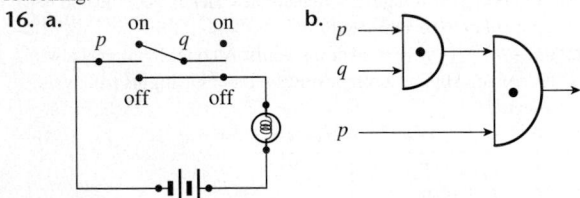

b.

17. Let d: I attend to my duties; r: I am rewarded; ℓ: I am lazy. Symbolic argument:

(1) $d \to r$ (2) $\ell \to \sim r$
(2) $\ell \to \sim r$ (3) ℓ
(3) ℓ (4) $\therefore \sim r$ Direct reasoning

(1) $d \to r$
(4) $\dfrac{\sim r}{\therefore \sim d}$ Indirect reasoning

Conclusion: I do not attend to my duties.

18. Let o: This is organic food; h: This is healthy food; s: This is an artificial sweetener; p: This is a prune. Symbolic argument:

(1) $o \to h$ (3) $p \to o$
(2) $s \to \sim h$ (1) $o \to h$
(3) $p \to o$ (4) $\dfrac{p \to o}{p \to h}$ Transitive
           (2') $\dfrac{h \to \sim s}{p \to \sim s}$ Law of contraposition
            Transitive

Conclusion: No prune is an artificial sweetener.

19. Let s: This is a square; r: This is a rectangle; q: This is a quadrilateral; p: This is a polygon. Symbolic argument:

(1) $s \to r$
(2) $\dfrac{r \to q}{\therefore \ s \to q}$ Transitive
(3) $\dfrac{q \to p}{\therefore \ s \to p}$ Transitive

Conclusion: All squares are polygons.

20.

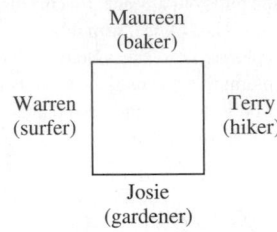

Chapter 3 The Nature of Numeration Systems

3.1 Early Numeration Systems, page 108

7. a. Roman, Egyptian **b.** Roman, Babylonian **c.** Egyptian, Roman, Babylonian **d.** Egyptian, Roman, Babylonian **e.** Roman, Babylonian **f.** Roman **9.** $\frac{1}{200}$ **11.** 100,010 **13.** $\frac{1}{12}$ **15.** $\frac{1}{100}$ **17.** 1,000,001 **19.** 1,997 **21.** 709 **23.** 2,001 **25.** 400,000 **27.** 9,712 **29.** 24 **31.** 671 **33.** 28 **35.** 25 **37.** One; the only one known to be going to St. Ives is myself. **39. a.** ∩∩∩∩∩∩ ||||| **b.** LXXV **c.** ▼ ◁ ▼▼▼▼▼ **41. a.** 𝟗𝟗𝟗𝟗𝟗∩∩| **b.** DXXI **c.** ▼▼▼▼▼▼▼◁◁◁◁▼ **43. a.** 𝄢𝄢| **b.** MMI **c.** ◁◁◁▼▼▼◁◁▼ **45.** 133 **47.** 𝟗∩∩∩∩∩∩∩∩ ||||||| || **49.** ◁◁◁◁◁ **51.** ◁◁▼▼▼▼▼ **53. a.** MDCLXVI **b.** MCDXLIV **55. a.** 9 **b.** 8 **c.** 99 **d.** 89 **e.** 999 **57.** ◁◁ by ◁▼◁◁

3.2 Hindu–Arabic Numeration System, page 113

1.

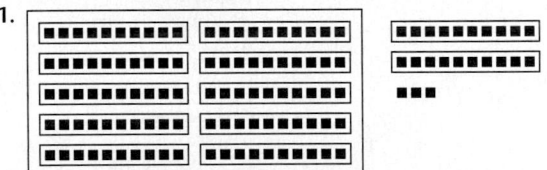

3. Let X represent ■■■■■■■■ of the larger groups.

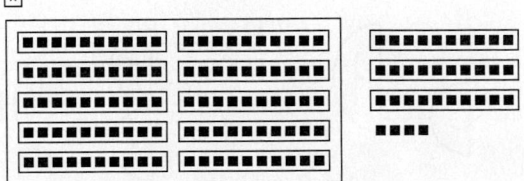

5. $b^n = b \cdot b \cdot \cdots \cdot b$, where there are n factors of b (for any counting number n) **7.** 5 units **9.** 5 thousandths **11. a.** 100,000 **b.** 1,000 **13. a.** 0.0001 **b.** 0.001 **15. a.** 5,000 **b.** 500 **17. a.** 0.06 **b.** 0.00009 **19.** 10,234 **21.** 521,658 **23.** 7,000,000.03 **25.** 500,457.34 **27.** 20,600.40769 **29.** $7 \times 10^5 + 2 \times 10^4 + 8 \times 10^3 + 4 \times 10^2 + 7 \times 10^0$ **31.** $2 \times 10^1 + 7 \times 10^0 + 5 \times 10^{-1} + 7 \times 10^{-2} + 2 \times 10^{-3}$ **33.** $5 \times 10^2 + 2 \times 10^1 + 1 \times 10^0$ **35.** $2 \times 10^6 + 3 \times 10^5 + 5 \times 10^3 + 6 \times 10^2 + 8 \times 10^1 + 1 \times 10^0$

37. $5 \times 10^3 + 2 \times 10^2 + 4 \times 10^1 + 5 \times 10^0 + 5 \times 10^{-1}$
39. $1 \times 10^5 + 1 \times 10^{-3}$ **41.** $8 \times 10^0 + 5 \times 10^{-5}$
43. $5 \times 10^4 + 7 \times 10^3 + 2 \times 10^2 + 8 \times 10^1 + 5 \times 10^0 + 9 \times 10^{-1}$
$+ 3 \times 10^{-2} + 6 \times 10^{-3} + 1 \times 10^{-4}$ **45.** 3,201
47. 5,001,005 **49.** 8,009,026

51. **53.**

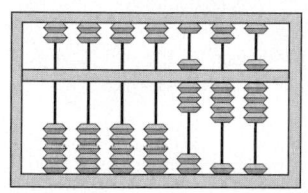 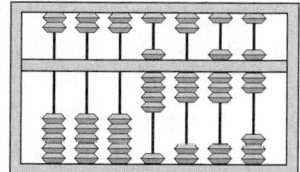

55. **57.**

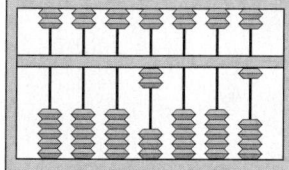 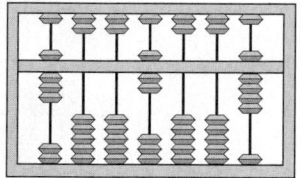

59. 22, 23, 24, 30, 31, 32, 33, 34, 40, . . .

3.3 Different Numeration Systems, page 118

7. a. 13 **b.** 23_{five} **c.** $10_{thirteen}$ **d.** 15_{eight} **e.** 1101_{two}
f. 11_{twelve} **9. a.** $6 \times 8^2 + 4 \times 8^1 + 3 \times 8^0$
b. $5 \times 12^3 + 3 \times 12^2 + 8 \times 12^1 + 7 \times 12^0 + 9 \times 12^{-1}$
11. a. $1 \times 2^5 + 1 \times 2^4 + 1 \times 2^2 + 1 \times 2^1 + 1 \times 2^0 + 1 \times 2^{-1}$
$+ 1 \times 2^{-4}$
b. $5 \times 6^3 + 4 \times 6^2 + 1 \times 6^1 + 1 \times 6^0 + 1 \times 6^{-1} + 2 \times 6^{-3}$
$+ 3 \times 6^{-4}$
13. a. $3 \times 4^5 + 2 \times 4^4 + 3 \times 4^3 + 2 \times 4^{-1}$
b. $2 \times 5^5 + 3 \times 5^4 + 4 \times 5^3$
15. 343 **17.** 13.75 **19.** 116 **21.** 11.625 **23.** 807
25. 66 **27.** 351.125 **29.** 3,042 **31.** 21310_{four}
33. $2E7_{twelve}$ **35.** 1147_{eight} **37.** 1001100111_{two}
39. 2214_{seven} **41.** $28T3_{twelve}$ **43.** 1030_{four} **45.** 3720_{eight}
47. 7 weeks, 3 days **49.** 4 ft, 7 in. **51.** 3 gross, 5 doz, 8 units
53. $84 = 314_{five}$ so you would need 8 coins
55. $954_{twelve} = 1,360$ **57.** $44 = 62_{seven}$; 6 weeks and 2 days
59. $29 = 15_{twenty\text{-}four}$; 1 day and 5 hours

3.4 Binary Numeration System, page 123

3. 39 **5.** 167 **7.** 13 **9.** 11 **11.** 29 **13.** 27
15. 99 **17.** 184 **19.** 1101_{two} **21.** 100011_{two}
23. 110011_{two} **25.** 1000000_{two} **27.** 10000000_{two}
29. 1100011011_{two} **31.** 68 79 **33.** 69 78 68 **35.** HAVE
37. STUDY **39.** 101_{two} **41.** 1101_{two} **43.** 10_{two}
45. 100011_{two} **47.** The computer is correct because there must
be either war or peace. **49. a.** 101_{two} **b.** 110_{two}
51. a. $001\ 110\ 111_{two}$ **b.** $110\ 010\ 100_{two}$
53. a. $101\ 111\ 000\ 000_{two}$ **b.** $000\ 100\ 011\ 010\ 000_{two}$
55. a. 3_{eight} **b.** 1_{eight} **57.** 400567_{eight} **59.** 453127_{eight}

3.5 History of Calculating Devices, page 134

11. a. 27 **b.** 63 **c.** 54 **13. a.** 243 **b.** 432 **c.** 504
15. ENIAC, UNIVAC, Cray, Altair, and Apple *Answers to
Problems 17–33 vary.* **17.** Aristophanes developed a finger-
counting system about 500 B.C. **19.** Babbage built a calculating
machine in the 19th century. **21.** Berry helped Atanasoff build the
first computer. **23.** Cray developed the first supercomputer.

25. Engelbart invented the computer mouse. **27.** Jobs was the
cofounder of Apple Computer and codesigned the Apple II computer.
29. Leibniz built a calculating device in 1695. **31.** Napier
invented a calculating device to do multiplication (in 1617).
33. Wozniak codesigned the Apple II computer. **35.** yes; speed,
complicated computations **37.** yes; speed, repetition **39.** yes;
repetition **41.** yes; ability to make corrections easily **43.** yes
(but not completely); it can help with some of the technical aspects
45. yes; speed, complicated computations, repetition **47.** E
49. D

Chapter 3 Review Questions, page 137

1. Answers vary. The position in which the individual digits are listed
is relevant; examples will vary. **2.** Answers vary. Addition is
easier in a simple grouping system; examples will vary.
3. Answers vary. It uses ten symbols; it is positional; it has a place-
holder symbol (0); and it uses 10 as its basic unit for grouping.
4. Answers vary. Should include finger calculating, Napier's rods,
Pascal's calculator, Leibniz' reckoning machine, Babbage's difference
and analytic engines, ENIAC, UNIVAC, Atanasoff's and Eckert and
Mauchly's computers, and the dispute they had in proving their position in the history of computers. Should also include the role and impact of the Apple and Macintosh computers, as well as the
supercomputers (such as the Cray). **5.** Answers vary. Should include illegal (breaking into another's computer; adding, modifying, or
destroying information; copying programs without authorization or
permission) and ignorance (assuming that output information is correct, or not using software for purposes for which it was intended).
6. Answers vary. **a.** the physical components (mechanical,
magnetic, electronic) of a computer system **b.** the routine programs,
and associated documentation in a computer system **c.** the process
of creating, modifying, deleting, and formatting text and materials
d. a device connected to a computer that allows the computer to communicate with other computers using electronic cables or phone lines
e. electronic mail—that is, messages sent along computer modems
f. Random-Access Memory or memory where each location is uniformly accessible, often used for the storage of a program and data
being processed **g.** an electronic place to exchange information with
others, usually on a particular topic **7.** 10^9
8. $4 \times 10^2 + 3 \times 10^1 + 6 \times 10^0 + 2 \times 10^{-1} + 1 \times 10^{-5}$
9. $5 \times 8^2 + 2 \times 8^1 + 3 \times 8^0$
10. $1 \times 2^6 + 1 \times 2^3 + 1 \times 2^2 + 1 \times 2^1$
11. 4,020,005.62
12. $1 \times 2^4 + 1 \times 2^3 + 1 \times 2^2 + 1 \times 2^0 = 29$
13. $1 \times 2^6 + 1 \times 2^5 + 1 \times 2^4 + 1 \times 2^3 + 1 \times 2^1 + 1 \times 2^0 = 123$
14. $1 \times 3^2 + 2 \times 3^1 + 2 \times 3^0 = 17$
15. $8 \times 12^2 + 2 \times 12^1 + 1 \times 12^0 = 1,177$

16.
```
     0  r.1
  2) 1  r.1
  2) 3  r.0
  2) 6  r.0
  2)12
12 = 1100_two
```

17.
```
      0  r.1
   2) 1  r.1
   2) 3  r.0
   2) 6  r.1
   2)13  r.0
   2)26  r.0
    2)52
52 = 110100_two
```

18.
```
         0  r.1
     2)  1  r.1
     2)  3  r.1
     2)  7  r.1
     2) 15  r.1
     2) 31  r.0
     2) 62  r.1
     2) 125 r.0
     2) 250 r.0
     2) 500 r.1
     2)1001 r.1
      2)2003
2,003 = 11111010011_two
```

19. $11110100001001000000_{two}$

20. a.

$$\begin{array}{r} 0 \text{ r.9} \\ 12)\overline{9} \text{ r.2} \\ 12)\overline{110} \text{ r.}E \\ 12)\overline{1,331} \end{array}$$

$1,331 = 92E_{twelve}$

b.

$$\begin{array}{r} 0 \text{ r.4} \\ 5)\overline{4} \text{ r.0} \\ 5)\overline{20} \text{ r.0} \\ 5)\overline{100} \end{array}$$

$100 = 400_{five}$

CHAPTER 4 THE NATURE OF NUMBERS

4.1 Natural Numbers, page 146

1. $\mathbb{N} = \{1, 2, 3, 4, 5, \ldots\}$ **3.** Subtraction is defined in terms of addition: $m - n = x$ means $m = n + x$. **5.** For an operation $\circ$ and elements a and b in a set S, $a \circ b = b \circ a$. **7.** For operations $\circ$ and $\otimes$ and elements a b, and c in S, $a \otimes (b \circ c) = (a \otimes b) \circ (a \otimes c)$. **9.** The associative property changes the grouping, and the commutative property changes the order. **11. a.** $4 + 4 + 4$ **b.** $3 + 3 + 3 + 3$ **13. a.** $184 + 184$ **b.** $2 + 2 + \cdots + 2$ (a total of 184 terms) **15. a.** $y + y + y + \cdots + y$ (a total of x terms) **b.** $x + x + \cdots + x$ (a total of y terms) **17.** commutative **19.** associative **21.** commutative **23.** commutative **25.** commutative **27.** commutative **29.** None are associative. **31.** Answers vary; dictionary definitions are often circular. **33. a.** $-i$ **b.** -1 **c.** 1 **d.** $-i$ **e.** i **f.** -1 **35.** Answers vary; yes **37.** Answers vary; **a.** yes **b.** yes **39.** yes; answers vary **41. a.** $\square$ **b.** $\circ$ **c.** yes; no, $\circ \cdot \triangle \neq \triangle \cdot \circ$ **d.** yes **43.** Check: $a \downarrow (b \rightarrow c) = (a \downarrow b) \rightarrow (a \downarrow c)$. Try several examples; it holds. **45. a.** $5 \times (100 - 1) = 500 - 5 = 495$ **b.** $4 \times (90 - 2) = 360 - 8 = 352$ **c.** $8 \times (50 + 2) = 400 + 16 = 416$ **47.** yes **49.** yes; even + even = even **51.** yes; odd $\times$ odd = odd **53.** yes; let a and b be any elements in S. Then, $a \odot b = 0 \cdot a + 1 \cdot b = b$ and we know that b is an element of S. **55.** It is commutative but not associative: $2 \otimes 3 = 10$ and $3 \otimes 2 = 10$; try other examples. $(2 \otimes 3) \otimes 4 \neq 2 \otimes (3 \otimes 4)$ **57.** yes **59.** Answers vary; $0 + 1 + 2 - 3 - 4 + 5 - 6 + 7 + 8 - 9 = 1$

4.2 Prime Numbers, page 158

7. a. prime; use sieve **b.** not prime; $3 \cdot 19$ **c.** not prime; only 1 divisor **d.** prime; check primes under 45 **9. a.** prime; use sieve **b.** prime; use sieve **c.** not prime; $3^2 \cdot 19$ **d.** not prime; $3 \cdot 23 \cdot 29$ **11. a.** T **b.** F **c.** T **d.** F **13. a.** F **b.** T **c.** T **d.** T **15.** Use the sieve of Eratosthenes; you only need to cross out the multiples of primes up to 17. The list of primes is: 2, 3, 5, 7, 11, 13, 17, 19, 23, 29, 31, 37, 41, 43, 47, 53, 59, 61, 67, 71, 73, 79, 83, 89, 97, 101, 103, 107, 109, 113, 127, 131, 137, 139, 149, 151, 157, 163, 167, 173, 179, 181, 191, 193, 197, 199, 211, 223, 227, 229, 233, 239, 241, 251, 257, 263, 269, 271, 277, 281, 283, and 293. **17. a.** $2^3 \cdot 3$ **b.** $2 \cdot 3 \cdot 5$ **c.** $2^2 \cdot 3 \cdot 5^2$ **d.** $2^4 \cdot 3^2$ **19. a.** $2^3 \cdot 3 \cdot 5$ **b.** $2 \cdot 3^2 \cdot 5$ **c.** $3 \cdot 5^2$ **d.** $3 \cdot 5^2 \cdot 13$ **21.** prime **23.** prime **25.** $13 \cdot 29$ **27.** $3 \cdot 5 \cdot 7$ **29.** prime **31.** $3^2 \cdot 5 \cdot 7$ **33.** $3^4 \cdot 7$ **35.** $19 \cdot 151$ **37.** 12; 360 **39.** 3; 2,052 **41.** 1; 252 **43.** 15; 1,800 **45.** The least common multiple of 6 and 8 is 24, so the next night off together is in 3 weeks and 3 days. **47.** Answers vary. **49.** 1, 3, 7, 9, 13, 15, 21, 25, 31, 33, 37, 43, 49, 51, 63, 67, 69, 73, 75, 79, 87, 93, and 99 **51.** Let $2n + 1$, $2n + 3$, and $2n + 5$ be any 3 consecutive odd numbers. If $n = 1$, we have the prime triplets 3, 5, and 7. If $n > 1$, then one of the three numbers must be divisible by 3. Suppose the first is divisible by 3; then they are not prime triplets. If the first is not divisible by 3, then dividing it by 3 leaves a remainder of 1

or 2. If it leaves a remainder of 1, then the middle number is divisible by 3. If it leaves a remainder of 2, then the last one is divisible by 3. In all cases, the numbers will not be prime triplets. **53. a.** yes **b.** yes **c.** yes **55.** $2 \cdot 3 \cdot 5 \cdot 7 \cdot 11 \cdot 13 + 1 = 30,031$, which is not prime, since $30,031 = 59 \cdot 509$ **57.** Answers vary; some other possibilities are: 2, 17, 37, 101, 197, 257, 401, 577, 677, 1297, 1601, 2917, 3137, 4357, 5477, 7057, 8101, 8837. A computer could also be used to answer this question. **59.** $2^{5-1}(2^5 - 1) = 2^4(2^5 - 1) = 16(31) = 496$. The proper divisors of 496 are 1, 2, 4, 8, 16, 31, 62, 124, and 248. The sum of these numbers is 496.

4.3 Integers, page 166

1. If $x = 0$, then $x + 0 = 0 + x = x$. To add nonzero integers x and y, look at the signs of x and y:

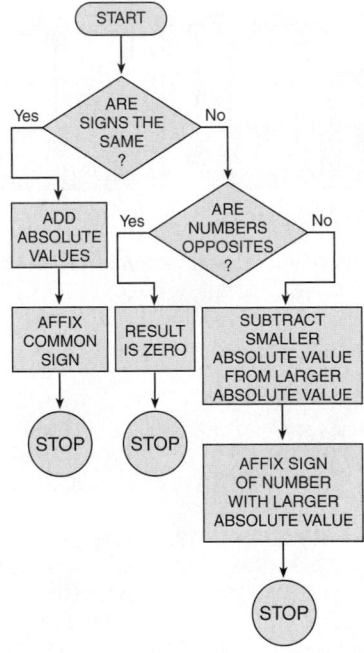

3. If $x = 0$, then $x \cdot 0 = 0 \cdot x = 0$. To multiply nonzero integers x and y, look at the signs of x and y:

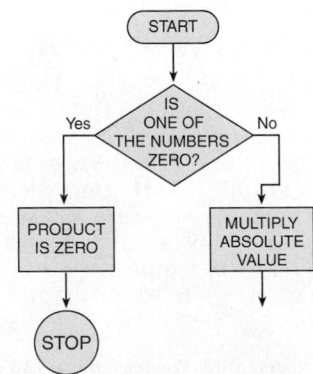

(continues on page A9)

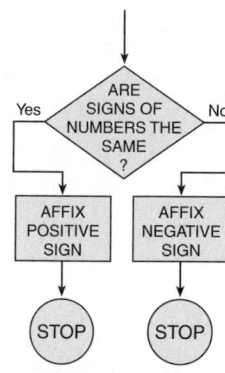

5. $0 \div 5$ means $\dfrac{0}{5}$, which is equal to 0; on the other hand, $5 \div 0$ is not defined. **7. a.** 30 **b.** 30 **c.** 0 **d.** -30 **e.** 0
9. a. 8 **b.** -2 **11. a.** 10 **b.** -4 **13. a.** -7 **b.** 12
15. a. 3 **b.** -3 **17. a.** -18 **b.** -20 **19. a.** -70
b. 70 **21. a.** -3 **b.** 7 **23. a.** 132 **b.** -1
25. a. -56 **b.** -75 **27. a.** -4 **b.** 4 **29. a.** 150
b. -19 **31. a.** 5 **b.** 10 **33. a.** -12 **b.** 15
35. a. 0 **b.** -8 **37. a.** 0 **b.** -23 **39. a.** -7 **b.** 0
41. a. 8 **b.** -2 **43. a.** 6 **b.** 7 **45. a.** 1 **b.** 22
47. a. 4 **b.** 16 **49. a.** 44 **b.** 16 **51. a.** 1 **b.** 1
c. 1 **d.** 1 **e.** 1 **53. a.** 1 **b.** -1 **c.** -1 **d.** 1 **e.** -1
55. Answers vary. **a.** An operation $\star$ is commutative for a set S if $a \star b = b \star a$ for all elements a and b in S. **b.** Yes, addition is commutative for the integers (try several examples). **c.** Not commutative for subtraction since $7 - 5 = 2$, but $5 - 7 \neq 2$. **d.** Yes, multiplication is commutative for integers. **e.** Not commutative for division since $8 \div 4 = 2$, but $4 \div 8 \neq 2$. **57.** Use the fact that the integers are closed for addition and that the opposite of any integer is also an integer. From this and the definition of subtraction, the desired results follows.
59. a. 12345668765433
b. Search for patterns:
$$1 \times 9 = 9$$
$$12 \times 99 = 1{,}188$$
$$123 \times 999 = 122{,}877$$
$$1{,}234 \times 9{,}999 = 12{,}338{,}766$$
By patterns: 12,345,668,765,433

4.4 Rational Numbers, page 174

7. a. $\frac{1}{5}$ **b.** $\frac{1}{4}$ **9. a.** 2 **b.** 2 **11. a.** 3 **b.** $\frac{2}{3}$
13. a. $\frac{1}{8}$ **b.** $\frac{1}{3}$ **15. a.** $\frac{6}{35}$ **b.** $\frac{3}{20}$ **17. a.** $\frac{17}{21}$ **b.** $\frac{7}{8}$
19. a. $\frac{-92}{105}$ **b.** $\frac{5}{6}$ **21. a.** $\frac{1}{9}$ **b.** $\frac{71}{63}$ **23. a.** $\frac{20}{7}$ **b.** $\frac{-7}{27}$
25. a. $\frac{8}{15}$ **b.** $\frac{7}{10}$ **27. a.** $\frac{10}{21}$ **b.** $\frac{1}{20}$ **29. a.** 1 **b.** -1
31. a. 36 **b.** -25 **33. a.** $\frac{18}{35}$ **b.** $\frac{4}{45}$ **35. a.** $\frac{1}{12}$ **b.** $\frac{-53}{48}$
37. a. 20 **b.** 6 **39. a.** $\frac{-5}{43}$ **b.** $\frac{28}{33}$ **41. a.** $\frac{-3}{4}$ (*Note:* Use distributive property) **b.** $\frac{-3}{4}$ **43.** $\frac{2{,}137}{10{,}800}$ **45.** $\frac{971}{3{,}060}$
47. $\frac{10{,}573}{13{,}020}$
49. Given any two rationals $\frac{a}{b}$ and $\frac{c}{d}$; show that $\frac{a}{b} - \frac{c}{d}$ is rational.
Now, $\frac{a}{b} - \frac{c}{d} = \frac{ad - bc}{bd}$ by the definition of subtraction. Also, ad and bc are integers and bd is a nonzero integer since a, c are integers and b, d are nonzero integers and the set of integers is closed for multiplication. Finally, $ad - bc$ is an integer because the integers are closed for subtraction. Therefore, $\frac{ad - bc}{bd}$ is a rational by the definition of a rational number. **51.** The set of integers, $\mathbb{Z}$, is closed for addition,

subtraction, and multiplication, but not for division since $4 \div 5$ is not an integer. **53.** Yes; answers vary. **55.** Answers vary; $\frac{3}{4} = \frac{1}{2} + \frac{1}{4}$ **57.** Answers vary; $\frac{67}{120} = \frac{1}{3} + \frac{1}{8} + \frac{1}{10}$ or $\frac{1}{2} + \frac{1}{20} + \frac{1}{120}$ **59.** It checks.

4.5 Irrational Numbers, page 182

7. true **9. a.** 30 **b.** 36 **c.** 807 **d.** 169 **11. a.** a
b. xy **c.** $4b$ **d.** $40w$ **13. a.** irrational; 3.162 **b.** irrational;
5.477 **c.** irrational; 9.870 **d.** irrational; 7.389 **15. a.** rational;
13 **b.** rational; 20 **c.** irrational; 23.141 **d.** irrational; 22.459
17. a. rational; 32 **b.** rational; 44 **c.** irrational; 1.772
d. irrational; 1.253 **19. a.** $10\sqrt{10}$ **b.** $20\sqrt{7}$ **c.** $8\sqrt{35}$
d. $21\sqrt{10}$ **21. a.** $\frac{1}{2}\sqrt{2}$ **b.** $\frac{1}{3}\sqrt{3}$ **c.** $\frac{1}{5}\sqrt{15}$ **d.** $\frac{1}{7}\sqrt{21}$
23. a. $\frac{1}{2}\sqrt{2}$ **b.** $-\frac{1}{3}\sqrt{3}$ **c.** $\frac{2}{5}\sqrt{5}$ **d.** $\frac{1}{2}\sqrt{10}$
25. a. simplified **b.** $x + 2$ **27. a.** $10\sqrt{2}$ **b.** 0
29. a. $3 + \sqrt{5}$ **b.** $2 - \sqrt{3}$ **31. a.** $1 - 3\sqrt{x}$ **b.** $-3 - \sqrt{x}$
33. a. $\frac{2x}{5y}\sqrt{y}$ **b.** $\frac{1}{4x}\sqrt{5xy}$ **35.** $\frac{1 - \sqrt{19}}{6}$ **37.** $6 + \sqrt{37}$
39. 24 ft **41.** 10 ft **43.** $\sqrt{8}$ in. or $2\sqrt{2}$ in. **45.** 200 ft
47. 15 ft **49.** Answers vary; 1.2323323332 . . .
51. Answers vary; 0.0919919991 . . . **53.** Solution comes from Pythagorean theorem. **a.** 2-in. square **b.** same **c.** 7-in. square
d. 9-in square **e.** 15-in. square
55.

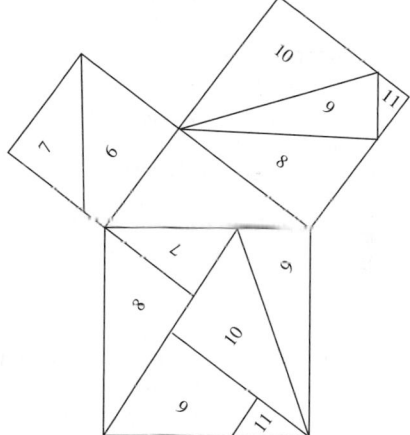

57. Answers vary; any number can be written as the sum of three triangular numbers. He must have meant 3 or fewer triangular numbers because to represent 2, for example, we write $1 + 1$. **59.** Consider the second diagonal: 1, 3, 6, 10, 15, 21, ... If we take the sum of any 2 adjacent terms, we obtain square numbers: 4, 9, 16, 25, ... (*Note:* Don't forget that we start counting with the 0th diagonal.)

4.6 Groups, Fields and Real Numbers, page 192

5. Let $\mathbb{S}$ be any set, let $\circ$ be any operation, and let a, b, and c be any elements of $\mathbb{S}$. We say that $\mathbb{S}$ is a **group** for the operation of $\circ$ if the following properties are satisfied: 1. The set $\mathbb{S}$ is *closed* for $\circ$: $(a \circ b) \in \mathbb{S}$. 2. The set $\mathbb{S}$ is *associative* for $\circ$: $(a \circ b) \circ c = a \circ (b \circ c)$. 3. The set $\mathbb{S}$ satisfies the *identity* for $\circ$: There exists a number $\mathrm{I} \in \mathbb{S}$ so that $x \circ \mathrm{I} = \mathrm{I} \circ x = x$ for *every* $x \in \mathbb{S}$. 4. The set $\mathbb{S}$ satisfies the *inverse* for $\circ$: For *each* $x \in \mathbb{S}$, there exists a corresponding $x^{-1} \in \mathbb{S}$ so that $x \circ x^{-1} = x^{-1} \circ x = \mathrm{I}$, where I is the identity element in $\mathbb{S}$.
7. a. $\mathbb{N}, \mathbb{Z}, \mathbb{Q}, \mathbb{R}$ **b.** $\mathbb{Q}, \mathbb{R}$ **c.** $\mathbb{Q}', \mathbb{R}$ **d.** $\mathbb{Q}', \mathbb{R}$ **e.** $\mathbb{Q}, \mathbb{R}$
f. $\mathbb{Q}, \mathbb{R}$ **g.** $\mathbb{Q}, \mathbb{R}$ **h.** $\mathbb{N}, \mathbb{Z}, \mathbb{Q}, \mathbb{R}$ **i.** $\mathbb{Z}, \mathbb{Q}, \mathbb{R}$ **9. a.** 1.5
b. $0.\overline{7}$ **c.** 0.6 **d.** 1.8 **11. a.** $0.\overline{6}$ **b.** $2.1\overline{53846}$ **c.** 5
d. 1.09 **13. a.** $\frac{1}{2}$ **b.** $\frac{4}{5}$ **15. a.** $\frac{9}{20}$ **b.** $\frac{117}{500}$

17. a. $\frac{987}{10}$ **b.** $\frac{63}{100}$ **19. a.** $\frac{153}{10}$ **b.** $\frac{139}{20}$ **21. a.** 3.179
b. -5.504 **c.** 1.901 **d.** -6.31 **23. a.** -0.13112
b. -65.415 **c.** 12.4 **d.** 0.85 **25. a.** 7.46 **b.** -5.15
c. 72.6 **d.** 45.96 **27.** commutative **29.** distributive
31. commutative **33.** associative **35.** identity
37. $\mathbb{N}$ is not a group for $-$ since it is not closed. **39.** $\mathbb{W}$ is not a
group for $\times$ since the inverse property is not satisfied.
41. $\mathbb{Z}$ is not a group for $\times$ since the inverse property is not satisfied.
43. $\mathbb{Q}$ is not a group for $\times$ since there is no multiplicative inverse for
the rational number 0.
45. a.

$\times$	1	2	3	4
1	1	2	3	4
2	2	4	6	8
3	3	6	9	12
4	4	8	12	16

b.

$*$	1	2	3	4
1	2	2	2	2
2	4	4	4	4
3	6	6	6	6
4	8	8	8	8

   c. Operation $\times$: Not closed; associative; identity is 1; no inverse
   property; commutative
   Operation $*$: Not closed; not associative; no identity; no inverse property; not commutative
Answers for Problems 47 *and* 49 *may vary.* **47.** 2.5; $\frac{2\pi}{3}$
49. 4.$\overline{5}$; 4.545545554 . . .
51. $\left\{\frac{1}{2}, \frac{1}{3}, \frac{2}{3}, \frac{1}{4}, \frac{3}{4}, \frac{1}{5}, \frac{2}{5}, \frac{3}{5}, \frac{4}{5}, \frac{1}{6}, \frac{5}{6}, \frac{1}{7}, \frac{2}{7}, \ldots\right\}$ **53.** $a \circ b = ab$
55. $a \circ b = a + b + 1$ **57.** $a \circ b = 2(a + b)$
59. $a \circ b = a^2 + 1$

4.7 Discrete Mathematics, page 202

5. a. 3 **b.** 10 **7. a.** 5 **b.** 10 **9. a.** 12 **b.** 8
11. a. 8 **b.** 6 **13. a.** 8 **b.** 1 **15. a.** T **b.** F
17. a. T **b.** T **19. a.** T **b.** T **21. a.** 0, (mod 5)
b. 8, (mod 12) **23. a.** 3, (mod 4) **b.** 3, (mod 5)
25. a. 2, (mod 5) **b.** 2, (mod 8) **27. a.** 3, (mod 5)
b. 10, (mod 11) **29. a.** 4, (mod 7) **b.** 4, (mod 5)
31. a. 5, (mod 6) **b.** 3, (mod 7) **33. a.** 1, 3, (mod 4)
b. 2, (mod 9) **35. a.** 0, (mod 4) **b.** 2, (mod 4)
37. all x's (mod 2) **39. a.** Sunday **b.** Thursday **c.** Wednesday **d.** Monday **41.** Let $x =$ the number of miles to aunt's
house. Then the total mileage for the six round trips is $12x \equiv 8$, (mod
10). Solving, $x \equiv 4, 9$, (mod 10). The possible distances are 4, 14, 24,
34, . . . or 9, 19, 29, 39, ... **43.** She lives 14 miles from your
house. **45.** Answers vary; it is a group for addition modulo 7.
47.

+	0	1	2	3	4	5	6	7	8	9	10
0	0	1	2	3	4	5	6	7	8	9	10
1	1	2	3	4	5	6	7	8	9	10	0
2	2	3	4	5	6	7	8	9	10	0	1
3	3	4	5	6	7	8	9	10	0	1	2
4	4	5	6	7	8	9	10	0	1	2	3
5	5	6	7	8	9	10	0	1	2	3	4
6	6	7	8	9	10	0	1	2	3	4	5
7	7	8	9	10	0	1	2	3	4	5	6
8	8	9	10	0	1	2	3	4	5	6	7
9	9	10	0	1	2	3	4	5	6	7	8
10	10	0	1	2	3	4	5	6	7	8	9

$\times$	0	1	2	3	4	5	6	7	8	9	10
0	0	0	0	0	0	0	0	0	0	0	0
1	0	1	2	3	4	5	6	7	8	9	10
2	0	2	4	6	8	10	1	3	5	7	9
3	0	3	6	9	1	4	7	10	2	5	8
4	0	4	8	1	5	9	2	6	10	3	7
5	0	5	10	4	9	3	8	2	7	1	6
6	0	6	1	7	2	8	3	9	4	10	5
7	0	7	3	10	6	2	9	5	1	8	4
8	0	8	5	2	10	7	4	1	9	6	3
9	0	9	7	5	3	1	10	8	6	4	2
10	0	10	9	8	7	6	5	4	3	2	1

49. no **51.** no **53 a.** 8 **b.** 5 **55.** The possible check
digits are 0, 1, 2, . . . , 10. The check digit is 10. **57.** Assign
eleven teams the numbers 1 to 11. On the first day, all teams whose
sum is 1, (mod 11) play; day 2 matches those whose sum is 2,
(mod 11); . . . The last team plays the leftover numbered team so each
team plays every day. The schedule follows:

Day 1:	1–11;	2–10;	3–9:	4–8;	5–7	6–12
Day 2:	2–11;	3–10;	4–9;	5–8;	6–7;	1–12
Day 3:	3–11;	4–10;	5–9;	6–8;	1–2;	7–12
Day 4:	4–11;	5–10;	6–9;	7–8;	1–3;	2–12
Day 5:	5–11;	6–10;	7–9;	1–4;	2–3;	8–12
Day 6:	6–11;	7–10;	8–9;	1–5;	2–4;	3–12
Day 7:	7–11;	8–10;	1–6;	2–5;	3–4;	9–12
Day 8:	9–10;	1–7;	2–6;	3–5;	8–11;	4–12
Day 9:	9–11;	1–8;	2–7;	3–6;	4–5;	10–12
Day 10:	10–11;	1–9;	2–8;	3–7;	4–6;	5–12
Day 11:	1–10;	2–9;	3–8;	4–7;	5–6;	11–12

Note: The answer is not unique. In fact, there are 39,916,800
possible solutions.
59. The smallest result is 785.

4.8 Cryptography, page 207

1. 14-5-22-5-18-29-19-1-25-29-14-5-22-5-18-28
3. 25-15-21-29-2-5-20-29-25-15-21-18-29-12-9-6-5-28 **5.** ARE
WE HAVING FUN YET **7.** FAILURE TEACHES SUCCESS
9. Divide by 8. **11.** Subtract 2 and then divide by 20.
13. Divide by 2, then subtract 2, then divide by 4.
15. MOHOY .CQ MOHOYZ **17.** ZFREIOPEZFRLE WQOC
19. HUMPTY DUMPTY IS A FALL GUY. **21.** MIDAS HAD A
GILT COMPLEX. **23.** D UPEOTEHEQAUXUJGEWAUOEHE-
QHGHEXAUNTDDCRZEDN UULEHRQECEPCLLED UPEVUJE-
HERCRNUOXJGTAB **25.** ANYONE WHO SLAPS CATSUP,
MUSTARD, AND RELISH ON HIS HOT DOG IS TRULY A MAN
FOR ALL SEASONINGS. **27.** CRYSTAL-CLEAR AIR OF
ROCKY MOUNTAINS PROVIDES IDEAL ENVIRONMENT FOR
WEATHER STATION. **29.** $3,915 + 15 + 4,826 = 8,756$

Chapter 4 Review Questions, page 210

1. $-4 + 5(-3) = -4 - 15 = -19$
2. $\frac{4}{7} \cdot \frac{9}{9} + \frac{5}{9} \cdot \frac{7}{7} = \frac{36}{63} + \frac{35}{63} = \frac{71}{63}$
3. $30 = 2^1 \cdot 3^1 \cdot 5^1 \cdot 7^0 \cdot 11^0$
   $42 = 2^1 \cdot 3^1 \cdot 5^0 \cdot 7^1 \cdot 11^0$
   $99 = 2^0 \cdot 3^2 \cdot 5^0 \cdot 7^0 \cdot 11^1$
   l.c.m. $= 2^1 \cdot 3^2 \cdot 5^1 \cdot 7^1 \cdot 11^1 = 6,930$

$$\frac{7}{2\cdot 3\cdot 5}\cdot\frac{\mathbf{3\cdot 7\cdot 11}}{\mathbf{3\cdot 7\cdot 11}}=\frac{1{,}617}{2\cdot 3^2\cdot 5\cdot 7\cdot 11}$$

$$\frac{5}{2\cdot 3\cdot 7}\cdot\frac{\mathbf{3\cdot 5\cdot 11}}{\mathbf{3\cdot 5\cdot 11}}=\frac{825}{2\cdot 3^2\cdot 5\cdot 7\cdot 11}$$

$$\frac{5}{3^2\cdot 11}\cdot\frac{\mathbf{2\cdot 5\cdot 7}}{\mathbf{2\cdot 5\cdot 7}}=\frac{350}{2\cdot 3^2\cdot 5\cdot 7\cdot 11}$$

$$\frac{2{,}792}{2\cdot 3^2\cdot 5\cdot 7\cdot 11}=\frac{1{,}396}{3^2\cdot 5\cdot 7\cdot 11}=\frac{1{,}396}{3{,}465}$$

4. $-\sqrt{10}\cdot\sqrt{10}=-10$

5. $\left(\frac{11}{12}+2\right)+\frac{-11}{12}=2+\left(\frac{11}{12}+\frac{-11}{12}\right)=2$

6. $\frac{3^{-1}+4^{-1}}{6}=\frac{\frac{1}{3}+\frac{1}{4}}{6}=\frac{\frac{7}{12}}{6}=\frac{7}{12}\times\frac{1}{6}=\frac{7}{72}$

7. $\frac{-7}{9}\cdot\frac{99}{174}+\frac{-7}{9}\cdot\frac{75}{174}=\frac{-7}{9}\left(\frac{99}{174}+\frac{75}{174}\right)=-\frac{7}{9}$

8. $\frac{-3+\sqrt{3^2+4(2)(3)}}{2(2)}=\frac{-3+\sqrt{33}}{4}$ **9.** $\frac{8}{3}$ (*Note:* $2\frac{2}{3}$ is

mixed number form, but $\frac{8}{3}$ is reduced.) **10.** $\frac{16}{18}=\frac{2\cdot 8}{2\cdot 9}=\frac{8}{9}$

11. $\frac{100}{825}=\frac{25\cdot 4}{25\cdot 33}=\frac{4}{33}$ **12.** $\frac{184}{207}=\frac{23\cdot 8}{23\cdot 9}=\frac{8}{9}$

13. $\frac{1{,}209}{2{,}821}=\frac{3\cdot 13\cdot 31}{7\cdot 13\cdot 31}=\frac{3}{7}$ **14.** 0.375, rational

15. $2.\overline{3}$, rational **16.** $0.\overline{428571}$, rational **17.** 6.25, rational
18. $0.\overline{230769}$, rational **19.** 89 is prime **20.** 101 is prime
21. 349 is prime (check primes up to 17)
22. $1{,}001=7\cdot 11\cdot 13$ (use a factor tree)
23. $6{,}825=3\cdot 5^2\cdot 7\cdot 13$ (use a factor tree)

24. $\frac{x}{5}=2$, (mod 8) means $x=5\cdot 2=10\equiv 2$, (mod 8)

25. $2x\equiv 3$, (mod 7); consider the set $\{0, 1, 2, 3, 4, 5, 6\}$ and try each, one at a time, to find $2\cdot 5=10\equiv 3$, (mod 7), so $x\equiv 5$, (mod 7).
26. $2x^2+7x+1\equiv 0$, (mod 2); consider the set $\{0, 1\}$ and try
each: $x\equiv 1$, (mod 2) **27.** $49=7^2$
$$1{,}001=7^1\cdot 11^1\cdot 13^1$$
$$2{,}401=7^4$$
$$\text{g.c.f.}=7^1=7$$

28. l.c.m. $=7^4\cdot 11^1\cdot 13^1=343{,}343$
29. $1\odot 2=1\times 2+1+2=2+1+2=5;$
$3\odot 4=3\times 4+3+4=12+3+4=19;$
$5\odot 19=5\times 19+5+19=95+5+19=119$
30. $(1\downarrow 2)\uparrow(2\downarrow 3)=1\uparrow 2=2$ **31.** For $a\neq 0$, multiplication is defined as: $a\times b$ means $\underbrace{b+b+b+\cdots+b}_{a\text{ addends}}$.

If $a=0$, then $0\times b=0$. **32.** $a-b=a+(-b)$

33. $\frac{a}{b}=m$ means $a=bm$ where $b\neq 0$.

34. If we use the definition of division, $\frac{x}{0}=m$, then $0\cdot m=x$.

If $x\neq 0$, then there is no solution because $0\cdot m=0$ for all numbers m.

Also, if $\frac{0}{0}=m$, then $0=0\cdot m$ which is true for *every* number m.

Thus, $\frac{0}{0}=0$ checks, and $\frac{0}{0}=1$ checks, and since two numbers equal

to the same number must also be equal, we obtain the statement $0=1$;

thus, we say $\frac{0}{0}$ is indeterminate.

35. Answers vary; $34.1011011101111011\ldots$

36. A **field** is a set $\mathbb{R}$, with two operations $+$ and $\times$ satisfying the
following properties for any elements $a, b, c\in\mathbb{R}$:

	Addition, +	Multiplication, ×
Closure:	1. $(a+b)\in\mathbb{R}$	2. $ab\in\mathbb{R}$
Associative:	3. $(a+b)+c$ $=a+(b+c)$	4. $(a\times b)\times c$ $=a\times(b\times c)$
Identity:	5. There exists $0\in\mathbb{R}$ so that $0+a=$ $a+0=a$ for every element a in $\mathbb{R}$.	6. There exists $1\in\mathbb{R}$ so that $1\times a=$ $a\times 1=a$ for every element a in $\mathbb{R}$.
Inverse:	7. For each $a\in\mathbb{R}$, there is a unique number $(-a)\in\mathbb{R}$ so that $a+(-a)$ $=(-a)+a=0$	8. For each $a\in\mathbb{R}$, $a\neq 0$, there is a unique number $\frac{1}{a}\in\mathbb{R}$ so that $a\times\frac{1}{a}=\frac{1}{a}\times a=1$
Commutative:	9. $a+b=b+a$	10. $ab=ba$
Distributive for multiplication over addition:	11. $a\times(b+c)=a\times b+a\times c$	

37. Let a, b, and c be any elements in $\mathbb{N}$.
Closure: $a\not{D}b\in\mathbb{N}$ because the g.c.f. is the product of factors and $\mathbb{N}$ is closed for multiplication.
Associative: $(a\not{D}b)\not{D}c=a\not{D}(b\not{D}c)$ because the g.c.f. is the prod-uct of factors and $\mathbb{N}$ is associative for multiplication.
Identity: Look for I so that $a\not{D}\mathrm{I}=\mathrm{I}\not{D}a=a$, but there is no such number.
Inverse: Since there is no identity, there can be no inverse property.
$\mathbb{N}$ is not a group for $\not{D}$, and therefore cannot be a commutative group

38. $\text{BMI}=\dfrac{703w}{h^2}$ **a.** 27.5 **b.** 25.1 **c.** $w=155$ lb

39. You can use a sieve or prime factorizations. A door will be opened or closed by a tenant only if the door number can be divided evenly by the number of that tenant. For example, door 9 will be touched (opened or closed) by tenants 1, 3, and 9; door 10 by tenants 1, 2, 5, and 10. Thus, the only doors left open are those with an odd number of divi-sors. The open doors are the perfect squares: 1, 4, 9, 16, 25, 36, . . . , 841, 900, 961. There are 31 doors left open.

40. 8 ft = 96 in. and $\dfrac{96}{8}=12$, so 12 stairs are necessary. The total

length of the segments is $12+8=20$ ft, and the length of the
diagonal is $\sqrt{12^2+8^2}=\sqrt{208}=4\sqrt{13}$.

CHAPTER 5 THE NATURE OF ALGEBRA

5.1 Polynomials, page 222

9. $-x-8$, 1st degree binomial **11.** $3x-6y-4z$; 1st degree
trinomial **13.** $5x^2-7x-7$; 2nd degree trinomial
15. $-x^2-5x+3$; 2nd degree trinomial **17.** $x-31$; 1st degree
binomial **19.** $16x^2+9x-3$; 2nd degree trinomial
21. $-3x^2+15x-17$; 2nd degree trinomial **23. a.** x^2+5x+6
b. y^2+6y+5 **c.** $z^2+4z-12$ **d.** s^2+s-20
25. a. c^2-6c-7 **b.** $z^2+2z-15$ **c.** $2x^2-x-1$
d. $2x^2-5x+3$ **27. a.** $x^2+2xy+y^2$ **b.** $x^2-2xy+y^2$
c. x^2-y^2 **d.** a^2-b^2 **29. a.** $x^2+8x+16$ **b.** y^2-6y+9
c. $s^2+2st+t^2$ **d.** $u^2-2uv+v^2$ **31.** $6x^3+x^2-12x+5$
33. $5x^4-9x^3+13x^2+3x$ **35.** $-4x^3+25x^2-19x+22$
37. $7x^2$

39.

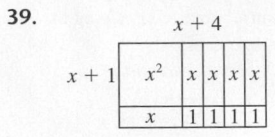

$(x + 1)(x + 4) = x^2 + 5x + 4$

41.

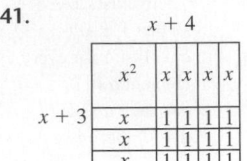

$(x + 3)(x + 4) = x^2 + 7x + 12$

43.

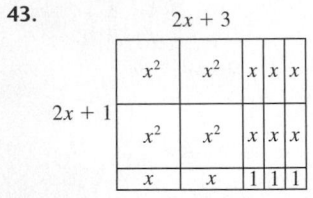

$(2x + 1)(2x + 3) = 4x^2 + 8x + 3$

45. $(x - 1)^3 = x^3 - 3x^2 + 3x - 1$
47. $(x + y)^6 = x^6 + 6x^5y + 15x^4y^2 + 20x^3y^3 + 15x^2y^4 + 6xy^5 + y^6$
49. $(x - y)^8 = x^8 - 8x^7y + 28x^6y^2 - 56x^5y^3 + 70x^4y^4 - 56x^3y^5$
$+ 28x^2y^6 - 8xy^7 + y^8$
51. $16x^4 - 96x^3y + 216x^2y^2 - 216xy^3 + 81y^4$
53. $91x^2y^{12} + 14xy^{13} + y^{14}$
55. $(6x + 2)(51x - 7) = 306x^2 + 60x - 14$
57. $(6b + 15)(10 - 2b) = 150 + 30b - 12b^2$
59. $(10x + y)(10x + z) = 100x^2 + 10xz + 10xy + yz$
$= 100x^2 + 10x(z + y) + yz$
$= 100x^2 + 10x(10) + yz$
$= 100x^2 + 100x + yz$
$= 100(x^2 + x) + yz$
$= 100[x(x + 1)] + yz$

5.2 Factoring, page 228

3. $2x(5y - 3)$ **5.** $2x(4y - 3)$ **7.** $(x - 3)(x - 1)$
9. $(x - 3)(x - 2)$ **11.** $(x - 4)(x - 3)$ **13.** $(x - 6)(x + 5)$
15. $(x - 7)(x + 5)$ **17.** $(3x + 10)(x - 1)$
19. $(2x - 1)(x - 3)$ **21.** $(3x + 1)(x - 2)$
23. $(2x + 1)(x + 4)$ **25.** $(3x - 2)(x + 1)$
27. $x(5x - 3)(x + 2)$ **29.** $x^2(7x + 3)(x - 2)$
31. $(x - 8)(x + 8)$ **33.** $25(x^2 + 2)$
35. $(x - 1)(x + 1)(x^2 + 1)$
37.

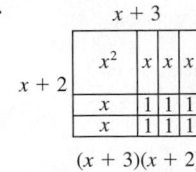

$(x + 3)(x + 2)$

39.

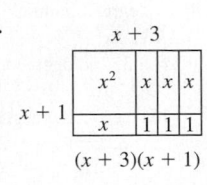

$(x + 3)(x + 1)$

41.

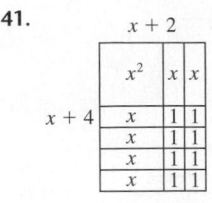

$(x + 2)(x + 4)$

43.

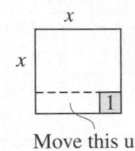

Move this unshaded piece to form area:
$(x - 1)(x + 1)$

45.

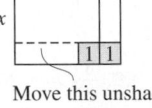

Move this unshaded piece to form area:
$(x - 1)(x + 2)$

47.

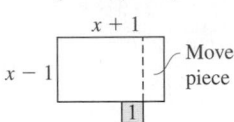

Move this unshaded piece to form area:
$(x - 2)(x + 1)$

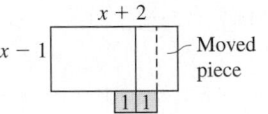

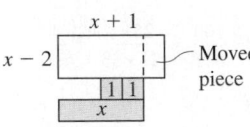

49. The dimensions are $x - 13$ feet by $x + 11$ feet.
51. The time is $2x - 1$ hours.   **53.** $(3x + 2)(2x + 1)$
55. $x^2(x - 3)(x + 3)(x - 2)(x + 2)$ **57.** $x^2(4y + 5z)(5y - 2z)$
59. Let $x - 1$, x, and $x + 1$ be the three integers. Then
$(x - 1)(x + 1) = x^2 - 1$ so the square of the middle integer is one
more than the product of the first and third.

5.3 Evaluation, Applications, and Spreadsheets, page 237

5. a. $+(2/3)*A1^2$ **b.** $+5*A1^2 - 6*A2^2$
7. a. $+12*(A1^2+4)$ **b.** $+(15*A1+7)/2$
9. a. $+(5 - A1)*(A1+3)^2$ **b.** $+6*(A1+3)*(2*A1 - 7)^2$
11. a. $+(1/4)*A1^2 - (1/2)*A1+12$
b. $+(2/3)*A1^2+(1/3)*A1 - 17$ **13. a.** $4x + 3$
b. $5x^2 - 3x + 4$ **15. a.** $\frac{5}{4}x + 14^2$ **b.** $(\frac{5}{4}x + 14)^2$

17. a. $\frac{x}{y}(z)$ **b.** $\frac{x}{yz}$ **19.** $A = 13$ **21.** $C = 8$

23. $E = 21$ **25.** $G = 19$ **27.** $I = 2$ **29.** $K = 4$
31. $M = 9$ **33.** $P = 20$ **35.** $R = 11$ **37.** $T = 17$
39. $V = 48$ **41.** $X = 100$ **45. a.** 8   **b.** 5   **c.** 9   **d.** 3.5

47.

	A	B	C	D	E
a. 1	1	3	4	5	6
b. 2	1	3	3	3	3
3					
4					
5					

49. a.

	A	B	C
1	7	0	5
2	2	4	6
3	3	8	1

b.

	A	B	C
1	107	100	105
2	102	104	106
3	103	108	101

51.

	A	B	C	D	E
1	NAME	SALES	COST	PROFIT	COMMISSION
2				+B2-C2	+.08*D2
3	Replicate Row	2 for Rows 3 to 21.			
4					
5					

53. Genotype: black, 42.25%; black (recessive brown), 45.5%; brown, 12.25%. Phenotype: black, 87.75%; brown, 12.25%
55. Genotypes and phenotypes are the same: red, 4%; pink, 32%; white, 64%.   **57.** $B = 50\%$ and $b = 50\%$ **59.** Answers vary. For the general population the result is FF is 49%, Ff is 42%, and ff is 9%; Free hanging is 91% and attached is 9%. In other words, F is 70% and f is 30%.

5.4 Equations, page 245

3. a. 15 **b.** 8   **5. a.** 32 **b.** -44 **7. a.** 0 **b.** 0
9. a. $X = \frac{1}{5}$ **b.** $C = 10$ **11. a.** $F = 1$ **b.** $G = 6$
13. a. $J = 15$ **b.** $K = 9$ **15. a.** $N = -6$ **b.** $P = -8$
17. $S = 8$ **19.** $U = 3$ **21.** $W = -15$ **23.** $Z = 14$
25. 0, 10   **27.** 11, -6 **29.** 0, 2, -2 **31.** $-\frac{3}{2}, \frac{3}{2}$
33. $\dfrac{-7 \pm \sqrt{41}}{2}$ **35.** $\dfrac{5 \pm \sqrt{37}}{2}$ **37.** $3 \pm \sqrt{2}$
39. $\dfrac{-5 \pm \sqrt{73}}{6}$ **41.** no real value   **43.** $4, -\frac{1}{3}$ **45.** $0, \frac{5}{6}$
47. 1.41, 2.83   **49.** 0.91, -2.33 **51.** $-41.54, -0.01$
53. 7.98, 3.02   **55. a.** 45.5 mg **b.** 20 mg   **57.** On the last step, dividing by $a + b - c$ is not allowed because $a + b = c$ so $a + b - c = 0$; cannot divide by 0.   **59. a.** $-13, 3$ **b.** 3

5.5 Inequalities, page 250

3.
5.
7.
9.
11.
13.
15.
17.
19. $x \le 7$ **21.** $y > 5$ **23.** $y > -8$ **25.** $t \ge 3$
27. $y \le -1$ **29.** $w \ge -4$ **31.** $y \ge 5$ **33.** $s > -3$
35. $b < 8$ **37.** $t \ge 3$ **39.** $v > 2$ **41.** $x \le -2$
43. $B < -\frac{3}{2}$ **45.** $D > -4$ **47.** $F > \frac{3}{5}$ **49.** $H < -\frac{17}{2}$

51. $J \le \frac{15}{4}$ **53.** any number less than 0   **55.** any number greater than -5 **57.** The number is -2.   **59.** The number is greater than -6.

5.6 Algebra in Problem Solving, page 259

3. $2(\text{NUMBER}) + 7 = 17$ **5.** $5(\text{NUMBER}) - (-10) = -30$
7. INTEGER + NEXT INTEGER = 117
9. INTEGER + NEXT INTEGER + THIRD CONSECUTIVE INTEGER = 105
11. VALUE OF HOUSE + VALUE OF LOT = 212,400
13. AMOUNT PAID FOR CORRECT PROBLEMS + AMOUNT OF FINES = 0
15. $10^2 + 15^2 = (\text{LENGTH OF GUY WIRE})^2$ **17.** DIST FROM J TO O + DIST FROM O TO P + DIST FROM P TO M = TOTAL DISTANCE
19. DIST FROM S TO A + DIST FROM A TO W + DIST FROM W TO D = TOTAL DISTANCE   **21.** DIST SHUTTLECRAFT TRAVELS + HEAD START = DIST ENTERPRISE TRAVELS   **23.** DIST OF SLOWER WALKER + HEAD START = DIST OF FASTER WALKER   **25.** DIST OF 1ST JOGGER + DIST OF 2ND JOGGER = TOTAL DISTANCE   **27.** The number is 9.
29. The number is 6.   **31.** The integers are 46 and 48.
33. The integers are 17, 18, 19, and 20.   **35.** The prices of the cabinets are $830 and $3,320.   **37.** The gambler started with $30; note, all that is necessary is to solve: $4(\text{ORIG AMT}) - 72 = 48$.
39. $2\sqrt{65}$, or approximately 16 ft   **41.** 250 mi
43. 12.5 sec   **45.** 6 min (or 0.1 hour)   **47.** 6 mph and 8 mph
49. The length of the shorter side is 5 cm.   **51.** The base is 3.6 ft and the height is 1.6 ft.   **53.** Need to obtain about 41.4% interest.
55. The height is approximately 10.5 ft.   **57.** There are 120 lilies.
59. The second monkey jumped 50 cubits into the air.

5.7 Ratios, Proportions, and Problem Solving, page 270

5. 20 to 1   **7.** 53 to 50   **9.** 18 to 1   **11. a.** yes **b.** yes
c. yes   **13. a.** yes **b.** no **c.** no   **15. a.** $>$ **b.** $<$ **c.** $>$
17. a. $<$ **b.** $>$ **c.** $>$ **d.** $<$ **19.** $A = 30$ **21.** $C = 10$
23. $E = 8$ **25.** $G = 21$ **27.** $I = 16$ **29.** $K = \frac{15}{2}$ or 7.5
31. $M = 4$ **33.** $P = 3$ **35.** $R = 9$ **37.** $T = \frac{1}{4}$ or 0.25
39. $V = \frac{5}{2}$ or 2.5   **41.** $Y = 22$ **43.** $0.91   **45.** 2 gallons
47. 27 minutes   **49.** 2,475 calories   **51.** 159.25   **53.** $\frac{2}{3}$
55. $3\frac{1}{3}$ ft or 3 ft 4 in.   **57.** $780   **59.** Since she started with 32 gallons and ended with 1 pt (of pure soft drink), the amount of soft drink served was 31 gal, 3 qt, and 1 pt.

5.8 Percents, page 279

1. $\frac{3}{4}$; 75%   **3.** $\frac{2}{5}$; 0.4   **5.** $0.\overline{3}$; $33\frac{1}{3}\%$ **7.** $\frac{17}{20}$; 85%
9. 0.375; 37.5%   **11.** $\frac{6}{5}$; 1.2   **13.** $\frac{1}{20}$; 5%   **15.** $0.1\overline{6}$; $16\frac{2}{3}\%$
17. $\frac{2}{9}$; $0.\overline{2}$ **19.** $\frac{7}{40}$; 17.5%   **21.** $\frac{1}{400}$; 0.25%
Estimates in Problems 23–27 may vary.   **23. a.** 9,500 **b.** 8.56

25. a. 200 **b.** 200 **27. a.** 40 **b.** 8,100

29. $\dfrac{15}{100} = \dfrac{A}{64}$; 9.6 **31.** $\dfrac{14}{100} = \dfrac{21}{W}$; 150

33. $\dfrac{P}{100} = \dfrac{10}{5}$; 200% **35.** $\dfrac{P}{100} = \dfrac{4}{5}$; 80%

37. $\dfrac{P}{100} = \dfrac{9}{12}$; 75% **39.** $\dfrac{35}{100} = \dfrac{49}{W}$; 140

41. $\dfrac{120}{100} = \dfrac{16}{W}$; $13\frac{1}{3}$ **43.** $\dfrac{33\frac{1}{3}}{100} = \dfrac{12}{W}$; 36 **45.** $\dfrac{6}{100} = \dfrac{A}{8,150}$; $489

47. 19.8 million **49.** $10.86 **51.** 5% **53.** 40%
55. The tax withheld is $2,624. **57.** 90%
59. The old wage was $1,250 and the new wage is $1,350.

5.9 Modeling Uncategorized Problems, page 287

7. A **9.** A **11.** A **13.** B **15.** D **17.** C
19. A **21.** A **23.** B **25.** D **27.** B **29.** B
31. C **33.** D **35.** Answers vary. **37. a.** $2,817.70
b. $459.73 **c.** $741.50 **d.** $333.68 **e.** $3,062.39
39. Standard Oil bldg., 1,136 ft; Sears, 1,454 ft. **41.** 10,100
43. 70 **45.** 198 hours **47.** $x = 2\ell - \frac{1}{30}c\ell - 4$ **49.** The
selling price of the house was $224,000. **51.** Cut off 2 in.; the box
is $8 \times 9 \times 12$ in. **53.** The wind speed is 21 mph. **55.** $2,350
with 30 additional persons **57.** The reading would not change.
59. a. 2, 5, 8, 10, 13, . . . **b.** No; $2 \cdot 8 = 16$ and 16 is
not sacred.

Chapter 5 Review Questions, page 291

1. a. Algebra refers to a structure as a set of axioms that forms the
basis for what is accepted and what is not.
  b. The four main processes are *simplify* (carry out all operations
  according to the order of operations agreement and write the
  result in a prescribed form), *evaluate* (replace the variable(s)
  with specified numbers and then simplify), *factor* (write the
  expression as a product), and *solve* (find the replacement(s) for
  the variable(s) that make the equation true).
2. a. $(x-1)(x^2 + 2x + 8) = (x-1)(x^2) + (x-1)(2x) + (x-1)(8)$
$= x^3 - x^2 + 2x^2 - 2x + 8x - 8$
$= x^3 + x^2 + 6x - 8$
  b. $x^2(x^2 - y) - xy(x^2 - 1) = x^4 - x^2y - x^3y + xy$
3. a. $x^3 - y^3 = 2^3 - 3^3 = -19$
  b. $(x-y)(x^2 + xy + y^2) = (2-3)(2^2 + 2 \cdot 3 + 3^2) = -19$
4. a. $3x^2 - 27 = 3(x^2 - 9) = 3(x-3)(x+3)$
  b. $x^2 - 5x - 6 = (x-6)(x+1)$
5. a. $8x - 12 = 0$
$8x = 12$
$x = \frac{12}{8} = \frac{3}{2}$
  b. $8x - 12 = -2x^2$
$2x^2 + 8x - 12 = 0$
$x^2 + 4x - 6 = 0$
$x = \dfrac{-4 \pm \sqrt{16 - 4(1)(-6)}}{2}$
$= \dfrac{-4 \pm \sqrt{40}}{2}$
$= \dfrac{-4 \pm 2\sqrt{10}}{2}$
$= \dfrac{2(-2 \pm \sqrt{10})}{2}$
$= -2 \pm \sqrt{10}$

6.

$x^2 + 5x + 6 = (x+3)(x+2)$

7. a. $2x + 5 = 13$ **b.** $3x + 1 = 7x$
$2x = 8$ $\qquad 1 = 4x$
$x = 4$ $\qquad \frac{1}{4} = x$

8. a. $\dfrac{2x}{3} = 6$ **b.** $2x - 7 = 5x$
$2x = 18$ $\qquad -7 = 3x$
$x = 9$ $\qquad -\frac{7}{3} = x$

9. a. $\dfrac{P}{100} = \dfrac{3}{20}$ **b.** $\dfrac{25}{W} = \dfrac{80}{12}$
$P = 15$ $\qquad W = 3.75$

10. $x^2 = 4x + 5$
$x^2 - 4x - 5 = 0$
$(x-5)(x+1) = 0$
$x = 5, -1$

11. $4x^2 + 1 = 6x$
$4x^2 - 6x + 1 = 0$
$x = \dfrac{6 \pm \sqrt{36 - 4(4)(1)}}{2(4)}$
$x = \dfrac{6 \pm \sqrt{20}}{8} = \dfrac{6 \pm 2\sqrt{5}}{8} = \dfrac{3 \pm \sqrt{5}}{4}$

12. $3 < -x$ **13.** $2 - x \geq 4$
$x < -3$ $\qquad -x \geq 2$
$\qquad\qquad x \leq -2$

14. $3x + 2 \leq x + 6$ **15.** $14 > 5x - 1$
$2x \leq 4$ $\qquad 15 > 5x$
$x \leq 2$ $\qquad 3 > x$
$\qquad\qquad x < 3$

16. a. $\dfrac{2}{3} \,\underline{\quad}\, \dfrac{67}{100}$ **b.** $\dfrac{3}{4} \,\underline{\quad}\, \dfrac{75}{100}$
$2(100) \,?\, 3(67)$ $\qquad 3(100) \,?\, 4(75)$
$200 < 201$ $\qquad 300 = 300$
  c. $\dfrac{23}{27} \,\underline{\quad}\, \dfrac{92}{107}$ **d.** $0.05 \,\underline{\quad}\, 0.1$
$23(107) \,?\, 27(92)$ $\qquad 0.05 \,?\, 0.10$
$2,461 < 2,484$ $\qquad\qquad <$
  e. $0.99 \,\underline{\quad}\, 0.909$
$0.990 \,?\, 0.909$
$\qquad >$

17. 1 is to 5 as 2.5 is to how much? $\dfrac{1}{5} = \dfrac{2.5}{x}$; this means
$x = 5(2.5) = 12.5$. $12\frac{1}{2}$ cups of flour.
18. Let x, $x + 2$, $x + 4$, and $x + 6$ be the four consecutive even
integers. Then
$x + (x+2) + (x+4) + (x+6) = 100$
$4x + 12 = 100$
$4x = 88$
$x = 22$ The integers are 22, 24,
26, and 28.

19. $(T + t)^2 = T^2 + 2Tt + t^2$
Genotypes: 27% tall: $T^2 = (0.52)^2 = 0.2704$;
50% tall (recessive short):
$2Tt = 2(0.52)(0.48) = 0.4992$;
23% short: $t^2 = (0.48)^2 = 0.2304$
Phenotypes: 77% tall: $0.2704 + 0.4992 = 0.7696$;
23% short: 0.2304

20. $\left(\begin{smallmatrix}\text{NEW YORK TO}\\\text{CHICAGO}\end{smallmatrix}\right)$ + (CHICAGO TO SF) + (SF TO HONOLULU) = 4,980

(CHICAGO TO SF − 1,140)   (CHICAGO TO SF + 540)

Let $x =$ DISTANCE FROM CHICAGO TO SF

$(x - 1,140) + x + (x + 540) = 4,980$
$3x - 600 = 4,980$
$3x = 5,580$
$x = 1,860$

The distance from New York to Chicago is 720 mi, from Chicago to San Francisco is 1,860 mi, and from San Francisco to Honolulu is 2,400 mi.

CHAPTER 6 THE NATURE OF GEOMETRY

6.1 Geometry, page 301

9. **11.**

13. **15.**

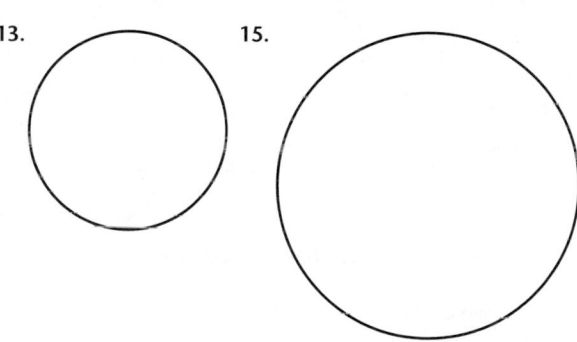

17.

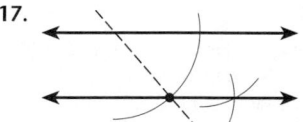

19.

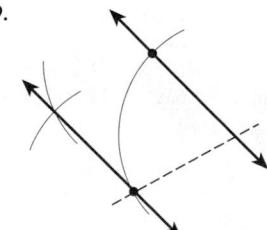

21. symmetric   **23.** symmetric   **25.** symmetric
27. symmetric   **29.** $\overset{\bullet}{P} \quad \overset{\bullet}{Q}$ **31.** $\leftarrow\!\!\!\overset{\bullet}{P}\!\!\!-\!\!\!\overset{\bullet}{Q}$
33. $\overset{\bullet}{P}\quad\overset{\bullet}{Q}$ **35.** $\leftarrow\!\!\overset{\bullet}{P}\!-\!\overset{\bullet}{Q}$
37. not symmetric   **39.** symmetric   **41.** not symmetric
43. not symmetric   **45.** not symmetric   **47.** rotation

Answers for Problems 49–53 may vary. If you have difficulty with this type of spatial visualization, try finding a die or other cube around your house and use self-adhesive notes on the faces.

49. **51.** **53.**

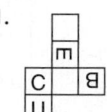

55. C   **57.** C   **59.** (1) letters with no symmetry; (2) letters with horizontal line symmetry; (3) letters with vertical line symmetry; (4) letters with symmetry around both vertical and horizontal lines

6.2 Polygons and Angles, page 311

9. a. quadrilateral **b.** pentagon   **11. a.** triangle **b.** hexagon
13. a. quadrilateral **b.** heptagon   **15. a.** T **b.** T
17. a. T **b.** F   **19. a.** F **b.** F   **21. a.** yes **b.** yes
c. yes **d.** yes **e.** yes   **23. a.** no **b.** no **c.** no **d.** no
e. no   **25.**

27.   **29.**

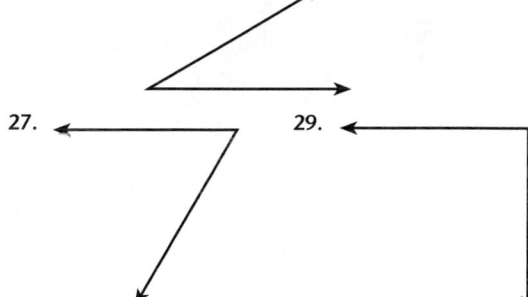

31. a. acute **b.** right   **33. a.** right **b.** obtuse
35. a. acute **b.** acute   **37.** adjacent and supplementary
39. adjacent and supplementary   **41.** vertical   **43.** alternate interior angles   **45.** F   **47.** T   **49.** F
51. $m\angle 1 = m\angle 3 = m\angle 5 = m\angle 7 = 115°$; $m\angle 2 = m\angle 4 = m\angle 6 = m\angle 8 = 65°$ **53.** $m\angle 1 = m\angle 3 = m\angle 5 = m\angle 7 = 153°$; $m\angle 2 = m\angle 4 = m\angle 6 = m\angle 8 = 27°$ **55.** $m\angle 1 = m\angle 3 = m\angle 5 = m\angle 7 = 163°$; $m\angle 2 = m\angle 4 = m\angle 6 = m\angle 8 = 17°$
57. $x = 132.5°$; $y = 47.5°$
59.

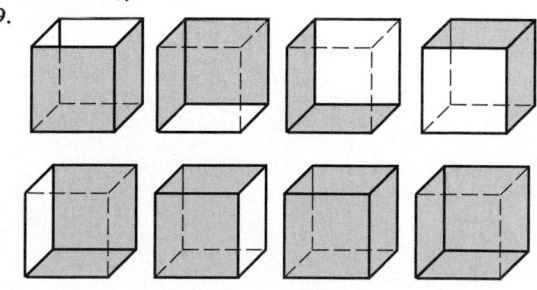

6.3 Triangles, page 317

5. $\overline{AB} \simeq \overline{ED}$; $\overline{AC} \simeq \overline{EF}$; $\overline{CB} \simeq \overline{FD}$; $\angle A \simeq \angle E$; $\angle B \simeq \angle D$; $\angle C \simeq \angle F$
7. $\overline{RS} \simeq \overline{TU}$; $\overline{RT} \simeq \overline{RT}$; $\overline{ST} \simeq \overline{UR}$; $\angle SRT \simeq \angle UTR$; $\angle S \simeq \angle U$; $\angle STR \simeq \angle URT$ **9.** $\overline{JL} \simeq \overline{PN}$; $\overline{LK} \simeq \overline{NM}$; $\overline{JK} \simeq \overline{PM}$; $\angle J \simeq \angle P$; $\angle L \simeq \angle N$; $\angle K \simeq \angle M$ **11.** 88°   **13.** 145°   **15.** 56°
17. 75°   **19.** 80°   **21.** 100°

23. 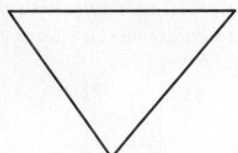 **25.**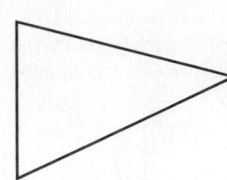

27.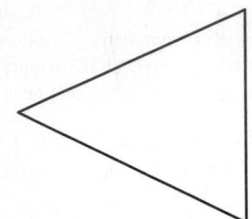

29. An isosceles right triangle is a right triangle whose legs have the same length. **31.** 20° **33.** 21° **35.** 6°
37. 60°; 60°; 60° **39.** 50°; 60°; 70° **41.** 13°; 53°; 114°

Answers to Problems 43–50 may vary.

43. 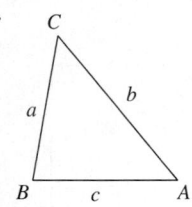 **45.**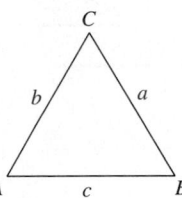

47. 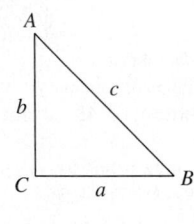 **49.**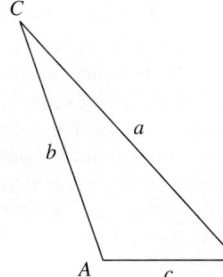

51. 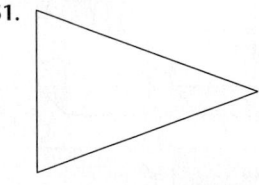 **53.**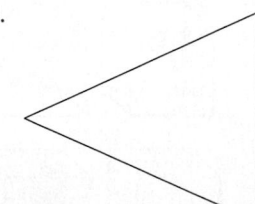

59. 540°

6.4 Similar Triangles, page 322

3. similar **5.** not similar **7.** similar
9. $m\angle A = m\angle A' = 25°$; $m\angle B = m\angle B' = 75°$; $m\angle C = m\angle C' = 80°$
11. $m\angle A = m\angle D = 38°$; $m\angle B = m\angle E = 68°$; $m\angle C = m\angle F = 74°$
13. $m\angle A = m\angle A' = 54°$; $m\angle B = m\angle B' = 36°$;
$m\angle C = m\angle C' = 90°$ **15.** $|\overline{AC}| = |\overline{AB}| = 11$; $|\overline{A'C'}| = |\overline{A'B'}| = 22$; $|\overline{BC}| = 5$; $|\overline{B'C'}| = 10$ **17.** $|\overline{GH}| = 14$; $|\overline{HI}| = 12$; $|\overline{GI}| = 16$;
$|\overline{DF}| = 20$; $|\overline{DE}| = 17.5$; $|\overline{EF}| = 15$ **19.** $|\overline{AB}| = 10$; $|\overline{AC}| = 6$;
$|\overline{BC}| = 8$; $|\overline{B'C'}| = 5$; $|\overline{A'B'}| = 6.25$; $|\overline{A'C'}| = 3.75$
21. $\sqrt{80}$ or $4\sqrt{5}$ **23.** 4 **25.** $\frac{16}{3}$ **27.** $\frac{8}{3}$ **29.** 22

31. 7.8 **33.** 13.7
35. $m\angle BMA = m\angle BMC$ Right angles
     $m\angle A = m\angle C$ Isosceles triangle property
     $\triangle ABM \sim \triangle CBM$ Similar triangle theorem
37. $m\angle ACT = m\angle TCO$ Given that they are bisectors
     $m\angle ATC = m\angle OTC$ Given
     $\triangle CAT \sim \triangle COT$ Similar triangle theorem
39. The lake is 125 ft long. **41.** The height of the building
is 24 ft. **43.** The building is 45 ft tall. **45.** The bell tower is
70 ft tall. **47.** The tree is 17 ft tall. **49.** The tree is 6 ft 8 in.
tall. **51.** Denver to New Orleans is 1,080 mi and Chicago to
Denver is 1,020 mi. **53.** The footbridge is 21 ft above the base.

6.5 Right-Triangle Trigonometry, page 332

5. Rope *A* would form a right triangle. **7.** *b* **9.** *a*
11. $\frac{a}{c}$ **13.** $\frac{b}{c}$ **15.** $\frac{a}{b}$ **17.** 0.8290 **19.** 0.8746
21. 0.5878 **23.** 0 **25.** 0.4452 **27.** 3.7321 **29.** 30°
31. 45° **33.** 60° **35.** 56° **37.** 37°
39. $\sin A = \frac{12}{13}$; $\cos A = \frac{5}{13}$; $\tan A = \frac{12}{5}$

41. $\sin A = \dfrac{\sqrt{35}}{6} \approx 0.9860$; $\cos A = \frac{1}{6} \approx 0.1667$;
$\tan A = \dfrac{\sqrt{35}}{1} \approx 5.9161$ **43.** $\sin A = \dfrac{1}{\sqrt{5}} \approx 0.4472$;
$\cos A = \dfrac{2}{\sqrt{5}} \approx 0.8944$; $\tan A = \frac{1}{2} = 0.5000$

45. $\sin A = \frac{1}{2} = 0.5000$; $\cos A = \dfrac{\sqrt{3}}{2} \approx 0.8660$; $\tan A = \dfrac{1}{\sqrt{3}} \approx$
0.5774 **47.** $\sin A = \dfrac{\sqrt{2}}{3} \approx 0.4714$; $\cos A = \dfrac{\sqrt{7}}{3} \approx 0.8819$;
$\tan A = \dfrac{\sqrt{2}}{\sqrt{7}} \approx 0.5345$ **49.** The height is 109 ft.

51. The distance is 199 m. **53.** The top of the ladder is 12 ft 7 in.
55. The chimney stack is 1,251 ft tall. **57.** $\cos 30° = \dfrac{\sqrt{3}}{2}$,
$\sin 30° = \frac{1}{2}$, and $\tan 30° = \dfrac{\sqrt{3}}{3}$ **59.** 12 in.

6.6 Golden Rectangles, page 339

7. 1, 3, 4, 7, 11, 18, 29, 47, 76, 123; the ratios are 3, 1.33, 1.75, 1.57,
1.64, 1.61, 1.62, 1.62, 1.62; it is τ **9.** Ratio of *s* to $\frac{1}{2}b$ is 1.62;
b to *h* is 1.57; these are both about the same as τ. **11.** Ratio is
1.875; this is close to τ. **13. a.** $5'' \times 3''$ **b.** 1.67 (about the
same) **15. a.** 10 in. by 8 in. **b.** 1.25 (not too close)
17. Answers vary. **19. a.** 2 **b.** 8 **c.** 144
d. 1, 2, 1.50, 1.67, 1.60, 1.63, 1.62, 1.62, 1.62, 1.62, 1.62; close to τ
21. $\frac{2}{1} = 2$; $\frac{3}{2} = 1.5$; $\frac{5}{3} \approx 1.67$; $\frac{8}{5} = 1.6$; $\frac{13}{8} \approx 1.63$; $\frac{21}{13} \approx 1.62$; close to τ
23. 3.1 ft; 8.1 ft **25.** 15 cm **27.** The length-to-width ratio
will remain unchanged if $\dfrac{L}{W} = \sqrt{2}$. **29.** $x = \dfrac{1 \pm \sqrt{5}}{2}$
b. They are negative reciprocals.

6.7 Projective and Non-Euclidean Geometries, page 344

5. Euclidean **7.** Euclidean **9.** elliptic **11.** hyperbolic
13. It is a Saccheri quadrilateral. **15.** It is not a Saccheri quadrilateral. **17.** B **19.** C **21.** Their sum is greater than 180°.
23. a. A great circle is a circle on a sphere with a diameter equal to
the diameter of the sphere. **b.** ℓ is a line (a great circle), but *m* is not.
c. yes **25.** Lines are great circles, and the circle labeled *m* is not a

great circle.   **27.** True; answers vary.   **29.** Just down from the North Pole there is a parallel that has a circumference of exactly one mile. If you begin anywhere on the circle that is a parallel one mile north of this, the conditions are satisfied.

Chapter 6 Review Questions, page 348

1. a. 0   **b.** 8 (corners of cube)   **c.** 24 (2 such pieces on each of the 12 edges)   **d.** 24 (4 on each face)   **e.** 8 (interior unpainted pieces)   **2. a.** Flag is symmetric; picture is not (if you include the flagpole).   **b.** no   **c.** no   **d.** yes   **3. a.** corresponding angles   **b.** adjacent angles and supplementary angles   **c.** vertical angles   **d.** alternate interior angles   **e.** alternate exterior angles   **f.** vertical angles   **g.** ℓ_1 and ℓ_2 are parallel (given); ℓ_3 is horizontal.
4. a. 41°   **b.** 131°   **c.** 41°
5. $(3x + 20) + (2x - 40) + (x - 16) = 180$
$$6x - 36 = 180$$
$$6x = 216$$
$$x = 36$$

6.
$$\frac{h}{10} = \frac{10}{h + 10}$$
$$h(h + 10) = 100$$
$$h^2 + 10h - 100 = 0$$
$$h = \frac{-10 \pm \sqrt{10^2 - 4(1)(-100)}}{2}$$
$$\approx 6.18, -16.18 \quad \text{Disregard negative value.}$$
The width is about 6 in.
7. $13^2 = x^2 + 12^2$
$$169 - 144 = x^2$$
$$25 = x^2$$
$$x = \pm 5 \quad \text{Disregard negative value.}$$
The other leg is 5 in.
8. a. sin 59° ≈ 0.8572   **b.** tan 0° = 0   **c.** cos 18° ≈ 0.9511
c. tan 82° ≈ 7.1154
9. Use the definition of the tangent ratio:
$$\tan 12° = \frac{160}{d} \quad \text{Let } d \text{ be the distance to the lighthouse.}$$
$$d \tan 12° = 160 \quad \text{Multiply both sides by } d.$$
$$d = \frac{160}{\tan 12°} \quad \text{By calculator: } \boxed{160} \div \boxed{\tan} \boxed{12} \boxed{=}$$
$$d \approx 753 \quad \text{Display: } 752.7408175$$
$$\text{By table: Use 0.2126 for tan 12°.}$$
10. A Saccheri quadrilateral is a quadrilateral $ABCD$ with base angles A and B right angles and with sides AC and BD with equal lengths. If the summit angles C and D are right angles, then the result is Euclidean geometry. If they are acute, then the result is hyperbolic geometry. If they are obtuse, the result is elliptic geometry.

CHAPTER 7 THE NATURE OF MEASUREMENT

7.1 Perimeter, page 358

7. a. ─────────────────────

Chapter 7 Review Questions, page 393

1. a. ──────────────────────
b. 1.1 cm   **c.** Answers vary; $\frac{1}{2}$ in.; 1.5 cm   **2. a.** First find the distance around the semicircle: $C = \frac{1}{2}\pi d = 4\pi \approx 12.56637061$. The distance around is $10 + 7 + 4\pi + 7 + 6$. To the nearest inch, the perimeter is 43 in.   **b.** First find the area of the semicircle:

b. ─────   **c.** ─────   **9.** A   **11.** B   **13.** A   **15.** C
17. B   **19.** B   **21.** C   **23.** C   **25.** C   **27.** A
29. C   **31.** A   **33.** A   **35. a.** 3 cm   **b.** 3.4 cm   **c.** 1 in.
d. $1\frac{3}{8}$ in.   **37.** 18 in.   **39.** 9 dm   **41.** 15.08 m
43. 397.08 ft   **45.** 46 ft   **47.** 257.08 cm   **49.** 35.71 in.
51. 11.14 cm   **53.** 300 m   **55.** The sides are 26 in., 39 in., and 52 in.   **57.** probably smaller   **59.** The approximate measurements of the ark are 158 m long, 26 m wide, and 16 m high.

7.2 Area, page 366

5. a. 5 cm²   **b.** 7 cm²   **c.** 12 cm²   **7.** 6 cm²   **9.** C
11. B   **13.** A   **15.** B   **17.** C   **19.** 15 in.²
21. 1,196 m²   **23.** 100 mm²   **25.** 7,560 ft²
27. 136.5 dm²   **29.** 6,600 cm²   **31.** 28 in.²
33. 314.2 in.²   **35.** 78.5 in.²   **37.** 307.9 in.²   **39.** 7.6 cm²
41. A; $2.86/ft² for Lot A and $3.19/ft² for Lot B.   **43.** 22.2 acres
45. 216 in.²   **47.** $93\frac{1}{2}$ in.²   **49.** 20 pounds are necessary;
$117   **51.** $s = 11.5; A \approx 20$ ft²   **53.** $s = 230$;
area of triangle ≈ 9,591.66; area of figure ≈ 52,792   **55.** 172 in.²
57. 483 in.²   **59.** 79 in.²

7.3 Surface Area, Volume, and Capacity, page 377

7. 3,125 cm²   **9.** 780 cm²   **11.** 35,000 cm²
13. 36π cm² ≈ 113.1 cm²   **15.** 1.25π ft² ≈ 3.9 ft²
17. 7.80 in.²   **19.** 60 cm³   **21.** 125 ft³   **23.** 8,000 cm³
25. 24 ft³   **27.** 96,000 cm³   **29. a.** 2 c   **b.** 16 oz
31. a. 13 oz   **b.** 380 ml   **33. a.** 25 ml   **b.** 75 ml   **c.** 70 ml
35. B   **37.** A   **39.** C   **41.** B   **43.** B   **45.** 4.3 gal
47. 2.5 L   **49.** 9 L   **51.** 500 L   **53. a.** 45.375 ft³
b. 26.375 ft³   **55.** 2.5 yd³   **57.** 112 kl
59. a. 560,000 people/mi²   **b.** 11,071 mi²; this is less than the size of the state of Maryland.   **c.** about 5.4 acres per person

7.4 Miscellaneous Measurements, page 388

5. a. centimeter   **b.** meter   **7. a.** liter   **b.** kiloliter
9. a. Celsius   **b.** Celsius   **11.** C   **13.** C   **15.** B
17. C   **19.** A   **21.** B   **23.** C   **25.** B   **27.** B
29. A   **31.** 0.000063 kiloliter; 0.00063 hectoliter; 0.0063 dekaliter; 0.063 liter; 0.63 deciliter; 6.3 centiliter; **63 milliliter**
33. 0.0035 kiloliter; 0.035 hectoliter; 0.35 dekaliter; **3.5 liter;** 35 deciliter; 350 centiliter; 3,500 milliliter   **35.** 0.08 kiloliter; 0.8 hectoliter; **8 dekaliter;** 80 liter; 800 deciliter; 8,000 centiliter; 80,000 milliliter   **37.** 0.31 kiloliter; **3.1 hectoliter;** 31 dekaliter; 310 liter; 3,100 deciliter; 31,000 centiliter; 310,000 milliliter
39. a. 0.001   **b.** 0.000001   **c.** 1,000   **d.** 100   **41. a.** 1,000
b. 10   **c.** 0.001   **d.** 1,000,000   **43.** 71 cm²   **45.** 2,827 cm²
47. 30 ft³   **49.** 600 cm³   **51.** $S = 452$ in.²; $V = 905$ in.³
53. The area is increased six-fold.   **55.** The area is increased nine-fold.   **57.** The area is increased four-fold.
59. Answers vary; use the formula $F = \frac{9}{5}C + 32$ in the column headed Fahrenheit.

$A = \frac{1}{2}\pi(4)^2 = 8\pi \approx 25.13274123$. The area of the trapezoid is $A = \frac{1}{2}(8)(7 + 13) = 80$. The area of the entire figure is $8\pi + 80 \approx 105.1327412$. To the nearest square inch, the area is 105 in.².

3. a. $V = \ell wh$
$= 2(3)(5)$
$= 30$ The volume is 30 dm^3.
b. Since 1 L = dm^3, we see from part **a** that the box holds 30 liters.
c. BASE + 2(SIDE) + 2(FRONT) = $2 \cdot 3 + 2(2 \cdot 5) + 2(3 \cdot 5) = 56$ dm^2
4. a. $S = 4\pi r^2 = 4\pi(1 \text{ ft})^2 = 4\pi$ ft^2 **b.** 4π ft$^2 \approx 12.6$ ft^2
c. $V = \frac{4}{3}\pi r^3 = \frac{4}{3}\pi(1 \text{ ft})^3 = \frac{4}{3}\pi$ ft^3 **d.** $\frac{4}{3}\pi \approx 4.188790205$; the volume is 4.2 ft^3. **5. a.** feet and meters **b.** Answers vary; 40°C; 100°F **c.** one-thousandth **d.** capacity **e.** 10 km = 1,000,000 cm
6. a. 2,400 ft^2 **b.** 7,200 ft^3 **c.** 53,856 gal
7. 11 ft $\times$ 16 ft $= 3\frac{2}{3}$ yd $\times 5\frac{1}{3}$ yd
$= \frac{11}{3} \times \frac{16}{3}$ yd^2
$= \frac{176}{9}$ yd^2
≈ 19.6 yd^2; You need to purchase 20 square yards.
8. a. The 6-in. pizza is about \$0.13/in.^{2}; the 10-in. and 12-in. pizzas are about \$0.10/in.^{2}, and the 14-in. pizza is about \$0.09/in.^{2}. The best value is the large size. **b.** Answers vary; the size is 201 in.^{2}; price for a large should be about \$0.09 per square inch; I would price it at \$17.95 (\$16.95 to \$18.25 is acceptable, calculate $201 \times 0.09 \approx 18.09$).
9. a.

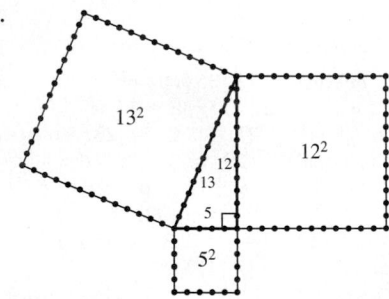

$5^2 + 12^2 = 25 + 144 = 169 = 13^2$ **b.** $25^2 + 312^2 = 313^2$, so the answer is yes, it is possible to form such a triangle.
10. Use the diagram to give a geometric justification of the distributive law.
Area of large rectangle is $a(b + c)$ and is the same as the area I and II.
Area of rectangle I: ab
Area of rectangle II: ac
Area of large rectangle is $ab + ac$.

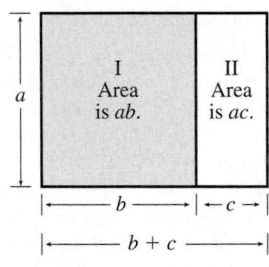

CHAPTER 8 THE NATURE OF GROWTH

8.1 Exponential Equations, page 404

9. a. $6 = \log_2 64$ **b.** $2 = \log 100$ **c.** $p = \log_n m$
11. a. $-1 = \log \frac{1}{10}$ **b.** $2 = \log_6 36$ **c.** $n = \log_t s$ **13. a.** 1
b. 3 **c.** -5 **15. a.** 1 **b.** 2 **c.** -3 **17. a.** 1 **b.** 3
c. -6 **19. a.** $\log_4 5$ **b.** $\log_5 8$ **c.** $\log_6 4.5$ **21. a.** $\log 15$
b. $\log 2.5$ **c.** $\log 45$ **23. a.** 0.63 **b.** 2.00 **c.** 3.00
25. a. 4.85 **b.** 2.00 **c.** 0.00 **27. a.** -1.38 **b.** -0.49
c. -1.27 **29. a.** 2.30 **b.** 4.61 **c.** 6.91 **31.** 3.4650
33. 1.2851 **35.** 0.8044 **37.** -7.4804 **39.** 1.1073
41. 0.4809 **43. a.** 7 **b.** 5 **45. a.** -2 **b.** $-\frac{4}{3}$
47. a. $\ln 4 \approx 1.386294361$ **b.** $x = \ln 25 \approx 3.218875825$
49. a. $x = \frac{1}{5}\log 5 \approx 0.1397940001$
b. $x = \frac{1}{3}\log 0.45 \approx -0.115595829$
51. a. $x = \frac{1}{2}(1 - \ln 3) \approx -0.049306144$

b. $x = \frac{1}{5}(1 - \ln 25) \approx -0.443775165$ **53.** $x = 3$
55. $x \approx 8.665302427$
57. $P = P_0 e^{rt}$ **59.** One hour, 18 min.
$$\frac{P}{P_0} = e^{rt}$$
$$rt = \ln \frac{P}{P_0}$$
$$t = \frac{1}{r}\ln \frac{P}{P_0}$$

8.2 Logarithmic Equations, page 412

5. F **7.** F **9.** T **11.** F **13.** F **15.** F **17.** F
19. F **21. a.** 23 **b.** 3.4 **c.** x **23. a.** x **b.** x
25. a. 4 **b.** 3 **27. a.** 3 **b.** -3 **29. a.** e^2 **b.** e^3
31. a. 9.3 **b.** 109 **33. a.** $\ln 6$ **b.** $\ln 4$ **c.** $\ln \frac{2}{243}$
35. a. $\log(x - 3)$ **b.** $\log(x - 3)$ **c.** $\ln(x - 2)$ **37. a.** 4
b. 10^4 **39. a.** 4 **b.** 4 **41. a.** $\frac{1}{2}$ **b.** $\sqrt{10}$ **43. a.** 2
b. 2 **45.** e^{10} **47.** 3, -3 **49.** $\frac{e^{1.8}}{4.8}$ **51.** $\frac{10^{12}}{8}$
53. 15 **55. a.** Could expect to sell 3,572 items. **b.** Could expect to sell 4,746 items. **c.** Need to spend about \$4,200.
57. a. 3.89 **b.** 8.8 **c.** $10^{23.8}$ **d.** $10^{1.5M+11.8}$ **59. a.** 80
b. $N = 80(1 - e^{-t/62.5})$

8.3 Applications of Growth and Decay, page 420

7. a. 1.6% **b.** 678,662 **d.** 811,078 **9. a.** 0.7%
b. 502,112 **d.** 541,917 **11. a.** 0.006% **b.** 365,362
d. 365,608 **13. a.** 0.07% **b.** 273,041 **d.** 275,270
15. 896,716 **17.** -0.023104906 **19.** The half-life is about
28 years. **21.** 6.7 **23.** An earthquake with magnitude 6 is 100 times stronger than an earthquake of magnitude 4.
25. 30 dB **27.** 107 dB **29.** 114 dB **31.** $10^{24.25}$ ergs
33. a. Growth rate of about 1.6%; predict the 2004 population to be 4,364,000. **b.** Growth rate of about -0.27%; predict the 2004 population to be 3,358,000. **c.** Growth rate of about 1.07%; predict the 2004 population to be about 3,838,000. **35.** The artifact is about 10,500 years old. **37.** The elapsed time is about 7 years.
39. 63 times more energy **41.** The artifact is about 1,300 years old. **43.** 1.41 times more energy **45.** 31.6 times more energy **47.** 7.2% lost; yes **49.** The cookie is about 105°F. **51.** 37°C **53.** 5,313 ft **55.** The satellite will operate for about 128 days. **57.** In 1988, Shroud was about 662 years old. **59.** When $x = 75$, $y \approx 0.16$ or no O-ring failures; when $x = 32$, $y \approx 4.80$ or 5 O-ring failures.

Chapter 8 Review Questions, page 423

1. $\log 100 + \log\sqrt{10} = 2 + \frac{1}{2}$
$= 2\frac{1}{2}$
2. $\ln e + \ln 1 + \ln e^{542} = 1 + 0 + 542$
$= 546$
3. $\log_8 4 + \log_8 16 + \log_8 8^{2.3} = \log_8 64 + 2.3$
$= 2 + 2.3$
$= 4.3$
4. $10^{\log 0.5} = 0.5$
5. $\ln e^{\log 1,000} = \ln e^3$
$= 3$
6. a. $\log 8.43 \approx 0.93$ **b.** $\log 9,760 \approx 3.99$
7. a. $\ln 2 \approx 0.69$ **b.** $\ln 0.125 \approx -2.08$

8. a. $\log_2 10 \approx 3.32$ **b.** $\log_\pi \dfrac{1}{\pi} = \log_\pi (\pi)^{-1} = -1$

9. $10^x = 85$
$x = \log 85$
≈ 1.929418926

10. $e^x = 500$
$x = \ln 500$
≈ 6.214608098

11. $435^x = 890$
$x = \log_{435} 890$
≈ 1.1178328654

12. $e^{3x+1} = 45$
$3x + 1 = \ln 45$
$x = \dfrac{\ln 45 - 1}{3}$
≈ 0.9355541633

13. $\log_6 x = 4$
$x = 6^4$

14. $2^{3x-1} = 6$
$3x - 1 = \log_2 6$
$x = \dfrac{\log_2 6 + 1}{3}$

15. $10^{2x} = 5$
$2x = \log 5$
$x = \dfrac{\log 5}{2}$

16.
$$\log(x + 1) = 2 + \log(x - 1)$$
$$\log(x + 1) - \log(x - 1) = 2$$
$$\log \dfrac{x + 1}{x - 1} = 2$$
$$\dfrac{x + 1}{x - 1} = 10^2$$
$$x + 1 = 100x - 100$$
$$101 = 99x$$
$$x = \dfrac{101}{99}$$

17. $3 \ln \dfrac{e}{\sqrt[3]{5}} = 3 - \ln x$
$\ln \dfrac{e^3}{5} + \ln x = 3$
$\ln \dfrac{e^3 x}{5} = 3$
$\dfrac{e^3 x}{5} = e^3$
$x = 5$

18. $A = P(1 + i)^x$
$\dfrac{A}{P} = (1 + i)^x$
$x = \log_{(1+i)} \dfrac{A}{P}$

19. $A = A_0 e^{-0.1t}$
$\dfrac{A}{A_0} = e^{-0.1t}$
$-0.1t = \ln \dfrac{A}{A_0}$
$t = -10 \ln \dfrac{A}{A_0}$
$= -10 \ln 0.5$
≈ 6.93 days
About 7 days

20. $A = A_0 e^{r(14/12)}$
$\dfrac{A}{A_0} = e^{r(14/12)}$
$r = \dfrac{12}{14} \ln 2$
≈ 0.59
Equation is $A = A_0 e^{0.59t}$, where A_0 is the number of teens infected and t is the number of years after 1992.

CHAPTER 9 THE NATURE OF FINANCIAL MANAGEMENT

9.1 Interest, page 438

7. C; interest rate is not stated, but you should still recognize a reasonable answer. **9.** B is the most reasonable. **11.** A **13.** D
15. B **17.** $1,500 **19.** $4,200 **21.** $1,028.25
23. $16,536.79 **25.** $1,132,835.66 **27.** $1,661.44
29. $165,298.89 **31. a.** $28,940.57 **b.** $29,222.27
c. $29,367.30 **d.** $29,465.60 **e.** $29,513.58 **f.** $29,515.22
g. $29,515.24 **h.** $22,800 **33.** $1,548.13 **35. a.** $1.18
b. $1.96 **c.** $3.06 **d.** $2.43 **e.** $940.99 **f.** $21,956.37
g. $43,912.74 **h.** $25,092.99 **37.** 4 years, 61 days
39. 7 years, 49 days **41.** $1,220.19 **43.** $11,023.18

45. $5,219.84 **47.** $279,064 **49.** $13,055.60
51. $P = \$101,500$ **53. a.** about $19,000 per person
b. about $10,845 per second **55.** $P = \$5,000,000$
57. a. 3 years, 3 months **b.** 10 years **c.** The interest rate is about 14%.

59. $Y = \left(1 + \dfrac{r}{n}\right)^n - 1$

9.2 Installment Buying, page 447

7. B **9.** B **11.** B **13.** B **15.** B **17.** B
19. C **21.** A **23.** B (it should be the least expensive)
25. 15% **27.** 18% **29.** 8% **31. a.** $4.50 **b.** $3.75
c. $3.95 **33. a.** $37.50 **b.** $36.88 **c.** $36.58
35. $14,181 **37.** $18,975 **39.** 23.4% **41.** 21.1%
43. a. $344.17 **b.** $349.69 **c.** 4.9% **d.** 0% is better
45. a. $711.67 **b.** $704.79 **c.** 4.9% **d.** 2.5% is better
47. a. $11,430 **b.** $2,430 **c.** 0.09 **d.** 17.5%
49. a. $23,443.20 **b.** $9,093.20 **c.** 0.1584181185 **d.** 31.0%
51. a. $8,886.48 **b.** $2,088.48 **c.** 0.08702
d. 17.0% **53.** 8% add-on rate; APR is about 15.7%
55. 11% add-on rate; APR is about 21.6% **57.** previous balance method, $45; adjusted balance method, $40.50; average daily balance method, $43.35 **59.** $67.58

9.3 Sequences, page 459

7. a. geometric **b.** $r = 2$ **c.** 32 **9. a.** arithmetic
b. $d = 10$ **c.** 35 **11. a.** Fibonacci-type **b.** $s_1 = 5, s_2 = 15$
c. 35 **13. a.** geometric **b.** $r = \frac{1}{5}$ **c.** $\frac{1}{5}$
15. a. geometric **b.** $r = 3$ **c.** 27 **17. a.** none of the classified types **b.** differences are $-2, 1, -2, 1, -2, \ldots$ **c.** 5
19. a. geometric **b.** $r = 2$ **c.** 96 **21. a.** both arithmetic and geometric **b.** $d = 0$ or $r = 1$ **c.** 10 **23. a.** Fibonacci-type
b. $s_1 = 3, s_2 = 6$ **c.** 24 **25. a.** geometric **b.** $r = \frac{3}{2}$ **c.** $\frac{81}{2}$
27. a. geometric **b.** $r = 4^{-1}$ or $\frac{1}{4}$ **c.** 4 **29. a.** arithmetic
b. $d = \frac{1}{10}$ **c.** $\frac{3}{5}$ **31. a.** arithmetic **b.** $d = \frac{1}{12}$ **c.** $\frac{11}{12}$
33. a. $0, 3, 6$ **b.** arithmetic; $d - 3$ **35. a.** $1, 0, -1$
b. arithmetic; $d = -1$ **37. a.** $0, -10, -20$ **b.** arithmetic; $d = -10$ **39. a.** $0, \frac{1}{2}, \frac{2}{3}$ **b.** neither **41. a.** $1, 3, 6$
b. neither **43. a.** $-1, 1, -1$ **b.** geometric, $r = -1$
45. a. $\frac{2}{3}, \frac{2}{3}, \frac{2}{3}$ **b.** both; $d = 0$ or $r = 1$ **47. a.** $-2, 3, -4$
b. neither **49.** -200 **51.** 625 **53.** $3, 1, \frac{1}{3}, \frac{1}{9}, \frac{1}{27}$
55. $1, 2, 3, 5, 8$ **57.** It is Fibonacci.
59. 1st number: x
2nd number: y
3rd number: $x + y$
4th number: $x + 2y$
5th number: $2x + 3y$
6th number: $3x + 5y$
7th number: $5x + 8y$
8th number: $8x + 13y$
9th number: $13x + 21y$
10th number: $21x + 34y$
SUM: $55x + 88y = 11(5x + 8y)$

9.4 Series, page 468

5. 30 **7.** 150 **9.** -1 **11.** 12 **13.** 90 **15.** 10
17. 33 **19.** 2 **21.** $\frac{3}{2}$ **23.** $-\frac{40}{3}$ **25.** 25 **27.** 15
29. 110 **31.** 10,000 **33.** 5,050 **35.** $n(n + 1)$
37. 11,500 **39.** 2,030 **41.** 120 **43.** 56; 64
45. 9,330 **47.** 3,828 blocks **49.** 1,024,000,000 present after 10 days **51.** 31 games **53.** 364 games **55.** 100 cm
57. 1,500 revolutions **59.** 50 ft

9.5 Annuities, page 476

7. $1,937.67   **9.** $2,026.78   **11.** $41,612.93
13. $74,517.97   **15.** $15,528.23   **17.** $18,294.60
19. $170,413.86   **21.** $540,968.11   **23.** $56,641.61
25. $59,497.67   **27.** $99,386.46   **29.** $38,754.39
31. $2,204.31   **33.** $252,057.07   **35.** $1,193.20
37. $1,896.70   **39.** $7,493.53   **41.** $8,277.87
43. $2,261.06   **45.** $489.00   **47.** $915,239.29
49. $27,298.98   **51.** $2,880.97   **53.** $1,666,454.24
55. $12,716.35   **57.** $1,216,445.09

9.6 Amortization, page 483

3. m is the amount of a periodic payment (usually a monthly payment); n is the number of payments made each year; t is the number of years; r is the annual interest rate; A is the future value; and P is the present value   **5.** $m = 50, r = 0.06, t = 5; P = \$2,586.28$
7. $m = 150, r = 0.05, t = 30; P = \$27,942.24$ **9.** $m = 150$,
$r = 0.08, t = 30; P = \$20,442.52$ **11.** $m = 1,050, r = 0.06$,
$t = 30; P = \$175,131.20$ **13.** $P = 14,000, r = 0.10, t = 5$;
$m = \$297.46$ **15.** $P = 150,000, r = 0.08, t = 30; m = \$1,100.65$
17. $P = 150,000, r = 0.10, t = 30; m = \$1,316.36$
19. $P = 260,000, r = 0.09, t = 30; m = \$2,092.02$
21. $6,882.42   **23.** $10,827.33   **25.** $5,429.91
27. $12,885.18   **29.** $7,407.76   **31.** $9,299.39   **33.** $6.38
35. $148.31   **37.** $430.73   **39.** $73.35   **41.** $780.54
43. $10,186.47   **45.** $117,238   **47.** $367,695.71
49. The annuity is the better choice.   **51.** $12,462,210.34
53. $206,029.43   **55.** $1,510.92; $543,931.20; $213,046.20
57. price range of $175,322.31 to $204,386.77   **59. a.** increases
b. decreases   **c.** increases

9.7 Summary of Financial Formulas, page 486

7. present value   **9.** annuity   **11.** amortization   **13.** future value; $1,404.93   **15.** present value; $5,674.27   **17. a.** future value   **b.** $1,347.85   **19. a.** present value   **b.** $5,536.76
21. a. sinking fund   **b.** $1,670.92   **23. a.** future value
b. $27,081.62   **25. a.** present value of an annuity   **b.** $5,756.94
27. a. amortization   **b.** $1,028.61   **29. a.** annuity
b. $23,193.91   **31. a.** present value   **b.** $7,215.46
33. a. annuity   **b.** $21,867.63   **35. a.** present value
b. $165,134.88   **37. a.** amortization   **b.** $1,317.40
39. a. annuity   **b.** $228,803.04   **41. a.** future value
b. $884.12   **43. a.** future value   **b.** $9,506.04
45. a. present value   **b.** $3,655.96   **47. a.** present value
b. $148,348   **49. a.** present value   **b.** $500,420   **51.** Take $10,000 now and $45,000 in one year.   **53.** $292,918.40
55. $179,986.40, total; $112,932 savings   **57.** $726,624
59. A 22-yr loan increases the monthly payment to $3,474.80. This loan has total payments of $917,347.20 with total interest of $498,847.20. The interest savings is $227,776.80.

Chapter 9 Review Questions, page 491

1. A sequence is a list of numbers having a first term, a second term, and so on; a series is the indicated sum of the terms of a sequence. An arithmetic sequence is one that has a common difference, $a_n = a_1 + (n - 1)d$; a geometric sequence is one that has a common ratio, $g_n = g_1 r^{n-1}$; and a Fibonacci-type sequence, $s_n = s_{n-1} + s_{n-2}$, is one that, given the first two terms, the next is found by adding the previous two terms. The sum of an arithmetic sequence is
$A_n = n\left(\dfrac{a_1 + a_n}{2}\right)$ or $A_n = \dfrac{n}{2}[2a_1 + (n - 1)d]$, and the sum of a

geometric sequence is $G_n = \dfrac{g_1(1 - r^n)}{1 - r}$.   **2.** Answers vary; a good procedure is to ask a series of questions. **Is it a lump-sum problem?** If it is, then what is the unknown? If FUTURE VALUE is the unknown, then it is a *future value* problem. If PRESENT VALUE is the unknown, then it is a *present value* problem. **Is it a periodic payment problem?** If it is, then is the periodic payment known? If the PERIODIC PAYMENT IS KNOWN and you want to find the future value, then it is an *ordinary annuity* problem. If the PERIODIC PAYMENT IS KNOWN and you want to find the present value, then it is a *present value of an annuity* problem. If the PERIODIC PAYMENT IS UNKNOWN and you know the future value, then it is a *sinking fund* problem. If the PERIODIC PAYMENT IS UNKNOWN and you know the present value, then it is an *amortization* problem.   **3. a.** arithmetic; $a_n = 5n$ **b.** geometric; $g_n = 5 \cdot 2^{n-1}$
c. Fibonacci-type; $s_1 = 5, s_2 = 10, s_n = s_{n-1} + s_{n-2}, n \geq 3$
d. none of these (add 5, 10, 15, 20, . . .); 55, 80   **e.** geometric; $g_n = 5 \cdot 10^{n-1}$ **f.** none of these (alternate terms); 5, 50

4. a. $\displaystyle\sum_{k=1}^{3}(k^2 - 2k + 1)$
$$= (1^2 - 2(1) + 1) + (2^2 - 2(2) + 1) + (3^2 - 3(2) + 1) = 5$$
b. $\displaystyle\sum_{k=1}^{4}\dfrac{k-1}{k+1} = \dfrac{0}{2} + \dfrac{1}{3} + \dfrac{2}{4} + \dfrac{3}{5} = \dfrac{43}{30}$ or $1.4\overline{3}$

5. parents, grandparents, great-grandparents, . . . ; that is 2, 4, 8, 16, ... Find G_{10} where $g_1 = 2$ and $r = 2$; $G_{10} = \dfrac{2(1 - 2^{10})}{1 - 2} = 2,046$.
Thus, there are a minimum of 2,046 people.   **6.** There will be 72 divisions in 24 hours; $g_{73} = 2^{10} \cdot 2^{72} = 2^{82}$
7. $18,579(1 + 0.05) = \$19,507.95$. You should offer $19,500 for the car.   **8. a.** $I = Prt = 13,500(0.029)(2) = 783$;
$A = P + I = 13,500 + 783 = 14,283$; monthly payment is $14,283 \div 24 = \$595.13$. The total interest is $783, and the monthly payment is $595.13.   **b.** APR $= \dfrac{2Nr}{N + 1} = \dfrac{2(24)(0.029)}{25} = 0.05568$;
the APR is 5.568%.   **9.** $A = 48(353.04) = 16,945.92$;
$I = A - P = 16,945.92 - 11,450.00 = 5,495.92$; also,
$$I = Prt$$
$5,495.92 = 11,450(r)(4)$ *48 months is 4 years, so t = 4.*
$1,373.98 = 11,450r$ *Divide both sides by 4.*
$0.12 \approx r$ *Divide both sides by 11,450.*
Finally, APR $= \dfrac{2Nr}{N + 1} = \dfrac{2(48)r}{49} \approx 0.235$. The APR is about 23.5%.

10. The adjusted balance method is most advantageous to the consumer.

PREVIOUS BALANCE METHOD
$I = Prt$
$= 525(0.09)(\frac{1}{12})$
$= 3.9375$
The finance charge is $3.94.

ADJUSTED BALANCE METHOD
$I = Prt$
$= (525 - 100)(0.09)(\frac{1}{12})$
$= 3.1875$
The finance charge is $3.19.

AVERAGE DAILY BALANCE
$I = Prt$
$= \dfrac{525 \times 7 + 425 \times 24}{31}(0.09)(\frac{31}{365})$
$= 3.421232877$
The finance charge is $3.42.

11. $A = \$1,000,000; t = 50; r = 0.09; n = 12$; Thus,
$$P = A\left(1 + \dfrac{r}{n}\right)^{-nt} = 1,000,000\left(1 + \dfrac{0.09}{12}\right)^{-(50)12} \approx 11,297.10$$
Deposit $11,297.10 to have a million dollars in 50 years.

12. a. Amortization; where $P = 154,000$, $n = 12$, $r = 0.08$, and $t = 20$. Thus,

$$m = \frac{P\left(\frac{n}{r}\right)}{1 - \left(1 + \frac{r}{n}\right)^{-nt}} = \frac{154,000\left(\frac{0.08}{12}\right)}{1 - \left(1 + \frac{0.08}{12}\right)^{-(12)(20)}} \approx 1,288.117706$$

(by calculator)

The monthly payments are $1,288.12.

b. $80\%(\text{PRICE OF HOME}) = \$154,000$

$\text{PRICE OF HOME} = \$192,500$

c. Future value continuous compounding:

$A = Pe^{rt}$

$\quad = \$192,500e^{0.04(20)}$

$\quad = \$428,416.63$

13. Present value; where $A = \$100,000$; $r = 0.064$; $t = 3\frac{4}{12} = \frac{40}{12}$; $n = 12$;

thus, $P = A\left(1 + \frac{r}{n}\right)^{-nt} = 100,000\left(1 + \frac{0.064}{12}\right)^{-(12)(40/12)} \approx 80,834.49$

14. present value; $A = \$420,000$, $t = 30$, $n = 1$, $r = 0.05$; $P = \$97,178.53$ **15.** annuity; $m = \$20,000$, $t = 29$, $n = 1$, $r = 0.05$; $A = \$1,246,454.24$ **16.** present value of an annuity; $m = \$20,000$, $t = 29$, $n = 1$, $r = 0.05$; $P = \$302,821.47$

17. present value; $A = \$420,000$, $t = 30$, $n = 1$, $r = 0.05$, $P = \$97,178.53$; You need to set aside $97,178.53.

18. $I = Prt$

$\quad = \$200,000(0.05)(1)$

$\quad = \$10,000$

19. You can compare present values of each or the future values of each. It is easier to compare present values.

Cash value: $300,000 Annuity: $302,821.47 (see Problem 16)

Present value of cash: $97,178.53

Total: $400,000

Since $400,000 is worth more than $300,000, take the installments.

20. Compare present values.

Cash value: $300,000 Present value of annuity:

$m = \$20,000$, $t = 29$, $n = 1$, $r = 0.10$

$187,392.12

Present value of cash:

$A = \$420,000$, $t = 30$, $n = 1$, $r = 0.10$

$24,069.59

Total: $211,461.71

Since $300,000 is worth more than $211,461.71, take the one-time payment.

CHAPTER 10 THE NATURE OF SET THEORY AND COUNTING

10.1 Sets, Subsets, and Venn Diagrams, page 504
5. $\{2, 6, 8, 10\}$ **7.** $\{2, 3, 5, 6, 8, 9\}$ **9.** $\{3, 4, 5\}$
11. $\{1, 2, 3, 4, 5, 6, 7\}$ **13.** $\varnothing$ **15.** $\{1, 2, 3, 4, 5, 6\}$
17. $\{1, 2, 3, 4, 5, 7\}$ **19.** $\{5\}$ **21.** $\{5, 6, 7\}$
23. $\{1, 2, 4, 6\}$ **25.** A or $\{1, 2, 3, 4\}$ **27.** $A \cap B$
29. 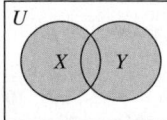 **31.**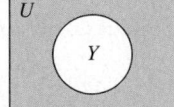

33. 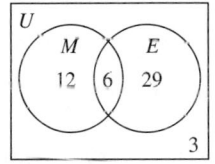 **35.**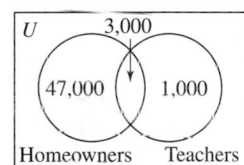

37. True; every element of the first set is also in the second.
39. False; m is an element of the second set, but $\{m\}$ is not. The correct statement would be $\{m\} \subset \{m, a, t, h\}$. **41.** False; the two sets are equal, so one cannot be a proper subset of the other. The correct statement would be $\{m, a, t, h\} \subseteq \{h, t, a, m\}$. **43.** True; math and history are certainly high school subjects, and there is at least one high school subject, say, English, that is not in the first set.
45. False; blue is a color of the rainbow but $\{blue\} \neq blue$. A correct statement would be blue $\in$ {colors of the rainbow} or $\{blue\} \subseteq$ {colors of the rainbow}. **47.** False; correct statements would be $1 \in \{1, 2, 3, 4, 5\}$ or $[1] \subseteq \{1, 2, 3, 4, 5\}$.
49. True **51.** False; the number zero and the empty set are very different. **53.** False; the set containing the empty set is not the same as the empty set. **55.** We work this problem without Venn diagrams to show an alternate method of solution.
$|B \cup S| = |B| + |S| - |B \cap S|$
$\qquad = 50 + 36 - 14$
$\qquad = 72$
Yes, they can get by with two buses.
57. 3 people **59.** 51,000 booklets are needed.

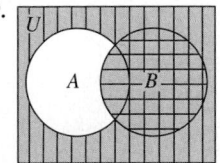 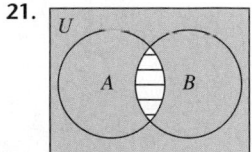

10.2 Combined Operations with Sets, page 511

3. $\{3, 5\}$ **5.** $\{7\}$ **7.** $\{5, 6, 7\}$ **9.** $\{1, 2, 4, 6, 7\}$
11. $\overline{X} \cup \overline{Y}$ **13.** $\overline{X \cap Y}$ **15.** $\overline{X} \cup Y$ **17.** $\overline{X} \cup \overline{Y}$
19. 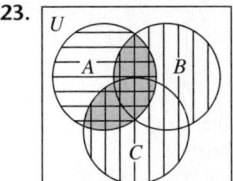 **21.**

23. 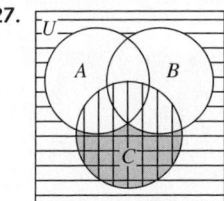 **25.**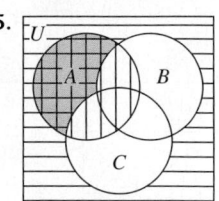

27. 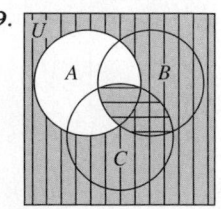 **29.**

31.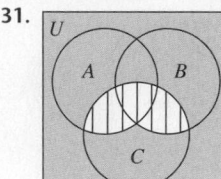

33.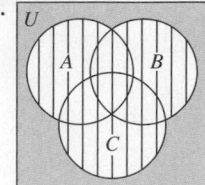

35.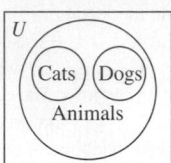

37.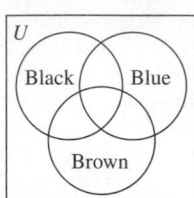

39. *Note:* We interpret white, black, or Hispanic to mean both parents are of that race. I: 31,560,000
II: 210,400,000   III: 23,670,000
IV: 5,260,000   V: 1,315,000
VI: 2,630,000   VII: 0
VIII: 11,835,000

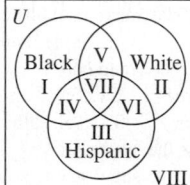

41. $\overline{A \cup B}$ **43.** $A \cap (B \cup C)$

45. Steffi Graf, II; Michael Stich, V; Martina Navratilova, III; Stefan Edberg, V; Chris Evert Lloyd, VI; Boris Becker, V; Evonne Goolagong, II; Pat Cash, V; Virginia Wade, II; John McEnroe, IV

47.

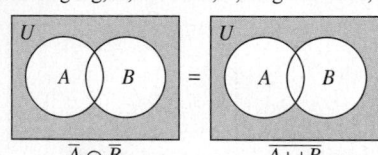

49.

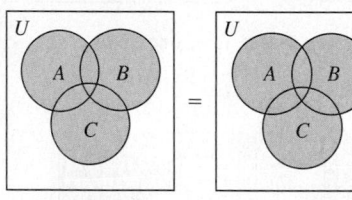

51.

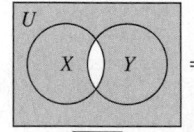 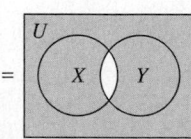

53.

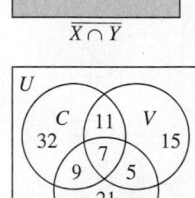

$C = \{$like comedies$\}$
$V = \{$like variety$\}$
$D = \{$like drama$\}$
No, there were 102 persons polled, not 100.

55.

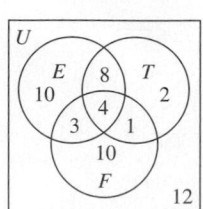

$E = \{$favor Prop. 8$\}$
$T = \{$favor Prop. 13$\}$
$F = \{$favor Prop. 5$\}$
b. 12   **c.** 7

57.

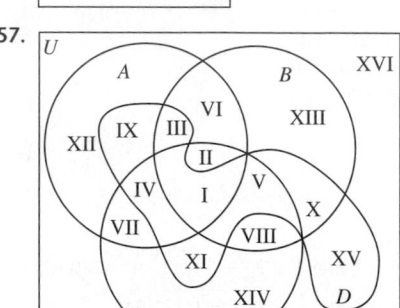

59. a. region 32   **b.** region 2

10.3 Permutations, page 520

3. 72   **5.** 3,024   **7.** 120   **9.** 42   **11.** 100   **13.** 60

15. 1   **17.** 1   **19.** 2,520   **21.** $\dfrac{n!}{(n-4)!}$ **23.** 120

25. 25   **27.** 336   **29.** 90   **31.** $\dfrac{n!}{(n-5)!}$

33. $\dfrac{x!}{(x-y)!}$ **35.** 60   **37.** 19,958,400   **39.** 129,729,600

41. 3,326,400   **43.** 778,377,600   **45.** 210   **47.** 600
49. 40,320   **51.** 47,045,520   **53.** 60   **55.** 3,125
57. 362,880   **59.** 210

10.4 Combinations, page 525

1. 9   **3.** 84   **5.** 1   **7.** 22,100   **9.** 1   **11.** 35

13. 1,225   **15.** $\dfrac{g!}{h!(g-h)!}$ **17.** $\dfrac{k!}{(k-4)!}$

19. $\dfrac{m!}{n!(m-n)!}$ **21.** 4,200   **23.** 1,260   **25.** 105

27. 792   **29.** 4   **31.** 1   **33.** 1,287   **35.** 24
37. combination; $_{30}C_5$; 142,506   **39.** combination; $_{20}C_2$; 190
41. combination; $_{31}C_3$; 4,495   **43.** neither; distinguishable permutation; 1,260   **45.** permutation; $_{10}P_2$; 90
47. permutation; $_6P_6$; 720   **49.** combination; $_{12}C_{10}$; 66
51. permutation; $_7P_7$; 5,040   **53.** neither;
$2 \cdot 2 \cdot 2 \cdot \cdots \cdot 2 = 2^{10}$; 1,024   **55.** neither, $5 \cdot 5 \cdot 5 \cdot \cdots \cdot 5 = 5^{10}$; 9,765,625   **57.** $_nC_5$ **59.** $_{16}P_2 \cdot {}_{19}P_2 \cdot {}_{31}C_6$; approximately $6.043394448 \times 10^{10}$

10.5 Counting without Counting, page 533

3. 15   **5.** 786,240   **7.** 21   **9.** 17,100,720   **11.** 1,296
13. 31   **15.** 127   **17.** 2,730   **19.** $41^7 \approx 1.9465 \times 10^{11}$
21. 135   **23.** 27,405   **25.** 120   **27.** 286   **29.** 125!
31. 360   **33.** 14,400   **35.** 24   **37.** 720   **39.** 66
41. Subsets: $\{a, b\}$; Arrangements: $(a, b), (b, a)$ **43.** Subsets: $\{a, b, c, d\}$; 24 arrangements   **45.** 810,000,000
47. 3,486,784,401   **49.** 26!   **51.** Count the number of kidney

beans that will fit into one cubic inch. Suppose this number is 15. Then, calculate the volume of the jar; the cylinder is about 1,176 in.^{3} Finally, multiply to estimate the number of beans to be 17,671.
53. About a half a year; more precisely, 30 weeks working five days per week, 8 hours per day. **55.** 0 **57.** 524,800
59. 6,161,805

10.6 Binomial Theorem, page 540

3. 8 **5.** 28 **7.** 56 **9.** 12 **11.** 1 **13.** 153
15. $x^4 + 4x^3 + 6x^2 + 4x + 1$
17. $x^4 + 16x^3 + 96x^2 + 256x + 256$
19. $a^7 + 7a^6b + 21a^5b^2 + 35a^4b^3 + 35a^3b^4 + 21a^2b^5 + 7ab^6 + b^7$
21. $x^9 - 9x^8 + 36x^7 - 84x^6 + 126x^5 - 126x^4 + 84x^3 - 36x^2 + 9x - 1$
23. $x^5 - 5x^4y + 10x^3y^2 - 10x^2y^3 + 5xy^4 - y^5$
25. $x^6 - 12x^5 + 60x^4 - 160x^3 + 240x^2 - 192x + 64$
27. $16x^4 + 96x^3y + 216x^2y^2 + 216xy^3 + 81y^4$ **29.** 462
31. 1,001 **33.** 1,820 **35.** 2,016 **37.** -16
39. F; $(a + b)^3 = a^3 + 3a^2b + 3ab^2 + b^3$ **41.** F; look at the coefficient of H^3T^2 **43.** $x^{15} - 15x^{14}y + 105x^{13}y^2 - 455x^{12}y^3$
45. $x^{12} - 24x^{11}y + 264y^{10}y^2 - 1{,}760x^9y^3$
47. $a^{15}b^{15} - 30a^{14}b^{15} + 420a^{13}b^{15} - 3{,}640a^{12}b^{15}$
49. $z^{22} + 55z^{20}k + 1{,}375z^{18}k^2 + 20{,}625z^{16}k^3$
51. 128 **53.** 10 **55.** 35 **57.** 210

10.7 Rubik's Cube and Instant Insanity, page 543

3. a.  **b.**  **c.**

d.  **5. a.**  **b.**

c.  **d.**  **7.**

9. **11.**  **13.**

15.  **17.** T^2 or $(T^{-1})^2$ **19.** $U^{-1}T^{-1}$
21. L **23.** BF **25.** no
27. 41,472 **29.**

Chapter 10 Review Questions, page 545

1. a. $A \cup B = \{1, 3, 5, 7, 9\} \cup \{2, 4, 6, 9, 10\}$
$= \{1, 2, 3, 4, 5, 6, 7, 9, 10\}$
b. $A \cap B = \{1, 3, 5, 7, 9\} \cap \{2, 4, 6, 9, 10\} = \{9\}$
2. a. $\overline{B} = \{2, 4, 6, 9, 10\}$
$= \{1, 3, 5, 7, 8\}$ You need to look at the universal set, U, to obtain this result.
b. $|\varnothing| = 0$
3. a. $|U| = 10$ **b.** $|A| = 5$
4. a. $\overline{A \cup B} = \underline{\{1, 3, 5, 7, 9\}} \cup \{2, 4, 6, 9, 10\}$
$= \overline{\{9\}}$ Intersection first, then complement
$= \{1, 2, 3, 4, 5, 6, 7, 8, 10\}$
b. $\overline{A} \cup \overline{B} - \overline{\{1, 3, 5, 7, 9\}} \cup \overline{\{2, 4, 6, 9, 10\}}$
$= \{2, 4, 6, 8, 10\} \cup \{1, 3, 5, 7, 8\}$ Complements first, then union
$= \{1, 2, 3, 4, 5, 6, 7, 8, 10\}$
5. $\overline{A} \cap (B \cup A) = \{2, 4, 6, 8, 10\} \cap \{1, 2, 3, 4, 5, 6, 7, 9, 10\}$
$= \{2, 4, 6, 10\}$
6. $\overline{(A \cup B) \cap A} = \overline{\{1, 2, 3, 4, 5, 6, 7, 9, 10\} \cap \{1, 3, 5, 7, 9\}}$
$= \overline{\{1, 3, 5, 7, 9\}}$
$= \{2, 4, 6, 8, 10\}$
7. a. 25 **b.** 15 **c.** 14 **d.** 26
8. a. **b.**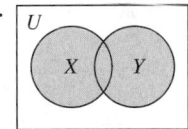

c.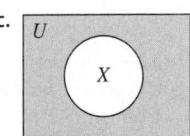

9. a. 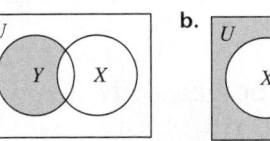 **b.**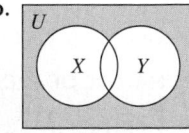

10. a. $8! - 3! = 40{,}320 - 6 = 40{,}314$ **b.** $8 - 3! = 8 - 6 = 2$
c. $(8 - 3)! = 5! = 120$ **d.** $\left(\dfrac{8}{2}\right)! = 4! = 24$ **d.** $\dbinom{8}{2} = 28$
11. Answers for **a–d** are found using Pascal's triangle: **a.** $_5C_3 = 10$
b. $_8P_3 = 3! \cdot {}_8C_3 = 6 \cdot 56 = 336$ **c.** $_{12}P_0 = 1 \cdot 1 = 1$
d. $_{14}C_4 = 1{,}001$ **e.** $_{100}P_3 = \dfrac{100!}{97!} = 100 \cdot 99 \cdot 98 = 970{,}200$
12. $_{12}C_3 = 220$; they can form 220 different committees.
13. $_5P_5 = 5! = 120$; they can form 120 different lineups at the teller's window. **14.** happy: $\dbinom{5}{1, 1, 2, 1} = \dfrac{5!}{1!1!2!1!} = 60$

college: $\begin{pmatrix} 7 \\ 1, 1, 2, 2, 1 \end{pmatrix} = \frac{7!}{2!2!} = 1,260$ **15.** If you draw three balls, then to have at least one red, you would need to have: 1 red and 2 whites or 2 red and 1 white or 3 red and 0 white $= \begin{pmatrix} 4 \\ 1 \end{pmatrix}\begin{pmatrix} 6 \\ 2 \end{pmatrix} + \begin{pmatrix} 4 \\ 2 \end{pmatrix}\begin{pmatrix} 6 \\ 1 \end{pmatrix} + \begin{pmatrix} 4 \\ 3 \end{pmatrix} = 100$

16. a. $A \cup (B \cap C) \neq (A \cup B) \cap C$; disproved

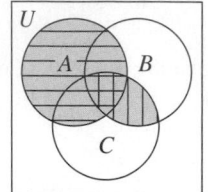 $\neq$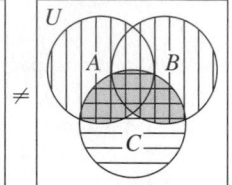

b. $(A \cup B) \cap C = (A \cap C) \cup (B \cap C)$; proved

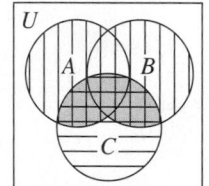 $=$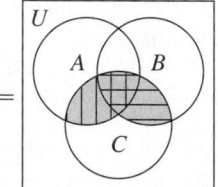

17. Draw a Venn diagram; begin with the innermost part first: 15 have all three.
Next, use the information for two overlapping sets to fill in 2, 20, and 10.
Fill in the regions in each circle not yet completed to fill in 5, 5, and 3.
Finally, total all the numbers to see that 10 of the students have none of these items.

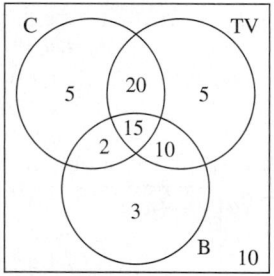

18. Fundamental counting principle; each item can be included or not included, and there are 8 items, so the number of possibilities is $2^8 = 256$. The claim is correct. **19.** The number of arrangements is $_{10}P_{10} = 10! = 3,628,800$. This is almost 10,000 years, so that day will never come for the members of this club. **20. a.** $24^5 = 7,962,624$ **b.** Divide by 60 to change to minutes; divide again by 60 to change to hours; then divide by 24 to change to days. The result is more than 92 days (nonstop).

CHAPTER 11 THE NATURE OF PROBABILITY

11.1 Introduction to Probability, page 560
5. a. $\frac{1}{4}$ **b.** $\frac{1}{4}$ **c.** $\frac{1}{2}$ **7. a.** $\frac{1}{2}$ **b.** $\frac{1}{3}$ **c.** $\frac{2}{3}$ **9.** about 0.05
11. 0.19 **13.** P(royal flush) $\approx 0.000001539077169$
15. P(four of a kind) ≈ 0.0002400960384 **17.** P(flush) $\approx$ 0.00196540155 **19.** P(three of a kind) ≈ 0.02112845138
21. P(one pair) ≈ 0.42256902761 **23. a.** P(five of clubs) $= \frac{1}{52}$
b. P(five) $= \frac{1}{13}$ **c.** P(club) $= \frac{1}{4}$ **25. a.** P(five and a jack) $= 0$
b. P(five or a jack) $= \frac{2}{13}$ **27. a.** $\frac{1}{12}$ **b.** $\frac{7}{12}$ **c.** $\frac{1}{12}$
29. a. $\frac{1}{4}$ **b.** $\frac{3}{4}$ **31.** P(five) $= \frac{1}{9}$ **33.** P(seven) $= \frac{1}{6}$
35. P(nine) $= \frac{1}{9}$ **37.** P(four *or* five) $= \frac{7}{36}$
39. P(eight *or* ten) $= \frac{2}{9}$ **41.** Player A can spin a 5 or 3 and C can

spin 2 or 4; look at the sample space:

C\A	4	2
5	A wins	A wins
3	C wins	A wins

Pick A: $P(A$ winning$) = \frac{3}{4}$
43. Player D can spin a 1, 7, or 8 and E can spin a 4, 5, or 6.

D\E	4	5	6
1	E	E	E
7	D	D	D
8	D	D	D

$P(D) = \frac{2}{3}$; $P(E) = \frac{1}{3}$; so pick D.
45. Player D can spin a 1, 7, or 8 and F can spin a 2, 3, or 9.

D\F	2	3	9
1	F	F	F
7	D	D	F
8	D	D	F

$P(F) = \frac{5}{9}$; $P(D) = \frac{4}{9}$, so pick F.
47. Player C can spin a 2 or 4, and F can spin a 2, 3, or 9. If both players spin a 2, play again.

C\F	2	3	9
2	Tie	F	F
4	C	C	F

$P(C) = \frac{2}{5}$; $P(F) = \frac{3}{5}$; pick F.
49. Answers vary; the answers here are the theoretical probabilities, but the problem requests empirical probabilities.
a. P(two) $= 0.0278$ **b.** P(three) $= 0.0556$ **c.** P(four) $= 0.0833$
d. P(five) $= 0.1111$ **e.** P(six) $= 0.1389$ **f.** P(seven) $= 0.1667$
g. P(eight) $= 0.1389$ **h.** P(nine) $= 0.1111$ **i.** P(ten) $= 0.0833$
j. P(eleven) $= 0.0556$ **k.** P(twelve) $= 0.0278$ **51.** yes
53. P(earthquake) $= \frac{870}{8,766} \approx 10\%$ **55. a.** $\frac{2}{9}$ **b.** $\frac{1}{9}$
57. a.

	1	2	3	4
1	(1, 1)	(1, 2)	(1, 3)	(1, 4)
2	(2, 1)	(2, 2)	(2, 3)	(2, 4)
3	(3, 1)	(3, 2)	(3, 3)	(3, 4)
4	(4, 1)	(4, 2)	(4, 3)	(4, 4)

b. P(two) $= \frac{1}{16}$ **c.** P(three) $= \frac{1}{8}$ **d.** P(four) $= \frac{3}{16}$ **e.** P(five) $= \frac{1}{4}$
f. P(six) $= \frac{3}{16}$ **g.** P(seven) $= \frac{1}{8}$ **59.** P(brain) $= \frac{2}{3}$

11.2 Mathematical Expectation, page 568

7. B **9.** B **11.** B **13.** $E = \$0.08$ **15.** $E = \$7.20$
17. $E = 0.125$; A fair price would be to pay \$0.25 for two plays of the game. **19.** \$1.50 **21.** They should play. **23.** $E = \$6,415$
25. $E = \$6,875$; They should dig the well because the expectation is positive. **27.** $E \approx \$0.02$; yes, you should play the game.
29. \$0.05 **31.** The expected number of tardies is 1.82.
33. They should not purchase them. **35.** $-\$0.05$
37. $-\$0.05$ **39.** $-\$0.05$ **41.** $-\$0.05$ **43.** $-\$0.05$
45. fair game **47.** not a fair game; don't play
49. You should be willing to pay \$2.84 to play the game.
51. The player should continue. **53.** $E = \$1.50$ **55.** \$500.00

11.3 Probability Models, page 580

7. C **9.** B **11.** B **13.** $\frac{1}{5}$ **15.** 0.995 **17.** $\frac{4}{5}$ **19.** $\frac{15}{16}$
21. odds of 15 to 1 **23.** Odds in favor are about 5 to 1.
25. $P(\#1) = \frac{1}{3}$; $P(\#2) = \frac{1}{16}$; $P(\#3) = \frac{2}{5}$; $P(\#4) = \frac{5}{12}$; $P(\#5) = \frac{1}{2}$. *Note:*
At a horse race, odds are determined by track betting and the sum of the probabilities is not necessarily 1. **27.** $\frac{33}{34}$ **29.** $\frac{4}{11}$ **31.** 1
33. $\frac{1}{3}$ **35.** $\frac{1}{2}$ **37.** $\frac{4}{51}$ **39.** $\frac{4}{17}$ **41.** $\frac{26}{51}$ **43.** About 51.5% of the plates have repetitions. **45.** 36.1% **47. a.** 0.3125 **b.** 0.675
c. 0.389 **d.** 0.70 **49. a.** 0 **b.** 1 **51. a.** $\frac{1}{5}$ **b.** $\frac{1}{5}$ **c.** 0

53. $\dfrac{P(\overline{E})}{P(E)} = \dfrac{\frac{f}{n}}{\frac{s}{n}} = \dfrac{f}{n} \cdot \dfrac{n}{s} = \dfrac{f}{s}$ = odds against **55. a.** $\frac{2}{7}$

b. $\frac{5}{7}$ **c.** $\frac{9}{34}$ **d.** $\frac{25}{34}$ **e.** $\frac{5}{17}$ **f.** $\frac{12}{17}$ **g.** $\frac{2}{7}$

h. *First selection*   *Second selection*

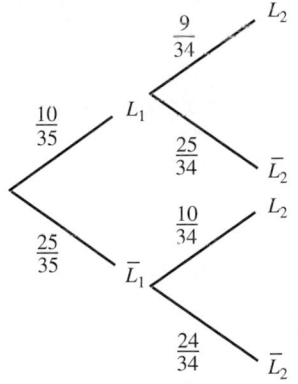

11.4 Calculated Probabilities, page 592

7. $\frac{1}{2}$ **9.** $\frac{5}{6}$ **11.** $\frac{1}{12}$ **13.** $\frac{2}{3}$ **15.** $\frac{4}{9}$ **17.** $\frac{11}{12}$ **19.** $\frac{1}{3}$
21. $\frac{5}{9}$ **23.** $\frac{35}{36}$ **25.** no **27.** yes **29.** yes **31.** $\frac{1}{6}$
33. $\frac{1}{2}$ **35.** $\frac{11}{36}$ **37. a.** 0.000455 **b.** 0.01229 **c.** 0.01229
d. 0.1538 **e.** 0.0237 **39.** $-\$0.08$ **41.** $-\$1.42$
43. $-\$0.92$ **45.** 2.7% **47.** $\frac{1}{32}$ **49. a.** $\frac{25}{64}, \frac{5}{14}$ **b.** $\frac{9}{64}, \frac{3}{28}$
c. $\frac{15}{32}, \frac{15}{28}$ **d.** $\frac{15}{64}, \frac{15}{56}$ **51. a.** play **b.** play **53.** $\frac{1}{38}$
55. 0.71 **57.** $-\$0.46$ **59.** $-\$0.55$

11.5 The Binomial Distribution, page 599

3. 0.132 **5.** 0.128 **7.** 0.016 **9.** 1.701×10^{-4}
11. 0.3125 **13.** 0.3125 **15.** 0.161 **17.** 0.3125
19. 0.044 **21.** 0.0107 **23.** 0.296 **25.** 0.222
27. 0.656 **29.** 0.0001 **31.** 0.9477 **33.** 0.512
35. 0.096 **37.** 0.488 **39.** 0.2109 **41.** 0.4019
43. 0.3025 **45.** 0.1699 **47.** 0.1028 **49.** 0.2903

51. 0.0837 **53.** The smallest number of missiles is 16.
55. The probability of denying the claim is 0.1114265. **57.** 0.75
59. 0.3125

Chapter 11 Review Questions, page 602

1. $P(\text{defective}) = \frac{4}{1,000} = 0.004$ **2.** The possible primes are 2, 3, and 5, so $P(\text{prime}) = \frac{3}{6} = \frac{1}{2}$. **3.** Look at Figure 11.3. $P(\text{eight}) = \frac{5}{36}$
4. $P(\text{jack or better}) = \frac{16}{52} = \frac{4}{13}$ **5.** $P(\text{ace}) = \frac{3}{51} = \frac{1}{17}$ **6.** Look at Figure 11.3. **a.** $P(5 \text{ on at least one of the dice}) = \frac{11}{36}$
b. $P(5 \text{ on one die or 4 on the other}) = \frac{20}{36} = \frac{5}{9}$ **c.** $P(5 \text{ on one die and 4 on the other}) = \frac{2}{36} = \frac{1}{18}$ **7.** $P(E) = 0.01$; $P(\overline{E}) = 1 - P(E) = 1 - 0.01 = 0.99$ **8.** $P(E) = \frac{9}{10}$; $P(\overline{E}) = 1 - \frac{9}{10} = \frac{1}{10}$;
Odds in favor $= \dfrac{P(E)}{P(\overline{E})} = \dfrac{\frac{9}{10}}{\frac{1}{10}} = \frac{9}{10} \cdot \frac{10}{1} = \frac{9}{1}$ or 9 to 1

9. Let E be the event. Given $f = 1,000$ and $s = 1$,
$$P(E) = \frac{s}{s + f} = \frac{1}{1 + 1,000} = \frac{1}{1,001}$$
10. $E = P(\text{one}) \cdot 12 = \frac{1}{6}(12) = 2$; the expected value is \$2.
11. a. $P(\text{orange} \mid \text{orange}) = \frac{3}{5}$ **b.** $P(\text{orange} \mid \text{orange}) = \frac{2}{4} = \frac{1}{2}$
12. $P(\overline{A \cup B}) = 1 - P(A \cup B) = 1 - [P(A) + P(B) - P(A \cap B)]$
$= 1 - P(A) - P(B) + P(A \cap B) = 1 - P(A) - P(B) + P(A)P(B)$
$= 1 - \frac{2}{3} - \frac{3}{5} + \frac{2}{3} \cdot \frac{3}{5} = \frac{2}{15}$ **13.** $P(X = x) = \dbinom{n}{x}(0.001)^x(0.999)^{n-x}$
14. $P(X = 3) = \dbinom{5}{3}(0.85)^3(0.15)^2 \approx 0.138$
15. $P(X = 4) = \dbinom{4}{4}(0.85)^4(0.15)^0 \approx 0.522$
16. List the sample space:

your selection ⌐   ⌐cat left behind

M M ⎫ two successes (M) out of 3 possibilities;
M F ⎬ $P(\text{male}) = \frac{2}{3}$
F M ⎭
F F } Vet rules this out, so delete this from the sample space.

17. $P(X \geq 1) = 1 - P(X = 0) = 1 - \dbinom{4}{0}(0.35)^0(0.65)^4 \approx 0.821$

18. Write out the possible sample spaces. The best possibility is with die C:

				D				
	5	5	5	5	1	1	1	1
2	5	5	5	5	2	2	2	2
2	5	5	5	5	2	2	2	2
2	5	5	5	5	2	2	2	2
C 2	5	5	5	5	2	2	2	2
2	5	5	5	5	2	2	2	2
2	5	5	5	5	2	2	2	2
6	6	6	6	6	6	6	6	6
6	6	6	6	6	6	6	6	6

We see that there are 64 possibilities and that die C wins 40 of those times; thus, $P(C \text{ wins}) = \frac{40}{64} = \frac{5}{8}$
19. $P(\text{winning}) = \frac{1}{5}$. It is tempting to say 1 out of 25, but after you pick a key, one of five cars will fit that key.
20. $P(\text{winning a particular car}) = \frac{1}{5} \cdot \frac{1}{5} = \frac{1}{25}$

CHAPTER 12 THE NATURE OF STATISTICS

12.1 Frequency Distributions and Graphs, page 616

5. a.

Salary	Tally	Frequency
$60,000	|	1
$50,000	|	1
$35,000	|	1
$30,000	||	2
$25,000	|||	3
$20,000	|	1
$18,000	||	2
$16,000	|	1
$14,000	|	1

b. (in thousands)

1	4 6 8 8
2	0 5 5 5
3	0 0 5
4	
5	0
6	0

c.

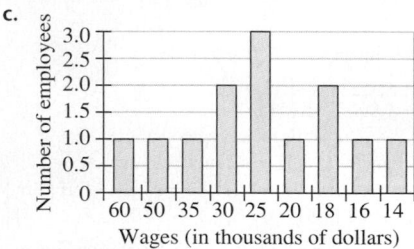

d.

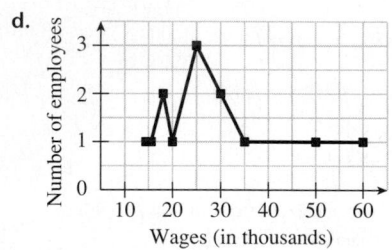

7.

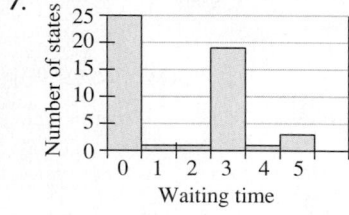

9.

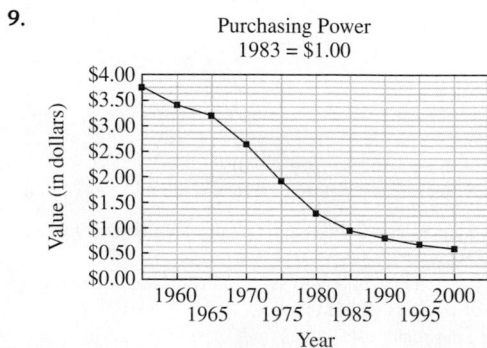

11.

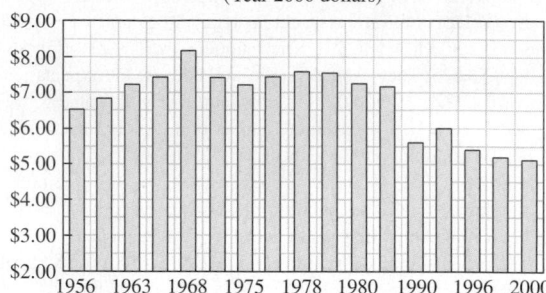

13. They seem to cross in about 1958.

15.

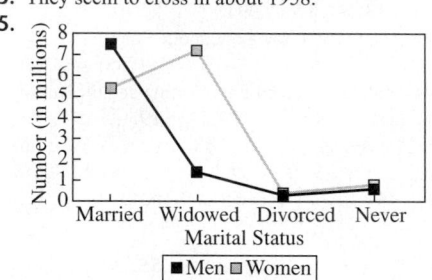

17. a. Gonorrhea cases by race and sex are graphed. The categories are black males, black females, white males, and white females.
b. Black females and white males are almost the same in 1981 (acceptable answer), but the only one that shows the same on the graph is white males and white females in 1986. **c.** The number of cases in black females was increasing between 1984 and 1986.
d. 100,000 cases **19.** B **21.** 1965 and 1990
23. 30%; NBC **25.** Answers vary. **27.** more imports
29. $90 billion **31.** more imports **33.** $56.8 billion
35. 25 times **37.** maybe DUI (illegal if under 18 yr old) **39.** no
41.

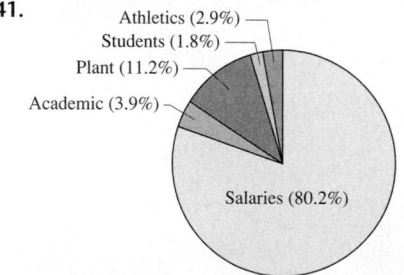

43.

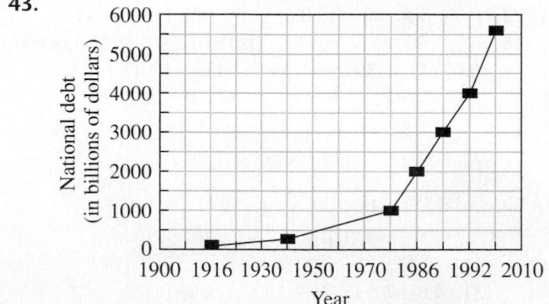

45. Answers vary; graph is meaningless without a scale.
47. The graph is appropriate and accurate. **49.** The advertisement says that the car is 57 in. wide on the outside, but a full 5 ft wide

across on the inside. **51.** 3.4 persons **53.** 2036
55. The line graph at the right begins at 50 rather than 0.
57.

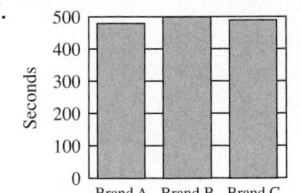

59. The graph in Figure 12.25b

12.2 Descriptive Statistics, page 630

13. mean = 19; median = 19; no mode **15.** mean = 767;
median = 767; no mode **17.** mean = 11; median = 9; no mode
19. mean = 82; median = 81; no mode **21.** mean = 6;
median = 2.5; mode = 1 **23.** range = 4; var. = 2.5; $s \approx 1.58$
25. range = 4; var. = 2.5; $s \approx 1.58$ **27.** range = 24; var. = 93.5;
$s \approx 9.67$ **29.** range = 25; var. = 83; $s \approx 9.11$
31. range = 21; var. = 62.86; $s \approx 7.93$ **33.** median = 432.51;
$Q_1 = 427.48$; $Q_3 = 442.28$

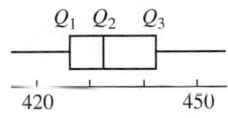

35. mean = 13; median = 11; mode = 10 **37.** mean = 68; median = 70; mode = 70; range = 50 **39.** mean $\approx$ 366; median =
365; no mode **41.** A **43.** C **45. a.** mean = 105.25
b. median = 12 **c.** mode = 42 **47.** 5 **49.** 12 **51.** 15
53. mean = 19,200; range = 4,000; $s \approx 1,483$ **57. a.** $\bar{x} = 6.5$;
H.M. = 4.7 **b.** $\bar{x} = 56.5$ mph; H.M. = 56.1

12.3 The Normal Curve, page 640

7.

	Cumulative
0	5%
1	16%
2	45%
3	79%
4	94%
5	99%
6 or more	100%
$\bar{x} = 2.62$	
median = 3	
mode = 3	

9.

	Cumulative
0	1%
1	12%
2	47%
3	68%
4	88%
5	94%
6	97%
7	99%
8	100%
$\bar{x} = 2.94$	
median = 3	
mode = 2	

11.

	Cumulative
2	0.28
6	0.40
8	0.75
12	0.81
16	1.00

$\bar{x} = 7.84$
median = 8
mode = 8

For Problems 13–23, *remember a normal curve is symmetric about the
mean.*

13. 41.92% **15.** 49.25% **17.** 49.99% **19.** 49.01%
21. 17.72% **23.** 48.68% **25. a.** 34 people **b.** 33 people

27. 91.92%

29.

		Cumulative
155		0.1%
160	1	2.3%
165	7	15.9%
170	17	50.0%
175	17	84.1%
180	7	97.7%
185	1	99.9%
190		100.0%

31. 60 or above **33.** 25 **35.** Cut off is 87. **37.** C

39.

		Cumulative
A	87^+	6%
B	80 86	22%
C	70–79	78%
D	65–69	94%
F	64^-	100%

41. Answers vary; the mode is the largest. See Figure 12.39.
43. $x = 98.0675$ **45.** $x = 68.919$ **47.** 0.1587
49. 0.4207 **51.** 0.0228 **53.** 0.3085
55. a. 50% **b.** 12.4 oz

57.

-4	0.00013
-3	0.00443
-2	0.05399
-1	0.24197
0	0.39894
1	0.24197
2	0.05399
3	0.00443
4	0.00013

59.

20	0.0004
30	0.0054
40	0.0242
50	0.0399
60	0.0242
70	0.0054
80	0.0004

12.4 Correlation and Regression, page 648

7. no **9.** yes **11.** no **13.** yes **15.** yes **17.** no
19. $r = -0.765$, not significant

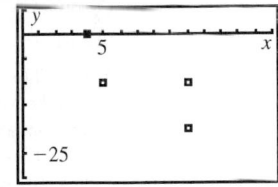

21. $r = -0.954$, significant at 5%

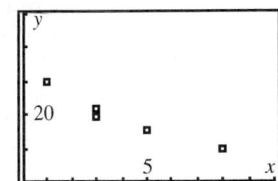

23. $r = -0.890$, significant at 5%

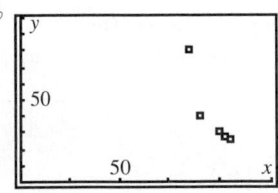

25. $y' = -1.95x + 4.146$ **27.** $y' = -2.7143x + 30.0571$
29. $y' = -2.380x + 270.00$ **31.** D **33.** A **35.** C

37. $y' = 6; r = 0$ **39.** $y' = x + 2; r = 1$ **41.** $r = 0.8830$, significant at 5% **43.** $r = 0.8421$, significant at 1%
45. $y' = 2.4545x + 6.0909$

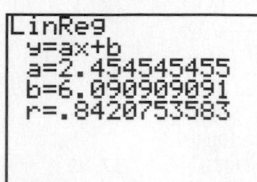 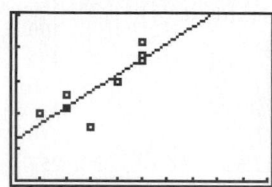

47. $r = 0.358$, no significant correlation **49.** $r = -0.936$, significant at 1%

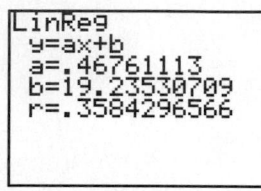 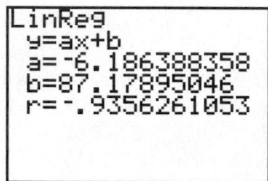

51. $y' = 0.468x + 19.235$ **53.** $y' = -6.186x + 87.179$

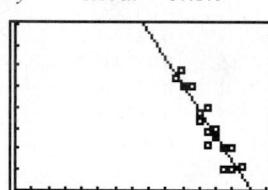

55. Time 27:31; the difference is 4 hr 29 min. **57.** Time is 44 hr 21 min. **59.** $r = 0.380$, significant at the 5% level

12.5 Sampling, page 655

5. B **7.** B **9.** D **11.** A **13.** D **15.** Survey AT&T's customers by randomly choosing from a list of AT&T's customers. **17.** Survey from a random sample of people whose names are randomly chosen from a local list of registered voters who are party members. **19.** Survey a selection of people whose names are randomly chosen from a list of all union members. **21.** Survey a random selection of students chosen from a list of all students. **23.** Survey a selection of people whose names are randomly chosen from a list of all customers. **25.** (1) You accept that 72 is the mean, and it is the mean. (2) You accept that 72 is the mean, and it is not the mean. This is Type II error (accept a false conclusion). (3) You do not accept that 72 is the mean, and it is the mean. This is Type I error (reject a true conclusion). (4) You do not accept that 72 is the mean, and it is not the mean.

Chapter 12 Review Questions, page 659

1. a. Heads: ⫫⫫ ⫫⫫ ⫫⫫ ||| (18); Tails ⫫⫫ ⫫⫫ ⫫⫫ ⫫⫫ || (22)
b.

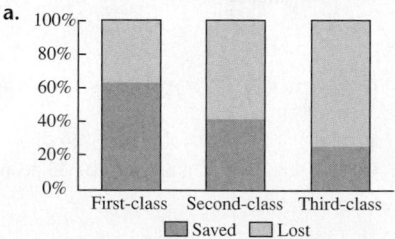

2. a.

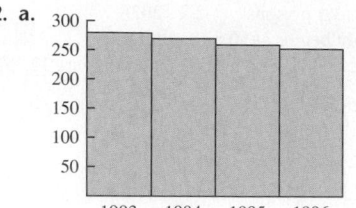

b.

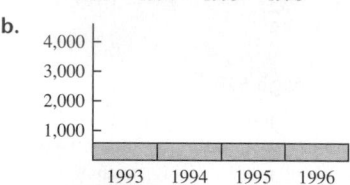

c.

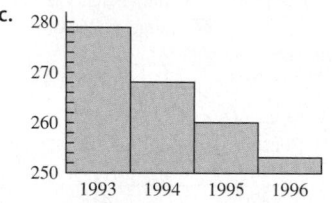

d. Answers vary; impressions can be greatly influenced by using faulty or inappropriate scale (or even worse, no scale at all).

3. mean $= \dfrac{5 + 21 + 21 + 25 + 30 + 40}{6} = \dfrac{142}{6} = 23\frac{2}{3}$

4. The median is the mid value or the mean of the midvalues; $\dfrac{21 + 25}{2} = 23$. **5.** The mode is the most frequently occurring value, namely, 21. **6.** The range is $40 - 5 = 35$.

7. The standard deviation is found:

Score	*(deviation from the mean)*^{2}
5	348.44
21	7.11
21	7.11
25	1.78
30	40.11
40	266.78

$$\frac{344.4 + 7.11 + 7.11 + 1.78 + 40.11 + 266.78}{6 - 1} = \frac{671.328}{5}$$

$$\approx 134.266$$

The standard deviation is $\sqrt{134.266} \approx 11.59$

8. mean $= 11$; median $= 12$; mode $= 12$; mode is the most appropriate measure **9.** mean $= 79$; median $= 74$; no mode; mean is the most appropriate measure **10.** mean $= \$64,741.37$; median $= \$68,345.50$; no mode; the median is the most appropriate measure

11. a.

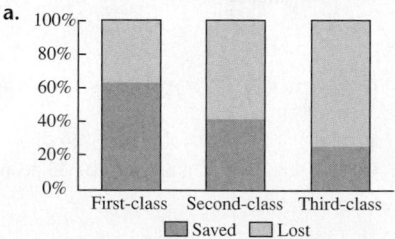

Survival rate of first-, second-, and third-class passengers.

b.

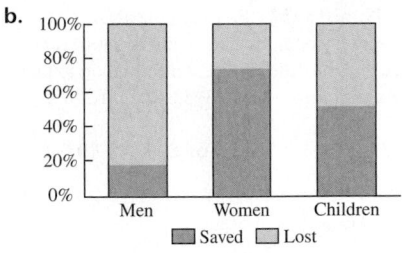

Survival rates of men, women, and children.

12. a. 1982 **b.** 1994 **c.** $41,500

13. a.

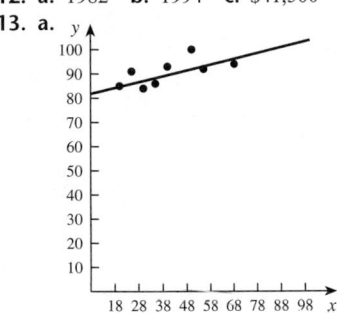

$$m = \frac{n(\Sigma xy) - (\Sigma x)(\Sigma y)}{n(\Sigma x^2) - (\Sigma x)^2}$$

$$= \frac{8(29,677) - (323)(725)}{8(14,899) - 104,329}$$

$$\approx 0.2180582655$$

$$b = \frac{\Sigma y - m(\Sigma x)}{n}$$

$$= \frac{725 - (0.2180582655)(323)}{8}$$

$$\approx 81.82089753$$

The best-fitting line is $y' = 0.22x + 81.8$.

b. $n = 8$ $\Sigma x^2 = 14,899$
$\Sigma x = 323$ $\Sigma y^2 = 65,907$
$\Sigma y = 725$ $(\Sigma x)^2 = 104,329$
$\Sigma xy = 29,677$ $(\Sigma y)^2 = 525,625$

$$r = \frac{n\Sigma xy - (\Sigma x)(\Sigma y)}{\sqrt{n(\Sigma x^2) - (\Sigma x)^2} \sqrt{n(\Sigma y^2) - (\Sigma y)^2}}$$

$$= \frac{8(29,677) - (323)(725)}{\sqrt{8(14,899) - 104,329} \sqrt{8(65,907) - 525,625}}$$

≈ 0.6582620404; not significant at 1% level or 5% level

14. $314 - 266 = 48$; since the standard deviation is 16 days, we note this is 3 standard deviations greater than the mean. From the standard normal curve, this should happen about 0.001 (0.1%) of the time. This means that approximately 1 in 1,000 pregnancies will have a 314-day duration.

15. a.

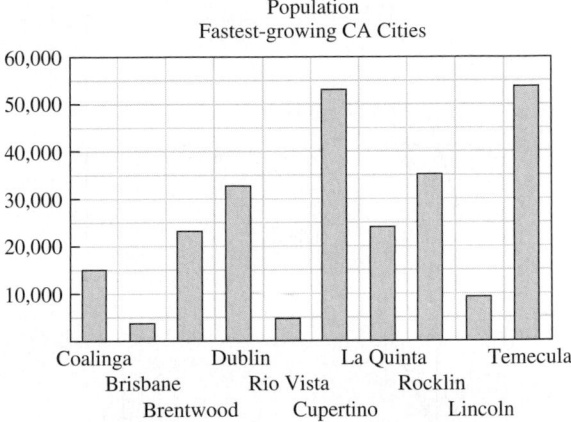

b. mean $= \dfrac{255,585}{10} = 25,558.5$; median $= 23,675$; no mode

CHAPTER 13 THE NATURE OF GRAPHS AND FUNCTIONS

13.1 Cartesian Coordinates and Graphing Lines, page 673
1. a. Aphrodite **b.** Maxwell Montes **c.** Atalanta Planitia
d. Rhea Mons **e.** Lavina Planitia **3.** The y-intercept of a (non-vertical) line is the point where the line crosses the y-axis. The slope is the steepness of the line. **5.** Horizontal lines have slope 0; vertical lines have no slope.

Ordered pairs in Problems 7–17 may vary.

7 $(0, 5), (1, 6), (2, 7)$ **9.** $(0, 5), (1, 7), (-1, 3)$
11. $(0, -1), (1, 0), (2, 1)$ **13.** $(0, 1), (1, -1), (-1, 3)$
15. $(0, 1), (1, 3), (2, 5)$ **17.** $(0, 2), (2, \frac{1}{2}), (4, -1)$

19.   **21.**

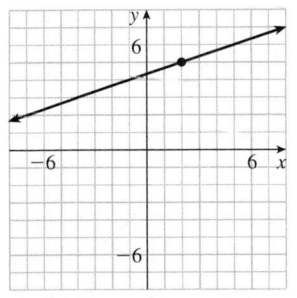 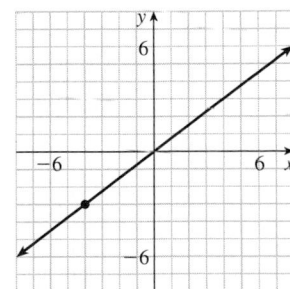

23.   **25.**

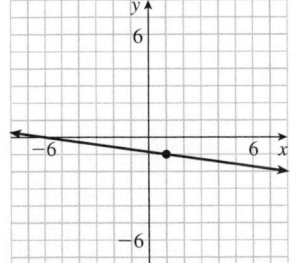 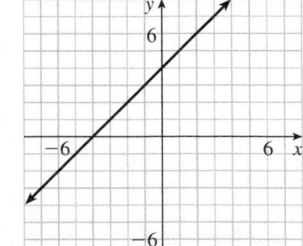

27.

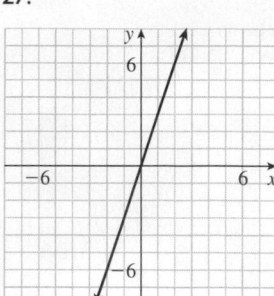

29.

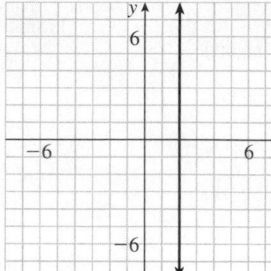

31.

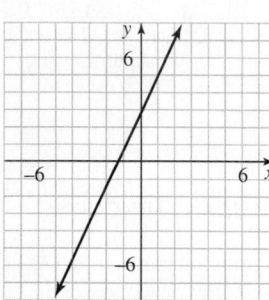

33.

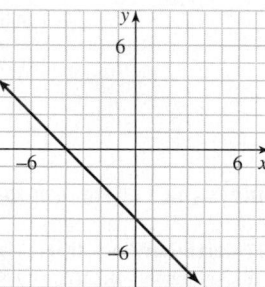

35.

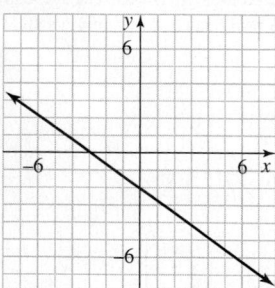

37.

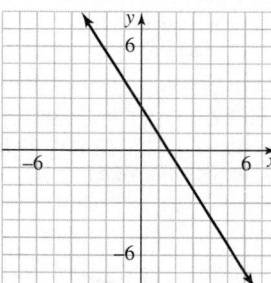

39.

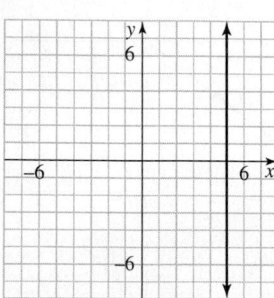

41.

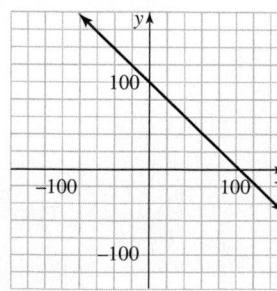

43.

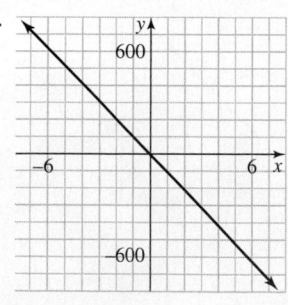

45. C **47.** A **49.** E **51. a.** Declining at a rate that is nearly linear, but you can use slopes to prove that it is not linear. (Looking at a graph does not provide sufficient information to make a decision.) **b.** Answers vary; $y = -0.14x + 5.8$ **c.** In 2005, predict $y = 0.9$.

53. a. $y = 0.35x + 17$ **b.** 800 **c.** 10,000

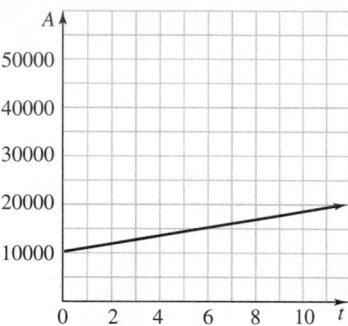

55. a.

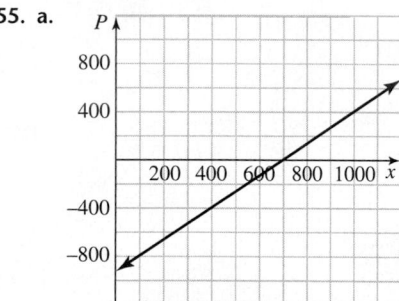

b. loss of $850 **c.** 1.25; it is the profit increase corresponding to each unit increase in number of items sold

57. $y = 0.35x + 17$

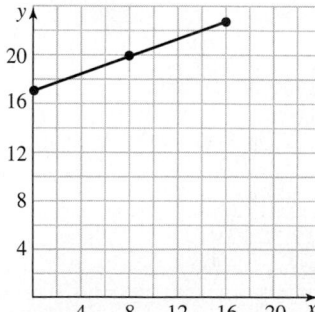

The projected population is 22.3 million.

59. Given $y - k = m(x - h)$. Since the line passes through (x_1, y_1) and (x_2, y_2) the slope is $m = \dfrac{y_2 - y_1}{x_2 - x_1}$. Let $(h, k) = (x_1, y_1)$ be the known point. Then, by substitution, $y - y_1 = \left(\dfrac{y_2 - y_1}{x_2 - x_1}\right)(x - x_1)$.

13.2 Graphing Half-Planes, page 676

3. F; the boundary is $2x + 5y = 2$ **5.** F; $(0, 0)$ is not a test point because it lies on the boundary. **7.** T **9.** T **11.** T

13.

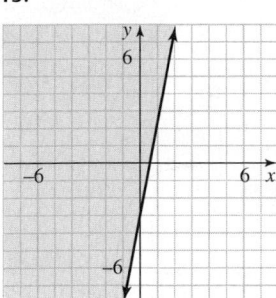

15.

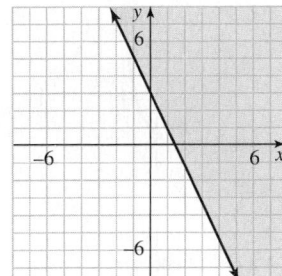

29.

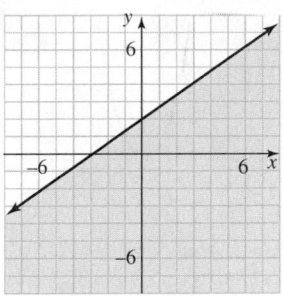

17.

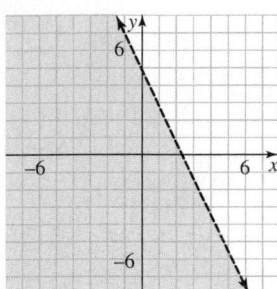

19.

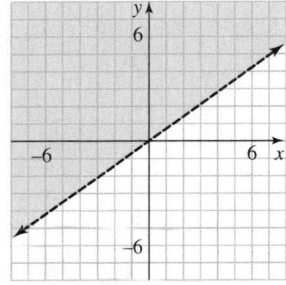

13.3 Graphing Curves, page 681

5.

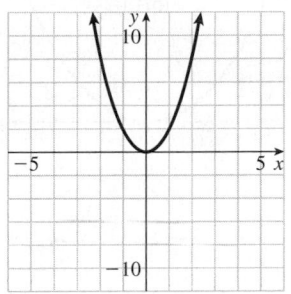

7.

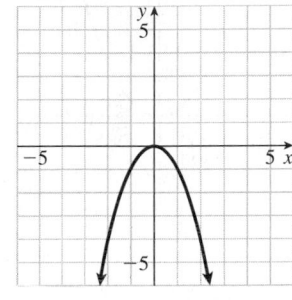

21.

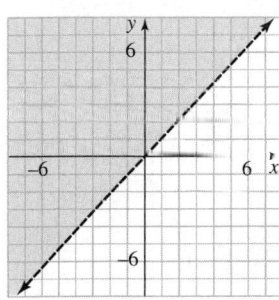

23.

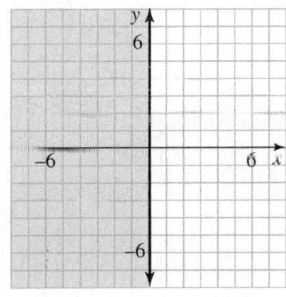

9.

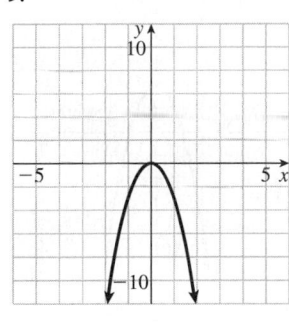

11.

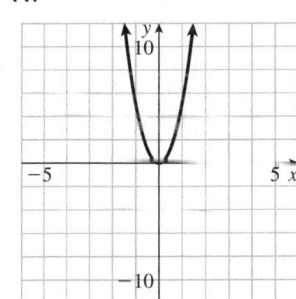

25.

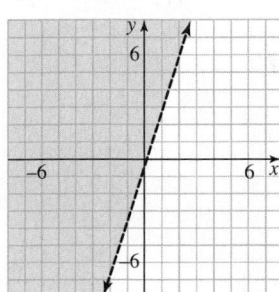

27.

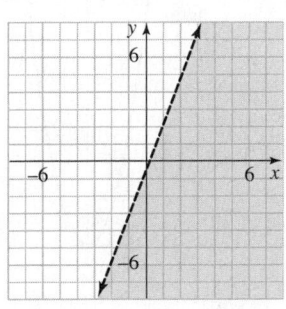

13.

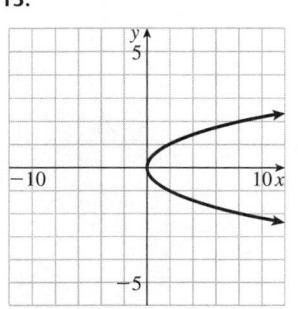

15.

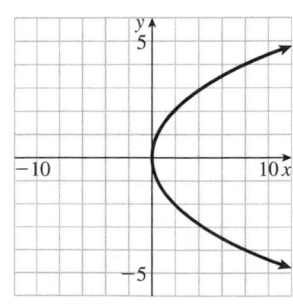

17.

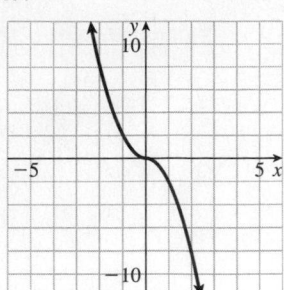

19.

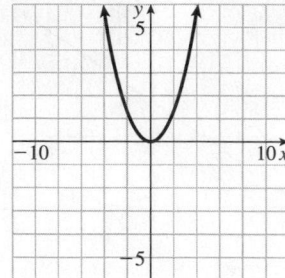

33.

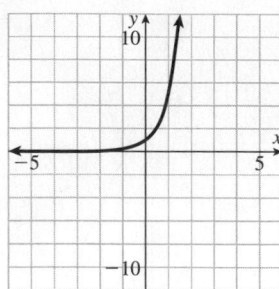

35.

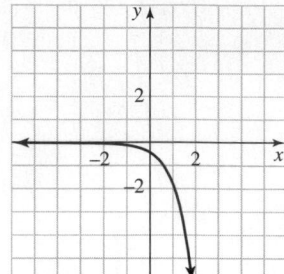

21.

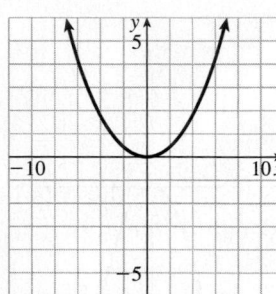

23.

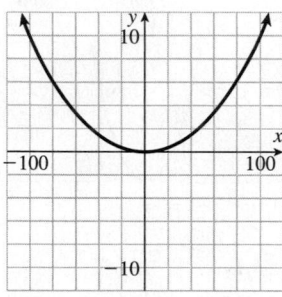

37.

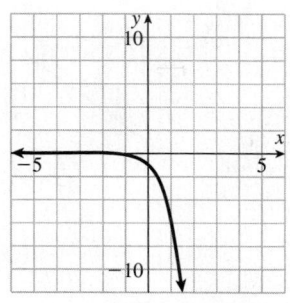

39.

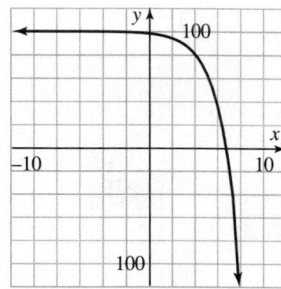

25.

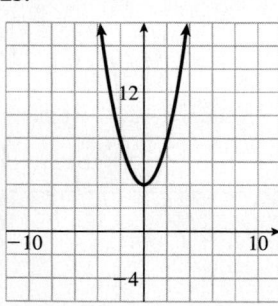

27.

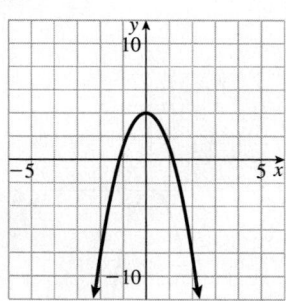

41.

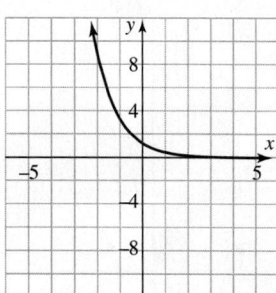

43.

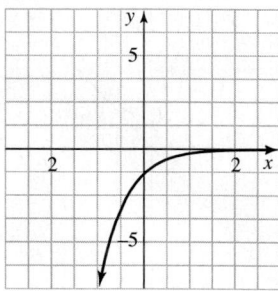

29.

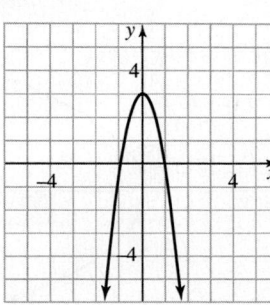

31.

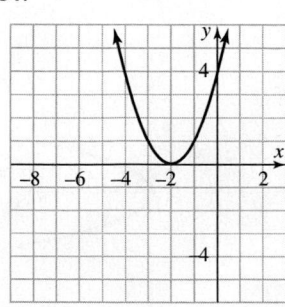

45. a.

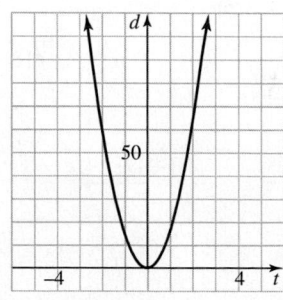

b. upward

c.

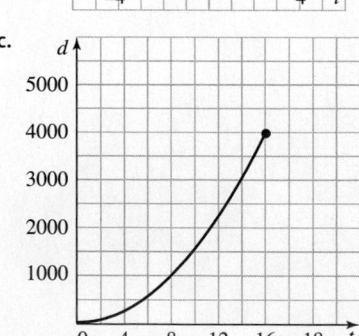

47.

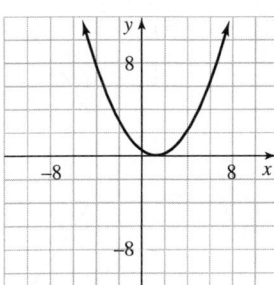

49.

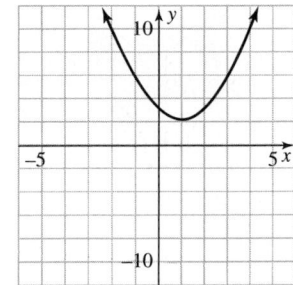

51.

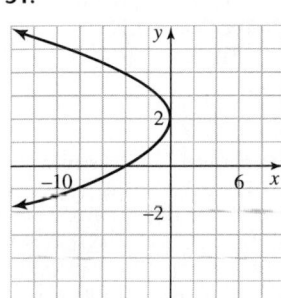

53.

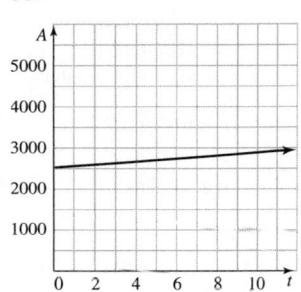

55.

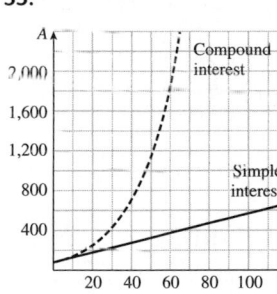

57.

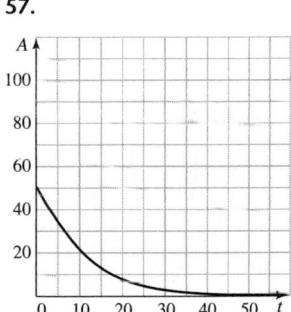

59.

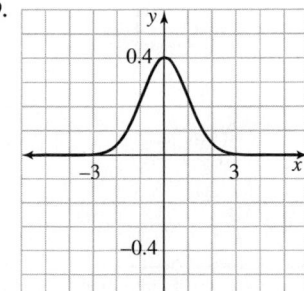

15.

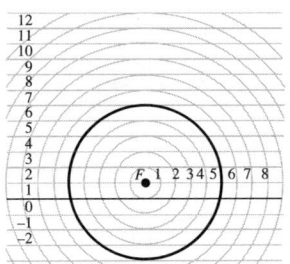

17.

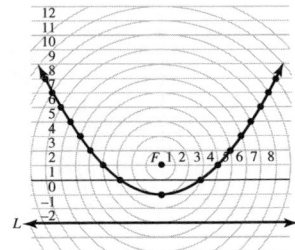

19.

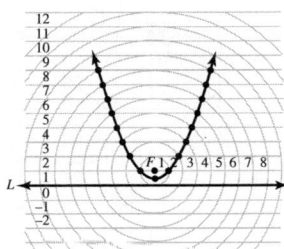

21.

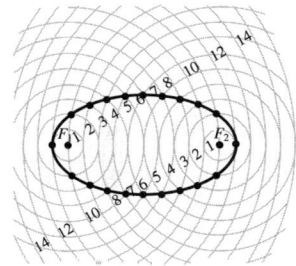

23.

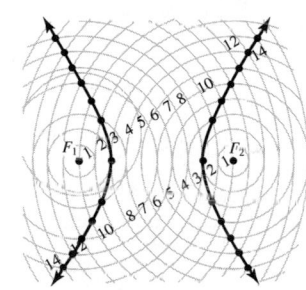

25. a. line **b.** ellipse **c.** parabola
27. a. parabola **b.** hyperbola **c.** cllipse
29. **31.**

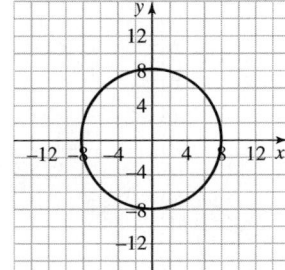

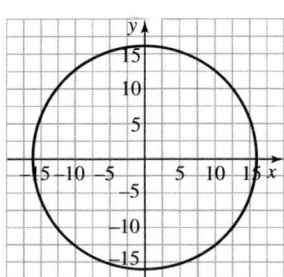

13.4 Conic Sections, page 694

7. a. Both variables are first-degree. **b.** One variable is first-degree; the other is second-degree. **c.** Both variables are second-degree and both have the same sign (in general form). **d.** Both variables are second-degree and they have opposite signs (in general form).
9. A and C have opposite signs. **11.** $A = 0$ and $C \neq 0$ or $A \neq 0$ and $C = 0$ **13.** $A = C = 0$

33.

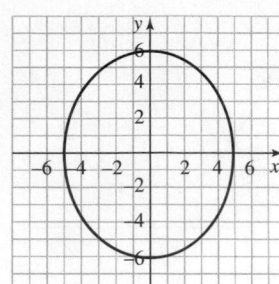

35.

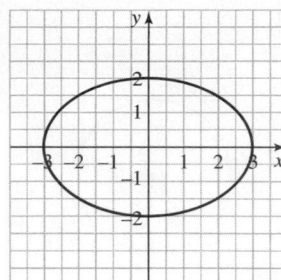

37.

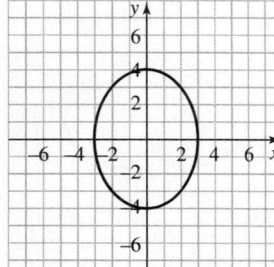

39.

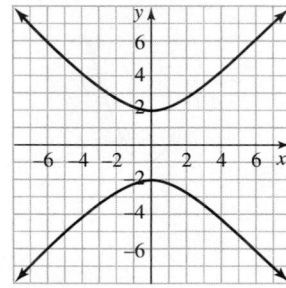

41.

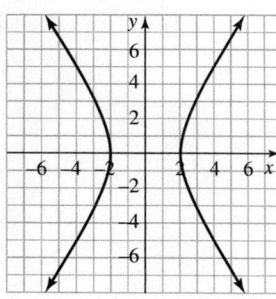

43.

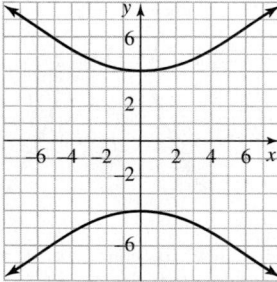

45.

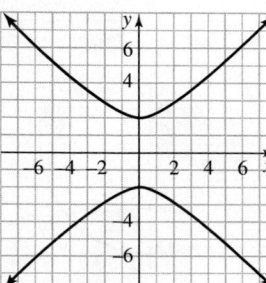

47.

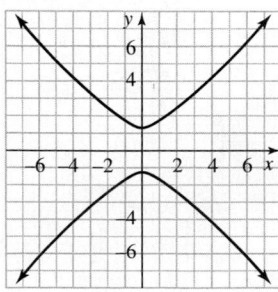

49.

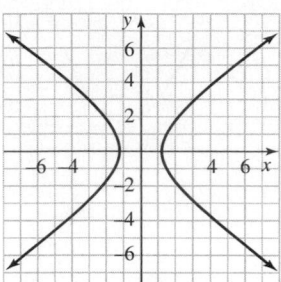

51.

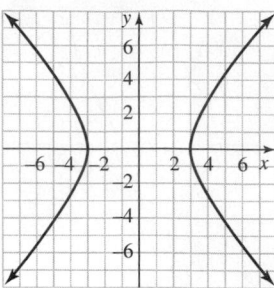

53. aphelion = 1.5×10^8 mi; perihelion = 1.3×10^8 mi
55. Eccentricity is about 0.053.   **57.** The focus is 2.25 m from the
vertex on the axis of the parabola.   **59.** The focus is 8 in. from the
vertex on the axis of the parabola.

13.5 Functions, page 701

3. function   **5.** function   **7.** not a function   **9.** function
11. function   **13.** not a function   **15. a.** $f(4) = 12$
b. $f(6) = 14$ **c.** $f(-8) = 0$ **d.** $f(\frac{1}{2}) = 8\frac{1}{2}$ **e.** $f(x) = x + 8$
17. a. $M(4) = 17$ **b.** $M(6) = 37$ **c.** $M(-8) = 65$
d. $M(\frac{1}{2}) = 1\frac{1}{4}$ **e.** $M(x) = x^2 + 1$ **19. a.** $g(4) = 7$
b. $g(6) = 11$ **c.** $g(-8) = -17$ **d.** $g(\frac{1}{2}) = 0$ **e.** $g(x) = 2x - 1$
21. a. 8   **b.** -16 **c.** $p - 7$ **23. a.** -1 **b.** -31
c. $3a - 1$ **25. a.** 5   **b.** -2 **c.** $\frac{3}{2}$ **27.** not a function;
domain: $-1 \le x \le 1$; range: $-3 \le y \le 3$ **29.** not a function;
domain: $x \ge -3$; range: $\mathbb{R}$ **31.** function; domain: $-2 \le x \le 3$;
range: $-8 \le y \le 4$
33. quadratic;   **35.** logarithmic;

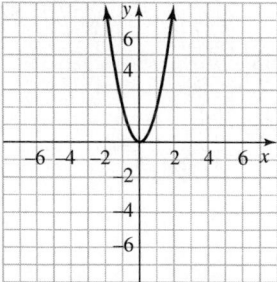

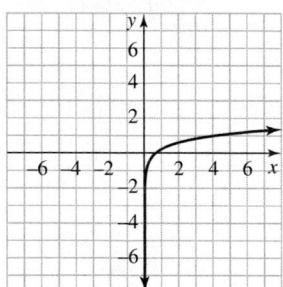

37. probability;

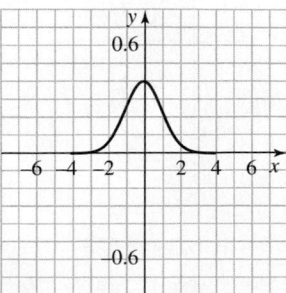

39. 3 **41.** $3x^2 + 3xh + h^2$ **43.** $\dfrac{-1}{x(x+h)}$ **45. a.** 64
b. 96 **c.** 128 **d.** 256 **e.** 512 **47. a.** 1,430 **b.** 1,050
c. 670 **d.** 290 **e.** 100 **49. a.** $P(10) = 4$
b. $P(-10) = 0$ **c.** $P(100) = 25$ **51.** $y = 3x - 5$; domain: $\mathbb{R}$
53. $y = \sqrt{5 - x}$; domain: $x \le 5$ **55. a.** $\{(a, e), (e, a), (b, e),$
$(e, b), (b, c), (c, b), (c, e), (e, d), (d, e)\}$; it is not a function
b. $\{(a, e), (e, b), (b, c), (c, e), (d, e)\}$; it is a function
57. They could toss a rock into the well and listen for when it hits the
bottom. 1 mi = 5,280 ft so $5,280 = 16t^2$. Solve for t to find that
$t \approx 18$ seconds. **59.** $A = s^2$ and $P = 4s$, so $A = \left(\dfrac{P}{4}\right)^2$.

Chapter 13 Review Questions, page 704

1. Solve for y: $5x - y = 15$
$\qquad\qquad\qquad\quad 5x - 15 = y$
The y-intercept is -15; the
slope is 5; rise of 5 and run
of 1.

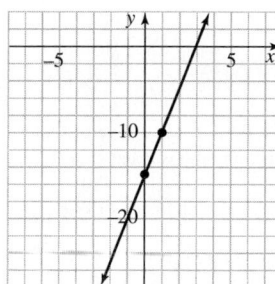

2. $y = -\frac{4}{5}x - 3$
The y-intercept is -3 and the slope is $-\frac{4}{5}$; rise of -4, and run of 5.

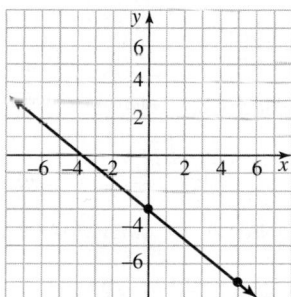

3. Solve for y: $2x + 3y = 15$
$\qquad\qquad\qquad\quad 3y = -2x + 15$
$\qquad\qquad\qquad\quad\ \ y = -\frac{2}{3}x + 5$
The y-intercept is 5; the slope is $-\frac{2}{3}$; rise of -2 and run of 3.

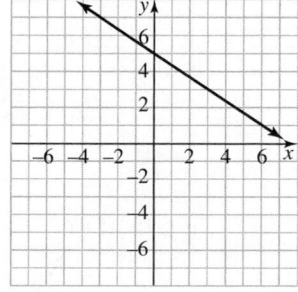

4. Solve for y: $\quad x = -\frac{2}{3}y + 1$
$\qquad\qquad\qquad 3x = -2y + 3$
$\qquad\qquad\qquad 2y = -3x + 3$
$\qquad\qquad\qquad\ \ y = -\frac{3}{2}x + \frac{3}{2}$
y-intercept is $\frac{3}{2}$; the slope is $-\frac{3}{2}$; rise of -3 and run of 2.

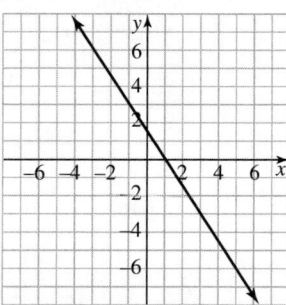

5. $x = 150$; recognize this as a vertical line; watch the scale.

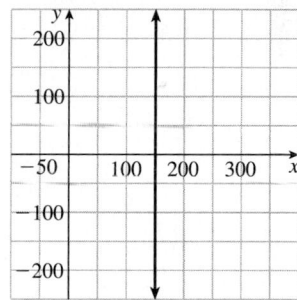

6. $x < 3y$
Graph the boundary $x = 3y$

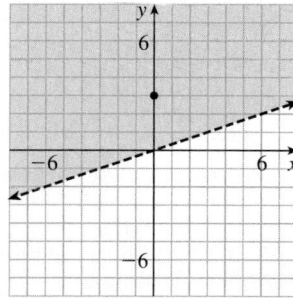

Test point: $(0, 3)$
$x < 3y$
$0 < 3(3)$ is true
Shade the half-plane on the same side as the test point.

7. $y = 1 - x^2$

x	y
1	0
-1	0
2	-3
3	-8

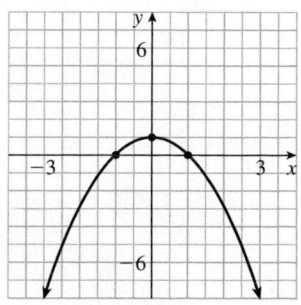

8. $y = -2^x$

x	y
0	-1
1	-2
2	-4
3	-8
-1	$-1/2$
-2	$-1/4$
-3	$-1/8$

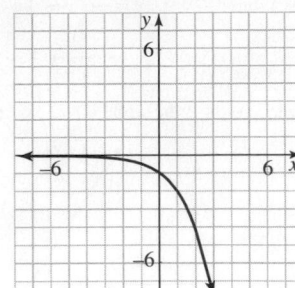

9.   **10.**

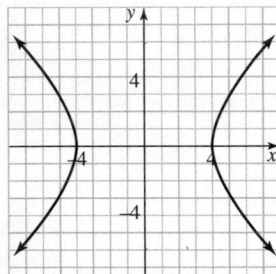   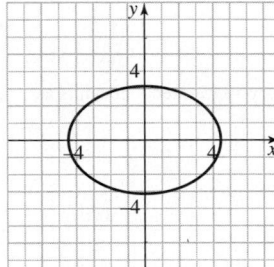

11.   **12.**

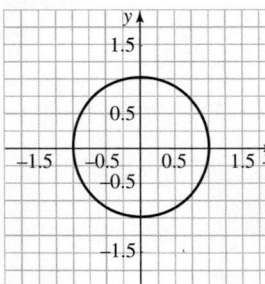   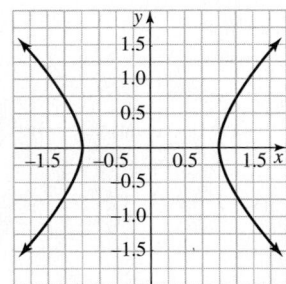

13. Yes, it is a function because each x-value (namely, 4, 5, and 6) is associated with exactly one y-value.   **14.** $f(6) = 3(6) + 2 = 20$
15. $g(0) = 0^2 - 3 = -3$ **16.** $F(10) = 5(10) + 25 = 75$
17. $m(10) = 5$ **18.** $A = 1.7P$; label the horizontal axis P and the vertical axis A. The intercept is 0 and the slope is 1.7—that is, a rise of 1.7 and a run of 1 (or a rise of 17 and a run of 10). Note that since P represents an amount of money, it is nonnegative.

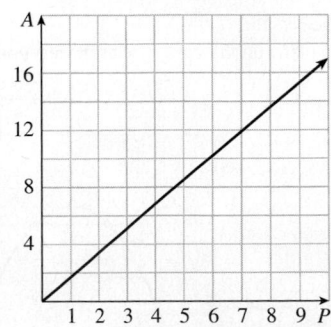

19. Form a table of values for $y = 128t - 16t^2$:

t	0	1	2	3	4	5	6	7	8
y	0	112	192	240	256	240	192	112	0

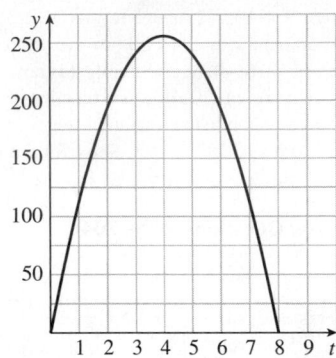

20. Form a table of values for $P = 53{,}000e^{0.013t}$; label the horizontal axis t and the vertical axis P (in units of thousands)

t	0	10	20	30	40	50
P	53,000	60,358	68,737	78,280	89,147	101,524

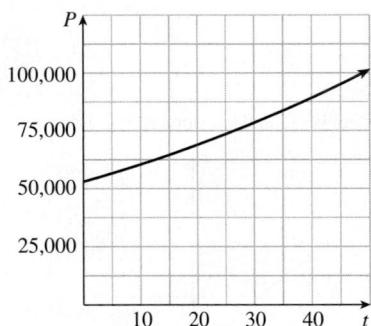

CHAPTER 14 THE NATURE OF MATHEMATICAL SYSTEMS

14.1 Systems of Linear Equations, page 713
7. $(4, -1)$ **9.** inconsistent

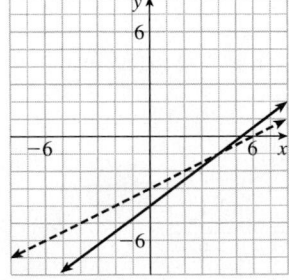   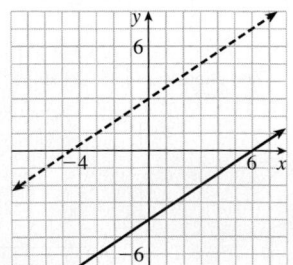

11. $(-1, 1)$ **13.** $(1, 1)$

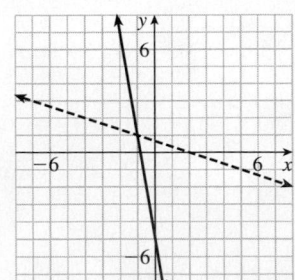   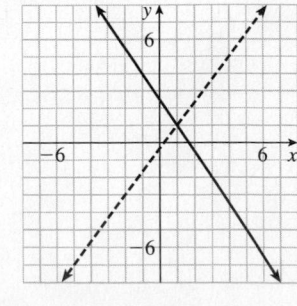

15. $(-5, -3)$ **17.** $(-1, -2)$ **19.** $(2, 10)$ **21.** $(9, -4)$
23. $(4, -1)$ **25.** $(6, -2)$ **27.** $(525, 35)$ **29.** $(5, -2)$
31. dependent   **33.** $(s_1, s_2) = (3, 4)$ **35.** $(u, v) = (5, -2)$
37. $(\frac{3}{5}, \frac{1}{2})$ **39.** $(5, -3)$ **41.** $(3, 1)$ **43.** $(3, 1)$
45. $(4, -1)$ **47.** $(\frac{1}{3}, -\frac{2}{3})$ **49.** $(1, 1)$ **51.** inconsistent
53. $(8, 2)$ **55.** $(-\frac{8}{5}, -\frac{21}{5})$ **57.** 2,601 days after the cat's
birth—namely, June 2, 1999
59. $\left(\dfrac{a + \sqrt{a^2 - 4}}{2}, \dfrac{a - \sqrt{a^2 - 4}}{2}\right), \left(\dfrac{a - \sqrt{a^2 - 4}}{2}, \dfrac{a + \sqrt{a^2 - 4}}{2}\right)$

14.2 Problem Solving with Systems, page 722

1. The box has 31 nickels and 56 dimes.   **3.** There are
18 nickels.   **5.** The plane's speed in still air is 175 mph.
7. The equilibrium point for the system is (50, 250 000).
9. The optimum price for the items is $3.   **11.** Debra Winger was
born in 1955.   **13.** Mixture **a** has 7 lb of micoden.
15. Mixture **c** has $0.5p$ L of bixon.   **17.** Mixture **a** contains $33\frac{1}{3}\%$
bixon.   **19.** There are 40 oz of the base metal.   **21.** Mix 45 gal
milk with 135 gal of cream.   **23.** Both rates are the same if you
drive 160 miles.   **25.** The equilibrium point is (2, 10000).
27. There are 14 dimes and 28 quarters.   **29.** The wind speed
is 22 mph.   **31.** Kristy McNichol was born in 1967.
33. Add $\frac{1}{3}$ gal water.   **35.** NY is 47,831 sq mi and CA is
156,361 sq mi.   **37.** There are 37 quarters.   **39.** There are
30 dimes.   **41.** The robbery included 14 $5 bills, 70 $10 bills, and
31 $20 bills.   **43.** The wind speed is 22.5 mph.
45. Mix together 28.5 gal of milk with 1.5 gal of cream.
47. Mix together 38.46 mg of 12% aspirin with 61.54 mg of 25%
aspirin.   **49.** The Sears Tower is 1,454 ft and the Standard Oil
building is 1,136 ft.   **51.** The Verrazano-Narrows bridge is
4,260 ft.   **53.** The equilibrium point is $2 for 40,000 items.
55. Replace 9 qt of the liquid with antifreeze.   **57. a.** 125 items
would be supplied; 75 would be demanded.   **b.** No items would be
supplied at $200.   **c.** No items would be demanded at $400.
d. The equilibrium price is $\frac{700}{3} \approx \$233.33$ **e.** The number of items
produced at the equilibrium price is $\frac{250}{3} \approx 83$.

14.3 Matrix Solution of a System of Equations, page 732

5. false; it is 4×5 **7.** true   **9.** true   **11.** false; multiply
row 4 by $\frac{1}{3}$ and add to row 2   **13.** false; RowSwap ([A],1,2)

15. a. $\begin{cases} 6x + 7y + 8z = 3 \\ x + 2y + 3z = 4 \\ y + 3z = 4 \end{cases}$ **b.** $\begin{cases} x_1 = 3 \\ x_2 + 2x_3 = 4 \end{cases}$ **c.** $\begin{cases} x_1 = 32 \\ x_2 = 27 \\ x_3 = -5 \\ 0 = 3 \end{cases}$

17. $\begin{bmatrix} 1 & 0 & 2 & \vdots & -8 \\ -2 & 3 & 5 & \vdots & 9 \\ 0 & 1 & 0 & \vdots & 5 \end{bmatrix}$ **19.** $\begin{bmatrix} 1 & 4 & 3 & \vdots & \frac{6}{5} \\ 7 & -5 & 3 & \vdots & 2 \\ 12 & 0 & 1 & \vdots & 4 \end{bmatrix}$
RowSwap([B], 1, 2)   *Row(1/5, [A], 1)

21. $\begin{bmatrix} 1 & 3 & -5 & \vdots & 6 \\ 0 & 13 & -14 & \vdots & 20 \\ 0 & 5 & 1 & \vdots & 3 \end{bmatrix}$ **23.** $\begin{bmatrix} 1 & 5 & 3 & \vdots & 2 \\ 0 & -7 & -7 & \vdots & 0 \\ 0 & -13 & -8 & \vdots & -6 \end{bmatrix}$
*Row+(3, [B], 1, 2)   *Row+(-2, [A], 1, 2)
                         Row +(-3, [Ans], 1, 3)

25. $\begin{bmatrix} 1 & 5 & -3 & \vdots & 5 \\ 0 & 1 & 3 & \vdots & -5 \\ 0 & 2 & 1 & \vdots & 5 \end{bmatrix}$ **27.** $\begin{bmatrix} 1 & 3 & -2 & \vdots & 0 \\ 0 & 1 & -4 & \vdots & 8 \\ 0 & 3 & 6 & \vdots & 1 \end{bmatrix}$
*Row(1/3, [B], 2)   *Row+(-1, [A], 3, 2)

29. $\begin{bmatrix} 1 & 0 & 12 & \vdots & 27 \\ 0 & 1 & -2 & \vdots & -5 \\ 0 & 0 & -2 & \vdots & -4 \end{bmatrix}$ **31.** $\begin{bmatrix} 1 & 0 & -26 & -8 & \vdots & 8 \\ 0 & 1 & 5 & 2 & \vdots & 0 \\ 0 & 0 & -1 & -2 & \vdots & 5 \\ 0 & 0 & -13 & -3 & \vdots & 7 \end{bmatrix}$
*Row+(-3, [B], 2, 1)   *Row+(-5, [D], 2, 1)
*Row+(2, [Ans], 2, 3)   *Row+(-1, [Ans], 2, 3)
                           *Row+(-2, [Ans], 2, 4)

33. $\begin{bmatrix} 1 & 0 & -4 & \vdots & -5 \\ 0 & 1 & 3 & \vdots & 6 \\ 0 & 0 & 1 & \vdots & 1.5 \end{bmatrix}$ **35.** $\begin{bmatrix} 1 & 0 & -8 & 2 & \vdots & 8 \\ 0 & 1 & -1 & 3 & \vdots & 2 \\ 0 & 0 & 1 & 0 & \vdots & 5 \\ 0 & 0 & -2 & 1 & \vdots & 6 \end{bmatrix}$
*Row(1/8, [B], 3)   *Row(1/2, [D], 3)

37. $\begin{bmatrix} 1 & 0 & 0 & \vdots & 7 \\ 0 & 1 & 0 & \vdots & -7 \\ 0 & 0 & 1 & \vdots & 3 \end{bmatrix}$ **39.** $\begin{bmatrix} 1 & 0 & 0 & 2 & \vdots & 24 \\ 0 & 1 & 0 & 2 & \vdots & -8 \\ 0 & 0 & 1 & 0 & \vdots & 2 \\ 0 & 0 & 0 & 1 & \vdots & 9 \end{bmatrix}$
*Row+(-4, [B], 3, 2)   *Row+(8, [D], 3, 1)
*Row+(3, [Ans], 3, 1)   *Row+(-4, [Ans], 3, 2)
                           *Row+(1, [Ans], 3, 4)

41. $(5, -3)$ **43.** $(3, 1)$ **45.** $(4, -1)$ **47.** $(4, -1)$
49. $(2, 0, 1)$ **51.** $(3, 2, 5)$ **53.** $(2, -3, -1)$
55. $(2, -3, 2)$ **57.** $(1, 2, -\frac{1}{2})$ **59.** Produce 4 units of candy I,
5 units of candy II, and 6 units of candy III.

14.4 Inverse Matrices, page 745

9. a. $\begin{bmatrix} 2 & 4 & 2 \\ 6 & -2 & 4 \\ 2 & 2 & 5 \end{bmatrix}$ **b.** $\begin{bmatrix} -6 & -1 & -2 \\ 3 & -7 & -3 \\ 4 & -7 & -2 \end{bmatrix}$

11. a. $\begin{bmatrix} 20 & 21 & 34 \\ 29 & 16 & 15 \\ 7 & 48 & 5 \end{bmatrix}$ **b.** $\begin{bmatrix} -1 & 37 & 32 \\ 4 & 14 & 45 \\ 27 & 9 & 28 \end{bmatrix}$

13. a. $\begin{bmatrix} 14 & 14 \\ -7 & 7 \end{bmatrix}$ **b.** $\begin{bmatrix} 1 & 0 & 0 & 0 \\ 0 & 1 & 0 & 0 \\ 0 & 0 & 1 & 0 \\ 0 & 0 & 0 & 1 \end{bmatrix}$

15. a. not conformable   **b.** not conformable

17. $\begin{cases} x + 2y + 4z = 13 \\ -3x + 2y + z = 11 \\ 2x + z = 0 \end{cases}$

19. $[A][B] = [B][A] = \begin{bmatrix} 1 & 0 \\ 0 & 1 \end{bmatrix}$

21. $\begin{bmatrix} 2 & 7 \\ 1 & 4 \end{bmatrix}$ **23.** $\begin{bmatrix} 9 & -4 & -2 \\ -18 & 9 & 4 \\ -4 & 2 & 1 \end{bmatrix}$ **25.** $\begin{bmatrix} 1 & 0 & -1 & 0 \\ 0 & \frac{1}{2} & 0 & 0 \\ -2 & 0 & 2 & 1 \\ 0 & 0 & 1 & 0 \end{bmatrix}$

27. $(3, 2)$ **29.** $(5, -4)$ **31.** $(38, 21)$ **33.** $(3, -2)$
35. $(3, -5)$ **37.** $(-4, 1)$ **39.** $(3, 1)$ **41.** $(-2, 2)$
43. $(1, -8)$ **45.** $(5, 6, 1)$ **47.** $(-26, 52, 15)$
49. $(88, -176, -38)$ **51.** $(1, 1, 1)$ **53.** $(2, -4, -1)$

55. a. $P = \begin{bmatrix} 1 & 0 & 0 & 0 & 0 \\ -1 & 1 & 0 & 0 & 0 \\ 1 & -2 & 1 & 0 & 0 \\ -1 & 3 & -3 & 1 & 0 \\ 1 & -4 & 6 & -4 & 1 \end{bmatrix}$

b. It is the same as the Pascal's matrix except that the signs of the
terms alternate.

57. $[A]^3 = \begin{bmatrix} 2 & 4 & 1 & 3 \\ 4 & 2 & 3 & 4 \\ 1 & 3 & 0 & 1 \\ 3 & 4 & 1 & 2 \end{bmatrix}$

For example, the U.S. can talk to Cuba through two intermediaries in one way, namely, U.S. to Mexico to Russia to Cuba.

59. a. Riesling costs 24; Charbono, 25; and Rosé, 51   **b.** Outside bottling, 230; produced and bottled at winery, 520; estate bottles, 280 **c.** It is the total cost of production of all three wines.

14.5 Systems of Inequalities, page 749

3. a.   **b.**

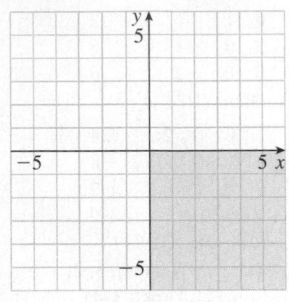

5.   **7.**

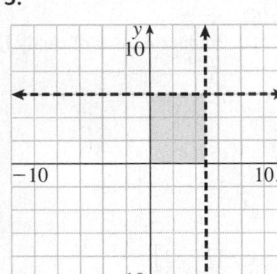   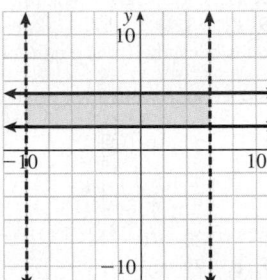

9.   **11.**

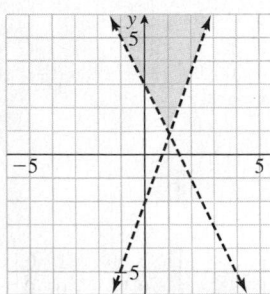   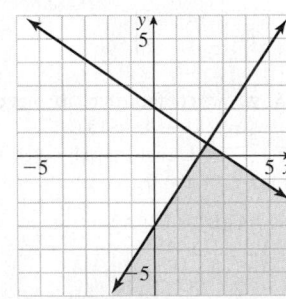

13.   **15.**

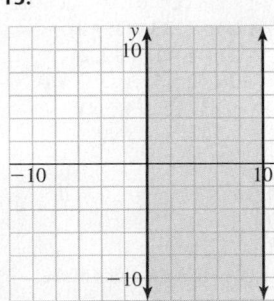   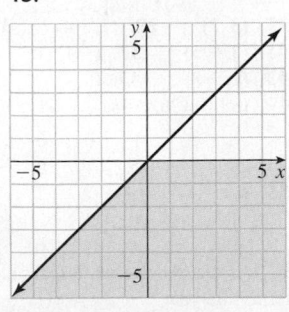

17.   **19.**

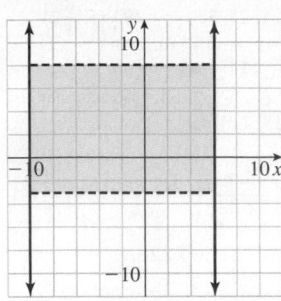   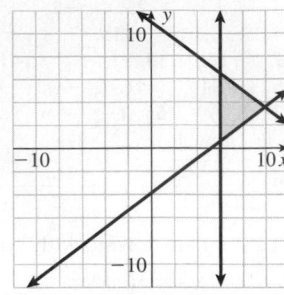

21.   **23.**

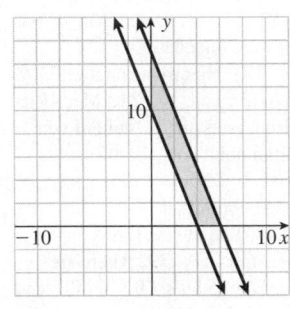   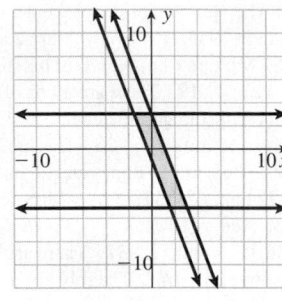

25.   **27.**

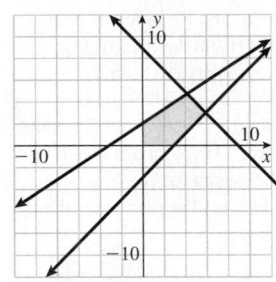   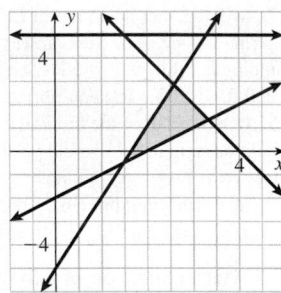

29.

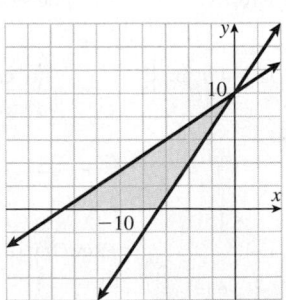

14.6 Modeling with Linear Programming, page 755

3. a. no **b.** yes **c.** no **d.** no **e.** no **f.** no
5. $(0, 0), (0, \frac{9}{2}), (5, 2), (6, 0)$ **7.** $(0, 0), (0, 4), (2, 3), (4, 0)$
9. $(0, 0), (0, 4), (4, 4), (6, 2), (6, 0)$ **11.** $(0, 0), (0, \frac{8}{5}), (\frac{24}{13}, \frac{16}{13})$,
$(\frac{8}{3}, 0)$ **13.** $(50, 0), (\frac{200}{7}, \frac{60}{7}), (8, 24), (0, 40)$ **15.** $(3, 2), (5, 5)$,
$(7, 5), (\frac{10}{3}, \frac{4}{3})$ **17.** maximum $W = 190$ at $(5, 2)$
19. maximum $P = 500$ at $(2, 3)$ **21.** minimum $A = -12$ at $(0, 4)$

23. Let x = number of units of food A
y = number of units of food B
Minimize $C = 0.29x + 0.15y$
Subject to $\begin{cases} x \geq 0, y \geq 0 \\ 10x + 5y \geq 200 \\ 2x + 5y \geq 100 \\ 3x + 4y \geq 20 \end{cases}$

25. Let x = amount invested in stock (in millions of dollars)
y = amount invested in bonds (in millions of dollars)
Maximize $T = 0.12x + 0.08y$
Subject to $\begin{cases} x \geq 0, y \geq 0 \\ x \leq 8, y \geq 2 \\ x + y \leq 10 \\ x \leq 3y \end{cases}$

27. The maximum profit P = \$14,300 is achieved with all 100 acres planted in corn.   **29.** The minimum cost of \$1.19 is obtained with 17 oz of Corn Flakes and no Honeycombs.

Chapter 14 Review Questions, page 758

1. $[C][B] - 3[D] = \begin{bmatrix} 2 & 0 \\ 1 & 2 \\ -1 & 1 \end{bmatrix} \begin{bmatrix} 2 & -1 & 0 \\ 1 & 0 & 1 \end{bmatrix} - 3 \begin{bmatrix} 0 & 1 & 0 \\ -1 & 0 & 0 \\ 0 & 1 & -1 \end{bmatrix}$

$= \begin{bmatrix} 4 & -2 & 0 \\ 4 & -1 & 2 \\ -1 & 1 & 1 \end{bmatrix} - \begin{bmatrix} 0 & 3 & 0 \\ -3 & 0 & 0 \\ 0 & 3 & -3 \end{bmatrix} = \begin{bmatrix} 4 & -5 & 0 \\ 7 & -1 & 2 \\ -1 & -2 & 4 \end{bmatrix}$

2. $[A][B][C] = \begin{bmatrix} 1 & 0 \\ 2 & -1 \end{bmatrix} \begin{bmatrix} 2 & -1 & 0 \\ 1 & 0 & 1 \end{bmatrix} \begin{bmatrix} 2 & 0 \\ 1 & 2 \\ -1 & 1 \end{bmatrix}$

$= \begin{bmatrix} 2 & -1 & 0 \\ 3 & -2 & -1 \end{bmatrix} \begin{bmatrix} 2 & 0 \\ 1 & 2 \\ -1 & 1 \end{bmatrix} = \begin{bmatrix} 3 & -2 \\ 5 & -5 \end{bmatrix}$

3. $[B]$ is a 2×3 matrix and $[A]$ is 2×2 so these matrices are not conformable for multiplication.

4. $[C][A][B] = \begin{bmatrix} 2 & 0 \\ 1 & 2 \\ -1 & 1 \end{bmatrix} \begin{bmatrix} 1 & 0 \\ 2 & -1 \end{bmatrix} \begin{bmatrix} 2 & -1 & 0 \\ 1 & 0 & 1 \end{bmatrix}$

$= \begin{bmatrix} 2 & 0 \\ 5 & -2 \\ 1 & -1 \end{bmatrix} \begin{bmatrix} 2 & -1 & 0 \\ 1 & 0 & 1 \end{bmatrix} = \begin{bmatrix} 4 & -2 & 0 \\ 8 & -5 & -2 \\ 1 & -1 & -1 \end{bmatrix}$

5. $\begin{bmatrix} 2 & 1 & | & 1 & 0 \\ -\frac{3}{2} & -\frac{1}{2} & | & 0 & 1 \end{bmatrix} \rightarrow \begin{bmatrix} 1 & \frac{1}{2} & | & \frac{1}{2} & 0 \\ -\frac{3}{2} & -\frac{1}{2} & | & 0 & 1 \end{bmatrix} \rightarrow \begin{bmatrix} 1 & \frac{1}{2} & | & \frac{1}{2} & 0 \\ 0 & \frac{1}{4} & | & \frac{3}{4} & 1 \end{bmatrix}$

$\rightarrow \begin{bmatrix} 1 & \frac{1}{2} & | & \frac{1}{2} & 0 \\ 0 & 1 & | & 3 & 4 \end{bmatrix} \rightarrow \begin{bmatrix} 1 & 0 & | & -1 & -2 \\ 0 & 1 & | & 3 & 4 \end{bmatrix}$ Inverse is $\begin{bmatrix} -1 & -2 \\ 3 & 4 \end{bmatrix}$.

6. $\begin{bmatrix} 1 & 3 & 3 & | & 1 & 0 & 0 \\ 1 & 4 & 3 & | & 0 & 1 & 0 \\ 1 & 3 & 4 & | & 0 & 0 & 1 \end{bmatrix} \rightarrow \begin{bmatrix} 1 & 3 & 3 & | & 1 & 0 & 0 \\ 0 & 1 & 0 & | & -1 & 1 & 0 \\ 0 & 0 & 1 & | & -1 & 0 & 1 \end{bmatrix}$

$\rightarrow \begin{bmatrix} 1 & 0 & 3 & | & 4 & -3 & 0 \\ 0 & 1 & 0 & | & -1 & 1 & 0 \\ 0 & 0 & 1 & | & -1 & 0 & 1 \end{bmatrix} \rightarrow \begin{bmatrix} 1 & 0 & 0 & | & 7 & -3 & -3 \\ 0 & 1 & 0 & | & -1 & 1 & 0 \\ 0 & 0 & 1 & | & -1 & 0 & 1 \end{bmatrix}$

Inverse is $\begin{bmatrix} 7 & -3 & -3 \\ -1 & 1 & 0 \\ -1 & 0 & 1 \end{bmatrix}$.

7. Graph $2x - y = 2$ by writing $y = 2x - 2$; y-intercept is -2 and slope is $2 = \frac{2}{1}$. Graph $3x - 2y = 1$ by writing $y = \frac{3}{2}x - \frac{1}{2}$; y-intercept is $-\frac{1}{2}$ and slope is $\frac{3}{2}$. The intersection point appears to be $(3, 4)$.

Check:
$2(3) - 4 = 6 - 4 = 2$
$3(3) - 2(4) = 9 - 8 = 1$

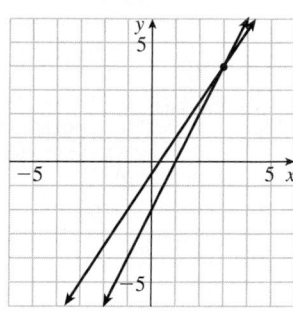

8. $2\begin{cases} x + 3y = 3 \\ 4x - 6y = -6 \end{cases} + \begin{cases} 2x + 6y = 6 \\ 4x - 6y = -6 \end{cases}$
$6x = 0$
$x = 0$ If $x = 0$, then $0 + 3y = 3$
or $y = 1$.
The solution is $(0, 1)$.

9. Substitute the first equation into the second: $5x + 2(1 - 2x) = 1$
$5x + 2 - 4x = 1$
$x = -1$
If $x = -1$, then $y - 1 - 2(-1) = 1 + 2 = 3$; the solution is $(-1, 3)$.

10. $\begin{bmatrix} 1 & 1 & 1 & | & 2 \\ 1 & 2 & -2 & | & 1 \\ 1 & 1 & 3 & | & 4 \end{bmatrix} \rightarrow \begin{bmatrix} 1 & 1 & 1 & | & 2 \\ 0 & 1 & -3 & | & -1 \\ 0 & 0 & 2 & | & 2 \end{bmatrix} \rightarrow \begin{bmatrix} 1 & 0 & 4 & | & 3 \\ 0 & 1 & -3 & | & -1 \\ 0 & 0 & 2 & | & 2 \end{bmatrix}$

$\rightarrow \begin{bmatrix} 1 & 0 & 4 & | & 3 \\ 0 & 1 & -3 & | & -1 \\ 0 & 0 & 1 & | & 1 \end{bmatrix} \rightarrow \begin{bmatrix} 1 & 0 & 0 & | & -1 \\ 0 & 1 & 0 & | & 2 \\ 0 & 0 & 1 & | & 1 \end{bmatrix}$ The solution is $(-1, 2, 1)$.

11. Let $[A] = \begin{bmatrix} 2 & 1 \\ -\frac{3}{2} & -\frac{1}{2} \end{bmatrix}$, $[X] = \begin{bmatrix} x \\ y \end{bmatrix}$, $[B] = \begin{bmatrix} 13 \\ 10 \end{bmatrix}$;
the solution is $[X] = [A]^{-1}[B]$.
We found the inverse in Problem 5;
then $[X] = \begin{bmatrix} -1 & -2 \\ 3 & 4 \end{bmatrix} \begin{bmatrix} 13 \\ 10 \end{bmatrix} = \begin{bmatrix} -33 \\ 79 \end{bmatrix}$. The solution is $(-33, 79)$.

12.

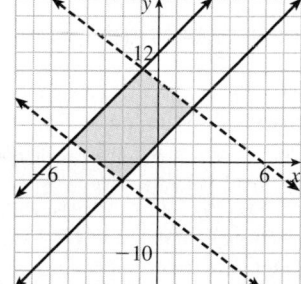

13. Let x = number of bars of product I.
y = number of bars of product II.

	A	B
Product I:	$3x$	$5x$
Product II:	$4y$	$7y$
Total:	33	56

Thus,
$\begin{cases} 3x + 4y = 33 \\ 5x + 7y = 56 \end{cases}$
The solution to this system is $(7, 3)$. This means she uses 7 bars of Product I and 3 bars of Product II.

14. Let x, y, and z be the number of products of I, II, and III that can be manufactured. This leads to the following system:

$$\begin{cases} 2x + 2y + 3z = 1{,}250 \\ x + 2y + 2z = 900 \\ x + y + 2z = 750 \end{cases}$$

The solution to this system is (100, 150, 250). This means the number of product I to be manufactured is 100; product II, 150; and product III, 250.

15. Let x = number of acres of corn; y = number of acres of wheat. Maximize profit, $P = 2.10(100)x + 2.50(40)y = 210x + 100y$. The constraints are:

$$\begin{cases} x \geq 0 \\ y \geq 0 \\ x + y \leq 500 \\ 100x + 40y \leq 18{,}000 \\ 120x + 60y \leq 24{,}000 \end{cases}$$

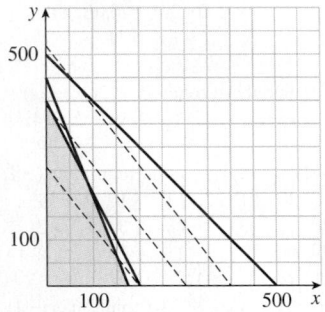

The corner points are (100, 200); (0, 400); and (180, 0).

Corner point	Values of objective function
(100, 200)	41,000
(0, 400)	40,000
(180, 0)	37,800

This means the farmer will maximize the profit if 100 acres of corn and 200 acres of wheat are planted (200 acres left unplanted).

CHAPTER 15 THE NATURE OF NETWORKS AND GRAPH THEORY

15.1 Euler Circuits and Hamiltonian Cycles, page 770

7. not an Euler circuit; traversable **9.** not an Euler circuit; not traversable **11.** not an Euler circuit; not traversable **13.** Hamiltonian cycle; $A \to B \to C \to D \to A$ **15.** Hamiltonian cycle; $A \to B \to C \to G \to H \to F \to E \to D \to A$ **17.** Hamiltonian cycle; $A \to B \to C \to D \to A$

Possible paths in Problems 19–23 may vary.

19. not traversable **21.** not traversable **23.** traversable **25.** not an Euler circuit; not traversable **27.** not an Euler circuit; not traversable **29.** not possible **31.** not possible **33.** *Answers may vary.* $1 \to 2 \to 3 \to 10 \to 11 \to 12 \to 4 \to 5 \to 14 \to 13 \to 19 \to 18 \to 17 \to 9 \to 8 \to 7 \to 16 \to 20 \to 15 \to 6 \to 1$ **35.** Transform the problem into a network; it is not traversable since there are more than two odd vertices. **37.** There are two odd vertices, so it is traversable; begin at either Queens or Manhattan. **39.** Yes; 2 odd vertices. **41.** No; there are 8 odd vertices. **43. a.** NYC to Boston to Washington, D.C., to NYC; 892 mi; forms a

loop without including Cleveland. **b.** NYC to Boston to Washington, D.C., to Cleveland to NYC; 1,513 mi **45.** Brute force (same as 44b); 4,627 mi. **47.** NYC $\to$ B $\to$ DC $\to$ C $\to$ A $\to$ NYC; 2,699 mi. **49.** A tetrahedron has four triangular faces; the sum of the measures of the angles on each face is 180°, so the sum of the measures of the face angles of a tetrahedron is 720°.
51. a. Room 1: 1 path; room 2: 2 paths; room 3: 3 paths; room 4: 5 paths **b.** 89 **c.** 377 **53. a.** Start at the top (or at the right) at one of the two odd vertices. **b.** The chance that you will choose one of the two odd vertices at random is small. **55.** The result is a twisted band twice as long and half as wide as the band with which you began. **57.** Two interlocking pieces, one twice as long and half as wide as the other **59.** One edge; one side; after cutting, you will have a single loop with a knot.

15.2 Trees and Minimum Spanning Trees, page 781

5. not a tree **7.** not a tree **9.** not a tree **11.** a tree
13.

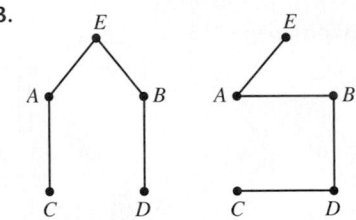

15.

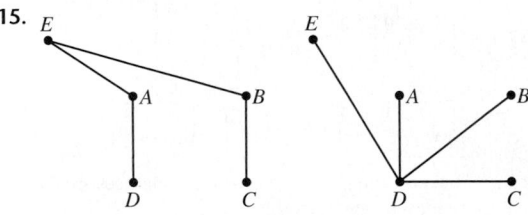

17.

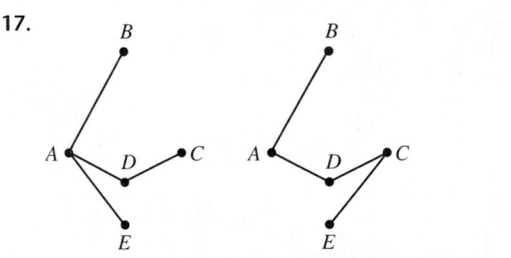

19.

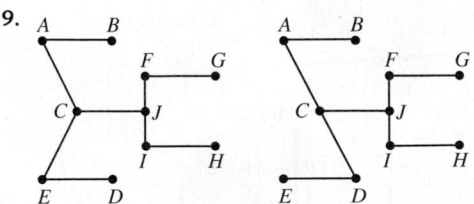

21.

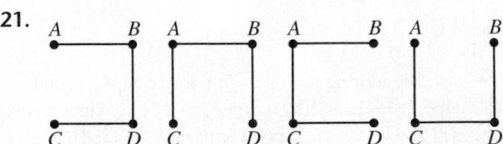

23.

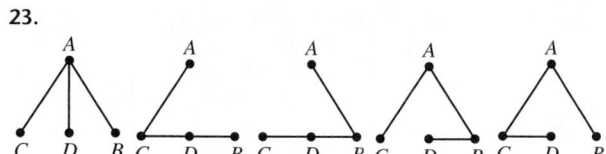

25.

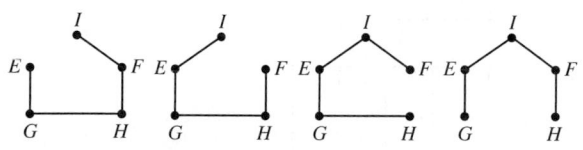

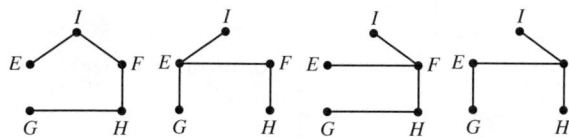

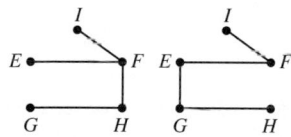

27. a tree

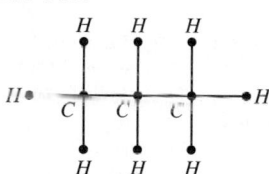

29. a tree

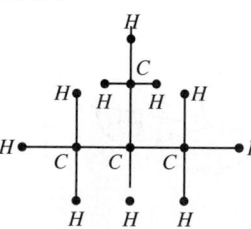

31. not a tree

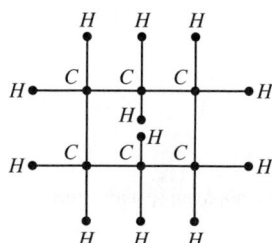

33.

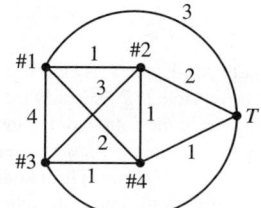

Minimum value, 60

35.

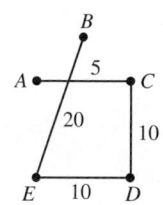

Minimum value, 45

37.

Minimum value, 180

39.

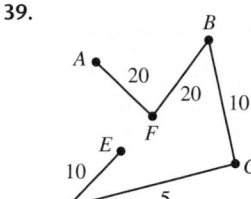

Minimum value, 65

41.

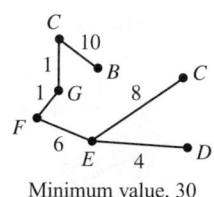

Minimum value, 30

43. 65   **45.** No, because some sites lead you back to previously visited sites, which completes a circuit.

47.

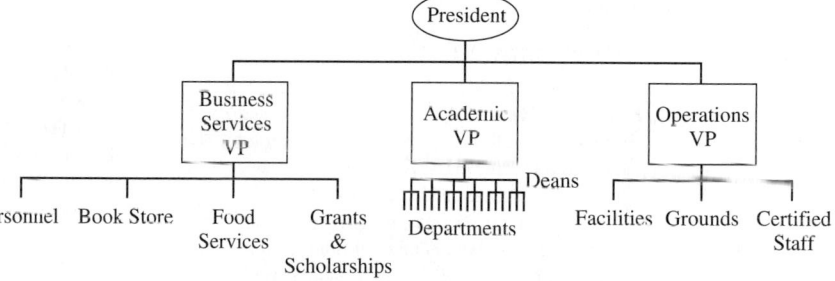

49. 49   **51.** $38,500   **53.** The minimum spanning tree connects San Francisco to Oakland (smallest distance, 13 miles); Oakland to San Jose (next smallest distance, 46 miles); Manteca to Merced (54 miles); Merced to Fresno (56 miles); Santa Rosa to San Francisco (58 miles); Oakland to Manteca (66 miles); and then complete the graph with Merced to Yosemite Village (56 miles). The minimum spanning tree is 375 miles.

55. a.

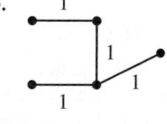

b.

c. The minimum cost is $4,000,000.

57. a.

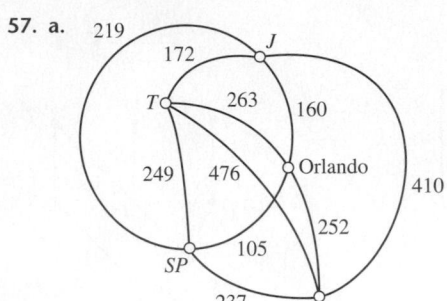

b.

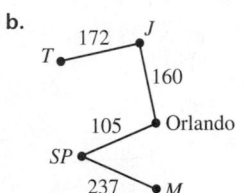

c. The minimum distance is 674 miles. The minimum cost is $57,290.

59. a. If a graph is a tree with n vertices, then the number of edges is $n - 1$.   **b.** There are no edges; yes, the property holds because $1 - 1 = 0$.   **c.** If there are two vertices and two or more edges, then there is a circuit, which is not permitted.   **d.** If there are three vertices, then it takes two edges to form a tree. If there are three or more edges, then there is a circuit.

15.3 Topology and Fractals, page 791

3. A and F; B and G; C and D   **5.** A and F are simple closed curves.   **7.** C, E, F, G, H, I, J, K, L, M, N, S, T, U, V, W, X, Y, and Z are the same class; A, D, O, P, Q, and R are another class; B is a third class   **9.** A, C, E, and F are the same class; B and D are the same class; G is different.   **11.** B and C are inside; A is outside.
13. A and B are outside; C is inside.   **15. a.** It will cross an odd number of times for $\overline{AX}$ and $\overline{CX}$ and an even number of times for $\overline{BX}$.
b. It will cross an odd number of times for $\overline{BX}$ and $\overline{CX}$ and an even number of times for $\overline{AX}$.   **c.** It will cross an odd number of times for $\overline{AX}$ and $\overline{CX}$ and an even number of times for $\overline{BX}$.   **d.** If $\overline{PX}$ crosses the curve an even number of times, then P is outside; an odd number of times, P is inside (for some point X that is outside).
21. Answers vary; construct a fractal curve.   **23.** Construction varies.   **25. a.** Answers vary (there are two possibilities).
b. Answers vary (there are 364 possibilities).   **29. a.** No, it will now fit the left hand.   **b.** Yes

Chapter 15 Review Questions, page 796

1. Classify the vertices; they are all odd (degree 3), so there are more than 2 odd vertices. Thus it is not traversable.   **2.** Classify the vertices; the outside corners and the corner at the bottom of the inner rectangle are even; the intersecting lines at the top show an even vertex, and the two vertices at the top of the inner rectangle are odd (degree 5). Start at either of these odd vertices, and end at the other to show that the network is traversable.   **3.** There are 7 rooms, including the exterior. There are two rooms with an odd number of doors, so begin in either of these rooms and end in the other—it is possible.

4. Answers vary.   **a.**

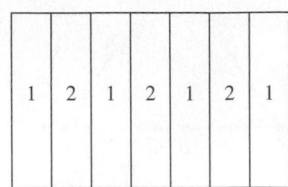

b.

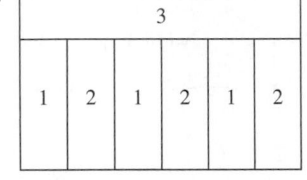

c.

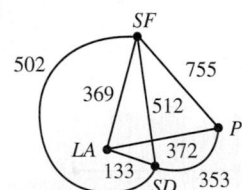

d. No such map exists.

5. a. Using the formula from Example 7 in Section 15.1, we see there are 9 cities, so there are $\dfrac{8 \cdot 7 \cdot 6 \cdot 5 \cdot 4 \cdot 3 \cdot 2 \cdot 1}{2} = 20{,}160$ different routes.
b. SF → LA → SD → P → SF; $369 + 133 + 353 + 755 = 1{,}610$
SF → LA → P → SD → SF; $369 + 372 + 353 + 502 = 1{,}596$
SF → SD → LA → P → SF; $502 + 133 + 372 + 755 = 1{,}762$
SF → SD → P → LA → SF; $502 + 353 + 372 + 369 = 1{,}596$
SF → P → SD → LA → SF; $755 + 353 + 133 + 369 = 1{,}610$
SF → P → LA → SD → SF; $755 + 372 + 133 + 502 = 1{,}762$
It looks like either of the routes shown in boldface is the best brute-force route.
6. a. SF → LA → SD → LA; does not include Phoenix, so this method does not give a best route.
b. The weighted graph is shown:

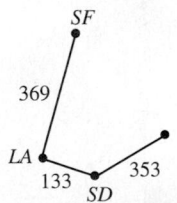

The minimum spanning tree is:

The minimum distance is $369 + 133 + 353 = 855$ miles.
7. a. Yes, there are many possibilities, for example, Minnesota, Wisconsin, Iowa, Illinois, Indiana, Michigan, Ohio, and Pennsylvania.
b. Actually no, but if you use the ferry from Wisconsin to Michigan, then the answer is yes: Iowa, Minnesota, Wisconsin, Michigan, Ohio, Indiana, Illinois, and Iowa.

8. a. $A \to B \to G \to L \to K \to F \to A$; the cost of this trip is 20.
b. $M \to H \to I \to N \to O \to J \to E \to D \to C \to B \to G \to L \to K \to F \to A \to B$; the cost of this trip is 49.
c. There are 15 vertices, so the number of possibilities is
$$\frac{14 \cdot 13 \cdot \cdots \cdot 3 \cdot 2 \cdot 1}{2} = 4.3589 \times 10^{10}.$$
d.

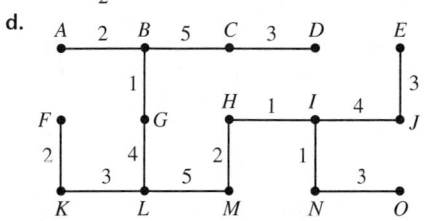

The cost of this trip is 39.

9. a. No, because there are more than two odd vertices.
b. $A \to B \to C \to D \to I \to E \to F \to G \to H \to A$

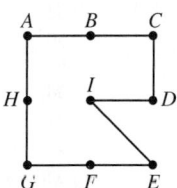

c. Answers vary; here is one possibility.

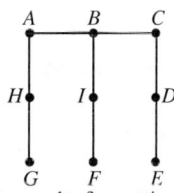

10. There are two edges and two sides; the result after cutting is two interlocking loops.

CHAPTER 16 THE NATURE OF VOTING AND APPORTIONMENT

16.1 Voting, page 811

13. 22   **15. a.** "(CBA)" means that the voter ranks three candidates in the order of C first, A next, and candidate B last.   **b.** It means that in the election there were 5 voters who ranked the candidates in the order CAB.   **17. a.** (ABC)   **b.** 8 voters   **19.** 19 votes
21. a. "(DACB)" means the voter picks D as the first choice, followed by A, then C, with B in last place.   **b.** It means that in the election there were 5 voters who ranked the candidates in the order DACB.   **23. a.** (BCDA)   **b.** 0   **25. a.** 6   **b.** 24   **c.** 120
27. $n!$ **29.** A wins   **31.** A wins   **33.** A wins
35. snacks   **37.** Howard Dean   **39.** no winner
41. A wins   **43.** A wins   **45.** no winner   **47.** C wins
49. C wins   **51. a.** B   **b.** plurality vote   **53. a.** C
b. plurality vote   **55.** 120   **57.** C wins   **59.** The Borda count is A: 2,540; B: 3,480; C: 2,880; D: 3,820; E: 3,780; D wins

16.2 Voting Dilemmas, page 827

11. a. California Teachers Association   **b.** no   **13. a.** There is no majority winner; A wins plurality.   **b.** A   **c.** no
15. a. A wins   **b.** C wins   **c.** yes   **17. a.** A   **b.** There is no majority winner; B wins plurality. This violates the Condorcet criterion.   **19. a.** A wins   **b.** B wins   **c.** yes   **21.** There is no

majority winner; C wins plurality. This violates the Condorcet criterion.   **23.** C wins; yes   **25.** A is the Condorcet candidate.
27. A wins; no   **29.** A wins; no   **31.** E wins
33. E wins; no   **35.** M wins; yes   **37.** Lillehammer wins; yes
39. a. Beijing wins   **b.** no   **41. a.** Chirac and Le Pen
b. Answers vary.   **c.** Answers vary.   **43.** Betty wins; no
45. Betty; yes, none   **47.** Dave wins; no   **49.** D; no
51. a. no winner   **b.** A, C, or D could win depending on the way they are paired.   **c.** Yes, both of these methods violate the condition of decisiveness.   **53. a.** B wins   **b.** yes   **55. a.** no majority; C wins plurality   **b.** B and C tie   **c.** B wins   **d.** C wins; yes
57. Answers vary.   **59.** If the 90% vote is spread out evenly over the ten serious candidates, then a 10% vote for the radical candidate is theoretically enough to win the plurality vote.

16.3 Apportionment, page 847

7. 3; 4   **b.** 3.5   **c.** 3.46   **d.** 4; 4   **9. a.** 1; 2   **b.** 1.5   **c.** 1.41
d. 1; 2   **11. a.** 2; 3   **b.** 2.5   **c.** 2.45   **d.** 2; 3
13. a. 1,695; 1,696   **b.** 1,695.5   **c.** 1,695.50   **d.** 1,695; 1,695
15. 6,500   **17.** 126   **19.** 120,833.33   **21.** 184,000

	Year	d	Manhattan	Bronx	Brooklyn	Queens	Staten Island
23.	1800	10,125	6.02	0.20	0.59	0.69	0.49
25.	1900	429,750	4.30	0.47	2.72	0.36	0.16
27.	1990	915,500	1.63	1.32	2.51	2.13	0.41

29. a. 11,600   **b.** 3.02, 1.81, 1.03, 4.14   **c.** $3 + 1 + 1 + 4 = 9$
d. 10,000   **31. a.** 80,000   **b.** 1.69, 2.89, 1.48, 3.95
c. $1 + 2 + 1 + 3 = 7$ **d.** 65,000   **33. a.** 11,600   **b.** 3.02, 1.81, 1.03, 4.14   **c.** $4 + 2 + 2 + 4 = 12$ **d.** 13,000
35. a. 80,000   **b.** 1.69, 2.89, 1.48, 3.95   **b.** $2 + 3 + 2 + 4 = 11$
d. 110,000

37.

CT	DE	GA	ME	MD	MA	NH	NJ
6.41	1.59	2.23	1.99	2.61	8.62	10.21	3.83	4.97

NY	NC	PA	RI	SC	VT	VA
9.17	10.65	11.69	1.86	6.72	2.30	20.16

39.

CT	DE	GA	ME	MD	MA	NH	NJ
6	1	2	1	2	8	10	3	4

NY	NC	PA	RI	SC	VT	VA
9	10	11	1	6	2	20

Total is 96.

41. $n = 10; d = \dfrac{25,000}{10} = 2,500$ $D = 3,125$

	q	Lower quota	Upper quota	Q	Adams' plan
North	8,700	3.48	3	4	2.78	3
South	5,600	2.24	2	3	1.79	2
East	7,200	2.88	2	3	2.30	3
West	3,500	1.40	1	2	1.12	2
TOTAL: 25,000			8	12		10

43. $n = 10; d = \dfrac{25,000}{10} = 2,500$

	q	Lower quota	Hamilton's plan
North	8,700	3.48	3	4 (#2)
South	5,600	2.24	2	2
East	7,200	2.88	2	3 (#1)
West	3,500	1.40	1	1
TOTAL: 25,000			8	10

45. $n = 10$; $d = \dfrac{25,000}{10} = 2,500$; $D = 2,500$ (no modification necessary)

	q	a Lower quota	b Upper quota	$\sqrt{ab}$	HH's plan
North	8,700	3.48	3	4	3.46	4
South	5,600	2.24	2	3	2.45	2
East	7,200	2.88	2	3	2.45	3
West	3,500	1.40	1	2	1.41	1
TOTAL: 25,000			8	12		10

47. $n = 26$; $d = \dfrac{62,000}{26} \approx 2,384.62$ $D = 2,210$

	q	Lower quota	Upper quota	Q	Jefferson's plan
North	18,200	7.63	7	8	8.24	8
South	12,900	5.41	5	6	5.84	5
East	17,600	7.38	7	8	7.96	7
West	13,300	5.58	5	6	6.02	6
TOTAL: 62,000			24	28		26

49. $n = 26$; $d = \dfrac{62,000}{26} \approx 2,384.62$; $D = 2,384.62$ (no modification necessary)

	q	Round nearest	Webster's plan
North	18,200	7.63	8	8
South	12,900	5.41	5	5
East	17,600	7.38	7	7
West	13,300	5.58	6	6
TOTAL: 62,000			26	26

51. $n = 16$; $d = \dfrac{62,000}{16} = 3,875$; $D = 4,425$

	q	Lower quota	Upper quota	Q	Adams' plan
North	18,200	4.70	4	5	4.11	5
South	12,900	3.33	3	4	2.92	3
East	17,600	4.54	4	5	3.98	4
West	13,300	3.43	3	4	3.01	4
TOTAL: 62,000			14	18		16

53. $n = 16$; $d = \dfrac{62,000}{16} = 3,875$

	q	Lower quota	Upper quota	Hamilton's plan
North	18,200	4.70	4	5	5 (#1)
South	12,900	3.33	3	4	3
East	17,600	4.54	4	5	5 (#2)
West	13,300	3.43	3	4	3
TOTAL: 62,000			14	18	16

55. $n = 16$; $d = \dfrac{62,000}{16} = 3,875$; $D = 3,875$ (no modification necessary)

	q	a Lower quota	b Upper quota	$\sqrt{ab}$	HH's plan
North	18,200	4.70	4	5	4.47	5
South	12,900	3.33	3	4	3.46	3
East	17,600	4.54	4	5	4.47	5
West	13,300	3.43	3	4	3.46	3
TOTAL: 62,000			14	18		16

57. Standard divisor: $\dfrac{16,630,000}{475} \approx 35,010.53$ Modified divisor: 34,720

Region	Number	Std. quota	Lower quota	Upper quota	Modified quota	Jefferson's plan
N	1,820,000	51.98	51	52	52.42	52
NE	2,950,000	84.26	84	85	84.97	84
E	1,760,000	50.27	50	51	50.69	50
SE	1,980,000	56.55	56	57	57.03	57
S	1,200,000	34.28	34	35	34.56	34
SW	2,480,000	70.84	70	71	71.43	71
W	3,300,000	94.26	94	95	95.05	95
NW	1,140,000	32.56	32	33	32.83	32
TOTAL: 16,630,000			471	479		475

59. Standard divisor: $d = \dfrac{16,630,000}{475} \approx 35,010.53$

Modified divisor: 35,010.53 (no modification necess.)

Region	Number	Std quota	Round	Webster's plan
N	1,820,000	51.98	52	52
NE	2,950,000	84.26	84	84
E	1,760,000	50.27	50	50
SE	1,980,000	56.55	57	57
S	1,200,000	34.28	34	34
SW	2,480,000	70.84	71	71
W	3,300,000	94.26	94	94
NW	1,140,000	32.56	33	33
TOTAL:	16,630,000		475	475

16.4 Apportionment Paradoxes, page 854

5. State A violates the quota rule.   **7.** State A violates the quota rule.   **9.** State D violates the quota rule.   **11.** State A illustrates the Alabama paradox.   **13.** State B illustrates the Alabama paradox.   **15.** State C illustrates the population paradox.   **17.** State C illustrates the population paradox.   **19.** State A illustrates the new states paradox.   **21.** State B illustrates the new states paradox.   **23. a.** 62.78 **b.** A: 199.43; B: 72.55; C: 12.93; D: 15.08 **c.** A: 199, 200; B: 72, 73; C: 12, 13; D: 15, 16 **d.** modified quota is 62.4; A: 200, B: 72, C: 13, D: 15 **e.** no
25. a. 203.3 **b.** Uptown (U): 83.52; Downtown (D): 16.48
c. U: 84; D: 16 **d.** U: 83; D: 17; new: 12 **e.** yes   **27.** yes
29. Adams' or Webster's plan gives the correct apportionment of the horses.

Chapter 16 Review Questions, page 858

1. There is a total of 45 votes, so a majority is 23; there is no majority. The plurality vote goes to C.   **2.** There is a total of $3! = 6$ arrangements, and 3 are shown here, so there are 3 possibilities that received no votes.
3. Here are the rankings:

    A: 15; B: 12; C: 18

Eliminate B for the second round of votes:

    A: 15 + 12 = 27

    C: 18

A wins the majority in this round, so A is the winner.
4. A over B: 15; B over A: 18 + 12 = 30; B wins 1 point.

    A over C: 15 + 12 = 27; C over A: 18; A wins 1 point.

    B over C: 15 + 12 = 27; C over B: 18; B wins 1 point.

    A has 1 point, B has 2 points, and C has 0 points, so B is the winner.

5. (1) The majority criterion is not violated because there is no majority winner. (2) We look for a Condorcet candidate by finding the one-on-one pairings:

A with B: **B wins**

A with C: **A wins**

B with C: **B wins**

	A	B	C
A	—	B	A
B	B	—	B
C	A	B	—

B is a Condorcet candidate. We see that the Hare method (Problem 3) violates the Condorcet criterion.

6. Here are the results (in percents):

A: $22 + 23 = 45$

B: $15 + 29 = 44$

C: $7 + 4 = 11$

Candidate A has a plurality.

7. We look for a Condorcet candidate by finding the one-on-one pairings:

A over B: $22 + 23 + 4 = 49$

B over A: $15 + 29 + 7 = 51$; **B wins**

A over C: $22 + 23 + 15 = 60$

C over A: $29 + 7 + 4 = 40$; **A wins**

B over C: $22 + 15 + 29 = 66$

C over B: $23 + 7 + 4 = 34$; **B wins**

	A	B	C
A	—	B	A
B	B	—	B
C	A	B	—

B is the Condorcet winner.

8. Three points for a first-place vote, 2 points for a second-place vote, and 1 point for a third-place vote.

David Carr (Fresno State);	$34(3) + 60(2) + 58(1) = 280$
Eric Crouch (Nebraska),	$162(3) + 98(2) + 88(1) = 770$
Ken Dorsey (Miami),	$109(3) + 122(2) + 67(1) = 638$
Dwight Freeney (Syracuse);	$2(3) + 6(2) + 24(1) = 42$
Rex Grossman (Florida);	$137(3) + 105(2) + 87(1) = 708$
Joey Harrington (Oregon);	$54(3) + 68(2) + 66(1) = 364$
Bryant McKinnie (Miami);	$26(3) + 12(2) + 14(1) = 116$
Julius Peppers (North Carolina);	$2(3) + 10(2) + 15(1) = 41$
Antwaan Randle El (Indiana);	$46(3) + 39(2) + 51(1) = 267$
Roy Williams (Oklahoma);	$13(3) + 36(2) + 35(1) = 146$

The winner was Eric Crouch with 770 Borda points.

9. The votes are:

A: 7

B: 4

C: 5

D: 1

The least first-place votes were for D, so we delete D from the second round of votes, with these results:

A: 7

B: $4 + 1 = 5$

C: 5

There is a tie for last place. The Hare method fails to pick a winner unless we specify some way of breaking a tie. For example, if we have a runoff between just B and C, we find

B: $7 + 4 + 1 = 12$

C: 5

B is the winner, so for the next round we eliminate C, with these results:

A: $7 + 5 = 12$

B: $4 + 1 = 5$

We declare A the winner.

10. Here are the pairwise point counts.

A over B: $7 + 5 = 12$; B over A: $4 + 1 = 5$; A wins 1 point.

A over C: $7 + 1 = 8$; C over A: $5 + 4 = 9$; C wins 1 point.

A over D: $7 + 5 = 12$; D over A: $4 + 1 = 5$; A wins 1 point.

B over C: $7 + 4 + 1 = 12$; C over B: 5; B wins 1 point.

B over D: $5 + 4 = 9$; D over B: $7 + 1 = 8$; B wins 1 point.

C over D: $5 + 4 = 9$; D over C: $7 + 1 = 8$; C wins 1 point.

The final point count is: A, 2 points; B, 2 points; C, 2 points; D, 0 points. There is no pairwise winner without specifying a method of breaking a tie. If we have a runoff among A, B, and C, we find the following preferences:

A: 7

B: $4 + 1 = 5$

C: 5

We declare A the winner.

11. If B pulls out of the race, we have the following preferences:

(ADC)	(CAD)	(CDA)	(DAC)
7	5	4	1

The first vote is

A: 7

C: $5 + 4 = 9$

D: 1

C has a majority, so C is the winner. If we compare this with the result of Problem 9 we see that this result does violate the irrelevant alternatives criterion.

12. The standard divisor is $\dfrac{790}{100} = 7.9$. **13.** The standard quotas are EA: 11.39; MC: 27.22; M: 33.92; P: 16.84; SR: 10.63. The lower and upper quotas are EA: 11, 12; MC: 27, 28; M: 33, 34; P: 16, 17; SR: 10, 11. **14.** Adams' plan (modified divisor is 8.15): EA: 12; MC: 27; M: 33; P: 17; SR: 11. **15.** Jefferson's plan (modified divisor is 7.67): EA: 11; MC: 28; M: 34; P: 17; SR: 10. **16.** Hamilton's plan: EA: 11; MC: 27; M: $33 + 1 = 34$ (#1); P: $16 + 1 = 17$ (#2); SR: $10 + 1 = 11$ (#3). **17.** Webster's plan (modified divisor is 8): EA: 11; MC: 27; M: 34; P: 17; SR: 11. **18.** HH's plan: EA: 11; MC: 27; M: 34; P: 17; SR: 11. **19.** No, not from the given information. We can see that the data do not violate the quota rule, and without more information, we cannot check to see whether any of the other paradoxes are illustrated. **20.** $d = 0.348\overline{2}$; Downtown: 70; Fairground: 29; Columbus Square: 28; Downtown West; 27; Peabody: 26

CHAPTER 17 THE NATURE OF CALCULUS

17.1 What Is Calculus? page 870

5. $\frac{1}{3}$ **7.** 1 **9.** π

11.

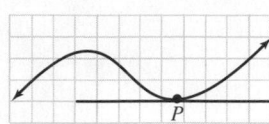

13.

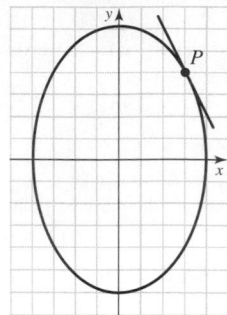

15.

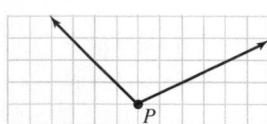

No single tangent line exists.

17. $\frac{2}{3}$ **19.** $\frac{1}{2}$ **21.** $\frac{3}{2}$ **23.** 20   **25.** 4 or 5   **27.** 3

17.2 Limits, page 876

3. 0, 2, 0, 2, 0   **5.** $\frac{4}{3}, \frac{7}{4}, 2, \frac{13}{6}, \frac{16}{7}$ **7.** 8,000   **9.** $\frac{2}{3}$ **11.** 5
13. 4   **15.** 0   **17.** 0   **19.** 0.7   **21.** 0   **23.** ∞
25. -3 **27.** 3   **29.** The sequence is 12, 6, 3, $\frac{3}{2}, \frac{3}{4}$; there is
0.75 mg of the drug present. At the end of n hours, there is $24(\frac{1}{2})^n$ mg
present.

17.3 Derivatives, page 886

1. a. 5.71   **b.** 15   **c.** 40   **d.** 40   **3. a.** 59 mph   **b.** 18 mph
c. 36 mph   **d.** 38 mph   **5.** Let $ be trillions of dollars.
a. 0.198 $/yr   **b.** 0.255 $/yr   **c.** 0.307 $/yr   **d.** 0.348 $/yr
e. 0.371 $/yr   **f.** It is changing at the rate of $371 billion per year.
7.   **9.**

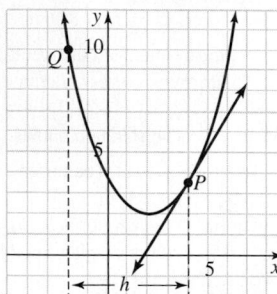

slope is 3/2

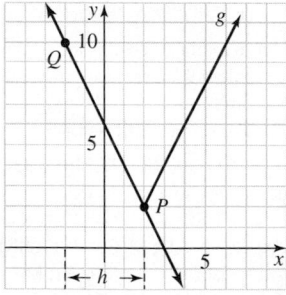

slope is -2

11.

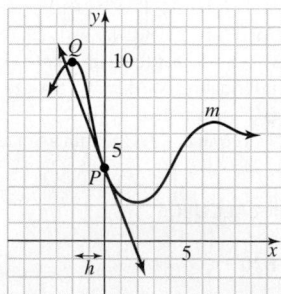

slope is -2

13. a. 0   **b.** 0   **15. a.** -25 **b.** -15 **17.** x^2
19. $-6e^{-6x}$ **21.** $2 - 6x$ **23.** $16x - y - 32 = 0$
25. $4x - y = 0$ **27. a.** The height of the tower is 176 ft.
b. The velocity is $-32t + 96$.   **29. a.** 8,900   **b.** 6,200
c. 5,930   **d.** $5,900 + 30h$

17.4 Integrals, page 895

1. 40   **3.** 24   **5.** 19.5   **7.** $6x + C$ **9.** $\frac{1}{2}x^2 + 5x + C$
11. $\frac{3}{2}x^2 + 4x + C$ **21.** $242\frac{2}{3}$ **23.** 82.5   **25.** $2e - 2$
27. 76   **29.** 58,965.3

Chapter 17 Review Questions, page 897

1. $\lim\limits_{n \to \infty} \dfrac{1}{n} = 0$

2. $\lim\limits_{n \to \infty} \dfrac{3n^4 + 20}{7n^4} = \lim\limits_{n \to \infty} \dfrac{\dfrac{3n^4}{n^4} + \dfrac{20}{n^4}}{\dfrac{7n^4}{n^4}} = \lim\limits_{n \to \infty} \dfrac{3 + \dfrac{20}{n^4}}{7} = \dfrac{3}{7}$

3. $\lim\limits_{n \to \infty} (2n + 3) = \infty$ **4.** $\lim\limits_{n \to \infty} \left(1 + \dfrac{1}{n}\right)^n = e$
5. The series illustrates the idea of a tangent line ("instantaneous
growth rate"). The first shows 5 measurements; the second, 20 measurements; and the third, 40 measurements.

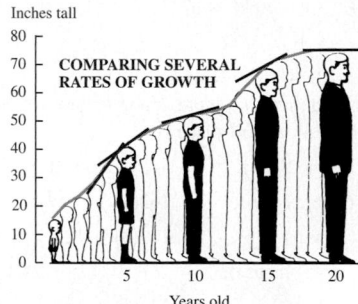

Each point on the curve has a tangent which indicates
rate of growth at that point. The steepest tangent shown
here occurs at about age three: the boy's growth was
most rapid then.

Note this one measures the growth rate at an instant. The closer the
measurements are together, the better the approximation at a particular
point.
6. The main ideas of calculus are limits, derivatives, and integrals.
(1) $\lim\limits_{n \to \infty} a_n = L$ means that the sequence a_n becomes closer and closer
to the number L as n becomes larger and larger. (2) The derivative
illustrates the idea of a tangent line. (3) The integral is used to find the
area under a curve. Answers vary.

7. $\lim\limits_{h \to 0} \dfrac{f(x + h) - f(x)}{h} = \lim\limits_{h \to 0} \dfrac{[6 - 4(x + h)^2] - [6 - 4x^2]}{h}$

$\qquad = \lim\limits_{h \to 0} \dfrac{6 - 4x^2 - 8xh - 4h^2 - 6 + 4x^2}{h}$

$\qquad = \lim\limits_{h \to 0} \dfrac{-8xh - 4h^2}{h}$

$\qquad = \lim\limits_{h \to 0} (-8x - 4h) = -8x$

8. $\displaystyle\int_{-1}^{0} (x - 4x^3)\, dx = \left(\tfrac{1}{2}x^2 - x^4\right)\Big|_{-1}^{0} = \tfrac{1}{2}$

9. $\displaystyle\int_{1}^{2} e^x\, dx = e^2 - e \approx 4.67$

10. Total amount used is $\int_0^t 32.4e^{0.048x}\,dx = \dfrac{4{,}050}{6}(e^{6t/125} - 1)$

Solve

$$670 = \frac{4{,}050}{6}(e^{6t/125} - 1)$$

$$670 = 675(e^{6t/125} - 1)$$

$$\frac{670}{675} = e^{6t/125} - 1$$

$$\frac{1{,}345}{675} = e^{6t/125}$$

$$\frac{6t}{125} = \ln\left(\frac{1{,}345}{675}\right)$$

$$t = \frac{125}{6}\ln\left(\frac{1{,}345}{675}\right)$$

$$\approx 14.363$$

This says that the oil reserves will be depleted in 2014.

Epilogue Problem Set, page E18

3. a. social science **b.** natural science **c.** natural science
d. humanities **e.** social science   **5. a.** social science
b. natural science **c.** social science **d.** social science
e. natural science **f.** natural science   **7.** all natural science
9. 1.46×10^{10} **11.** Venus and Neptune have the smallest eccentricity and therefore have the most circular orbits.   **13.** We expect the ostrich egg to weigh 2.89 oz.   **15.** genotype $SS = 0.3025$; genotype wS or $Sw = 0.495$; genotype $ww = 0.2025$
17. 0.45 cm²   **19.**

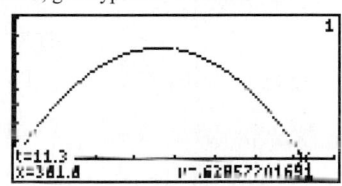

$$X_{1_T} = (64\cos 60)\,T$$
$$Y_{1_T} = (64\sin 60)\,T - 4.9\,T^2$$
$$\text{Tmin} = 0$$
$$\text{Tmax} = 12$$
$$\text{Tstep} = .1$$

Xmin = 0	Ymin = -25
Xmax = 400	Ymax = 200
Xscl = 50	Yscl = 25

21. a. 0.0996 **b.** Let each unit be $\sqrt{10^{16}} = 10^8$.

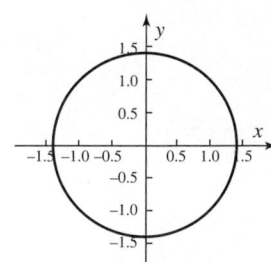

23. 24
25. a.

	A	B	C	D
a	(1, 2)	(3, 1)	(4, 1)	(2, 2)
b	(2, 3)	(4, 4)	(3, 2)	(1, 1)
c	(2, 4)	(3, 3)	(4, 4)	(1, 3)
d	(3, 1)	(2, 2)	(4, 3)	(1, 4)

b.

	A	B	C	D	
a	(1, 2)	(3, 1)	(4, 1)	[(2, 2)]	unstable
b	(2, 3)	(4, 4)	[(3, 2)]	(1, 1)	
c	(2, 4)	[(3, 3)]	(4, 4)	(1, 3)	
d	[(3, 1)]	(2, 2)	(4, 3)	(1, 4)	

or

	A	B	C	D	
a	[(1, 2)]	(3, 1)	(4, 1)	(2, 2)	stable
b	(2, 3)	(4, 4)	(3, 2)	[(1, 1)]	
c	(2, 4)	(3, 3)	[(4, 4)]	(1, 3)	
d	(3, 1)	[(2, 2)]	(4, 3)	(1, 4)	

27.

```
Y1■8sin (360*261
.626X)+4sin (720
*261.626X)
Xmin=0    Ymin=-15
Xmax=.02  Ymax=15
Xscl=.001 Yscl=1
```

29. There are 6 possibilities.
The matrix of their stated choices is:

	A	B	C
a	(3, 1)	(1, 3)	(2, 2)
b	(2, 2)	(3, 1)	(1, 3)
c	(1, 3)	(2, 2)	(3, 1)

Here are the six possibilities:

	A	B	C	
a	[(3, 1)]	(1, 3)	(2, 2)	
b	(2, 2)	[(3, 1)]	(1, 3)	stable; each school has its first
c	(1, 3)	(2, 2)	[(3, 1)]	choice

	A	B	C	
a	[(3, 1)]	(1, 3)	(2, 2)	a would rather be with C, and C
b	(2, 2)	(3, 1)	[(1, 3)]	with a; unstable
c	(1, 3)	[(2, 2)]	(3, 1)	

	A	B	C	
a	(3, 1)	[(1, 3)]	(2, 2)	
b	[(2, 2)]	(3, 1)	(1, 3)	
c	(1, 3)	(2, 2)	[(3, 1)]	c would rather be with B, and B with c; unstable

	A	B	C	
a	(3, 1)	[(1, 3)]	(2, 2)	
b	(2, 2)	(3, 1)	[(1, 3)]	
c	[(1, 3)]	(2, 2)	(3, 1)	stable; each person has his or her first choice

	A	B	C	
a	(3, 1)	(1, 3)	[(2, 2)]	a prefers B, but B is happier with c
b	[(2, 2)]	(3, 1)	(1, 3)	b prefers C, but C is happier with a
c	(1, 3)	[(2, 2)]	(3, 1)	c prefers A, but A is happier with b stable

	A	B	C	
a	(3, 1)	(1, 3)	[(2, 2)]	
b	(2, 2)	[(3, 1)]	(1, 3)	b would rather be with A and A
c	[(1, 3)]	(2, 2)	(3, 1)	with b; unstable

CREDITS

This page constitutes an extension of the copyright page. We have made every effort to trace the ownership of all copyrighted material and to secure permission from copyright holders. In the event of any question arising as to the use of any material, we will be pleased to make the necessary corrections in future printings. Thanks are due to the following authors, publishers, and agents for permission to use the material indicated.

PROLOGUE

P1: © Gianni Dagli Orti/CORBIS. **P2:** © Bettmann/CORBIS. **P5:** © Scala/Art Resource, NY.

CHAPTER 1

8: Dale Seymour Publications, Palo Alto, CA. **10:** © Tony Freeman/PhotoEdit. **15:** "The Thoth Maneuver," by Clifford A. Pickover, *Discover*, March 1996, p. 108. Clifford Pickover/ © 1996. Reprinted with permission of *Discover* Magazine. Nenad Jakesevic and Sonja Lamut/ © 1996. Reprinted with permission of *Discover* Magazine. **19:** *B.C.* reprinted by permission of Johnny Hart and Creators Syndicate. **20 top:** Tom Henderson, *The Saturday Evening Post* © 1960. **20 bottom:** Quotation in News Clip from *Everybody Counts*, A Report to the Nation on the Future of Mathematics Education, National Research Council, Washington, D.C. 1989, p. 3. **22:** *B.C.* reprinted by permission of Johnny Hart and Creators Syndicate. **25 top:** Illustration "How Big Is the Cosmos?" adapted from *The Universe*, Life Nature Library. **25 bottom:** *B.C.* reprinted by permission of Johnny Hart and Creators Syndicate. **27:** © NASA/Stock Boston. **28 all:** Courtesy Texas Instruments Incorporated. **33:** Courtesy of Anaheim Public Information. **34:** DENNIS THE MENACE® used by permission of Hank Ketcham and © by North American Syndicate. **35:** *Graffiti* reprinted by permission of Newspaper Enterprise Association, Inc. **37 left:** © Fredrik Bodin/Stock Boston. **37 right:** © John Coletti/Stock Boston. **38 top:** David Gilkey. Courtesy of the *Daily Camera*, Boulder, Colorado. **39 top:** © Rafael Macia/Photo Researchers.

CHAPTER 2

57: Courtesy of Vermont Agency of Transportation. **62:** Robert Mankoff from *Saturday Review*, June 11, 1977. Copyright © 1977 by Robert Mankoff. **63:** PIXIES © 1972. Reprinted by permission of United Feature Syndicate, Inc. **75:** Sidney Harris *American Scientist* magazine © 1977. Reprinted by permission. **83:** *Beetle Bailey* © 1974 by King Feature Syndicate, Inc. **83:** Problems 55 and 56 adapted from "Solving the Mystery," by Francis R. Curicio and J. Lewis McNeece, *The Mathematics Teacher*, November 1993. pp. 682–685. **89:** Problem 7 reprinted from *Introduction to Logic*, 2/e, by Irving Copi. Copyright © 1961 by the Macmillan Company. **90:** Problem 10, by Gurney Mentes in *Games*, August 1996, p. 43. **97:** Problem G4 from "Ask Marilyn," by Marilyn vos Savant, *Parade Magazine*, October 31, 1993. Reprinted with permission from Parade, copyright © 1993.

CHAPTER 3

103: Alan Dunn, 1952, appeared in *The New Yorker Magazine*. **106:** © Erich Lessing/PhotoEdit. **111:** *Margarita Philosophica Nova*, 1512. © Museum of the History of Science, Oxford. **114:** NASA. **120:** Bob Schochet. Reprinted with permission. **125:** Reprinted by permission of *Datamation* magazine. Copyright © by Technical Publishing Company. A Division of Dunn-Donnelley Publishing Corporation, 1969. All rights reserved. **126 top:** Courtesy of International Business Machines. **126 bottom:** Courtesy of Pickett Industries. **127 top and middle:** Courtesy of International Business Machines. **127 bottom:** © SPL/Photo Researchers. **129:** AP/Wide World Photos. **132:** *A Computer Glossary* © 1968 International Business Machines Corporation.

CHAPTER 4

145: © Mark Antman/The Image Works **148:** Adapted from the Vanishing Leprechaun © 1968 by the W. A. Elliott Company, Toronto. All rights reserved under the Universal Copyright Convention. **157:** Courtesy of Donald B. Gillies, University of Illinois. **161:** *B.C.* reprinted by permission of Johnny Hart and Creators Syndicate. **168:** *B.C.* reprinted by permission of Johnny Hart and Creators Syndicate. **177:** *B.C.* reprinted by permission of Johnny Hart and Creators Syndicate. **178:** Illustrations in Historical Note courtesy of the British Library, London. **179:** © Myrleen Ferguson Cate/PhotoEdit. **180:** © John Maher/Stock Boston. **181:** © 1979 United Feature Syndicate, Inc. Reprinted by permission. **182:** Problem 7 from *OMNI Magazine*, March 1995, "Games" department, p. 104. **186:** Courtesy of Patrick J. Boyle. **195:** Courtesy of Karl J. Smith. Image on the cover courtesy of Professor Leon D. Harmon, Case Western Reserve University, Cleveland, Ohio. **204:** *Games Magazine*, December 1985. **206:** *The Mathematics Teacher*, January, 1990, p. 70. **207:** *Science News*, May 7, 1994. **211:** From "Ask Marilyn," by Marilyn vos Savant, *Parade Magazine*, April 15, 1990. Reprinted with permission from Parade, copyright 1990. **212:** Courtesy S. Forseth and A. P. Troutman, The National Council of Teachers of Mathematics.

CHAPTER 5

232: From *A Computer Glossary* © 1968 International Business Machines Corporation. **241:** Tom Sampson from Black Lake High School. **271:** *PEANUTS* reprinted by permission of United Feature Syndicate, Inc. 1979. **285:** *PEANUTS* reprinted by permission of United Feature Syndicate, Inc. 1972.

CHAPTER 6

296: "The Meaning of Life," copyright 1971. Reprinted by permission of *Saturday Review* and Henry Martin. **299 top:** M. C. Escher's "Drawing Hands" © 2003 Cordon Art–Baarn–Holland. All rights reserved. **299 bottom left:** Courtesy of National Oceanic and Atmospheric Administration. **299 bottom center:** © Stock Montage. **299 bottom right:** Courtesy of Chrysler Corporation. **301 left:** Scott Morris, *OMNI Magazine*, April 1992, p. 94. **301 right:** Advertisement by Daniel Wade Arthur, from *The Los Angeles Times*. **303 top left:** © Susan Van Etten/ Stock Boston. **303 top right:** © Tim Zurowski/CORBIS. **303 left center:** © 1982 Asian Art Museum of San Francisco. All rights reserved. **303 right center:** Da Vinci's Dodecahedron Courtesy of the Metropolitan Museum of Art, Whittesley Fund. **303 bottom left:** © Robert Landau/CORBIS. **303 right:** © Dorling Kindersley, Ltd. **304:** Reprinted by permission of Tribune Media Services. © Tribune Media Services, Inc. All rights reserved. Reprinted with permission. **305:** Mirror Image by Diane Dawson, from *Games Magazine*, May/June 1979. **313:** Problem 60 adapted from "Mathematical Recreations," by A. K. Dewdney, *Scientific American*, July 1990, p. 118. Copyright 1990 by Scientific American, Inc. All rights reserved. Illustration by Slim Films. **330:** Reprinted by permission of United Feature Syndicate, Inc. 10/26/73. **333:** © Richard T. Nowitz/CORBIS. **341:** Adapted from "Angles of Elevation at the Pyramids of Egypt," in *The Mathematics Teacher*, February 1982, pp. 124–127. **335:** Dale Seymour Publications. **336:** *La Parade* by Georges Seurat. The Metropolitan Museum of Art, bequest of Stephen C. Clark, 1960. (60.101.17) Photograph © 1989 The Metropolitan Museum of Art. **337:** © N. Brandt/Photo Researchers. Copyright VEB Bibliographisches Institute Leipzig. Reprinted by permission. **338 bottom:** © Susan Van Etten/Stock Boston. **340:** Courtesy of Opera del Duomo, Siena. **341 top:** *B.C.* reprinted by permission of Johnny Hart and Creators Syndicate. **341 center:** Courtesy of FOTO Marburg/Art Resource, NY. **343:** Courtesy of Donald M. Welch. **346:** M. C. Escher's "Circle Limit III" © 2003 Cordon Art–Baarn–Holland. All rights reserved. **349:** Adapted from "The Amazing Mirror Maze" by Walter Wick in *Games* magazine, September/October 1981, p. 25.

CHAPTER 7

354: *Graffiti* reprinted by permission of Newspaper Enterprise Association, Inc. **360:** Musée du Louvre, by M. C. Huzeville, Paris. **390:** From *The Metrics Are Coming! The Metrics Are Coming!* by R. Cardwell. Copyright 1975 by Dorrance and Company. Reprinted by permission. **394:** Courtesy of Round Table Pizza.

CHAPTER 8

422: © David Lees/ CORBIS. **424:** Photograph of Aquarium of Americas from author's collection.

CHAPTER 9

441: MasterCard is a registered trademark of MasterCard International, Inc. **442:** AP/Wide World Photos. **446:** VISA is a registered trademark of the Bank of America, N.T. & S.A. **454:** Courtesy of Dale Seymour Publications, Palo Alto, CA. **460:** Problem 59 from "Fibonacci Forgeries" by Ian Stewart, *Scientific American,* May 1995, p. 104. Illustration by Johnny Johnson. Copyright © 1995 by Scientific American, Inc. All rights reserved. **464:** *PEANUTS* 1967 reprinted by permission of United Feature Syndicate, Inc. **489:** Courtesy of the Fields Institute and the University of Toronto.

CHAPTER 10

498: Assessment question from the 1987 Examination of the California Assessment Program (along with analysis from 1989) in *A Question of Thinking: A First Look at Students' Performance on Open-ended Questions in Mathematics.* **504:** From the 1987 Examination of the California Assessment Program. **505 left:** © Robert Daemmrich. **505 right:** Problem 60 from *Math Puzzles and Logic Problems,* © 1995 by Dell Magazines, a division of Penny Marketing Limited Partnership, reprinted by permission of Dell Magazines. **508:** Frontispiece of a book written by Augustus De Morgan in 1838. **509:** David Weintraub/Stock Boston. **526:** Reprinted by permission of *The Saturday Evening Post,* copyright 1976. **528 top:** Copyright National Museum of American Art, Washington, D.C./Art Resource, NY. **528 bottom:** © Sandra Lord/Photonight. **529:** Text of Little Ceasar's Pizza Pizza television advertisement. **542:** © Bettmann/CORBIS. **545:** Advertisement courtesy of Wendy International, Inc. **548:** Avenger Puzzle, Douglas Engel, Inglewood, Colorado. Photo appeared in *Games* magazine, June/July 1987.

CHAPTER 11

581: © Time Inc. Reprinted by permission. **562:** *B.C.* reprinted by permission of Johnny Hart and Creators Syndicate. **563:** Problem 59 by Clifford A. Pickover from *Discover,* March 1997, p. 94. Reprinted with permission of Discover Magazine, published by Disney Magazine Publishing Inc. **570:** Courtesy of Reader's Digest Inc. **572:** The Bettmann Archive. Copyright Corbis. **581:** AP/Wide World Photos. **582 top:** © Kathleen Olson. **582 bottom:** Problem 46 adapted from the DEAR ABBY column by Abigail Van Buren. Copyright 1974. Distributed by Universal Press Syndicate. Reprinted by permission. All rights reserved. **591:** Guest essay from *Innumeracy: Mathematical Illiteracy and Its Consequences,* by John Paulos, pp. 54–55. Copyright © 1988 John Paulos. **594 right:** Keno pay-offs courtesy of the California Lottery. **594 left:** AP/Wide World Photos.

CHAPTER 12

619: Blood alcohol concentration prepared by the Department of Motor Vehicles in cooperation with the California Highway Patrol. **620 bottom :** Advertisement courtesy of Anacin. **620 top:** Graph from U.S. Forest Service. **621 top:** Advertisement copyright 1971 by SAAB-Scania of America, Inc. **621 bottom:** Problem 60 adapted from "Caveat Emptor" by Robert Leighton in *Games,* January 1988, p. 38. **622:** *PEANUTS* 1961 reprinted by permission of United Feature Syndicate, Inc. **633:** Problem 59

from "Instances of Simpson's Paradox," by Thomas R. Knapp in the *College Mathematics Journal,* July 1985, pp. 209–211. **634:** Cartoon by David Pascal. **648 top:** Graph in Problem 5 by Slim Films from "Diet and Cancer," by Leonard A. Cohen, *Scientific American,* November 1987, p. 44. Copyright 1987 by Scientific American, Inc. All rights reserved. **648 bottom:** Graph in Problem 6 by Patricia J. Wynne from "How Dinosaurs Ran," by R. P. McNeill Alexander, *Scientific American,* April 1991, p. 132. Copyright 1991 by Scientific American, Inc. All rights reserved. **650:** Problems 54–59 are based on an article, "Data Analysis and the Hard Rock 100," by Marny Frantz and Sylvia Lazarnick in *The Mathematics Teacher,* April 1997, pp. 274–276. **651:** Problem 60 based on data from "The Correlation Coefficient and Influential Data Points," by Donna J. Dessart, *The Mathematics Teacher,* March 1997, pp. 242–246. **653:** © CORBIS. **661:** News Clip taken from DEAR ABBY column by Abigail Van Buren. Copyright 1973 distributed by Universal Press Syndicate. Reprinted with permission. All rights reserved.

CHAPTER 13

673: Courtesy of NASA. **695:** © Topham/The Image Works. **696:** *B.C.* reprinted by permission of Johnny Hart and Creators Syndicate. **702:** *B.C.* reprinted by permission of Johnny Hart and Creators Syndicate.

CHAPTER 14

714: Problem 56 from "When Does a Dog Become Older Than Its Owner?" by Anne Larson Quinn and Karen R. Larson, *The Mathematics Teacher,* December 1996, pp. 734–737. **731:** Courtesy of Patrick J. Boyle. **751:** AT&T Archives. Reprinted with permission of AT&T. **759–760:** Problem G57 adapted from R. E. Kohn, "A Mathematical Programming Model for Air Pollution Control," *Science and Mathematics,* June 1969, pp. 487–499.

CHAPTER 15

773: Quotation in Problem 53 by Claudia Zaslavsky in *Africa Counts* (Prindle, Weber, and Schmidt, 1973) from *On the Trail of the Bushongo* by Emil Torday. **784:** California-Nevada driving distances map copyright 2002 AAA. **787:** From the author's collection. **788:** Courtesy of the University of Illinois. **789 left:** © IBM Research/Peter Arnold, Inc. **789 right:** From *Fractal Vision: Put Fractals to Work for You.* © 1992 SAMS Publishing, a division of Prentice-Hall Computer Publishing. **790 bottom:** M. C. Escher's Circle Limit I © 2003 Cordon Art–Baarn–Holland. All rights reserved. **790 top:** Guest essay "What Good Are Fractals?" from "Geometrical Forms Known As Fractals Find Sense in Chaos" by Jeanne McDermott in the *Smithsonian,* December 1983, pp. 110–116. **793:** Dale Seymour Publications. Palo Alto, CA. **794:** Photo by Tibor Hirsch. **795:** Guest essay " Chaos" by Jack Wadhams.

CHAPTER 16

802: © AFP/ CORBIS.

CHAPTER 17

867: News Clip illustration by Hank Iken from "Plateau Uplift and Climatic Change," by William F. Ruddman and John E. Kutzbach, *Scientific American,* March 1991, p. 68. Copyright 1991 by Scientific American, Inc. All rights reserved. **873:** © David Young-Wolff/ PhotoEdit. **884:** Courtesy of Paramount's Great America.

EPILOGUE

E8: © David Lees/ CORBIS. **E13:** © Scala/Art Resource, NY. **E14 top:** © Scala/Art Resource, NY. **E14 bottom:** © Historical Picture/Archive/ CORBIS. **E15:** Perspective in Art, *Mathematics* by David Bergamini. Copyright Time Inc. Reprinted by permission. **E16:** Music score from The Selmer Bandwagon, 1961, 9(2).

Index